ADVANCES IN CERAMICS • VOLUME 10

STRUCTURE AND PROPERTIES OF MgO AND Al$_2$O$_3$ CERAMICS

Volume 1   Grain Boundary Phenomena in Electronic Ceramics
Volume 2   Physics of Fiber Optics
Volume 3   Science and Technology of Zirconia
Volume 4   Nucleation and Crystallization in Glasses
Volume 5   Materials Processing in Space
Volume 6   Character of Grain Boundaries
Volume 7   Additives and Interfaces in Electronic Ceramics
Volume 8   Nuclear Waste Management
Volume 9   Forming of Ceramics

ADVANCES IN CERAMICS • VOLUME 10

STRUCTURE AND PROPERTIES OF MgO AND Al_2O_3 CERAMICS

Edited by

W. D. Kingery
Massachusetts Institute of Technology
Cambridge, Massachusetts

The American Ceramic Society, Inc.
Columbus, Ohio

Proceedings of an international symposium held at the Massachusetts Institute of Technology in Cambridge, Massachusetts on June 13-16, 1983.

The June 1983 meeting at the Massachusetts Institute of Technology and the subsequent editing of this volume have been supported by the Office of Basic Energy Sciences, Materials Science Program, Division of Physical Research, U.S. Department of Energy under Contract No. DE-AC02-76ER02390.

Library of Congress Cataloging in Publication Data

Structure and properties of MgO and Al_2O_3 ceramics.

(Advances in ceramics, ISSN 0730-9546 ; v. 10)
 "Proceedings of an international symposium held at the Massachusetts Institute of Technology in Cambridge, Massachusetts on June 13-16, 1983"—T.p. verso.
 Includes bibliographies and index.
 1. Magnesia—Congresses. 2. Aluminum oxide—Congresses. 3. Ceramics—Congresses. I. Kingery, W. D.
TP889.S77   1984   666   84-11106
ISBN 0-916094-62-6

Coden: ADCEDE

©1984 by The American Ceramic Society, Inc. All rights reserved.

No part of this book may be reproduced, stored in a retrieval system, or transmitted in any form or by any means, electronic, mechanical, photocopying, microfilming, recording, or otherwise, without written permission from the publisher.

Printed in the United States of America.

Foreword

This volume consists of papers presented at a meeting of about one hundred participants held at the Massachusetts Institute of Technology, June 13-16, 1983. It represents the best current research on the two most studied ceramic oxides and thus serves as an indicator of the current status of ceramic science, from which the reader can identify levels of success achieved in different directions. It will be clear, for example, that the tools and techniques for studying surface structure and grain boundaries have only recently become really effective. As a result, the sorts of questions posed about surfaces and boundaries have not reached the level of sophistication and detail as those applied to the structure and properties of the bulk crystal.

Magnesium oxide and aluminum oxide are wide-band-gap ionic materials with relatively simple crystal structures that have generally been regarded and used as "model" materials, serving as paradigms for other materials that are sometimes more interesting but often more difficult to study. One use of this volume will be to evaluate the fruits of this approach and identify areas in which this reductionist methodology of focusing on model materials has led to a broad understanding that can be applied to a wide range of ceramics and, perhaps, to areas in which the approach has concealed technologically important and interesting complexities.

Structure-property relationships are basic to our whole approach to ceramic science and technology. The present level of understanding of the structure and composition of crystals, point defects, dislocations, grain boundaries, and surfaces in MgO and Al2O3 is implicit in the research results which are reported. Since our present understanding of the structure-property relationships is stronger for MgO and for Al2O3 than for other ceramic materials, this volume should provide a guide for approaching oncoming studies of new and different ceramics which lie outside the family of wide-band-gap ionic materials.

While we shall let the record speak for itself, it is quite clear, and should be gratifying for the sponsors of modern research in ceramics, that the level of questions now posed—as compared with those posed in the recent past—clearly demonstrates the enormous contribution that scientific studies of ceramic materials have provided for the technology which has advanced in a parallel way.

W. David Kingery
Massachusetts Institute of Technology

Contents

Electrical Properties of α-Al$_2$O$_3$... 1
    F. A. Kröger

Polarization-Depolarization Studies of Defects in Oxides....... 16
    J. H. Crawford, Jr.

Luminescence and Photoconductivity in MgO and
α-Al$_2$O$_3$ Crystals... 25
    G. P. Summers

Optical and Mössbauer Spectra of Transition-Metal-Doped
Corundum and Periclase... 46
    R. C. Burns and V. M. Burns

Calculated Point Defect Formation, Association, and
Migration Energies in MgO and α-Al$_2$O$_3$..................... 62
    W. C. Mackrodt

Calculations of Entropies and of Absolute Diffusion
Rates in Oxides... 79
    A. M. Stoneham, M. J. L. Sangster, and D. K. Rowell

Experimental and Calculated Values of Defect Parameters
and the Defect Structure of α-Al$_2$O$_3$.......................... 100
    F. A. Kröger

Low Atomic Number Impurity Atoms in Magnesium
Oxide: Hydrogen and Carbon.. 119
    F. Freund, B. King, R. Knobel, and H. Kathrein

Defect Association in MgO... 139
    T. A. Yager and W. D. Kingery

Semiempirical Analysis of the Electronic Processes of
F-Type Centers in α-Al$_2$O$_3$.................................. 152
    S. Choi and T. Takeuchi

Defects Produced by Shock Conditioning:
An Overview.. 157
    R. F. Davis, Y. Horie, R. O. Scattergood, and H. Palmour III

Surfaces of Magnesia and Alumina...................................... 176
    P. W. Tasker

Point Defects and Chemisorption at the {001}
Surface of MgO.. 190
    E. A. Colbourn and W. C. Mackrodt

Surface Electronic Structure of MgO and Al$_2$O$_3$ Ceramics...... 205
   V. E. Henrich

Calcium Segregation to MgO and α-Al$_2$O$_3$ Surfaces........... 217
   R. C. McCune and R. C. Ku

Dislocations in α-Al$_2$O$_3$..................................... 238
   A. H. Heuer and J. Castaing

Dislocations in MgO..................................... 258
   Y. Moriyoshi

Properties of Grain Boundaries in Rock Salt Structured Oxides 275
   D. M. Duffy and P. W. Tasker

Comparison of the Calculated Properties of High-Angle (001) Twist Boundaries in MnO, FeO, CoO, and NiO with MgO... 290
   D. Wolf

Grain-Boundary Structure in Al$_2$O$_3$........................... 303
   C. B. Carter and K. J. Morrissey

Electron Diffraction and Microscopy Studies of the Structure of Grain Boundaries in NiO....................... 324
   J. Eastman, F. Schmückle, M. D. Vaudin, and S. L. Sass

Boundary Structure Observed in MgO Bicrystals............. 347
   S. Kimura, E. Yasuda, N. Horiai, and Y. Moriyoshi

Grain-Boundary Microstructures in Alumina Ceramics........ 357
   D. S. Phillips and Y. R. Shiue

Solute Segregation at Grain Boundaries in Polycrystalline Al$_2$O$_3$...................................... 368
   C. W. Li and W. D. Kingery

Oxygen Diffusion in MgO and Al$_2$O$_3$....................... 379
   Y. Oishi and K. Ando

Secondary Ion Mass Spectrometric Analysis of Oxygen Self-Diffusion in Single-Crystal MgO....................... 394
   H. I. Yoo, B. J. Wuensch, and W. T. Petuskey

Grain-Boundary and Lattice Diffusion of ^{51}Cr in Alumina and Spinel... 406
   V. S. Stubican and J. W. Osenbach

Understanding Defect Structure and Mass Transport
in Polycrystalline Al$_2$O$_3$ and MgO via the Study of
Diffusional Creep.. 418
    R. S. Gordon

Lattice-, Grain-Boundary, Surface-, and Gas-Diffusion
Constants in Magnesium Oxide............................ 438
    J. M. Vieira and R. J. Brook

Oxygen Surface Diffusion and Surface Layer Thickness
in MgO... 464
    Y. Ikuma and W. Komatsu

Oxygen Diffusion in Undoped and Impurity-Doped
Polycrystalline MgO.. 474
    S. Shirasaki, S. Matsuda, H. Yamamura, and H. Haneda

Migration Energies of Interstitials and Vacancies in MgO...... 490
    C. Kinoshita, K. Hayashi, and T. E. Mitchell

Sintering of Shock-Conditioned Materials................... 506
    H. Palmour III, T. M. Hare, A. D. Batchelor, K. Y. Kim, K. L. More, and T. T. Fang

Role of Al$_2$O$_3$ in Sintering of Submicrometer
Yttria-Stabilized ZrO$_2$ Powders........................... 526
    R. C. Buchanan and D. M. Wilson

Vaporization from Magnesia and Alumina Materials.......... 541
    T. Sata and T. Sasamoto

Vaporization of Magnesium Oxide in Hydrogen............... 553
    V. Bheemineni and D. W. Readey

Influence of MgO Vaporization on the Final-Stage
Sintering of MgO-Al$_2$O$_3$ Spinel......................... 562
    M. Matsui, T. Takahashi, and I. Oda

Solid-State Sintering: The Attainment of High Density........ 574
    S. Wu, E. Gilbart, and R. J. Brook

Initial Sintering of MgO in Several Water Vapor Pressures...... 583
    O. J. Whittemore and J. A. Varela

Model of Interactions between Magnesia and Water........... 592
    E. Longo, J. A. Varela, C. V. Santilli, and O. J. Whittemore

Effects of Crystallinity in Initial-Stage Sintering.............. 601
    J. G. Dash and R. Stone

Effect of Magnesium Chloride on Sintering of Magnesia....... 610
   K. Hamano, Z. Nakagawa, and H. Watanabe

Final-Stage Sintering of MgO.............................. 619
   C. A. Handwerker, R. M. Cannon, and R. L. Coble

Processing of Narrow Size Distribution Alumina.............. 644
   T. R. Gattuso and H. K. Bowen

Sintering of α-Al$_2$O$_3$ in Gas Plasmas........................ 656
   D. L. Johnson, W. B. Sanderson, J. M. Knowlton, and E. L. Kemer

Effect of Magnesia on Grain Growth in Alumina.............. 666
   R. D. Bagley and D. L. Johnson

Use of Solid-Solution Additives in Ceramic Processing........ 679
   M. P. Harmer

Mechanical Behavior of Alumina: A Model Anisotropic Brittle Solid... 697
   A. G. Evans and Y. Fu

Effects of Temperature and Stress on Grain-Boundary Behavior in Fine-Grained Alumina.......................... 720
   J. D. Fridez, C. Carry, and A. Mocellin

High Creep Ductility in Alumina Containing Compensating Additives.................................... 741
   W. R. Cannon

Crack Healing in Al$_2$O$_3$, MgO, and Related Materials........... 750
   T. K. Gupta

Fracture Toughness of Single-Crystal Alumina............... 767
   M. Iwasa and R. C. Bradt

Effect of MgO Dopant Dispersing Method on Density and Microstructure of Alumina Ceramics.................... 780
   A. Cohen, C. P. Van der Merwe, and A. I. Kingon

Overview of Current Understanding of MgO and Al$_2$O$_3$
   Defects in MgO and Al$_2$O$_3$............................... 791
      J. H. Crawford

   Dislocations in Ceramics................................ 799
      A. H. Heuer

   Structure of MgO and Al$_2$O$_3$ Surfaces..................... 801
      V. Henrich

   Grain Boundaries in MgO and Al$_2$O$_3$...................... 803
      W. D. Kingery

Mechanical Properties of MgO and Al$_2$O$_3$................. 818
R. M. Cannon

Sintering and Grain Growth in Alumina and Magnesia....... 839
R. L. Coble, H. Song, R. J. Brook, C. A. Handwerker,
and J. M. Dynys

Electrical Properties of α-Al$_2$O$_3$

F. A. KRÖGER

University of Southern California
Department of Materials Science
University Park MC 0241
Los Angeles, CA   90089-0241

The properties of α-Al$_2$O$_3$ are always determined by the presence of impurities, which act as acceptors (e.g., Mg, Fe, Co, V, Ni) and donors (e.g., H, Ti, Si, Zr, Y). Conductivities (σ) are of a mixed ionic–electronic character, the relative contribution of each varying with oxygen pressure and temperature. For acceptor-dominated Al$_2$O$_3$, $\sigma = \sigma_i$ at low p_{O_2}, and $\sigma = \sigma_h$ at high p_{O_2}; for donor-dominated Al$_2$O$_3$, $\sigma = \sigma_e$ at low p_{O_2}, and $\sigma = \sigma_i$ at high p_{O_2}. Compensation of acceptors by donors leads to material with almost intrinsic properties from which basic disorder parameters can be obtained. Analysis of the experimental data leads to positions of dopant energy levels and of oxidation–reduction constants.

Alpha aluminum trioxide, α-Al$_2$O$_3$, with an enthalpy of formation of 1674.4 kJ/mol, is one of the most strongly bonded compounds in existence. Energies of formation of point defects are correspondingly large. Table I shows computed energies of disorder per defect for various disorder mechanisms.[1,2] The energy of electronic disorder is equally large,[3] the optical band gap being $\approx$9.9 eV at 25°C, giving an energy of formation of electrons and holes by intrinsic disorder of $\approx$5 eV per electron or hole.

Calculation on this basis of the concentration of defects formed by disorder at 1600°C gives values well below 1 ppm. The same applies to defects formed by nonstoichiometry. Since impurity levels of the purest Al$_2$O$_3$ available are $\geq$1 ppm per impurity, it is obvious that the properties of α-Al$_2$O$_3$ up to 1600°C are impurity dominated. Indeed, intrinsic properties have so far never been observed and have only been deduced from the properties of doped, compensated material.

Electrical Conductivity

Measurements of electrical conductivity are often complicated by the presence of barriers at the contacts. These complications can be eliminated in DC measurements by measuring potentials on non-current-carrying probes or in AC measurements by employing sufficiently high frequencies. For Al$_2$O$_3$ the bulk resistivities are so high that AC measurements give practically the same results as DC measurements with potential measurement on current-carrying contacts.[4] Correct results require, however, that surface and gas-phase conduction are eliminated with the aid of a biased volume guard.[5–8] Conductivity isotherms as a function of oxygen pressure p_{O_2} in general are U-shaped, indicating contributions by different species at high and low p_{O_2}. Emf measurements on cells of the type

$$O_{2,I}, Pt|Al_2O_3|Pt, O_{2,II} \tag{1}$$

make it possible to determine the degree to which ions and electrons or holes

Table I. Computed Energies of Formation per Defect for Various Types of Disorder

	Energies					
	Ref. 1				Ref. 2	
	Empirical potentials		Nonempirical potentials			
Disorder type	eV	kJ/mol	eV	kJ/mol	eV	kJ/mol
Schottky	4.18	402	5.14	495	5.7	549
Cation Frenkel	5.22	502	7.09	682	10.0	962
Anion Frenkel	3.79	365	8.27	796	7.0	674
Interstitial	4.54	437	10.44	1005	11.5	1107

contribute to the conductivity:

$$E = \frac{RT}{nF} \int_I^{II} t_i \, d \ln p_{O_2} \tag{2}$$

with $t_i = \sigma_i/\sigma$.

In general $t_i = f(p_{O_2})$. Its value at a given p_{O_2} can be found by measuring the emf of cell I with $p_{O_2,I}$ constant and $p_{O_2,II}$ variable. Then

$$(t_i)_{p_{O_2,II}} = \frac{nF}{RT}\left(\frac{\delta E}{\delta \ln p_{O_2,II}}\right)_{p_{O_2,I}} \tag{3}$$

Effects of surface and/or gas-phase conduction are again eliminated with the aid of a biased volume guard.[6] Then $\sigma_i = t_i\sigma$ and $\sigma_{el} = t_{el}\sigma = (1 - t_i)\sigma$.

Two types of behavior have been observed. In one, encompassing Al_2O_3:Mg,[9,10] Fe,[11,12] Co,[13] or V,[14] the conductivity is mainly ionic at low p_{O_2} and electronic at high p_{O_2}. Electronic conduction increasing with p_{O_2} indicates that holes are the carriers involved: the dopants mentioned act as acceptors. The same type of behavior is generally found in undoped Al_2O_3 and is attributed to the presence of Mg and/or Fe as an impurity.[15] Typical isotherms of conductivities σ, σ_i, and $\sigma_{el} \equiv \sigma_h$ and emf of a concentration cell based on Mg-doped crystals are shown in Fig. 1(a,b). In the second type of behavior, the conductivity is mainly ionic at high p_{O_2} and electronic at low p_{O_2} (with electrons the carriers of the electronic conductivity). This behavior is found in Al_2O_3 doped with Ti,[16] Y,[17] H,[14,15,18] and Si;[19] these dopants act as donors. Typical isotherms of conductivity, emf, t_i, and partial conductivities σ_i and σ_e are shown in Fig. 2(a,b). Donor behavior is expected for Ti_{Al} and Si_{Al}, Ti and Si having more valence electrons than Al; donor activity in the case of H can be accounted for if the hydrogen replaces oxygen (H$^-$ at an O^{2-}, H$_O'$). In fact, three H species were found: H$_O'$, H$_i^{\cdot}$, and H$_i'$, the former two being ionized donors and the last an ionized acceptor. The donors dominate with [H$_O'$] $\gg$ [H$_i^{\cdot}$] and [H$_i'$]. The interstitial species are mobile and give rise to proton conduction.[18] The presence of H$_O'$ has also been linked to a long phosphorescence.[20–22]

Yttrium, being isoelectronic with aluminum, presents a problem. The donor activity of Y_{Al} is attributed to the large size of the Y^{3+} ion with a correspondingly large repulsive energy.[17] Ionization either of the Y^{3+} itself or of an adjacent O^{2-} with formation of Y^{4+} or O$^-$ — both smaller ions — reduces the stress and thus should stabilize the ionized state relative to the nonionized one. This has the effect

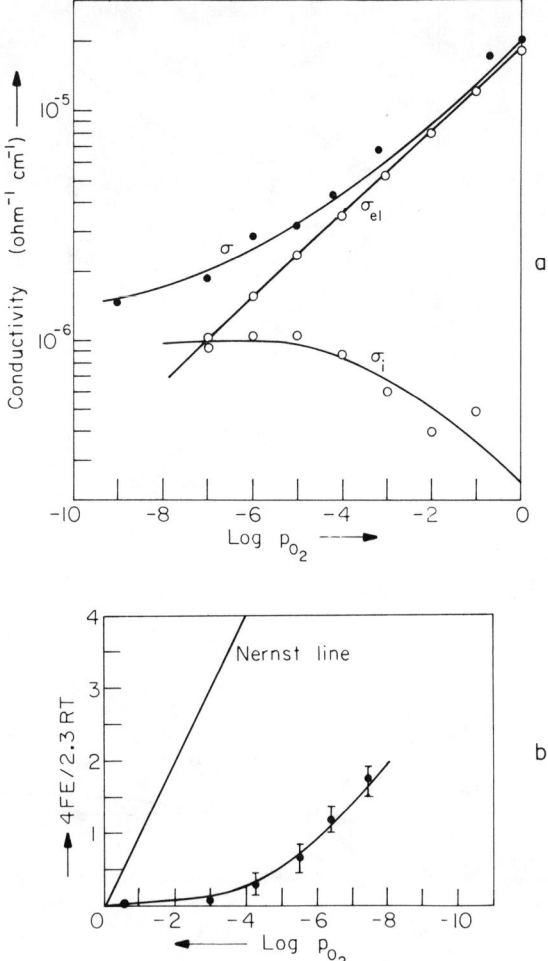

Fig. 1. (a) Total and partial electronic and ionic conductivities at 1600°C as $f(p_{O_2})$ for Al$_2$O$_3$:Mg, 60° with c axis. (b) Emf for an oxygen concentration cell with Al$_2$O$_3$:Mg as the electrolyte with p_{O_2} at one side equal to 10^5 Pa.

of raising the Y_{Al}^x level in the band picture. Energy considerations indicate O^{2-} next to Y_{Al}^x as the probable donor.[17]

The amount of acceptors present in undoped Al$_2$O$_3$ can be estimated by indiffusion of known concentrations of donors, noting the concentration at which the material becomes donor dominated. Such a "titration" was done successfully with Ti as the donor[23] (Fig. 3(a,b)). This procedure is correct if the Fermi level position changes sharply close to equivalence. However, this does not imply that donors D or acceptors A can only affect the defect structure when [D] > [A] or [A] > [D]. In fact, Ti donors have been found to affect creep in Al$_2$O$_3$: Fe + Ti in air at concentration [Ti] ≃ 0.2[Fe].[24] The explanation is that in air $[Fe_{Al}^x] > [Fe_{Al}']$, with Ti starting to affect the defect structure when $[Ti_{Al}^·] = [Fe_{Al}']$.

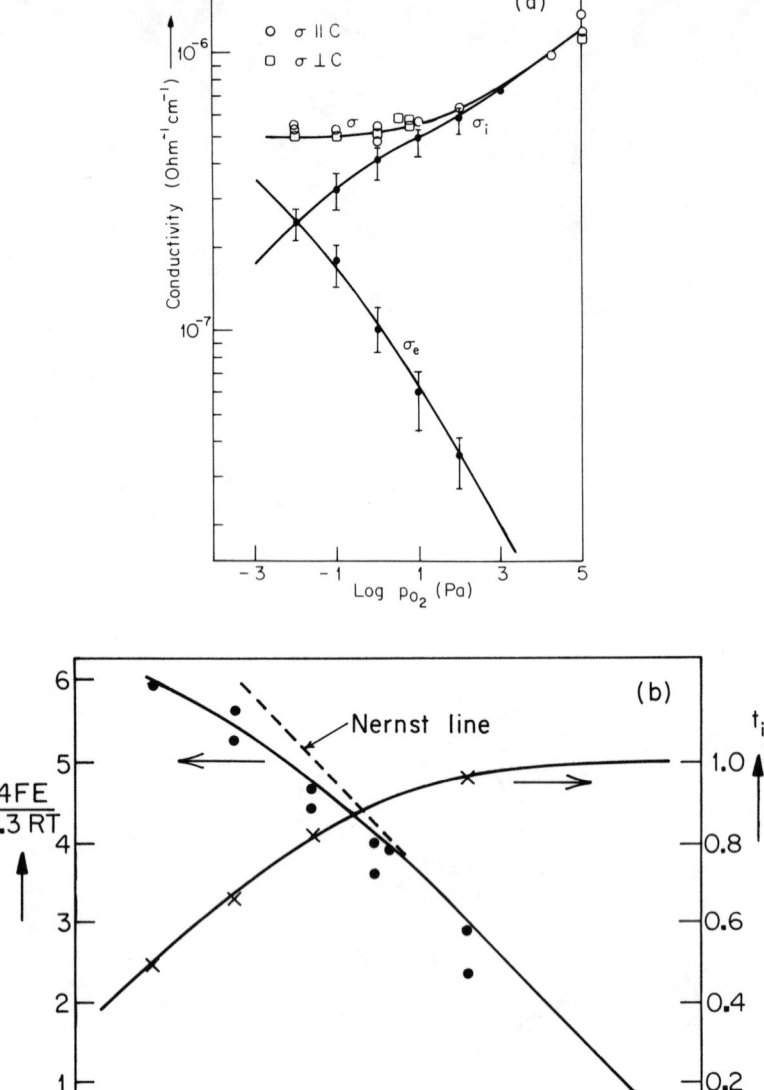

Fig. 2. (a) Total and partial electronic and ionic conductivities as $f(p_{O_2})$ for Al_2O_3:Ti at 1600°C. (b) Emf for an oxygen concentration cell based on Al_2O_3:Ti with p_{O_2} at one side equal to 10^5 Pa, and ionic transference number t_i (p_{O_2}) derived from it.

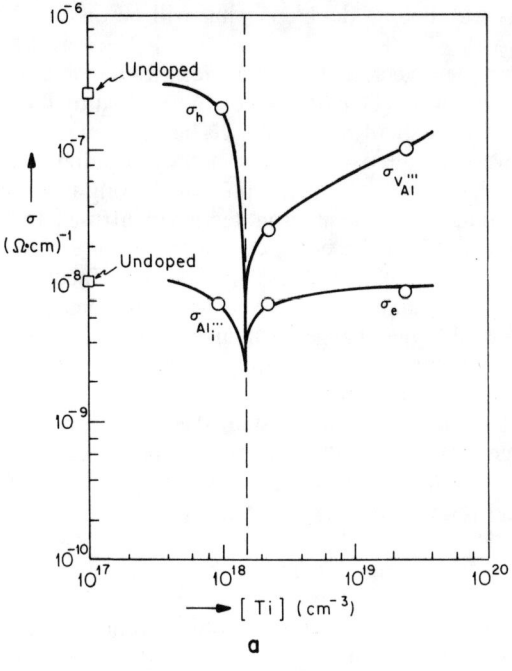

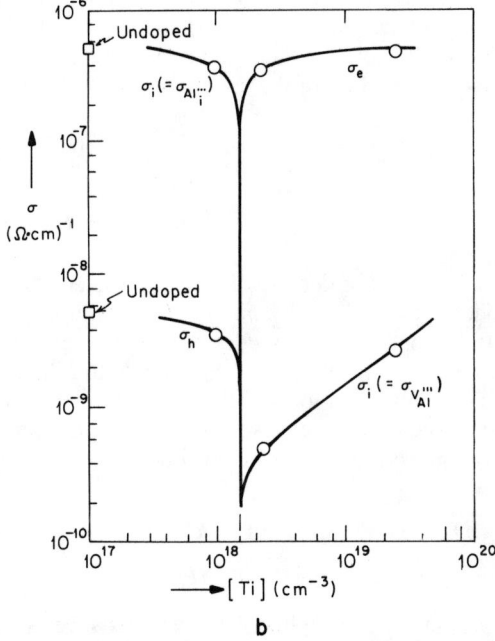

Fig. 3. Conductivities of Al_2O_3:A + Ti as $f([Ti])$ at high p_{O_2} (a) and low p_{O_2} (b).

The complications surrounding double doping will be discussed in detail in our second paper.[46]

Attempts to compensate acceptor-dominated Al_2O_3 with Si—therewith proving the donor activity of this element—are complicated by the low solubility of SiO_2 in Al_2O_3 and by its small diffusivity.[19] The latter was the reason that long annealing times were required to achieve homogenization with dissolution of second-phase particles of SiO_2 originally present. During this time, acceptor impurities from the furnace wall (mainly iron) tended to diffuse into the material, thus reducing the effect of Si on the conductivity and even causing an increase of conductivity at long times of anneal; the material remained acceptor dominated (Fig. 4). Only when the samples were protected from contamination by a screen of undoped Al_2O_3 could type change be induced[19] (Fig. 5).

Conductivity isotherms can be determined under two different conditions. In one the temperature is varied slowly, allowing equilibrium between the crystal and the surrounding atmosphere to remain established. Since the degree of oxidation at constant p_{O_2} in general varies with temperature, this procedure leads to variation of the composition, and the temperature dependence of conductivity includes variations in composition, of internal disorder, and of mobility. On the other hand, if the temperature is varied rapidly (or if the measurements are extended to low temperatures where diffusivities get small), the crystals do not change their composition by evolution or uptake of oxygen, and the only effects observed are those due to variations in internal disorder and mobility. For ionic conductivity under nonequilibrium conditions, the mobility variations dominate, the temperature dependence of σ_i giving the activation energy of the mobility of the ionic defect involved. For electronic conductivity the variation of the electron and hole mobilities μ_e or μ_h are negligible relative to the effects due to internal disorder. The temperature dependence of σ_e or σ_h under such conditions give the position of the Fermi level E_F relative to the edges of the conduction band E_c or valence band E_v:

$$\sigma_e = c_e q \mu_e = \frac{N_c \mu_e q}{1 + \exp(E_c - E_F)/kT} \simeq N_c \mu_e q \exp-(E_c - E_F)/kT \qquad (4)$$

$$\sigma_h = c_h q \mu_h = \frac{N_c \mu_e q}{1 + \exp(E_F - E_v)/kT} \simeq N_v \mu_h q \exp-(E_F - E_v)/kT \qquad (5)$$

where N_c and N_v are the effective density of states in the conduction and valence band. If a band model applies, $N = 2(2\pi m^* kT/h^2)^{3/2}$ with $m^* = m_e^*$ for N_c and $m^* = m_h^*$ for N_v. If a small polaron model applies, N_c and N_v are constants. Positions of donor or acceptor levels E_D and E_A relative to E_c and E_v can be found if the concentrations of neutral and ionized donors or acceptors are known, either from other experiments such as optical absorption or electron spin resonance or by implication from titration experiments. The level positions are found by using the relations[25]

$$[D^\cdot]/[D^\times] = \{g(D^\cdot)/g(D^\times)\} \exp(E_D - E_F)/kT \qquad (6)$$

$$[A']/[A^\times] = \{g(A')/g(A^\times)\} \exp(E_F - E_A)/kT \qquad (7)$$

with the g's being the statistical weights of the various species.

Relatively large electron mobilities of 3 cm^2/s[26,27] or 100 cm^{-2}/s[28] suggest that electrons are large polarons (band model). Recent results indicate that holes are also large polarons,[10] these being $\simeq 0.4$ eV more stable than small polarons.[29] Level positions determined in this manner are *thermal* levels, valid for the tem-

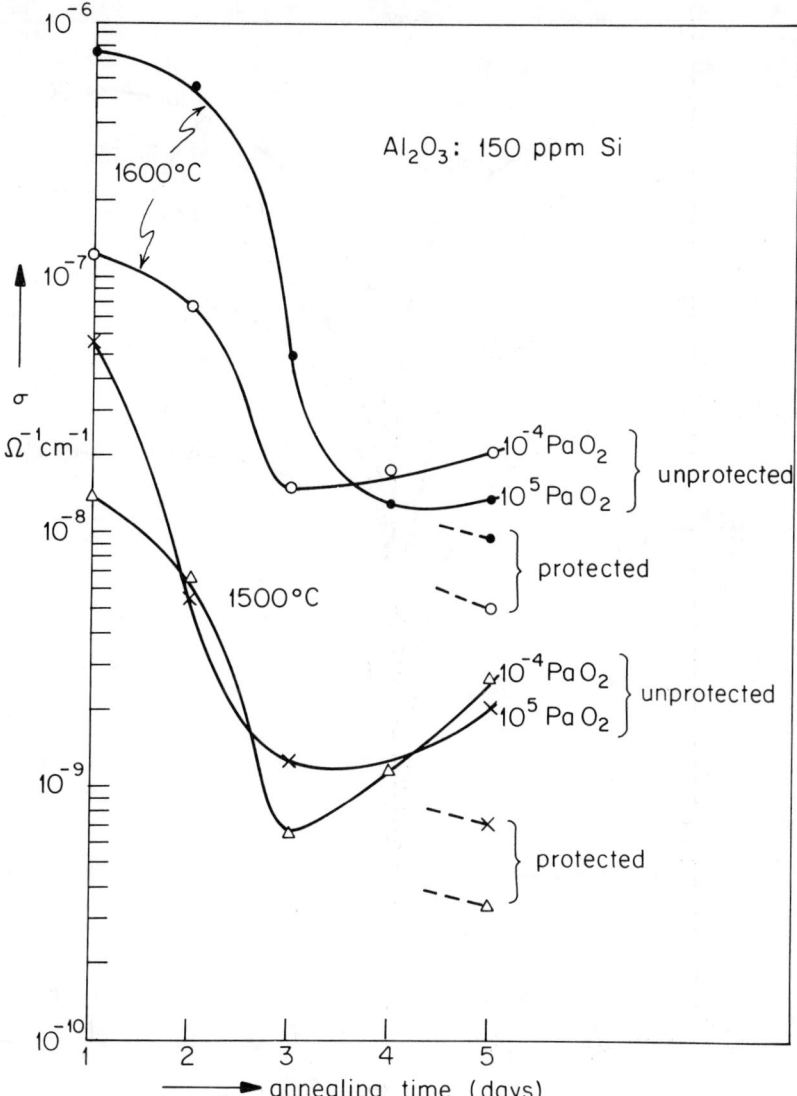

Fig. 4. Conductivity at high and low p_{O_2} of Al_2O_3: 150-ppm Si as a function of time of annealing at 1600°C for an unprotected sample; the material remains acceptor dominated.

perature at which the Fermi level position was determined. Variation of the level position with temperature should give rise to a variation of E_F with T, i.e., to a curvature of log $(\sigma_{el})_{noneq}$ against $1/T$. Thermal level positions can also be determined from the rate of thermal release of electrons or holes trapped after their formation by irradiation; this has been done for the Mg'_{Al} level.[10] They are also found from a study of defect concentrations as $f(T)$ on doubly doped material as

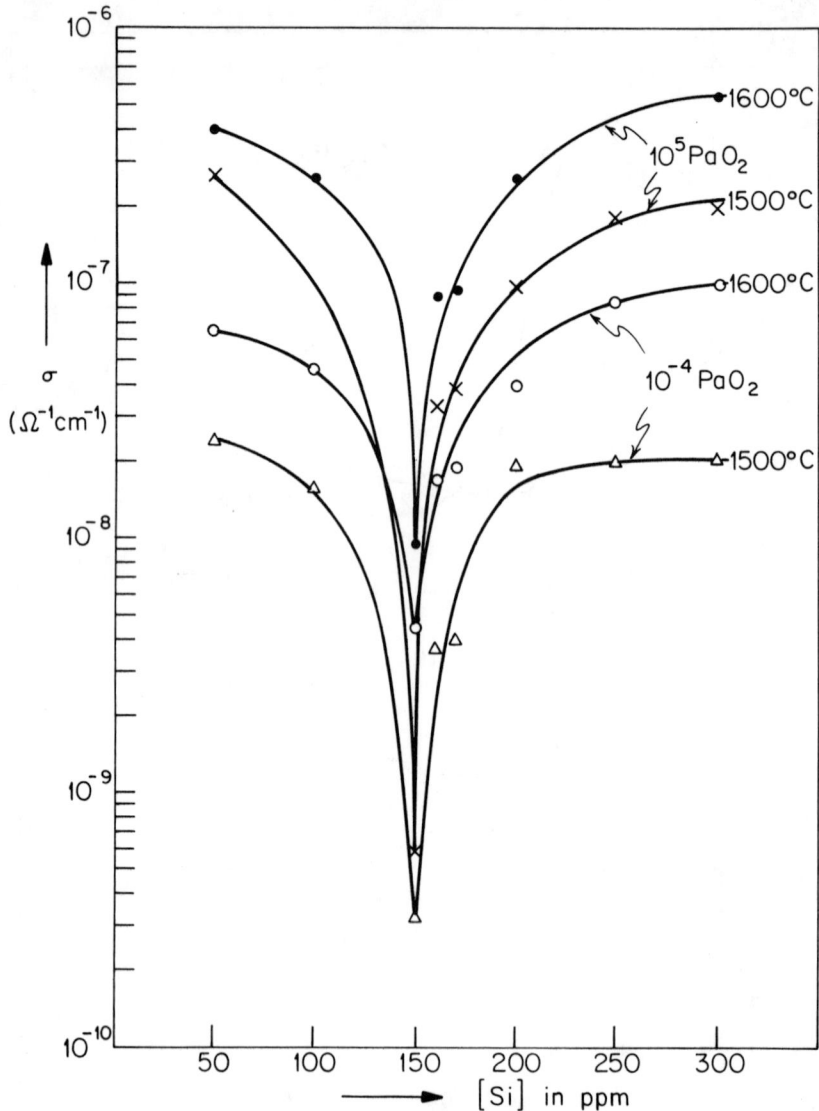

Fig. 5. Conductivity at Al_2O_3:A + Si as f[Si] at low and high p_{O_2} for materials preannealed for 216 h at 1600°C in air with protection against indiffusion of impurities.

has been done for Al_2O_3:Co + V, giving the position of the V_{Al}^x level 0.7 eV below the Co'_{Al} level.[30]

Optical level positions can be determined by studies of optical absorption and/or luminescence involving electron-transfer transitions. In general, level separations determined by absorption are larger, and those determined by luminescence are smaller than thermal separations. Such studies have been performed for Al_2O_3

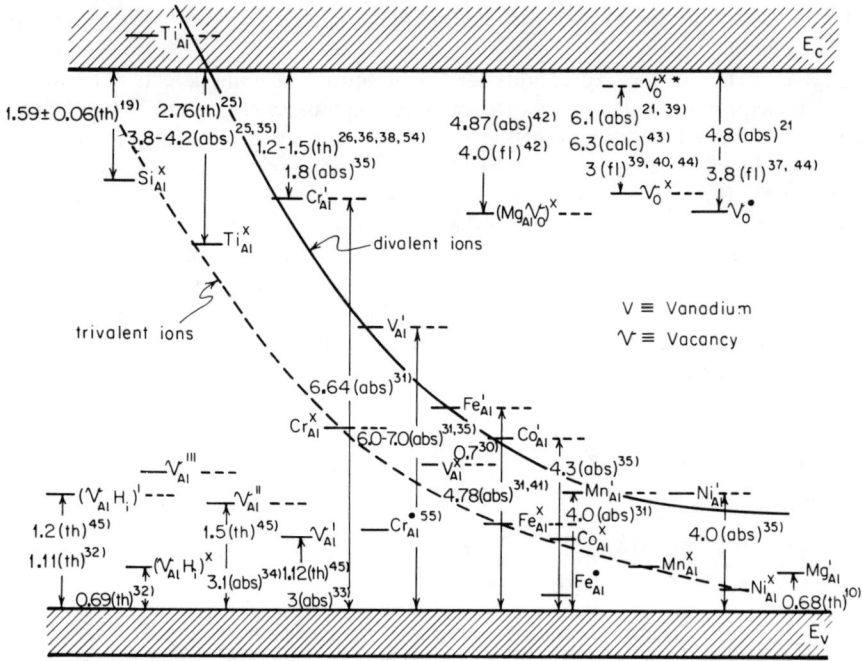

Fig. 6. Optical and thermal energy levels of native defects and dopants. Levels are marked according to the state existing when the levels are occupied by an electron.

doped with transition elements[31] and for Al_2O_3 with native defects.[21,32–43] Figure 6 shows schematically the positions of various levels in the forbidden gap.

A recent paper[56] places the V_O levels below the valence band and even suggests the neutral center to be an electron trap. This result is in disagreement with experimental results which show that deficiency of oxygen raises, not lowers, the Fermi level as would be the case if the proposed level positions would be correct.

Table II. Parameters in the Expressions for the Mobilities of Ionic Defects $\mu_i = \mu_i^\circ \exp(-H(\mu_i)/kT)$ (cm²/V·s), Assuming the Mobile Defects to be Majority Species

Material	Defect	μ_i° (cm²/V·s)	$H(\mu_i)$ (eV)	(kJ/mol)	Ref.
Al_2O_3:Co	$Al_i^{\cdots}$ (or $V_O^{\cdot\cdot}$)	1.22×10^7	3.97 ($\perp c$)	382	13, 15
Al_2O_3:Mg	$Al_i^{\cdots}$ (or $V_O^{\cdot\cdot}$)	2.2×10^9	4.84 ($\|c$)	466	10, 9
		2.27×10^8	4.39 ($\perp c$)	422	
Al_2O_3:A	$Al_i^{\cdots}$ (or $V_O^{\cdot\cdot}$)		3.6 (60°c)	346	15
Al_2O_3:A*	$Al_i^{\cdots}$ (or $V_O^{\cdot\cdot}$)	2×10^8	4.74 (60°c)	456	15
Al_2O_3:Ti	V_{Al}''' (or O_i'')	5.2×10^3	3.78	364	16
Al_2O_3:Y	V_{Al}''' (or O_i'')		3.45	332	17
Al_2O_3:Si	V_{Al}''' (or O_i'')		3.58	345	19

A computation for the oxygen vacancy in $SrTiO_3$ indeed gives levels close to those of the conduction band.[58]

Parameters of the expressions for the mobilities of ionic defects determined from nonequilibrium variations of σ_i with temperature are given in Table II. The data are obtained by assuming that the mobile species are the major ones: for acceptor-doped materials either $Al_i^{\cdots}$ or $V_O^{\cdot\cdot}$ and for donor-doped materials V_{Al}''' or O_i''. Since self diffusion of Al is much faster than that of O, it is likely that Al defects are the active carriers in all cases. If the mobile species are minority species, e.g., $Al_i^{\cdots}$ with $[Al_i^{\cdots}] \ll [A'] \simeq 2[V_O^{\cdot\cdot}]$ or V_{Al}''' with $[V_{Al}'''] \ll [D^{\cdot}] \simeq 2[O_i'']$, the parameters μ_i^0 and $H(\mu_i)$ found are not simply the preexponential and the enthalpy of the mobility but contain parameters of the equilibrium constants of the disorder processes linking variations of the majority species to those of the minority species.[46] In donor-dominated material

$$\mu_i^0 \equiv 0.94(K_{F,O}^{3/2})_0(K_S)_0^{1/2}[D^{\cdot}]^{-1/2}$$

$$H(\mu_{V_{Al}'''}) = H(\mu_i) + \tfrac{1}{2}H_{F,O} - \tfrac{1}{2}H_S$$

In acceptor-dominated material

$$\mu_i^0 \equiv 0.94(K_S)_0^{1/2}(K_{F,Al})_0^{3/2}$$

$$H(\mu_{Al_i^{\cdots}}) = H(\mu_i) + \tfrac{1}{2}H_S - H_{F,Al}$$

K_S, $K_{F,Al}$, and $K_{F,O}$ are the equilibrium constants of Schottky, Frenkel, and anti-Frenkel disorder, the H's being the corresponding enthalpies and the K_0's the preexponentials. The values for $Al_i^{\cdots}$ are slightly anisotropic; those for V_{Al}''' are isotropic.

Combination of the results of σ as a function of temperature under equilibrium and nonequilibrium conditions leads to expressions for oxidation–reduction of doped and undoped Al_2O_3. For results, see Ref. 46. As mentioned earlier, compensation of acceptors by donors can be used to change the position of the Fermi level at a given p_{O_2} and temperature from a low to a high position in the forbidden gap, causing at the same time a change in the concentrations of ionic defects V_{Al}''' and $Al_i^{\cdots}$ responsible for ionic conductivity. Near the compensation point the material has virtually intrinsic properties. Compensation can and has been achieved by stable donors such as Ti, Y, and Zr. With Ti and Zr, the properties studied were initial sintering[47–49] and creep rate[50] — both believed to be dependent on the concentration of ionic species, V_{Al}''' and $Al_i^{\cdots}$. With Y, ionic and electronic conductivity were studied.[17] In such experiments, a series of samples with different concentrations of donors must be made and studied. Only in one case (Al_2O_3:Zr^{50}) was the compensation point closely approached. More convenient is compensation by hydrogen as a donor. This element has the advantage of diffusing easily in and out of single-crystal samples even at relatively low temperature.[14,15] The donor concentration can be changed reversibly by changing either p_{H_2} or T; exact compensation is easily achieved, and type conversion can be studied in one sample. The only complication is that proton conduction occurs in addition to the normal ionic and electronic conductivity. Nevertheless, it proved possible to show that in exactly compensated samples σ_e, σ_h, $\sigma_{V_{Al}}$, and σ_{Al} are of the same order of magnitude (Fig. 7). Figure 8 shows values of $(\sigma_{i,I}\sigma_{i,II})^{1/2}$ and $(\sigma_e\sigma_h)^{1/2}$ as $f(T)$. With estimated values of the mobilities, values of disorder constants could be determined.[51] Note that the electronic conductivities are orders of magnitude smaller than those reported by Kitazawa and Coble,[52] indicating that these authors must have misinterpreted their data.

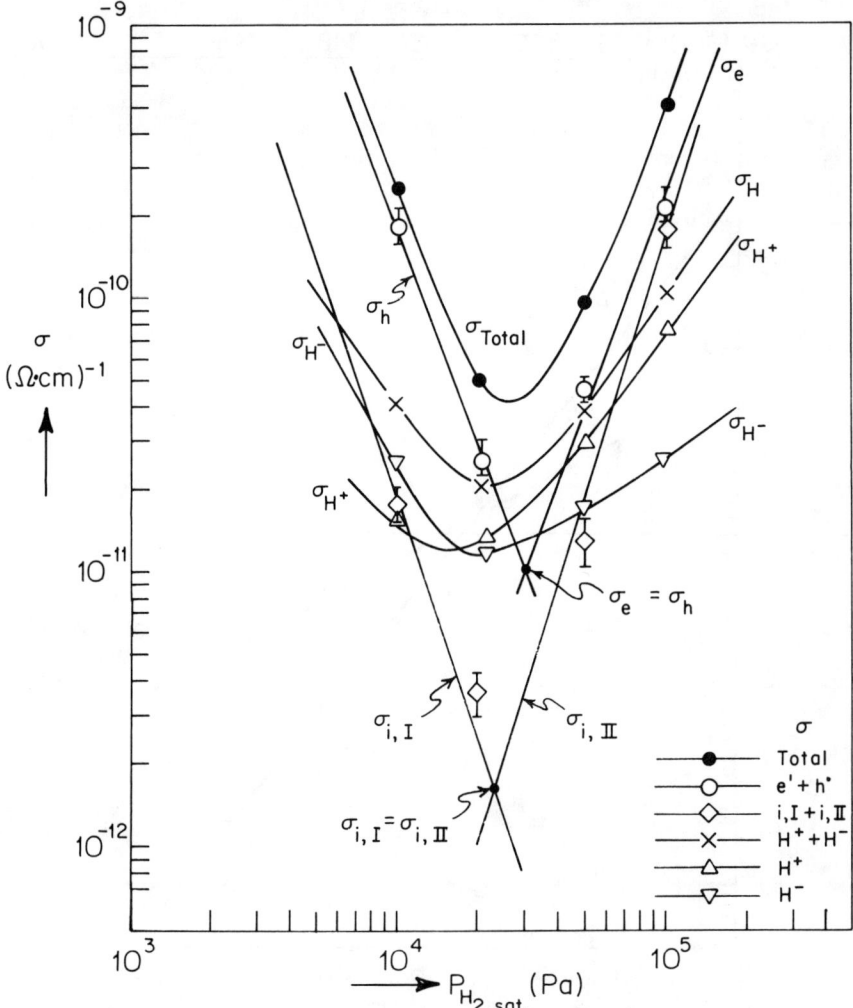

Fig. 7. Conductivity at $T_{measd} = 1450°C$ for Al_2O_3 loaded with hydrogen by annealing at $T_{anneal} = 1350°C$ at various H_2 pressures (Ref. 51).

As seen in Fig. 9, polycrystalline material with grain size ≥ 10 μm shows conductivities identical to those of single crystals. At smaller grain sizes some samples show larger electronic conductivities but smaller ionic conductivities, indicating that grain boundaries may be preferential conduction paths for electrons but not for ions. The fast grain boundary diffusion observed for oxygen therefore must involve neutral species, probably O_i^x.[53] The fact that we do not observe contributions from a large ionic grain boundary conductivity is not in conflict with an observation by Lessing and Gordon,[57] who found that doping with iron changes limitation of creep rate by Al bulk diffusion to limitation by grain boundary diffusion of a species first believed to be oxygen but later identified with aluminum.[24] The latter identification cannot be correct. Since $[Al_i^{...}]_{bulk} \propto [Al_i^{...}]_{gb} \propto [Fe]^n$

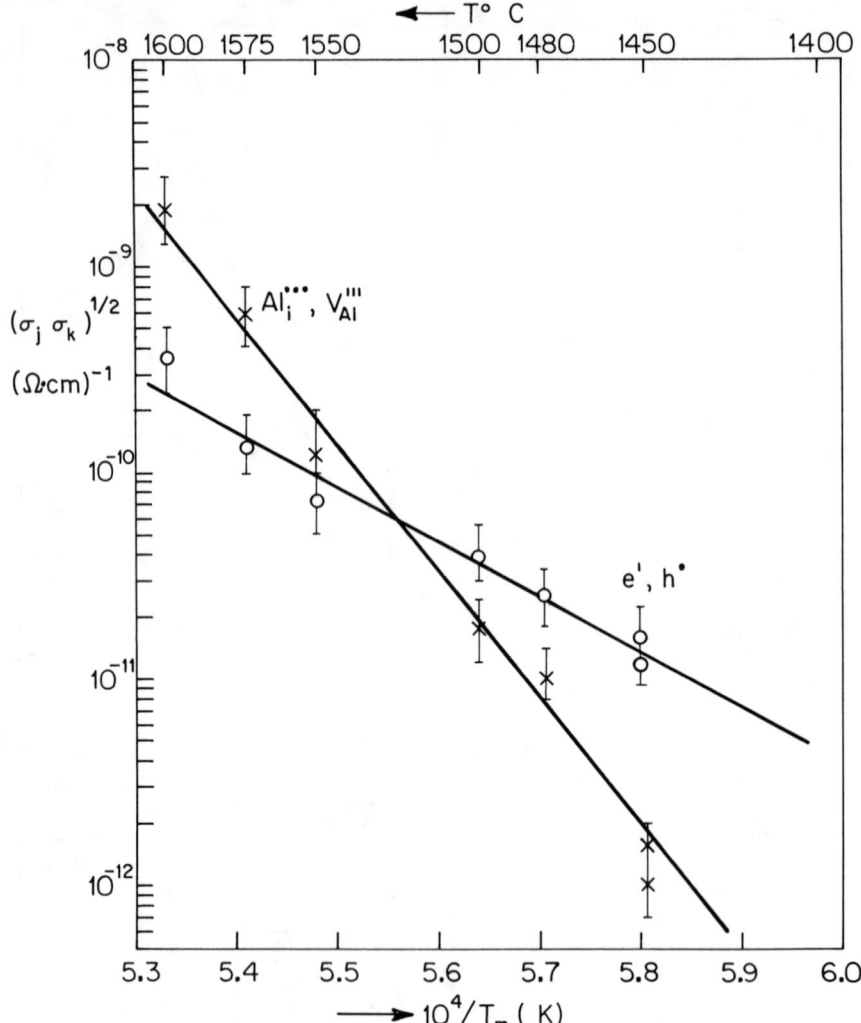

Fig. 8. Values of $(\sigma_{i,\text{I}}\sigma_{i,\text{II}})^{1/2}$ and $(\sigma_e\sigma_h)^{1/2}$ as $f(T)$ (Ref. 51).

with the same value of n for bulk and grain boundary, there is no reason to expect that an increase of [Fe] will cause a change from limitation by bulk diffusion to limitation by grain boundary diffusion. This objection does not hold if the results by Lessing and Gordon are interpreted in the original way as a change from limitation by bulk diffusion of $Al_i^{\cdots}$ to grain boundary diffusion by oxygen. Such a change is to be expected because $[O_i] \propto [Fe]^m$ with $m = 0$ for O_i^x, but $0 < m < n$ even for O_i''—the species with the highest possible charge. For O_i^x as the diffusing species, the increase of the creep rate with increasing [Fe] should be limited to the effect originating from the elimination of the rate limitation by Al; for O_i' or O_i'' with $m > 0$, the increase should continue.

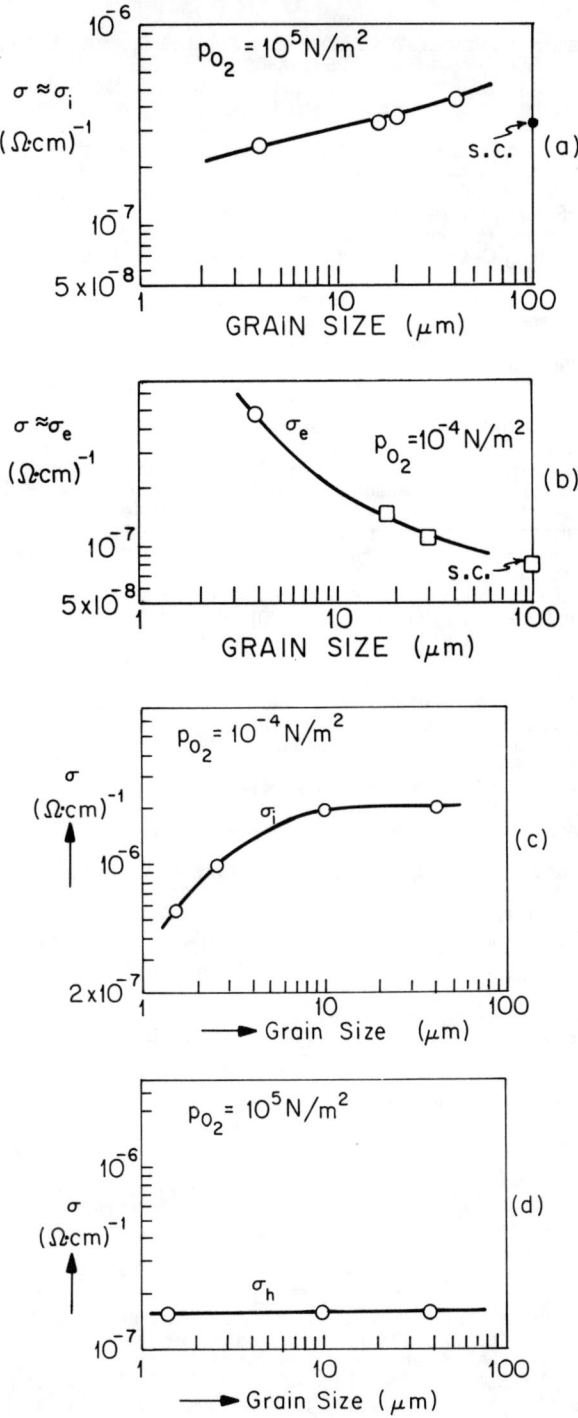

Fig. 9. Ionic and electronic conductivities of polycrystalline Al$_2$O$_3$ doped with Ti (a, b) or Fe (c, d) as a function of grain size.

References

[1] C. R. A. Catlow, R. James, W. C. Mackrodt, and R. F. Stewart, "Defect Energetics in α-Al$_2$O$_3$ and Rutile TiO$_2$," *Phys. Rev. B*, **25** [22] 1006–26 (1982).

[2] D. J. Dienes, D. O. Welch, C. R. Fischer, R. D. Hatcher, O. Lazareth, and M. Samberg, "Shell Model Calculations: Some Point Defect Properties in α-Al$_2$O$_3$," *Phys. Rev. B*, **11** [8] 3060–70 (1975).

[3] W. R. Strehlow and E. L. Cook, "Energy Band Gaps in Elemental and Binary Compound Semiconductors and Insulators," *J. Phys. Chem., Ref. Data* **2** [1] 163–93 (1973).

[4] H. M. Kizilyalli and P. R. Mason, "D.C. and A.C. Electrical Conduction in Single Crystal Alumina," *Phys. Status Solidi A*, **36A** [2] 499–508 (1976).

[5] A. J. Moulson and P. Popper, "Problems Associated with the Measurement of Volume Resistivity of Insulating Ceramics at High Temperature," *Proc. Br. Ceram. Soc.*, **10**, 41–50 (1968).

[6] J. Yee and F. A. Kröger, "Measurements of Electromotive Force in Al$_2$O$_3$—Pitfalls and Results," *J. Am. Ceram. Soc.*, [4] 189–91 (1973).

[7] O. T. Özkan and A. J. Moulson, "Electrical Conductivity of Single Crystal and Polycrystalline Aluminum Oxide," *J. Appl. Phys.*, **3** [6] 983–86 (1971).

[8] R. J. Brook, J. Yee, and F. A. Kröger, "Electrochemical Cells and Electrical Conduction of Pure and Doped Al$_2$O$_3$," *J. Am. Ceram. Soc.*, **54** [9] 444–51 (1971).

[9] S. K. Mohapatra and F. A. Kröger, "Defect Structures of α-Al$_2$O$_3$ Doped with Magnesium," *J. Am. Ceram. Soc.*, **60** [3–4] 141–48 (1977).

[10] H. A. Wang, C. H. Lee, F. A. Kröger, and R. T. Cox, "Point Defects in α-Al$_2$O$_3$:Mg Studied by Electrical Conductivity, Optical Absorption, and EPR," *Phys. Rev. B*, **27** [6] 3821–41 (1983).

[11] J. Pappis and W. D. Kingery, "Electrical Properties of Single and Polycrystalline Alumina at High Temperatures," *J. Am. Ceram. Soc.*, **44** [9] 459–64 (1961).

[12] B. V. Dutt and F. A. Kröger, "High Temperature Defect Structures of Iron-Doped α-Alumina," *J. Am. Ceram. Soc.*, **58** [11–12] 474–76 (1975).

[13] B. V. Dutt, J. P. Hurrell, and F. A. Kröger, "High-Temperature Defect Structure of Cobalt-Doped α-Alumina," *J. Am. Ceram. Soc.*, **58** [9–10] 420–27 (1975).

[14] M. M. El-Aiat and F. A. Kröger, "D.C. Conductivity of α-Al$_2$O$_3$ Doped with Vanadium and Hydrogen," (Chemical Metallurgy—A Tribute to Carl Wagner) *Proc. Symp. Metallurg. Soc. AIME*, 483–92 (1981).

[15] S. K. Mohapatra, S. K. Tiku, and F. A. Kröger, "Defect Structure of Unintentionally Doped α-Al$_2$O$_3$," *J. Am. Ceram. Soc.*, **62** [1–2] 50–57 (1979).

[16] S. K. Mohapatra and F. A. Kröger, "Defect Structure of α-Al$_2$O$_3$ Doped with Titanium," *J. Am. Ceram. Soc.*, **60** [9–10] 381–87 (1977).

[17] M. M. El-Aiat and F. A. Kröger, "Yttrium, an Isoelectronic Donor in α-Al$_2$O$_3$," *J. Am. Ceram. Soc.*, **65** [6] 280–83 (1982).

[18] M. M. El-Aiat and F. A. Kröger, "Hydrogen Donors in α-Al$_2$O$_3$," *J. Appl. Phys.*, **53** [5] 3658–67 (1982).

[19] C. H. Lee and F. A. Kröger, "Electrical Conductivity of Polycrystalline Al$_2$O$_3$ Doped with Silicon"; to be published in *Journal American Ceramic Society*.

[20] H. W. Lehmann and H. Gunthard, "Luminescence and Absorption Studies on Sapphire with Flash Light Excitation," *J. Phys. Chem. Solids*, **25** [9] 941–50 (1964).

[21] K. H. Lee and J. H. Crawford, Jr., "Luminescence of the F Center in Sapphire," *Phys. Rev. B*, **19** [6] 3217–21 (1979).

[22] B. T. Jeffries, R. Gonzales, Y. Chen, and G. P. Summers, "Luminescence in Thermochemically Reduced MgO: The Role of Hydrogen," *Phys. Rev. B*, **25** [3] 2077–80 (1982).

[23] M. M. El-Aiat, L. D. Hou, S. K. Tiku, H. A. Wang, and F. A. Kröger, "High Temperature Conductivity and Creep of Polycrystalline Al$_2$O$_3$ Doped with Fe and/or Ti," *J. Am. Ceram. Soc.*, **64** [3] 174–82 (1981).

[24] Y. Ikuma and R. S. Gordon, "Effect of Doping Simultaneously with Iron and Titanium on the Diffusional Creep of Polycrystalline Al$_2$O$_3$," *J. Am. Ceram. Soc.*, **66** [2] 139–47 (1983).

[25] S. K. Tiku and F. A. Kröger, "Energy Levels of Donor and Acceptor Dopants and Electron and Hole Mobilities in α-Al$_2$O$_3$," *J. Am. Ceram. Soc.*, **63** [1–2] 31–32 (1980).

[26] R. C. Hughes, "Generation, Transport, and Trapping of Excess Charge Carriers in Czochralski-Grown Sapphire," *Phys. Rev. B*, **19** [10] 5318–28 (1979).

[27] S. Dasgupta, "The Hall Effect and Magnetoresistance of Alumina Single Crystals," *J. Phys. D*, **8** [15] 1822–26 (1975).

[28] B. A. Green and M. V. Davis, "Temperature-Dependent Hall Effect Measurements in Alumina," *Trans. Am. Nucl. Soc.*, **16** [1] 75–76 (1973).

[29] E. A. Colbourn and W. C. Mackrodt, "Optical, Electrical, and Polaron Energy Levels in α-Al$_2$O$_3$, "*Solid State Commun.*, **40** [3] 265–67 (1981).

[30] C. J. Delbecq, S. A. Marshall, and P. H. Yuster, "Photo and Thermal Reactions in α-Al$_2$O$_3$:V$_2$O$_3$:Co$_2$O$_3$," *Bull. Am. Phys. Soc.*, **24** [3] 375 (1979); "Optical and Thermal Stimulation of Reactions in α-Al$_2$O$_3$ Containing V^{3+}, V^{4+}, Co^{3+}, and Co^{4+}," *Phys. Status Solidi B*, **99B** [1] 377–86 (1980).

[31] H. H. Tippins, "Charge Transfer Spectra of Transition Metal Ions in Corundum," *Phys. Rev. B*, **1** [1] 126–35 (1970).

[32] D. W. Cooke, H. E. Roberts, and C. Alexander, Jr., "Thermoluminescence and Emission Spectra of UV Grade Al_2O_3 from 90 to 500°K," *J. Appl. Phys.*, **49** [6] 3451–57 (1978).

[33] K. H. Lee, G. E. Holmberg, and J. H. Crawford, Jr., "Hole Centers in γ-irradiated Oxidized Al_2O_3," *Solid State Commun.*, **20** [3] 183–85 (1976).

[34] S. Govinda, "Coloration and Luminescence in α-Al_2O_3 Single Crystals Irradiated with X-rays at Room Temperature," *Phys. Status Solidi A*, **32A** [2] K95–100 (1975).

[35] I. A. Burlyaev, E. B. Perchik, Yu V. Tipunin, and Yu. K. Shalabutov, "Optical and Electrical Properties and the Energy Level Diagrams of Corundum," *Sov. Phys. Solid State*, **11** [5] 1152–53 (1969).

[36] R. W. Klaffky, B. H. Rose, A. W. Goland, and G. J. Dienes, "Reduction-Induced Conductivity of Al_2O_3: Experiment and Theory," *Phys. Rev. B*, **21** [8] 3610–34 (1980).

[37] P. A. Kulis, M. J. Springis, I. A. Tale, and J. A. Valbis, "Recombination Luminescence in Single Crystal Al_2O_3," *Phys. Status Solidi A*, **53A** [1] 113–19 (1979).

[38] A. Niklas and B. Sujak, "Thermoluminescence of a Ruby Crystal Colored by X-rays," *Acta Phys. Polon. A*, **48** [2] 291–305 (1975).

[39] J. D. Brewer, B. T. Jeffries, and G. P. Summers, "Low-Temperature Fluorescence in Sapphire," *Phys. Rev. B*, **22** [10] 1900–6 (1980).

[40] B. T. Jeffries, G. P. Summers, and J. H. Crawford, Jr., "F-Center Fluorescence in Neutron-Bombarded Sapphire," *J. Appl. Phys.*, **61** [7] 3984–86 (1980).

[41] J. B. Blum, H. L. Tuller, and R. L. Coble, "Temperature Dependence of the Iron Acceptor Level in Aluminum Oxide," *J. Am. Ceram. Soc.*, **65** [8] 379–82 (1982).

[42] P. A. Kulis, M. J. Springis, I. A. Tale, V. S. Vainer, and A. Valbis, "Impurity-Associated Colour Centres in Mg- and Cr-Doped Al_2O_3 Single Crystals," *Phys. Status Solidi B*, **104B** [2] 719–25 (1981).

[43] S. Y. La, R. H. Bartram, and R. T. Cox, "Luminescence of the F Center in Sapphire," *J. Phys. Chem. Solids*, **34** [6] 1079–86 (1973).

[44] K. H. Lee and J. H. Crawford, Jr., "Additive Coloration of Sapphire," *Appl. Phys. Lett.*, **33** [4] 273–75 (1978).

[45] S. Kawamura and B. S. H. Royce, "Thermally Stimulated Current Studies of Electrons and Hole Traps in Single Crystal Al_2O_3," *Phys. Status Solidi A*, **50A** [2] 669–77 (1978).

[46] F. A. Kröger, "Experimental and Calculated Values of Defect Parameters and the Defect Structure of α-Al_2O_3"; these proceedings.

[47] R. D. Bagley, I. B. Cutler, and D. L. Johnson, "Effect of TiO_2 on Initial Sintering of Al_2O_3," *J. Am. Ceram. Soc.*, **53** [3] 136–41 (1970).

[48] R. J. Brook, "Effect of TiO_2 on the Initial Sintering of Al_2O_3," *J. Am. Ceram. Soc.*, **55** [2] 114–15 (1972).

[49] F. A. Kröger, "Defect Models for Sintering and Densification of Al_2O_3:Ti and Al_2O_3:Zr," *Journal American Ceramic Society*, **67** [6] 390–92 (1984).

[50] M. P. Harmer and R. J. Brook, "Influence of ZrO_2 Additions on the Kinetics of Hot Pressing of Al_2O_3"; for abstract see *Bull. Am. Ceram. Soc.*, **60** [3] 385 (1981).

[51] M. M. El-Aiat and F. A. Kröger, "Determination of the Parameters of Native Disorder in α-Al_2O_3," *J. Am. Ceram. Soc.*, **65** [3] 162–66 (1982).

[52] K. Kitazawa and R. L. Coble, "Electrical Conduction in Single Crystal and Polycrystalline Al_2O_3 at High Temperature," *J. Am. Ceram. Soc.*, **57** [6] 245–50 (1974).

[53] H. A. Wang and F. A. Kröger, "Chemical Diffusion in Polycrystalline Al_2O_3," *J. Am. Ceram. Soc.*, **63** [11–12] 613–19 (1980).

[54] G. E. Arkhangelskii, Z. L. Morgenshtern, and N. B. Nenstruen, "On the Nature of Color Centres in Ruby," *Phys. Status Solidi*, **29** [2] 831–36 (1968).

[55] M. A. Brown, "The Stabilization of Cr^{5+} in Al_2O_3," *J. Phys. C*, **10** [24] 4939–43 (1977).

[56] S. Choi and T. Takeuchi, "Electronic States of F-type Centers in Oxide Crystals: a New Picture," *Phys. Rev. Lett.*, **50** [19] 1474–77 (1983).

[57] P. A. Lessing and R. S. Gordon, "Creep of Polycrystalline Alumina, Pure and Doped with Transition Metal Impurities," *J. Mater. Sci.*, **12** [11] 2291–2303 (1977).

[58] M. O. Selme and P. Pêcheur, "A Tight Binding Model of the Oxygen Vacancy in $SrTiO_3$," *J. Phys. (C)*, **16** [13] 2559–68 (1983).

---

*Work supported by the United States Department of Energy under Contract No. AS03-76-SF001113, Project Agreement AT03-76 ER 71027.

Polarization–Depolarization Studies of Defects in Oxides

J. H. Crawford, Jr.

University of North Carolina at Chapel Hill
Department of Physics and Astronomy
Chapel Hill, NC 27514

Since its introduction by Bucci and coworkers in the early 1960s, thermally stimulated depolarization (TSD), initially known as ionic thermoconductivity (ITC), has proved a powerful means of investigating defects in solids. Usually confined to dipolar complexes in ionic crystals in which one component can move by thermally activated jumping, it can give information on the magnitude of the dipole moment and the concentration of dipoles as well as the activation energy and the mode of thermally activated reorientation. Such techniques can also be applied to Maxwell–Wagner or interfacial polarization involving ionic or electron (hole) migration within domains or regions of conducting material (precipitate particles), which are embedded in a lower conducting matrix. In the case of MgO and Al_2O_3, these methods have not been extensively exploited. Exceptions are the studies of relaxation of interfacial polarization in regions of high $[Li]^0$ center concentration in MgO:Li crystals quenched after an oxidizing heat treatment and undoped, irradiated MgO crystals. The origin of the TSD peaks in these materials has not been firmly established. The value of such studies and future potential for TSD studies of dipole reorientation in ionic oxides will be discussed.

Perhaps the main bottleneck to unraveling defect-related properties in MgO and Al_2O_3 crystals and ceramics and their homologs is an inadequate knowledge of the parameters associated with defect migrations. Ionic conductivity and diffusion measurements are experimentally difficult because of a variety of reasons, most of which are associated with the inconveniently high temperatures at which they must be carried out. Indeed, the very feature that makes these materials so useful, namely their high-temperature stability, also makes it difficult to investigate their defect-sensitive behavior, and it is quite obvious that any experimental approach that would permit defect motion studies at a lower temperature would be most helpful. Thermally stimulated depolarization (TSD) and thermally stimulated polarization (TSP) are techniques that are very sensitive to the spatial displacement of electrically charged defects and, hence, can give indications of the motion of ionic defects at lower temperatures than those that are required for the more conventional methods. Therefore, this paper is essentially a plea for the exploitation of TSD and TSP techniques in studies of both crystalline and polycrystalline magnesia, alumina, and spinel.

In the early 1960s Bucci and coworkers[1,2] introduced the TSD technique, which they termed ionic thermoconductivity (ITC), a technique that has proved to be an exceptionally powerful one for characterizing dipolar complexes in solids. The experimental method is quite straightforward: An electric field is applied to the solid specimen at a temperature called the polarization temperature (T_p) at which the rate of reorientation of the reorientable dipoles of interest is high. The crystal is then cooled with the field applied to a temperature low enough to freeze in the polarization. The field is then removed, and the crystal is warmed at a linear rate during which the displacement current due to relaxation is monitored. As thermal activation becomes sufficient for reorientation of the dipoles, the displacement current increases exponentially, reaches a maximum at a temperature T_M and then declines to zero as the polarization is exhausted and the dipoles resume their completely random arrangement. The area under this TSD peak is a measure of the polarization, which for a single set of dipoles is given by

$$P = \frac{\mu^2 N_D E}{3kT_p} \qquad (1)$$

where P is the polarization per unit area, μ is the dipole moment, N_D is the dipole concentration, and E is the applied polarizing electric field. Hence, if the dipole moment is known, P yields directly the concentration of dipoles. In the absence of interference from other relaxation peaks, the activation energy E_d for the dipole reorientation can be obtained from the shape of the TSD peak and the value of T_M. Under the assumption of noninteracting dipoles, which is reasonable for dilute dipole solutions, the temperature dependence of I_d, the depolarization current, is given by

$$I_d = \frac{AN_d\mu^2}{3kT_p} \tau^{-1} \exp\left\{-\frac{1}{b}\int_{T_0}^{T} \tau^{-1}(T')\,dT'\right\} \qquad (2)$$

where A is the area of the specimen, b is the heating rate (dT/dt), and

$$\tau = \tau_0 \exp[E_d/kT] \qquad (3)$$

with E_d representing the activation energy and τ_0 the reciprocal frequency factor that includes the entropy of activation. At early times during the heating, the exponential in Eq. (2) is nearly unity and the value of E_d can be obtained in good approximation from the slope of the $\ln I_d$ vs $1/T$ plot. A more precise determination of both E_d and τ_0 can be obtained from a computer fit of the TSD peak to the complete TSD equation. Experimentally, the essential ingredients are a sensitive electrometer for measuring currents as small as 10^{-15} A and a high-impedance specimen chamber with provision for rapid cooling, linear heating rates of a few degrees per minute, and imposition of a polarizing field of a few thousand V/cm. Thermally stimulated release of polarization has the advantage of much greater sensitivity than the AC dielectric relaxation method, which can also be used to examine dipole polarization processes.

The TSD approach has been applied with excellent results to impurity-vacancy and impurity-interstitial dipoles in alkali halides,[1-3] silver halides,[3] alkaline earth fluorides,[4] and more recently to CeO_2 doped with CaO and Y_2O_3.[5,6] Information has been gained on not only the activation energies of dipole reorientation but their structure,[7] e.g., whether the defect occupies the nearest or next-nearest available position, as well. Broadening of the peaks due to interaction of dipoles has been studied in detail,[8,9] and complex dipoles associated with defect clustering have been studied.[10] The vast literature is too extensive to be reviewed here.

Although not so extensively used, TSP can also be a valuable approach. Here the field is applied at low temperature where the dipoles are immobile. A displacement current peak is reached as each set of dipoles is saturated. Finally, the current begins to increase exponentially as ionic or electronic conductivity begins to take over. TSP can help distinguish between various processes, particularly when used in conjunction with TSD. Van Turnhout has discussed the advantages of this combination.[11]

TSD can also be exceedingly valuable in a study of Maxwell–Wagner or interfacial polarization, which comes about when a conducting phase is isolated within a poorly conducting phase. Usually AC conductivity measurements are used to investigate the conductivity of the encapsulated conductor. However, because it is applicable at a much lower temperature and exhibits a high sensitivity, TSD (and TSP) are very helpful here as well. Best results are often obtained where the AC and TSD measurements are used together. For ionic processes, the activation energy of the conductivity E_σ is related to the migration activation energy of the mobile charged species though it may contain an additional term due to the energy of association of the defect (vacancy or interstitial) to a complex, viz,

$$E_\sigma = E_M + E_A \tag{4}$$

in the low-temperature range where association is prevalent. Hence, in polycrystalline specimens for which the grain-boundary material usually has a much lower conductivity than the center of the grains, one can from thermally stimulated relaxation obtain insight into certain types of defect motion.

Normally one would expect that electronic interfacial polarization would be temperature dependent only through the electron or hole mobility and that no frozen-in polarization would be possible. However, if the inclusion is semiconducting instead of a metallic conductor, frozen-in polarization is possible by cooling to a temperature at which all electrons (holes) are condensed out onto donors (acceptors). It turns out that such a situation may occur in MgO:Li[12] and possibly in Al_2O_3:Mg.[13]

In the following paragraphs some recent work in ceria will be cited that is illustrative of what might be experimentally possible for the oxides of interest here. Also, studies of what appear to be semiconducting inclusions in MgO:Li as indicated by TSD relaxation peaks will be discussed, and finally, some work on MgO crystals containing cation vacancies before and after exposure to UV radiation will be considered.

Nowick and coworkers[5,6,14–16] have made extensive studies of the electrical conductivity and TSD processes in CeO_2 doped with Y_2O_3. This oxide crystallizes in the flourite structure, which is rather open, allowing the oxygen vacancies to migrate with a rather small activation energy (~0.7 eV). They have studied polycrystalline specimens with Y^{3+} contents ranging from 0.05 to 6 mol%. The Y^{3+} impurity ions are compensated by oxide vacancies, two for one, to conserve electrical neutrality. Three types of TSD peaks are observed. In the absence of any precautions, the TSD spectrum is dominated by a large peak near $-10°C$ that has an amplitude of $\sim 10^{-9}$ A. This enormous peak (the upper curve, Fig. 1) is far too large to be caused by dipole reorientation. Since this peak parallels the AC conductivity measurement, showing the same trend in activation energy with composition, it is concluded that it is associated with Maxwell–Wagner type relaxation due to conductivity within the grains which is higher than that of the grain-boundary material. They were able to relate the grain conductivity to the value of T_M of the relaxation peak and thereby extend the conductivity measurements downward by 3.5 orders of magnitude on the same $\ln \sigma T$ vs $1/T$ curve, thereby giving increased confidence in the conductivity activation energy of 0.955 eV in the association range, i.e., $E_\sigma = E_M + E_A$.

Since establishment of the interfacial polarization for this large, high-temperature TSD peak requires migration of charge a considerable distance across the grain, a substantial time is required to polarize to saturation, a time which scales with the dielectric relaxation time τ_R given by

$$\tau_R = \frac{\varepsilon\varepsilon_0}{\sigma} \quad (5)$$

where ε_0 is the permittivity of free space and ε is the dielectric constant. Consequently, to look for relaxation processes associated with other processes that occur at much shorter times and lower temperatures, Wang and Nowick[5] were able to show that polarization for short times suppressed the high-temperature conductivity process in favor of two additional relaxation peaks, as can be seen in the middle curve of Fig. 1. A similar effect could be achieved by inserting insulating

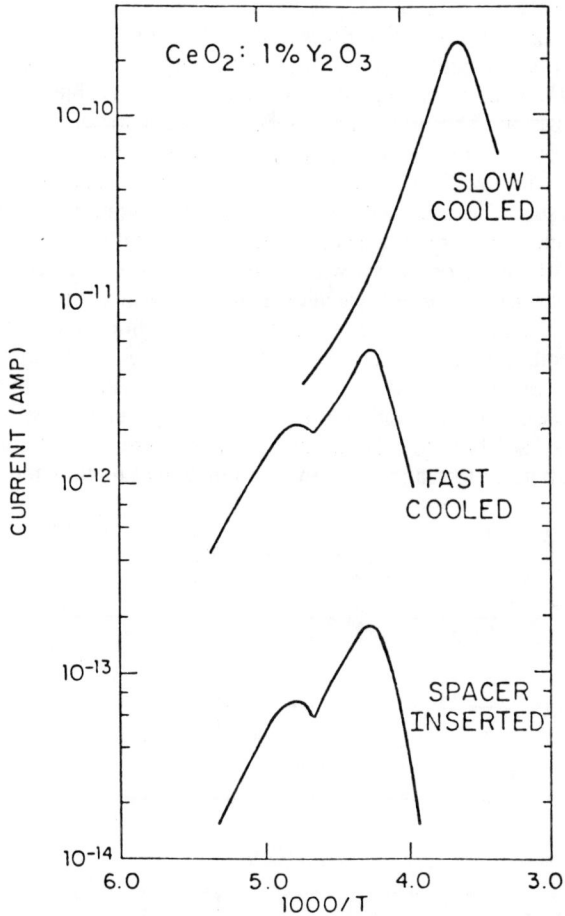

Fig. 1. Comparison of TSDC peaks obtained by three methods for a CeO_2:1% Y_2O_3 sample. The upper curve is obtained by the conventional method, middle curve by fast cooling, and lowest curve by insertion of a dielectric spacer. Polarization in each case was at room temperature (after Wang and Nowick, Ref. 6).

spacers between the specimen and the electrodes, since this arrangement also reduced long-range charge transport and suppressed the high-temperature peak, as shown by the bottom curve of Fig. 1. The question then arises as to the cause of these two peaks at lower temperature. The lowest temperature peak (peak 1) has a maximum at $T_M = -64°C$ which is independent of Y_2O_3 content. The analysis of the peak yields $E_d = 0.64$ eV and $\tau_0 = 2 \times 10^{-14}$ s with a single relaxation governed by these parameters giving an acceptable fit to the data. A similar relaxation with these parameters was obtained for CaO-doped CeO_2. Thus, it is concluded that this relaxation is associated with the reorientation of the Y^{3+}–oxide vacancy dipolar complex, which should have a reciprocal frequency factor of this magnitude.

The next highest peak, peak 2, exhibits a T_M that is displaced toward lower temperatures with increasing Y^{3+} concentration up to 2.5% but moves to higher values above 2.5%. At 2.5%, peak 1 disappears and only peak 2 is visible. This peak is completely missing in CeO_2 doped with CaO. The shift in T_M corresponds to the increase in σ up to 2.5% followed by a decline at higher values, which suggests that this relaxation is associated with long-range migration of a defect, presumably an oxygen vacancy, which at the same time is able to relax a configurational polarization. The model proposed to account for these observations is a random array of Y^{3+} ions on Ce^{4+} sites, half of which are associated with O^{2-} vacancies. To minimize the Coulomb energy, each Y^{3+}–oxygen vacancy complex would tend to surround itself with Y^{3+} ions and the situation is reversed for each isolated Y^{3+} ion, approximating a super lattice. Polarization would come about by the formation of *wrong pairs*, shown schematically in Fig. 2, and it would relax by the migration of an oxygen vacancy from the wrongly placed complex to a neighboring Y^{3+} ion, thus erasing that unit of polarization. In essence, the process erasing local polarization is the process that is responsible for charge transport over long distances. Such a process is excluded from CeO_2 doped with Ca^{2+} since the simplest complex is neutral rather than singly negative, as in Y^{3+}-doped CeO_2.

Let us now see how these experimental approaches might apply to MgO or Al_2O_3. It is important to realize that energies of defect motion and interaction are

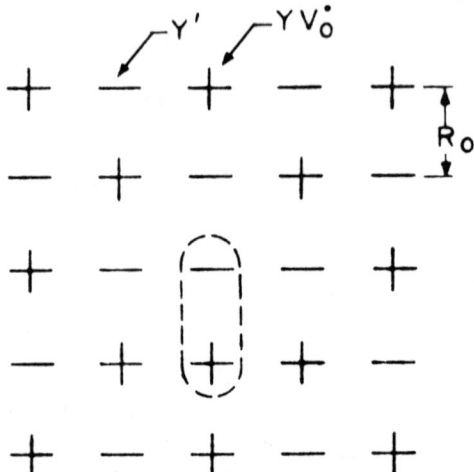

Fig. 2. Model of an array of (YV$\ddot{\text{o}}$) and Y′ defects showing a *wrong pair* (after Wang and Nowick, Ref. 6).

expected to be larger in magnesia and alumina than in ceria because of the closer packing and smaller dielectric constants (~10 vs ~25) in the former. Hence, the temperatures at which relaxation times for similar processes would be comparable would also be higher in the former materials. This would entail a certain loss of sensitivity for TSD, since the saturation polarization of a set of dipoles for a given applied field is inversely proportional to temperature. Even so, however, one would expect a measurable effect due to dipolar complex reorientation, e.g., Li–O^{2-} vacancies in MgO, Mg–O^{2-} vacancies in Al_2O_3, or perhaps Fe^{3+}–Mg^{2+} vacancies in MgO, at T_m's well above room temperature. Any difficulty would likely be associated with maintaining the necessary high-impedance isolation of the specimen for measurement purposes rather than maintaining the magnitude of the effect itself. Other candidates for study would be the various hydrogen-containing centers in MgO and Al_2O_3, several of which are dipolar in character, and here the peak temperatures would be expected to fall well below room temperature.

In order for any *wrong-pair* relaxation domains to exist, incomplete compensation of defect charge by an impurity is necessary; i.e., two (or more) impurity ions are needed to compensate the charge associated with a vacancy. This condition obtains in MgO for Li^+ and in Al_2O_3 for Li^+ or Mg^{2+}. Therefore, the solubility limit, either the quenched-in or the equilibrium, of these impurities might be explored by searching for *wrong-pair* TSD peaks. Since both dipole reorientation and *wrong-pair* relaxation do not require single crystals in order to be observed, some of the more restrictive problems associated with specimens with desired dopings are eased. Indeed, it is even possible in principle to make such measurements on polycrystalline ceramics. At issue, of course, is the migration energy of the oxygen vacancy with attendant complications associated with charge state. Thus it would be valuable to study specimens after various reducing treatments. A point to remember is that Pells[17] reports that the activation energy for F_2 center formation (two coupled anion vacancies in an unknown charge state) from isolated F centers (anion vacancies containing two electrons) is only 0.6 eV. Hence, this relaxation may occur at rather low temperatures under appropriate conditions. Another point to remember is that the isolated oxygen vacancy charge state itself is uncertain, since Choi[18] has pointed out that removing an O^- ion from an O^{2-} site leaves the remaining electron more tightly bound than when the site is occupied. Such a situation should have important implications for oxide ion vacancy-impurity complex formation in these materials.

Another important potential application of TSD involves Maxwell–Wagner polarization associated with ionic conductance in polycrystalline specimens of MgO, Al_2O_3, and $MgAl_2O_4$. The great sensitivity of this method would permit measurements of activation energy for conduction at temperatures well below the usual range with a considerable gain in experimental convenience.

We now turn attention to electronic relaxation processes. Anneals of either MgO:Li[19] or Al_2O_3:Mg crystals[13,20] in an oxidizing atmosphere followed by rapid cooling result in the introduction of [Li]^0 or [Mg]^0 centers (Li_{Mg}^x or Mg_{Al}^x in the Kröger and Vink notation). Studies of charge release from these centers, created in unoxidized MgO:Li crystals by X rays or γ rays well below room temperature, indicate that their thermal ionization energy as acceptors, e.g., [Li]$^0 \rightarrow$ [Li]$^-$ + hole, is 0.6–0.7 eV. TSD measurements carried out on MgO:Li crystals[12,21] reveal a very large peak located at 250 K when polarized in this vicinity and a second smaller peak when polarized at a higher temperature (Fig. 3). The 250 K peak is much too large to be attributed to dipole reorientation. The relaxation activation energy is 0.69 eV, which corresponds to the expected value of [Li]^0 center

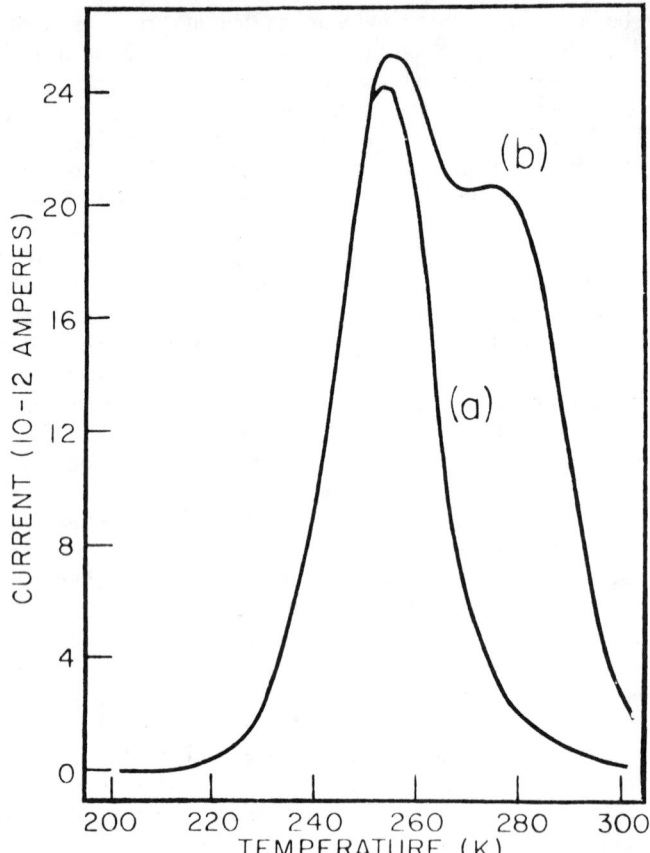

Fig. 3. TSD curve for MgO:Li annealed in static air (a) polarized at 250 K for 2000 s and (b) polarized at 280 K for 180 s (after D. J. Eisenberg, Ref. 21).

ionization energy. Therefore, it is reasonable to conclude that the large peak is associated with interfacial polarization involving holes, with the temperature dependence entering through thermally activated hole release. The magnitude of the effect is too great to be associated with electrode polarization or even spherical semiconducting regions. Analysis of the dielectric loss curves suggests that the conducting regions are oriented filaments such as might result from high [Li]^0 concentrations along edge dislocations.[22] It should be pointed out that more recent measurements of the dielectric behavior in MgO:Li crystals[23] cast doubt on this conclusion, and the actual distributions of [Li]^0 acceptor are still very much up in the air. Nevertheless, some type of isolation of conducting regions must be involved, since measurement of the current–voltage characteristic[24] also shows that the conductivity is not a uniform bulk effect. Indeed, the I–V curve is highly nonlinear (Fig. 4). In fact, one has here a high-voltage switch with switching occurring at ±8000 V/cm. A possible interpretation of these results is field-assisted tunneling from one filamentary semiconducting inclusion to another.

In another study of MgO, Krokoszinski et al.[25] examined the effect of UV irradiation on the polarization behavior. Before irradiation, they observed several

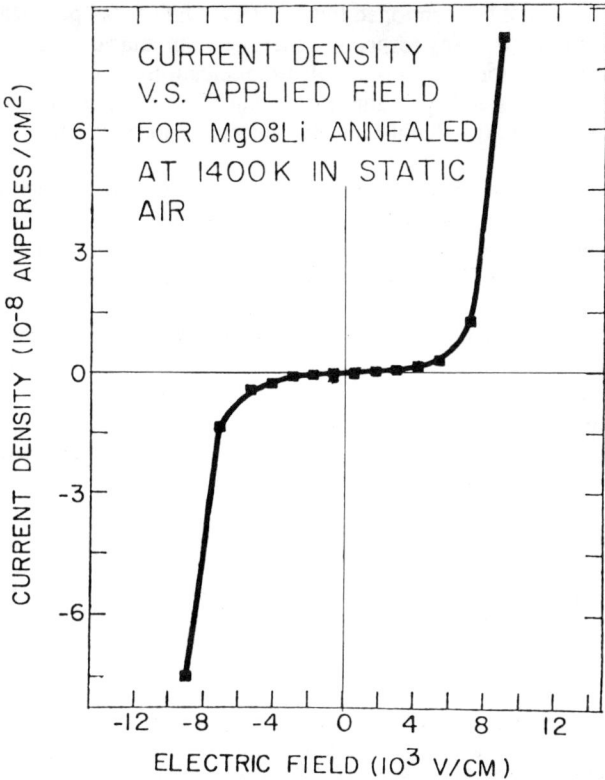

Fig. 4. *I–V* characteristic of MgO:Li annealed in static air (after D. J. Eisenberg, Ref. 21).

weak low-temperature TSD peaks (200–250 K) that appeared to behave as dipole relaxation peaks and one large high-temperature (~400 K) peak that had characteristics expected for interfacial polarization. The large high-temperature peak was insensitive to UV irradiation, which incidentally produced *V*-band coloration. The low-temperature peaks, on the other hand, were much affected, one of them increasing by a factor of 100. This enhancement decayed with *V*-band coloration, leading to the conclusion that the low-temperature peaks must be associated in some way with one of the *V* centers present.

Although the models invoked to account for these observations are speculative, the results are real and indicate that TSD together with other electrical measurements is a potentially valuable experimental approach to a study of impurity centers of variable charge states and electronic conductivity in these ionic oxides. In particular, it would be of great interest to learn whether Al_2O_3:Mg, about which we have heard at this conference,[13,25] exhibits the same relaxational characteristics and whether other alkali metal ions in MgO behave the same way as Li^+. The possibility of a high-field switch is itself quite interesting and suggests certain applications. Therefore, a deeper study of these various processes is warranted. For example, if dislocations are indeed involved in the formation of semiconducting domains, deformation by bending the crystals around appropriate crystallographic axes could result in dislocation arrangements, which might give rise to controlled, anisotropic *I–V* characteristics.

Finally, it should be mentioned that electronic interfacial polarization, and its relaxation, is similar in many respects to thermally stimulated conductivity (TSC), a technique that monitors the thermal release of electrons from traps.[26,27] The same experimental approach and apparatus are used in TSC as in TSP. This experimental approach can also yield valuable insights into the nature of defect and impurity states in oxides.

In summary, it appears from work on other oxides as well as MgO:Li and UV-irradiated MgO that thermally stimulated reorientation of dipoles and charge release from impurity centers in alumina, magnesia, and spinel can be investigated effectively by TSD and TSP methods, which have been applied effectively in other ionic systems. An obvious advantage of this approach for these high-temperature materials derives from the greater sensitivity to relaxation processes at lower temperatures than that of the conventional AC or DC methods. Finally, such techniques may be applied to processes involving charge release from defect centers where such centers act as donors or acceptors. Hence, TSD and TSP may prove to be very effective tools in studying defect and carrier kinetics in ionic oxide crystals and polycrystalline specimens.

Acknowledgment

The author acknowledges support from the Department of Energy under Contract No. DE-A505-78ER5866.

References

[1] C. Bucci and R. Fieschi, *Phys. Rev. Lett.,* **12**, 16 (1964).
[2] C. Bucci, R. Fieschi, and G. Guidi, *Phys. Rev.,* **148**, 816 (1966).
[3] For a review of early TSD results on alkali halides and silver halides, see: J. H. Crawford and L. M. Slifkin, *Ann. Rev. Mat. Sci.,* **1**, 139 (1971).
[4] For a review of early work on alkaline-earth fluorides see the review of: J. H. Crawford, "Impurity-Defect Dipole Reorientation in Alkaline Earth Fluorides"; p. 26 in the Symposium on Physics of Condensed Matter. Edited by E. Laredo. Universidad Simon Bolivar, Caracas, Venezuela, 1980.
[5] D. Y. Wang and A. S. Nowick, *Phys. Status Solidi A,* **A73**, 165 (1982).
[6] D. Y. Wang and A. S. Nowick, *J. Phys. Chem. Solids,* in press.
[7] G. E. Matthews and J. H. Crawford, *Phys. Rev. B,* **B15**, 55 (1977).
[8] W. van Weperen and H. W. den Hartog, *Phys. Rev. B,* **B18**, 2857 (1978).
[9] E. Laredo, M. Puma, and D. R. Figueroa, *Phys. Rev. B,* **B19**, 2224 (1979).
[10] R. Cappelletti, E. Okuna, and J. H. Crawford, *Phys. Status Solidi,* **47**, 617 (1978).
[11] J. Van Turnhout, "Thermally Stimulated Discharge of Electrets"; in Electrets. Edited by G. M. Sessler. Springer-Verlag, Berlin, 1978.
[12] D. J. Eisenberg, L. S. Cain, K. H. Lee, and J. H. Crawford, *Appl. Phys. Lett.,* **33**, 479 (1978).
[13] H. A. Wang, C. H. Lee, F. A. Kröger, and R. T. Cox, *Phys. Rev. B,* **B27**, 3821 (1983).
[14] H. L. Tuller and A. S. Nowick, *J. Electrochem. Soc.,* **122**, 255 (1975).
[15] D. Y. Wang, D. S. Park, J. Griffith, and A. S. Nowick, *Solid State Ionics,* **2**, 95 (1981).
[16] D. Y. Wang and A. S. Nowick, *Solid State Ionics,* **5**, 551 (1981).
[17] A. Y. Stathopoulos and G. P. Pells, *Philos. Mag, [Part] A,* **A47**, 381 (1983).
[18] S. I. Choi and T. Takeuchi, Phys. Rev. Lett., **50**, 1474 (1983).
[19] Y. Chen, H. T. Tohver, J. Narayan, and M. M. Abraham, *Phys. Rev. B,* **B16**, 5535 (1977).
[20] R. T. Cox, *J. Phys. Colloq.,* **34** [C-9] 333 (1973).
[21] D. J. Eisenberg; Ph.D. Thesis, University of North Carolina, Chapel Hill, NC; unpublished work.
[22] J. H. Crawford and D. J. Eisenberg, *J. Physique, Colloq.,* **41** [C-6] 394 (1980).
[23] M. Puma, A. Lorincz, J. F. Andrews, and J. H. Crawford, *J. Appl. Phys.,* **53**, 4546 (1982).
[24] Y. Chen, R. H. Kerrohan, J. L. Boldu, M. M. Abraham, D. J. Eisenberg, and J. H. Crawford, *Solid State Commun.,* **33**, 441 (1980).
[25] H. J. Krokoszinski, K. Bärner, and F. Freund, *J. Chem. Phys.,* **72**, 2616 (1980).
[26] F. A. Kröger; this conference.
[27] W. Mallard and J. H. Crawford, *J. Appl. Phys.,* **43**, 2060 (1972).
[28] G. S. White, K. H. Lee, and J. H. Crawford, *Semicond. Insul.,* **5**, 123 (1980).

Luminescence and Photoconductivity in MgO and α-Al$_2$O$_3$ Crystals

G. P. Summers

Oklahoma State University
Department of Physics
Stillwater, OK 74078

Elucidation of several unsolved problems in luminescence and charge motion in MgO and α-Al$_2$O$_3$ has come about because of a study of the effects of thermochemical reduction at high temperatures. Evidence is presented that substitutional negative hydrogen ions, i.e., [H]$^+$ centers, which are found to be thermally stable to high temperatures in thermochemically reduced MgO, act as electron traps and determine the lifetime of the 2.3-eV F center luminescence near room temperature. [H]$^+$ ions also determine the spectral dependence of the luminescence in MgO:Mg over the temperature range 6–300 K and enable infrared and red radiation to excite 2.3-eV luminescence at low temperatures. By analogy with the results for MgO, it is tentatively proposed that the same electron trap enables the interconversion of different charge states of anion vacancies in α-Al$_2$O$_3$ to occur. Photoconductivity in reduced α-Al$_2$O$_3$ over the temperature range 10–300 K is described, and the effect of C_2 site symmetry on the luminescence of anion vacancies in α-Al$_2$O$_3$ is discussed. Photoconductivity from positively charged anion vacancies, F$^+$ centers, is discussed in relation to recent theoretical results.

Although luminescence from anion vacancies in alkaline earth oxides[1–10] and sapphire[11–15] has been studied for many years, one problem that has remained unsolved until recently was the cause of the long lifetime observed near room temperature for the F center in thermochemically reduced samples. An F center consists of an oxygen ion vacancy that has trapped two electrons. In sapphire, this problem has not yet been completely resolved, although by analogy with MgO and CaO we can infer the cause quite confidently.[16] We now have very strong evidence that dissolved hydrogen ions, which are introduced by very high-temperature reduction of the sample, are intimately involved in determining the details of the luminescence spectra observed at temperatures in the range 4–400 K,[17–19] and thus the motion of charge at all temperatures.

Protons are present in alkaline earth oxides under normal circumstances as OH$^-$ ions, which can be detected by their infrared absorptions in the 3000–4000-cm^{-1} region.[1] However, when MgO or CaO crystals are thermochemically reduced at high temperature, protons appear in the anion sublattice and these give rise to local mode absorptions in the region of $\sim$1000 cm^{-1}. These modes have been attributed to substitutional negative hydrogen ions[20] (protons surrounded by two electrons at oxygen ion sites), which we refer to as [H]$^+$ ions, because they have a net positive charge with respect to the lattice.

In MgO, both the F$^+$ (anion vacancy with one electron) and F center absorptions occur at 5.0 eV (250 nm).[2] When a sample of thermochemically reduced MgO is illuminated with 5.0-eV light, two luminescence bands are possible.[7,21]

One of these is at 3.2 eV and decays very rapidly once the excitation is removed. This band has been assigned to a $^2T_{1u} \rightarrow\ ^2A_{1g}$ transition of the F^+ center. The other band is at 2.3 eV and has a lifetime in the temperature range 5–300 K that is sample dependent and can range from a fraction of a second to several minutes. In α-Al$_2$O$_3$, the F^+ center has two distinct absorption bands at 4.8 and 5.4 eV. A third band near 6.0 eV is observed in the excitation spectrum of the 3.8-eV F^+ center luminescence.[12,13,22] The lifetime of the luminescence is approximately 7 ns.[12] The F center has its main absorption band at 6.1 eV.[13,14] Excitation into this band produces a luminescence at 3.0 eV,[11,13,14] which, like MgO, has a sample dependent lifetime near room temperature.[15] The lifetime can range from the intrinsic value of 34 ms to many seconds. At temperatures in the 5–50 K range, the lifetime varies in a way that reflects the site symmetry.[15]

Excitation of the F center in α-Al$_2$O$_3$ produces photoconductivity which is largest at low temperature. The photoconductivity is strongly temperature dependent in samples showing long phosphorescence near room temperatures, due to the presence of an electron trap that probably is [H]$^+$ ions. Photoconductivity from the F center in MgO is not easy to detect[23,24] in thermochemically reduced samples, although other experimental and theoretical results suggest that it should occur even at low temperatures. Also, surface and electrode effects are problematical. F center photoconductivity is readily observable in CaO,[25] however. More surprisingly, because of the positive charge with respect to the lattice, F^+ centers in both CaO[25] and SrO[26] show photoconductivity. These results suggest that the F^+ center photoconductivity in MgO should be observable above room temperature, but so far as we know, this has not yet been detected. Recent theoretical results by Choi,[27] which are reported at this conference, indicate that photoconductivity should be observed from F^+ centers in α-Al$_2$O$_3$. We have some evidence that this is indeed the case,[16] although more work is needed in this area too.

In this article we first briefly review the evidence for the existence of [H]$^+$ ions in MgO. Then we give an interpretation of the effect these ions have on the spectral dependence and lifetime of the F center luminescence in MgO, including infrared conversion effects. The temperature dependence of F center luminescence and photoconductivity in α-Al$_2$O$_3$ is reviewed, and the photoconductivity from F^+ centers is discussed.

Substitutional [H]$^+$ Ions in MgO:Mg

The thermochemically reduced samples of alkaline-earth oxides which are discussed below, were prepared at Oak Ridge National Laboratory by Y. Chen. The main evidence for the presence of substitutional [H]$^+$ ions has been presented elsewhere.[20] A brief review is given here as an introduction to the luminescence and photoconductivity results.

Hydrogen occurs in MgO and presumably other oxide crystals because the starting powder from which the crystals are grown invariably contains a certain amount of hydroxide. This is certainly the case for CaO. A large variation in hydrogen content is possible, depending on how damp the powder becomes before the crystals are grown. For example, high concentrations of hydrogen can be produced by presoaking MgO powder in water. Alkaline earth oxide crystals are usually grown by arc fusion and often appear cloudy due to the presence of cavities of hydrogen gas,[28] which provide a ready source of hydrogen ions in subsequent thermochemical reduction.

In the case of α-Al$_2$O$_3$, crystals of which are usually grown by the Czochralski or variations of the Bridgman process, reducing conditions are often present during growth. The possibility of hydrogen ions being formed from hydroxyl ions in the starting material obviously exists, although less is known directly than in the case of MgO.

The reduction process in oxides has been commonly called "additive coloration" due to its apparent similarity to the process used for alkali halides. In the case of oxides, however, it is more aptly called "subtractive coloration"[29] since the experimental evidence is that the stoichiometric imbalance occurs by the removal of oxygen from the sample rather than the absorption of cations from the vapor. The samples discussed here were heated to a temperature between 2000 and 2400 K under several atmospheres of cation vapor in a tantalum bomb, before being quenched to room temperature. In the case of α-Al_2O_3 crystals, these were colored during growth and were used as received from the manufacturer.

Figure 1(a) shows the infrared absorption spectrum at room temperature of a crystal of thermochemically reduced MgO (MgO:Mg), which had been treated to contain a large concentration of $[H]^+$ ions. Three fundamental absorptions are observed at 1053, 1032, and 1024 cm.$^{-1}$ The sample had been heated to 1900 K to remove anion vacancies, which also indicates the high thermal stability of $[H]^+$ ions. The ions can be moved by irradiating the sample with high-energy electrons. Figure 1(b,c) shows the effect of cumulative doses up to 2×10^{17} electrons·cm^{-2}. At the same time the anion vacancy absorption at 5.0 eV increased, and new infrared absorptions in the region of 3400 and 2400 cm^{-1} appeared. Similar results are obtained for CaO. Figure 2 shows the effect of an electron flux of $\sim 3 \times 10^{17}$ electrons·cm^{-2} on the absorptions of F and F^+ centers in CaO.

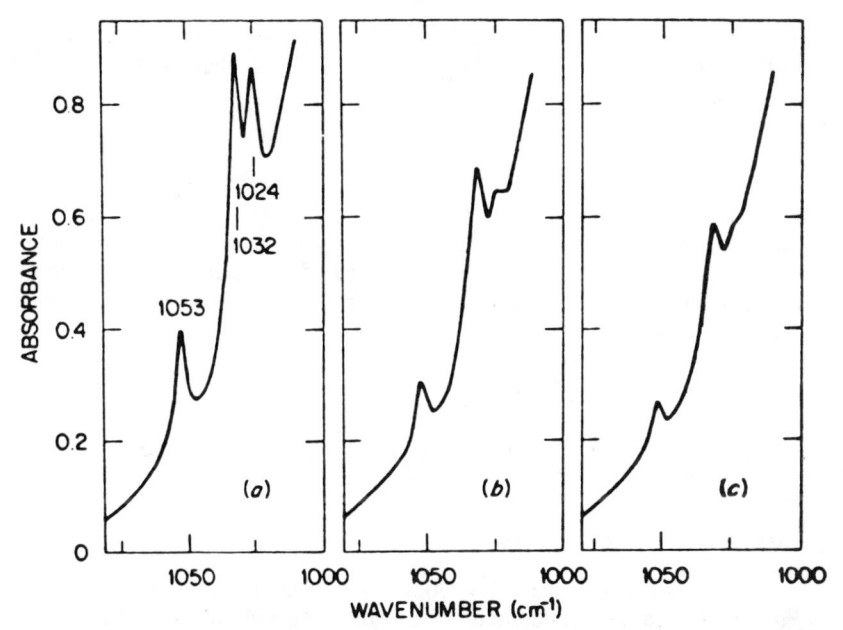

Fig. 1. Infrared absorption spectra for $[H]^+$ ions in MgO:Mg prior to electron irradiation (curve a) and after irradiations with cumulative doses of 5×10^{16} electrons·cm^{-2} (curve b) and 2×10^{17} electrons·cm^{-2} (curve c).

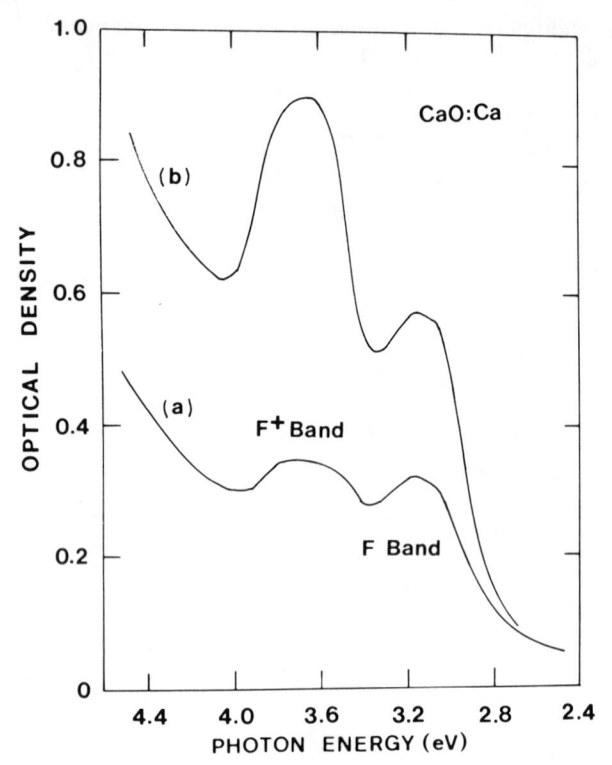

Fig. 2. Optical absorption spectra for anion vacancies in CaO:Ca after heating to 1773 K for 1 h (curve a) and after irradiation with ~3 × 10^17 electrons·cm^{-2} (curve b).

These results enable us to draw several conclusions. First, the [H]$^+$ ions are located at oxygen ion sites. Second, the cross section for displacement by an incident electron is large. We estimate it to be ~1 × 10^8 barns in the case of [H]$^+$ ions in MgO and ~2 × 10^6 barns in CaO. Third, we can derive a formula by which the concentration of [H]$^+$ ions, $n(\text{H}^+)$, can be estimated from the local mode absorption. In MgO we find

$$n(\text{H}^+) = 5.5 \times 10^{16} \sum \alpha(\text{H}^+) \tag{1}$$

where $\sum \alpha(\text{H}^+)$ is the sum of the peak absorption coefficients at 1053, 1032, and 1024 cm^{-1}; in CaO we find

$$n(\text{H}^+) = 1.2 \times 10^{16} \sum \alpha(\text{H}^+) \tag{2}$$

where the sum is over the peak absorption coefficients at 911 and 880 cm^{-1}.

Following high-temperature anneals of MgO:Mg and CaO:Ca, there were no optical absorptions in the region 0.5–6.0 eV which could be attributed to an electronic transition of [H]$^+$ ions. This result suggests that either the oscillator strength is unexpectedly small (<10^{-3}) or no transition of the [H]$^+$ ion occurs in this region. Wilson[30] has recently calculated that the main electronic transition of the $[H]^+$ ion in MgO occurs from a ground state that is depressed into the valence band and that may be as large as 9.6 eV.

Luminescence in MgO:Mg

In this section evidence is presented that substitutional $[H]^+$ ions in MgO:Mg are the main electron traps responsible for (i) the spectral dependence of the luminescence between 6 and 300 K, (ii) the lifetime of the 2.3-eV F center luminescence near room temperature, and (iii) infrared conversion effects below 250 K. Most of the data discussed here were obtained from four samples of MgO:Mg that contained a representative range of concentrations of anion vacancies and $[H]^+$ ions. These samples are labeled MgO-I, MgO-II, MgO-III, and MgO-IV. Their characteristics are given in Table I.

Table I shows that MgO-II contained a large concentration of anion vacancies (4.1×10^{18} cm^{-3}) and a relatively small concentration of $[H]^+$ ions (1.3×10^{17} cm^{-3}). Figure 3 shows the luminescence spectrum excited in this sample by 5.4-eV light over the temperature range 5–290 K. Because the 3.2-eV F^+ band is weak, the shape of the 2.3-eV F band can be seen at all temperatures without further analysis. The 2.3-eV band is slightly asymmetric with a half-width at 5 K of 0.68 eV, which increases to 0.72 eV at 290 K. The peak position does not change significantly between 5 and 290 K. Figure 3 also shows the excitation spectrum of the 2.3-eV band. This band has a peak position of 5.0 eV and a half-width at 77 K of 0.77 eV. One other characteristic of samples with a much lower concentration of $[H]^+$ ions than that of anion vacancies is that the intensity of the 3.2-eV band does not vary between 100 and 300 K. In contrast to Fig. 3, Fig. 4 shows the luminescence spectrum of MgO-I, which contained a large concentration of both $[H]^+$ ions and anion vacancies. At low temperatures, the 2.3-eV band is small and the 3.2-eV band is large. When the temperature increased above 200 K, the intensity of the 2.3-eV band rapidly increased, while the 3.2-eV band decreased in an approximately one-to-one proportion.

The results of Figs. 3 and 4 can be best understood by a consideration of the thermoluminescence glow curves of MgO:Mg. Figure 5 shows these curves for MgO-I and MgO-III. The luminescence at 40, 260, and 470 K is similar to the 2.3-eV photoluminescence band (Fig. 3). Because of its correlation with the $[H]^+$ ion concentration in the samples, we attribute the 260 K peak to release of electrons trapped at $[H]^0$ ions, i.e., $[H]^+$ ions that have trapped electrons photoexcited from F centers. Below 260 K, these electrons are permanently trapped. Therefore, when a sample such as MgO-I, which contains comparable concentrations of $[H]^+$

Table I. Characteristics of Thermochemically Reduced MgO Samples

Sample	α_F (cm^{-1})*	n_F (cm^{-3})†	α_H (cm^{-1})‡	n_H (cm^{-3})§
MgO-I	330	1.6×10^{18}	11	3.0×10^{18}
MgO-II	820	4.1×10^{18}	0.50	1.3×10^{17}
MgO-III	110	5.5×10^{17}	0.20	5.4×10^{16}
MgO-IV	750	3.7×10^{18}	0.60	1.6×10^{17}

*Absorption coefficient at 5.0 eV.
†Calculated from $n_F = 5.0 \times 10^{15} \alpha_F$ (see Ref. 18).
‡Absorption coefficient at 1053 cm^{-1}.
§Calculated from $n_H = 2.7 \times 10^{17} \alpha_H$ where α_H is the absorption coefficient at 1053 cm^{-1} (see Ref. 18).

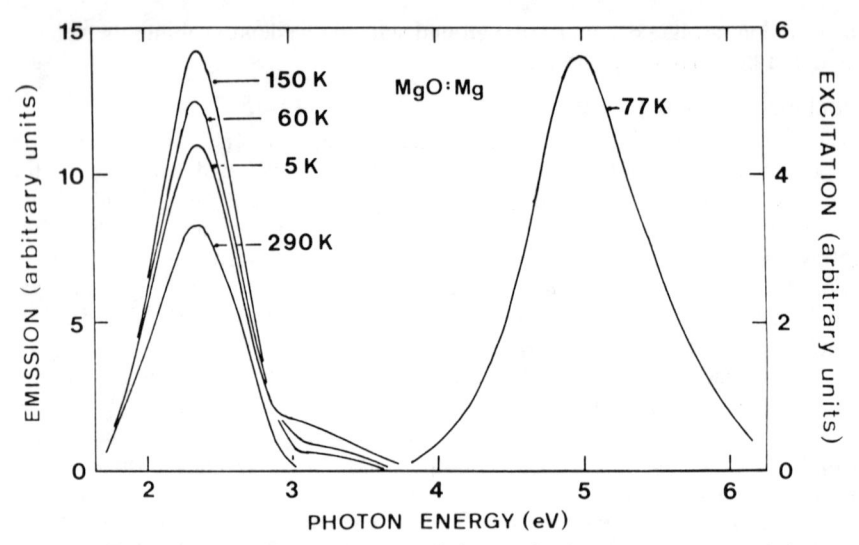

Fig. 3. Temperature dependence of the emission spectrum and the excitation spectrum of the 2.3-eV emission band at 77 K for sample MgO-II (Table I). This sample contained a large concentration of anion vacancies and a much smaller concentration of [H]$^+$ ions. For the emission spectra, E_{ex} = 5.4 eV, and for the excitation spectrum, 2.1 < E_{em} < 2.6 eV.

ions and anion vacancies, is excited by 5.4-eV light at low temperatures, many F centers are converted to F^+ centers and the luminescence occurs mainly at 3.2 eV. As the temperature increases above ~200 K, the transferred electrons are able to return to the anion vacancies, causing mainly 2.3-eV F center luminescence. In a sample such as MgO-II in which the anion vacancy concentration is dominant, few excited electrons are trapped and the luminescence is expected to occur mainly from F centers at all temperatures (Fig. 3).

Evidence is now presented that the concentration of [H]$^+$ ions in a sample determines the lifetime of the 2.3-eV luminescence at 260 K. First, we show a correlation between the concentration of [H]$^+$ ions and this lifetime in a series of different samples, and second, we show how the lifetime can be varied by changing the F center to [H]$^+$ ion concentrations in a single sample. Figure 6 shows the luminescence decay of the F band in MgO-I, MgO-II, and MgO-III. In these samples, the [H]$^+$ ion concentration decreases from sample I to sample III, Table I, and the lifetime decreases correspondingly. The [H]$^+$ ion concentration in MgO-III is about 60 times smaller than in MgO-I. We find that, in general, if the ratio of the concentrations [H]$^+$/F center is about unity or larger, a luminescence lifetime in excess of 200 s can be expected, where we have taken the time over which the intensity fills to 0.1 of the initial value as a reasonable measure of the lifetime.

By thermally annealing MgO-II, we could decrease the concentration of anion vacancies present without affecting the concentration of [H]$^+$ ions. Figure 7 shows the result of measuring the luminescence lifetime at 260 K following these anneals. There was an unexplained decrease in the lifetime following the first anneal at 1400 K, but for the following anneals the trend became clear. As the ratio H$^+$/F center increased due to the removal of the anion vacancies, the luminescence

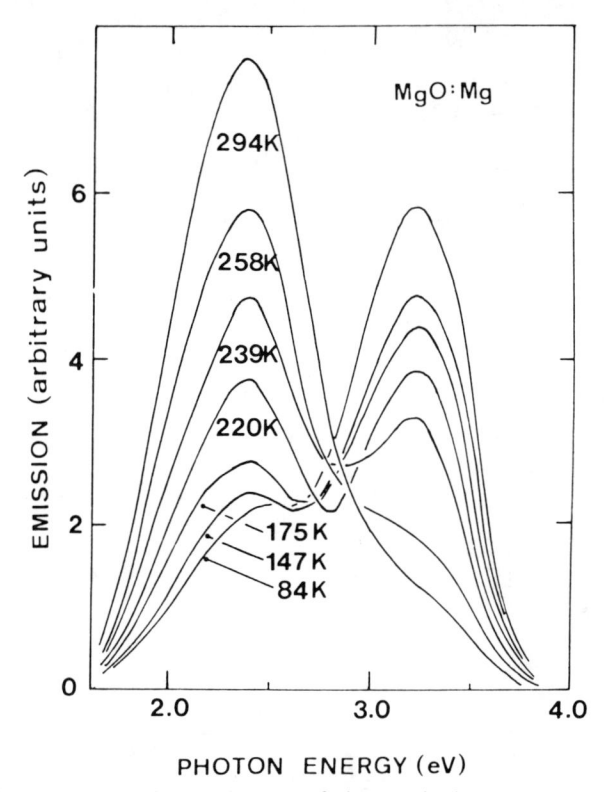

Fig. 4. Temperature dependence of the emission spectrum in sample MgO-I (Table I). This sample contained a relatively large concentration of both anion vacancies and [H]⁺ ions.

lifetime increased markedly. After MgO-II had been heated to 1825 K, the lifetime increased by more than 1 order of magnitude (Fig. 8).

The decay of the 2.3-eV luminescence does not follow first-order kinetics. After an initial transient period, the decay is approximately second order, which would be expected from our model for the results discussed above. This model is summarized in Fig. 9. Absorption of a photon by an F center excites an electron from the $^1A_{1g}$ ground state to a $^1T_{1u}$ excited state. There is a high probability that the electron will escape into the conduction band, from which it can be trapped elsewhere in the crystal. We have proposed that the major effective traps are [H]⁺ ions. Above ~240 K, [H]⁰ ions become thermally unstable, and the outermost electron is released into the conduction band with an activation energy of 0.56 eV. It can either return to an F^+ center, resulting in 2.3-eV luminescence, or be retrapped at another [H]⁺ ion, thereby delaying further the return to an F^+ center. It is this successive capture at [H]⁺ ions that leads to the long lifetime.

The experimental data presented above and other recent results prompted a new look by Wilson at the theoretical calculations[5] of the electronic structure of the F center in MgO. The status of these calculations had been less satisfactory than for CaO,[9] mainly because of the greater differences in the relative sizes and polarizabilities of Mg^{2+} and O^{2-} ions in MgO compared to those of Ca^{2+} and O^{2-}

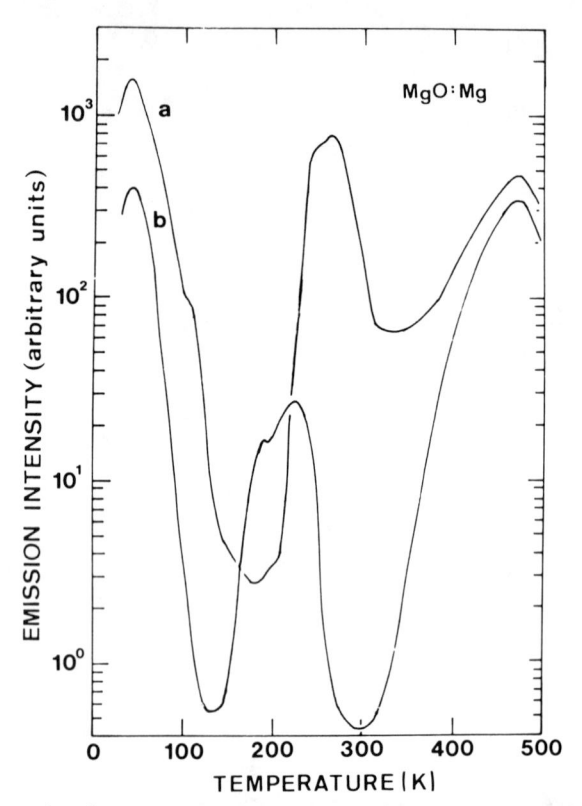

Fig. 5. Thermoluminescence glow curves for sample MgO-I (curve a) and sample MgO-III (curve b). MgO-I contained about 60 times the concentration of [H]$^+$ ions compared to MgO-III (see Table I). The heating rate was ~5 K·min^{-1}.

in CaO. In the recent calculations, the electronic polarization energy was treated more precisely than before. Figure 10 shows the calculated configuration coordinate diagram for the luminescence states for an A_{1g} relaxation of the surrounding Mg^{2+} ions. The important result of the calculation is that it predicts that the $^3A_{1g}$ state has almost the same energy as the $^3T_{1u}$ state for an A_{1g} relaxation. At the lowest temperatures, an excited electron would be expected to fall into the $^3A_{1g}$ state, which is in accord with ODMR results.[4] Mixing is expected of the $^3A_{1g}$ and $^1T_{1u}$ states which is only ~0.05 eV higher. Therefore, the calculation predicts that at most temperatures the 2.3-eV luminescence band is due mainly to a $^1T_{1u} \rightarrow {^1A_{1g}}$ transition and that emission does not occur from a $^3T_{1u}$ state as is the case in CaO. In MgO the $^3T_{1u} \rightarrow {^1A_{1g}}$ transition energy is predicted to ~2.9 eV. So far, no confirmed identification of this transition has been given.

Finally, we describe an experiment in MgO:Mg in which low-energy light is converted into 2.3-eV F center luminescence. Figure 11 shows the results of an experiment on MgO-IV. The sample was cooled to 260 K and illuminated with 5.4-eV light until the 2.3-eV intensity reached a maximum. The excitation was then removed and the sample rapidly cooled to 77 K, which quenched the luminescence. The sample was then illuminated with a 4-s pulse of light. Figure 11

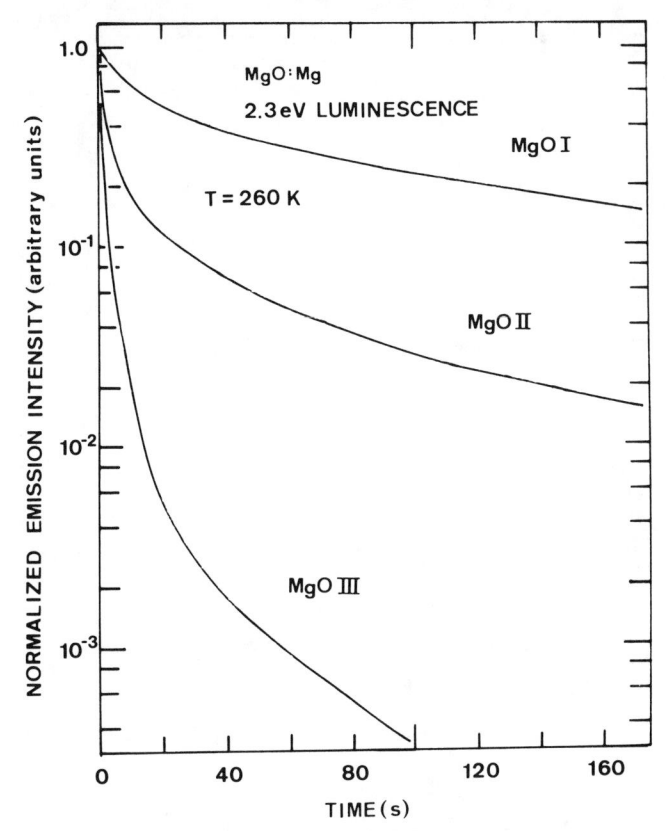

Fig. 6. Decay at 260 K of the 2.3-eV luminescence intensity in MgO:Mg. The concentration of [H]$^+$ ions decreases from sample MgO-I through MgO-III (Table I).

shows the effect for $\lambda > 500$ nm. The 2.3-eV emission was instantly restored, after which it decayed again. This enhancement could be repeated many times. A completely satisfactory explanation of this effect has not yet been found. We propose, however, that the low-energy radiation optically excites electrons from [H]^0 ions into the conduction band. Some of these are immediately captured by F^+ centers, giving 2.3-eV luminescence, while others are captured at shallow traps, from which they can leak to F^+ centers. This shallow trap is responsible for the 40 K glow peak (Fig. 5). The nature of this low-temperature electron transfer is uncertain, although there is theoretical[19] evidence that some kind of tunneling process might be involved.

F Center Luminescence in α-Al$_2$O$_3$

In contrast to the cubic symmetry in MgO, anion vacancies in α-Al$_2$O$_3$ have only C_2 symmetry (Fig. 12). The effect of this low symmetry is observable in several optical properties such as the absorption spectrum of the F^+ center.[12] We present here evidence for the effect on the luminescence of the F center. Figure 13 shows the absorption spectra of samples of α-Al$_2$O$_3$ from several different manufacturers. Three of the samples were reduced during growth, which introduced the

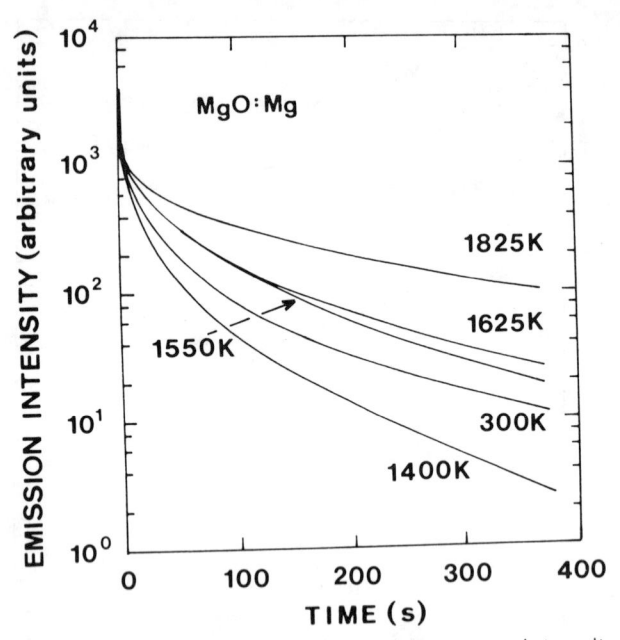

Fig. 7. Decay at 260 K of the 2.3-eV luminescence intensity in sample MgO-II after annealing at different temperatures.

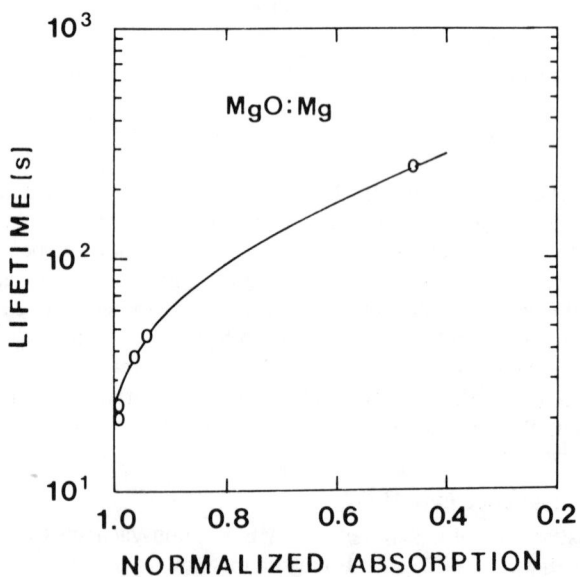

Fig. 8. Lifetime of the 2.3-eV luminescence vs the normalized absorption of the 5.0-eV anion vacancy absorption band. The data were obtained by annealing sample MgO-II (Table I) at different temperatures, resulting in a loss of anion vacancies but no loss of [H]$^+$ ions.

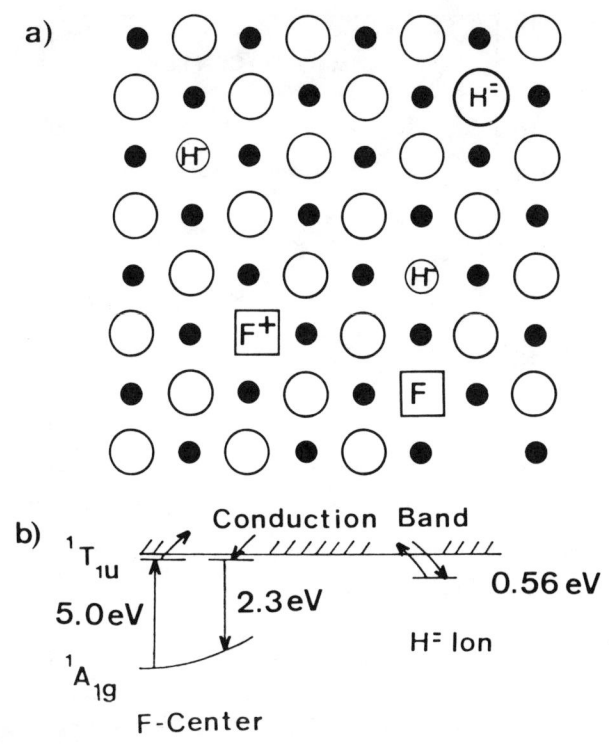

Fig. 9. Schematic representation of the model for the effect of substitutional hydrogen ions on the 2.3-eV F center luminescence in MgO:Mg. (a) Schematic MgO lattice containing anion vacancies and hydrogen ions. Negative hydrogen ions (H⁻ ions) are referred to as [H]⁺ ions in the text because of their charge with respect to the lattice. (b) Energy level diagram for the F center and [H]⁰ ion (see text).

F band at 6.1 eV. The Linde sample showed no 6.1-eV band initially, and the spectrum shown in Fig. 13 was obtained following neutron irradiation. The bands at 4.8 and 5.4 eV are due to F^+ centers.

Figure 14 shows the luminescence excited by unpolarized 6.1-eV light in reduced α-Al$_2$O$_3$ at 20 K. The luminescence is polarized with the maximum intensity for light with the electric vector perpendicular to the crystallographic c axis, $\mathbf{E}_{\perp c}$. This means that the emission is polarized with the electric vector along the axis of the F center. The ratio of the intensities for $\mathbf{E}_{\perp c}$ and $\mathbf{E}_{\|c}$, which was ~1.7 for a $\langle 1\bar{1}02 \rangle$ direction of emission, was found to be independent of temperature up to 55 K. The band centroid, which is at 2.97 eV for both perpendicular and parallel polarizations, is almost independent of temperature up to 300 K, whereas the zeroth moment decreased by about 20% over the same range.

The full width at half-maximum intensity, FWHM, is 0.36 eV at 10 K and increases to 0.44 eV at 300 K following a hyperbolic cotangent relation, with an effective interaction energy $\hbar\omega$, of 345 cm^{-1}. When linear vibronic coupling is assumed, the Huang–Rhys factor, S, can be determined by using the formula $S = \langle E^2 \rangle / (\hbar\omega)^2$, where $\langle E^2 \rangle$ is the second moment of the band. $\langle E^2 \rangle$ was found to

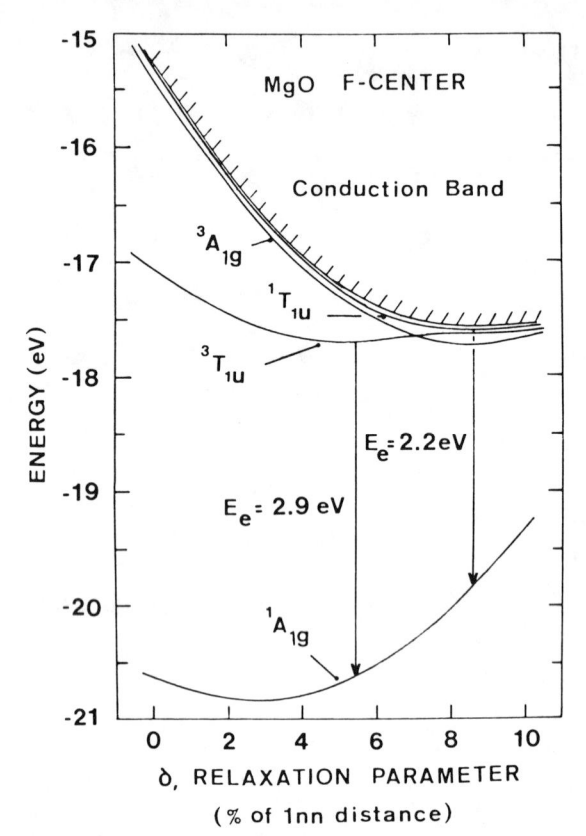

Fig. 10. A_{1g} configuration coordinate curves for the relaxed F center in MgO (after T. M. Wilson; see Ref. 19).

be 0.027 (eV)², giving a value of $S \approx 15$. When the same effective vibronic frequencies for both the emitting and the ground state are assumed, the Stokes shift, ΔE_s, can be calculated from $\Delta E_s = \hbar\omega(2s - 1)$. A value for ΔE_s of 1.23 eV is obtained, which is much less than the observed value of 3.1 eV.

At temperatures between 50 and 200 K, the luminescence decay after pulse excitation follows closely a single exponential with a lifetime of 34 ms. Below 50 K, however, the decay is more complicated and can best be analyzed in terms of the sum of two exponentially decreasing components with temperature-dependent lifetimes, as shown in Fig. 15. Although the total intensity is constant between 10 and 50 K, most of the intensity is in the fast component at low temperatures whereas at 50 K the whole intensity is in the slower component.

The experimental results can be interpreted in terms of a model for the electronic structure of the luminescence states, which is shown schematically in Fig. 16. Absorption of a 6.1-eV photon raises an electron from the ^1A ground state to a ^1p-like excited state. Since there appears to be only one F center absorption band, the local symmetry is apparently not effective in splitting the p-like state, possibly because it is very diffuse, unlike the analogous state in the F^+ center. A lifetime of the order of tens of milliseconds indicates that the 3.0-eV

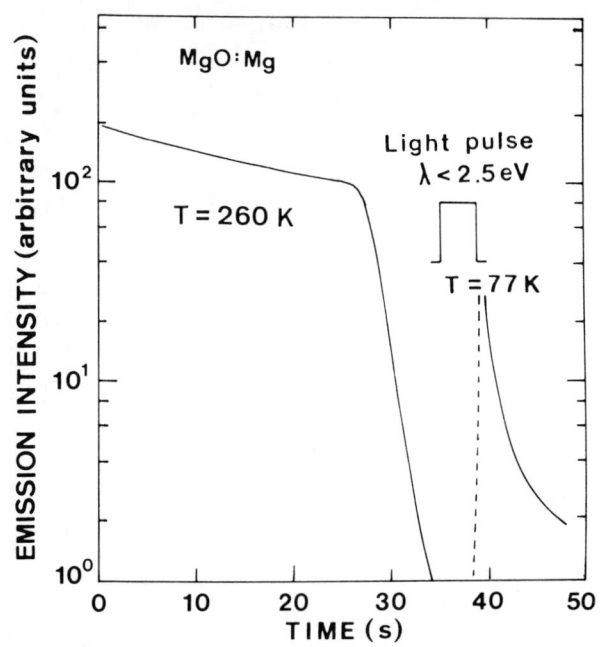

Fig. 11. Excitation process for the 2.3-eV luminescence in MgO:Mg. The decaying 2.3-eV luminescence intensity at 260 K was quenched by cooling the sample rapidly to 77 K after ~30 s. A 4-s pulse of light with $\lambda < 2.5$ eV re-excites the luminescence.

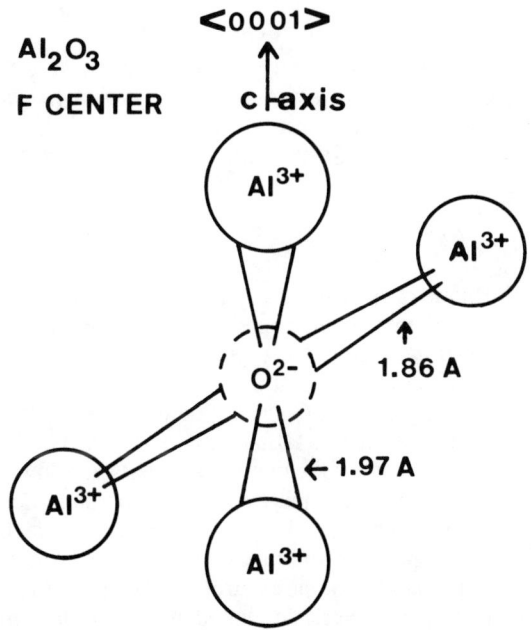

Fig. 12. Positions of the four nearest-neighbor Al^{3+} ions surrounding an oxygen ion vacancy in α-Al_2O_3. An anion vacancy has C_2 symmetry.

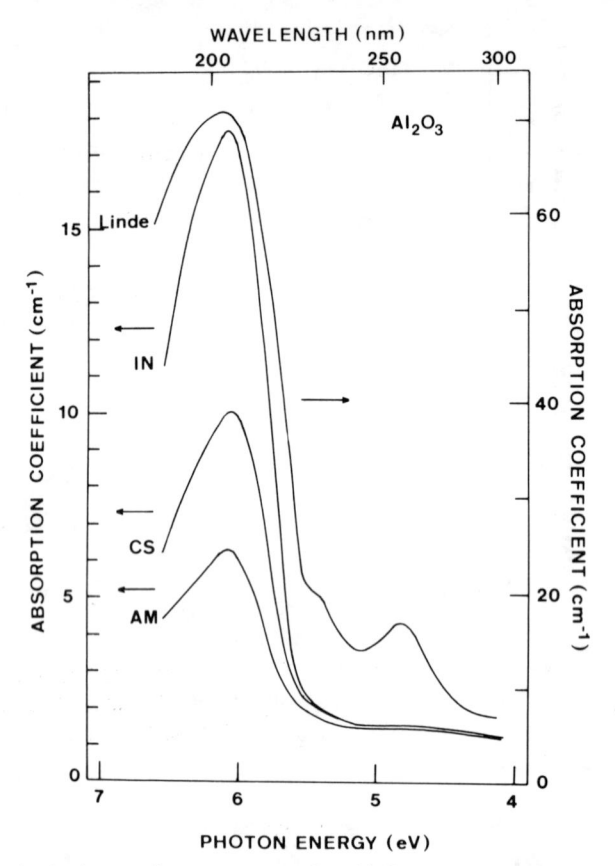

Fig. 13. Optical absorption spectra of α-Al$_2$O$_3$ at room temperature. The IN (Insaco, Inc.), CS (Crystal Systems, Inc.), and AM (Adolf Meller, Inc.) samples were thermochemically reduced during growth. The Linde sample was irradiated near room temperature with ~5 × 10^{17} neutrons. Notice the different absorption scales.

transition is spin forbidden, as is the case, for example, for the F center luminescence in CaO. Since the observed Stokes shift is much larger than that calculated from the Huang–Rhys factor, we suggest that the ^{3}p-like emitting state is ~1.6 eV below the ^{1}p-like state and that the electron falls into the ^{3}p-like state as a result of a radiationless decay. The C_2 crystal field is expected to split the ^{3}p-like state into three components with 1A, 1B, and 2B character. Because of the nature of the polarized emission, we place the two B character states lowest. The splitting between these two states was estimated from the temperature dependence of the luminescence lifetime between 20 and 50 K. Actually, the electronic processes involved in emission are probably more complicated than indicated in Fig. 16 because there is a high probability of the electron escaping into the conduction band even at low temperatures. However, the photoconductivity lifetime is quite short at these low temperatures, which indicates that the long-lived luminescence component is not due to electrons leaking away from a trap.

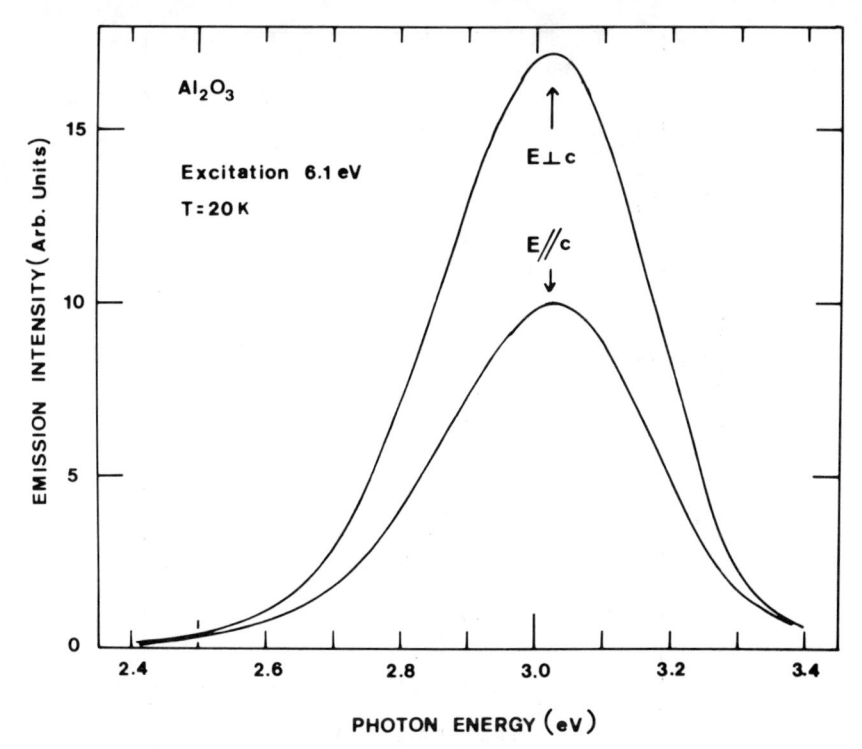

Fig. 14. Polarized luminescence spectrum from thermochemically reduced α-Al$_2$O$_3$ at 20 K; E_{ex} = 6.2 eV.

Photoconductivity in α-Al$_2$O$_3$

Our main interest in measuring photoconductivity in oxides has been to produce additional information about the electronic structure of point defects. In the case of α-Al$_2$O$_3$, one particular problem which we discuss here, concerns the process whereby F centers are converted to F^+ centers by optical excitation. This interconversion is both sample and temperature dependent. In those samples where it is efficient, there is also long-lived phosphorescence observed from the F center near room temperature. We call those samples in which interconversion occurs type I and those in which it is absent or very small type II.

Figure 17 shows the interconversion process in a type I sample at room temperature, and the insert, the same effect at 10 K. Curve a shows the optical spectrum following irradiation with a 6.4-eV light, and curve b, the optical spectrum following irradiation with a 5.0-eV light. Curves c and d show the spectrum following an anneal at 500°C and following X irradiation, respectively. A 6.4-eV irradiation reduces the 6.1-eV F band and increases the 4.8- and 5.4-eV F^+ bands and vice versa. A sample that has been stored in the dark at room temperature for several weeks shows only the 6.1-eV band (see insert, Fig. 17). The interconversion process is about 4 times more efficient at 10 K than at room temperature and is not fully reversible. The interpretation of the conversion of F to F^+ centers is that the 6.1-eV light excites an electron into the conduction band from which it is trapped, thereby leaving behind an F^+ center. Several electron traps are in-

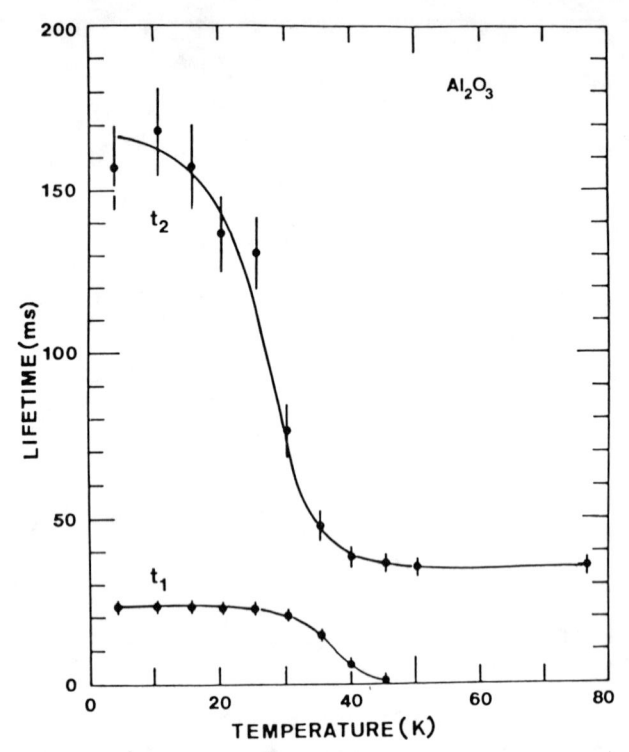

Fig. 15. Temperature dependence of the lifetimes of the two exponential components in the decay of the 3.0-eV luminescence of thermochemically reduced α-Al$_2$O$_3$.

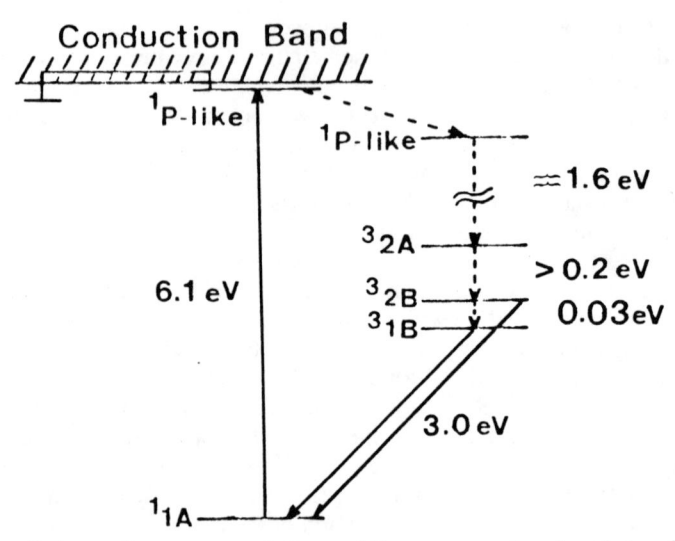

Fig. 16. Schematic representation of the energy levels of the F center in α-Al$_2$O$_3$.

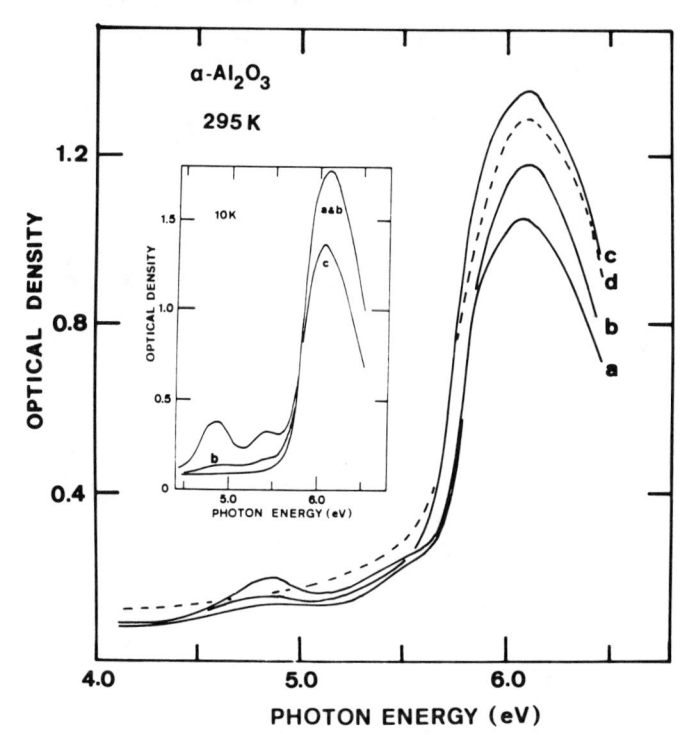

Fig. 17. Optical absorption spectra of thermochemically reduced α-Al$_2$O$_3$ at 295 K after bleaching with 6.4-eV light (curve a); after bleaching with 5.0-eV light (curve b); after heating to 500°C for 30 min (curve c); and after irradiation with 1 Mrad of 70-kV X rays (curve d). Insert: optical spectra at 10 K after several months at 300 K (curve a); after bleaching with 5.0-eV light (curve b); and after bleaching with 6.4-eV light (curve c).

volved, but the main one is thermally unstable at temperatures below 295 K. Our results for MgO strongly suggest that this trap is an [H]$^+$ ion, which was produced by the reducing condition under which the sample was grown. Presumably, those samples containing small concentrations of [H]$^+$ ions are those in which the conversion is very inefficient, type II.

The motion of the electrons can be observed in the photoresponse of the samples (Fig. 18). Curves a and b show the photoresponse for a type I and a type II sample, respectively. The photoresponse is given as $(\eta\omega_0)_\varepsilon$, where η is the number of free carriers produced per incident photon and ω_0 is the mean range per unit applied electric field. In terms of the measurable quantities

$$(\eta\omega_0)_\varepsilon = (I/N)_\varepsilon(d/eE)$$

where I is the photocurrent, N is the incident photon flux, d is the thickness of the sample, E is the applied electric field, and e is the electronic charge. Using the data in Fig. 18 and the absorption coefficients of the samples, we can calculate the photoresponse per absorbed photon. We find 3.2×10^{-9} cm$^2 \cdot$V^{-1} for the type I sample and 5.7×10^{-8} cm$^2 \cdot$V^{-1} for the type II sample. The mean range is larger in the type II sample because it contains fewer of the electron traps. Our data can

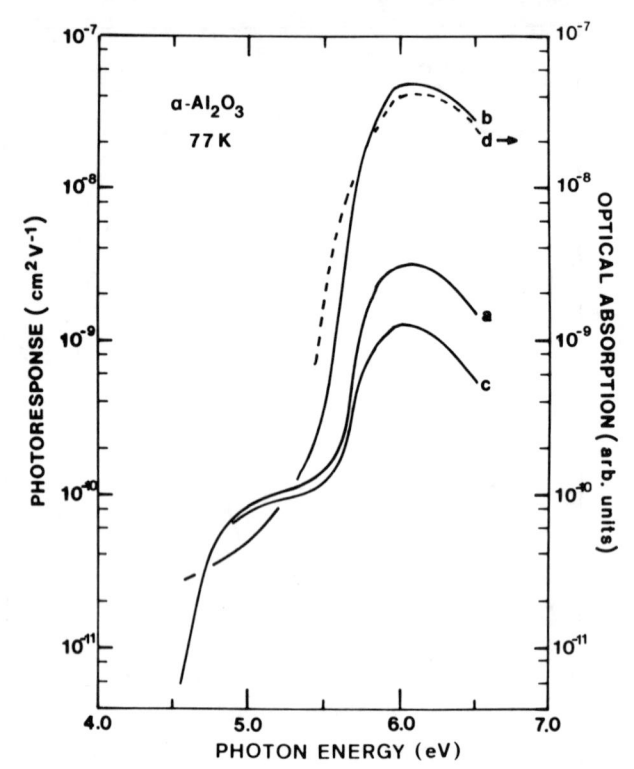

Fig. 18. Photoresponse of thermochemically reduced α-Al$_2$O$_3$ at 77 K. Type I sample after one scan (curve a), type II sample (curve b), and type I sample after six scans (curve c). Spectrum of light absorbed the type II sample at 77 K.

be compared to those of Hughes[31] for high-purity Linde samples. He finds values in the $(8-30) \times 10^{-8}$ cm$^2 \cdot$V^{-1} range for carriers excited by a pulse of X rays. Photons (6.1 eV) produce an immeasurable photoresponse in these samples.

Figure 19 shows the temperature dependence of the photoresponse over the range 10–300 K. In all samples the photoresponse is highest at the lowest temperatures and decreases with increasing temperature. In the type I sample, however, a rapid increase occurs around 240 K that is not seen in the type II sample. At the same time, the lifetime of the photoconductivity increases from <300 μs at low temperatures to many seconds, which shows that the increase in photoresponse is due to thermal extension of the range, ω_0, as an electron trap becomes emptied. This is the trap that we believe allows the conversion of F centers to F^+ centers to occur.

Thermal release of electrons from the trap can be seen in the glow curves for the samples (Fig. 20). The peak near 255 K has been known for a long time. The emission has a peak energy near 3.0 eV and is similar to F center photoluminescence. Figure 20 shows that the trap is present in a much smaller concentration in type II samples, as expected. In the case of the neutron-irradiated sample, the size of the glow curve is due, we think, to the formation of anion

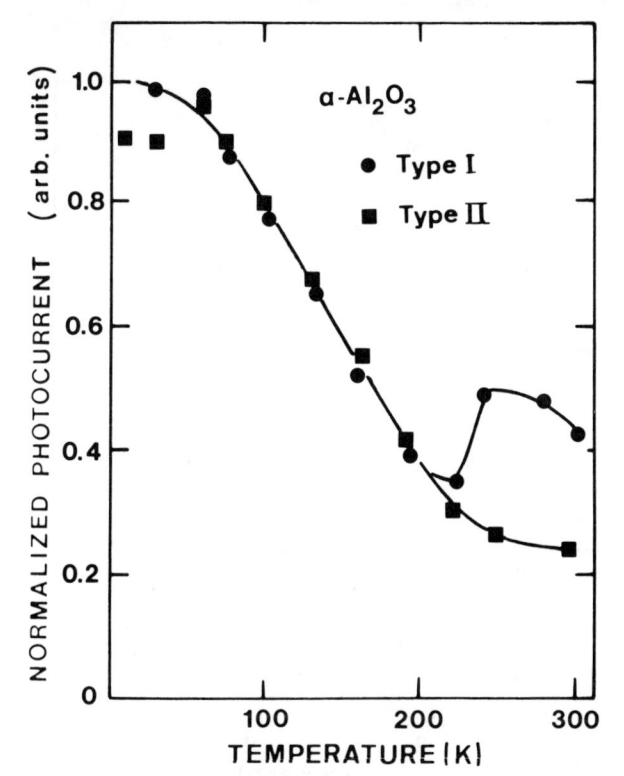

Fig. 19. Temperature dependence of the normalized photoresponse of the 6.1-eV band in thermochemically reduced α-Al$_2$O$_3$.

vacancies rather than the introduction of the trap. We note that in MgO [H]$^+$ ions are much more thermally stable than anion vacancies.

The presence of the trap produces long-lived phosphorescence from F centers at temperatures above 240 K in type I samples, which is not seen in type II samples. The activation energy for release of electrons from the trap can be measured from the glow curves, the photoconductivity or the phosphorescence. In each case an activation energy of 0.73 ± 0.03 eV is obtained, with a pre-exponential frequency factor of ~10^{12} s^{-1}.

Although we strongly suspect that the electron trap involved is an [H]$^+$ ion, we have so far not been able to confirm this. No infrared absorption in the region from 2000 cm^{-1} to the *restrahlen* edge could be detected in the type I samples that could be attributed to [H]$^+$ ions. It is possible that no local mode of [H]$^+$ ions would be detectable outside the *restrahlen* in α-Al$_2$O$_3$ because of the relatively small mass of the Al^{3+} ion.

Finally, we give a brief discussion of photoconductivity from F^+ centers. F^+ centers are positively charged with respect to the lattice, and from a simple viewpoint it would seem unlikely that an optically excited F^+ center would lead to release of a free carrier into the conduction band. However, photoconductivity has been observed from F^+ centers in both CaO[25] and SrO.[26] In the case of SrO, an attempt was made to determine the sign of the charge carriers, and although

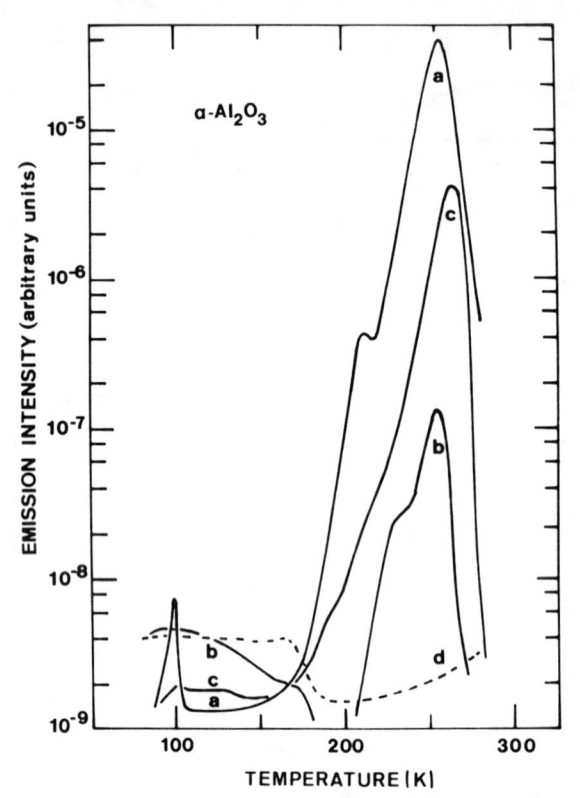

Fig. 20. Thermoluminescence glow curves of α-Al$_2$O$_3$. Type I (curve a), type II (curve b), sample irradiated with $\sim 10^{17}$ reactor neutrons (curve c), and sample untreated (curve d).

the type of experiment used was difficult, the evidence was that they were electrons. A subsequent theoretical calculation by Wilson[32] put the ground state of the F^+ center in SrO in the forbidden gap of the host and the excited states close to the conduction band, in agreement with the experimental result. The raising of the electronic structure was due mainly to a large polarization energy. There has been discussion in the literature for some time,[33] however, that, because the free O^{2-} ion is unstable (i.e., the ionization potential is negative), the site energy at an oxygen vacancy would put the ground state of the F^+ center below the top of the host valence band. Choi will discuss this effect in detail for α-Al$_2$O$_3$ in another conference paper. Photoconductivity could occur then from the F^+ center in α-Al$_2$O$_3$[34] by a sort of negative U effect.[35] It is possible that the photoconductivity at 5.0 eV in Fig. 18 originates from F^+ centers. Alternatively, the photoconductivity could be due to optical release of electrons from the electron trap, which we have identified tentatively as $[H]^+$.

Concluding Remarks

Although there is now considerable experimental and theoretical data available concerning luminescence and photoconductivity from defects in MgO and

α-Al$_2$O$_3$, there are still some important questions to be answered. We mention a few here. In particular, the details of the luminescence at temperatures below 100 K in MgO:Mg still need to be elucidated. A first step would be the identification of the electron trap responsible for the 40 K glow peak, which is introduced by high-temperature thermochemical reduction. Second, more theoretical calculations of the electronic structure of defects in all oxides need to be made. At the moment there are no theoretical results for the F center in α-Al$_2$O$_3$, for example. Third, more experimental data are needed concerning photoconductivity from F^+ centers in order to determine unambiguously the sign of the charge carriers and thus the mechanism by which the photocurrent is produced.

Acknowledgments

The author acknowledges the many contributions of his collaborators, especially Y. Chen, R. Gonzalez, and T. Wilson. J. H. Crawford has provided many stimulating discussions, and his interest in this work is greatly appreciated. This research was supported in part by the U.S. Department of Energy under Contract No. EY-76-S-05-4837.

References

[1] A. E. Hughes and B. Henderson; in Point Defects in Solids. Edited by J. H. Crawford and L. Slifkin. Plenum, New York, 1972.
[2] Y. Chen, J. Kolopus, and W. A. Sibley, *Phys. Rev.*, **186** [3] 865–70 (1969).
[3] L. A. Kappers, R. L. Kroes, and E. B. Hensley, *Phys. Rev. B*, **B1** [10] 4151–7 (1970).
[4] P. Edel, B. Henderson, Y. Merle d'Aubigne, R. Romestain, and L. A. Kappers, *J. Phys. C*, **12** [21] 5245–53 (1979).
[5] T. M. Wilson and R. F. Wood, *J. Phys., Colloq.*, **37** [C7] 190–6 (1976).
[6] B. D. Evans, J. C. Cheng, and J. C. Kemp, *Phys. Lett. A*, **27A** [8] 506–7 (1968).
[7] B. Henderson, Y. Chen, and W. A. Sibley, *Phys. Rev. B*, **B6** [10] 4060–8 (1972).
[8] C. Escribe and A. E. Hughes, *J. Phys. C*, **4** [5] 2537–49 (1971).
[9] R. F. Wood and T. M. Wilson, *Phys. Rev. B*, **B15** [8] 3700–13 (1977).
[10] P. Edel, B. Henderson, and R. Romestain, *J. Phys. C*, **15** [7] 1569–80 (1982).
[11] H. W. Lehmann and H. H. Gunthard, *J. Phys. Chem. Solids*, **25** 941–50 (1964).
[12] B. D. Evans and M. Stapelbroek, *Phys. Rev. B*, **B18** [12] 7089–98 (1978).
[13] B. G. Draeger and G. P. Summers, *Phys. Rev. B*, **B19** [2] 1172–7 (1979).
[14] K. H. Lee and J. H. Crawford, *Phys. Rev. B*, **B19** [6] 3217–21 (1979).
[15] J. D. Brewer, B. T. Jeffries, and G. P. Summers, *Phys. Rev. B*, **B19** [2] 3217–21 (1979).
[16] B. T. Jeffries, J. D. Brewer, and G. P. Summers, *Phys. Rev. B*, **B24** [10] 6074–82 (1981).
[17] B. T. Jeffries, R. Gonzalez, Y. Chen, and G. P. Summers, *Phys. Rev. B*, **B25** [3] 2077–80 (1982).
[18] Y. Chen, R. Gonzalez, O. E. Schow, and G. P. Summers, *Phys. Rev. B*, **B27** [2] 1276–82 (1983).
[19] G. P. Summers, T. M. Wilson, B. T. Jeffries, Y. Chen, and M. M. Abraham, *Phys. Rev. B*, **B27** [2] 1283–91 (1983).
[20] R. Gonzalez, Y. Chen, and M. Mostoller, *Phys. Rev. B*, **B24** [12] 6862–9 (1981).
[21] B. Henderson, S. E. Stokowski, and T. C. Ensign, *Phys. Rev.*, **183** [3] 826–31 (1969).
[22] K. H. Lee and J. H. Crawford, *Phys. Rev. B*, **B15** [8] 4065–70 (1977).
[23] W. T. Peria, *Phys. Rev.*, **112** [2] 423–33 (1958).
[24] R. W. Roberts and J. H. Crawford, *J. Nonmetals*, **2** 133–40 (1974).
[25] J. Feldott and G. P. Summers, *Phys. Rev. B*, **B16** [4] 1722–9 (1977).
[26] J. Feldott and G. P. Summers, *Phys. Rev. B*, **B15** [4] 2295–303 (1977).
[27] S. Choi and T. Takeuchi; private communication.
[28] A. Briggs, *J. Mater. Sci.*, **10** [5] 729–46 (1975).
[29] K. H. Lee and J. H. Crawford, *Appl. Phys. Lett.*, **33** [4] 273–5 (1978).
[30] T. M. Wilson; private communication.
[31] R. C. Hughes, *Phys. Rev. B*, **B19** [10] 5318–28 (1979).
[32] J. Feldott, G. P. Summers, T. M. Wilson, H. T. Tohver, M. M. Abraham, Y. Chen, and R. F. Wood, *Solid State Commun.*, **25** [10] 839–42 (1978).
[33] J. C. Kemp and V. I. Neeley, *Phys. Rev.*, **132** [1] 215–23 (1963).
[34] J. H. Crawford, *Bull. Am. Phys. Soc.*, **26** [3] 235 (1981).
[35] P. W. Anderson, *Phys. Rev. Lett.*, **34** [15] 953 (1975).

Optical and Mössbauer Spectra of Transition-Metal-Doped Corundum and Periclase

ROGER G. BURNS AND VIRGINIA MEE BURNS

Massachusetts Institute of Technology
Department of Earth, Atmospheric, and Planetary Sciences
Cambridge, MA   02139

Natural and synthetic periclases and corundums colored by transition-metal cations have been extensively studied by absorption spectroscopy in the visible region. While assignments of the optical spectra of single-cation-doped MgO and Al_2O_3 crystals are generally well understood, ambiguities remain over interpretations of the electronic spectra of mixed-cation assemblages occurring in V^{3+}-bearing rubies, blue sapphires, and iron-doped periclases and magnesiowustites. Mössbauer spectral measurements of iron-bearing MgO and Al_2O_3 crystals reveal coexisting Fe^{2+} and Fe^{3+} ions, providing evidence that $Fe^{2+} \rightarrow Fe^{3+}$ intervalence transitions contribute to the optical spectra of periclases and that $Fe^{2+} \rightarrow Ti^{4+}$ charge transfer is responsible for the dark blue color of sapphire.

The minerals periclase and corundum are the naturally occurring forms of MgO and α-Al_2O_3, respectively. They often host and are colored by transition-metal ions; ruby and sapphire are specific names for colored varieties of corundum, while magnesiowustite refers to MgO–FeO solid solutions. In nature periclase is formed at relatively high temperatures from the metamorphism of dolomites and magnesian limestones.[1] It often occurs in contact aureoles surrounding marble in association with calcite, forsterite, monticellite, and other calc-silicates. Corundum is a widespread mineral occurring in a variety of high-temperature rock types, including pegmatites, syenites, anorthosites, schists, gneisses, and marbles.[1] It also occurs in meteorites and has been reported in cosmic dust. The extreme hardness and resistance of corundum to weathering leads to its occurrence in placer deposits. The world's finest rubies from Burma occur in marbles associated with gneisses and mica schists. However, the actual deposits mined are residual alluvial or stream placer deposits formed by intense chemical weathering of the source rocks. The sapphires from Sri Lanka were formed in gneisses or marbles but are mined from alluvium.

The colors of deep red ruby and dark blue sapphire, highly valued as gemstones on account of their rarity, beauty, and extreme hardness, have attracted the attention of mankind since antiquity and are the focus of considerable mineralogical and spectroscopic research. Small quantities of Fe and Ti in blue sapphire and Cr^{3+} in ruby are well-known to be responsible for the colors of these gems.[2] Crystal growth studies have enabled a variety of ions to be incorporated into the periclase and corundum structures, with the result that numerous spectroscopic techniques have been applied to several transition-metal-doped MgO and Al_2O_3

phases. Thus, in addition to visible-region (optical) absorption and diffuse reflectance spectroscopies, Mössbauer effect and electron paramagnetic resonance measurements have been used to identify the valencies, site symmetries, electronic structures, and magnetic states of transition-metal cations in host MgO and Al_2O_3 crystal structures. In general, assignments of absorption bands in visible-region spectra of single-cation-doped MgO and Al_2O_3 crystals are relatively straightforward, but interpretations of spectra of phases containing assemblages of transition-metal ions remain controversial. Some of these problems are addressed in this paper.

Geochemically, iron and to lesser extents Ti, Cr, and Mn are the predominant transition elements on the surfaces of the Earth, Moon, Mars, and other terrestrial planets. Iron cations present in oxide phases structurally related to corundum (e.g., hematite, ilmenite) contribute to the visible-region reflectance spectral profiles of planetary surfaces.[3,4] This has led to laboratory measurements of optical spectra of transition-metal-bearing oxide and silicate minerals at elevated and low temperatures, simulating conditions on sunlit planetary surfaces, aimed at interpreting remote-sensed reflectance spectra measured through telescopes focused on the hot (e.g., Mercury, Moon) and cold (e.g., Mars, asteroids) exteriors of planetary objects.[5] Iron is also a major constituent of the Earth's interior where it may occur in dense structure types including periclase, corundum, and perovskite (isochemical with corundum) in the Lower Mantle at depths below 650 km.[6] Therefore, much of the current mineral spectroscopic research is focused on optical and Mössbauer spectral measurements of iron-bearing minerals at very high pressures as well as at elevated temperatures aimed at understanding geophysical properties of the Earth's interior.[7] Such high P and T spectral data for transition-metal-bearing MgO and Al_2O_3 phases are also reviewed here.

Crystal Structures

Although the crystal structures of MgO and Al_2O_3 are well-known, it is useful to summarize here structural features that are important for interpreting optical spectral data. The periclase structure consists of a cubic closest-packed lattice of O^{2-} anions, in which all octahedral sites are filled by Mg^{2+} cations. Unit cell parameters are the following: $Fm3m$; $a = 421.2$ pm; $Z = 4$. The [MgO_6] octahedra are undistorted and Mg^{2+} ions are centrosymmetric in the coordination site with all Mg–O distances equal to 210.6 pm. Each octahedron shares all its edges with 12 adjacent [MgO_6] octahedra, and next-nearest-neighbor Mg–Mg distances are 298 pm. Substitution of larger Fe^{2+} cations (octahedral ionic radius 77 pm) for smaller Mg^{2+} ions (72 pm) leads to a small expansion of the cubic cell parameter to 424 pm in the magnesiowustite $Mg_{0.75}Fe_{0.25}O$ (denoted as Wu_{25}).[8] The Ni^{2+} (70 pm) and Co^{2+} (73.5 pm) cations are accommodated in doped-MgO crystals without much change in the unit cell parameters. Because the cubic periclase structure contains cations in regular centrosymmetric octahedral sites, dichroism is not observed in transition-metal-doped MgO crystals; positions and intensities of absorption bands in optical spectra are unaffected by the direction of propagation of polarized light through the crystals. Such properties do not apply to corundum host crystals, however.

The corundum structure is based on a hexagonal closest-packed lattice of O^{2-} anions, in which two-thirds of the octahedral sites are filled by Al^{3+} cations. Unit cell parameters of this trigonal α-Al_2O_3 phase (referred to hexagonal axes) are the following: $R\bar{3}c$; $a = 475.8$ pm; $c = 1299.1$ pm; $Z = 6$. In corundum, pairs of

Al³⁺ ions are stacked along the c axis, and each Al³⁺ is surrounded by a trigonally distorted octahedron of oxygen ions. These pairs of adjacent Al³⁺ ions share three oxygens located at the vertices of an equilateral triangle between the two cations, and the Al–O distances are 197 pm to these three shared oxygens. The other three oxygens of each [AlO₆] octahedron, which also surround unfilled octahedral sites above and below the Al³⁺ ion pairs parallel to the c axis, lie closer to each Al³⁺ and these three Al–O distances are 186 pm. Therefore, the Al³⁺ cations are not centrally located in the [AlO₆] octahedra. The Al–Al distance between pairs of face-shared [AlO₆] octahedra is 265 pm parallel to the c axis. Each [AlO₆] octahedron also shares edges with three neighboring [AlO₆] octahedra, and the Al atoms lie in puckered planes approximately normal to the c axis. The Al–Al distances across edge-shared octahedra are 279 pm at about 80° to the c axis. Although all Al³⁺ sites are crystallographically equivalent in the corundum structure, there are two magnetically inequivalent sites.[9] All cations in a given (0001) plane of edge-shared octahedra are equivalent but magnetically inequivalent to cations in adjacent (0001) planes separating the face-shared octahedra.

Most of the cations substituting for Al³⁺ cations (octahedral ionic radius 53 pm) in transition-metal-bearing Al₂O₃ have significantly larger ionic radii (e.g., Ti³⁺, 67 pm; V³⁺, 64 pm; Cr³⁺, 61.5 pm; Mn³⁺, 65 pm; Fe³⁺, 64.5 pm; Fe²⁺, 77 pm; Ti⁴⁺, 60.5 pm), exceptions being low-spin Co³⁺ (52.5 pm) and Ni³⁺ (56 pm). Their larger radii are manifested by larger unit cell parameters and metal–oxygen and metal–metal distances of the transition-metal sesquioxides. Pertinent data[10–13] for Ti₂O₃, V₂O₃, Cr₂O₃, Fe₂O₃, and FeTiO₃ phases, the spectra of which are discussed later, are summarized in Table I. As a result of occupying compressed Al³⁺ sites in the host corundum structure, the positions of absorption bands in optical spectra of transition-metal-doped Al₂O₃ crystals generally occur at

Table I. Crystallographic Data for Corundum and Related Transition-Metal Sesquioxides

Compound (mineral)	Cell parameters (pm) a	c	Metal–oxygen distances (pm)	Metal–metal distances (pm) ∥c	⊥c	Ref.
Al₂O₃ (corundum)	475.8	1299.1	186 and 197	265	279	10
Al₂O₃:Cr³⁺ (ruby)	476.1	1299.5	185.6 and 197.1	265.5	279.2	11
Cr₂O₃ (eskolaite)	496.1	1359.9	197 and 202	265	289	10
Fe₂O₃ (hematite)	503.4	1374.2	194.5* and 211.6*	289§	297§	12
FeTiO₃ (ilmenite)	508.8	1408.0	215* and 203* 214† and 192†	294‡	300§ 299¶	13
Ti₂O₃	514.9	1364.2	201† and 208†	259¶	299¶	10
V₂O₃	495.2	1400.2	196 and 206	270	288	10

*Fe–O distances.
†Ti–O distances.
‡Fe–Ti distances.
§Fe–Fe distances.
¶Ti–Ti distances.

higher energies relative to their own sesquioxide phases. In addition, the occurrence of transition-metal cations in noncentrosymmetric distorted (point symmetry C_3) octahedral sites in the trigonal corundum structure leads to dichroism and, hence, polarization dependence of positions and intensities of absorption bands, in polarized visible-region spectra.

Optical Spectra: Corundum

General Information

The optical spectra of transition-metal-doped corundums have been extensively studied both theoretically and experimentally, and early work is reviewed by Hush and Hobbs.[14] In a classic paper, McClure[15] described the polarized spectra and discussed band assignments of Al_2O_3 crystals doped with individual Ti^{3+}, V^{3+}, Cr^{3+}, Mn^{3+}, Fe^{3+}, Co^{3+}, and Ni^{3+} cations. Although many refinements and several additions have been made subsequently to McClure's spectral data and interpretations,[16-18] his results for single-cation-doped Al_2O_3 crystals remain a bench mark in transition-metal-bearing corundum optical spectroscopy. Visible-region absorption bands for Al_2O_3 crystals doped with single-transition-metal cations are summarized in Table II.

Measurements of mixed-cation assemblages in natural and synthetic transition-metal-bearing corundums have been much more controversial, particu-

Table II. Energies of Absorption Bands in Optical Spectra of Transition-Metal-Bearing Corundum Structures*

Oxide (mineral)	Positions of absorption bands (cm^{-1})	Δ_o (cm^{-1})	Ref.
Ti^{3+}:Al_2O_3	18 450 ($\parallel$, $\perp$); 20 300 ($\parallel$, $\perp$)	19 500	15, 31
V^{3+}:Al_2O_3	17 510 ($\parallel$); 24 930 ($\parallel$) 17 420 ($\perp$); 25 310 ($\perp$)	17 500	15
Cr^{3+}:Al_2O_3 (ruby)	18 450 ($\parallel$); 25 200 ($\parallel$)	18 150	15
Cr_2O_3 (eskolaite)	16 600; 21 700	16 600	20–22
Mn^{3+}:Al_2O_3	18 750 ($\perp$); 20 600 ($\parallel$)	19 470	15
Fe^{3+}:Al_2O_3 (yellow sapphire)	9 450 ($\parallel$, $\perp$); 14 350 ($\perp c$) 18 700 ($\perp$, $\parallel$); 22 220 ($\perp$, $\parallel$) 25 600 ($\perp$, $\parallel$); 26 700 ($\perp$, $\parallel$)	14 300	32, 35, 36
Fe_2O_3 (hematite)	11 600; 18 500; 22 200	14 000	44
Co^{3+}:Al_2O_3	15 740 ($\parallel$); 23 170 ($\parallel$) 15 380 ($\perp$); 22 800 ($\perp$)	18 300	15, 16
Ni^{3+}:Al_2O_3	16 800 ($\parallel$); ≈20 000 ($\parallel$) 16 300 ($\perp$); ≈20 000 ($\parallel$)	18 000	15, 16
Co^{2+}:Al_2O_3	9 100 ($\perp$); 21 400 ($\parallel$); 22 800 ($\parallel$, $\perp$) (7 700; 16 000; 17 200, 18 300: spin-forbidden)	12 300	16
Ni^{2+}:Al_2O_3	18 000 ($\parallel$, $\perp$); 25 000 ($\parallel$,$\perp$) (16 200; 22 700: spin-forbidden)	10 700	16

*$\parallel$ and $\perp$ refer to light polarized parallel and perpendicular to the corundum c axis.

larly for V^{3+}–Cr^{3+}-doped rubies and Fe–Ti-bearing sapphires. In addition, recent spectral measurements of rubies and sapphires have been extended to elevated temperatures or high pressures. It is these problems for sapphire and the high-P or -T spectral data of ruby that are summarized in the following sections.

Ruby

Because of its importance in ruby-laser technology, the spectra of Cr^{3+}-doped Al_2O_3 have been studied extensively under a variety of experimental conditions. Summarized here are temperature- and pressure-induced variations, as well as composition dependencies, of the visible-region spectra. Typical room-temperature, 1-atm (9.8 × 10^4 Pa) optical spectra of a ruby single crystal measured in polarized light are illustrated in Fig. 1. Two broad intense bands occurring around 18 000 and 25 000 cm^{-1} represent, respectively, the spin-allowed $^4A_2 \rightarrow {}^4T_2(F)$ and $^4A_2 \rightarrow {}^4T_1(F)$ transitions in octahedrally coordinated Cr^{3+} ions, while the weak peaks at 14 430 and 15 110 cm^{-1} originate from spin-forbidden transitions $^4A_2 \rightarrow {}^2E(G)$ and $^4A_2 \rightarrow {}^2T_1(G)$. The polarized spectra portray the dichroism expected for Cr^{3+} in corundum's trigonally distorted (symmetry C_3) octahedral site, with small but significant differences of band maxima and intensities for light polarized parallel (Fig. 1(a)) and perpendicular (Fig. 1(b)) to the c axis. In the $\mathbf{E}_{\|c}$ spectrum, band maxima are at 18 300 cm^{-1} ($^4A_2 \rightarrow {}^4A_1(^4T_2)$) and 24 800 cm^{-1} ($^4A_2 \rightarrow {}^4A_2(^4T_1)$), and in the $\mathbf{E}_{\perp c}$ spectra, they occur at 18 100 cm^{-1} ($^4A_2 \rightarrow {}^4E(^4T_2)$) and 24 500 cm^{-1} ($^4A_2 \rightarrow {}^4E(^4T_1)$). Strictly speaking, the $^4A_2 \rightarrow {}^4A_1$ transition in the $\mathbf{E}_{\|c}$ spectrum is symmetry forbidden by group theoretical selection rules,[15] and the presence of the absorption band at 18 300 cm^{-1} is attributed to excitation of allowed vibrational modes during the electronic transition within the Cr^{3+} ions. Such vibronic coupling is accentuated by increased temperature. The energy of the absorption band around 18 000 cm^{-1} represents the crystal field splitting parameter, Δ_0, used in thermodynamic discussions of octahedral Cr^{3+} ions.[19]

The spectra of ruby illustrated in Fig. 1 also show that regions of minimum absorption occur at 21 000 cm^{-1} (blue) and beyond 16 000 cm^{-1} (red), the "window" in the red dominating the human eye's perception of ruby's color in transmitted and reflected light.[20] The optical spectrum of eskolaite or chromia, Cr_2O_3,[20,21] however, has the intense spin-allowed transitions centered at 16 600 and 21 700 cm^{-1} and a "window" of a minimum absorption near 20 000 cm^{-1} (green), resulting in its green color. Numerous studies have been made of the compositional variations of the optical spectra of the Al_2O_3–Cr_2O_3 solid solution series.[20–22] Absorption bands show a red shift (move to lower energies) with increasing Cr concentration due to expansion of the octahedral sites as larger Cr^{3+} replace Al^{3+} ions in the corundum structure. The color change from red to green appears between 20 and 40 mol% Cr_2O_3[20] and is influenced by temperature.

Measurements of ruby at elevated temperatures[15,23] have revealed that there is a general broadening, intensification (integrated areas), and shift of band maxima to lower energies between room temperature and 900°C, the effects being most pronounced for the 18 000-cm^{-1} band (particularly in the $\mathbf{E}_{\|c}$ spectrum), which moves toward 16 000 cm^{-1} at 900°C. Above 500°C, the color of a typical ruby becomes green.[20] The effect of increased pressure on ruby optical spectra is to cause intensification and a blue shift of the spin-allowed transitions at high pressures.[24–26] For example, in unpolarized spectra the $^4A_2 \rightarrow {}^4T_2$ transition moves from 18 020 to 19 880 cm^{-1} at 46 GPa, and the $^4A_2 \rightarrow {}^4T_1$ transition shifts from 24 690 to 25 910 cm^{-1} at 32 GPa.[26] On the other hand, the ruby R_1 fluorescence

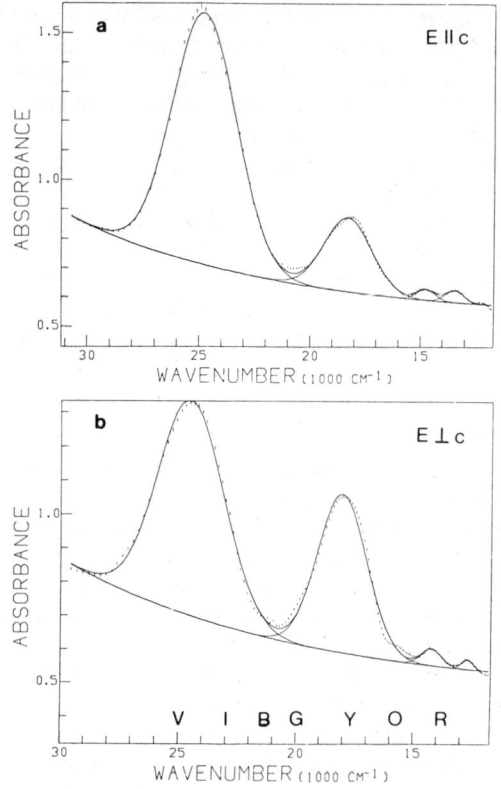

Fig. 1. Optical absorption spectra of ruby in light polarized (a) parallel and (b) perpendicular to the c axis (Ref. 23). The natural ruby crystal contains 0.3 wt% Cr_2O_3.

line representing one peak of the spin-forbidden $^4A_2 \rightarrow {}^2E(G)$ transition doublet shows a red shift, decreasing from 14 390 cm^{-1} at 1 atm (9.80 × 10^4 Pa) to about 13 600 cm^{-1} at 112 GPa.[26] This pressure-induced shift of the ruby fluorescence spectrum is utilized as an in situ pressure gauge in experiments using the diamond anvil cell.[27]

"Alexandrite Effect"

The color of ruby discussed earlier is dominated by the absorption minimum in the red region, and the color changes with rising temperature and increasing Cr^{3+} content of Al_2O_3 crystals. However, the type of incident light can also influence the color perceived by the eye in certain transition-metal-doped corundums. This phenomenon, known as the "alexandrite effect,"[28] refers to the behavior of Cr^{3+}-bearing chrysoberyl (Al_2BeO_4), which appears green in daylight but red in incandescent light. The positions and intensities of absorption bands in optical spectra of alexandrite[29] are polarization dependent as a consequence of Cr^{3+} ions occurring in distorted, low-symmetry six-coordination sites in the orthorhombic chrysoberyl structure. By analogy with Cr^{3+}- and V^{3+}-doped corundums, the optical spectra of alexandrite are bimodal with absorption maxima in the violet-blue and yellow-orange regions,[29] so that "windows" of comparably low absorption occur in the red and green regions. Because daylight is equally rich in all wavelengths in the visible region, alexandrite transmits equal proportions of red and green light. However, because the eye is most sensitive to green light, alexandrite is perceived to be green. Under incandescent light, which is richer in low-energy red wavelengths, alexandrite transmits more red light swamping out the green so that the gem appears red. The optical spectra of V^{3+}-doped Al_2O_3 crystals[15] are remarkably similar to those of the mineral alexandrite[29] with respect to positions and intensities of absorption bands and "windows," so that vanadium-bearing corundums show the alexandrite effect. Indeed, V^{3+}-doped corundums are used widely as imitations of the rare gem alexandrite. Because Cr^{3+}-doped Al_2O_3 crystals show red shifts of absorption maxima and minima with increasing Cr^{3+} concentration,[20-22] the alexandrite effect is also displayed by some natural rubies containing significant concentrations of both vanadium and chromium.[30,43]

Sapphire Blue Problem

The cause of the dark blue coloration in sapphire has been a long-standing problem in mineralogy and solid-state chemistry.[31-42] Analyses of natural sapphires, coupled with crystal growth studies, established early that both Fe and Ti must be present but in only low concentrations to produce the intense blue color.[43] Natural and synthetic corundums containing only Fe^{3+} ions have yellow or pale green-blue colors,[43] while synthetic Ti^{3+}-doped Al_2O_3 crystals are rose colored.[31] Laboratory investigations have also revealed that heat treatment of natural and synthetic sapphires in oxidizing or reducing atmospheres induce color changes[37,42] and that variations exist for sapphires originating from similar localities.[43] Numerous optical spectral measurements have been brought to bear on the origin of the dark blue color of sapphire,[31-42] but considerable confusion exists in the literature over assignments of absorption spectra in the visible and nearby ultraviolet and infrared regions.[37,42]

Representative optical spectra of various sapphires are illustrated in Fig. 2. It is apparent from the spectra of natural blue sapphires (Fig. 2(b)) that absorption minima in the violet-indigo and blue-green regions located between sharp peaks at 25 680 and 22 220 cm^{-1} and broad bands spanning 17 800–14 200 cm^{-1} are responsible for the color of blue sapphires. The latter feature is relatively less intense in the spectra of natural yellow sapphires (Fig. 2(a)) believed to contain negligible Ti. The spectra of synthetic Ti^{3+}-doped Al_2O_3 (Fig. 2(c)) show absorption maxima at 18 450 and 20 300 cm^{-1}, with the suggestion of a weak broad band centered at about 12 500 cm^{-1}. In a synthetic Fe–Ti-doped Al_2O_3 (Fig. 2(d)), band maxima occur at 17 400 cm^{-1} ($\mathbf{E}_{\perp c}$) and 14 000 cm^{-1} ($\mathbf{E}_{\|c}$), with prominent shoulders near 12 500 and 20 300 cm^{-1}. The 18 450 and 20 300-cm^{-1} bands (Fig. 2(c)) represent crystal field transitions within Ti^{3+} ions,[15,31] and the weaker band near 12 500 cm^{-1} may represent a $Ti^{3+} \rightarrow Ti^{4+}$ intervalence charge-transfer transition.[71] There is also general agreement[17,18,32,35,37] that the peaks in yellow and blue sapphire spectra clustered at 22 200 cm^{-1} and around 26 000 cm^{-1} represent the spin-forbidden $^6A_1 \rightarrow {}^4A_1$, $^4E(G)$ and $^6A_1 \rightarrow {}^4T_2$, $^4E(D)$ transitions in octahedrally coordinated Fe^{3+} ions, intensified by exchange interactions[9] between adjacent Fe^{3+} ion pairs in the corundum structure.[17] Debate centers over origins of the band maxima at 17 400 and 14 200 cm^{-1} and contributions to the broad absorption envelope spanning 9000–20 000 cm^{-1}. It is apparent from Table II that most transition-metal-doped Al_2O_3 crystals have absorption bands in this region. In natural corundums, however, only Cr, Fe, Ti, and V cations occur in sufficiently high concentrations[43]

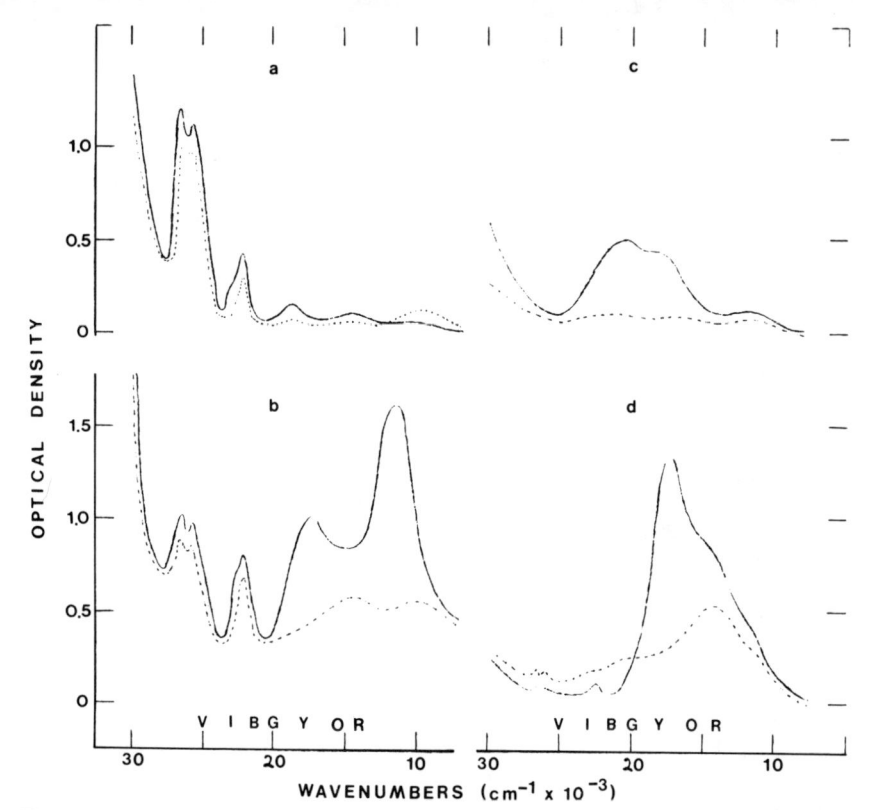

Fig. 2. Optical spectra of natural and synthetic sapphires in light polarized perpendicular (—) and parallel (---) to the c axis: (a) natural yellow sapphire (based on Refs. 32, 35, 42); (b) natural dark blue sapphire (based on Refs. 32, 38, 39, 42); (c) synthetic Ti-doped Al_2O_3 (based on Refs. 15, 31); (d) synthetic Fe-Ti-doped Al_2O_3 (based on Ref. 31).

to contribute to the optical spectra. Studying the spectra of synthetic Fe- and Fe–Ti-doped Al_2O_3 crystals confines the problem to transitions involving only iron and titanium cations.

The Fe^{3+} crystal field peaks near 26 000 and 22 200 cm^{-1} imply that weaker contributions from the spin-forbidden $^6A_1 \rightarrow {}^4T_1(G)$ and $^6A_1 \rightarrow {}^4T_2(G)$ transitions must contribute to the optical spectra at lower energies. Studies of natural yellow sapphires[32,35] and Fe^{3+}-doped Al_2O_3 crystals[17,18,44] suggest that they occur around 9450, 14 350, and 18 700 cm^{-1}. However, an additional intense broad band appears at 11 500 cm^{-1} in flux-grown Al_2O_3 crystals[17] and in many blue sapphires;[32,38,39] this band appears to originate from Fe^{2+} ions, the presence of which are made possible by charge compensation by F^- or OH^- anions replacing O^{2-} ions in the corundum structure. This 11 500-cm^{-1} band has been assigned to the spin-allowed $^5T_{2g} \rightarrow {}^5E_g$ crystal field transition within octahedral Fe^{2+} ions,[32-34] perhaps intensified by exchange interactions with neighboring Fe^{3+} ions,[39,40] and to a $Fe^{2+} \rightarrow Fe^{3+}$ intervalence charge-transfer transition between edge-shared cations perpendicular to the c axis.[38] The corresponding $Fe^{2+} \rightarrow Fe^{3+}$ intervalence transition across face-shared octahedra parallel to the c axis was suggested[38] to occur at 9700 cm^{-1}, overlapping with the $^6A_1 \rightarrow {}^4T_1(G)$ crystal field transition in Fe^{3+}. An inverse temperature dependence of the bands at 11 500 and 9700 cm^{-1}, which are intensified in low-temperature spectra,[38] supports their assignments to Fe$^{2+} \rightarrow$ Fe^{3+} intervalence transitions. The broad bands around 17 400 and 14 200 cm^{-1} also intensify at lower temperatures,[38] leading to suggestions[35-38] that they represent Fe$^{2+} \rightarrow$ Ti^{4+} intervalence transitions across edge-shared ($\perp c$) and face-shared ($\| c$) octahedra, respectively.[71] The loss of intensity of these bands when Fe–Ti-doped Al$_2$O$_3$ and some natural blue sapphires are heated in oxygen[37] or air[42] is attributed to oxidation of Fe^{2+} to Fe^{3+} ions and the removal of Fe$^{2+} \rightarrow$ Ti^{4+} intervalence charge-transfer transitions near 17 400 and 14 300 cm^{-1}. The latter confine the "windows" of minimum absorption to the blue and indigo-violet portion of the spectrum, resulting in dark blue sapphires. The absence of titanium and/or conversion of Fe^{2+} cations to Fe^{3+} ions eliminates Fe$^{2+} \rightarrow$ Ti^{4+} charge transfer, resulting in yellow or pale blue-green sapphires.

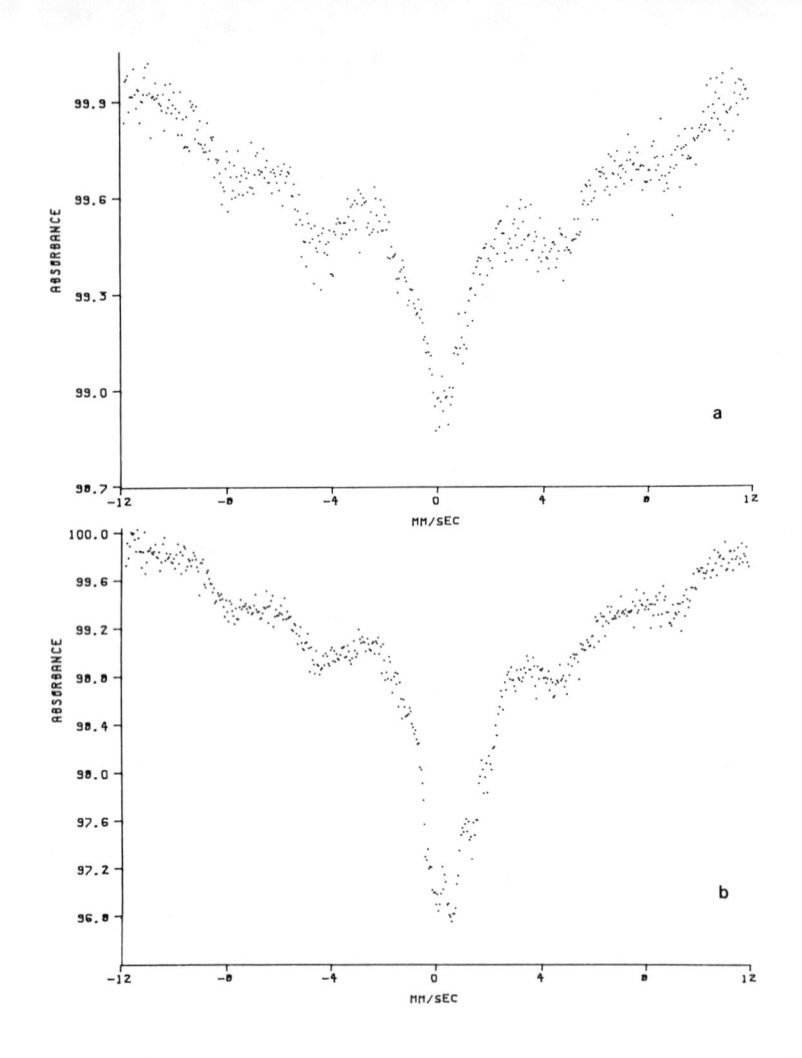

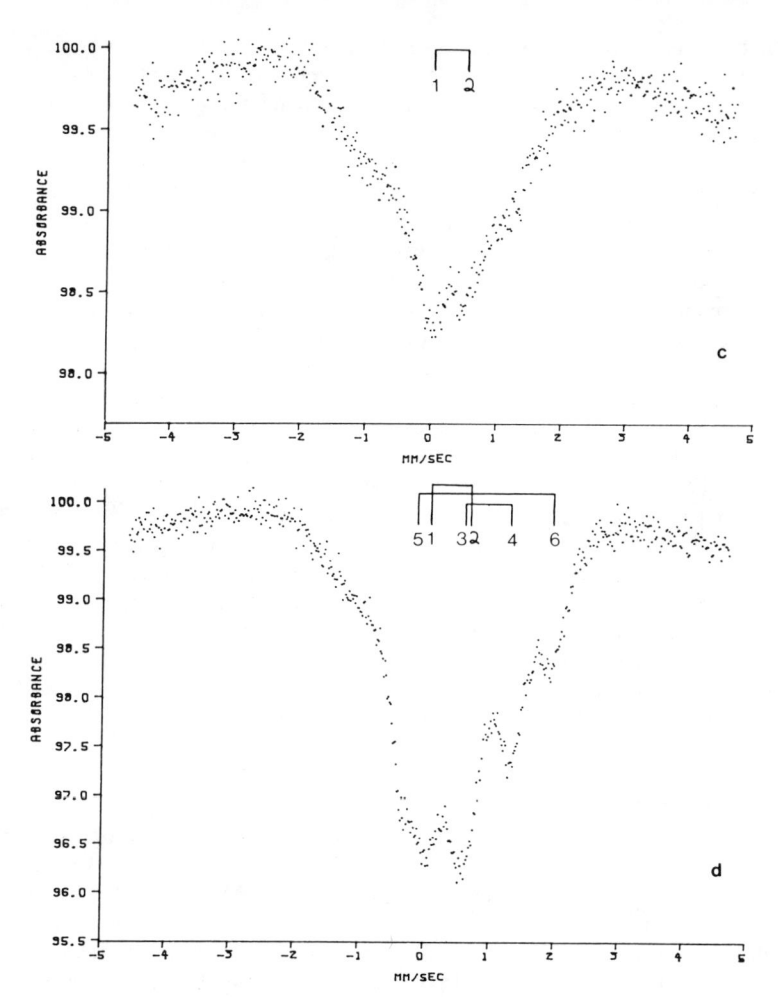

Fig. 3. Mössbauer spectra of synthetic ^{57}Fe- and ^{57}Fe–Ti-doped Al$_2$O$_3$ crystals: (a, b) expanded velocity scale; (c, d) higher resolution in the ±5 mm/range. The crystals were grown by the laser-heated floating zone process from a sintered mixture containing 10 g of Al$_2$O$_3$ and 140 mg of Fe$_2$O$_3$ (enriched in the Mössbauer-active isotope ^{57}Fe) without (Fig. 3(a, c)) and with (Fig. 3(b, d)) 30 mg of TiO$_2$ added.

Mössbauer Spectra: Fe in Al$_2$O$_3$

Since the stable isotope ^{57}Fe, constituting 2.19% of natural iron, is one that exhibits the Mössbauer effect, the valence and coordination symmetry of iron cations in sapphires can be studied by Mössbauer spectroscopy. However, although numerous Mössbauer studies have been made of hematite, data are scant for iron-bearing corundum[45–47] because the amount of iron present in natural sapphires is too low to obtain well-resolved Mössbauer spectra. The Mössbauer spectra of crystals containing less than 0.1% Fe are complicated by spin-lattice relaxation effects, which produce magnetic hyperfine structure in dilute paramagnetic Fe^{3+}

ions in the nonmagnetic Al_2O_3 structure.[48,72] One report[42] stated that only Fe^{3+} ions were detected in three sapphires before and after heat treatment. However, co-existing Fe^{2+} and Fe^{3+} ions were found in synthetic flux-grown Al_2O_3 crystals doped with 2% Fe.[17]

In an attempt to monitor changes of crystal chemistry of iron induced by titanium in blue sapphires, we had synthesized for us Fe- and Fe–Ti-doped Al_2O_3 crystals grown by the laser-heated floating zone process by J. S. Haggerty,[49] using >90% isotopically enriched $^{57}Fe_2O_3$ starting material. Examples of the Mössbauer spectra of sapphires grown by this technique are shown in Fig. 3. Measurements with an expanded velocity range (Fig. 3(a)) revealed magnetic hyperfine structure for both Fe- and Fe–Ti-doped Al_2O_3 crystals probably due to spin-lattice relaxation effects.[48,72] High-resolution spectra measured within a smaller velocity range (Figs. 3(c,d)) show significant differences between the Fe–Ti–Al_2O_3 (blue) and Fe–Al_2O_3 (colorless) sapphires. At least three doublets may be resolved in the spectrum of the blue Fe–Ti–Al_2O_3, two of which (peaks 3,4 and 5,6) represent octahedral Fe^{2+} cations and the third (peaks 1,2) octahedral Fe^{3+} cations. The Mössbauer parameters of peaks 3,4 are comparable to those of Fe^{2+} in ilmenite (Table III), indicating that clusters of Fe^{2+}–Ti^{4+} pairs in adjacent face-shared octahedra occur in this blue sapphire. However, since the proportion of Fe^{2+} ions estimated from the peak areas exceeds the Ti content of the blue sapphire, the second Fe^{2+} doublet probably originates from Fe^{2+} surrounded by Fe^{3+} ions only located in face-shared ($\|c$) and edge-shared ($\perp c$) octahedra. These preliminary Mössbauer measurements support the interpretation that intense blue coloration in sapphires is due to $Fe^{2+} \rightarrow Ti^{4+}$ intervalence charge transfer.[71]

Optical Spectra: Periclase

General Information

Numerous experimental and theoretical spectral studies have been made of transition-metal-doped MgO, including crystal field spectra of Ni^{2+}, Co^{2+}, Fe^{2+}, Mn^{2+}, Fe^{3+}, and Cr^{3+} ions in the periclase structure. Many of the early optical spectral data were reviewed by Hush and Hobbs.[14] Some data for doped MgO crystals and end-member binary oxide phases are summarized in Table IV.[53–65] Measurements have also been made of pressure-induced changes of crystal field spectra of MgO crystals hosting Ti^{3+}, Cr^{3+}, Ni^{2+}, Co^{2+}, and Fe^{2+} ions.[24,66] Surprisingly, high-temperature optical spectra of transition-metal-bearing MgO have not been studied. Most of the recent spectral measurements have centered on iron-doped periclases[54–57] and magnesiowustites,[67,68] triggered by the belief that the periclase structure-type occurs in the Earth's lower mantle and that the presence of iron in this phase influences geophysical properties of the Earth's interior.[7]

(2) Magnesiowustite and Fe-Doped MgO

The spectra of iron-bearing periclases contain a broad asymmetric band in the near-infrared region with maxima around 10 000 and 11 600 cm^{-1} and sharper peaks in the visible region at 21 300 and 26 200 cm^{-1}.[54–57,66–68] The two maxima at 10 000 and 11 600 cm^{-1} represent spin-allowed $^5T_{2g} \rightarrow {}^5E_g$ crystal field transitions in octahedral Fe^{2+} cations, the excited ^5E_g level of which shows Jahn–Teller splitting. Spin-forbidden $^5T_{2g} \rightarrow {}^3T_{1g}(H)$ and $^5T_{2g} \rightarrow {}^3T_{2g}(H)$ transitions in Fe^{2+} are responsible for the peaks at 21 300 and 26 200 cm^{-1}, respectively. In MgO crystals doped with low concentrations of Fe, sharp intense peaks around 9350 and 9500 cm^{-1} assigned to vibrational transitions are observed,[54–56] the intensities of which increase nonlinearly with iron concentrations above 4300 ppm due to the

Table III. Mössbauer Parameters for Iron in Corundum and Periclase Structures*

Oxide	Fe²⁺ Isomer shift	Fe²⁺ Quadrupole splitting	Fe³⁺ Isomer shift	Fe³⁺ Quadrupole splitting	Ref.
Fe:Al₂O₃	1.05	1.09	0.4	0.54	46
Fe:corundum	1.05	1.09	0.35	0.48	47
Fe:Al₂O₃			0.396		48
(Fe₀.₇₇Al₀.₂₃)₂O₃			0.38	0.20	50
(Fe₀.₃Cr₀.₇)₂O₃			0.43	0.10	50
(Fe₀.₃V₀.₇)₂O₃			0.53	0.10	50
Fe:Ti₂O₃	1.0	1.0	0.4	0.5	51
FeTiO₃ (ilmenite)	1.07	0.71	0.35	0.67	52
Fe–Ti:Al₂O₃	1.0† / 1.0‡	0.7† / 1.9‡	0.3§	0.6§	this work
Fe:Al₂O₃			0.3¶	0.5¶	this work
Fe:MgO	1.00	0	0.37	0.74	69
(Fe₀.₁Mg₀.₉)O	1.08	0.34			50
Fe₀.₉O	1.08	0.30			50

*Isomer shifts relative to Fe foil standard; parameters are in units of mm/s.
†Peaks 3, 4 (Fig. 3(b))
‡Peaks 5, 6 (Fig. 3(b))
§Peaks 1, 2 (Fig. 3(b))
¶Peaks 1, 2 (Fig. 3(a))

Table IV. Energies of Absorption Bands in Optical Spectra of Transition-Metal-Bearing Periclase Structures*

Oxide	Positions of absorption bands (cm⁻¹)	Δ_o (cm⁻¹)	Ref.
Cr³⁺:MgO	16 200; 22 700; 29 700 (14 319)	16 200	53
Fe²⁺:MgO	10 040; 11 628 (21 511; 26 422)	10 040	54–57
Co²⁺:MgO	8470; 18 700; 19 600 (9080; 17 200; 24 600; 28 500)	9 000	58–61
CoO	8333; 17 240; 18 520 (13 000; 16 340; 18 250–18 700; 20 080–20 600; 21 010)	8 500	61
Ni²⁺:MgO	8600; 14 800; 24 500 (13 400; 21 550; 25 950; 28 300; 34 500)	8 500	62–64
NiO	8756; 14 080; 24 100 (15 550; 21 550; 26 450)	8 750	65

*Spin-forbidden transitions appear in parentheses.

emergence of a broad band around 9000 cm⁻¹.[54,57] The latter feature was assigned to a crystal field transition in Fe²⁺ ions in noncubic sites[54] in which some Fe³⁺ ions were postulated to be present in adjacent edge-shared octahedra. Electron paramagnetic resonance spectral measurements have demonstrated coexisting Fe²⁺ and Fe³⁺ in iron-doped MgO,[55] but the weak spin-forbidden Fe³⁺ crystal field peaks could not be detected in optical spectra.[55] The intensity of the 9000-cm⁻¹ band diminished considerably in absorption spectra of Fe-doped MgO crystals heated in reducing atmospheres,[54,57] leading to an alternative suggestion that it represents an Fe^{2+}–Fe^{3+} intervalence transition.[57] Another broad band situated at 13 200 cm^{-1} in as-grown, nonreduced Fe-doped MgO crystals was also assigned to $Fe^{2+} \rightarrow Fe^{3+}$ intervalence charge transfer.[57] The presence of Fe^{3+} cations was also suggested to intensify the Fe^{2+} crystal field bands at 10 000 and 11 600 cm^{-1} by exchange interactions.[57]

The effect of pressure on the optical spectra of Fe^{2+}-doped periclase[66] showed that there are blue shifts (to higher energies) of about 55 and 72 cm^{-1}/GPa, respectively, for the Fe^{2+} crystal field bands near 10 000 and 11 600 cm^{-1} up to 5 GPa. Measurements of four synthetic magnesiowustites containing 9–56 mol% FeO[67] showed similar pressure-induced shifts of the Fe^{2+} crystal field bands. However, the same measurements indicated that the absorption edge in the ultraviolet shifted dramatically into the visible region, obscuring the Fe^{2+} crystal field bands beyond pressures of 20 GPa.[67] Such pressure-induced opacity by these oxygen $\rightarrow$ metal charge-transfer absorption bands has important implications to electrical conductivity and radiative heat-transfer properties of the Earth's interior.[7] However, more recent shock-wave measurements of the visible-region spectra of two synthetic magnesiowustites Wu_{14} and Wu_{26} up to 41 GPa[68] demonstrated that magnesiowustites have considerable lower opacities at short wavelengths (i.e., smaller pressure-induced shifts of the oxygen $\rightarrow$ iron absorption edge) than those inferred from diamond cell measurements,[67] suggesting that the presence of Fe^{3+} ions in iron-doped periclase crystals profoundly affect oxygen $\rightarrow$ iron charge-transfer bands in the ultraviolet region.

Mössbauer Spectra: Fe in MgO

Mössbauer-effect studies have demonstrated that Fe^{2+} and Fe^{3+} coexist in synthetic Fe-doped MgO[54,69] and magnesiowustites.[70] The Mössbauer spectra of magnesiowustites indicated discrete clusters of Fe^{3+} ions[58] with octahedral/tetrahedral ratios of about 3:1. Similar clustering was not observed for coexisting Fe^{2+} ions. Such cation occupancies have important implications in defect structures of iron-doped periclases.

Conclusions

This survey of the optical spectra of transition-metal-bearing corundums and periclases indicates that the electronic spectra of single-cation-doped Al_2O_3 and MgO crystals are well understood, but problems exist in the assignment of spectral features in mixed-cation assemblages. The optical spectra of Cr^{3+} ions in ruby have been thoroughly investigated for a variety of compositions and at elevated temperatures and pressures. Measurements of compositional variations and temperature-induced changes of the optical spectra other than transition-metal-doped Al_2O_3 crystals are needed, as are studies of electronic spectra of doped-MgO crystals at high temperatures.

Many of the problems of assignments of absorption spectra in the visible region involve iron-bearing MgO and Al_2O_3 phases, and valuable information has been derived from Mössbauer spectral measurements. Preliminary Mössbauer data for synthetic sapphires indicate that Fe^{2+} and Fe^{3+} ions coexist in Fe–Ti-doped blue sapphires, supporting the origin of their dark blue colors to $Fe^{2+} \rightarrow Ti^{4+}$ intervalence transitions in the visible region.

Acknowledgments

We thank J. S. Haggerty for synthesizing [57]Fe-doped sapphires by the laser-heated floating zone process. The research is supported by a grant from the National Science Foundation (Grant No. EAR83-13585) and the National Aeronautics and Space Administration (Grant No. NSG-7604).

References

[1] W. A. Deer, R. A. Howie, and J. Zussman; pp. 2, 11 in Rock-Forming Minerals, Non-Silicates, Vol. 5. Longmans, London, 1962.
[2] B. M. Loeffler and R. G. Burns, "Shedding Light on the Color of Gems and Minerals," Am. Sci., 64, 636–47 (1976).
[3] R. G. Burns and D. J. Vaughan, "Polarized Electronic Spectra"; pp. 39–72 in Chapter 2 of Infrared and Raman Spectroscopy of Lunar and Terrestrial Minerals. Edited by C. Karr. Academic Press, New York, 1975.
[4] J. B. Adams, "Interpretation of Visible and Near-Infrared Diffuse Reflectance Spectra of Pyroxenes and Other Rock-Forming Minerals"; pp. 91–116 in Chapter 4 of Infrared and Raman Spectroscopy of Lunar and Terrestrial Minerals. Edited by C. Karr. Academic Press, New York, 1975.
[5] T. B. McCord and J. B. Adams, "Use of Ground-Based Telescopes in Determining the Composition of the Surfaces of Solar System Objects"; pp. 893–922 in The Soviet-American Conference on Cosmochemistry of the Moon and Planets. NASA SP-370, Part 2, 1977.
[6] L. G. Liu, "Phase Transformations and the Constitution of the Deep Mantle"; pp. 117–202 in The Earth: Its Origin, Structure, and Evolution. Edited by M. W. McElhinny. Academic Press, London, 1979.
[7] R. G. Burns, "Electronic Spectra of Minerals at High Pressures: How the Mantle Excites Electrons"; pp. 223–46 in High-Pressure Researches in Geosciences. Edited by W. Schreyer. E. Schweizerbart'sche Verlagsbuchhandlung, Stuttgart, 1982.
[8] B. Simons, "Composition-Lattice Parameter Relationships of the Magnesiowüstite Solid Solution Series"; pp. 376–80 in the Annual Report of the Geophysical Laboratory of the Carnegie Institute, Washington, D.C., Yearbook 79, 1980.
[9] R. Bramley and M. B. McCool, "EPR of Fe^{3+} Pairs in α-Al_2O_3," J. Phys. C. Solid State Phys., 9, 1793–1803 (1976).
[10] R. E. Newnham and Y. M. de Haan, "Refinement of the α-Al_2O_3, Ti_2O_3, V_2O_3, and Cr_2O_3 Structures," Z. Kristallogr., 117, 235–37 (1962).
[11] L. W. Finger and R. M. Hazen, "Crystal Structure and Compression of Ruby to 46 kbar," J. Appl. Phys., 49, 5823–26 (1978).
[12] R. L. Blake, R. E. Hessevick, T. Zoltai, and L. W. Finger, "Refinement of the Hematite Structure," Am. Mineral., 51, 123–29 (1966).
[13] G. Shirane, S. J. Pickart, B. Nathans, and Y. Ishikawa, "Neutron-Diffraction Study of Antiferromagnetic $FeTiO_3$ and Its Solid Solutions with α-Fe_2O_3," J. Phys. Chem. Solids, 10, 35–43 (1959).
[14] N. S. Hush and R. J. M. Hobbs, "Absorption Spectra of Crystals Containing Transition Metal Ions," Prog. Inorg. Chem., 10, 259–486 (1968).
[15] D. S. McClure, "Optical Spectra of Transition-Metal Ions in Corundum," J. Chem. Phys., 36, 2757–79 (1962).
[16] R. Muller and H. H. Gunthard, "Spectroscopic Study of the Reduction of Nickel and Cobalt Ions in Sapphire," J. Chem. Phys., 44, 365–73 (1966).
[17] J. J. Krebs and W. G. Maisch, "Exchange Effects in the Optical-Absorption Spectrum of Fe^{3+} in Al_2O_3," Phys. Rev., B.: Solid State, 4, 757–69 (1971).
[18] K. Eigenmann, K. Kurtz, and H. H. Gunthard, "The Optical Spectrum of α-Al_2O_3: Fe^{3+}," Chem. Phys. Lett., 13, 54–7 (1972).
[19] R. G. Burns, "Crystal Field Effects in Chromium and its Partitioning in the Mantle," Geochim. Cosmochim. Acta, 39, 857–64 (1975).
[20] C. P. Poole, Jr., "The Optical Spectra and Color of Chromium Containing Solids," J. Phys. Chem. Solids, 25, 1169–82 (1964).
[21] A. Neuhaus, "Uber die Ionenfarben der Kristalle und Minerale am Beispiel der Chromfarbungen," Z. Kristallogr., 113, 195–233 (1960).
[22] D. Reinen, "Ligand-Field Spectroscopy and Chemical Bonding in Cr^{3+}-Containing Oxidic Solids," Struct. Bonding (Berlin), 6, 30–51 (1969).
[23] K. M. Parkin and R. G. Burns, "High Temperature Crystal Field Spectra of Transition Metal-Bearing Minerals: Relevance to Remote-Sensed Spectra of Planetary Surfaces," Proc. Lunar Planet. Sci. Conf., 11th, 731–55 (1980).
[24] H. G. Drickamer and C. W. Frank; 220 pp. in Electronic Transitions and the High Pressure Chemistry and Physics of Solids. Chapman and Hall, London, 1973.
[25] E. S. Gaffney and T. J. Ahrens, "Optical Absorption Spectra of Ruby and Periclase at High Shock Pressures," J. Geophys. Res., 78, 5942–53 (1973).

[26] T. Goto, T. J. Ahrens, G. R. Rossman, and Y. Syono, "Absorption Spectra of Cr^{3+} in Al_2O_3 under Shock Compression," *Phys. Chem. Mineral.*, **4**, 253–64 (1979).

[27] H. K. Mao, P. M. Bell, J. W. Shaner, and D. J. Steinberg, "Specific Volume Measurements of Cu, Mo, Pd, and Ag and Calibration of the Ruby R_1 Fluorescence Pressure Gauge from 0.06 to 1 Mbar," *J. Appl. Phys.*, **49**, 3276–83 (1978).

[28] W. B. White, R. Roy, and J. M. Crichton, "The Alexandrite Effect": An Optical Study," *Am. Mineral.*, **52**, 867–71 (1967).

[29] E. F. Farrell and R. E. Newnham, "Crystal Field Spectra of Chrysoberyl, Alexandrite, Peridot, and Sinhalite," *Am. Mineral.*, **50**, 1972–81 (1965).

[30] K. Schmetzer, H. Bank, and E. Gubelin, "The Alexandrite Effect in Minerals: Chrysoberyl, Garnet, Corundum, and Fluorite," *Neues Jahrb. Mineral. Abh.*, **138**, 147–64 (1980).

[31] M. G. Townsend, "Visible Charge Transfer Band in Blue Sapphire," *Solid State Commun.*, **6**, 81–3 (1968).

[32] G. Lehmann and H. Harder, "Optical Spectra of Di- and Trivalent Iron in Corundum," *Am. Mineral.*, **55**, 98–105 (1970).

[33] G. H. Faye, "On the Optical Spectra of Di- and Trivalent Iron in Corundum: A Discussion," *Am. Mineral.*, **56**, 344–8 (1971).

[34] G. Lehmann, "On the Optical Spectra of Di- and Trivalent Iron in Corundum: Reply," *Am. Mineral.*, **56**, 344–50 (1971).

[35] J. Ferguson and P. E. Fielding, "The Origins of the Colours of Yellow, Green, and Blue Sapphires," *Chem. Phys. Lett.*, **10**, 262–5 (1971); *Aust. J. Chem.*, **25**, 1371–85 (1972).

[36] K. Eigenmann and H. H. Gunthard, "Valence States, Redox Reactions, and Biparticle Formation of Fe and Ti Doped Sapphire," *Chem. Phys. Lett.*, **13**, 58–61 (1972).

[37] K. Eigenmann, K. Kurtz, and H. H. Gunthard, "Solid State Reactions and Defects in Doped Verneuil Sapphire. III. Systems α-Fe_2O_3: Fe, α-Al_2O_3: Ti, and α-Al_2O_3: (Fe, Ti)," *Helv. Phys. Acta*, **45**, 452–80 (1972).

[38] G. Smith and R. G. J. Strens, "Intervalence-Transfer Absorption in Some Silicate, Oxide, and Phosphate Minerals"; pp. 583–612 in The Physics and Chemistry of Rocks and Minerals. Edited by R. G. J. Strens. Wiley, New York, 1976.

[39] L. V. Nikolskaya, V. M. Terekhova, and M. I. Samoilovich, "On the Origin of Natural Sapphire Color," *Phys. Chem. Minerals*, **3**, 213–24 (1978).

[40] G. Smith, "Evidence for Absorption by Exchange-Coupled Fe^{2+}-Fe^{3+} Pairs in the Near Infra-red Spectra of Minerals," *Phys. Chem. Minerals*, **3**, 375–83 (1978).

[41] B. G. Granadchikova, L. V. Nikol'skaya, and M. I. Samoilovich, "Nature of the Colours of Natural and Synthetic Sapphires and a Rapid Way of Identifying Them," *Soviet Phys. Dokl.*, **24**, 328–9 (1979).

[42] K. Schmetzer and H. Bank, "Explanations of the Absorption Spectra of Natural and Synthetic Fe- and Ti-Containing Corundums," *Neues Jahrb. Mineral. Abh.*, **139**, 216–25 (1980).

[43] K. Schmetzer and H. Bank, "The Colour of Natural Corundum," *Neues Jahrb. Mineral., Monatsh.*, **2**, 59–68 (1981).

[44] R. V. Morris and H. V. Lauer, "Optical Spectra and Band Assignments for the Hematite (α-Fe_2O_3)-Corundum (α-Al_2O_3) Series," *Lunar Planet. Science*, **XIV**, 524–5 (1983).

[45] G. Shirane, D. E. Cox, and S. L. Ruby, "Mössbauer Study of Isomer Shift, Quadrupole Interaction, and Hyperfine Field in Several Oxides Containing ^{57}Fe," *Phys. Rev.*, **125**, 1158–65 (1962).

[46] W. Triftshauser and D. Schroeer, "Investigations of Charge States in Co^{57}-Doped Oxides Using the Mössbauer Effect and Delayed-Coincidence Techniques," *Phys. Rev.*, **187**, 491–8 (1969).

[47] K. Jonas, K. Solymar, and J. Zoldi, "Some Applications of Mössbauer Spectroscopy for the Quantitative Analysis of Minerals and Mineral Mixtures," *J. Mol. Struct.*, **60**, 449–52 (1980).

[48] G. K. Wertheim and J. P. Remeika, "Mössbauer Effect Hyperfine Structure of Trivalent ^{57}Fe in Corundum," *Phys. Rev.*, **10**, 14–5 (1964).

[49] J. S. Haggerty, "Production of Fiber by a Floating Zone Fiber Drawing Technique," NAS 3-14328, Final Rept. (1972).

[50] G. Shirane, D. E. Cox, and S. L. Ruby, "Mössbauer Study of Isomer Shift, Quadrupole Interaction, and Hyperfine Field in Several Oxides Containing ^{57}Fe," *Phys. Rev.*, **125**, 1158–65 (1962).

[51] C. Blaauw, F. Leenhouts, and F. van der Woude, "Mössbauer Effect of ^{57}Fe in Ti_2O_3," *J. Phys.*, **12**, C6:607–10 (1976).

[52] T. C. Gibb, N. N. Greenwood, and W. Twist, "The Mössbauer Effect of Natural Ilmenites," *J. Inorg. Nucl. Chem.*, **31**, 947–54 (1969).

[53] W. Low, "Paramagnetic Resonance and Optical Absorption Spectra of Cr^{3+} in MgO," *Phys. Rev.*, **105**, 801–5 (1957).

[54] K. W. Blazey, "Optical Absorption of MgO:Fe," *J. Phys. Chem. Solids*, **38**, 671–5 (1977).

[55] F. A. Modine, E. Sonder, and R. A. Weeks, "Determination of the Fe^{2+} and Fe^{3+} Concentration in MgO," *J. Appl. Phys.*, **48**, 3514–8 (1977).

[56] N. B. Manson, J. T. Gourley, E. R. Vance, D. Sengupta, and G. Smith, "The $^5T_{2g} \to {}^5E_g$ Absorption in MgO:Fe^{2+}," *J. Phys. Chem. Solids*, **37**, 1145–8 (1976).

[57] G. Smith, "Evidence for Optical Absorption by Fe^{2+}-Fe^{3+} Interactions in MgO:Fe," *Phys. Status Solidi A,* **61A**, Kl 91–Kl 95 (1980).

[58] W. Low, "Paramagnetic and Optical Spectra of Divalent Cobalt in Cubic Crystalline Fields," *Phys. Rev.,* **109**, 256–65 (1958).

[59] D. Reinen, "Farbe und Konstitution bei anorganischen Feststoffen. II. Die Lichabsorption des oktaedrisch koordinierten Co^{2+}-Ions in der Mischkristallreihe $Mg_{1-x}Co_xO$ und anderen oxidischen Wirtsgittern," *Monatsh. Chem.,* **96**, 730–9 (1965).

[60] R. Pappalardo, D. L. Wood, and R. C. Linares, Jr., "Optical Absorption Study of Co-Doped Oxide Systems," *J. Chem. Phys.,* **35**, 2041–59 (1961).

[61] G. W. Pratt, Jr., and R. Coelho, "Optical Absorption of CoO and MnO Above and Below the Neel Temperature," *Phys. Rev.,* **116**, 281–6 (1959).

[62] W. Low, "Paramagnetic and Optical Spectra of Divalent Nickel in Cubic Crystalline Fields," *Phys. Rev.,* **109**, 247–55 (1958).

[63] O. Schmitz-DuMont, H. Gossling, and H. Brokopf, "Die Lichabsorption des zweiwertigen Nickels in oxydischen Koordinationsgittern," *Z. Anorg. Allg. Chem.,* **300**, 159–74 (1959).

[64] R. Pappalardo, D. L. Wood, and R. C. Linares, Jr., "Optical Absorption Spectra of Ni-Doped Oxide Systems," *J. Chem. Phys.,* **35**, 1460–78 (1961).

[65] G. R. Rossman, R. D. Shannon, and R. W. Waring, "Origin of the Yellow Color of Complex Nickel Oxides," *J. Solid State Chem.,* **39**, 277–87 (1981).

[66] T. J. Shankland, A. G. Duba, and A. Woronow, "Pressure Shifts of Optical Absorption Bands in Iron-Bearing Garnet, Spinel, Olivine, Pyroxene, and Periclase," *J. Geophys. Res.,* **79**, 3273–82 (1974).

[67] H. K. Mao, "Charge Transfer Processes at High Pressure"; pp. 573–81 in Physics and Chemistry of Minerals and Rocks. Edited by R. G. J. Strens. Academic Press, London, 1976.

[68] T. Goto, T. J. Ahrens, G. R. Rossman, and Y. Syono, "Absorption Spectrum of Shock-Compressed Fe^{2+}-Bearing MgO and the Radiative Conductivity of the Lower Mantle," *Phys. Earth Planetary Interiors,* **22**, 277–88 (1980).

[69] H. R. Leider and D. N. Pipkorn, "Mössbauer Effect in MgO:Fe^{2+}; Low-Temperature Quadrupole Splitting," *Phys. Rev.,* **165**, 494–500 (1968).

[70] G. A. Waychunas, "Mössbauer, X-Ray, Optical, and Chemical Study of Cation Arrangements and Defect Association in Fm3m Solid Solutions in the System Periclase-Wustite-Lithium Ferrite"; Ph.D. Thesis, U.C.L.A., 479 pp., 1981.

[71] R. G. Burns, "Intervalence Transitions in Mixed-Valence Minerals of Iron and Titanium," *Ann. Rev. Earth Planet. Sci.,* **9**, 345–83 (1981).

[72] J. Hess and A. Levy, "Response of the Mössbauer Spectrum of Paramagnetic Fe^{3+} in Al_2O_3 to Nuclear Dipole Fields," *Phys. Rev. B: Condens. Matter.,* **B22**, 5068–78 (1980).

Calculated Point Defect Formation, Association, and Migration Energies in MgO and α-Al$_2$O$_3$

W. C. Mackrodt

I.C.I. plc., New Science Group
The Heath, Runcorn
Cheshire WA7 4QE, England

Calculated energies for the formation, migration, and association of point defects in MgO and α-Al$_2$O$_3$ are reviewed. A brief outline of the theoretical basis for these calculations is given, and where possible, comparisons are made with experimental data.

The traditional role of theory in both solid-state physics and materials science has been largely to provide a framework for (a) assessing existing experimental information and (b) provoking new investigation. However, over the past decade or so, a third component has emerged: it involves the *direct calculation* of solid-state properties, many of which are either inaccessible to experiment or obtainable only with extreme difficulty. The computed point defect structures of ionic materials, including oxides, are a good illustration of this comparatively recent innovation, which could have important implications for the future development of ceramic and other technologies. Accordingly, this paper provides a survey of atomistic calculations for the simplest of ceramic materials, namely, MgO and α-Al$_2$O$_3$, the structure and properties of which are the central theme of the present conference. In no way is this survey to be regarded as comprehensive, for that would require much greater coverage than is appropriate here; however, it is intended to present a self-consistent description of the point defect structures of MgO and α-Al$_2$O$_3$, in that all the results derive from the same well-defined theoretical methods and computational procedures. The objective is to show that a diversity of point defect properties can be calculated, often to experimental accuracy, by means of a relatively simple theoretical approach; this suggests a possible extension of the methods to systems of wider technological significance.

Theoretical Basis

The theoretical basis for the calculations reported here has been discussed in full recently.[1,2] Two features, in particular, deserve comment, for ultimately they determine the extent to which the methods used for MgO and α-Al$_2$O$_3$ can be applied to a wider range of ceramics. They are the derivation of interatomic potentials and the treatment of lattice relaxation. Both relate to the physical model invoked for these materials, which can be thought of as essentially ionic; that is to say, electron transfer from cations to anions leads to *effective* ionic charges that are very close to integral values. This has often been assumed to be the case,[2] and recent ab initio quantum mechanical calculations confirm this, for they indicate that in MgO, for example, the net ionic charges are approximately $\pm 1.99|e|$, based on a Mulliken population analysis of cluster wave functions.[3] By comparison, identical calculations for FeO suggest a value closer to $\pm 1.70|e|$ in keeping with the somewhat more covalent bonding of the transition-metal oxides.[4] An important consequence of this high degree of ionicity is that interatomic potentials can be represented, to a good approximation, by *two-body interactions*. These are much simpler to construct than higher-order (angle-dependent) terms and are relatively straightforward to implement in lattice simulation codes, which are used to evaluate energies and local atomic configurations.[1]

The derivation of interatomic potentials, whether two-body or otherwise, is a ubiquitous problem in the physics of condensed matter. For ionic and semiionic materials, *particularly in relation to lattice defect calculations*, it is essentially twofold, in that it concerns both the short-range (repulsive) interaction between ions and the polarization of the lattice in the presence of excess local charge. The most convenient, and as yet most accurate, way of dealing with the latter is by means of the "shell model" introduced by Dick and Overhauser some time ago.[5] It has been discussed on a number of occasions in the present context.[2,6] Despite its simplicity, it has proved to be remarkably successful in accounting for polarization effects in both defective and nondefective lattices[2,7] and has been used throughout the calculations reported here. Short-range interactions, on the other hand, have been approached in a number of different ways.[2] It has been argued elsewhere[6,7] that, of these, procedures based on the electron-gas approximation offer a number of advantages, notably that (a) they are essentially nonempirical and hence do not rely on structural information, which may not always be available, (b) they relate directly to the electronic structure of the material, (c) impurity potentials can be derived in a way identical to those for the host lattice, so that impurity-host interactions can be distinguished for different lattices, and (d) potentials for different charge states of ions, including the polaron hole, can be calculated in a consistent way. For these and other reasons,[8] all the calculations reported here are based on modified electron-gas potentials of the type discussed by Mackrodt and Stewart[9] and collated recently by Colbourn et al.[10]

A feature of ionic materials, as distinct from metals and covalent solids, is that lattice relaxation plays a major role in determining defect structures and energies. The most convenient way of dealing with this from a computational point of view is that developed by Norgett[11] in his Hades procedures, from which other codes have been derived.[1] The essential strategy is to partition the solid surrounding a defect into a finite proximate region, in which the lattice relaxation is calculated explicitly by solving the appropriate elastic equations of the defective crystal, and an infinite outer region in which the ion displacements are estimated on the basis of the Mott–Littleton approximation.[12] The derivation of the relevant equations and their method of solution have been discussed recently by Catlow and the author.[1] The whole procedure involves a number of *approximations*, and it is, perhaps, worthwhile mentioning *two of these*, for they have not been emphasized sufficiently yet could be significant in comparisons between computed and experimental defect energies. When the kinetic and potential energies, respectively, are written as K and V, the total energy, $E_L(T)$, of a *nondefective* lattice at a temperature T, is given by

$$E_L(T) = K_L(T) + V_L(T) \tag{1}$$

and that of a *defective* lattice by

$$E_D(T) = K_D(T) + V_D(T) \tag{2}$$

Hence, the defect energy, $\Delta E(T)$, is given by

$$\Delta E(T) = [K_D(T) - K_L(T)] + [V_D(T) - V_L(T)] \tag{3}$$

However, in most treatments two assumptions are made. The first is that

$$[K_D(T) - K_L(T)] \approx 0 \tag{4}$$

and the second that

$$\Delta E(T) \approx \Delta E(0) \tag{5}$$

In other words, the effects of anharmonicity and lattice expansion are neglected: both could be important in certain circumstances. As a result, we would expect *absolute* energies such as the formation energy of defects to be calculated less accurately than energy *differences* such as those involved in the association and migration of defects.

Calculated Defect Energies

As mentioned earlier, all the calculations reported here are based on shell-model, electron-gas potentials.[9] In some cases, there are small differences between the present results and those reported earlier.[9,13–15] These arise solely from an improved method for implementing the short-range potentials in the lattice relaxation procedures.[16] Furthermore, recent investigations have shown that, while calculations for the most symmetric defects in cubic lattices can be made to converge to ±0.01 eV, the results for asymmetric defects in noncubic materials are generally no better than ±0.1 eV. For consistency, therefore, all energies are reported here to the nearest tenth of a volt.

Intrinsic Defects

Table I lists the calculated formation energies of intrinsic defects in MgO and α-Al$_2$O$_3$. For both materials Schottky disorder is predicted, with effective formation energies *per defect* of 3.8 and 5.0 eV, respectively: the level of intrinsic defects, therefore, is expected to be extremely low. Frenkel disorder, on the other hand, would require cation interstitials to be approximately 4–5 eV more stable than those given in Table I, while anti-Frenkel disorder would necessitate a lowering in energy of about 10–12 eV for anion interstitials. The latter, in particular,

Table I. Calculated Formation Energies of Intrinsic Defects in MgO and α-Al$_2$O$_3$

Defect	Energy (eV) MgO	Energy (eV) α-Al$_2$O$_3$
Cation vacancy	25.4	61.0
Anion vacancy	22.8	21.7
Cation interstitial	−12.8	−47.0
Anion interstitial	−5.5	−5.2
Vacancy pair	45.6	
Schottky*	7.5 (3.8)†	25.2 (5.0)
Frenkel	13.0 (6.5)	14.0 (7.0)
Anti-Frenkel	17.6 (8.6)	16.5 (8.3)

*Lattice energy: MgO, −40.7 eV; α-Al$_2$O$_3$, −161.9 eV.
†Energy per defect in parentheses.

would be difficult to accommodate within the present theoretical framework. Furthermore, identical calculations for other oxides ranging from Li_2O, through CaO and SrO, to CeO_2 and ThO_2, suggest quite strongly that the type of disorder is determined by cation size. For the smallest cations, such as Li^+, Frenkel defects predominate the oxide, whereas anti-Frenkel defects seem to be associated with large cations such as Ce^{4+}, U^{4+}, and Th^{4+}. Vacancy pairs in MgO are found to be strongly bound, by 2.6 eV, so that the effective formation energy is reduced to 4.9 eV. However, this is still greater than the effective formation energy per defect of free vacancies. A comparable interaction of about -3 eV would be expected for α-Al_2O_3.

Dopant Energies and Interactions

Whether by accident or design, impurities play a major role in determining the defect properties of most materials, for their effects are numerous, far reaching, and varied. An understanding of "model" systems such as MgO and α-Al_2O_3, therefore, would seem to be an essential precursor to the investigation of a wider (and more complex) range of ceramic materials. Here we are concerned solely with *cation* impurities and, in particular, with dopant energies and interactions, with the order of magnitude of these energies, and the influence that dopant-defect interactions have on the mode of solution. Throughout, we concentrate on those generic features that might reasonably apply elsewhere.

In principle, dopants can enter a lattice either as interstitials, as substitutions, or, in some cases, as both; and it is the potential diversity of the different modes of solution that is often the root of the problem, but where calculations might prove most useful. Table II illustrates this for a divalent oxide such as MgO in the *absence* of dopant-defect interactions, which inevitably complicate the situation still further. We distinguish three types of dopant, namely, *subvalent,* in which the dopant charge is less than that of the host cation, *isovalent,* in which the charges are equal, and *supervalent,* in which the dopant charge is greater than that of the host. Typical examples are the following: subvalent, Li^+:MgO and Mg^{2+}:α-Al_2O_3; isovalent, Ca^{2+}:MgO and Ti^{3+}:α-Al_2O_3; supervalent, Sc^{3+}:MgO and Si^{4+}:α-Al_2O_3. We also distinguish three limiting situations. They are the high-temperature infinite-dilution limit, in which there are no defect-dopant

Table II. Possible Modes of Solution for the Doping of Divalent Oxides (MO) in the Absence of Oxygen Exchange

Monovalent (X_2O)

$X_2O \rightleftharpoons 2X_i^\cdot + O_i''$
$X_2O + M_M^x \rightleftharpoons 2X_i^\cdot + V_O^{\cdot\cdot} + MO$
$X_2O + 2M_M^x + O_O^x \rightleftharpoons 2X_M' + V_O^{\cdot\cdot} + 2MO$
$X_2O + M_M^x \rightleftharpoons X_i^\cdot + X_M' + MO$

Divalent (XO)

$XO \rightleftharpoons X_i^{\cdot\cdot} + O_i''$
$XO + M_M^x \rightleftharpoons X_i^{\cdot\cdot} + V_M'' + MO$
$XO + M_M^x \rightleftharpoons X_M^x + MO$

Trivalent (X_2O_3)

$X_2O_3 \rightleftharpoons 2X_i^{\cdot\cdot\cdot} + 3O_i''$
$X_2O_3 + 3M_M^x \rightleftharpoons 2X_i^{\cdot\cdot\cdot} + 3V_M'' + 3MO$
$X_2O_3 + 3M_M^x \rightleftharpoons 2X_M^\cdot + V_M'' + 3MO$

interactions, the intermediate-temperature low-concentration limit, in which the dopant is present largely as defect dimers and trimers, and the low-temperature high-concentration limit, in which there is appreciable aggregation or clustering of defects.

We begin by considering the high-temperature infinite-dilution limit where the resulting disorder is determined solely by the relative energies of the basic defects. Tables III and IV list the calculated heats of solution for typical 1+ to 4+ oxides in MgO and α-Al$_2$O$_3$, based on interstitial and substitution modes, respectively. In the absence of dopant interactions, the compensating defects simply follow the basic disorder, namely, vacancy in both cases. An interesting point that does emerge from these results is that interstitial dopants represent a highly unfavorable mode of solution, irrespective of cation size or charge. Indeed, it is only for Li$^+$ in α-Al$_2$O$_3$ that an interstitial mechanism is at all favorable. From this we conclude that the (cation) doping of oxides, quite generally, leads to *lattice substitution* in the absence of defect interactions and that the only exception to this might be Li$^+$.

We turn now to defect-dopant interactions and, in particular, to the interaction of impurities with vacancies in MgO. In the *absence* of oxygen exchange, calculations predict that Na$^+$ will be compensated by oxygen vacancies. As shown in Table V, the corresponding dimers are strongly bound with interaction energies *per vacancy* close to 1.5 eV. Trimers appear to be only slightly less stable. This suggests that the concentration of free vacancies will remain low even at high temperatures and that the calculated heats of solution will be reduced by over 2 eV from the value at infinite dilution. However, it is important to note that the calculated enthalpy for the reaction

$$\tfrac{1}{2}O_2(g) + V_O^{\cdot\cdot} \rightarrow O_O^{\times} + 2h^{\cdot} \tag{6}$$

is -1.2 eV; if incorporation of oxygen into Na$^+$:MgO is allowed, electron holes are the preferred defects rather than oxygen vacancies. We note, without exploring the ramifications, the interaction energy of $+0.7$ eV between Na$'_{Mg}$ and a cation vacancy. This implies a severe restriction on the migration of Na$^+$ and suggests that

Table III. Calculated Heats of Solution for Interstitial Doping of MgO and α-Al$_2$O$_3$ in the Absence of Defect Interactions

Dopant oxide	Anion interstitial compensation	Cation vacancy compensation
	MgO	
Na$_2$O	12.1	5.1
MgO	19.5*	11.9†
α-Al$_2$O$_3$	26.4	14.9
TiO$_2$	41.6	26.4
	α-Al$_2$O$_3$	
Li$_2$O	15.1	3.5
MgO	19.3	11.2
α-Al$_2$O$_3$	52.3*	14.0†
TiO$_2$	39.4	23.2

*Energy for interstitial disorder.
†Frenkel energy.

the dissolution of Na₂O in MgO will be limited despite the low enthalpy of solution.

Doping by *isovalent* cations is determined essentially by size effects and, in particular, by the way in which the elastic strain due to the impurity can be accommodated. In the presence of neighboring vacancies, this strain can be relieved by a movement of the impurity away from the normal cation site toward the vacancy with a lowering of the lattice energy. It might be appropriate to refer to the resulting defect as a "strain-compensated dimer" by analogy with the more common charge-compensated defect.[13] Table V lists the calculated association energies of Ca^{2+}, Sr^{2+}, and Ba^{2+} with a {110} cation vacancy in MgO. Two points emerge from this: the first is the magnitude of the interaction, which can be a few tenths of an eV; the second is the clear size effect shown by this series of impurities. This was noted originally in MgO[13] and confirmed subsequently for other systems.[18,19]

Supervalent impurities give rise to cation vacancies in MgO, which can associate to form charge-compensated dimers and trimers. Some typical calculated

Table IV. Calculated Heats of Solution for the Substitutional Doping of MgO and α-Al₂O₃ in the Absence of Defect Interactions

Dopant oxide	Heats of solution (eV) per dopant cation	
	Interstitial compensation	Vacancy compensation
	MgO	
Na₂O	5.1	3.0
MgO		
α-Al₂O₃	6.0	2.2
TiO₂	7.3	3.5
	α-Al₂O₃	
Li₂O	6.8	5.9
MgO*	3.6	3.2
α-Al₂O₃		
TiO₂	7.6	3.5

*For compensation by magnesium interstitials, the calculated heat of solution is 3.0 eV.

Table V. Calculated Association Energies of Cation Impurities with Vacancies in MgO

Defect	Association energy (eV)
$Li'_{Mg}-V^{\bullet\bullet}_O$	−1.4
$Na'_{Mg}-V^{\bullet\bullet}_O$	−1.3
$Na'_{Mg}-V''_{Mg}$	+0.7
$Ca^{x}_{Mg}-V''_{Mg}$	−0.2
$Sr^{x}_{Mg}-V''_{Mg}$	−0.3
$Ba^{x}_{Mg}-V''_{Mg}$	−0.4
$Al^{\bullet}_{Mg}-V''_{Mg}\{100\}$	−0.9
$Sc^{\bullet}_{Mg}-V''_{Mg}$	−0.8
$Y^{\bullet}_{Mg}-V''_{Mg}$	−1.0
$In^{\bullet}_{Mg}-V''_{Mg}$	−1.0
$Ti^{\bullet\bullet}_{Mg}-V''_{Mg}$	−2.0
$Ge^{\bullet\bullet}_{Mg}-V''_{Mg}$	−1.5

binding energies are given in Table V. They range from 0.8 eV for Sc^{3+} to 2 eV for Ti^{4+}, which suggests that four-valent impurities will be present largely as dimers, whereas doping by three-valent cations will lead to free vacancies, isolated impurities, dimers, and trimers with consequent implications for defect-related properties ranging from ionic conduction to space-charge layer formation at surfaces. Present calculations cannot sensibly distinguish cations of similar size such as Sc^{3+}, Cr^{3+}, and Fe^{3+}, for example, all of which are predicted to bind to vacancies by just over 0.8 eV; however, as is the case for the divalent impurities reported here, grosser size effects do emerge, not only with regard to binding energies but in the case of Al^{3+}, an alternative orientation of the dimer. While all the other supervalent impurities form {110} vacancy dimers, calculations suggest that Al^{3+} forms a {100} dimer. The reason for this is that the local distortions surrounding the vacancy and the impurity are "in phase" in the {100} configuration thereby leading to the maximum degree of relaxation, whereas in the {110} configuration, they are "out of phase" so that the resulting lattice stabilization is limited.[15]

Defect-impurity interactions in α-Al_2O_3 have not been examined as widely as they have in MgO; however, both Ti^{4+} and Mg^{2+} have been considered in some detail, for they are the two impurities that seem to have received closest experimental scrutiny. We recall in Table IV the calculated heat of solution of TiO_2 in α-Al_2O_3 based on oxygen interstitial and aluminum vacancy compensation: they are 7.6 and 3.5 eV, respectively. Thus, vacancies are favored by about 4 eV. To assess the influence of defect-impurity interactions, James[17] has investigated various interstitial and vacancy clusters involving $Ti^{·}_{Al}$; his calculated association energies and resulting heats of solution are collected in Table VI. From this we note first that the binding energy of these aggregates is appreciable, in excess of 1.5 eV per impurity ion, and second that cation vacancies remain the favored mode of compensation in the absence of entropic considerations, although the difference between the two modes of solution is now only 1.7 eV. However, in any event, calculations strongly suggest that extensive clustering will occur even at high temperatures and that the effects of this need to be included in a proper interpretation of experimental data.

For the most part, the defect structures referred to in the preceding discussion have been relatively straightforward in that calculations have favored a particular type of disorder or mode of solution for impurities. We now consider two examples for which a more complex situation, involving more than one type of disorder, is predicted. Table VII lists various defect reactions for the dissolution of Li_2O in MgO together with the calculated enthalpies. The less favorable modes of solution involving Li^+ substitution compensated by magnesium interstitials and Li^+ inter-

Table VI. Calculated Association Energies and Heats of Solution of Ti^{4+} Defect Aggregates in α-Al_2O_3[17]

Defect aggregate	Association energy (eV)	Heat of solution (eV)
$Ti^{·}_{Al}$–O''_i–$Ti^{·}_{Al}$	3.1	3.5
$(Ti^{·}_{Al}$–$V''''_{Al}) + 2Ti^{·}_{Al}$	1.9	2.9
$(Ti^{·}_{Al}$–V''''_{Al}–$Ti^{·}_{Al}) + Ti^{·}_{Al}$	3.5	2.3

$(Ti^{·}_{Al}$–V''''_{Al}–$Ti^{·}_{Al})$
    |
  $Ti^{·}_{Al}$
	5.2	1.8

stitials compensated by magnesium vacancies and oxygen interstitials have been omitted, for the calculated enthalpies are all 5 eV per impurity or greater. However, as Table VII indicates, even without these unfavorable modes of solution, calculations suggest that the defect structure of Li$_2$O in MgO could consist of free anion vacancies, Li$^+$ substitutions, Li$^+$ interstitials, vacancy dimers and trimers, interstitial complexes, free holes, and V_{Li} centers. The complex, (Li$_i^·$–V$_{Mg}^{''}$–Li$_i^·$), has been discussed previously.[15] It is formed by the association of an interstitial, Li$_i^·$, and a substitution, Li$_{Mg}'$, which moves off center to give a "dumb-bell"-like defect oriented in the {111} direction. The rotation barrier in the {110} plane is calculated to be about 0.5 eV. Clearly, entropic considerations are important in this case, as are the temperature and oxygen partial pressure. What the calculations here suggest, though, is that the disorder is likely to be complicated.

Our second example relates to the doping of α-Al$_2$O$_3$ by magnesia, which has been examined recently by James.[17] With reference to Table III, magnesium interstitials compensated by either oxygen interstitials or aluminum vacancies appear to be highly energetic; however, as shown in Table VIII, the three alternative modes, namely, magnesium substitution compensated by both oxygen vacancies and aluminum interstitials and the self-compensating mode, (2Mg$_{Al}'$ + Mg$_i^{··}$), are

Table VII. Calculated Heats of Solution for Li$_2$O:MgO

Defect process	Calculated enthalpy (eV) per Li$^+$ ion
Li$_2$O + 2Mg$_{Mg}^x$ + O$_O^x$ ⇌ 2Li$_{Mg}'$ + V$_O^{··}$ + 2MgO	2.7
⇌ (Li$_{Mg}'$–V$_O^{··}$) + Li$_{Mg}'$ + 2MgO	2.0
⇌ (Li$_{Mg}'$–V$_O^{··}$–Li$_{Mg}'$) + 2MgO	1.3
Li$_2$O + Mg$_{Mg}^x$ ⇌ Li$_{Mg}'$ + Li$_i^·$ + MgO	2.0
⇌ (Li$_i^·$–V$_{Mg}^{''}$–Li$_i^·$) + MgO	0.8
½O$_2$(g) + V$_O^{··}$ ⇌ O$_O^x$ + 2h$^·$	−1.2*
Li$_2$O + 2Mg$_{Mg}^x$ + ½O$_2$ ⇌ 2Li$_{Mg}'$ + 2h$^·$ + 2MgO	2.2
⇌ V$_{Li}'$ + Li$_{Mg}'$ + h$^·$ + 2MgO	~1.5–2
⇌ 2V$_{Li}'$ + 2MgO	~1–1.5

*The calculated enthalpy for this reaction is based on the following defect energies: formation energy of a large polaron hole, h, 5.4 eV; sum of ionization potentials of lattice oxygen, −8.2 eV; dissociation energy of O$_2$(g), 5.2 eV.

Table VIII. Calculated Heats of Solution of MgO:α-Al$_2$O$_3$ after James[17]

Defect Process	Calculated enthalpy (eV)
MgO + ⅔Al$_{Al}^x$ ⇌ ⅔Mg$_{Al}'$ + ⅓Mg$_i^{··}$ + Al$_2$O$_3$	3.0
⇌ ⅓(Mg$_{Al}'$–Mg$_i^{··}$–Mg$_{Al}'$) + Al$_2$O$_3$	2.2
MgO + Al$_{Al}^x$ + ½O$_O^x$ ⇌ Mg$_{Al}'$ + ½V$_O^{··}$ + ½Al$_2$O$_3$	3.2
⇌ ½(Mg$_{Al}'$–V$_O^{··}$) + ½Mg$_{Al}'$ + ½Al$_2$O$_3$	2.5
⇌ ½(Mg$_{Al}'$–V$_O^{··}$–Mg$_{Al}'$) + ½Al$_2$O$_3$	1.9
MgO + Al$_{Al}^x$ ⇌ Mg$_{Al}'$ + ⅓Al$_i^{···}$ + ⅓Al$_2$O$_3$	3.6
½O$_2$(g) + V$_O^{··}$ ⇌ O$_O^x$ + 2h$^·$	3.6–4.4*

*Calculated value taken from Ref. 14. The two values correspond to large and small polaron models, respectively, for the free hole.

predicted to be much more favorable, with an energy difference of only 0.6 eV between the three. In the absence of defect-impurity interactions, the self-compensation mode is lowest in energy, but the difference between this and vacancy compensation appears to be small. The inclusion of clustering to form interstitial trimers, $(Mg'_{Al}-Mg''_{I}-Mg'_{Al})$, and vacancy dimers, $(Mg'_{Al}-V''_{O})$, and trimers, $(Mg'_{Al}-V''_{O}-Mg'_{Al})$, reduces both the enthalpy of solution (by about 1 eV) and the difference in energy between the various modes, although now vacancy aggregates are thought to be the most stable. However, the important point is that calculations suggest that the disorder associated with the dissolution of magnesia in alumina is likely to be complicated. Unlike MgO, the oxidation of vacancies to form holes is predicted to be endothermic by about 4 eV so that the level of electronic disorder would be expected to be very low in this case.

Ion Migration

Ion migration is a sensitive probe in the study of defect structures; it is also implicated in a number of important nonequilibrium processes in solids. It would seem to be advantageous, therefore, to have a *quantitative* description of the characteristic mechanisms and energies in simple oxides such as MgO and α-Al_2O_3 prior to the investigation of a wider range of ceramic materials.

The discussion of the preceding sections suggests that ion migration in MgO takes place by a *vacancy mechanism*. In the case of free vacancies, the process is quite straightforward in that it involves a jump in the {110} direction, as shown in Fig. 1(a), with the corresponding saddle point configuration in 1(b). The calculated activation energies are 2.1 and 2.4 eV for cation and anion vacancies, respectively. Vacancy pairs, on the other hand, can migrate either by reorientation or by a "dissociative mechanism"; these are illustrated in Fig. 2(a,b). Of the two, the dissociative mechanism is calculated to be the more facile with an overall migration energy of 2.5 eV corresponding to a cation vacancy jump of the type shown in 2(b). The anion vacancy jump re-forming the vacancy pair requires 1.4 eV.

The migration of impurities by a vacancy mechanism is also a two-stage process involving the direct exchange of the impurity and the migration of a bound cation vacancy as shown in Fig. 3; either can determine the migration energy. Table IX lists the calculated activation energies for a series of 2+, 3+, and 4+

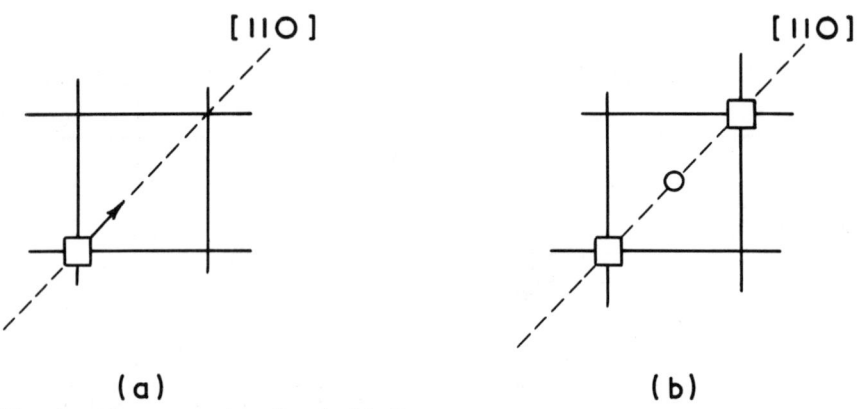

Fig. 1. Vacancy migration in MgO.

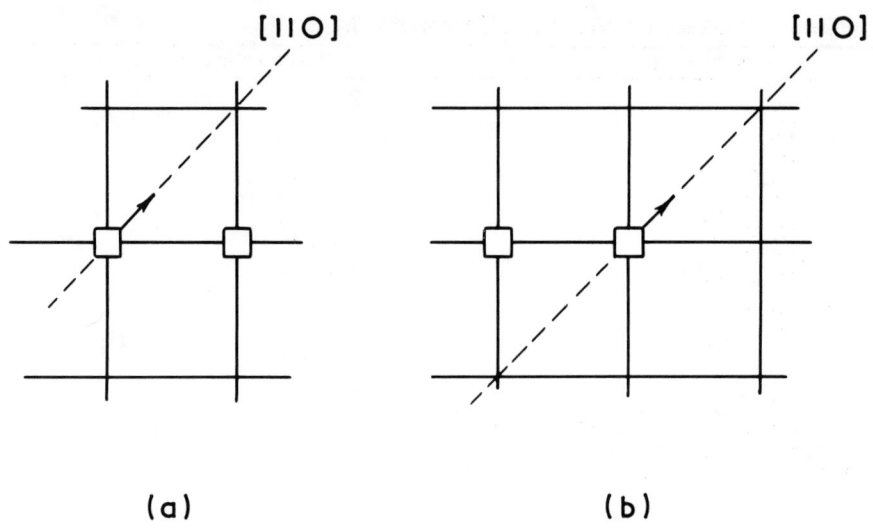

Fig. 2. Vacancy pair migration in MgO.

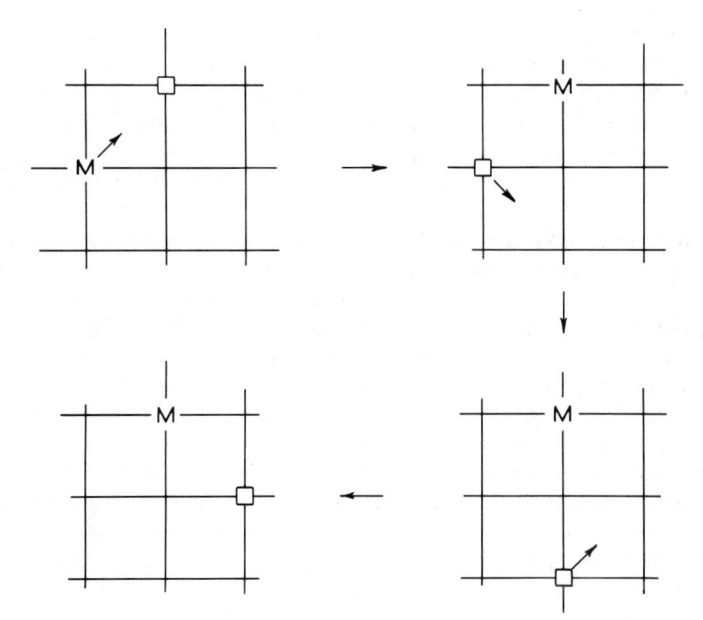

Fig. 3. Impurity dimer migration in MgO.

Table IX. Calculated Migration Energies in MgO

Migrating species	Activation energy (eV)
V''_{Mg}	2.1
$V^{\cdot\cdot}_O$	2.4
$(V''_{Mg}-V^{\cdot\cdot}_O)$	2.5 (V''_{Mg})*
$(Ca^x_{Mg}-V''_{Mg})$	2.5 (Ca^x_{Mg})
$(Sr^x_{Mg}-V''_{Mg})$	2.7 (V''_{Mg})
$(Ba^x_{Mg}-V''_{Mg})$	2.6 (V''_{Mg})
$(Al^{\cdot}_{Mg}-V''_{Mg})$	2.8 $(Al^{\cdot}_{Mg})$
$(Sc^{\cdot}_{Mg}-V''_{Mg})$	3.4 $(Sc^{\cdot}_{Mg})$
$(Y^{\cdot}_{Mg}-V''_{Mg})$	3.3 $(Y^{\cdot}_{Mg})$
$(In^{\cdot}_{Mg}-V''_{Mg})$	3.5 $(In^{\cdot}_{Mg})$
$(Ge^{\cdot\cdot}_{Mg}-V''_{Mg})$	$4.0 < E_m < 4.5$ $(Ge^{\cdot\cdot}_{Mg})$†

*The parentheses include the defect responsible for the migration energy.
†Numerical convergence problems arise here for the saddle-point transition state so that the migration energy, E_m, can only be estimated.

impurity-vacancy dimers in MgO based on this two-stage process. Migration energies are predicted to increase with both charge and size, from 2.5 eV for Ca^{2+} to ~4.5 eV for Ge^{4+}. A rough "rule-of-thumb" seems to be ~2.5 eV for divalent, ~3.5 eV for trivalent, and ~4.5 eV for quadrivalent impurities, respectively. Size effects, on the other hand, appear not to be quite so clear-cut, as witnessed by the calculated energies for Ba^{2+} and Y^{3+}. For the most part, the direct exchange of the impurity determines the migration energy, with two exceptions, namely, $(Sr^x_{Mg}-V''_{Mg})$ and $(Ba^x_{Mg}-V''_{Mg})$, for which the migration of the bound vacancy is found to be the more activated by 0.4 and 0.7 eV. Overall, then, calculations suggest that ion migration in MgO is not a facile process, requiring in excess of 2 eV, whatever the ion.

Vacancy migration in α-Al_2O_3 is somewhat more complicated to deal with than in MgO due, quite simply, to the increased complexity of the corundum structure and the number of different routes that are possible. For cation vacancies, James has identified six unique jumps, of which the most facile that contributes to ion migration has an activation energy of 1.8 eV.[17] It is illustrated in Fig. 4 and corresponds to a cation jump across an anion plane between two sites that do not share an edge of the trigonal bipyramid consisting of four cation (C_1–C_2–C_3–C_3) and two interstitial (O_1–O_2) sites surrounding an anion. We note that this particular jump has components both parallel to and perpendicular to the c axis that would lead to *isotropic* conductivity. James has also identified five anion jumps, and here the lowest energy path is along a shared edge between two filled octahedra, as shown in Fig. 5. The migration energy is calculated to be 1.8 eV, and once again this route has components parallel and perpendicular to the c axis. Thus, calculations predict that ionic conductivity in α-Al_2O_3 will be isotropic and that the Arrhenius energies for cation and anion diffusion will be practically identical. So far there have been no comparable studies of impurity migration.

Comparison with Experiment

We turn now to a comparison with existing data, for it is only in relation to experiment that the prospective utility of calculations of this type can be assessed. There seems to be widespread agreement that Schottky disorder predominates in

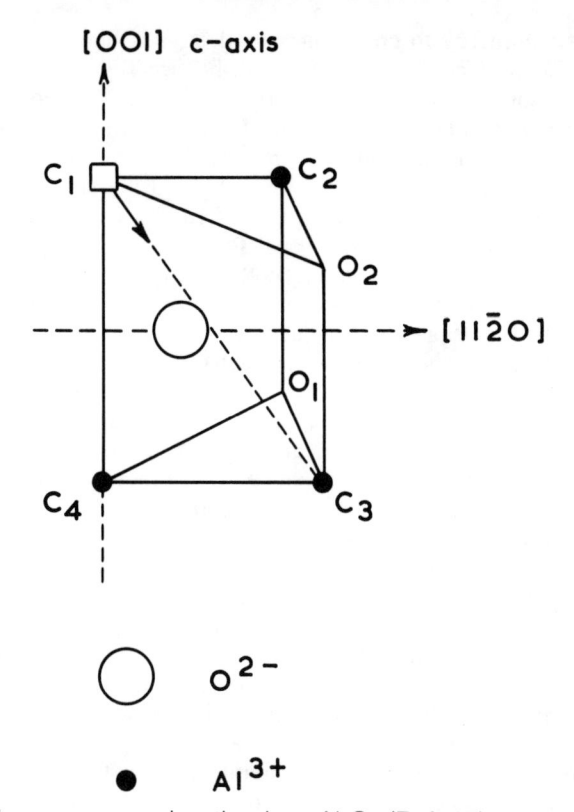

Fig. 4. Cation vacancy migration in α-Al$_2$O$_3$ (Ref. 17).

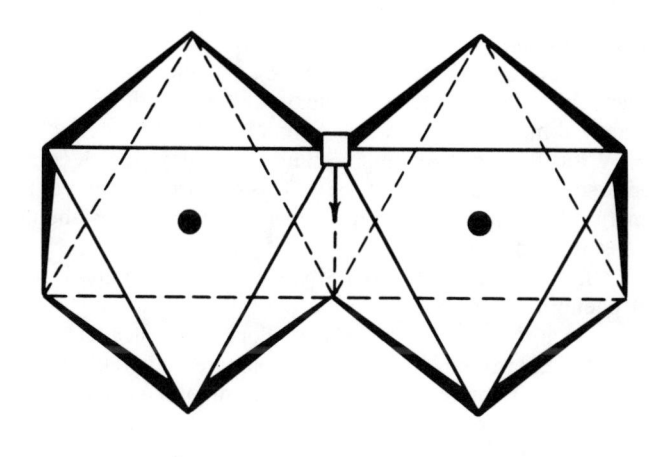

Fig. 5. Anion vacancy migration in α-Al$_2$O$_3$ (Ref. 17).

MgO and that the energy to create intrinsic defects is high.[9,20-22] Wuensch has suggested a value in excess of 4.5 eV,[21] and this accords with calculations. For α-Al$_2$O$_3$ the position is less clear. In an early study based on theoretical methods, which were broadly similar to those outlined here, Dienes et al.[23] concluded that Schottky defects were the major defects in α-Al$_2$O$_3$, with a calculated energy per defect of 5.7 eV, in good agreement with that given in Table I. Mohapatra and Kröger[24] found support for this view of the disorder from an analysis of data on doped α-Al$_2$O$_3$, although the energy per defect they suggested was about 2 eV lower than the theoretical value. Subsequently, Phillips et al.,[25] from what seems to have been a very careful examination of dislocation loops surrounding precipitated rutile in star sapphire, concluded that oxygen interstitials rather than aluminum vacancies are the compensating defects for Ti^{4+} in α-Al$_2$O$_3$; and this *in the absence of defect-impurity interactions* points to anti-Frenkel disorder. However, these results are not in stark contradiction to calculation, for we note two points in particular. First, our results (and those of Dienes et al.[23]) refer to pure α-Al$_2$O$_3$; and, as Phillips et al.[25] have pointed out, Schottky defects could still be the major disorder in the pure material. Second, our calculations suggest very appreciable defect-impurity interactions for Ti^{4+} in alumina. So far, we have considered only the simplest type of clusters and shown that the energy per Ti^{4+} is reduced more for oxygen interstitials than for aluminum vacancies. It seems quite feasible that more extensive clustering, which might reasonably be expected in the vicinity of precipitated rutile, would favor oxygen interstitial compensation altogether, but this we leave for future investigation. The role that clustering is thought to play is further emphasized by the calculated defect structure of magnesium-doped alumina. Here defect association is predicted to favor oxygen vacancy compensation, in preference to aluminum interstitials. This view seems to be supported by Pleska et al.[26] on the basis of precipitation studies of MgO in α-Al$_2$O$_3$. We conclude, therefore, that, as far as the major disorder in the *pure* materials is concerned, there is no experimental evidence in direct contradiction to our view that Schottky defects predominate the defect equilibria in both MgO and α-Al$_2$O$_3$.

With regard to quantitative details of defect-impurity association, here the data are sparse; however, two reports, which are available, support our calculated energies. Weber et al.[27] have deduced a value of 0.88 eV for the association energy of the Cr vacancy dimer in MgO, while Sempolinski and Kingery[28] obtained a value of 0.73 eV for the corresponding scandium complex, which compares with our calculated association energy of 0.8 eV. This confirms a general point made earlier that energy differences based on comparable calculations should be more reliable than absolute energies.

Data for ion mobility are relatively profuse, at least by comparison with other defect-related phenomena, so that here we have the opportunity for a detailed confrontation between theory and experiment. For cation vacancies in MgO, the calculated migration energy is 2.1 eV, which compares favorably with values of 2.28 ± 0.21 and 2.2 eV reported recently by Sempolinski and Kingery[28] and Duclot and Deportes,[29] respectively. However, we note that higher values have been reported, including one of 2.76 ± 0.08 eV by Wuensch et al.[30] In the case of oxygen migration, the level of agreement is similar. With the assumption of a free vacancy mechanism, the calculated energy is 2.4 eV compared with 2.71 eV found by Oishi and Kingery,[31] 2.61 eV by Hasimoto et al.,[32] and 2.42 eV by Shirasaki and Hama.[33] With reference to closed-shell impurities, there is adequate correspondence for calcium and strontium. Thus, Wuensch and Vasilos[34] obtained a value of 2.76 eV for Ca^{2+}; calculations give 2.5 eV. Likewise, Mortlock and Price[35] reported 2.91 ± 0.13 and 3.09 ± 0.19 eV for Sr^{2+}; the predicted value is 2.7 eV. The discrepancy for barium on the other hand, seems to be much greater. Harding[36] has deduced energies of 1.85 ± 0.07 and 3.38 ± 0.05 eV, depending on solute penetration, whereas calculations suggest a value close to 2.6 eV. However, Wuensch[21] has cast some doubt on the interpretation of the data as truly representative of lattice migration. Of the trivalent impurities considered here, we can make a direct comparison for yttrium but can only speculate in the case of scandium. For Y^{3+} we calculate an activation energy of 3.3 eV, which accords with that of 3.10 ± 0.07 eV reported by Berard[37] from tracer measurements. For Sc^{3+} the predicted value of 3.4 eV seems to be a reasonable candidate for the migration energy of the putative impurity associate, which Sempolinski and Kingery[28] have suggested as responsible for the change in slope of their ionic conductivity plots for Sc-doped MgO. From their graphical data, we deduce an energy of approximately 3 eV, which lends some support to our postulate. The only four-valent impurity that appears to have been investigated is Ge^{4+}, for which Harding[38] has obtained a migration energy of 4.0 ± 0.2 eV. While the calculated value here is less exact than for other impurities, due to numerical convergence problems for the saddle-point transition state, our estimate of $4.0 < E_m < 4.5$ is of the right order of magnitude. Overall, then, calculations seem to provide a satisfactory account of migration energies in MgO, with discrepancies of about 0.2 eV from the reported experimental values.

There have been numerous studies involving defect migration in both "pure" and doped, single-crystal and polycrystalline α-Al_2O_3; interpretation of the data, however, seems to have been somewhat problematic, although this is not altogether surprising in view of the diversity of opinion as to the basic disorder. Here we can consider only a selected part of this data for comparison with theory. From measurements of oxygen self-diffusion coefficients, Oishi and Kingery[39] deduced an Arrhenius energy of 6.59 ± 1.08 eV, which they attributed to intrinsic diffusion. More recently, Reddy and Cooper[40] reported a value of 6.37 ± 0.43 eV, but favored an extrinsic mechanism wherein their Arrhenius energy was interpreted as *one-third* of the Schottky energy plus the migration energy. Subsequently, Reed and Wuensch[41] obtained a much larger temperature coefficient of 8.15 ± 0.30 eV, which they also interpreted in terms of extrinsic control by silicon impurity. Now on the assumption of equally dependable data, it is not unreasonable to conclude that these results refer to at least two different mechanisms, one for the lower values, which for convenience might be averaged to 6.5 eV, and the other for the higher value of 8.2 eV. From a simple analysis, it can be shown that there are two possible interpretations for these energies;[42] both involve defect-impurity association. They are based on the calculated free vacancy migration energy of 1.8 eV and attribute the lower energy to intrinsic diffusion and the higher to extrinsic control. The first assumes anti-Frenkel disorder with a formation energy of 9.4 eV and an *effective* association energy of oxygen interstitials to silicon of 3 eV. The second assumes Schottky disorder, with a formation energy of 23.5 eV and an *effective* association energy of a cation vacancy to Si^{4+} of 2.2 eV. Calculations favor the second interpretation, for the predicted formation energies are 16.5 and 25.0 eV, respectively; however, it is important to emphasize that anti-Frenkel disorder is consistent with the oxygen diffusion data.

The migration of aluminum defects in α-Al_2O_3 seems to have been studied less extensively. Paladino and Kingery[43] reported self-diffusion coefficients of aluminum in α-Al_2O_3, from the temperature dependence of which they deduced an Arrhenius energy of 4.94 eV. Calculations, which predict a value of 6.8 eV, suggest that this is rather too low for intrinsic diffusion in the bulk; however, it could correspond to intrinsic behavior enhanced by grain boundary migration. Writing

$$Q(\text{intrinsic/grain}) = E'_m + \tfrac{1}{5}E_s \tag{7}$$

in which E'_m is the grain boundary migration energy, we find $E'_m = 0.24$ eV for $E_s = 23.5$ eV (cf. oxygen diffusion) and $E'_m = 0$ for $E_s = 25.0$ eV (calculation), which is consistent with the notion that migration at grain boundaries is substantially less than in bulk lattice. Alternatively, the energy obtained by Paladino and Kingery[43] could refer to extrinsic control involving four-valent impurities, in which case the effective association energy is ~3 eV, assuming the calculated migration energy for V'''_{Al}. Later, Kitazawa and Coble[44] reported an activation energy of 2.47 eV for ionic conduction in α-Al$_2$O$_3$, which they ascribed to an extrinsic mechanism. This value could be attributed to the migration energy of a free vacancy, in which case the calculated value is in error by about 0.8 eV. It would also suggest that the association energy that might be attributed to Paladino and Kingery's results is 2.47 eV, which is close to that deduced for the oxygen diffusion data. On the other hand, the difference of 0.8 eV might reasonably represent the association energy of an aluminum vacancy to some unspecified impurity. More recently, Mohapatra and Kröger[45] examined the conductivity of Ti-doped alumina and obtained an activation energy of 3.77 eV. Even if allowance for uncertainties in the calculated value is made, this would seem to be rather too high for the migration energy of a free aluminum vacancy. However, if we assume the calculated value of 1.8 eV, this suggests an association energy of ~2 eV, which accords with both previous deductions and the dimer formation energy of 1.9 eV given in Table VI.

Summary, Conclusions, and Prospects

This review has been concerned with calculated defect energies in MgO and α-Al$_2$O$_3$. Of necessity, it has been limited in scope and has concentrated solely on the formation and migration of lattice defects and their association with closed-shell impurities. Where possible, comparison has been made with experiment, although uncertainties in the interpretation of the data have restricted this comparison to a few examples, largely for MgO.

The principal conclusions to emerge from calculations are the following:

(i) Schottky disorder predominates the defect equilibria in the pure materials.

(ii) The migration energy of the host lattice is close to 2 eV in both MgO and α-Al$_2$O$_3$, while the ionic conductivity in alumina is predicted to be close to isotropic.

(iii) The migration energies of *closed-shell* impurities in MgO range from ~2.5 eV for divalent cations to ~4.5 eV for four-valent ions.

(iv) Association energies in MgO are <1 eV for trivalent and ~2 eV for four-valent cations. This suggests that trivalent impurities will give rise to a multiplicity of defects including free vacancies, dimers, and trimers, whereas four-valent impurities will be present largely as associates.

(v) Association energies in α-Al$_2$O$_3$ are large; for four-valent impurities these are in excess of 2 eV. In the case of Ti^{4+}-doped alumina, substantial clustering is expected with the possibility that oxygen interstitials might contribute to the charge-compensating disorder.

(vi) The defect structures of both Li-doped MgO and Mg-doped α-Al$_2$O$_3$ are complex.

Finally, we consider future prospects. With regard to MgO and α-Al$_2$O$_3$, these include a more complete description of electronic defects in these materials than has hitherto been given and an extension of the present calculations to transition-metal and covalent impurities such as silicon and carbon. This must necessarily involve quantum mechanical methodology in procedures that include both electronic structure calculation and defect lattice simulation; such procedures are in development. With regard to calculations for a wider class of ceramics, the present restriction is simply that the materials be predominantly ionic; however, with improvements in interatomic potentials this restriction is likely to be eased. A range of other oxides, including CaO, BaO, Fe$_2$O$_3$, ThO$_2$, CeO$_2$, and Li$_2$O, has been investigated already by the methods described in the present review, and more recently these have been extended to include the simplest nitrides and carbides. In view of the difficulties associated with the experimental investigation of ceramic materials and the ambiguities of interpreting the data, calculations would seem to have a role to play in the future development of both this and other technologies.

References

[1] C. R. A. Catlow and W. C. Mackrodt; Chapter 1 in Computer Simulation of Solids. Edited by C. R. A. Catlow and W. C. Mackrodt. Springer-Verlag, Berlin–Heidelberg–New York, 1982.

[2] C. R. A. Catlow, M. Dixon, and W. C. Mackrodt; Chapter 10 in Computer Simulation of Solids. Edited by C. R. A. Catlow and W. C. Mackrodt. Springer-Verlag, Berlin–Heidelberg New York, 1982.

[3] E. A. Colbourn and W. C. Mackrodt, "A Theoretical Study of CO Chemisorption at {001} Surfaces of Non-Defective and Doped MgO"; to be published in Surf. Sci.

[4] J. B. Goodenough; Chapter 6 in Solid State Chemistry. Edited by C. N. R. Rao. Marcel Dekker, Inc., New York, 1974.

[5] B. G. Dick and A. W. Overhauser, "Theory of Dielectric Constants of Alkali Halide Crystals," Phys. Rev., **112**, 90 (1958).

[6] W. C. Mackrodt; Chapter 5 in Mass Transport in Solids. Edited by F. Beniere and C. R. A. Catlow. Plenum Press, New York, 1983.

[7] W. C. Mackrodt; Chapter 12 in Computer Simulation of Solids. Edited by C. R. A. Catlow and W. C. Mackrodt. Springer-Verlag, Berlin–Heidelberg–New York, 1982.

[8] J. Kendrick and W. C. Mackrodt, "Interatomic Potentials for Ionic Materials from First-Principle Calculations," Solid State Ionics, **8**, 247 (1983).

[9] W. C. Mackrodt and R. F. Stewart, "Defect Properties of Ionic Solids: II Point Defect Energies Based on Modified Electron-Gas Potentials," J. Phys. C, **12**, 431 (1979).

[10] E. A. Colbourn, J. Kendrick, and W. C. Mackrodt, "Tables of Interatomic Potentials Based on the Electron-Gas Approximation I. Magnesium Oxide and Cation Dopants." I.C.I. plc Corporate Laboratory Rept., CL-R/81/1637/A; unpublished work.

[11] M. J. Norgett, "A General Formulation of the Problem of Calculating the Energies of Lattice Defects in Ionic Crystals." A.E.R.E. Rept., AERE-R.7650, 1974.

[12] N. F. Mott and M. T. Littleton, "Conduction in Polar Crystals. I Electolytic Conduction in Solid Salts," Trans. Faraday Soc., **34**, 485 (1938).

[13] W. C. Mackrodt and R. F. Stewart, "Defect Properties of Ionic Solids: III The Calculation of the Point-Defect Structure of the Alkaline-Earth Oxides and CdO," J. Phys. C, **12**, 5015 (1979).

[14] C. R. A. Catlow, R. James, W. C. Mackrodt, and R. F. Stewart, "Defect Energies in α-Al$_2$O$_3$ and Rutile TiO$_2$," Phys. Rev. B: Condens. Matter, **B25**, 1006 (1982).

[15] E. A. Colbourn and W. C. Mackrodt, "The Calculated Defect Structure of Bulk and {001} Surface Cation Dopants in MgO," J. Mater. Sci., **17**, 3021 (1982).

[16] J. Kendrick, E. A. Colbourn, and W. C. Mackrodt, "The Calculated Defect Structure of Planar and Non-planar Surfaces of MgO and FeO," Radiat. Eff., **73**, 259 (1983).

[17] R. James, "Disorder and Non-stoichiometry in Rutile and Corundum Structured Metal Oxides"; Ph.D. Thesis, University of London, 1979, and A.E.R.E. Rept., T.P. 814, 1979.

[18] V. Butler, C. R. A. Catlow, B. E. F. Fender, and J. H. Harding, "Dopant Ion Radius and Ionic Conductivity in Cerium Dioxide," Solid State Ionics, **8**, 109 (1983).

[19] E. A. Colbourn and W. C. Mackrodt, "The Calculated Defect Structure of Thoria," J. Nucl. Mater., **118**, 50 (1983).

[20] G. Ramani and K. J. Rao, "Formation Energies of Schottky Defects in Alkaline Earth Oxides," *J. Solid State Chem.*, **16**, 63 (1975).

[21] B. J. Wuensch, "On the Interpretation of Lattice Diffusion in Magnesium Oxide"; in Mass Transport Phenomena in Ceramics. Edited by A. R. Cooper and A. H. Heuer. Plenum, New York, 1975.

[22] C. R. A. Catlow, I. D. Faux, and M. J. Norgett, "Shell and Breathing Shell Model Calculations for Defect Formation Energies and Volumes in Magnesium Oxide," *J. Phys. C*, **9**, 419 (1976).

[23] G. J. Dienes, D. O. Welch, C. R. Fischer, R. D. Hatcher, O. Lazareth, and M. Samberg, "Shell-Model Calculation of Some Point-Defect Properties in α-Al$_2$O$_3$," *Phys. Rev. B: Condens. Matter*, **B11**, 3060 (1975).

[24] S. K. Mohapatra and F. A. Kröger, "The Dominant Type of Atomic Disorder in α-Al$_2$O$_3$," *J. Am. Ceram. Soc.*, **61**, 106 (1978).

[25] D. S. Phillips, T. E. Mitchell, and A. H. Heuer, "Precipitation in Star Sapphire: III Chemical Effects Accompanying Precipitation," *Philos. Mag.* [Part A], **42A**, 417 (1980).

[26] B. J. Pleska, T. E. Mitchell, and A. H. Heuer, "Dislocation Studies in Doped Sapphire Deformed by Basic Slip"; Unpublished work reported by A. H. Heuer, "The Role of MgO in the Sintering of Alumina," in *J. Am. Ceram. Soc.*, **62**, 317 (1979).

[27] G. W. Weber, W. R. Bitler, and V. S. Stubican, "Diffusion of ^{51}Cr in Cr-Doped MgO," *J. Phys. Chem. Solids*, **41**, 1355 (1980).

[28] D. R. Sempolinski and W. D. Kingery, "Ionic Conduction and Magnesium Vacancy Mobility in Magnesium Oxide," *J. Am. Ceram. Soc.*, **63**, 664 (1980).

[29] M. Duclot and C. Deportes, "Influence des Impuretes sur la Conductivite Cationique dans l'oxyde de Magnesium Monocristallin," *J. Solid State Chem.*, **31**, 377 (1980).

[30] B. J. Wuensch, W. C. Steele, and T. Vasilos, "Cation Self-Diffusion in Single-Crystal MgO," *J. Chem. Phys.*, **58**, 5258 (1973).

[31] Y. Oishi and W. D. Kingery, "Oxygen Diffusion in Periclase Crystals," *J. Chem. Phys.*, **33**, 905 (1960).

[32] H. Hasimoto, M. Hama, and S. Shirasaki, "Preferential Diffusion of Oxygen along Grain Boundaries in Polycrystalline MgO," *J. Appl. Phys.*, **43**, 4828 (1972).

[33] S. Shirasaki and M. Hama, "Oxygen-Diffusion Characteristics of Loosely-Sintered Polycrystalline MgO," *Chem. Phys. Lett.*, **20**, 361 (1973).

[34] B. J. Wuensch and T. Vasilos, "Impurity Cation Diffusion in Magnesium Oxide"; p. 95 in Mass Transport in Oxides. Edited by J. B. Wachtman, Jr., and A. D. Franklin. U.S. National Bureau of Standards, Special Publication No. 296, 1968.

[35] A. J. Mortlock and D. M. Price, "Diffusion of Sr^{2+} in Single Crystal MgO," *J. Chem. Phys.*, **58**, 634 (1973).

[36] B. C. Harding, "The Diffusion of Barium in Magnesium Oxide," *Philos. Mag.*, **16**, 1039 (1967).

[37] M. F. Berard, "Yttrium Impurity Diffusion in MgO Single Crystals," *J. Am. Ceram. Soc.*, **54**, 58 (1971).

[38] B. C. Harding, "Cation Diffusion in Single Crystal MgO up to 2500°C," *Phys. Status Solidi B*, **56**, 645 (1973).

[39] Y. Oishi and W. D. Kingery, "Self-Diffusion of Oxygen in Single Crystal and Polycrystalline Aluminium Oxide," *J. Chem. Phys.*, **33**, 480 (1960).

[40] K. P. R. Reddy and A. R. Cooper, "Diffusion of Oxygen in Sapphire," *Am. Ceram. Soc. Bull.*, **55**, 402 (1976); *Am. Ceram. Soc. Bull.*, **57**, 306 (1978).

[41] D. J. Reed and B. J. Wuensch, "Ion-Probe Measurement of Oxygen Self-Diffusion in Single Crystal Al$_2$O$_3$," *J. Am. Ceram. Soc.*, **63**, 83 (1980).

[42] W. C. Mackrodt; unpublished work.

[43] A. E. Paladino and W. D. Kingery, "Aluminium Ion Diffusion in Aluminium Oxide," *J. Chem. Phys.*, **37**, 957 (1962).

[44] K. Kitazawa and R. L. Coble, "Electrical Conduction in Single-Crystal and Polycrystalline Al$_2$O$_3$ at High Temperatures," *J. Am. Ceram. Soc.*, **57**, 245 (1974).

[45] S. K. Mohapatra and F. A. Kröger, "Defect Structure of α-Al$_2$O$_3$ Doped with Titanium," *J. Am. Ceram. Soc.*, **60**, 381 (1977).

Calculations of Entropies and of Absolute Diffusion Rates in Oxides*

A. M. Stoneham

B424.4, A.E.R.E.
Theoretical Physics Division
Harwell, Didcot, Oxon. OX11 ORA U.K.

M. J. L. Sangster and D. K. Rowell

University of Reading
J. J. Thomson Physical Laboratory
Whiteknights, Reading, RG6 2AF U.K.

We survey methods for the quantitative calculation of vibrational entropies and diffusion jump rates. These methods exploit the same combination of interatomic forces and the shell model, well-known from calculations of defect energies and other defect properties. As in the calculations of energies, it is essential to treat lattice relaxation and polarization, and this is done by using a general-purpose computer code. We make two applications for detailed comparison with experiment. The first is of formation entropies for Fe^{3+}/Mg^{2+}–vacancy dimers in MgO, where our predictions agree well with measurements of Gourdin, Kingery, and Driear. The second predictions are of Mg^{2+} and Fe^{3+} diffusion rates. Theoretically, rate theory and dynamical theory give rather different results, and these are not easily correlated with the unexpectedly high Mg^{2+} vacancy mobility reported by Sempolinski and Kingery. The outstanding theoretical problems appear to arise in the approximations used to evaluate statistical mechanical expressions rather than in the defect calculations. The discrepancies are accentuated in the cases we treat because, at the normal saddle point, there is a relatively soft mode for Mg^{2+}; for Fe^{3+}, the same mode is unstable, giving a bifurcated jump path.

In the theory of defects and impurities in crystals, considerable advances have been made during the last decade, particularly in the calculation of internal energies or enthalpies of formation, motion, and aggregation. These calculations involve evaluations of the energies of static defect configurations, and the advances have exploited two major developments: the availability of general and flexible computer codes (such as HADES)[1] for obtaining fully relaxed defect configurations from assumed interatomic potential models and the systematic development and verification of such potential models.[2] For ionic crystals, these interatomic potentials are generally described within the framework of the shell model, which has a long-established record of success in the field of lattice dynamics. Coulombic and short-range interactions between polarizable ions are included, an essential feature being the coupling of polarization to the short-range interactions. In the following section we shall briefly review the potential models for oxides used in the calculations to be presented here.[3] We then present the results for the relevant static calculations of defect energies.

More recently attention has been turned to the role of dynamical properties of defects. Knowledge of these properties is required for the calculation of vibrational contributions to entropies. Along with the enthalpies mentioned above, these terms provide a complete description of the thermodynamic variables for the model systems, since the force constants needed for the dynamical calculations are just derivatives of the interatomic potentials evaluated for relaxed configurations. The methods used for the calculation of the vibrational properties are by now routine, but with the availability of large and fast computers, it is now feasible to approach the problem in several ways. These approaches are discussed in later sections, where we shall consider two main problems quantitatively. The first (Vibrational Properties of Defects) estimates the concentrations of several different configurations of Fe^{3+}–vacancy dimers in MgO.[4,5] These results can be compared with the experimental data of Gourdin, Kingery, and Driear,[5] who show the $\langle 110 \rangle$ configuration to be 0.28 eV less stable than the $\langle 100 \rangle$ configuration for which there is an intervening O^{2-} ion. The second calculation (in Absolute Diffusion Rates) estimates the absolute diffusion rates for the migration of Mg^{2+} and Fe^{3+} ions in MgO by the vacancy mechanism. The basic diffusion jumps are shown in Fig. 1

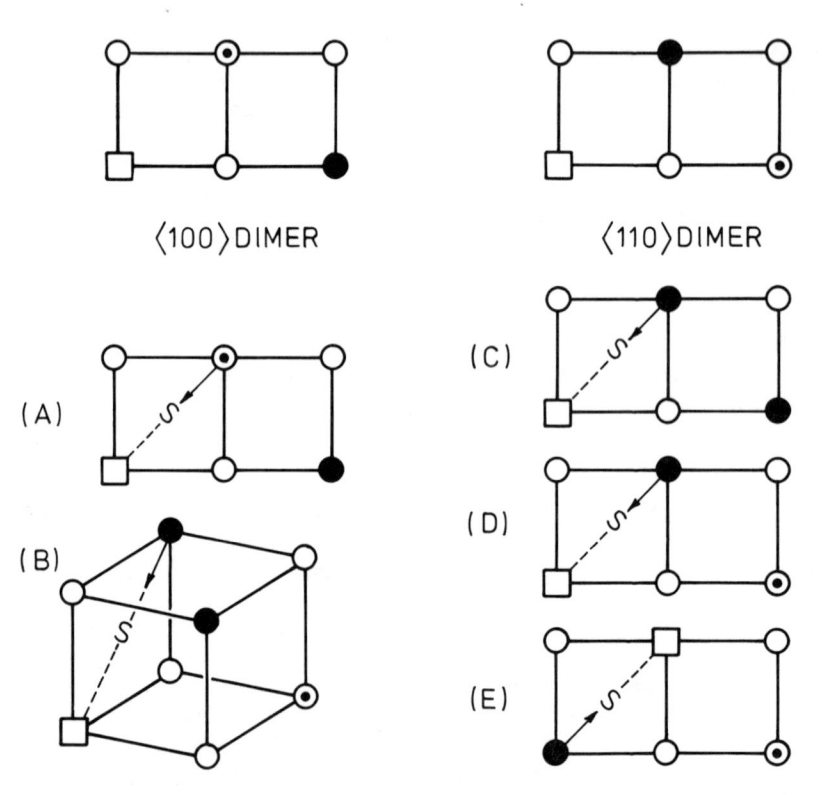

Fig. 1. Atomic configurations of dimers and the geometries of the jumps between them. The labeling (A)–(E) corresponds to that of Table II. Saddle points are labeled S for the jump indicated by an arrow and broken line. Oxygen ions are denoted ○; Mg^{2+} ions ●; Fe^{3+} ions ⊙; Mg vacancies □.

and are the standard atom–vacancy interchanges. At the saddle point, the diffusing ion may lie out of the plane of the diagram. We shall find that the saddle point does indeed lie out of the plane for Fe^{3+}, but for the self-diffusion of Mg in MgO, the lowest energy saddle point lies in the plane shown. Our predictions for Mg motion can be compared with the mobility measurements of Sempolinski and Kingery.[6]

Interatomic Potentials

Sangster and Stoneham[7] have obtained potentials for a range of metal oxides with the rock-salt structure including both alkaline-earth and transition-metal oxides. For the problems considered here, use is made of the model potentials for MgO and FeO. An important feature of these potentials is that O^{2-}–O^{2-} interactions and the parameters that describe the oxygen polarizability are taken to be the same in all cases. While these assumptions, and particularly that of crystal-independent polarizability for O^{2-}, lack full justification (and indeed go against other conclusions such as those of Tessman, Kahn, and Shockley[8]), for progress to be made in impurity calculations it seems essential to us that an unambiguous prescription of this type be provided. The potential models successfully reproduce static and dynamic properties of the perfect crystals, as well as intrinsic defect energies (Rowell and Sangster[9]), and correctly predict occurrences and symmetries of off-center divalent substitutional impurity ions in alkaline-earth oxide host crystals (Sangster and Stoneham[7]). An ionic picture is assumed throughout. This does not require that covalency is zero but merely (Catlow et al.;[10] Catlow and Stoneham[11]) that it is represented adequately within the empirical components of the potential.

Details of the potentials for MgO and FeO are given in Table I. We have treated Fe^{3+} impurities in two ways (see Stoneham and Sangster[12] and Sangster[13]): (i) simply by addition of an extra charge to the ion core or (ii) by an additional adjustment of the Fe–O interaction to take some account of the reduced ionic radius for the ion in its trivalent state.

These rather cavalier treatments appear to incorporate the more important of the required changes. The symmetries of off-center displacements of Fe impurity ions in SrO are correctly calculated to be [110] for Fe^{3+} ions and [111] for Fe^{2+} ions (Rowell and Sangster[14]). Very recently Müller and Berlinger[15] have examined the

Table I. Parameters for Interionic Potentials in MgO and FeO*

(a) Ion polarizabilities α and shell charges Y[†]

Ion	α (Å^3)[‡]	Y ($|e|$)
O^{2-}	2.466	−2.8107
Mg^{2+}	0	(0)
Fe^{2+}	2.010	3.372

(b) Short-range potentials $\phi(r) = B \exp(-r/\rho) - C/r^6$

	B (eV)	ρ (Å)[‡]	C (eV · Å^6)
O^{2-}–O^{2-}	22764.3	0.1490	20.37
Mg^{2+}–O^{2-}	1275.2	0.3012	
Fe^{2+}–O^{2-}	1249.1	0.3079	

*Ion charges are ±2$|e|$, and short-range interactions between cations are ignored.
[†]The polarization force constant is $(Y|e|)^2/\alpha$.
[‡]1 Å = 0.1 nm.

effects of uniaxial [100], [111], and [110] stresses on EPR fine-structure lines of trivalent impurities in MgO. They analyze their results in terms of local strain coefficients and find their experimentally based estimates to be in excellent accord with values calculated with the above assumptions (Sangster[13]). Thus, while the potential models contain naive features, there is an accumulating body of evidence that their use produces physically reasonable results.

In the calculations reported here, no adjustment for change in ionic radius is made; i.e., Fe^{3+} is treated simply as Fe^{2+} carrying an extra charge. We intend to investigate the effects of ionic radius corrections in the future.

Static Calculations of Defect Energies

The results presented in this section were all obtained by using the Harwell HADES package (Norgett[1,16,17]). This package has been used extensively for efficient calculation of the relaxed configurations of defects in crystals from provided interaction potentials and has been reviewed fully elsewhere (see, e.g., Catlow and Mackrodt[2]). Energies of equilibrium configurations may be found, as can those when certain ions are held at nonequilibrium positions (e.g., saddle points), and all other ions are allowed to relax and polarize. In all the calculations reported here, around 100 ions of the crystal were treated explicitly (region I of HADES), the remainder of the crystal being treated as a dielectric continuum.

Formation energies, binding energies, and activation energies obtained from some relaxed equilibrium and saddle-point configurations are listed in Table II. For the vacancy–Fe^{3+} dimer, the $\langle 100 \rangle$ configuration is energetically favored relative to the $\langle 100 \rangle$ geometry by 0.27 eV. This value compares well with the value obtained by Gourdin and Kingery[4] (0.28 eV). As a result of the limited number of ions treated explicitly, the results of the calculation depend on the choice of origin

Table II. Static Defect Energies

Formation energies*	
Isolated substitutional Fe^{3+} and Mg^{2+} vacancy	−4.73
Dimer with $\langle 110 \rangle$ orientation	−5.34
Dimer with $\langle 100 \rangle$ orientation	−5.61
Dimer binding energies[†]			
Dimer orientated along $\langle 110 \rangle$	0.61	(0.69)	[0.85]
Dimer orientated along $\langle 100 \rangle$	0.88	(0.92)	[1.13]
Activation energies for various processes[‡]	
(A) $\langle 110 \rangle$ Fe^{3+}–vacancy interchange	2.697
(B) $\langle 110 \rangle$ Mg^{2+} motion to give another $\langle 110 \rangle$ dimer	2.365
(C) Mg^{2+} motion (no Fe^{3+} present)	
$T = 0°C$ lattice parameter	2.196
$T = 1400°C$ lattice parameter	1.903
(D) $\langle 100 \rangle$ Mg^{2+} motion to give $\langle 110 \rangle$ dimer	2.228
(E) $\langle 110 \rangle$ Mg^{2+} motion to give $\langle 100 \rangle$ dimer[§]	1.958

*In units of eV.
[†]Experimental values of Gourdin, Kingery, and Driear[5] are bracketed; the earlier predictions of Gourdin and Kingery[4] are in square brackets. Units are eV.
[‡]Here $\langle 100 \rangle$ and $\langle 110 \rangle$ specify the dimer orientation initially. The notation (A), (B), etc. corresponds to that in Fig. 1. The units are eV.
[§]Reverse of process (D).

for the defect system. The value tabulated for the ⟨110⟩ dimer was obtained with the vacancy at the origin: if the origin is taken at the midpoint of the dimer, the energy is increased by 0.01 eV. For the ⟨100⟩ dimer, the origin was taken at the intervening oxygen.

The activation energy for an Mg^{2+}–vacancy interchange (with no Fe^{3+} ion in the vicinity) is 2.2 eV; when there is an Fe^{3+} near neighbor, as shown in Fig. 1(B), the energy barrier is lowered by about 0.1 eV. The saddle point for an Mg^{2+} ion jump between the two dimer configurations (Fig. 1(C,D)) is about 2.3 eV above the equilibrium energy of the ⟨100⟩ dimer configuration and ~2.0 eV above the equilibrium energy of the ⟨110⟩ dimer configuration. Although we have not carried out calculations for all the necessary configurations and although the differences in activation energies are sufficiently small that we should take more careful consideration of effects such as choice of origin (e.g., by extending region I), it is reasonable to infer the following from Table II:

(a) For a ⟨110⟩ dimer, the jump with lowest activation (and hence the most likely jump) is that of Mg^{2+} to form a dimer oriented along ⟨100⟩.

(b) The direct exchange of Fe^{3+} and vacancy, inverting the ⟨110⟩ dimer, has a relatively high energy and is therefore less likely.

(c) Starting from a dimer oriented along ⟨100⟩, two jumps have very similar energies: the Mg^{2+} jump to give a ⟨110⟩ dimer and the jump that separates the vacancy and impurity further. This latter jump (not calculated explicitly) will have an energy close to that for the motion of the Mg^{2+} vacancy in isolation.

For the Mg^{2+}–vacancy interchange we have calculated the energies of relaxed configurations as the Mg^{2+} ion is moved from the saddle point to the equilibrium configurations. The results are shown in Fig. 2. In later sections, we shall use the curvatures at the saddle point and the minimum to investigate diffusion rates. We have also examined the variation in energy for displacements of the Mg^{2+} around

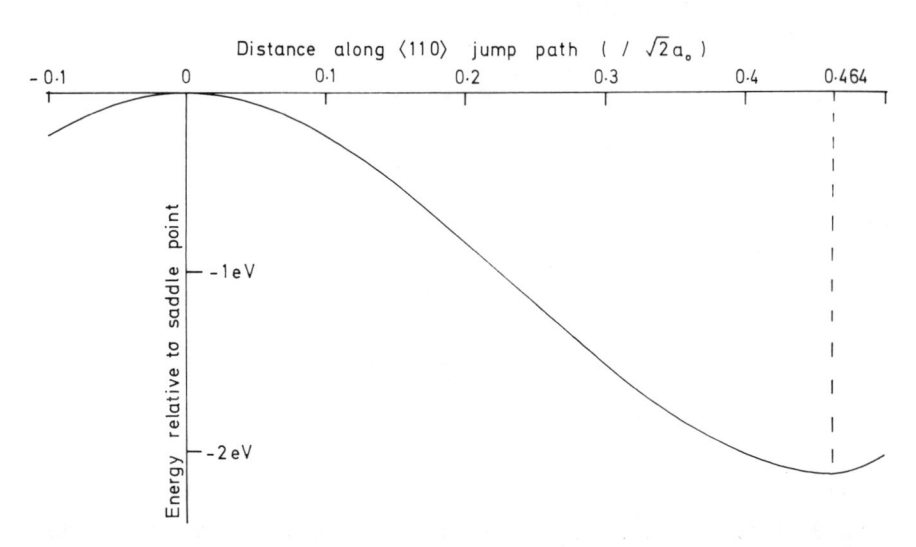

Fig. 2. Energy along jump path for Mg^{2+} diffusion (as in Fig. 1(C)). The curve shows the total energy for the fully relaxed system as a function of position of Mg^{2+} ion along the ⟨110⟩ jump path.

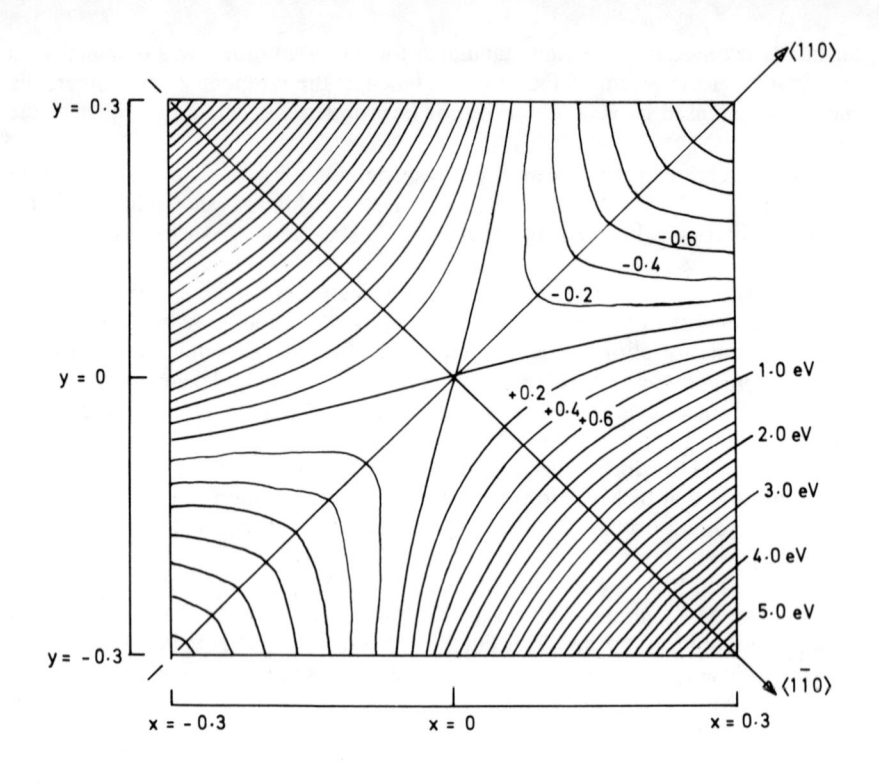

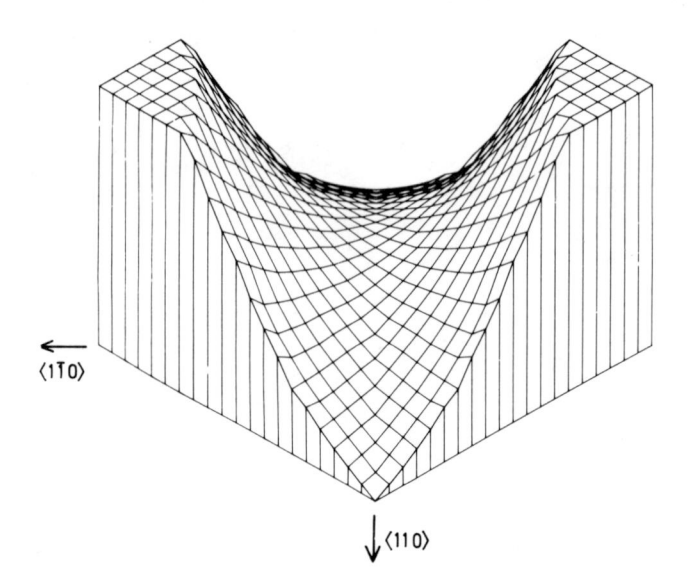

Fig. 3. Energy surface around saddle point for Mg^{2+} diffusion. (Perfect lattice sites are at $x,y = \pm 0.5$.) Both contour plots (at 0.2 eV intervals) and isometric plots are shown. The curvature normal to the jump path at the saddle point is small, giving a relatively low-frequency motion.

the saddle point and in the planes of Fig. 1. These results are shown in Fig. 3 both as contour plots and as isometric plots of energy surfaces. As in Fig. 2 (part of which is a cut on Fig. 3), the crystal atoms, other than the jumping Mg^{2+} ion, are allowed to relax and polarize fully. Displacements out of the plane of Fig. 1 are not shown, but the energy increases as the Mg^{2+} ion is moved out of the plane. Thus, we have the classic "saddle-point" energy surface. These figures will be considered further in later sections.

For the Fe^{3+}–vacancy interchange, the saddle point lies out of the plane of Fig. 1. Minimum energies are at $\pm 0.238 a_0$ from the plane, where a_0 is the MgO nearest-neighbor distance. The energy surface around the saddle points is shown in Fig. 4, again as energy contours and as an isometric plot. As in previous figures, the energy for a point in the plane shown is the energy for a relaxed configuration of the crystal when the migrating atom is held at this point. Note that the plane chosen for Fig. 4 is orthogonal to that for Fig. 3. For an Fe^{3+} ion jump, the lowest energy path follows the $\langle 110 \rangle$ valley until the ion is at a position $(0.1, 0.1, 0)a_0$ from the midpoint. It then follows a curved trajectory at almost constant energy through one of the two saddle points. The implications for calculations of diffusion rates will be considered later.

It is also of interest to know how these energies change with lattice parameter. As Gillan[18] has observed recently, these derivatives of energies with respect to lattice parameter determine the *volumes* of formation and activation. These are related to the pressure dependence of formation and activation energies (and hence require knowledge of the compressibility of the model system). Viewing the change in lattice parameter alternatively as a consequence of a temperature change, the same information, together with the thermal expansion coefficient of the model system, provides the temperature dependence of the energies. As an illustrative example, of which use will later be made, we calculate the dependence of the formation energies for the two configurations of the Fe^{3+}–vacancy dimer already considered on lattice parameter. The formation energies are calculated at expansions of ± 1.5 and $\pm 2.5\%$, a fairly extreme range, and are fitted to polynomials:

$$E = E_0 + c_1 x + c_2 x^2 + c_3 x^3 + c_4 x^4 \tag{1}$$

with x equaling $100(a - a_0)/a_0$, a being the nearest-neighbor distance in the expanded (or contracted) crystal, and a_0 being the equilibrium value at room temperature (0.2106 nm). Values for E_0 are as in Table II, and the polynomial coefficients are given in Table III. Table III also lists the room-temperature volumes of formation (Ω_f) for the two configurations as fractions of the molar volume (Ω_m). These are obtained from the result (Finnis and Sachdev[19]).

$$\Omega_f = -KV \frac{\partial E}{\partial V} = -\frac{1}{3} Ka \frac{\partial E}{\partial a} \tag{2}$$

from which, at room temperature ($a = a_0 = 0.2106$ nm)

$$\Omega_f / \Omega_m = -K 100 c_1/(6 a_0^3) \tag{3}$$

In these expressions K is the isothermal compressibility, and it is important that the value taken is that predicted by the model potentials. We have used the relation with the elastic moduli:

$$K = 3/(c_{11} + 2 c_{12}) \tag{4}$$

with the model values for c_{11} and c_{12} given by Sangster and Stoneham:[12] $c_{11} = 374$ and $c_{12} = 157$ GPa. At the room-temperature equilibrium spacing, the volumes of

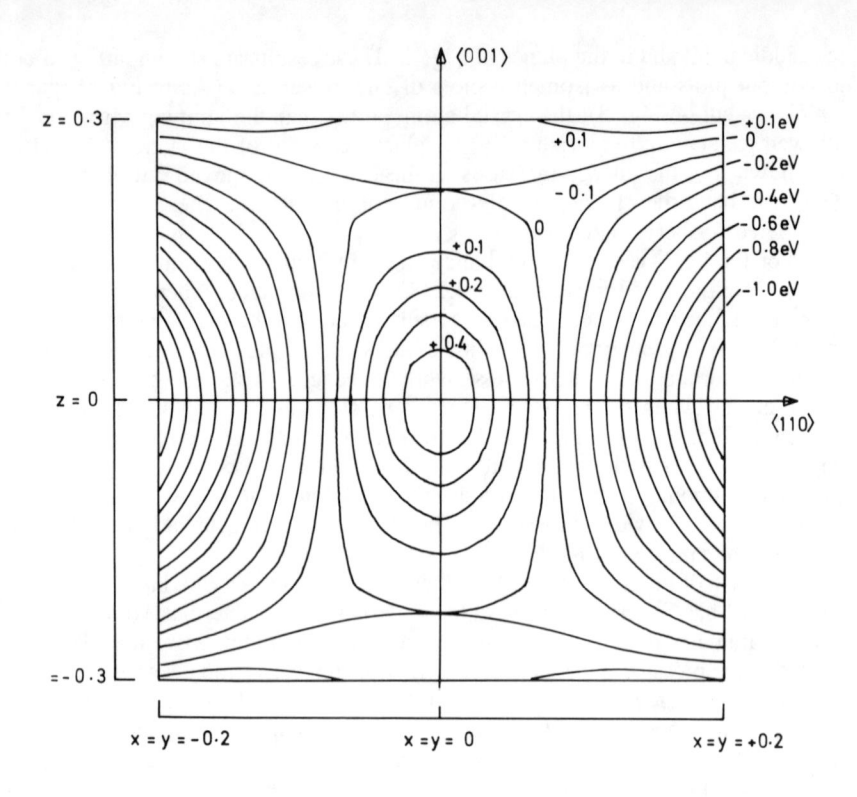

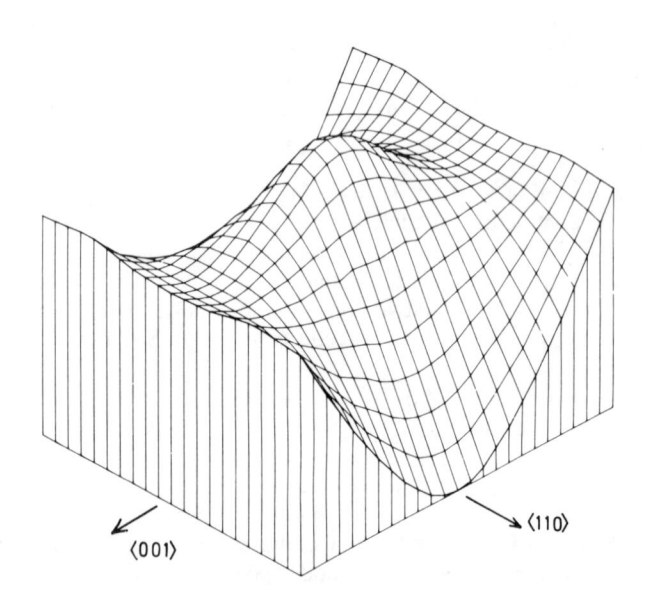

Fig. 4. Energy surface around saddle point for Fe^{3+} diffusion. Both contour plots (at 0.1-eV intervals) and isometric plots are shown. Note the different orientation from that of Fig. 3. The low-frequency motion of Fig. 3 is now unstable, so that the jump path now bifurcates near the saddle point.

Table III. Polynomial Coefficients for Formation Energy*

Configuration	c_0	c_1	c_2	c_3	c_4	Ω_f/Ω_m
110	−5.342 48	0.086 20	−0.021 40	−0.000 21	−0.000 06	−0.107
100	−5.611 52	0.055 54	−0.016 66	0.000 25	−0.000 03	−0.069

*Coefficients (in eV) of polynomial for formation energy for Fe^{3+}-vacancy dimers (Eq. (1)) and formation volume Ω_f as fraction of molar volume Ω_m.

formation of both dimers are negative. They change sign at (linear) expansions of 1.92% for the ⟨110⟩ configuration and 1.71% for the ⟨100⟩ configuration; i.e., $\partial E/\partial a = 0$ at these expansions. The formation energy for the ⟨110⟩ configuration falls below that for the ⟨100⟩ configuration at a (linear) compression of 5.60%, which could be achieved by a pressure of around 50 GPa (Perez-Albuerne and Drickamer[20]).

Vibrational Properties of Defects

We encounter the vibrational properties of defects in three main contexts. First, the defect vibrations may be observed directly, either by infrared or by Raman spectroscopy or as sidebands to electronic transitions. We shall not discuss these aspects specifically in this paper. Second, the vibrational energies enter into defect thermodynamic properties. In particular, the defect formation entropy, in the high-temperature limit, can be written

$$S/k = \ln\left(\prod_i^{3N}\omega_i \Big/ \prod_j^{3N}\omega_j'\right) \quad (5)$$

where frequencies ω_i refer to the perfect lattice and ω_j' to the defective lattice. It is these entropies that determine defect concentrations and that are the subject of this section. Third, the vibrational properties determine transport behavior. In Absolute Diffusion Rates, as an example, we shall discuss the effective frequency found in the pre-exponential factor of Vineyard theory, which involves frequencies with the defect both at its normal lattice site and at the relaxed saddle point.

Methods for Defect Dynamics

Calculations of the vibrations of lattices with defects involve both the loss of the translational symmetry exploited in perfect crystals and the important effects resulting from polarization and distortion of the lattice near the defect. For practical calculations, including this relaxation, and on the basis of the same interatomic potentials as the HADES calculations of energies, we have three main methods: (1) the supercell method (see Harding, Masri, and Stoneham;[21] Harding and Stoneham[22]), (2) the embedded crystallite method (see Jacobs and Gillan[23]), and (3) the Green's function method (see Jacobs and Gillan[23]). These are not the only methods, but they are the only ones for which serious quantitative calculations have been made, including lattice relaxation. Jacobs and Gillan give extensive references to previous work using Green's function and embedded crystallite methods.

Our present calculations use the *supercell* (large unit cell) *method*. Here one creates a superlattice of defects, thus restoring translational symmetry with a larger unit cell. The phonon modes and frequencies can be obtained by using the standard approaches for crystal lattice vibrations. Given a similar calculation for the defect-free lattice (with a corresponding large unit cell) the modes, entropy, and other features of defect dynamics follow directly.

The advantage here is that one can use the existing, relatively robust, and tested standard codes for lattice dynamics. There are no real restrictions on defect or host symmetry. The method has indeed been used in several demanding cases, including electron-hole excitation in UO_2 (Harding, Masri, and Stoneham[21]) and the dynamics of off-center Li in KCl (Sangster and Stoneham[24]). The worries about supercell methods refer to technical points that have not been verified fully, though it should be emphasized that these worries have not been shown significant in any case and that similar difficulties do not prove severe (except in special cases) in the parallel and long-established electronic structure calculations.[25] The first obvious issue is the size of the large unit cell; only 24 ions per cell were used in early work, whereas 64 or 128 per cell are practical now. The second is whether it is acceptable to take as lattice relaxation the results of a HADES calculation on an isolated defect, with some adjustment at the cell boundaries to recover periodicity. In principle, with use of the PLUTO code, it is possible and preferable to calculate relaxations in the periodic structure. The third question is whether charged defects cause problems. Since both the other main approaches do have difficulties with charged defects arising from boundaries between a "defect" region and that region in which it is embedded, we note specifically there is no evidence for such problems with supercell methods. The early Harding, Masri, and Stoneham calculations considered cells containing one hole, one electron, and one electron–hole pair; the results allowed no precise check of the point at issue nor were there hints of difficulties. The periodic boundary conditions (at least if implemented fully in the lattice relaxation) eliminate the possible awkward cell–cell boundary terms, and certainly the Coulomb potential of the array of net charges contains no obviously difficult features. The fourth point concerns a working approximation, checked more fully in the electronic case (Evarestov[25]) — the replacement of a calculation of all phonon frequencies with that of those near $\mathbf{q} = 0$ only. In essence, one relies on the effective averaging over the Brillouin zone, resulting from the "folding back" of phonon branches in the supercell. This working assumption is probably satisfactory, at least for entropies, especially if (following Evarestov) one chooses the supercell shape sensibly. Only one specific check of the use of $\mathbf{q} = 0$ frequencies was made in the course of calculating Hugoniots (Harding and Stoneham;[26] J. H. Harding, private communication), where checks showed very good agreement with a direct summation over a mesh.

Embedded crystallite method: Here one calculates the entropy exploiting the force constant matrices and harmonic vibrations of a perturbed (defective) region only, this defect region being embedded in a matrix of ions that do not vibrate. Finite cluster models may be regarded as a special case of the embedded crystallite method.

This approach is quite simple to use, and one imagines it will converge with crystallite size about as fast as the supercell method converges with cell size. Special care must be taken with charged defects to avoid an entropy contribution from the crystallite "surface." Such corrections are inconvenient, rather than serious obstacles. General codes for this approach remain to be written, though a substantial part of such a code has been developed by J. H. Harding and R. Ball for high-temperature self-consistent phonon calculations.

Green's function methods: Here the change in force constants is combined with the well-tried Green's function methods (see, e.g., Maradudin et al.[27]), in which the entropy is expressed in terms of perfect-lattice Green's functions and of force constant changes for ions in an inner defect region. These force constant changes incorporate lattice relaxation effects.

The approach is (as Jacobs and Gillan observe) very laborious to apply. Its convergence is, however, very good, so that the method is especially suitable for reference calculations as a test for the more easily applied methods. The Green's function methods also have the advantage of showing continuum limits clearly. Again, while there are problems with charged-defect entropies associated with the surface of the defect region, these are mainly a practical inconvenience.

In summary, for general defect applications, the natural choice is the large unit cell method, where an important factor is the existence of tested and robust codes capable of handling arbitrary symmetry. For special cases, either as reference examples or with unusual features, the extra effort of the Green's function may prove worthwhile.

Calculations of Defect Entropies: Fe^{3+}–Vacancy Dimers in MgO

We have performed supercell calculations with up to 64 ions/cell. In cases where the defect is not neutral (as for the Fe^{3+}–vacancy dimers), the excess charge is screened out in effect by omitting a divergent term in the Coulomb summations. We have made predictions for both formation entropies and for the absolute rates discussed in the next section. Thus, the supercell method has been used for frequencies both at the equilibrium geometry and at the saddle-point geometry.

For vibrational entropies, we have used the full temperature-dependent expression, for which the entropy for a mode with energy $\hbar\omega$ is

$$S/k = X/[\exp(X) - 1] - \ln[1 - \exp(-X)] \tag{6}$$

with $X = \hbar\omega/kT$. In practice, the high-temperature expansion ($X \ll 1$) is respectable above 2000 K for MgO. We have always used results for the correct temperature (usually 1400°C) in making comparison with experiment. However, our present results are at constant volume, and we have not yet estimated the effects of thermal expansion on the entropies.

Results are summarized in Tables IV and V. We note two technical points first of all. One is that the temperature dependence of the entropy is almost independent of which defect is involved. For the perfect lattice, the Mg^{2+} vacancy and the Fe^{3+} substitutional, one finds differences between 1000 and 2000 K of 2.045 k per atom, with a spread of only 0.00062 k (0.04 k per supercell) from case to case. There is no sign of any effect of defect charge. The second technical point is that the differences between the values obtained from the 64- and 54-site supercells are almost independent of defect. At 1400°C, for the $\langle 110 \rangle$ dimer, the $\langle 100 \rangle$ dimer, and

Table IV. Defect Entropies for MgO*

	1000 K	2000 K
Perfect MgO lattice	5.105 (5.188)	7.149 (7.233)
Mg^{2+} vacancy	5.130	7.174
Substitutional Fe^{3+}	5.144	7.189
$\langle 110 \rangle$ dimer	5.183 (5.276)	7.228 (7.321)
$\langle 100 \rangle$ dimer	5.173 (5.250)	7.218 (7.305)

*Units are k ion. Values are quoted for a 64-site supercell, with values for a 54-site supercell quoted in brackets for some cases only. The full finite temperature expression (Eq. (6)) has been used.

Table V. Entropies and Dimer Concentrations*

	Entropy (k)	Reaction constant Theory	Reaction constant Experiment
$\langle 110 \rangle$ dimer	0.896	2027	1438
$\langle 100 \rangle$ dimer	0.243	3434	3546

*The reaction constants K measure [dimer]/[Fe^{3+}][vacancy], as given in Eqs. (7) and (8), and are calculated here for 1400°C. The entropies are those defined in Eq. (9). Experimental results (which may not be strictly comparable) are from the table on p. 2079 of Gourdin, Kingery, and Driear.

the perfect solid, we find respective differences of 0.093, 0.086, and 0.084 k per atom. Even after multiplication by the number of atoms per cell, the spread is only 0.5 k. The differences presumably reflect the sampling of the phonon modes and the way this is affected by the size and shape of the supercell. The approximate constancy of the size dependence suggests that calculations for the entropy of the perfect solid will often give a good approximation when results are extrapolated to larger clusters. To the extent this approximate independence holds, one would use values of the perfect crystal $\Delta S_C(N)$ entropy for supercells of size N to extrapolate values $\Delta S_D(N)$ for the defect supercells by using $\Delta S_D(\infty) \approx \Delta S_D(N) + [\Delta S_C(\infty) - \Delta S_C(N)]$. We plan to check this point; if verified, the result will extend the supercell method very usefully. In any case, our present results give us confidence in comparing entropies of different defects for a chosen supercell.

We may use our calculated entropies to study three processes: the formation of $\langle 110 \rangle$ and $\langle 100 \rangle$ dimers from the constituent Fe^{3+} impurity and Mg^{2+} vacancy and the interconversion of the $\langle 110 \rangle$ and $\langle 100 \rangle$ dimers. Table V gives results for the 64-site supercell and a temperature of 1400°C ($kT = 0.1441$ eV). For the reaction constants K we define

$$K_{\langle 110 \rangle} = 12 \exp(\Delta S_{\langle 110 \rangle}/k) \exp(-\Delta E_{\langle 110 \rangle}/kT) \qquad (7)$$
$$K_{\langle 100 \rangle} = 6 \exp(\Delta S_{\langle 100 \rangle}/k) \exp(-\Delta E_{\langle 100 \rangle}/kT) \qquad (8)$$

where the entropy ΔS is given by

$$\Delta S_{dimer} = (N-1)S_{dimer} + NS_{perfect} - NS_{Fe} - (N-1)S_{vac} \qquad (9)$$

with the various S_i corresponding to entropies per ion (as in Table IV) and N the number of sites (64). The energies ΔE are given in Table II. The relative populations of $\langle 110 \rangle$ and $\langle 100 \rangle$ dimers follow from the ratios of the reaction constants.

Our first conclusion is that 37% of the dimers should be in the $\langle 110 \rangle$ configuration. This may be compared with 29% observed by Gourdin, Kingery, and Driear.[5] Our prediction would be slightly worse if we used the 54-site result, giving 44% $\langle 110 \rangle$ dimers. Our second conclusion is that the reaction constants are broadly consistent with those observed. Gourdin et al. strictly obtain the K by ignoring vibrational entropy and by using the observed energies which themselves were fitted to a set of reaction data without vibrational entropies. There is thus a possible difference in definition between the experimental and theoretical values. The theory values involve only calculated quantities (both energies and entropies) without adjustable parameters, and the agreement must be considered satisfactory.

Absolute Diffusion Rates

We now turn to the remaining problem in the quantitative prediction of defect processes, namely, the absolute rates of diffusion processes. Many of the techniques are the same as those we have described for defect entropies and energies. We shall use the same potentials; any phonon frequencies and eigenvectors will be calculated by using equivalent supercells, and averages over the Brillouin zone will exploit the same use of a small wave vector **q** and of folding back, as mentioned earlier. Technically, the main new feature is that we shall calculate phonons for constrained nonequilibrium situations, notably for the saddle point. At the saddle point, for example, there will be at least one unstable mode, and its frequency will be purely imaginary. Our methods obtain such imaginary frequencies and their associated eigenvectors in exactly the same way as for the real frequencies; in all cases frequencies are obtained either in the harmonic approximation (by dynamical supercell calculations) or with use of second derivatives of total energy surfaces.

Given the energy surfaces and the harmonic vibrational properties for any chosen stage of a defect's diffusive jump, how can one calculate the diffusion rate? It will emerge that our main limitation in calculating rates lies in the approximations used to simplify the statistical mechanical expressions for jump rates in thermal equilibrium and not, so far as we can tell, in any limits on the accuracy of the atomistic modeling. Methods for calculating absolute diffusion rates fall into a few broad categories. For light interstitials, like hydrogen or muons, or for the motion of self-trapped electronic carriers, one has variants of small polaron theory (Holstein;[28] Flynn and Stoneham;[29] Catlow et al.;[10] Stoneham[30]). At the other extreme, the full dynamical theory of Schober and Stoneham[31] can be used for heavy-atom resonances. For the vacancy systems of interest here, one has three main options: molecular dynamics, reaction rate theory, and the dynamical theory. We shall only make some general remarks about molecular dynamics, whereas we shall use the other two approaches in quantitative estimates.

Reaction Rate and Dynamical Theories

Reaction rate theory stems from the work of Eyring on chemical reaction rates (Glasstone et al.[32]) and was developed for applications to solids principally by Wert and Zener[33] and Vineyard.[34] The second approach, the dynamical theory, was proposed by Rice[35] and Slater.[36] The former theory makes more direct use of equilibrium statistical mechanics, whereas the latter stresses the atomistic jump process and employs random-variable arguments. A concise discussion of some of the limitations of the theories has been given by Glyde;[37] a penetrating discussion of some of the subtleties is given by Landauer.[38] Flynn[39] gives an excellent survey of the theories.

In either theory the jump frequency is expressed as

$$\Gamma = \Gamma_0 \exp(-E_A/kT) \tag{10}$$

where E_A is an energy of activation. The reaction rate (or Vineyard) argument is based on equilibrium statistical mechanics and adopts a many-body normal mode approach in which the crossing between equilibrium sites is identified with the passage through zero of the coordinate of the normal mode that is unstable at the saddle point. For the simple hopping mechanism that we consider, the dynamical theory identifies the crossing with the passage of a "reaction coordinate" through the midpoint of the equilibrium sites. The simplest choice of reaction coordinate would be one in which the jumping atom moves from an equilibrium site toward the saddle point: more realistic choices would involve accompanying adjustments of the positions and polarizations of surrounding atoms.

In both theories the energy of activation E_A in Eq. (10) is obtained from activation energies and volumes calculated by applying the static methods of the section Static Calculations of Defect Energies to equilibrium and (constrained) saddle-point configurations. Of main interest here is the pre-exponential factor Γ_0. Vineyard theory assumes that the energy for configurations around the saddle point may be adequately represented within the harmonic approximation and leads to the result

$$\Gamma_0 = \left(\prod_{i=1}^{3N} \nu_i\right) \Big/ \left(\prod_{i=1}^{3N-1} \nu_i'\right) \tag{11}$$

that is, the ratio of the product of the normal mode frequencies (ν_i) at the equilibrium configuration and the product of the frequencies (ν_i') of the *stable* normal modes at the saddle-point configuration. (N is the number of atoms in the crystal.) On the other hand, in the dynamical theory, the pre-exponential factor can be considered as an "attempt frequency," the frequency of approaches of the jumping atom toward the potential barrier that it is required to surmount. As indicated above, there is ambiguity in the precise definition of the coordinate that should be taken in estimates of this frequency: this will be considered further in the section Numerical Results.

Experimental Rates and Simple Estimates

For the systems of interest to us, the experimental quantity observed is either a diffusion constant D or a mobility μ, these being related by the Nernst–Einstein relation $\mu = (Ze/kT)D$ for a defect of charge Ze. The diffusion constant is related to the jump frequency Γ by $D \equiv \frac{1}{6}\Gamma d^2$ in cubic crystals, with d being the jump distance. When the jump frequency is written in the form of Eq. (10), i.e.,

$$\Gamma = \Gamma_0 \exp(-E_A/kT)$$

there are already some hidden assumptions. For example, at constant pressure, one will expect a temperature dependence of E_A from thermal expansion. Thus, if $E_A \equiv E_{A0} - \alpha kT$, then one could write Γ_0 as $\nu_0 \exp(\alpha)$, i.e., $\Gamma \equiv (\nu_0 \exp(\alpha)) \exp(-E_{A0}/kT)$, and one must take care to decide whether one is calculating ν_0 or Γ_0. It is usually Γ_0 that is measured and ν_0 that is calculated.

Sempolinski and Kingery[6] have measured the mobility of Mg^{2+} vacancies in MgO, from which one can calculate $\Gamma_0 = 2650$ THz. This is a large value, 2 orders of magnitude above lattice vibrational frequencies. We shall note it as it stands, though one must recall that measurements over restricted temperature ranges (here 1200–1600°C only) may contain deceptive features, especially if changes of mechanism or detrapping from a variety of traps is involved. We may combine the experimental Γ_0 with our calculations of the temperature dependence of the activation energy (Table II) to estimate ν_0. Including only the effects of the (experimental) lattice parameter change, we find $E_A \simeq E_{A0} - 2.43kT$, so that $\nu_0 \equiv \Gamma_0 \exp(-2.43)$. The resulting value, $\nu_0 \simeq 234$ Thz, is much reduced but is still high.

It is useful to start by making primitive estimates of ν_0. The first uses the approach frequency picture implicit in the dynamical theory. The particle makes excursions in the direction of the jump saddle point at irregular intervals determined by its random vibrational motion about the equilibrium site. Clearly, the particle cannot make systematic excursions of large amplitude in a chosen direction more often than at intervals which average out to about the period of the most rapid lattice vibrations, i.e., ν_{max}^{-1}. Thus, one anticipates $\nu_0 \lesssim \nu_{max}$. For MgO, if there are no local modes, this means $\nu_0 \lesssim 20$ THz. This rule can be broken, e.g., if successive jumps are correlated or if there is some cooperative phenomenon (de Boer[40]). But any case where $\nu_0 > \nu_{max}$ needs explanation. A second estimate puts a bound on Γ. Suppose the moving particle, of mass M, experienced no barrier whatsoever to motion from site to site. In thermal equilibrium, the mean kinetic energy is $\frac{1}{2}Mv^2 = \frac{3}{2}kT$, and the transit time is d/v. Thus, v/d is an upper bound on Γ; i.e., $\Gamma \leq (3kT/Md^2)^{1/2}$. For the Mg jump in MgO at 1400°C, with d as the Mg–Mg separation in MgO, this means $\Gamma \leq 4.4$ THz.

Another estimate of ν_0 is provided by the usual continuum approximation for the dynamical theory (Flynn[39]):

$$\nu_0 = \left(\frac{3}{5}\right)^{1/2} \nu_D \tag{12}$$

where ν_D is the Debye frequency, related to the Debye temperature θ_D by $h\nu_D = k\theta_D$. For MgO, θ_D is 941 K and hence $\nu_D = 19.6$ THz, and $\nu_0 = 15.2$ THz.

We also note that all jump rates calculated by molecular dynamic simulations appear to be consistent with Γ less than the maximum harmonic lattice frequency. These computer "experiments" in which the coupled equations of motion of atoms are solved numerically to yield trajectories for a system of modest sized, periodically repeated supercells clearly provide detailed information on individual jump events. There are, however, two problems. The first is that the number of time steps per jump is prohibitive except for idealized systems (or very high temperatures). Special techniques are needed to enhance the sampling in phase space around saddle configurations. The second problem is that some transits past the saddle point are followed by a very rapid return. One has to decide (see, e.g., McCombie and Sachdev[41]) whether these transits are jumps or merely spurious crossings. This problem is most severe when the energy surface is flat along the jump path at the saddle point. To the best of our knowledge, the problems we shall discuss later are of different origin, being most severe when the energy surface is flat *normal* to the jump path at the saddle point.

Numerical Results

We have two methods for estimating the frequencies required in the reaction rate and dynamical theories: (a) from the curvatures of the energy surfaces (e.g., Figs. 2–4) and (b) from the supercell normal mode frequencies. In expressing the curvatures as effective frequencies we adopt the following procedure. First we express the quadratic dependence of the energy changes for small displacements of the jumping atom from its equilibrium or saddle point position in terms of a force constant K

$$\Delta E = \frac{1}{2} K X^2 \tag{13a}$$

and then we obtain an effective frequency from

$$(2\pi\nu_{eff})^2 = K/m \tag{13b}$$

with m simply as the mass of the ion that makes the jump. It should be recalled however that, in the calculations leading to Figs. 2–4, the neighboring ions are allowed to relax, and therefore an increased mass should strictly be associated with displacement, leading to a reduction in the effective frequency. In Table VI (first column), we list the effective frequencies obtained in this way for displacements of a range of diffusing ions from their equilibrium positions toward the potential

Table VI. Dynamical Theory Predictions

	Shells and cores relaxed (THz)	Shells alone relaxed (THz)
(a) Divalent ions		
Alkaline earths		
Mg^{2+}	8.24	10.70
Ca^{2+}	6.14	8.64
Transition metals		
Mn^{2+}	5.65	7.12
Fe^{2+}	5.55	6.98
Co^{2+}	5.78	7.39
Ni^{2+}	5.72	7.15
(b) Effect of charge state		
Fe^{2+}	5.55	6.98
Fe^{3+}	5.92	8.48
(c) Simple models		
Debye model	15.2	
Maximum frequency	21.6	

*All results are in THz. In calculating frequencies from energy surfaces, we have used only the mass of the diffusing particle; an effective mass, weighted for relaxation of the neighbors, would reduce values somewhat. The results for Fe^{3+} do not include a factor of 2 associated with the double saddle point.

barrier (Fig. 1(A,C); displacements are always in the plane of these diagrams). These provide estimates (or in fact overestimates for the reason just stated) of the attempt frequency in the dynamical theory with one choice of reaction coordinate, namely, that in which positions of neighbors are adjusted to give a constrained minimum in the energy. The effective frequencies vary little with atomic species, the main variation being due to their different masses. (There is also some evidence of an ion size effect, most clearly seen from the corresponding force constants: for Ca^{2+} the force constant is 10% lower than for Mg^{2+}.) Also from the single case calculated, there is little dependence on charge state despite the significant changes in equilibrium configuration. In the last column of the table we list the effective frequencies for a second choice of reaction coordinate, that in which cores of neighbors are held fixed but polarization (displacement of shells) is allowed. In this case the mass used in the above procedure for extracting effective frequencies is correct. The larger force constants (and effective frequencies) from this choice of reaction coordinate arise, of course, from the constraints imposed on the neighbors. It should be noted that, with either choice of reaction coordinate, the effective frequencies lie well below those given by the Debye model.

In addition to the above ambiguity in the choice of reaction coordinate, a further ambiguity arises when the jump path bifurcates, giving separate saddle points as with Fe^{3+} (Fig. 4). Should the coordinate be the combination of displacements that, linearly extrapolated, takes the atom to the ultimate saddle point, or should it be along the initial part of the jump path?

In Table VII effective frequencies, from curvatures of fully relaxed energy surfaces, are given for Mg^{2+} and Fe^{3+} ions in equilibrium and saddle point configurations. Frequencies for displacements both along and normal to the jump path are included. Some normal mode frequencies for the saddle-point configurations

Table VII. Effective Frequencies from Curvatures of Energy Surfaces (from Eq. (13)).*

Direction	Mg^{2+}	Fe^{3+}
(a) Displacements from equilibrium configuration		
(110)	8.24	5.92
($1\bar{1}0$)	11.0	7.2
(001)	10.8	7.5
(b) Displacements from saddle-point configuration		
(110)	7.64i (8.61i)	0.91i (3.27i)
($1\bar{1}0$)	12.71	8.34
(001)	2.98 (2.04)	4.35

*Values in brackets in part (b) are frequencies for normal modes consisting mainly of the assigned displacements. Units are in THz.

are also given for comparison. The eigenvectors for these normal modes show that they are dominated by the displacements indicated: strict comparison is not, of course, possible since the displacements considered in the energy calculations are not normal modes.

The low effective frequency for (001) displacements of Mg^{2+} from the saddle point is of particular interest. If Mg^{2+} is replaced by Fe^{3+} (i.e., we hold an Fe^{3+} ion at the midpoint of its positions before and after a jump and allow relaxation of the neighbors), then this "soft" displacement becomes unstable. This causes the double saddle-point surface, Fig. 4, with the saddle points displaced in the (001) direction.

Using the normal mode frequencies from the 64-site supercell calculations for one small **q** value (in the 110 direction) with both equilibrium and saddle-point configurations, we have estimated the pre-exponential factor ν_0 from the Vineyard expression (Eq. (11)). We obtain 32.91 THz for the Mg^{2+} jump and $2 \times 1.21 = 2.42$ THz for the Fe^{3+} jump (the factor 2 coming from the pair of saddle points). One is immediately struck by two features of the ν_0 estimated for Mg^{2+} and Fe^{3+} motion: (a) the large difference between them and (b) the relation to the spectrum of normal mode frequencies for the host lattice (maximum frequency ~20 THz); for Mg^{2+} our estimate for ν_0 appears anomalously high, and that for Fe^{3+}, anomalously low.

From our discussion in the previous section, we do not expect ν_0 to exceed significantly the maximum lattice frequency. On the other hand, it is clear that the Vineyard expression (Eq. (11)) imposes no such limit. Suppose we idealize and represent the $3N - 1$ frequencies for normal modes orthogonal to the jump path by a single frequency $\bar{\nu}$ at the equilibrium point and $\bar{\nu}'$ at the saddle point. The expression for ν_0 becomes

$$\nu_0 = \nu^*(\bar{\nu}/\bar{\nu}')^{3N-1} \tag{14}$$

with ν^* the frequency for the (equilibrium configuration) normal mode that takes the atom along the jump path. The upper bound on ν_0 should certainly apply to ν^*, but the remaining factor can be as large or as small as we please: it depends only on the relative flatness of the energy surface in directions orthogonal to the jump at the equilibrium and saddle-point configurations. Increased flatness at the saddle point will enhance ν_0 and vice versa.

The effective frequencies in Table VII for displacements normal to the $\langle 110 \rangle$ jump direction may be taken as an indicator of the ratio $\bar{\nu}/\bar{\nu}'$. It can be seen that, for both the Mg^{2+} and Fe^{3+} jumps, the ratio is greater than unity due to the (001) effective frequencies and that the enhancement is much greater in the Mg^{2+} case. While this does not account for the extreme difference found between the Vineyard pre-exponential factors, it does indicate how a value above the lattice cutoff can be given by the Vineyard expression. Indeed examination of the sample of normal mode frequencies used in our estimate of ν_0 for the Mg^{2+} jump shows that the high value arises from a low frequency (2.04 THz) at the saddle point. The eigenvector for this mode is dominated by a (001) displacement of the jumping Mg^{2+} ion. Thus, both the normal mode analysis and examination of the energy surfaces at equilibrium and saddle points lead to the same explanation for the high frequency factor Γ_0 from the Vineyard theory for the Mg^{2+} jump.

We believe that the Vineyard result is unphysical in this case and that this comes about because the quasi-harmonic approximation used in the derivation of Eq. (11) is inapplicable for such flat energy surfaces; i.e., the harmonic expansion of the energy about the saddle point is insufficient. Only if the constrained hypersurface containing the saddle point has a deep minimum at the saddle point with crossings all in close vicinity of the saddle point will harmonic expansion be appropriate and the Vineyard expression reasonable. In Harding's[42] calculation on Na_2S, for example, no special problems with the Vineyard formalism were apparent. For cases like our Mg^{2+} jump, the integration over the hypersurface cannot be approximated by using harmonic expansions and would have to be carried out numerically.

Finally, ignoring the above caveat, we estimate the absolute jump frequency from the modified expression of the previous section as

$$\Gamma_{th} = \nu_0 \exp(2.43) \exp(-E_A^0/kT)$$

with E_A^0 the activation energy at 0 K and ν_0 as our estimate from the Vineyard theory (32.91 THz). Extrapolation of the activation energies in Table II gives $E_A^0 = 2.261$ eV, and hence at 1400°C we obtain

$$\Gamma_{th} = 57.2 \text{ MHz}$$

as compared with Sempolinski and Kingery's value of

$$\Gamma_{exp} = 356.2 \text{ MHz}.$$

The prediction is lower than the experimental value by a factor 6.2. Since both pre-exponential factor and activation energy in Γ_{th} are predicted without introducing adjustable parameters, the agreement is respectable. We are left with two problems, however. One is that the various theoretical approaches give distinctly different values for the pre-exponential factors for Fe^{3+} and Mg^{2+} motion: the dynamical theory predicts that Mg^{2+} jumps some 30% faster (39% when cores and shells are relaxed and 26% when shells alone are relaxed) whereas Vineyard theory predicts a factor 14 times larger for Mg^{2+} motion. Second, both our calculations (Table II) and the high measured mobility suggest that extrapolation of $D(T)$ or $\mu(T)$ to high temperature will present problems. Extrapolation using the experimental data will not include the significant temperature dependence of the activation energy, which is likely, and extrapolation using Vineyard theory may be sensitive to temperature-induced changes in the soft-mode frequency and in the energy surfaces near the saddle point.

Conclusions

In this paper we have combined some new calculations of defect concentrations and diffusion rates with a survey of the methods available for quantitative calculations. The key step in these recent developments was the realization in 1980 that supercell methods, using existing codes, offered a practical and flexible technique for defect dynamics. Clearly, there are technical points one would like verified in detail, and equally, comparisons with a full range of oxide data would be highly desirable. Nevertheless it is clear that there is already gratifying success, one that (subject to the qualifications just made) promises to match the value of the long-established calculations of defect energies.

To measure the degree of success, one must look at the results in context. Most calculations like ours will be used in combination with experimental data in a variety of ways. The first way is as a guide in deriving reaction constants and populations when there is a relatively complicated series of defect reactions in which experiment alone gives incomplete data. The entropies and energies are the key quantities, and present experience suggests theory will be able to guide and to fill gaps effectively. The position is less clear for internal friction, where defect dynamics are involved rather than defect populations. Even for relatively simple defects, like vacancy-impurity pairs, one may wish to know rather many relative rates for different types of jump (see e.g., Lidiard[43]). Usually the ratios are unknown and so are chosen arbitrarily to maximize convenience. In principle, the methods discussed in Absolute Diffusion Rates could give quantitative guidance, but this is a long way from being demonstrated.

A second use of theory and experiment together is as a check of models and theories. The apparent success of the energy and entropy terms, together with the many other properties modeled in the same way, suggests that the interatomic forces are adequately handled. Far less satisfactory are the theories that use some statistical mechanical argument to relate calculations to experimental jump rates. Indeed, our results suggest caution in extrapolating jump rates at all far from the original data. Even the simple phenomenology, based on the Arrhenius equation, should be used cautiously, for naive extrapolation may predict (as it does for Mg in MgO) that the jump rate with a diffusion barrier will ultimately exceed the rate when no barrier is present. Naive extrapolation may also suggest jumps occurring more rapidly than those of the highest frequency lattice vibrations; no cases with $\Gamma > \nu_{max}$ have been observed to our knowledge, either in experiments or in molecular dynamics, but cases where Γ_0 or ν_0 exceed ν_{max} should be extrapolated with care.

A third use of theory lies in the insight it can give into jump processes and to features not easily accessible experimentally. One feature identified from our Vineyard theory calculations, for instance, showed systematic behavior giving both an anomalously high ν_0 from soft phonons for Mg motion and a bifurcated jump path from the corresponding unstable phonons for Fe^{3+}.

The picture which emerges is that entropies and jump rates can be predicted by using the approaches and models which work well for defect energies. While such predictions will never be so straightforward as energy calculations and while many technical points remain to be checked, one can be cautiously optimistic of these approaches and of their value in understanding oxide defect behavior.

---

*Printed with permission to publish by the United Kingdom Atomic Energy Authority. Copyright remains the property of the UKAEA.

References

[1] A. B. Lidiard and M. J. Norgett, "Point Defects in Ionic Crystals"; p. 385 in Computational Solid State Physics. Edited by F. Herman, N. W. Dalton, and T. R. Koehler. Plenum, New York, 1974.

[2] C. R. A. Catlow and W. C. Mackrodt (Editors); Computer Simulation of Solids. Springer, New York, 1982.

[3] A. M. Stoneham; Handbook of Interatomic Potentials: I, Ionic Crystals. A.E.R.E. Rept. R9598, 1981.

[4] W. H. Gourdin and W. D. Kingery, "The Defect Structure of MgO Containing Trivalent Cation Solutes: Shell Model Calculations," *J. Mater. Sci.*, **14**, 2053 (1979).

[5] W. H. Gourdin, W. D. Kingery, and J. Driear, "The Defect Structure of MgO Containing Trivalent Cation Solutes: The Oxidation-Reduction Behavior of Iron," *J. Mater. Sci.*, **14**, 2074 (1979).

[6] D. R. Sempolinski and W. D. Kingery, "Ionic Conductivity and Magnesium Vacancy Mobility in Magnesium Oxide," *J. Am. Ceram. Soc.*, **63**, 11 (1980).

[7] M. J. L. Sangster and A. M. Stoneham, "Calculations of Off-Centre Displacements of Divalent Substitutional Ions in MgO, CaO, SrO, and BaO from Model Potentials," *Philos. Mag.*, [Part B], **43**, 597 (1981).

[8] J. R. Tessman, A. H. Kahn, and W. Shockley, "Electronic Polarizabilities of Ions in Crystals," *Phys. Rev.*, **92**, 890 (1953).

[9] D. K. Rowell and M. J. L. Sangster, "Calculation of Defect Energies and Volumes in Some Oxides," *Philos. Mag.*, [Part A],**44**, 613 (1981).

[10] C. R. A. Catlow, W. C. Mackrodt, M. J. Norgett, and A. M. Stoneham, "Basic Atomic Processes of Corrosion I," *Philos. Mag.*, **35**, 177 (1977).

[11] C. R. A. Catlow and A. M. Stoneham, "Ionicity in Solids," *J. Phys. C*, **16**, 4321 (1983).

[12] A. M. Stoneham and M. J. L. Sangster, "Multiple Charge States of Transition Metal Impurities," *Philos. Mag.*, [Part B], **43**, 609 (1981).

[13] M. J. L. Sangster, "Relaxations and Their Strain Derivatives around Impurity Ions in MgO" *J. Phys. C*, **14**, 2889 (1981).

[14] D. K. Rowell and M. J. L. Sangster, "Off-Centre Displacements of Trivalent Impurity Ions in SrO," *J. Phys. C*, **15**, L357 (1982).

[15] K. A. Müller and W. Berlinger, "Superposition Model for Sixfold Coordinated Cr^{3+} in Oxide Crystals"; to be published in *J. Phys. C*.

[16] M. J. Norgett, "A General Formulation of the Problem of Calculating the Energies of Lattice Defects in Ionic Crystals," A.E.R.E. Report R7650 (1974).

[17] C. R. A. Catlow and M. J. Norgett, "Lattice Structure and Stability of Ionic Materials," A.E.R.E. Rept. M2936, 1976.

[18] M. J. Gillan, "Volume of Formation of Defects in Ionic Crystals," *Philos. Mag.*, [Part A], **43**, 301 (1981).

[19] M. W. Finnis and M. Sachdev, "Vacancy Formation Volumes in Simple Metals," *J. Phys. F,* **6**, 965 (1976).

[20] E. A. Perez-Albuerne and H. G. Drickamer, "Effect of High Pressure on the Compressibility of Seven Crystals Having the NaCl or CsCl Structure," *J. Chem. Phys.*, **43**, 1381 (1965).

[21] J. H. Harding, P. Masri, and A. M. Stoneham, "Thermodynamic Properties of UO_2: Electronic Contributions to the Specific Heat," *J. Nucl. Mater.*, **92**, 73 (1980).

[22] J. H. Harding and A. M. Stoneham, "Vibrational Entropies of Defects in Solids," *Philos. Mag.*, [Part B], **43**, 43 (1980).

[23] M. J. Gillan and P. W. M. Jacobs, "On the Entropy of a Point Defect in an Ionic Crystal," *Phys. Rev. B*, **B28**, 759 (1983).

[24] M. J. L. Sangster and A. M. Stoneham, "Off-Center Li in KCl: Infrared Absorption and Behavior under Pressure," *Phys. Rev. B*, **B26**, 1026 (1982).

[25] R. A. Evarestov and V. A. Lovchikov, "Use of Symmetry-Adapted Orbitals in the Large Unit Cell Approach in Solids," *Phys. Status Solidi B*, **93**, 469 (1979). This paper cites earlier references, of which the most relevant are: A. M. Dobrotvorskii and R. A. Evarestov, *Phys. Status Solidi B*, **66**, 83 (1974) and R. A. Evarestov, *Phys. Status Solidi B*, **72**, 569 (1975).

[26] J. H. Harding and A. M. Stoneham, "The Calculation of Hugoniots in Ionic Solids"; to be published in *J. Phys. C*.

[27] A. A. Maradudin, E. W. Montroll, G. H. Weiss, and I. P. Ipatova, "Theory of Lattice Dynamics in the Harmonic Approximation," 2nd edition. Academic Press, New York and London, 1971.

[28] T. Holstein, "Small Polaron Motion in Molecular Crystals," *Ann. Phys. (New York)*, **8**, 343 (1959).

[29] C. P. Flynn and A. M. Stoneham, "Quantum Theory of Diffusion with Application to Light Interstitials in Metals," *Phys. Rev. B*, **B1**, 3966 (1970).

[30] A. M. Stoneham, "The Motion of Muons in Solids," *Hyperfine Interact.*, **6**, 211 (1979).

[31] H. R. Schober and A. M. Stoneham, "Motion of Interstitials in Metals: Quantum Tunnelling at Low Temperatures," *Phys. Rev. B*, **B26**, 1819 (1982).

[32] S. Glasstone, K. J. Laidler, and H. Eyring, "The Theory of Rate Processes," McGraw-Hill, New York, 1941.

[33] C. A. Wert and C. Zener, "Interstitial Atomic Diffusion Coefficients," *Phys. Rev.*, **76**, 1169 (1949).
[34] G. H. Vineyard, "Frequency Factors and Isotope Effects in Solid State Rate Processes," *J. Phys. Chem. Solids*, **3**, 121 (1957).
[35] S. A. Rice, "Dynamical Theory of Diffusion in Crystals," *Phys. Rev.*, **121**, 804 (1958).
[36] N. B. Slater; The Theory of Unimolecular Reactions. Cornell University Press, Ithaca, NY, 1959.
[37] H. R. Glyde, "Rate Processes in Solids," *Revs. Mod. Phys.*, **39**, 373 (1967).
[38] R. Landauer, "Stability and Relative Stability in Nonlinear Driven Systems," *Helv. Phys. Acta*, **56**, 847 (1983).
[39] C. P. Flynn; Defects and Diffusion in Solids. Oxford University Press, London, 1972.
[40] J. H. de Boer, "Steady State Processes Involving Lattice Rearrangement," *Disc. Farad. Soc.*, **23**, 171 (1957).
[41] C. W. McCombie and M. Sachdev, "Theories of Classical Thermal Hopping of a Defect," *J. Phys. C,* **C8**, L413 (1975).
[42] J. H. Harding, "The Physics of Superionic Conductors and Electrode Materials"; pp. 27–40 in Thermodynamic and Transport Properties of Superionic Conductors and Electrode Materials. Edited by J. W. Perram. Plenum, New York, 1983.
[43] A. B. Lidiard; Physics of Crystals with the Fluorite Structure. Edited by W. Hayes. Oxford University Press, London, 1973.

Experimental and Calculated Values of Defect Parameters and the Defect Structure of α-Al$_2$O$_3$*

F. A. Kröger

University of Southern California
Department of Materials Science
Los Angeles, CA 90089-0241

Comparison of experimental and theoretical values of the enthalpies of oxidation–reduction of Al$_2$O$_3$ doped with donors or acceptors gives reasonable agreement for models involving O_i'', V_{Al}''', $V_O^{\cdot\cdot}$, or $Al_i^{\cdot\cdot\cdot}$ if empirical potentials are used in the computation. Fair agreement is also found for the enthalpies of all four possible mechanisms of ionic disorder. Experimental values of the exponents n and m in $[j] \propto [\text{dopant}]^n p_{O_2}^m$ also fit all models. Results of microstructure studies point to anti-Frenkel disorder as the dominant disorder mechanism. Experimental and calculated values for the enthalpies of the mobilities of ionic defects agree with computed values if empirical potentials are used in the computation of disorder enthalpies. Electrons and holes are large polarons. The thermal band gap at 0 K is ≈ 9.7 eV with $\Delta = (E_g^\circ)_{opt} - (E_g^\circ)_{th} \approx 0.65$ eV. Oxygen diffusion in the bulk involves $V_O^{\cdot\cdot}$; that at grain boundaries is due to migration of neutral interstitial oxygen atoms, $O_i^\times$. Effects of double doping by donors and acceptors are discussed on the basis of the anti-Frenkel model.

The defect structure of α-Al$_2$O$_3$ and parameters of defect formation reactions can be obtained by analyzing experimental results of properties dependent on the presence of defects or by theoretical calculations. In this paper, results obtained in these two ways are compared, paying attention in particular to the following aspects: (1) ionic and electronic disorder; (2) dominant defects in crystals doped with aliovalent atoms acting as donors or acceptors; (3) oxidation–reduction processes; (4) defect models; (5) type of electronic carriers — small or large polarons; (6) mobilities of ionic defects; (7) electronic energy levels of defects — the band gap; (8) double doping with donors and acceptors.

Discussion

Experimental data are available for ionic and electronic conductivities as a function of oxygen fugacity, temperature, and type and concentration of aliovalent dopants for self-diffusion of oxygen and aluminum, for chemical diffusion involved in changing the stoichiometry, and for the rates of sintering, creep, and densification in hot-pressing. If self-diffusion involves migration of charged species, as is certainly the case for aluminum, the constants of tracer self-diffusion

$$D_j^* = f_j D_j[j] = f_j D_{\text{self},j} \tag{1}$$

and of ionic conductivity

$$\sigma_i = c_j z_j q \mu_j = N[j] z_j q \mu_j \tag{2}$$

with
$$\mu_j = D_j z_j q / kT \quad \text{(Einstein relation)}$$

are related through the Nernst–Einstein relation:

$$\sigma_i = ND_j[j]z_j^2 q^2 / kT = D_j^* N z_j^2 q^2 / f_j kT = D_{\text{self},j} N z_j^2 q^2 / kT \tag{3}$$

Here square brackets indicate concentrations in mole fractions, N is the number of molecules Al_2O_3 per cm^3, D_j and μ_j are, respectively, the diffusivity and mobility of the defect j, q is the electronic charge, z_j is the number of charges per defect, and f_j is a correlation coefficient.

For bulk diffusion, $D_{Al}^* \gg D_O^*$.[1-5] For surface and grain-boundary diffusion, $D_O^* \geq D_{Al}^*$.[2,4] As seen in Fig. 1, values of D_j^* calculated from σ_i by using Eq. (3) are close to those measured directly for D_{Al}^* but are orders of magnitude larger than those measured for D_O^*(bulk), indicating that ionic bulk conductivity involves migration of Al defects, $Al_i^{\cdots}$ or V_{Al}'''. Figure 1 shows the same correspondence with D_{Al}^* for D's calculated from creep rates. In creep as in sintering, oxygen migrates rapidly via grain boundaries, the creep rate being determined by the slower migration of Al in the bulk.[7]

The agreement between calculated D_{Al}^* and values deduced from measurements on different samples cannot be expected to be exact; absolute values depend on the presence or absence of dopants or impurities. Moreover, donor and acceptor dopants cause D_{Al}^* to be determined by $[V_{Al}''']$ and $[Al_i^{\cdots}]$, respectively, which should have somewhat different mobilities.

The species involved in the migration of oxygen may be charged (O_i'' or $V_O^{\cdot\cdot}$) or neutral ($O_i^\times$). As seen in Fig. 1, there is a one decade spread in the values found for diffusion in the bulk. According to Reddy,[4] the spread of values for the same undoped Al_2O_3 in experiments by him and Reed and Wuensch[3] must be attributed to the difference in the oxygen fugacity during the diffusion runs, that in Reddy's experiments (which give a large D_O) being much larger than that in the experiments by Reed and Wuensch. Undoped Al_2O_3 invariably contains acceptor impurities that dominate its properties, causing a large concentration of charged oxygen vacancies, $V_O^{\cdot\cdot}$. Since the concentration of $V_O^{\cdot\cdot}$ decreases with increasing p_{O_2}, $V_O^{\cdot\cdot}$ cannot be the species dominating oxygen diffusion if Reddy's assumption is correct: oxygen diffusion must involve migration of O_i. The concentration of O_i depends on p_{O_2} and temperature, and on doping when it is charged, but is independent of doping when it is neutral. The observation of almost the same D_O in samples with widely different impurity content[5,8] and the absence of an effect of doping with Mg^4 thus suggest that $O_i^\times$ is the migrating species.

There is a conflict, however, with the observation that a sample doped with Ti has a significantly smaller D_O,[4] while recent work by Castaing and Heuer (presented at this conference) shows a positive effect of Mg and a negative effect of Ti, indicating that the migrating species is charged. Donor doping should increase the concentration of O_i'' and decrease the concentration of $V_O^{\cdot\cdot}$, Mg having the opposite effect. Thus, a decrease of D_O by Ti and an increase by Mg can only be accounted for if $V_O^{\cdot\cdot}$ is the migrating defect in undoped (acceptor dominated) Al_2O_3.

Evidently some of the pieces of information must be in error. We are inclined to believe that the dopant effects are real and that $V_O^{\cdot\cdot}$ is the major mobile species, at least at low and moderate oxygen pressures. At extremely large oxygen activities present during a fast color change observed in Al_2O_3:Mg at $T < 900°C$,[8] $O_i^\times$ appears to be the major mobile species.

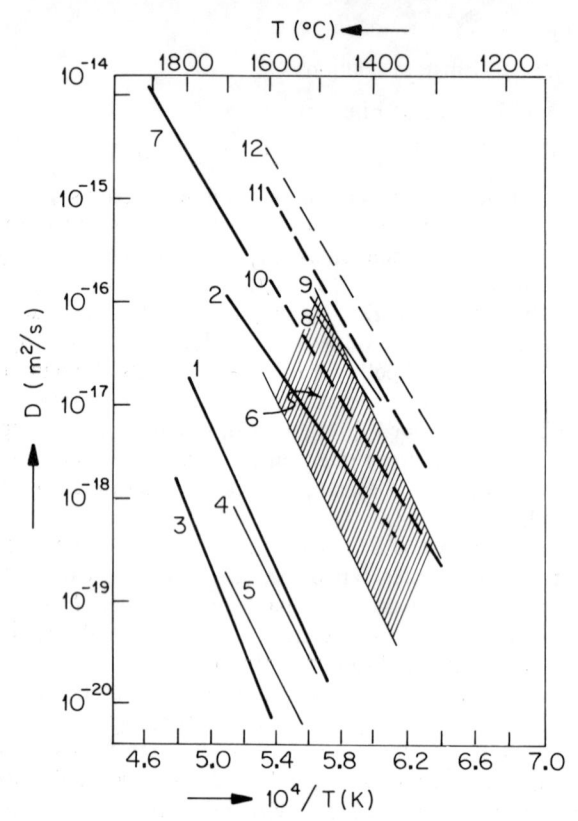

Fig. 1. Self-diffusion coefficients obtained from measurements of diffusion (using radioactive tracers or heavy isotopes), ionic conductivity, and creep: (1) D_O from diffusion of ^{18}O in single crystals;[2] (2) D_O from diffusion of ^{18}O in polycrystalline material;[2] (3) D_O from diffusion of ^{18}O in single crystal;[3] (4) D_O from diffusion of ^{18}O in single crystals of $Al_2O_3 \pm Mg$;[4] (5) D_O from diffusion of ^{18}O in single crystals of Al_2O_3:500 ppm Ti;[4] (6) D_O from diffusion of ^{18}O in polycrystalline Al_2O_3;[4] (7) D_{Al} from diffusion of Al in single and polycrystalline Al_2O_3;[1] (8) D from Nabarro–Herring creep in undoped polycrystalline Al_2O_3;[6] (9) D from Nabarro–Herring creep of Al_2O_3:500 ppm Ti (grain size 4 μm);[6] (10) D from σ_i for sample of curve 8 in air;[6] (11, 12) D from σ_i for $Al_2O_3 \pm 3$ wt % Fe, $p_{O_2} = 10^{-4}$ Pa.[6]

The lack of excess ionic conductivity in polycrystalline material at small grain sizes[6,10] and the dependence of chemical diffusion on oxygen fugacity[11] indicate that neutral O_i^x is also the species involved in the fast grain-boundary diffusion of oxygen at normal oxygen pressures.[11]

In any case, charged oxygen defects will be present and may be involved in disorder and charge compensation of dopants. The fact that for bulk diffusion and conduction $D_{Al}^* \gg D_O^*$ does not necessarily imply that the concentrations of Al defects are larger than those of the oxygen defects, provided the mobilities of the Al defects are considerably larger than those of the O defects. Investigations of the microstructure of Al_2O_3 doped with Ti[12] or Mg[13] in fact point to O_i'' and $V_O^{\cdot\cdot}$

rather than V'''_{Al} and $Al_i^{...}$ as the major charged ionic species.

Oxidation–reduction of donor- or acceptor-doped Al_2O_3 according to the reactions shown in Table I can be accounted for equally well on the basis of models with oxygen or aluminum defects dominant.[14] This is demonstrated in Table II, which shows experimental values of the enthalpies of oxidation–reduction reactions compared to theoretical values computed by Catlow et al.[15] Unfortunately, there is a wide spread in some of the computed values, related to the types of repulsion potentials used. Computed values in Table II are given separately for holes as small or large polarons and for corrected and uncorrected values in the electronic energies of Ref. 15 (the value of the band gap computed in this reference being ≈ 3 eV too small). Experimental results[9] as well as computation[18] indicate that holes are large polarons, large polarons being ≈ 0.4 eV more stable than small polarons.[18]

Further experimental input on the basis of which a choice between the various models can possibly be made concerns the values of exponents in the expression

$$[j] \propto [\text{dopant}]^n p_{O_2}^m \tag{4}$$

Table III gives expected values of n and m for donor-dominated Al_2O_3 with either $[D^.] \approx 2[O''_i]$ or $[D^.] \approx 3[V'''_{Al}]$ with $[D]_{total} \approx [D^.]$ expected at high p_{O_2} or $[D]_{total} \approx [D^\times]$ expected at low p_{O_2}. Values concerning dominant defects are underlined. Experimental values for n are not available; those for m of σ_e are in good agreement with the O''_i model.

Table IV gives values of n and m for acceptor-dominated Al_2O_3. The $Al_i^{...}$ model appears to be slightly favored. Creep results by Hollenberg and Gordon, on

Table I. Various Defect Reactions and the Corresponding Mass Action Relations Using Symbols Proposed by Kröger and Vink*

	Reaction	Mass action relation†
(a)	$\tfrac{3}{4}O_2(g) + 3e' \rightleftharpoons \tfrac{3}{2}O_O^\times + V'''_{Al}$	$K_{ox,V(Al)} = [V'''_{Al}][e']^{-3} p_{O_2}^{-3/4}$
(b)	$\tfrac{1}{2}O_2(g) + V_i^\times + 2e' \rightleftharpoons O''_i$	$K_{ox,O_i} = [O''_i][e']^{-2} p_{O_2}^{-1/2}$
(c)	$\tfrac{3}{4}O_2(g) + Al_i^{...} \rightleftharpoons \tfrac{3}{2}O_O^\times + Al_{Al}^\times + 3h^.$	$K_{ox,Al_i} = [h^.]^3[Al_i^{...}]^{-1} p_{O_2}^{-3/4}$
(d)	$\tfrac{1}{2}O_2(g) + V_O^{..} \rightleftharpoons O_O^\times + 2h^.$	$K_{ox,V(O)} = [h^.]^2[V_O^{..}]^{-1} p_{O_2}^{-1/2}$
(e)	$\tfrac{3}{4}O_2(g) + 3Ti_{Al}^\times \rightleftharpoons 3Ti_{Al}^. + V'''_{Al} + \tfrac{3}{2}O_O^\times$	$K_{ox,V(Al)}^{Ti} = [Ti_{Al}^.]^3[V'''_{Al}][Ti_{Al}^\times]^{-3} p_{O_2}^{-3/4}$
(f)	$\tfrac{1}{2}O_2(g) + 2Ti_{Al}^\times + V_i^\times \rightleftharpoons 2Ti_{Al}^. + O''_i$	$K_{ox,O_i}^{Ti} = [Ti_{Al}^.]^2[O''_i][Ti_{Al}^\times]^{-2} p_{O_2}^{-1/2}$
(g)	$Ti_{Al}^\times \rightleftharpoons Ti_{Al}^. + e'$	$K_d = [Ti_{Al}^.][e'][Ti_{Al}^\times]^{-1}$
(h)	$A^\times \rightleftharpoons A' + h^.$	$K_a = [A'][h^.][A^\times]^{-1}$
(i)		$K_{ox,V(Al)}^{Ti} = K_{ox,V(Al)} K_d^3$
(j)		$K_{ox,O_i}^{Ti} = K_{ox,O_i} K_d^2$
(k)	$O_O^\times + V_i^\times \rightleftharpoons O''_i + V_O^{..}$	$K_{F,O} = [O''_i][V_O^{..}]$
(l)	$2Al_{Al}^\times + 3O_O^\times \rightleftharpoons 2Al_i^{...} + 3O''_i$	$K_I = [Al_i^{...}]^2[O''_i]^3$
(m)	$0 \rightleftharpoons 2V'''_{Al} + 3V_O^{..}$	$[V'''_{Al}]^2[V_O^{..}]^3 = K_S$
(n)	$Al_{Al}^\times + V_i^\times \rightleftharpoons Al_i^{...} + V'''_{Al}$	$[Al_i^{...}][V'''_{Al}] = K_{F,Al}$
(o)	$0 \rightleftharpoons e' + h^.$	$K_i = [e'][h^.]$
(p)	$K_{F,Al} = K_i^3 K_{ox,V(Al)} K_{ox,Al_i}^{-1}$	$K_{F,O} = K_i^2 K_{ox,O_i} K_{ox,V(O)}^{-1}$
(q)	$K_S = K_i^6 K_{ox,V(Al)}^2 K_{ox,V(O)}^{-3}$	$K_I = K_i^6 K_{ox,O_i} K_{ox,Al_i}^{-2}$
(r)	$A + V'''_{Al} \rightleftharpoons A'_{Al} + 2e'$	$K_A = [A'_{Al}][e']^2[V'''_{Al}]^{-1} a_A^{-1}$
(s)	$D + V'''_{Al} \rightleftharpoons D^._{Al} + 3e'$	$K_D = [D^._{Al}][e']^3[V'''_{Al}]^{-1} a_D^{-1}$

*Ref. 16.
†Square brackets indicate concentrations in numbers of defects for molecule Al_2O_3.

Table II. Enthalpies for Various Native Oxidation Processes Determined from Experimental Data on Donor- and Acceptor-Doped Al_2O_3 Compared with Theoretical Values

	Experimental values					Computed values[15] based on			
	Al_2O_3:Ti		Al_2O_3:Fe or Mg			Empirical potentials		Nonempirical potentials	
	kJ/mol	eV	kJ/mol	eV		kJ/mol	eV	kJ/mol	eV
$H_{ox,V(Al)}$	−862	−9.0[¶]				−639	−6.64	−56.8	−0.59
H_{ox,O_i}	−578	−6.0[¶]				(−1293)[‡]	(−13.44)[‡]	(−712)[‡]	(−7.4)[‡]
						−368	−3.82	728	7.56
						(−811)[‡]	(−8.42)[‡]	(285)[‡]	(12.96)[‡]
H_{ox,Al_i}			626	6.51[17]		514	5.34	292	3.03
			335	3.48[9]		(1176)[§]	(12.22)[§]	(956)[§]	(9.93)[§]
$H_{ox,V(O)}$			390	4.05[17]		192;* 113[†]	2.0;* 1.17[†]	426;* 346[†]	4.43;* 3.6[†]
			210	2.18[9]		(632*)[§] (552[†])[§]	(6.57*)[§] (5.74[†])[§]	(866*)[§] (789[†])[§]	(9)*[§] (8.2)[†][§]

*With holes as small polarons.
[†]With holes as large polarons.
[‡]Electronic energies of Ref. 15 in error.
[§]Hole energies of Ref. 15 in error.
[¶]Ref. 14.

Table III. Values of n and m in $[j] \propto [D]^n p_{O_2}^m$

	$[D]_{total} \simeq [D]^\times$				$[D]_{total} \simeq [D']$				Experiment	
	$[D']\simeq 2[O_i'']$		$[D']\simeq 3[V_{Al}''']$		$[D']\simeq 2[O_i'']$		$[D']\simeq 3[V_{Al}''']$			
Species	n	m	n	m	n	m	n	m	n	m
O_i''	2/3	1/6	3/4	3/16	1	0	1	0		
V_{Al}'''	1	1/4	1	−3/16	3/2	0	1	0	0.16±0.02[6]	
e'	1/3	−1/6	1/4	1/4	1/2	−1/4	1/3	−1/4	−0.16±0.02[6]	
$[V_{Al}'''][e']$	4/3	1/2	1	0	2	−1/4	4/3	−1/4	−1/6 (low p_{O_2}); −1/4 (high p_{O_2})[12]	

Table IV. Values of n and m in $[j] \propto [A]^n p_{O_2}^m$

			$[A]_{total}=[A^\times]$				
	$[A']\simeq 2[V_O^{\cdot\cdot}]$		$[A']\simeq 3[Al_i^{\cdot\cdot\cdot}]$		Experiment		
Species	n	m	n	m	n	m	
$V_O^{\cdot\cdot}$	2/3	−1/6	1/2	−1/8	0.7 ± 0.03*	−3/16[†]	−4.8/16* −0.19±0.05[‡]
$Al_i^{\cdot\cdot\cdot}$	1	−1/4	3/4	−3/16		3/16[†]	0.17±0.03[‡]
$h^\cdot$	1/3	1/6	1/4	3/16			
$[Al_i^{\cdot\cdot\cdot}][h^\cdot]$	4/3	−1/12	1	0			

*Reference 20.
[†]Reference 19.
[‡]Reference 6.

Table V. Experimental and Theoretical Values of the Enthalpies per Defect of Interstitial and Anti-Frenkel Disorder

	Experimental values				Theoretical values					
	Ref. 14		Ref. 21		Empirical potentials[‡]		Nonempirical potentials[‡]		Ref. 22	
Enthalpies per defect	kJ/mol	eV	kJ/mol	eV	kJ/mol	eV	kJ/mol	eV	kJ/mol	eV
$1/5 H_I$	843*	8.76*	630±135	6.6±1.4	437–513	4.54–5.33	1005	10.44	1100	11.5
	1027[†]	10.67[†]								
$1/2 H_{F,O}$	517*	5.37*	530±116	5.5±1.2	365–391	3.79–4.06	796	8.27	673	7
	607[†]	6.31[†]								
$1/5 H_S$	564*	5.87*	630±135	6.6±1.4	402–452	4.18–4.70	495	5.14	550	5.7
	672[†]	6.99[†]								
$1/2 H_{F,Al}$	685*	7.11*	790±170	8.2±1.8	502–623	5.22–6.47	682	7.09	960	10
	830[†]	8.63[†]								

*Using Al_2O_3:Fe data from Ref. 17.
[†]Using Al_2O_3:Mg data from Ref. 9.
[‡]Reference 15.

the other hand, are best accounted for on the basis of the $V_O^{\cdot\cdot}$ model.[17]

On the basis of these results, all possibilities remain open, with either O_i'' or V_{Al}''' dominant in donor-dominated Al_2O_3 and $V_O^{\cdot\cdot}$ or $Al_i^{\cdot\cdot\cdot}$ dominant in acceptor-dominated Al_2O_3. As a result, none of the four possible disorder mechanisms involving these defects may be discarded with certainty.

In Table V experimental values for the enthalpy per defect formed in these four types of disorder are compared with values computed by Catlow et al.[15] and Dienes et al.[22] The agreement is reasonable for all types of disorder, but best for the theoretical results of Ref. 15 computed with empirical potentials. Experimental values of the preexponential factors of the disorder constants are given in Ref. 21. Comparing the four types of native ionic disorder, the present author favors anti-Frenkel disorder because it is in agreement with the microstructure observations by Phillips et al.[12] and Pletka et al.[13] and because it fits the general notion that, in asymmetrical compounds, the major disorder is the disorder mechanism involving the ions with the smaller charge and coordination number: F^- in CaF_2, O^{2-} in UO_2, and therefore O^{2-} in Al_2O_3.

Finally, when empirical potentials are used, the computed values are lowest for this mechanism. However, the other mechanisms cannot be ruled out with certainty. It is even possible that several mechanisms produce defects with comparable concentrations. Note that, since the experimental values of the parameters for different models were deduced from the same set of data, only those for one model can be correct. Once a choice has been made regarding the major types of defects and disorder, the values for what now are believed to be minority species and mechanisms must be discarded.

If mobile defects are majority species, analysis of the temperature dependence of ionic conductivity measured under conditions where the composition of the crystal is not changed (i.e., with rapid temperature changes) leads to expressions for the defect mobilities as a function of temperature, as given in Table II of Ref. 23. If the mobile defects are minority species, however, which is the case in all crystals if anti-Frenkel disorder is the main disorder mechanism, in acceptor-dominated Al_2O_3 if Schottky disorder is dominant, and in donor-dominated crystals if interstitial disorder is dominant, the true mobility of the mobile Al defects is related to μ_i of Table II of Ref. 23 (now called μ_{rep}), arrived at under the assumption that the mobile defects are majority species, by corrections resulting from the internal disorder processes by which the concentrations of major and minor species are related,[21] in a manner dependent on whether or not equilibrium for these processes is maintained. If no equilibrium is maintained, the observed activation energy of σ_i is the activation energy of the true ionic mobility, but the value of the preexponential is modified in the ratio between the oxygen and aluminum defects. When equilibrium for the internal disorder processes is maintained, the activation energy contains contributions from these processes. For donor-dominated material with anti-Frenkel disorder or interstitial disorder dominant, the relation between $H(\mu_{rep})$ and $H(\mu_{V_{Al}})$ is

$$H(\mu_{rep}) = H(\mu_{V_{Al}'''}) - 3/2H_{F,O} + 1/2H_S \tag{5}$$

For acceptor-dominated material with anti-Frenkel disorder or Schottky disorder dominant, the relation between $H(\mu_{rep})$ and $H(\mu_{Al_i})$ is

$$H(\mu_{rep}) = H(\mu_{Al_i^{\cdot\cdot\cdot}}) + H_{F,Al} - 1/2H_S \tag{6}$$

H_S, $H_{F,Al}$ and $H_{F,O}$ being the enthalpies of Schottky, Frenkel, and anti-Frenkel disorder. No corrections are involved at all for cation Frenkel disorder, for donor-dominated material with Schottky disorder dominant, and for acceptor-dominated material with interstitial disorder dominant, i.e.,

$$H(\mu_{rep}) = H(\mu_{V_{Al}}) \tag{7}$$

and

$$H(\mu_{rep}) = H(\mu_{Al_i}) \tag{8}$$

In Table VI experimental values for μ_{rep} are compared with values of the right-hand side of Eqs. (5 and 6). There is reasonable agreement for acceptor-dominated material using disorder enthalpies computed either with empirical or with nonempirical potentials for all types of disorder. For donor-dominated material, there is good agreement for Schottky or Frenkel disorder and fair agreement for interstitial and anti-Frenkel disorder, provided the disorder energies are computed by using empirical potentials, indicating that empirical potentials are preferable — a conclusion reached earlier in the interpretation of oxidation–reduction of Al_2O_3:Ti.[14] It is still uncertain, however, whether the values arrived at with or without maintenance of internal disorder equilibrium should be used.

Analysis of the conductivity of exactly compensated Al_2O_3 led not only to the parameters of ionic disorder but also to those of electronic disorder:[21]

$$\sigma_e \sigma_h = 3.3 \times 10^7 \times /: 7.7 \times 10^5 \exp\{-(10 \pm 2.1) \text{ eV}/kT\}$$
$$(\Omega^{-2} \cdot cm^{-2}) \tag{9}$$

with an activation energy of 10 ± 2.1 eV or 962 ± 202 kJ/mol. The symbol $\times/:$ stands for "multiply or divide."

Assuming electrons and holes to be large polarons, this expression is to be compared with the theoretical one for the band model:

$$\sigma_e \sigma_h = (1.0 \times /: 3.5) 10^7 \mu_e \mu_h (m_e^* m_h^*/m^2)^{3/2} \exp\{-(E_g^\circ + 0.3) \text{ eV}/kT\}$$
$$(\Omega^{-2} \cdot cm^{-2}) \tag{10}$$

with μ_e and μ_h the mobilities, m_e^* and m_h^* the effective masses of e' and $h^\cdot$, and m the free electron mass; E_g° is the thermal gap extrapolated to $T = 0$ K; thus $(E_g^\circ)_{th} = 9.7 \pm 2$ eV. This value is to be compared with an optical value $(E_g^\circ)_{opt} = 10.35$ eV, the gap decreasing with temperature with 1.5×10^{-3} eV/deg.[30] Since $(E_g^\circ)_{opt}$ should be smaller than $(E_g^\circ)_{th}$, the medium value of the latter has to be preferred; i.e., $(E_g^\circ)_{th} \simeq 9.7$ eV, giving $\Delta = (E_g^\circ)_{opt} - (E_g^\circ)_{th} \simeq 0.65$ eV. The experimental value of the preexponential of $\sigma_e \sigma_h$ is $\simeq 3.3 \times 10^7 \ \Omega^{-2} \cdot cm^{-2}$, in good agreement with the theoretical value if $\mu_e \simeq \mu_h \simeq 1$ cm^2/V·s and $m_e^* = m_h^* \approx m$. The reported value for $(E_g^\circ)_{th}$ is ≈ 3 eV larger than the one computed by Catlow et al.[15]

Finally, let us consider the defect structure of Al_2O_3 doped with donors D and acceptors A. Dependent on the concentrations of these dopants, the material will be donor or acceptor dominated, but the transition from one regime to the other does not necessarily occur at the point where $[D] = [A]$.[32] Oxygen pressure plays an essential role through its effect on stoichiometry and on the charge of the dopant centers.[23] Effects to be expected will be discussed on the basis of a model in which anti-Frenkel disorder is assumed to be the dominant ionic disorder mechanism. Defect reactions, and mass action relations, are given in Table I. Values of various equilibrium constants as used in the calculations are given in Table VII. The values are based on experimental data but have been modified to yield a consistent picture.

Table VI. Experimental Values of the Parameters of Reported Mobilities of Ionic Defects $\mu_{rep} = \mu^\circ_{rep} \exp\{-H(\mu_{rep})/kT\}$ and Computed Values of the Corresponding Enthalpies

Dopant	μ°_{rep} (cm²/V·s)	$H(\mu_{rep})$ eV	$H(\mu_{rep})$ kJ/mol	Ref.	Empirical potentials eV	Empirical potentials kJ/mol	Nonempirical potentials eV	Nonempirical potentials kJ/mol	Eq.	Dominant disorder mechanism
Acceptors										
Co	1.22 × 10⁷	3.97 (⊥c)	382	24, 25						
Mg	2.2 × 10⁹	4.84 (∥c)	466	9, 26	4.65−10.45 +10.44 =4.64 or 4.65 (no Eq.)	466	4.65−12.85 +14.18 =5.98 or 4.65 (no Eq.)	575	6	Anti-Frenkel or Schottky
A	2.27 × 10⁸	4.39 (⊥c) 3.6 (60°C)	422 346	9, 26 25	4.65	447	4.65	447	8	Interstitial or Frenkel
A*	2 × 10⁸	4.74 (60°C)	456	25						
Donors										
Ti	5.2 × 10³	3.78	364	27	3.65+10.45 −11.37 =2.73 or 3.65 (no Eq.)	263	3.65+12.85 −24.81 =−8.31 or 3.65 (no Eq.)	−200	5	Interstitial or Anti-Frenkel
Y		3.45	332	28	3.65	351	3.65	351	7	Frenkel or Schottky
Si		3.58	345	29						

*Using values of H_S, $H_{F,Al}$, and $H_{F,O}$ from Ref. 15 and $H(\mu) = H(D) − kT$ from Ref. 22.

Table VII. Values of Various Equilibrium Constants Used in the Calculations

Reaction in Table I	Equilibrium constants	Values of K at 1600°C Experimental	As used
d	$K_{ox,V(O)}$	10^9 cm^{-3}·Pa$^{-1/2}$ ‡	10^6 Pa$^{-1/2}$
h	K_a	1.2×10^8 cm^{-3} §	10^9 cm^{-3}
g	K_d	3.7×10^{12} cm^{-3} §	3.7×10^{13} cm^{-3}
o	K_i	7.4×10^{22} cm^{-6} ‡	7.4×10^{21} cm^{-6}
k	$K_{F,O}$	1.2×10^{32} cm^{-6} ¶	1.2×10^{28} cm^{-6}
n	$K_{F,Al}$		1.6×10^{23} cm^{-6} *
m	K_S		2×10^{65} cm^{-15} †

*Estimated, $<K_{F,O}$
†Estimated, $<K_{F,O}^{5/2}$
‡Reference 17.
§Reference 31.
¶Reference 21.

Table VIII. Sets of Species Dominating the Dopant and Charge Balance Equations in Various Domains

Domain	A	D	Neutrality Condition
I	$A^\times$	$D^{\cdot}$	$A', V_O^{\cdot\cdot}$
II	A'	$D^{\cdot}$	$A', V_O^{\cdot\cdot}$
III	A'	$D^\times$	$A', V_O^{\cdot\cdot}$
IV	$A^\times$	$D^{\cdot}$	$O_i'', V_O^{\cdot\cdot}$
V	$A^\times$	$D^{\cdot}$	$A', D^{\cdot}$
VI	A'	$D^{\cdot}$	$A', D^{\cdot}$
VII	A'	$D^\times$	$D^{\cdot}, O_i''$
VIII	$A^\times$	$D^{\cdot}$	$D^{\cdot}, O_i''$
XI	A'	$D^{\cdot}$	$D^{\cdot}, O_i''$
X	A'	$D^\times$	$A', D^{\cdot}$
XI	A'	$D^{\cdot}$	$O_i'', V_O^{\cdot\cdot}$

Solutions for defect concentrations are obtained in the usual way by approximating the neutrality condition and the balance equations for donors and acceptors by their dominant members, leading to expressions of the type

$$[j] = 2^r \prod_s K_s^n [A]^m [D]^p p_{O_2}^q \tag{11}$$

$$[j] = 2^r \prod_s K_s^n [A]^m a_D^p p_{O_2}^q \tag{12}$$

$$[j] = 2^r \prod_s K_s^n a_A^m a_D^p p_{O_2}^q \tag{13}$$

Here the K's are equilibrium constants of various reactions involved, [A] and [D] are the concentrations of A and D in Al_2O_3, and a_A and a_D are activities of A and D in an adjacent phase with which the Al_2O_3 is in equilibrium; p_{O_2} is the partial pressure of oxygen in the atmosphere.

Table IX. Values of Exponents in Eq. (11)*

Domain	Species	r	[A]	[D]	K_a	K_d	$K_{ox,V(O)}$	$K_{F,O}$	$K_{F,Al}$	K_S	K_i	p_{O_2}
I	$A^{\cdot}$, $2V_O^{\cdot\cdot}$	1/3			2/3		−1/3					−1/6
	$h^{\cdot}$	−1/3	2/3		1/3		1/3					1/6
	$D^{\times}$	1/3	1/3		−1/3		−1/3					−1/6
	$Al_i^{\cdot\cdot\cdot}$	−1	−1/3	1	1		−1/2		1			1/4
II	$h^{\cdot}$	−1/2	1				1/2			−1/2		1/4
	$A^{\times}$	−1/2	1/2		−1		1/2					1/4
	$D^{\cdot}$	−1/2	3/2				1/2					1/4
III	$h^{\cdot}$ as in II		1/2	1								
IV	$h^{\cdot}$						1/2	1/4				1/4
	$A^{\prime}$		1		1	−1	−1/2	−1/4				−1/4
	$D^{\times}$						−1/2	−1/4				−1/4
V	$h^{\cdot}$		1	1	1							
	$V_O^{\cdot\cdot}$		2	−2	2		−1	1				−1/2
	$O_i^{\prime\prime}$		−2	2	−2		1					1/2
	$V_{Al}^{\prime\prime\prime}$		−3	3	−3		3/2			1/2		3/4
	$D^{\times}$		−1	2								
VII	$h^{\cdot}$	1/3		−1/3		−1	1/3	1/3			1	1/6
	$D^{\cdot}$	1/3		2/3		−1/3	1/3	1/3			1/3	1/6
						2/3					−2/3	
VIII	$h^{\cdot}$	1/2		−1/2			1/2	1/2				1/4
IX	$h^{\cdot}$	1/2		−1/2			1/2	1/2				1/4
	$A^{\times}$	1/2	1	−1/2	−1		1/2	1/2				1/4
	$D^{\cdot}$	−1/2		3/2			−1/2	−1/2				−1/4
X	$h^{\cdot}$		1	−1		−1	−1/2				1	
	$V_O^{\cdot\cdot}$		2	−2		−2	−1				2	−1/2

*Values that are not included are zero.

Table X. Values of Exponents in Eq. (12)*

Domain	Species	r	[A]	$a_D K_D$	K_a	K_d	$K_{ox,V(O)}$	$K_{F,O}$	K_S	K_i	p_{O_2}
I	D·	−1/3	1/3	1	1/3		11/6		1/2	−4	11/12
II	D·	−1/2	1/2	1			2		1/2	−4	1
	h·	−1/2	1/2				1/2				1/4
III	D×			1		−1	3/2				3/4
	h·	−3/2	3/2				3/2				3/4
IV	h·						1/2				1/4
V	D·		1/2	1/2	1/2		3/4	1/4	1/4	−2	3/8
	e′		−1/2	1/2			3/4		1/4	−1	3/8
VI	h·			−1			3/2		−1/2	4	−3/4
VII	h·	1/3		−1/3			−1/6	1/3	−1/6	4/3	−1/12
VIII	h·	1/3		−1/3			−1/6	1/3	−1/6	4/3	−1/12
IX	h·	1/3		−1/3			−1/6	1/3	−1/6	4/3	−1/12
	A×	1/3	1	−1/3			−1/6	1/3	−1/6	4/3	−1/12
X	h·		1	−1	−1		−3/2		−1/2	4	−3/4

*Values that are not indicated are zero.

Table XI. Values of Exponents in Eq. (13)*

Domain	Species	r	$a_A K_A$	$a_D K_D$	$K_{ox,V(O)}$	$K_{F,O}$	K_S	K_i	p_{O_2}
I, II, III	h·	−1/3	1/3		5/6		1/6	−2/3	5/12
IV	h·				1/2	1/4			1/4
V, VI, X	e′		−1/2	1/2					
VII, VIII, IX, XI	h·	1/3		−1/3	−1/6	1/3	−1/6	4/3	−1/12
XII	h·		1/2		3/4		1/4	−1	3/8

*Values that are not indicated are zero.

Table VIII gives different sets of approximations. Values of exponents for defect concentrations according to Eqs. (11), (12), and (13) are given, respectively, in Tables IX–XI. Each set of approximations characterizes a different domain in isothermal plots of two of the three variables [A] or a_A, [D] or a_D, and p_{O_2}, the third variable being kept constant. Figures 2–6 show such plots for $T = 1600°C$, respectively, with p_{O_2} and [D] as variables, [A] constant, [A] and [D] as variables with p_{O_2} constant (large or small), and a_A and a_D as variables, p_{O_2} constant (and large). Only the central range of p_{O_2} in Figs. 2 and 3 is of practical importance. The extremes are added to demonstrate the general behavior to be expected. Similar figures can be constructed on the basis of other disorder processes. Domain boundaries were determined by equating different expressions for the concentration of the same defect in adjacent domains, using for Figs. 2, 4, and 5 the data of Table IX, for Fig. 3 the data of Table X, and for Fig. 6 the data of Table XI.

In general, at each boundary only one dominant species changes. The only exceptions are the boundaries II–X and V–IX in Fig. 2, V–IX and V–X in Fig. 4, and II–IX and V–X in Fig. 5.

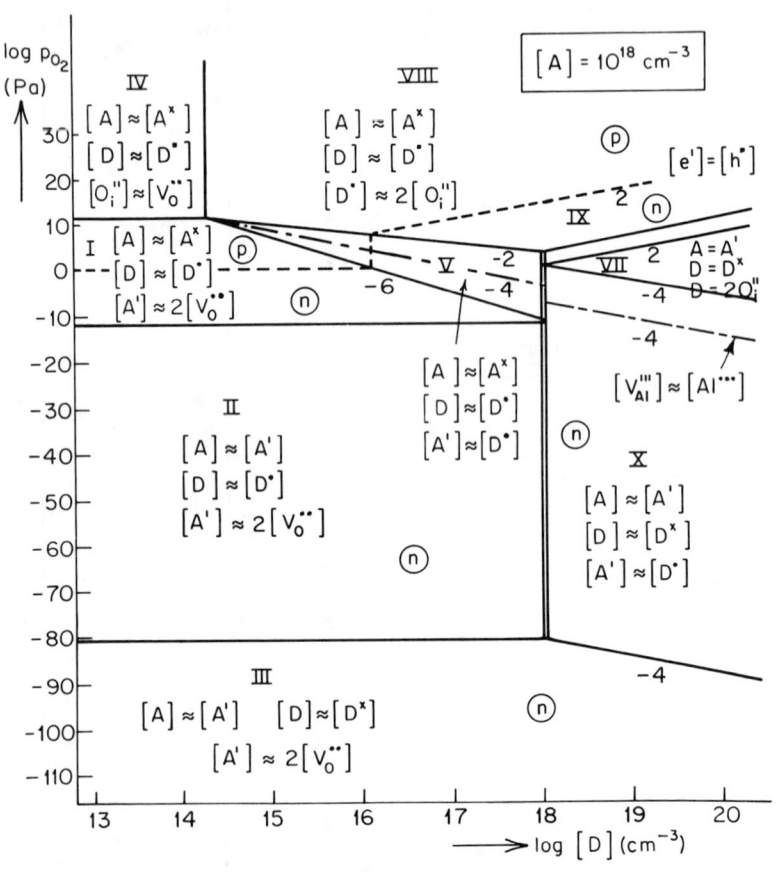

Fig. 2. Defect domains as a function of p_{O_2} and [D] at [A] = 10^{18} cm^{-3} and $T = 1600°C$; dashed lines indicate the loci $[e'] = [h^·]$ and $[V'''_{Al}] = [Al_i^{···}]$; n and p indicate regions where $[e'] >$ and $< [h^·]$, respectively.

Comparison of Fig. 2 with Fig. 3 and of Figs. 4 and 5 with Fig. 6 shows that this situation arises because domain VI characterized by $[A'] = [A] = [D^{\cdot}] = [D]$ collapses to a line in plots as a function of concentration. The collapse of domain VI causes titration steps in the defect concentrations at the double boundaries of Figs. 2, 4, and 5. The occurrence of such steps for concentrations plotted as a function of dopant concentration is clearly demonstrated in an isothermal section through Fig. 2 at $p_{O_2} = 10^2$ Pa, as shown in Fig. 7.

Dopants affect the defect structure whenever they dominate the neutrality condition. Therefore, in range V of Figs. 2, 4, and 5, D affects the defect structure although $[D] < [A]$. The reason is that $[D] = [D^{\cdot}] \simeq [A']$, but $[A] \simeq [A^\times] > [A']$. Similarly, A affects the defect structure in domains X of Figs. 4 and 5 although $[A] < [D]$. Creep experiments on $Al_2O_3:Fe + Ti$ in air by Ikuma and Gordon[32] fall in domain V of Fig. 4 in the region where it is adjacent to IX; remember, however, that the temperature of their experiments was not the same as that of Fig. 4, so there is no exact correspondence. As demonstrated by

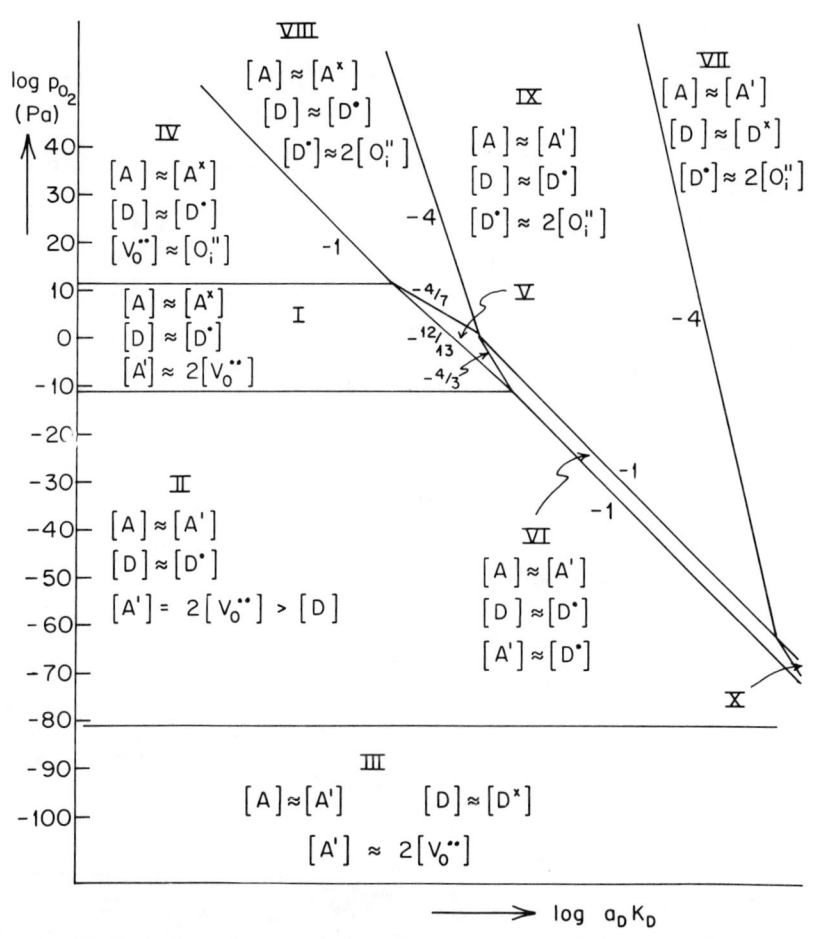

Fig. 3. Defect domains as a function of p_{O_2} and donor activity a_D at $[A] = 10^{18}$ cm^{-3} and $T = 1600°C$.

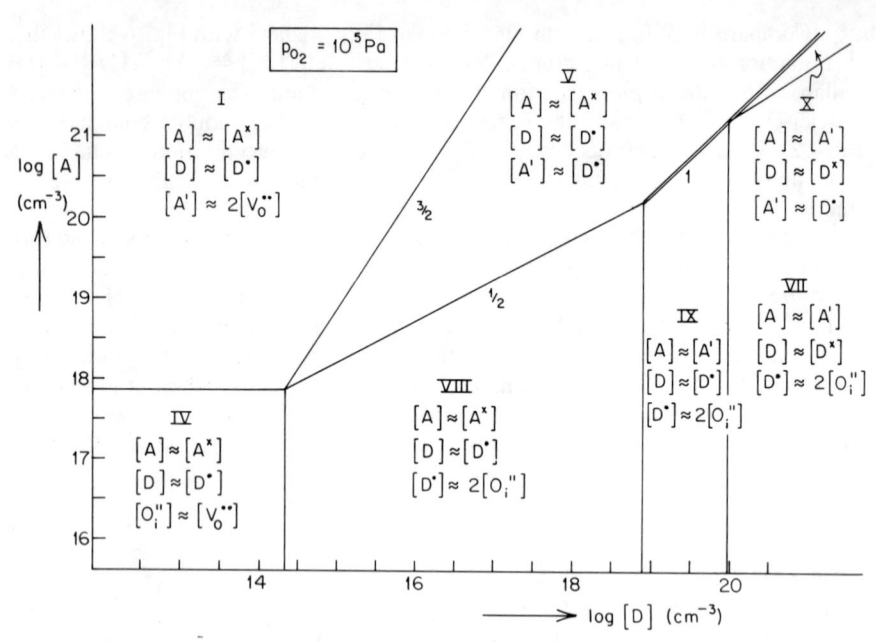

Fig. 4. Defect domains as a function of [A] and [D] at $p_{O_2} = 10^5$ Pa and $T = 1600°C$.

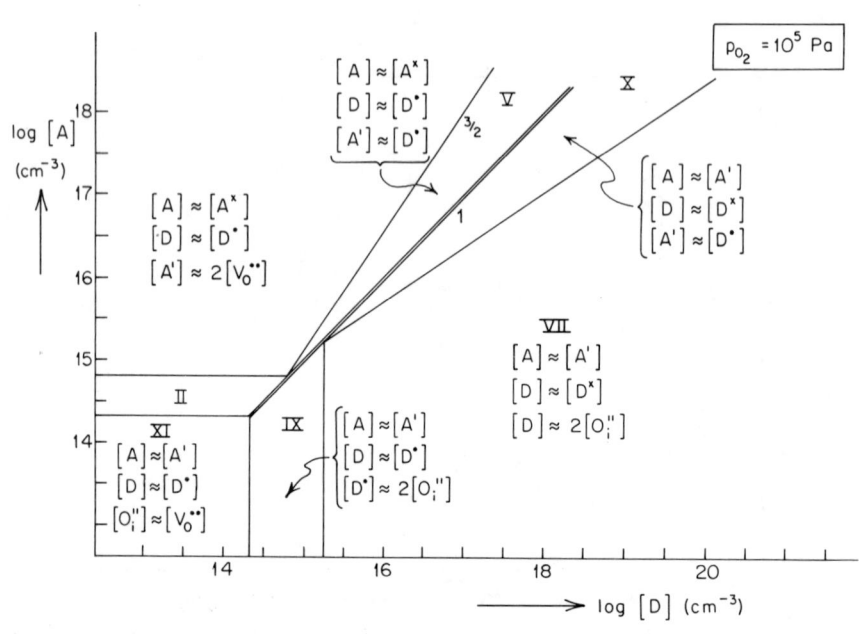

Fig. 5. Defect domains as a function of [A] and [D] at $p_{O_2} = 10^{-5}$ Pa and $T = 1600°C$.

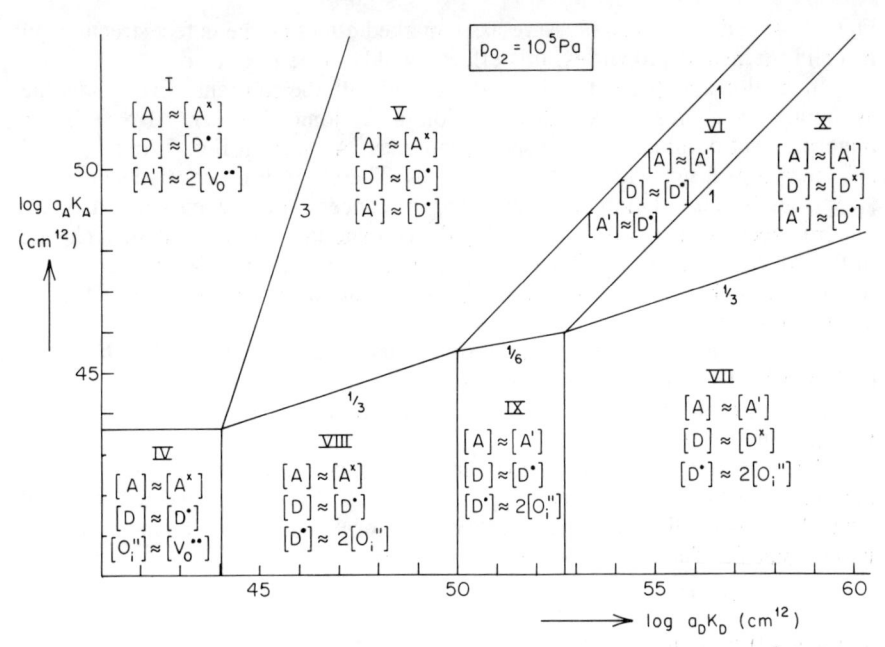

Fig. 6. Defect domains as a function of a_A and a_D at $p_{O_2} = 10^5$ Pa and $T = 1600°C$.

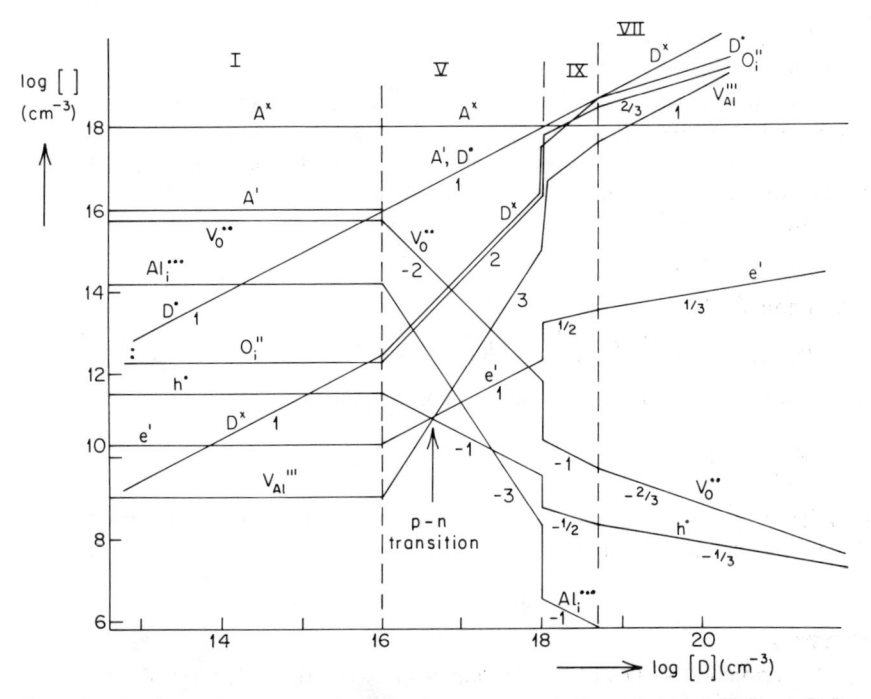

Fig. 7. Defect concentration isotherms as a function of [D] at [A] = 10^{18} cm^{-3}, $p_{O_2} = 10^2$ Pa, and $T = 1600°C$.

Figs. 2, 4, and 5, oxygen pressure has a marked effect on the defect structure. In modeling transport properties, this effect should not be neglected.[23]

In the domains presented so far, calculated with the constants given under the assumption that anti-Frenkel disorder dominates ionic disorder, Al defects are minorities but dominate the transport properties. As seen in Fig. 7, in range VII $[V'''_{Al}]$ increases faster with [D] than $[O''_i]$ does. Therefore, it is conceivable that the Al species become dominant at large dopant concentrations even when they are minority species in undoped and weakly doped material. A relatively small change in the constant will bring about this change already in range V. The combination of the higher charge and the high concentration may also cause pairing to become important.

Figure 2 also shows the loci of points where $[e'] = [h^\cdot] = K_i^{1/2}$ and $[V'''_{Al}] = [Al_i^{\cdots}] = K_{F,Al}^{1/2}$. For the values of the constants as used, the p–n transition occurs in the center of domain I, runs vertically through V, and varies proportional with $[D]^2$ in range VIII.

The vertical trajectory through boundary V occurs closer to equivalence when the p–n transition in boundary I approaches the I–II boundary and will occur at equivalence when the two boundaries coincide or when the p–n boundary moves into domain II. The p–n boundary is given by

$$(p_{O_2})_{p,n} = 4K_i^3[A]^{-2}K_a^{-2}K_{ox,V(O)}^{-2} \tag{14}$$

and the I–II boundary by

$$(p_{O_2})_{I,II} = 4[A]^{-2}K_a^4 K_{ox,V(O)}^{-2} \tag{15}$$

and thus

$$\frac{(p_{O_2})_{p,n}}{(p_{O_2})_{I,II}} = K_i^3 K_a^{-6} \tag{16}$$

In our calculation $K_i > K_a^2$; the two boundaries coincide when $K_i = K_a^2$, and the p–n boundary moves into domain II when $K_i < K_a^2$. Only a relatively small variation in the values of K_i or K_a presently used are required to bring this about. Experiments indicate that this is the case in Al_2O_3:Fe + Ti[6] and Al_2O_3:Mg, Fe + H.[33]

Equality of $[V'''_{Al}]$ and $[Al_i^{\cdots}]$ occurs in domain V at

$$(p_{O_2})_{V_{Al}=Al_i} = K_S^{-2/3} K_{F,Al}^{2/3} K_a^4 K_{ox,V(O)}^{-2}[A]^4[D]^{-4} \tag{17}$$

and in domain X at

$$(p_{O_2})_{V_{Al}=Al_i} = K_S^{-2/3} K_{F,Al}^{2/3} K_i^4 K_d^{-4} K_{ox,V(O)}^{-2}[A]^4[D]^{-4} \tag{18}$$

This leads to a step at the equivalence point equal to

$$\frac{(p_{O_2})_V}{(p_{O_2})_X} = (K_a K_d K_i^{-1})^4 \tag{19}$$

A variation of this nature was observed in Al_2O_3:Mg, Fe + H, but the value of the slope could not be determined.[33]

Summary

Experimental values of the enthalpies of oxidation–reduction of donor- and acceptor-doped Al_2O_3 are compared with computed values. Reasonable agreement is obtained for models involving O''_i, V'''_{Al}, $V_O^{\cdot\cdot}$, or $Al_i^{\cdots}$ provided empirical repulsive potentials are used in the computations. Good agreement is also found for the enthalpies of all four possible types of ionic disorder.

Experimental values of n and m in [j] $\propto$ [dopant]$^n p_{O_2}^m$ fit both models. Anti-Frenkel disorder is preferred in view of evidence from the microstructure and from general physical considerations. Expressions for experimental mobilities of ionic defects agree for all models with computed values if empirical potentials are used in the computation. Oxygen diffusion in the bulk involves $V_O^{\cdot\cdot}$ and that at grain boundaries involves O_i^x. Electrons and holes are large polarons. The band gap at $T = 0$ K is $(E_g^\circ)_{th} \simeq 9.7$ eV with $\Delta = (E_g^\circ)_{opt} - (E_g^\circ)_{th} \simeq 0.65$ eV. The complications involved in double doping with donors and acceptors are considered in detail.

References

[1] A. E. Paladino and W. D. Kingery, "Aluminum Ion Diffusion in Aluminum Oxide," *J. Chem. Phys.*, **37** [5] 957–62 (1962).

[2] Y. Oishi and W. D. Kingery, "Self-Diffusion of Oxygen in Single Crystal and Polycrystalline Aluminum Oxide," *J. Chem. Phys.*, **33** [2] 480–6 (1960).

[3] D. J. Reed and B. J. Wuensch, "Ion Probe Measurements of Oxygen Self-Diffusion in Al₂O₃," *Am. Ceram. Soc. Bull.*, **56** [3] 298 (1977); *J. Am. Ceram. Soc.*, **63** [1–2] 88–92 (1980).

[4] K. P. R. Reddy, "Oxygen Diffusion in Close-Packed Oxides"; Ph.D. Thesis, Case Western Reserve University, Cleveland, OH, 1975. K. P. R. Reddy and A. R. Cooper, *J. Am. Ceram. Soc.*, **65** [12] 634–8 (1982).

[5] Y. Oishi, K. Ando, and Y. Kubota, "Self-Diffusion of Oxygen in Single Crystal Alumina," *J. Chem. Phys.*, **73** [3] 1410–2 (1980).

[6] M. M. El-Aiat, L. D. Hou, S. K. Tiku, H. A. Wang, and F. A. Kröger, "High-Temperature Conductivity and Creep of Polycrystalline Al₂O₃ Doped with Fe and/or Ti," *J. Am. Ceram. Soc.*, **64** [3] 174–82 (1981).

[7] A. E. Paladino and R. L. Coble, "Effect of Grain Boundaries on Diffusion-Controlled Processes in Aluminum Oxide," *J. Am. Ceram. Soc.*, **46** [3] 133–6 and 460 (1963).

[8] Y. Oishi, K. Ando, and K. Matsuhiro, "Self-Diffusion Coefficient of Oxygen in Vapor-Grown Single Crystal Alumina," *Yogyo Kyokai Shi*, **85** [10] 522–4 (1977).

[9] H. A. Wang, Ch. H. Lee, F. A. Kröger, and R. T. Cox, "Point Defects in α-Al₂O₃: Mg Studied by Electrical Conductivity, Optical Absorption, and ESR," *Phys. Rev. B: Condens. Matter*, **27** [6] 3821–41 (1983).

[10] L. D. Hou, S. K. Tiku, H. A. Wang, and F. A. Kröger, "Conductivity and Creep in Acceptor-Dominated Polycrystalline Al₂O₃," *J. Mater. Sci.*, **14** [8] 1877–89 (1979).

[11] H. A. Wang and F. A. Kröger, "Chemical Diffusion in Polycrystalline Al₂O₃," **63** [11–12] 613–9 (1980).

[12] D. S. Phillips, T. E. Mitchell, and A. H. Heuer, "Precipitation in Star Sapphire III: Chemical Effects Accompanying Precipitation," *Philos. Mag.*, [Part] A, **42A** [3] 417–32 (1980).

[13] B. J. Pletka, T. E. Mitchell, and A. H. Heuer, "Dislocation Sub-Structures in Doped Sapphire (α-Al₂O₃) Deformed by Basic Slip," *Acta Metall.*, **30** [1] 147–56 (1982). A. H. Heuer, "The Role of MgO in the Sintering of Alumina," *J. Am. Ceram. Soc.*, **62** [5–6] 317–9 (1979).

[14] F. A. Kröger, "Oxidation–Reduction and the Major Type of Ionic Disorder in α-Al₂O₃," *J. Am. Ceram. Soc.*, **66** [10] 730–32 (1983).

[15] C. R. A. Catlow, R. James, M. C. Mackrodt, and R. F. Stewart, "Defect Energetics in α-Al₂O₃ and Rutile TiO₂," *Phys. Rev. B: Condens. Matter*, **25** [2] 1006–26 (1982).

[16] F. A. Kröger; p. 14 in The Chemistry of Imperfect Crystals, Vol. 2. 2nd Revised Edition, North-Holland, 1974.

[17] S. K. Mohapatra and F. A. Kröger, "The Dominant Type of Atomic Disorder in α-Al₂O₃," *J. Am. Ceram. Soc.*, **61** [3–4] 106–9 (1978).

[18] E. A. Colbourn and W. C. Mackrodt, "Optical, Electrical, and Polaron Energy Levels in α-Al₂O₃," *Solid State Commun.*, **40** [3] 265–7 (1981).

[19] B. V. Dutt and F. A. Kröger, "High-Temperature Defect Structure of Iron-Doped α-Alumina," *J. Am. Ceram. Soc.*, **58** [11–12] 474–6 (1975).

[20] W. Raja Rao and I. B. Cutler, "Effect of Iron Oxide on the Sintering Kinetics of Al₂O₃," *J. Am. Ceram. Soc.*, **56** [11] 588–93 (1973).

[21] M. M. El-Aiat and F. A. Kröger, "Determination of the Parameters of Native Disorder in α-Al₂O₃," *J. Am. Ceram. Soc.*, **65** [3] 162–6 (1982).

[22] G. J. Dienes, D. O. Welch, R. D. Hatcher, O. Lazareth, and M. Samberg, "Shell Model Calculations in Some Point Defect Properties in α-Al₂O₃," *Phys. Rev. B: Condens. Matter*, **11** [8] 3060–70 (1975).

[23] F. A. Kröger, "Electrical Properties of α-Al₂O₃"; these proceedings.

[24] B. V. Dutt, J. P. Hurrell, and F. A. Kröger, "High-Temperature Defect Structure of Cobalt-Doped α-Alumina," *J. Am. Ceram. Soc.*, **58** [9–10] 420–7 (1975).

[25] S. K. Mohapatra, S. K. Tiku, and F. A. Kröger, "Defect Structure of Unintentionally Doped α-Al$_2$O$_3$ Crystals," *J. Am. Ceram. Soc.*, **62** [1–2] 50–7 (1979).
[26] S. K. Mohapatra and F. A. Kröger, "Defect Structure of α-Al$_2$O$_3$ Doped with Magnesium," *J. Am. Ceram. Soc.*, **60** [3–4] 141–8 (1977).
[27] S. K. Mohapatra and F. A. Kröger, "Defect Structure of α-Al$_2$O$_3$ with Titanium," *J. Am. Ceram. Soc.*, **60** [9–10] 281–7 (1977).
[28] M. M. El-Aiat and F. A. Kröger, "Yttrium, an Isoelectronic Donor in α-Al$_2$O$_3$," *J. Am. Ceram. Soc.*, **65** [6] 280–3 (1982).
[29] C. H. Lee and F. A. Kröger, "Electrical Conductivity and the Defect Structure of α-Al$_2$O$_3$:Si"; to be published in *J. Am. Ceram. Soc.*
[30] W. R. Strehlow and E. L. Cook, "Energy Band Gaps in Elemental and Compound Semiconductors and Insulators," *J. Phys. Chem. Ref. Data*, **2** [1] 163–93 (1973).
[31] S. K. Tiku and F. A. Kröger, "Energy Levels of Donor and Acceptor Dopants and Electron and Hole Mobilities in α-Al$_2$O$_3$," *J. Am. Ceram. Soc.*, **63** [1–2] 31–2 (1980).
[32] Y. Ikuma and R. S. Gordon, "Effect of Doping Simultaneously with Iron and Titanium on the Diffusional Creep of Polycrystalline Al$_2$O$_3$," *J. Am. Ceram. Soc.*, **66** [2] 139–47 (1983).
[33] M. M. El-Aiat and F. A. Kröger, "Hydrogen Donors in α-Al$_2$O$_3$," *J. Appl. Phys.*, **53** [5] 3658–67 (1982).

---

*Work supported by the United States Department of Energy under Contract No. AS03-76-SF001113, Project Agreement AT03-76 ER 71027.

Low Atomic Number Impurity Atoms in Magnesium Oxide: Hydrogen and Carbon

FRIEDEMANN FREUND* AND BRUCE KING

Arizona State University
Department of Physics
Tempe, AZ 85287

ROLF KNOBEL

Mineralogisches Institut
Universität zu Köln D-5000 Köln-41
Federal Republic of Germany

HENDRIK KATHREIN

Bayer AG
Uerdingen
Federal Republic of Germany

Magnesium oxide single crystals, grown by arc fusion, contain nonnegligible concentrations of dissolved hydrogen and carbon. They derive from traces of H_2O and CO/CO_2 component taken up in solid solution during crystal growth. The H_2O yields an equimolal concentration of extrinsic cation vacancies and OH^- ions. Fully OH^--compensated cation vacancies convert to a large extent into molecular H_2 and peroxy ions, O_2^{2-}. The H_2 may be lost, leaving O_2^{2-} as excess oxygen. The thermal decay of O_2^{2-} above 530°C generates positive holes, both vacancy-bound O^- states (V^- centers) and highly mobile unbound O^- states, which diffuse toward the surface. They release oxygen, set up a surface charge, and also cause autoxidation of transition-metal impurities in the bulk. Solute carbon does not occur as carbonate ions, CO_3^{2-}, but as CO_2^{2-} and CO^- with the carbon being very reduced, essentially zerovalent. The solute carbon is endowed with an unusually low activation energy of diffusion, 22.5 ± 2.5 kJ·mol^{-1}, and a strong tendency for subsurface segregation, even at low temperatures. Two distinct subsurface carbon concentration profiles have been recognized. One extends 1–2 μm into the bulk and is probably controlled by surface strain relaxation with long-range, dipolar strain–strain interaction between the solute carbon species. The other profile is confined to a narrow subsurface zone, approximately 5–10 nm wide. It is probably controlled by surface charges and the associated subsurface space charge layer.

For many technical applications of ceramic materials, it is important to know their impurity content and impurity distribution. Medium to high atomic number impurities are relatively easy to analyze. The analysis of low atomic number elements, however, in particular H and C, requires special techniques that are often not available in laboratories where the physical and chemical properties of ceramic materials are studied. As a result of this deficiency, the presence of hydrogen and carbon impurities in refractory oxides and silicates has received comparatively

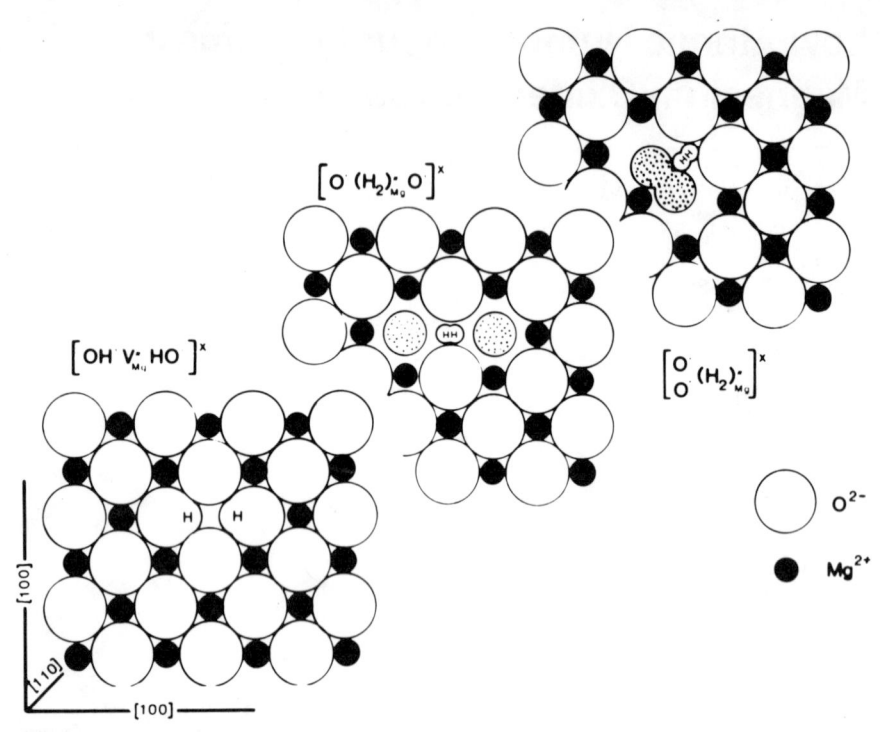

Fig. 1. Charge-transfer conversion of two OH⁻ ions facing each other across a cation vacancy in MgO into an H_2 molecule and a peroxy anion, O_2^{2-} (idealized (100) section; lattice relaxation omitted).

little attention or has been ignored all together, while transition-metal impurities have been the subject of many excellent studies.[1-3]

In this report we shall review the progress that has been made over the past 5–6 years in studying specifically the dissolution mechanism of H_2O and CO_2 molecules in densely packed, structurally simple oxides like MgO. We shall point out the basic principle common to both dissolution mechanisms, the physical and chemical interactions among the solute species, and their influence upon certain material properties. We shall also include some very recent results that may help to better understand subsurface segregation and diffusion of the solute carbon for which a large body of experimental data is now available.

The MgO structure is particularly simple inasmuch as the number of anion sites (face-centered cubic close packing of O^{2-}) equals the number of cations. The cations occupy all available octahedrally coordinated sites but leave the tetrahedrally coordinated sites empty. The latter are the only available interstitial sites.

H_2O Dissolution

When a H_2O molecule is dissolved in a given volume of MgO, one oxygen is added and, hence, one extrinsic cation vacancy, $V''_{Mg'}$ is created.[†] The missing divalent cation is charge-wise compensated by the two protons forming hydroxyl ions (OH⁻) on O^{2-} lattice sites, OH˙. The preferred configuration will be the fully

OH⁻-compensated vacancy, neutral to the surrounding lattice, $[\text{OH}^·V''_{\text{Mg}}\text{HO}^·]^×$. This defect is expected to be oriented parallel to the [100] direction of the MgO structure, as shown at the bottom left-hand side of Fig. 1.

In an infrared (IR) study of high-purity MgO and CaO single crystals containing traces of dissolved "water," it was found[5,6] that the intensity of the IR signal of this particular defect MgO was unexpectedly low in comparison to the intensity of the half-compensated cation vacancy, $[\text{OH}^·V''_{\text{Mg}}]'$. At the same time, a weak sharp band was observed near 4150 cm⁻¹. Such a band has been repeatedly reported for H_2O-rich glasses, where it was assigned to a combination band of the Si–OH and O–H stretching modes.[7,8] In the case of MgO and CaO, however, no combination mode yielded an absorption band around 4150 cm⁻¹.

The appearance of this IR band rather indicated the presence of H_2 molecules on some specific, noncentrosymmetric lattice sites. Two questions at once arose: (1) how can we understand the formation of H_2 molecules, and (2) where can we expect lattice sites in the cubic structure of MgO and CaO that would be sufficiently asymmetric to render H_2 molecules IR active?

One possibility for the formation of H_2 would be that traces of transition-metal impurities such as Fe^{2+} interact with the OH⁻, yielding Fe^{3+} and H_2. In the cases studied, this reaction can be ruled out on the basis of the internal redox equilibria, for instance Fe^{2+}/Fe^{3+}, evaluated by electron paramagnetic resonance (EPR) spectroscopy.[9] In addition, during an earlier mass spectroscopic study[10] of the thermal decomposition of ultrahigh-purity $Mg(OH)_2$, molecular H_2 was shown to evolve in unexpectedly large quantities, besides H_2O, from the freshly formed, finely divided MgO. The underlying mechanism is by now rather well understood.

As illustrated by Fig. 1, the H_2 formation occurs when two OH⁻ ions are adjacent to an Mg^{2+} vacancy in the above-mentioned $[\text{OH}^·V''_{\text{Mg}}\text{HO}^·]^×$ configuration. The two protons reduce themselves to H_2 by taking over electrons from two oxygens such that two O⁻ states are formed that in turn dimerize to give the diamagnetic peroxy ion, O_2^{2-}, which will be aligned parallel to the [110] direction:

$$[\text{OH}^·V''_{\text{Mg}}\text{HO}^·]^× \rightleftharpoons [^{O^·}_{O^·}(H_2)''_{\text{Mg}}]^× \tag{1}$$

We shall call this particular reaction a charge-transfer (CT) reaction. It has been studied for both H_2O- and D_2O-doped MgO single crystals and powders by a variety of techniques.[11] Theoretical calculations[12] indicate that the probability to form $(H_2 + O_2^{2-})$ according to Eq. (1) depends upon the O–O distance across the cation vacancy: if the O–O distance is large, as in CaO, less $(H_2 + O_2^{2-})$ will be formed. This agrees with the IR results.[6]

The observed IR activity of the H_2 can be explained as follows: if the axis of the H_2 molecule is oriented perpendicular to the axis of the O_2^{2-} ion (see Fig. 1 upper right), we expect a weak bonding between the σ^b orbital of H_2 to the π_{xy} orbital of O_2^{2-}. This would then render the H_2 molecule asymmetric and, hence, IR active even in the cubic MgO and CaO structure.

It is to be noted that the volume of the O_2^{2-} ion is much smaller than that of the two O^{2-} for which it substitutes. Hence, we expect a rather large lattice relaxation to occur in the vicinity of this defect (not shown in Fig. 1). Probably this lattice relaxation contributes favorably to the energy balance of Eq. (1).

The H_2 can be removed from the cation vacancy side without perturbing the local charge balance:

$$[^{O^·}_{O^·}(H_2)''_{\text{Mg}}]^× \rightleftharpoons [^{O^·}_{O^·}V''_{\text{Mg}}]^× + (H_2)^×_i \tag{2}$$

where index i stands for interstitial. Strain relaxation at the surface of a crystal containing interstitial H_2 molecules will set up a driving force for surface segregation.[13] Unlike cationic impurities that are known to segregate to the surface and to remain there on account of their electric charge,[14–16] molecular H_2 may be entirely removed from the crystal and enter the gas phase:

$$(H_2)_i^x \rightleftharpoons H_2(g) \tag{3}$$

Combining Eqs. (2) and (3) leads to a loss of H_2 from the MgO, which originally contained traces of dissolved "water." The evolution of molecular H_2 can be experimentally demonstrated as shown in Fig. 2 taken from an earlier paper.[10]

The remaining oxide will be nonstoichiometric, $MgO_{1+\delta}$, with excess oxygen in the form of peroxy ions and an equimolal concentration of cation vacancies. Alternatively, one may say that a Mg deficiency has been introduced by the formation of excess Mg^{2+} vacancies, $Mg_{1-\delta}O$. The important point to note here is that, via the H_2O dissolution mechanism, relatively large amounts of excess oxygen can be introduced as if the MgO were heated to extremely high temperatures in an O_2 atmosphere.[17,18]

Though the value of δ may be small in absolute numbers, the presence of peroxy anions in the structure can affect various physical and catalytical properties. This becomes immediately apparent if we consider the thermal stability of O_2^{2-}. Bulk peroxides such as BaO_2, SrO_2, and CaO_2 decompose around 600–700°C by releasing oxygen;[19] for instance, $BaO_2 \rightleftharpoons BaO + \frac{1}{2}O_2$.

In the MgO structure a similar peroxy decay reaction starts around 550°C. Two cases must be distinguished. If the O_2^{2-} ion is located at the MgO surface, oxygen is released. An anion vacancy is left behind. Together with the associated cation vacancy, this gives a neutral Schottky pair:

$$\begin{bmatrix} O^{\cdot} \\ O^{\cdot} \end{bmatrix} V_{Mg}'' \Big]^x \rightleftharpoons [O^x V_O^{\cdot\cdot} V_{Mg}'']^x + O(g) \tag{4}$$

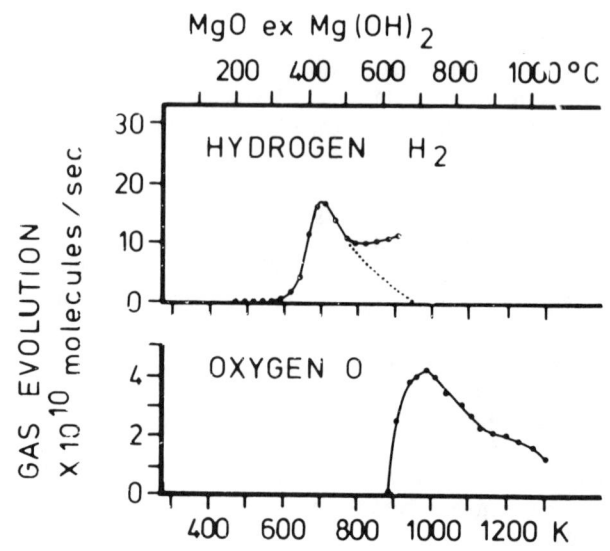

Fig. 2. Release of molecular H_2 from finely divided MgO between 300 and 550°C, followed by the release of atomic O between 550 and 750°C due to the thermal decay of O_2^{2-} anions (Ref. 10).

The release of atomic oxygen from finely divided MgO is shown in the lower part of Fig. 2. It starts sharply, rises to a maximum around 700°C, and then decays.[10]

If, however, the O_2^{2-} is located inside the MgO crystal, the reaction is more complex. It begins by the dissociation of the O_2^{2-} ion into two paramagnetic O^- states:

$$\begin{bmatrix} O^{\cdot} \\ O^{\cdot} \end{bmatrix} V_{Mg}'' \Big]^x \rightleftharpoons [O^{\cdot}V_{Mg}''']' + [O^{\cdot}] \tag{5}$$

One O^- state will remain bound to the cation vacancy for partial charge compensation, forming the V^- center, which has been extensively studied by electron paramagnetic resonance (EPR) spectroscopy.[20-22] The other O^- state becomes unbound and represents nothing else but a defect electron or positive hole in the O^{2-} matrix, often designated as $e\cdot$ or h. The unbound O^- is difficult to detect by EPR spectroscopy, because it usually has very short relaxation times, suggesting very fast exchange by electron hopping between stationary oxygen positions.[23] It may not be quenchable because it readily recombines with the V^- centers according to Eq. (5). If an unbound O^- state or hole encounters a transition-metal cation, for instance Fe^{2+} on an Mg^{2+} lattice site, oxidation leads to Fe^{3+}:

$$[Fe_{Mg}^x]^x + [O^{\cdot}] \rightleftharpoons [Fe_{Mg}^{\cdot}] + [O^x]^x \tag{6}$$

Figure 3 shows the spontaneous autoxidation of transition-metal trace impurities in MgO, which starts at the temperature at which Eq. (5) is activated.[9] Only the increase of the Fe^{3+} concentration and the decrease of the Mn^{2+} and Cr^{3+} concentrations are shown, as studied by EPR spectroscopy. Mn^{2+} probably oxidizes to Mn^{3+}. Cr^{3+}, the stable oxidation state in the octahedral environment of an Mg^{2+} lattice site, represents an excess positive charge and would thus repel the

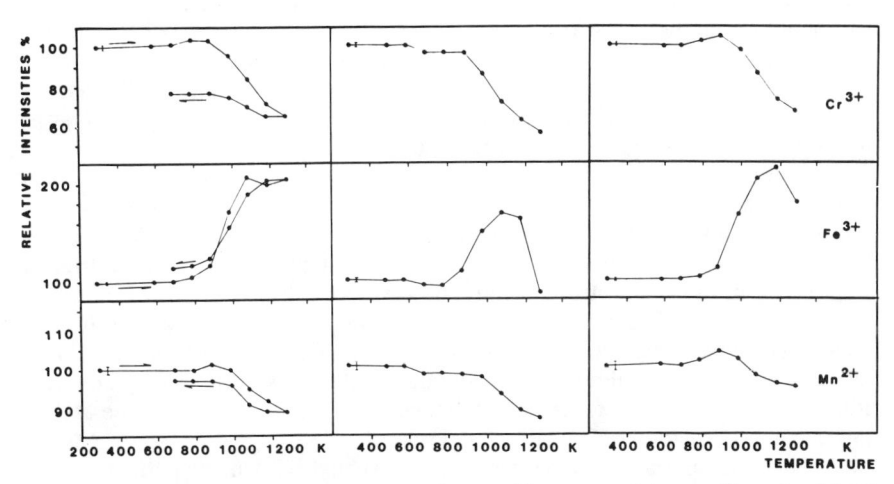

Fig. 3. Reversible autoxidation of transition-metal impurities in MgO caused by unbound O^- states (positive holes or defect electrons in the O^{2-} matrix) released during the thermal decay of O_2^{2-} anions (after Ref. 9).

hole. If, however, the Cr^{3+} is associated with a nearby Mg^{2+} vacancy, the interaction becomes attractive and oxidation of Cr^{3+} to Cr^{4+} may occur:

$$[Cr_{Mg}^{.}O^{\times}V_{Mg}^{''}]' + [O^{.}]' \rightleftharpoons [Cr_{Mg}^{.}O^{\times}V_{Mg}']^{\times} + [O^{\times}]^{\times} \quad (7)$$

An important point evidenced by Fig. 3 is that the internal redox reactions involving transition-metal impurities are largely reversible. In the case shown in Fig. 3, stepwise cooling with a 15-min holding time at each temperature suffices to rereduce Fe^{3+} to Fe^{2+}. In other words, Fe^{3+} releases its trapped hole and reconverts to Fe^{2+}, while the unbound hole is attracted by the excess negative charge of the V^- center, forming again a peroxy anion. Such internal equilibria suggest that transition-metal impurities in MgO tend to stabilize a certain fraction of peroxy anions that derive from the H_2O dissolution mechanism.

Figure 3 further shows that another type of self-reduction of Fe^{3+} and of other higher oxidation state transition-metal impurities occurs above 800–900°C. The details are still poorly understood. Maybe trimers form similar to the V_I centers, O_3^{3-} on (111) surfaces of MgO, which are stable up to 1200°C.[24] Maybe even diamagnetic tetramers, O_4^{4-}, exist in the bulk near Mg^{2+} vacancy pairs, remaining stable up to still higher temperatures.[25]

CO₂ Dissolution

When a CO_2 molecule is dissolved in MgO, two oxygens are added to the system and two extrinsic cation vacancies, V_{Mg}'', are created. Until recently it was unknown that structurally densely packed oxides like MgO can take up CO_2 in solid solution. Bulk chemical analyses of arc-fusion-grown, so-called high-purity MgO single crystals,[26] perfectly transparent and free of any inclusions, lead to mean carbon concentrations of the order of a few hundred wt-ppm, typically 300 wt-ppm C, corresponding to more than 1000 C atoms per 10^6 oxygen atoms.[27] Upon heating, CO_2 evolves in a peculiar temperature and time-dependent manner.[28]

With nuclear reaction profiling techniques, in particular the $^{12}C\,(d,p)$-^{13}C reaction,[27,29-31] and photoelectron spectroscopy,[32] the subsurface segregation and diffusion of the solute carbon-bearing species was studied. Due to limited space, we cannot repeat the experimental details here. We shall only give a brief account of our present understanding of the dissolution mechanism and diffusion mechanism of CO_2 in MgO, which draws additional support from other studies using a variety of techniques such as DC conductivity,[33] AC conductivity,[34] IR spectroscopy,[5,6] dilatometry, [35] and magnetic susceptibility.[36]

Figure 4 illustrates schematically the two possibilities how a CO_2 molecule can become incorporated in the dense O^{2-} packing. The MgO structure will undergo a rearrangement of the local electron density. In case (a) the charge of one O^{2-} would spread over the CO_2 molecule, yielding the well-known planar trigonal carbonate ion, CO_3^{2-}. Spectroscopically, however, one does not find any indication of CO_3^{2-} groups in MgO single crystals containing a few hundred ppm of C as mentioned above[5,6] and evolving CO_2 upon heating. Therefore, the CO_2 dissolution mechanism seems to follow the pathway depicted by cases (b) and (c) in Fig. 4. It involves a charge transfer from two lattice O^{2-} onto the CO_2 molecule and the intermediate formation of a CO_4^{4-} complex.

When the carbon is moved off-center and when it is bound more tightly to two O^-, this complex splits into a peroxy ion O_2^{2-} and the bent configuration CO_2^{2-}. The peroxy ion may be viewed as a dissolved oxygen atom (see Eq. (4)), while the CO_2^{2-} formally represents a dissolved CO molecule: $CO_2^{2-} \rightleftharpoons O^{2-} + CO(g)$. Thus,

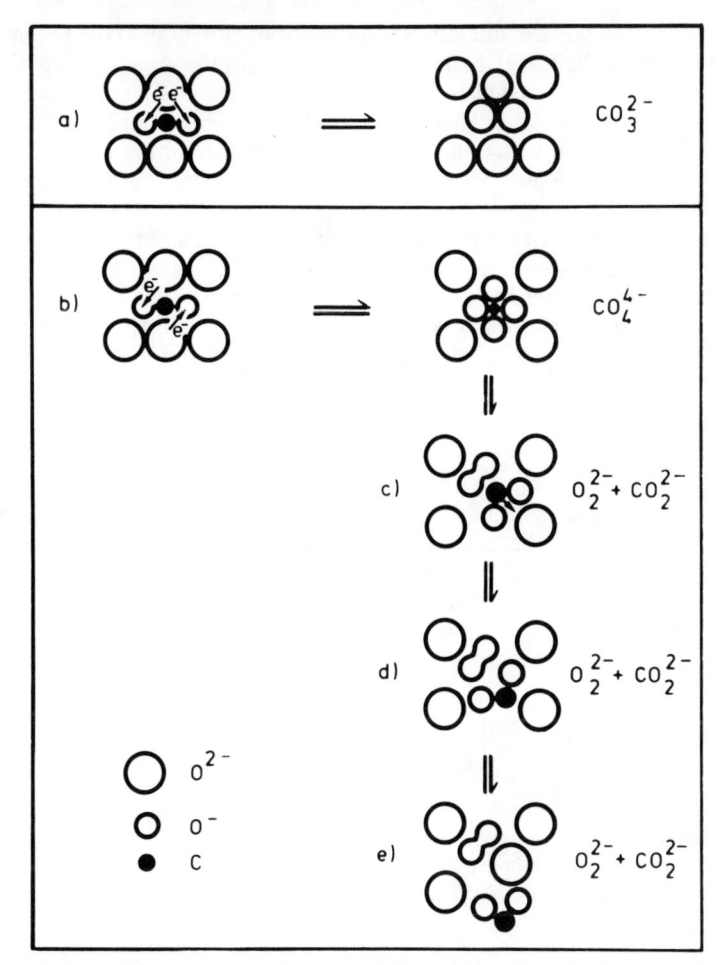

Fig. 4. Schematic representation of the CO₂ dissolution mechanism: (a) interacting electronically with one lattice O^{2-} to form CO_3^{2-}; (b) interacting electronically with two lattice O^{2-} to form CO_4^{4-}; (c) splitting of the CO_4^{4-} into a peroxy anion O_2^{2-} and the bent configuration CO_2^{2-}; (d) jumping of the C atom over the energy barrier that is lowered by the smallness of the two ligand O^-; (e) jumping two steps further after an electron exchange between an O^{2-} and an O^- and another diffusional jump of the C atom.

we can say that, upon solid-state dissolution, the CO₂ molecule splits in a strongly oxidized species and a strongly reduced species, O_2^{2-} and CO_2^{2-}, respectively. This is analogous to the H₂O dissolution mechanism, where OH^- pairs rearrange to give an O_2^{2-} ion plus molecular H₂.

Figure 4(d,e) depicts the essence of the diffusion mechanism of the C atom via the CO_2^{2-} configuration. Because of the smallness of the O^- to which the C atom is bonded, a local contraction zone is created. This drastically lowers the activation energy barrier for a diffusional jump of the C atom, as indicated by the arrow in Fig. 4(c). In the theory of diffusion this corresponds to a negative activation volume.[37,38]

The next step will be an electron transfer from an O^{2-} onto the O^- in such a way as to reorient the C–O$^-$ bond. Figure 4(e) shows the stage reached after one such O^{2-}–O^- charge transfer and an additional diffusional jump of the C atom away from the O_2^{2-} ion. Thus, the motion of the C atom is coupled to the motion of the O^- state, which is nothing but an electronic state of the oxygen, a defect electron, or positive hole.[23] We shall return to this question further below.

If the CO_2 component is incorporated into the MgO structure during crystal growth — for instance from the melt[26] or by decomposition of magnesium carbonates and sintering — a more or less homogeneous solid solution will be formed, where entropy is expected to favor a spatially uniform distribution of O_2^{2-} (with its associated cation vacancy) and of CO_3^{2-} (with the C atom residing on a cation vacancy site). Remember that two cation vacancies are introduced per dissolved CO_2 molecule:

$$\begin{bmatrix} O^\cdot & & O^\cdot \\ & C''_{Mg} & \\ O^\cdot & & O^\cdot \end{bmatrix}^x \rightleftharpoons \begin{bmatrix} O^\cdot \\ V''_{Mg} \\ O^\cdot \end{bmatrix}^x + \begin{bmatrix} O^\cdot \\ C''_{Mg} \\ O^\cdot \end{bmatrix}^x \tag{8}$$

The C diffusion mechanism as depicted by Fig. 4 is still too simplistic, because it implies that the C atom only occupies cation vacancy sites. In such a case the availability and/or mobility of the cation vacancies would be the rate-limiting step. In reality, however, a very high mobility of the segregating C-bearing species was found with an activation energy of the order of 22.5 kJ/mol.[27,32] The underlying reason is that the C atom also enters the interstitial sites:

$$\begin{bmatrix} O^\cdot \\ C''_{Mg} \\ O^\cdot \end{bmatrix}^x \rightleftharpoons [V''_{Mg}]'' + \begin{bmatrix} O^\cdot \\ C_i^x \\ O^\cdot \end{bmatrix}^{\cdot\cdot} \tag{9}$$

Experimentally, this reaction can be studied by considering the OH^- ions jointly present as traces in the MgO. Since the cation vacancy V''_{Mg} is negatively charged with respect to the perfect lattice, it attracts either one proton, yielding the half-compensated defect $[OH\cdot V''_{Mg}]'$, or two protons, giving the fully compensated defect introduced above as the source defect for H_2 formation.

Thus, we expect that the IR intensity of the absorption bands associated with these defects can be used to monitor Eq. (9). This is indeed the case. For instance, the formation of the half-OH-compensated defect may be written as an exchange reaction between one interstitial proton, $[OH_i]^\cdot$, and the C atom in Eq. (9):

$$\begin{bmatrix} O^\cdot \\ C''_{Mg} \\ O^\cdot \end{bmatrix}^x + [OH_i]^\cdot \rightleftharpoons [OH\cdot V''_{Mg}]' + \begin{bmatrix} O^\cdot \\ C_i^x \\ O^\cdot \end{bmatrix}^{\cdot\cdot} \tag{10}$$

This has been experimentally demonstrated.[5,6] Figure 5 shows the intensity variations of the IR signal assigned to $[OH\cdot V''_{Mg}]'$ as a function of temperature. (These intensity variations are fully reversible. They can be studied, if the crystal is quenched to liquid nitrogen temperature. Furthermore, it is to be noted that the IR signal is split into four equidistant components of decreasing intensity as a result of strong lattice strains, originating from nearby CO_3^{2-} configurations.[6] These latter are bent with an OĈO angle of the order of 130° and, hence, dipolar with respect to their charge and associated strain distribution.)

As shown by Figure 5, the IR intensity of the $[OH\cdot V''_{Mg}]'$ defect increases approximately 4-fold between 530 and 800°C, which is believed to be caused by the exchange reaction as given in Eq. (10). (The reason for the decrease of the IR intensity above 800°C is still unclear. It correlates with variations of the internal

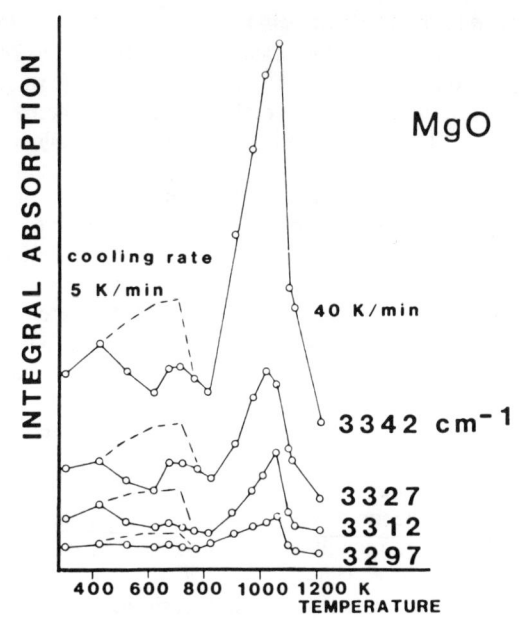

Fig. 5. Intensity variation of the infrared signal assigned to the half-OH$^-$-compensated cation vacancy defect in MgO. The IR signal is split into four components due to strain interaction with nearby C atoms (Ref. 6).

redox reactions as monitored by EPR, using traces of Fe^{2+}/Fe^{3+} or similar transition-metal impurities as local probes[9] as shown by Fig. 3.)

The situation is rendered still more complex by the fact that the CO$_2^{2-}$ configuration with the C atom residing on an interstitial site carries a double positive charge with respect to the perfect lattice, as clearly indicated in Eqs. (9) and (10). This is unfavorable. Electrical conductivity measurements[33] have provided evidence that concomitantly with or subsequent to the transition of the C atom from a cation vacancy to an interstitial site, i.e., above 530°C, the CO$_2^{2-}$ configuration releases a positive hole:

$$[V''_{Mg}]'' + \begin{bmatrix} O^{\cdot} \\ O^{\cdot} C_i^x \end{bmatrix}^{\cdot\cdot} \rightleftharpoons [O^{\cdot}V''_{Mg}]' + [O^{\cdot}C_i^x] \qquad (11)$$

Probably most of the O$^-$ states thus generated become trapped by the cation vacancies, forming the same paramagnetic V^- center as discussed above in relation with the thermal decay of O$_2^{2-}$ according to Eq. (5). Others contribute to the electronic conductivity of the MgO together with the unbound O$^{\cdot}$ states that appear on the right-hand side of Eq. (5). Therefore, the electronic conductivity of annealed MgO single crystals tends to increase sharply above 530°C as a result of these internal defect reactions.[33]

The species CO$^-$, which forms by the dissociation of CO$_2^{2-}$, is also paramagnetic. It probably represents the most mobile C-bearing species in the system liable to surface segregation. Before discussing in more detail the surface and subsurface segregation of C in MgO, we shall first take a look at the electronic structure of the CO$_2^{2-}$ configuration.

Electronic Structure of the CO_2^{2-} Species

The CO_2^{2-} ion contains 18 valence electrons as compared to 24 electrons in the case of the carbonate ion, CO_3^{2-}. The CO_2^{2-} derives from a CO_2 molecule by adding two more electrons. Figure 6 shows the molecular orbital scheme of the linear CO_2 molecule.[39] It is diamagnetic, and the highest occupied MO's are the essentially nonbonding π_{xy} levels.

When two more electrons are added, they can enter only the antibonding π_{xy}^* levels.[39] By bending, the degeneracy of π_{xy}^* orbitals is removed and two new states are created that have the symmetry a_1 and b_1. The electron density of these MO's is depicted in Fig. 6: in the case of the more stable a_1, the electron is strongly concentrated on the C atom and has essentially C 2s character, while in the case of b_1, it spreads over the O^- ligands in a manner that is antibonding with respect to the $C-O^-$ bonds but bonding with respect to the O^--O^- bond. (Note the sign of the electron density lobes in Fig. 6.)

The most stable configuration of CO_2^{2-} will be diamagnetic when both electrons enter the a_1 level and undergo spin pairing (Fig. 6, center). The relative position of a_1 and b_1, however, varies as a function of the $O\hat{C}O$ angle:[39] at small angles, the b_1 level may become more stable than a_1 on account of its O^--O^- bonding character. Hence, there will be a crossover at which the diamagnetic CO_2^{2-} becomes paramagnetic (Fig. 6, right).

Such a situation arises when the C atom residing on a cation vacancy site jumps to a nearest interstitial site, as shown in Fig. 6, top. Alternatively, one may say that during this jump the C atom passes through the plane of three oxygens. This corresponds to a transient CO_3^{4-} configuration, the electron density of which is strongly concentrated on the oxygen ligands.[39]

Recently, two temperature regions of transient paramagnetism have been identified by magnetic susceptibility measurements.[36,40] One occurs between 530 and 800°C. It is probably related to the release of positive holes in conjunction with the formation of V^- centers and CO^-, as described by Eqs. (5) and (11). Another one occurs between 300 and 530°C and is probably due to the spin decoupling in the CO_2^{2-} configuration during transition of the C atoms from cation vacancy sites to nearest interstitial sites, as depicted in Fig. 6. This transition, which precedes the actual dissociation according to Eq. (11), may be viewed as the formation of an associated pair $[V''_{MgO} \cdot C_i^x]^\times$.

During electrical conductivity measurements,[41] a conductivity maximum was found around 350°C that showed a pronounced hysteresis behavior as a function of the annealing time at room temperature. At the time when these measurements were made, the occurrence of this conductivity maximum was quite puzzling. On the basis of the susceptibility data, it now appears possible that this conductivity maximum is linked to the incipient transition of C atoms of CO_2^{2-} from cation vacancy sites to the interstitial sites according to Eq. (9). During this transition, the CO_2^{2-} configuration may release positive holes according to Eq. (11) that would contribute to the density of mobile charge carriers and cause a p-type conductivity.

Alternatively, one may suggest that, during the passage of the C atom through the plane of the three oxygens shown in Fig. 6 (top), the electron-rich configuration CO_3^{4-} releases electrons into the conduction band or some impurity-related conduction levels in the band gap, which would lead to an n-type conductivity. Since an n-type conductivity seems to prevail at these low temperatures,[33] we presently favor the latter mechanism.

The peculiarity of the behavior of MgO in the temperature regions 300–530°C and above 530°C is evidenced by many other measurements. The dielectric loss

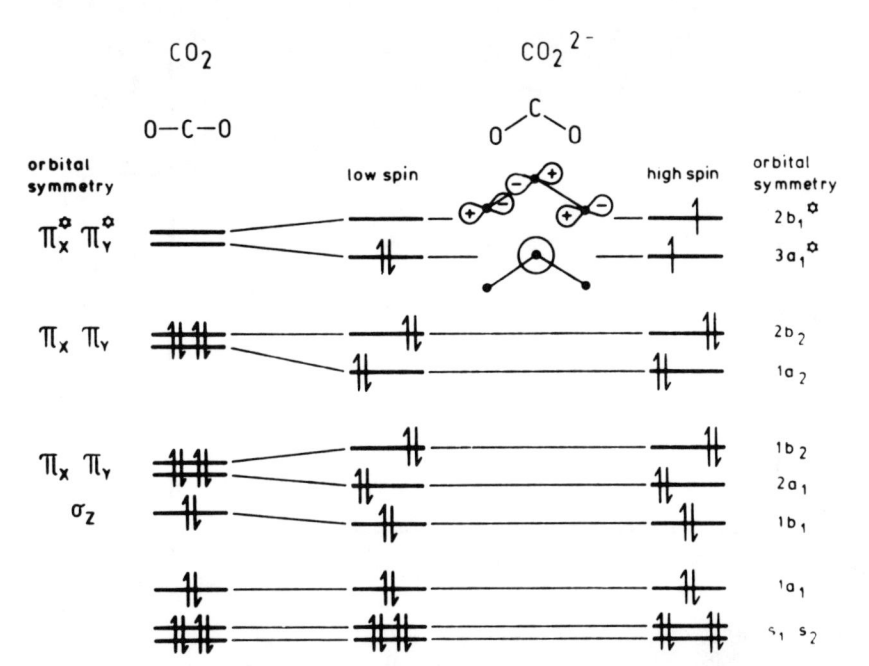

Fig. 6. Molecular orbital scheme for the 16-electron CO_2 molecule (left) and for two possible spin states of the 18-electron CO_2^{2-} configuration (center and right) with the electron density distribution in the highest occupied levels.

and the thermal expansion are both anomalous.[34,35] The intensity variation of the IR signal assigned to the half-OH$^-$-compensated cation vacancy between 530 and 800°C has already been discussed above. Between 300 and 530°C, it strongly depends upon the cooling rate of the crystal, as indicated in Fig. 5: in slowly cooled MgO crystals, the IR intensity is high but low in rapidly quenched crystals.[6] Also, the subsurface carbon concentration profiling by means of the $^{12}C(d,p)^{13}C$ technique also reveals strange anomalies:[27] in crystals slowly cooled from temperatures between 300 and 530°C, the subsurface C concentration is high but low in crystals that were more rapidly quenched. While the IR data point at rapid short-range diffusion of the type discussed above in connection with Eq. (6) occurring in the bulk, the (d,p) data indicate that some mobile C-bearing species are subject to fast long-range subsurface segregation.

Surface Charges and Subsurface Segregation

Frenkel[42] first pointed out that pure ionic crystals with Schottky disorder should, in thermal equilibrium, possess a charged surface and a region of space charge of the opposite sign beneath the surface. This surface dipole region arises from the fact that the free energies of formation of the anion and cation vacancies differ, which leads to different equilibrium concentrations near the surface and, hence, a potential difference between surface, subsurface region, and bulk. The space charge distributions for pure alkali halides and alkali halides doped with divalent cations have been calculated.[43–45]

Kingery[1] has applied this concept to heterovalent impurity segregation, grain-boundary diffusion, and electrostatic potentials in ceramic materials. For instance,

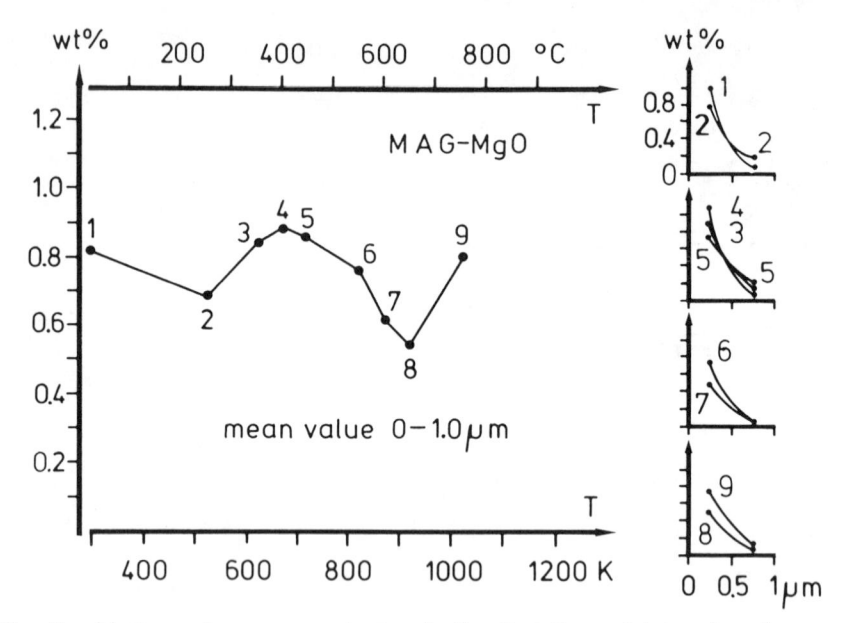

Fig. 7. Mean carbon concentration in the 0–1.5-μm thick subsurface region of an MgO single crystal as a function of temperature after a 30-min heating in ultrahigh vacuum (right, the subsurface carbon concentration profiles) as analyzed by the $^{12}C(d,p)^{13}C$ technique (Ref. 27).

Sc^{3+} in MgO represents a positively charged defect, $(Sc_{Mg}^{·})$, and will thus be attracted by a negative space charge on the MgO surface. Indeed, the experimentally determined Sc^{3+} subsurface concentration profiles in MgO agree rather well with the calculated width of the space charge layer.[2,3] On the contrary, in the case of Ca^{2+} in MgO, the surface segregation is strain induced.[15,16] Thus, both surface charges and strain relaxation need to be considered.

In the case of the solute carbon, both driving forces contribute to the surface segregation. The bent CO_2^{2-} configuration is inherently dipolar and so is the species CO^- as mentioned above. As long as the C atom resides on a cation vacancy site, the defect is neutral, but when it enters the interstitial sites, C-bearing species are created that, in addition to their strain dipole character, carry a double or a single positive charge, CO_2^{2-} or CO^-, respectively, as outlined by Eqs. (5) and (7).

Experimentally, by $^{12}C(d,p)^{13}C$ profiling, one finds very pronounced subsurface carbon concentration profiles extending at least 2 μm from the surface into the bulk.[27,29–31] Thus far, the depth resolution of this technique has not been better than 500 nm, allowing determination of hardly more than three points on the above-mentioned profiles. A recent report[46] on an argon ion backscattering technique, using a time-of-flight detection device, will probably lead to a better than 10-fold improvement of the depth resolution.

Figure 7 shows a series of carbon concentration profiles in an MgO single crystal containing approximately 1000 at.-ppm (0.03 wt%) of carbon, heated under ultrahigh-vacuum conditions in steps, held for 30 min at the indicated temperatures, and then quenched to liquid nitrogen temperature for analysis. The solid line indicates the mean carbon concentration in the analyzed layer, 0–2 μm, which is the depth range analyzed by the $^{12}C(d,p)^{13}C$ technique. The steepness of the profiles, i.e., the temperature-dependent carbon concentration in each layer, varies in a characteristic manner. The rise between 200 and 400°C is due to the increase of the carbon mobility, which allows more carbon to diffuse into the subsurface zone within the 30-min annealing time. The decrease above 400°C is due to losses of CO_2, CO, and other C-bearing gases from the MgO surface. Eventually, above 530°C the C atoms residing on cation vacancies are transferred onto interstitial sites, as discussed above, and mobilized. Thus, the supply of mobile C-bearing species all through the bulk suddenly increases, which manifests itself in an increase of the carbon concentration in the subsurface region due to an increased flux of C atoms from the bulk toward the surface.

At still higher temperatures (not shown in Fig. 7), the subsurface C concentration decreases again, because the solubility in the bulk increases, leading to a retrodiffusion of C atoms from the subsurface region into the bulk. This retrodiffusion has been studied by measuring the kinetic $^{13}C/^{12}C$ isotope fractionation that occurs after each step-wise temperature increment.[47]

According to our present understanding, this 0–2-μm deep subsurface C concentration profile is controlled by the strain relaxation acting between surface and bulk. The dipolar nature of the local strains associated with the CO_2^{2-} and CO^- configurations allows for the information to be transmitted over relatively large lattice distances. Dielectric loss factor measurements indeed indicate that the dipolar CO_2^{2-} and/or CO^- defects in MgO seem to form ordered domains by strain–strain interactions, leading to a weakly ferroelectric behavior.[34]

Carbon can be depleted from topmost subsurface layers by a heat treatment in air at 620°C. When the layers are cooled to about 200°C, a flat profile is obtained. The profile changes isothermally over several hours, indicating that the diffusion of the C atoms occurs in waves: the mean C concentration in the near-surface 0–0.5-μm layer starts to oscillate over several hours with a periodicity of about 45 min, while the mean C concentration in the next layer below, 0.5–1 μm, oscillates to a lesser degree, with a periodicity that is twice as long.[27] Details of this behavior are not yet understood. Similar 45-min oscillations were noted during the isothermal CO_2 evolution,[28] during DC conductivity and calorimetric measurements. These recurring observations suggest that the MgO surface, if placed far from equilibrium, approaches the equilibrium state in an oscillating manner.

The mean subsurface C concentrations, averaged over 0.5 μm (500 nm) as analyzed by the $^{12}C(d,p)^{13}C$ technique, may be as high as 50–100 times the mean bulk concentration. By photoelectron spectroscopy (XPS), the subsurface enrichment in the topmost 5–10-nm layer was found to rise to 500–1000 times the mean bulk concentration.[32] It appeared that this very high carbon concentration at or immediately beneath the surface is not due to contamination from the vacuum but to segregation of C-bearing solute species from the bulk.

Figure 8 shows in a somewhat schematic way the mean subsurface carbon concentrations as analyzed within the distinguishable layers between 0–1.5 μm by XPS and nuclear reaction profiling. The extremely high carbon value that was derived from the XPS data may be subject to an experimental uncertainty as high as 50%, due to the special calibration procedure used.[32] The C concentration may correspond to as much as two carbon atoms per unit cell MgO. The insert shows how the mean carbon concentration within this thin layer (expressed as carbon-to-oxygen ratio) varies with temperature. This is in agreement with theoretical expectations[14,15] and with observations made for the Ca^{2+} segregation.[16]

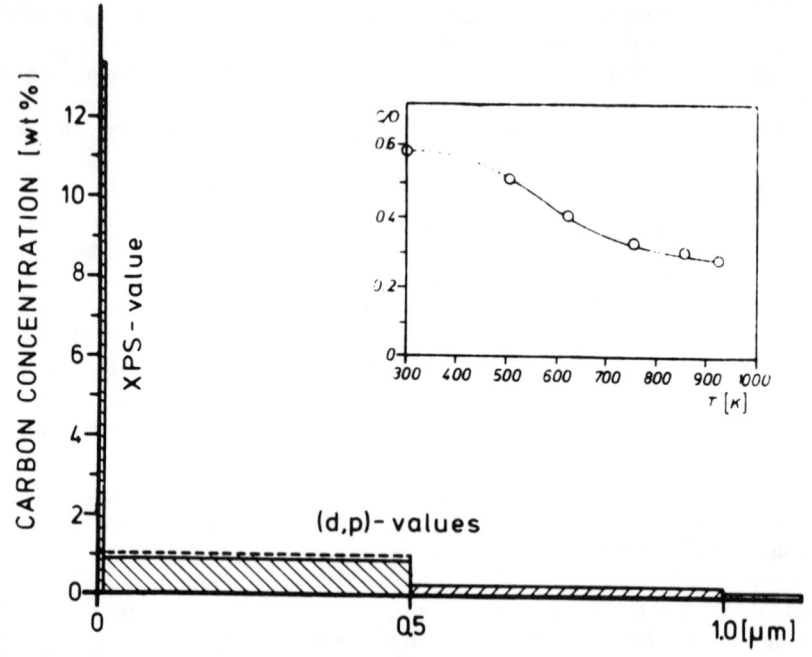

Fig. 8. Schematic representation of the subsurface carbon concentration as obtained by the XPS and (d,p) techniques. Note the extremely pronounced carbon segregation into the topmost 5–10 nm thick subsurface layer. Insert: Variation of the carbon-to-oxygen ratio as a function of temperature as derived from the XPS data (Ref. 32).

Discussion

If the C-bearing species segregate in such large numbers into the subsurface layer as shown by Fig. 8, two major problems arise that deserve special attention: stoichiometry and charge balance.

It is quite obvious that, on the basis of the available analytical data, the stoichiometry of the subsurface layer of an annealed MgO crystal is no longer correctly described as "MgO." This leads to the question of how the valency of the matrix constituents, Mg^{2+} and O^{2-}, will respond to the influx of the impurity species, mainly CO_2^{2-} and CO^-. Alternatively, one may ask what the valency state of the carbon is that segregates to the surface.

The molecular orbital scheme in Fig. 6 suggests that, if the CO_2^{2-} is present in its diamagnetic configuration with the a_1 level fully occupied, the two added electrons are strongly localized on the C atom. This raises the electron density at the C atom, making it essentially zerovalent. At least the carbon is not strongly positive, as one might suspect on the basis of a simple ionic model, assuming C^{4+} ions. Indeed, the XPS data indicate that the carbon in MgO is nearly zerovalent or at most slightly positive.[32] In earlier publications,[27-31] the solute carbon has therefore been called "atomic."

The electron density that is found on the C atoms is taken away from the O^{2-} ions in such a way as to oxidize them to the O^- state. The O^- is electronically quite different from the O^{2-}. It resembles a F atom and thus can tie a very covalent bond to the C atom. Experimentally, we expect that the electron density difference between the highly ionic O^{2-} of the MgO matrix and the O^- bonded to C should be measurable. Indeed, the XPS data provide strong evidence that the oxygen in the near-surface region of MgO exists in two distinctly different valency states, which may well be assigned to O^{2-} and O^-. It has been reported [32] that, in MgO, the C 1s signal is split into two components separated by about 2 eV and that the intensity of the O 1s component, which may be assigned to O^-, correlates with the 1s signal, the time-dependent intensity variations of which were used to determine the carbon diffusion coefficient.

In the mineral olivine, $(Mg,Fe)SiO_4$, the lattice oxygens are all bonded to Si. The Si–O bond is much more covalent than the Mg–O bond. Hence, the electron density difference between the oxygens bonded to C and those bonded to Si should be less. This trend is indeed confirmed by the XPS analysis,[48] which yields two O 1s components separated by not more than 1.5 eV.

In order to characterize the deviation of the stoichiometry of the MgO subsurface from the 1:1 stoichiometry of the perfect compound, a specific symbol for O^- should be used, for instance, $O^{\cdot}$ as suggested earlier.[25] By placing the symbol for essentially zerovalent carbon at the end of the formula, separated by a vertical bar, one can write the composition of the C-rich subsurface layer as $MgO_{1-x}O^{\cdot}_x/C_{\alpha x'}$, where α may adopt values between 0.5–1.0, depending upon the ration of the CO_2^{2-} and CO^- species in this layer. Probably CO^- prevails over CO_2^{2-} because the local density of carbon atoms is such that they can only be on interstitial sites. Therefore, α should be close to 0.5.

If we now turn to the H_2O dissolution mechanism, we may use the same concept of a nonstoichiometric MgO containing a trace δ of dissolved H_2O. If nothing but OH^- ions form, the solid solution may be written as $MgO_{1-\delta}(OH)_{2\delta}$. If, on the contrary, all OH^- ions were to convert into $(H_2 + O_2^{2-})$, the formula becomes $MgO_{1-\delta}O^{\cdot}_{2\delta}/(H_2)_\delta$. This changes into $MgO_{1-\delta}O^{\cdot}_{2\delta}$, if all H_2 molecules were lost from the solid. Rewriting this formula leads to $MgO_{1+\delta}$, where δ is the excess oxygen. This excess oxygen is in both cases balanced by an equal molar concentration of cation vacancies introduced by the CO_2 and H_2O dissolution mechanisms. Therefore, charge neutrality within each crystallite is maintained.

We shall consider next the charge balance inside the crystallite, i.e., between the surface–subsurface and the bulk. Each C atom that segregates toward the surface introduces a charge unbalance because it is bound to at least one O^- state. The O^- states migrate via the O^{2-} sublattice as a positive hole. They are positively charged with respect to the perfect lattice. If the subsurface C concentration profiles that extend 1–2 μm into the bulk are strain induced,[27] the region of the crystal where they exist necessarily becomes positively charged. The complemental negative charges remain in the bulk as half or fully uncompensated cation vacancies. It has been argued above that the more narrow subsurface C profile studied by XPS[32] is probably induced and controlled by the space charge layer associated with a surface charge.[36,49] This would be analogous to the subsurface profiles of Ca^{2+} in NaCl[45] and of Fe^{3+} and Sc^{3+} in MgO.[2,3] We are thus led to the conclusion that the presence of such positive surface layers implies a negatively charged MgO surface, irrespective of whether the subsurface segregation layer is made up of heterovalent cation impurities or of dissolved CO/CO_2 component.

If this is true, the question arises whether the MgO surface will always be negatively charged. In order to discuss this issue, we have to turn to the thermal stability of the peroxy ions, namely, $[O^{\cdot}_O V''_{Mg}]^\times$, which derive from the dissolution mechanisms of both H_2O and CO_2, as shown in Eqs. (2) and (3). Figure 2 demonstrates that the O_2^{2-} ions thermally disproportionate above 550°C into O^{2-} plus O, the latter evolving as a gas from the surface. However, through Eqs. (4) and (5), a clear distinction was made between the decay of those O_2^{2-} ions that exist at the surface of the crystallites and of those that are in the bulk. Obviously, in any sizeable crystallite the majority of the O_2^{2-} ions or similar, yet not fully characterized, peroxy configurations will be in the bulk. Those can only decay by releasing an unbound positive hole or O^- state according to Eq. (5). Being highly mobile, these unbound O^- states have a chance to reach the surface. At the surface, they may dimerize and form surface O_2^{2-} ions. Such surface O_2^{2-} ions are not charge compensated by nearby cation vacancies. Hence, all these O_2^{2-} ions carry two positive charges with respect to the bulk. If they disproportionate by releasing an oxygen atom, both charges are permanently deposited at the surface:

$$\begin{bmatrix} O^\cdot \\ O^\cdot \end{bmatrix} \Rightarrow [O^\times V_O^\cdot]^\cdot + O(g) \qquad (12)$$

If the outermost MgO surface was negatively charged before this series of events, it will now change sign and become positively charged. Temperature-dependent reversals of the sign of the surface charges have been predicted on the basis of thermodynamic arguments,[44] but in the case considered here, the cause is chemical.

Figure 9 shows normalized subsurface concentration profiles calculated for mobile positive holes (unbound O^- states) in MgO generated by the decay of peroxy ions according to Eq. (5), assuming that cation vacancies and V^- centers (the vacancy-bound O^- states) are still immobile.[49] Increasing the temperature increases the width of the O^- depletion zone near the surface. This is equivalent to a negatively charged space charge layer beneath a now positively charged surface (Fig. 9(A)). Increasing the charge carrier concentration decreases the width and thus increases the field gradient (Fig. 9(B)).

Several consequences of the change in surface polarity and of the buildup of a positive surface charge may be considered. (1) The unbound O^- states that exist in the bulk will be repelled from the surface. (2) Confined to the bulk, they may be consumed by oxidizing transition-metal impurities, as predicted by Eqs. (6) and (7) and as shown in Fig. 3. (3) Positively charged defect species such as CO_2^{2-} and/or CO^- that were contained in the preexisting subsurface layer of opposite polarity will undergo radical reactions. When O^- states are captured, they may give CO_2 and CO molecules that evolve as gases from the surface.[40] (4) Under the influence of the newly established positive surface potential the solute CO_2^{2-} and/or CO^- species will experience a force that causes them to retrodiffuse into the bulk. This retrodiffusion can be monitored by measuring the kinetic $^{13}C/^{12}C$ isotope fractionation:[47] since the ^{12}C isotopes retrodiffuse faster than the heavier ^{13}C isotopes, the latter become enriched relative to ^{12}C in the CO_2 gas molecules, which evolve from the surface.

One important aspect has still been disregarded in this discussion: the fact that hydrogen and carbon-bearing solute species will jointly be present in the MgO, because any MgO grown under realistic conditions, either by arc fusion or by sintering, will be contaminated at least on a trace level by dissolved H_2O and CO/CO_2 components. Hence, it is not unexpected that hydrogen and the C-bearing species interact with each other chemically in the near-surface region. Their cross-reactions lead to a large variety of organic molecules that also evolve from the

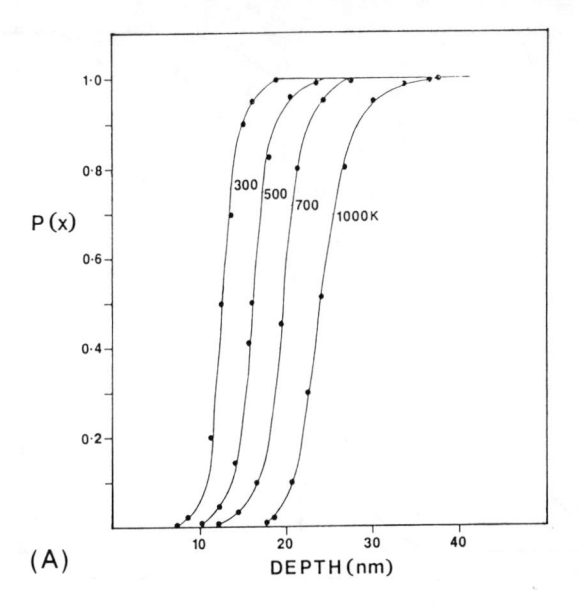

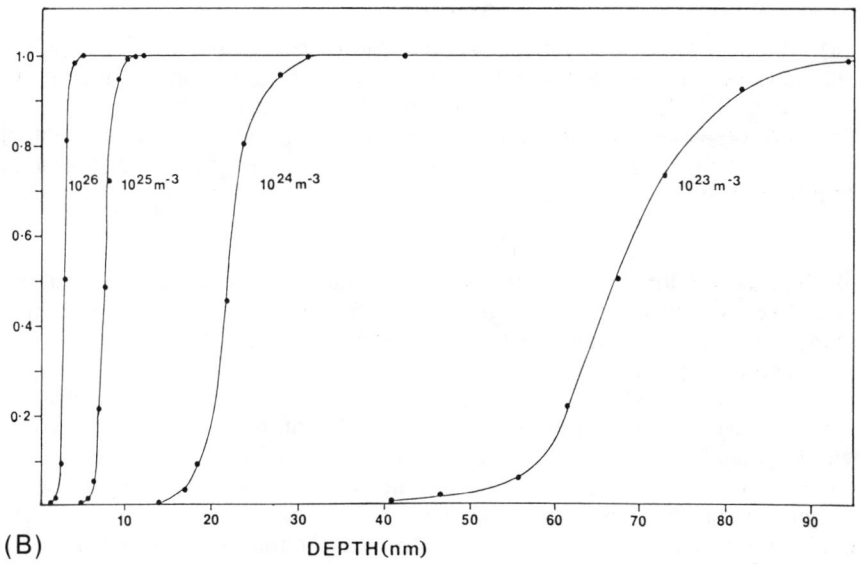

Fig. 9. Calculated normalized subsurface space charge distribution, assuming a positively charged MgO surface (due to oxygen release after peroxy decay), immobile cation vacancies, but mobile unbound holes, viz, O^- states (Ref. 49).

MgO surface.[40,50] The mechanisms and kinetics of these in part very complex reactions are still poorly understood.

If the positive surface charge remains positive and the temperature is increased up to the range where cation vacancies become mobile, a new situation arises: now, the negatively charged defects, cation vacancies, V^- centers, and the

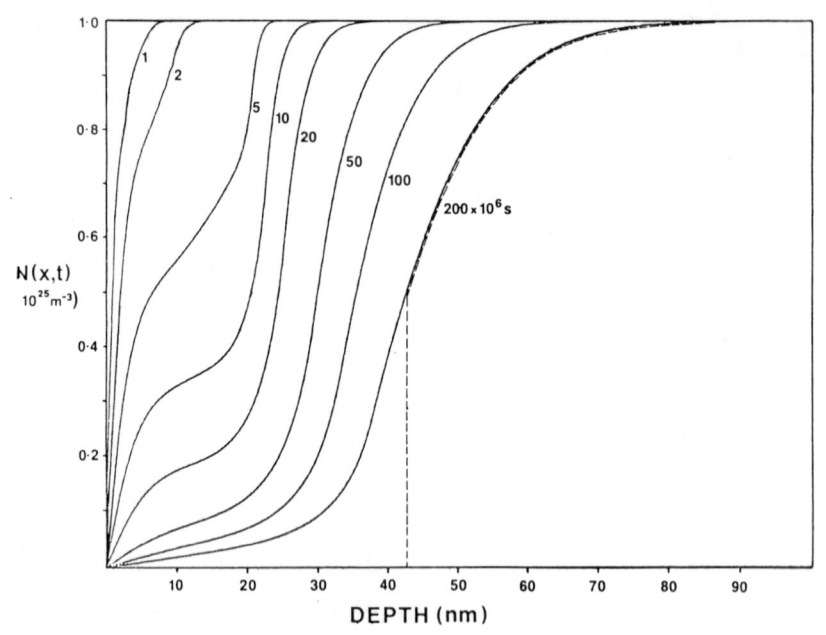

Fig. 10. Normalized subsurface concentration profiles of bound holes, viz, cation vacancies with one O$^-$, assuming that the cation vacancies are mobilized at higher temperatures, migrate to the surface, and anneal out at the surface while cations diffuse inward with a diffusion coefficient of 10^{-24} m$^2 \cdot$s. The dashed line represents the unbound hole concentration reached after 2×10^8 s (Ref. 49).

like will start to diffuse and to drift toward the surface under the influence of the field. Every cation vacancy, V''_{Mg}, that reaches the surface annihilates two positive charges. Thus, positive holes that still exist in the bulk may come to the surface, maintaining the positive potential.

Figure 10 shows a set of calculated subsurface vacancy concentration profiles[49] that are expected to develop as a function of time, assuming a constant surface potential, a reasonable vacancy concentration of 10^{25} m^{-3}, and a vacancy diffusion coefficient of 10^{-24} m$^2 \cdot$s^{-1}. Chemically speaking, this process describes nothing else but the evaporation of the remaining excess oxygen and the annealing of the excess cation vacancies that were originally introduced into the system by the dissolution of traces of H_2O and CO_2. Eventually, the MgO would purify itself by expelling all remaining traces of impurities that derive from solute H_2O and CO_2. However, this stage of ideal purity and of a truly intrinsic defect equilibrium will never be reached because, as the temperature increases, the equilibrium solubility for impurities also increases. This prevents the ultimate self-purification.

Acknowledgments

This work was supported by the Deutsche Forschungsgemeinschaft through several grants to F.F. and in part by the Gas Research Institute (Grant No. 5082-260-0707).

References

[1] W. D. Kingery, "Plausible Concepts Necessary and Sufficient for Interpretation of Ceramic Grain-Boundary Phenomena: I, Grain-Boundary Characteristics, Structure, and Electrostatic Potential," *J. Am. Ceram. Soc.*, 57 [1–8] (1974); "II, Solute Segregation, Grain-Boundary Diffusion, and General Discussion," *J. Am. Ceram. Soc.*, 57, 74–83 (1974).

[2] J. R. H. Black and W. D. Kingery, "Segregation of Aliovalent Solutes Adjacent Surfaces in MgO_3," *J. Am. Ceram. Soc.*, 62, 176–8 (1979).

[3] U. M. Chiang, A. F. Hendricksen, W. D. Kingery, and D. Finello, "Characterization of Grain-Boundary Segregation in MgO," *J. Am. Ceram. Soc.*, 64, 385–9 (1981).

[4] F. A. Kröger; The Chemistry of Imperfect Crystals. North-Holland, Amsterdam, 1964.

[5] H. Wengeler and F. Freund, "Atomic Carbon in Magnesium Oxide: VII, Infrared Analysis of OH^- Containing Defects," *Mater. Res. Bull.*, 15, 1747–53 (1980).

[6] F. Freund and H. Wengeler, "The Infrared Spectrum of OH^- Compensated Defect Sites in C-Doped MgO and CaO Single Crystals," *J. Phys. Chem. Solids*, 43, 129–45 (1982).

[7] R. H. Stolen and G. E. Walrafen, "Water and Its Relation to Broken Bond Defects in Fused Silica," *J. Chem. Phys.*, 64, 2623–31 (1976).

[8] J. Stone and G. E. Walrafen, "Overtone Vibrations of OH^- Groups in Fused Silica Optical Fibers," *J. Chem. Phys.*, 76, 1712–22 (1982).

[9] H. Kathrein, J. Nagy, and F. Freund, "O^--Ions and Their Relation to Traces of H_2O and CO_2 in Magnesium Oxide: An EPR Study"; to be published in *J. Phys. Chem. Solids*.

[10] R. Martens, H. Gentsch, and F. Freund, "Hydrogen Release during the Thermal Decomposition of Magnesium Hydroxide to Magnesium Oxide," *J. Catal.*, 44, 366–72 (1976).

[11] F. Freund, R. Martens, and H. Wengeler, "A Deuterium-Hydrogen Fractionation Mechanism in Magnesium Oxide," *Geochim. Cosmochim. Acta*, 46, 1821–9 (1982).

[12] J. M. André, E. G. Fripiat, and D. P. Vercauteren, "Quantum Mechanical Approach to the Chemisorption of Molecular Hydrogen on Defect Magnesium Oxide Surfaces," *Theor. Chim. Acta*, 43, 239–51 (1977).

[13] P. Wynblatt and R. C. Ku, "Surface Energy and Solute Strain Energy Effects in Surface Segregation," *Surf. Sci.*, 65, 511–31 (1977).

[14] C. Lea and M. P. Seah, "Kinetics of Surface Segregation," *Philos. Mag.*, 35, 213–28 (1977).

[15] E. A. Colbourn, W. C. Mackrodt, and P. W. Tasker, "The Segregation of Calcium Ions at the Surface of Magnesium Oxide: Theory and Calculation," *J. Mater. Sci.*, 18, 1917–24 (1983).

[16] R. C. McCune and P. Wynblatt, "Calcium Segregation to a Magnesium Oxide (100) Surface," *J. Am. Ceram. Soc.*, 66, 111–7 (1983).

[17] C. M. Osburn and R. W. Vest, "Electrical Properties of Single Crystals, Bicrystals and Polycrystals of MgO," *J. Am. Ceram. Soc.*, 54, 428–35 (1971).

[18] D. R. Sempolinski and W. D. Kingery, "Ionic Conductivity and Magnesium Vacancy Mobility in Magnesium Oxide," *J. Am. Ceram. Soc.*, 63, 644–99 (1980).

[19] N. G. Vannerberg, "Peroxides, Superoxides, and Ozonides of the Metals of Groups Ia, IIa, and IIb," *Prog. Inorg. Chem.*, 4, 125–95 (1962).

[20] A. M. Glass and T. M. Searle, "Reactions between Vacancies and Impurities in Magnesium Oxide: I, Mn^{4+} and OH-Ion Impurities," *J. Chem. Phys.*, 46, 2092–101 (1967).

[21] A. C. Tomlinson and B. Henderson, "Some Studies of Defects in Calcium Oxide: I, Impurity Effects," *J. Phys. Chem. Solids*, 30, 1793–9 (1969).

[22] M. M. Abraham, Y. Chen, and W. P. Unruh, "Formation and Stability of V^- and V_{Al} Centers in MgO," *Phys. Rev. B*, B9, 1842–52 (1974).

[23] J. E. Wertz and B. Henderson; Defects in the Alkaline Earth Oxides. Taylor & Francis, London, 1977.

[24] E. G. Derouane, U. Indovina, A. B. Walters, and M. Boudart, "Paramagnetic Defects at the Surface of Magnesium Oxide," pp. 703–6 in the Proceedings of the 7th International Symposium on the Reactivity of Solids. Chapman and Hall, London, 1972.

[25] F. Freund, "Mechanism of the Water and Carbon Dioxide Solubility in Oxides and Silicates and the Role of O^-," *Contrib. Mineral. Petrol.*, 76, 474–82 (1981).

[26] C. T. Butler, B. J. Sturm, and R. B. Quincy, "Arc-Fusion Growth and Characterization of High Purity MgO Crystals," *J. Cryst. Growth*, 8, 191–202 (1971).

[27] H. Wengeler, R. Knobel, H. Kathrein, F. Freund, G. Demortier, and G. Wolff, "Atomic Carbon in Magnesium Oxide Single Crystals—Depth Profiling, Temperature- and Time-Dependent Behavior," *J. Phys. Chem. Solids*, 43, 59–71 (1982).

[28] R. Knobel and F. Freund, "Atomic Carbon in Magnesium Oxide: IV, Carbon Dioxide Evolution under Argon and Oxygen," *Mater. Res. Bull.*, 15, 1247–53 (1980).

[29] F. Freund, G. Debras, and G. Demortier, "Carbon Content in Magnesium Oxide Single Crystals Grown by the Arc-Fusion Method," *J. Cryst. Growth*, 38, 227–80 (1977).

[30] F. Freund, G. Debras, and G. Demortier, "Carbon Content of High-Purity Alkaline Earth Oxide Single Crystals Grown by Arc Fusion," *J. Am. Ceram. Soc.*, 61, 429–34 (1978).

[31] F. Freund, H. Kathrein, H. Wengeler, and G. Demortier, "Atomic Carbon in Magnesium Oxide: I, Carbon Analysis by the $^{12}C(d,p)^{13}C$ Method," *Mater. Res. Bull.*, 15, 1011–18 (1980).

[32] H. Kathrein, H. Gonska, and F. Freund, "Subsurface Segregation and Diffusion of Carbon in Magnesium Oxide," *Appl. Phys., [Part] A*, 30, 33–41 (1983).

[33] H. Kathrein and F. Freund, "Electrical Conductivity of Magnesium Oxide Single Crystals below 1200 K," *J. Phys. Chem. Solids,* **44**, 177–84 (1983).

[34] F. Freund, "The O$^-$ State, Hydrogen and Carbon in Solid Solution in Refractory Oxides"; to be published in *High Temp. High Pressure.*

[35] H. Wengeler and F. Freund, "Atomic Carbon in Magnesium Oxide: III, Anomalous Thermal Expansion Behavior," *Mater. Res. Bull.,* **15**, 1241–5 (1980).

[36] F. Freund, R. Knobel, H. Kathrein, and H. Wengeler, "Pre-irradiation Defects in "Pure" MgO Associated with Hydrogen, Carbon and Peroxy Configurations"; to be published in the Proceedings of the 2nd International Conference on Radiation Effects in Insulators.

[37] A. M. Brown and M. F. Ashby, "Correlation for Diffusion Constants," *Acta Metall.,* **28**, 1085–101 (1980).

[38] C. G. Sammis, J. C. Smith, and G. Schubert, "A Critical Assessment of Estimation Methods for Activation Volume," *J. Geophys. Res.,* **86**, B11, 10707–18 (1982).

[39] P. W. Atkins and M. C. R. Symons; The Structure of Inorganic Radicals. Elsevier, Amsterdam, 1967.

[40] R. Knobel, "Formation and Evolution of Gaseous H-C-N-O Compounds from MgO and Silicate Surfaces (in German)"; Ph.D. Thesis, Univ. Köln, 1983.

[41] H. Kathrein, U. Knipping, and F. Freund, "Atomic Carbon in Magnesium Oxide: VI, Electrical Conductivity," *Mater. Res. Bull.,* **15**, 1393–99 (1980).

[42] J. Frenkel; p. 36 in the Kinetic Theory of Liquids. Oxford University Press, Oxford, New York, 1946.

[43] K. Lehovec, "Space-Charge Layer and Distribution of Lattice Defects at the Surface of Ionic Crystals," *J. Chem. Phys.,* **21**, 1123–8 (1953).

[44] K. L. Kliewer and J. S. Koehler, "Space Charge in Ionic Crystals: I, General Approach with Application to NaCl," *Phys. Rev. A,* **A140**, 1226–40 (1965).

[45] K. L. Kliewer, "Space Charge in Ionic Crystals: II, The Electron Affinity and Impurity Accumulation," *Phys. Rev. A,* **A140**, 1241–6 (1965).

[46] J. P. Thomas, J. Fallavier, D. Ramdane, N. Chevarier, and A. Chevarier, "High Resolution Depth Profiling of Light Elements in High Atomic Mass Materials"; 6th International Conference on Ion Beam Analysis. Tempe, AZ, May 23–7, 1983.

[47] F. Pineau and F. Freund; unpublished results.

[48] G. Oberheuser, H. Kathrein, G. Demortier, H. Gonska, and F. Freund, "Carbon in Olivine Single Crystals Analyzed by the ^{12}C(d,p)^{13}C Method and by Photoelectron Spectroscopy," *Geochim. Cosmochim. Acta,* **47**, 1117–29 (1983).

[49] B. King and F. Freund, "Surface Charges and Subsurface Space Charge Layers in Oxides Containing Traces of Dissolved Water"; to be published in *Phys. Rev. B.*

[50] F. Freund, R. Knobel, G. Oberheuser, G. C. Maiti, and R. G. Schaefer, "Atomic Carbon in Magnesium Oxide: V, Hydrocarbon Evolution," *Mater. Res. Bull.,* **15**, 1385–91 (1980).

---

*Also associated with the Mineralogisches Institut.

†Throughout this paper, the Kröger (Ref. 4) nomenclature will be used where a dot ($\cdot$), prime ($'$), and cross ($\times$) designate positive, negative, and neutral charges with respect to the ideal occupancy of the lattice points by the divalent cations, Mg^{2+}, and the divalent anions, O^{2-}. V stands for vacancy and i for interstitial.

Defect Association in MgO

T. A. Yager

Hewlett-Packard Laboratories
Palo Alto, CA 94304

W. D. Kingery

Massachusetts Institute of Technology
Department of Materials Science and Engineering
Cambridge, MA 02139

High-temperature EPR spectroscopy with a CO_2 laser heat source was used to study the defect structure of MgO single crystals doped with up to 4300 ppm of Fe below 1200°C. Mass action relations incorporating Debye–Hückel activity coefficient corrections were developed to describe the association of isolated iron and vacancies. Kinetic studies indicated that the room-temperature defect structure of liquid nitrogen quenched samples consists of many uniformly dispersed clusters, dimers, trimers, and unassociated Fe^{3+} but very few isolated vacancies. In samples of low dopant levels quenched from 1100°C, Fe^{3+} and vacancies aggregate to form 3-clusters below 600°C, reducing the mobility of defects. Above 600°C, clusters dissociate, liberating mobile dimers and vacancies.

The physical properties of MgO, including diffusivity,[1] electrical conductivity,[2] dielectric loss,[3] magnetic susceptibility,[4] optical absorption,[5] and mechanical strength,[6] have been related to the structure and chemical state of aliovalent solutes. Iron has proved to be a useful solute for studying trivalent substitutional defect reactions in MgO.[7] With use of Kröger–Vink notation, the solution, dimer association, trimer association, and precipitate reactions can be written, respectively, as

$$Fe_2O_3 \xrightleftharpoons{MgO} 2Fe_{Mg}^{\cdot} + V_{Mg}'' + 3O_O^{\times} \qquad (1)$$

$$Fe_{Mg}^{\cdot} + V_{Mg}'' \rightleftharpoons (Fe_{Mg}^{\cdot}\text{–}V_{Mg}'')' \qquad (2)$$

$$(Fe_{Mg}^{\cdot}\text{–}V_{Mg}'')' + Fe_{Mg}^{\cdot} \rightleftharpoons (Fe_{Mg}^{\cdot}\text{–}V_{Mg}''\text{–}Fe_{Mg}^{\cdot})^{\times} \qquad (3)$$

$$Mg_{Mg}^{\times} + 2Fe_{Mg}^{\cdot} + V_{Mg}'' + 4O_O^{\times} \rightleftharpoons MgFe_2O_4(ppt) \qquad (4)$$

Association and clustering processes were found to occur sufficiently rapidly that local high-temperature structures were not maintained in the room-temperature material after quenching.[8] Thus, the defect structure of quenched MgO does not correspond to either the equilibrium high-temperature or room-temperature structure and partially consists of metastable point defects, associates, and clusters. This paper analyzes the thermodynamic defect structure and kinetic reactions of Fe^{3+}-doped MgO below 1200°C.

Experimental Procedure

Single crystals of 4N (99.99% pure) MgO doped in the melt with 310, 2300, and 4300 ppm of iron were obtained commercially.* All samples were cleaved along {100} planes to produce dimensions of approximately 4.8 × 4.8 × 1 mm. A controlled-atmosphere molybdenum-wound tube furnace was used to fully oxidize and homogenize all dopants. The samples were slowly heated to 1600°C in flowing O_2, cooled, and held at 1100°C for 24 h, and then quenched in liquid nitrogen.

A high-temperature EPR spectrometer with a CO_2 laser heat source,[9] capable of heating to 1200°C and allowing a liquid-nitrogen quench within the spectrometer, was used to monitor defect concentrations. Equilibrium measurements were performed in the spectrometer by heating the sample to 1200°C for 0.5 h to allow the apparatus to reach a stable thermal state. The sample temperature was then lowered at a rate sufficiently slow to allow short-range equilibrium association to occur, and the EPR spectra were recorded at 25°C intervals.[10]

Kinetic measurements were performed in the spectrometer by heating the sample to 1200°C for 15 min and cooling to 1100°C for 0.5 h, followed by a quench. The sample was next heated to the appropriate isothermal anneal temperature, and the EPR spectra were recorded at appropriate time intervals. This procedure was repeated on the same sample to conduct several isothermal anneals.[11]

Phase equilibria measurements were performed in the controlled-atmosphere tube furnace by reheating the samples in flowing O_2 for a sufficient period of time to allow equilibration, near the temperature of phase separation. Following each of the many heat treatments per sample, the room-temperature EPR spectra were recorded, and the presence of a broad ferrimagnetic resonance peak provided evidence of a second phase.[10] After this investigation, the crystals were analyzed by quantitative atomic absorption spectroscopy.[†]

Results and Analysis

Simple Association

The EPR peak attributed to D⟨100⟩ was often obscured by broadening effects and overlapping of the D⟨110⟩ peak. The D⟨110⟩ peak provided more reproducible results and was therefore used in this analysis. According to Eq. (2), the equilibrium constant for dimer formation is given by

$$K_{D\langle 110\rangle} = \frac{A_{D\langle 110\rangle}}{A_{Fe'_{Mg}} A_{V''_{Mg}}} = K^0_{D\langle 110\rangle} \exp\left(\frac{E_{D\langle 110\rangle}}{k_b T}\right) \quad (5)$$

The dilute solution approximation at low concentrations or high temperatures assumes the activity equals the mole fraction. A manipulation and substitution of terms yields

$$\ln\left(\frac{I_{D\langle 110\rangle}}{I_{Fe'_{Mg}}}\right) = \frac{E_{D\langle 110\rangle}}{k_b T} + \ln\left(\frac{a_{Fe'_{Mg}}[V''_{Mg}] K^0_{D\langle 110\rangle}}{a_{D\langle 110\rangle}}\right) \quad (6)$$

where the last term can be considered constant at low solute concentration levels.

Figure 1 is a plot of Eq. (5) in a sample containing 310 ppm of Fe. A value of $E_{D\langle 110\rangle}$ equal to 0.81 eV was calculated from the straight line found at high temperatures. Figures 2 and 3 show similar plots for crystals containing 2300 and 4300 ppm of Fe. Since the low solute level approximation is less valid for these samples, more curvature and a lower slope were measured.

Phase Equilibria

Table I gives the analyzed Fe concentration and the phase separation temperature for the samples studied. These data, along with the earlier data of Roberts

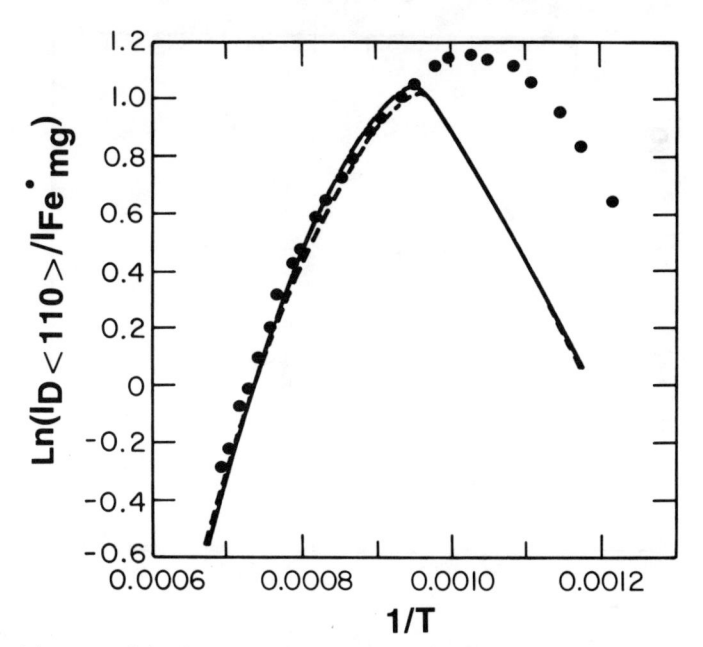

Fig. 1. Measured (solid points) and calculated (with (solid line) and without (dashed line) Debye–Hückel corrections) ratios of dimer and isolated iron ion intensities for a sample containing 310 ppm of Fe.

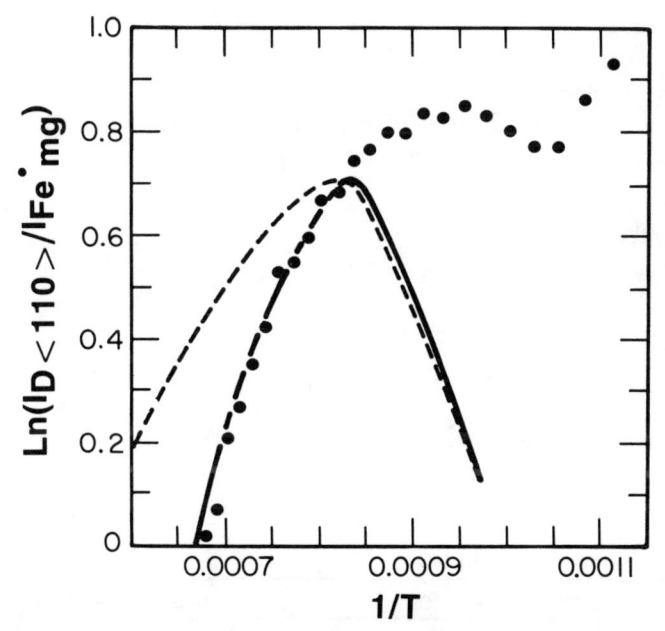

Fig. 2. Measured (solid points) and calculated (with (solid line) and without (dashed line) Debye–Hückel corrections) ratios of dimer and isolated iron ion intensities for a sample containing 2300 ppm of Fe.

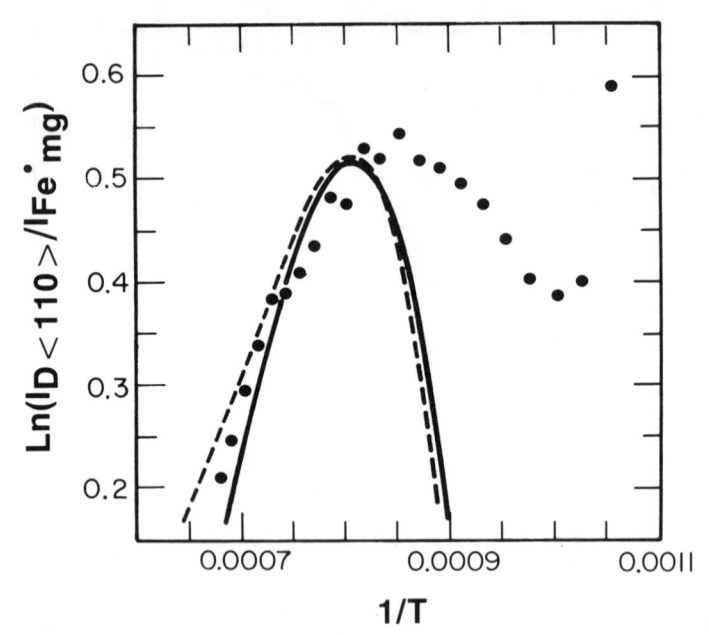

Fig. 3. Measured (solid points) and calculated (with (solid line) and without (dashed line) Debye–Hückel corrections) ratios of dimer and isolated iron ion intensities for a sample containing 4300 ppm of Fe.

and Merwin,[12] are shown in Fig. 4 as a plot of ln (mole fraction on phase boundary) against $1/T$.

Equilibrium Analysis

A numerical iterative computer program was used to theoretically analyze the data in Figs. 1–4. The following assumptions were made:[10]

$$\frac{A_{\text{dimer}}}{A_{\text{Fe}_{\text{Mg}}^{\cdot}} A_{V_{\text{Mg}}^{''}}} = K_{D\langle 110 \rangle}^{0} \exp\left(\frac{E_{D\langle 110 \rangle}}{k_b T}\right)$$

$$\frac{A_{\text{trimer}}}{A_{\text{dimer}} A_{\text{Fe}_{\text{Mg}}^{\cdot}}} = \frac{K_{D\langle 110 \rangle}^{0}}{24} \exp\left(\frac{1.4(\text{eV})}{k_b T}\right)$$

$$[\text{Fe}_T] = [\text{dimer}] + 2[\text{trimer}] + [\text{Fe}_{\text{Mg}}^{\cdot}]$$

$$[V_T] = 1/2[\text{Fe}_T] = [\text{dimer}] + [\text{trimer}] + [V_{\text{Mg}}^{''}]$$

$$K_{\text{ppt}} = \frac{A_{\text{MgFe}_2\text{O}_4}}{A_{\text{Fe}_{\text{Mg}}^{\cdot}}^{2} A_{\text{Mg}_{\text{Mg}}^{x}} A_{V_{\text{Mg}}^{''}} A_{\text{O}_{\text{O}}^{x}}} = \frac{1}{[\text{Fe}_{\text{Mg}}^{\cdot}]^2 f_{\text{FeMg}}^{2}} (1 - [V_{\text{Mg}}^{''}]) f_{V_{\text{Mg}}^{''}} [V_{\text{Mg}}^{''}]$$

Table I. Phase Equilibria Data

Analyzed Fe concentration (ppm)	Temperature of phase separation (°C)
250	772
2340	922
4590	975

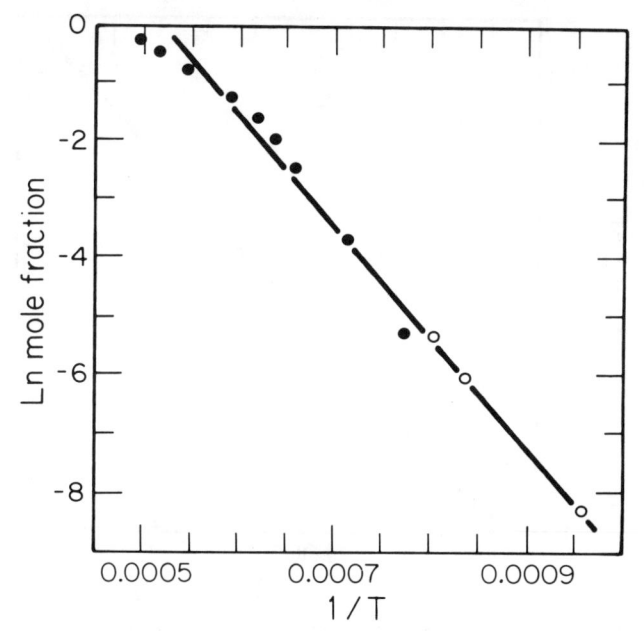

Fig. 4. Heat of solution plot for precipitation of MgFe$_2$O$_4$ from MgO. Solid points are data from Roberts and Merwin (Ref. 12).

$$= K_{ppt}^0 \exp\left(\frac{E_{ppt}}{k_b T}\right)$$

The dotted lines in Figs. 1–3 were calculated by assuming that the activity coefficient for all defects equals unity while the solid lines were calculated by using Debye–Hückel activity coefficient corrections. Since the ordinate axes in Figs. 1–3 have only relative meaning, these calculated curves were adjusted in the verticle direction to best fit the data. Good agreement of the calculated curves incorporating the Debye–Hückel corrections was found over a wide temperature and concentration range.

Values incorporating the Debye–Hückel corrections derived from this analysis are given in Table II. The phase separation data from Fig. 4 corrected for the defect structure is given in Fig. 5 as calculated ln (K_{ppt}) vs $1/T$. The straight line

Table II. Quantitative Evaluation of the Defect Structure of MgO:Fe^{3+} below 1200°C

Reaction	K^0	$-E$ (eV)	$-E/\text{Fe}_{Mg}^{\cdot}$
$\text{Fe}_{Mg}^{\cdot} + V_{Mg}'' = D\langle 110\rangle$	0.8†	0.85‡	0.85‡
$D\langle 110\rangle + \text{Fe}_{Mg}^{\cdot} = T\langle 100\rangle$	0.03*	1.4*	1.13*
$2D\langle 110\rangle + \text{Fe}_{Mg}^{\cdot} = 3\text{-cluster}$	8×10^{-6}*	2.38	1.36
$\text{Mg}_{Mg}^x + 2\text{Fe}_{Mg}^{\cdot} + V_{Mg}'' + 4O_0^x$ $= \text{MgFe}_2\text{O}_4(ppt)$	8.1×10^{-6}‡	3.83‡	1.91‡

*Approximated value.
†Based entirely on computer curve fit.
‡Based partially on computer curve fit.

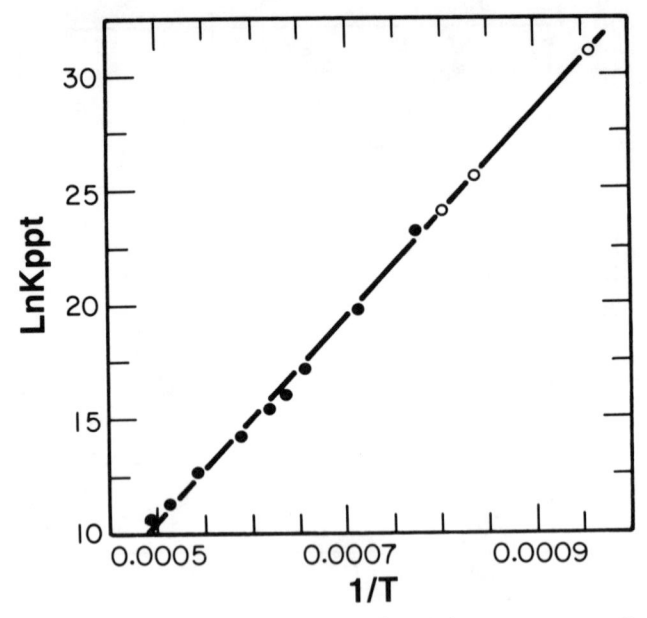

Fig. 5. Heat of solution data corrected for defect structure. Solid points are data from Roberts and Merwin (Ref. 12).

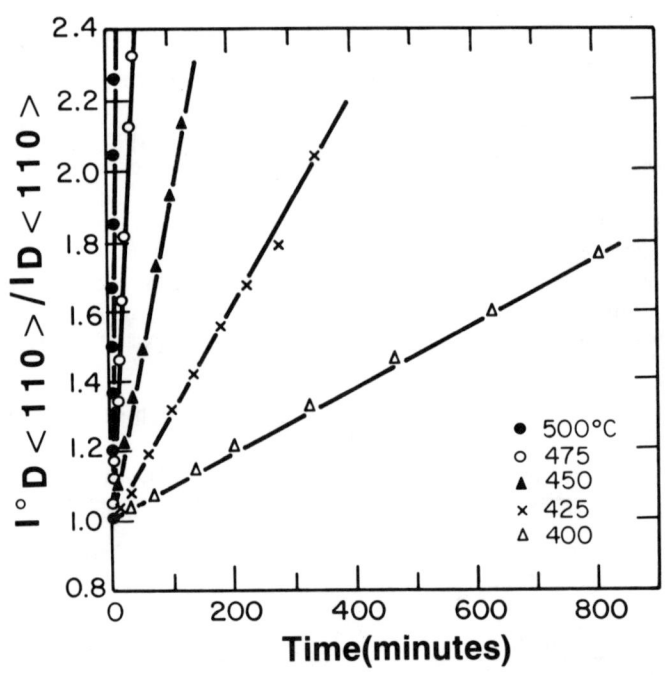

Fig. 6. Normalized second-order plot for D⟨110⟩ decay during isothermal anneals between 400 and 500°C in MgO containing 310 ppm of Fe.

that resulted provides additional evidence for the calculated defect structure. At low temperatures, the deviation of the calculated curves from the measured data in Figs. 1–3 implies that kinetic limitations prevented precipitation. This is reasonable, considering the long diffusion distances necessary for phase separation.

Kinetic Relaxation

A normalized second-order kinetic decay plot for the $D\langle 110\rangle$ peak in the 310-ppm sample below 500°C is shown in Fig. 6. Assuming that $[D\langle 110\rangle] \propto [D\langle 110, 100\rangle]$, one can write the second-order reaction as[11]

$$\frac{d[D\langle 110, 100\rangle]}{dt} = -k[D\langle 110, 100\rangle]^2 \tag{7a}$$

or by integration and manipulation of terms

$$\frac{I^0_{D\langle 110\rangle}}{I_{D\langle 110\rangle}} = k[D^0\langle 110, 100\rangle]t + 1 \tag{7b}$$

The straight lines that resulted confirm second-order decay kinetics. Plots for either first- or third-order kinetics were not linear. Figure 7 shows a similar plot for the same sample between 500 and 600°C. Curvature in this plot is explained below as the result of achieving a metastable state.

The intensity of the $Fe^{\cdot}_{Mg}$ peak was found to decay at a rate proportional to the decay of the $D\langle 110\rangle$ peak. Therefore

$$\frac{d[Fe^{\cdot}_{Mg}]}{dt} = B\frac{d[D\langle 110, 100\rangle]}{dt} \tag{8}$$

or

$$\frac{[Fe^{\cdot}_{Mg}]}{[Fe^{\cdot 0}_{Mg}]} = \frac{B[D^0\langle 110, 100\rangle]}{[Fe^{\cdot 0}_{Mg}]}\left(\frac{1}{k[D^0\langle 110, 100\rangle]t + 1}\right)$$
$$+ \left(1 - \left(\frac{B[D^0\langle 110, 100\rangle]}{[Fe^{\cdot 0}_{Mg}]}\right)\right) \tag{9}$$

This is plotted in Fig. 8 from which a slope of 0.48 and an intercept of 0.52 were obtained. The straight line obtained suggests the interaction of $Fe^{\cdot}_{Mg}$ and dimers during the relaxation reaction.

The 3-Cluster

The second-order decay of dimers implies the reaction of two dimers to form the initial cluster. The plausible reaction

$$2 \text{ dimers} + Fe^{\cdot}_{Mg} = \text{3-cluster} \tag{10}$$

may follow a two-step process

$$\text{dimer} + Fe^{\cdot}_{Mg} = \text{trimer} \tag{11}$$

$$\text{dimer} + \text{trimer} = \text{3-cluster} \tag{12}$$

and would follow second-order kinetics so long as Eq. (11) is rate limiting.[11] According to Eq. (10), the value of B in Eq. (8) should equal 0.5. Thus, from Eq. (9) and the experimental values obtained for the slope and intercept, $[\text{dimers}^0] \simeq [Fe^{\cdot 0}_{Mg}]$ in the sample containing 310 ppm of Fe, suggesting that $[V''^0_{Mg}] \simeq 0$.

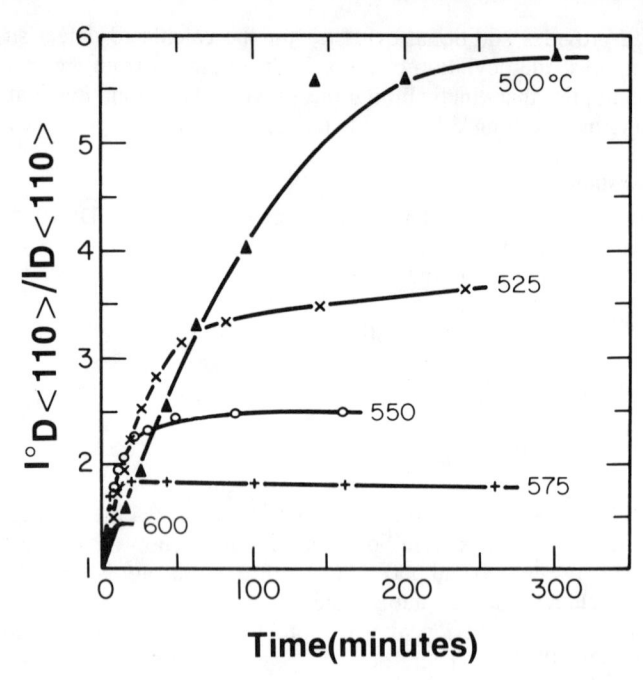

Fig. 7. Normalized second-order plot for D⟨110⟩ decay during isothermal anneals between 500 and 600°C in MgO containing 310 ppm of Fe.

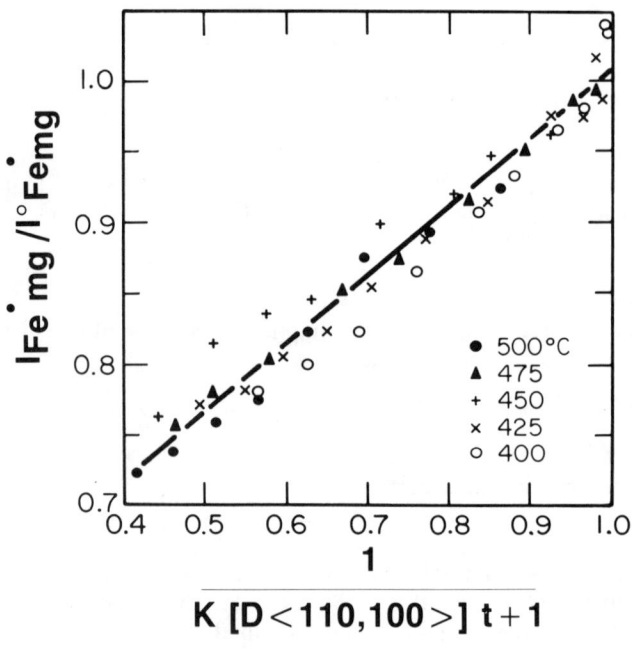

Fig. 8. Kinetic decay of Fe$_{Mg}^{\bullet}$ during isothermal anneals between 500 and 600°C in MgO containing 310 ppm of Fe.

Assuming an activity coefficient of 1, one can write the equilibrium constant for Eq. (10) as

$$\frac{[\text{3-cluster}]_{eq}}{[\text{dimer}]_{eq}^2 [\text{Fe}'_{Mg}]_{eq}} = K^0_{\text{3-cluster}} \exp\left(\frac{E_{\text{3-cluster}}}{k_b T}\right) \quad (13)$$

The intensity ratio at which the curves in Fig. 7 leveled off provided a measure of the metastable equilibrium. With

$$[\text{Fe}'_{Mg}]_{eq} = [\text{dimers}^0] - 1/2([\text{dimers}^0] - [\text{dimers}]_{eq})$$

and

$$[\text{3-cluster}]_{eq} = 1/2([\text{dimer}^0] - [\text{dimer}]_{eq})$$

a manipulation of terms yields

$$\ln\left[\frac{1 - \left(\frac{I^{eq}_{D\langle 110\rangle}}{I^0_{D\langle 110\rangle}}\right)}{\left(\frac{I^{eq}_{D\langle 110\rangle}}{I^0_{D\langle 110\rangle}}\right)^2 \left(1 + \left(\frac{I^{eq}_{D\langle 110\rangle}}{I^0_{D\langle 110\rangle}}\right)\right)}\right] = \frac{E_{\text{3-cluster}}}{k_b T} + \ln\left(D^0\langle 110,100\rangle\right) K^0_{\text{3-cluster}} \quad (14)$$

A plot of Eq. (14) is shown in Fig. 9. A straight line resulted from which the equilibrium constant was derived:

$$K_{\text{3-cluster}} = \frac{6.9 \times 10^{-15}}{[D^0\langle 110,100\rangle]^2} \exp\left(\frac{2.4(\text{eV})}{k_b T}\right) \quad (15)$$

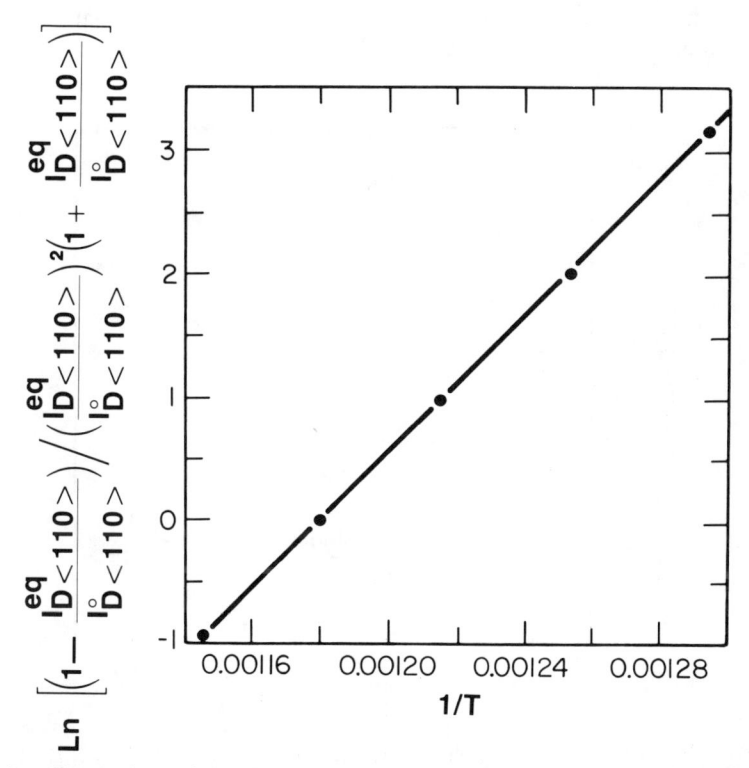

Fig. 9. Formation of the 3-cluster during isothermal anneal of MgO containing 310 ppm of Fe.

If $[D^0\langle 110, 100\rangle] \simeq 30$ ppm is approximated (10% of the total iron concentration), then

$$K_{\text{3-cluster}} \simeq 8 \times 10^{-6} \exp\left(\frac{2.4(\text{eV})}{k_b T}\right)$$

Kinetic Relaxation Processes During the Quench

The relative concentration of defect centers in samples quenched from 1100°C was determined by using the approximation from Eq. (8)

$$[i] = \frac{C}{W} \int\int I(H) dt_1 dt_2$$

where C is a constant, W is the sample weight, $I(H)$ is the EPR signal intensity as a function of the applied magnetic field, and dt_1 and dt_2 were the time constants of the double integrator. The magnetic field was increased at a constant rate so that $dH \propto dt$. Figure 10 shows a plot of the relative double-integrated intensity of $I_{\text{Fe}_{\text{Mg}}^\bullet}$ and $I_{D\langle 110\rangle}/I_{\text{Fe}_{\text{Mg}}^\bullet}$ for MgO crystals containing 310, 2300, and 4300 ppm of Fe. The constant C was adjusted so that the maximum value of each curve equaled 1. This plot shows that $[\text{Fe}_{\text{Mg}}^\bullet]$ levels off with increasing concentration, while the ratio of dimers to isolated iron increases. According to mass action relations, both $[\text{Fe}_{\text{Mg}}^\bullet]$ and $[\text{dimers}]/[\text{Fe}_{\text{Mg}}^\bullet]$ should continuously increase. This deviation suggests higher solute-containing samples show a greater extent of relaxation during the quench, where the more mobile dimer is preferentially eliminated from the system.

EPR kinetic relaxation curves at 450 and 600°C for a crystal containing 2300 ppm of Fe are shown in Figs. 11 and 12, respectively. At 450°C, the rate of

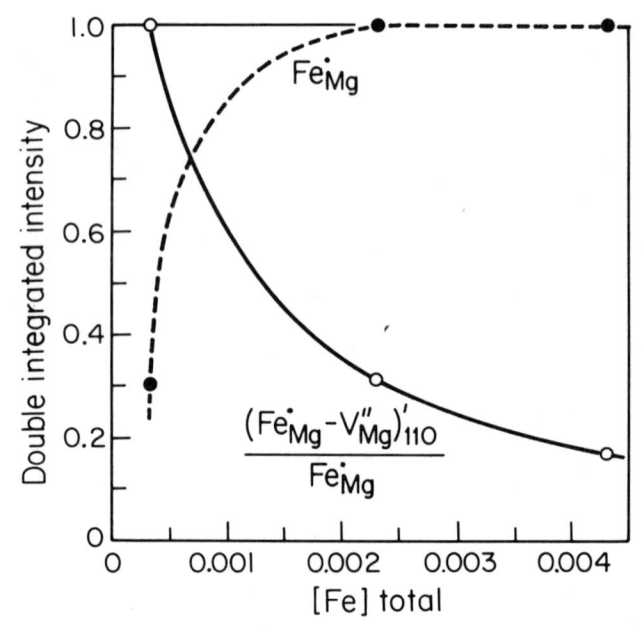

Fig. 10. Plot of relative double-integrated intensity of the $\text{Fe}_{\text{Mg}}^\bullet$ peak and relative double-integrated intensity of the $D\langle 110\rangle$ peak divided by the $\text{Fe}_{\text{Mg}}^\bullet$ peak vs the total iron concentration of quenched samples.

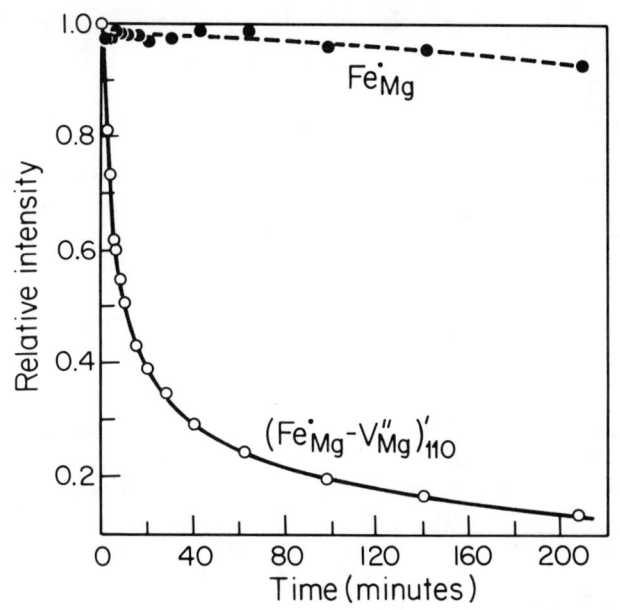

Fig. 11. Plot for isothermal anneal at 450°C in MgO containing 2300 ppm of Fe.

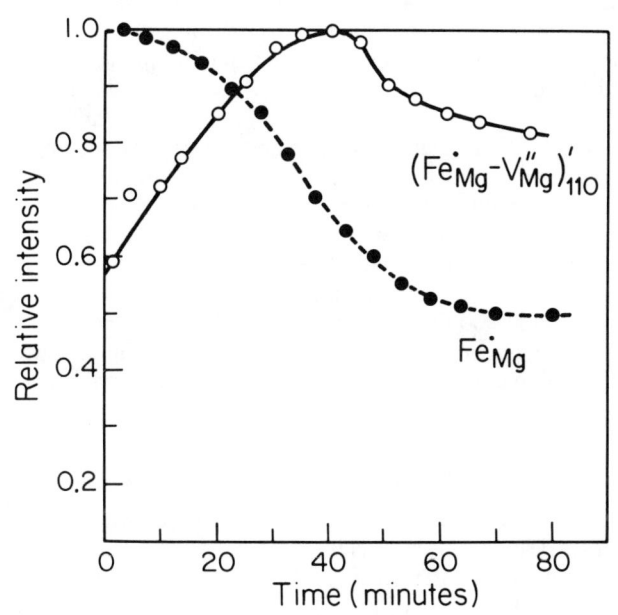

Fig. 12. Plot for isothermal anneal at 600°C in MgO containing 2300 ppm of Fe.

dimer decay is rapid; however, the rate of $Fe_{Mg}^{·}$ decay is negligible. With application of Eq. (9)

$$\frac{0.5[D^0\langle 110,100\rangle]}{[Fe_{Mg}^{·0}]} \simeq 0$$

or $[Fe_{Mg}^{·0}] \gg [\text{dimers}^0]$, confirming the preferential reduction of dimers during the quench at higher solute concentrations. At 600°C, the isolated iron decay is continuous, whereas the dimer concentration first increases and then slowly decreases. The large number of metastable clusters, formed during the quench, dissociate at 600°C to liberate trimers, mobile dimers, and vacancies. The vacancies combine to form simple associates, which then form more stable clusters or precipitates.

Discussion

A quantitative evaluation for the defect structure of Fe-doped MgO is shown in Table II. The column labeled $-E/Fe^{3+}$ gives the total energy change per iron moving from an isolated position into the product. A comparison of the various reactions shows that the energy per $Fe_{Mg}^{·}$ increases monotonically with increasing size of the cluster or associate. A difference of 0.55 eV exists between the energy change per $Fe_{Mg}^{·}$ for the 3-cluster and that for the precipitate, and therefore, several other species of metastable clusters could exist having energies between 1.36 and 1.91 eV/$Fe_{Mg}^{·}$.

A comparison of the present study with the published literature shows excellent agreement with the association energies calculated by using HADES.[13] Magnetic measurements found an average of six Fe^{3+} ions per cluster in 0.9% Fe^{3+} doped in MgO following a water quench,[4] confirming the rapid association and clustering during the quench. The unusual behavior found at 600°C has been related in this study to the dissociation of clusters formed during the cooling process. Below 600°C, defects aggregate to form clusters and their mobility is reduced. Above 600°C, clusters dissociate and the defects become more mobile. Etching techniques combined with microscopy found impurities to precipitate on dislocations and low-angle grain boundaries at 600°C and above.[8,14] This behavior was related to the mechanical strength of MgO,[15,16] where changes in the mechanical behavior were not produced below 600°C.[15] A study of the critical resolved shear stress as a function of temperature in iron-doped MgO showed a maximum hardening produced at 600°C, while higher temperatures resulted in softening due to overaging.[6] The agreement of this study with the literature shows the significance that metastable clusters play in dominating the physical properties of MgO at low temperatures.

Notation

A_i = activity of i.
a_i = EPR signal concentration proportionality constant, where $I_i = a_i[i]$.
$D\langle uvw\rangle$ = dimer oriented in the $\langle uvw\rangle$ direction.
$D\langle 110, 100\rangle$ = a dimer with either a $\langle 110\rangle$ or a $\langle 100\rangle$ orientation.
E_i = energy for the formation of i.
f_i = activity coefficient for defect i.
I_i = peak to peak spectral intensity attributed to defect center i.
K_i = equilibrium constant for the formation of i.
k_b = Boltzmann's constant.
T = absolute temperature.

t = time.
^{0} = a superscript "0" indicating $t = 0$ for kinetic equations or a preexponential value for equilibrium equations.

Acknowledgments

This research was conducted at the Massachusetts Institute of Technology with the support of the U.S. Department of Energy under Contract No. DE-AC02-76ER02390.

---

*W. and C. Spicer, Ltd., Cheltenham, England.
†Analysis was performed by W. Carrera at M.I.T.

References

[1] B. J. Wuensch; p. 211 in Mass Transport Phenomena in Ceramics. Edited by A. R. Cooper and A. H. Heuer. Plenum Press, New York, 1975.
[2] D. R. Sempolinski, "The High Temperature Electrical Properties of Single Crystalline Magnesium-Oxide"; Sc.D. Thesis in Materials Science, M.I.T., 1979.
[3] J. M. Driear, "Modification of the Permittivity of Magnesium-Oxide by Aliovalent Solutes and Dislocation"; Ph.D. Thesis in Materials Science, M.I.T., 1980.
[4] K. N. Woods and M. E. Fine, "Nucleation and Growth of Magnesioferrite in MgO Containing 0.9% Fe^{3+}," *J. Am. Ceram. Soc.*, **52** [4] 186–8 (1969).
[5] W. H. Gourdin, W. D. Kingery, and J. Driear, "The Defect Structure of MgO Containing Tivalent Cation Solutes: The Oxidation-Reduction Behavior of Iron," *J. Mater. Sci.*, **14**, 2074–82 (1979).
[6] R. W. Davidge, "Distribution of Iron Impurity in Single-Crystal Magnesium Oxide and Some Effects of Mechanical Properties," *J. Mater. Sci.*, **2** [4] 339–46 (1967).
[7] T. A. Yager, "Complex Clustering of Trivalent Cationic Solutes in MgO"; Ph.D. Thesis in Materials Science, M.I.T., 1980.
[8] T. A. Yager and W. D. Kingery, "Kinetic Reactions in Quenched MgO: Fe^{3+} Crystals," *J. Am. Ceram. Soc.*, **65** [1] 12–5 (1982).
[9] T. A. Yager and W. D. Kingery, "Laser-Heated High-Temperature EPR Spectroscopy," *Rev. Sci. Instrum.*, **51** [4] 464–6 (1981).
[10] T. A. Yager and W. D. Kingery, "The Equilibrium Defect Structure of Iron-Doped MgO in the Range 600–1200°C," *J. Mater. Sci.*, **16**, 489–94 (1981).
[11] T. A. Yager and W. D. Kingery, "The Kinetics of Clustering Reactions in Iron-Doped MgO," *J. Mater. Sci.*, **16**, 483–8 (1981).
[12] H. S. Roberts and H. E. Merwin, "The System $MgO-FeO-Fe_2O_3$ in Air at One Atmosphere," *Am. J. Sci.*, **21**, 147–57 (1931).
[13] W. H. Gourdin and W. D. Kingery, "The Defect Structure of MgO containing Trivalent Cationic Solutes: Shell Model Calculations," *J. Mater. Sci.*, **14**, 2053–73 (1979).
[14] D. H. Bowen and F. J. P. Clarke, "Impurity Precipitates in Magnesium Oxide," *Philos. Mag.*, **8**, 1257–68 (1963).
[15] R. J. Stokes, "Thermal-Mechanical History and the Strength of Magnesium Oxide Single Crystals: II, Etch Pit and Electron Transmission Studies," *J. Am. Ceram. Soc.*, **49** [1] 39–46 (1966).
[16] B. Ilschner and B. Reppich; Defects and Transport in Oxides. Edited by M. S. Seltzer and R. I. Jaffee. Plenum Press, New York, 1974.

Semiempirical Analysis of the Electronic Processes of *F*-Type Centers in α-Al$_2$O$_3$

SANG-IL CHOI

University of North Carolina
Department of Physics and Astronomy
Chapel Hill, NC 27514

TAKAO TAKEUCHI

State University of New York
Department of Physics
Agricultural and Technical College
Alfred, NY 14802

A semiempirical analysis of the electronic processes of F and F^+ centers in α-Al$_2$O$_3$ has been carried out with the help of our previously reported calculation.

In a recent article,[1] we reported an $x\alpha$ SCF calculation of the F^+ center of α-Al$_2$O$_3$, in which the lattice polarization was not incorporated. In the absence of the lattice polarization, the O$^-$ ion vacancy represents a deep electron trap largely because of the negative electron affinity of the O$^-$ ion, and the F^+ center may be viewed as a hole trapped by the electrons localized in the deep trap.

The important question to be addressed is what happens to this model if the lattice polarization is included? As the O$^-$ ion is removed, leaving one of the O^{2-} electrons behind, the equilibrium positions of the neighboring cations (Al^{3+}) are expected to shift away from the vacancy while the anions (O^{2-}) are expected to move in, accompanied by the polarization of each ion; the new equilibrium is obtained such that the total free energy is minimized. Although the effect of the vacancy on the electronic states of the rest of the crystal is often described by the polarizable ions and the dielectric constants,[2] such a simplified description may not be adequate if the vacancy represents an attractive center for an electron on the neighboring oxygen ions since the electrons on the oxygen ions may be transferred to the vacancy. In this case, at least a self-consistent calculation should be carried out for all the electrons of the neighboring oxygen ions and the electron left behind in the vacancy, in addition to the polarized ion description of the rest of the crystal.

In this article we examine semiempirically the effect of the lattice polarization on the F-type center electronic states in α-Al$_2$O$_3$.

Some Relevant Theoretical Results

In 1963 Kemp and Neeley[3] reported their calculation of the electronic states of the center in MgO, which was based on the LCAO (linear combination of atomic orbitals) method for the wave functions of the F-center electron. The result of including the lattice distortion and ion polarization was to raise by 6.2 eV the F-center energy level (the one-electron orbital energy), which was below the top of the highest valence band. Kemp and Neeley concluded that 6.2 eV was *not* enough to raise the F-center ground-state energy level above the maximum of the valence band.

Wilson and Wood[4,5] reported the calculations of the energies of the two-electron F centers of MgO, CaO, and SrO, in which the lattice distortion and ion polarization is shown to lower the ground-state energy levels by a fraction of 1 eV and the optical ionization energies by approximately 1 eV.

In a recent article, Selme and Pêcheur reported tight-binding model calculations of the oxygen vacancy in $SrTiO_3$.[6] This work shows the lowest energies of the singly occupied and doubly occupied F centers as well as the unoccupied to lie close to the minimum (bottom) of the conduction band instead of the valence band. In their work, the oxygen vacancy is represented by the *infinite* potential energy which automatically excludes electrons from the vacancy, leading to high energy levels. In either MgO or α-Al_2O_3, the O^{2-} vacancy is an attractive site for the electrons on the neighboring oxygen ions and certainly cannot be represented by an infinite potential energy in any realistic theoretical work of MgO and α-Al_2O_3.

The carrying out of a reliable self-consistent calculation involving all the relevant electrons and the ions seems to be an enormous task at this time. A reasonable semiempirical theory could be useful and is presented below.

Effects of the Lattice Distortion and Polarization

Our previous $x\alpha$ SCF calculation[1] without the lattice distortion and polarization effect showed that a doubly occupied energy level 6 eV below the top of the highest valence band has the electrons strongly localized in the vacancy, and also a singly occupied level is at the top of the valence band with the electron density concentrated in the vacancy and on its neighboring oxygen ions. These two levels are separated by approximately 6 eV, as shown in Fig. 1. Before we proceed to estimate the effects of the lattice polarization, we should emphasize that our above mentioned calculation is a self-consistent calculation for all electrons of 12 nearest-neighbor oxygen ions — except the 1s electrons — and one extra electron.

As the ions are allowed to adjust to the "oxygen ion vacancy," the oxygen ions are expected to shift their equilibrium positions toward the vacancy while the aluminum ions move away. Unlike the case in which a true O^{2-} vacancy is created and is kept free of electrons, we have more than one electron in the vacancy and, as a result, expect a smaller lattice relaxation and polarization energies.

The experimental data on the F-type centers in α-Al_2O_3 are summarized as follows: (i) the F center has a broad optical absorption peak at 6.1 eV with the width of approximately 1 eV,[7,8] and an emission band is excited by this absorption at 3.0 eV of a long lifetime ($\sim$36 ms); (ii) the F^+ center has the absorption peaks at 4.8 and 5.4 eV, which excite an emission band at 3.8 eV with a short lifetime;[7,8] (iii) the F-center absorption at 6.1 eV is accompanied by photoconductivity even at 10 K as well as at room temperature;[9,10] (iv) there is no conclusive evidence of the F^+-center photoconductivity; (v) optical absorption in the 6.1-eV band can bleach F-center absorption and increase the F^+-center bands at 4.8 and 5.4 eV;[9,10] (vi) photons of 4.8 and 5.4 eV bleach the F^+-center absorptions and increase the F-center absorption;[9,10] (vii) photons of energy as low as 4 eV bleach the F^+-center absorptions, but to a lesser degree;[9] (viii) the photobleaching processes are sample dependent;[10] (ix) the optical band gap of a pure crystal is approximately 10 eV.[11]

From (i) one can conclude that the ground state of the F center is at least 6.5 eV below the conduction band — i.e., the optical ionization energy of the F

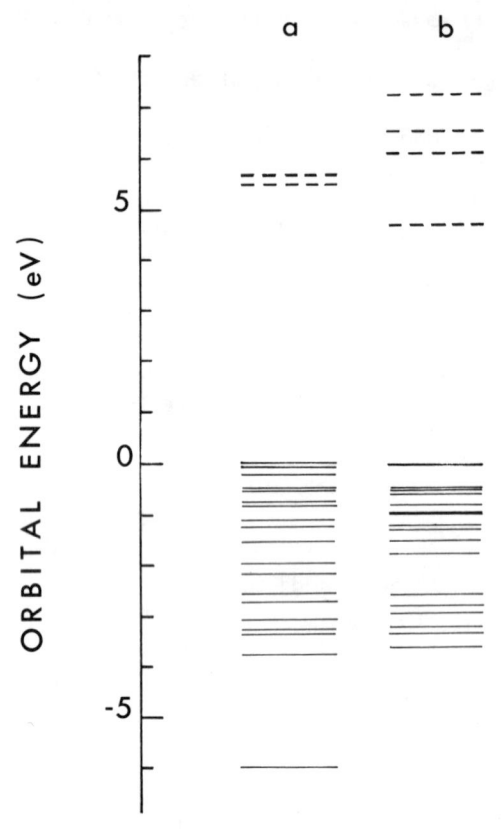

Fig. 1. One-electron energy levels (orbital energies) of the F^+ center by the $x\alpha$ SCF cluster model calculation; a and b denote the symmetry of the orbitals with respect to the 2-fold axis; dashed lines are unoccupied orbitals. The highest occupied a orbital is singly occupied while all others are doubly occupied.

center is at least 6.5 eV — and (iii) shows that the F center excited by the 6.1-eV band light autoionizes, as depicted in Fig. 2, indicating that the lattice relaxation lowers the ionization potential by at least 1 eV and that the thermal ionization energy of the F center is at least 5.5 eV if the 1-eV reduction is accepted; the calculations of Wilson and Wood on some alkaline earth oxides suggest that 1 eV is a reasonable estimate.[4,5]

For the F^+ center, (ii) and (iv) lead us to believe that the optical ionization potential is greater than 6 eV and that no autoionization occurs from the two excited states, 4.8 and 5.4 eV above the ground state. This is rather surprising if one expected that the relaxation and polarization, due to the extra charge after the removal of one electron from the F^+ center, would reduce the ionization energy even further to bring the F^+-center ground state close to the conduction band. One reasonable explanation for the seemingly large ionization potential, optical and thermal, of the F^+ center is that the bare vacancy is not created when an electron is removed from the F^+ center; that is, the ionization of the F^+ center does not lead to the O^{2-} vacancy devoid of electrons.

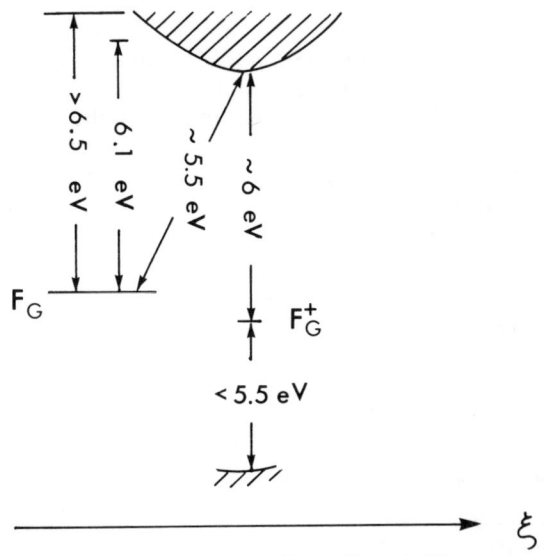

Fig. 2. Ground-state energy levels of the F and F^+ centers relative to the conduction band and the highest valence band; the upper shaded region is the conduction band, and the lower one indicates the highest valence band, with lattice distortion represented by ξ.

This picture can be easily understood from the model of Choi and Takeuchi.[1] In the absence of the lattice relaxation and the long range polarization, the electronic structure of the F^+ center consists of a doubly occupied deep trap, 6 eV below the top of the highest valence band, and a localized hole near the top of the band. As the lattice relaxation and polarization are turned on, these energy levels are expected to shift upward: the singly occupied level (the hole) shifts a few electron volts into the gap between the valence band and the conduction band, while the deep trap level moves into the valence band to become a resonance state. The deep trap electrons are the F^+-center electrons in the sense that these are mostly the electrons to be found in the vacancy. Elevating the electron in the singly occupied level to the conduction band still leaves those resonance state electrons in the vacancy and two holes bound to the vacancy.

Unlike the F center, optical absorption due to the transition of an electron from the valence band to the singly occupied state is possible for the F^+ center. Such a transition would create a hole in the valence band, which may move away, leaving an F center behind; this provides a simple mechanism for the F^+-to-F photobleaching mentioned in (iv) above. The energy needed to excite a valence band electron to the singly occupied level is difficult to estimate, but a rough estimate places it in the energy range known to be effective for the photobleaching. This number is obtained as the difference between the energy band gap and the vertical electron affinity of the F^+ center, i.e., 10 eV minus the vertical electron affinity of F^+. The vertical electron affinity of the F^+ center is equal in magnitude to the ionization energy of the F center with a proper adjustment for the lattice relaxation and is estimated to be *greater* than 4.5 eV. Then the minimum photon energy to excite a valence band electron and convert the F^+ center to the F center is *less* than 5.5 eV.

Summary

The available experimental data on α-Al$_2$O$_3$ strongly indicate that removing an electron from an F^+ center does not lead to a bare O^{2-} vacancy. This unusual phenomenon can be given a reasonable explanation on the basis of our previous calculation modified by the effect of the lattice relaxation and polarization. Photo-excitation of an electron from the valence band to the singly occupied energy level can convert the F^+ center to the F center; a rough estimate of the photon energy is given.

References

[1] Sang-il Choi and T. Takeuchi, *Phys. Rev. Lett.,* **50**, 1474 (1983).
[2] For example, J. S. Markham; F-Centers in Alkali Halides. Academic Press, New York, 1966.
[3] J. C. Kemp and V. I. Neeley, *Phys. Rev.,* **132**, 215 (1963).
[4] T. M. Wilson and R. F. Wood, *Phys. Rev. B: Condens. Matter,* **16**, 4594 (1977).
[5] R. F. Wood and T. M. Wilson, *Solid State Commun.,* **16**, 545 (1975); *Phys. Rev. B: Condens. Matter,* **15**, 3700 (1977).
[6] M. O. Selme and P. Pêcheur, *J. Phys. C: Solid State Phys.,* **16**, 2559 (1983).
[7] K. H. Lee and J. H. Crawford, Jr., *Phys. Rev. B: Condens. Matter,* **19**, 3217 (1979).
[8] B. D. Evans, *Bull. Am. Phys. Soc.,* **23**, 412 (1978).
[9] B. G. Draeger and G. P. Summers, *Phys. Rev. B: Condens. Matter,* **19**, 1172 (1979).
[10] B. J. Jeffries, J. D. Brewers, and G. P. Summers, *Phys. Rev. B: Condens. Matter,* **24**, 6074 (1981).
[11] E. T. Arakawa and M. W. Williams, *J. Phys. Chem. Solids,* **29**, 735 (1968).

Defects Produced by Shock Conditioning: An Overview

R. F. Davis, Y. Horie, R. O. Scattergood, and H. Palmour III
North Carolina State University
Materials Engineering and Civil Engineering Departments
Raleigh, NC 27650

Shock waves in crystalline solids cause extensive plastic deformation and the creation of a variety of defects. Current theoretical descriptions of these processes are restricted to monolithic solids and some less extensive analyses of extended porous materials. Though shock activation of ceramic powders has been shown to enhance sinterability, there are few studies that treat analytically the interaction of shock waves with powder assemblages and the resultant generation of defects. This paper reviews the currently available knowledge of shock-induced generation of defects in powders, examines deficiencies in that knowledge, and discusses these topics in terms of shock conditioning (activation) of oxide ceramic particulates like Al_2O_3.

The essence of materials science and engineering is the close relationship between the processing of materials and their properties. In recent years the strict requirements for the designs and applications of new and difficult-to-fabricate products from unconventional materials have produced a corresponding need to develop and employ unusual processing techniques, including dynamic compaction and shock conditioning of powders.

In dynamic compaction, a high-pressure shock wave encounters a pre-compacted powder mass, and if the flow stress of the powder mass is exceeded, densification occurs as the compressive shock wave progresses through the mass.[1] Since large plastic strains are required to produce a high degree of densification, localized defect generation and associated plastic flow mechanisms at the shock front must play a critical role in the compaction process. In many cases, however, direct dynamic compaction does not produce a strong, well-bonded material, and a combination of shock and conventional processing must be used. Shock conditioning (or shock activation) of a powder requires the generation within powder grains of a very high defect density via the passage of a shock wave through a powder compact. The normally friable, shock-compacted mass is then milled and conventionally sintered or hot-pressed. One of the aims of shock conditioning is to improve the sintering kinetics and hot compaction processes through the effects on these of a high defect density. In this context, particle size distribution (morphological) changes due to shock conditioning cannot be overlooked. While there exists a large body of experimental information and some theoretical models for the shock-induced defect generation mechanisms in homogeneous materials, the mechanisms in powder materials are less well documented and virtually no basic understanding currently exists.

On a macroscopic scale, the shock process comprises a precise, orderly, and comprehensible sequence of events. However, this sequence evolves rapidly. The very intense pressure pulse of the shock itself may rise to the full amplitude of 1 to >10 GPa in a time not exceeding a few nanoseconds, and a typical experiment runs its entire course in only about 1 μs.

If one seeks fundamental understanding of the basic mechanisms, wants to generalize such processes over a wide range of material choices, or attempts to produce articles of a given material in reliable, useful forms and shapes, it will be necessary to supplement the externally determined macroscale knowledge of the shock event itself with some very detailed microstructural characterizations and predictions. The first systematic investigation of the substructural changes induced by the passage of shock waves is described in the classic paper by Smith.[2] During the past 20 years, this activity has steadily continued, especially in metallurgy but also in ceramics as a consequence of the growing realization that the shock-wave parameters play an important role in the determination of these substructures and associated mechanical properties. Several review papers[1,3-7] have also appeared during that time.

This chapter briefly reviews the basic theory of the interaction between shock waves and materials as well as several models for dislocation generation by shock waves. The various defects that have been observed in shocked materials are described.

Fundamental Background

Shock-Wave Phenomena

Shock waves in solids are created by, e.g., a sudden release of energy from detonating explosives or high-speed impact of a solid body on another.

They are finite amplitude disturbances, propagating at a supersonic speed. Across a foward-facing front there exists a near-discontinuous change of the physical quantities like pressure, density, and temperature. The word "shock" often refers only to this discontinuous front, which is related to the fact that at high pressures the velocity of sound increases with increasing pressure and that this steepening is kept in balance by dissipative mechanisms like viscosity.

In this article we shall be mostly concerned with the nature of the shock transition, but we need to be aware that in practice the relief from the shocked state and the interactions of shock waves with one another or materials boundaries are equally important in the production and annihilation of defects.

A mathematical description of the shock wave or the shock transition is customarily restricted to one-dimensional waves. Two- and three-dimensional problems received considerably less attention because of the complexity and difficulty of their analysis and experiments.

Let there be a plane-steady shock front, L, moving through a tube of solid material with unit cross-sectional area indicated by $ABCD$ in Fig. 1. If the materials behave normally in compression, i.e., if the compressibility decreases with increasing pressure, the shock front will be essentially a discontinuity in stress, density, material velocity, and internal energy, which travels at supersonic velocity with respect to the material ahead.

The conservation of mass is expressed by observing that in a short time, dt, the shock front with a velocity of U advances a distance, Udt, while the matter that was initially at the plane L has advanced at a velocity of u_1 only a distance $u_1 dt$. Therefore, the material, $\rho_0 U dt$, encompassed by the shock during dt has been compressed into the space, $(U - u_1) dt$, at density ρ_1. If mass is to be conserved, we must have

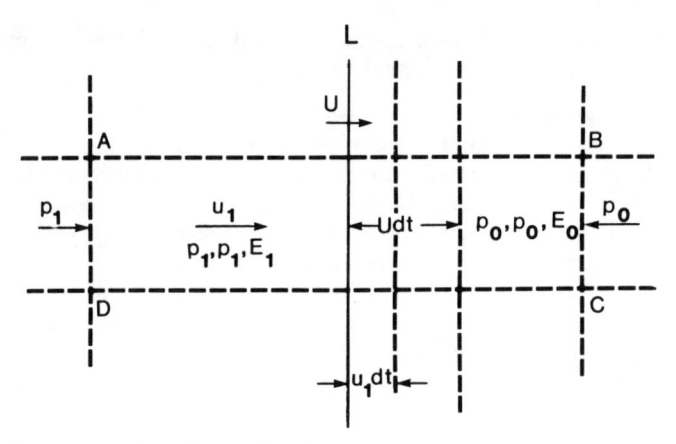

Fig. 1. Progress of a plane shock wave.

Fig. 2. Receding shock wave (reproduced with permission from: Courant and Friedrichs, Supersonic Flow and Shock Waves, Interscience, New York, 1948).

$$\rho_0 U = \rho_1 (U - u_1) \tag{1}$$

The expression of Newton's second law is obtained by noting that the force acting on matter contained in the flow tube $ABCD$ of unit area is, from the left, across the plane AD in the shocked state, p_1, and, from the right, p_0. The net force is $p_1 - p_0$ to the right. In the time interval, dt, the mass of material encompassed by the shock is $\rho_0 U dt$, and this is all accelerated to the velocity u_1. The increase in momentum of the system is then $\rho_0 U u_1 dt$, and the rate of change of momentum is found by dividing by the time interval, dt; thus

$$p_1 - p_0 = \rho_0 U u_1 \tag{2}$$

The expression for conservation of energy in the shock transition is derived in a similar way. The surface AD moves a distance, $u_1 dt$, during the time, dt, so the work done on mass contained in the unit area flow tube is $p_1 u_1 dt$. This must be equal to the sum of the increases of kinetic and internal energy of the system. The increase in kinetic energy is $\frac{1}{2}(\rho_0 U dt) u_1^2$, and the increase in internal energy is $(E_1 - E_0)\rho_0 U dt$, where E_1 and E_0 are internal energies per unit mass of shocked and unshocked material, respectively. Equating the work done to the gain in energy produces the energy conservation equation:

$$p_1 u_1 = \frac{1}{2}\rho_0 U u_1^2 + \rho_0 U(E_1 - E_0) \tag{3}$$

By combining Eqs. (1) and (2) and setting $1/\rho = V$, where V is the volume per unit mass, i.e., the specific volume, one obtains

$$U = V_0 \left(\frac{p_1 - p_0}{V_0 - V_1}\right)^{1/2} \tag{4}$$

and

$$u_1 = [(p_1 - p_0)(V_0 - V_1)]^{1/2} \tag{5}$$

The velocities may be eliminated from the energy equation to give the Rankine–Hugoniot relation

$$E_1 - E_0 = \frac{1}{2}(V_0 - V_1)(p_1 + p_0) \tag{6}$$

Equations (1), (2), and either (3) or (6) must be satisfied by material parameters on the two sides of the shock front. They can be shown to hold for any geometry, independent of the curvature of the shock front.[8] An illustration of the shock-wave concept is given in Fig. 2.

If the specific internal energy of a material is regarded as a function of its pressure and volume, then Eq. (6) represents the locus of all (p_1, V_1) states attainable by propagating a steady shock wave into a material having an initial state (p_0, V_0).* This locus is defined as the Rankine–Hugoniot curve centered about (p_0, V_0) as shown in Fig. 3.

The shock transition is a dissipative process, giving rise to entropy production. This entropy production can be shown to be third order in the compression

$$S_1 S_0 = -\frac{1}{12 T_0}\left(\frac{\delta^2 p}{\delta V^2}\right)_{S_0, V_0}(V_0 - V_1)^3 + \ldots \tag{7}$$

where S_0 is the entropy of the initial state. The corresponding temperature change can be calculated by use of the equation[9]

$$\frac{dT}{dV} = -T\frac{\gamma}{V} + \left\{\frac{dp}{dV}(V_0 - V) + (p - p_0)\right\}\frac{1}{2C_v} \tag{8}$$

where γ is the Grüneisen constant and C_v the specific heat. For a small compression, it can be approximated by the isentropic temperature rise that

$$T = T_0 \exp\left\{-\int_{V_0}^{V}\frac{\gamma}{V}dV\right\} \tag{9}$$

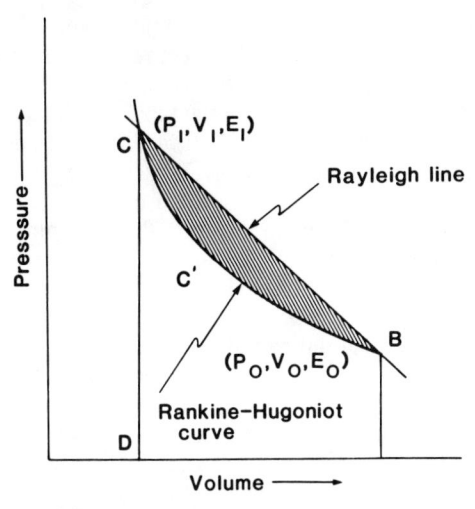

Fig. 3. Schematic of Rankine–Hugoniot curve.

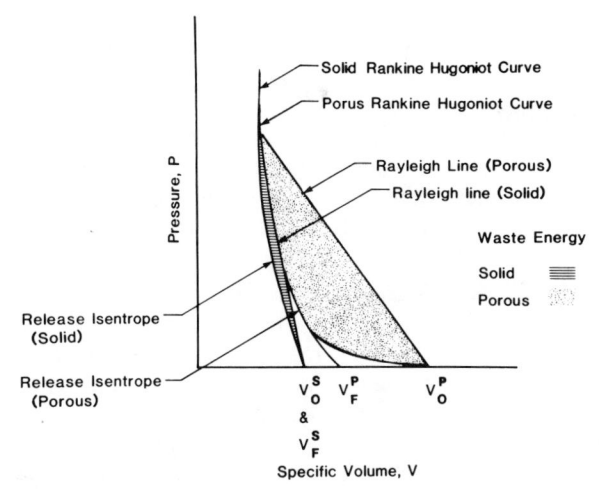

Fig. 4. Comparison of waste energies in solid vs porous materials.

In the preceding development it was assumed that for a given shock pressure only a single shock wave would propagate into an undisturbed medium. The shear strength had been neglected, and the material was treated like a fluid. The concept of a single "hydrostatic shock pressure" is not correct for materials that have a finite yield strength. It is a reasonable approximation only when the stresses are very much higher than the yield strength. Hence, the pressure, p_1, in the Eqs. (2–6) should properly be relaced by σ_x, the stress component acting normal to the wave front. Except by inference from mechanical equations of state, the stress components parallel to the wave front, σ_y and σ_z, are not known.

The differences between the shock behavior of a porous or powder material and its solid counterpart are numerous. Microscopically, the shock front, being perturbed by the pore and particle structure, is no longer uniform or steady. But if shock-wave measurements are done with specimens that are large compared with its particle and pore sizes and the steady state is maintained behind the shock front, the jump Eqs. (1), (2), and (3) or (6) will retain their usual meaning. However, the dissipative mechanisms such as frictional surface heating and particle fragmentation in distended materials may easily invalidate the energy conservation equation within the time scale of currently practiced measurement techniques.

For a steady shock in a distended material, the Hugoniot curve as defined by Eq. (6) with V_0^p replacing V_0: E_0 is unchanged if the energy associated with the porosity is ignored (usual assumption). This equation defines a Hugoniot curve on the high-pressure side of the Hugoniot curve for the solid counterpart. Upon passage of the shock, the material expands along a release isentrope to the final volume V_F^p at ambient pressure. The ratio V_F^p/V_0^p is a measure of the compaction.

An illustrative description of the compaction of a porous material, in terms of the energy absorbed, is shown schematically in Fig. 4. The major point to be observed is that waste energy represented by the area between the Rayleigh line and the release isentrope is considerably greater for the porous material as compared to that for a solid of the same material. This increase in waste energy in the powder material, as compared to a solid of the same material, accounts for the significant temperature rise in shock-compacted powders.

The requirement for providing a preliminary basis for the investigation of the behavior of powders under shock-loading conditions is the information on the equations of state for the powder material. However, it should be noted that, although adequate models may be developed in terms of the temperature rise and the pressures developed, these may not prove to be the most important considerations for the application of shock-wave compression to powdered compacts. Indeed, it may not be the quantity of energy but its distribution, as influenced by particle size, initial green density, etc., that exerts the controlling influence on the degree of compaction and its resultant properties.

Structural Changes in Shock-Loaded Samples
Materials Responses

The macroscopic (continuum) approach to shock-wave treatment of a solid is expressed in the three jump conditions involving p, ρ, and E. The material itself is treated as being homogeneous and possessing continuum properties, whereas even in solids, the actual processes by which real materials are able to absorb energy are known to be discrete, strongly orientation dependent, and inhomogeneous. In particulate masses, though some statistical averaging will occur, the elementary processes on a grain-by-grain basis must be locally discrete and quite inhomogeneous.

If one is to achieve a basic understanding of the effects of the shock compaction and/or conditioning on changes in the physical properties of a bulk solid or powdered compact, it will be necessary to conduct as detailed an analysis as possible on the mechanisms of materials responses to shock loading.

The present level of our understanding about the shock response of a material is still incomplete, but in solids considerable progress has been achieved in recent years.

Initially most solids behave elastically. Then the maximum shear stress in a plane shock is given by

$$\tau = \frac{3(1-2\nu)}{2(1+\nu)}p = \frac{1}{2}(\sigma_x - \sigma_y) = \frac{1}{2}(\sigma_x - \sigma_z) \qquad (10)$$

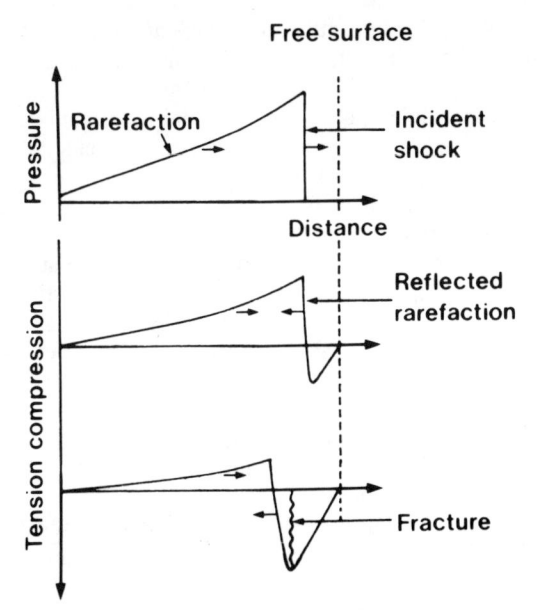

Fig. 5. Successive stages in the reflection of a finite shock pulse at a free surface resulting in Hopkinson fracture.

where ν is the Poisson's ratio. This high shear stress at the compression front causes initial inelastic phenomena such as plastic flow and fracture. Failures in the subsequent unloading are also expected.

In the case of distended materials, they will be further complicated by factors like particle and/or pore geometries and surface phenomena. A characteristic example is that of Hopkinson fracture caused by the reflection of a shock at a particle-free surface, as shown in Fig. 5. Other particulate-related phenomena are discussed in a recent paper by Palmour et al.[1] Some notable examples are (1) fracture and surface heating (or even melting) via particle–particle collision and sliding between particles, (2) stress concentration around particle contact points, and (3) the perturbation of a shock front. Very few papers have addressed such aspects of shock-wave propagation in distended materials. Clearly, in order to develop a physical description of the sequence of events that can occur in shock loading, one must employ various idealized, but plausible models. Inevitably, the prediction of structural changes in shock-loaded materials and their correlation with experiments will require a combination of approximate analytical solutions and a numerical simulation of shock-wave propagation by a technique such as finite difference codes.

Dislocation Generation

The dislocation substructures generated by shock loading depend on a number of shock-wave and material parameters. Among the former parameters, pressure is by far the most important. As the pressure is increased, so is the dislocation density. Murr and Kuhlmann-Wilsdorf[10] have shown that this density varies as the square root of pressure. This dependence breaks down at very high pressures (>100 GPa) as a result of shock heating.

The duration of the loading pulse does appear to have some effect on the dislocation density,[11,12] but the data are conflicting.[13] However, an increased duration does allow more time for dislocation reorganization into structures that more closely resemble those obtained after conventional deformation.

In the past 25 years, several models have been presented to account for the effects of the interaction of a shock wave with a crystalline lattice. The initial model was that of Smith,[14] which described the formation of networks of dislocations at the shock front to alleviate the elastic strain at the shock interface. This model, however, required that the dislocation network move at supersonic velocities with the progressing shock wave. Hornbogen[15] modified the Smith model by introducing the idea of the generation of loops by the shock forces, the edge components of which moved with the shock front, leaving screw segments behind in the lattice. The Hornbogen model again requires dislocation motion at supersonic velocities, and while the substructure in Fe at low shock pressures is indeed made up of straight screw dislocation arrays, consistent with the model's predictions, there are a wide variety of shock-induced dislocation substructures that cannot be reconciled with the model.[6]

More recently, Meyers[16] has presented a model that revises the original Smith concept. Meyers envisions the homogeneous nucleation of a dislocation network to relieve the elastic strain at the shock interface, but once the dislocation wall is formed, the shock wave moves on, leaving the dislocation wall in the lattice. The progressing shock wave again rebuilds the local stress at the shock interface such that another dislocation wall is nucleated. In this manner, the shock wave creates an intermittent succession of dislocation walls as it progresses through the lattice.

The essentially different and perhaps more physically reasonable feature of this model compared to that of the earlier ones is that supersonic dislocation velocities are not required. Intermittent generation of dislocation walls periodically relieves the buildup of the shock-front stresses, and dislocation motion is confined to the transient region behind the shock front where it can progress at subsonic velocities. In actuality, the mechanism envisaged by Meyers must be a rather complex one, involving dislocation generation and the subsequent dynamics of propagation and interaction of the dislocation stresses with the shock-front stresses, thereby producing a partial, intermittent cancellation of the latter.

A highly simplified schematic of this model is shown in Fig. 6. As the shock wave penetrates into the material, the initially cubic lattice is effectively distorted into a monoclinic lattice. Figure 6(*b*) shows the wave as the front coincides with the first dislocation interface. In Fig. 6(*c*), the front has moved ahead of the interface; additional layers of dislocation are shown in Fig. 6(*d*).

Recent experimental results[17,18] indicate that the rarefaction part of the wave plays only a minor role in dislocation generation. The principal reason is that this portion of the wave enters into a material that is already highly dislocated. This is consistent with Meyers' predictions[19] "that, if a prestrained material is shock loaded, part of the stresses at the shock front could be accommodated by existing dislocations; in this case the number of new dislocations that would be generated at the front would be reduced. The same argument can be extended to the rarefaction portion of the wave; it can accommodate the stresses by the movement of the existing dislocations." But here again there are experiments that indicate that "effects relating the stress state and strength of metals during shock compression occur that fall outside of the behavior in normal plastic processes."[20]

The various models have in common the fact that dislocation generation is assumed to occur in a homogeneous solid. The particulate nature of the medium

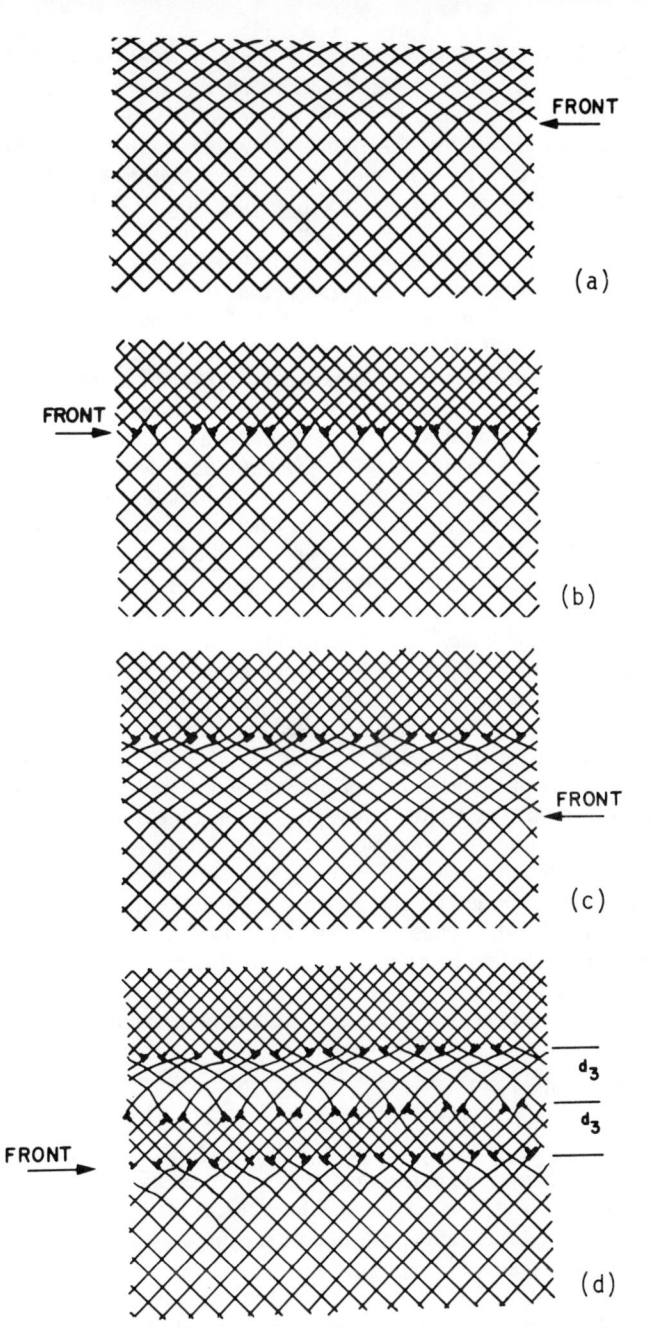

Fig. 6. Progress of a shock front according to a model by Meyers (from Ref. 6).

in the shock compaction or shock conditioning of powders is ignored. However, this factor may have a significant influence on the shock wave itself, as was discussed in an earlier section, and more importantly on the mechanisms of dislocation and defect generation. X-ray diffraction measurements on shocked MgO powders by Heckel and Youngblood[21] have shown that under similar conditions more damage occurs in powder material than in a single crystal. Furthermore, as powder size becomes finer, measurements by Prümmer and Ziegler[22] on Al_2O_3 have shown that higher degrees of internal residual strain can be retained.

Dislocation generation mechanisms such as the model proposed by Meyers must be operative in powder materials as well as in homogeneous solids because the generation is a result of incompatibilities at the shock front itself. Additional generation mechanisms may occur in powders, however, and the current authors believe that particle—particle contacts can play an important role in dislocation generation and fracture processes. As the shock wave passes through the powder mass, particles will accelerate and undergo a complicated variety of mutual sliding contact and impact events. While there is no available information on these events for powders per se, the processes that occur when a sharp indentor contacts the flat surface of a brittle solid have been extensively studied and are well documented.[23] An analog of these events could certainly occur during the shocking of powders, for example, when a sharp asperity on one particle meets the flank of a second particle. Contact loads under a static or sliding indentor produce localized plastic flow in brittle materials. Furthermore, particle contacts under impact conditions, either direct or oblique (with a sliding component), also produce localized plastic flow,[24] and TEM observations have disclosed extensive dislocation generation and even local melting at the impact site.[25,26] Plastic zone formation due to such contact events also leads to the initiation and propagation of radial and lateral cracks.[23]

In the case of powder grains, it is very likely that the fracture events will cause shock-induced comminution of the powder mass. The various processes mentioned are shown schematically in Fig. 7. Similar contact events will of course play a critical role in particle bonding and dynamic compaction processes, with intense, localized thermal heating caused by energy deposition mechanisms related to impact and sliding friction being a principal factor here. The effect of the particle medium on dislocation and defect damage production remains an outstanding problem, and it represents an area where further experimental and theoretical work is needed.

Point Defect Generation

All experimental investigations conducted to date[27–34] indicate that shock loading produces a high density of point defects. Direct quantitative evidence of vacancies and vacancy-type defects was first observed by Murr et al.[34] in shock-loaded Mo. Although very small vacancy loops accounted for only a small portion of the shock-induced vacancies, the majority exist as single vacancies or small clusters difficult to resolve by conventional TEM.

The primary source of point defects is the nonconservative motion of jogs generated by the intersection of screw or mixed dislocations. Figure 8 shows the direction of motion of dislocations under the effect of the residual shear stresses. As they move, these stresses decrease; however, in the process, the dislocations intersect each other and generate jogs. The nonconservative motion of these jogs produces strings of either vacancies or interstitials. These can also occur as dislocation loops when observed in the TEM. The subject has been treated in detail by Hirth and Lothe[35] and by Cohen and Weertman.[36] The significant conclusion of

(a)

(b)

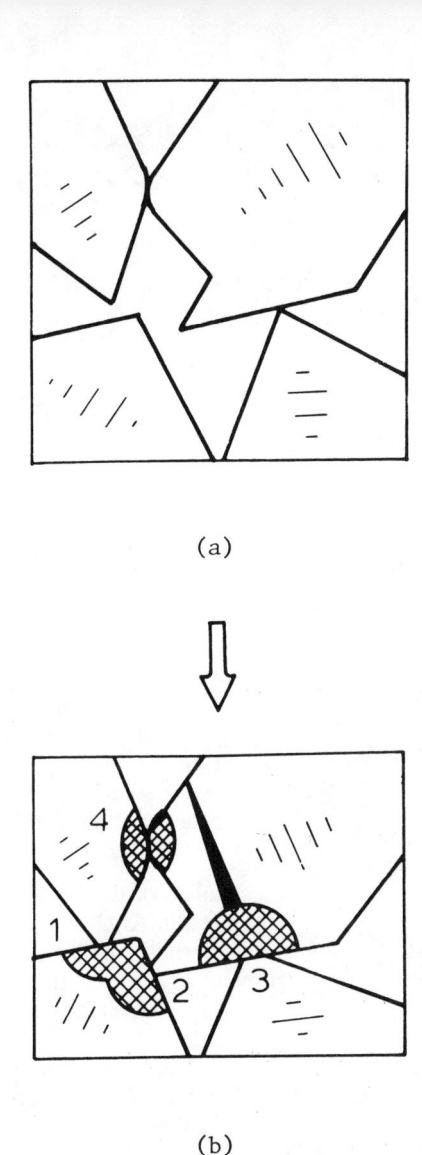

Fig. 7. Schematic diagram of the various contact events that might occur between powder particles during shock loading. The unloaded state in (a) becomes the loaded state in (b) due to the passage of a shock wave. Sections 1 and 2 are events where a sharp asperity on one particle causes a plastic zone (cross-hatched) to be produced upon impact on the flank of an adjacent particle. A similar event occurs in section 3 for the case where the sharp asperity is initially in contact with the other particle. This plastic zone has also caused a crack (shaded) to form in the upper grain. The event at section 4 is primarily a sliding contact between rounded asperities, presumably causing less plastic flow. Some deformation will also occur under the tips of the sharp asperities, but for clarity this is not shown. In all cases, the plastic zone formation will be accompanied by localized heating.

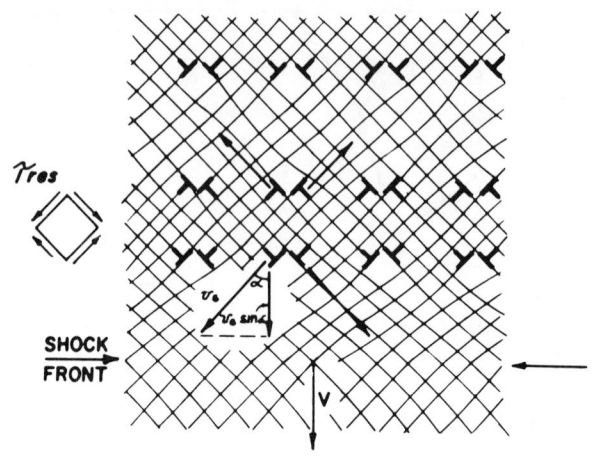

Fig. 8. Movement of dislocations generated at the shock front (from Ref. 32).

this theory is that, as the velocity of the dislocation line increases, the effectiveness of jogs as barriers to dislocation motion decreases. They are effectively dragged along by the dislocation, generating a large number of point defects in the process. At the high shear–stress levels that can exist during shock loading, dislocation velocities can be very high, even if not supersonic, and a large density of point defects as one component in the damage substructure can be expected.

Deformation Twinning

Twinning is a very common mechanism of deformation in metals subjected to shock loading. It has also been observed in Al_2O_3 under similar conditions, as described below. It is apparent that materials which do not easily twin under conventional deformation at low temperatures can be made to do so under shock loading. The applied pressure, pulse duration, stacking fault energy, existing substructure, crystallographic orientation, and grain size all play a role in creating a susceptibility to twinning. The crystallographic and morphological features of shock-induced twins do not differ greatly from those formed by conventional means. A decrease in temperature or an increase in strain rate tends to favor twinning over slip by dislocation motion.[37] However, the mechanism(s) responsible for the nucleation and growth of these twins continues to be a point of debate. The motion of Shockley partials produced at Cottrell–Lomer locks;[36] homogeneous nucleation of dislocations;[35] and the cooperative movement of dislocations on parallel planes[6] have been proposed for fcc metals.

Phase Transformations

The occurrence of a phase transformation in metals during the passage of a shock wave has been noted by many investigators. A very detailed review has been presented by Duvall and Graham;[38] thus, this subject will not be covered in this paper.

Experimental Observations of the Effects of Explosive Shocking of Ceramics

The majority of the *post explosionem* research conducted on ceramic materials has been concerned with residual stress buildup during compaction or shock

(a)

(b)

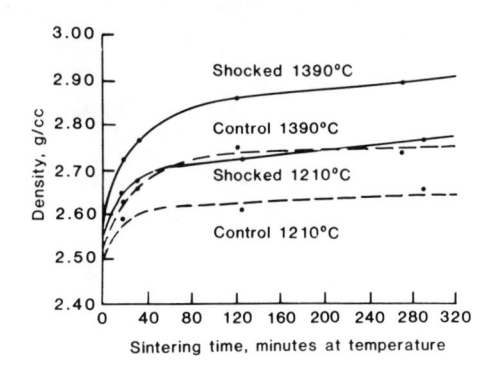

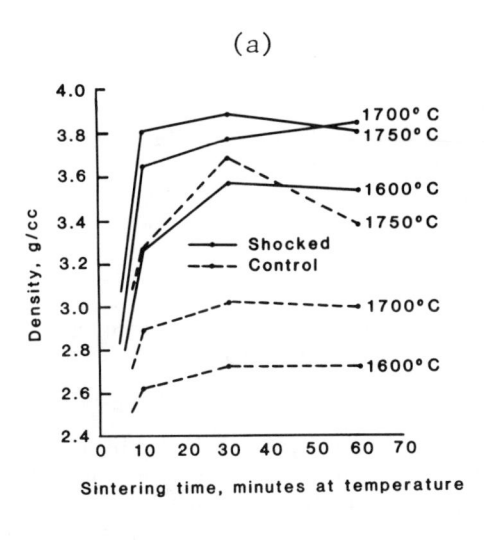

Fig. 9. Effects of shock conditioning of (a) MgO and (b) Al$_2$O$_3$ on subsequent densification (from Ref. 39).

conditioning. Unfortunately, very little knowledge concerning the defect substructure of ceramics after shock loading exists at the present time. A brief review of this research is presented below.

Bergmann and Barrington[39] have investigated the effects of explosive shocking on a range of materials including MgO and Al$_2$O$_3$. Fourier analysis of the line broadening in the X-ray diffraction pattern of explosively treated MgO and Al$_2$O$_3$ powders showed that the broadening in these two materials was caused primarily by strain and to a lesser extent by the reduction in crystallite size. This was in agreement with the small increase in the surface area of the materials after shocking. Exactly the opposite effect was noted in SiC.

The final product of a shock conditioned and subsequently sintered material may be similar to the material that is only sintered or hot-pressed. However, since one can often get a substantial boost in densification from the release of the

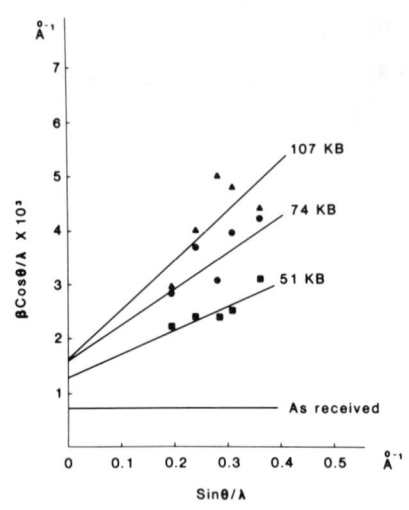

Fig. 10. Hall–Williamson plots of X-ray line broadening for explosively shocked alumina (after Kim, Ref. 42). With increasing levels of nominal shock pressure, the calculated residual strains are 0.116, 0.177, and 0.231%, respectively.

considerable internal energy imparted during shock conditioning, the temperature necessary for densification may be much reduced with resultant reduction in average final grain size at equal or higher densities and with improved mechanical and other properties.

The MgO and Al_2O_3 powders shock conditioned by Bergmann and Barrington[39] were found to be unusually responsive to sintering operations. When such powders were cold-pressed by conventional means into compacts and then fired, the resulting sintered material was considerably denser than controls made from unshocked material (see Fig. 9). The substantially higher sintering activity of shocked ceramic powders is believed to be related to the introduction of larger numbers of defects into the crystal lattice by the shock treatment. Since crystallite reduction rather than lattice strain was produced in SiC and Si evaporation occurred at the high sintering temperatures (1900–2500°C), only neck growth without densification was observed in SiC. (Recent sintering studies of shocked ceramic powders are discussed in a companion experimental paper[40] in this volume.)

Heckel and Youngblood[21] have also conducted full Fourier analysis of the total line broadening of the X-ray diffraction patterns from explosively shocked MgO and Al_2O_3 powders. Their results as well as TEM observations of the size of the shocked and unshocked powder by Greenham and Richards[41] again confirm that the line broadening in these two materials was caused primarily by lattice strain. In addition, the values of the microstrain were used to calculate the approximate postshocked dislocation densities of 3×10^{11} cm^{-2} for both materials.

Other more recent investigators, including Prümmer and Ziegler,[22] Kim,[42] Morosin et al.,[43] and Horie et al.,[44] have employed a variety of X-ray methods in obtaining and analyzing diffraction data to estimate both residual strain and crystallite (subgrain) size for shocked ceramic powders. The results obtained, though subject to experimental scatter (as well as much variability regarding the specifics

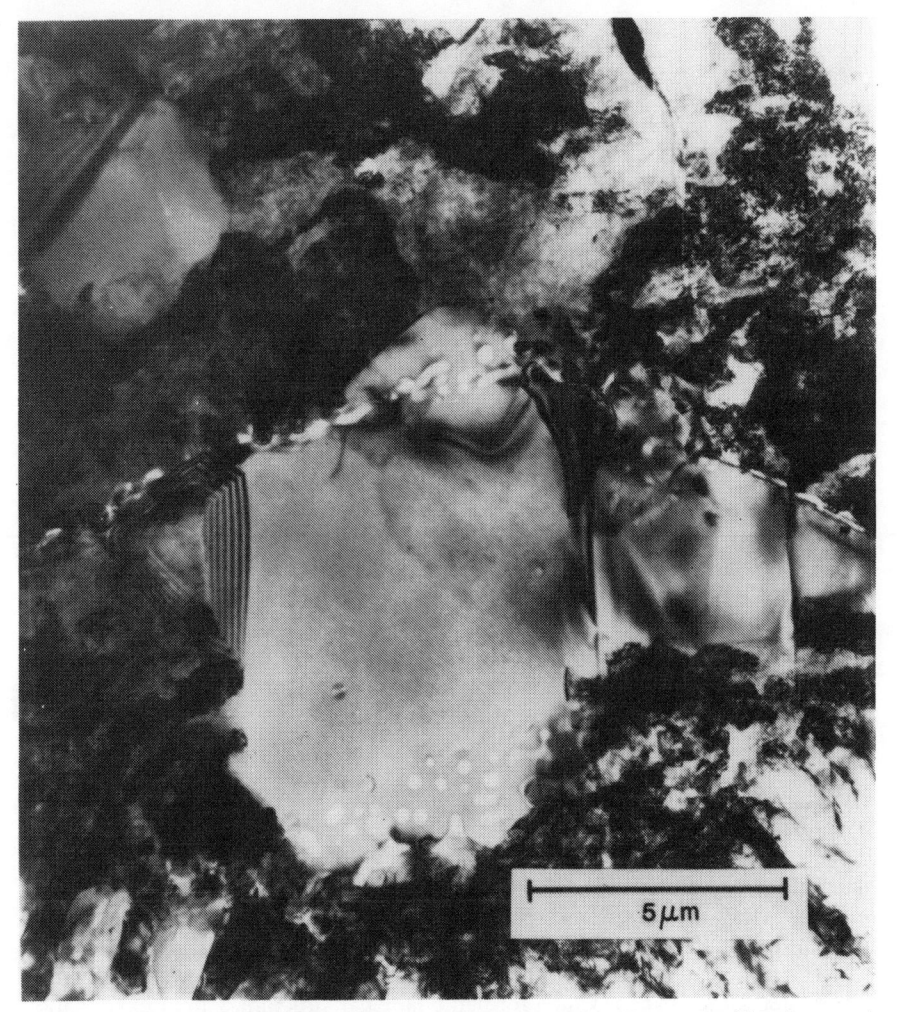

Fig. 11. Extensive plastic deformation and localized recrystallization and grain growth in central region of cylinder of dynamically compacted alumina (after Hoenig and Yust, Ref. 47). Note the well-rounded porosity at the top and bottom of large recrystallized grain and the region of columnar grains at bottom of the figure. Both features are indicative of locally high temperatures, near T_m.

of the prior shock treatments employed), have been in generally good agreement, with some residual strains of >0.25% having been reported. The X-ray line-broadening data have commonly been given in the form of Hall–Williamson plots[45] (see Fig. 10), from which the residual strains may be calculated, e.g., by the methods of Ziegler.[46]

The most informative investigations concerning the resulting substructure of dynamically compacted (3.6–7.5 GPa) ceramic powders have been the TEM studies of Hoenig and Yust[47] and Yust and Harris.[48] The central core region of a rod-shaped Al_2O_3 sample (precompacted particle size range of 10–177 μm) con-

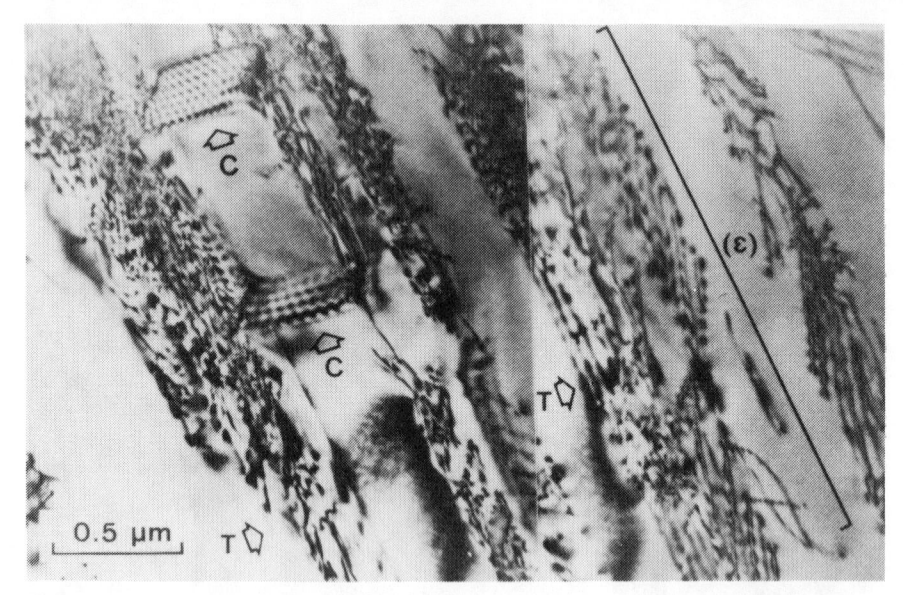

Fig. 12. The differently misoriented lattice bands are separated by parallel dislocation arrays lying on the basal plane (after Hoenig and Yust, Ref. 47). Basal plane twins are indicated by T, microcracks or crack precursors by C, and a dislocation group by the bracket marked "ε." The basal plane is tilted 40° to the plane of the photograph.

tained a densely faulted and strained polycrystalline matrix having some recrystallized strain-free grains (see Fig. 11) and evidence of columnar grain growth at crack boundaries as a result of shocking. The final average grain size of all the material in the central core was reduced below 10 μm, which again confirmed previous X-ray investigations (see above) which showed that all particles in the original distribution were broken down.

The principal substructural features in the shocked material were differently misoriented lattice bands, which formed a lamellar structure. Closer examination of the structure showed that the bands were comprised primarily of dislocation arrays on the basal planes along with basal-plane twins. Typical features of the dislocation substructure can be seen in the micrograph in Fig. 12. The fringes present along some of the bands are suggestive of sharp, crystallographic boundary structures, and crack precursors are present between some of the active slip regions. Further observations on sections of different orientation showed that considerable dislocation activity exists on slip systems normal to the basal plane.

A more detailed analysis of the Burgers vector geometry of the array dislocations (denoted by the "ε" bracket in Fig. 12) was carried out. The arrays have a large fraction of $\langle 11\bar{2}0 \rangle$ type dislocations with differently oriented Burgers vectors. At first sight, the dislocation substructure appears consistent with the intermittent, homogeneous nucleation of dislocation bands proposed by Meyers in the model mentioned earlier. However, most of the differently oriented $\langle 11\bar{2}0 \rangle$ dislocations in the lamellar arrays appear to be formed from the reaction of a single set of $\langle 10\bar{1}0 \rangle$ type dislocations. The reactions can be rationalized on the basis of Frank's rule; however, elastic anisotropy may play a role, and the issue remains in part unresolved. The most important result of the analysis lies in the fact that the Meyers model is not unambiguously confirmed. If the dislocation substructure consisted of several different sets of independently generated dislocations, the confirmation of the model would be more certain. In addition some features of the substructure suggest a loop-generating mechanism rather than the intermittent formation of the lamellar dislocation arrays that are envisioned in the Meyers model.

Practical Considerations

In addition to its thorough reviews of the fundamental scientific and/or engineering aspects of shock waves, materials responses, characterizations of resultant defects, densification processes et al. involved in the dynamic compaction or shock conditioning of powder materials, Ref. 7 also considered known or anticipated technical advantages and benefits. It also described problem areas and discussed some of the pertinent economic and safety issues.

Among those identified were the following: (a) deficiencies in real time knowledge of shock-wave events, particularly in particles and particle assemblages; (b) the need for more extensive (and intensive) characterization of shocked materials (and their unshocked controls); together with (c) intensified efforts in modeling, scaleup, and extrapolation studies; (d) better methods for instrumenting and monitoring (or of simulating) shock waves and their interactions with particulate materials; (e) effective methods for reducing and/or eliminating undesirable side effects in dynamic compaction (including coarse and fine cracking, residual internal strains, and the formation of excessive gradients in terms of strain and thermodynamic heating effects and in resultant microstructural development); (f) better diagnostic computer codes.[7] Many of these interesting issues are now being explored in a coordinated, multilaboratory experimental program.[49]

In Ref. 1, Palmour et al. considered some postulates about the statistical nature of the shocking and subsequent sintering of ceramic powders (see also Ref. 7). They proposed two desirable shock-conditioning situations: case I, powder shock compacted to some intermediate and uniform green density, free of gross flaws, and thus highly sinterable; and case II, powder shock conditioned but still friable and capable of being reconstituted by normal processing methods to yield sound compacts suitable for subsequent sintering.

To them has now been added an undesirable situation: case III, powder shocked, but *overcompacted,* yielding material that is nominally friable but containing both a population of gross flaws and a population of strongly aggregated particle clusters that are quite resistant to subsequent comminution, thus creating many morphologically dictated anomalies during sintering.[44] These points are considered in greater detail in a companion experimental paper.[40]

Summary

Although a number of materials have been dynamically compacted or shock conditioned, a relatively small number of investigations concerned with the characterization of the resulting defect substructure have been conducted. This is particularly true for ceramic materials where only the study by Yust and coworkers exists as the definitive work for Al_2O_3. Even more importantly, essentially no direct evidence exists that directly links the type and concentration of defects with increased reactivity of shock-conditioned powders, though most assuredly, there is a direct connection for most materials.

Acknowledgments

This work was supported as a part of the DARPA Dynamic Synthesis and Consolidation Program, under Lawrence Livermore National Laboratory Subcontract No. UCRL-6853501.

References

[1] H. Palmour III, V. D. Linse, and M. Spriggs; Sintering Theory and Practice. Edited by D. Kolar, S. Pejovnik, and M. M. Ristic. Elsevier, Amsterdam, 1982.
[2] C. S. Smith, *Trans. AIME*, **212**, 574 (1958).
[3] G. E. Dieter; p. 409 in Response of Metals to High Velocity Deformation. Edited by P. W. Shewmon and V. F. Zackay. Interscience, New York, 1961.
[4] W. C. Leslie; p. 571 in Metallurgical Effects at High Strain Rates. Edited by R. W. Rohde, B. M. Butcher, J. R. Holland, and C. H. Karnes. Plenum, New York, 1973.
[5] L. Davison and R. A. Graham, *Phys. Repts.*, **55**, 255 (1979).
[6] M. A. Meyers and L. E. Murr; p. 487 in Shock Waves and High Strain Rate Phenomena in Metals. Plenum, New York, 1981.
[7] National Materials Advisory Board; "Dynamic Compaction of Metal and Ceramic Powders," Rept. No. 394, Oct, 1982.
[8] W. Bleakney and A. H. Taub, *Rev. Mod. Phys.*, **21**, 584 (1979).
[9] M. H. Rice, R. G. McQueen, and J. M. Walsh, *Solid State Phys.*, **6**, 593, (1958).
[10] L. E. Murr and D. Kuhlmann-Wilsdorf, *Acta Metall.*, **26**, 654 (1978).
[11] S. LaRouch et al., *Metall. Trans., A*, **12A** (1981).
[12] R. N. Wright et al.; p. 703 in Ref. 6.
[13] A. S. Appleton and J. S. Waddington, *Acta Metall.*, **12**, 681 (1963).
[14] C. S. Smith, *Trans. Met. Soc. AIME*, **574** (1958).
[15] E. Hornbogen, *Acta Metall.*, **10**, 978 (1962).
[16] M. A. Meyers, *Scr. Metal.*, **12**, 21 (1978).
[17] K. C. Hsu, C. Y. Hsu, L. E. Murr, and M. A. Meyers; p. 433 in Ref. 6.
[18] B. Kazmi and L. E. Murr; p. 733 in Ref. 6.
[19] M. A. Meyers, *Scr. Metal.*, **12**, 21 (1978).
[20] D. E. Grady et al.; Sandia Rept. SAND82-1627C, 1982.
[21] R. W. Heckel and J. L. Youngblood, *J. Am. Ceram. Soc.*, **51**[7] 398–401 (1968).
[22] R. Prümmer and D. Ziegler, *Ber. Dtsch. Keram. Ges.*, **51**, 343 (1974).
[23] B. Lawn and R. Wilshaw, *J. Mater. Sci.*, **10**, 1049 (1975).
[24] A. G. Evans; p. 1 in Treatise on Materials Science and Technology, Vol. 16. Edited by C. M. Preece. Academic Press, New York, 1979.
[25] B. J. Hockey, S. M. Wiederhorn, and H. Johnson; p. 379 in Fracture Mechanics of Ceramics, Vol. 3. Edited by R. C. Bradt, D. P. H. Hasselman, and F. F. Lange. Plenum, New York, 1977.
[26] C. S. Yust and R. S. Crouse, *Wear*, **51**, 193 (1978).
[27] H. Kressel and N. Brown, *J. Appl. Phys.*, **38**, 138 (1967).
[28] D. C. Brillhart, R. M. DeAngelis, A. G. Preban, J. B. Cohen, and P. Gordon, *Trans. AIME*, **239**, 836 (1967).
[29] M. F. Rose and T. L. Berger, *Philos. Mag.*, **17**, 1121 (1968).
[30] D. L. Styris and G. E. Duvall, *High Temp.-High Press.*, **2**, 477 (1970).
[31] M. J. Ginsberg and D. E. Grady, *Bull. Am. Phys. Soc.*, **17**, 477 (1970).
[32] A. Christou, *Philos. Mag.*, **26**, 97 (1972).
[33] J. J. Dick and D. L. Styris, *J. Appl. Phys.*, **46**, 1602 (1975).
[34] L. E. Murr, O. T. Inal, and A. A. Morales, *Acta Metall.*, **24**, 261 (1976).
[35] J. P. Hirth and J. Lothe; p. 689 in Theory of Dislocations, 2nd ed. McGraw-Hill, New York, 1968.
[36] J. B. Cohen and J. Weertman, *Acta Metall.*, **11**, 997 (1963); 1368 (1963).
[37] D. Peckner; p. 49 in The Strengthening of Metals. Reinhold, New York, 1964.
[38] D. E. Duvall and R. A. Graham, *Rev. Med. Phys.*, **49**, 523 (1977).
[39] O. R. Bergmann and J. Barrington, *J. Am. Ceram. Soc.*, **49**[9] 502–7 (1966).
[40] H. Palmour III, T. M. Hare, A. D. Batchelor, K. Y. Kim, K. L. More, and T. T. Fang, "Sintering of Shock-Conditioned Materials"; this proceedings.
[41] A. C. Greenham and B. P. Richards, *Trans. Br. Ceram. Soc.*, **69**, 115 (1970).
[42] K. Y. Kim, "Densification of Shock-Conditioned Ceramic Powder by Rate Control Sintering"; Ph.D. Dissertation, North Carolina State University, 1983.
[43] (a) R. A. Graham, E. K. Beauchamp, B. Morosin, E. Venturi, D. M. Webb, and L. W. Davison, "Shock Activated Sintering"; pp. 258–297 in First Quarterly Rept., DARPA Dynamic Synthesis and Consolidation Program. Edited by C. L. Cline. Lawrence Livermore National Laboratory, Rept. UCID-19663, Aug 1982. (b) B. Morosin; to be published in Ref. 44.
[44] Y. Horie, H. Palmour III, and J. K. Whitfield, "Shock Compaction and Sintering Behavior of Selected Ceramic Powders (Appendix II)"; pp. 130–216 in Second Quarterly Rept., DARPA

Dynamic Synthesis and Consolidation Program. Edited by C. L. Cline. Lawrence Livermore National Laboratory, Rept. UCID-19663-83-1, March 1983.

[45]G. K. Williamson and W. H. Hall, "X-ray Line-Broadening from Filled Aluminum and Tungsten," *Acta Metall.*, **1** [22] (1953).

[46]G. Ziegler, "Structural and Morphological Investigations of Ceramic Powders and Compacts," *Powder Metall. Int.*, **8**, [2] (1978).

[47]C. L. Hoenig and C. S. Yust, *Am. Ceram. Soc. Bull.*, **60** [11] 1175 (1981).

[48]C. S. Yust and L. A. Harris; p. 881 in Ref. 6.

[49]First Quarterly Rept., DARPA Dynamic Synthesis and Consolidation Program. Edited by C. L. Cline. Lawrence Livermore National Laboratory, Rept. UCID-19663, Aug 1982.

---

*The measurement of any pair of variables used in Eqs. (1–6), when coupled with the known initial conditions, is sufficient to define a point on the shock-wave loci, or Hugoniot curve. A particularly useful pair of variables is the force p and the specific volume V. However, in Fig. 6, the combination of force and particle velocity u_1 are used.

Surfaces of Magnesia and Alumina

P. W. TASKER

A.E.R.E. Harwell
Theoretical Physics Division
Oxfordshire, England OX11 ORA

The surface properties of ceramic materials determine many important processes including catalysis and corrosion. Computer simulation methods can complement experiment in probing the surface structure and predict the behavior of impurities and defects. Calculations on the surface structure and energy of the flat surfaces of alumina and magnesia are discussed and compared with the available experimental data. Steps are common surface defects, and their properties have been calculated for MgO. The calculations predict a modified step geometry and a short-range repulsive interaction between the steps. Composition of the surface will, in general, differ from the bulk. A detailed calculation on the segregation of calcium to the magnesia surface is discussed and compared with experiment. The different behavior of high surface area materials is discussed.

The surfaces of crystalline ceramics affect many important properties including shape, mechanical strength, chemical behavior including adsorption, and even electrical characteristics. They have a dominant contribution in many processes such as catalysis, corrosion, electrode behavior, and sintering. However, the structure and composition of the surface region is still often poorly understood. In particular, the surface may differ in structure and will almost certainly differ in composition from a simple termination of the bulk crystal. Because of this, much experimental and theoretical effort has been expended in investigating and characterizing surface composition and behavior. A valuable theoretical technique that has proved its value in investigating a wide range of defect processes in ionic materials is static lattice computer simulation.[1] In this paper we describe the application of this method to the study of surface structure. We have studied the surfaces of magnesium oxide and α-alumina as typical examples of ceramic materials. In Method of Calculation, we briefly describe the principles of the calculation. The section Perfect Surfaces discusses the calculations and experimental observations on the perfect flat surfaces of magnesia and alumina. Most real surfaces are, of course, defective, and one of the most important surface irregularities are steps. Calculations on the stepped surfaces of magnesia are described in Stepped Surfaces. Changes in composition in the surface region are discussed in Surface Segregation with a calculation of impurity segregation in magnesium oxide.

Method of Calculation

In a static lattice computer simulation, the crystal is described as an assembly of atoms with the perfect bulk structure. The atoms interact with each other and are held in equilibrium by a potential that describes the variation of energy with interatomic separation. The defect is introduced and destroys the equilibrium of the crystal. The atoms are then relaxed with an efficient numerical method until no forces act on them.[1] In the surface calculations described here, the crystal maintains two-dimensional periodicity parallel to the surface. It is constructed from a stack of crystal planes that terminate in a free surface on one side and is matched to bulk crystal on the other side.[2] The ions in the stack are relaxed by using a variable metric, quasi-Newton–Raphson method.[3,4] The surface energy is calculated by subtracting the energy of an assembly of bulk ions from the energy of a crystal block with a surface. The size of the explicitly relaxed surface region is varied to ensure that the surface energy and structure have converged to values independent of size. For an ionic crystal, the interatomic potential consists of a short-range part and a long-range Coulomb interaction. The long-range part cannot be summed directly in real space but must be handled by an Ewald type method. We use the form derived by Parry.[5] Details of the crystal stacks and the lattice summation methods have been given elsewhere[2,6] and are implemented in the Harwell MIDAS code.

Since a static lattice calculation relaxes the defective crystal to mechanical equilibrium but contains no dynamical effects, the results strictly apply to absolute zero temperature with neglect of the zero-point vibration. The energies can be extrapolated to higher temperature through calculation of the entropy. If the defect entropy is small, the calculated results may be directly comparable with experiments at moderate temperatures. In many cases, the structural information obtained is valid to higher temperatures, at least qualitatively. The accuracy of the static lattice calculations depends on the validity of the potentials used to describe the interatomic interactions. These are in three parts. First, the Coulomb potential is dominant in ionic materials. In the crystals studied here, the full nominal ionic charges of Mg^{2+}, Al^{3+}, and O^{2-} have been used. Second, a short-range potential that is repulsive at very close range is essential for crystal equilibrium. This may be calculated by a quantum mechanical procedure or determined empirically.[7,8] In these calculations, we have used empirical potentials obtained by Sangster and Stoneham[9] for MgO and by Catlow et al.[10] for Al_2O_3. In both cases, the potentials are described by a simple analytic form and the parameters determined by fitting to bulk elastic and dielectric properties. The third component of the potential is the ionic polarizability. This is described by the shell model[11] in which each ion is treated as a core and shell of opposite charge coupled harmonically. Displacements of the shell introduce dipole moments on the ion site. The short-range potential acts between shells, thus coupling the interatomic and intraatomic potentials. The shell parameters are obtained by fitting to the dielectric properties.[8]

The potential parameters were not modified for interactions in the surface, but the change in environment produces a change in effective ion polarizabilities used in the calculation. Although, in reality, some change in the interactions between ions in the surface is likely, such modifications are probably small for ionic materials. The approximation in neglecting them means that calculations presented here can be used predictively.

Perfect Surfaces

The low energy and, hence, the most common surfaces of a crystal are generally those of low Miller index. These planes are closest packed with large interplanar spacings. However, in ionic materials, other constraints also apply. If the Madelung energy of the crystal is not to diverge with increasing crystal size, the crystal must not only be electrically neutral, but must also not have a net dipole

Table I. Surface Energies for Magnesium Oxide*

	This work	Other theoretical[†,‡]	Experimental[‡]
MgO (001)	1.16	1.17, 1.45	1.04, 1.2, 1.15
MgO (110)	2.92	2.98, 3.89	

*Units are $J \cdot m^{-2}$.
[†]Ref. 15.
[‡]Ref. 16.

moment.[12] Thus, for a rock-salt crystal, the (001) and (110) planes are neutral and thus potentially low energy surfaces, but the (111) direction is polar and the crystal cannot terminate in this direction without restructuring. This would necessarily lead to a rather high surface energy. The criterion of no dipole moment often helps to determine the likely low-energy surfaces. For example, (111) fluorite crystal surfaces must terminate with anions while the (100) surface will not be stable as a simple bulk termination.[13] Observations on UO_2 confirm this with the (100) face being observed but with only a half coverage of ions in the surface plane.[14] This is exactly sufficient to remove the dipole moment. While the possible surfaces are rather simple for magnesia, this general principle has been of help in determining the likely surfaces for calculation for α-alumina.

Magnesia Surfaces

Magnesium oxide has the rock-salt structure and cleaves naturally to produce (001) surfaces. The (110) face is also electrically neutral and is likely to be of relatively low energy. The (111) surface can be neglected since its dipolar nature makes it unstable for a macroscopic crystal. The simulation of the surfaces gives a value of $1.16 \; J \cdot m^{-2}$ for the (001) surface and $2.92 \; J \cdot m^{-2}$ for the (110) face. This agrees well with previous theoretical values[15] and experimental estimates.[16] The values from this and other work are summarized in Table I. The low value for the (001) face agrees qualitatively with the ease of cleavage along this plane. Only small relaxations take place in the calculation, particularly on the (001) face. The anions move out slightly and the cations inward, producing a "rumpled" surface, as was found for the corresponding alkali halide surfaces.[2] However, much smaller displacements are calculated for the more highly charged ionic oxides. The calculated surface still closely resembles the bulk termination. This seems to be a common property of the lowest energy ionic crystal surfaces in contrast to the behavior in semiconductors. Experimentally, the (001) surface has been studied by low-energy electron diffraction (LEED).[17–19] This has also shown very small displacements from the ideal geometry. The structure is generally characterized by the displacements of ions in the outer layer. If ε_1 is the outward displacement of the anion and ε_2 is the outward displacement of cations, the structure is defined by[20]

$$\text{relaxation} = \tfrac{1}{2}(\varepsilon_1 + \varepsilon_2)$$

and

$$\text{rumple} = \varepsilon_1 - \varepsilon_2$$

The experimental and theoretical estimates of these quantities are summarized in Table II. Our calculations are in very good agreement with the observations. The differences with the other theoretical values indicate the dependence on the details of the interatomic potential. However, the experimental measurement is made by scattering electrons off the ion cores, while the cores in the calculation are a

Table II. Theoretical and Experimental Structure Determinations for MgO (001)

	Relaxation (%)	Rumple (%)
Experimental	0 to −3	0 to +5*
	0 ± 0.75	2 ± 2[†]
Other theory	0	5–9[‡]
	0	7[§]
	0	3[¶]
	−2	2
This work	−0.7	2.5

*Ref. 18.
[†]Ref. 19.
[‡]Ref. 20.
[§]Ref. 21.
[¶]Ref. 22.

function of the shell model and given a charge to reproduce the correct ion polarizability. There is not an exact correspondence between the measurement and the calculated quantity. Furthermore, the magnitude of the rumple is only ~0.05 Å (~0.005 nm). Thus, all the calculations are in reasonable agreement both with each other and with experiment. The calculations from this work do show that it is not necessary to include changes in potential at the surface in order to predict the very small rumplings observed. All the calculations indicate that the ionic model reproduces the behavior of the surface rather well. The only anomalous feature of the MgO (001) face is the observation of an apparent phase transition. Some recent electron diffraction results indicate that the cleaved surface is flat but metastable.[23,24] It undergoes an irreversible reconstruction to the rumpled state on being heated. This cannot easily be explained by a simple ionic model, and electron–phonon coupling has been invoked to interpret it.[25] If these experiments do reflect the properties of the perfect planar MgO surface and are not influenced by surface defects, they indicate that the surface properties are more complicated than originally deduced. However, the majority of measurements on vacuum-cleaved MgO and the calculation by static lattice simulation agree well. Both indicate the lowest energy (001) surface is very similar in structure to a simple bulk termination but with a slight rumpling in the ion positions.

Alumina Surfaces

The most stable form of alumina, α-alumina, has the corundum structure. This consists of hexagonal close-packed oxygen ions with aluminum cations in two-thirds of the octahedral interstices. The more complex structure than that of magnesia leads to a large number of low-index and potentially low-energy surfaces. In the calculations described here, we have considered the basal plane (0001) and four other surfaces. At right angles to the basal plane are the surfaces related to the hexagonal prism. The prism face is indexed ($10\bar{1}0$) and the diagonal of the prism is indexed ($11\bar{2}0$). The primitive unit cell is rhombohedral, and faces related to this should have the smallest two-dimensional unit cells. The rhombohedral faces considered are ($10\bar{1}1$) and ($10\bar{1}2$). Alumina has no simple cleavage planes such as (001) in magnesia, though the corundum-structured transition-metal oxides do cleave along ($10\bar{1}2$).[26,27]

Figures 1–4 show the structure of four of the surfaces calculated. This is illustrated in two ways. First, a schematic diagram shows the stacking sequence of

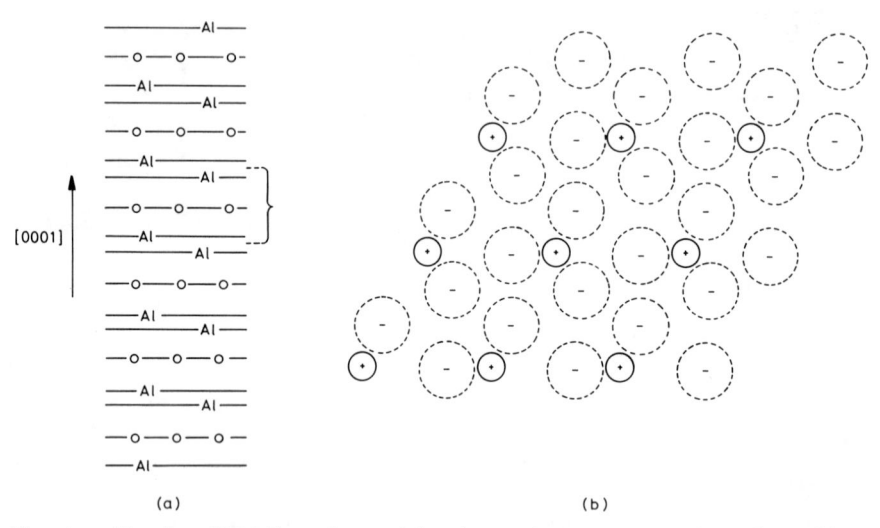

Fig. 1. Alumina (0001) surface: (a) schematic representation of stacking sequence; (b) structure of surface plane.

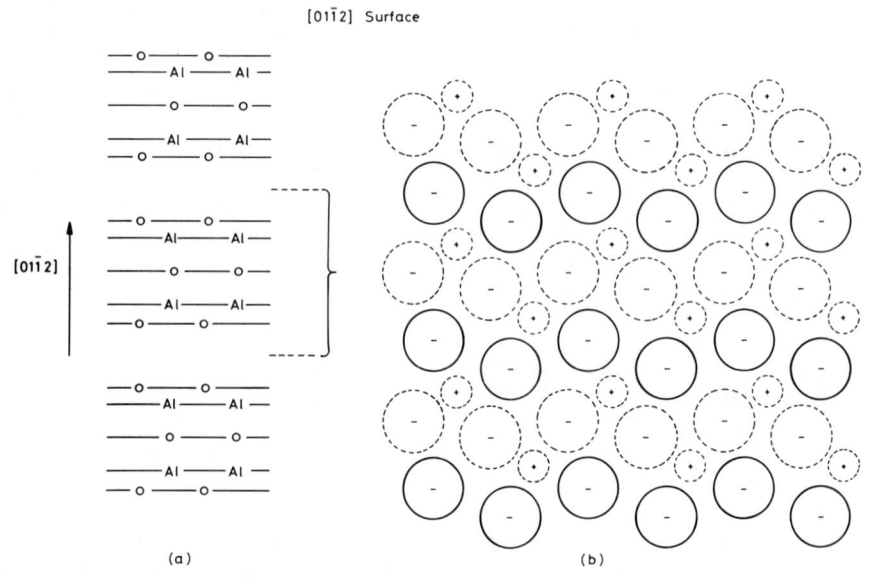

Fig. 2. Alumina (01$\bar{1}$2) surface: (a) schematic representation of stacking sequence; (b) structure of surface plane.

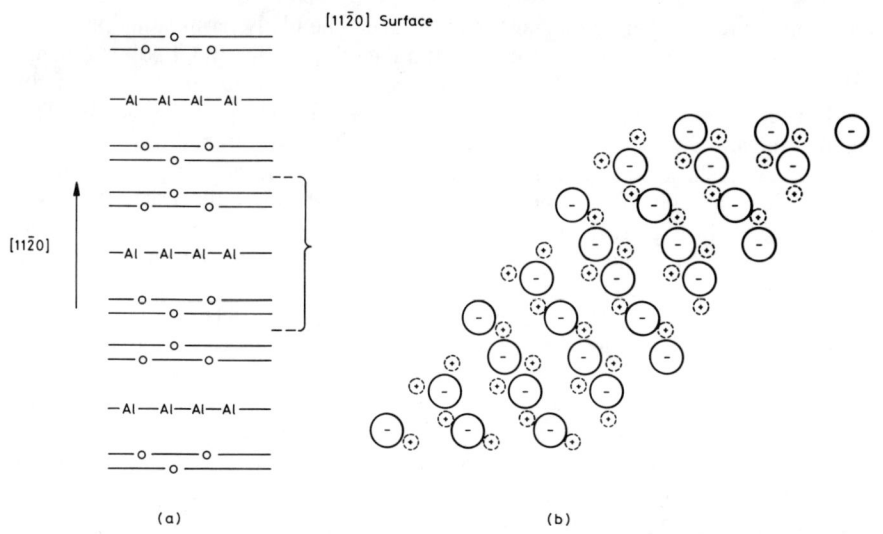

Fig. 3. Alumina (11$\bar{2}$0) surface: (a) schematic representation of stacking sequence; (b) structure of surface plane.

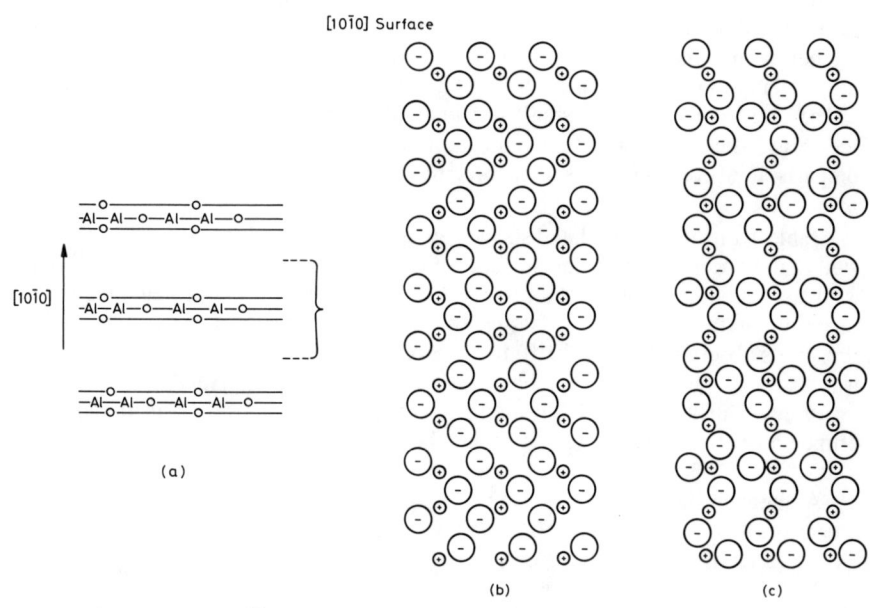

Fig. 4. Alumina (10$\bar{1}$0) surface: (a) schematic representation of stacking sequence; (b) structure of surface plane unrelaxed; (c) structure of equilibrated surface plane.

planes in a direction perpendicular to the surface. The likely terminating plane for the low-energy face can be determined from the criterion that no dipole moments can be left across the block. This usually gives the same result as the commonly used criterion of breaking the minimum number of anion–cation bonds. The figures show the repeat unit perpendicular to the surface that has no dipole moment. Also shown on each figure is the structure of the surface plane with the ion positions for the lowest energy configuration. Figure 1 illustrates the basal plane surface. In Fig. 1(a), the stacking sequence is shown as planes of oxygen ions interleaved by planes of cations. Since the cation planes are puckered, they are shown as two planes. Cleavage between the cation planes produces a surface with no dipole moment. The structure is shown in Fig. 1(b). The surface layer of oxygen ions has aluminum ions above it in one-third of the potential sites. The surface ($10\bar{1}2$) is shown in Fig. 2, where the stacking sequence in Fig. 2(a) shows the surface plane must be oxygen ions. The structure is shown in Fig. 2(b) with the aluminum ions below the surface and the oxygen ions of the next anion plane down shown with broken lines. This surface is the same as that designated (047) and illustrated by Henrich.[28] Figure 3 shows the hexagonal prism related face ($11\bar{2}0$), where again the surface must consist of oxygen ions. The surface anions are not quite coplanar, as indicated in Fig. 3(a). The cations from the plane below are also shown. The next lowest plane of anions would fill up the gaps in the structure, as drawn in Fig. 3(b). Figure 4 shows the prism face ($10\bar{1}0$) in which the surface is composed of a plane of anions and cations in stoichiometric ratio. As indicated in Fig. 4(a), the plane is not quite flat with anions both above and below the mean plane position. Figure 4(b) shows the surface structure before relaxation. For the other surfaces described here, the surface relaxations are modest and involve mainly small perpendicular relaxation. However, this prism face shows the most relaxation, and the final structure is shown in Fig. 4(c). A comparison of Fig. 4(b,c) shows that the coordination of the surface cations has altered.

Table III lists the calculated surface energies for the surfaces studied. They are listed in the order of increasing surface unit cell, since the smaller unit cells imply less dense packing of the crystal planes. The surface energies are calculated before relaxation of the structure and after equilibrium has been attained. The larger relaxations observed for the ($10\bar{1}0$) face are reflected in the large decrease in energy on relaxation. The ($10\bar{1}2$) face has the lowest surface energy and, in particular, a very low surface energy before the crystal relaxations are considered. Experimentally, this surface seems to be the most stable in corundum-structured oxides.[26–28]

A measurement of the surface energy of polycrystalline alumina gave an energy of 0.9 $J \cdot m^{-2}$ at 1850°C.[29] If we assume a surface energy at absolute zero of 2.6 $J \cdot m^{-2}$ from these calculations, then the surface entropy would be 8×10^{-4} $J \cdot m^{-2} \cdot K^{-1}$. Measurements on stoichiometric uranium dioxide give a temperature coefficient of 6.3×10^{-4} $J \cdot m^{-2} \cdot K^{-1}$, which is comparable.[30] Therefore, although a direct comparison between calculated and experimental surface energies is not possible, the reported values are not inconsistent. The (0001) surface of corundum has been studied by LEED,[31] and the (1 × 1) observed pattern was interpreted in terms of the simple structure illustrated in Fig. 1. When the surface was heated, more complex structures related to supercells on the surface were formed. Similar behavior has been observed on the (0001) surface of α-Fe_2O_3.[32] The larger cells may be related to impurities or changes in stoichi-

Table III. Surface Energies Calculated for the Faces of α-Alumina

Surface	Unrelaxed surface energy (J·m^{-2})	Relaxed surface energy (J·m^{-2})
(0001)	6.53	2.97
(10$\bar{1}$1)	6.41	3.27
(10$\bar{1}$2)	3.55	2.57
(11$\bar{2}$0)	5.17	2.65
(10$\bar{1}$0)	6.87	2.89

ometry or may indicate that the (1 × 1) cell is metastable. The current calculations have only considered the primitive unit cell on the surface, but further studies could consider the stability of larger cell structures. The (10$\bar{1}$2) surface studied by LEED for Ti_2O_3 showed no signs of restructuring. In our calculations, this has the lowest energy and thus is also the most stable surface.

Stepped Surfaces

The perfect flat surfaces described in the previous section are an idealization. A real surface is rarely flat on an atomic scale, though substantial regions of terrace may exist. One of the most common forms of irregularity is a surface step. The structure and properties of steps may have a disproportionate importance in determining surface properties. For example, they are believed to provide active sites for catalysis and to be vital to the understanding of the mechanism of evaporation and crystallization. They may be particularly important in surface diffusion and, hence, affect processes such as sintering. The modified surface defect sites may be important in determining the surface space charge.

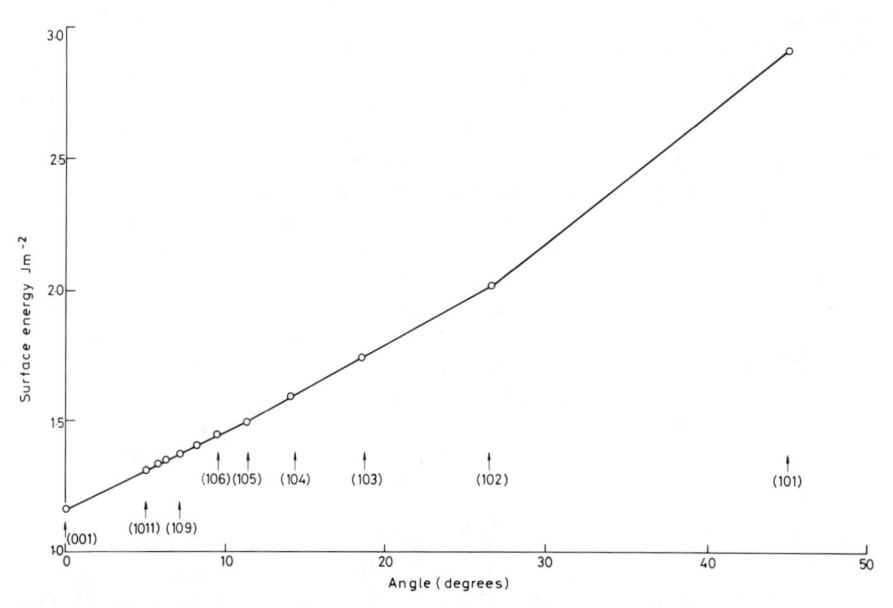

Fig. 5. Variation of surface energy with angle for MgO (10n) surfaces (Ref. 37).

In magnesium oxide, surface steps have been observed and shown to be monatomic.[33] Symmetry arguments show that an unrelaxed surface step would have a Madelung energy at an ion site of 90% of the bulk value while a kink site is bound by half the bulk energy.[34] However, there have been very few detailed calculations including lattice relaxation. Tsang and Falicov[35] have studied a stepped (1011) face in NaCl, and Colbourn and Mackrodt[36] have studied indentation in the flat surface bounded by steps. When a series of regularly stepped surfaces is studied with differing interstep separation, not only can the step energy and structure be determined but also the step–step interactions. This has been done for magnesium oxide and nickel oxide,[37] and the magnesia results are discussed below. The series of surfaces indexed $(10n)$ can be considered as (001) terraces with a monatomic step every n interatomic separations. They can also be characterized by an angular deviation from (001) by $\theta = \arctan(n^{-1})$. The series for MgO from (1011) to (101) has been studied and the calculated surface energies as a function of angle are shown in Fig. 5. This figure illustrates the large variation of surface energy with angle found in ionic crystals. This is in contrast to the behavior in metals where surface energies are much more isotropic. The surface steps are at right angles to the low-energy terraces in this series, and so the surface energy can be related to the terrace energy by

$$E_s = E_{001} \cos\theta + E_{step} \frac{\sin\theta}{a} + E_{int} \frac{\sin\theta}{a} \qquad (1)$$

where E_{001} is the surface energy of the (001) terrace (1.16 J·m^{-2}), E_{step} is the step energy, and E_{int} is the step–step interaction energy. At low angle or high index, n, a plot of $E_s - E_{001}\cos\theta$ vs $\sin\theta$ is linear, indicating that the interaction terms are negligible. For magnesia the step–step interaction can be neglected for separations greater than six interatomic distances. From the linear region, we deduce the step energy as 3.62×10^{-10} J·m^{-1}. Another way of expressing it is as 0.49 eV per molecule unit along the step. This is comparable with the surface energy of 0.64 eV per molecule unit in the flat surface. The lattice relaxations are substantial, as indicated by the large relaxation energy for the step. The step energy calculated before relaxation is over twice as large at 8.16×10^{-10} J·m^{-1}. The relaxed geometry is shown in Fig. 6, where an obtuse step angle is the main result. This is rather different from the results obtained by Tsang and Falicov.[35] Although some difference may be due to the different material that they studied, the improved potentials and computational procedures of the present work probably account for much of the discrepancy. Similar structural results were obtained for the small steps around the surface cavities studied by Colbourn and Mackrodt.[36]

At separations shorter than six interatomic units, the step–step interactions become important. The interaction function is shown in Fig. 7. The steps not only have positive energy but interact repulsively, as suggested by Blakely and Schwoebel.[38] However, the interaction is rather short range and is screened by the relaxation of the intervening ions. The function illustrated in Fig. 7 is not well represented by an inverse power of the separation but fitted rather accurately by an exponential function

$$E_{int}(R) = 2.86 \times 10^{-10} \; (\text{J·m}^{-1}) \; \exp(-R/3.730 \; (\text{Å})) \qquad (2)$$

In conclusion, surface steps on magnesia show much greater relaxation from the ideal geometry than from the perfect flat surface. This manifests itself in a greatly reduced step energy and is rather short range for the step–step interaction function, which is repulsive. The calculated geometry is a much more realistic

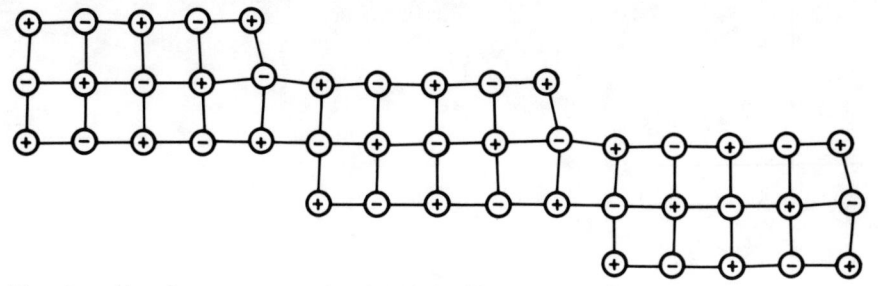

Fig. 6. Atomic structure of a (105) MgO surface (Ref. 37).

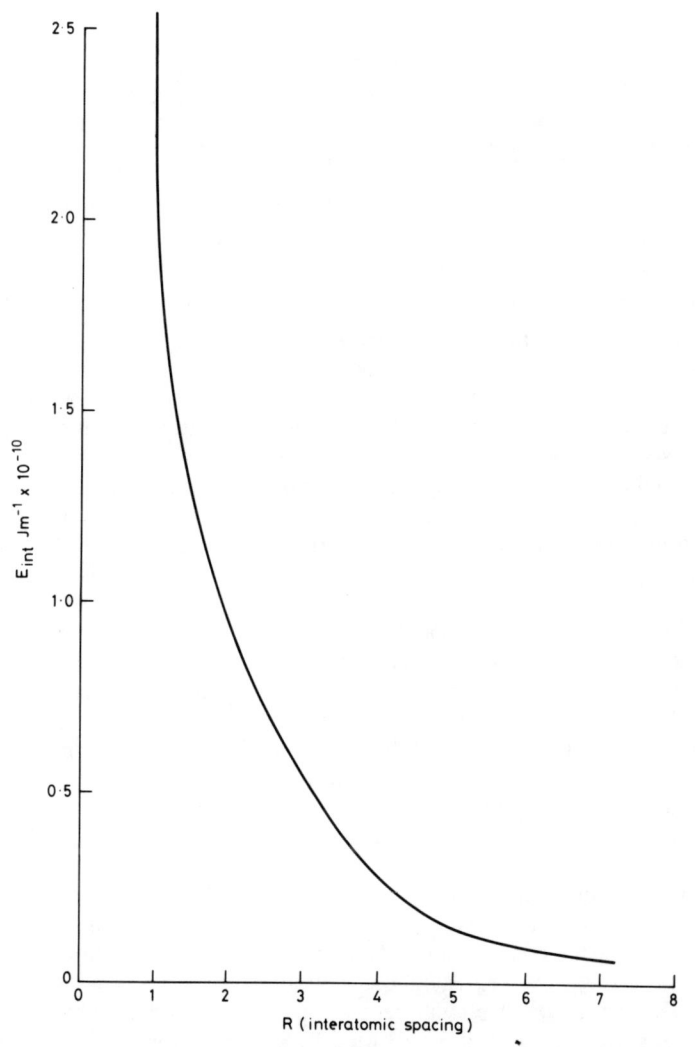

Fig. 7. Step–step interaction energy as a function of interstep spacing for monatomic steps on an (001) MgO surface (Ref. 37).

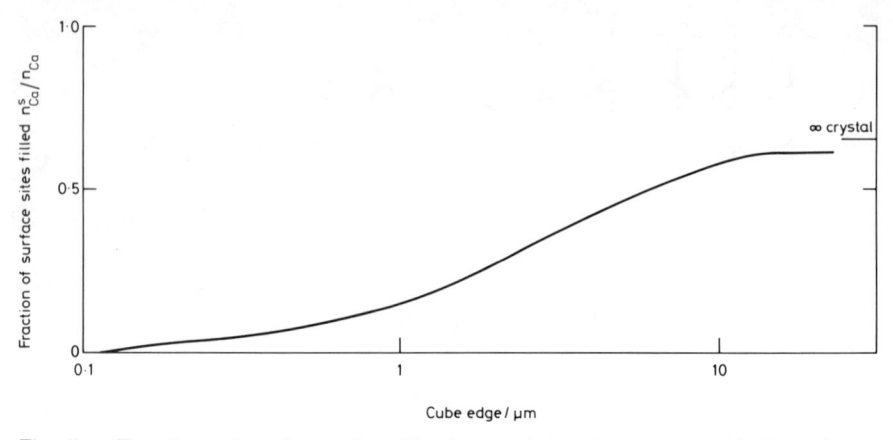

Fig. 8. Fraction of surface sites filled by calcium ions on an MgO surface as a function of crystal size at 1000 K and a constant average enthalpy of segregation assumed at −0.8 eV. The total calcium concentration is assumed to be 220 ppm.

starting point for discussing step–point defect interactions, diffusion, and adsorption than calculations based on an ideal configuration.

Surface Segregation

Surfaces differ from ideal terminations of the bulk crystal not only in structure but also in composition. Impurities dissolved in the crystal may have a lower energy at surface sites than at bulk sites. Under equilibrium conditions, segregation occurs with an enhanced concentration of impurity or dopant in the surface. This modifies surface properties and is a potential control for many processes such as catalysis and corrosion. Recently, experimental investigation of oxides has begun to yield information on the segregation behavior.[39,43] The calculation of impurity substitution energies in the bulk and the surface can give important information on the propensity of different elements to segregate.[44] At high levels of segregation, complete layers of segregant can be formed.[45] The theoretical methods are most powerful when the point defects and layers of segregants are both considered. Then the variation of enthalpy of segregation with concentration can be investigated. This has been done recently for the simple case of calcium dissolved in magnesium oxide.[46] In considering isovalent substitutions, we need not be concerned with space–charge effects. Ion size and the consequent strain are the important factors. The results of the calculation are summarized in Table IV. This shows the energies of substitution of calcium singly, as clusters and as a complete plane near the (001) MgO surface. The calculations were done with a nonempirical potential calculated by the electron–gas approximation.[46] The results show that, at all coverages, the surface ion is more stable than the bulk so that surface segregation will occur. However, intercalcium repulsions reduce the stability of the ions in the surface layer with increasing concentration. Thus, the enthalpy of segregation varies from −1.08 eV at very low segregant concentrations to −0.42 eV for a saturated surface. This agrees very well with recent experimental studies, which obtained an enthalpy of segregation of −0.52 eV for 20% surface coverage.[43] The variation of coverage with temperature can be determined by a statistical–mechanical treatment. If the crystal is large, and we recognize only surface sites and bulk sites but

Table IV. Energies per Substituted Calcium Ion at an (001) MgO Surface from Colbourn et al.*

Plane	Single substance	(110) Dimer	(110) Trimer	Pentamer	Complete plane
1	3.54	3.57	3.56	3.63	4.20
2	4.61	4.59	4.54	4.61	4.74
Bulk	4.62	4.60	4.62	4.71	4.72

*Units are eV (Colbourn et al., Ref. 46).

no intermediates, and if the solution is assumed to be regular, the ratio of any two components in the surface of a multicomponent system is

$$\frac{n_1^s}{n_2^s} = \frac{n_1}{n_2} \exp(-\Delta G/kT) \tag{3}$$

where n_i^s is the number of atoms of component i in the surface and n_i is the total number of atoms of component i. ΔG is the difference in the free energies of segregation of components 1 and 2. For a two-component mixture, this is the same as the well-known McLean–Langmuir form:

$$\frac{X^s}{1 - X^s} = \frac{X}{1 - X} \exp(-\Delta G/kT) \tag{4}$$

where X is the concentration of impurity. The calculations show that ΔG in the above expressions is dependent on the surface concentration X^s. The calculations also indicate that, for the (001) magnesia surface, the perturbed segregant concentration is limited to the outermost atomic layer of the surface. This is confirmed experimentally in the ion scattering results of McCune and Wynblatt.[43] The calculations for the complete layer discussed above assumed that the surface unit cell remained as the primitive $p(1 \times 1)$. A calculation of the surface phonons shows an anomaly for the segregated surface with a phonon of imaginary frequency.[47] This indicates some additional restructuring, which is difficult to find a priori in a static lattice calculation. Further calculation has shown that the stable configuration has twice the unit cell size, $c(\sqrt{2} \times \sqrt{2})$, with the lattice strain moving the surface calcium ions out of plane. The substitution energies are only lowered by ~0.1 eV and do not modify the previous conclusions. This shows the power of surface vibration calculations not only in determining surface entropy,[48] but also in discovering larger scale restructuring that is hard to predict by other methods. In the case of heavily segregated magnesium oxide, we predict a segregation-induced surface restructuring.

The statistics of segregation discussed above are only valid for large crystals. The situation for small crystal sizes is illustrated in Fig. 8. The degree of segregation varies with crystal size due to the insufficiency in the number of impurity ions to fill all the surface sites for high surface area materials. In the example shown, the crystal size must be larger than 10 μm before the large crystal value is approached. Thus, the segregation and surface properties of a material will depend on how finely it is divided. This is important particularly in catalysis where high surface area materials are used.

Summary

In this paper we have described the application of static lattice simulation calculations to the study of the surface properties of magnesia and alumina. Very little theoretical work has been done on the hexagonal alumina. Surface energies for the low index faces are reported here. The surface structures do not deviate far from simple bulk terminations and can be deduced from the criterion of eliminating dipole moments across the crystal. The largest relaxations are observed for the prism face ($10\bar{1}0$) where the ion coordination in the surface changes. The lowest energy surface is calculated as ($10\bar{1}2$), which agrees with observation. Magnesia shows little surface relaxation in its lowest energy (001) face. The surface energy and structure agree well with experiment. Steps on the surface can be studied by considering the series of vicinal surfaces indexed ($10n$). This enables the positive step energy to be deduced and shows that step–step interactions are repulsive. These are well described by an exponential function. The much larger atomic relaxations around the step indicate that calculations of step behavior based on the ideal geometry will not be valid. Finally, surface segregation with impurities can be calculated, and the example has been given of calcium in magnesium oxide. This is predicted to strongly segregate, in agreement with experiment. When the material is very finely divided, different behavior is predicted with lower apparent levels of segregation.

References

[1] C. R. A. Catlow and W. C. Mackrodt (Editors); Computer Simulation of Solids. Springer-Verlag, Berlin, 1982.
[2] P. W. Tasker, *Philos. Mag.*, [*Part A*], **A39**, 119 (1979).
[3] R. Fletcher and M. D. J. Powell, *Comput. J.*, **6**, 163 (1963).
[4] R. Fletcher and M. J. Norgett, *J. Phys. C*, **3**, L190 (1970).
[5] D. E. Parry, *Surf. Sci.*, **49**, 433 (1975); *ibid.*, **54**, 195 (1976).
[6] P. W. Tasker and T. J. Bullough, *Philos. Mag.*, [*Part A*], **A43**, 313 (1981).
[7] W. C. Mackrodt and R. F. Stewart, *J. Phys. C*, **12**, 431 (1979).
[8] C. R. A. Catlow, M. Dixon, and W. C. Mackrodt; Computer Simulation of Solids. Edited by C. R. A. Catlow and W. C. Mackrodt. Springer-Verlag, Berlin, 1982.
[9] M. J. L. Sangster and A. M. Stoneham, *Philos. Mag.*, **43**, 597 (1981).
[10] C. R. A. Catlow, R. James, W. C. Mackrodt, and R. F. Stewart, *Phys. Rev. B*, **B25**, 1006 (1982).
[11] B. G. Dick and A. W. Overhauser, *Phys. Rev.*, **112**, 90 (1958).
[12] P. W. Tasker, *J. Phys. C*, **12**, 4977 (1979).
[13] P. W. Tasker, *Surf. Sci.*, **87**, 315 (1979).
[14] T. N. Taylor and W. P. Ellis, *Surf. Sci.*, **107**, 249 (1981).
[15] W. C. Mackrodt and R. F. Stewart, *J. Phys. C*, **10**, 1431 (1977).
[16] M. P. Tosi, *Solid State Phys.*, **16**, 1 (1964).
[17] C. G. Kinniburgh, *J. Phys. C*, **8**, 2382 (1975); **9**, 2695 (1976).
[18] M. Prutton, J. A. Walker, M. R. Welton-Cook, R. C. Felton, and J. A. Ramsey, *J. Phys. C*, **12**, 5271 (1979).
[19] M. R. Welton-Cook and W. Berndt, *J. Phys. C*, **15**, 5691 (1982).
[20] M. R. Welton-Cook and M. Prutton, *Surf. Sci.*, **74**, 276 (1978).
[21] R. N. Barnett and R. Bass, *J. Chem. Phys.*, **67**, 4620 (1977).
[22] A. J. Martin and H. Bilz, *Phys. Rev. B*, **B19**, 6593 (1979).
[23] T. Goth, S. Marakami, K. Kinosita, and Y. Murata, *J. Phys. Soc. Jpn.*, **50**, 2063 (1981).
[24] Y. Murata, *J. Phys. Soc. Jpn.*, **51**, 1932 (1982).
[25] A. Fujimori and N. Tsuda, *Surf. Sci.*, **100**, L445 (1980).
[26] R. L. Kurtz and V. E. Henrich, *Phys. Rev. B*, **B25**, 3563 (1982).
[27] R. L. Kurtz and V. E. Henrich, *Phys. Rev. B*, **B26**, 6682 (1982).
[28] V. E. Henrich; unpublished work.
[29] W. D. Kingery, H. K. Bowen, and D. R. Uhlmann; Introduction to Ceramics, 2nd ed. Wiley, New York, 1976.
[30] E. N. Hodkin and M. G. Nicholas, *J. Nucl. Mater.*, **47**, 23 (1973).
[31] C. C. Chang, *J. Appl. Phys.*, **39**, 5570 (1968).
[32] R. C. Kurtz and V. E. Henrich; *Surf. Sci.*, **129**, 345 (1983).
[33] G. Lehmfuhl and Y. Uchida, *Ultramicroscopy*, **4**, 275 (1979).
[34] J. Magill, J. Bloem, and R. W. Ohse, *J. Chem. Phys.*, **76**, 6227 (1982).
[35] T. W. Tsang and L. M. Falicov, *Phys. Rev. B*, **B12**, 2441 (1975).
[36] E. A. Colbourn and W. C. Mackrodt, *Solid State Ionics*, **8**, 221 (1983).
[37] P. W. Tasker and D. M. Duffy, *Surf. Sci.*, **137**, 91 (1984).
[38] J. M. Blakely and R. L. Schwoebel, *Surf. Sci.*, **26**, 321 (1971).

[39] J. R. Black and W. D. Kingery, *J. Am. Ceram. Soc.,* **62** [3–4] 176–78 (1979).
[40] W. D. Kingery, T. Mitamura, J. B. Van der Sande, and E. L. Hall, *J. Mater. Sci.,* **14**, 1766 (1979).
[41] M. W. Roberts and R. St. C. Smart, *Surf. Sci.,* **108**, 271 (1981).
[42] A. Cimino, B. A. De Angelis, G. Minelli, T. Persini, and P. Scarpini, *J. Solid State Chem.,* **33**, 403 (1980).
[43] R. C. McCune and P. Wynblatt, *J. Am. Ceram. Soc.,* **66** [2] 111–17 (1983).
[44] E. A. Colbourn and W. C. Mackrodt, *J. Mater. Sci.,* **17**, 3021 (1982).
[45] P. W. Tasker; Proceedings of the NATO ASI on Mass Transport in Solids. Edited by F. Beniere and C. R. A. Catlow. Plenum Press, New York, 1983.
[46] E. A. Colbourn, W. C. Mackrodt, and P. W. Tasker, *J. Mater. Sci.,* **18**, 1917–26 (1983).
[47] P. Masri and P. W. Tasker; to be published in *Surface Science*.
[48] P. W. Tasker, *Solid State Ionics,* **8**, 233 (1983).

Point Defects and Chemisorption at the {001} Surface of MgO

E. A. Colbourn and W. C. Mackrodt

ICI plc, New Science Group
The Heath, Runcorn
Cheshire WA7 4QE, England

Calculated point defect energies and the chemisorption of carbon monoxide and hydrogen at the {001} surface of MgO are reviewed. Differences between the bulk and surface are identified with particular reference to impurity cations and electron-transfer processes. It is shown that the dissociative chemisorption of hydrogen requires the presence of surface defects and that trapped holes are likely sites for reversible adsorption whereas anion vacancies lead to irreversible trapping. The binding of carbon monoxide to various planar and nonplanar cation sites is examined and the net charge transfer shown to be from the adsorbate to the surface in every case.

The properties of ceramic surfaces are of interest in many areas of technology, including refractories, corrosion control, catalysis, electronics, and lubrication. Experimental information has accumulated in these and other areas of materials science, although only a fraction of this is at a detailed atomic level. Consequently, understanding and predictive capability have been restricted. For example, it is only comparatively recently that the crystal structure of free planar surfaces of divalent fcc oxides has been determined,[1] while the corresponding information for even the simplest grain boundaries remains uncertain.[2] Furthermore, the structure of surface irregularities such as steps, corners, and kinks, which are implicated in the reactive properties of these materials, seems unlikely to be determined in the foreseeable future. However, recent advances in theoretical methods and computational procedures suggest that this situation might change as the reliability and scope of calculations improve. The present paper, then, is concerned with these calculations for the very simplest of ceramic surfaces, namely, the {001} surface of MgO. In particular, it is concerned with point defect structures and energetics and the chemisorption of simple molecules at planar and nonplanar surfaces. Our emphasis throughout is on aspects that might be relevant in a wider context than just MgO and on features that are likely to prove elusive to determine by experiment alone.

Theoretical Methods

The calculations reported here involve the lattice simulation of defective and nondefective surfaces and ab initio molecular orbital calculations: the details of both have been described previously.[3,4] The salient features of lattice simulation are also reviewed elsewhere in this volume.[5] In the present context, we draw attention to the following points. First, interatomic potentials for the surface are assumed to be identical to those for the bulk. This can be justified, in part, by reference to recent electronic structure calculations, which indicate that the ionicity of surface ions is very similar to that of the bulk lattice.[6] The second is that the neglect of anharmonicity and lattice expansion effects is likely to introduce greater inaccuracies in surface calculations than those for the bulk, so that we reemphasize the point made elsewhere[5] that calculated energy differences are more reliable than absolute energies. However, we stress that all defect energies, whether they refer to planar or nonplanar surfaces, are calculated with respect to the fully relaxed (and polarized) nondefective surface.[7] Our model, therefore, is that for a real surface;[8] furthermore, numerical and other procedures used in surface calculations reported here parallel exactly those for the bulk, so that even relatively minor differences between the bulk and the surface can be identified.

To investigate the chemisorption of small molecules at the {001} surface of MgO, we have adopted a finite cluster approach. Here the surface is simulated by a relaxed cluster of six surface ions in a point ion field that is adjusted to give the surface Madelung potential at the adsorption site. From molecular orbital calculations of the adsorbate plus cluster (and point ion field), we deduce chemisorption energies and equilibrium geometries. Our calculations all involve extended basis sets of the type proposed by Dunning[9] and Veillard,[10] with the inclusion of polarization functions in certain important cases. They are mostly at the SCF, near Hartree–Fock level, although more extensive calculations, including electron correlation, have been carried out for the dissociative interaction of hydrogen with surface defects.

Lattice Calculations

All the calculations reviewed here are based on shell-model, electron-gas potentials of the type discussed by Mackrodt and Stewart.[11] They involve an improved method for implementing those potentials reported by Kendrick et al.[12] and modifications to the Mott–Littleton procedure described by Duffy et al.[13] Nonplanar irregularities are represented by cavities and adatom clusters of various sizes; these have been considered in detail recently by Colbourn and Mackrodt.[14]

Surface Topography

Lattice structure is a rudimentary feature of crystalline materials; it is also highly influential in determining the nature and energetics of point defects and associated properties. We begin, therefore, with the topography of the {001} surface of MgO. Tasker considers the planar surface and periodic structures in detail elsewhere in this volume.[15] Here we concentrate on nonplanar irregularities; as noted previously, they are the putative source of chemical reactivity.[16] Experimental studies, mainly LEED, indicate that there is very little distortion of the planar nondefective {001} surface of MgO (and other divalent fcc oxides) from the perfect lattice termination.[1] The calculated relaxation, defined as the change in the distance between the two topmost layers, and rumpling, defined as the outward displacement of anions minus the outward displacement of cations, are 1.97 and 0.68% of a lattice spacing, respectively, in good agreement with experiment.[1] In contrast, nonplanar surfaces have been found to exhibit appreciable relaxation from the perfect lattice structure.[14] To illustrate this, we refer to three basic types of irregularity; they are step or ledge sites, shown as A ... A''' in Fig. 1, corner sites, shown as B ... B'', and cavity sites, shown as C' ... C'''. The detailed lattice relaxation at these sites, corresponding to zero strain, is shown in Figs. 2–4, in which the movement of ions, indicated by arrows, is drawn to scale.

At ledge sites (Fig. 2), we see that the greatest relaxations are those at A and A'. At A there is a small movement out of the surface but a more noticeable lateral

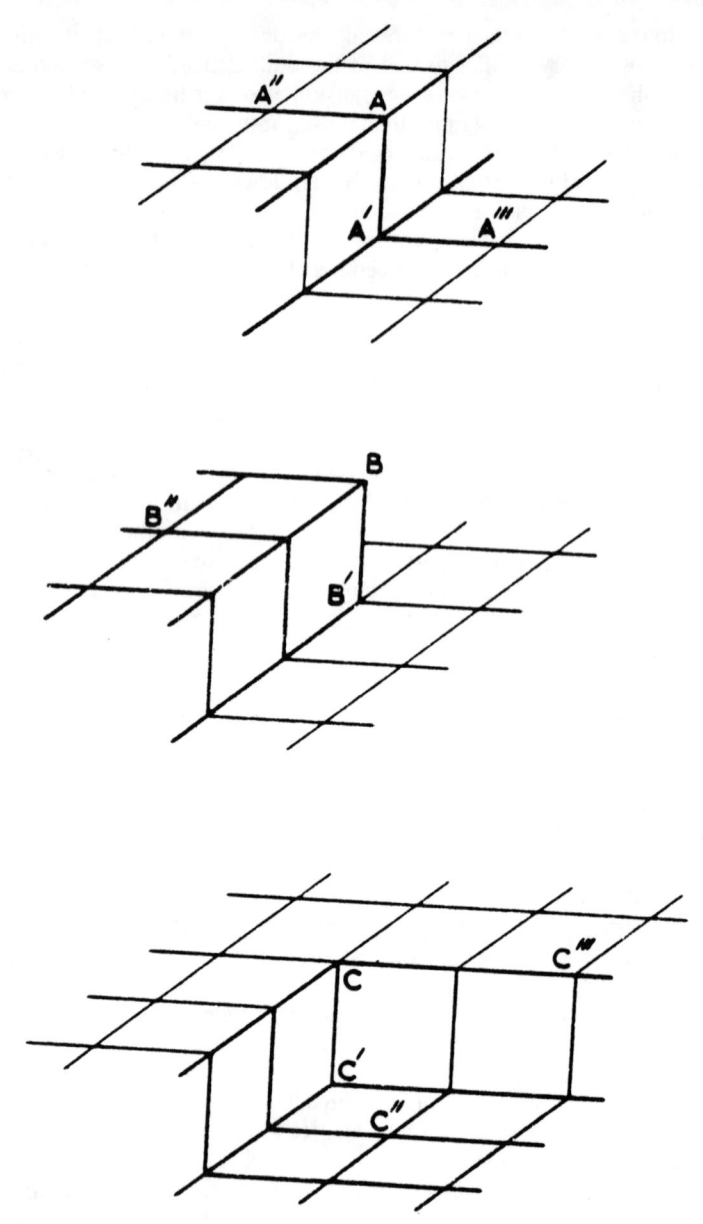

Fig. 1. Nonplanar irregularities at the {001} surface (Ref. 14).

displacement away from the perfect lattice position toward A″. Ions at A′, on the other hand, move both upward and outward, and these trends hold for anions and cations alike. At corner sites (Fig. 3), we find a marked relaxation of ions at both B and B′. At B the predominant displacement is in the x–y plane with an appreciable contraction from the perfect lattice position. At B′, the lateral movement (in the x–y plane) is small, although that away from the surface (in the z direction) is

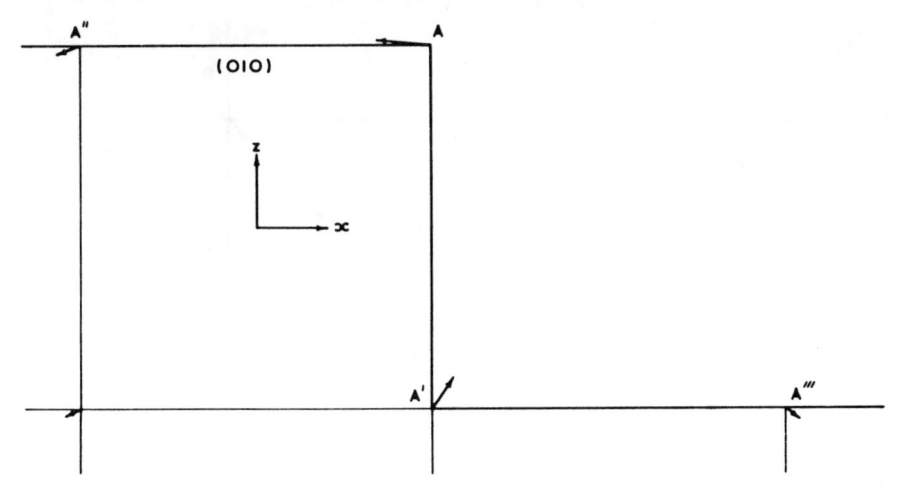

Fig. 2. Relaxed structure of step sites at the {001} surface (Ref. 14).

large, of the order of 20–25% of a lattice spacing. At cavity sites (Fig. 4) the relaxation is less than at corner sites, since the nearest neighbor coordination at C and C' is that of the planar surface, with only next-nearest neighbors removed. Nevertheless, the relaxation is still greater than 0.2 Å (0.02 nm). Overall, then, at irregularities of this type there is a marked "rounding" of the discontinuities with consequent changes in the surface area, and it seems likely that the surfaces of other ceramic materials will exhibit similar effects. Furthermore, the magnitude of these relaxations, which in the case of corner sites is ~0.5 Å (~0.05 nm), suggests that they need to be taken into account when chemical reactions at surfaces are considered, particularly in relation to potential energy profiles. We note, for example, that 0.5 Å (0.05 nm) is half the standard CH bond length and three-quarters that of H_2, so that inaccuracies in surface geometry of this order of magnitude could lead to highly unreliable conclusions about surface reactivity.

Energies of Intrinsic Defects

It is the accepted view that vacancies predominate the lattice disorder in MgO;[5] accordingly, we consider their formation, migration, and association energies at the {001} surface compared with those of the bulk. They are given in Table I. Previously, it was suggested that the formation energy of free vacancies at the {001} surface is ~0.5 eV less than in the bulk. The reason for this was thought to be a substantially reduced long-range polarization energy due to the interaction of *charged* defects at the surface with induced dipoles in the rest of the lattice. Recently, however, this has been shown not to be the case.[13] Although the volume over which the surface interaction is integrated is half that of the bulk, the electrostatic field due to a surface charge is approximately twice that due to the corresponding charge in the bulk, so that the total long-range polarization energy is practically the same in the two cases. Differences in defect energy between the bulk and surface, then, are due largely to differences in the short-range polarization and elastic interactions. As a result, we now find vacancy energies at the planar {001} surface to be 0.3–0.4 eV less than in the bulk. However, it is unlikely that energy differences of this magnitude will lead to an appreciable concentration of free vacancies at the surface. The neutral pair association energy is less than that given

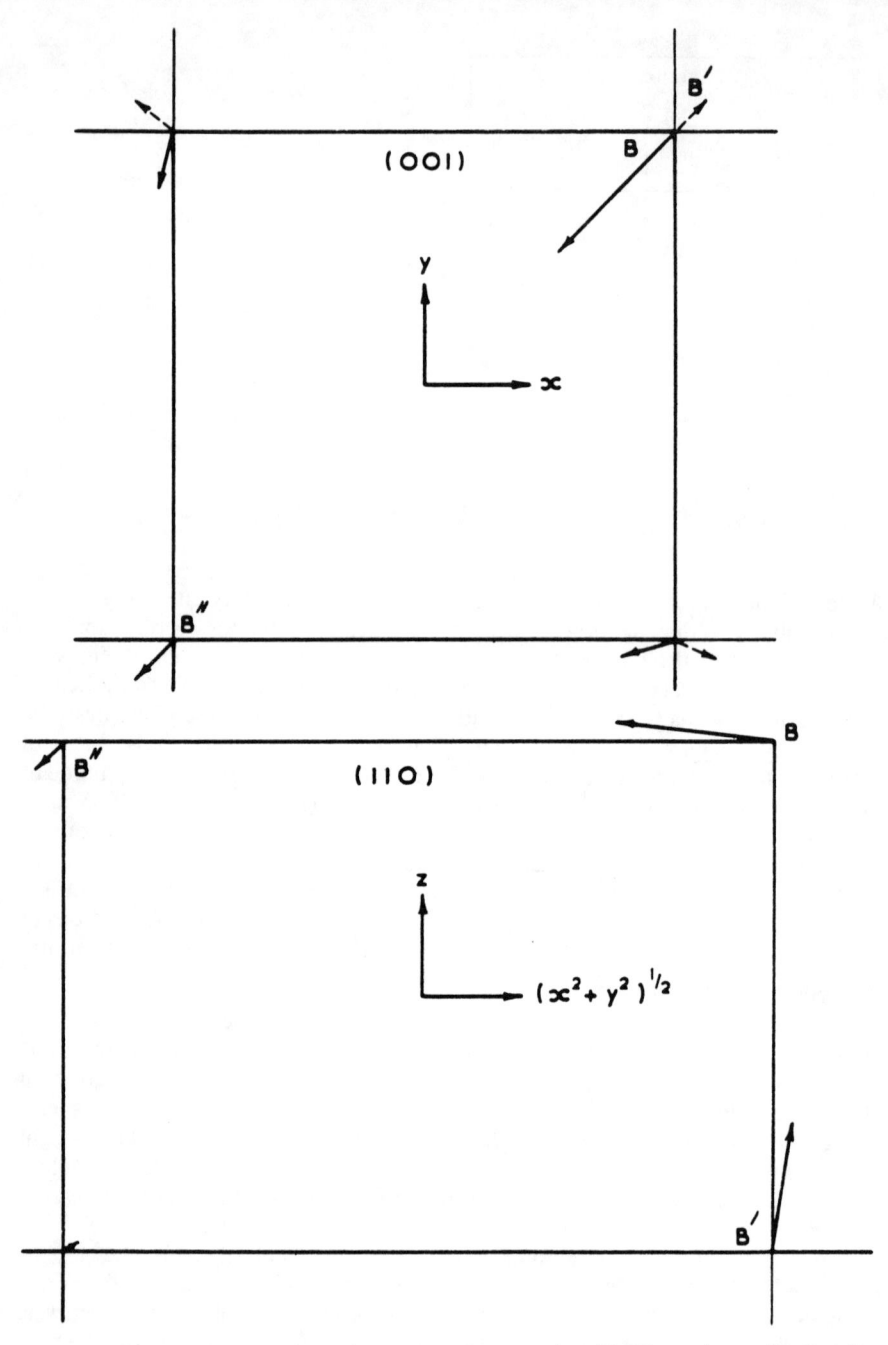

Fig. 3. Relaxed structure of corner sites at the {001} surface (Ref. 14).

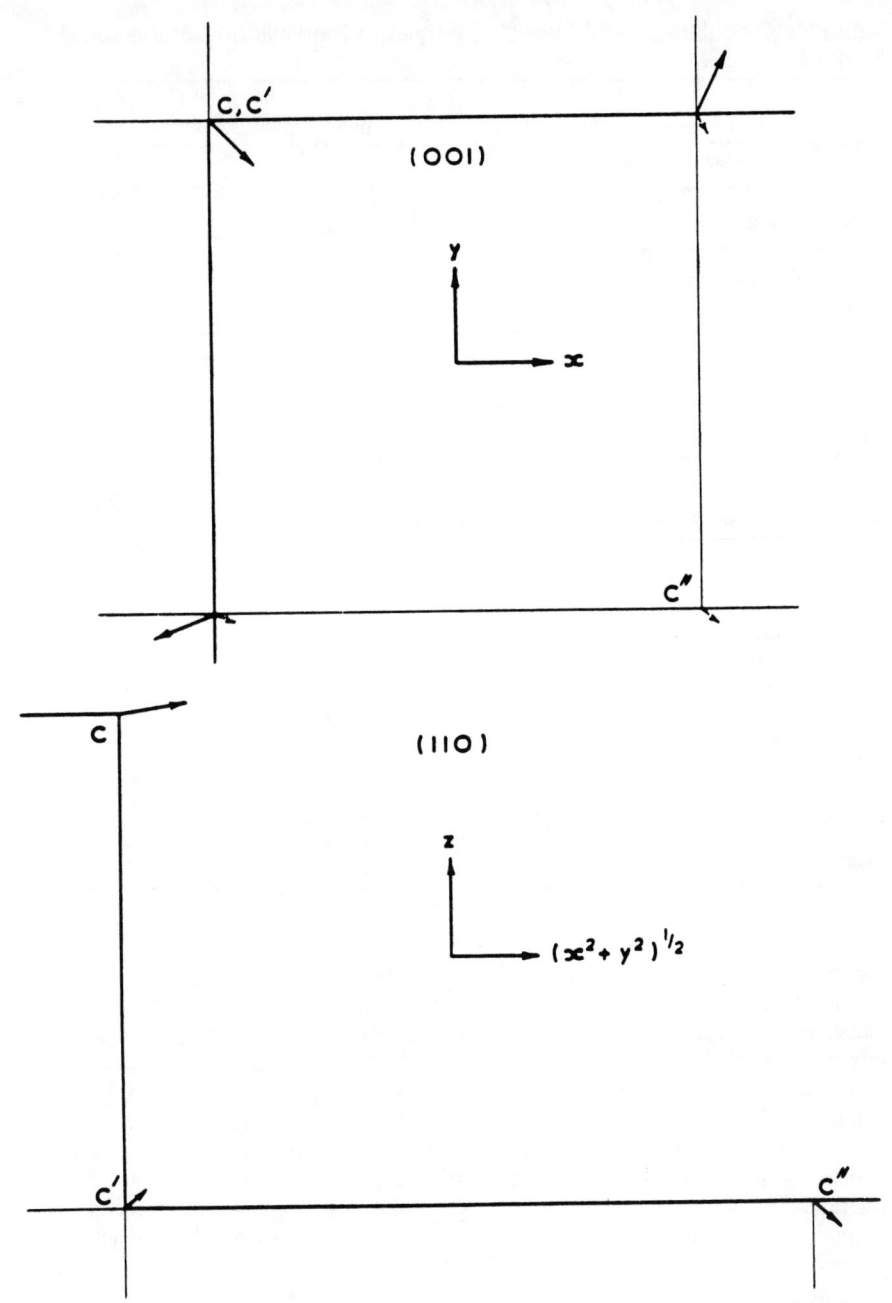

Fig. 4. Relaxed structure of cavity sites at the {001} surface (Ref. 14).

Table I. Calculated Point Defect Energies in the Bulk and at the {001} Surface of MgO

Defect	Energy (eV) Bulk	{001} Surface
Cation vacancy	25.4	24.9
Anion vacancy	22.8	22.5
Vacancy pair	45.6	43.8
Pair association	2.6	3.7
Cation vacancy migration	2.1	1.5
Anion vacancy migration	2.4	0.9

Table II. Calculated Cation Substitution Energies in the Bulk and at the Planar {001} Surface of MgO

Substituent	Energy (eV) Bulk	{001} Surface
Li^+	16.2	16.1
Na^+	18.7	17.7
Ca^{2+}	4.6	3.5
Sr^{2+}	7.9	5.6
Ba^{2+}	11.6	7.7
Al^{3+}	−29.0	−28.5
Sc^{3+}	−22.4	−22.1
Y^{3+}	−17.6	−18.0
In^{3+}	−17.8	−18.2
Si^{4+}	−69.9	−69.8
Ti^{4+}	−59.9	−58.8
Ge^{4+}	−60.0	−58.8

previously (5.2 eV),[14] though it is still greater than in the bulk. Migration energies, on the other hand, are unaffected by this change; for both cation and anion vacancies, we find a substantial reduction in the activation energy for surface migration, such that anions are the more mobile species, unlike those in the bulk. Two points emerge from these calculations that might reasonably apply more generally. The first is that the formation of vacancies is slightly more facile at surfaces than in the bulk, although the magnitude of the difference is likely to depend on the material in question. In the case of MgO, we note that different interatomic potentials lead to quite large variations in energy differences. The second point is that vacancy migration, at least, is an appreciably less activated process at surfaces than in the bulk.

Impurities

In view of the low level of intrinsic disorder in MgO and α-Al_2O_3, impurities play a major role in determining the defect equilibria in these materials. It is quite possible that they control the surface properties to an even greater extent. However, for this to be the case, there must be an adequate driving force for the concentration (or depletion) of impurities at the surface. While the basic rules that govern this are clear, calculations provide quantitative details that are instructive.

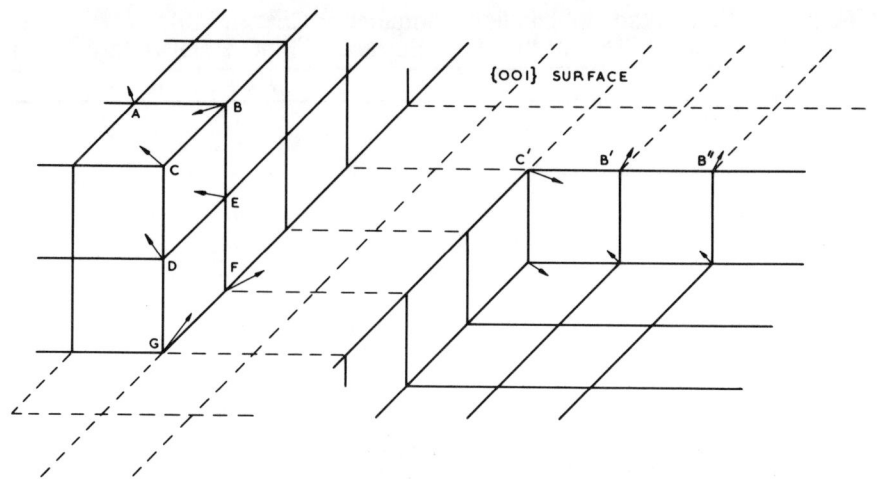

Fig. 5. Nonplanar sites for Ca^{2+} substitution at the {001} surface (Ref. 14).

There are two essential features of the surface that determine the driving force: they are the reduced Madelung potential and a small, though significant, contraction of the lattice at the surface. From this we deduce that isovalent impurities mostly favor the surface: smaller ions substitute the surface layer itself with an inward relaxation of the lattice while larger ions relax outward from the surface to reduce the elastic strain. In either case the energy is reduced by comparison with that of the bulk. Likewise, impurities with a lower charge than that of the host cation substitute at the surface preferentially, by virtue of electrostatic and elastic considerations; *both* favor the surface, irrespective of ion size. For impurities with a higher charge, on the other hand, electrostatic and elastic effects are in opposition, for the former favors bulk substitution and the latter the surface, and this can lead to a potentially complicated pattern of behavior. Table II illustrates these points in detail for a series of monovalent to tetravalent impurities in MgO. From this we note, in particular, the magnitude of the energy differences between the bulk and surface. In the extreme case of Ba^{2+} substitution, it is 3.9 eV, but differences of 1 eV or more are predicted for Na^+, Ca^{2+}, Sr^{2+}, and Ti^{4+}. This is comparable to the interaction energy of impurity-defect associates,[5] which suggests that the same defect-equilibria considerations should apply. It is important to stress that the energies given in Table II correspond to the limit of infinite dilution and that they do not determine surface segregation directly. Furthermore, in the case of aliovalent impurities, the energies of the appropriate dimers and trimers need to be considered. However, they give a strong indication as to which impurities are likely to concentrate at the surface. In the case of MgO, these are Na^+, Ca^{2+}, Sr^{2+}, Ba^{2+}, and Ti^{4+}, and this seems to be borne out in practice.[17]

It could be argued that "real" surfaces contain irregularities of the type described previously and that these might provide the most favorable sites for impurity substitution, rather than the planar surface. We have examined this possibility in the case of Ca^{2+} for various nonplanar sites shown in Fig. 5; the corresponding energies are given in Table III. For all the surface sites considered, the substitution energy is found to be lower than that in the bulk, which suggests that at infinite dilution, at least, Ca^{2+} ions will concentrate at the surface irre-

Table III. Calculated Substitution Energies for Ca^{2+} at Various Nonplanar Irregularities* at the {001} Surface of MgO Relative to the Bulk

Site	Energy difference (eV)[†]
Planar	1.1
A	0.1
B	0.7
C	0.7
D	0.5
E	0.3
F	0.6
G	0.8
C'	1.8
B'	1.1
B"	1.2

*The nonplanar sites A...B" listed here are shown in Fig. 5.
[†]Energy difference equals E(bulk) $-$ E(surface).

spective of the topography. However, this may not be a general rule, either for MgO [14] or for other materials. We find sites in the surface plane itself, such as C', B', and B", to be lowest in energy and also close to the planar value. This is particularly important in relation to more extensive calculations of surface segregation,[18] described elsewhere in this volume by Tasker.[15] The enthalpy of segregation, which can be compared with experiment, is deduced from calculations of a planar monolayer, and this is shown to be a reasonable model for a real surface.

Electron-transfer processes involving impurities play an important role in corrosion and catalysis, both of which are essentially surface phenomena. However, they are often discussed in terms of bulk energy levels, for the corresponding surface values are mostly inaccessible to direct experimental measurement. Calculations, therefore, have an obvious utility here, particularly in relation to a wider range of ceramic materials. In Table IV, we list calculated thermal excitation energies for four electron-transfer processes in the bulk and at the {001} surface of MgO. We see, first of all, that there are differences of more than 1 eV between the bulk and surface, particularly for the two impurities. For the ionization of a magnesium vacancy and the formation of a small polaron hole, we find a decrease in energy of about 1 eV; for the two impurities, on the other hand, we find increases

Table IV. Thermal Excitation Energies for Electronic Processes at the {001} Surface of MgO

Process	Excitation energy (eV) Bulk	{001} Surface
$O_O^x \rightarrow O_O^. + e_c'^*$	6.1	5.1
$V_{Mg}'' + O_O^x \rightarrow V_{Mg}''\text{--}O_O^. + e_c'$	4.6	3.4
$Fe_{Mg}^x \rightarrow Fe_{Mg}^. + e_c'$[†]	3.2[‡]	4.5
$Cu_{Mg}' \rightarrow Cu_{Mg}^x + e_c'$	2.1	3.8

*Conduction band edge taken to be -0.6 eV at the bulk and the surface. O_O^x ionization potential calculated along the lines given in Ref. 11.
[†]For Fe and Cu free-ion ionization potentials taken from Ref. 25.
[‡]Experimental value is 3.31 ± 0.5 eV reported by Sempolinski et al (Ref. 19).

of 1.3 and 1.5 eV, respectively, for copper and iron. The most significant point, however, is that calculations predict a change in the donor–acceptor properties of V''_{Mg}, Fe^x_{Mg}, and Cu'_{Mg} at the surface from that in the bulk. If we take the experimental band gap of 6.8 ± 0.5 eV reported by Sempolinski et al.,[19] V''_{Mg} is below the Fermi level in the bulk but is at it in the surface, whereas Fe^x_{Mg} and Cu'_{Mg} are above the Fermi level in the bulk but below it at the surface. Changes of this type could have important implications in the design and application of ceramic materials.

Chemisorption

Chemisorption is an essential feature of the majority of reactive processes at surfaces, including those involved in catalysis and corrosion. While phenomenological information is profuse, atomic details remain sparse, particularly with regard to the nature of the chemisorbed state, the role of the surface defect structure, and the influence of impurities. However, here as elsewhere,[5] calculations might have a useful role to play in both the interpretation of data and the provision of inaccessible information. To illustrate this potential utility in a wide area of ceramic science, we consider the chemisorption of hydrogen and carbon monoxide at the {001} surface of MgO.[20,21] In the case of hydrogen, we concentrate on dissociative chemisorption, for this is believed to be an essential precursor to a number of important processes. For carbon monoxide, on the other hand, our primary concern is with the influence of surface topography, impurity effects, chemisorption energies, and electron transfer.

Dissociative Chemisorption of Hydrogen

In view of the bond strength of hydrogen, ~435 kJ/mol, a prerequisite for the dissociative chemisorption on MgO is the formation of stable OH and MgH groups in various charge states.

We begin, therefore, by considering the interaction of a hydrogen *atom* with O^{2-}, O^-, Mg^{2+}, and Mg^+ ions at the planar {001} surface. As mentioned earlier, our approach is based on ab initio cluster calculations, which, in this case, include the effects of electron correlation. The results are given in Table V. We see that only OH^-_s (the surface OH^-_s group) is bound; in all other cases the interaction is *totally repulsive*. The calculated bond strength of OH^-_s is 524.6 kJ/mol, of which approximately 127.5 kJ/mol is due to electron correlation. This compares with a value of 489.9 kJ/mol for molecular OH^-.[20] Hence, we predict that the homolytic splitting of H_2 at the {001} surface of MgO given by

$$H_2 + (MgO)_s \rightleftharpoons OH^{2-}_s + MgH^{2+}_s$$

$$H_2 + 2O^{2-}_s \rightleftharpoons 2OH^{2-}_s$$

$$H_2 + 2Mg^{2+}_s \rightleftharpoons 2MgH^{2+}_s$$

Table V. Calculated Binding Energy of a Hydrogen Atom at Nondefective Sites at the {001} Surface of MgO

Surface site	Product	Interaction energy (kJ/mol)
O^x_O	OH^{2-}_s	Unbound
Mg^x_{Mg}	MgH^{2+}_s	Unbound
O'_O	OH^-_s	524.6
Mg'_{Mg}	MgH^+_s	Unbound

does not lead to bound states. For the alternative dissociative processes, involving the formation of OH_s^-, namely:

$$H_2 + (MgO)_s \rightleftharpoons OH_s^- + MgH_s^+ \quad \text{(heterolytic splitting)}$$
$$H_2 + 2O_s^{2-} \rightleftharpoons 2OH_s^- + 2e_c^- \quad \text{(ionized splitting)}$$

the enthalpies of chemisorption are calculated to be in excess of 10^3 kJ/mol, due primarily to the ionization potential of hydrogen (1312.5 kJ/mol). We conclude, therefore, that the dissociative chemisorption of hydrogen cannot take place at a nondefective {001} surface of MgO. Furthermore, the OH_s^- group is only slightly more bound at irregularities such as those described earlier (~563 kJ/mol at the corner site, B), so that this conclusion applies to both planar and nonplanar surfaces.

We turn now to defective surfaces and, in particular, to defects that might lead to the formation of the OH_s^- group. In effect, we have already considered the self-trapped hole, O^-; to this we add the V^- center and an anion vacancy. We use the same finite cluster approach here for the point defects and the hydrogenated products, $H_O^- - OH_s^-$ and $V_{Mg}'' - OH_s^-$, in which H_O^- is a hydride ion, H^-, trapped at an O^{2-} vacancy. The calculated chemisorption energies are given in Table VI. In each case we find dissociative chemisorption to be exothermic with a wide range of reversibilities depending on the type of defect involved. Furthermore, the three processes are all calculated to be *nonactivated*. Experimentally, it has been suggested that the surface V^- center promotes o/p conversion and H_2/D_2 exchange,[22,23] and from this it can be inferred that hydrogen (and deuterium) is dissociatively chemisorbed at these defects and that the barrier, if it exists at all, must be small, although no exact data are available for comparison.

We conclude from these results that trapped holes in the form of O^- ions in oxides more generally are likely sites for the reversible chemisorption of hydrogen, whereas anion vacancies lead to the irreversible trapping of hydrogen in the form of hydride ions and surface OH^- groups.

Table VI. Calculated Energies for the Dissociative Chemisorption of Hydrogen at Point Defects in the {001} Surface of MgO

Process	Energy (kJ/mol)
$H_2 + O_O^* \rightleftharpoons OH_s^- + H$	−93.2
$H_2 + V^-$ center $\rightleftharpoons V_{Mg}'' - OH_s^- + H$	−230.7
$H_2 + V_O^{\cdot\cdot} \rightleftharpoons H_O^- - OH_s^-$	−650.0

Table VII. Calculated Binding Energies, Equilibrium Separations, and Charge Transfer for CO Chemisorption on Planar and Nonplanar Mg^{2+} Ions at the {001} Surface of MgO

Site	Binding energy (kJ/mol)	Equilibrium separation (Å)	Net charge transfer from CO (|e|)
Planar	37.0	2.49	0.064
Ledge/step (A)	39.3	2.38	0.030
Corner (B)	60.6	2.38	0.025

Carbon Monoxide

There have been several reports of the chemisorption of CO on MgO.[24] A common feature of this work seems to have been the formation of paramagnetic species, which have been presumed to be the result of electron transfer from the surface to the adsorbate. There has been some speculation as to the nature of the active sites,[24] although further details remain elusive. We begin by considering the chemisorption of CO at the planar {001} surface, based on ab initio cluster methods referred to previously. An extensive examination of the interaction hypersurface, described in detail elsewhere,[6] indicates that CO binds to an Mg^{2+} ion in the nondefective {001} surface, through the carbon atom in a perpendicular configuration. The binding energy is calculated to be 37 kJ/mol at an equilibrium separation from the surface of 2.49 Å (0.249 nm). Furthermore, we find a net electron transfer *from* CO to the surface of $0.06|e|$, based on a Mulliken population analysis. Our calculations suggest that CO is less bound through the oxygen atom and at other positions on the surface and that it is not bound at an O^{2-} ion nor through its π-electron system, parallel to the surface. With regard to nonplanar irregularities, we have examined the chemisorption of CO at ledge or step sites, A, and at corner sites, B; the results are collected in Table VII, together with those for the planar surface. At the ledge we find a binding energy of 39.3 kJ/mol at an equilibrium distance of 2.38 Å (0.238 nm) from an Mg^{2+} ion at A, although this seems to be relatively insensitive to the orientation. At corner sites, which have threefold coordination, the binding energy is 60.6 kJ/mol at 2.38 Å (0.238 nm), and once again this is insensitive to the exact orientation. In both cases we find a net charge transfer from CO to the surface, $0.03|e|$ at A and $0.02|e|$ at B. As

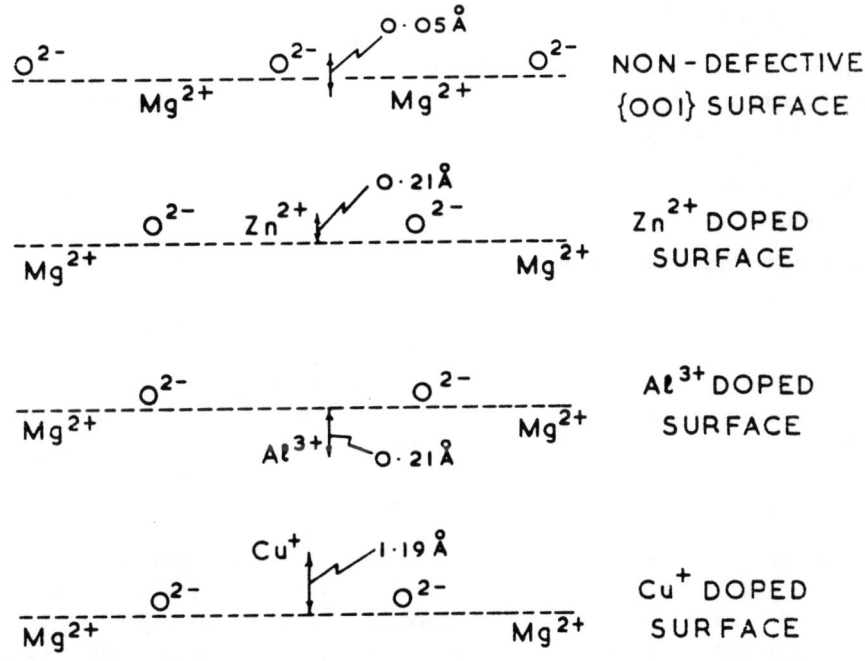

Fig. 6. Relaxed structure of doped {001} surfaces of MgO (Ref. 20).

Table VIII. Calculated Binding Energies, Equilibrium Separations, and Charge Transfer for CO Chemisorption at Impurity Ions in a Fully Relaxed {001} Surface of MgO

Impurity	Binding energy (kJ/mol)	Equilibrium separation (Å)	Net charge transfer from CO ($|e|$)
Li^+	26.8	2.38	0.155
Na^+	21.7	2.65	0.039
Cu^+	47.2	2.12	0.012
Cu^{2+}	39.0	2.38	0.060
Zn^{2+}	44.7	2.38	0.036
Al^{3+}	33.9	2.48	0.122

before, CO is unbound at O^{2-} ions in these positions. We conclude, therefore, that chemisorption of CO at *nondefective* surfaces of MgO is heteroenergetic in the range 40–60 kJ/mol and that such charge transfer that does occur is from CO to the surface.

We have also examined the chemisorption of CO at a number of nontransition-metal, mono-, di-, and trivalent impurities at the {001} surface. Here the relaxation of the defective surface is particularly important, for as Fig. 6 shows, impurity ions can move both inward and outward from the surface plane. Lattice simulation procedures were used to determine the fully relaxed surface geometries, which formed the basis for the appropriate cluster calculations. The calculated binding energies, equilibrium separations, and charge transfer for CO chemisorption at these impurity ions are given in Table VIII. As shown, we find energies in the range 21–47 kJ/mol, at equilibrium separations between 2.12–2.65 Å (0.212–0.265 nm), and a net charge transfer from CO to the surface in every case. We conclude, then, that CO chemisorbs preferentially at cation sites in MgO with energies in the range 20–60 kJ/mol with no appreciable charge transfer either to the nondefective surface or to impurity ions of the type we have examined. In the event that charge transfer occurs to give paramagnetic species such as $(CO)_2^-$, we are inclined to the view proposed by Cordischi et al.[24] that transition-metal ions at the surface are responsible.

Summary, Conclusions, and Prospects

This paper has been concerned largely with the calculation of point defect and chemisorption energies at the {001} surface of MgO, based on lattice simulation and ab initio molecular orbital methods. The point defect structure of surfaces is a largely unexplored area of ceramic science, and the principal concern here has been to identify differences between the bulk and the surface, with an emphasis on those features that might prove difficult to elucidate from experiment alone. Ab initio cluster methods have been used to investigate the dissociative chemisorption of hydrogen on MgO and the detailed energetics of carbon monoxide binding at specified lattice sites at the surface. There is experimental information on both, which the present calculations help to clarify. The principal conclusions are the following: (i) The formation energies of intrinsic defects at surfaces that are at equilibrium are similar to, though slightly lower than, the corresponding bulk values; migration energies, on the other hand, appear to be appreciably lower. (ii) Isolated impurities distinguish between the bulk and surface by virtue of both charge and size effects. Mono- and divalent impurities seem to favor the surface and tetravalent cations the bulk, while trivalent dopants exhibit a mixed preference. (iii) Thermal excitation energies for electronic processes at surfaces differ from those in the bulk with predicted variations in excess of 1 eV. (iv) The dissociative chemisorption of hydrogen on predominantly ionic oxides requires the presence of surface defects. Trapped hole states involving O$^-$ ions are likely sites for reversible chemisorption, whereas anion vacancies lead to irreversible binding. (v) The chemisorption of carbon monoxide on MgO is reversible and heteroenergetic, with binding energies in the range 20–60 kJ/mol. There is only limited charge transfer, and that is from CO to the surface. Chemisorption on transition-metal ions seems the most probable source for the reported paramagnetic species found experimentally.

In view of the difficulties associated with the investigation of insulator surfaces, calculations seem likely to play an increasing role in the science and technology of ceramic surfaces. A substantial part of this will be the provision of structural, thermodynamic, and mechanistic information inaccessible to experiment. In the case of MgO, a comprehensive account of defect-associated properties is in prospect, including features such as the structure and stability of nonplanar surfaces, the segregation of impurities, surface diffusion, adatom effects, surface reconstruction, and vibrational properties. Improvements in interatomic potentials will lead inevitably to the treatment of more complex materials, and with this, theoretical approaches to chemisorption and surface reactivity will move away from "model" calculations to those that reflect the genuine properties of real systems.

References

[1] M. R. Welton-Cook and W. Berndt, "A LEED Study of the MgO (100) Surface: Identification of a Finite Rumple," *J. Phys. C*, **15**, 5691 (1982).
[2] D. Wolf and R. Benedek, "Energy of (001) Coincidence Twist Boundaries in Oxides with NaCl Structure"; Advances in Ceramics, Vol. 1. Edited by L. M. Levinson. The American Ceramic Society, Inc., Columbus, OH, 1981.
[3] C. R. A. Catlow and W. C. Mackrodt; Chapter 1 in Computer Simulation of Solids. Edited by C. R. A. Catlow and W. C. Mackrodt. Springer-Verlag, Berlin, 1982.
[4] V. R. Saunders and M. F. Guest; Atmol 3 Documentation. S.E.R.C. Rutherford Laboratory, 1976.
[5] W. C. Mackrodt, "Calculated Point Defect Formation, Migration, and Association Energies in MgO and α-Al$_2$O$_3$"; these proceedings.
[6] E. A. Colbourn and W. C. Mackrodt; unpublished results.
[7] W. C. Mackrodt and R. F. Stewart, "Defect Properties of Ionic Solids: I. Point Defects at the Surfaces of Face-Centred Cubic Crystals," *J. Phys. C*, **10**, 1431 (1977).
[8] A. M. Stoneham, "Ceramic Surfaces: Theoretical Studies," *J. Am. Ceram. Soc.*, **64**, 54 (1981).
[9] T. H. Dunning and P. J. Hay; Chapter 1 in Methods of Electronic Structure Theory. Edited by H. F. Schaefer. Plenum, New York, 1977.
[10] A. Veillard, "Gaussian Basis Set for Molecular Wavefunctions Containing Second-Row Atoms," *Theor. Chim. Acta*, **12**, 405 (1968).
[11] W. C. Mackrodt and R. F. Stewart, "Defect Properties of Ionic Solids: II. Point Defect Energies Based on Modified Electron-Gas Potentials," *J. Phys. C*, **12**, 431 (1979).
[12] J. Kendrick, E. A. Colbourn, and W. C. Mackrodt, "The Calculated Defect Structure of Planar and Non-Planar Surfaces of MgO and FeO," *Rad. Eff.*, **73**, 259 (1983).
[13] D. Duffy, P. W. Tasker, E. A. Colbourn, and W. C. Mackrodt; unpublished results.
[14] E. A. Colbourn and W. C. Mackrodt, "Irregularities at the {001} Surface of MgO: Topography and Other Aspects," *Solid State Ionics*, **8**, 221 (1983).
[15] P. W. Tasker, "The Surfaces of Magnesia and Aluminum"; these proceedings.
[16] S. Ferrer, J. M. Rojo, M. Salmeron, and G. A. Somorjai, "The Role of Surface Irregularities (Steps, Kinks) and Point Defects on the Chemical Reactivity of Solid Surfaces," *Philos. Mag.*, [Part A], **A45**, 261 (1982).
[17] E. A. Colbourn and W. C. Mackrodt, "The Calculated Defect Structure of Bulk and {001} Surface Cation Dopants in MgO," *J. Mater. Sci.*, **17**, 3021 (1982).

[18] E. A. Colbourn, W. C. Mackrodt, and P. W. Tasker, "The Segregation of Calcium Ions at the Surface of Magnesium Oxide: Theory and Calculation," *J. Mater. Sci.*, **18**, 1917 (1983).

[19] D. R. Sempolinski, W. D. Kingery, and H. L. Tuller, "Electron Conductivity of Single Crystalline Magnesium Oxide," *J. Am. Ceram. Soc.*, **63**, 669 (1980).

[20] E. A. Colbourn and W. C. Mackrodt, "Theoretical Aspects of H_2 and CO Chemisorption on MgO Surfaces," *Surf. Sci.*, **117**, 571 (1982).

[21] E. A. Colbourn, J. Kendrick, and W. C. Mackrodt, "Defective Non-Planar Surfaces of MgO," *Surf. Sci.*, **126**, 550 (1983).

[22] J. H. Lunsford, "A Study of Irradiation-Induced Active Sites on Magnesium Oxide Using Electron Paramagnetic Resonance," *J. Phys. Chem.*, **68**, 2312 (1964).

[23] D. D. Eley and M. A. Zammitt, "Spin Centres and Catalysis on Magnesia," *J. Catal.*, **21**, 377 (1971).

[24] (a) J. H. Lunsford and J. P. Jayne, "Study of CO Radicals on Magnesium Oxide with Electron Paramagnetic Resonance Effects," *J. Chem. Phys.*, **44**, 1492 (1966). (b) A. Zecchina and F. S. Stone, "Reflectance Spectra of Carbon Monoxide Adsorbed on Alkaline Earth Oxides," *J. Chem. Soc., Faraday Trans. 1*, **74**, 2278 (1978). (c) D. Cordischi, V. Indovina, and M. Occhiuzzi, "Formation of Polymeric Radical Ions by Adsorption of CO on High Surface Area CoO–MgO," *J. Chem. Soc., Faraday Trans. 1*, **76**, 1147 (1980).

[25] R. C. Weast; Handbook of Physics and Chemistry, 60th ed. CRC Press, Baton Rouge, LA, 1981.

Surface Electronic Structure of MgO and Al$_2$O$_3$ Ceramics

Victor E. Henrich

Yale University
Applied Physics
New Haven, CT 06520

The current state of our understanding of the electronic structure of MgO and Al$_2$O$_3$ surfaces is summarized. The small amount of information that is available for Al$_2$O$_3$ surfaces is consistent with surface electron structure being essentially the same as that of the bulk. Electron spectroscopic measurements performed on the (100) cleavage face of single-crystal MgO indicate a valence band structure similar to that of the bulk but with a reduction of the optical band gap at the surface by 1.5–2 eV. This effect and other changes in the electronic structure of surface cations result from the presence of strong electric fields at the surface of ionic insulators.

MgO and Al$_2$O$_3$ are two of the most important ceramic materials, and as such, most of their bulk and some of their surface properties have been rather thoroughly investigated. But studies of the electronic properties of their surfaces have lagged behind. This is partly a practical problem, since both materials are excellent insulators and the surface-sensitive electron-spectroscopic techniques that have been used so successfully with metals and semiconductors are difficult to apply. But our lack of knowledge is also partially a motivational problem. Since MgO and Al$_2$O$_3$ have band gaps of 8–9 eV, they are highly insulating compared to most other materials, including transition-metal oxides; they are thus often relegated to the role of substrates or supports, and surface scientists have concentrated their efforts more on the metallic and semiconducting components of systems, where electrical conduction is important.

This neglect of the surface electronic structure of MgO and Al$_2$O$_3$ is unfortunate, since all materials interact with their environment via their surface ions. Mackrodt has discussed the very important problem of chemisorption on Al$_2$O$_3$ and MgO, where the electronic charge distribution on the surface ions plays an important role.[1] This paper will consider what is currently known about the surface electronic structure of Al$_2$O$_3$ and MgO. The discussion will be concerned primarily with studies of nearly perfect single-crystal surfaces, since only on such surfaces is it possible to compare experimental results with theoretical calculations. The atomic geometry of both perfect and defect MgO and Al$_2$O$_3$ surfaces is considered in the section titled Geometry of MgO and Al$_2$O$_3$ Surfaces, and the electronic structure of Al$_2$O$_3$ and MgO surfaces is discussed in sections termed Surface Electronic Structure of Al$_2$O$_3$ and Electronic Structure of MgO (100).

Geometry of MgO and Al$_2$O$_3$ Surfaces

The bulk crystal structures of MgO (cubic rock salt) and Al$_2$O$_3$ (trigonal corundum) are fundamentally different. A common feature of both lattices, how-

Rocksalt (100)

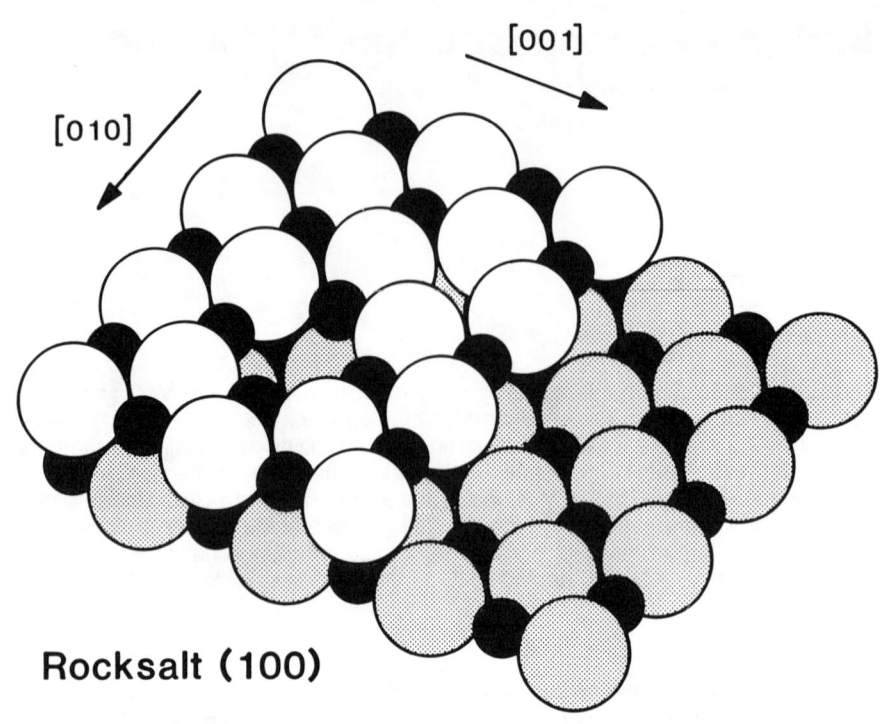

Fig. 1. Model of the rock-salt (100) surface. Large circles are O anions and small circles are metal cations. A {100} step to another (100) terrace is shown, as are both missing anion and missing cation point defects.

ever, is the octahedral O–ligand coordination of the metal cations. In MgO, the coordination is exactly octahedral, while the octahedron is distorted in Al_2O_3. This similar cation ligand coordination is apparent in their surface geometries.[2] The MgO lattice is by far the simpler one, and MgO is found to cleave preferentially along the cubic (100) face; other crystal planes have much higher surface energies and may even be unstable.[3,4] The rock-salt (100) surface is shown in Fig. 1. A surface step and point defect have also been included; these will be discussed below. Both cations and anions on the (100) surface are 5-fold coordinated with nearest-neighbor ligands.

Al_2O_3 does not exhibit any good cleavage plane. Several stable surface planes occur in other corundum structure oxides,[2] and two of them are shown in Figs. 2 and 3. The cleavage plane for the transition-metal oxides Ti_2O_3[5] and V_2O_3[6] is shown in Fig. 2. This plane indexes nearly as (047) in binary–bisectrix–trigonal notation and exactly as ($10\bar{1}2$) in hexagonal notation. This is also a common orientation for sapphire substrates that are used for epitaxial growth of semiconductors, superconductors, etc. In the bulk corundum structure, one-third of the lattice sites that are surrounded by O octahedra are empty, and the (047) surface corresponds to cleavage along the plane containing those vacancies. Although this surface is quite different from that of rock salt (100), the local ligand coordination of the surface cations is also 5-fold. Adjacent O octahedra are inclined alternately with respect to the macroscopic surface plane. (Follow a line of uppermost,

Corundum (047) [100]

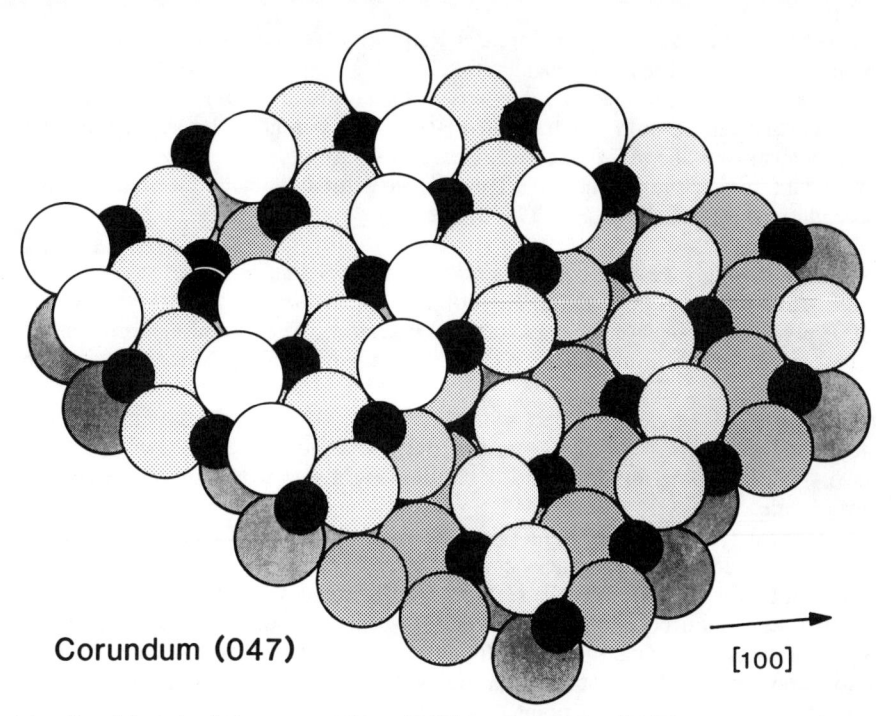

Fig. 2. Model of the corundum (047) surface. An {047} step to another (047) terrace and an O vacancy defect are shown.

Corundum (001)

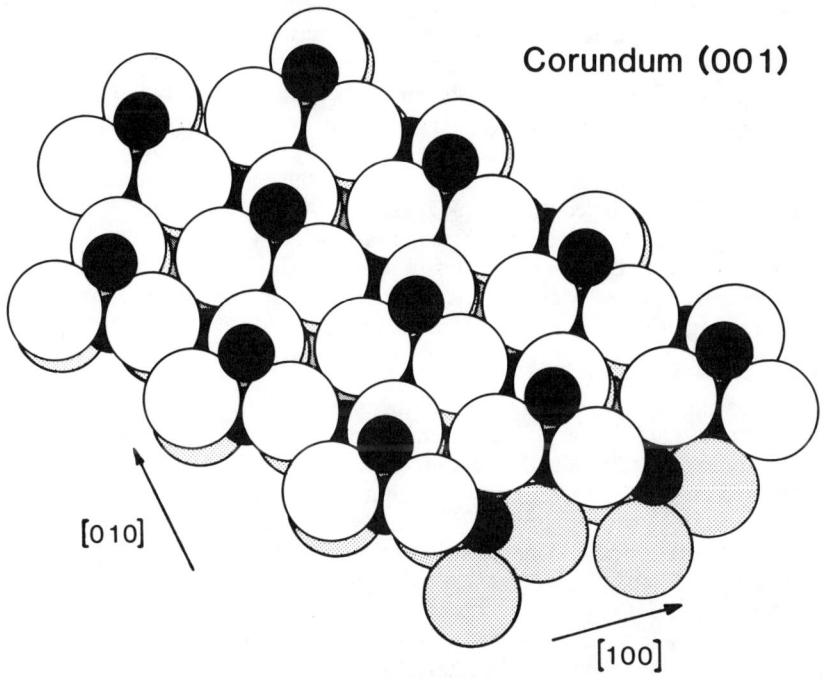

[010] [100]

Fig. 3. Model of the corundum (001) surface.

unshaded O ions vertically in Fig. 2.) Point defects and steps have also been included in Fig. 2 for later consideration.

Figure 3 shows the atomic geometry of the corundum (001) surface, which is an important one in both Al_2O_3 and α-Fe_2O_3.[7] This is a nonpolar surface in spite of its outermost plane of 3-fold coordinated cations. (Opposing cleavage faces have exactly the same atomic structure.) This surface is fundamentally different than the corundum (047) and the rock-salt (100) due to the much lower ligand coordination of the surface cations.

Any real surface will contain a myriad of steps and point defects; some of these are shown in Figs. 1 and 2. A {100} step of one atom height down to another (100) plane is shown for the rock-salt (100) surface. The ligand environment of both cations and anions is different at step edges than on terraces, with the coordination reduced to 4-fold. A double-height step would not exhibit any new types of sites. A single atom step on the corundum (047) surface, having an {047} step face, is shown in Fig. 2. As in the rock-salt structure, the step edge cations have 4-fold coordination. Here also a double-height step would introduce no new surface sites.

Several types of point defects are also shown in Figs. 1 and 2. The most common defect on oxide surfaces consists of an O vacancy,[2] and one such defect is shown on the corundum (047) surface. Two of the three cations adjacent to the vacancy have their coordination reduced to 4-fold, with the third cation (lying in the plane below) now having five O ligands. An important difference between the 4-fold coordinated cations at an O vacancy and those at step edges is that removing a surface O ion to create a point defect greatly reduces the screening between adjacent cations. Charge neutrality requires that one or two electrons not normally on the cations be associated with the cations at defect sites. Due to the reduced cation–cation screening, this charge may be covalently shared by two or more cations. Studies of defects on the surfaces of transition-metal oxides show a profoundly different electronic structure at defect sites than at steps.[2]

Two types of point defects are shown on rock salt (100) in Fig. 1. A missing O ion, referred to as a surface F (or F_s) center, creates a symmetric cluster of four 4-fold coordinated cations in the surface plane and one 5-fold cation in the plane below. More cations are thus available to share the charge associated with the defect than those on the corundum (047) surface. A missing surface cation is also shown in Fig. 1. Such a defect is referred to as a surface V (or V_s) center, in analogy with bulk notation. Both types of defect sites are shown since, as discussed in Electronic Structure of MgO(100), certain features in the electron energy loss spectra for MgO(100) could arise from either type of defect.

Surface Electronic Structure of Al_2O_3

Schematic energy level diagrams of the bulk electronic structure of MgO and Al_2O_3 are presented in Fig. 4. The valence bands of both oxides consist primarily of O(2p) orbitals, with a small admixture of cation 3s and 3p orbitals.[8] The cation orbitals and the bonding O(2p) orbitals comprise the lower half of the band, with the upper half consisting primarily of nonbonding O(2p) orbitals. The lower portion of the conduction bands consists primarily of cation 3s and 3p orbitals.[8] The bulk optical band gap is about 7.8 eV in MgO[9] and 9.5 eV in Al_2O_3;[10] thermal band gaps are 1–3 eV smaller.[11,12]

The electronic structure that is of primary importance for the interaction of a surface with its surroundings, whether at solid/solid, solid/liquid, or solid/gas interfaces, is that of the outer few atomic layers. Experimental determinations of

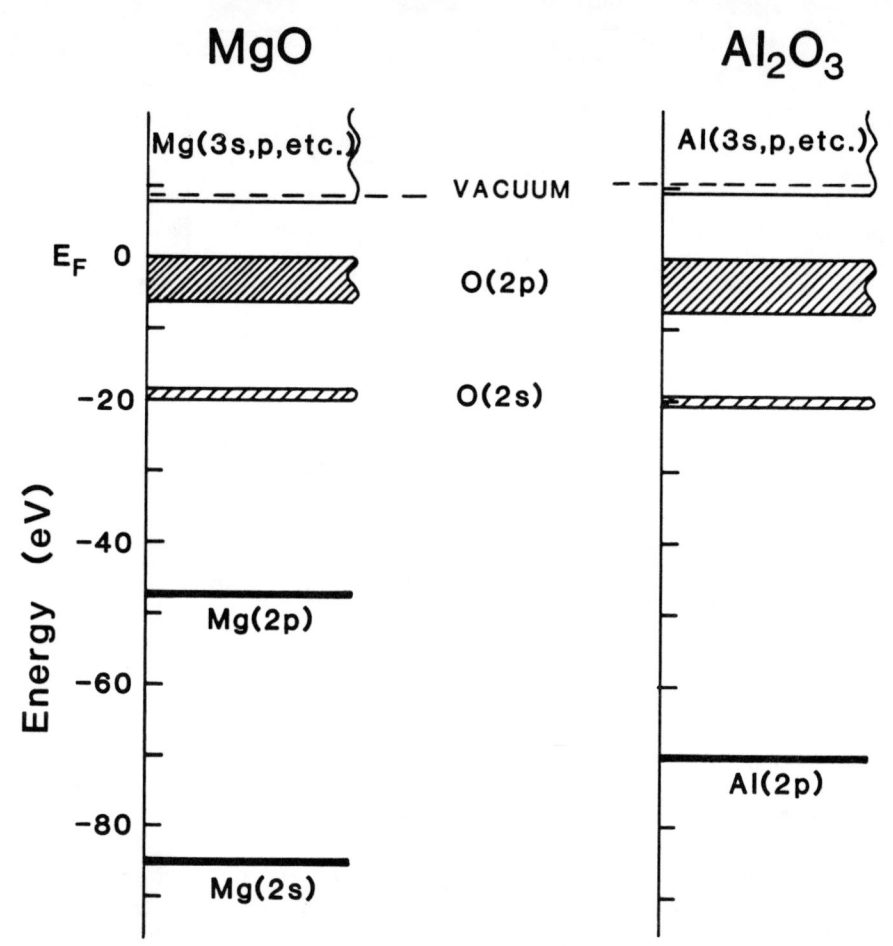

Fig. 4. Schematic energy level diagrams of the bulk electronic structure of MgO and Al_2O_3.

this structure require the use of probes that are sensitive to only the first few angstroms of the crystal. The most commonly used surface probes — ultraviolet and X-ray photoemission spectroscopy (UPS, XPS), low-energy electron diffraction (LEED), Auger and electron energy loss (ELS) spectroscopies, etc. — utilize the extremely short mean free paths (as small as a few Å) of electrons in the energy range of 10–1000 eV. All of the results discussed here have employed those techniques, usually in combination with an ultrahigh-vacuum (UHV) environment to ensure surface cleanliness.

The most direct surface electronic structure information is generally obtained from UPS and XPS, where electrons on surface ions are photoexcited into vacuum and their kinetic energies are measured with an electron spectrometer. A problem with these spectroscopies is that there is a net flux of electrons out of the sample surface, which results in a positive surface charge if the sample does not have sufficient conductivity to allow electrons to flow to the sample surface from the support. At best, the positive surface charge shifts the UPS or XPS spectra so that

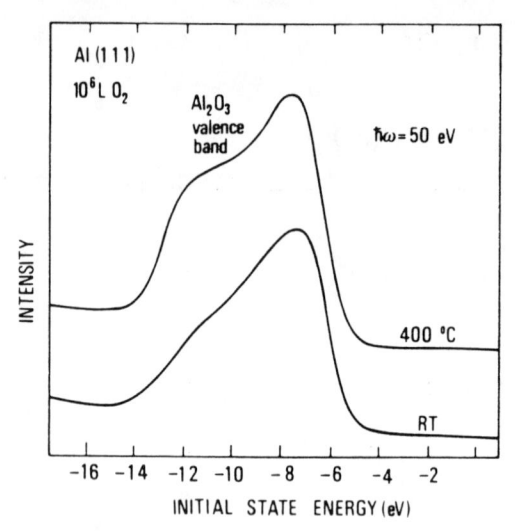

Fig. 5. Photoemission spectrum, using 50-eV photon energy, of the valence band of Al$_2$O$_3$ grown on the (111) surface of an Al single crystal exposed to 10^6 L of oxygen. The spectrum of the crystalline oxide was obtained by heating the sample to 400°C (Ref. 14).

absolute band positions cannot be measured; at worst, it eliminates the spectra entirely. Al$_2$O$_3$ and MgO are such good insulators that no reliable UPS spectra from single-crystal surfaces have yet been reported. XPS spectra from both oxides have been obtained, but use of an electron flood gun was necessary to minimize charging.[13] Thus, absolute binding energies could not be obtained.

UPS spectra of the valence band region of a thin α-Al$_2$O$_3$ layer grown on Al have been measured by Bianconi et al.,[14] and their data are shown as the 400°C curve in Fig. 5. Since the photon energy used is close to the minimum in the electron mean free path in solids, the spectra originate from only the first few angstroms of the surface. The O(2p) valence band is seen to consist of two main peaks, with a total band width of 7–8 eV. This is only slightly larger than the theoretically predicted values of about 6.5 eV.[12] The peak near the upper edge of the valence band presumably corresponds to the O(2p) nonbonding orbitals, and the other peak should consist of the bonding O(2p) orbitals, with some Al (3s) and (3p) admixture.[8] There are thus no observable differences between this surface valence band spectrum and that for bulk Al$_2$O$_3$. This is in agreement with measurements on a number of transition-metal oxides, where the density of occupied states at the surface looks essentially like that of the bulk.[2] Since the absolute energy location of the valence band cannot be determined, it is not possible to determine the position of the Fermi level, E_F, at the surface.

One of the ways in which the empty density of surface states above E_F can be examined is by electron energy loss spectroscopy.[15] ELS spectra are somewhat more difficult to interpret than UPS or XPS spectra in that they yield the joint density of both filled and empty states. ELS measurements have been made recently on thin films of α-Al$_2$O$_3$ by Olivier and Poirier.[16] Their spectra are consistent with optical reflectance measurements performed on single-crystal Al$_2$O$_3$, which measure bulk band structure parameters.[10] There is thus no evidence

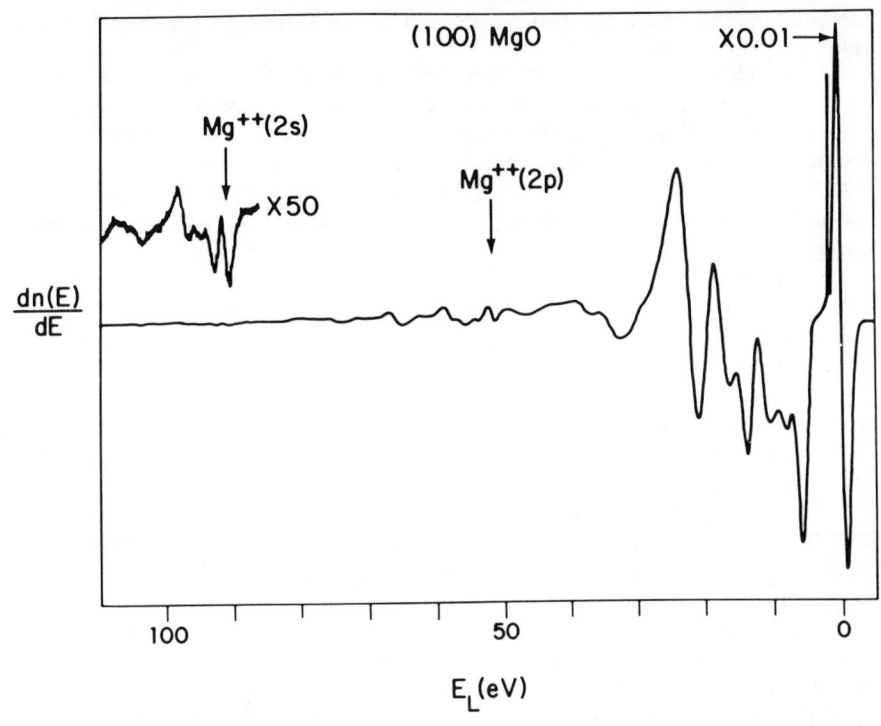

Fig. 6. Electron energy loss spectrum, $dn(E)/dE$ vs E_L, for MgO(100) with $E_p = 200$ eV. Arrows indicate thresholds for excitation from the $Mg^{2+}(2p)$ and $Mg^{2+}(2s)$ core levels.

to date to indicate that the electronic structure of Al_2O_3 surfaces is different than that of the bulk.

The measurements made to date on Al_2O_3 surfaces are crude compared to those that have been performed on semiconductor or metal surfaces. No measurements have been made on surfaces whose geometric structure has been characterized by LEED. No work has been done on either point defects or steps on Al_2O_3 surfaces. On the basis of the behavior of corundum transition-metal oxides, point defects on Al_2O_3 would be expected to have a major effect on electronic structure.[2] No theoretical work has been done on any of the electronic properties of Al_2O_3 surfaces. In short, we know little about the details of the surface electronic structure of Al_2O_3, and a great deal more work, both theoretical and experimental, is clearly necessary.

Electronic Structure of MgO(100)

The (100) surface of MgO is better understood than are Al_2O_3 surfaces. LEED measurements on UHV-cleaved single-crystal surfaces have shown that the surface geometry is essentially a termination of the bulk lattice.[17] No UPS spectra are available for MgO, but XPS measurements indicate a two-peaked valence band having a width of 6.5 eV, very similar to that for Al_2O_3 and other oxides.[13] (It should be noted that XPS spectra are somewhat less surface sensitive than are the 50-eV UPS spectra shown in Fig. 5.) There is thus no clear evidence that the structure of filled electronic states on MgO(100) is measurably different from the bulk structure.

The spectrum of empty electronic states on MgO(100) surfaces and the properties of point defects have been studied by ELS by Henrich et al.[15] and by Underhill and Gallon.[18] Electron energy loss spectra exhibit peaks in $n(E)$ corresponding to transitions from maxima in the initial, filled density of states to maxima in the final, empty density of states.[15] Figure 6 shows the first 100-eV range of the ELS spectrum for MgO(100), taken with a primary electron beam energy of 200 eV.[15] (The ordinate in Fig. 6 is the first derivative of the electron energy distribution, so maxima in the joint density of states appear as maximum negative slopes in $dn(E)/dE$.) Due to the short electron mean free path, such spectra are extremely surface sensitive. The features in Fig. 6 can be correlated with those of the energy level diagram for MgO in Fig. 4. Below 25 eV, the loss spectra arise from O-to-Mg interionic transitions, with the exception of a bulk plasmon loss feature at 22.4 eV. Between 50 and 90 eV, the loss spectrum arises from intraionic Mg(2p) → Mg(3s) etc. levels. Above 90 eV, transitions originating with the Mg(2s) core level occur.

By varying the primary electron energy used in an ELS experiment, it is possible to separate the surface from the bulk electronic structure. Those measurements have been reported previously,[15,19] and they will only be summarized here. Transitions originating with the narrow Mg(2p) and Mg(2s) core levels yield directly the density of empty final states in the conduction band of MgO (with some matrix element effects). Those spectra indicate that the resulting excited state of the bulk Mg^{2+} ions looks very nearly like that of a free Mg^{2+} ion. Surface Mg^{2+} ions give rise to a spectrum whose peaks are shifted relative to those of the bulk ions by the strong surface electric field that is generated by the gradient of the Madelung potential at the surface. The magnitude of the electric field present at the site of a surface Mg ion was found from the Stark shift to be about 2.9×10^8 V/cm, compared to the value of 1.9×10^8 V/cm calculated for the Madelung electric field at the center of a Mg^{2+} ion on an unrelaxed MgO(100) surface.[19]

Subsequent shell model calculations by Martin and Bilz[20] conclude that a small surface rumpling and polarization of the surface O ions will result in a 30% increase in the surface electric field, more nearly in agreement with the above experimental results. Associated with this increase in electric field is a 5% increase in the ionicity of the surface O and Mg ions. Recent He diffraction measurements from MgO(100), however, suggest that the surface ions may be slightly less ionic than bulk ions.[21] The two experiments may be sensitive to somewhat different surface properties, but no clear consensus has yet been reached on the charge on the surface ions for MgO(100).

A different class of transitions occurs for loss energies below 20 eV. The spectra in this region arise from O-to-Mg transitions, which are no longer localized on a single ion. Figure 7 shows the low-energy region of the ELS spectrum for primary electron energies from 100 to 2000 eV.[15] The lower primary energies are more surface sensitive, and the spectra taken with 100 eV electrons exhibit primarily the surface electronic structure while those taken with 2000-eV electrons are dominated by the bulk electronic structure. The only feature that varies strongly with primary electron energy corresponds to a transition at 6.2 eV, whose amplitude change identifies it as being of surface origin.

That the loss peak at 6.2 eV involves surface ions is also obvious since its energy is 2 eV less than that of the bulk band gap. The locations of features in the

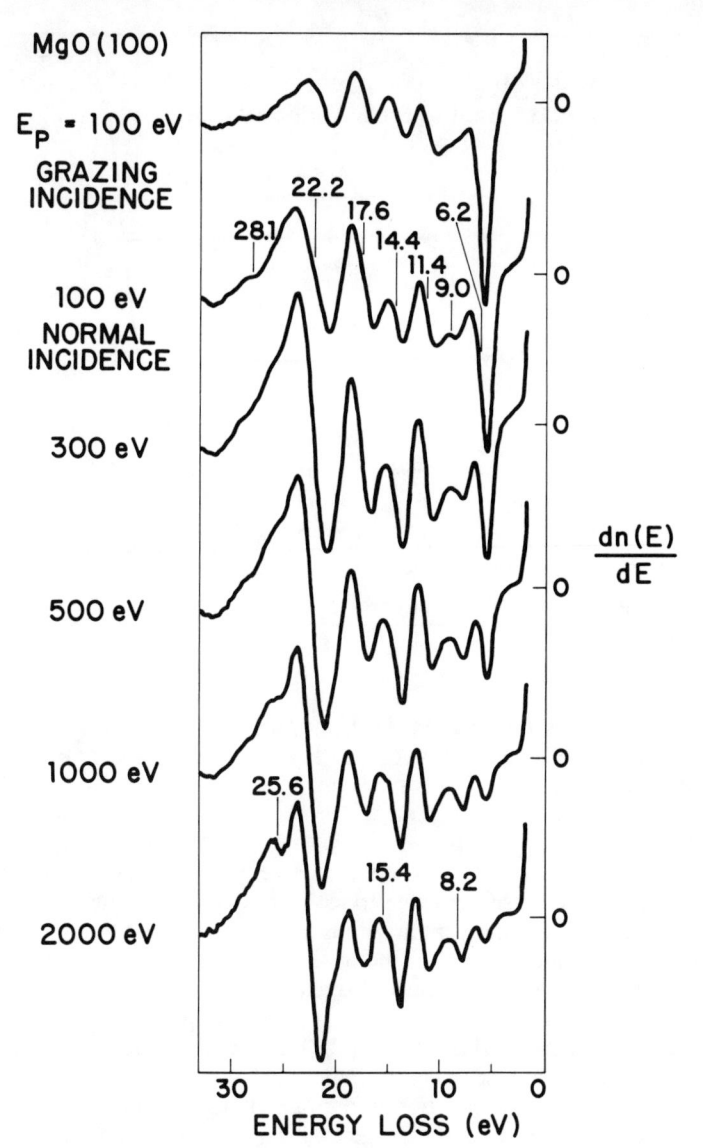

Fig. 7. ELS spectra for annealed MgO(100) for 100 eV $\leq E_p \leq$ 2000 eV.

ELS spectra are compared with the energy loss function, $Im(-1/\varepsilon)$, computed from ultraviolet reflectance measurements, in Fig. 8. Reflectance spectra measure the bulk electronic structure only. All features are in excellent agreement except for the 6.2-eV surface loss peak. The optical band gap at the surface of MgO(100) thus appears to be 1.5–2 eV smaller than the bulk band gap.

The surface electronic structure of MgO has been considered theoretically by two groups. Lee and Wong[22] have performed Green's function and LCAO calculations of the surface energy band structure of MgO(100). They find a surface-state band that extends about 0.5 eV above the top of the bulk O(2p) valence band,

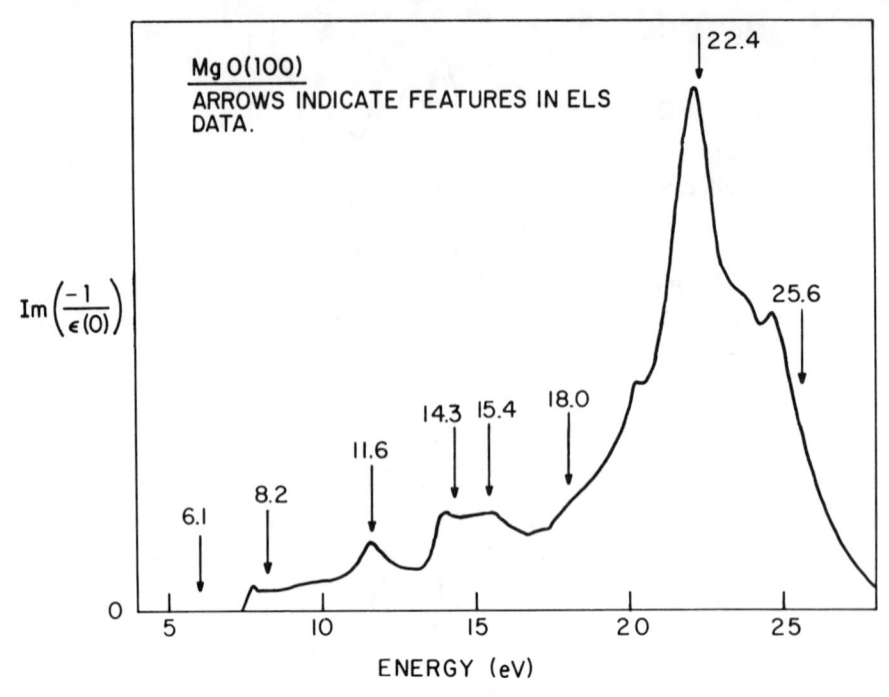

Fig. 8. Energy loss function for MgO computed from dielectric constant measurements of Roessler and Walker (Ref. 9). Arrows indicate location of observed ELS peaks.

while the Mg(3s) surface band lies entirely above the bulk conduction band minimum. No direct transitions involving surface states exist in their model for energies less than 7.6 eV, and no indirect transitions have energies less than 7.3 eV. Their calculation thus cannot account for the observed loss peak at 6.2 eV.

Satoko et al.[23,24] have performed discrete variational $X\alpha$ cluster calculations for both bulk and surface MgO clusters. They find that a peak in the surface density of states occurs 2 eV below the bottom of the bulk Mg(3s) conduction band, which is in agreement with the ELS spectra. In those calculations, the 2-eV shift results from a balance between a reduction in the Madelung potential near the surface, differences between cation–anion charge transfer for surface and bulk ions, and a polarization of the wave function of the surface ions due to the potential gradient at the surface. The first and third of these are also included in the Stark effect interpretation of the surface shift of intraionic ELS transitions discussed above. It thus seems well established that strong electric fields at the surface of MgO alter the electronic orbital structure of surface ions. Defect lattice calculations by Kendrick et al.[11] also show a 1-eV narrowing of the thermal band gap of MgO(100) at the surface.

Electron energy loss measurements have also identified a defect surface state on MgO(100).[15,18,25] Figure 9 shows the ELS spectrum from a UHV-cleaved MgO(100) surface (a) before and (b) after exposure to 10^6 L of O_2. The clean, cleaved surface spectrum in Fig. 9(a) exhibits a peak at about 2.3 eV that was absent for the annealed surface shown in Fig. 7. This peak is generally present on

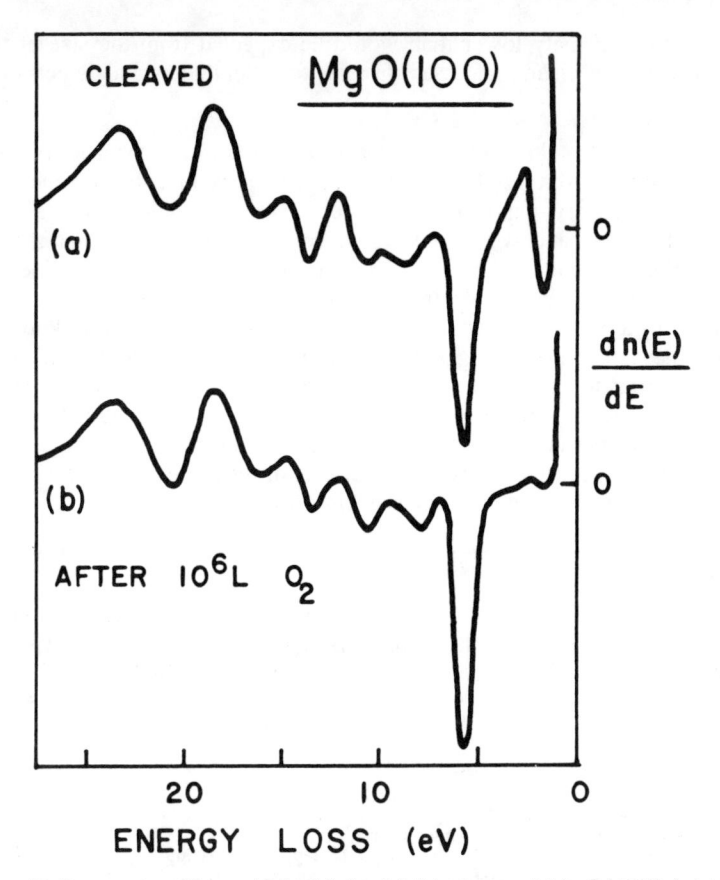

Fig. 9. ELS spectra (E_p = 100 eV) for UHV-cleaved MgO(100) (a) before and (b) after exposure to 10^6 L of O_2.

cleaved surfaces, disappears when the surfaces are annealed, and reappears when the sample is subjected to electron bombardment and, in some instances, also as a result of ion bombardment. That it is clearly a transition involving a surface defect state is shown by its sensitivity to O_2 adsorption, Fig. 9(b), and to its behavior under electron irradiation and annealing. (Note that O_2 exposure does *not* affect the intrinsic surface-state transition at 6.2 eV in Fig. 9.)

Although the 2.3-eV transition is known to involve a defect surface state, it is not yet clear just what type of defect is involved. The reducibility of oxide surfaces suggests an F_s center, Fig. 1, as a likely candidate.[15,25] Such centers have been identified on MgO powders by spin-resonance techniques.[26] But this defect has been considered theoretically in great detail, and its transition energy should be very close to that of the bulk F^+ center, about 5 eV.[27,28] An alternative possibility is a V_s center, Fig. 1.[18] The transition energy for bulk V^- centers has been measured as 2.35 eV, and the surface V center would again be expected to have a roughly equal transition energy. The V_s^- center might be depopulated upon O_2 adsorption by the formation of an O_2^- ion. The magnitude of the 2.3-eV ELS feature is also puzzling, since the LEED patterns for MgO(100) surfaces suggest a surface defect density lower than would be expected from the size of the loss peak. More work is clearly necessary in order to better understand defects on MgO.

Summary

Both experimental and theoretical work on the surface properties of wide-gap insulators such as MgO and Al_2O_3 has lagged far behind that on metal, semiconductor, and narrow-gap insulator surfaces. The information that is available is in bits and pieces, and we do not yet have a complete picture of the surface structure of either oxide. UPS and ELS measurements on polycrystalline α-Al_2O_3 are consistent with a surface electronic structure similar to that of the bulk, but the position of E_F at the surface and the nature of surface defects are not known. XPS and ELS measurements on MgO(100) surfaces indicate a valence band electronic structure similar to that of the bulk, but the optical band gap at the surface is reduced by 1.5–2 eV. Theoretical calculations for MgO(100) predict this band-gap narrowing based partially upon the strong electric fields that exist at MgO surfaces. Core-level ELS spectra also exhibit the effects of these electric fields in the Stark shifting of the excited states of surface Mg^{2+} ions.

Acknowledgments

The author is indebted to G. Dresselhaus, R. L. Kurtz, and H. J. Zeiger for invaluable contributions to some of the work reported here. This work was partially supported by National Science Foundation Grant No. DMR 82-02727.

References

[1] W. C. Mackrodt; these proceedings.
[2] V. E. Henrich, *Prog. Surf. Sci.*, **14**, 175 (1983).
[3] P. Mark, *J. Phys. Chem. Solids*, **29**, 689 (1968).
[4] V. E. Henrich, *Surf. Sci.*, **57**, 385 (1976).
[5] R. L. Kurtz and V. E. Henrich, *Phys. Rev. B*, **B25**, 3563 (1982).
[6] R. L. Kurtz and V. E. Henrich, *Phys. Rev. B*, **28**, 6699 (1983).
[7] R. L. Kurtz and V. E. Henrich, *Surf. Sci.*, **129**, 345 (1983).
[8] J. A. Tossell, *J. Phys. Chem. Solids*, **36**, 1273 (1975).
[9] D. M. Roessler and W. C. Walker, *Phys. Rev.*, **159**, 733 (1967).
[10] E. T. Arakawa and M. W. Williams, *J. Phys. Chem. Solids*, **29**, 735 (1968).
[11] J. Kendrick, E. A. Colbourn, and W. C. Mackrodt; to be published in *Radiation Effects*.
[12] E. A. Colbourn and W. C. Mackrodt, *Solid State Commun.*, **40**, 265 (1981).
[13] S. P. Kowalczyk, F. R. McFeely, L. Ley, V. T. Gritsyna, and D. A. Shirley, *Solid State Commun.*, **23**, 161 (1977).
[14] A. Bianconi, R. Z. Bachrach, S. B. M. Hagstrom, and S. A. Flodström, *Phys. Rev. B*, **B19**, 2837 (1979).
[15] V. E. Henrich, G. Dresselhaus, and H. J. Zeiger, *Phys. Rev. B*, **B22**, 4764 (1980).
[16] J. Olivier and R. Poirier, *Surf. Sci.*, **105**, 347 (1981).
[17] M. R. Welton-Cook and M. Prutton, *Surf. Sci.*, **74**, 276 (1978).
[18] P. R. Underhill and T. E. Gallon, *Solid State Commun.*, **43**, 9 (1982).
[19] V. E. Henrich, G. Dresselhaus, and H. J. Zeiger, *Phys. Rev. Lett.*, **36**, 158 (1976); **38**, 872 (E) (1977).
[20] A. J. Martin and H. Bilz, *Phys. Rev. B*, **B19**, 6596 (1979).
[21] K. H. Rieder, *Surf. Sci.*, **118**, 57 (1982).
[22] V.-C. Lee and H.-S Wong, *J. Phys. Soc. Jpn.*, **45**, 895 (1978).
[23] C. Satoko, M. Tsukada, and H. Adachi, *J. Phys. Soc. Jpn.*, **45**, 1333 (1978).
[24] M. Tsukada, H. Adachi, and C. Satoko, *Prog. Surf. Sci.*, **14**, 113 (1983).
[25] V. E. Henrich and R. L. Kurtz, *J. Vac. Sci. Technol.*, **18**, 416 (1981).
[26] A. E. Hughes and B. Henderson; p. 381 in *Point Defects in Solids*, Vol. I. Edited by J. H. Crawford, Jr., and L. M. Slifkin. Plenum, New York, 1972.
[27] R. R. Sharma and A. M. Stoneham, *J. Chem. Soc., Faraday Trans. II*, **72**, 913 (1976).
[28] H. A. Kassim, J. A. D. Matthew, and B. Green, *Surf. Sci.*, **74**, 109 (1978).

Calcium Segregation to MgO and α-Al$_2$O$_3$ Surfaces

R. C. McCune and R. C. Ku

Ford Motor Co.
Dearborn, MI 48121

The segregation of calcium to MgO (100) and various surfaces of single and polycrystalline α-Al$_2$O$_3$ has been determined by use of Auger electron spectroscopy (AES) and low-energy ion scattering (LEIS) on specimens equilibrated at elevated temperatures under ultrahigh vacuum. The segregation enthalpy, ΔH_s, for Ca to MgO (100) between 1323 and 1723 K was found to range between -44 and -80 kJ/mol, with a maximum solute occupation of surface sites approaching 50% below about 1273 K. The segregation of Ca to α-Al$_2$O$_3$ surfaces exhibited transient and irreversible behavior. The maximum observed occupation of surface cation sites was found to approach 9% at a temperature of 1723 K for a polycrystalline specimen. Extimates of ΔH_s were in the range -90 to -190 kJ/mol over temperatures from 1623 K to approximately 1900 K.

Solute segregation at interfaces in ceramic materials has received considerable attention in recent years because of its role in such phenomena as sintering, densification, fracture, and electrical behavior.[1-5] Much of the current interest is a consequence of the application of relatively new surface sensitive spectroscopies to grain-boundary surfaces produced by predominantly intergranular fracture[3] and of the development of microanalytical techniques for the study of intact interfaces.[6,7]

This work focuses on segregation of calcium at free surfaces of MgO and α-Al$_2$O$_3$ (hereafter referred to as Al$_2$O$_3$) measured by Auger electron spectroscopy (AES) and low-energy ion scattering (LEIS). While free-surface composition may be of interest in its own right in such areas as catalysis,[8] corrosion,[9] and tribology,[10] its relevance to MgO and Al$_2$O$_3$ ceramics stems from a traditional interest in grain-boundary phenomena. Studies of free-surface composition, structure, and chemical state, utilizing various surface-sensitive probes, should provide an indication of solute behavior at grain boundaries where structures and energetics are generally more complex.

Descriptions of our approach to these surface studies may be found in prior work.[4,11] The measurement of surface concentration of the solute as a function of temperature permits the determination of an enthalpy for segregation, ΔH_s, in the limiting case of regular solution behavior. Estimates of ΔH_s, which should reflect the internal energy difference in substituting a solute atom for a solvent atom at the surface, can be made from thermodynamic and physical property data for the solute and solvent materials in their pure states. Recent theoretical assessments of the segregation enthalpy for Ca^{2+} in MgO[12-14] from appropriate crystal lattice–defect pair potentials provide an additional comparison with experimental data. A simplified version of these approaches has also been advanced recently for prediction of solute segregation tendencies in divalent transition-metal oxide solid solutions.[15]

Theoretical Background

Although a conceptual framework for solute segregation has existed since the time of Gibbs,[16] a number of statistical models based on regular solution behavior have been developed recently.[17,18] In part, these approaches stem from a number of experimental difficulties that limit the utility of the Gibbs formalism. The Gibbs adsorption equation[19] relates the specific surface work of the solution, γ, to the excess concentration of species i, Γ_i, at the interface, by an equation of the form

$$d\gamma = -S^s dT - \sum_i \Gamma_i d\mu_i \tag{1}$$

where S^s is the specific interfacial excess entropy and μ_i is the chemical potential for species i. Despite the usual practice of confining measurements to a single temperature (Gibbs isotherm), there remains a formidable problem of establishing the specific surface work γ as a function of solute concentration and surface enrichment. Linford[20] has discussed these problems at length for the case of crystalline surfaces.

In the statistical models advanced by McLean[17] for grain boundaries in metals and by Defay et al.[18] for liquids, the surface mole fraction ratio for components A and B, X_A^s/X_B^s (where A is taken to be the solute), is related to the bulk mole fraction ratio X_A^b/X_B^b by an equation of the form

$$\frac{X_A^s}{X_B^s} = \frac{X_A^b}{X_B^b} \exp(-\Delta H_s/RT) \tag{2}$$

where ΔH_s is the enthalpy of segregation. Since Eq. (2) is the outcome of a regular solution equilibrium where only the entropy of mixing at bulk and surface sites is considered in the reaction, excess interfacial entropy terms associated with real solutions are incorporated in an expanded form of Eq. (2) such that

$$\frac{X_A^s}{X_B^s} = \frac{X_A^b}{X_B^b} \exp(-\Delta H_s/RT) \exp(\Delta S_s/R) \tag{3}$$

or

$$\frac{X_A^s}{X_B^s} = \frac{X_A^b}{X_B^b} \exp(-\Delta G_s/RT) \tag{4}$$

with

$$\Delta G_s = \Delta H_s - T\Delta S_s \tag{5}$$

The segregation enthalpy, ΔH_s, may thus be extracted from a logarithmic plot of surface mole fraction vs inverse temperature.

In the original McLean approach,[17] the segregation enthalpy was interpreted to arise solely from elastic strain relaxation accompanying substitution of a solute atom from the bulk with one at an interface, the magnitude of the strain energy being determined from linear elasticity.

In metals, the specific surface work difference, $\Delta\gamma$, for the constituents[21] and the pairwise bond enthalpies have been recognized as contributing to ΔG_s. This has permitted assessment of segregation in alloys, based on quasi-chemical theory.[23] A number of recent treatments of solute segregation in metals recognize contributions from all three components.[19,22-25]

The McLean treatment,[17] exemplified by Eq. (2), is similar in several respects to gas adsorption behavior in the Langmuir isotherm regime.[26] In particular, the adsorption behavior is predicated on a finite number of noninteracting sites, all of which have the same characteristic energy for adsorption of an atom or molecule. Furthermore, the energy for adsorption is taken to be independent of coverage, which may only approximate the situation for high levels of coverage where substantial solute–solute interactions can occur.

The situation is more complex in ceramics because of the nature of the strong pairwise bonds between nearest neighbors and the possibility of various charge states for solutes, vacancies, and associated lattice defects. Considering only substitutions on the cation sublattice in binary oxides, the situation may be partitioned into two domains. In the first case, the solute cation is isovalent with the host cation and segregation may be treated in terms of solute strain, surface work difference, and chemical attraction or repulsion of the constituents.[4] The monolayer or McLean[17] model is expected to be a reasonable approximation for dilute solutions in this instance. For aliovalent solutes, on the other hand, there exists the possibility for interactions with the interfacial space charge layer,[27] which may develop as the consequence of unequal formation energies of complementary point defects in the vicinity of the interface.[28] In this case, the spatial distribution of effectively charged solute and lattice defects must be consistent with Poisson's equation. Yan et al.[27] have shown that the combination of elastic and electronic contributions results in a coupled problem in terms of predicting solute spatial distribution; that is, there will be a synergism between effectively charged solutes that segregate in response to elastic effects and solutes that migrate under the influence of space charge alone. Blakely and Danyluk[29] have suggested that the solute distribution profile may adopt either of two forms depending upon its electrical sign relative to that of the surface charge. In cases where the solute has the same effective sign as the surface charge, it may replace an equivalent unit of charge and reside at the physical surface. When the solute carries a charge opposite to that of the boundary charge, however, it may be expected to accumulate in the subsurface region within a depth determined by the Debye length for the material. In both situations, it may be possible to realize an interfacial enrichment of solute when the appropriate "dividing surface" is chosen in the Gibbsian analysis.[4]

The sign of the boundary charge will, in general, be dependent upon total aliovalent impurity content, temperature, and the extent to which coupled defects are formed.[30] Kingery[2] has used experimental methods to arrive at the sign of boundary charge in several ceramic materials. The ramifications of aliovalent solute distribution at interfaces are crucial for surface spectroscopies, since a static surface measurement utilizing a technique such as LEIS may understate the situation for subsurface enrichment of solute. Although there have been several experimental studies of aliovalent solute segregation in apparent response to the boundary charge,[31–33] none of these has had sufficient depth sensitivity to establish whether the solute was located at the physical boundary or in the subsurface region.

In ionic materials, such as the alkali and silver halides, the magnitude of the space charge may be estimated for both intrinsic and extrinsic defects from formation energies.[27–30,34] For MgO and Al_2O_3 however, the relatively high energies for intrinsic defect formation[35,36] ensure that space charge behavior will be dominated by aliovalent impurities. In MgO, for example, trivalent solutes (e.g., Sc^{3+}, Cr^{3+}, and Fe^{3+}) are attracted to the near-surface region[32,33] in response to the negative surface charge[2] presumably associated with condensation of the charge-compensating vacancies at the interface. In polycrystalline Al_2O_3 doped with MgO, Kingery[2] has reported a net positive boundary charge, suggesting that an effectively negative-charged impurity, such as Ca^{2+}, would be attracted to the discontinuity. Tiku and Kröger[37] also suggested that a divalent impurity will accumulate at interfacial regions in Al_2O_3.

In accounting for space charge attraction in the monolayer regime of the McLean expression (Eq. (2)), it may be possible to adjust ΔH_s for boundary accumulated solute by an amount $-q\phi_\infty$, where q is the effective charge on the solute in its substitutional site (e.g., -1 for Ca^{2+} in Al_2O_3) and ϕ_∞ is the bulk potential referenced to a null surface potential. (As indicated by Blakely and Danyluk,[29] the surface potential may alternatively be expressed as $\phi_s = -\phi_\infty$, with the reference null potential at the crystal interior.) A positive boundary charge leads, therefore, to a net negative value of ϕ_∞,[37] and the Ca^{2+} impurity will experience a more negative contribution to ΔH_s. For aliovalent solutes distributed within the space charge region, however, the electrical potential will be position dependent, and solution of the Poisson equation is required to predict the spatial variation of solute concentration.

The segregation of Ca to MgO (100) is an example of the isovalent case, and ΔH_s should contain terms corresponding to lattice strain, surface work difference, and a net negative chemical interaction contribution suggested by clustering tendencies of the binary system in contrast to compound formation.[38] Estimates of ΔH_s from tabulated physical property data are about -59 kJ/mol.[11] Recent calculations of segregation enthalpy for highly enriched surface layers of CaO on MgO using pair potentials[13,14] indicate this to be a reasonable estimate.

For Ca segregation in Al_2O_3, the principal driving force should result from the large misfit strain of the Ca^{2+} cation in Al^{3+} substitutional sites. Estimates of the solute strain energy are about -134 kJ/mol.[4] In the CaO-Al_2O_3 system,[39] there is a strong tendency for compound formation, which should, in principle, act to reduce the magnitude of ΔH_s. Tabulated values of specific surface work, γ, for CaO and Al_2O_3[21] do not suggest major contributions to ΔH_s based on $\Delta \gamma$, and as indicated previously, an attraction to the surface may be expected from the apparent space charge introduced by the divalent impurity.[37] This should increase the magnitude of ΔH_s in the absence of significant quantities of other impurities.

Experimental Procedures

Details of the experimental approach have been described in the prior work.[11] Specimens (≤ 0.5-mm thick) of the particular crystal or polycrystal are affixed to metal foil heating strips (Ta for MgO; Re for Al_2O_3), which are then mounted on a movable manipulator that is used in ultrahigh-vacuum chambers of either an Auger electron spectrometer (AES)* or low-energy ion scattering spectrometer (LEIS).† For AES measurements, an electron beam of minimum diameter ~ 10 μm, energy 3 keV, and current 1 μA excited emission of Auger electrons.[40] The Auger current for a given transition was measured in the derivative mode with a 4–6 eV modulation amplitude; the Auger electron energies determined by a cylindrical mirror type electrostatic analyzer operated in a phase-locked detection mode. LEIS utilizes measurement of the energy distribution of noble gas ions reflected from the surface of the specimen. Application of binary elastic scattering kinematics permits assignment of peaks in the energy distribution to appropriate atom concentrations in the surface layer. In principle, LEIS assesses the composition of the outermost surface of the target.[41,42] In this work,[4] He^+ ions at an energy of 500 eV were generated by an ion gun producing an approximate 0.5-mm diameter beam, at currents of $\sim 4.0 \times 10^{-8}$ A. Energy analysis was performed with a cylindrical mirror type electrostatic analyzer. In both modes of

Table I. Major Impurities in MgO Crystals*

Sample	Ca	Fe	Si	Al	Ni	Cr	Mn	Na	K	Ti
MgO-I	200	90	50	30	40	20	24	110	NA	22
MgO-II	180	200	NA	70	43	NA	20	NA	<100	<10
MgO-III	220	106	14	33	5	24	23	35	66	NA

*Units are wt ppm. NA means not analyzed.

Table II. Summary of Free-Surface Segregation in α-Al$_2$O$_3$

Material	Orientation	[Ca] (wt ppm)	Analysis mode	Maximum observed X^s_{Ca}/X^s_{Al}	Maximum surface enrichment X^s_{Ca}/X^b_{Ca}	Temp (K)	$-\Delta H_s$ (kJ/mol)
Linde Cz Sapphire	(10$\bar{1}$2)	26±3	LEIS	0.02–0.05	1348	1673	
Baikowski Verneuil Sapphire	(10$\bar{1}$0)	12±2	LEIS	0.003–0.007	511	1673	
Baikowski Verneuil Sapphire	Random	50±5	LEIS	0.01–0.024	369	1673	
Meller Substrate Sapphire	(10$\bar{1}$0)	40±5 ~40	LEIS AES	0.009–0.022 0.011–0.015	423 300	1723 1723	188 156
Lucalox*	Polycrystal	39±5	AES	0.096	1936	1723	184

*General Electric Co., Schenectady, NY.

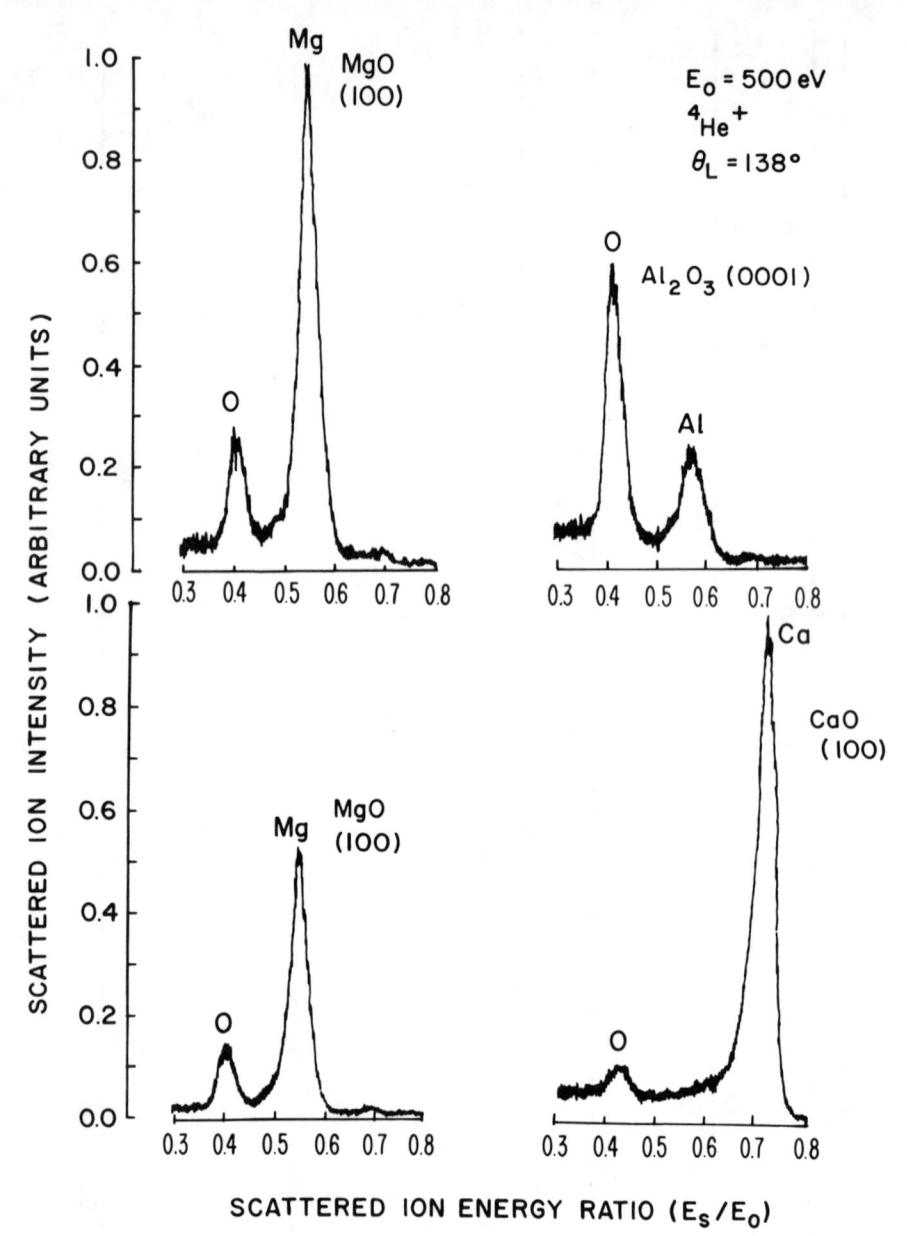

Fig. 1. Comparison of ^{4}He$^+$ LEIS spectra for MgO (100), Al$_2$O$_3$ (0001), and CaO (100) at a beam energy of 0.5 keV. Scales were adjusted to permit use of MgO (100) as a reference in both cases.

analysis, the specimen surface composition could be monitored at temperature, usually with some associated distortion in relative peak intensities, or peak shapes. Where comparisons were made between measurements at elevated temperature and room temperature on the same surface, the elevated temperature composition could

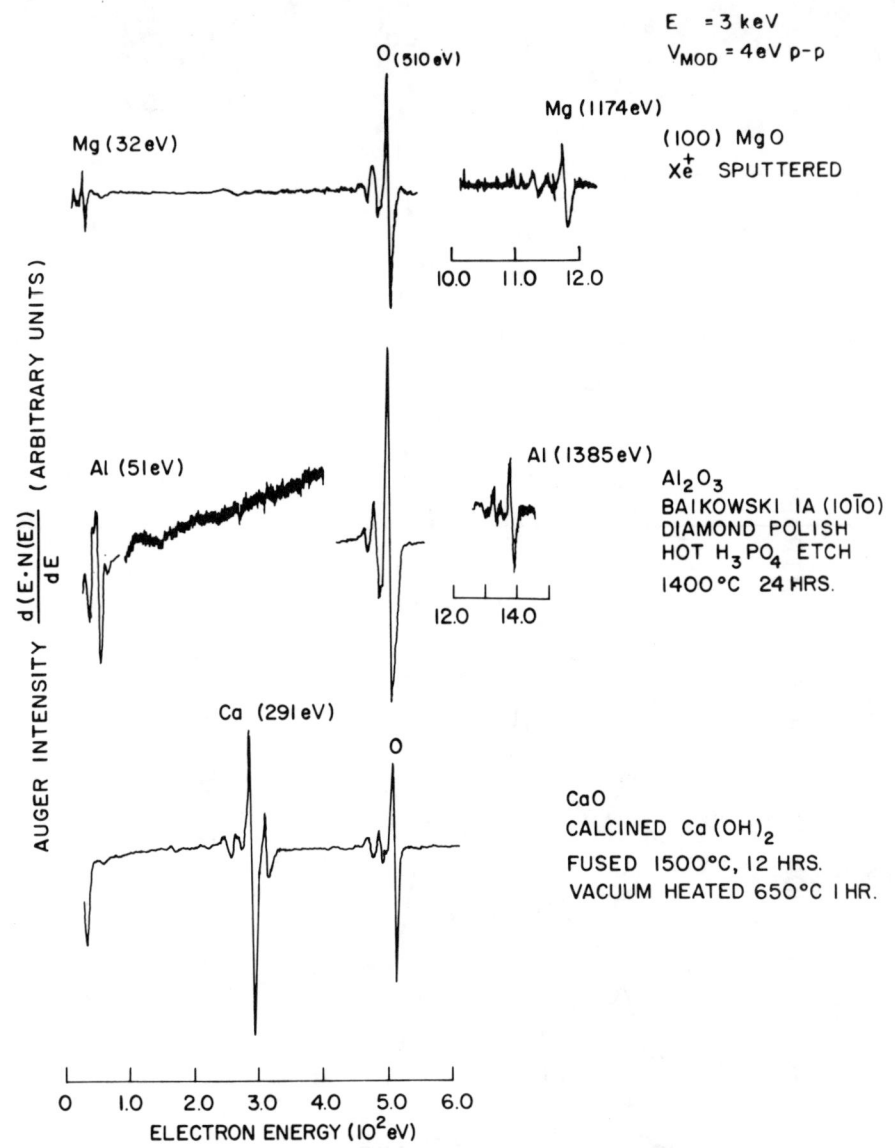

Fig. 2. AES spectra for reference materials as indicated.

be maintained by quenching the crystal by cessation of current through the heating strip. The quench rate for most specimens, estimated by loss of incandescence of the crystal, was on the order of 100°C/s. Measurements on Al_2O_3 specimens were only on quenched samples, while results for MgO contain analyses in both modes.

MgO (100) crystals obtained from several lots of material, believed to have been originally grown by arc fusion, were prepared by cleavage to provide an $0.8 \times 0.8 \times 0.05$ cm specimen. Chemical analyses for the three crystals employed are presented in Table I. Inductively coupled plasma (ICP) emission spectroscopy and atomic absorption were utilized for impurity analyses. A variety of

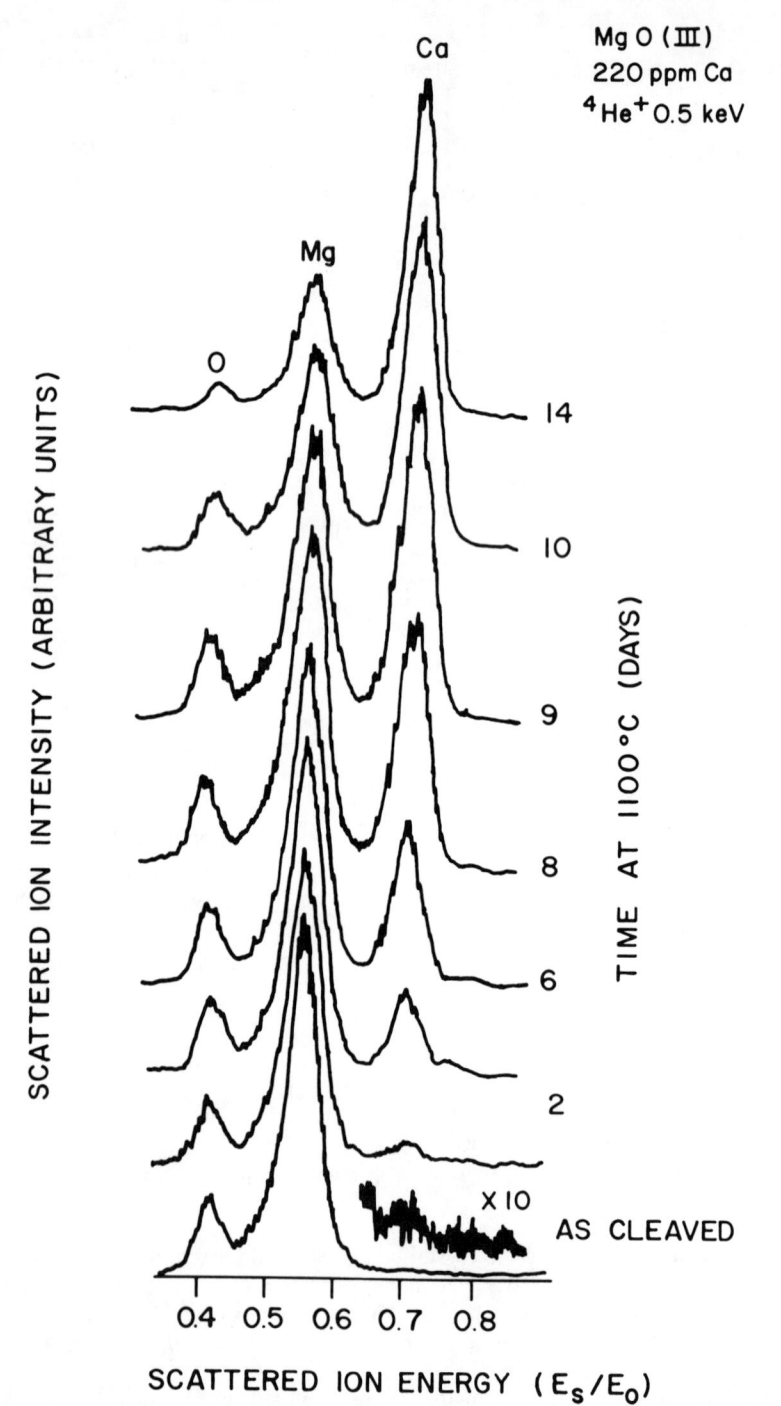

Fig. 3. LEIS spectra for a (100) surface of MgO at 1100°C, showing evolution of segregated calcium-rich surface. Spectra have been normalized to the largest peak in each case.

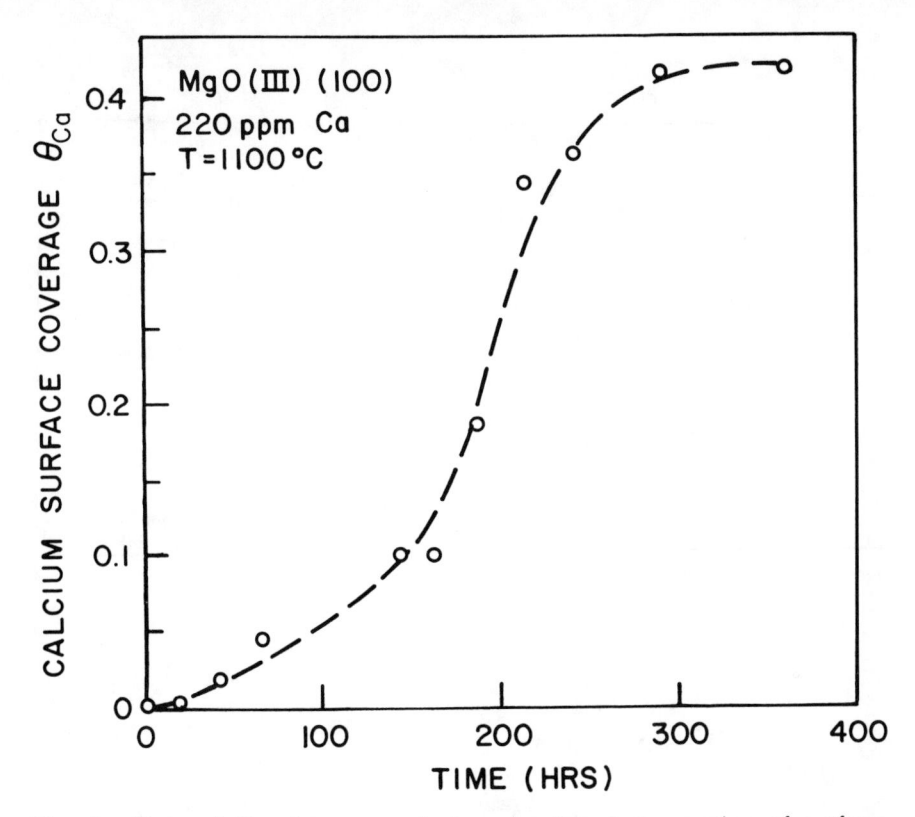

Fig. 4. Data of Fig. 3 represented as fractional occupation of surface cation sites as a function of time after the onset of heating.

single-crystal and polycrystalline Al_2O_3 was studied, with bulk calcium concentrations determined by optical emission spectroscopy.[‡] Results are reported in Table II, which also summarizes maximum observed surface enrichments. Because Al_2O_3 surfaces of known orientation are not readily produced by cleavage, it was necessary in most instances to prepare surfaces by a combination of diamond wafering, diamond polishing (through 1-μm abrasive), followed by chemical etching in hot orthophosphoric acid.[43] Since this last step leaves residual aluminum phosphate on the surface,[44] a further etch in boiling $3HCl/1HNO_3$, followed by deionized water rinsing, was employed. Final cleaning was by sputter ion etching in the appropriate spectrometer prior to equilibration. The chemical cleanliness of the surface could then be established before being heated. Once surface cleanliness had been ensured spectroscopically, the specimen was heated by passing a current through the heater strip. The temperature of the heater strip adjacent the specimen was measured by an optical pyrometer. The surface composition was monitored until no further change could be observed at a given temperature. The rate of surface segregation slowed as the concentration increased.[45] The strip temperature was changed and the process repeated. For Ca segregation to MgO (100), the composition was determined on both ascending and descending temperatures.

Quantification of spectroscopic intensities was performed by calibration of relative cation/oxygen sensitivities on clean surfaces of MgO, Al_2O_3, and CaO. A

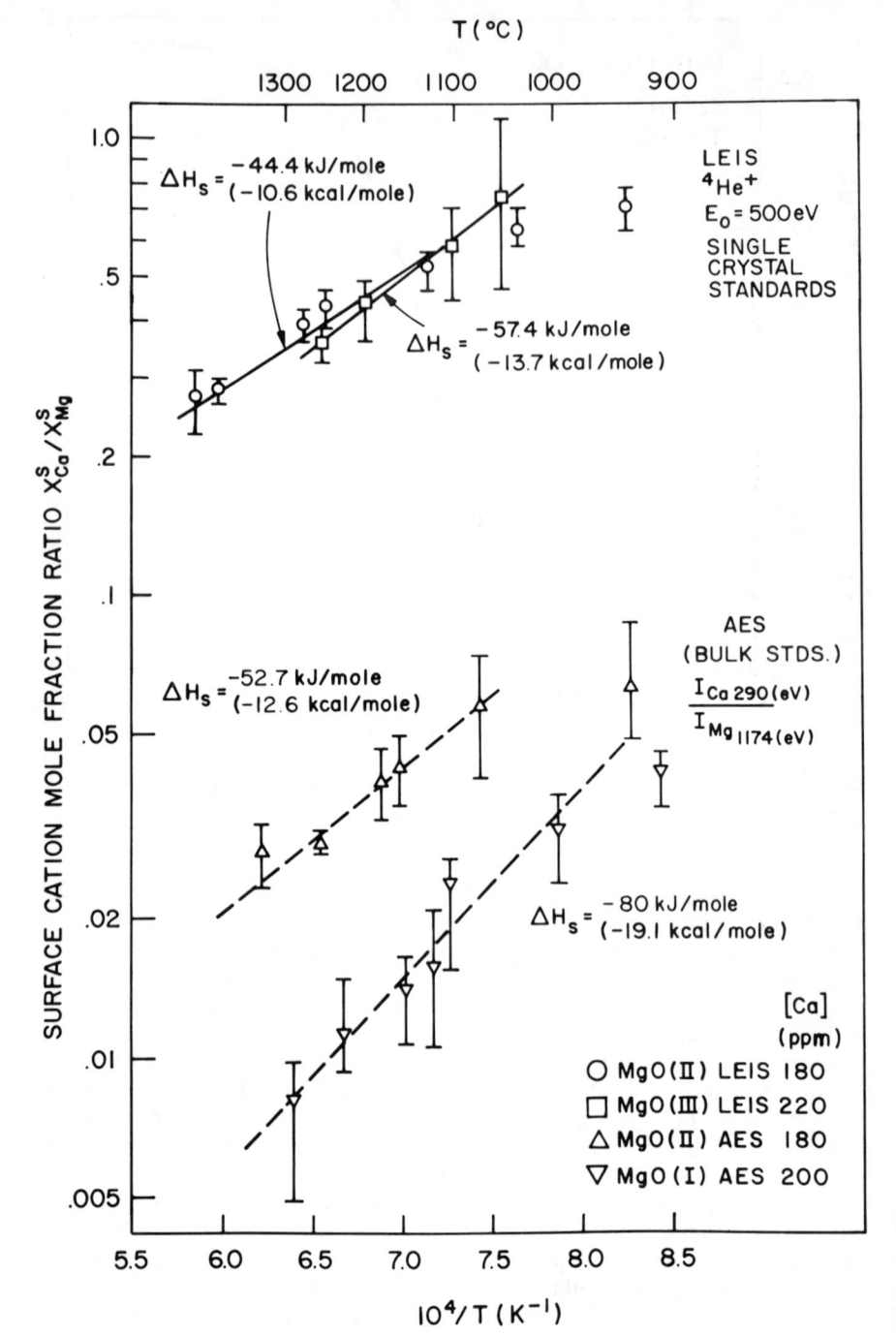

Fig. 5. Surface cation mole fraction ratios for calcium segregated to MgO (100) on several crystals. LEIS measurements are reported in terms of single-crystal references. The AES measurements are reported for calcined $Ca(OH)_2$ as the reference material.

single-crystal CaO (100)[§] surface was utilized for LEIS calibration, whereas calcined Ca(OH)$_2$ was utilized as the CaO standard for AES measurements.

Figure 1 illustrates the comparative LEIS spectra for the reference oxide single-crystal surfaces, from which relative sensitivities may be deduced either by ratio to the common oxygen present or by direct comparison of peak heights, normalized by the respective areal atom densities. This latter procedure yields a lower relative Ca/Mg sensitivity and consequently higher surface enrichments than those previously reported.[11] The relative Ca/Mg sensitivity as measured from (100) single crystals is approximately 2.6, agreeing with the calculated value of ~2.7 from tabulated scattering cross sections and known instrument factors.[11]

Figure 2 illustrates typical AES spectra from clean oxide standards from which sensitivities for each cation relative to the common oxygen, normalized for appropriate stoichiometry, were obtained. Values for the Ca/Mg and Ca/Al relative sensitivities obtained in this manner were somewhat higher than sensitivities deduced from published tabulations.[46]

Results

Ca Segregation to MgO (100)

The evolution of the Ca-enriched surface layer can be followed from the LEIS spectra in Fig. 3, where the specimen was equilibrated at a strip temperature of 1373 K. Figure 4 illustrates the fractional surface coverage for LEIS values, based on the single-crystal standards. Agreement with kinetics proposed for surface enrichment in alloys, wherein a constant enrichment ratio was employed,[45] was poor. Use of a variable enrichment ratio method, proposed by Brailsford,[47] may be helpful in interpretation of these kinetics, although results now are preliminary.

Surface cation mole fraction ratios for calcium, X^s_{Ca}/X^s_{Mg}, are shown for MgO in Fig. 5, along with apparent segregation enthalpies deduced from linear fits to the data represented in this form. The LEIS data were calibrated in terms of single-crystal CaO and MgO standards, while AES results are reported for MgO (100) and calcined Ca(OH)$_2$ references, in the absence of AES data from CaO (100). The discrepancy in magnitude of AES values from MgO-I and MgO-II is believed to result from disruptive effects of the electron beam, which was used at a higher current density in the case of MgO-I, causing greater apparent loss of the segregant by electron-induced desorption.[11,48]

Ca Segregation to Surfaces of Al$_2$O$_3$

Determination of calcium segregation at surfaces of Al$_2$O$_3$ has been generally less straightforward than for MgO. Table II summarizes maximum observed surface mole fraction ratios, (X^s_{Ca}/X^s_{Al}), maximum enrichment ratio, (X^s_{Ca}/X^b_{Ca}), and enthalpy of segregation, ΔH_s, where such estimates were possible.

Conclusions from the studies of calcium segregation to Al$_2$O$_3$ surfaces can be summarized. Virtually all Al$_2$O$_3$ specimens examined (including several not indicated in Table II) showed some indication of surface enrichment in calcium following high-temperature annealing of the spectroscopically clean surface. The extent of calcium segregation for a given sample was not simply related to the measured bulk concentration. In many specimens, the surface calcium concentration had a nonuniform spatial dependence and was time dependent, suggesting loss of solute by volatilization.[45] Measurements of bulk calcium concentration before and after equilibration at temperatures in excess of 1873 K indicate, however, no apparent loss of bulk solute.

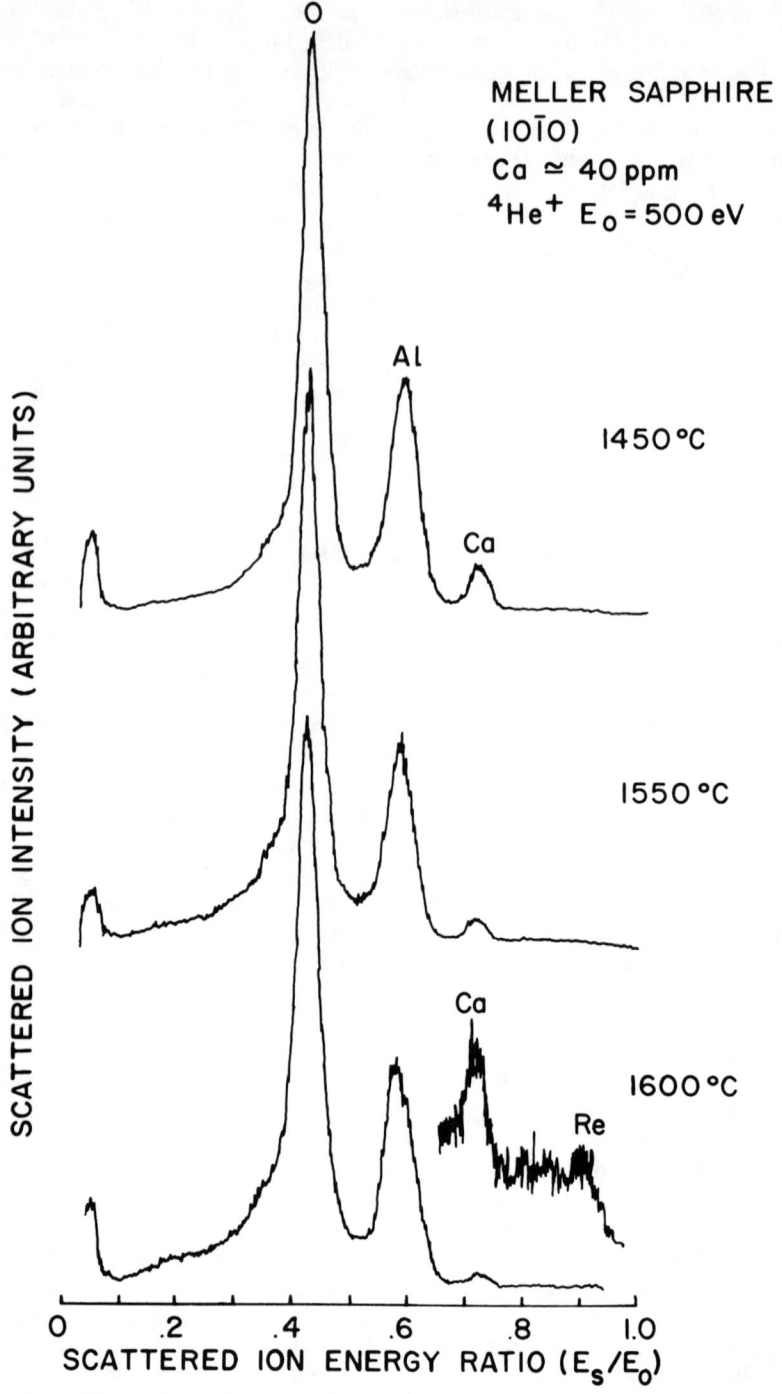

Fig. 6. LEIS spectra from a single-crystal Al$_2$O$_3$ surface following equilibration at several temperatures, indicating magnitude of calcium segregation observed.

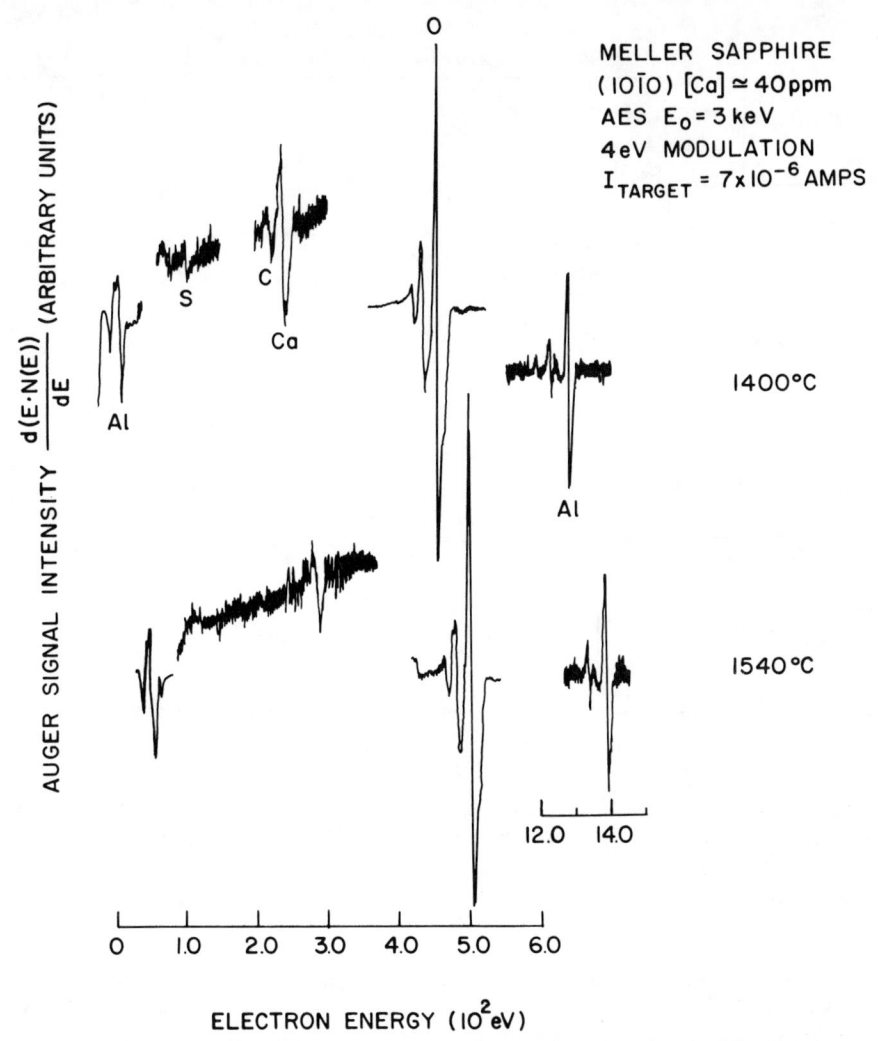

Fig. 7. AES spectra from a single-crystal Al$_2$O$_3$ surface after equilibration at temperatures indicated.

The specimen designated Meller ($1\bar{0}10$) in Table II provided the most consistent results with respect to segregation, as observed in both LEIS and AES. It should be pointed out, however, that considerable variation in bulk composition of the original 2.5 × 2.5 cm wafer was observed, with values ranging from 12–90 ppm, the average calcium concentration being ~40 ppm. Figures 6 and 7 show LEIS and AES spectra for separate specimens of this material following equilibration at several temperatures. A decrease in surface concentration of calcium with increasing temperature was observed, although neither sample exhibited reversibility of surface enrichment. Comparison of surface mole fraction data X^s_{Ca}/X^s_{Al}, using LEIS and AES for these specimens is shown in Fig. 8, where the calcined Ca(OH)$_2$ reference was used for calibration.

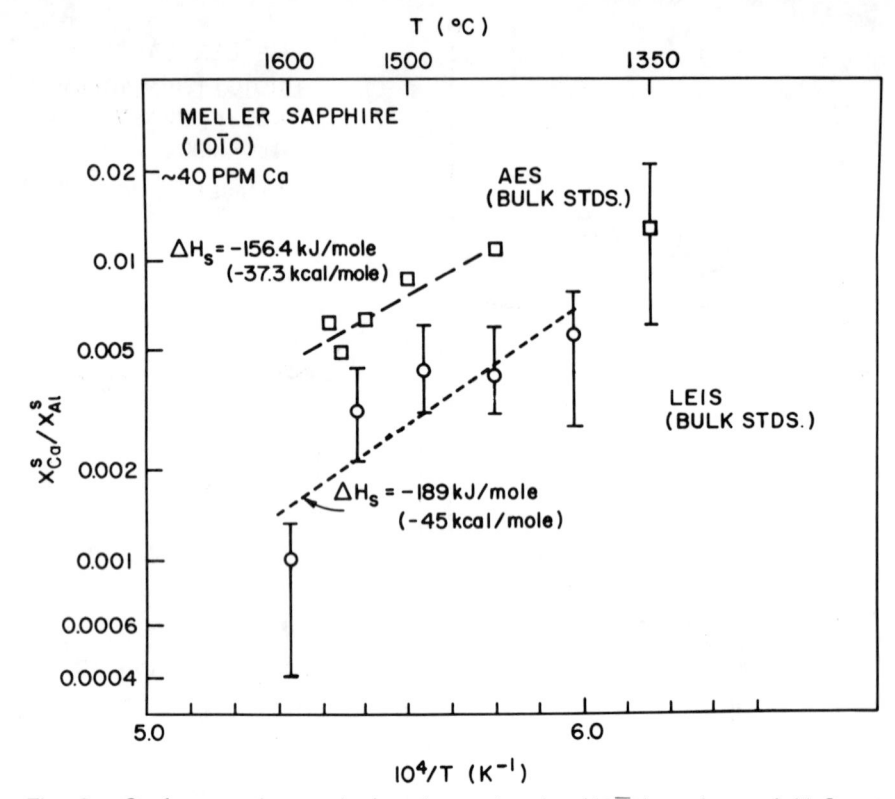

Fig. 8. Surface cation mole fraction ratios for (10$\bar{1}$0) surface of Al$_2$O$_3$ at various temperatures, and estimated values of segregation enthalpy, ΔH_s. Calcined Ca(OH)$_2$ was utilized as the reference material for both AES and LEIS values.

The temperature dependence of surface cation mole fraction ratio for segregation of Ca at the surface of a MgO-doped Al$_2$O$_3$, is shown in Fig. 9, where values were determined from AES using bulk oxide standards. The temperature dependences of grain-boundary Ca segregation reported by Johnson and Stein[49] and by Jupp et al.[50] for polycrystalline aluminas are shown for comparison. Heats of segregation calculated from surface atomic compositions reported in these works are also indicated. Least-squares linear fits to these data points produce segregation enthalpies somewhat different from those reported in the original works.[49,50]

Discussion

Ca Segregation to MgO (100)

Segregation enthalpies for various specimens of MgO, containing between 180 and 220 wt ppm, of calcium lie between −44 and −80 kJ/mol in the temperature range from approximately 1323 to 1723 K. Calculated values for heat of substitution of Ca to pure CaO layers on MgO indicate a range of −52 to −72 kJ/mol,[13,14] and estimates from combinations of solute strain energy and

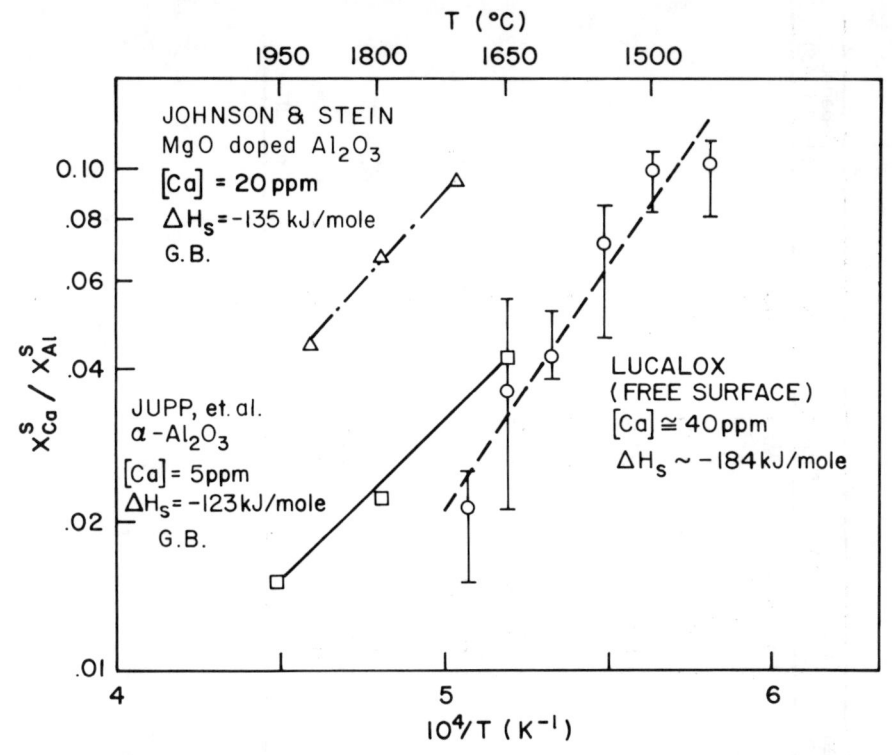

Fig. 9. Surface cation mole fraction for calcium segregated to the surface of a MgO-doped Al_2O_3 at various temperatures, as determined by AES. Grain-boundary segregation results for calcium in a similar material (Ref. 49) and a higher purity alumina (Ref. 50) are shown for comparison. ΔH_s was calculated by least-squares fits to the data as presented.

specific surface work difference[11] indicate a value near −59 kJ/mol when temperature dependence of the bulk elastic properties or surface work is not taken into account. The most recent LEIS calibrations from single crystals of MgO and CaO suggest a limiting surface coverage approaching 50% of the cation sites below about 1323 K.

Use of the bulk mole fraction ratios determined from Table I and the McLean Eq. (2) suggest that segregation is in excess of that predicted from regular solution theory. In light of the reasonable agreement for segregation enthalpy, the enhancement may arise from additional entropic terms, suggested by Eq. (3).

The discrepancy between absolute surface coverages determined by LEIS and AES may be understood in terms of the relative sensitivity of the techniques. In general, AES samples a depth determined by the inelastic mean free path (IMFP) for the emitted Auger electrons.[51] When the surface mole fraction ratio is determined from relative peak intensities, the matrix will contribute from a depth determined by the IMFP, while the total contributions from the segregated monolayer will be recorded. The matrix contribution thus "dilutes" the true mole fraction ratio, since its contribution occurs from a greater depth. Compensation can be made if the spatial distribution of the solute is known.[51] In LEIS, on the other hand,

Table III. Segregation at α-Al$_2$O$_3$ Grain Boundaries

Material	[Ca] (wt ppm)	Anneal temp (K)	X^s_{Ca}/X^s_{Al} from stated at.% concentration	Enrichment X^s_{Ca}/X^b_{Ca} (×10³)	Technique	Ref.	Reported $-\Delta H_s$ (kJ/mol)
Lucalox[†]	5–15	‡	0.5–0.14	9.8–19.6	AES	55	
Lucalox[†]	10–15	‡	0.16	7.5–11.3	XPS*	56	
Lucalox[†]	50	‡	0.02	0.313	XPS		
Linde A + 0.1% MgO	65	2173	0.01	0.126	XPS	57	
Linde A + 0.25% MgO	110	2173	0.02	0.143	XPS		
Al$_2$O$_3$ + 0.1% MgO	45	2123	0.176	2.6	AES	59	
Lucalox[†]	36 (est)	1973	0.093	0.85–3.3	AES	49	121
		2073	0.067	0.63–2.5			
		2173	0.044	0.40–1.6			
Al$_2$O$_3$ High purity	5	1823	0.042	6.3	AES	50	127
		2073	0.085	3.5			
		2223	0.060	2.4			
Al$_2$O$_3$	36 (est)	2073	0.012	0.28	XPS	59	
Lucalox[†]	39	1723	0.096	1.9	AES	this work	187 free surface

*X-ray Photoelectron Spectroscopy
[†]General Electric Co., Schenectady, NY.
[‡]Information unavailable.

the intensities arise from scatter events at the outermost surface[41] and should reflect true first-layer composition.

The high surface enrichment deduced by LEIS suggests that surface compositions may greatly exceed the bulk solubility.[38] In such situations, the surface structures may differ from those of the original (100) MgO cleavage surface.

Ca Segregation to Al$_2$O$_3$ Surfaces

The segregation of calcium at grain boundaries in high-purity Al$_2$O$_3$ or MgO-doped Al$_2$O$_3$ is a widely reported phenomenon. Interest in such segregation has stemmed, in part, from studies of grain-boundary composition in MgO-doped Al$_2$O$_3$,¶ where MgO, present in amounts typically near 0.1 wt%, promotes densification and minimizes discontinuous grain growth.[52] There has also been interest with regard to potential effects of calcium segregation on fracture toughness of sintered aluminas.[50,53,54]

A number of determinations of calcium segregation in Al$_2$O$_3$ are summarized in Table III. Surface mole fraction ratios were estimated from reported atomic concentrations (balance assumed to be Al$_2$O$_3$) at the surface. The enrichment ratio was deduced from the stated bulk concentration by using wt ppm in the absence of stated units. Unfortunately, a lack of commonality in measurement technique, quantification, and reporting of values makes any comparison difficult.

In general, results obtained in this study are in an order of magnitude agreement with grain-boundary enrichments of calcium reported in the literature, although our equilibration temperatures were generally lower in an attempt to minimize volatilization. Where possible, linear segments were estimated from surface mole fraction ratio, X_{Ca}^s/X_{Al}^s, as a function of inverse temperature, such that enthalpies of segregation can be deduced. The range for these values (Table II) is between -92 and -189 kJ/mol. Values from grain-boundary studies of -121[49] and -127 kJ/mol[50] have been reported. The calculated elastic strain energy is about -134 kJ/mol.[4] While the data as measured from free surfaces agree with this trend, lack of reversibility, in most cases, limits the applicability of a model based on an equilibrium solute exchange between surface and bulk.

The transient nature of the surface segregation of calcium appears to be at least partly dependent on the mechanical processing and thermal history of the crystal. It has been shown, for example, that subsurface damage introduced by mechanical polishing[60,61] may be lost during annealing above 1500°C. Thus, entrained mechanical damage, which may enhance diffusion at temperatures below approximately 1500°C,[62] is annealed out during equilibration. The loss of mechanically induced defects by annealing could slow the rate of segregation at lower temperatures during cooling, despite the fact that these pathways were available during heating.

In general, the bulk concentration of calcium, itself, was not an adequate predictor of maximum surface concentration. This can be seen by comparison of maximum surface enrichment values, X_{Ca}^s/X_{Ca}^b, for Meller (1010) and the MgO-doped polycrystalline Al$_2$O$_3$,¶ where the latter showed a 6-fold increase in surface enrichment at 1450°C over that from the single crystal, despite similar bulk concentrations of calcium. Both overall purity and grain-boundary diffusion of solute may also increase calcium segregation in the MgO-doped Al$_2$O$_3$.

Comparison of LEIS- and AES-determined values for surface mole fraction ratio on a similar material (Fig. 9) suggests appreciably different behavior than the monolayer segregation observed for calcium on MgO. For calcium segregation in this particular specimen, LEIS produces a lower set of values than comparable

AES measurements on a similar (but not identical) sample using the calcined Ca(OH)$_2$ reference material. Use of single-crystal reference materials for LEIS (Fig. 1) tends to increase the values for the surface mole fraction ratios, to approaching those determined by AES. Agreement between the modes of analysis, in this case, however, precludes the possibility of assigning a depth dependence on the basis of differing measurements in two static techniques. The implication is that calcium does not as readily absorb at free surfaces in Al$_2$O$_3$ as in MgO and may have consequently greater enrichments in the subsurface region, giving rise to a relatively greater AES intensity when electron inelastic mean free paths are considered.

As indicated by Table II, calcium segregation measured at free surfaces is generally less than for comparable grain-boundary enrichments. As indicated, poor equilibration kinetics may play an appreciable role; however, the possible loss of segregant by volatilization[45] should also be considered. In general, experiments conducted at higher temperatures (>1573 K) with Al$_2$O$_3$ were commensurate with more rapid volatilization of solute than experiments conducted at lower temperatures with MgO.

Summary and Conclusions

Measurements of free-surface composition utilizing energetic particles on oxide surfaces under conditions of ultrahigh vacuum and high temperature are naturally subject to a number of uncertainties. In particular, this work has not attempted to address issues of solute and solvent sublimation, faceting, and redox reactions, all of which can occur in experiments of this type. Since the aim was to obtain a static measurement of the surface monolayer composition in the McLean approximation,[17] no effort was made to determine compositional depth profiles. The approach appears to be satisfactory for the Ca/MgO system, but may be only an approximation for Ca/Al$_2$O$_3$. In any equilibration experiment, there are issues of kinetics, and this would be particularly crucial in these segregation experiments, since the greatest apparent concentrations should occur at low temperatures where solute diffusion is slow. Surface damage introduced during specimen preparation may enhance diffusion in some instances, but the effect should be transient. Finally, the limits of spatial resolution of the surface analytical techniques used does not permit an accurate assessment of when solute or second-phase precipitation has occurred. This has been shown to be a significant factor in surface compositional measurements.[63] Surface diffraction probes may be useful in interpreting these phenomena.

Measurements of free-surface composition may offer an attractive alternative to grain-boundary segregation studies. Such measurements should be particularly useful in materials where solute diffusivities permit equilibration at temperatures where sublimation or differential sublimation is not a critical concern. Measured enthalpies of segregation appear to be in reasonable agreement with other measurements for grain boundaries as well as with calculated estimates.

Acknowledgments

The authors thank W. C. Leslie of the University of Michigan for critical review of the manuscript. W. T. Donlon and T. J. Whalen have provided useful comments and suggestions. P. Wynblatt of Carnegie-Mellon University was instrumental in initiating this work, and M. D. Wiggins assisted in data collection. This work is based, in part, on research conducted by R. C. McCune for the Ph.D. degree, Department of Materials and Metallurgical Engineering, at the University of Michigan.

References

[1] J. H. Westbrook; pp. 263–84 in Science of Ceramics, Vol. 3. Edited by G. H. Stewart. Academic Press, London, 1967.

[2] W. D. Kingery, "Plausible Concepts Necessary and Sufficient for Interpretation of Ceramic Grain-Boundary Phenomena: I, Grain-Boundary Characteristics, Structure, and Electrostatic Potential," *J. Am. Ceram. Soc.*, **57** [1] 1–8 (1974); "II, Solute Segregation, Grain-Boundary Diffusion, and General Discussion," *J. Am. Ceram. Soc.*, **57** [2] 74–83 (1974).

[3] W. C. Johnson, "Grain-Boundary Segregation in Ceramics," *Metall. Trans. A*, **8A** [9] 1413–22 (1977).

[4] P. Wynblatt and R. C. McCune; pp. 83–95 in Surfaces and Interfaces in Ceramic and Ceramic-Metal Systems. Edited by J. A. Pask and A. G. Evans. Plenum, New York, 1981.

[5] W. D. Kingery; pp. 1–22 in Advances in Ceramics, Vol. 1. Edited by L. M. Levinson. American Ceramic Society, Columbus, OH, 1981.

[6] J. B. Vander Sande and E. L. Hall, "Applications of Dedicated Scanning Transmission Electron Microscopy to Non-Metallic Materials," *J. Am. Ceram. Soc.*, **62** [5–6] 246–54 (1979).

[7] B. Bender, D. B. Williams, and M. R. Notis, "Investigation of Grain-Boundary Segregation in Ceramic Oxides by Analytical Scanning Transmission Electron Microscopy," *J. Am. Ceram. Soc.*, **63** [9–10] 542–6 (1980).

[8] P. Menon and T. S. R. Prasada Rao, "Surface Enrichment in Catalysis," *Catalysis Reviews*, **20** [1] 97–120 (1979).

[9] M. C. Kung and H. H. Kung, "The Surface Cation Densities of Iron Oxide–Chromium Oxide Solid Solutions," *Surf. Sci.*, **104** [1] 253–69 (1981).

[10] K. Miyoshi, D. H. Buckley, and M. Srinivasan, "Tribological Properties of Sintered Polycrystalline and Single-Crystal Silicon Carbide," *Am. Ceram. Soc. Bull.*, **62** [4] 494–500 (1983).

[11] R. C. McCune and P. Wynblatt, "Calcium Segregation to a Magnesium Oxide (100) Surface," *J. Am. Ceram. Soc.*, **66** [2] 111–7 (1983).

[12] P. W. Tasker, "Computer Simulation of Ionic Crystal Surfaces," T. P. 849, AERE Harwell, Oxfordshire, England, May, 1980.

[13] D. Wolf, "Formation Energy of Point Defects in Free Surfaces and Grain Boundaries in MgO," Presented at the 4th Europhysical Topical Conference on Lattice Defects in Ionic Crystals. Dublin, 1982.

[14] E. A. Colbourn, W. C. Mackrodt, and P. W. Tasker, "The Segregation of Calcium Ions at the Surface of Magnesium Oxide: Theory and Calculations," *J. Mater. Sci.*, **18** [7] 1917–24 (1983).

[15] S. Y. Liu and H. H. Kung, "Surface Cation Ratios of Binary Oxide Solid Solutions," *Surf. Sci.*, **110** [2] 504–22 (1981).

[16] J. W. Gibbs; p. 219 in The Scientific Papers of J. Willard Gibbs, Vol. 1. Dover, New York, 1961.

[17] D. McLean; Grain Boundaries in Metals. Oxford University Press, London, 1957.

[18] R. Defay, I. Prigogine, A. Bellemans, and D. H. Everett; p. 158 in Surface Tension and Adsorption. Wiley, New York, 1966.

[19] P. Wynblatt and R. C. Ku; pp. 115–36 in Interfacial Segregation. Edited by W. C. Johnson and J. M. Blakely. American Society for Metals, Metals Park, OH, 1979.

[20] R. G. Linford; pp. 1–152 in Solid State Surface Science, Vol. 2. Edited by M. Green. Marcel Dekker, New York, 1979.

[21] S. H. Overbury, P. A. Bertrand, and G. A. Somorjai, "The Surface Composition of Binary Systems. Prediction of Surface Phase Diagrams of Solid Solutions," *Chem. Rev.*, **75** [5] 547–60 (1975).

[22] F. L. Williams and D. Nason, "Binary Alloy Surface Compositions from Bulk Alloy Thermodynamics Data," *Surf. Sci.*, **45** [2] 377–408 (1974).

[23] A. R. Miedema, "Surface Segregation in Alloys of Transition Metals," *Z. Metallkunde*, **69** [7] 455–61 (1978).

[24] F. F. Abraham and C. R. Brundle, "Surface Segregation in Binary Solid Solutions: A Theoretical and Experimental Perspective," *J. Vac. Sci. Technol.*, **18** [2] 506–19 (1981).

[25] M. P. Seah, "Quantitative Prediction of Surface Segregation," *J. Catal.*, **57** [3] 450–7 (1979).

[26] I. Langmuir, "The Adsorption of Gases on Plane Surfaces of Glass, Mica and Platinum," *J. Am. Chem. Soc.*, **40** [9] 1361–1403 (1918).

[27] M. F. Yan, R. M. Cannon, and H. K. Bowen, "Space Charge, Elastic Field, and Dipole Contributions to Equilibrium Solute Segregation at Interfaces," *J. App. Phys.*, **54** [2] 764–78 (1983).

[28] J. Frenkel; p. 37 in Kinetic Theory of Liquids. Dover, New York, 1955.

[29] J. M. Blakely and S. Danyluk, "Space Charge Regions at Silver Halide Surfaces: Effects of Divalent Impurities and Halogen Pressure," *Surf. Sci.*, **40** [1] 37–60 (1973).

[30] K. L. Kliewer and J. S. Koehler, "Space Charge in Ionic Crystals: I, General Approach with Application to NaCl," *Phys. Rev. A*, **140** [4A] 1226–40 (1965).

[31] Y. T. Tan, "Impurity Distribution in AgBr," *Surf. Sci.*, **61** [1] 1–9 (1976).

[32] J. R. H. Black and W. D. Kingery, "Segregation of Aliovalent Solutes Adjacent Surfaces in MgO," *J. Am. Ceram. Soc.*, **62** [3–4] 176–8 (1979).

[33] Y. M. Chiang, A. F. Henriksen, W. D. Kingery, and D. Finello, "Characterization of Grain-Boundary Segregation in MgO," *J. Am. Ceram. Soc.*, **64** [7] 385–9 (1981).

[34] K. Lehovec, "Space-Charge Layer and Distribution of Lattice Defects at the Surface of Ionic Crystals," *J. Chem. Phys.*, **21** [7] 1123–8 (1953).

[35] E. A. Colbourn and W. C. Mackrodt, "The Calculated Defect Structure of Bulk and {001} Surface Cation Dopants in MgO" *J. Mater. Sci.*, **17** [10] 3021–38 (1982).

[36] C. R. A. Catlow, R. James, W. C. Mackrodt, and R. F. Stewart, "Defect Energetics in α-Al_2O_3 and Rutile TiO_2," *Phys. Rev. B*, **B25**, [2] 1006–26 (1982).

[37] S. K. Tiku and F. A. Kröger, "Effects of Space Charge, Grain Boundary Segregation, and Mobility Differences Between Grain Boundary and Bulk on the Conductivity of Polycrystalline Al_2O_3," *J. Am. Ceram. Soc.*, **63** [3–4] 183–9 (1980).

[38] R. C. Doman, J. B. Barr, R. N. McNally, and A. M. Alper, "Phase Equilibria in the System CaO-MgO," *J. Am. Ceram. Soc.*, **46** [7] 313–6 (1963).

[39] M. Rolin and P.-H. Thanh, "Phase Diagrams of Mixtures Not Reacting with Molybdenum $Ca(AlO_2)_2$-, $NaAlO_2$-, and La_2O_3-Al_2O_3," *Rev. Hautes Temp. Refract.*, **2** [2] 175–85 (1965).

[40] J. T. Grant, "Surface Analysis with Auger Electron Spectroscopy," *Appl. Surf. Sci.*, **13** [1] 35–62 (1982).

[41] J. A. Van den Berg and D. G. Armour, "Low Energy Ion Scattering (LEIS) and the Compositional and Structural Analysis of Solid Surfaces: I," *Vacuum*, **31** [6] 259–70 (1981).

[42] E. Taglauer and W. Heiland, "Surface Analysis with Low Energy Ions," *Appl. Phys.*, **9** [4] 261–75 (1976).

[43] J. A. Champion and M. A. Clemence, "Etch Pits in Flux-Grown Corundum," *J. Mater. Sci.*, **2** [2] 153–9 (1967).

[44] D. J. Barber and N. J. Tighe, "Observations of Dislocations and Surface Features in Corundum Crystals by Electron Transmission Microscopy," *J. Res. Natl. Bur. Stds.*, **69A** [3] 271–80 (1965).

[45] C. Lea and M. P. Seah, "Kinetics of Surface Segregation," *Philos. Mag.*, **35** [1] 213–22 (1977).

[46] L. E. Davis, N. C. MacDonald, P. W. Palmberg, G. E. Riach, and R. E. Weber; p. 13 in the Handbook of Auger Electron Spectroscopy, 2nd ed. Physical Electronics Industries, Eden Prairie, MN, 1976.

[47] A. D. Brailsford, "Surface Segregation Kinetics in Binary Alloys," *Surf. Sci.*, **94** [1] 387–402 (1980).

[48] C. G. Pantano and T. E. Madey, "Electron Beam Damage in Auger Electron Spectroscopy," *Appl. Surf. Sci.*, **7** [1–2] 115–41 (1981).

[49] W. C. Johnson and D. F. Stein, "Additive and Impurity Distributions at Grain Boundaries in Sintered Alumina," *J. Am. Ceram. Soc.*, **58** [11–12] 485–8 (1975).

[50] R. S. Jupp, D. F. Stein, and D. W. Smith, "Observations on the Effect of Calcium Segregation on the Fracture Behavior of Polycrystalline Alumina," *J. Mater. Sci.*, **15** [1] 96–102 (1980).

[51] L. Marchut and C. J. McMahon, Jr.; pp. 183–217 in Electron and Positron Spectroscopies in Materials Science and Engineering. Edited by O. Buck, J. K. Tien, and H. L. Marcus. Academic Press, New York, 1979.

[52] R. L. Coble, "Sintering Crystalline Solids: II, Experimental Tests of Diffusion Models in Powder Compacts," *J. Appl. Phys.*, **32** [5] 793–9 (1961).

[53] A. W. Funkenbusch and D. W. Smith, "Influence of Calcium on the Fracture Strength of Polycrystalline Alumina," *Metall. Trans., A*, **6A** [12] 2299–301 (1975).

[54] G. De With and N. Hattu, "Influence of CaO Doping on the Fracture Toughness of Hot-Pressed Al_2O_3," *J. Mater. Sci.*, **16** [3] 841–4 (1981).

[55] H. L. Marcus and M. E. Fine, "Grain-Boundary Segregation in MgO-Doped Al_2O_3," *J. Am. Ceram. Soc.*, **55** [11] 568–70 (1972).

[56] R. I. Taylor, J. P. Coad, and R. J. Brook "Grain-Boundary Segregation in Al_2O_3," *J. Am. Ceram. Soc.*, **57** [12] 539–40 (1974).

[57] R. I. Taylor, J. P. Coad, and A. E. Hughes, "Grain-Boundary Segregation in MgO-Doped Al_2O_3," *J. Am. Ceram. Soc.*, **59** [7–8] 374–5 (1976).

[58] J. G. J. Peelen; pp. 443–53 in Materials Science Research, Vol. 10. Edited by G. C. Kuczynski. Plenum Press, New York, 1975.

[59] P. E. C. Franken and A. P. Gehring, "Grain-Boundary Analysis of MgO-Doped Al_2O_3," *J. Mater. Sci.*, **16** [2] 384–8 (1981).

[60] B. J. Hockey, "Plastic Deformation of Aluminum Oxide by Indentation and Abrasion," *J. Am. Ceram. Soc.*, **54** [5] 223–31 (1971).

[61] A. Reisman, M. Berkenblit, J. Cuomo, and S. A. Chan, "The Chemical Polishing of

Sapphire and MgAl Spinel," *J. Electrochem. Soc.,* **118** [10] 1635–7 (1971).
[62]D. J. Reed and B. J. Wuensch, "Ion-Probe Measurement of Oxygen Self-Diffusion in Single Crystal Al$_2$O$_3$," *J. Am. Ceram. Soc.,* **63** [1–2] 88–92 (1980).
[63]W. D. Kingery, W. L. Robbins, A. F. Henriksen, and C. E. Johnson, "Surface Segregation of Aluminum (Spinel Precipitation) in MgO Crystals," *J. Am. Ceram. Soc.,* **59** [5–6] 239–41 (1976).

---

*Model 545, Physical Electronics Industries, Eden Praire, MN.
†Model 525, 3M Corp., St. Paul, MN.
‡Coors Spectrochemical Laboratory, Golden, CO.
§W. & C. Spicer, Ltd., Cheltenham, Gloucester, England.
¶Lucalox, General Electric Co., Schenectady, NY.

Dislocations in α-Al$_2$O$_3$

A. H. HEUER

Case Institute of Technology
Department of Metallurgy and Materials Science
Case Western Reserve University
Cleveland, OH 44106

J. CASTAING

Laboratoire de Physique des Matériaux
C.N.R.S. Bellevue
92195 Meudon Cedex, France

Perfect dislocations, with **b** = $\frac{1}{3}\langle 11\bar{2}0\rangle$, $\frac{1}{3}\langle 10\bar{1}1\rangle$, and $\langle 10\bar{1}0\rangle$, and partial dislocations, with **b** = $\frac{1}{9}\langle 11\bar{2}0\rangle$, $\frac{1}{3}\langle 10\bar{1}0\rangle$, $\frac{1}{3}\langle 10\bar{1}1\rangle$, and $\frac{1}{3}\langle 0001\rangle$, have been observed and studied in α-Al$_2$O$_3$. We consider in this paper the energy of these several types of dislocations, their relation to the crystal structure of α-Al$_2$O$_3$, and their occurrence during crystal growth, plastic deformation, and irradiation with energetic particles. Finally, the importance of dislocations to high-temperature processes such as precipitation and diffusion are discussed.

Dislocations in α-Al$_2$O$_3$ have been of interest for more than 25 years. The early work dealt with both crystal growth[1] and crystal plasticity.[2] Concern with plastic deformation of sapphire, the single-crystal form of α-Al$_2$O$_3$, has continued to the present time and has occupied the attention of a small group of dedicated researchers. Dedication is necessary when deformation of sapphire is studied because of the extreme experimental conditions — high temperatures or high hydrostatic confining pressures — needed to induce dislocation motion, considering the large Peierls barrier of the α-Al$_2$O$_3$ lattice.

Dislocations also form and multiply in α-Al$_2$O$_3$ during irradiation with energetic particles[3,4] and during precipitation from solid solution.[5] For comparison purposes, dislocation climb and other types of recovery processes during high-temperature deformation can often be considered in parallel with other high-temperature diffusion-controlled properties such as sintering and grain growth, and studies of pipe diffusion can provide insight into other types of short-circuit diffusion processes, such as grain-boundary diffusion.

In the following, we examine in detail those characteristics of dislocations in α-Al$_2$O$_3$ that are involved in these various processes. After a brief description of the crystallographic properties of dislocations, which is derived solely from consideration of the crystal structure, we review the more common types of dislocations, i.e., those with the most frequently observed Burgers vectors. We next deal with dislocation interactions, particularly dislocation dissociation and dislocation reactions. Finally, examples of the role of dislocations on several kinetic processes in α-Al$_2$O$_3$ will be given.

Dislocations and the Corundum Lattice

Crystal Structure

The crystallography of α-Al_2O_3 has been described in detail by Kronberg,[2] who carefully distinguished the hexagonal and rhombohedral descriptions of α-Al_2O_3, and the morphological and structural cells, both of which have been used in previous work. In this paper, we use the (now conventional) structural hexagonal unit cell; it is twice as high as the morphological cell and rotated by 180° (when referred to hexagonal axes).[2] (This rotation has sometimes been overlooked by previous workers.)

Figures 1 and 2 show the structure of α-Al_2O_3. It can be viewed as an hcp sublattice of oxygen ions, with ⅔ of the octahedral interstices being filled with Al ions in an ordered array. Off-center displacements of the Al ions occur and increase

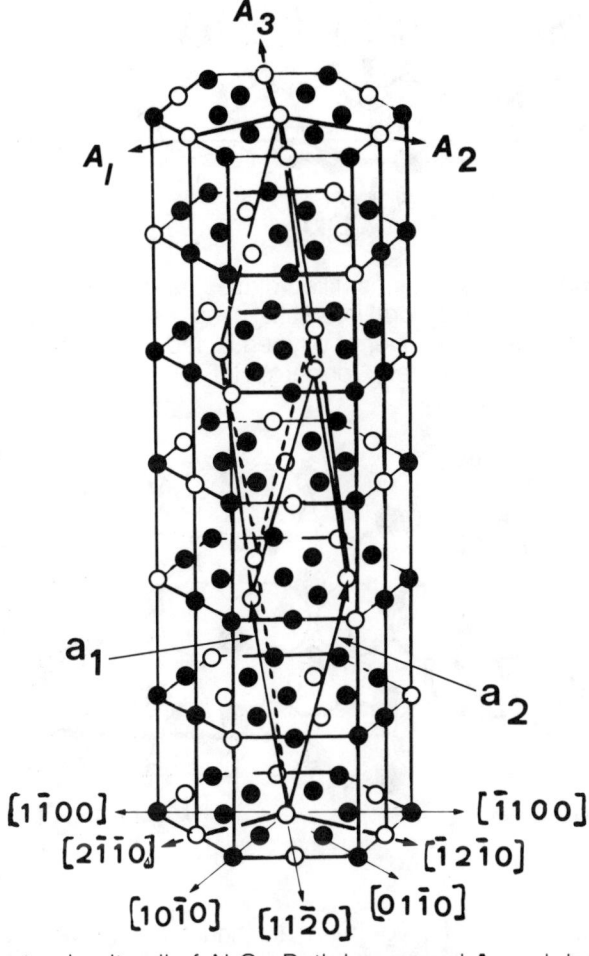

Fig. 1. Structural unit cell of Al_2O_3. Both hexagonal A_i and rhombohedral a_i basis vectors are shown. Oxygen ions are not indicated; Al ions and unoccupied octahedral sites are shown as filled and open circles, respectively. The staggered displacement of Al ions along the [0001] axis is not shown.

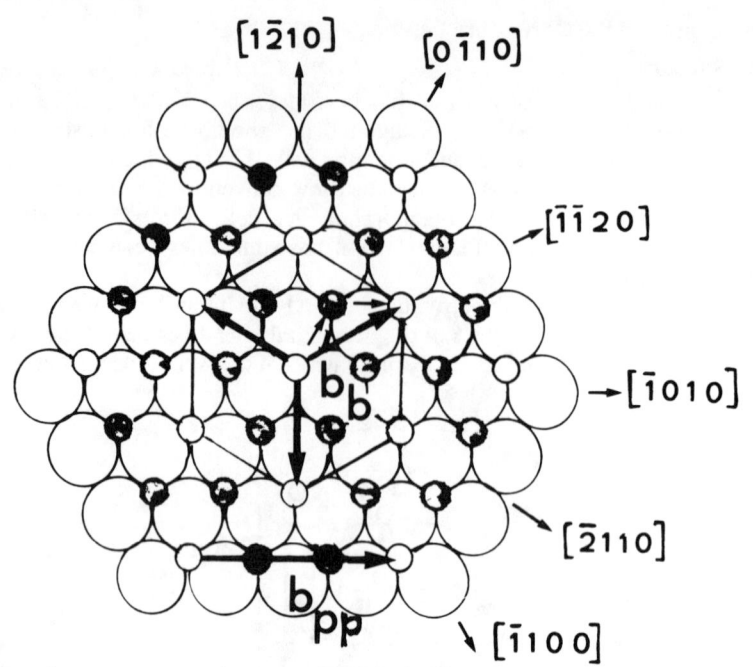

Fig. 2. Arrangement of Al ions (filled circles) and unoccupied sites (small open circles) in the octahedral interstitial sites of the hcp oxygen sublattice (large open circles). One layer of oxygen is shown together with the basal hexagonal cell vectors (after Kronberg; Ref. 2).

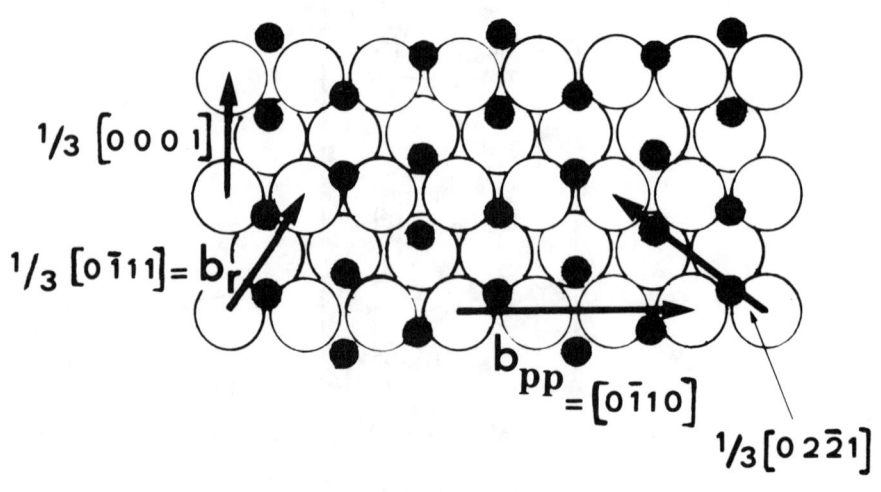

Fig. 3. Arrangement of oxygen and aluminum ions viewed along [$\bar{2}110$] in the prism slip plane. Several possible Burgers vectors for dislocations are also shown.

Table I. Burgers Vectors in Al$_2$O$_3$ and Corresponding Elastic Energy of Dislocations*

b		b (nm)	μb^3 (eV)	μb^2 (J/nm)	
$\mathbf{b}_b$ =	⅓⟨11$\bar{2}$0⟩	0.475	100	34	Perfect, basal, observed
$\mathbf{b}_r$ =	⅓⟨10$\bar{1}$1⟩	0.512	126	39	Perfect, rhombohedral, observed
	⅓⟨$\bar{2}$021⟩	0.698	319	73	Perfect
$\mathbf{b}_{pp}$ =	⟨10$\bar{1}$0⟩	0.822	555	101	Perfect, prism plane, observed
	⅓⟨21$\bar{3}$1⟩	0.844	564	107	Perfect
	⅓⟨$\bar{1}$012⟩	0.908	702	124	Perfect
	[0001]	1.297	2045	252	Perfect, observed
	⅓⟨10$\bar{1}$0⟩	0.274	21	11	Partial, observed
	⅓[0001]	0.432	76	29	Partial, observed
	⅑⟨11$\bar{2}$0⟩	0.158	3.7	3.8	Partial, observed

*μ = 150 GPa; 1 eV = 1.6 × 10^{-19} J.

their separation by 18% along [0001] (Fig. 3); this has been explained[2] as arising from the electrostatic interactions between the positively charged Al ions. As the packing of Al ions can be described by an ....abcabc ....sequence and that of the close-packed oxygen layers by the *ABAB* sequence, the minimum repeat distance along [0001] corresponds to six planes of the hcp sublattice (Fig. 1). The hexagonal parameters defined in Figs. 1 and 2 are c = 1.297 nm and a = 0.475 nm, with c/a = 2.73. For the oxygen hcp sublattice alone, c/a = 1.58, slightly smaller than the ideal value for a hard sphere model of 1.63 and presumably also a result of the electrostatic binding that causes the "puckering" of the Al sublattice.[2]

Burgers Vectors

The shortest repeat distances of the crystal lattice are the possible Burgers vectors, **b**, for perfect dislocations. They are shown in Figs. 1–3 and are listed in Table I, following Snow and Heuer.[6] Table I also includes values of the parameters μb^3, proportional to the elastic energy of a screw dislocation per repeat distance along the dislocation line, and μb^2, the elastic energy per unit length of dislocation (μ is the shear modulus). To obtain dislocation line energies, these values have to be multiplied by $(1/4\pi)(\ln R/b)$ ($\approx$0.55 if R, the outer cutoff radius,[7] is taken as 1000b). The values are appropriate for isotropic crystals; they are satisfactory in this case for α-Al$_2$O$_3$ because of its nearly isotropic elastic properties. (The isotropic approximation for α-Al$_2$O$_3$ has been confirmed, for example, by the very small difference between line energies computed by using anisotropic and isotropic elasticity theory for dislocations with **b** = ⅓⟨11$\bar{2}$0⟩ and ⟨10$\bar{1}$0⟩.[8])

It is not surprising that the "basal" dislocation, $\mathbf{b}_b$, with the shortest Burgers vector (⅓⟨11$\bar{2}$0⟩; Table I) has been observed in the greatest number of experiments (Table II). However, other dislocations with larger Burgers vectors have also been found, viz., the rhombohedral dislocation, $\mathbf{b}_r$ = ⅓⟨10$\bar{1}$1⟩, and the prism plane* dislocation, $\mathbf{b}_{pp}$ = ⟨10$\bar{1}$0⟩ (Table I).

Table II. Observations of Dislocations in Al_2O_3 for Various Burgers Vectors

Burgers vector	Grown-in (Ref.)	Network (Ref.)	Plasticity (Ref.)	Irradiation damage (Ref.)	Precipitate (Ref.)
b_b	Barber 1966 (21) Caslavsky 1972 (22) Watanabe 1976 (23) Wada 1980 (24)	Barber 1966 (21)	Pletka 1974 (15) Mitchell 1976 (13) Pletka 1982 (17) Phillips 1982 (14) Phillips 1982 (16) Lagerlof 1983 (48) Barber 1965 (21) Wiederhorn 1973 (27) Kotchick 1980 (19) Cadoz 1982 (20) Becher 1971 (53) Yust 1981 (54) Hockey 1971 (26)	Rechtin 1979 (3) Barber 1968 (29)	Phillips 1980 (5)
b_r		Gulden 1967 (35) Hockey 1981 (28)	Bayer 1967 (31) Tressler 1974 (32) Firestone 1976 (33) Cadoz 1976 (34) Gooch 1973 (36) Becher 1971 (53)	Rechtin 1979 (3) Howitt 1981 (4) Barber 1968 (29)	Phillips 1980 (5)
b_{pp}	Watanabe 1976 (23)		Gooch 1973 (36) Bilde-Sorensen 1976 (37) Cadoz 1977 (18) Cadoz 1980 (8) Cadoz 1982 (20) Phillips 1982 (38) Cadoz 1978 (40) Hockey 1975 (43) Yust 1981 (54)		
$1/3\langle 10\bar{1}0\rangle$		Hockey 1981 (28)	Mitchell 1976 (13) Phillips 1982 (14) Phillips 1982 (38) Rivière 1980 (42)	Howitt 1981 (4)	Phillips 1980 (5)
$1/3[0001]$		Hockey 1981 (28)		Rechtin 1979 (3) Howitt 1981 (4) Stathopoulos 1983 (44)	Phillips 1980 (5)
$1/9[11\bar{2}0]$		Hockey 1981 (28)	Rivière 1980 (42)		

The ⅓⟨$\bar{2}021$⟩ dislocation has never been observed. This is probably due to the fact that it has a large self-energy (Table I), and although it is a translation vector of the α-Al$_2$O$_3$ lattice, it does not correspond to any simple vector in the hcp oxygen sublattice (Fig. 3). The prism plane dislocation has an even higher line energy, but it corresponds to a close-packed direction of the oxygen sublattice (Figs. 2 and 3) and furthermore can decrease its energy by dissociation. This will be discussed in more detail below, along with further discussion of the properties of $\mathbf{b}_b$, $\mathbf{b}_r$, and $\mathbf{b}_{pp}$ dislocations.

The [0001] dislocation has been observed only in very restricted circumstances. Its large energy (Table I) can be substantially reduced only if the logarithmic term is small, i.e., if R can be reduced. This is the case for whiskers grown along the [0001] direction. The elastic equilibrium of these whiskers requires the presence of axial screw dislocations with Burgers vectors several times [0001],[9] which actually take the form of hollow tunnels.[10] Except for this case, the other perfect dislocations in Table I with Burgers vectors larger than 0.822 nm have never been observed in α-Al$_2$O$_3$.

Dislocation Properties
Basal Dislocations

Basal dislocations with $\mathbf{b}_b$ = ⅓⟨$11\bar{2}0$⟩ Burgers vector have the lowest energy of any perfect dislocation in α-Al$_2$O$_3$ (Table I) and are the most easily introduced by high-temperature plastic deformation; basal slip has the lowest critical resolved shear stress at all temperatures above 700°C.[11,12]

The crystallographic structure of edge dislocations with this Burgers vector can be described most simply by considering the stacking along ⟨$11\bar{2}0$⟩ (Fig. 3). Two different slices of crystal, which differ only in their arrangement of Al ions, are sufficient to construct the crystal structure. One of these slices is shown in Fig. 3, and an edge dislocation can be formed by adding two "extra half-plane" slices (Fig. 4(a)).

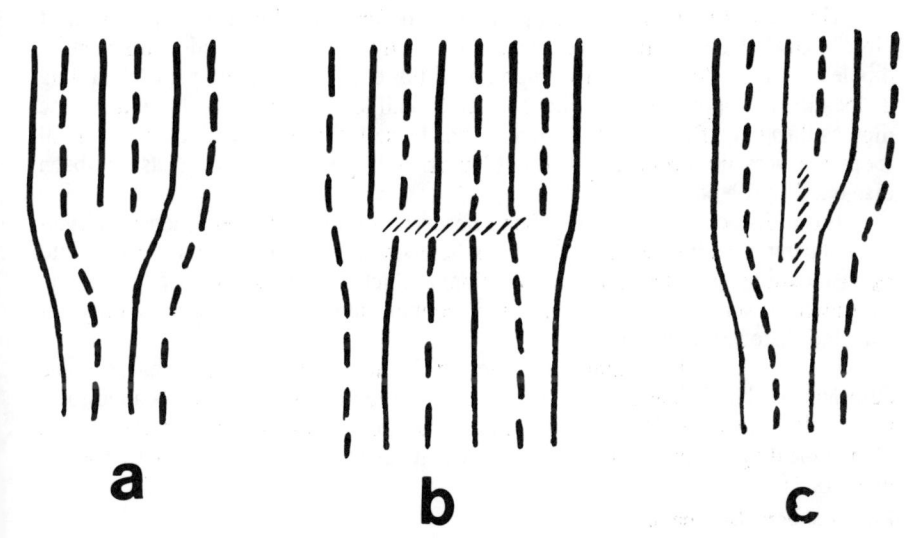

Fig. 4. Edge dislocation with $\mathbf{b}_b$ or $\mathbf{b}_r$ Burgers vector. The perfect dislocation (a) can dissociate by glide (b) or by climb (c); the two partials having noncolinear vectors.

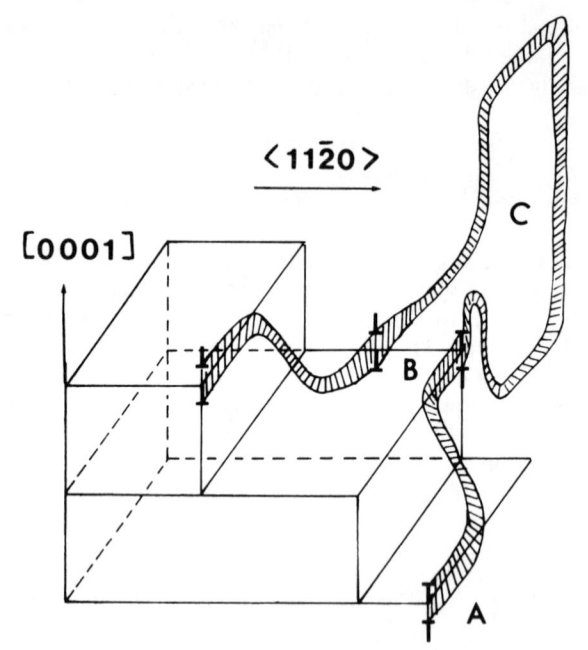

Fig. 5. Edge dislocation dipole formed during basal slip. Part A corresponds to elastic equilibrium. Parts B and C are formed by self-climb at the tip of the dipole (Refs. 13, 14). At B, dislocations are in the basal plane, and at C, they almost form a prismatic loop. There is no systematic constriction of the stacking fault ribbon.

The dislocations found in sapphire after deformation by basal slip consist of "free" basal dislocations, predominantly of edge character,[15] edge dislocation dipoles, formed during slip by trapping of the edge portions of parallel gliding dislocations of opposite sign (Fig. 5),[15] and small loops, formed by breakup of the dipoles through a fluctuation of their widths by self-climb.[†] The dipoles and small loops are very important in work hardening and dynamic recovery, as has been discussed by Pletka et al.[15,17,25]

Basal dislocations have also been observed to form a three-dimensional network during prism plane slip,[18-20] due to the decomposition of b_{pp} dislocations into b_b dislocations (see Dislocation Reactions). Such networks are also present in as-grown crystals,[21-24] which may also contain subboundaries with readily resolvable dislocation structures.[21]

In addition to these examples where dislocations are related to macroscopic deformation, basal dislocations have been observed in indented and abraded samples,[26] after crack healing following microfracture at various temperatures[27,28] and after annealing of irradiation damage, where dislocation loops react to form a network.[3,29]

Rhombohedral Dislocations

Knowledge of rhombohedral dislocations is limited compared to that of basal or prism plane dislocations (Table II). This results mainly from the difficulty of introducing such dislocations by plastic deformation and avoiding twinning;[30] it can be done by testing in tension[31,32,36] or in shear,[34] but there are inherent difficulties in using such techniques with brittle materials.

Dislocations with $\mathbf{b}_r = \frac{1}{3}\langle 10\bar{1}1\rangle$ have an energy slightly higher than that of basal dislocations (Table I). Their Burgers vector corresponds to a *structural* rhombohedral unit cell vector (Fig. 1) and is the shortest repeat distance out of the basal plane (Fig. 3). Careful study of Fig. 3 reveals that an edge dislocation with this Burgers vector can also be created by inserting two extra half-plane slices, giving rise to the configuration already shown in Fig. 4(*a*).

Rhombohedral dislocations form and multiply during pyramidal slip, but they do not lie in a clearly defined pyramidal slip plane, probably due to extensive climb (the temperature of deformation needs to be above $0.8T_M$ to activate pyramidal slip); rather, the dislocations form a three-dimensional network with many dislocation reactions.[33]

Rhombohedral dislocations are sometimes present in low-angle grown-in subgrain boundaries in α-Al$_2$O$_3$;[35] more significantly, they are prominant components in the dislocation substructures arising from irradiation damage in α-Al$_2$O$_3$ (see next section) and are also prominant following precipitation from TiO$_2$-doped ("star") sapphire,[5] as will be discussed in High-Temperature Processes.

Irradiation of α-Al$_2$O$_3$ at various temperatures has been performed with neutrons,[29] various ions,[3] and electrons.[4] The large number of Frenkel-pair point defects produced during irradiation either (i) mutually recombine, (ii) annihilate at fixed sinks, e.g., grain boundaries and grown-in dislocations, or (iii) aggregate to form dislocation loops and voids. The loops that form initially are faulted; they lie either on (0001) and have a ⅓[0001] Burgers vector or on $\{10\bar{1}0\}$ and have a $\frac{1}{3}\langle 10\bar{1}0\rangle$ Burgers vector. When the loops grow to a sufficient size, they tend to unfault, thus eliminating their stacking fault; this can be accomplished by the nucleation of an additional partial dislocation, which glides across the loop to give a perfect $\mathbf{b}_r$ dislocation according to the dislocation reaction

$$\frac{1}{3}\langle 10\bar{1}0\rangle + \frac{1}{3}[0001] \rightarrow \frac{1}{3}\langle 10\bar{1}1\rangle \tag{1}$$

(Basal dislocations are noticeably absent in irradiated samples of Al$_2$O$_3$, because their self-energy is only slightly smaller than that of rhombohedral dislocations and they cannot accommodate any lattice strain with a component out of the basal plane.)

Finally, $\mathbf{b}_r$ dislocations were observed as part of an interfacial misfit dislocation network following spontaneous crack healing and had also formed via Eq. (1).[28]

It appears, therefore, that rhombohedral dislocations may play their most important part in processes other than those involving plastic deformation in Al$_2$O$_3$; their absence in crystals deformed by prism plane slip implies a very high Peierls barrier compared to that of prism plane dislocations, whose glissile motion may be substantially assisted by dissociation.

Prism Plane Dislocations

From an energetic point of view, the occurrence of prism plane dislocations with $\mathbf{b}_{pp} = \langle 10\bar{1}0\rangle$ is rather unexpected. The energy of this dislocation is 3 times that of basal dislocations (Table I), into which they can decompose with a 33% decrease in energy according to the dislocation reaction

$$\langle 10\bar{1}0\rangle \rightarrow \frac{1}{3}\langle 2\bar{1}\bar{1}0\rangle + \frac{1}{3}\langle 11\bar{2}0\rangle \tag{2}$$

The stability of $\mathbf{b}_{pp}$ dislocations during prism plane slip probably comes from the fact that the Burgers vector corresponds to a close-packed direction of the oxygen sublattice (Figs. 2 and 3), with a consequent lowering of the Peierls barrier, coupled with the dissociation into three colinear partials (to be discussed below) that apparently have a high mobility.[20]

Therefore, prism plane slip occurs along the $\mathbf{b}_{pp}$ direction, rather than along the $\mathbf{b}_r$ direction, which is also possible for this slip plane (Figs. 2 and 3) and is in preference to the alternative slip system $\{10\bar{1}0\}\langle11\bar{2}0\rangle$. Furthermore, the usual experimental conditions for nonbasal plastic deformation of α-Al_2O_3, i.e., compression perpendicular to [0001], provide a much larger Schmid factor for $\mathbf{b}_{pp}$ dislocations than for $\mathbf{b}_r$ dislocations.

A schematic drawing of an edge dislocation with a $\mathbf{b}_{pp}$ Burgers vector can be constructed by noting that the stacking along $\langle10\bar{1}0\rangle$ is made up of three different "slices" of crystal,[39] representing a 12-layer sequence.[4] An edge dislocation then contains *three* "extra half-planes," each of thickness $\frac{1}{3}\langle10\bar{1}0\rangle$ (Fig. 6(a)).

Prism plane dislocations, sometimes in dipoles or as isolated loops, are observed after small strains in samples deformed by prism plane slip; they have dominant edge character.[18,40] After large strains, they disappear, since several dislocation reactions leading to their elimination (discussed in Dislocation Reactions) become very effective. $\langle10\bar{1}0\rangle$ dislocations have also been observed around microhardness identations, but they are on $\{11\bar{2}3\}$ slip planes or are associated with basal microtwins.[43] In the latter case, different sets of $\langle10\bar{1}0\rangle$ dislocations (i.e., $[10\bar{1}0]$ and $[0\bar{1}10]$) were present on opposite faces of wedge-shaped twin interfaces, and TEM contrast analysis demonstrated they were not twinning dislocations, but perfect accommodation dislocations.[43]

Partial Dislocations, Twins and Stacking Faults

The dissociation of basal and prism dislocations introduced by plastic deformation has been studied in detail by weak-beam transmission electron microscopy (Table II). They have been found to dissociate into partials, always with a $\frac{1}{3}\langle10\bar{1}0\rangle$ Burgers vector. This vector is a translation vector of the oxygen sublattice (Figs. 2 and 3); the resulting stacking fault only involves the Al sublattice.

We consider *basal* dislocations first. These can in principle separate to give a conventional dissociated dislocation (Fig. 4(b)); however, such *glide*-dissociated dislocations have been observed only after low-temperature deformation.[11] On the other hand, *climb* dissociation (Fig. 4(c)) has been frequently observed after high-temperature deformation for edge dislocation dipoles collapsing to faulted dipoles[13] and for a few isolated dislocations[14] according to the dislocation reaction (Fig. 2)

$$\tfrac{1}{3}[11\bar{2}0] \rightarrow \tfrac{1}{3}[10\bar{1}0] + \tfrac{1}{3}[01\bar{1}0] \tag{3}$$

The faulted dipoles form by annihilation of the inner partials of the constituent dislocations.[13,14,16] Faulted and unfaulted dipoles, as well as strings of faulted and unfaulted loops formed by dipole breakup, are therefore typical components of the dislocation debris in both pure and doped α-Al_2O_3 after basal slip.[13-15]

Prism plane dislocations are even more prone to dissociation, given their large Burgers vectors; they dissociate into three colinear partials (Fig. 6):

$$\langle1\bar{1}00\rangle \rightarrow \tfrac{1}{3}\langle1\bar{1}00\rangle + \tfrac{1}{3}\langle1\bar{1}00\rangle + \tfrac{1}{3}\langle1\bar{1}00\rangle \tag{4}$$

There is a substantial decrease of energy for this reaction, provided the stacking fault energy is reasonably low. Glide dissociation via this reaction (Fig. 6(b)) may have been observed,[37] but a climb component could also have been present but not detected.[20,42] Certainly, after high-temperature deformation, dissociation occurs

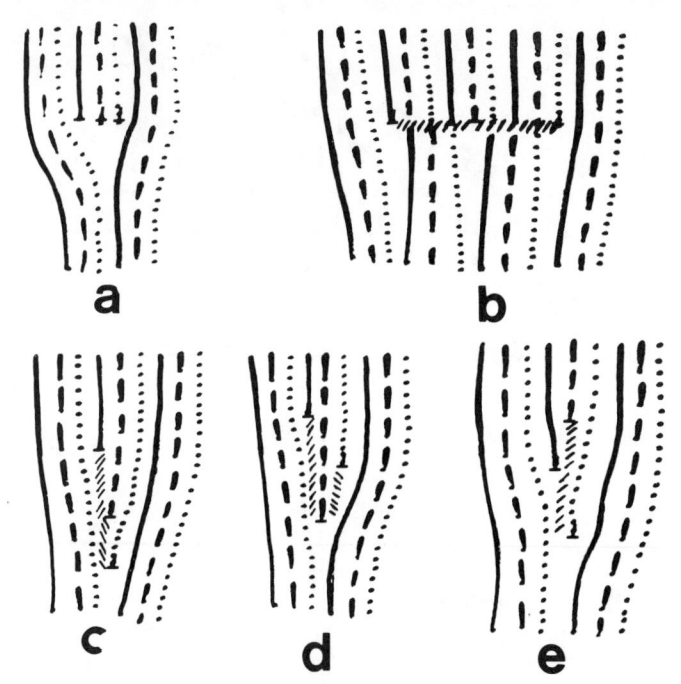

Fig. 6. Edge dislocation with a prism plane $\langle 10\bar{1}0 \rangle$ Burgers vector. The perfect dislocation (a) may dissociate into three colinear partials in a glissile configuration (b) or in three different climb configurations (c), (d), and (e).

mostly by climb (Fig. 6(c–e)).[39,42] Different configurations are possible, giving different combinations of three stacking faults. The upper fault in each of Fig. 6(c–e) corresponds to the removal of one layer (vacancy fault), and the lower fault involves insertion of an extra layer (i.e., an interstitial fault). The configurations of Fig. 6(c, e) are energetically equal, and it has been suggested that they are favored over that of Fig. 6(d).[39]

$\frac{1}{3}\langle 10\bar{1}0 \rangle$ partial dislocations are also involved in the formation of basal twins in α-Al$_2$O$_3$. The basal twin law for α-Al$_2$O$_3$ is[41,43] $K_1 = (0001)$; $\eta_1 = \langle 10\bar{1}0 \rangle$; $K_s = \{10\bar{1}1\}$; $\eta_2 = \langle \bar{1}012 \rangle$; $s = 0.635$.

Crystallographically, such twinning can be described as a rotation of the lattice by 180° around [0001]. This requires that for mechanical twins, the macroscopic shear deformation be along $[10\bar{1}0]$, $[\bar{1}100]$, or $[0\bar{1}10]$ and not the reverse directions. However, Kronberg[2] pointed out that the crystal structure of α-Al$_2$O$_3$ does not permit a twinned structure to be generated by a unidirectional shear along η_1, which is the conventional way of generating deformation twins. Kronberg[2] suggested that two sets of quarter-partials, with $\mathbf{b} = \frac{1}{9}\langle 11\bar{2}0 \rangle$ and movement in "synchroshear," lead to a classically unpredicted twinning interface; this interface was actually a glide plane rather than a true mirror but was in accord with macroscopic experimental data.

Hockey,[43] however, on the basis of his observations that the $\langle 10\bar{1}0 \rangle$ accommodation dislocations present at twin interfaces were all of one sign on each interface and glissile on $\{0001\}$, suggested a different model. Each face of a mechanical twin

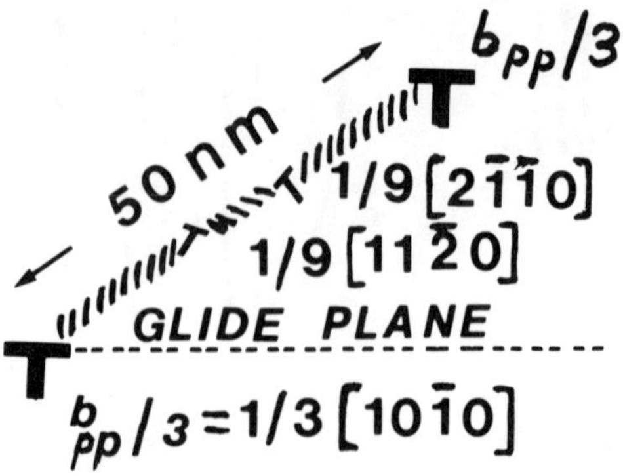

Fig. 7. Representation of a dissociated prism plane dislocation with a stacking fault at 26° to the glide plane (Refs. 39, 42).

can be likened to a simple shear boundary. Displacements of ⅓[$\bar{1}$010], ⅓[01$\bar{1}$0], and ⅓[1$\bar{1}$00], i.e., *opposite* to the macroscopic twinning shear but on every other basal plane, will result in a twinned structure in agreement with macroscopic observation. Hockey[43] therefore suggested that passage of ⅓[$\bar{1}$010] partial dislocation on every other basal plane above the midplane of the twin, and ⅓[01$\bar{1}$0] partials on every other basal plane below the midplane of the twin, is the mechanism by which basal twins are generated. This model is essentially one of a hypothetical double-ended pole dislocation source emitting ⅓[$\bar{1}$010] dislocations at one end and ⅓[01$\bar{1}$0] dislocations at the other. Although different from the usual pole model of twinning suggested for metals, this model satisfies all the macroscopically observed features of basal twinning and results in a twin interface, which is a true mirror plane. Rhombohedral twinning is dealt with in Appendix II.

Finally, Hockey[43] also demonstrated that the [10$\bar{1}$0] dislocations associated with {11$\bar{2}$3} slip bands around microhardness indents were glide dissociated, but he did not determine the Burgers vectors of the partial dislocations bounding the stacking faults.

Dissociation of *rhombohedral* dislocations into partials with the *morphological* rhombohedral cell vectors as Burgers vector is also possible, according to the reaction

$$\tfrac{1}{3}[0\bar{1}11] \rightarrow \tfrac{1}{6}[\bar{2}021] + \tfrac{1}{6}[2\bar{2}01] \quad (5)$$

This dissociation has never been observed; however, the TEM studies of such dislocations[3–5,29] have not been exhaustive, and this reaction cannot be discounted at this time.

Further dissociation of ⅓⟨10$\bar{1}$0⟩ partials into quarter-partials, this time with the introduction of a fault in the oxygen lattice, was predicted by Kronberg[2] according to the reaction

$$\tfrac{1}{3}[10\bar{1}0] \rightarrow \tfrac{1}{9}[2\bar{1}\bar{1}0] + \tfrac{1}{9}[11\bar{2}0] \quad (6)$$

Reaction (6) has been observed during the dissociation of the inner partial of a dissociated prism plane dislocation (Fig. 7)[39,42] and within dislocation networks produced by crack healing.[28] In both of these examples, the fault configuration was unstable, implying a very high stacking fault energy; for this reason and because of its rarity, we shall not discuss this dissociation any further.

One of the most interesting characteristics of almost all dislocation studies reported to date is the observation that dislocation dissociation often occurs by *climb* rather than by glide. Similar results have been found in other brittle compounds (for a review, see Ref. 12), whenever dislocations experience high temperatures. It is probably only surprising because of its infrequent (or unrecognized) occurrence in metallic systems. Climb dissociation is a very natural process by which the self-energy of a dislocation can be decreased, since it leads to a configuration of minimum elastic interactions between the partial dislocations (see Appendix I). (In fact, a similar low-energy configuration exists for perfect dislocations making up a polygon wall.[47]) Climb dissociation only requires climb over limited distances (Figs. 4–7), which occurs readily at the high temperatures of deformation usually employed ($>0.7\ T_M$). Whenever a dislocation is immobilized, a climb-dissociated configuration is expected, especially if the dislocation was dissociated by glide during its motion and if the stacking fault energy is reasonably isotropic.[39]

Quite different types of partial dislocations are introduced during irradiation with energetic particles, due to condensation of excess point defects. As already mentioned, this leads to the production of faulted loops with a ⅓[0001] Burgers vector in the basal plane and to faulted loops with a Burgers vector of ⅓⟨10$\bar{1}$0⟩ in the {10$\bar{1}$0} prism plane. From the maximum size of the irradiation-induced faulted loops, *stacking fault energies* γ have been deduced[4] whose magnitudes are in agreement with similar estimates obtained from the width of dissociated dislocations on {10$\bar{1}$0}, {1$\bar{2}$10}, and {10$\bar{1}$2} planes and introduced by high-temperature plastic deformation.[39] All γ values fall within the range 0.15–0.7 J/m², suggesting that the stacking fault energy in Al₂O₃ is fairly isotropic for planes parallel to [0001] or at small angles to [0001],[39] with a mean value of 0.3 J/m². A calculation based on electrostatics gives a value of 0.5 J/m² for the ⅓[0$\bar{1}$10] basal fault; however, values 10 times larger than this were found for faults on {1$\bar{2}$10} and {10$\bar{1}$0} prism planes.[45,46] This discrepancy is probably due to the inadequacy of the Coulomb potential used; an alternative explanation is that precipitation of point defects onto actual faults occurs, giving the impression that the fault energy is low and rather isotropic.

Finally, the trigonal symmetry of α-Al₂O₃ gives rise to a singular type of partial dislocation in the faulted interfacial misfit dislocation network around spontaneously healed microcracks.[28] Consider reaction (1) again, but this time change the sign of the ⅓[0001] partial

$$\tfrac{1}{3}\langle 10\bar{1}0\rangle + \tfrac{1}{3}[000\bar{1}] \rightarrow \tfrac{1}{3}\langle 10\bar{1}\bar{1}\rangle \tag{7}$$

⟨10$\bar{1}\bar{1}$⟩ is *not* a lattice vector in α-Al₂O₃, and the region in the dislocation network bounded by such dislocations is of necessity faulted. TEM contrast analysis confirmed[28] that the dislocation networks in nonbasal healed microcracks could always be described in terms of two sets of ⅓⟨10$\bar{1}$0⟩ partials cross-linked by a set of ±⅓[0001] partials, and nodes described by *both* reactions (1) and (7) were found.

Dislocation Reactions

Dislocation reactions are the first step in the formation of subboundaries and in the annihilation of dislocations. Since there are three basal, three rhombohedral, and three prism plane Burgers vectors, many possible reactions must be considered, and indeed, many have been observed. For example, reactions between basal dislocations are common:[20–22,36]

$$\tfrac{1}{3}[1\bar{2}10] + \tfrac{1}{3}[\bar{2}110] \rightarrow \tfrac{1}{3}[\bar{1}\bar{1}20] \tag{8}$$

All other reactions involve Burgers vectors of different types. The following reaction

$$\tfrac{1}{3}[\bar{1}101] + \tfrac{1}{3}[1\bar{2}10] \rightarrow \tfrac{1}{3}[0\bar{1}11] \tag{9}$$

has been observed after both deformation[34,36] and precipitation experiments.[5] These reactions obviously lead to a decrease in energy. The situation is somewhat different for the decomposition of prism plane dislocations already mentioned (reaction (2)), which has been observed within prism plane slip bands:[20,40]

$$\langle 10\bar{1}0 \rangle \rightarrow \tfrac{1}{3}\langle 2\bar{1}\bar{1}0 \rangle + \tfrac{1}{3}\langle 11\bar{2}0 \rangle \tag{2}$$

The gain in energy is, at first sight, so favorable that it virtually forbids the existence of $\mathbf{b}_{pp}$ dislocations! Mechanistically, this is a very easy reaction for dislocations in screw orientation at high temperatures, since the product basal dislocations are very mobile (for other orientations of the dislocation line, cross-slip or climb must occur). However, an alternative and efficient way to decrease the energy of $\mathbf{b}_{pp}$ dislocations, which competes with reaction (2), is the dissociation into three colinear partials (reaction (4)). On the assumption that the energy of the triply dissociated $\langle 10\bar{1}0 \rangle$ dislocation is equal to that of two basal dislocations resulting from reaction (2), a value of γ in good agreement with experiment was obtained.[39]

Reaction (2) plays an important role in the work-hardening/recovery process during high-temperature prism plane slip,[20] in that the product basal dislocations readily form a three-dimensional network, which leads to very high work-hardening rates.[20] In fact, basal dislocations can be formed from prism plane dislocations by an even more efficient process, the reaction of prism plane and basal dislocations to give a basal dislocation:

$$\langle 1\bar{1}00 \rangle + \tfrac{1}{3}\langle \bar{2}110 \rangle \rightarrow \tfrac{1}{3}\langle 1\bar{2}10 \rangle \tag{10}$$

which, of course, also leads to a very large reduction in energy. (Actually, both basal and prism plane dislocations are climb dissociated when they are static, and the energy reduction also includes the reduction in stacking fault area.) Reactions (2) and (10) are very efficient in creating a large number of basal dislocations during prism plane slip. In fact, after a few percent deformation at high temperatures, the specimens contain no prism plane dislocations at all![18,20] On the other hand, after low-temperature prism plane slip under hydrostatic pressure,[11] specimens do not contain basal dislocations, reactions (2) and/or (10) apparently being inhibited.

High-Temperature Processes

We deal in this section with a few ways in which study of dislocations in Al_2O_3 relates to processes other than plastic deformation and radiation damage,[‡] either directly or indirectly. In particular, we wish to deal with the role of dislocations in accommodating volume changes during a high-temperature precipitation reaction[5] and to describe studies of the kinetics of dislocation annihilation, which can provide information on both lattice and pipe diffusion in α-Al_2O_3.[48]

Dislocations play two roles in high-temperature precipitation reactions. Unless the lattice parameters of parent and matrix are particularly well matched (e.g., with misfits <0.01) and the particles are coherent with the matrix, the interface must be semicoherent or incoherent. In the latter case, the structure is similar to

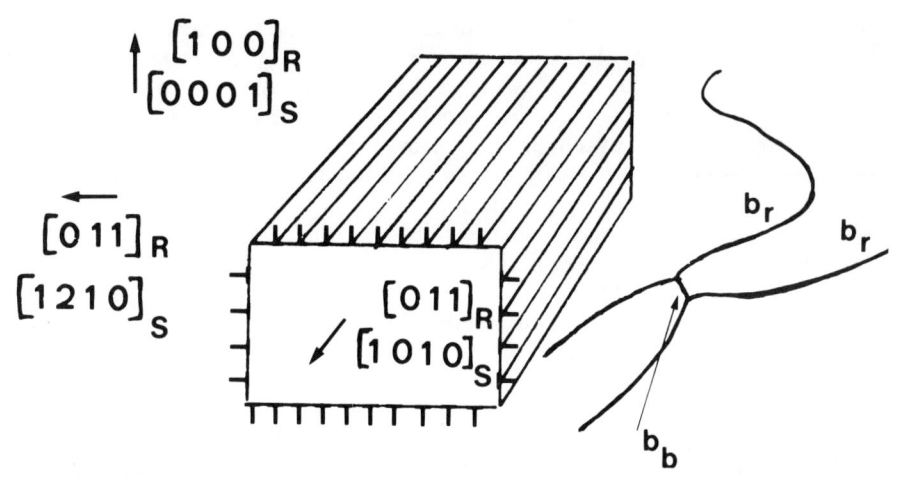

Fig. 8. Precipitate of rutile (R) in a sapphire matrix (S) showing interface dislocations and surrounding rhombohedral dislocations ($\mathbf{b}_r$) with some short basal segments ($\mathbf{b}_b$). The scheme is according to Ref. 5.

that of a high-angle grain boundary. A semicoherent interface consists of an array of evenly spaced dislocations. If the dislocation spacing is s, the misfit strain ε accommodated by the array of dislocations is b/s; any additional misfit must be accommodated elastically. (If the Burgers vector does not lie in the plane of the interface, then it is the component of b resolved in the plane of the interface that is effective in relieving the misfit strain.) In the case of the needle-like rutile precipitates in α-Al$_2$O$_3$, where the precipitate/matrix habit planes are (0001) and $\{11\bar{2}0\}$, alternate $\frac{1}{3}[1\bar{1}00]$ and $\frac{1}{3}[10\bar{1}0]$ dislocations spaced 5.5 nm apart and $\frac{1}{3}[0001]$ dislocations spaced 7 nm apart, respectively, were thought to be present to relieve the lattice strain, although arrays of $\mathbf{b}_r$ dislocations with these same spacings could not be discounted (Fig. 8).[5]

A more subtle need for dislocations during precipitation reactions is to carry away (or bring in) the excess (deficit) volume when the precipitate is less (more) dense than the matrix. In metallic systems, this is usually accomplished by conventional plastic flow. In star sapphire, the oxygen sublattice of the rutile precipitate is 9.3% less dense than the oxygen sublattice in α-Al$_2$O$_3$; rutile is not as close packed as α-Al$_2$O$_3$. The "excess" Al$_2$O$_3$ that must be removed so that rutile precipitates can form "plate out" on basal planes as climbing interstitial loops with rhombohedral Burgers vectors (Fig. 8); the $\mathbf{b}_r$ dislocations subsequently react via Eq. (9) to form a network. In fact, the area of climbed-out loop per unit volume of precipitated rutile can be used as a very precise measure of the density of the Ti-doped Al$_2$O$_3$ solid solution *prior to precipitation*. This, and precise lattice parameter measurements of the solid solution, permitted Phillips et al.[5] to conclude that the charge compensating defect for Ti^{4+} was an oxygen interstitial (probably in a defect cluster) and not the commonly accepted aluminum vacancy.

Finally, studies of dislocation annihilation permit measurements of lattice and pipe diffusion in ways that may be simpler than tracer measurements and that permit determination of diffusivity to lower temperatures than are possible with tracer techniques. This approach, first used by Narayan and Washburn[49] for MgO, has been recently exploited by Lagerlof et al.[48] in α-Al$_2$O$_3$, which had been deformed by basal slip at elevated temperatures (>1200°C). As mentioned in the section on Basal Dislocations, basal edge dislocation dipoles are prominent constituents of the dislocation debris. The dipoles can become single-ended or double-ended elongated loops by climb or by cross-slip of screw components of the individual dislocations, and they further break up into small circular or nearly circular (aspect ratio ≤ 5) prismatic loops. The formation of prismatic loops occurs by self-climb, a process controlled by pipe diffusion (D_p) kinetics. These small prismatic loops are also unstable and can coarsen by glide and self-climb or, more importantly, can shrink by pure climb, a process controlled by lattice diffusion (D_l) kinetics. For both self-climb and pipe diffusion, and pure climb and lattice diffusion, stoichiometric quantities of Al and O must be added to or taken away from the dislocation line, and the process in each case is controlled by the slower species (which may not be the same in the two climb processes).

Study of a single dipole breaking up, or a single prismatic loop shrinking, yields a measure of D_p or D_l, respectively. This has been accomplished by Lagerlof et al.[48] by characterizing the dislocation substructure in a thin foil by TEM, annealing the foil for a given time at a given temperature, and studying the changes induced in the now well-characterized dislocation debris. To date, lattice diffusion kinetics have been measured at temperatures between 1200 and 1500°C for undoped sapphire and sapphire doped with either Mg^{2+} or Ti^{4+}; the results are displayed in Fig. 9. Because $D_l^{Al} > D_l^O$, we expect the dislocation annihilation kinetics to yield data on D_l^O; it is gratifying to see that the results are in excellent agreement with the extrapolation of three tracer studies (Fig. 9). D_l^O measurements as low as 10^{-23} m^2/s have been made without any major difficulty.

The measurements on the doped crystals are consistent with the notions that the oxygen vacancy concentrations (assuming oxygen diffuses by a vacancy mechanism) are, respectively, increased and decreased by the solutes Mg^{2+} and Ti^{4+}. The best fit diffusivities for the new data of Fig. 9 are

$$D_{\text{undoped}} = 2.9 \times 10^{-4} \exp\left[-\left(\frac{5.8 \text{ (eV)}}{kT}\right)\right] \text{ (m}^2\text{/s)}$$

$$D_{\text{Mg-doped}}^{2+} = 1.13 \exp\left[-\left(\frac{6.3 \text{ (eV)}}{kT}\right)\right] \text{ (m}^2\text{/s)}$$

$$D_{\text{Ti-doped}}^{4+} = 2.2 \times 10^{-7} \exp\left[-\left(\frac{5.1 \text{ (eV)}}{kT}\right)\right] \text{ (m}^2\text{/s)}$$

The work of Lagerlof et al.[48] on pipe diffusion is preliminary at this time, but initial data indicate $D_p \sim 10^3 D_l$.

Summary and Conclusions

The various perfect and partial dislocations possible in α-Al$_2$O$_3$ have been described and interpreted in terms of the crystal structure and discussed in terms of their energetics. Their role in crystal plasticity and their formation during irradiation with energetic particles are reviewed, and their importance to various high-temperature processes such as precipitation are emphasized. Finally, their utility in novel determinations of bulk and short-circuit diffusivities are demonstrated.

It can be concluded that knowledge of dislocations in Al$_2$O$_3$ is at a tolerably advanced state; this knowledge can now be exploited in both traditional and unconventional ways in the advancement of ceramic science.

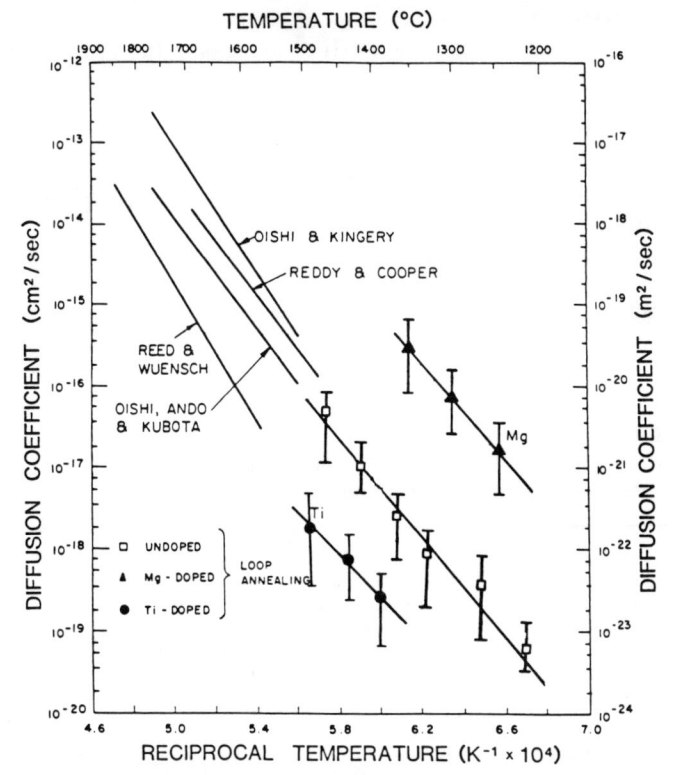

Fig. 9. Diffusion coefficients D vs $1/T$ derived from loop annealing studies (Ref. 48). Tracer data for oxygen diffusion are also included: figure after (Ref. 48).

Acknowledgments

A. H. H.'s research on deformation of sapphire has been supported by the NSF, while that of J.C. is supported by the CNRS. Both authors acknowledge the contributions both of the scholars Averroes and Maimonides of Cordoba, Spain (b. 1126, d. 1198 and b. 1135, d. 1204, respectively), whose work provided the inspiration for the written version of this work during the authors' visit to Cordoba in Holy Week, 1983, and of the authors' Cleveland and Bellevue colleagues for pleasant and profitable collaboration over a number of years. Finally, A.H.H. acknowledges the Alexander von Humboldt Foundation for a Senior Scientist Award, which made possible his sabbatical leave at the Max-Planck-Institut für Metallforschung, Stuttgart, FRG, where this paper was written.

Appendix I

Glide vs Climb Dissociation

The calculation of equilibrium dissociation is generally made by letting the partial dislocations move in their glide plane. The energy minimum under such a condition gives the dissociation width. In this Appendix, we consider the case where no such restrictive condition is applied and the stacking fault energy γ is isotropic. We treat edge partial dislocations with colinear Burgers vectors for simplicity of exposition, and without loss of generality, the conclusions are valid for mixed dislocations. According to Hirth and Lothe (p. 117 in Ref. 7), the interaction force per unit length is

$$\frac{F(r)}{L} = \frac{\mu b^2}{2\Pi(1-\nu)r}$$

where **b** is the partial Burgers vector, r is the distance between partials, and ν is Poisson's ratio. The first partial dislocation is taken as the origin, and polar coordinates (r,θ) are used. Note that $F(r)$ does not depend on θ. This is well-known for screw dislocations, where the dissociation plane is determined by the fault energy anisotropy. The equilibrium distance obtained by letting

$$\frac{F(r)}{L} = \gamma$$

is the same for glide, climb, or "intermediate" dissociation.

We now determine the configuration of minimum energy at fixed r. The interaction energy is given by Hirth and Lothe[7] as

$$\frac{W_{12}}{L} = \frac{\mu b^2}{2\Pi(1-\nu)} \ln\left(\frac{r}{r_a}\right) - \frac{\mu b^2 \sin^2\theta}{2\Pi(1-\nu)r^2}$$

where r_a is the core radius. It is clear the W_{12} is maximum for $\theta = 0$ or Π (glide dissociation) and minimum for $\theta = \Pi/2$ or $3\Pi/2$ (climb dissociation).

Appendix II

Rhombohedral Twinning in α-Al₂O

Rhombohedral twinning is an important deformation mode in α-Al_2O_3.[41,50] Atom movements during deformation twinning were considered by Scott,[50] but not in terms of twinning dislocations. In addition, steps at a rhombohedral twin interface in the isostructural hematite (α-Fe_2O_3) were imaged by using high-resolution electron microscopy by Bursill and Withers.[51] Both Scott[50] and Bursill and Withers[51] proposed structures for the twin interface. Scott's structure did not contain a true mirror plane, as discussed by Fortuneé,[52] but was a faulted plane that gave the correct crystallographic orientation between twin and matrix.

The rhombohedral twin in α-Al_2O_3 is actually a symmetrical twin boundary obtained after a 64.789° rotation about $[11\bar{2}0]$. This rotation leaves the oxygen sublattice undisturbed but introduces faults into the cation sublattice. In the reference crystal, a row of unoccupied octahedral interstices ("holes") are aligned parallel to the boundary plane; in the rotated crystal, the cation plane parallel to the boundary plane contains alternate cations and empty holes.[52] Clearly, a pair of adjacent holes at the boundary represents an electrostatic charge "problem," which can be removed by a further 180° rotation of the cation sheets about the c axis of the rotated crystal. Three distinct positions are possible for the last rotation axis and correspond to the axis passing through either the A, B, or C position of the cation stacking sequence at the boundary plane. One of these positions gives the twin boundary structure of Bursill and Withers and is a true mirror plane; a second gives Scott's structure, and the third gives still another twin boundary structure that has not previously been described. In all three structures, the cations in the boundary plane are probably in tetrahedral coordination. Fortuneé noted that the "shuffle" needed to bring the cations to the tetrahedral positions was minimum in Bursill and Wither's structure, which made it the most likely. This description of twinning in terms of two rotations arises from the trigonal symmetry of α-Al$_2$O$_3$ and is unrelated to the actual atom motions that are involved in the twinning.

A description of nucleation and/or growth of rhombohedral twins in α-Al$_2$O$_3$ in terms of twinning dislocations still does not exist. The "twinning dislocations" described by Bursill and Withers derived from steps in the twin plane and do not correspond to the types of twinning dislocations usually envisaged in other systems. In particular, their structural model for a twin dislocation at a rhombohedral twin boundary was claimed to show twin dislocations parallel to $[10\bar{1}2]$ and with $b = 0.74$ Å (0.074 nm) (sic). We have drawn a Burgers circuit around their structural model of a twin dislocation (their Fig. 10; actually a boundary containing a step), using their twin interface model (their Fig. 9) as the reference lattice. The closure failure can be thought of as a dislocation with $\mathbf{b} = \frac{1}{3}\langle 10\bar{1}l \rangle$ (it lies on the basal plane in a $[11\bar{2}0]$ projection) and must correspond to the displacement introduced by the step itself. There is no assurance, however, that the dislocations associated with such steps are involved in rhombohedral twinning. We thus conclude that further theoretical and experimental work on the nature and role of twinning dislocations in rhombohedral twinning in α-Al$_2$O$_3$ remains an important task for future work.

References

[1] G. W. Sears and R. C. de Vries, "Morphological Development of Aluminum Oxide Crystals Grown by Vapor Deposition," *J. Chem. Phys.*, **39** [11] 2837–45 (1963).
[2] M. L. Kronberg, "Plastic Deformation of Single Crystals of Sapphire: Basal Slip and Twinning," *Acta Metall.*, **5** [9] 507–24 (1957).
[3] M. D. Rechtin, "A Transmission Electron Microscopy Study of the Defect Microstructure of Al$_2$O$_3$ Subjected to Ion Bombardment," *Rad. Eff.*, **42**, 129–44 (1979).
[4] D. G. Howitt and T. E. Mitchell, "Electron Irradiation Damage in α-Al$_2$O$_3$," *Philos. Mag., [Part A]*, **A44** [1] 229–38 (1981).
[5] D. S. Phillips, A. H. Heuer, and T. E. Mitchell, "Precipitation in Star Sapphire," *Philos. Mag., [Part A]*, **A42** [3] 285–432 (1980).
[6] J. D. Snow and A. H. Heuer, "Slip Systems in Al$_2$O$_3$," *J. Am. Ceram. Soc.*, **56** [3] 153–7 (1973).
[7] J. P. Hirth and J. Lothe; Theory of Dislocations. Wiley, New York, 1982.
[8] J. Cadoz, J. Castaing, P. Veyssière, and G. Faivre, "Study of the Motion of a $\langle 10\bar{1}0 \rangle$ Dislocation in $(1\bar{2}10)$ in α-Al$_2$O$_3$," *Electron Microsc.*, **4**, 408–11 (1980); Proceedings of the International Conference of High Voltage Electron Microscopy, Sept. 1–3, Antwerp, 1980.
[9] R. D. Dragsdorf and W. W. Webb, "Detection of Screw Dislocations in α-Al$_2$O$_3$ Whiskers," *J. Appl. Phys.*, **29** [5] 817–9 (1958).
[10] D. J. Barber, "Electron Microscopy and Diffraction of Aluminum Oxide Whiskers," *Philos. Mag.*, **10** [103] 75–94 (1964).
[11] J. Castaing, J. Cadoz, and S. H. Kirby, "Deformation of Al$_2$O$_3$ Single Crystals between 25°C and 1800°C Basal and Prismatic Slip," *J. Phys. Colloq. C3*, **42** [6] 43–7 (1981).
[12] T. Bretheau, J. Castaing, J. Rabier, and P. Veyssière, "Dislocation Motion and High Temperature Plasticity of Binary and Ternary Oxides," *Adv. Phys.*, **28** [6] 829–1014 (1979).
[13] T. E. Mitchell, B. J. Pletka, D. S. Phillips, and A. H. Heuer, "Climb Dissociation of Dislocations in Sapphire (α-Al$_2$O$_3$)," *Philos. Mag.*, **34** [3] 441–51 (1976).
[14] D. S. Phillips, T. E. Mitchell, and A. H. Heuer, "Climb Dissociation of Dislocations in Sapphire (α-Al$_2$O$_3$) Revisited: Crystallography of Dislocation Dipoles," *Philos. Mag., [Part A]*, **45** [3] 371–85 (1982).
[15] B. J. Pletka, T. E. Mitchell, and A. H. Heuer, "Dislocation Structures in Sapphire Deformed by Basal Slip," *J. Am. Ceram. Soc.*, **57** [9] 388–93 (1974).
[16] D. S. Phillips, B. J. Pletka, A. H. Heuer, and T. E. Mitchell, "An Improved Model of Break-up of Dislocation Dipoles into Loops: Application to Sapphire α-Al$_2$O$_3$," *Acta Metall.*, **30**, 491–8 (1982).
[17] B. J. Pletka, T. E. Mitchell, and A. H. Heuer, "Dislocation Substructures in Doped Sapphire (α-Al$_2$O$_3$) Deformed by Basal Slip," *Acta Metall.*, **30**, 147–56 (1982).
[18] J. Cadoz, D. Hokim, M. Meyer, and J. P. Rivière, "Observation of Dislocations Associated with Prism Plane Slip in Alumina Single Crystals," *Rev. Phys. Appl.*, **12**, 473–81 (1977).
[19] D. M. Kotchick and R. E. Tressler, "Deformation Behavior of Sapphire via the Prismatic Slip System," *J. Am. Ceram. Soc.*, **63** [7–8] 429–34 (1980).

[20] J. Cadoz, J. Castaing, D. S. Phillips, A. H. Heuer, and T. E. Mitchell, "Work-Hardening and Recovery in Sapphire α-Al$_2$O$_3$ Undergoing Prism Plane Deformation," *Acta Metall.*, **30**, 2205–18 (1982).

[21] D. J. Barber and N. J. Tighe, "Electron Microscopy and Diffraction of Synthetic Corundum Crystals," *Philos. Mag.*, **11**, [111] 495–512 (1965); **14** [129] 531–44 (1966).

[22] J. L. Caslavasky and C. P. Gazzara, "Dislocation Behavior in Sapphire Single Crystals," *Philos. Mag.*, **26** [4] 961–75 (1972).

[23] K. Watanabe and Y. Sumiyoshi, "Relationship between Habit and Etch Figures of Corundum Crystals Grown from Molten Cryolite Flux," *J. Cryst. Growth*, **32**, 316–26 (1976).

[24] K. Wada and K. Hoshikawa, "Growth and Characterization of Sapphire Ribbon Crystals," *J. Cryst. Growth*, **50**, 151–9 (1980).

[25] B. J. Pletka, A. H. Heuer, and T. E. Mitchell, "Work-Hardening in Sapphire," *Acta Metall.*, **25**, 25–33 (1977).

[26] B. J. Hockey, "Plastic Deformation of Aluminum Oxide by Indentation and Abrasion," *J. Am. Ceram. Soc.*, **54**, 223–31 (1971).

[27] S. M. Wiederhorn, B. J. Hockey, and D. E. Roberts, "Effect of Temperature on the Fracture of Sapphire," *Philos. Mag.*, **28** [4] 783–96 (1973).

[28] B. J. Hockey, "Crack Healing in Brittle Materials"; pp. 637–58 in Fracture Mechanics in Ceramics, Vol. 6. Edited by R. C. Bradt, A. G. Evans, D. P. H. Hasselman, and F. F. Lange. Plenum, New York, 1983.

[29] D. J. Barber and N. J. Tighe, "Neutron Damage in Single Crystal Aluminum Oxide," *J. Am. Ceram. Soc.*, **51** [11] 611–7 (1968).

[30] W. D. Scott and K. K. Orr, "Rhombohedral Twinning in Alumina," *J. Am. Ceram. Soc.*, **66** [1] 27–32 (1983).

[31] P. D. Bayer and R. E. Cooper, "A New Slip System in Sapphire," *J. Mater. Sci.*, **2**, 301–2 (1967).

[32] R. E. Tressler and D. J. Barber, "Yielding and Flow of C-Axis Sapphire Filaments," *J. Am. Ceram. Soc.*, **57** [1] 13–9 (1974).

[33] R. F. Firestone and A. H. Heuer, "Creep Deformation of 0° Sapphire," *J. Am. Ceram. Soc.*, **59** [1–2] 24–9 (1976).

[34] J. Cadoz and P. Pellissier, "Influence of Three-Fold Symmetry on Pyramidal Slip of Alumina Single Crystal," *Scr. Metall.*, **10**, 597–600 (1976).

[35] T. D. Gulden, "Direct Observation of Nonbasal Dislocations in Sintered Alumina," *J. Am. Ceram. Soc.*, **50** [9] 472–5 (1967).

[36] D. J. Gooch and G. W. Groves, "Non-basal Slip in Sapphire," *Philos. Mag.*, **28** [3] 623–37 (1973).

[37] J. B. Bilde-Sorensen, A. R. Tholen, D. J. Gooch, and G. W. Groves, "Structure of the $\langle 01\bar{1}0 \rangle$ Dislocation in Sapphire," *Philos. Mag.*, **33** [6] 877–89 (1976).

[38] D. S. Phillips and J. L. Cadoz, "Climb Dissociation of $\langle 10\bar{1}0 \rangle$ Dislocations in Sapphire (α-Al$_2$O$_3$)," *Philos. Mag.*, [Part A], **46** [4] 583–95 (1982).

[39] K. P. D. Lagerlof, T. E. Mitchell, A. H. Heuer, J. P. Rivière, J. Cadoz, J. Castaing, and D. S. Phillips, "Stacking Fault Energy in Sapphire (α-Al$_2$O$_3$)"; to be published in *Acta Metall.*

[40] J. Cadoz, J. Castaing, and J. Philibert, "Dislocations in Alumina Deformed by Prismatic Slip"; pp. 606–7 in Proceedings of the 9th International Congress of Electron Microscopy, Vol. 1. Toronto, 1978.

[41] A. H. Heuer, "Deformation Twinning in Corundum," *Philos. Mag.*, **13**, 379–93 (1966).

[42] J. P. Rivière, J. Cadoz, and J. Philibert, "Climb Dissociation of [$1\bar{1}00$] Dislocations in $(11\bar{2}0)$ Prismatic Slip in Sapphire"; p. 410 in EUREM 80, Proceedings of the 7th European Conference of Electron Microscopy, Vol. 1. Edited by P. Brederoo and G. Boom, The Hague, Aug. 24–9, 1980.

[43] B. J. Hockey, "Pyramidal Slip on $\{11\bar{2}3\}$ $\langle \bar{1}100 \rangle$ and Basal Twinning in Al$_2$O$_3$"; pp. 167–79 in Deformation of Ceramic Materials. Edited by R. C. Bradt and R. E. Tressler. Plenum, New York, 1975.

[44] A. Y. Stathopoulos and G. P. Pells, "Damage in the Cation Sublattice of α-Al$_2$O$_3$ Irradiated in a HVEM," *Philos. Mag.*, [Part A], **47** [3] 381–94 (1983).

[45] J. Rabier, P. Veyssière, and J. Grilhé, "Energy Computations of Planar Defects in Some Oxides," *J. Phys. Colloq. C9*, **34**, 373–7 (1973).

[46] J. Rabier; Thesis, University of Poitiers, France, 1979.

[47] M. L. Kronberg, "Polygonization of a Plastically Bent Sapphire Crystal," *Science (Washington, D.C.)*, **122**, 599–600 (1955).

[48] K. P. D. Lagerlof, B. J. Pletka, T. E. Mitchell, and A. H. Heuer, "Deformation and Diffusion in Sapphire α-Al$_2$O$_3$," *Rad. Eff.*, **74**, 87–107 (1983).

[49] J. Narayan and J. Washburn, "Self-Diffusion in Magnesium Oxide," *Acta Metall.*, **21**, 533–8 (1973).

[50] W. D. Scott, "Rhombohedral Twinning in Aluminum Oxide"; pp. 151–66 in Deformation of Ceramic Materials. Edited by R. C. Bradt and R. E. Tressler. Plenum, New York, 1975.

[51] L. A. Bursill and R. L. Withers, "Twinning Dislocations in Haematite Iron Ore," *Philos. Mag.*, [Part A], **40**, 213–32 (1979).

[52] R. Fortunée; M.S. Thesis, Case Western Reserve University, Cleveland, OH, 1981.

[53]P. F. Becher, "Deformation Substructure in Polycrystalline Alumina," *J. Mater. Sci.*, **6**, 275–80 (1971).
[54]C. S. Yust and L. A. Harris, "Observation of Dislocations and Twins in Explosively Compacted Alumina"; pp. 881–94 of Chapter 50 in Shock Waves and High Strain Rate Phenomena in Metals. Plenum, New York, 1981.

---

*We use the rather clumsy term "prism plane" because of the importance of dislocations with a $\langle 10\bar{1}0 \rangle$ Burgers vector during prism plane slip in sapphire and so that confusion could be avoided with "prismatic loops," which in dislocation jargon always connote loops whose Burgers vectors do not lie in the plane of the loop.

†Self-climb is the conservative motion of an edge dislocation away from its glide surface, which occurs by pipe diffusion such that positive climb occurs in some regions and negative climb in others (Ref. 13).

‡The relations between dislocations and plastic flow and dislocations and irradiation damage are so obvious and have been alluded to sufficiently in previous sections that nothing further need be said here. Likewise, the intimate relation between dislocations and the structure of low-angle grain boundaries needs no comment here. The dislocation-like character of high-angle grain boundaries is dealt with in other papers in these proceedings.

Dislocations in MgO

Yusuke Moriyoshi and Takayasu Ikegami

National Institute for Research in Inorganic Materials
Tsukuba Science City
Ibaraki 305, Japan

Dislocation-related phenomena in MgO such as high-temperature creep, thermal behavior of dislocations, dislocation motion, dislocations and impurities, and dislocations and cathodoluminescence are reviewed. Also, a trial to observe a dislocation core with high-resolution electron microscopy is briefly described.

Since dislocations in crystals were first postulated by Orowan, Taylor, and Polanyi, both theoretical and experimental studies of dislocations have been extensive. Dislocations in crystals are involved with various phenomena including mechanical properties, solid-state reactions, crystal growth, electrical properties of semiconductors, and catalysis. These have been studied by many workers. However, we still have many things to investigate. In comparison with metallic systems, dislocations in ceramics are more complex. Two extra half-planes of atoms are required for ceramics in order to maintain electrical neutrality, and dislocation jogs have an effective charge. Interactions between dislocations and lattice defects, solute segregation and precipitation along dislocations, the dislocation velocity under stress, and many other things still remain to be studied in detail. MgO is a typical ceramic with a simple crystal structure, so it is natural that many workers have studied dislocation phenomena in MgO.

If we search a commercial index* by the key words "dislocation" and "MgO," a large number of papers can be found for the past 5 years. These papers include many dislocation-related phenomena such as creep, dislocation motion, fracture, diffusion, cathodoluminescence, dislocation theories, and so forth. MgO is an ionic material with the simple face-centered cubic structure with (100) cleavage faces. At low temperature, slip takes place in the (110) $\langle 110 \rangle$ systems, whereas at higher temperature, it occurs in both the (100) $\langle 110 \rangle$ and (110) $\langle 110 \rangle$ systems. The Burgers vectors are $a/2\langle 110 \rangle$. Because of its high solubility in acids, it is easy to make thin films by chemical thinning for transmission electron microscopy (TEM). Therefore, basic studies of dislocations in MgO have been conducted by using TEM. These have been discussed in excellent review papers by Kingery,[1] Heuer,[2] Hobbs,[3] Mitchell,[4] and Vander Sande.[5]

In this paper, a brief review of progress during the past 5 years and our recent experimental data concerning MgO, observed with TEM, high-temperature creep, dislocation and diffusion, dislocation motion, dislocation and impurities, and high-resolution TEM observations are presented.

MgO Samples for Dislocation Studies

Observation of dislocations in MgO has been done both by etch pits and by TEM. The former is effective for observing the macroscopic characteristics of dislocations such as the dislocation density and slip bands. A dislocation etch pit technique for MgO was first introduced by Stokes et al.,[6] who developed a solution of 5 parts of NH$_4$Cl saturated aqueous solution with 1 part of concentrated H$_2$SO$_4$ and 1 part of distilled H$_2$O as an etchant. The etchant has been widely used by many workers to measure dislocation densities as well as to study the dynamic behavior of dislocations. Sangual[7] has investigated the effect of temperature and concentration of inorganic and organic acids on the etching rate for (100), (110), and (111) crystal faces. Harada[8] has proposed a new etchant, FeCl$_3$·6H$_2$O solution, to reveal dislocations on the (100) faces of the crystal. He successfully distinguished the different shapes between edge and screw slip bands in the etchant of 5% FeCl$_3$·6H$_2$O solution for 20 min at 45°C. These etch-pit techniques are easy and convenient to study dislocation behavior in MgO; however, because of optical resolution, it is difficult to clarify dislocation configurations such as the dislocation arrangements within slip bands and dislocation interactions at networks. Thus, TEM would be best for such precise studies. Recently, the 0.2-nm lattice fringe was resolved. However, TEM observations are not without their problems.

Chemical thinning with H$_3$PO$_4$ and ion bombardment[9] are general methods for the fabrication of MgO thin foils. The former, developed by Washburn et al.,[10] is useful for single crystals but not for samples with precipitates that are insoluble in acids. The latter is suitable for both single crystals and polycrystals, but it is important that the thickness of the mechanically polished samples before ion bombardment be less than 100 μm, because dislocations are introduced by the mechanical polishing to about 30 μm in depth. Also, selective ion etching can occur in polycrystalline samples and those with precipitates. A demerit of ion bombardment is that not only does it take a long time to obtain an electron transparent foil, but that a good foil for high-resolution observations is not always obtained. Radiation damage by the electron beam is another problem, which has been recently reviewed by Hobbs.[3] A high-resolution microscope equipped with a cold stage and ultrahigh-vacuum system needs to be developed. The preparation of suitably thin foils is extremely important; further work is absolutely necessary on this.

Dislocations and High-Temperature Creep

Steady-state creep of polycrystal MgO under low stress has been elucidated by the Nabarro-Herring[11] and Coble[12] diffusion models in which strain rate is inversely proportional to the square of grain size and to the cube of grain size, respectively. On the basis of these models, modified theories,[13,14] deformation mechanism maps,[15] and many other studies[16-18] have been reported.

The creep theory associated with dislocation climb has been developed by Weertman.[19] Bilde-Sorensen[20] studied compressional creep of polycrystal MgO at 1300–1400°C under 2.5–5.5 kg/mm^2. As a result, it has been shown that glide is the principal cause of deformation in creep, but the rate-limiting process is diffusion controlled. Huether and Reppich[21] investigated dislocation structures generated during compression creep of MgO single crystals, using both the etch-pit technique and TEM. Clauer and Wilcox[22] studied in detail dislocation structures during ⟨110⟩ tensile creep in MgO single crystals. They discussed in detail the creep substructures and creep mechanisms. Moriyoshi et al.[23] also studied dislocation structures developed during ⟨100⟩ tensile creep in MgO single crystals at 1250–1450°C in air, using TEM (Fig. 1). Consequently, it was suggested that oxygen ion diffusion in MgO is the rate-determining process at stresses lower than 200 kg/cm^2, and dislocation glide begins to take place at stresses higher than 300 kg/cm^2, with screw dislocations gliding at lower stress than edge dislocations.

As shown in Fig. 2, dislocation loops with the Burgers vector of $a/2\langle 110\rangle$ result from the breaking of dislocation dipoles by a diffusion process. Narayan et al.[24] and Narayan[25,26] indicated from the observation of MgO electric field treated at high temperatures for a long period of time dislocation loops with both $a/2\langle 110\rangle$ and $a\langle 100\rangle$ generated on (110) and (100) planes, respectively. They analyzed in detail the nature of these dislocation loops and concluded they were of the vacancy type. The formation of the loops was suggested to result from clustering of vacancies generated in the electric field. Recently, Kinoshita et al.[27] reported from the study of electron irradiation in a high-voltage TEM that dislocation loops generated in MgO were of an interstitial type.

Dislocations and Oxygen Diffusion Coefficients

Both cation and anion diffusion coefficients in MgO are very important in understanding sintering, solid-state reactions, high-temperature creep, etc. Oishi and Kingery[28] established a method of measuring oxygen diffusion coefficients in oxides and reported values for O^{2-} ions in MgO. The diffusion coefficients of Mg^{2+} ions in MgO have been reported by Wuensch et al.[29] and Lindner et al.[30]

In polycrystals, we must pay attention to different diffusivities along grain boundaries and in the bulk. In this case, Shirasaki et al.[31] reported a method to calculate oxygen diffusion coefficients in polycrystalline materials.

Dislocations are easily introduced into MgO by mechanical treatments such as polishing on sandpaper, crushing in an agate mortar, and cutting with a diamond wheel, as shown in Fig. 3. The density of dislocation introduced by such mechanical treatments is $10^{10}-10^{12}/cm^2$. As oxygen diffusion coefficients are usually measured in such crushed samples with high dislocation densities, it is of interest to know the relationship between oxygen diffusion coefficient and dislocation density. For this, several experiments have been carried out. Moriyoshi, Haneda, and Shirasaki[32] reported that oxygen diffusion coefficients in MgO single crystals, in which one had a high dislocation density and the other was chemically polished (low dislocation density), are independent of dislocation density. In this study, however, impurity concentrations in the crystals were high. Recently, Oishi and Ando[33] measured oxygen diffusion coefficients in very high-purity MgO single crystals in which one was crushed, another was as-cleaved, and a third was chemically polished after cleavage. In the results, plotted against reciprocal temperatures in Fig. 4, there are break points in the data for as-cleaved and chemically polished samples. At low temperature, all the curves have approximately similar slopes and the chemically polished sample has the lowest oxygen diffusion coefficient. From this evidence, they concluded that the origin of extrinsic oxygen diffusion in MgO is not impurities but dislocations introduced during sample preparation. High diffusion coefficients in a sample crushed and subsequently treated at 1700°C can be related to enhanced diffusion through polygonized dislocation networks. These data show that mechanically introduced dislocations play an extremely important role in diffusion in MgO. However, as vacancies and interstitial ions are also introduced by dislocation motion, it is uncertain whether rapid pipe diffusion along dislocation cores or point defects introduced by dislocation motion are the cause of oxygen diffusion enhancement.

Thermal Behavior of Dislocations in MgO

Dislocation climb takes place by diffusion processes. The rate-determining step in dislocation climb can be estimated from the change of dislocation loop radii with time. Narayan and Washburn[34] indicated that the diffusion coefficients ob-

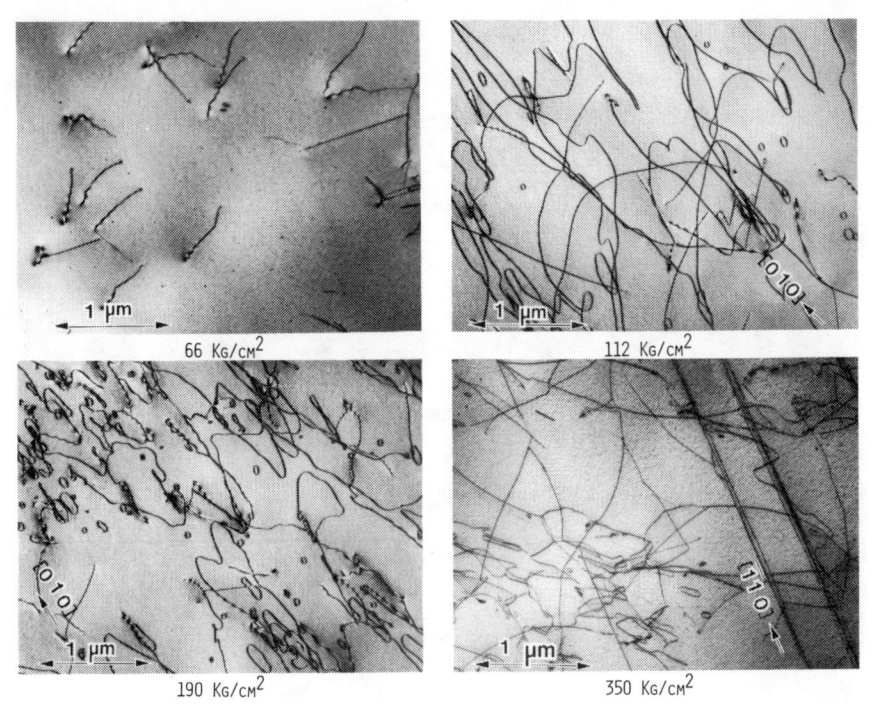

Fig. 1. Dislocation structures during [100] tensile creep of MgO single crystals at 1400°C in air (Ref. 23).

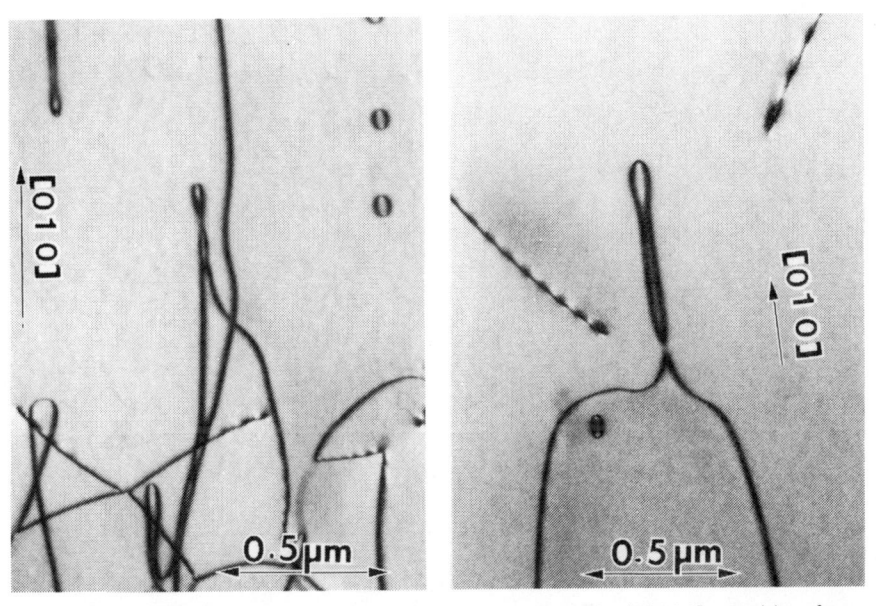

Fig. 2. Dislocation loops with the Burgers vector of $a/2[011]$ resulting from the breaking of dislocation dipoles (Ref. 23).

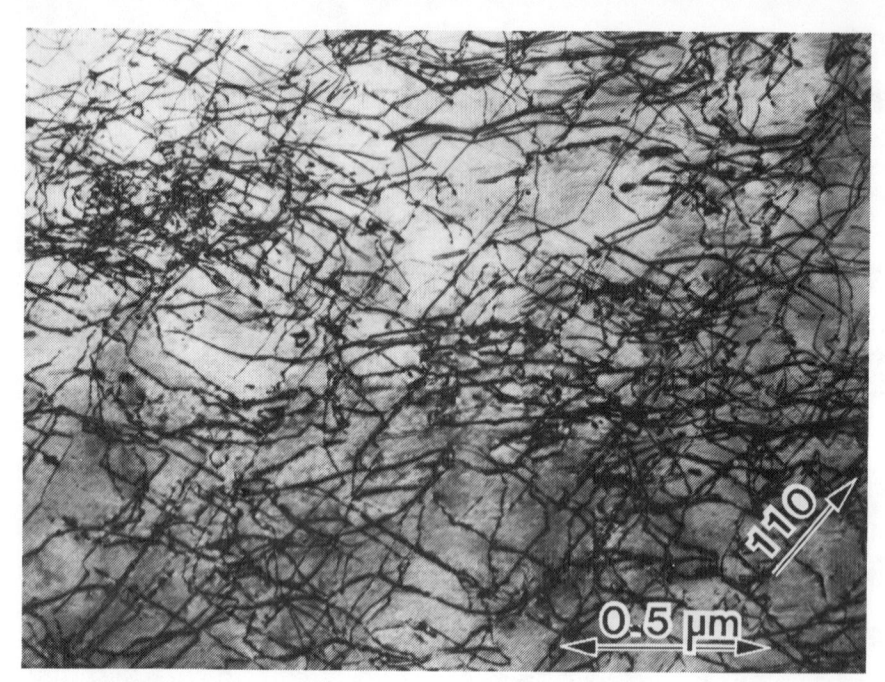

Fig. 3. Dislocations in MgO single crystals introduced during polishing on a diamond disk.

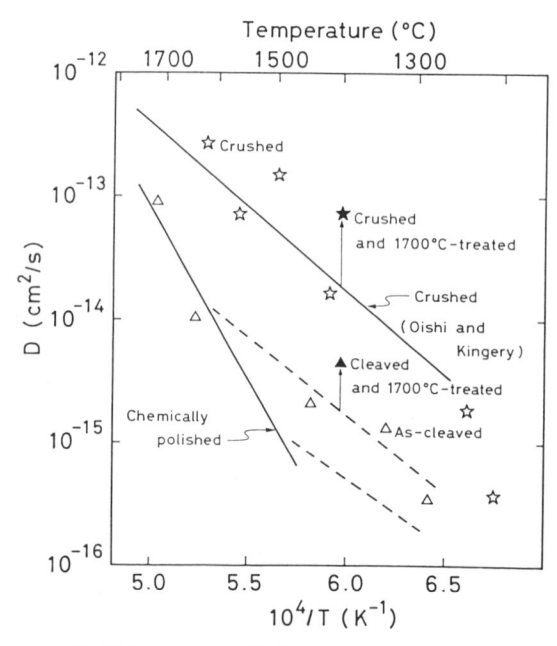

Fig. 4. Oxygen self-diffusion coefficients crushed and cleaved as compared with those for chemically polished commercial MgO (Ref. 33).

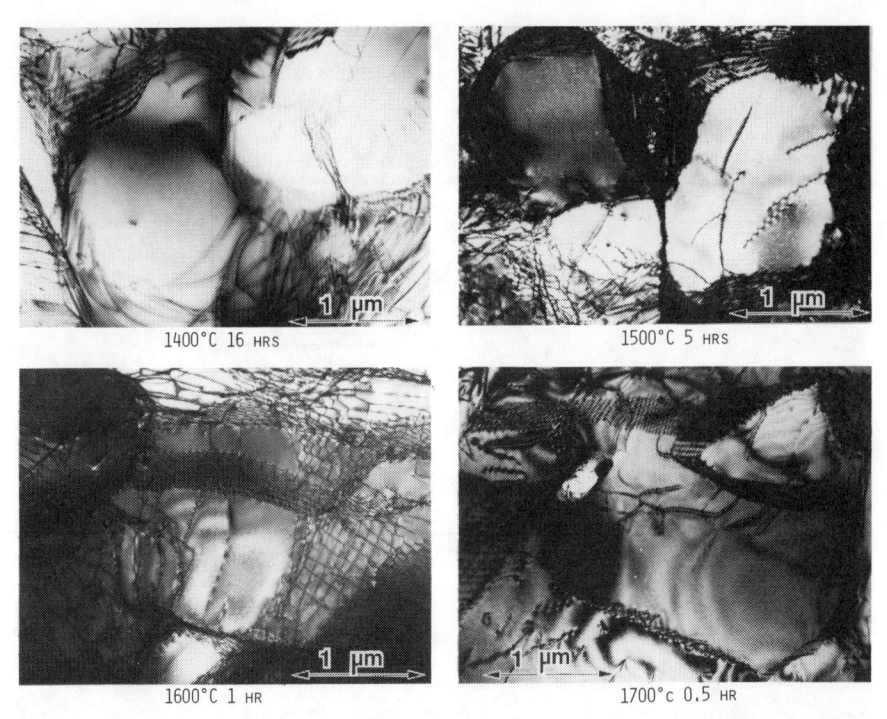

Fig. 5. Polygonized dislocations in a MgO single crystal heat-treated at various temperatures (Ref. 36).

tained from the change of the radii agree fairly well with those of oxygen diffusion data reported by Oishi and Kingery.[28]

When plastically deformed MgO is annealed, edge dislocation dipoles break up into rows of prismatic dislocation loops by pipe diffusion. Narayan and Washburn[35] reported a diffusion coefficient, $D_p = 3.0 \times 10^{-3} \exp(-67200 \pm 5000/RT)$, for this process. Moriyoshi et al.[36] also estimated the diffusion coefficients during polygonization processes of dislocations in MgO single crystals heat treated at high temperature. That is, when MgO with high dislocation density is heat treated at a given temperature and time, dislocations tend to form subgrain boundaries by so-called polygonization processes as shown in Fig. 5. The diffusion coefficient estimated from the polygonization rate agrees fairly well with that obtained by Oishi and Kingery.[28] The character of dislocations at subgrain boundaries has been determined from the out-of-contrast conditions corresponding to $\mathbf{b} \cdot \mathbf{g} = 0$ and $\mathbf{b} \cdot \mathbf{g} \times \mathbf{u} = 0$ elsewhere. A typical example of subgrain boundary dislocations is shown in Fig. 6. At the nodes of the dislocation network, the following reactions[37] take place: $a/2[1\bar{1}0] + a/2[110] = a[100]$ and $a/2[110] + a/2[101] = a/2[211]$. Two intersecting glide planes, $(\bar{1}10)$ and (101), are at an angle of 120° to each other and the resultant dislocation is parallel to [111]. The glide plane of the resultant dislocation is (211). As the glide planes of (100) and (211) are not favorable for MgO, the dislocations are sessile. Clauer and Wilcox[22] have reported that the annealing of MgO single crystals for 30 min at 2000°C in an Ar atmosphere can remove dislocations introduced during machin-

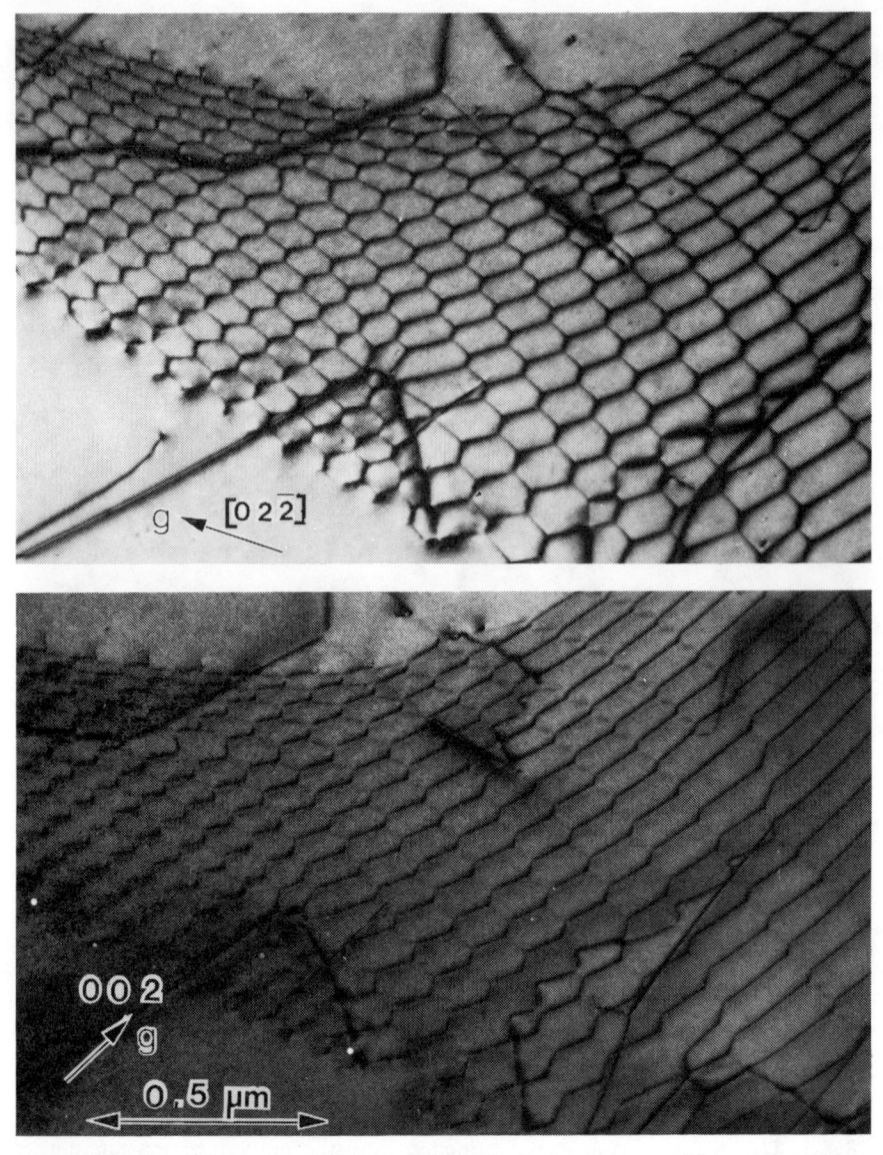

Fig. 6. Typical dislocation network in a MgO single crystal heat-treated at 1900°C for 3 h under vacuum (Ref. 36).

ing. However, this seems very doubtful since in our experiments the dislocations were polygonized by heat treatments at high temperature.

The weak-beam method developed by Cockayne et al.[38] is a powerful tool for observing grain-boundary dislocations in crystals. Its application to the study of dislocations in ceramics was reviewed by Mitchell,[4] and particularly, its application to the study of grain-boundary dislocations in ceramics was recently reviewed in detail by Carter and Sass.[39]

Dislocation Motion in MgO

Many attempts have been made to observe and study dislocation motion in MgO by using electron microscopy and etch-pit techniques.[40-43] These experimental investigations have reported data based on dislocation positions before and after various mechanical treatments. However, dynamic aspects of dislocation phenomena such as velocities, multiplication, and interaction of dislocations have not been studied thoroughly and conclusively due to the lack of direct and continuous observation of dislocation motion. The reasons are related to the brittle properties of MgO thin foils.

The first attempts at in situ observation of dislocation motion in MgO were made independently by Appel, Bethge, and Messerschmidt[44] and Moriyoshi, Kingery, and Vander Sande.[45] Appel et al.[44] recorded photographically the change of the dislocation structure during relaxation of deformation, but since in the range of plastic deformation very fast dislocation motion took place, it could not be recorded by the photographic arrangement available even at the low deformation rate. On the other hand, Moriyoshi et al.[45] recorded dislocation motion on videotape and analyzed the dislocation images filmed with a 16-mm movie camera at a speed of 24 frames s^{-1}. The results in both papers are approximately in good agreement.

In general, it is said from conventional etch-pit techniques that edge dislocations move faster than screw dislocations.[42,43] However, Moriyoshi et al.[45] indicated that the motion of dislocation was discontinuous; that is, the motion was thought to be hindered by obstacles such as impurity precipitation, jogs, and forest dislocations. The average dislocation velocity depends strongly on the number as well as the kind of obstacles that exist in the path of the moving dislocations. This interaction explains the difference between the edge and screw dislocation velocities measured in the etch-pit technique. From data in the continuous observation method described above, the minimum dislocation velocities are calculated as 2.3×10^{-4} and 2.4×10^{-4} cm/s for edge and screw dislocations, respectively; the two calculated values are approximately equal, and because of the limitation of the time interval (1/24 s) associated with the 16-mm film analyzed, these velocities are lower limits. With the data obtained from the motion of U-shaped dislocations, the stress value is estimated to be about 7×10^7 N/m^2 for both edge and screw dislocations. These results suggest that the dislocation mobility in MgO might be controlled by the Peierls mechanism rather than by the Peierls mechanism with double kink. Puls and Norgett[46] calculated atomistically the core configuration and energies of the two symmetric configurations of an edge dislocation in MgO. They indicated that the Peierls stress derived from the energy calculation ranged between 162 and 383 MPa. The Frank–Read mechanism for dislocation multiplication was not observed in these experiments; the cross-slip mechanism for dislocation multiplication occurs, as is observed in other alkali halide type crystals.

From in situ observation, it was also shown that impurities in MgO play an important role in dislocation motion. Moriyoshi et al.[47] pointed out the following results from extending MgO bicrystals with about 500 ppm of impurities by a tensile device in a 100-kV TEM. No impurity precipitates were observed at the grain boundary, and edge dislocations initially present at the boundary did not move under extension. Fracture in these samples always occurred not at the ground boundary but in the bulk crystal adjacent to a boundary. Even after fracture, the original edge dislocations remained in the same position as before extension. This is thought to result from the edge dislocations being pinned by

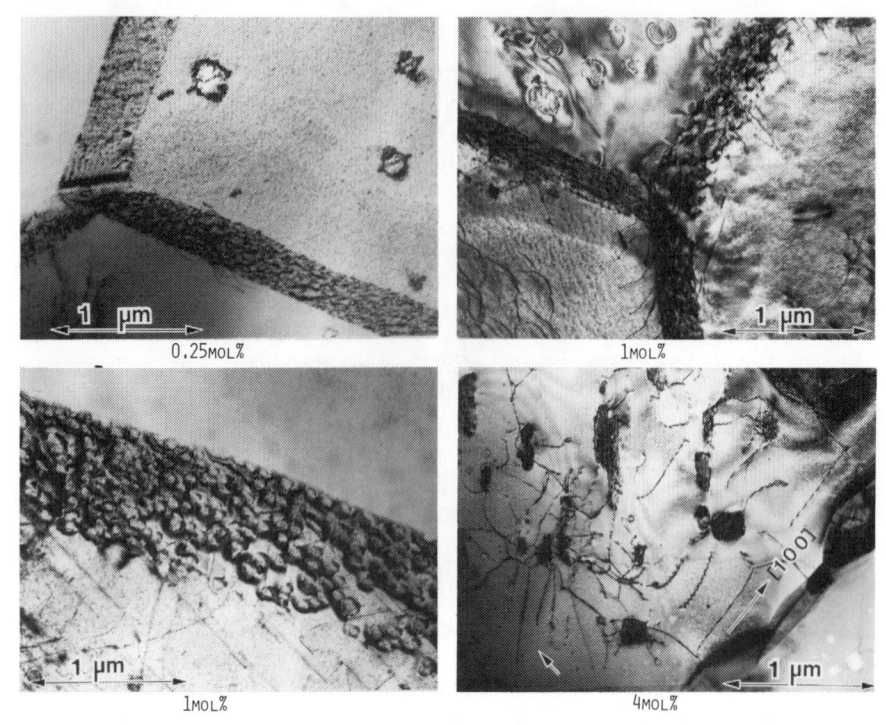

Fig. 7. Micrographs of MgO-containing Al_2O_3 sintered at 1700°C for 1 week in air and annealed at 1300°C for 1 week in air.

impurities and, therefore, being rendered immobile. The fact that not only edge dislocations at grain boundaries but also grown-in dislocations do not move under extension suggests that impurities pin these dislocations also. Kingery et al.[48] reported that the solute segregated impurity concentration near grain boundaries in MgO is higher than in bulk.

The fracture mainly occurs along the (100) plane with steps on the (010) plane, but sometimes fracture takes place along the (110) planes as well. It is evaluated as a process of crack formation in single crystals, where crack formation is believed to occur by a dislocation pile-up mechanism in which immobile grown-in dislocations and impurity solutes act as obstacles.

Dislocations and Impurities

The mechanical properties of MgO are much affected by impurities. The introduction of Fe^{3+} ions into MgO increases the strength while introduction of Fe^{2+} ions shows no effect; these phenomena are elucidated from the viewpoint of electrostatic energy.[41,49] Precipitation also affects the strength of MgO.[50,51] In general, the solubility behavior of impurities in oxides is more complex than that in metals where the major factor determining solubility is the strain energy introduced by solute atoms. In the case of MgO, much of the energy for the introduction of aliovalent ions is related to vacancy formation rather than strain energy, and the impurity solubility depends strongly on temperature. At low temperatures, extremely low impurity solubilities commonly cause impurity precipitation at dislocations and grain boundaries, as discussed in the paper by Kingery.[1]

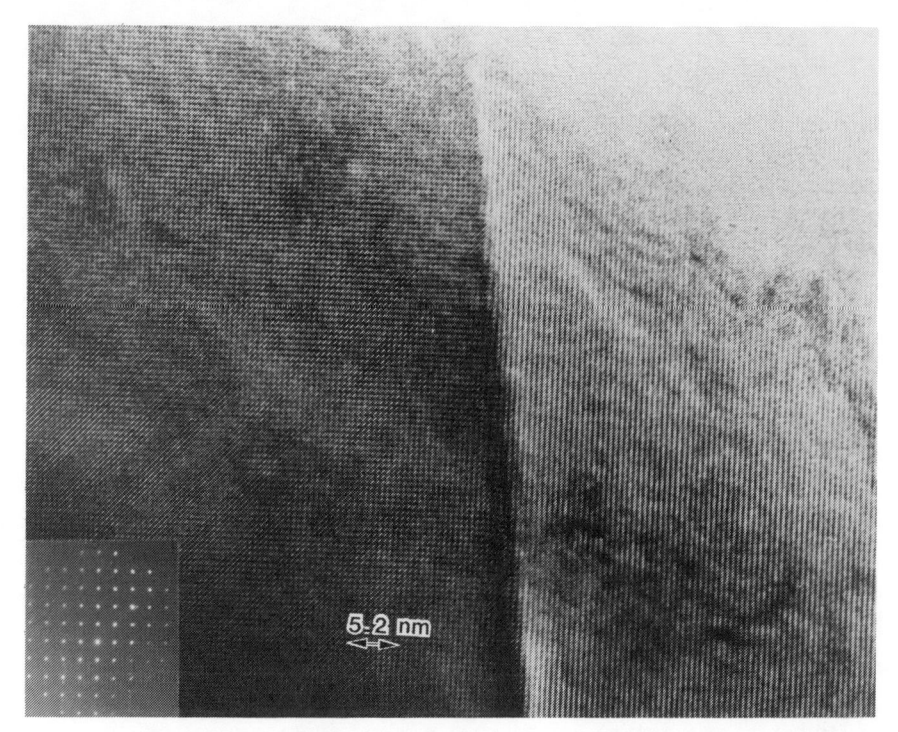

Fig. 8. High-resolution micrograph of a boundary in monoclinic ZrO_2 (Ref. 52).

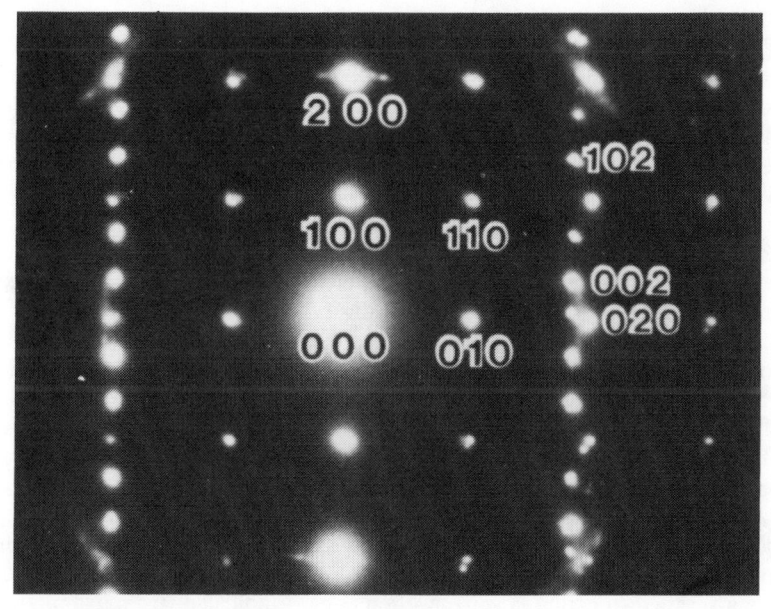

Fig. 9. Diffraction pattern of Fig. 8 (Ref. 52).

Figure 7 shows the microstructure of polycrystalline MgO containing 0.25–4 mol% Al_2O_3, which was sintered at 1700°C for 1 week in air, annealed at 1300°C for 1 week in air, and then quenched to room temperature. It can be seen from these micrographs that granular precipitation takes place at grain boundaries and along dislocations at grain boundaries at low concentrations of Al_2O_3. Precipitation occurs in the bulk at higher concentrations of Al_2O_3. Dislocations among precipitates are generated during cooling to relax a lattice strain due to different thermal expansions between precipitates and the bulk. These precipitates line up in $\langle 110 \rangle$ directions and give rise to an increase of the critical resolved shear stress as a function of particle size. An interesting interpretation of dislocation particle size relations has been given by Reppich.[50] He proposed a simple hardening model by describing the increase of the critical resolved shear stress (CRSS) of MgO caused by coherent stress-free magnesium ferrite particles. The comparison with experimental data indicated that the calculated CRSS agreed well with measured values.

The interface structure between precipitates and the bulk in the MgO-Al_2O_3 system has not been clarified, since it is difficult to prepare a satisfactory thin foil for observation with high-resolution TEM. Another problem is that the zone axes of the two grains adjacent to a grain boundary do not coincide. In this connection, recent observations of boundaries in ZrO_2 using high-resolution TEM are interesting.[52] As can be seen in Fig. 8, the lattice fit at a grain boundary is very good. Clearly from the diffraction pattern shown in Fig. 9, the zone axes are $\langle 010 \rangle$ and $\langle 001 \rangle$ for one grain and the other, respectively. We can clearly identify two-dimensional lattice fringes of about 0.52 nm at the left side of the figure and one-dimensional lattice fringes at the right side. It is interesting that there are stairlike steps at the grain boundary, and the lattice orientation is regular adjacent to the grain boundary. A similar micrograph is shown in Fig. 10.

Dislocations and Cathodoluminescence

Photoluminescence and cathodoluminescence from deformed MgO crystals have been studied by many workers. Chen et al.[53] suggested that defects such as vacancy complexes responsible for the photoluminescence were formed at the onset of plastic deformation. Velendnitskaya et al.[54] proposed, in a study to clarify the deformation mechanism by indentation of MgO, that the luminescence was caused by interstitial point defects created by indentation. On the other hand, Pennycook and Brown[55] suggested that the local dilation due to a dislocation is large enough to cause a decrease of as much as 4 eV in the band gap. Llopis et al.[56] deduced from the observation of cathodoluminescence from slip planes in deformed MgO that grown-in dislocations have less influence on the cathodoluminescence than dislocations originated during plastic deformation; that is, they indicated an influence of point defect complexes, resulting from plastic deformation on the cathodoluminescence. Their suggestion agrees fairly well with the results reported by Chen et al.[53] and Velendnitskaya et al.[54] Recently Chaudhri et al.[57] found that in MgO the intensity of luminescence from screw dislocation was markedly higher than from edge dislocations for all possible orientations of the specimen. They cast doubt upon the explanation by Pennycook and Brown,[55] in which cathodoluminescence is attributed to the reduction in band gap at dislocations due to dilatation, since screw dislocations give higher intensity cathodoluminescence than edge dislocations, in spite of the fact that the dilatation around a screw dislocation is zero. Melton et al.[58] indicated, from examining the luminescence of MgO during mechanical deformation in crystals of the highest

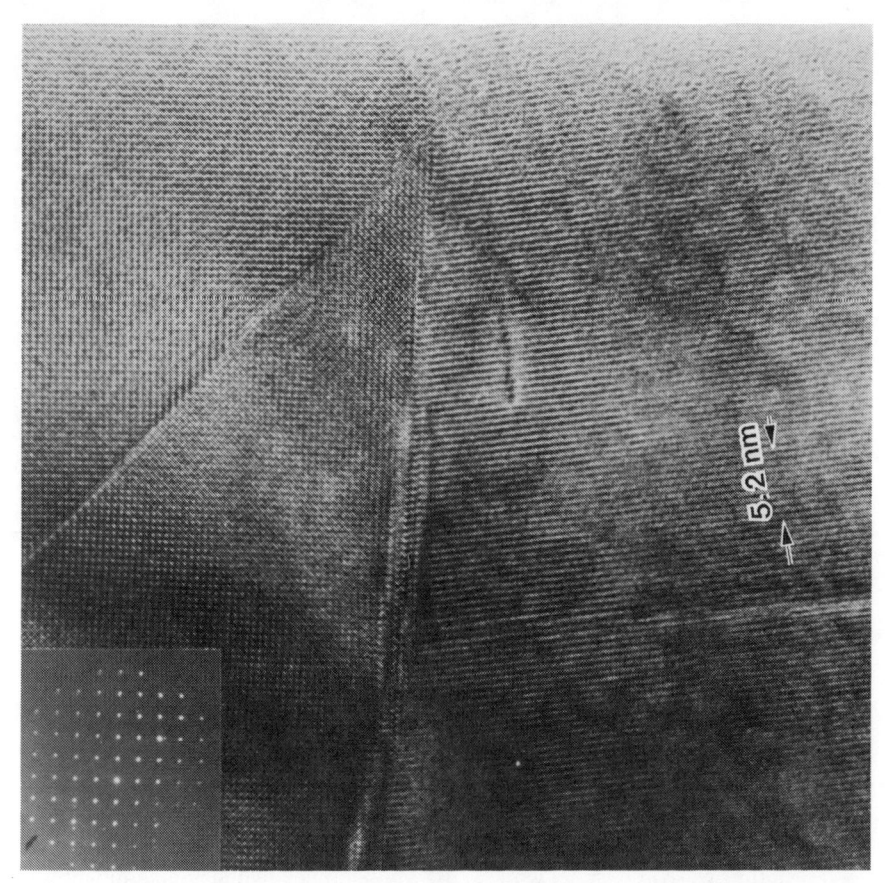

Fig. 10. High-resolution micrograph of monoclinic ZrO$_2$ (Ref. 52).

purity available as well as relatively impure crystals, that the luminescence data are essentially the same, even though there is a significant difference in yield stress. They also pointed out the importance of defect production by dislocation motion.

High-Resolution Observation with TEM

Atomic resolution micrographs have been taken with TEM. These techniques have been applied to ceramic materials, which were recently reviewed by Clarke.[59]

A recent ultrahigh-voltage transmission electron micrograph (UTEM) taken at 1 MV has a point resolution of 0.2 nm in vertical illumination with a specimen goniometer. With this UTEM, we are trying to observe dislocations in MgO; however, it is extremely difficult to obtain good micrographs, since thin specimens obtained by either ion thinning or chemical thinning are too thick for high-resolution observation of MgO. In general, a knife-edge fragment crushed in an agate mortar is used as a specimen for high-resolution observation; however, this is not suitable for MgO, because it has (100) cleavage faces and the lattice spacing between equivalent atomic planes of [200], 0.21 nm, is too small to resolve with even UTEM. From the viewpoint of clarifying atomic arrangements, observation

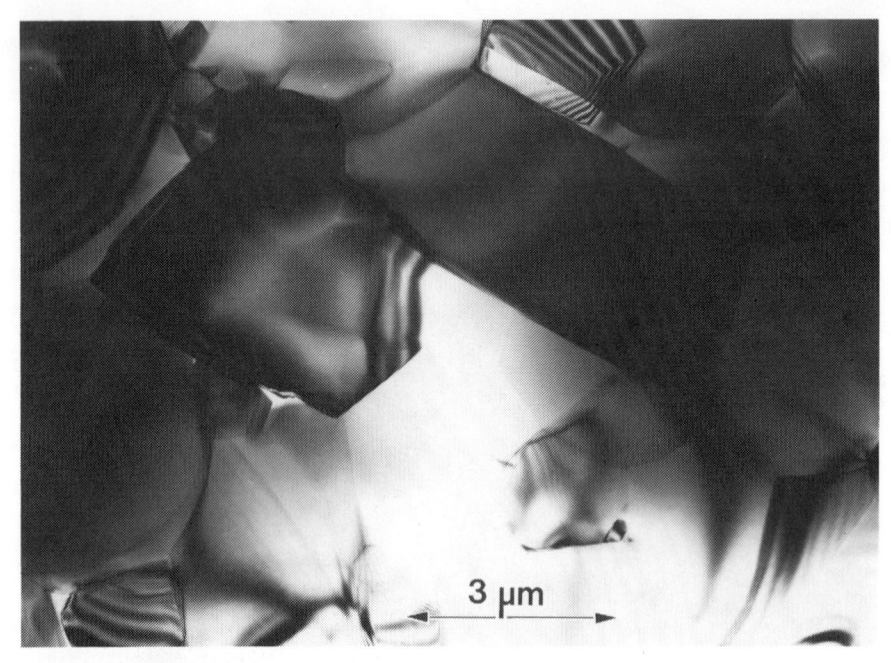

Fig. 11. Micrograph of system MgAl$_2$O$_4$-MgO.

from the $\langle 110 \rangle$ direction is most desirable, since oxygen atoms and magnesium atoms line up independently in this direction. However, it is extremely difficult to find a fragment with (110) planes for the reason mentioned above. Instead, a polycrystalline MgAl$_2$O$_4$ spinel with excess MgO thinned by ion bombardment was observed with UTEM.

Spinel powder was obtained by a conventional method. The resultant powder with 10% added MgO was hydrostatically pressed to a compact, which was then sintered at 1650°C for 24 h in air. The compact was mechanically polished to less than 100 μm in thickness, after which it was further thinned by Ar ion bombardment from both sides at a rate of about 0.2 μm/h. In order to prevent charging, a carbon coating was applied to the foil. The resultant thin foils were observed with the UTEM[†] operating at 1 MV.

Figure 11 is a typical micrograph of the two-phase region at low magnification. The average grain size is about 5 μm. Grains of spinel and MgO are indistinguishable. Figure 12 shows a typical high-resolution micrograph of spinel viewed from the $\langle 110 \rangle$ direction. The lattice fringe of 0.4 nm is clearly identified. Figure 13 also shows a high-resolution micrograph of spinel viewed from the $\langle 110 \rangle$ direction. A defect indicated by the arrow can be seen; however, its character is not clear. Figure 14 shows a high-resolution micrograph of the interface between a spinel and MgO grain. The micrograph was taken with the incident electron beam normal to the (110) plane of the spinel grain, as shown in the diffraction pattern inset. The lattice fringe spacing in the MgO grain is 0.3 nm, which agrees very well with the spacing between (110) planes of MgO. Careful observation in the interface of the micrograph indicates that there are dislocations along the interface. Unfortunately, a diffraction pattern of the MgO grain is not available; how-

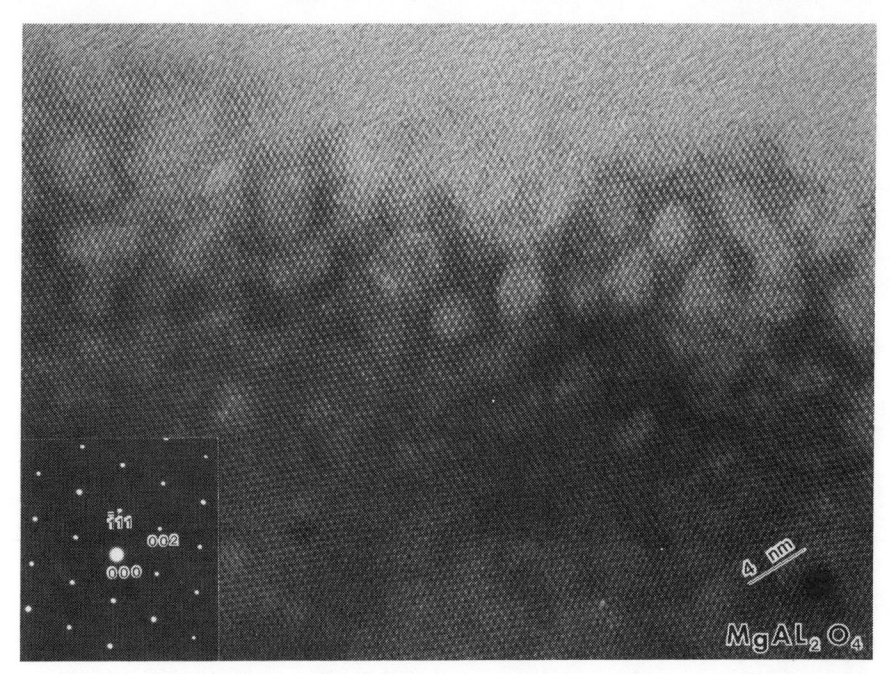

Fig. 12. High-resolution micrograph of MgAl$_2$O$_4$ spinel viewed from the [110] direction.

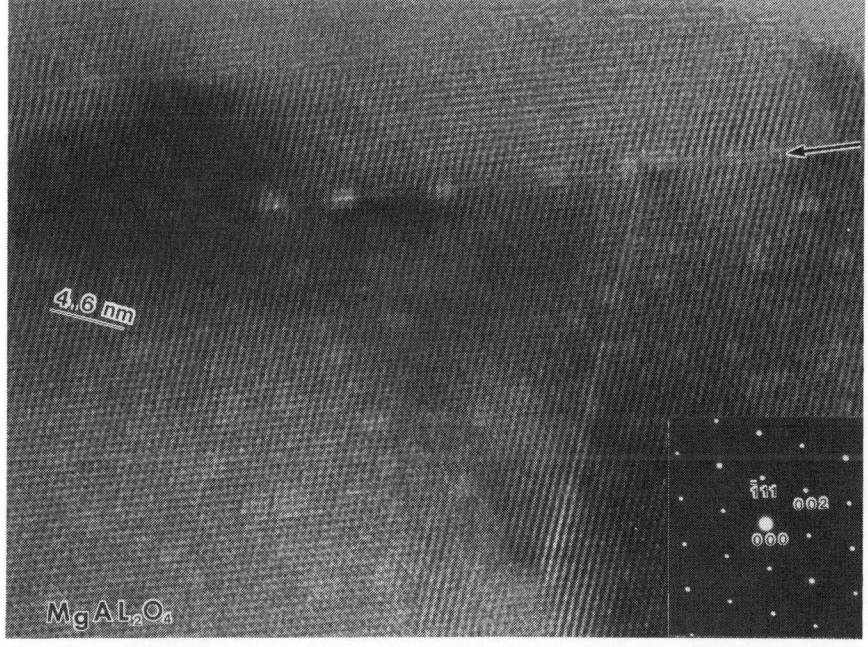

Fig. 13. High-resolution micrograph of MgAl$_2$O$_4$ spinel viewed from the [110] direction.

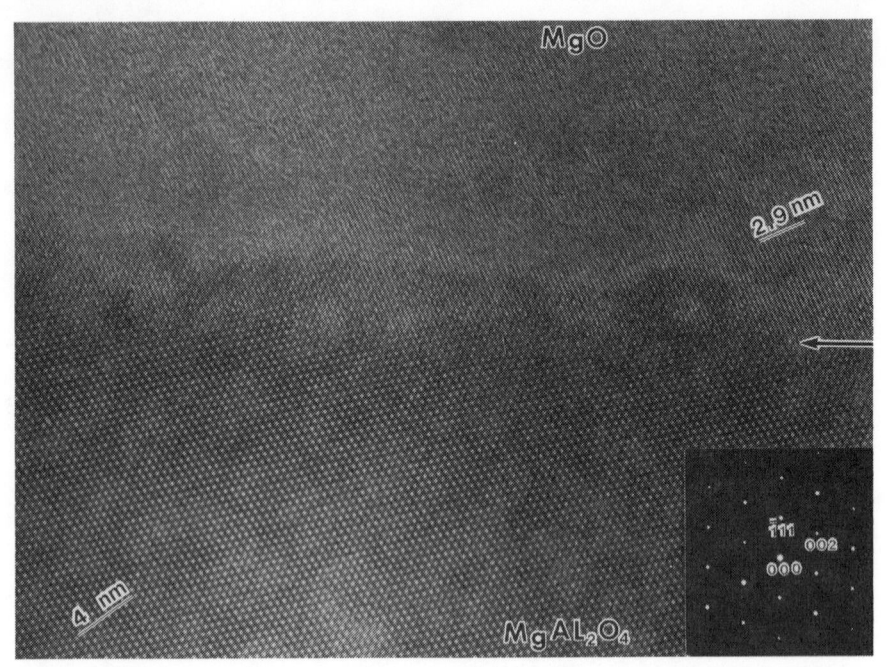

Fig. 14. High-resolution micrograph of grain boundary of system MgAl$_2$O$_4$-MgO viewed from the [110] direction of the MgAl$_2$O$_4$ grain.

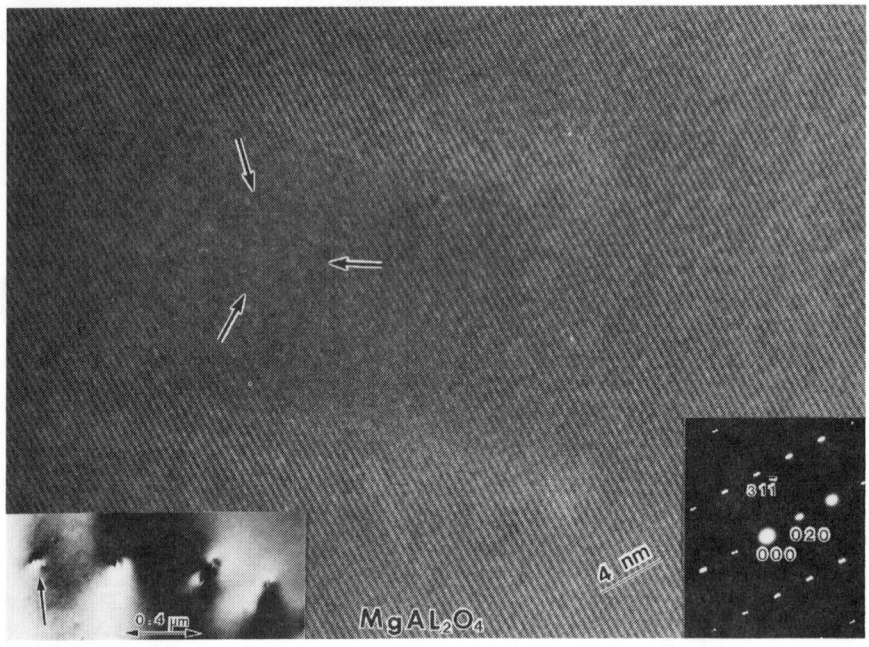

Fig. 15. High-resolution micrograph of the MgAl$_2$O$_4$ spinel viewed from the [103] direction. Three arrows indicate a dislocation position.

ever, presumably the zone axis of the MgO grain would coincide with that of the spinel grain.

Figure 15 shows a high-resolution micrograph of spinel viewed from the ⟨103⟩ direction with a low-magnification micrograph and diffraction pattern inset. Dislocations in the low-magnification image are grown-in dislocations. The position indicated by three arrows in the lattice image corresponds to that of the dislocation shown with a small arrow. It is very interesting that no lattice fringes exist in this region. This may be related to impurity segregation or precipitation along the dislocation that causes the lattice fringes to be irregular. However, more studies are necessary to understand this point.

Other Subjects

In addition to the subjects mentioned above, dislocations in MgO have been studied in relation to phenomena such as high-temperature strength,[60] internal friction,[61,62] dislocation theories,[63-67] and neutron irradiation.[68] Recently, relationships between dislocations and sintering[69,70] have been reported.

References

[1] W. D. Kingery, *J. Am. Ceram. Soc.*, **57** [1] 1–8 (1974).
[2] A. H. Heuer, *J. Am. Ceram. Soc.*, **62** [5–6] 226–35 (1979).
[3] L. W. Hobbs; pp. 267–78 in Ref. 2.
[4] T. E. Mitchell; pp. 254–67 in Ref. 2.
[5] J. B. Vander Sande and E. L. Hall; p. 246 in Ref. 2.
[6] R. J. Stokes, T. L. Johonston, and C. H. Li, *Philos. Mag.*, **3**, 718 (1958).
[7] K. Sangual, *J. Mater. Sci.*, **15**, 237 (1980).
[8] T. Harada, *J. Crystal Growth*, **44**, 635 (1978).
[9] G. Carter and J. S. Colligon; Ion Bombardment of Solid. Heinemann Educational Books Ltd., London, 1968.
[10] J. Washburn, G. W. Groves, A. Kelly, and G. K. Williamson, *Philos. Mag.*, **5**, 991 (1960).
[11] F. R. N. Nabarro; p. 75 in the Report of the Conference on Strength of Solids. University of Bristol, England, 1947. C. Herring, *J. Appl. Phys.*, **21**, 437 (1950).
[12] R. L. Coble, *J. Appl. Phys.*, **34**, 1679 (1963).
[13] R. S. Gordon, *J. Am. Ceram. Soc.*, **56** [3] 147–52 (1973).
[14] R. S. Gordon and J. D. Hodge, *J. Mater. Sci.*, **10**, 200 (1975).
[15] M. Ashby, *Acta Metall.*, **20**, 887 (1972).
[16] H. E. Evans and G. Knowles, *Acta Metall.*, **26**, 141 (1978); **25**, 963 (1977).
[17] J. D. Hodge, P. A. Lessing, and R. S. Gordon, *J. Mater. Sci.*, **12**, 1598 (1977).
[18] Y. Ikuma and R. S. Gordon, *J. Am. Ceram. Soc.*, **66** [2] 139–47 (1983).
[19] J. Weertman, *J. Appl. Phys.*, **26**, 1213 (1955); **28**, 362 (1957).
[20] J. B. Bilde-Sorensen, *Acta Metall.*, **21**, 1495 (1973); *J. Am. Ceram. Soc.*, **55** [12] 606–10 (1972).
[21] W. Huether and B. Reppich, *Philos. Mag.*, **28**, 362 (1973).
[22] A. H. Clauer and B. A. Wilcox, *J. Am. Ceram. Soc.*, **59** [3–4] 89–96 (1976). A. H. Clauer, M. S. Seltzer, and B. A. Wilcox, *J. Am. Ceram. Soc.*, **62** [1–2] 85–94 (1979).
[23] Y Moriyoshi, T. Ikegami, Y. Sekikawa, and S. Shirasaki, *Z. Phys. Chem.*, **NF119**, 239 (1980).
[24] J. Narayan, R. A. Weeks, and E. Sonder, *J. Appl. Phys.*, **49**, 5977 (1978).
[25] J. Narayan, *Philos. Mag.*, **37**, 457 (1978).
[26] J. Narayan, *Phys. Status Solidi*, **45**, 625 (1978).
[27] T. Kinoshita, K. Hayashi, K. Nakai, and S. Kitajima; 21st Meeting of Japan Ceramic Society, 1983.
[28] Y. Oishi and W. D. Kingery, *J. Chem. Phys.*, **33**, 480 (1960).
[29] B. J. Wuensch, W. C. Steele, and T. Vasilos, *J. Chem. Phys.*, **58**, 5258 (1972).
[30] R. Lindner and G. D. Parfitt, *J. Chem. Phys.*, **26**, 182 (1957).
[31] S. Shirasaki, I. Shindo, H. Haneda, M. Ogawa, and K. Manabe, *Chem. Phys. Lett.*, **50**, 459 (1977).
[32] Y. Moriyoshi, H. Haneda, and S. Shirasaki; 16th Meeting of Japan Ceramic Society, 1978.
[33] Y. Oishi and K. Ando, *J. Am. Ceram. Soc.*, **66** [4] C-60–C-62 (1983).
[34] J. Narayan and J. Washburn, *Acta Metall.*, **21**, 533 (1973).
[35] J. Narayan and J. Washburn, *Crystal Lattice Defects*, **3**, 91 (1972).
[36] Y. Moriyoshi, T. Ikegami, S. Matsuda, and S. Shirasaki, *Z. Phys. Chem. (Wiesbaden)*, **NF118**, 187 (1979).
[37] J. Narayan, *J. Am. Ceram. Soc.*, **56** [12] 644–47 (1973).

[38] J. H. Cockayne, I. L. F. Ray, and M. J. Whelan, *Philos. Mag.*, **20**, 1265 (1969).
[39] C. B. Carter and S. L. Sass, *J. Am. Ceram. Soc.*, **64** [6] 335–45 (1981).
[40] R. N. Singh; p. 217 in Deformation of Ceramic Materials. Plenum Press, New York and London, 1974.
[41] C. N. Ahlquist; p. 233 of Ref. 40.
[42] R. N. Singh and R. L. Coble, *J. Appl. Phys.*, **45**, 981 (1974).
[43] S. N. Valikovskii and E. M. Nadgornyi, *Sov. Phys. Solid State*, **17**, 1733 (1976).
[44] F. Appel, H. Bethge, and V. Messerschmidt, *Phys. Status Solidi A*, **A38**, 103 (1976).
[45] Y. Moriyoshi, W. D. Kingery, and J. B. Vander Sande, *J. Mater. Sci.*, **13**, 2507 (1978).
[46] M. P. Puls and M. J. Norgett, *J. Appl. Phys.*, **47**, 466 (1976).
[47] Y. Moriyoshi, W. D. Kingery, and J. B. Vander Sande, *J. Mater. Sci.*, **12**, 1062 (1977).
[48] W. D. Kingery, T. Mitamura, and J. B. Vander Sande, *J. Mater. Sci.*, **14**, 1766 (1979).
[49] J. J. Gilman, *J. Appl. Phys.*, **45**, 508 (1974).
[50] B. Reppich, *Acta Metall.*, **23**, 1055 (1975).
[51] H. Knoch and B. Reppich; p. 1061 in Ref. 50.
[52] Y. Moriyoshi, T. Ikegami, H. Yamamura, and A. Watanabe; ZrO_2 Symposium in Japan, 1982.
[53] Y. Chen, M. M. Abraham, T. J. Turner, and C. M. Nelson, *Philos. Mag.*, **32**, 99 (1975).
[54] M. A. Velendnitskaya, V. N. Rozhanski, L. F. Comolova, and G. V. Saparin, *Phys. Status Solidi A*, **A32**, 123 (1975).
[55] S. J. Pennycook and L. M. Brown, *J. Lumin.*, **18/19**, 905 (1979).
[56] J. Llopis, J. Piqueras, and L. Bru, *J. Materi. Sci.*, **13**, 1361 (1978).
[57] M. M. Chaudhri, J. T. Hagan, and J. K. Wells, *J. Mater. Sci.*, **15**, 1189 (1980).
[58] R. Melton, N. Danieley, and T. J. Turner, *Phys. Status Solidi A*, **A57**, 755 (1980).
[59] D. R. Clarke, *J. Am. Ceram. Soc.*, **62** [5–6] 236–46 (1979).
[60] F. Sato and K. Sumino, *J. Mater. Sci.*, **15**, 1625 (1980).
[61] M. Gabbay, C. Esnouf, and G. Fantozzi, *J. Phys. Lett.*, **39**, L271 (1978).
[62] M. Gabbay, C. Esnouf, and G. Fantozzi, *J. Phys.*, **37**, 561 (1976).
[63] A. Kröger, *J. Phys. Chem. Solids*, **41**, 741 (1980).
[64] M. P. Puls, *Philos. Mag., [Part] A*, **A41**, 353 (1980).
[65] U. Messerschmidt and F. Appel, *Krist. Tech.*, **14**, 1331 (1979).
[66] C. H. Woo and M. P. Puls, *Philos. Mag.*, **35**, 727 (1977).
[67] E. M. Nadgornyi and S. I. Zaitsev, *Mater. Sci. Eng.*, **52**, 69 (1982).
[68] G. F. Hurley, J. C. Kennedy, and F. W. Clinard, Jr., *J. Nucl. Mater.*, **103**, 761 (1982).
[69] C. S. Morgan and V. J. Tennery, *Mater. Sci. Res.*, **13**, 427 (1980).
[70] C. S. Morgan, *Phys. Sintering*, **5**, 31 (1973).

---

*Chemical Abstracts, Columbus, OH.
†Hitachi H-1250, Hitachi Ltd., Toyko, Japan.

Properties of Grain Boundaries in Rock Salt Structured Oxides

D. M. Duffy* and P. W. Tasker

Theoretical Physics Division
A.E.R.E. Harwell
Oxfordshire, England OX11 ORA

A number of coincident grain boundaries in nickel and magnesium oxide have been modeled by using an atomistic simulation technique. Stable structures were found for all the ⟨001⟩ and ⟨011⟩ tilt boundaries, but it was necessary to introduce Schottky defects into the interface of the $\Sigma = 5$, [001] twist boundary in order to stabilize the structure against dissociation into two free surfaces. The energies of intrinsic defects and impurities near tilt boundaries were calculated, and the resulting defect concentrations were estimated. Space charge layers, compensated by a net boundary charge, similar to those near general grain boundaries and surfaces in ionic crystals, were found to exist at coincidence grain boundaries. Planar defects orientated along polar directions in ionic crystals introduce a potential difference between the two halves of the crystal, which enhances the concentration of positively charged defects on one side of the interface and negatively charged defects on the other side. The potential resulting from this defect distribution cancels the boundary potential at some distance from the interface, but a potential barrier remains in the boundary region.

Solid-state diffusion along grain boundaries in ceramic materials plays an important role in the corrosion of metals and in many other technological processes.[1] The distribution of defects and impurities near grain boundaries also has a strong influence on the electrical characteristics of ceramics. There is much interest in developing devices (for example varistors) that exploit these properties.[2] Despite the importance of grain-boundary phenomena, the atomistic structure in oxide materials is poorly understood. Experimental techniques are probing the structure with increasing resolution,[3,4] and theoretical computer simulation can aid both in determining low-energy structures and in studying the defect properties.

The theoretical simulations require some two-dimensional periodicity in the boundary plane and so have been limited to coincidence boundaries. The first detailed simulations for oxides were carried out by Wolf and Benedek[13] for twist boundaries in MgO and NiO. They found the interfaces to be hardly bound. The stability of these boundaries will be discussed in more detail in ⟨001⟩ Twist Boundaries. We have also studied a series of tilt grain boundaries as a preliminary to the study of transport properties. Both magnesium and nickel oxide have been modeled, and the results are discussed in the next section. Grain boundaries may influence the material properties by producing a perturbed defect population. The energies for the formation of intrinsic defects and for impurity substitution near a variety of grain boundaries have been calculated and are discussed in Energies of Defects near Grain Boundaries. Under certain conditions, the defect distributions resulting from the calculated formation energies can be determined. Finally, we discuss the origin of a special type of space charge effect that is associated with grain boundaries orientated along particular crystallographic directions.

Although a comparison between grain boundary structures for MgO and NiO is presented in the next section, the majority of the structures and all the defect energies have been calculated for NiO only. The structure and lattice parameter for MgO are similar to those for NiO and thus so is the empirical potential.[5] We expect, therefore, that most of the results presented here will be applicable, at least qualitatively, to MgO as well as to NiO.

Structure and Energies of Coincidence Grain Boundaries

⟨001⟩ Tilt Boundaries

The relaxed structures and energies of seven coincident symmetric tilt boundaries, with rotation axes orientated along the [001] direction, have been calculated[6] by using the Harwell MIDAS code.[7] Empirical potentials for nickel oxide were used, and the ionic polarization was described by the shell model.[5] Four of the calculations were repeated by using the corresponding potentials for magnesium oxide.

A [001] projection of the calculated structure of the 36.9°, $\Sigma = 5$, (310)/[001] tilt boundary is represented in Fig. 1. Differences between the relaxed

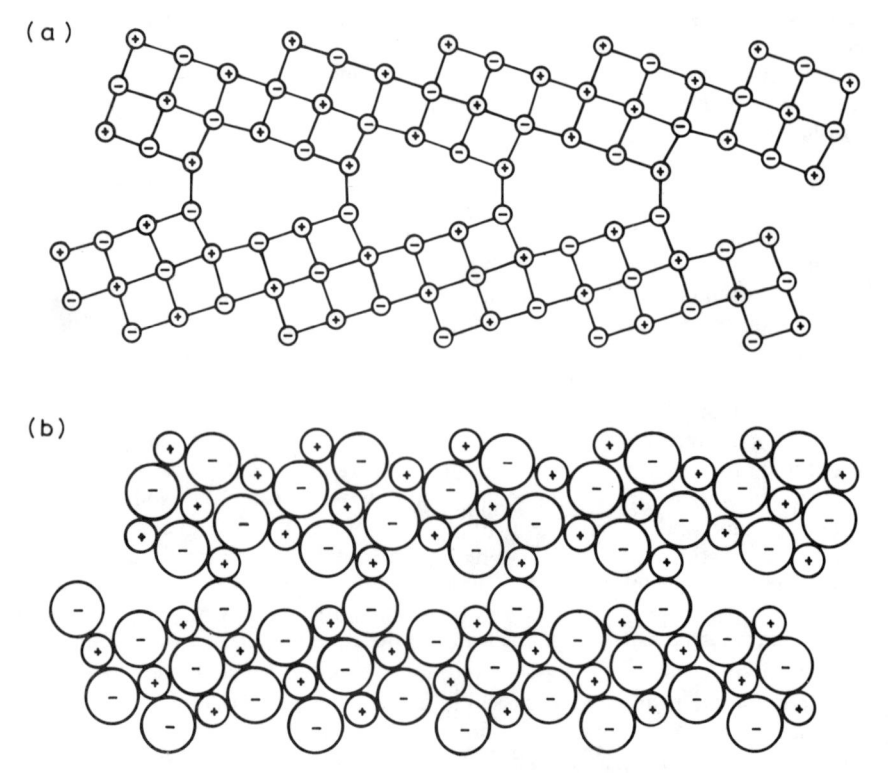

Fig. 1. Relaxed structure of the 36.9° (310)/[001] tilt boundary that (a) shows the [100] dislocation structure and (b) is drawn with Pauling's ionic radii.

structures of the boundaries in NiO and MgO are negligible (of the order of $0.01a$, where a is the lattice spacing). The steps on the surfaces of the two halves of the bicrystal line up to produce a symmetrical configuration with relatively large gaps in the boundary plane. This configuration is much less dense than that of the corresponding grain boundary in metals.[8,9] The structure can be resolved into an array of [100] dislocations. Isolated [100] dislocations are not stable in ionic crystals with the rock-salt structure as they tend to dissociate into dislocations with smaller Burgers vectors. However, the high angle of the boundary accommodates the large Burgers vector without excessive lattice strain. The relaxed configurations of all the $(n10)/[001]$ interfaces ($n = 2$–6) are analogous, but the spacing between the dislocations increases with the integer n. In each case, there is a parallel array of open pipes in the boundary core that may act as fast diffusion paths for defects.

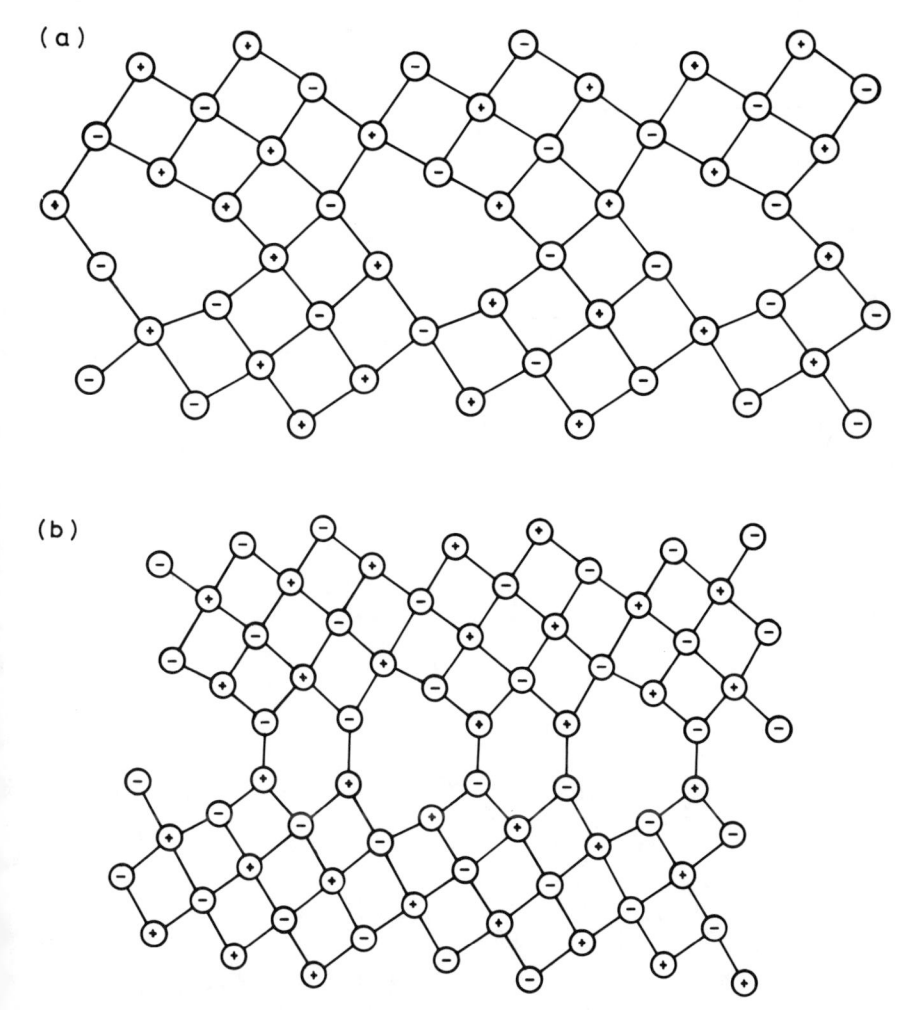

Fig. 2. Relaxed structure of the 67.4° (320)/[001] tilt boundary in (a) NiO and (b) MgO.

The calculated structure of the 67.4°, $\Sigma = 13$, (320)/[001] interface in NiO is represented in Fig. 2(a). This boundary, together with the (430)/[001] interface, can be resolved into an array of [110] dislocations. The (320)/[001] interface in MgO is shown in Fig. 2(b). This is the only case for which a significant difference was found between the structures of grain boundaries in the two oxides. In contrast to the case of the NiO interface, unlike ions are matched across the boundary in MgO but the configuration has a more open structure. A local minimum with this configuration was also found in NiO, but the energy was 0.3 J·m^{-2} higher than that of the structure shown in Fig. 2(a). The Coulomb energy is, therefore, lower in the MgO structure, but the short-range energy is higher. The main difference between the two potentials is that the Ni^{2+} ion is polarizable whereas Mg^{2+} is not. It appears that the local relaxations are insufficient in MgO to stabilize the configuration with the high Coulomb energy. The dislocation structure of the (320) interface in MgO is ambiguous, but it more closely resembles

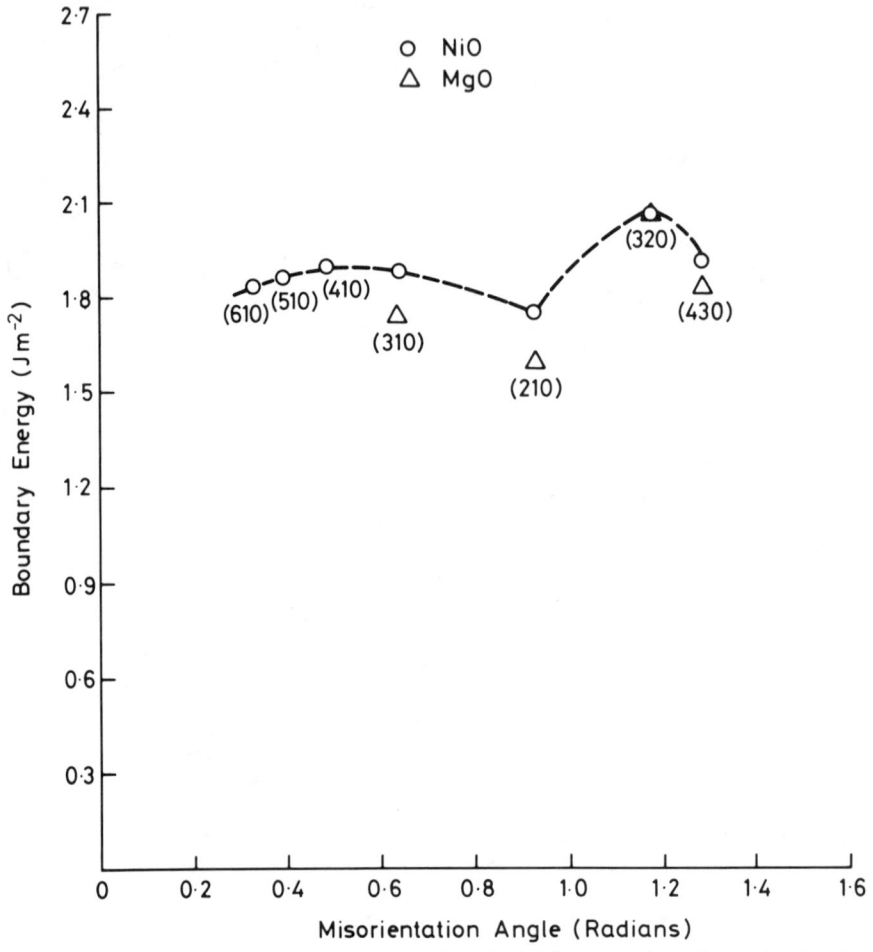

Fig. 3. Energy of ⟨001⟩ tilt boundaries in NiO as a function of misorientation angle. The corresponding energies in MgO are also shown.

Table I. Calculated Energies and Binding Energies for ⟨001⟩ Tilt Boundaries in NiO and MgO

Boundary plane	Misorientation angle (deg)	Σ	Boundary energy (J·m^{-2}) NiO	Boundary energy (J·m^{-2}) MgO	Binding energy (J·m^{-2}) NiO	Binding energy (J·m^{-2}) MgO
(610)	18.9	37	1.82		1.01	
(510)	22.6	13	1.86		1.08	
(410)	28.1	17	1.89		1.20	
(310)	36.9	5	1.88	1.76	1.58	1.72
(210)	53.1	5	1.75	1.61	2.07	2.43
(320)	67.4	13	2.04	2.05	2.37	
(430)	73.7	25	1.91	1.84	2.83	

an array of [100] dislocations than an array of [110] dislocations.

The interfacial energies of the grain boundaries are summarized in Table I and plotted against misorientation angles in Fig. 3. The interfacial energies were found to be slightly lower in MgO than in NiO. The plot follows a smooth curve except for a shallow cusp at the (210)/[001] interface, which corresponds to the transition between the two types of dislocations in NiO. It should be noted, however, that all points on the plot correspond to coincidence boundaries and may, therefore, be cusp points on the complete energy plot. The binding energies (defined as the energy required to dissociate the interface into two free surfaces) are relatively high (of the order of 1–2 J·m^{-2}); therefore, these boundaries are stable even at high temperatures.

⟨011⟩ Tilt Boundaries

The structures and energies of six ⟨011⟩ symmetric tilt boundaries in NiO have been calculated by the same procedure.[10] The interfacial energies are summarized in Table II and plotted against misorientation angles in Fig. 4. The energies are, in general, higher than for the ⟨001⟩ interfaces, but there is a very deep cusp at 70.5°, corresponding to the (1$\bar{1}$1)/(11$\bar{1}$) interface. The relaxed structure of this boundary is shown in Fig. 5(a), with the ions represented by circles of radii equal to the Pauling values. All nearest-neighbor distances in the boundary plane are very close to the ideal crystalline value, but the local environment of the ions is altered. This accounts for the small, but finite, interfacial energy. The 70.5°, (1$\bar{1}$1)/[011] interface is, in fact, a coherent twin that corresponds to a reversal of the fcc stacking sequence.

On the high-angle side of the cusp, the energy rises steeply for the $(n\bar{1}1)$ $(n = 2-4)$ boundaries. One layer of the [011] projection of the structure of the 129.5° (3$\bar{1}$1)/[011] interface is shown in Fig. 5(b). The alternate layer is displaced by half the boundary periodicity. The resulting configuration is relatively close packed, and there are no well-defined open pipes. The structures of the (2$\bar{1}$1) and (4$\bar{1}$1) interfaces are analogous. The (1$\bar{2}$2) and (1$\bar{3}$3) interfaces also have a close-packed configuration. Thus, in general, coincidence tilt boundaries with rotation axes orientated along the [011] direction have higher energies and denser packing than [001] tilt boundaries.

⟨001⟩ Twist Boundaries

In contrast to the results of the experimental evidence,[11,12] theoretical modeling of ⟨001⟩ coincident twist boundaries in MgO and NiO has not found stable

Table II. Calculated Energies of ⟨011⟩ Tilt Boundaries in NiO and MgO

Boundary plane	Misorientation angle (deg)	Σ	Boundary energy (J·m⁻²) NiO	MgO
(1$\bar{3}$3)	26.5	19	2.91	
(1$\bar{2}$2)	38.9	9	2.56	
(1$\bar{1}$1)	70.5	3	0.95	0.91
(2$\bar{1}$1)	109.5	3	2.56	2.44
(3$\bar{1}$1)	129.5	11	3.13	
(4$\bar{1}$1)	141.1	9	3.05	

bound structures.[13] It was felt that defects in the boundary plane could play an important role in the stability of such boundaries, and indeed, calculations have shown that isovalent impurities do increase the binding energy by a small amount.[12] However, more recent calculations on the Σ = 5, 36.9° [001] twist boundary have indicated that a very stable structure (with a binding energy of 1.5 J·m⁻²) is obtained by introducing Schottky defects into the boundary plane.[15]

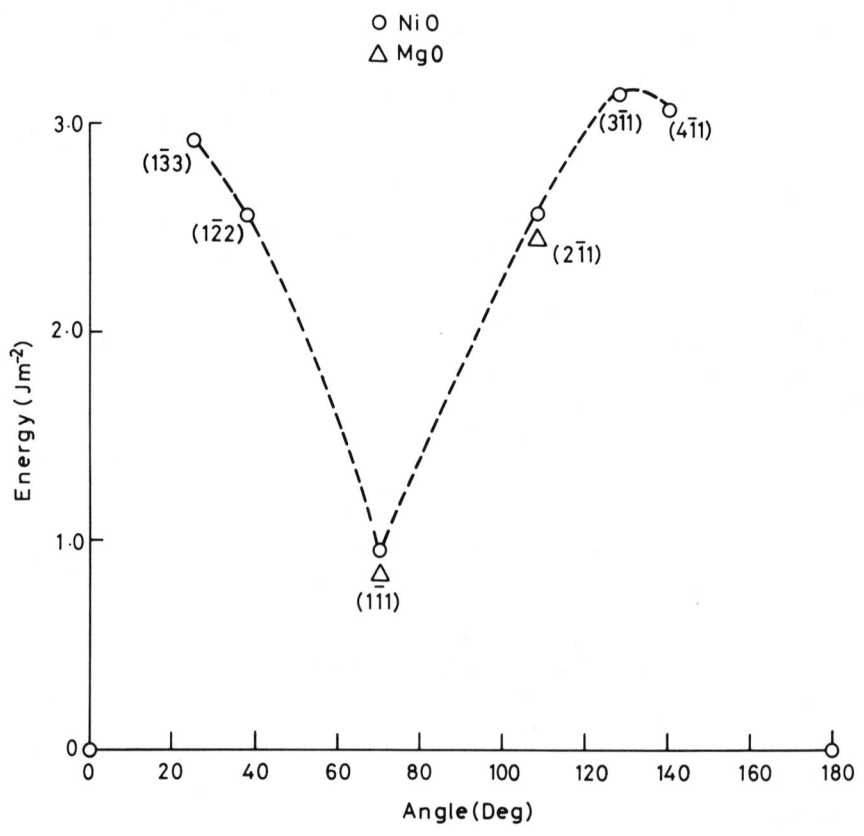

Fig. 4. Energy of ⟨001⟩ tilt boundaries in NiO and MgO as a function of misorientation angle.

(a)

(b)

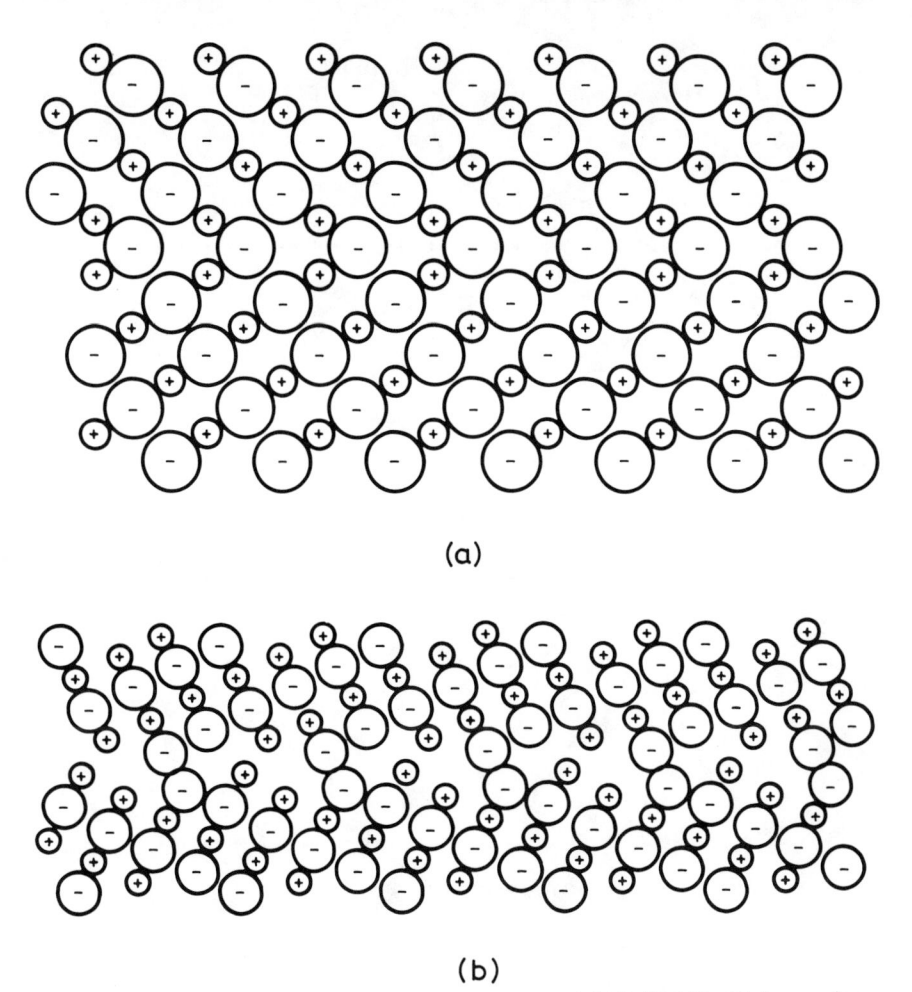

Fig. 5. (a) Relaxed structure of the 70.5° $(1\bar{1}1)/[011]$ tilt boundary. (b) Relaxed structure of the 129.5° $(3\bar{1}1)/[011]$ tilt boundary.

The starting point is the anticoincidence configuration, which has like ions facing each other across the boundary on the anticoincidence sites. All other ions are reasonably close to ions of the opposite charge. The anticoincidence ions in one plane are then removed and the remaining ions relaxed to equilibrium. The resulting structure is shown in Fig. 6. The solid lines represent the reconstructed boundary plane, and the broken lines are the planes above and below the boundary. The relative rotation between the planes represented by the broken lines is 36.9°, but the boundary phase has an intermediate angle. There is no significant dilation of the crystal across the interface, which is a further indication of the stability of this configuration. The Schottky defects have a negative energy of formation (-1.4 eV), and they reduce the interfacial energy of the boundary from 2.7 $J \cdot m^{-2}$ for the perfect, unstable configuration to 2.2 $J \cdot m^{-2}$ for the reconstructed stable structure. The predicted structure, therefore, should not be considered as "defective" but as the most stable configuration for the grain boundary.

(a)

(b)

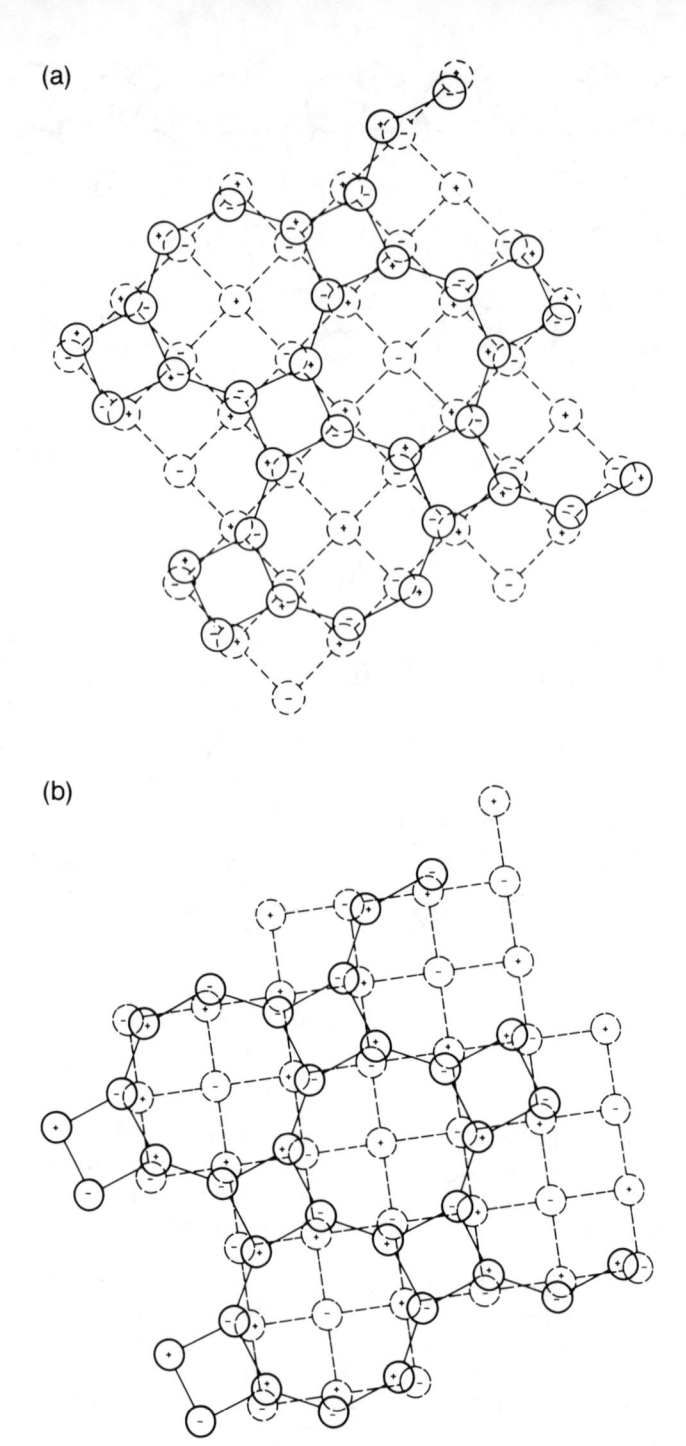

Fig. 6. Structure of a stable 36.9° [001] coincidence twist boundary in NiO. The broken lines in (a) and (b) indicate the structure of the lower and upper grains.

Energies of Defects near Grain Boundaries
Intrinsic Defects

The structural disorder around grain boundaries in oxides introduces a spread into the formation energy of defects. The resulting defect distributions modify the transport properties of the interface with respect to the bulk crystal and contribute to the enhanced diffusion coefficient of the grain-boundary region. The formation energies of point defects in perfect ionic crystals can be calculated by using the Harwell HADES code.[16] A similar program has been developed for the calculation of defect formation energies of defects near grain boundaries or other planar defects in ionic crystals. Here the perfect or reference configuration is the relaxed grain-boundary structure, calculated by using the MIDAS code. The crystal is divided into two regions. The inner region is a spherical volume of ions surrounding the point defect in which the ions are relaxed explicitly. The outer region, the rest of the crystal, is treated as a dielectric elastic continuum in which the ionic relaxations are calculated by the Mott–Littleton method.[17] The Coulomb energies of the ions are determined with a two-dimensional lattice-summation technique.[18,19]

The formation energies of a number of point defects near coincident tilt boundaries in NiO have been calculated by this method.[20] One boundary was chosen to represent each type of structure modeled previously (i.e., the (310)/[001], (320)/[001], (2$\bar{1}$1)/[011], and (1$\bar{2}$2)/[011] interfaces). The defect sites were selected by inspection of the Madelung potentials of the ions in the boundary. This represents the Coulomb energy of the defect in the unrelaxed configuration. The other components of the final defect energy, the short-range energy and the relaxation energy, also show significant variation around a given interface; therefore, several sites were selected in each case.

The lowest defect formation energies for cation and anion vacancies and holes (Ni^{3+} ions) are summarized in Table III. The highest energies are also included for cation vacancies to give a measure of the energy spread around a given boundary. At each interface, there are ionic sites for which the values of the defect formation energies are lower than the corresponding bulk values. The structure of some coincidence boundaries is relatively open; therefore, the interstitial energies should be significantly reduced. The formation energies of a cation and an anion interstitial in the 36.9°, (310)/[001] interface were, therefore, calculated and found to be reduced by 3.7 and 3.2 eV with respect to the corresponding bulk values. In spite of this large reduction, the cation and anion Frenkel defect energies are high (8.3 and 7.9 eV). Therefore, the concentration of interstitials at grain boundaries is negligible.

Table III. Energies of Intrinsic Defects near Tilt Boundaries in NiO*

Boundary	Cation defect energy min	Cation defect energy max	Vacancy interaction energy min	Vacancy interaction energy max	Anion defect energy	Vacancy interaction energy	Hole (Ni^{3+} ion) defect energy	Interaction energy
(310)/[001]	24.06	25.30	−0.36	+0.88	24.39	−0.30	−30.86	−0.25
(3$\underline{2}$0)/[001]	22.81	24.88	−1.61	+0.46	23.44	−1.25	−31.06	−0.45
(2$\underline{1}$1)/[011]	23.49	24.89	−0.95	+0.47	23.71	−0.98	−30.84	−0.23
(1$\underline{2}$2)/[011]	22.88	24.58	−1.54	+0.15	23.13	−1.56	−31.15	−0.54

*The interaction energy is defined as the difference between the energy at the boundary and the energy in the bulk (units are in eV).

A general grain boundary has a large number of steps and kinks, which makes it an effective source of defects. Such a boundary is equivalent to two free surfaces, back to back, and the difference between the formation energies of cation and anion vacancies leads to a space charge layer with a compensating boundary charge.[21,22] A coincidence boundary, on the other hand, cannot be treated as a vacancy source because the interstitial energies are high and there are no low-energy kink sites. Charged defects will, however, segregate to coincidence boundaries, and the concentration of positively charged defects is not necessarily equal to the concentration of negatively charged defects. Thus, the boundary may have a net charge that is compensated by a space charge layer.

The defect distributions around the $(2\bar{1}1)/[011]$ interface have been calculated for the case when the predominant defects are singly charged vacancies, compensated by holes.[20] In the situation where the defect concentration at the boundary is much higher than in the bulk, the boundary acts as an effective defect source, with defect formation energies of $-F_{INT}^v$ and $-F_{INT}^h$. Here, F_{INT}^v is the difference between the energies of a singly charged vacancy at the boundary and in the bulk (calculated to be -0.95 eV), and F_{INT}^h is the corresponding energy for holes (-0.23 eV). We have neglected the entropy contribution to the free energy because there have been no reliable estimates of this term. At 1000 K, the concentration of vacancies on the boundary was found to be enhanced by a factor of 1.2×10^3 with respect to the bulk, and the concentration of holes is enhanced by a factor of 0.8×10^3. The resultant boundary charge density is -1.9×10^{17} $e \cdot m^{-2}$, and the difference between the potential at the boundary and the bulk is 0.34 V. Thus, the low-energy sites for defect formation at coincidence grain boundaries introduce space charge layers similar to those formed at general grain boundaries and surfaces in ionic crystals.

Impurities

The structural disorder around a grain boundary also affects the energy of substitutional impurities. Impurities are known to segregate to both surfaces and grain boundaries in ionic crystals, and they can have a significant effect on the properties of the space charge layer.[23] Impurities with zero net charge segregate to grain boundaries because of size effects, but the boundary charge and the spread of the Madelung potentials are the dominating factors for aliovalent impurities.

The energies of neutral (Co^{2+}), singly charged (Al^{3+}), and doubly charged (Ce^{4+}) substitutional impurities at coincidence grain boundaries in NiO have been calculated by using the program discussed in the previous section. The results are summarized in Table IV. As with intrinsic defects, there are ionic sites at each boundary with negative segregation energies. Thus, all impurities will tend to segregate to grain boundaries in ionic crystals. The distribution of defects in the

Table IV.   Maximum Interaction Energies of Substitutional Defects at Tilt Boundaries in NiO*

Boundary	Co^{2+}	Al^{3+}	Ce^{4+}
(310)/[001]	−0.09	−0.18	−1.15
(320)/[001]	−0.22	−0.39	−1.73
$(2\bar{1}1)/[011]$	−0.12	−0.24	−0.41
$(1\bar{2}2)/[011]$	−0.06	−0.53	−0.74

*Units are eV.

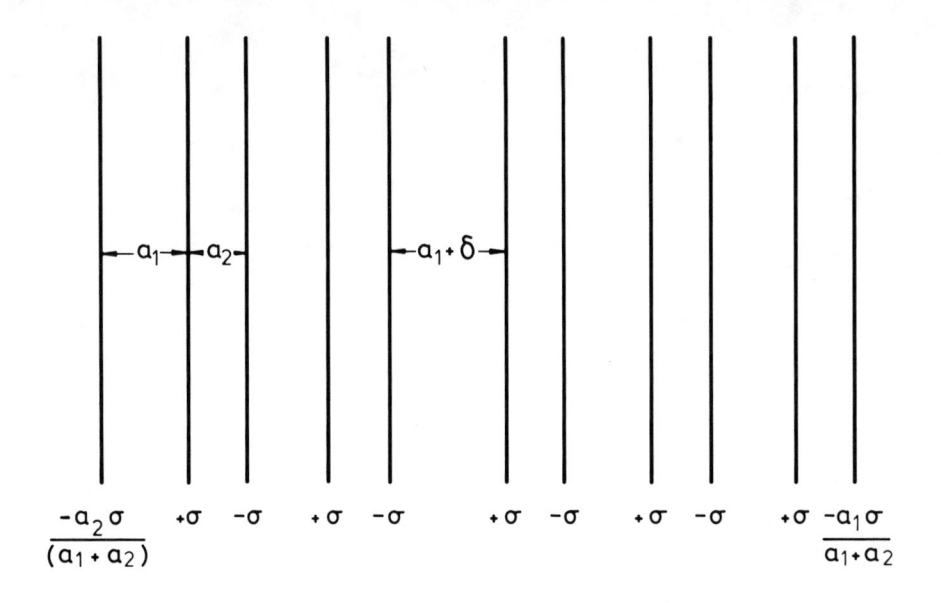

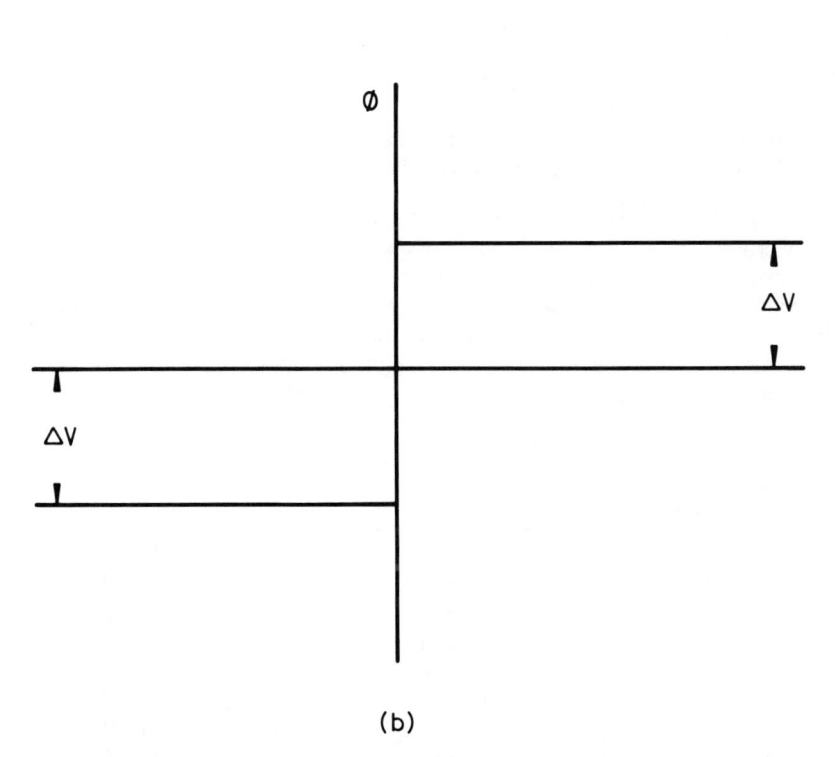

Fig. 7. Schematic representation of a tilt boundary orientated along a polar direction in (a) an ionic crystal and (b) the boundary potential.

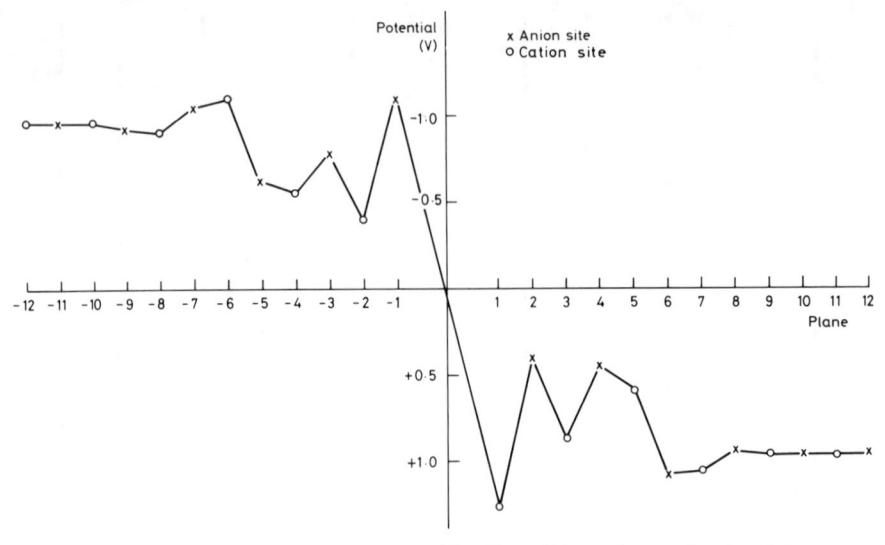

Fig. 8. Boundary potential of the $(3\bar{1}1)/[011]$ interface obtained from an atomistic calculation.

space charge region depends on the boundary charge, which is a function of temperature. As with free surfaces,[22] we expect the boundary charge to pass through zero at a finite temperature (the isoelectric temperature). Below the isoelectric temperature, the boundary has a positive charge; therefore, positively charged impurities will be repelled from the space charge layer. Above the isoelectric temperature, the opposite effect occurs. A more detailed calculation of impurity distributions near coincidence boundaries is published elsewhere.[24]

Dipolar Grain Boundaries

Certain crystallographic planes in ionic crystals contain only one type of ion and have, therefore, a net charge. A neutral stack of such planes has a net dipole moment that produces an electric field in the bulk of the crystal. Such a field is obviously not present in a real crystal, and it must, therefore, be removed by imposing suitable boundary conditions. Kummar and Yao[25] have noted that a dipole layer of charge density $\sigma a_1/(a_1 + a_2)$ and spacing $n(a_1 + a_2)/2$ cancels the

Table V. Interaction Energies of Vacancies near the $(3\bar{1}1)/[011]$ Tilt Boundary in NiO

Plane	Vacancy	Interaction energy (eV)
1	Cation	−1.66
−1	Anion	−1.69
4	Anion	+1.84
−4	Cation	+1.90
11	Cation	−1.87
−11	Anion	−1.88
12	Anion	+2.06
−12	Cation	+2.12

dipole moment of the crystal and thus quenches the field. Here σ is the charge density of the ionic planes, a_1 and a_2 are the interplanar distances, and n is the number of planes. The modified outer layers represent either reconstructed outer surfaces or tapered low-energy surfaces beyond the boundaries.

A grain boundary, or other planar defect, that is orientated along a charged direction in the crystal may introduce a net dilation (δ) across the interface. A continuum representation of such a defect is shown in Fig. 7(a). There is a net charge in the crystal on either side of the interface; therefore, any dilation will introduce a potential difference of $4\pi\sigma\delta a_1/(a_1 + a_2)\varepsilon$, where ε is the dielectric constant, across the boundary (Fig. 7(b)). The potential is equivalent to that of a dipole layer with charge density $a_1\sigma/(a_1 + a_2)$ and spacing δ in the boundary plane.

One example of such a planar defect is the $(3\bar{1}1)/[011]$ interface in NiO (Fig. 5(b)). In this case σ is $0.14\ e\cdot\text{Å}^{-2}$, $a_1 = a_2$, $\delta = 1$ Å, (0.01 nm), and $\varepsilon = 12.6$; therefore, the potential drop across the boundary can be estimated as 1.0 V. The continuum model is not an accurate representation of the detailed structure of the boundary; therefore, an atomistic calculation of the Madelung potentials of the ions in the relaxed structure was carried out by using MIDAS. The results are plotted in Fig. 8. This is the atomistic equivalent of the continuum plot in Fig. 7(b). The oscillations close to the boundary plane die out, and the potential settles down to a constant value of 0.95 V on one side of the boundary and -0.95 V on the other side. Thus, the potential drop calculated by the atomistic method is a factor of 2 larger than in the continuum approximation.

The boundary potential modifies the energy of charged defects by $\pm q\Delta V$ on either side of the boundary. Here q is the net charge of the defect, and ΔV is the magnitude of the potential introduced by the boundary. The energies of cation and anion vacancies near the $(3\bar{1}1)/[011]$ interface in NiO were calculated,[26] and the results are summarized in Table V. After a few atomic distances from the interface, the energies are modified by approximately ± 1.9 eV ($\pm q\Delta V$) with respect to the

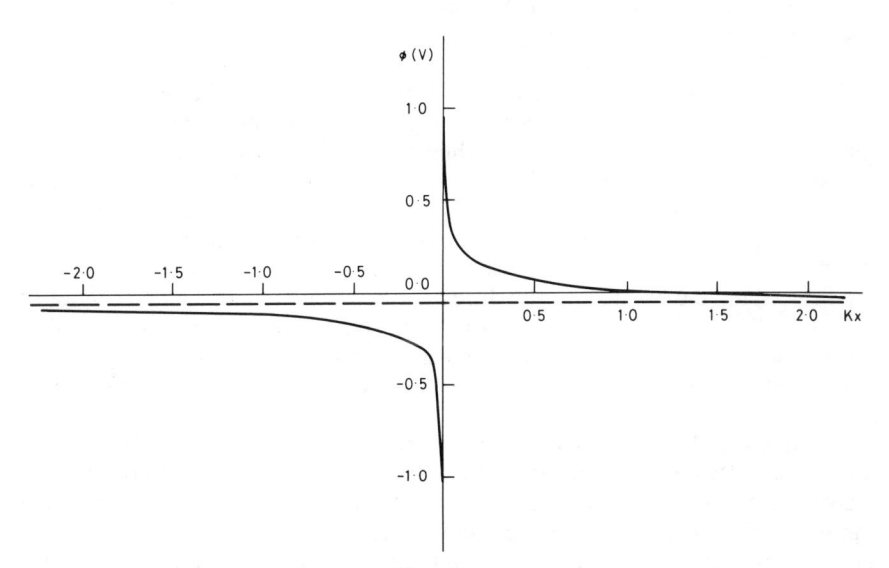

Fig. 9. Total potential near a $(3\bar{1}1)/[011]$ tilt boundary in NiO.

corresponding defect energies in a perfect crystal. The energies of all other charged defects (aliovalent impurities and holes) will suffer similar modifications.

The distribution of charged defects around dipolar grain boundaries can be calculated by minimizing the free energy and solving a Poisson's equation.[26] The problem is equivalent to the calculation of defect distributions near free surfaces.[22] The result is that the concentration of positively charged defects is enhanced on one side of the interface and depleted on the other side. The opposite effect occurs for negatively charged defects. The charge density resulting from the defect distribution introduces an additional potential that cancels the boundary potential at some distance from the interface. The total potential near a $(3\bar{1}1)/[011]$ tilt boundary in NiO is plotted in Fig. 9. There is incomplete cancellation between the defect potential and the boundary potential near the interface, and an asymmetric potential barrier of height 1 V (ΔV) remains. Such a barrier will introduce nonlinear electrical properties into the crystal. In addition, the defect distribution is such that the diffusion coefficient will be enhanced on one side of the interface but reduced on the other side. This inhomogeneity might be revealed by a tracer diffusion experiment parallel to the boundary. Thus, there are space charge layers associated with grain boundaries orientated along polar directions in ionic crystals even when such a boundary is not a source of defects. The resultant potential barrier has the same form as that associated with a general grain boundary, but it is asymmetric. It will, however, have a similar effect on the electrical characteristics across the boundary.

Conclusions

We have used an atomistic simulation technique to model coincidence grain boundaries in nickel and magnesium oxide and found that the dominating factor in determining the structure is the Coulomb interaction. The boundary configurations are, in general, more open than the corresponding configurations in metals; therefore, we expect strong enhancement of diffusion along grain boundaries in ionic crystals. In the case of the $\Sigma = 5$, [001] twist boundary in NiO, the stable structure has a lower density of ions in the boundary plane. The resulting configuration should have an enhanced, isotropic diffusion coefficient in the boundary plane.

The spread in the formation energy of defects around a given grain boundary in NiO was found to be considerable. The primary cause of the modification of defect energies is the spread in the Madelung potentials of the ions in the boundary plane; hence, the largest effects are exhibited for defects with a high net charge. For all types of defect, some sites with lower energies than the bulk values exist. The concentration of both intrinsic defects and impurities is, therefore, enhanced at grain boundaries. Under certain conditions, coincidence grain boundaries act as defect sources with energies of formation equal to the interaction energies between the defects and the boundary. The difference between the interaction energies of defects with opposite charges leads to a net charge on the boundary, with a compensating space charge layer. Similar space charge layers are expected at general grain boundaries, but the defect formation energies are different in this case.

Grain boundaries orientated along polar directions in ionic crystals introduce a potential difference between the crystal on the two sides of the interface. Such a potential enhances the concentration of positively charged defects on one side of the interface and depletes it on the other side. The opposite effect occurs for negatively charged defects. The potential resulting from the modified defect distribution cancels the boundary potential at some distance from the interface, but a potential barrier remains in the boundary region. This effect is independent of the space charge effects introduced by grain boundaries acting as defect sources.

Thus, grain boundaries in oxides have a strong influence on defect populations both in the boundary plane and in a finite region on either side of the interface. Transport properties will be modified accordingly, with the enhancement of diffusion and conductivity being pronounced in the boundary plane. Nonlinear electrical characteristics across the boundary may be introduced by the space charge layers and the resulting potential barriers.

Acknowledgments

D.M.D. acknowledges funding from the Materials Development Division of A.E.R.E. Harwell under an extramural research contract.

References

[1] A. Atkinson and R. I. Taylor, *Philos. Mag., [Part A]*, **A43** [4] 979 (1981).
[2] W. D. Kingery, "Grain Boundary Phenomena in Electronic Ceramics"; p. 1 in Advances in Ceramics, Vol. 1. Edited by M. Levinson. The American Ceramic Society, Columbus, OH, 1981.
[3] H. Schmid, M. Rühle, and N. L. Peterson; p. 177 in Surfaces and Interfaces in Ceramic and Ceramic-Metal Systems, Edited by J. Pask and A. Evans, Plenum, New York, 1981.
[4] M. Vaudin, M. Rühle, and S. L. Sass, *Acta Metall.*, **31**, 1109 (1983).
[5] M. J. L. Sangster and A. M. Stoneham, *Philos. Mag., [Part B]*, **B43**, 597 (1981).
[6] D. M. Duffy and P. W. Tasker, *Philos. Mag., [Part A]*, **A47**, 817 (1983).
[7] P. W. Tasker; AERE Rept. R.9130, 1978.
[8] G. Hasson, J. Y. Boos, I. Herbeuval, M. Biscondi, and G. Goux, *Surf. Sci.*, **31** [1] 115 (1972).
[9] D. A. Smith, V. Vitek, and R. C. Pond, *Acta Metall.*, **25** [5] 475 (1977).
[10] D. M. Duffy and P. W. Tasker, *Philos. Mag., [Part A]*, **A48**, 155 (1983).
[11] P. Chaudari and J. W. Matthews, *J. Appl. Phys.*, **42**, 3063 (1971).
[12] C. P. Sun and R. W. Balluffi, *Philos. Mag., [Part A]*, **A46**, 49 (1982).
[13] D. Wolf and R. Benedek; pp. 107–13 in Advances in Ceramics, Vol. 1. The American Ceramic Society, Columbus, OH, 1981.
[14] D. Wolf, Proceedings of the 84th Meeting of the American Ceramic Society; to be published in *J. Am. Ceram. Soc.*
[15] P. W. Tasker and D. M. Duffy, *Philos. Mag., [Part A]*, **A47**, L45 (1983).
[16] M. J. Norgett; AERE Harwell Rep. R.7650, 1974.
[17] N. F. Mott and M. J. Littleton, *Trans. Faraday. Soc.*, **34**, 485 (1938).
[18] D. Parry, *Surf. Sci.*, **49**, 433 (1975).
[19] D. Parry, *Surf. Sci.*, **54**, 195 (1976).
[20] D. M. Duffy and P.W. Tasker, *Philos. Mag., [Part A]* **50**, 143 (1984).
[21] J. D. Eshelby, C. W. A. Newey, P. L. Pratt, and A. B. Lidiard, *Philos. Mag.*, **3**, 75 (1958).
[22] K. L. Kliewer and J. S. Koehler, *Phys. Rev. A*, **A140**, 1226 (1965).
[23] M. F. Yan, R. M. Cannon, and H. K. Bowen, *J. Appl. Phys.*, **54**, 764 (1983).
[24] D. M. Duffy and P. W. Tasker, *Philos. Mag. [Part A]*, **50**, 155 (1984).
[25] J. T. Kummer and Y. Y. Yao, *Can. J. Chem.*, **45**, 421 (1967).
[26] D. M. Duffy and P. W. Tasker, *J. Appl. Phys.*, **56**, 971 (1984).

---

*Department of Physics, University of Reading, Reading, England.

Comparison of the Calculated Properties of High-Angle (001) Twist Boundaries in MnO, FeO, CoO, and NiO with MgO*

Dieter Wolf

Argonne National Laboratory
Materials Science and Technology Division
Argonne, IL 60439

The energy and structure of (001) coincident-site lattice (CSL) twist grain boundaries and of the free (001) surface in the transition-metal oxides with rock-salt structure (MnO, FeO, CoO, and NiO) are determined by means of a computer code developed in recent years. Boundaries with values of Σ, the inverse density of coincidence sites, ranging between 5 and 65 are considered. Our comparison with similar results for MgO confirms our earlier suggestion that the effective Van der Waals attraction between oxygen ions on opposite sides of the interface is mainly responsible for the rather weak cohesion of these bicrystals while Coulombic interactions are thought to play a minor role only. As for MgO, the comparison of the relaxed energies, E_Σ, of boundaries characterized by different values of Σ and the misfit angle θ suggests that no "special" boundaries (giving rise to "cusps" in the $E_\Sigma(\theta)$ curve) exist on (001) planes in the NaCl structure.

Bollmann's coincident-site lattice (CSL) model[1] has met with considerable success in the prediction of dislocation arrangements in high-angle grain boundaries.[2-5] According to his model, arrays of secondary grain-boundary (GB) dislocations are formed whenever the misfit angle, θ, deviates slightly from certain misorientations for which the inverse density, Σ, of CSL sites is particularly low. From this and other observations,[6-9] it is generally concluded that such "special" boundaries are also characterized by especially low energies; i.e., the GB energy vs θ curve shows "cusps" near special misorientations.

(001) twist boundaries in the fcc and NaCl structures represent ideal model systems in that (i) a fair amount of experimental information has become available in recent years[2-5,9,10] and (ii) they are easily amenable to computer simulation techniques. Recent computer calculations of the energy and structure of (001) twist boundaries both in fcc metals[11-13] and in MgO[14-17] employing lattice statics techniques have yielded the somewhat puzzling result that energy "cusps" near special misorientations may not exist for (001) twist boundaries. Moreover, for MgO it was found[14-17] that defect-free (001) twist boundaries whose geometry derives itself from the ideal-crystal positions of the ions in the CSL unit cell should be only marginally stable with respect to breaking up to form two free (001) surfaces. It was suggested that Coulomb interactions do not contribute to the cohesion of such bicrystals.[16] Instead, their rather weak cohesion observed in these calculations was attributed mainly to a relatively weak effective Van der Waals attraction between ions on opposite sides of the interface.[16,17] More recent calculations suggest that the rather strong cohesion of (001) bicrystals manufactured by pressure sintering[4,5]

may be due to impurities[18,19] or Schottky pairs[20] in these boundaries.

This apparent discrepancy between experiments and computer calculations concerning (i) the existence of "cusps" and (ii) the reasons for the stability of (001) twist boundaries in MgO represents probably one of the more puzzling aspects of grain-boundary research at this time. In this article, calculations similar to the ones for MgO are presented for the transition-metal oxides with NaCl structure, MnO, FeO, CoO, and NiO, and the results obtained are compared with the earlier ones for MgO.

Computational Procedure and Interionic Potentials

An iterative energy minimization procedure described earlier in some detail[17] is applied to determine the relaxed grain-boundary energy and structure. The computer program developed in recent years at Argonne employs techniques similar to those incorporated in Harwell's HADES program but with the advantage of allowing both bulk and interfacial properties to be considered.

The interionic potentials applied in our calculations are due to Catlow et al.,[21] who assumed an entirely ionic central-force model. The short-range repulsive cation–anion potential is assumed to be of the Born–Mayer type, according to

$$V_{+-}(r) = A_{+-}e^{-r/\rho_{+-}} \qquad (1)$$

while the oxygen–oxygen interaction is represented by the Buckingham form

$$V_{--}(r) = A_{--}e^{-r/\rho_{--}} - C_{--}/r^6 \qquad (2)$$

where C_{--} is an effective Van der Waals constant. Owing to the small cation sizes, the cation–cation interactions were neglected entirely.

The short-range potentials of Catlow et al.[21] are shown in Fig. 1, which illustrates that for all but extremely short distances the effective Van der Waals attraction between oxygen ions dominates over their Born–Mayer repulsion. The oxygen–oxygen interaction (Eq. (2)) was assumed to be identical for all four oxides considered;[21] i.e., the values for A_{--}, ρ_{--}, and C_{--} were assumed to be the same ones for all four oxides.[21] In contrast, A_{+-} and ρ_{+-} in Eq. (1) were assumed to depend on the particular oxide considered.

In addition to the Coulomb and short-range interactions, the ion polarizability was incorporated via the shell model. The same shell-model parameters for O^{2-} were assumed for all four compounds while the much smaller (and hence much less polarizable) cations were assumed to be unpolarizable.[21]

These potentials are very similar to the ones of Catlow et al.[22] for MgO, which were used successfully for point defect calculations[22] as well as more recently for grain-boundary studies.[14-17] The above potentials for the transition-metal oxides have been applied, with reasonable success, to investigate both their electronic conduction[21] and their defect structure and cation-transport properties.[23]

In contrast to the bulk of the crystal, ions near an interface are not arranged in well-defined shells.[13,17] To avoid undesirable cutoff effects, it is, therefore, imperative not to invoke any arbitrary cutoff radii, R_c, for the short-range potentials. Particularly, the longer range Van der Waals interaction in Eq. (2) may give rise to cutoff effects in the GB energy when terminated, for example, after the second nearest neighbors. To avoid such effects, it was found necessary to extend R_c^{+-} to $1.3a$ and $R_c^{--} = R_c^{++}$ to $2.6a$, where a is the lattice parameter (i.e., the edge of the unit cell). These values are substantially larger than the cutoff radii after the first and second nearest neighbors, respectively, for which these potentials were originally parametrized.[21] As a consequence, it was found necessary to slightly

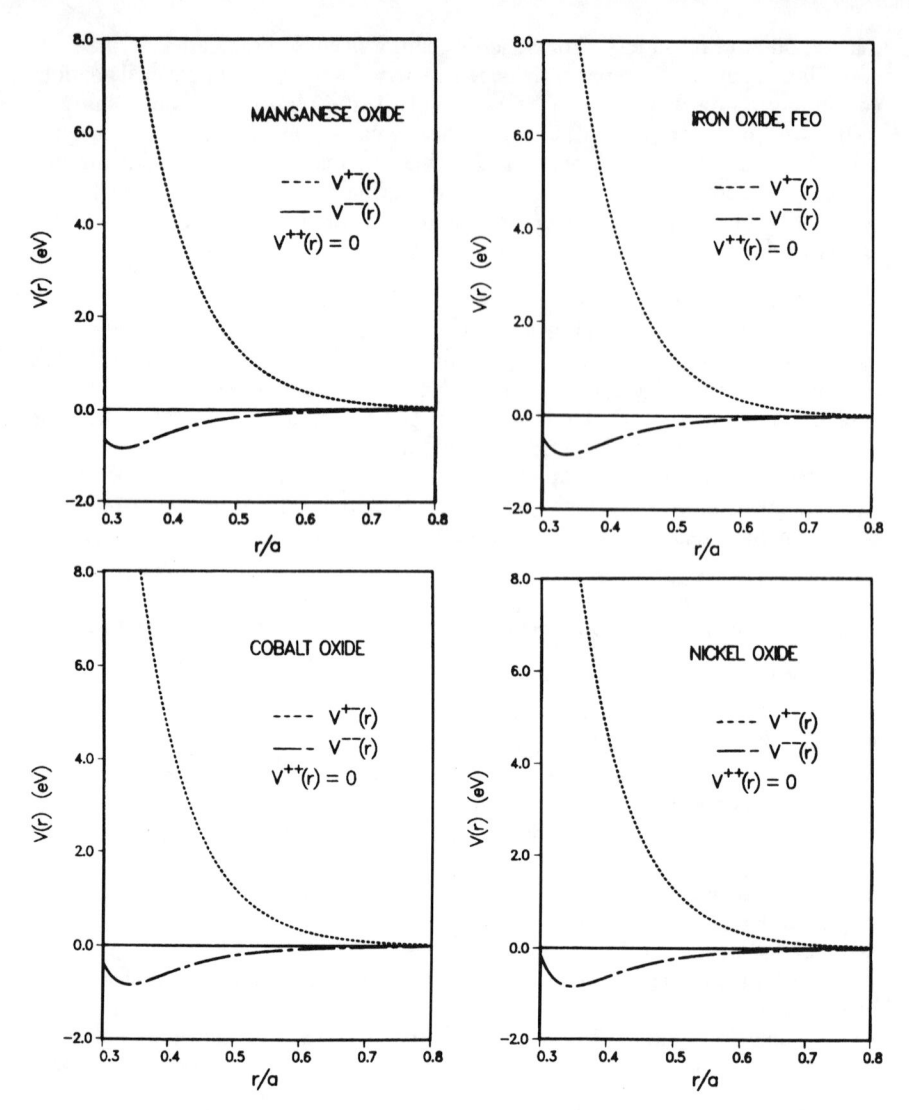

Fig. 1. Short-range interionic potentials for MnO, FeO, CoO, and NiO due to Catlow et al. (Ref. 21).

readjust the lattice parameters to ensure equilibrium of the crystals described by these slightly modified potentials (see also Table I). While the cohesive energy changes very little as a result of these adjustments, the energy of the free (001) surface depends more strongly on the lattice parameter and the cutoff radii.[17] However, neither the stability of (001) twist boundaries relative to the free surface nor their structure is changed significantly due to these adjustments.[17]

Free (001) Surface

When the two semiinfinite single crystals forming a (001) twist boundary are well separated from each other, the interfacial energy corresponds to the energy

Table I. Unrelaxed ($\sigma^\circ_{(001)}$) and Relaxed ($\sigma_{(001)}$) Free-Surface Energies Obtained by Means of the Potentials of Catlow et al. (Ref. 21), Which Were Modified Slightly in the Manner Described Above*

	a_0 (nm)	E_{coh} (eV)	$\sigma^\circ_{(001)}$ (erg/cm^2)	$\sigma_{(001)}$ (erg/cm^2)
MnO	0.226 37	−37.33	720	718
FeO	0.216 02	−39.43	947	945
CoO	0.213 70	−39.93	1004	1002
NiO	0.208 71	−40.95	1113	1112
NiO (Ref. 26)	0.208 00	−40.95	1080	1077
MgO[17] Pot. 1	0.211 06	−40.85	1226	1188
Pot. 2	0.217 03	−40.75	1116	1069

*For comparison, the results obtained by means of two rather different sets of short-range potentials for MgO are also shown (see Ref. 17); a_0 and E_{coh} denote the nearest-neighbor distance and the cohesive energy, respectively.

$2\sigma_{(001)}$ of the two free (001) surfaces formed.[13–17] Before investigating the properties of such bicrystals, we, therefore, first must consider this limiting case.

The unrelaxed and relaxed free-surface energies, $\sigma^\circ_{(001)}$ and $\sigma_{(001)}$, thus determined for the four transition-metal oxides and for the two different sets of short-range interionic potentials in MgO considered in Ref. 17 are listed in Table I. The corresponding structures and electrical polarizations for MnO and NiO are shown in Fig. 2. The more interesting features of these results are the following four:

First, for the four transition-metal oxides, the monotonous decrease in the lattice parameter from MnO to NiO results in a monotonous increase in the cohesive energy and a corresponding increase in $\sigma^\circ_{(001)}$ and $\sigma_{(001)}$. This type of correlation between σ and E_{coh} has been observed also for alkali–halide crystals.[24]

Second, the relaxation energies, $\sigma^\circ_{(001)}-\sigma_{(001)}$, are much smaller for the transition-metal oxides than for MgO. The reason for this difference lies in the different sets of shell-model parameters that take into account polarization effects. Thus, for MnO to NiO, the anion shell charge chosen is $Y^- = -2.0066\ e$ (Ref. 21), which results in an anion core charge of $+0.0066\ e$ (notice that for unpolarizable anions $Y^- = -2.0$ with a vanishing core charge). Hence, the O^{2-} ions in the transition-metal oxides described by the potentials of Catlow et al.[21] are not very polarizable; this is known to result in rather small relaxation energies. In contrast, shell charges of -2.62 and $-1.78\ e$, respectively, were chosen for the first[22] and second set[25] of MgO potentials considered, thus permitting a stronger polarization and hence relaxation of the oxygen ions. This interpretation was confirmed by calculations of $\sigma_{(001)}$ with and without use of the shell model. While in MnO through NiO, practically the same relaxed energies are obtained whether or not the shell model is used, in MgO the relaxation energy is found to be substantially larger for polarizable than for rigid ions.

Third, the comparison of the $\sigma^\circ_{(001)}$ and $\sigma_{(001)}$ values for NiO with those calculated earlier[26] by using the same set of interionic potentials (however, with the cutoff radii and lattice parameter for which these potentials were originally derived) illustrates the sensitivity of the surface energy to small changes in the short-range potentials.

Finally, the structures shown in Fig. 2 differ rather drastically from the ones obtained in Ref. 17 for MgO in that (i) none of the four transition-metal oxides shows the rumpling effect observed for MgO for the first set of short-range potentials[22] and (ii) an inward relaxation of the outermost lattice plane is observed

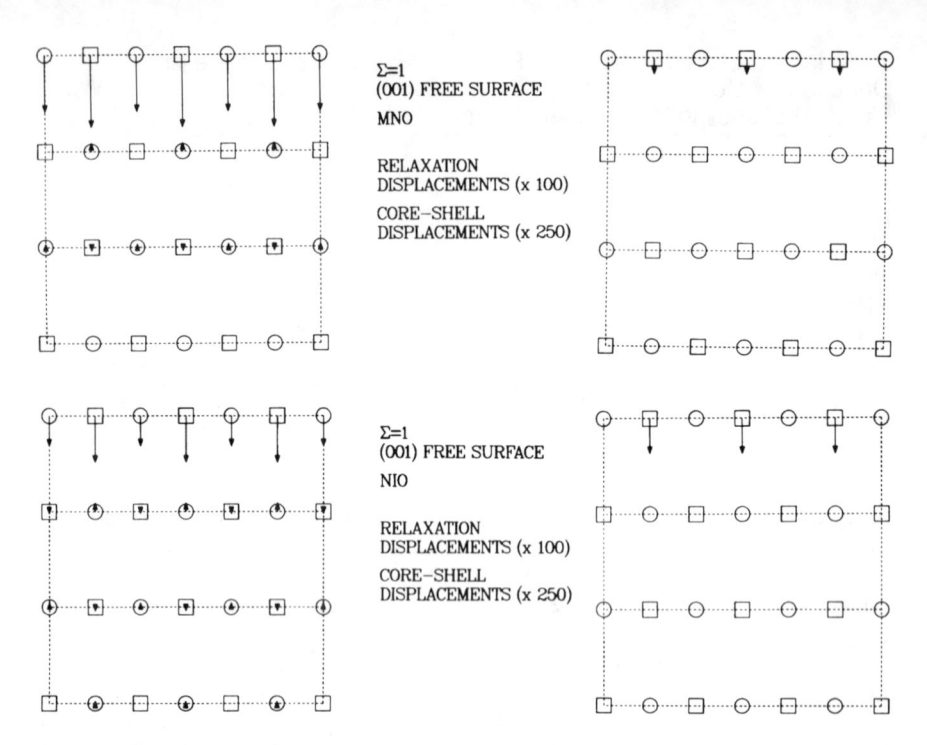

Fig. 2. Relaxed structure (left half) and electrical polarization (right half) of the free (001) surfaces of MnO and NiO. On the left the *unrelaxed* cation and anion positions are shown together with the relaxation displacements (mulitplied by a factor of 100). On the right the *relaxed* core positions together with the core-shell displacemens are illustrated. Anions are shown as squares and cations as circles. The relaxation and core-shell displacements for FeO and CoO (not shown) lie in between the ones shown here for MnO and NiO.

while in MgO no such net displacement was found. Similar to MgO, however, in all four transition-metal oxides, lattice relaxations and core-shell displacements are limited to essentially the two lattice planes nearest to the free surface.

Grain-Boundary Energies

As pointed out earlier[13–17] for grain-boundary calculations in ionic crystals, the volume change $\delta V_\Sigma = d/d_o - 1$ due to the introduction of the interface is an important variable. Here d_o denotes the ideal-crystal (001) interplanar lattice constant while d represents the interplanar lattice constant at the interface (see, e.g., Fig. 2 in Ref. 17).

The (001) twist-boundary energies as a function of δV_Σ for $\Sigma = 5, 13$, and 17 are shown in Fig. 3. These results may be summarized as follows: (i) For all four oxides the *unrelaxed* energies, $E°_\Sigma$, depend strongly on δV_Σ but are completely independent of Σ. (ii) Only weakly stable boundaries are predicted for all four oxides; i.e., for the $\Sigma = 5, 13$, and 17 boundaries the *relaxed* interfacial energy, E_Σ, drops only slightly below the stability limit $2\sigma_{(001)}$. (iii) The minima in E_Σ correspond to volume increases at the interface of typically about 80–100%.

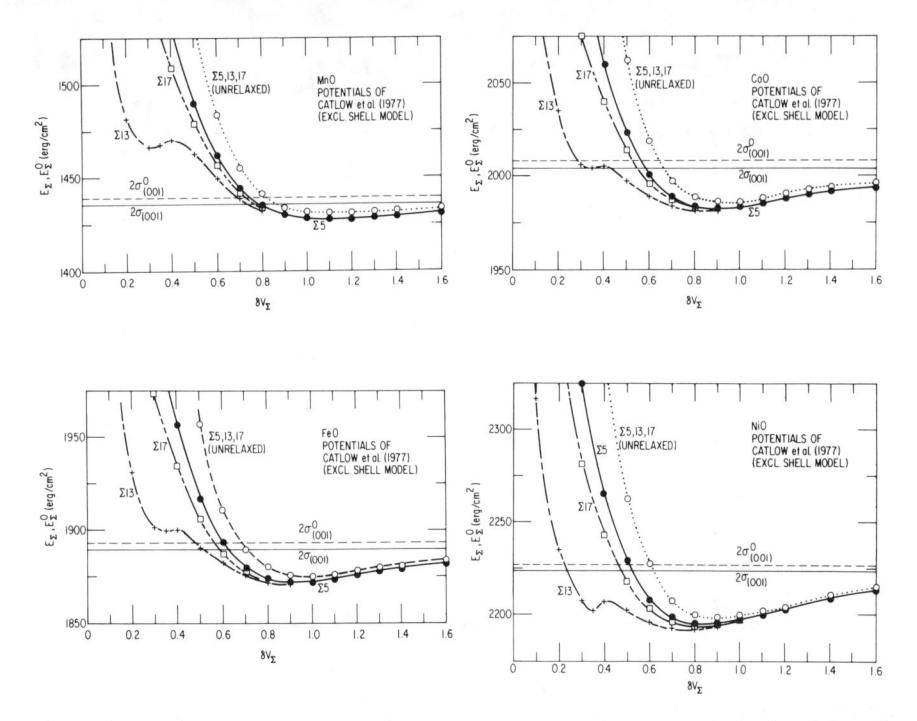

Fig. 3. Unrelaxed (dotted lines) and relaxed (solid and dashed lines) energies of the $\Sigma = 5$, 13, and 17 boundaries as a function of the volume increase at the interface due to the introduction of a grain boundary into the crystal (without shell model). For comparison, the unrelaxed and relaxed free-surface energies are also shown. All energies are in units of erg/cm².

(iv) The stability, $2\sigma_{(001)} - E_\Sigma$, increases slightly as one proceeds from MnO to NiO. (v) The $\Sigma = 13$ boundary shows two minima, a stable one for a rather large volume increase and a metastable one for smaller values of δV_Σ. In all oxides but NiO, the value of E_Σ at this metastable minimum lies above $2\sigma_{(001)}$. Hence, for NiO two configurations characterized by different δV_Σ values might exist.

All of these results were obtained for unpolarizable ions. However, as already mentioned in the preceding section, the application of the shell model changes the energies in the oxides described by the potentials of Catlow et al.[21] very little due to the small core charge of the O^{2-} ions. Also, in earlier work on MgO, it was shown that the properties of (001) CSL twist boundaries (such as the stability or the δV_Σ values at the minima in Fig. 3) are practically the same for both polarizable and rigid ions.[16,17] For that reason, the results presented in the remainder of this section have been obtained without the use of the shell model.

For boundaries with Σ values exceeding $\Sigma = 17$, the minima in the related E_Σ vs δV_Σ curves (similar to the ones shown in Fig. 3) were obtained by means of the "block-relaxation procedure" described in Ref. 17. The results thus obtained for $\Sigma \le 65$ are shown in Fig. 4. For comparison, our earlier results for MgO have also been included in Fig. 4. According to Fig. 4, MnO, FeO, CoO, and NiO show the same smooth $E_\Sigma(\theta)$ curve as MgO. Also, for angles of $\theta \gtrsim 23°$, the energy levels off and becomes independent of θ. These similarities between the transition-metal

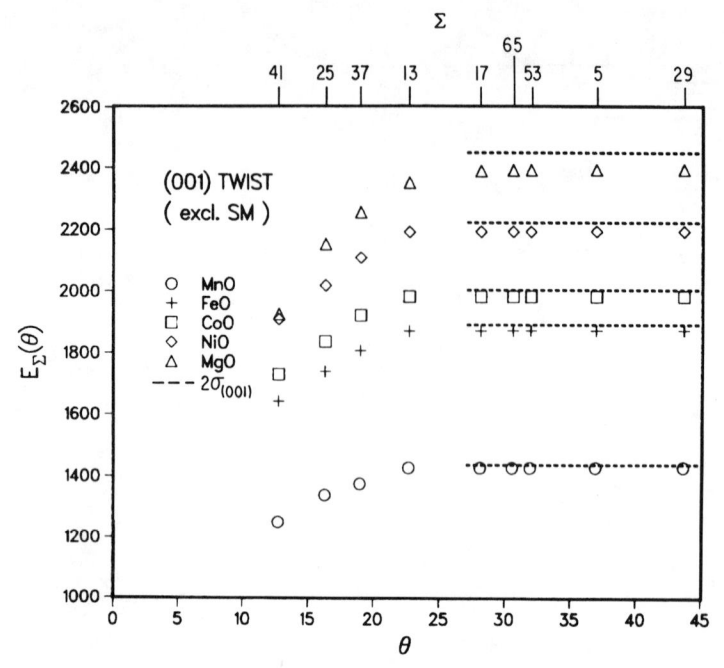

Fig. 4. Relaxed energies (in erg/cm^2) of (001) twist boundaries for $\Sigma \leq 65$ and for the free (001) surface for the four transition-metal oxides and for MgO as a function of the twist angle θ.

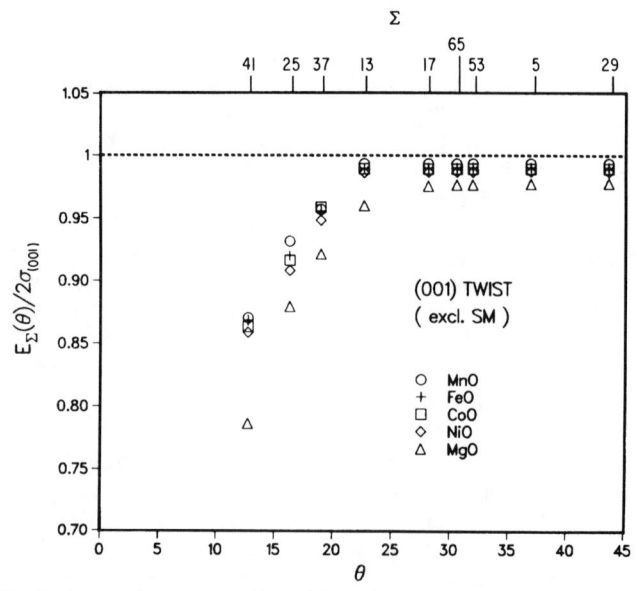

Fig. 5. Grain-boundary energies of Fig. 4 normalized to the energy, $2\sigma_{(001)}$, of the two free surfaces into which the bicrystal may be decomposed.

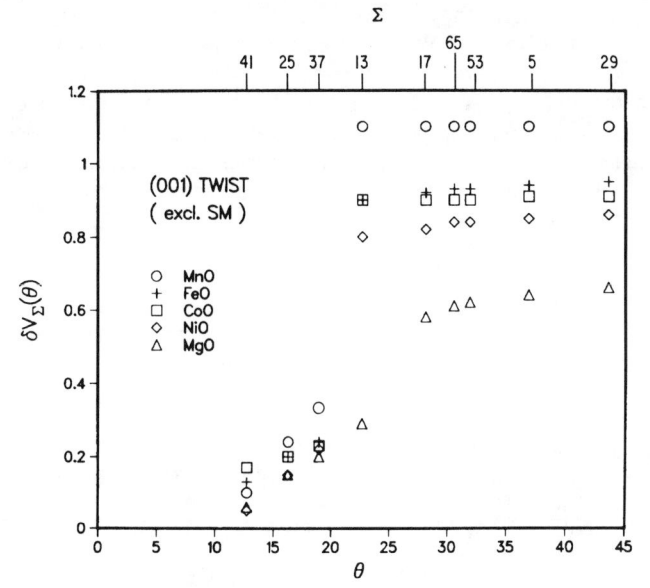

Fig. 6. Volume increase, $\delta V_\Sigma = d/d_0 - 1$, at the interface due to the introduction of a grain boundary into a crystal for the same (001) twist boundaries whose energies are shown in Figs. 4 and 5.

oxides and MgO become even more obvious when, as in Fig. 5, all boundary energies are normalized to the corresponding value of $2\sigma_{(001)}$.

The volume increase at the boundary obtained for boundaries with $\Sigma \leq 65$ is shown in Fig. 6, together with our earlier results for MgO. Similar to MgO, two regions may be distinguished according to Fig. 6: (i) For $\theta \lesssim 23°$ ("low-angle boundaries"), δV_Σ depends approximately linearly on θ. (ii) For $\theta \gtrsim 23°$ ("high-angle boundaries"), δV_Σ is practically independent of θ (or Σ).

Finally, the effect of in-boundary translations (parallel to the boundary plane) has been investigated for the $\Sigma = 5$ and $\Sigma = 13$ boundaries in NiO. Similar to our earlier results for MgO, it was found that E_Σ and $\delta(V_\Sigma)$ are almost entirely independent of the translation vector.

Discussion

As already pointed out for MgO,[16,17] two problems are of particular importance when the properties of (001) twist boundaries in compounds with rocksalt structure are discussed. These two are (i) the question of the stability (or existence) of these boundaries and (ii) the question whether or not there should be "cusps" in the $E_\Sigma(\theta)$ curve.

From our earlier calculations for MgO with two different sets of short-range interionic potentials, it was concluded that the prime reason for the weak stability of high-angle (001) twist boundaries is due to the rather weak effective Van der Waals attraction between oxygen ions on opposite sides of the interface. The same oxygen–oxygen interaction was used in the above calculations for all four transition-metal oxides. However, an increase in the stability of the high-angle boundaries was observed as one proceeds from MnO to NiO.

This apparent discrepancy, at first sight, is easily understood in terms of the total strength of the Van der Waals interaction between oxygen ions, which increases with decreasing lattice parameter if, as in the oxides considered here, all three parameters in Eq. (2) are identical for all four of them. Independent of cutoff radius, the total strength of the Van der Waals interaction is proportional to C_{--}/a_0^6, where $a_0 = a/2$ is the nearest-neighbor distance. The relative stability of high-angle twist boundaries, on the other hand, may be characterized by the "stability factor," Δ, defined by

$$\Delta = (2\sigma_{(001)} - E_{\Sigma 5})/2\sigma_{(001)} = 1 - E_{\Sigma 5}/2\sigma_{(001)} \quad (3)$$

where we have utilized the fact that the energy and the volume increase for the $\Sigma = 5$ boundary appears to be representative for all high-angle (001) twist boundaries (see Figs. 4–6). With the nearest-neighbor distances listed in Table I, values

Table II. Strength of the Effective Van der Waals Interaction, C_{--}/a_0^6, Stability Factor, Δ, and Grain-Boundary Free Volume, δV_Σ, for High-Angle (001) Twist Boundaries

	C_{--}/a_0^6 (eV)	$\Delta = 1 - E_{\Sigma 5}/2\sigma_{(001)}$ (%)	$\delta V_{\Sigma 5}$
MnO	0.151	0.65	1.10
FeO	0.200	0.97	0.94
CoO	0.214	1.06	0.90
NiO	0.246	1.27	0.85
MgO (Ref. 17)	0.328	2.25	0.64

of C_{--}/a_0^6 are readily determined. They are listed in Table II together with the values for Δ. For comparison, MgO has also been included in this table. According to these results, a steady increase in the strength of the Van der Waals interaction is accompanied by a steady increase in the stability factor, Δ, and a corresponding decrease in the grain-boundary free volume, δV_Σ. It ought to be pointed out, however, that in all five oxides considered here, *high-angle* (001) twist boundaries are only marginally stable in that the energy difference for an ion in the grain boundary and in the free surface is of the order of the thermal energy kT at temperatures around or only slightly above room temperature. Nevertheless, the above results confirm our earlier conclusion for MgO that the Van der Waals attraction between oxygen ions is mainly responsible for the weak cohesion of (001) bicrystals. The absence of Coulomb attractions between the two semiinfinite crystals that form the bicrystal is also thought to be the reason for the observed phenomenon that E_Σ is practically the same, independent of translations parallel to the boundary plane (see also Ref. 17).

Similar to our MgO results, Figs. 4–6 suggest the distinction between low-angle boundaries (for $\theta \lesssim 23°$) and high-angle boundaries (for $\theta \gtrsim 23°$). As illustrated by the properties of the $\Sigma = 13$ boundary (with $\theta = 22.62°$), however, the angle for which this separation can be made in a meaningful manner is somewhat arbitrary. According to Fig. 3, this boundary with its metastable configuration for lower δV_Σ values and its stable configuration for larger δV_Σ values represent a borderline case between high- and low-angle boundaries. In contrast to the transition-metal oxides, only one such configuration, namely the one corresponding to smaller volume increases at the boundary, has been found for MgO.[17] The

$\Sigma = 13$ boundary in MgO therefore behaves more like a low-angle boundary, and the discontinuity in the $\delta V_\Sigma(\theta)$ curve occurs for $\theta > 22.62°$. With decreasing Van der Waals attraction, however, the angle for this transition shifts toward lower values with the result that in CoO, FeO, and MnO only the minimum in E_Σ at the higher δV_Σ values yields a stable grain-boundary configuration.

The second question concerns the reasons for the existence or nonexistence of "cusps" in the $E_\Sigma(\theta)$ curve. According to Figs. 4 and 5, for Σ values up to $\Sigma = 65$, smooth $E_\Sigma(\theta)$ curves are obtained for all oxides considered here. Similarly smooth curves were also found for a number of alkali-halide (001) bicrystals.[27] Moreover, the fcc metals Al,[13] Cu, Ag, and Au[28] showed a similarly smooth behavior for Σ values up to $\Sigma = 73$ for a wide range of interatomic potentials. This leads one to suspect that smooth $E_\Sigma(\theta)$ curves for (001) twist boundaries might be a feature of the fcc structure in general. In particular, the observation of secondary grain-boundary dislocations near special misfit orientations in both fcc metals[2,3] and metal oxides with rock-salt structure[4,5] strongly suggests certain low-energy configurations near certain misfit angles. The reasons for this discrepancy are not understood at this time. Such reasons have been speculated about before[16,17] and will not be repeated here. It appears, however, that the resolution of this apparent conflict between experiments and calculations is one of the more challenging problems in the way of a better understanding of the physical nature of high-angle grain boundaries.

Grain-Boundary Structure

The structures of the perfect-coincidence $\Sigma = 5$ boundaries in NiO and MnO whose energies and free volumes were discussed above (see Fig. 3) are shown in Fig. 7. The left half of these figures represents a view on to the *two* lattice planes adjacent to the boundary $(x–y)$ plane while in the right half a view parallel to the boundary plane (i.e., the ion positions in the $x–z$ plane) is shown for the *four* planes nearest to the grain-boundary plane for the ions in the top quarter of the $x–y$ unit cell on the left. The displacements of the remaining ions are determined by symmetry. Anions are represented by squares and cations by circles. According to Fig. 7, the ion-relaxation displacements *parallel* to the boundary plane increase from MnO (via FeO and CoO, which are not shown) to NiO while simultaneously the displacements *perpendicular* to the boundary plane decrease slightly. Notice that all displacements are enhanced by a factor of 150.

As discussed in connection with Fig. 3, the $\Sigma = 13$ boundary in NiO may exist in a stable and a metastable configuration with free volumes of $\delta V_\Sigma = 0.8$ and 0.35, respectively. The corresponding structures are shown in Fig. 8. Although the energies associated with these two structures are rather similar (see Fig. 3), the structures are obviously very different in that the metastable configuration shows much larger relaxation displacements than the stable structure. The displacements for the latter are roughly the same as for the $\Sigma = 5$ boundary.

In MgO it was observed[17] that the magnitude of the relaxation displacements depends strongly on the twist angle in that a lower twist angle corresponds to larger relaxations (in both the structure and in the energy). The same phenomenon is observed for the transition-metal oxides although in the latter the relaxations are slightly less pronounced due to the greater rigidity of the O^{2-} ions as expressed in the shell-model parameters in the potentials of Catlow et al.[21] The dual character of the $\Sigma = 13$ boundary in NiO as both a low-angle and a high-angle boundary is consistent with these earlier observations in MgO. The smaller free volume and the much larger relaxation displacements associated with the metastable configuration

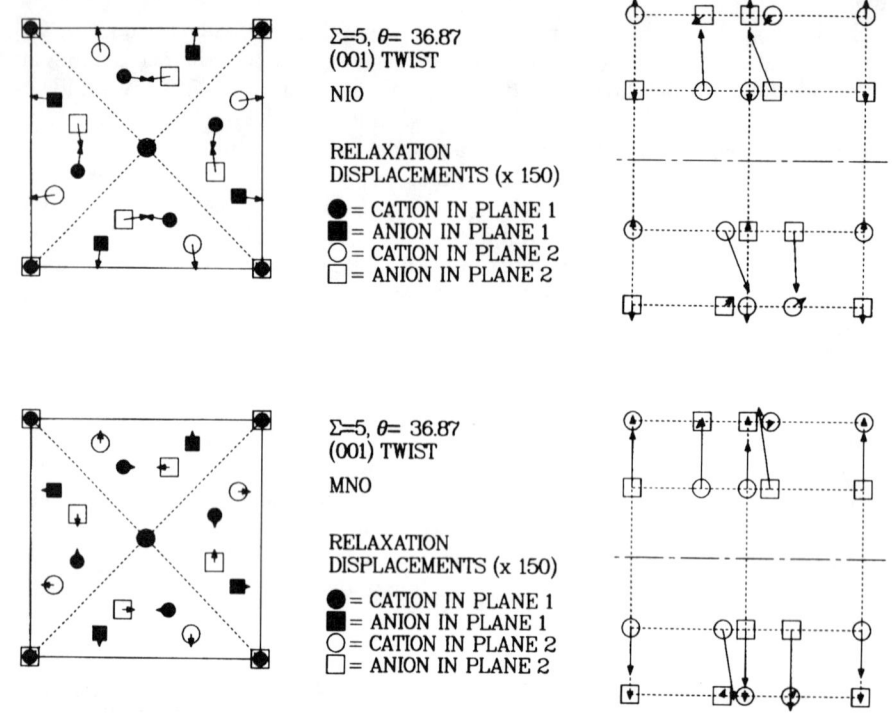

Fig. 7. Relaxed structure of the $\Sigma = 5$ perfect CSL boundary (i.e., without translation parallel to the boundary plane) in NiO and MnO. In the left half, the unrelaxed ion positions and relaxation displacements *parallel* to the boundary (x–y) plane are illustrated while the right half shows the relaxation displacements *perpendicular* to the boundary plane (i.e., in the x–z plane) for the ions in the top quarter of the CSL unit cell on the left. The relaxations for FeO and CoO (not shown) were found to lie in between those for MnO and NiO.

in Fig. 3 make it appear rather similar to a low-angle boundary. On the other hand, the larger free volume and the very small relaxation displacements associated with the stable configuration in Fig. 3 are characteristic for large-angle boundaries.[17] The $\Sigma = 13$ boundary in NiO thus represents a rather illustrative example for the different structural characteristics of low- and high-angle boundaries.

Concluding Remarks

There are two obvious limitations to our treatment of interfaces in metal oxides. First, we have chosen central-force interionic potentials although it is known that the Cauchy relation is violated for all oxides considered, thus suggesting the importance of many-body (covalent-type) interactions between the ions. However, to some degree the effective Van der Waals interaction in Eq. (2) does take into account a covalent (although nondirectional) attraction between oxygen ions, and this limitation may not be quite as severe as it appears at first sight.

The second limitation arises from our inability to consider CSL unit cells

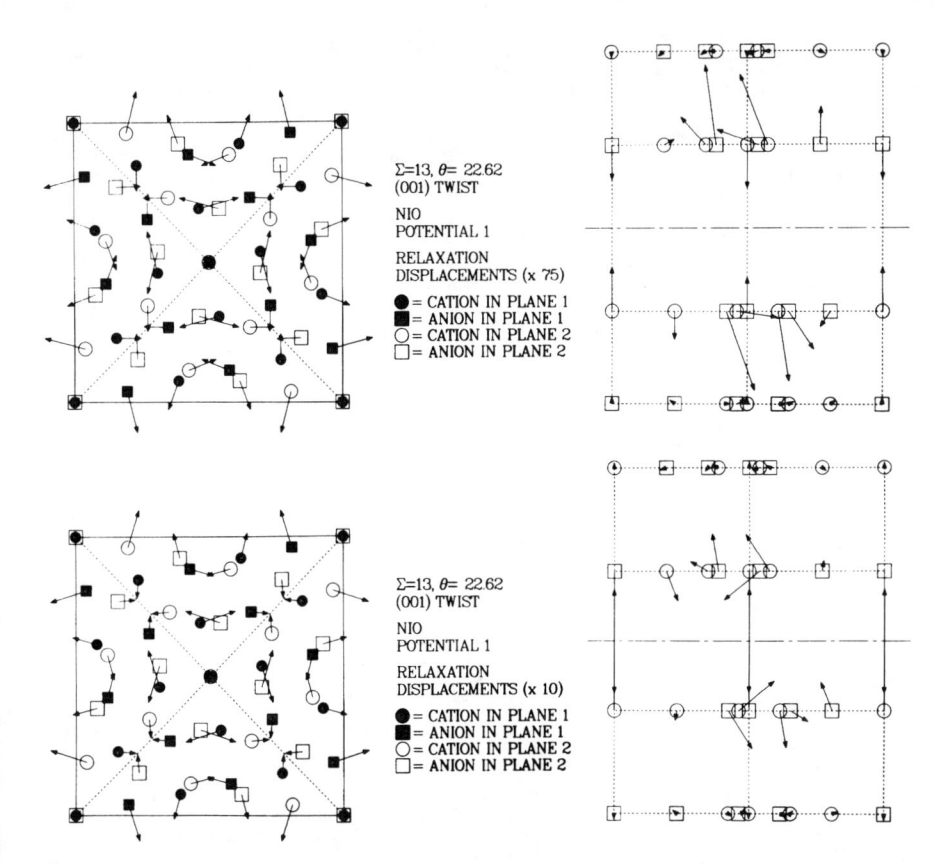

Fig. 8. Relaxed structure of the stable (top half) and metastable (bottom half) $\Sigma = 13$ perfect CSL boundary in NiO whose energy as a function of the volume increase at the boundary was discussed in connection with Fig. 3. Notice that all arrows in the top part represent 75 times the actual relaxation displacements while an enhancement factor of 10 was applied in the bottom part (for further details, see the caption to Fig. 7).

containing more than typically about 700 ions with presently available computers. In practice, we are thus limited to perfect CSL boundaries with only moderately high Σ values, typically $\Sigma \lesssim 65$. For a detailed analysis of the vicinity of certain "special" low-Σ orientations, much larger Σ values would have to be considered. At this time we are, therefore, limited to investigate only the bottoms of the hypothetical "cusps." On the other hand, since the boundary energy cannot exceed the energy of two free surfaces, the maximum depth of such "cusps" is governed by the difference $E_\Sigma - 2\sigma_{(001)}$. In the transition-metal oxides and in MgO, this difference is typically of the order of 50 erg/cm^2 (5×10^{-6} J/cm^2) from which one is tempted to conclude that, if such "cusps" for (001) twist boundaries exist at all, they cannot be very deep.

References

[1] W. Bollmann, Crystal Defects and Crystalline Interfaces. Springer-Verlag, New York, 1970.

[2] T. Schober and R. W. Balluffi, "Quantitative Observation of Misfit Dislocation Arrays in Low and High Angle Twist Grain Boundaries," *Philos. Mag.*, **21**, 109–23 (1970).

[3] R. W. Balluffi, Y. Komem, and T. Schober, "Electron Microscope Studies of Grain Boundary Dislocation Behavior," *Surf. Sci.*, **31**, 68–103 (1972).

[4] C. P. Sun and R. W. Balluffi, "Secondary Grain Boundary Dislocations in (001) Twist Boundaries in MgO. I. Instrinsic Structures," *Philos. Mag.*, [Part A], **A46**, 49–62 (1982).

[5] K. Y. Liou and N. L. Peterson; pp. 189–98 in Surfaces and Interfaces in Ceramic and Ceramic-Metal Systems. Edited by J. Pask and A. Evans. Plenum Press, New York, 1981.

[6] P. Chaudhari and J. W. Matthews, "Coincidence Twist Boundaries between Crystalline Smoke Particles," *J. Appl. Phys.*, **42**, 3063–6 (1971).

[7] H. Mykura, P. S. Bansal, and M. H. Lewis, "Coincidence Site Lattice Relations of MgO-CdO Interfaces," *Philos. Mag.*, [Part A], **A42**, 225–33 (1980).

[8] G. Herrmann, H. Gleiter, and G. Barö, "Investigation of Low Energy Grain Boundaries in Metals by a Sintering Technique," *Acta Metall.*, **24**, 353–9 (1976).

[9] H. Schmid, M. Rühle, and N. L. Peterson; pp. 177–88 in Surfaces and Interfaces in Ceramic and Ceramic-Metal Systems. Edited by J. Pask and A. Evans. Plenum Press, New York, 1981.

[10] P. D. Bristowe and S. L. Sass, "The Atomic Structure of a Large Angle (001) Twist Boundary in Gold Determined by a Joint Computer Modeling and X-Ray Diffraction Study," *Acta Metall.*, **28**, 575–88 (1980).

[11] P. D. Bristowe, "Computer Modeling of Grain Boundaries in Cubic Metals," *J. Phys. Colloq.* (Orsay Fr.), **43**, C6–33–42 (1982).

[12] P. D. Bristowe and A. G. Crocker, "The Structure of High Angle (001) CSL Twist Boundaries in fcc Metals," *Philos. Mag.*, [Part A], **A38**, 457–502 (1978).

[13] D. Wolf, "Effect of Interatomic Potential on the Calculated Energy and Structure of High-Angle Coincident-Site Grain Boundaries. I. (100) Twist Boundaries in Aluminum"; to be published in *Acta Metall.*, **32**, 245–58 (1984).

[14] D. Wolf, "On the Energy of (100) Coincidence Twist Boundaries in Transition Metal Oxides," *J. Phys. Colloq.* (Orsay, Fr.), **41**, C6–142–5 (1980).

[15] D. Wolf and R. Benedek, "Energy of (001) Coincidence Twist Boundaries in Oxides with NaCl Structure"; pp. 107–13 in Advances in Ceramics, Vol. 1. Edited by L. M. Levinson. The American Ceramic Society, Columbus, OH, 1981.

[16] D. Wolf, "On the Stability of (001) CSL Twist Boundaries in MgO: A Theoretical Study," *J. Phys. Colloq.* (Orsay, Fr.), **43**, C6–45–63 (1982).

[17] D. Wolf, "Energy and Structure of (001) Coincident-Site Twist Boundaries and the Free (001) Surface in MgO: A Theoretical Study," *J. Am. Ceram. Soc.*, **67**, 1–13 (1984).

[18] D. Wolf, "Effect of Fe^{2+} Substitutional Impurities on the Stability of a $\Sigma = 5$ (100) CSL Twist Boundary in MgO: A Theoretical Study"; pp. 36–43 in Advances in Ceramics, Vol. 6. The American Ceramic Society, Columbus, OH, 1983.

[19] D. Wolf, "Formation Energy of Point Defects in Free Surfaces and Grain Boundaries in MgO," *Radiat. Eff.*, **75**, 203–9 (1983).

[20] P. W. Tasker and D. M. Duffy, "On the Structure of Twist Grain Boundaries in Ionic Oxides," *Philos. Mag.*, [Part A], **A47**, L45–48 (1983).

[21] C. R. A. Catlow, W. C. Mackrodt, M. J. Norgett, and A. M. Stoneham, "The Basic Atomic Processes of Corrosion. I. Electronic Conduction in MnO, CoO, and NiO," *Philos. Mag.*, **35**, 177–87 (1977).

[22] C. R. A. Catlow, I. D. Faux, and M. J. Norgett, "Shell and Breathing Shell Model Calculations for Defect Formation Energies and Volumes in Magnesium Oxide," *J. Phys. C: Solid State Phys.*, **9**, 419–29 (1976).

[23] C. R. A. Catlow, W. C. Mackrodt, M. J. Norgett, and A. M. Stoneham, "The Basic Atomic Processes of Corrosion. II. Defect Structure and Cation Transport in Transition Metal Oxides," *Philos. Mag.*, **40**, 161–72 (1980).

[24] P. W. Tasker, "The Surface Energies, Surface Tensions, and Surface Structure of the Alkali Halide Crystals," *Philos. Mag.*, [Part A], **A39**, 119–36 (1979).

[25] W. C. Mackrodt and R. F. Stewart, "Defect Properties of Ionic Solids: II. Point Defect Energies Based on Modified Electron-Gas Potentials," *J. Phys. C: Solid State Phys.*, **12**, 431–49 (1979).

[26] D. Wolf; pp. 13–22 in Surfaces and Interfaces in Ceramic and Ceramic-Metal Systems. Edited by J. Pask and A. Evans. Plenum Press, New York, 1981.

[27] D. Wolf, "Properties of High-Angle (001) Twist Boundaries in Alkali-Halide Bicrystals: A Theoretical Investigation"; to be published in *Philos. Mag.*, [Part A], Aug., 823–44 (1984).

[28] D. Wolf, "Effect of Interatomic Potential on the Calculated Energy and Structure of High-Angle Coincident-Site Grain Boundaries: II, (100) Twist Boundaries in Cu, Ag, and Au," *Acta Metall.*, **32**, 735–48 (1984).

---

*Work supported by the U.S. Department of Energy.

Grain-Boundary Structure in Al$_2$O$_3$

C. B. Carter and K. J. Morrissey

Cornell University
Department of Materials Science and Engineering
Ithaca, NY 14853

The factors affecting the structure of grain boundaries in Al$_2$O$_3$ polycrystalline materials are discussed. Illustrations are presented on the basal twin ($\Sigma = 3$), the rhombohedral twin ($\Sigma = 8$), and two different $\Sigma = 13$ grain boundaries. The CSL/O lattice models are helpful in understanding the predominant grain-boundary plane, but it is suggested that kinetic rather than energetic factors may cause other facets to lie parallel to densely packed planes in the oxygen sublattice rather than parallel to planes containing either a high density of coincident lattice sites or O points. The DSC lattice can be used to predict the allowed grain-boundary dislocation Burgers vectors. The structures of the boundaries examined are shown to influence segregation to, diffusion along, and migration of the boundary. Illustrations of grain boundary/pore breakaway are presented for a $\Sigma = 3$ and a $\Sigma = 13$ boundary.

The present state of understanding of structure/property relations of grain boundaries in ceramics is similar to the situation that existed for semiconductors until quite recently. In the semiconductor field, many observations had been made on electrical properties of grain boundaries without knowing the structure: it is now realized that the structure is important. In the field of ceramics, observations have been made on anomalous grains and segregation to grain boundaries without knowing the structure of the grain boundaries involved. In this paper, it will be shown that many different types of special grain boundaries do exist in polycrystalline compacts of Al$_2$O$_3$ and that these boundaries will have properties that depend both on the type of grain boundary and on the particular plane it adopts. It has been suggested by other workers that a thin film of amorphous material is present at the interface between two grains and that this causes or is associated with the faceting of grain boundaries.[1,2] Such films are not present in the boundaries discussed here; a detailed criticism of such observations and their interpretation is presented elsewhere.[3]

In relating grain-boundary structure to properties such as migration of, segregation to, and diffusion along grain boundaries in Al$_2$O$_3$, it is important to know the structure of the particular interface. For example, some grain-boundary planes can be expected to show little segregation and to move very slowly relative to other facet planes; illustrations of slow-moving, clean facets would be the common basal plane on a basal twin and the common rhombohedral plane on a rhombohedral twin. If a glassy phase were present at the boundary, segregation to and diffusion along that interface would not depend on the structure of the interface, although migration of the interface would still be affected if particular densely packed Al$_2$O$_3$ facet planes developed.

The present paper is concerned with discussing factors which dictate the structure of special grain boundaries in Al$_2$O$_3$ and will be illustrated by four experimentally observed examples, namely, the basal twin, the rhombohedral twin, and two different boundaries for which $\Sigma = 13$. The notation and terminology used throughout this paper will follow that presently employed in the discussion of grain boundaries in metals and elemental semiconductors. This terminology is based on Bollman's O lattice theory,[4,5] which has been extended and elucidated by Smith and Pond,[6] Warrington,[7] and others. When two identical interpenetrating crystal lattices are rotated with respect to one another through certain angles, a large fraction of lattice sites in one crystal may be coincident with the same large fraction of lattice sites in the other crystal. This lattice of coincident sites is referred to as the coincident site lattice (CSL), and the inverse of the fraction of coincident lattice sites is defined to be Σ. A difficulty encountered with the CSL model is that the two adjoining grains are often translated relative to one another so that there are no coincident lattice sites. Wagner, Tan, and Balluffi,[8] therefore, proposed the generalized CSL in which the coincident lattice sites in the two grains need neither be identical nor be occupied by atoms. It is in this sense that the CSL model will be used in the present paper.

The CSL concept can be extended to include coincidence of all equivalent points in the two grains. Such points are known as O points and define the O lattice. The O lattice can in fact consist of a lattice of parallel lines or of parallel planes.[5] The grain boundary is now formed by inserting a plane through the two interpenetrating lattices and removing all the lattice points of one crystal on one side of the plane and conversely on the other side. A third lattice, which can be constructed, is defined by the three-dimensional array of displacement vectors by which one crystal can be moved relative to the other without changing the physical geometry of the boundary; this lattice is known as the DSC lattice (DSC indicates "displacement shift complete"). It can be described[6] as the coarsest lattice that can be constructed to contain both individual crystal lattices in the coincidence orientation.

Since the aim of this paper is to illustrate the relation between the structure and properties of grain boundaries in Al_2O_3, emphasis will be placed on using the results of structure studies rather than presenting a detailed crystallographic analysis. For full details of the TEM techniques used in this study the reader is, therefore, referred elsewhere (see e.g., Refs. 9–13).

Experimental Procedures

A wide range of high-purity grade polycrystalline materials have been examined and are broadly characterized in this paper as extruded and sintered or as hot-pressed. One of the sintered materials has been examined earlier by Carter, Kohlstedt, and Sass[9] and more recently by Hansen and Phillips.[1] Throughout this paper, emphasis is placed on the general features of certain faceted, special grain boundaries that can occur in any type of polycrystalline alumina. The materials are impure by metallurgical standards, containing up to 2000 ppm of dopants, but this fact will not affect the crystallographic arguments presented here although it will affect the local chemical composition and may cause certain facets to occur more frequently than might otherwise be the case.

It is not the intention of this program to characterize any individual manufacturer's product. The hot-pressed material had a typical grain size of $\sim$2–5 μm and, therefore, contained a large number of grains that on a subjective comparison tended more frequently to be faceted. The sintered material generally had a larger grain size of 10–20 μm, which was useful in the identification of facet planes along a particular grain boundary. This material also contained a large number of pores, and an amorphous phase was often observed at the triple point. The largest grain materials examined were several translucent aluminas taken from unused Na vapor lamp envelopes. Few special boundaries were identified in such materials and two reasons can be suggested for this, first, that since the grain size is large (typically > 30 μm), fewer boundaries are present in each sample, and second, since extensive grain growth has occurred, the presence of facets which was often used in this study to locate special grain boundaries was seldom detected, thus making any such boundaries even less readily identifiable.

One of the extruded and sintered materials was that examined by Hansen and Phillips,[1] who refer to it as a liquid-phase sintered material. During the present study many special, structured, high-angle grain boundaries have been observed in this material by the present authors, and preliminary reports have been given elsewhere.[10-12] It is important to realize that a commercial material can vary considerably from sample to sample and even within the same sample. Caution should, therefore, be used when observations on a particular material are interpreted where the chemical composition is critical to that interpretation: this is not the case where special structures of grain boundaries are being examined unless details of segregation and other properties are also being determined.

Samples of commercial Al_2O_3 containing small quantities of second-phase particles were cut by using a diamond saw and mechanically polished to a thickness of less than 75 μm. Disks were cut from these slices by using an ultrasonic drill* and thinned to perforation by using an ion miller† with an argon gas and a 5-kV operating voltage. The samples were coated with a thin layer of carbon to ensure electrical conductivity in the microscope and were examined with a transmission electron microscope‡ operating at 125 kV.

All of the materials examined contained K and Mg in second-phase particles, which consisted primarily of β'''-Al_2O_3.[14] Some Na was also present in such samples but was primarily found in large regions of amorphous phase along with Si and Ca. The hot-pressed material contained little amorphous material but did include ½–1-μm size particles, which contained intentionally added Ni.

Experimental Observations and Discussion

Grain-Boundary Structures

The simplest grain boundary in Al_2O_3 that can be used to illustrate the concepts of the CSL and DSC lattice is the mirror or glide basal twin discussed by Kronberg[15] and illustrated for the $11\bar{2}0$ projection in Fig. 1. The lattice sites are shown in Fig. 1(c) for the two crystals (represented by circles and stars for the upper and lower crystals, respectively). The CSL is then defined by the overlapping of the circles and stars. The work on grain boundaries in metals would suggest that preferred grain-boundary planes would be those that contain a high density of coincident lattice sites (or rather of O lattice points, which are equivalent in this situation). Thus, the predicted preferred grain-boundary planes are (0001), $\{11\bar{2}0\}$, $\{\bar{1}010\}$, and $\{\bar{1}012\}$, in that order. These planes have all been observed in this study and are each illustrated in Fig. 2. Statistics are limited at present, but in this study (0001) planes have been most frequently observed. This observation, however, does not "prove" that such a plane has the lowest energy, since as will be shown below, it is predicted to be the slowest moving facet and will, therefore, in time predominate, independent of energy considerations.

The DSC lattice is defined by vectors 1 and 2 and a third vector (e.g., ⅓[$1\bar{1}00$]) inclined to the plane of the figure. If one lattice is translated with respect

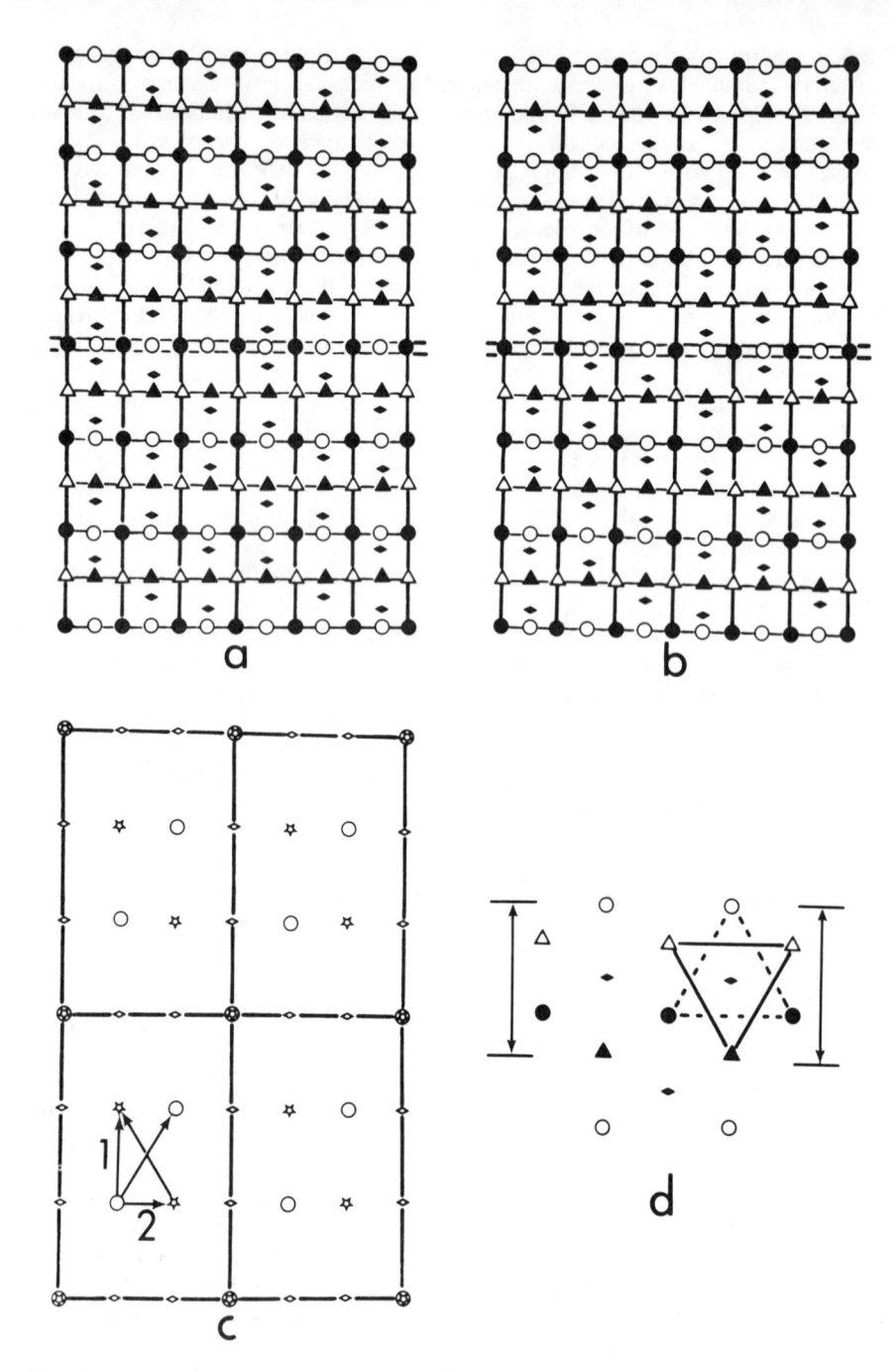

Fig. 1. Schematic representations of two basal twin boundaries in Al$_2$O$_3$. In a–c, the atoms are projected onto the (1$\bar{2}$10) plane, and the symbols are defined in the (0001) projection in d: (a) mirror twin, (b) glide twin, (c) generalized CSL, the DSC that consists of circles, stars, and open diamonds, and the four shortest DSC vectors in this plane.

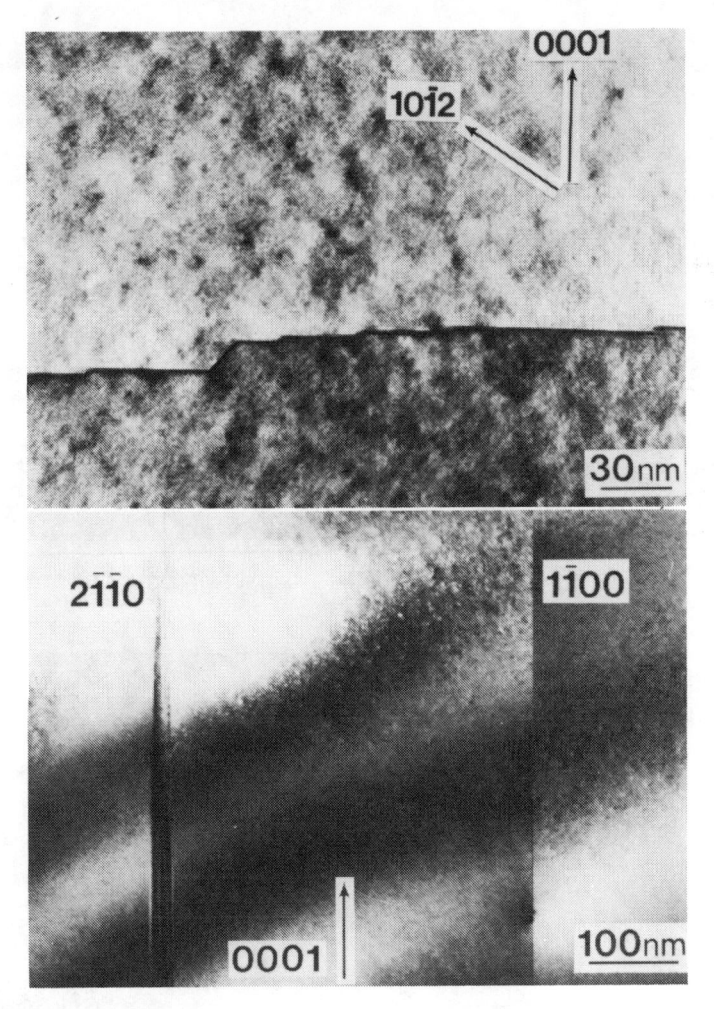

Fig. 2. Faceted basal twin boundaries in Al$_2$O$_3$.

to the other by such a vector, the CSL is reproduced but is itself shifted. Thus, the plane of the boundary will step. This result is an important one and will be found for less simple boundaries. Since a translation by a DSC vector does not change the structure of the boundary, such vectors are the Burgers vectors of the intrinsic grain-boundary dislocations. Any deviation from the exact CSL orientation will be accommodated by so-called secondary dislocations whose Burgers vectors are DSC vectors (usually, but not necessarily, the shortest).

The number of dislocations present in a basal twin interface can be large as illustrated in Fig. 3. This figure also emphasizes that grain-boundary dislocations can occur on any facet plane. Those illustrated in Fig. 3 are the $\{11\bar{2}3\}$ and $\{11\bar{2}0\}$ planes with (0001) plane being parallel to the electron beam. Dislocation interactions do occur and can be understood by the DSC construction: an example of a widely spaced dislocation network on a twin boundary is shown in Fig. 4. Dislocation Burgers vectors of $\frac{1}{3}\langle 10\bar{1}0\rangle$, $\frac{1}{3}[0001]$, and $\frac{1}{3}\langle 10\bar{1}1\rangle$ have all been

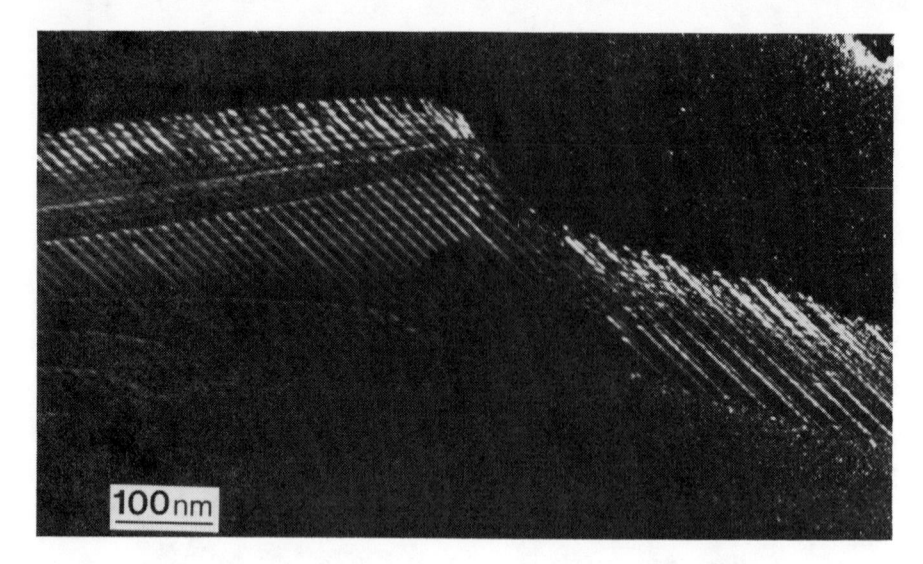

Fig. 3. High density of dislocations in a basal twin boundary.

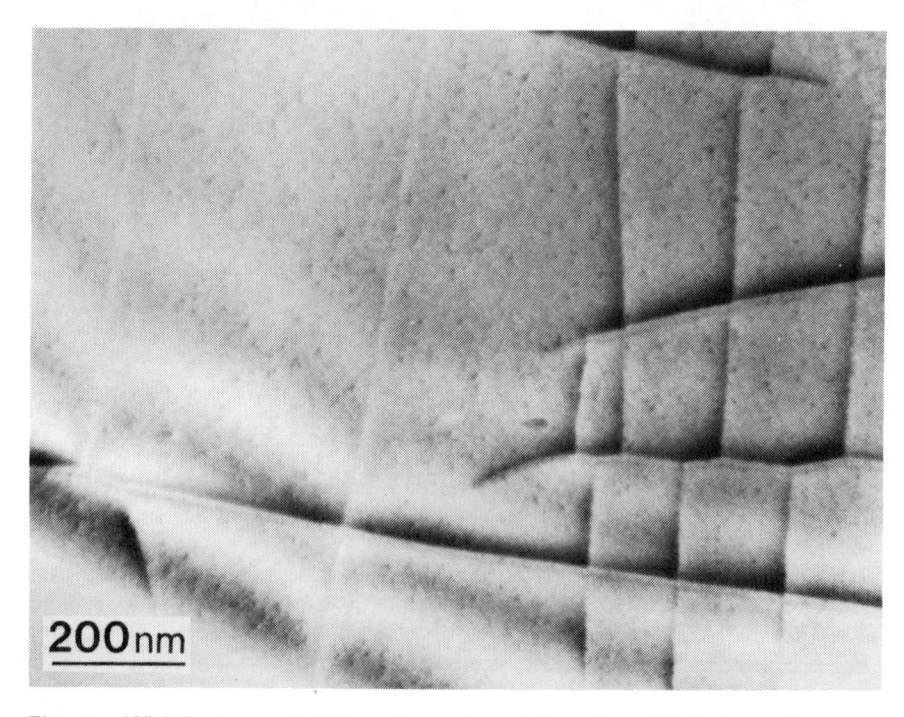

Fig. 4. Widely spaced dislocation network in a basal twin boundary.

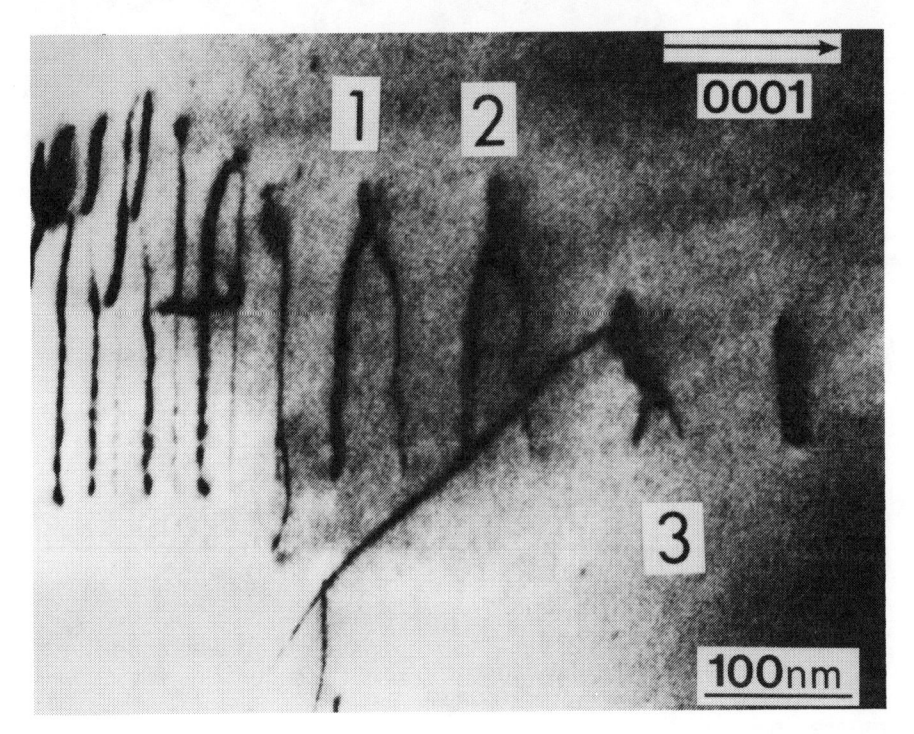

Fig. 5. Dissociation of lattice dislocations on entering a basal twin boundary.

experimentally observed for twin boundaries. The $\frac{1}{3}\langle 10\bar{1}1\rangle$ dislocation is special since it is a perfect dislocation both in the twin boundary and in one (but only one) of the grains. The final illustration of dislocations in a basal twin is shown in Fig. 5 where a $\frac{1}{3}[11\bar{2}0]$ lattice dislocation is seen to dissociate into $\frac{1}{3}[10\bar{1}0]$ and $\frac{1}{3}[01\bar{1}0]$ twin boundary dislocations. This process is predicted by the DSC model and is directly analogous to the dissociation of $\frac{1}{2}\langle 110\rangle$ dislocations in $\Sigma = 3$ twins in fcc metals. A particularly interesting feature of the dissociation shown in Fig. 5 is the fact that the twin plane is $11\bar{2}0$ and the dislocations are, therefore, dissociated by a pure climb mechanism which is related to that observed for lattice dislocations (see e.g., Ref. 16).

The next example is that of a near-CSL boundary, which can contain a special mirror plane but does involve distortion of the oxygen ion sublattice. The rhombohedral twin in Al_2O_3 has previously been investigated by Heuer[17] and Scott[18] and is essentially the same as the rhombohedral twin in Fe_2O_3 (hematite), which has been discussed by Bursill and Withers.[19] A CSL model can be constructed for this case and is shown in Fig. 6, where only the lattice sites are included by using the same projection as shown in Fig. 1. The boundary can be described as a near-CSL, $\Sigma = 8$. The row of coincident lattice sites on the left defines the $(1\bar{1}02)$ plane while those along the other $(1\bar{1}02)$ plane are actually near-CSL sites. The diffraction pattern from a rhombohedral twin viewed in this orientation is shown in Fig. 7 together with a strong-beam dark-field image recorded by using the common $(1\bar{1}02)$ reflection. The major facet lies parallel to the common $(1\bar{1}02)$ plane, as

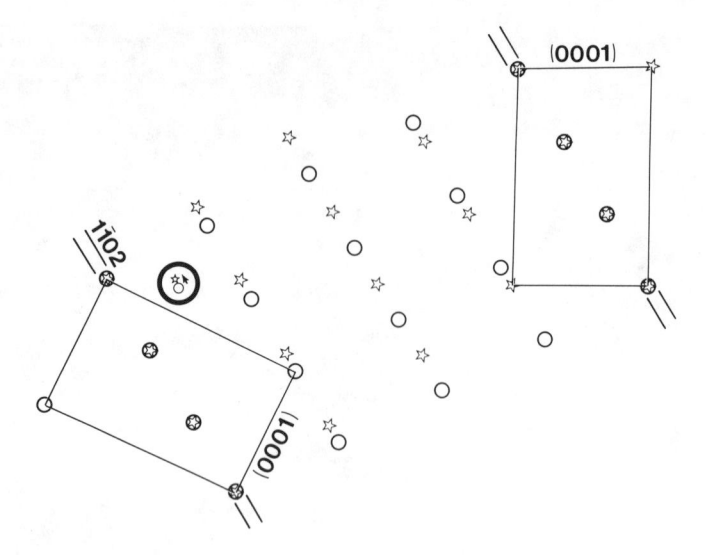

Fig. 6. Schematic illustration of the origin of the (near) CSL corresponding to the rhombohedral twin in Al_2O_3. For simplicity, only the relative orientations of the two unit cells and the position of the lattice sites are shown. The stars and circles identify the two crystals, and the superimposed motifs represent the CSL sites. The ($1\bar{1}02$) twin planes are defined by two close parallel lines.

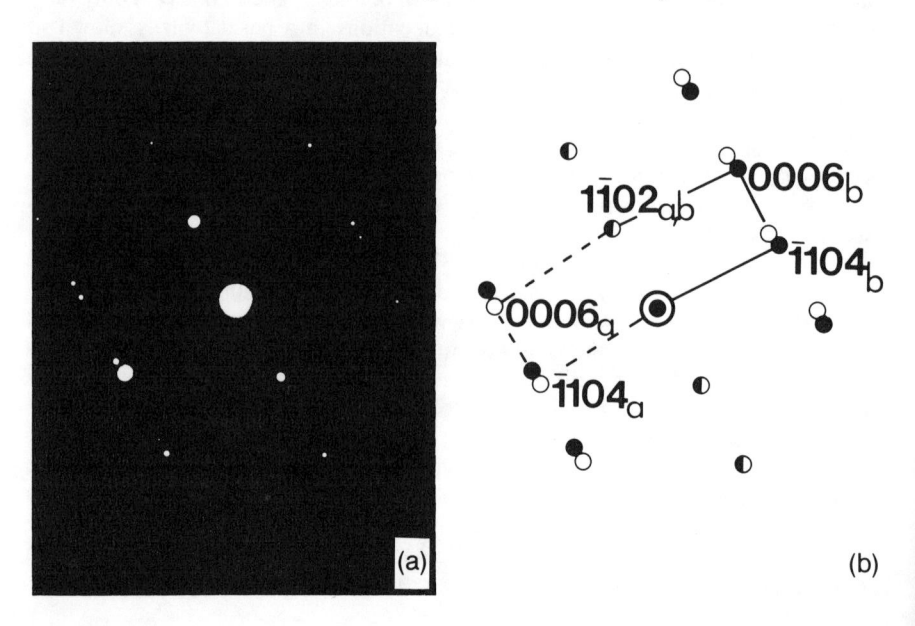

would be expected from Fig. 6, since it contains a high density of coincident lattice sites. (It is in fact an O plane for the mirror case.[5]) The other facet in Fig. 7 lies parallel to the basal plane in one grain. The basal plane also contains a high density of coincident lattice sites, as seen in Fig. 6, but this is actually not the second most favored plane according to the density of coincident sites arguments. This observation, therefore, implies that the basal plane facet is particularly favored even when a higher density of coincident lattice sites could be achieved. It will be suggested below that this result may be due to kinetic rather than energetic considerations.

The aluminum ions do not directly influence the type of allowed boundary plane, but they do influence the local atomic arrangement, as shown schematically in Fig. 8. These two configurations have been proposed by Bursill and Withers[19] for hematite and by Scott[18] for Al_2O_3; other variations are also possible, and the influence of impurity ion segregation which can accommodate local non-stoichiometry must be considered. The arrangement of the aluminum ions may, of course, influence the stability of particular boundary configuration and will probably also influence the choice of facet plane (cf. the basal twin).

The DSC lattice for the rhombohedral twin can be used to deduce the allowed Burgers vectors of the disloctions in the rhombohedral twin interface. The shortest DSC vector is shown by the arrow in Fig. 6 and connects two lattice points that are not in the plane of the figure (circled). When a rhombohedral twin contains such a dislocation, the boundary plane must step to the adjacent rhombohedral

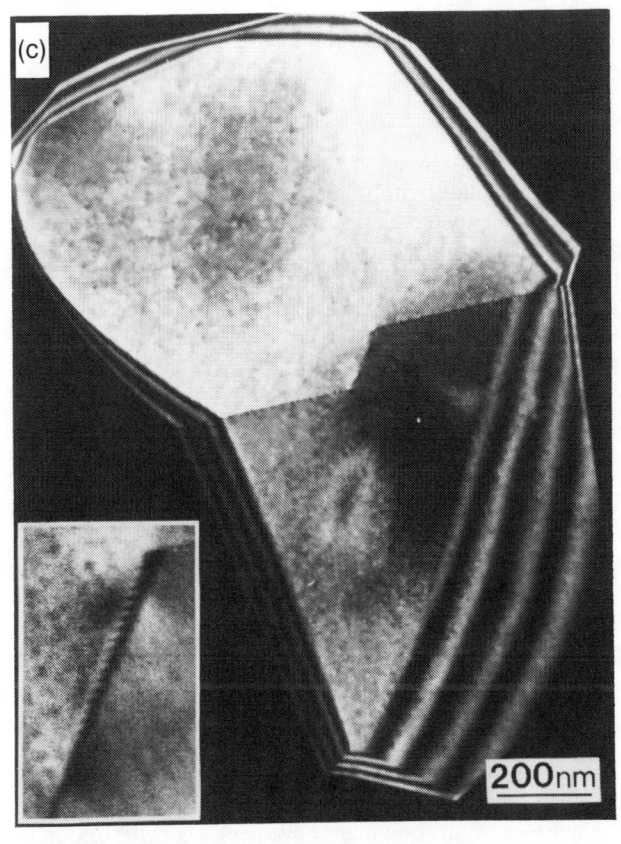

Fig. 7. (a) Diffraction pattern and (b) schematic of (a) for a rhombohedral twin in Al_2O_3. Open circles correspond to lower grain, and (c) shows a strong-beam image of the interface.

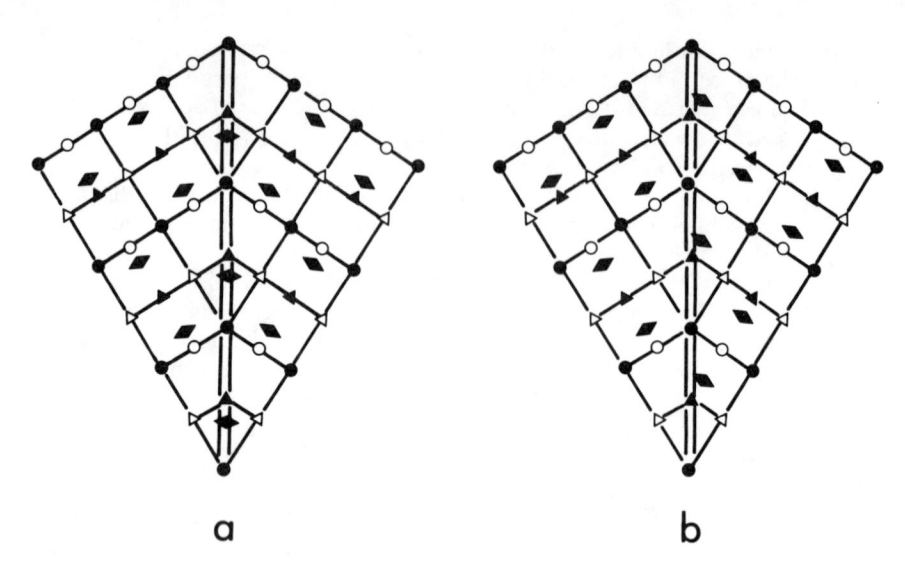

Fig. 8. Two of the possible configurations of the rhombohedral twin in Al_2O_3. The twin plane is defined by a pair of verticle lines (symbols are those used in Fig. 1).

plane which contains these new coincident lattice sites. Such a dislocation has been reported for a rhombohedral twin in hematite by Bursill and Withers,[19] who used a geometrical model to explain the observation of a step in a lattice-fringe image. Since the Burgers vector of such a dislocation is $(1/24)\langle 10\bar{1}1\rangle$, i.e., 0.064 nm, it will be very difficult to detect such dislocations by diffraction contrast experiments; however, the presence of such dislocations will influence the properties of the rhombohedral twin.

Thus far, it has now been shown that, in the case of the only two previously recognized special grain boundaries in Al_2O_3, the boundary plane can be explained by the CSL model with the proviso that densely packed planes in the oxygen sublattice appear to be particularly favored. The DSC lattice has been shown to correctly predict the observed Burgers vectors. These conclusions hold for the basal twin which has an exact CSL ($\Sigma = 3$) and the rhombohedral twin which has a near CSL ($\Sigma = 8$). The following two illustrations have been chosen to demonstrate that other exact CSL and near-CSL boundaries exist in polycrystalline Al_2O_3 and that the above general concepts apply to these boundaries also.

An exact CSL for which $\Sigma = 13$ can be formed by rotating one crystal with respect to the other through an angle of 27.8° about the [0001] axis. Such a boundary will be referred to as $\Sigma = 13$, [0001], 27.8° (Σ, axis, angle notation). The (0001) plane shown in Fig. 9 is a schematic to show the lattice sites superimposed in the same plane such that the coincident lattice sites can be identified. This type of special grain boundary can be identified relatively easily, since the $\langle 11\bar{2}0\rangle$ direction in one crystal is nearly parallel to the $\langle 10\bar{1}0\rangle$ direction in the other crystal. Five examples of this boundary have been characterized in this study, and one of these is illustrated in Fig. 10. This illustration is particularly interesting because the boundary separates into a $\Sigma = 39$ boundary plus a $\Sigma = 3$ basal twin along part of its length. The boundary lies exactly parallel to the common basal

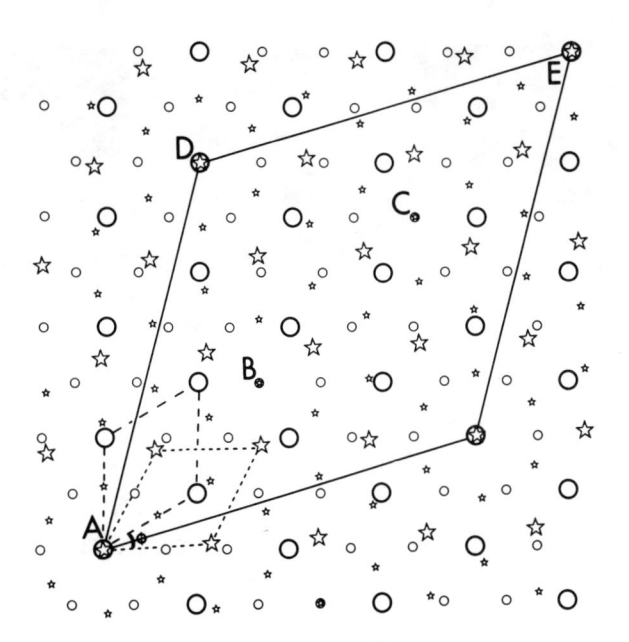

Fig. 9. Schematic representations of the $\Sigma = 13$ twist boundaries in Al_2O_3. The circles and stars define the oxygen ion sites in the two crystals; for this representation the sites are coplanar. The larger circles and stars define lattice sites in the two Al_2O_3 lattices. The two unit cells illustrate the rotation (ϕ) used to produce the boundary. The superimposed large motifs A, D, and E, are coincident lattice sites.

plane over part of its length, as shown in Fig. 10, but then facets over the remainder. When the special boundary changes from the $\Sigma = 13$ to the $\Sigma = 39$ configuration, the faceted structure shows no obvious change, which is consistent with the proposal that the facet planes in boundaries other than the basal twin are primarily dictated by the oxygen ion sublattice, which is unaffected by the addition of a mirror or glide basal twin.

The fourth type of special boundary to be discussed in this paper is illustrated in Fig. 11. It appears at first sight to be a "typical" high-angle boundary with fine fringes and faceting being apparent at the higher magnification. However, as shown by the diffraction pattern in Fig. 12, it is actually a special boundary formed by a rotation of 57.5° about the $\langle \bar{1}100 \rangle$ axis or by mirroring across the $(11\bar{2}3)$ plane. The simplest configuration for this twin interface can be understood by considering the $\bar{1}100$ projection shown in Fig. 12 and then constructing the mirror twin as illustrated in Fig. 13(A). This twin on the $(11\bar{2}3)$ plane is actually analogous to the $(10\bar{1}1)$ twin that has been observed in hcp metals.[20] The actual oxygen and aluminum ions along this boundary plane, which are shown as partially overlapping circles, would most likely occupy the mean sites and thus introduce a small distortion. This configuration is not the only one possible: shifting one lattice with respect to the other by any of the vectors shown in Fig. 13(B) would produce one of several other configurations each of which would not, however, contain a mirror plane. The grain boundary shown in Fig. 13 is actually a near-CSL, $\Sigma = 13$

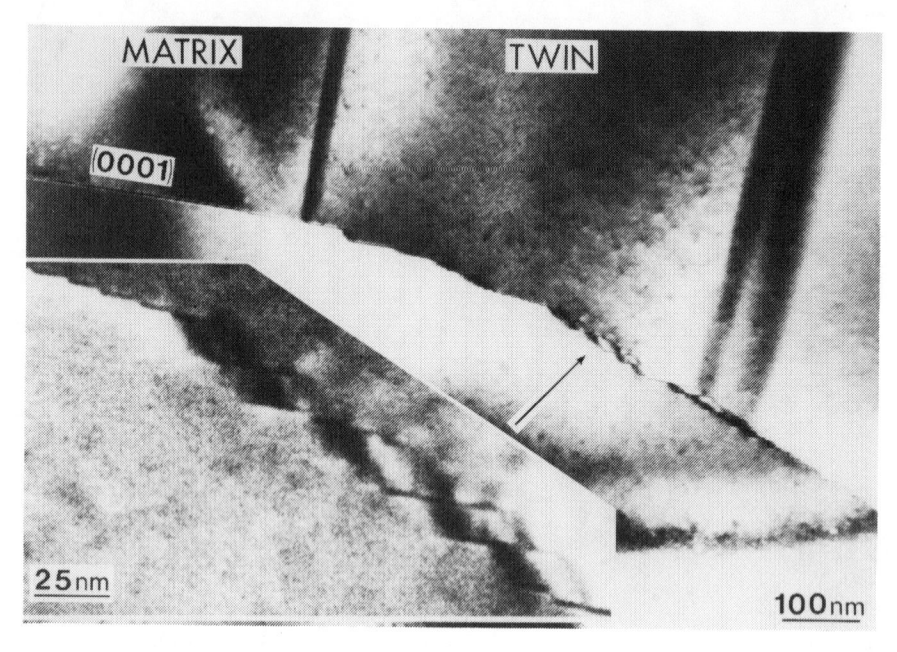

Fig. 10. $\Sigma = 13$ boundary in Al_2O_3 that has partly separated into a $\Sigma = 39$ plus a $\Sigma = 3$ boundary.

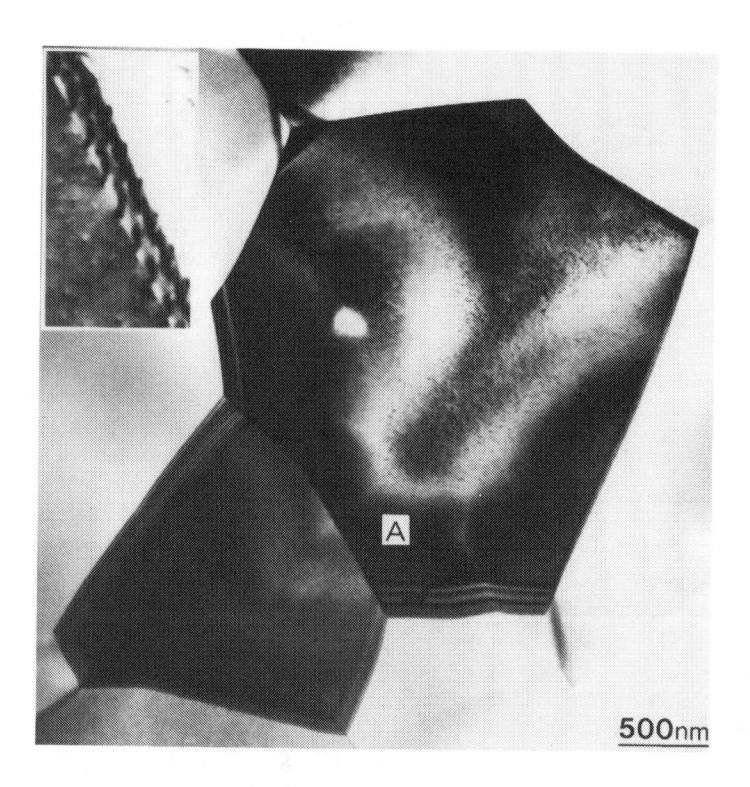

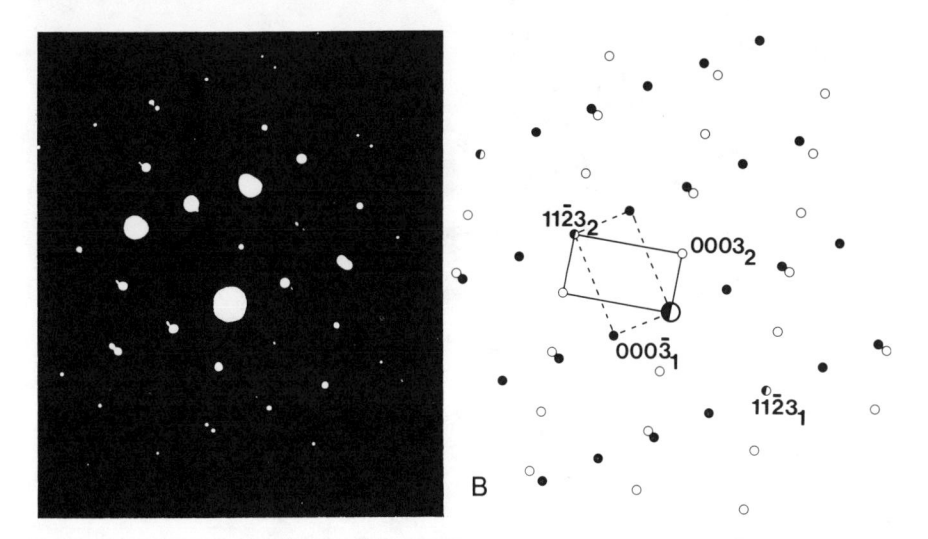

Fig. 11. $\Sigma = 13$, near-CSL boundary formed by a 57.5° rotation about the [$\bar{1}100$] direction or by twinning across the ($11\bar{2}3$) plane: (A) strong-beam image (insert is from area A); (B) diffraction pattern and schematic indexed for the twinning operation.

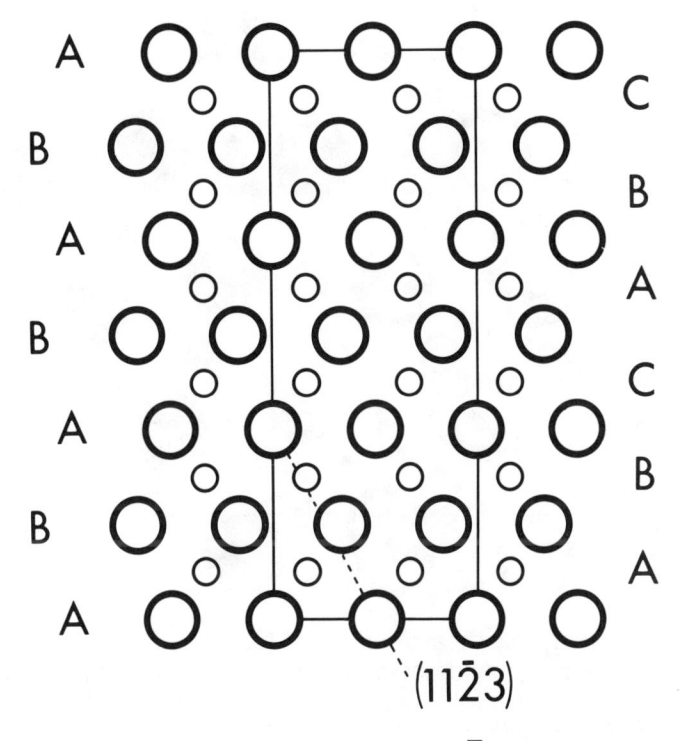

Fig. 12. View of the Al_2O_3 structure down the $\langle\bar{1}100\rangle$ direction. Large and small circles represent oxygen and aluminum ions, respectively.

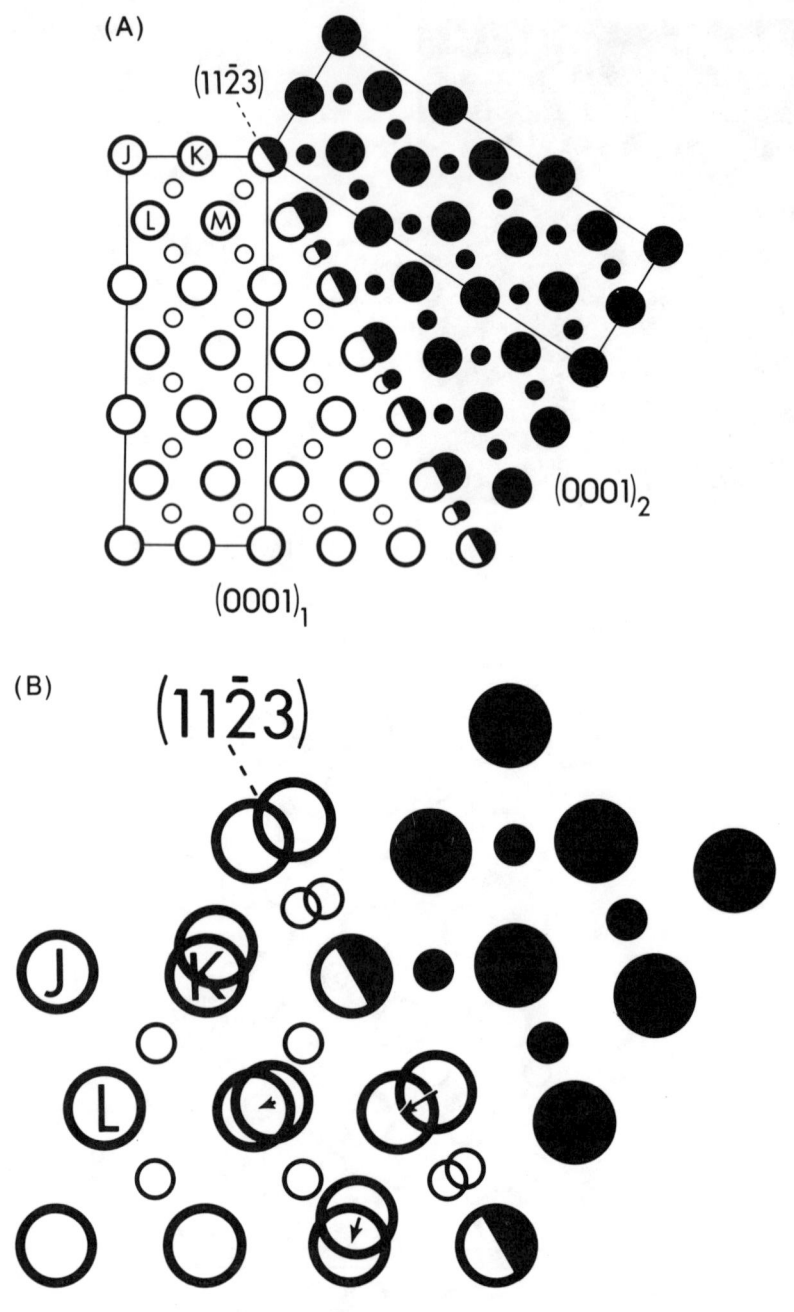

Fig. 13. (A) Mirror configuration of the (11$\bar{2}$3) twin viewed in the ⟨$\bar{1}$100⟩ direction. Overlapping circles indicate the distortion expected for this interface. Large and small circles represent oxygen and aluminum ions, respectively. (B) Illustration of how different small shifts of the two lattices could change the relative positions of the generalized coincident lattice sites in the (11$\bar{2}$3) twin.

boundary. Such a special angle/axis pair in Al_2O_3 has also been deduced by Fortunee[21] using computer simulation.

Structure/Property Relations for Special Boundaries

The properties of special grain boundaries in Al_2O_3 are very strongly related to the structure of the interface. In particular, it will now be briefly demonstrated that the migration and chemical composition (i.e., effects of impurity segregation) of special grain boundaries depends both on the type of boundary (i.e., $\Sigma = 3$, $\Sigma = 13$, etc.) and on the grain-boundary plane (e.g., compare the common (0001) to others for the basal twin).

The basal twin (glide or mirror geometry) is again a model case, as illustrated in Fig. 14, since only cations need be considered. In order for the $\{1\bar{1}00\}$ facet to move, only the aluminum ions need move and the required movement is shown by arrows. It should be noted that, in the plane shown, only 4 of the 12 Al^{3+} ions are required to change sites to translate the vertical segment a distance $a\langle 10\bar{1}0\rangle$ or equivalently to move the horizontal segment a distance of 13 Å (1.30 nm) and that this will most likely be a cooperative effect. It should also be recognized that this boundary motion will occur as a recrystallization process: it does not distort the oxygen ion sublattice. A similar model can be constructed for the rhombohedral twin, as shown in Fig. 15. The step shown in Fig. 15 is a basal facet which moves the boundary from one $\{\bar{1}012\}$ plane to a parallel one without the necessity of including dislocations. (A model incorporating dislocation dipoles can still be used). The presence of dislocation-free steps thus allows the grain boundary to move faster than would otherwise be possible (see e.g., Ref. 22). In the experimental observation of a rhombohedral twin shown in Fig. 7, the basal facet was 130-nm long: this observation and many similar observations on the basal twin and other special boundaries imply that although single steps are possible,[19] larger steps

a   b

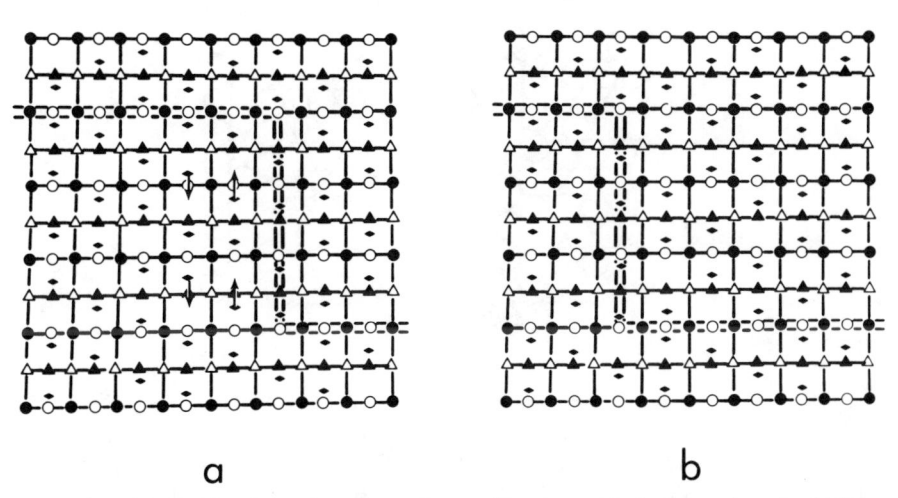

Fig. 14. Model for the migration of a perfect basal twin boundary in Al_2O_3. The arrows in (a) show the Al ion movement required to move the interface from that shown to the position in (b).

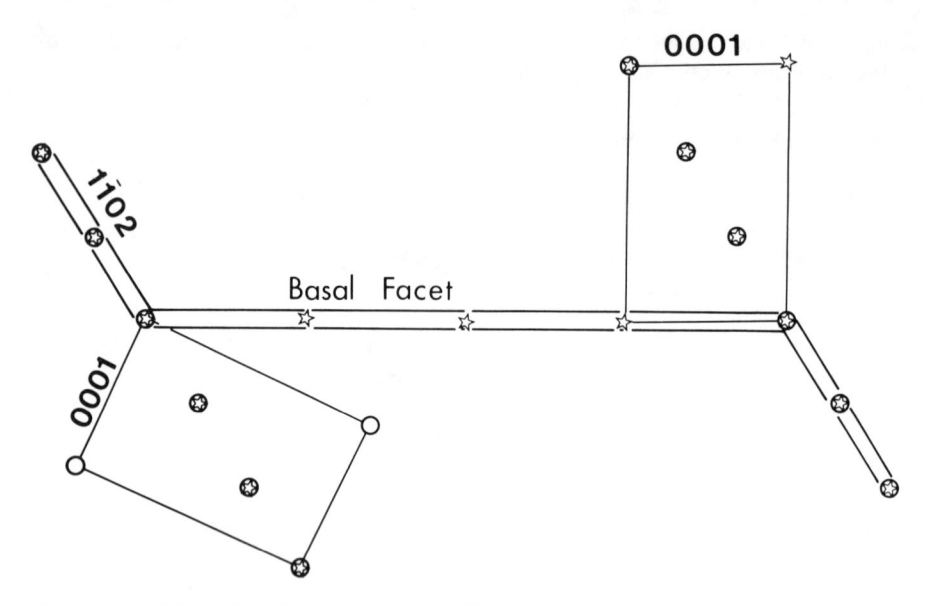

Fig. 15. Dislocation-free step on a ($1\bar{1}02$) rhombohedral twin plane that is proposed as a mechanism for the rapid migration of the rhombohedral twin.

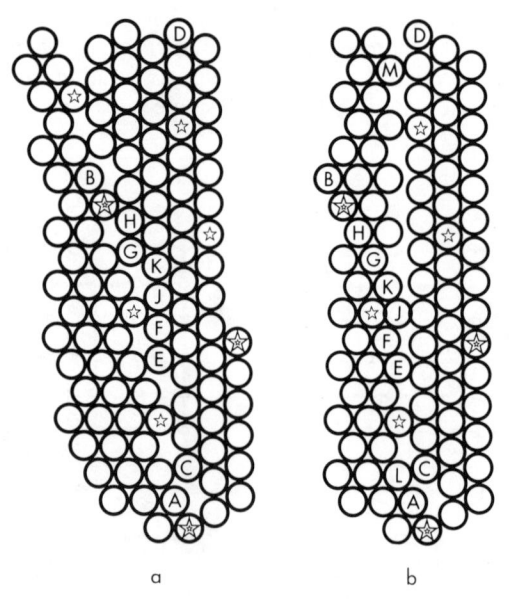

Fig. 16. Two possible configurations for a nonbasal facet on a $\Sigma = 13$, [0001], 27.8° boundary. If the interface is parallel to the rotation axis, then in (a) it will be a symmetric tilt boundary while in (b) it will be an asymmetric one: in such an asymmetric tilt boundary, the $\{10\bar{1}0\}$ plane of the left grain (LJ and KM) and the $\{11\bar{2}0\}$ plane of the right grain (CD) will be nearly parallel. A higher density of coincident lattice sites will lie on the boundary if in (b) it is not pure tilt.

are favored. A similar observation has been made in Mg–Al spinel,[23,24] and the proposed explanation is that the situation is analogous to the formation of superjogs on a dislocation rather than a series of single jogs or to the formation by polygonalization of a tilt boundary; i.e., a lowering of the strain energy is achieved by such special alignment (i.e., coalescence) of smaller facets.

When the migration of a nonbasal plane facet on a $\Sigma = 13$, [0001], 27.8° boundary such as that shown in Fig. 16 is considered, it can be appreciated that the packing of oxygen ions must be reduced at the boundary (cf. the $\Sigma = 3$ incoherent twin facets in Mg–Al spinel[23,24]) and that the Al^{3+} ion packing must be similarly reduced. Thus, migration of the boundary by the jumping of oxygen ions from one grain to the other (e.g., E to K) will be facilitated by this decrease in packing density. Also, as illustrated for the analogous $\Sigma = 7$, [0001], 22.8° boundary by Shackleford[25], this boundary facet will be a preferred site for large cation impurities, which are always present in polycrystalline Al_2O_3 (ionic radii for Al^{3+}, Mg^{2+}, Ca^{2+}, Na^+, and K^+ are 0.53 Å (0.053 nm), 0.72 Å (0.072 nm), 1.00 Å (0.100 nm), 1.02 Å (0.102 nm), and 1.38 Å (0.138 nm), respectively[25]). It is clear that such large cations must either diffuse along the boundary or be dragged along when it migrates. Segregation of large-ion impurities to the boundary must further decrease the Al^{3+} concentration at the boundary. Figure 16 actually illustrates a second reason for the more densely packed planes being favored in Al_2O_3 even when a plane with higher density of coincident lattice sites could be adopted; i.e., that densely packed planes will be favored by the migration process since they are the slower moving planes.

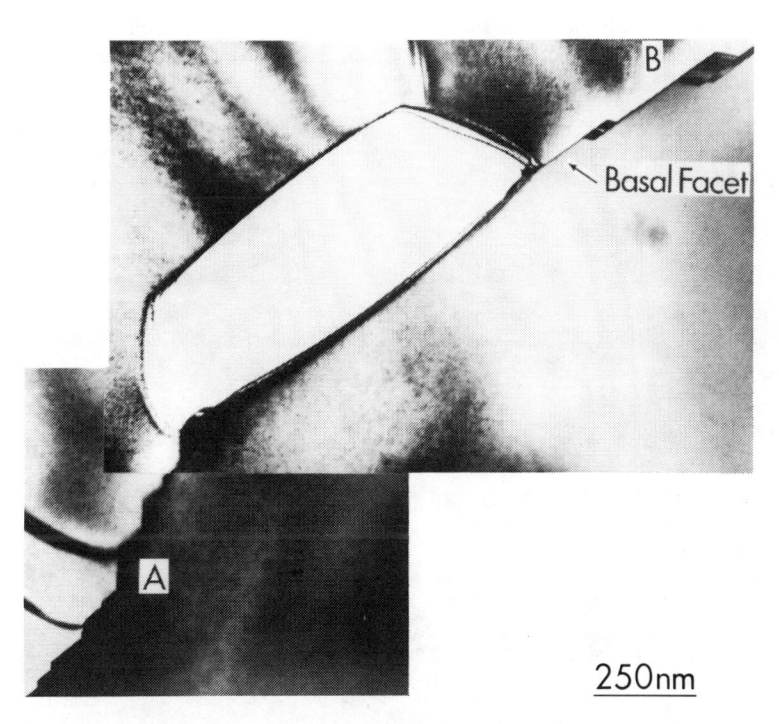

Fig. 17. $\Sigma = 13$, [0001], 27.8° boundary that may be in the process of breaking away from a pore.

The ease with which a certain special boundary can move may be directly related to the ability of that boundary to break away from a pore, second-phase particle, or small grain, and each of these situations will now be briefly discussed. A major difficulty in discussing the breakaway phenomena is that all the observations are made on static boundaries and the breakaway process is then inferred. Pores are identified in a transmission electron micrograph of Al_2O_3 by the steep sides of the holes rather than tapered sides produced by ion thinning. The correlation of the end of the pore with the long facet of the twin boundary is directly related to this breakaway. The boundary breaking away from the pore in Fig. 17 is a $\Sigma = 13$, [0001], 27.8° boundary and illustrates directly the relationship between the faceted nature of the boundary and the method of breakaway. Breakaway occurs by the facets moving along the boundary; i.e., the basal plane facet does not move forward, but rather the less densely packed facet moves along it. The second illustration of this phenomena is shown in Fig. 18 where the boundary is a $\Sigma = 3$ basal twin and breakaway has already occurred. The observations reported in Figs. 17 and 18 each show a faceted boundary, while in Fig. 19, which shows a grain boundary interacting with a second-phase particle, the boundary did not appear to be faceted on a detectable scale. This new boundary could not be identified as a special boundary and curves in much the same manner as envisaged by Reynen[27] and Sakarcan, Hsueh, and Evans[28] for pores. Clearly, the nature of the boundary interacting with a pore must be taken into account. The balance between surface diffusion and grain-boundary diffusion at a boundary/pore intersection has been discussed by Reynen[27] and will only be briefly illustrated here. If a facet is particularly mobile and the grain boundary is special (i.e., low Σ and, hence, low energy; see discussion of this assumption by Wolf[29]), it will be relatively easy for the grain boundary to become detached. The situation for a second-phase particle

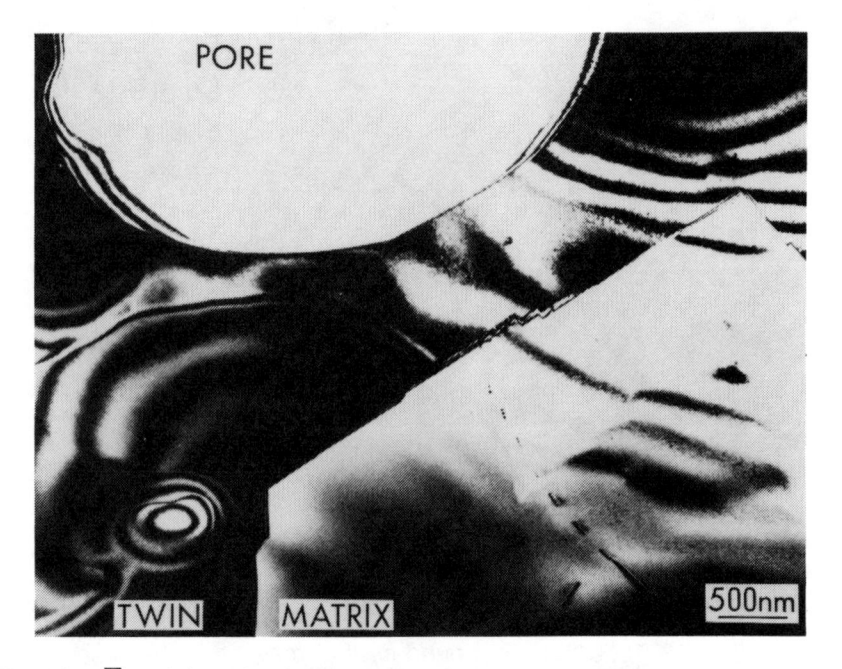

Fig. 18. $\Sigma = 3$ basal twin boundary that has broken away from a pore.

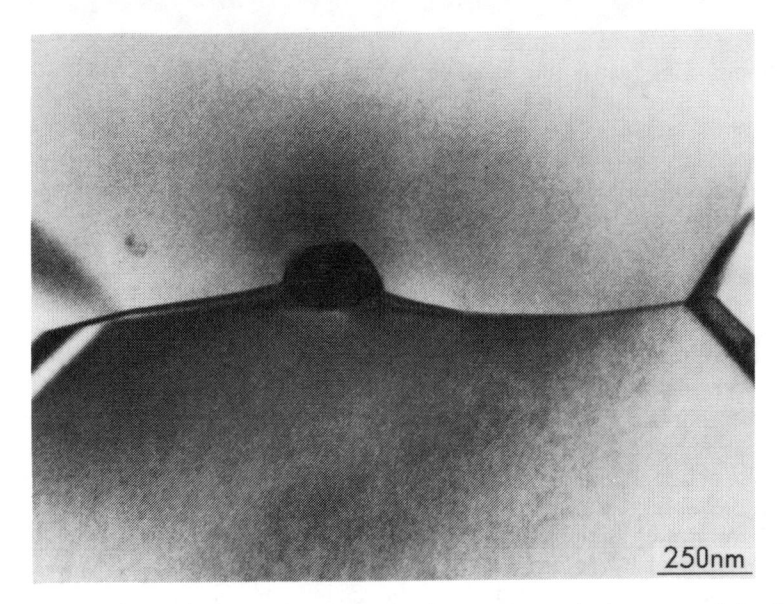

Fig. 19. High-angle boundary attached to a second-phase particle. EDS showed that this type of particle is Ni or NiO.

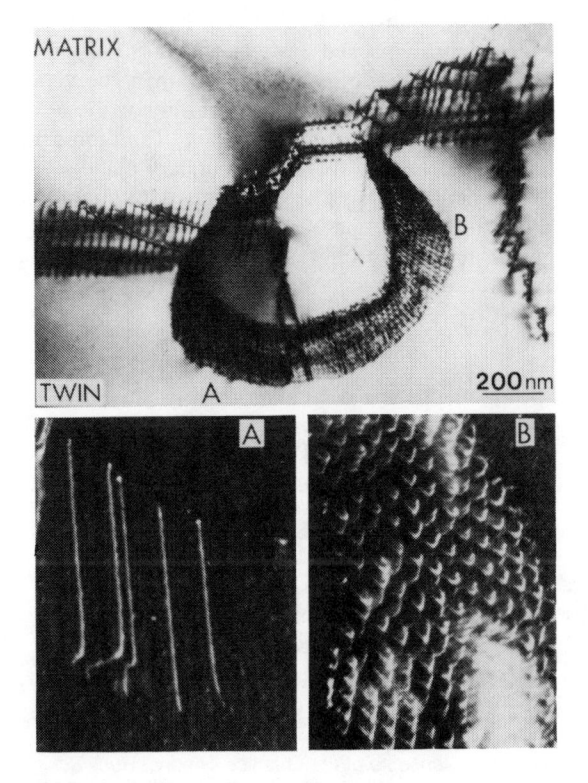

Fig. 20. Basal twin boundary interacting with a small grain.

or small grain is more complex because diffusion around the obstacle will also be anisotropic. This fact is illustrated in Fig. 20 by a $\Sigma = 3$ basal twin boundary interacting with a small grain. The twin boundary is faceted, and the boundary enclosing the small grain necessarily changes its character on different segments of the boundary: two different areas are shown in the enlargements and illustrate the change from tilt to twist character. Movement of the low-angle grain boundary requires long-range oxygen ion diffusion while movement on the basal twin only requires short-range hopping of the Al^{3+} ions although dislocations present on the twin boundary will slow its migration.

Conclusions

The concepts of the CSL, O lattice, and DSC lattice can be applied to Al_2O_3.

The primary plane adopted by special grain boundaries contains the highest number of coincident lattice sites, as might be predicted by analogy with nonionic materials, and the DSC lattice can be used to understand the Burgers vectors of dislocations observed in the grain boundaries.

The mirror or glide basal twin is unique since the oxygen ion sublattice is to a first approximation undisturbed by the presence of the grain boundary. In this case, the facet plane is dictated by the aluminum ion sublattice.

In the other boundaries examined, the preferred boundary plane is one where best matching of the oxygen ions occurs and other facets tend to lie parallel to densely packed planes in the oxygen ion sublattice in one crystal or the other, even though this contains a lower density of coincident lattice points than is possible.

The preference for densely packed oxygen ion planes may be associated with migration of the grain boundary, i.e., kinetic rather than energetic factors.

The structure of a grain boundary depends not only on the relative rotation of the two grains but also on the plane of the boundary. Therefore, properties of the grain boundary will also depend on the grain-boundary plane.

Examples of this dependence have been illustrated for migration of the boundary and by implication for segregation of impurities to the boundary. Specifically, a facet with an open structure will migrate faster than a densely packed one and will be associated with a high concentration of impurity ions.

Acknowledgments

The authors acknowledge the assistance of Ray Coles, who maintains the electron microscope facility of the Materials Science Center at Cornell. J. Porter kindly provided the hot-pressed material. This research program is supported by the Department of Energy under Contract No. DE-AC02-82ER12076.

References

[1]S. C. Hansen and D. S. Phillips, "Grain Boundary Microstructures in a Liquid-Phase Sintered Alumina (α-Al_2O_3)"; to be published in *Philos. Mag.*
[2]M. P. Harmer, "The Use of Solid Solution Additives in Ceramics Processing"; these proceedings.
[3]K. J. Morrissey and C. B. Carter, "On the Identification of Grain Boundary Liquid Phases in Al_2O_3"; unpublished work.
[4]W. Bollman; "Crystal Defects and Crystalline Interfaces." Springer-Verlag, New York, 1970.
[5]W. Bollman, "The Basic Concepts of the O-lattice Theory," *Surf. Sci.*, **31**, 1–11 (1972).
[6]D. A. Smith and R. C. Pond, "Bollman's O-lattice Theory; A Geometrical Approach to Interface Structure," *Int. Met. Rev.*, **No. 205** (1976).

[7]D. Warrington, "Formal Geometrical Aspects of Grain Boundary Structure," in Grain Boundary Structure and Kinetics," ASM, 1–12 (1979).

[8]W. R. Wagner, T. Y. Tan, and R. W. Balluffi, "Faceting of High-Angle Grain Boundaries in the Coincidence Lattice," *Philos. Mag.*, **29**, 895–904 (1974).

[9]C. B. Carter, D. L. Kohlstedt, and S. L. Sass, "Electron Diffraction and Microscopy Study of the Structure of Grain Boundaries in Al_2O_3," *J. Am. Ceram. Soc.*, **63** [11-12] 623–7 (1980).

[10]K. J. Morrissey and C. B. Carter, "Faceted Interfaces in Al_2O_3"; Paper No. 1-B-81 presented at the 83rd Annual Meeting of the American Ceramic Society. Washington, DC, 1981.

[11]K. J. Morrissey and C. B. Carter, "Faceted Twin Boundaries in Al_2O_3"; pp. 343–4 in the Proceedings of the International Conference on Electron Microscopy. Hamburg, 1982.

[12]K. J. Morrissey and C. B. Carter, "Dislocations in Twin Boundaries in Al_2O_3"; pp. 85–95 in Advances in Ceramics, Vol. 6. Edited by M. F. Yan and A. H. Heuer. The American Ceramic Society, Columbus, OH, 1983.

[13]C. B. Carter and S. L. Sass, "Electron Diffraction and Microscopy Techniques for Studying Grain Boundary Structure," *J. Am. Ceram. Soc.*, **64** [6] 335–45 (1981).

[14]K. J. Morrissey and C. B. Carter, "Analysis of Second-Phase Particles in Al_2O_3," *Mater. Sci. Res.*, **15**, 297–307 (1983).

[15]M. L. Kronberg, "Plastic Deformation of Single Crystals of Sapphire: Basal Slip and Twinning," *Acta Metall.*, **5**, 507–24 (1957).

[16]T. E. Mitchell, "Application of Transmission Electron Microscopy to the Study of Deformation in Ceramic Oxides," *J. Am. Ceram. Soc.*, **62** [5–6] 254–67 (1979).

[17]A. H. Heuer, "Deformation Twinning in Corundum," *Philos. Mag.*, **13**, 379–93 (1966).

[18]W. D. Scott, "Rhombohedral Twinning in Aluminum Oxide"; pp. 151–66 in Deformation of Ceramic Materials. Edited by R. C. Bradt and R. E. Tressler. Plenum, New York, 1965.

[19]L. A. Bursill and R. L. Withers, "Twinning Dislocations in Haematite Iron Ore," *Philos. Mag., [Part A]*, **A40**, 213–32 (1979).

[20]A. Dubertret and A. LeLann, "Development of a New Model for Atom Movement in Twinning," *Phys. Status Solidi A*, **A60**, 145–51 (1980).

[21]R. P. Fortunee, "Grain Boundary Structure in α-Al_2O_3"; M.S. Thesis, Case Western Reserve University, Cleveland, OH, 1981.

[22]J. P. Hirth, "Defect Structures in Grain Boundaries", *Acta Metall.*, **22**, 1023–31 (1974).

[23]T. M. Shaw and C. B. Carter, "Faceting of Twin Boundaries in Spinel," *Scr. Metall.*, **16**, 1431–6 (1982).

[24]K. M. Ostyn and C. B. Carter, "Structure of Interfaces in Cubic Oxides"; pp. 44–58 in Ref. 12.

[25]J. F. Shackleford, "A Canonical Hole Model of Grain Boundaries in Oxides", pp. 96–101 in Ref. 12.

[26]W. D. Kingery, H. K. Bowen, and D. R. Uhlmann; Introduction to Ceramics. Wiley, New York, 1976.

[27]P. Reynen, "The Impact of Sintering Theory on Practical Powder Metallurgy," *Mater. Sci. Res.*, **13**, 355–75 (1980).

[28]M. Sakarcan, C. H. Hsueh, and A. G. Evans, "Experimental Assessment of Pore Breakaway During Sintering," *J. Am. Ceram. Soc.*, **66** [6], 456–61 (1983).

[29]D. Wolf, "Effect of Fe^{2+} Substitutional Impurities on the Stability of a $\Sigma = 5$, (100) CSL Twist Boundary in MgO: A Theoretical Study"; pp. 36–43 in Ref. 12.

---

*Raytheon Ultrasonic Impact Grinder, Model 100, Londondary, NH.
†Technics Inc., Alexandria, VA.
‡Siemens 102, Siemens AG, Karlsruhe, Federal Republic of Germany.

Electron Diffraction and Microscopy Studies of the Structure of Grain Boundaries in NiO

J. Eastman, F. Schmückle, M. D. Vaudin, and S. L. Sass

Cornell University
Department of Materials Science and Engineering
Ithaca, NY 14850-0121

The structure of tilt grain boundaries in NiO was studied with an electron microscopy technique that is sensitive to the change in mean inner potential at the core of dislocations. Boundaries possessing a variety of rotation axes were examined. In each case, the boundary was observed to be faceted. It is suggested that in NiO a tilt boundary with an arbitrary rotation axis will be faceted. By using the experimentally determined rotation axis, angle, and boundary plane, Bollmann's O lattice theory was applied in an attempt to determine the geometry and dislocation content of the faceted boundaries. For certain of the boundaries, the predictions of the theoretical analysis agreed with all details of the observed geometry and dislocation content, while for other boundaries, disagreements were noted. The origin of the discrepancies is not understood at present. The magnitude of the expansion and the structural width of [001] twist boundaries in NiO were estimated by using a new electron diffraction technique. The average expansions were 2% for a $\theta = 7°$ boundary (θ is the misorientation angle) and 18% for a $\theta = 22°$ boundary. Comparison of results from NiO and Au $\sigma = 22°$ boundaries indicated that there are large differences in boundary expansion in these two materials.

In recent years a considerable amount of experimental information has been obtained about the structure of grain boundaries in metals, in particular, concerning the defects lying in the plane of the boundary. Electron microscopy techniques have been used to study the dislocation structure of the boundary.[1] For large-angle boundaries, where the period of the ordered structure can be small, electron and X-ray diffraction techniques have been used to detect periodicities in the boundary plane[2] and in certain cases to determine the boundary structure.[3] It is of considerable interest to obtain detailed information on the structure of boundaries in ceramic materials. The important questions that need to be answered about the structure of boundaries in ceramics include the following: (1) What is the atomic structure of large-angle boundaries? (2) What is the influence of nonstoichiometry and impurity segregation on boundary structure? (3) What is the frequency of occurrence of faceting of boundaries? What is the detailed structure of faceted boundaries, and in particular, for low-angle boundaries, what is the dislocation content of the facets? Can the observed facet structures be understood in terms of theoretical models of boundary structure? (4) What is the local expansion at a boundary and the structural width of a boundary? The present paper will be concerned with answering the last two questions. It is worthwhile to begin by discussing the reasons these questions are important.

It has been suggested that the presence of steps, or facets, in grain boundaries plays an important role in the mechanisms of grain-boundary sliding[4] and migration.[5,6] As the first step in relating faceting with a property such as grain-boundary sliding, it is necessary to possess information on the detailed structure of the boundary. While it is easier to study simple boundaries such as pure tilt and twist, it is more relevant to concentrate on boundaries with a mixed character when structure and properties are related. Since this is experimentally difficult to accomplish, tilt boundaries in NiO with increasingly high-index rotation axes were studied with the goal of extrapolating these results to the structure of completely arbitrary tilt boundaries. This study is the subject of the first part of this paper.

Information on the magnitude of any expansion at a grain boundary could help in the understanding of properties such as grain-boundary diffusion as well as the cohesive strength of boundaries. Experimental information about the structure of boundaries along the direction normal to the interface is relatively rare. Electron[7] and X-ray[8] diffraction techniques have recently been used to measure the structural width of grain boundaries, i.e., the extension of the region of large strain associated with a grain boundary in the direction normal to the interface. In the second part of this paper, it will be demonstrated that an electron diffraction technique can be used to estimate the expansion as well as the structural width of twist boundaries in NiO. The first results from a study using this approach will be discussed.

Faceting of Tilt Boundaries in NiO

Experimental Procedure

The NiO used in the faceting study was in the form of coarse-grain rods, which were produced with the Verneuil technique at Argonne National Laboratory, starting with 99.999% NiO powder. There is no information available at present about the concentration of impurities at the grain boundaries in these rods. The first step in the preparation of electron microscope specimens was to cut 3-mm diameter cylinders from bulk material with an ultrasonic cutting tool. Disks with thicknesses of 200–300 μm were then cut from the cylinders with a diamond saw. The disks were ground to a thickness of ~100 μm using an extremely careful polishing procedure, with the edges of the specimen protected by a hard wax. Specimens for the electron microscope were produced by ion thinning. The transmission electron microscopy studies were performed at either 100 or 125 kV.* The specimens were aligned in the microscope so that the dislocations associated with the grain boundary were exactly parallel to the electron beam. The grain boundary was imaged in the bright-field mode, using conditions that were as kinematical as possible. The observed contrast is mainly due to the change in mean inner potential at the dislocation core.[9] In this imaging mode, changes in image contrast can be due to variations of both the magnitude of the dislocation Burgers vector and its edge component.

Experimental Observations

Bright-field images of five different small-angle tilt grain boundaries with a range of rotation axes (**w**) and angles (θ) are shown in Fig. 1. These boundaries were imaged in the overfocused condition, and the black spots correspond to dislocations that are viewed in the end-on orientation. The boundary structure is also shown schematically in each case. It is seen that the boundaries are faceted and that the facet configuration is complex in some cases. For example, in Fig. 2, which is a shallow angle view along the $\theta = 12.5°$, **w** = [127] boundary in Fig. 1(e), a complicated double-facet structure can clearly be seen. The short facets have a length of ~7.0 nm and contain dislocations with a spacing of

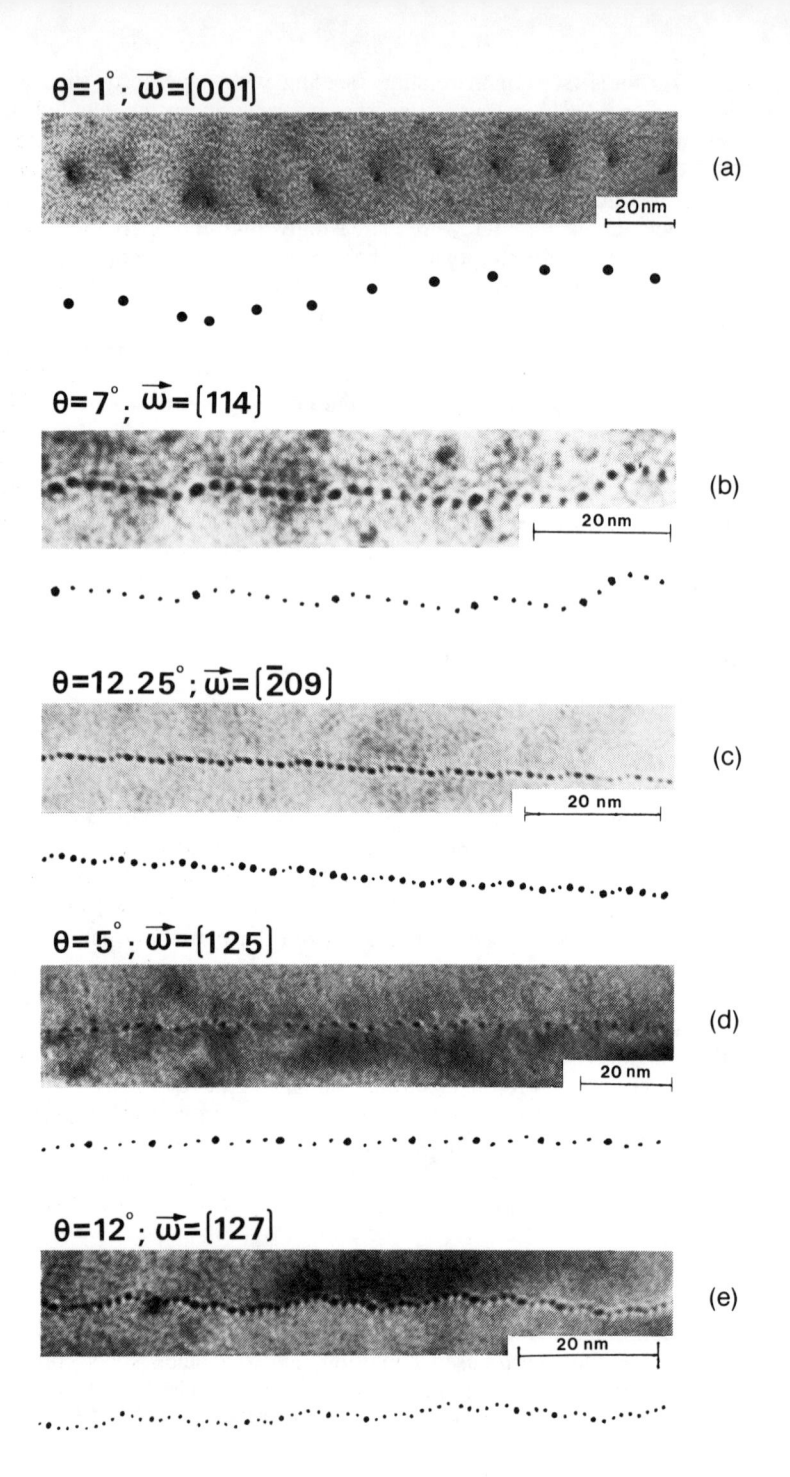

Fig. 1. (a–e) Electron micrographs and schematic diagrams of a variety of tilt boundaries in NiO imaged in the overfocused condition.

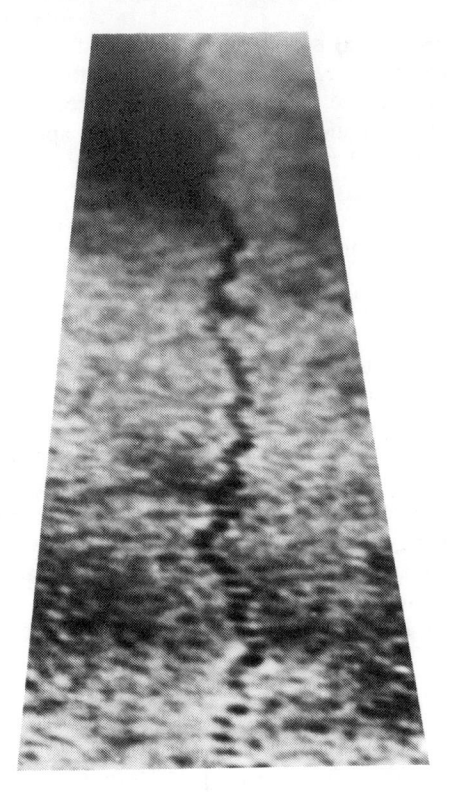

Fig. 2. View at a shallow angle along the boundary in Fig. 1(e) showing a complicated double-facet structure.

~1.2 nm. The large facets have a dimension of 25–30 nm, and it is seen that they are not arranged periodically.

These observations clearly demonstrate that tilt boundaries in NiO have a strong tendency to facet. An analysis based on the geometrical ideas of Bollmann[10] was used in an attempt to learn more about the structure of the observed boundaries. This analysis is the subject of the next section.

Analysis of the Facet and Dislocation Structure

For a naturally crystalline solid, perfect crystallinity is the lowest energy structure. Therefore, the energy of a grain boundary is dependent on its deviation from the ideal crystal configuration. In simple terms, regions of good match in a grain boundary have lower energy than regions of bad match. It is possible that a grain boundary may minimize its energy by faceting, in order to maximize the areas of good match in the interface.

For small angles of rotation, regions of bad match in the boundary are associated with dislocations. For a tilt boundary, the rotation axis lies in the boundary plane, and therefore, the boundary contains an array of dislocations aligned parallel to the rotation axis. By calculation of the distribution of points of good match (called the O lattice) for a given rotation axis and angle, the facet structure and detailed dislocation content of a grain boundary can be predicted.

This is accomplished by (a) postulating a combination of dislocation Burgers vectors that can accommodate the misorientation of the two crystals at the grain boundary, (b) calculating the corresponding O lattice, and (c) determining the predicted dislocation configuration. This follows the approach used by Vaudin, Rühle, and Sass[11] and will be carried out for three of the boundaries shown in Fig. 1.

$\vec{w} = [001]$ *Tilt Boundary* ($\theta = 1°$): Figure 1(a) is a micrograph of a [001] tilt boundary that is faceted with the long facet being the average ($1\bar{1}0$) plane of the two crystals separated by the boundary. The spacing, d, of the dislocations within each facet is ~17 nm.

Determination of the Appropriate Dislocation Burgers Vectors: The b lattice for a particular crystal structure is defined as the lattice of allowed dislocation Burgers vectors for that structure. A rotation about the [001] axis can be accommodated by dislocations whose Burgers vectors lie in the plane normal to the rotation axis (the b plane). Figure 3 shows all the possible Burgers vectors of the type $a/2\langle 110 \rangle$ in the NiO structure and the [001] rotation axis. It is seen that there are two linearly independent Burgers vectors in the b plane, (001), which corresponds to the case that Bollmann called a "special" boundary. The b net is the set of b lattice vectors that can accommodate the misorientation between the two crystals at the grain boundary. For this case, the b net is a planar square lattice with basis $a/2[110]_1$ and $a/2[1\bar{1}0]_1$, where the subscripts indicate in which crystal the vector lies.

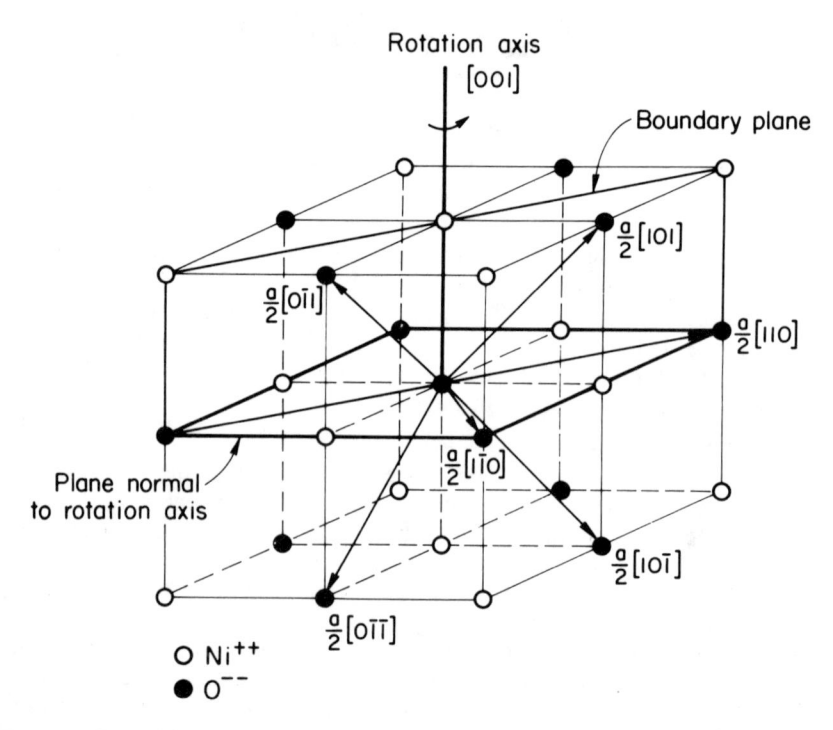

Fig. 3. Possible Burgers vectors of $(a/2)\langle 110 \rangle$ type in a NiO structure: [001] rotation axis, (001) b plane, and boundary plane are indicated with bold lines.

Determination of the O Lattice: The two crystal lattices, interpenetrating and initially coincident, are misoriented about the common [001] axis by an angle θ; this rotation is expressed by the transformation matrix R:

$$R = \begin{pmatrix} \cos\theta & -\sin\theta \\ \sin\theta & \cos\theta \end{pmatrix} \quad (1)$$

The distribution of regions of good match for the observed misorientation axis and angle is determined. It is recognized that these regions of good match occur when positions in the two crystals with common internal coordinates superimpose. Since the two interpenetrating crystals are related by a rotation, the points of best fit form lines parallel to the rotation axis. These lines constitute the O lattice, which is of interest because it is believed that a grain boundary will pass through as many regions of good match as possible in order to minimize boundary energy. The basis vectors $\mathbf{u}_x$ and $\mathbf{u}_y$, of the O lattice ($\mathbf{u}_z$ being parallel to the rotation axis) are given in terms of the base vectors of crystal 1 by the following expressions:

$$\begin{aligned} \mathbf{u}_x &= a/4[1 + \cot(\theta/2), \quad 1 - \cot(\theta/2)] \\ \mathbf{u}_y &= a/4[1 - \cot(\theta/2), \quad -1 - \cot(\theta/2)] \end{aligned} \quad (2)$$

which were derived by using the expression

$$\begin{pmatrix} \mathbf{u}_x \\ \mathbf{u}_y \end{pmatrix} = (I - R^{-1})^{-1} \begin{pmatrix} b_x \\ b_y \end{pmatrix} \quad (3)$$

where I is the identity matrix and b_x and b_y are the components of $\mathbf{b}$ in the b plane (001). Expressed in words, the O lattice vectors are derived from the b net vectors by a similarity relation defined as an expansion of $\tfrac{1}{2}\mathrm{cosec}(\theta/2)$ and a rotation clockwise by $90° - \theta/2$.[12] This operation assumes that a counterclockwise rotation takes crystal 1 into crystal 2 and that the b net is crystallographically aligned with crystal 1, which is the normal convention and will be used throughout this paper. The basis vectors $\mathbf{u}_x$, $\mathbf{u}_y$, and $\mathbf{u}_z$ define a square lattice of O lines. It is convenient at this point to introduce the concept of the median lattice,[13] which is identical in structure to crystal lattices 1 and 2 and lies halfway between them in orientation, i.e., $\theta/2$ counterclockwise from lattice 1. The subscript m will be used to denote vectors and planes in the median lattice. Equation (2) shows that the basis vectors $\mathbf{u}_x$ and $\mathbf{u}_y$ are in the $[110]_m$ and $[1\bar{1}0]_m$ directions, respectively, independent of the value of θ. The magnitudes of both basis vectors for a misorientation of 1° is 17 nm.

Prediction of the Boundary Dislocation Configuration: Wigner–Seitz cell walls, which locate the regions of worst fit, are constructed halfway between the O lines. The bicrystal containing the grain boundary is formed by passing the boundary plane through the two interpenetrating crystal lattices and discarding a different set of atoms from each side. The intersection of the boundary plane with the Wigner–Seitz cell network gives the predicted dislocation structure of the boundary. It can be seen that the regions of worst fit cut by the boundary plane are dislocations by noting that between one O line and the next, the two crystals rotate away from perfect crystal alignment by a vector $\mathbf{b}$. A dislocation of Burgers vector $\mathbf{b}$ causing this displacement is found between the two O lines. The choice of particular Burgers vectors dictates the geometry of the O lattice; for a different choice of $\mathbf{b}$, there would be a different O lattice.

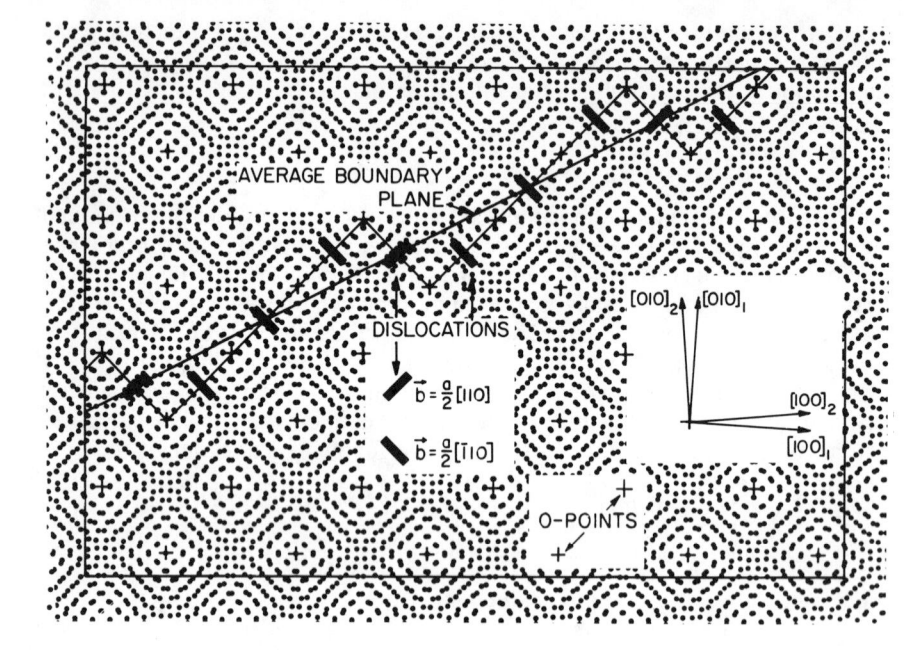

Fig. 4. Two superimposed and misoriented fcc (001) planes; faceted boundary is shown passing through O points.

It is now possible to predict the expected faceting, based on the premise that it is energetically favorable for a boundary to pass through an O element when this will result in only a small deviation from the average boundary plane. Figure 4 shows two superimposed square lattices ((001) fcc[†] planes) slightly misoriented, on which the points of best fit (O points) correspond to the O lines projected onto (001) and have been marked with a cross. As can be seen, if the average boundary plane is not $(110)_m$ or $(1\bar{1}0)_m$, the boundary will facet so as to pass through as many O points as possible, and dislocations will be found halfway between the O points, as indicated.

For the average grain-boundary plane that is 7° away from the $(1\bar{1}0)_m$ plane (see Fig. 1(a)), the boundary structure was determined by using the preceding analysis with the result shown in Fig. 5. It is seen that the dislocation spacing in the long facet is 17 nm, and the height of the facet is also 17 nm. Comparison of Fig. 5 with Fig. 1(a) shows excellent agreement for both the facet geometry and dislocation spacing in the facet. This is a simple boundary that in the past has been analyzed without using the O lattice theory. In the examples that follow, the

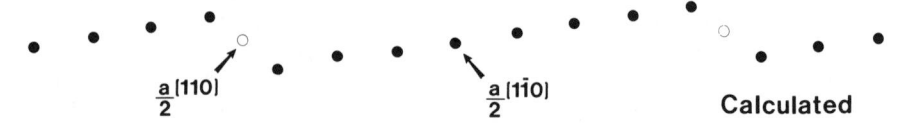

Fig. 5. Predicted structure, using O lattice theory, of the boundary in Fig. 1(a).

concepts that have been introduced will be found invaluable for the analysis of more complex boundaries.

$\vec{w} = [\overline{2}09]$ *Tilt Boundary* ($\theta = 12.25°$): Figure 1(c) is a micrograph of a $[\overline{2}09]$ tilt boundary that is regularly faceted with a period of 8.8 nm, although two irregularities were found within the total length of observed boundary. The misorientation angle, θ, was determined to be $12.25° \pm 0.25°$. Figure 6(a) is an enlargement of the image of part of the boundary and shows the detailed structure of a single boundary repeat unit. The black spots correspond to dislocations that are being viewed in the end-on orientation. There are typically six dislocations identically arranged in each repeat unit, with the circular images having lower contrast than the elongated images. The crystallography of the repeat unit is described in Fig. 6(b). It was observed that the long facet is along the $[010]_m$ direction and the average boundary plane contains the $[010]_2$ direction.

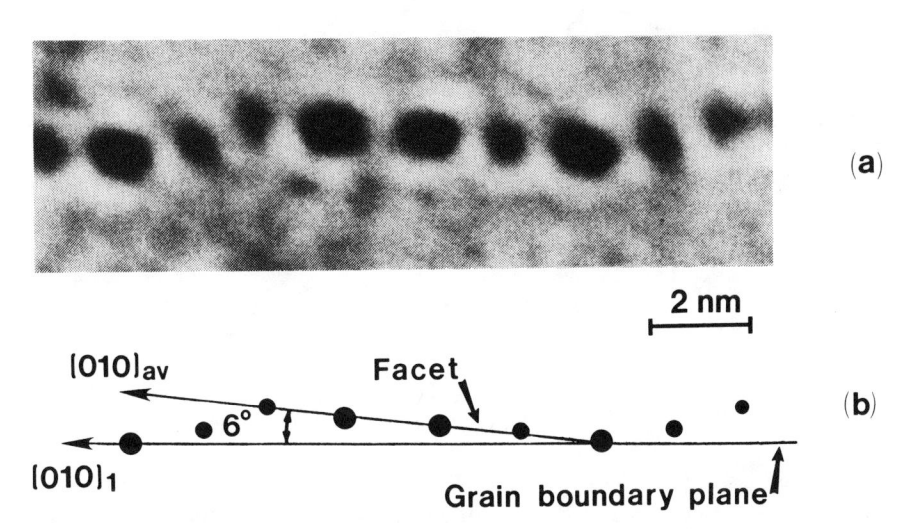

Fig. 6 (a) Enlargement of part of the **w** = $[\overline{2}09]$, $\theta = 12.25°$ tilt boundary in Fig. 1(c); the boundary repeat length is 8.8 nm. (b) Crystallography of one boundary repeat unit.

Since the observed contrast is believed to be due to the change in mean inner potential at the dislocation core, the variation in contrast along the facet corresponds to a variation in dislocation character and/or Burgers vector. This information will be useful when the predicted structure is compared with the observations.

Determination of the Appropriate Dislocation Burgers Vectors: Examination of Fig. 7, which is similar to Fig. 3 but has the $[\overline{2}09]$ rotation axis and its associated b plane normal to the rotation axis drawn in, shows that no b lattice vectors lie in this b plane. According to Bollmann, this is a "general" boundary. The b plane is approximated by steps or facets that contain $a/2\langle 110\rangle$ Burgers vectors. This forms a stepped b net. Any $(h0l)$ plane can be approximated by segments of (001) and $(\overline{1}01)$ planes, which contain the vectors $a/2[110]$, $a/2[1\overline{1}0]$, and $a/2[101]$. These vectors and combinations of them are therefore expected to be the Burgers vectors of the dislocations in the boundary. Since these three vectors have components parallel and perpendicular to the rotation axis and the boundary was observed to

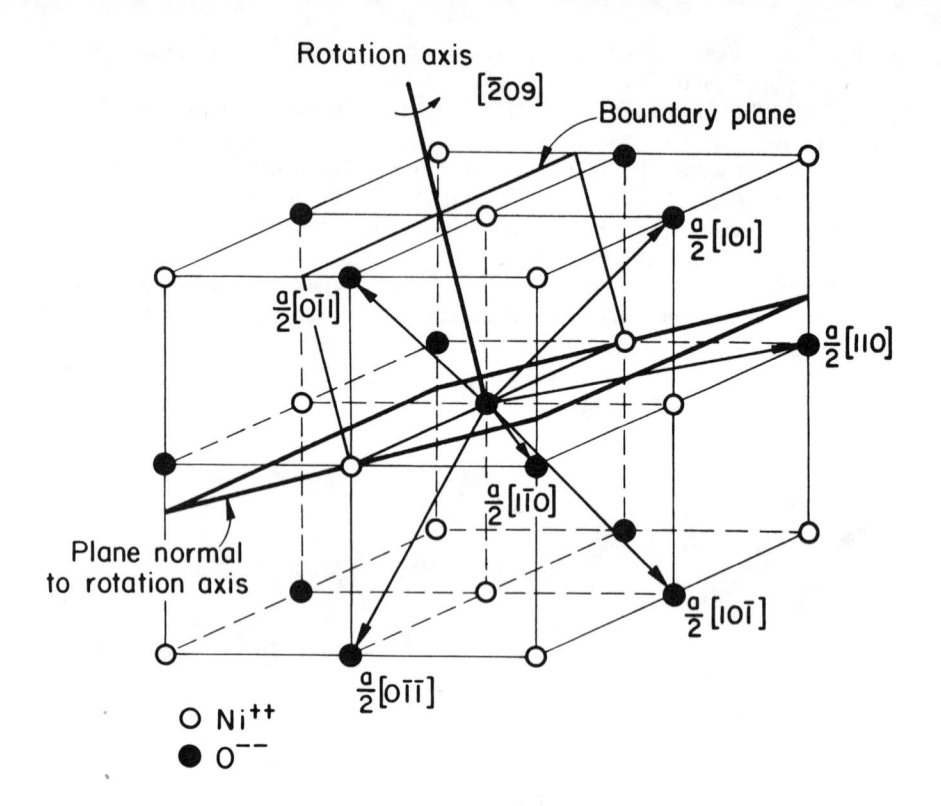

Fig. 7. Possible Burgers vectors of a/2⟨110⟩ type in a NiO structure. [$\bar{2}$09] rotation axis, ($\bar{2}$09) b plane, and boundary plane are indicated with bold lines.

have no long-range stress field, the following conditions, embodied in Frank's formula and implicit in the O lattice formalism,[10] must apply: the sum of the parallel components of the dislocations in one boundary period must be zero, and the sum of the perpendicular components must be normal to the average boundary plane. The stepped b net construction ensures that these conditions are met.

Different stepped or faceted b nets can be devised for the same b plane. A particular stepped b net predicts a unique dislocation network for a given boundary plane, and conversely, the predicted dislocation network that best agrees with the experimental observation corresponds with a unique stepped b net, from which Burgers vectors can be assigned to all the observed dislocations. One of the criteria that was used to choose the appropriate (001) and ($\bar{1}$01) steps needed to make up a given stepped b net was that of reducing to a minimum the volume enclosed between the b plane and the stepped surface.

Figure 8 shows a (010) projection of the cations of the NiO structure (equivalent to the b lattice), and the trace of the ($\bar{2}$09) plane is drawn in together with a possible stepped approximation. The planar b net was created by projecting the Burgers vectors embedded in the stepped b net onto the ($\bar{2}$09) plane. Analytically, the projected Burgers vectors $\mathbf{b}_p$ were found from

$$\mathbf{b}_p = P\mathbf{b} \qquad (4)$$

where

$$P = \begin{pmatrix} 0 & 1 & 0 \\ 9/85^{1/2} & 0 & 2/85^{1/2} \end{pmatrix}$$

for this case. This gives the components of each projected Burgers vector, which lie in the b plane $(\bar{2}09)$, in the orthonormal basis $[010]_1$ and $1/85^{1/2}[902]_1$ and hence establishes the b net.

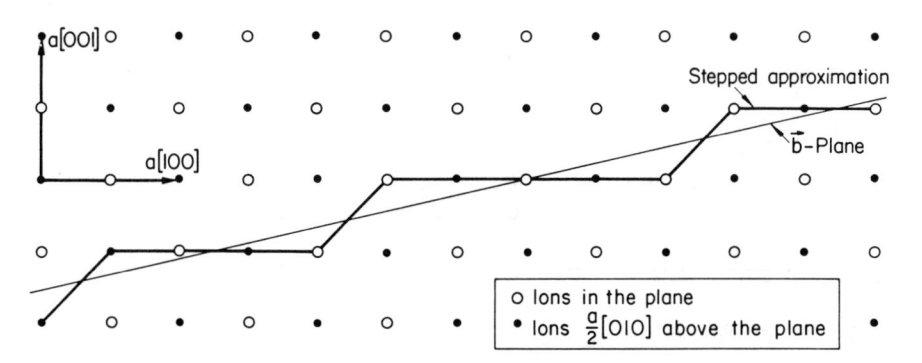

Fig. 8. Stepped approximation to the $(\bar{2}09)$ b plane in the fcc b lattice shown in (010) projection.

Determination of the O Lattice: The O lattice for this boundary is not a simple lattice as the unit cell contains several O lines. It was constructed graphically as before by using the similarity relation between the b net and the O lattice. The result is shown in Fig. 9 projected onto the $(\bar{2}09)$ b plane. Analytically, the

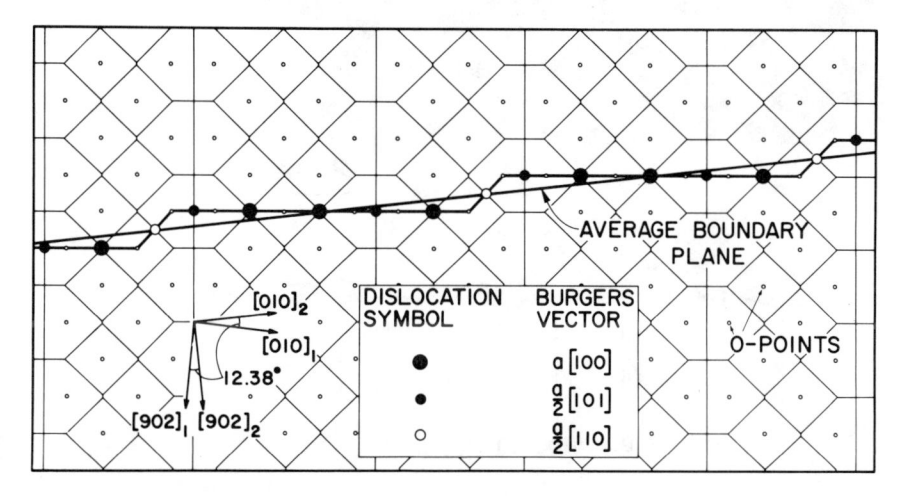

Fig. 9. Calculated O network for rotation axis of $[\bar{2}09]$. Wigner–Seitz cell walls, average boundary plane, and predicted dislocation spacing are indicated.

O vectors can be calculated in the b net basis by using Eq. 3 with **b** now replaced by $\mathbf{b}_p$.

Prediction of the Boundary Dislocation Configuration: The Wigner–Seitz cell walls are shown in Fig. 9 as bold lines bisecting each O vector, and the trace of the average boundary plane is also indicated. As can be seen, a small deviation from planarity allows the boundary plane to pass through a high density of O elements. Dislocations are found as indicated, and Burgers vectors are assigned to each dislocation by using the relationship between the b net and the O lattice.[‡] Figure 10(a) shows the predicted arrangement of the dislocations, and in Fig. 10(b) their Burgers vectors are superimposed on the observed facet. It is clear from Fig. 10(a) that the calculated spatial arrangement of the dislocations agrees with the observations to within 0.1 nm. Figure 10(c) is a perspective view of the boundary, indicating the arrangement of the dislocations and their associated Burgers vectors.

As noted previously, the variation in contrast could be due to a variation in the magnitude of the Burgers vector, and this interpretation of the contrast is

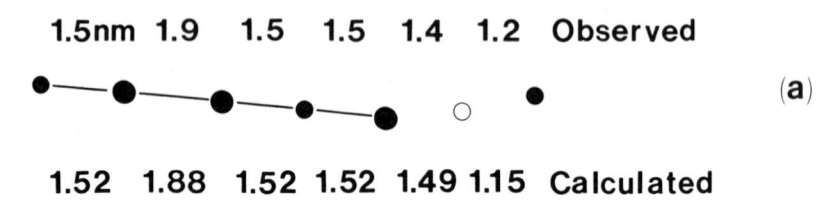

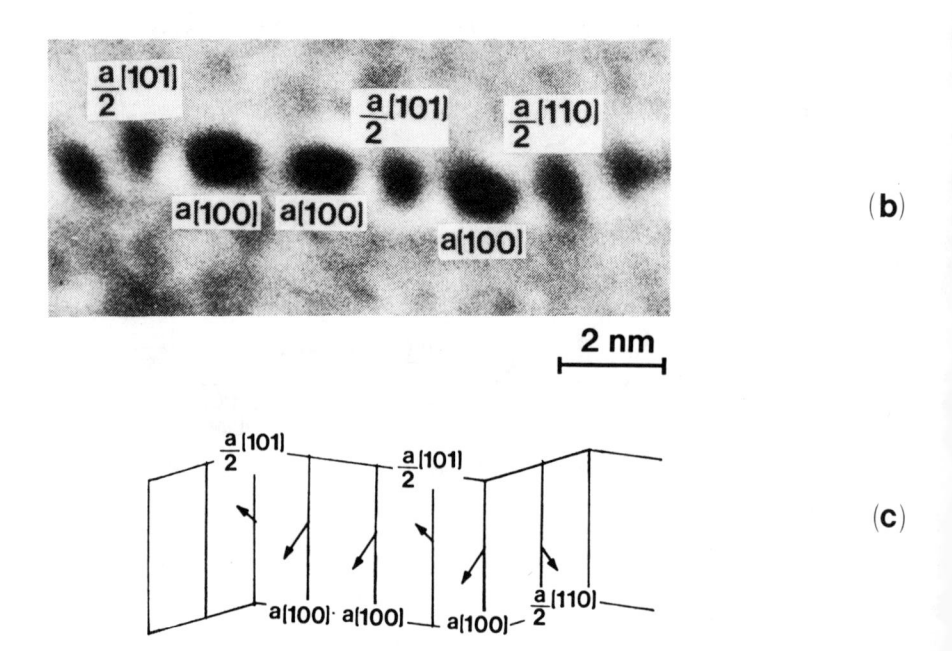

Fig. 10. Predicted structure for the boundary in Fig. 6: (a) dislocation arrangement; (b) Burgers vectors assigned to the observed dislocations; (c) perspective view of the faceted structure.

in good agreement with the calculated Burgers vectors; i.e., the intensities of the images scale qualitatively with the magnitude of the edge component of the assigned Burgers vectors. Further, it is believed that the shapes of the dislocation images are related to the dislocation core structure. In particular, the elongated appearance of the $a[100]$ dislocation suggests that the core of this dislocation consists of $(a/2)[110]$ and $(a/2)[1\bar{1}0]$ cores slightly separated along $[010]_m$, the long facet direction.

w = $[114]$ Tilt Boundary ($\theta = 7°$): Figure 1(b) is a micrograph of a [114] tilt boundary that has long facets, typically containing seven dislocations of identical contrast, separated by steps containing a single high-contrast dislocation. The boundary is not truly periodic within the observed area. Some of the long facets contain more than seven dislocations while others contain as few as five dislocations. Two large steps, containing three dislocations instead of the usual one, were also observed. The long facet plane in all cases is the $(1\bar{1}0)_m$. Figure 11(a) is an enlargement of the image of part of the boundary and Fig. 11(b) shows the boundary crystallography. The crystallography of the facet is described in Fig. 11(b). It was observed that the average boundary plane is 6° from $(1\bar{1}0)_m$.

Determination of the Appropriate Dislocation Burgers Vectors: Figure 12 shows that only one $(a/2)\langle 110 \rangle$ type Burgers vector ($(a/2)[1\bar{1}0]$) lies in the b plane for this case. This is therefore another general boundary. A stepped approximation to the (114) b plane can be constructed by using segments of (001) and (111) planes. Both of these planes contain $(a/2)[1\bar{1}0]$ and at least one other linearly independent $(a/2)\langle 110 \rangle$ type vector. This stepped approximation is shown in Fig. 13. Again, a minimum enclosed volume criterion was used in choosing the stepped b plane. The b plane was created by projecting the Burgers vectors lying in the stepped surface onto the (114) plane. Analytically, this was achieved by using Eq. (4) with

$$P = \begin{pmatrix} 1/2^{1/2} & -1/2^{1/2} & 0 \\ -2/3 & -2/3 & 1/3 \end{pmatrix}$$

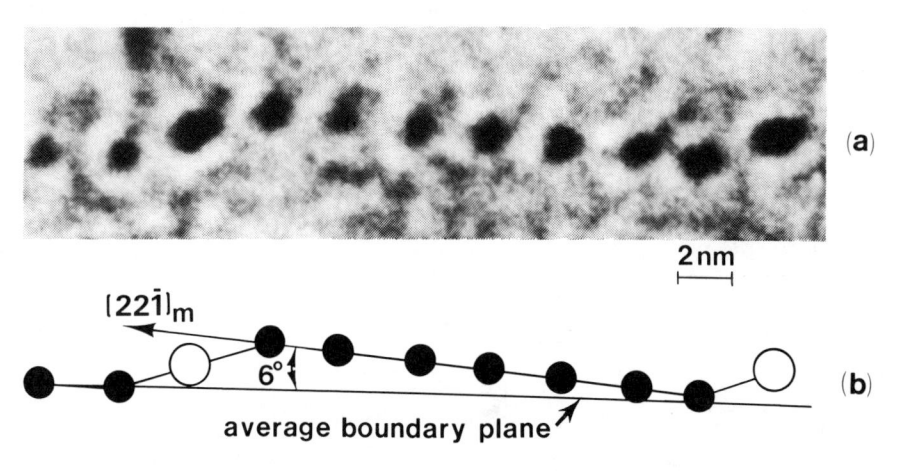

Fig. 11 (a) Enlargement of part of the **w** = [114], θ = 7° tilt boundary in Fig. 1(b). (b) Crystallography of one facet.

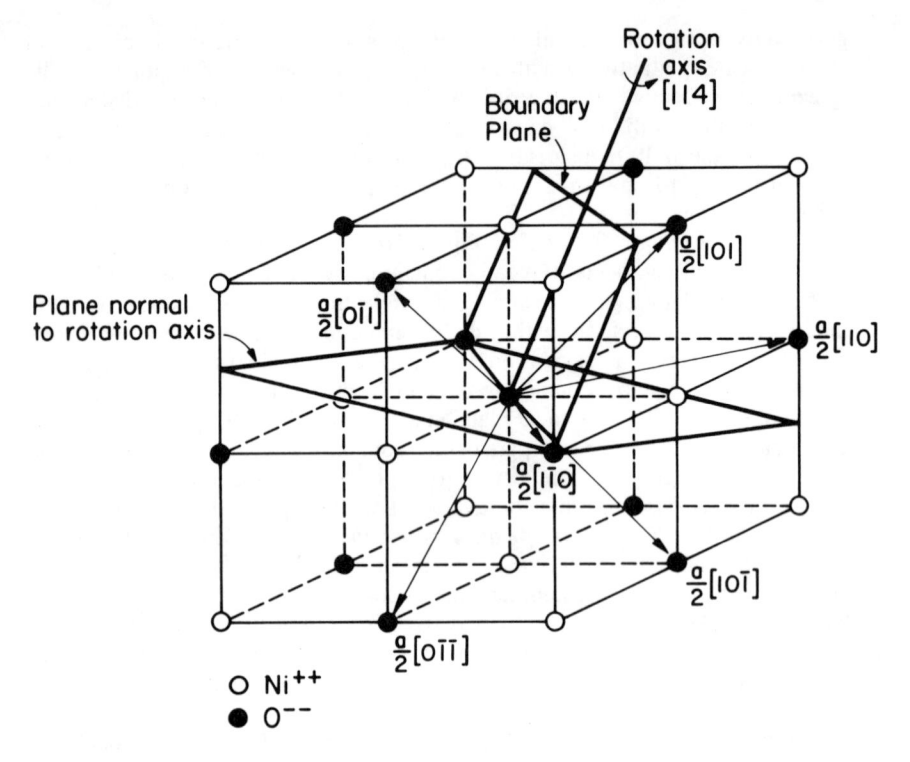

Fig. 12. Possible Burgers vectors of $a/2\langle 110\rangle$ type in a NiO structure. [114] rotation axis, (114) b plane, and boundary plane are indicated with bold lines.

This gives the components of each projected Burgers vector, which lie in the b plane (114), in the orthonormal basis $1/2^{1/2}[1\bar{1}0]_1$ and $1/3[\overline{221}]_1$.

Determination of the O Lattice: The O lattice was constructed by using Eq. (3), again with **b** replaced by $\mathbf{b}_p$. The result is shown projected onto the (114) b plane (in Fig. 14). As before, the O lattice has a large unit cell.

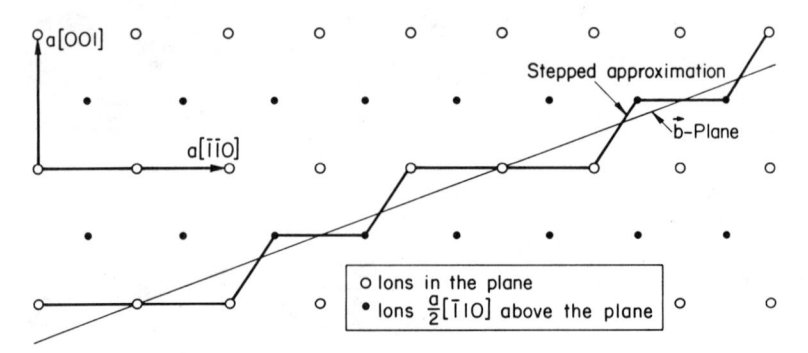

Fig. 13. Stepped approximation to the (114) b plane in the fcc b lattice shown in $(\bar{1}10)$ projection.

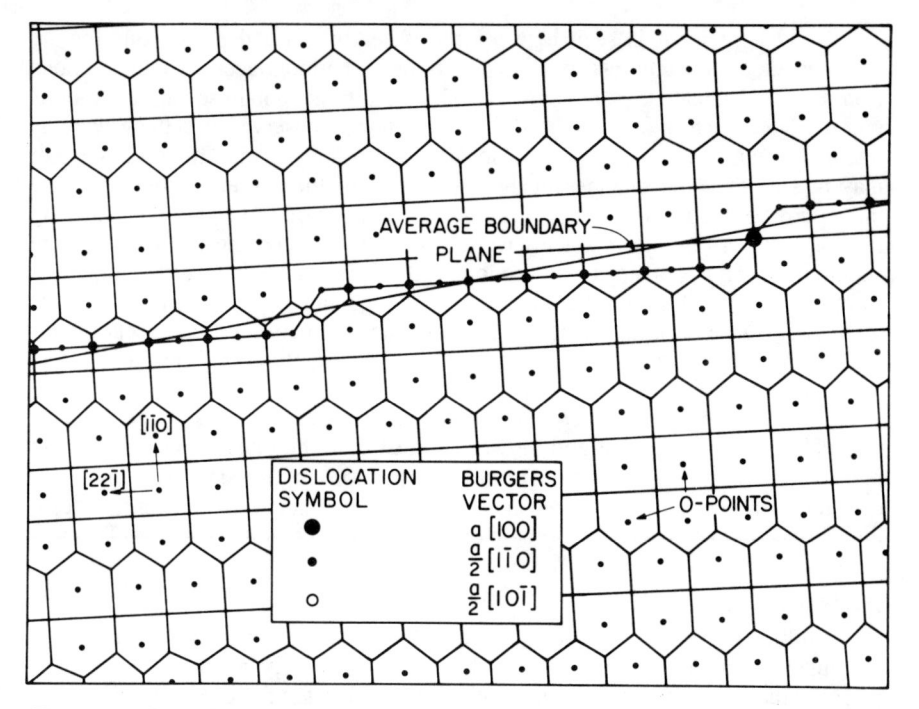

Fig. 14. Calculated O lattice for rotation axis of [114]. Wigner–Seitz cell walls, average boundary plane, and predicted dislocation spacing are indicated. The crystallographic directions are in the median lattice.

Prediction of the Boundary Dislocation Configuration: The Wigner–Seitz cell walls and the trace of the average boundary plane are shown in Fig. 14. The average boundary plane passes through a low density of O elements. A small deviation from planarity again allows the boundary plane to pass through a higher density of O elements. The resulting predicted dislocation structure is compared with the observations in Fig. 15.

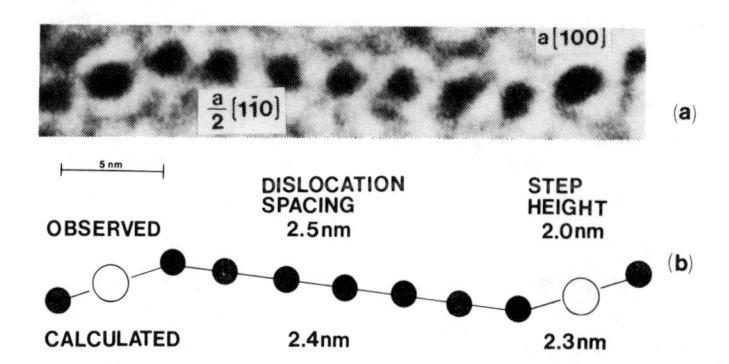

Fig. 15. Predicted structure for the boundary in Fig. 11: (a) dislocation arrangement; (b) Burgers vectors assigned to the observed dislocations.

Good agreement is seen between the observations and predictions for the spacings and image contrast of the dislocations in the long facets. The predictions from Fig. 14 indicate that the long facet should always contain seven dislocations with Burgers vectors of the type $(a/2)[1\bar{1}0]$. It was observed that there were as few as five and as many as nine dislocations in a long facet. This disagreement may be due to small changes in the orientation of the average boundary plane that could be accommodated by varying the number of dislocations in the long facet. More serious discrepancies were also noted. According to Fig. 14, there should be two types of steps between long facets: one containing a dislocation with Burgers vector $a[100]$ and the other containing a dislocation with Burgers vector $(a/2)[10\bar{1}]$. In the former case, the height of the step between long facets should be 2.3 nm, while in the latter case the step height should be 1.7 nm. The actual measured step height in all cases was 2.0 nm. The dislocation with Burgers vector $a[100]$ should show higher contrast, while the dislocation with Burgers vector $(a/2)[10\bar{1}]$ should show slightly lower contrast than the dislocations in the long facets. The observed contrast for the dislocation in the step was high and constant in all cases. Finally, the calculated structure suggests that, in the steps, three high contrast dislocations should exist for every two of low contrast. This was not observed.

Discussion

Bollmann's O lattice theory is based on the principle that a grain boundary can minimize its energy by maximizing the area of good fit in the boundary. The present study has shown that the relatively complex structure of certain general low-angle tilt boundaries in NiO can be understood in terms of this principle. The present work has also demonstrated that tilt boundaries in NiO with misorientation axes far from low index directions have faceted structures.

In order to carry out the faceting study, a new imaging technique was employed, which is sensitive to the change in mean inner potential at the dislocation core. Since the image contrast is both localized to a ~0.8 nm or smaller diameter region and dependent on the magnitude of the dislocation Burgers vector, it is possible to obtain detailed information on the dislocation content of the facets. In order to make progress in understanding the details of faceting in boundaries other than those of very low angle, it is necessary to use either the imaging technique employed in the present work or the lattice imaging technique. For a tilt boundary with a high index rotation axis, the latter technique is of only limited value since for this case there will be few, if any, low-order reflections available to contribute to the lattice image.

The low-angle tilt boundaries examined in this work fall into the categories, special and general, as defined by Bollmann. For the $[001]/1°$ boundary (special), the analysis is simple because the dislocation Burgers vectors do not have components parallel to the rotation axis. In the other cases (general), the analysis is more complex because many of the dislocation Burgers vectors have components parallel to **w**.

The $[\bar{2}09]/12.25°$ boundary is regularly faceted, and for most of its observed length, the boundary structure consists of a repeating unit of six dislocations. In the center of Fig. 1(c), there is a unit containing nine dislocations, and in a thicker part of the foil, there is a unit containing three dislocations; these are the only two deviations from uniform faceting in over 30 boundary units. The two deviating units can be combined to form two normal six dislocation units, and therefore their effects cancel out. Since the boundary is periodic, the Burgers vectors of the six dislocations in one boundary period sum to a vector **B**, which is rational and normal to **w**. Each six-dislocation repeat unit has identical dimensions, and therefore the boundary separates two crystals that are in a coincidence relationship; i.e., the vector **X** perpendicular to **w** describing one boundary period is a lattice vector in both crystals. For a noncoincidence misorientation, **X** would not be a lattice vector and it is expected that systematic variations of the dislocation spacing would occur, resulting in nonuniform facet dimensions.

Variations in facet dimension are seen in the [114]/7° boundary. Thus, either this boundary is not a coincidence boundary or the boundary period is larger than the portion of the boundary that was observed. The predicted structure was not identical with the observed structure. This suggests that a problem exists with the geometrical analysis used for this case. Two possible explanations for the discrepancies are as follows: (i) the wrong choice of Burgers vectors was used to form the stepped approximation to the (114) *b* plane (i.e., the wrong O lattice was obtained); (ii) the geometrical analysis used does not take other important factors into account that are not yet fully understood. For example, a quantitative calculation of the strain energy may be required to decide which particular configuration of O points gives the lowest energy.

Several combinations of Burgers vectors that gave different stepped approximations to the (114) *b* plane (thus, different O lattices) were examined. The one used in this paper gave the best agreement between predictions and observations. Although not discussed in this paper, the O lattice theory was also used to investigate the structure of the [127]/12.5° boundary (Fig. 1(*e*)) and the [125]/5° boundary (Fig. 1(*d*)). Again, significant discrepancies were found to exist between the predicted and observed structures. This suggests that the second explanation for the observed discrepancies is more likely.

Conclusions

General tilt grain boundaries in NiO show a strong tendency to facet. The observations of this study suggest that NiO tilt boundaries possessing arbitrary rotation axes will be faceted. In fact, it is likely that only special, symmetrical tilt boundaries will not be faceted.

Bollmann's O lattice theory can be used to analyze the structure of some low-angle tilt boundaries in NiO. Excellent agreement was found between the predictions of the analysis and the observations for a boundary having a rotation axis equal to [001] and $\theta = 1°$ and a boundary having a rotation axis equal to [$\bar{2}$09] and $\theta = 12.25°$. The first boundary is special because two linearly independent $(a/2)\langle 110 \rangle$ Burgers vectors lie in the *b* plane. The second boundary is not completely arbitrary because it is a small period coincidence boundary. For more arbitrary boundaries, the O lattice theory alone does not appear to fully explain the observed facet structures.

Measurement of the Expansion at [001] Twist Boundaries in NiO

Thin-Crystal Model as the Basis of a Diffraction Technique for Detecting the Expansion at a Twist Boundary[14]

Figure 16(*a*) is a schematic representation of a [001] twist boundary in an end-on orientation with the boundary plane parallel to the (002) planes of the crystals on both sides of the interface. Due to rearrangement of atoms in the boundary region, it is expected that there will be an expansion normal to the boundary. This means that the spacing of the (002) planes lying parallel to

Real Space .... Reciprocal Space

(002) planes

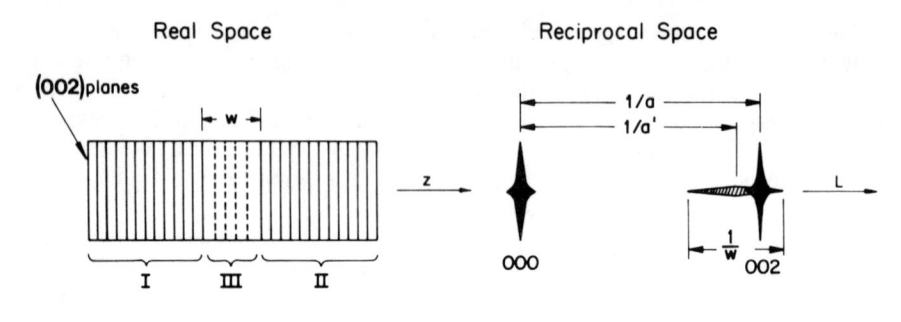

I, II = perfect crystal, plane spacing = a

III = grain boundary, plane spacing = a', width = w

(a)   (b)

Fig. 16. Thin-crystal model of grain boundary. (a) Schematic representation of a [001] twist boundary contained in a bicrystal; I and II are regions of perfect crystal with interplanar spacing, a; III is the grain-boundary region of width w and with interplanar spacing a'. (b) L direction in reciprocal space due to the bicrystal in part a, showing the extra reflection from the grain boundary.

the boundary is slightly larger in the vicinity of the boundary (region III) than in the undisturbed crystals (regions I and II).

As a first approximation to describe the expansion, it is useful to consider the grain-boundary region as being a thin crystal having a width w and a uniform interplanar spacing a', which is slightly larger than the spacing a in the perfect crystals. What are the consequences in reciprocal space of this model of the boundary? Particular attention is paid to the L direction passing through 000 in Fig. 16(b), since for this direction the diffraction intensity is influenced only by the component along L of the atomic displacements in the vicinity of the boundary. The model in Fig. 16(a) predicts that the perfect crystals will give rise to common 002 reflections at a distance $1/a$ from 000, while the scattering from the thin boundary region will be in the form of an extra reflection at a distance $1/a'$ from 000. The extra grain-boundary reflection is elongated because region III is thin and is displaced away from the 002 reflections toward 000, since $1/a > 1/a'$.

Thus, according to the thin-crystal model, the plane spacing a' in the boundary region can be determined by examining the interface edge-on in the electron microscope and measuring the position of the elongated grain-boundary reflection (termed a relrod) relative to the 002 reflection. The structural width of the boundary region can be determined from the length of the grain-boundary relrod.[7,8]

It is expected that a' will not be constant but vary along the z direction, since the stress field associated with a grain boundary is predicted to fall off smoothly with distance from the boundary plane.[15] It may be possible to determine this variation, $a'(z)$, by analysis of observations on the systematic row of reflections along the L direction through 000. This has not been done in the present paper, where, for simplicity, the grain-boundary region was treated as a thin crystal with a uniform plane spacing.

Specimen Preparation

In order to obtain grain boundaries suitable for the diffraction studies discussed above, NiO bicrystals were fabricated by a hot-pressing technique, which had been employed by Sun and Balluffi[16] for MgO and by Liou and Peterson[17] for NiO. The single crystals used for fabricating the bicrystals were cut from NiO single crystals that were grown from high-purity (99.999%) NiO with an arc melting technique at Argonne National Laboratory. Two such crystals were placed together with their cleaved (001) surfaces facing each other as shown in Fig. 17. The misorientation angle was set to the desired value, and then the crystals were hot-pressed for 24 h in air at 1450°C. During that time, a pressure of 14–20 MPa (2000–3000 psi) was applied by means of two SiC pistons. The load was then removed, and the bicrystal was annealed for an additional 48 h at a temperature of 1550°C, which results in a deviation from stoichiometry of $\sim 4 \times 10^{-3}$.[18] After the hot-pressing, the misorientation angle of the bicrystal was checked with the Laue back-reflection technique. Slices ~ 250-μm thick were then cut from the bicrystal with the boundary in an end-on orientation. These slices were further thinned by mechanical polishing to a thickness of ~ 70 μm, and the final thinning was done in an ion miller to obtain electron microscope specimens.

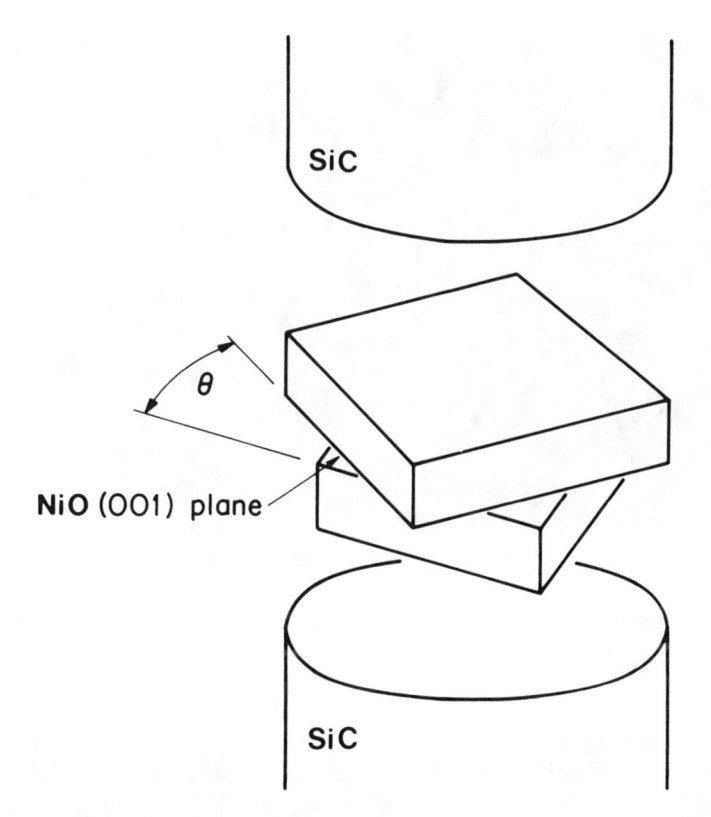

Fig. 17. Schematic drawing showing the procedure used for fabricating bicrystals of NiO. Two NiO single crystals are hot-pressed between SiC pistons in a furnace operating in air.

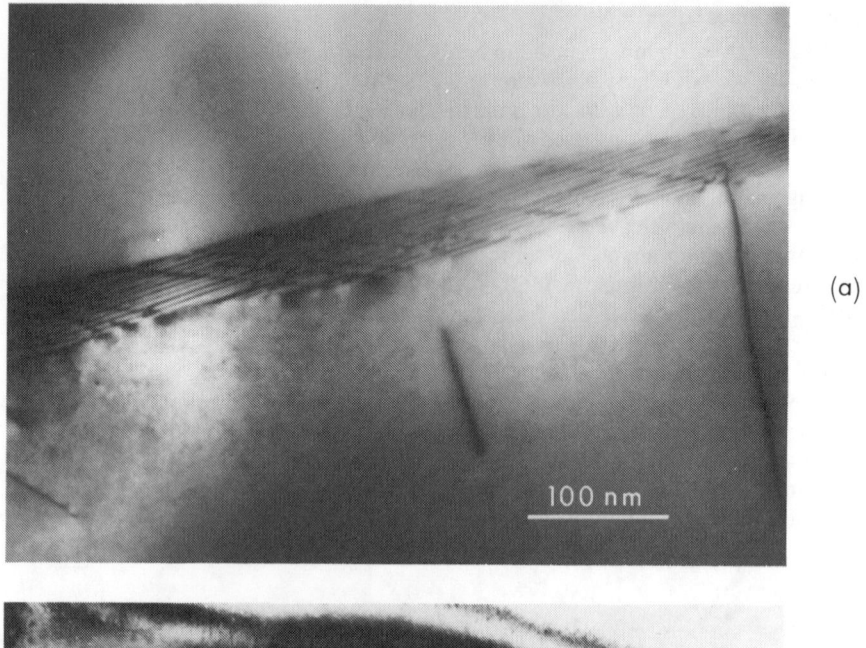

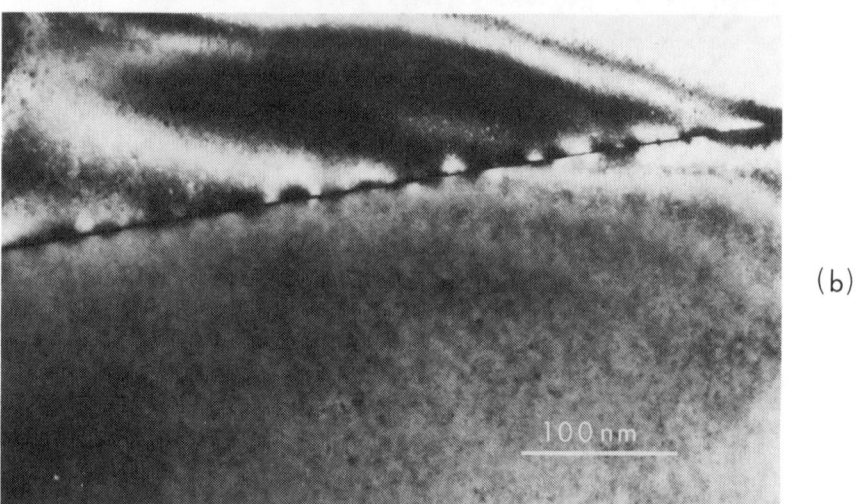

Fig. 18. Electron micrographs of [001] twist boundary in NiO ($\theta = 22°$): (a) boundary inclined ~30° to the incident beam; (b) same boundary viewed edge-on.

Electron micrographs from a 22° twist boundary fabricated by using this technique are shown in Fig. 18. In Fig. 18(a), the boundary is tilted ~30° away from its end-on orientation, and dislocations are clearly visible in the interface. Figure 18(b) shows the boundary in its end-on orientation, which is the position that was used for the diffraction studies. It can be seen that the boundary is free of precipitates and voids and is very straight.

Experimental Procedure and Observations

When taking an electron diffraction pattern, it is important to note that the grain-boundary region itself is only a small fraction of the total crystal volume exposed to the electron beam, and therefore, it is expected that the streaks due to the grain-boundary relrods will be quite weak. In fact, the streaks are so weak that they are not possible to observe directly on the viewing screen of the electron microscope, and a special procedure is required in order to make certain that they are recorded. The approach used is shown schematically in Fig. 19(a), for the relrod associated with the 002 reflection. In order to excite the relrod, it must be

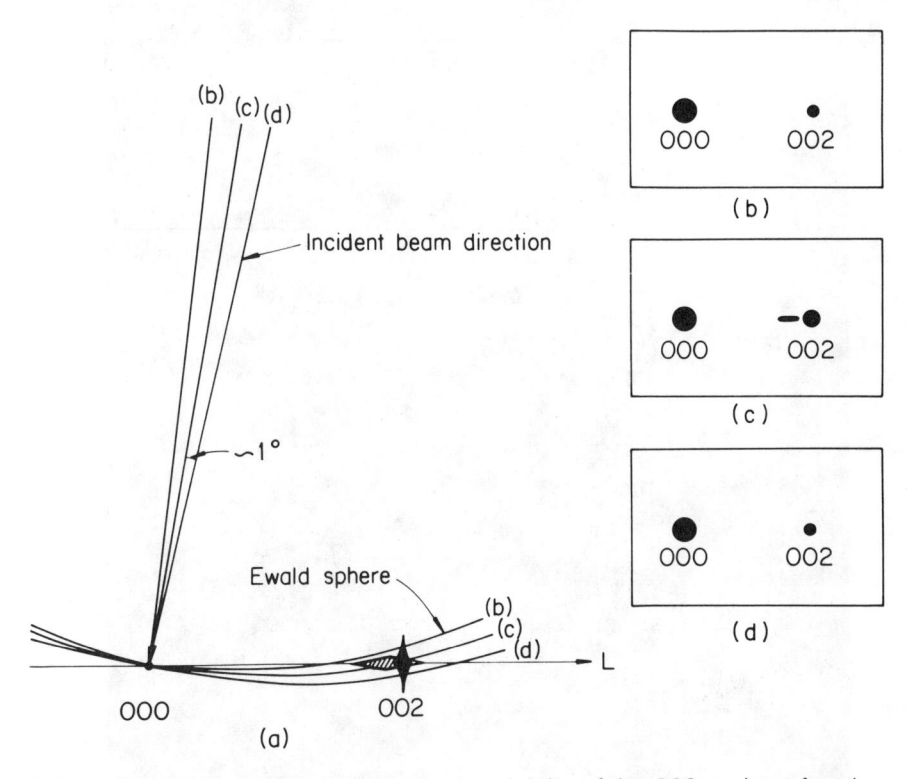

Fig. 19. (a) Diffraction geometry in the vicinity of the 002 region of reciprocal space. (b, c, d) Schematic diffraction patterns corresponding to the different orientations of the Ewald sphere in (a).

cut by the Ewald sphere, and this can be done in the following way. The boundary is slightly tilted away from its exact end-on position in order to strongly excite the 002 reflection. Then, the dark-field beam tilt controls are used to tilt the incident electron beam in fine steps along the L direction, and after each tilt step the diffraction pattern is recorded. Thus, the reciprocal space in the vicinity of the 002 reflection is scanned, and the grain-boundary streak will be recorded on one or several of the diffraction patterns taken in this series. Figure 19(b, c, d) shows schematically the type of observations that are expected with the grain-boundary streak imaged in Fig. 19(c). In order to record the streaks, it is necessary to use a well-defocused second condenser lens and exposure times of ~100 s.

Part of such a series of diffraction patterns taken from the 22° twist boundary shown in Fig. 18(b) is seen in Fig. 20. The 000, 002, and 004 reflections along the L direction are shown. The diffraction patterns shown differ by small changes in beam orientation. The grain-boundary relrod at the 002 spot is strongly excited in Fig. 20(c), where it can be clearly seen that its maximum is shifted toward the origin. The streak maximum moves when the Ewald sphere orientation is changed.

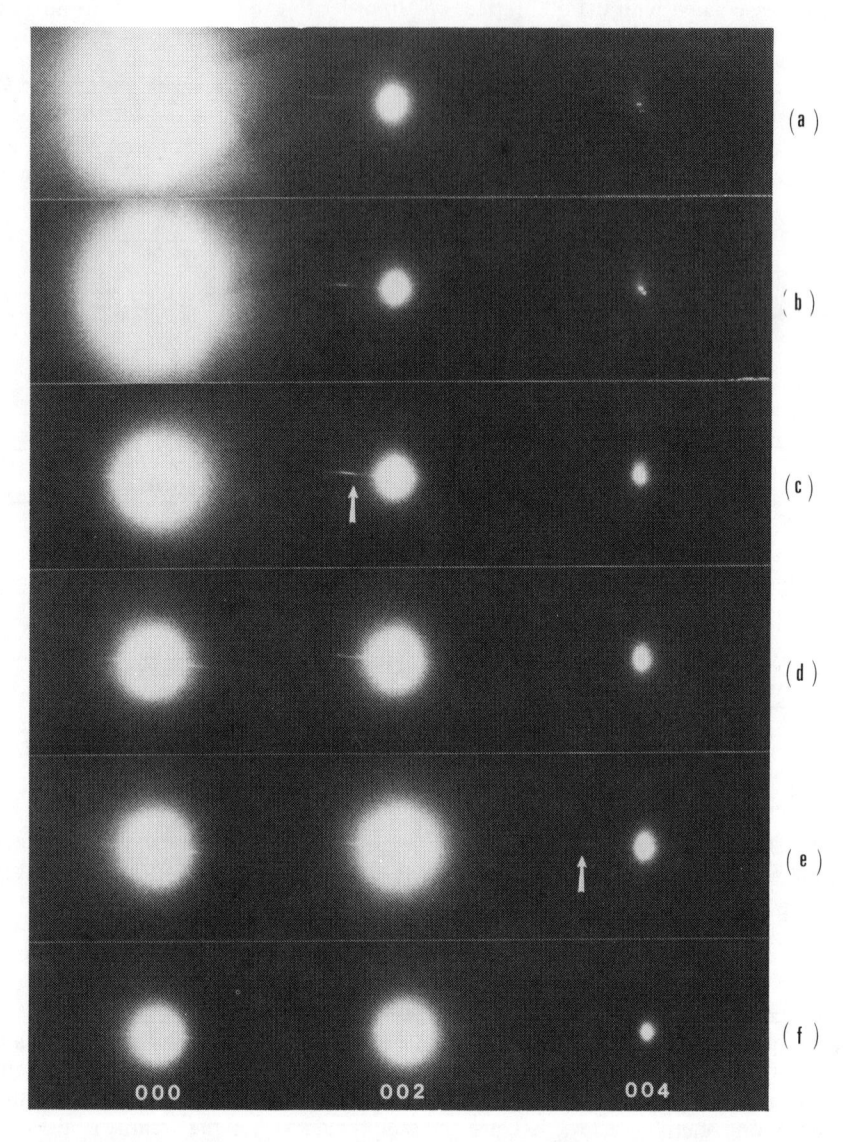

Fig. 20. Electron diffraction patterns from a long series taken on the 22° twist boundary shown in Fig. 18. The beam orientation changes in small steps from (a) to (f).

Table I. Structural Data from (001) Twist Boundaries Obtained by Electron Diffraction

Misorientation angle (deg)	Boundary width (nm)	Increase in plane spacing (%)
NiO, 22	0.5 ± 0.1	18 ± 5
NiO, 7	0.5 ± 0.1	2 ± 1
Au, 22	0.9 ± 0.4	5 ± 3

This is due to the Ewald sphere cutting the grain-boundary relrod at different points along its profile as the beam is tilted. A complete series of diffraction patterns shows that the midpoint of the grain-boundary relrod is displaced toward 000, in agreement with the expectation that there is a local expansion at the grain boundary. Further, it is even possible to excite the relrod at the 004 reflection, however, with a much lower intensity (see Fig. 20(e)).

From the length of these streaks, the grain-boundary width was determined to be 0.5 ± 0.1 nm, which corresponds to 2–3 (002) interplanar spacings. The increase in the average interplanar spacing at the grain boundary was estimated from the shift of the midpoint of the relrod relative to the corresponding matrix reflections and was found to be 18 ± 5%.* Observations were also made on a small angle [001] twist boundary. The results of all of the observations are summarized in Table I,[§] which also includes values obtained for a 22° [001] twist boundary in Au.[14] Comparison between the NiO and Au results on the same boundary suggests that there is a characteristic difference between the structure of grain boundaries in metals and ceramic materials in terms of both boundary width and expansion. To confirm this suggestion, more experimental evidence from boundaries with different misorientations and from different types of material is needed. In addition, it is of interest to determine whether point defects, stoichiometry, and segregation have an influence on the magnitude of the expansion at grain boundaries in ceramics. Experiments to examine these questions are in progress.

Conclusions

The expansion at twist boundaries and their structural width can be estimated using electron diffraction techniques. The magnitude of the expansion at twist boundaries in NiO was observed to increase with increasing misorientation angle. Measurements of the expansion at a 22° twist boundary in NiO and Au indicate that there is a significant difference in structure normal to the boundary in metals and ceramics.

Acknowledgments

This work was supported by the Department of Energy under Contract No. DE-AC02-81ER10456. The authors thank N. L. Peterson, Argonne National Laboratory, Argonne, IL, for providing the NiO specimens used in the faceting study and for allowing the use of the facilities in his laboratory for growing the NiO single crystals needed for the study of the expansion at grain boundaries.

References

[1]T. Schober and R. W. Balluffi, "Quantitative Observation of Misfit Dislocation Arrays in Low and High Angle Twist Grain Boundaries," *Philos. Mag.*, **21**, 109–23 (1970).
[2]S. L. Sass and P. D. Bristowe, "Diffraction Studies of the Atomic Structure of Grain Boundaries"; pp. 71–113 in Grain Boundary Structure and Kinetics. ASM, 1980.

[3]J. Budai, P. D. Bristowe, and S. L. Sass, "The Projected Atomic Structure of a Large Angle [001] $\Sigma = 5$ ($\theta = 36.9°$) Twist Boundary in Gold: Diffraction Analysis and Theoretical Predictions," *Acta Metall.*, **31**, 699–712 (1983).
[4]R. Raj and M. F. Ashby, "On Grain Boundary Sliding and Diffusional Creep," *Metall. Trans.*, **2**, 1113–27 (1971).
[5]R. W. Balluffi, "High Angle Grain Boundaries as Sources or Sinks for Point Defects"; pp. 279–329 in Grain Boundary Structure and Kinetics. ASM, 1980.
[6]D. A. Smith, C. M. F. Rae, and C. R. M. Grovenor, "Grain Boundary Migration"; pp. 337–77 in Grain Boundary Structure and Kinetics. ASM, 1980.
[7]C. B. Carter, A. M. Donald, and S. L. Sass, "The Study of Grain Boundary Thickness Using Electron Diffraction Techniques," *Philos. Mag.*, **41**, 467–76 (1980).
[8]J. Budai, W. Gaudig, and S. L. Sass, "The Measurement of Grain Boundary Thickness Using X-ray Diffraction Techniques," *Philos. Mag., [Part A]*, **40**, 757–67 (1979).
[9]M. Rühle and S. L. Sass, "The Detection of the Change in Mean Inner Potential at Dislocations in Grain Boundaries in NiO"; pp. 99–100 in 10th International Congress on Electron Microscopy. Hamburg, 1982.
[10]W. Bollmann; Crystal Defects and Crystalline Interfaces. Springer-Verlag, New York, 1970.
[11]M. Vaudin, M. Rühle, and S. L. Sass, "Faceting of Tilt Boundaries in NiO," *Acta Metall.*, **31**, 1106–16 (1983).
[12]J. J. Bacmann, G. Silvestre, M. Petit, and W. Bollman, "Partial Secondary Dislocations in Germanium Grain Boundaries, I. Periodic Network in a $\Sigma = 5$ Coincidence Boundary," *Philos. Mag., [Part A]*, **43**, 189–200 (1981).
[13]F. C. Frank; p. 124 in the Symposium on Plastic Deformation of Crystalline Solids. Carnegie Institute of Technology, 1950.
[14]P. Lamarre and S. L. Sass, "Detection of the Expansion of a Large Angle [001] Twist Boundary Using Electron Diffraction," *Scr. Metall.*, **17**, 1141–46 (1983).
[15]A. H. Cottrell; p. 94 in Dislocations and Plastic Flow in Crystals. Oxford University Press, London, 1953.
[16]C. P. Sun and R. W. Balluffi, "Secondary Grain Boundary Dislocations in [001] Twist Boundaries in MgO, I. Intrinsic Structures," *Philos. Mag., [Part A]*, **46**, 49–62 (1982).
[17]K.-Y. Liou and N. L. Peterson, "Observations on Grain Boundary Structures in Nickel Oxide," pp. 189–98 in Surfaces and Interfaces in Ceramic and Ceramic-Metal Systems. Materials Science Research, Vol. 14, Plenum Press, New York, 1981.
[18]A. Dominguez-Rodrigues and J. Castaing, "Deformation Plastique De L'Oxyde De Nickel Monocristallin," *Rev. Phys. Appl.*, **11**, 387–91 (1976).

---

*Siemens Elmiskop 102, Siemens AG, Karlsruhe, Federal Republic of Germany.
†For clarity, only one set of ions in the NiO structure is shown.
‡The positions of the dislocation lines predicted by the intersection of the boundary plane with the Wigner–Seitz cell walls can be found either by graphical methods, as in Fig. 9, or by computation. It is easily shown that the separation of two dislocations with Burgers vectors $\mathbf{b}_1$ and $\mathbf{b}_2$ is $\frac{1}{2}(I - R^{-1})^{-1}P(\mathbf{b}_1 + \mathbf{b}_2)$.
§These values were obtained under the assumption that the grain boundary is a thin crystal with a uniform plane spacing. A more detailed treatment shows that the relationships between the streak position and length and between the boundary expansion and width are more complicated than those predicted from the simple thin-crystal model. The results of this treatment will be published elsewhere.

Boundary Structure Observed in MgO Bicrystals

SHIUSHICHI KIMURA AND EIICHI YASUDA

Tokyo Institute of Technology
Research Laboratory of Engineering Materials
Yokohama, 227 Japan

NAOSHI HORIAI

Nippon Gakki Co., Ltd.
Iwatagun, Shizuoka, 438-01 Japan

YUSUKE MORIYOSHI

National Institute for Research in Inorganic Materials
Tsukuba Science City, Ibaraki, 305-31 Japan

Bicrystals with (001) symmetrical tilt boundaries were obtained by joining two single crystals at about 2480°C. The joined boundaries were observed with optical microscopy and high-voltage electron microscopy. The joining was mainly controlled by evaporation–condensation, which was confirmed by the appearance of many square patterns at the partially joined area. Many edge dislocations were observed at the grain boundary parallel to the ⟨001⟩ and ⟨011⟩ directions. Boundary energies were measured by thermal grooving. Relative boundary energies (σ_b/σ_s) for tilt angles below 23° followed the Read and Shockley model. The mean distance of the dislocations at the boundary calculated from the Read and Shockley model corresponded to electron microscope observations. The ratio σ_b/σ_s for high tilt angles was about 0.65, a value higher than that for metals.

Grain boundaries play an important role in controlling many physical, mechanical, chemical, and electrical properties of polycrystalline solids. Many reports of the boundary structures and properties from bicrystal experiments have been presented for metals[1-4] and also for ceramics,[5] especially for halides,[6] because specimens are easy to prepare. Models for the tilt boundary have been proposed by Read and Shockley[1] and R. W. Balluffi et al.[5] Most experimental studies of grain boundaries have used sintered or recrystallized polycrystalline materials. However, many impurities are preferentially concentrated at the boundary, and it is preferable to use synthesized bicrystals and/or tricrystals.

In this paper, synthetic bicrystals of MgO with (001) tilt boundaries were prepared from the joining of two single crystals, and the structures were observed by ultrahigh-voltage electron microscopy. The grain-boundary energies were measured by thermal grooving.

Experimental Procedure

Preparation of MgO Bicrystals

MgO single crystals contained mainly Al, Si, and Ca (total 1350 ppm). The crystals (1-cm cube) were cleaved on the (001) plane at first and then cut at a prescribed angle to (001) by diamond abrasive (No. 200). The cut surfaces were polished with 9-μm diamond powder on copper disks and finally with ¼-μm powder on tin disks. The polished surfaces were found by a surface roughness tester to have flatness below 1 μm and roughness below 100 Å (10 nm). The tilt angles were measured by an optical microscope to an accuracy of 0.1°. Two polished crystals with the same angle were held tightly together with cotton thread to maintain the symmetric tilt orientation before being heated.

Preliminary experiments indicated that the specimens should be heated at about 2480°C to join the MgO single crystals. The vapor pressure of MgO at about 2480°C is very high (sometimes the cover of the graphite crucible was brown due to the vapor pressure of MgO). To prevent the vaporization of the single crystals, they were packed in MgO powder (99.9% purity, 0.3-μm mean diameter) and then rubber pressed to 20 mm in diameter and 25 mm in height. The single crystals with MgO powder were set in a graphite cylinder, which was set in a furnace with a graphite heating element, as shown in Fig. 1.

The assembly was heated slowly (ca. 6°C/min) up to 1500°C to sinter the powder uniformly without cracks. From 1500 to 2480°C, it was heated quickly (11°C/min) to prevent the vaporization of MgO during the heating. After being held for 10 min at 2480°C, it was cooled slowly (9°C/min) to 2250°C and then the heating power supply shut off.

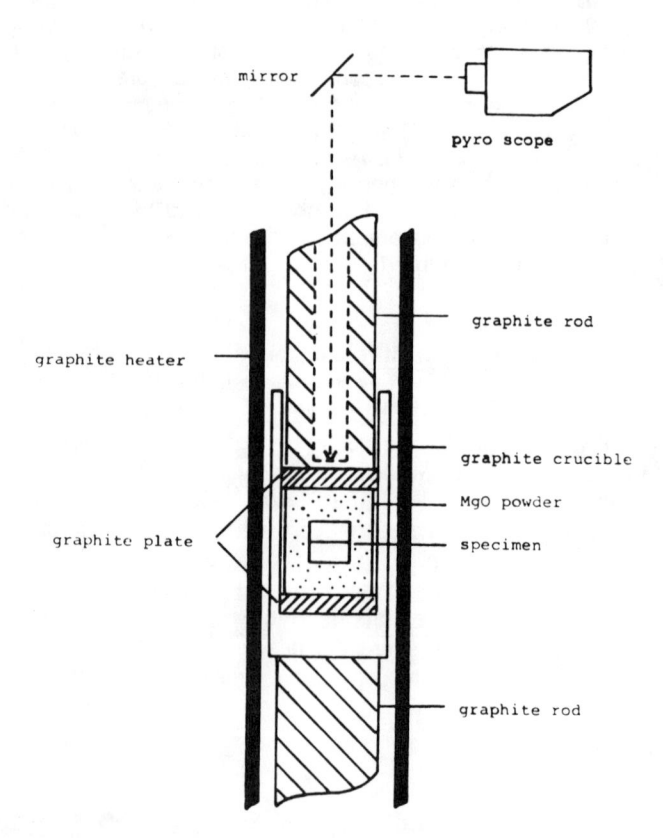

Fig. 1. Setting assembly of joining for MgO bicrystals.

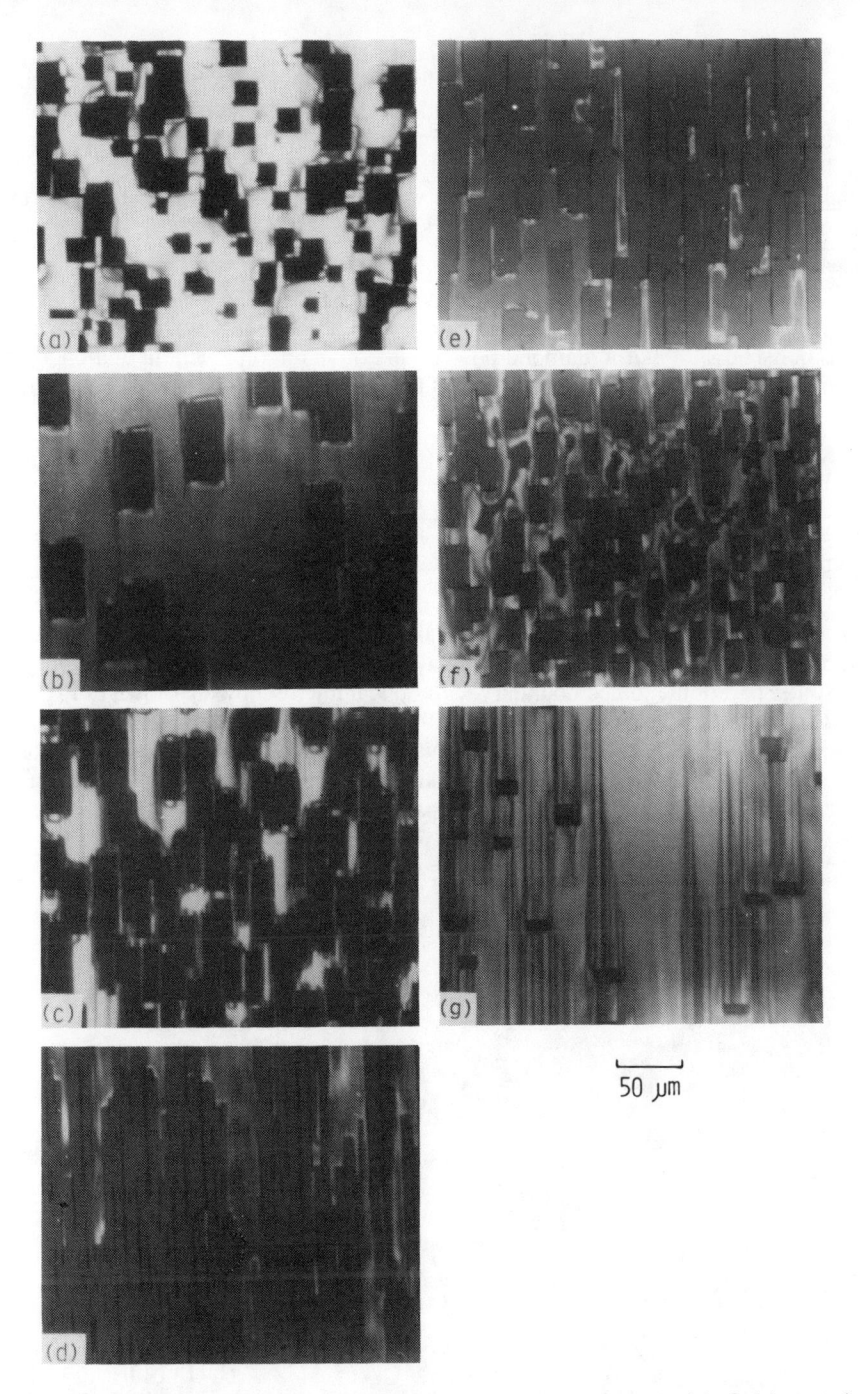

Fig. 2. Square patterns on the partially joined surface in MgO bicrystals (dark parts correspond to the joined parts): (a) $\theta = 0°$; (b) $\theta = 10.0°$; (c) $\theta = 20.2°$; (d) $\theta = 38.6°$; (e) $\theta = 14.7°$ near center; (f) between e and g; (g) $\theta = 14.7°$ far from the joined area.

Observations of the Grain Boundary

The bicrystals were separated from the surrounding polycrystalline MgO and the boundaries observed through the single-crystal layer with optical microscopy. Afterward, the bicrystals were cut perpendicular to the grain boundary, polished to less than 200-μm thickness, chemically thinned with hot orthophosphoric acid, and finally ion thinned. The thin sections were observed in a transmission electron microscope* operating at 100 and 1000 kV.

Measurement of Boundary Energy

The bicrystals were separated from the sintered MgO and then cut on the (100) plane perpendicular to the joined boundary with a diamond saw. The cut surface was polished with 1-μm diamond powder and olive oil to prevent hydration. The thermal grooving was performed at 1500°C in an MgO crucible made of the same material as the specimen. The width and depth of the groove were measured with a sensitivity of 1 μm and 500 Å (50.0 nm), respectively, by using an interferometer[†] (λ = 543 μm).

Results and Discussion

Observations of the Tilt Boundary

After being heated by the schedule described above, the interface could not be recognized with the naked eye because the greater part of the single-crystal surfaces was joined well. However, many rectangular particles were observed at the outer part of the joined boundary with optical microscopy, as shown in Fig. 2. In these photographs, the dark parts with rectangular shape are joined, and the white parts are noncontacted areas with Newton rings. Figure 2(*a*) shows portions of a zero tilt boundary on the (100) plane as squares. The size of the squares decreased away from the central joined part, as shown in Fig. 2(*e,f, g*), and also

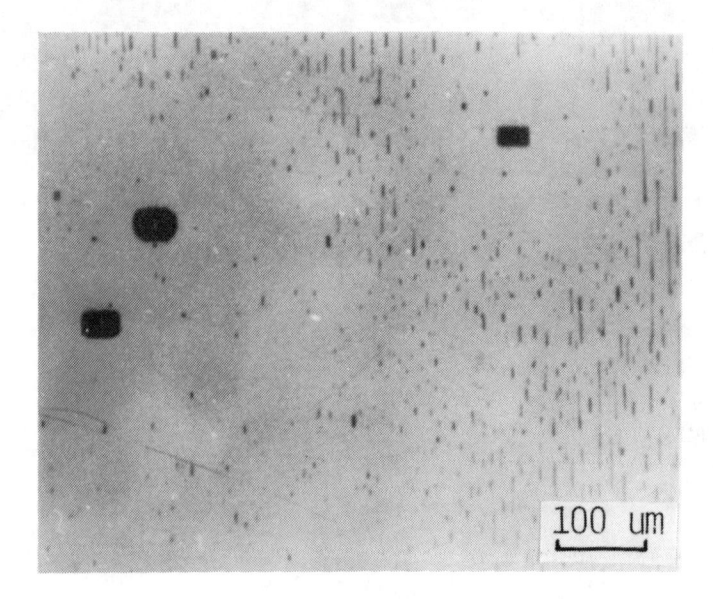

Fig. 3. Microstructure of the grain boundary around the perfectly joined area.

decreased with increasing cooling rate from 2480 to 2250°C. On the other hand, the aspect ratio of the rectangular particles increased with increasing tilt angle as shown in Fig. 2(a, b, c, d). Although the length of the shorter side decreased a little with increasing tilt angle, the length of the longer side which corresponds to the ⟨001⟩ direction, increased with increasing tilt angle.

These results suggest that the boundary formation of MgO is mainly controlled by vapor deposition. The preferred growth plane and direction should be (001) and ⟨001⟩, respectively. However, diffusion should also take part in the final stage, which is the disappearance of pores formed after the contact of rectangular particles at the interface, as shown in Fig. 3.

Observation of the microstructure by electron microscopy revealed many ellipsoidal pores of size 0.8×3 μm at a constant repeat distance, as shown in Fig. 4(a). If examined in detail, there were flat planes (facets) on the surface of the pores. Kinks could also be seen at the grain boundary, which indicated that there were regions of symmetric and asymmetric boundary. The same symmetric and asymmetric boundaries have also been reported for Si.[7] These structures result from the boundary formation process of vapor deposition. Around the pores, polycrystalline areas were observed; since many dislocations are present in this area, these are believed to be crystals of brucite produced during the chemical thinning. Figure 4(b) shows the (020) dark-field image of the 6.8° tilt boundary. We can recognize many crisscrossed edge dislocations at the grain boundary and parallel to the ⟨100⟩ direction as shown in Fig. 4(c, d). These suggest that two different edge dislocations are present. This will be discussed in the following section.

Boundary Energy

The grain boundary showed a hump after thermal grooving, as shown in Fig. 5. The width (W) and depth (d) of the grooves increased with heating period (t), as shown in Fig. 6. The width and depth agreed well with the reported value by Robertson.[8] The time exponent (m) of W and also d was about 0.25 ± 0.04. When bulk diffusion or evaporation–condensation is the dominant transport mechanism, the time exponent m should be 0.33 or 0.5, respectively. The obtained value of m indicates that thermal grooving of the tilt boundary is controlled by surface diffusion, as analyzed by Mullins.[9] After a long period of grooving (~20 h), the surface became roughened.

The measured value of the relative boundary energy (σ_b/σ_s) of a given (prescribed) tilt angle did not change during heat treatment from 10 to 80 h at 1500°C. The relative boundary energy increased with the tilt angle (θ) and approached a constant value of 0.65, as shown in Fig. 7. Dhalenne[10] reported that the relative boundary energy of NiO was also about 0.66 for a high-angle (001) tilt boundary. The value of 0.65 is higher than that of metals, showing that the grain-boundary energy of ceramics is closer to the surface energy than that of metals.[11]

Read and Shockley[1] proposed that the grain-boundary energy (E) would vary as a function of tilt angle (θ), based on the low-angle dislocation model, as shown in Eq. (1).

$$E = E_0\theta(A - \ln \theta) \tag{1}$$

where

$$E_0 = Ga/4\pi(1 - \mu) \tag{2}$$

$$A = 1 + \ln (a/2\pi r_0) \tag{3}$$

Fig. 4. TEM observation of the grain boundary: (a) bright-field image; (b) dark-field image; (c) crossed-edge dislocations on grain boundary in b; (d) parallel-edge dislocations on grain boundary.

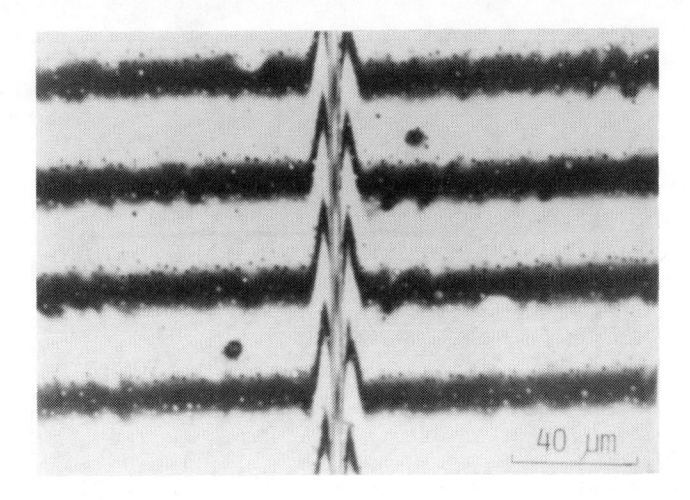

Fig. 5. Interferogram of a tilt grain boundary observed on MgO bicrystals.

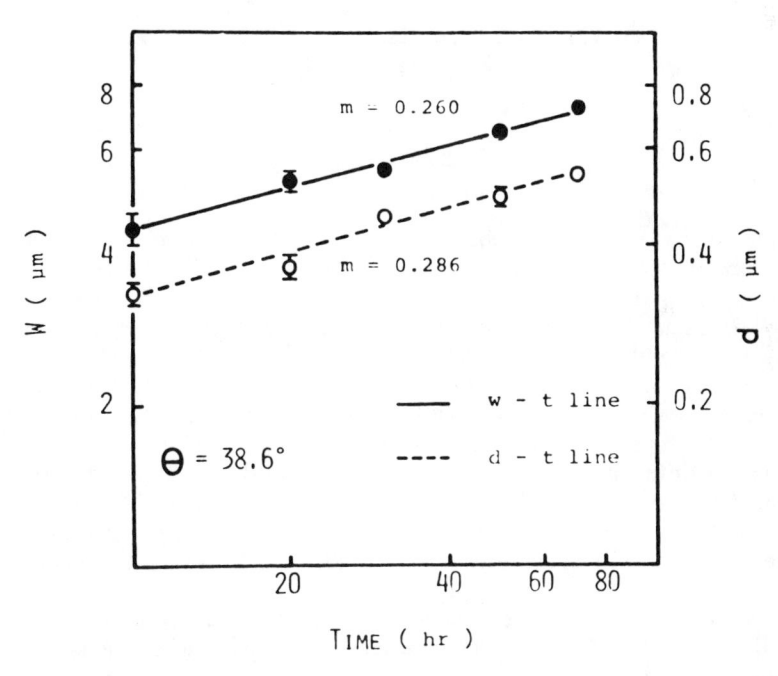

Fig. 6. Growth of width and depth of thermal groove annealed at 1500°C.

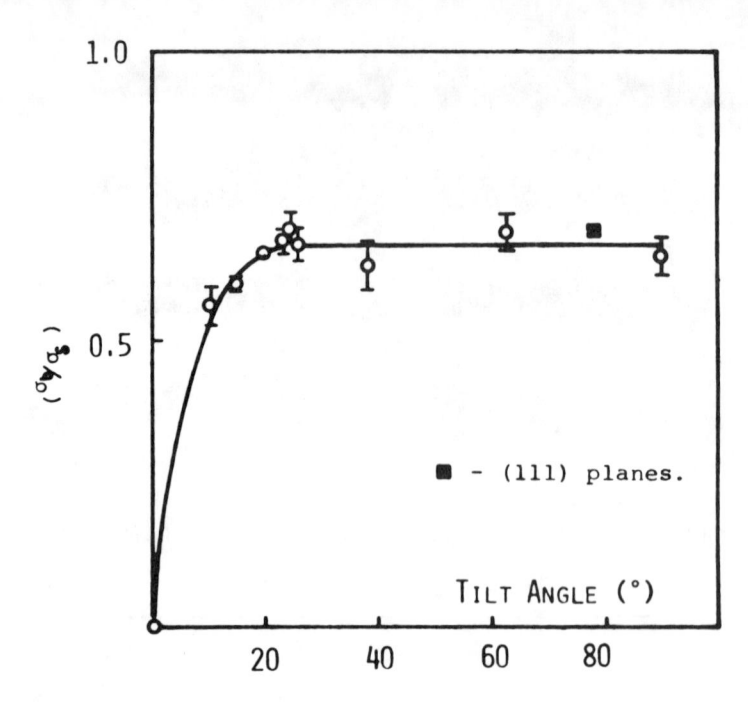

Fig. 7. Influence of the tilt angle θ upon the relative energy (σ_b/σ_s) of MgO at 1500°C.

G is the rigidity modulus, a is the lattice constant, μ is Poisson's ratio, and r_0 is the dislocation core radius. Plots of σ_b/σ_s vs ln θ for the (001) tilt boundary are shown in Fig. 8. The data can be represented by a straight line about 23° of tilt, beyond which the data deviate from linearity. Recently, Osenbach et al.[11] reported grain-boundary diffusion results in MgO bicrystals with tilt boundaries. They concluded that grain-boundary diffusion for low angles (<10°) could be related to the orientation of dislocations and that the mechanism was one of dislocation pipe diffusion. The upper limit of the low-angle tilt boundary behavior reported by Osenbach is different from our result. The difference should be derived from the measured properties. In metals, for example, Cu, the slope changed at about 5°.[12] The upper limit of the Read and Shockley model for ceramics appears to be larger than that for metals, perhaps because of the difference between the monatomic materials and ionic compounds.

For twist boundaries, the coincidence site lattice (CSL) theory has been applied successfully.[13] The CSL theory has been also applied to tilt boundaries; for example, a possible 36.9° symmetric tilt boundary structure in the rock-salt structure was suggested by Kingery on this basis.[14] A (001) tilt boundary model of MgO is illustrated in Fig. 9. We can see that two dislocations exist together at a point on the grain boundary. The repeat distance of the dislocations become shorter with an increase of the tilt angle as follows: 10°, 34 Å (3.4 nm); 15°, 22.5 Å (2.25 nm) 20°, 17.4 Å (1.74 nm); 30°, 12 Å (1.2 nm); 45°, 8.4 Å (0.84 nm). As can be seen in Fig. 4 (c,d), the shortest dislocation repeat distance is about 80 Å (8.0 nm), which agrees well with the calculated distance for the 6.8 tilt boundary (71 Å) (7.1 nm).

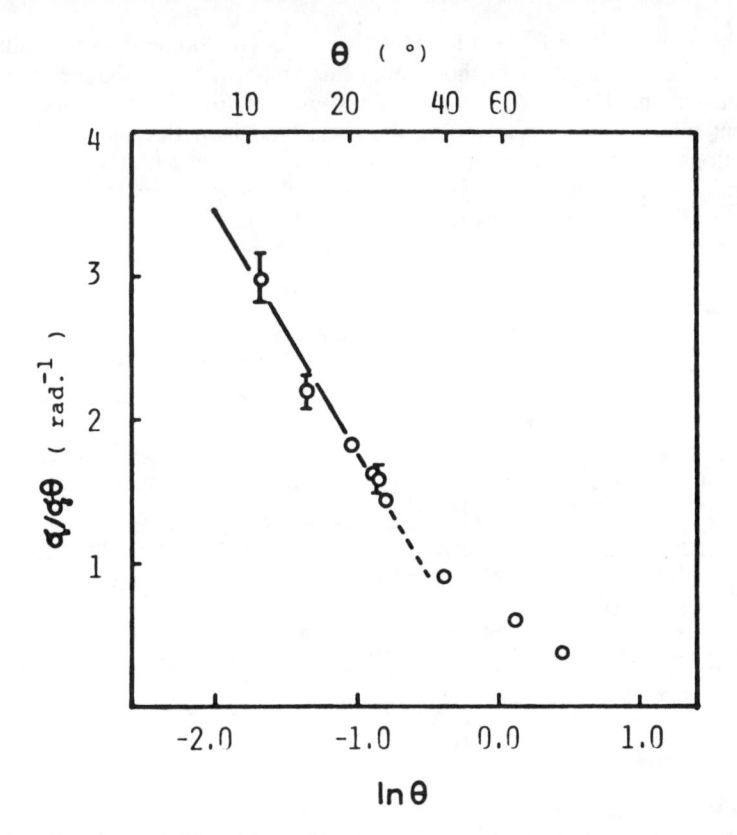

Fig. 8. Read and Shockley plot for symmetrical tilt boundary of MgO measured by thermal grooving at 1500°C.

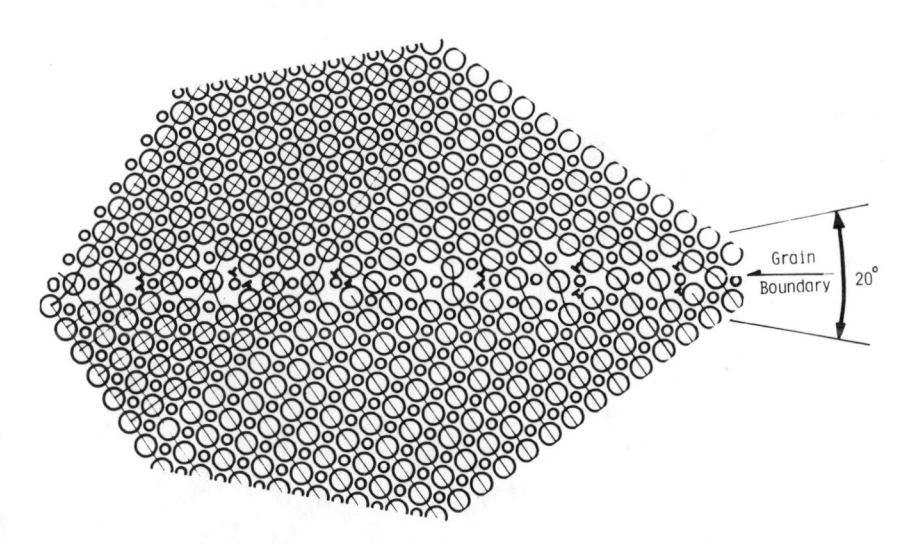

Fig. 9. Schematic illustration of symmetrical tilt boundary of MgO.

The value of A in Eq. (1) for MgO was calculated to be 0.08 from the data by using the least-squares method. With this value, the core diamter r_0 can be calculated to be 1.68 Å (0.168 nm). As shown in Fig. 4, dislocations of two different directions are observed on the grain boundary. Because of dislocation interactions, it is possible that the core diameter in the present (001) bicrystals of MgO may be larger than the single dislocation core diameter on the (011) plane for MgO (1.23 Å) (0.123 nm), as reported by Puls et al.[15]

References

[1] W. I. Read and W. Shockley, *Phys. Rev.*, **78**, 275–89 (1950).
[2] M. F. Ashby, F. Spaepen, and S. Williams, *Acta Metall.*, **26** 1647–63 (1961).
[3] J. C. M. Li, *J. Appl. Phys.*, **32**, 525–41 (1961).
[4] C. Fontaine and D. A. Smith; pp. 39–43 in Grain Boundary in Semiconductors. Edited by Pike et al. Elsevier, Amsterdam, 1982.
[5] R. W. Balluffi, P. D. Bristowe, and C. P. Sun, *J. Am. Ceram. Soc.*, **64** [1] 23–34 (1981).
[6] N. Fuschillo, M. L. Gimpl, and A. D. MacMaster, *J. Appl. Phys.*, **37**, 2044–51 (1966).
[7] B. Cunningham and D. Ast; pp. 21–6 in Grain Boundaries in Semiconductors. Edited by Pike et al. Elsevier, Amsterdam, 1982.
[8] W. M. Robertson, "Material Science Research"; pp. 49–60 in The Role of Grain Boundary and Surfaces in Ceramics, Vol. 3. Edited by Kriegel. Plenum, New York, 1966.
[9] W. W. Mullines, *J. Appl. Phys.*, **28** 333–9 (1957).
[10] G. Dhalenne, M. Dechamps, and A. Revcolevschi; pp. 13–9 in Grain Boundaries in Semiconductors. Edited by Pike et al. Elsevier, Amsterdam, 1982.
[11] J. W. Osenbach and V. S. Stubican, *J. Am. Ceram. Soc.*, **66** [3] 191–5 (1983).
[12] N. A. Gjostein and F. N. Rhines, *Acta Metall.*, **7**, 319–30 (1959).
[13] T. Schober and R. W. Balluffi, *Phys. Status Solidi B*, **B44**, 115–26 (1971).
[14] W. D. Kingery, *J. Am. Ceram. Soc.*, **57**, 1–8 (1974).
[15] M. P. Puls and M. J. Norgett, *J. Appl. Phys.*, **47**, 446–77 (1976).

---

*Hitachi H-1250, Hitachi Ltd., Tokyo, Japan.
†Nikon surface finish microscope, Nikon Co. Ltd., Tokyo, Japan.

Grain-Boundary Microstructures in Alumina Ceramics

D. S. Phillips and Y. R. Shiue
University of Illinois
Department of Metallurgy and Mining
Urbana, IL 61801

Grain-boundary microstructures in four alumina alloys, including both solid-state- and liquid-phase-sintered materials and a hot-pressed one, have been analyzed by TEM. Both liquid-phase-sintered doped materials and a hot-pressed undoped one showed readily identifiable basal and rhombohedral annealing twins. In contrast, a 2-year search covering over 200 boundaries in magnesia-doped solid-state-sintered material showed no twins. Those materials in which twins have been detected are invariably prone to exaggerated grain growth; the one in which they have not been detected is not. A preliminary experiment designed to measure the relative mobilities of twins and random boundaries is discussed.

The processing of dense ceramic bodies from powders of relatively pure (>99%) alumina alloys has received more attention than that of any other single oxide system. Existing technology permits fabrication of dense specimens of these materials by hot-pressing undoped powders[1] and by pressureless sintering with MgO[2] and $CaSiO_3$ additives. MgO-doped alumina appears to sinter in the solid state, while a representative $CaSiO_3$-doped material has been shown to contain a glassy grain-boundary phase indicative of liquid-phase sintering.[3]

Both undoped and $CaSiO_3$-doped materials are susceptible to exaggerated growth of a small fraction of the grains in the sample, leading to wide distributions of grain sizes and frequently to isolated pores trapped within the larger grains. MgO-doped sintered alumina, however, undergoes normal grain growth to quite large grain sizes (≥ 20 μm) and avoids pore detachment; their translucency makes these materials commercially important.

Because of this variety of grain growth behavior, alumina alloys seem an ideal system for investigating both differences in grain-boundary structure left from the several processing techniques mentioned and the possible importance of those differences in regulating the mobilities of grain boundaries in the products (i.e., propensity to exaggerated grain growth).

Experimental Procedure*

The grain size distributions in these materials were characterized by either scanning or transmission electron microscopy, depending on the average grain size. Specimens were prepared for TEM by argon ion thinning; a EM400T[†] microscope with an EDAX X-ray detection system was used.

An electron irradiation experiment using the Argonne National Laboratory HVEM was performed on McDanel 998* in order to measure the effectiveness of twins and "random" boundaries as vacancy sinks. This experiment was performed

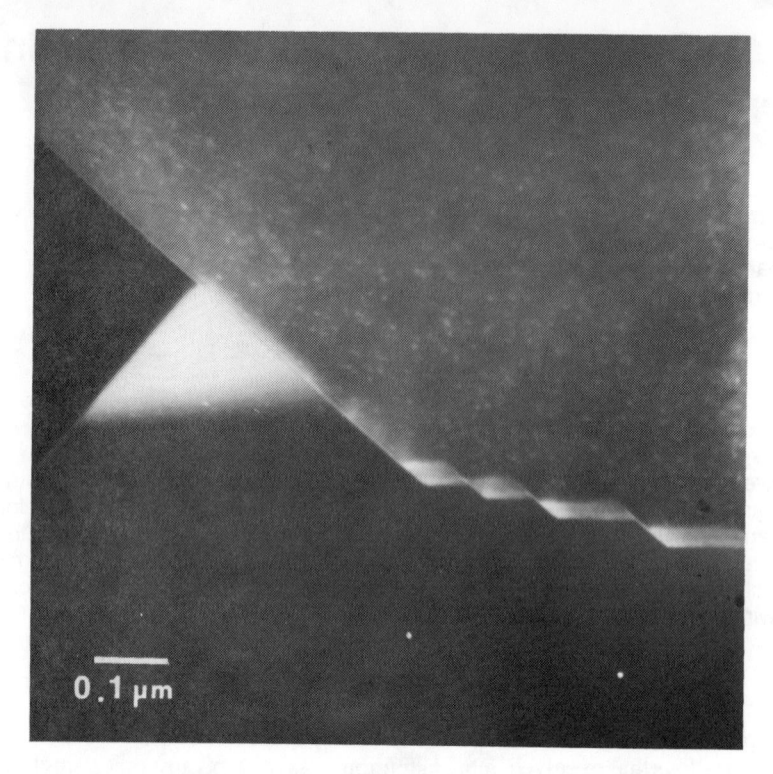

Fig. 1. Dark-field electron micrograph (from Ref. 3) formed by using a portion of the "amorphous halo" diffracted by the glass in McDanel 998. The boundary region is strongly faceted and wet by the glass.

at 800°C at a fluence of 1×10^4 C/cm^2; the results were analyzed by using the EM400T after slightly rethinning the specimen.

Results

Figure 1 is a dark-field micrograph showing a large triple-point glass pocket in the McDanel 998. It is evident that the glass does wet the adjacent boundary but that its distribution is not homogeneous. This boundary, along with numerous other glass-wet ones investigated, is faceted, in this case on ($\bar{1}012$) and ($11\bar{2}0$). Figure 2 is an energy-dispersed X-ray spectrum showing the glass to be rich in Ca, Si, and Al; the high Al concentration explains the large volume of glass present at the doping level given. The composition inferred from the spectrum with the assumption of a thin film limit is 40% Al_2O_3, 30% SiO_2, and 30% CaO by weight.

Basal annealing twins are particularly common in the liquid-phase-sintered materials. They sometimes but not always contain secondary grain-boundary dislocations, as does the one shown in Fig. 3. Figure 3(B) shows a diffraction pattern taken from the boundary region that identifies the boundary as a basal twin — each of the two (1012) reflections labeled is forbidden in one of the crystals. The plane of the boundary is ($11\bar{2}0$), and the termination of the boundary makes a near 90° angle with the two neighboring boundaries. A conventional **g·b** analysis

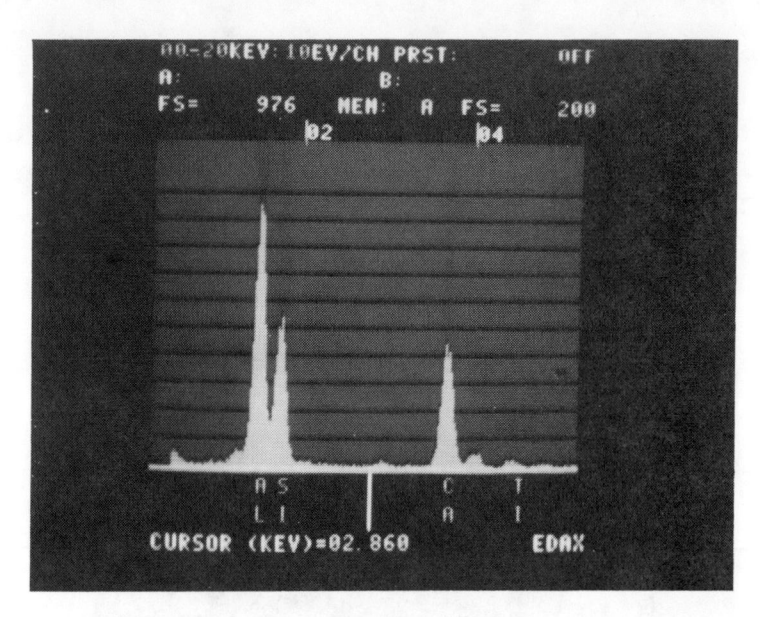

Fig. 2. Energy-dispersed X-ray analysis showing strong Al, Si, and Co and weaker Mg and Ti signals from a glass pocket in McDanel 998. The glass composition is roughly 40% Al_2O_3, 30% SiO_2, and 30% CaO by weight.

of the dislocations in the boundary shows that they have the three $⅓\langle 10\bar{1}0\rangle$ Burgers vectors.

A much smaller number of rhombohedral twins has been observed primarily in hot-pressed material; one is shown in Fig. 4 along with a diffraction pattern sufficient for its recognition. These twins lie predominantly on $(\bar{1}012)$ rhombohedral planes and invariably contain dislocations, usually of Burgers vector $⅓\langle 10\bar{1}1\rangle$.

Our qualitative observations of special boundaries in a variety of alumina alloys are summarized in Table I. Note that materials prone to exaggerated grain growth tend to contain both glassy phases and twin boundaries, but the absence of glass detected to date in the LANL hot-pressed material suggests that the latter correlation is the more important.

Figure 5 shows the results of the electron irradiation experiment. Small dislocation loops were formed by condensation of interstitials in the neighborhood of both twin (Fig. 5(A)) and random (Fig. 5(B)) boundaries, with no clearly denuded zone visible in either case. The loops near the twin boundary appear to be larger and less numerous than those near the random one. Voids formed by vacancy condensation are less visible in the figure and likewise indicate no denuded zone. An usually large void concentration however was noted at the random boundary.

A typical three-grain junction in MgO-saturated material is shown for comparison in Fig. 6. Note the 120° dihedral angle, suggesting that all three boundaries have nearly equal surface tensions. A search covering over 200 boundaries failed to detect a single annealing twin of either variety or any evidence of amorphous boundary phase in this material.

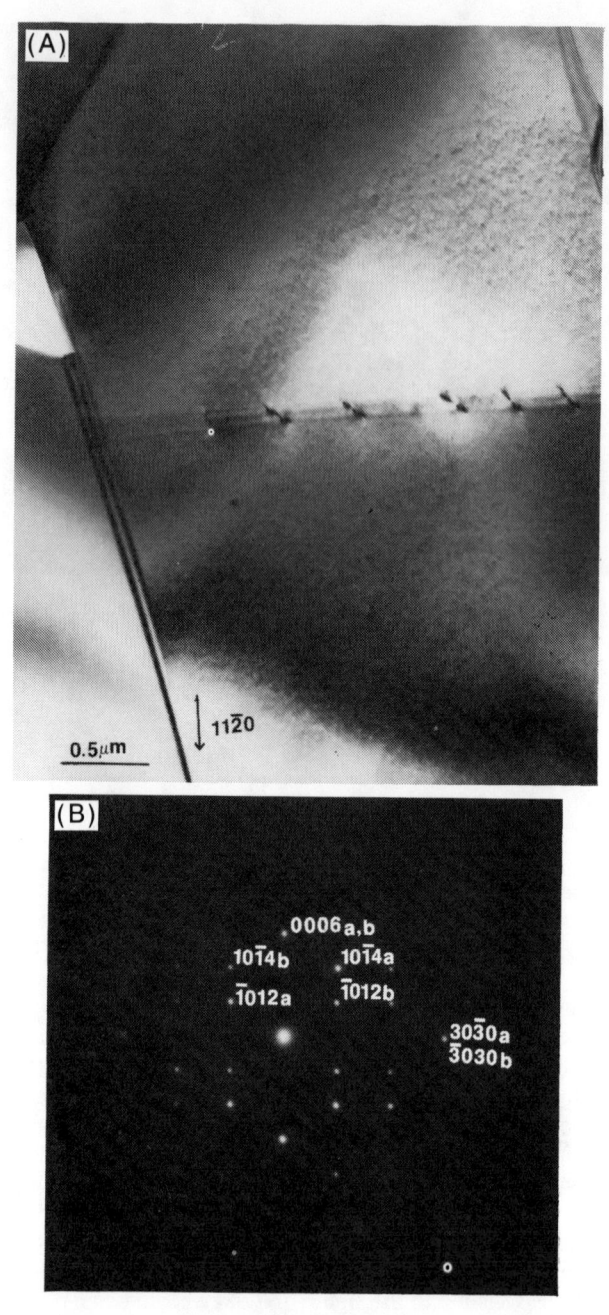

Fig. 3. (A) Bright-field micrograph showing a dislocated basal twin boundary in Feldmuhle implant material. Note that the neighboring boundary remains nearly straight at the termination of the twin. The presence of dislocations proves that the twin boundary is not wet by glass. (B) Selected-area diffraction pattern from basal twin boundary showing superimposed $\langle 1\bar{2}10 \rangle$ and $\langle \bar{1}2\bar{1}0 \rangle$ patterns diagnostic of a basal twin.

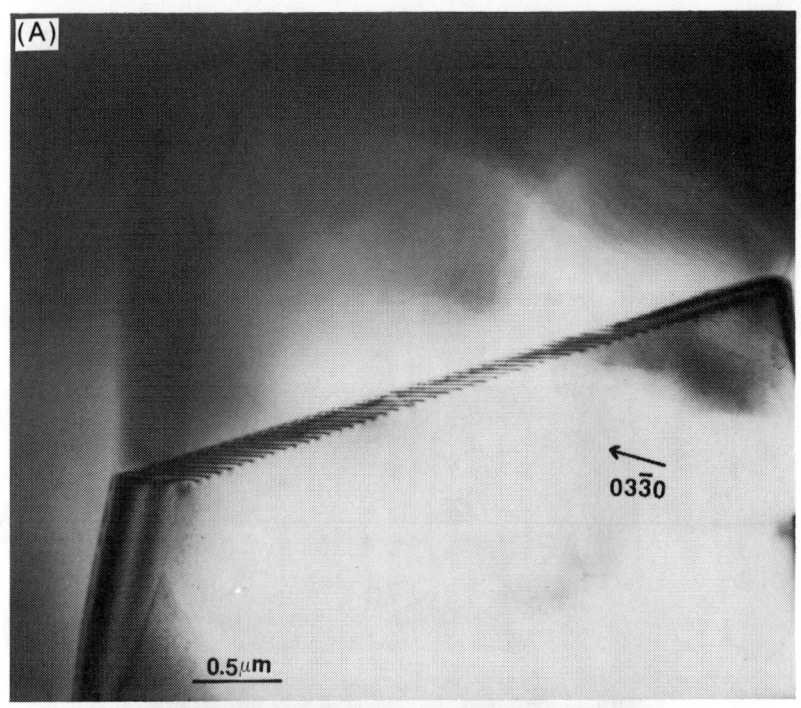

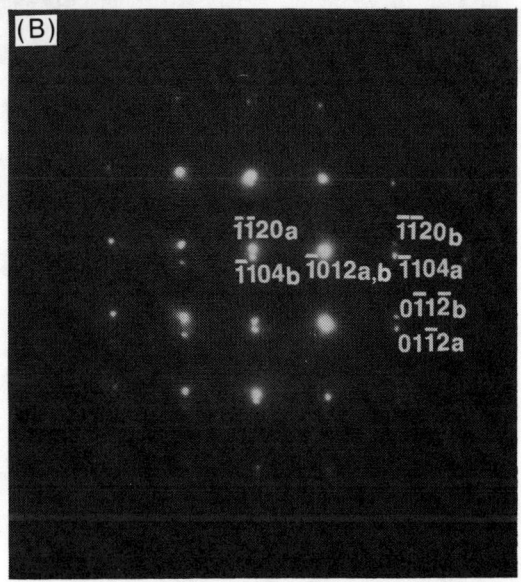

Fig. 4. (A) Bright-field micrograph showing a dislocated rhombohedral twin in laboratory hot-pressed material. Again the boundary has a lower than usual surface tension and is not wet by glass. (B) Selected-area electron diffraction pattern from rhombohedral twin boundary showing superimposed ⟨$\bar{2}201$⟩ and ⟨$22\bar{0}\bar{1}$⟩ patterns diagnostic of a rhombohedral twin.

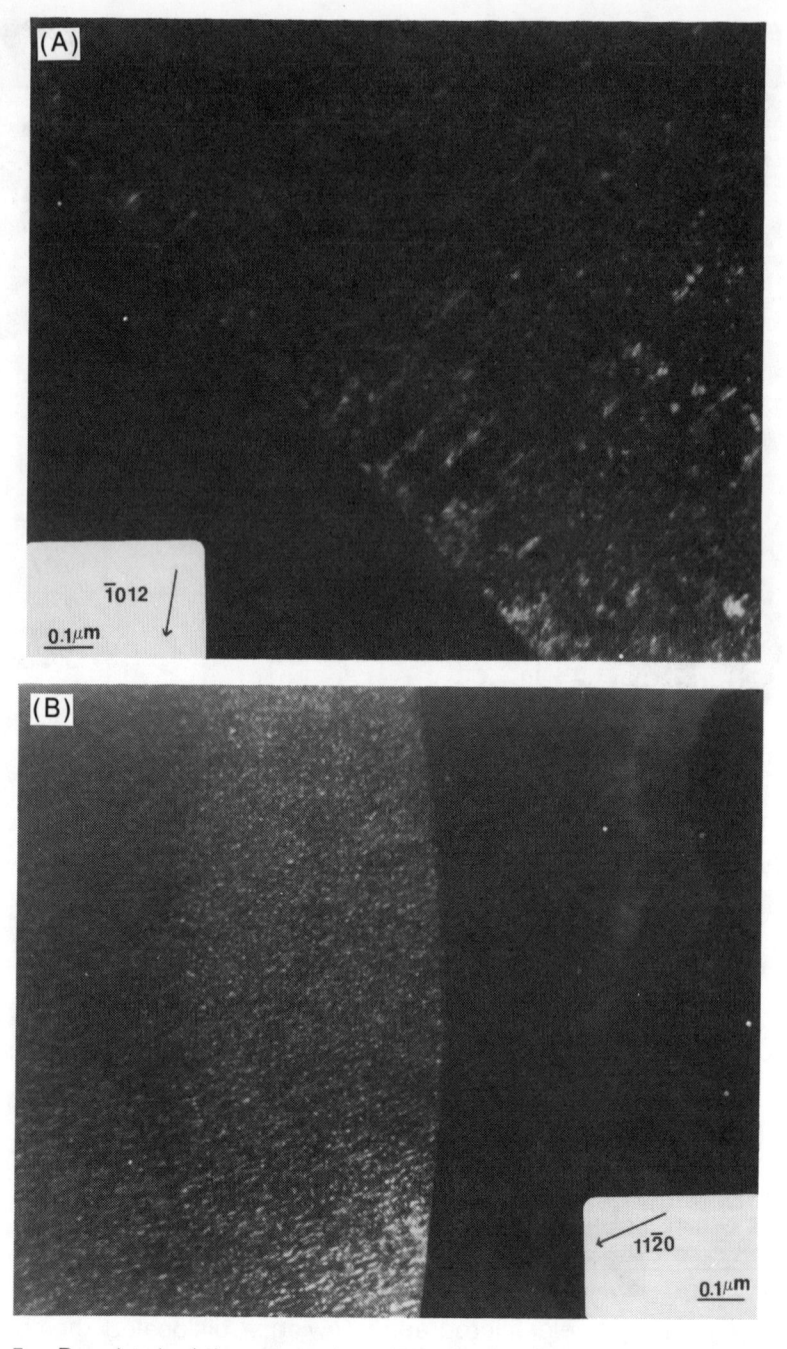

Fig. 5. Basal twin (*A*) and random (*B*) boundaries in McDanel 998 irradiated to 2×10^4 C/cm^2 at 800°C. Small dislocation loops are formed by coalescence of irradiation-induced point defects even in the boundary region, indicating that neither boundary can absorb point defects at the production rate obtained.

Fig. 6. Typical three-grain junction in MgO-saturated sintered alumina showing classic 120° dihedral angles. Neither twins nor glass nor exaggerated grain growth has been observed in this material.

Discussion

Observations in the CaSiO$_3$-doped material show two major anisotropies in surface tension in the faceting behavior of glass-wet boundaries and in the small surface tension of apparently clean twin boundaries. The remainder of the experimental program discussed here is designed to find which, if either, of these causes can be linked with the onset of exaggerated grain growth in alumina.

The presence of a continuously glassy phase is far from unique, and similar glasses are well-known in Si$_3$N$_4$ ceramics.[5,6] The chemical stability of alumina

Table I. Special Boundaries in a Variety of Alumina Alloys

Material	Detectable liquid?	Mode of growth AR*	Mode of growth AN*	Special boundaries?[‡]
McDanel 998	yes	exg[†]	exg[†]	sa, b
McDanel impl	no	nrm[†]	exg[†]	sa, b, r
Feldmuhle	yes	nrm[†]	exg[†]	sa, b
LASL HP	no	nrm[†]	exg[†]	sa, b, r
Lucalox	no	nrm[†]	nrm[†]	none

*AR, AN: as-received, annealed condition.
[†]exg, nrm: exaggerated, normal grain growth.
[‡]sa, b, r: small angle boundaries, basal twins, rhombohedral.

alloys however permits comparison of the observed glass composition with the established Al_2O_3-CaO-SiO_2 equilibrium diagram, Fig. 7 (from Ref. 7). The experimental composition is found to lie at a ternary eutectic, as described by Kingery.[8] This eutectic is *not* the first encountered on cooling from the bulk composition at the alumina corner as might at first be expected; this "overshoot"

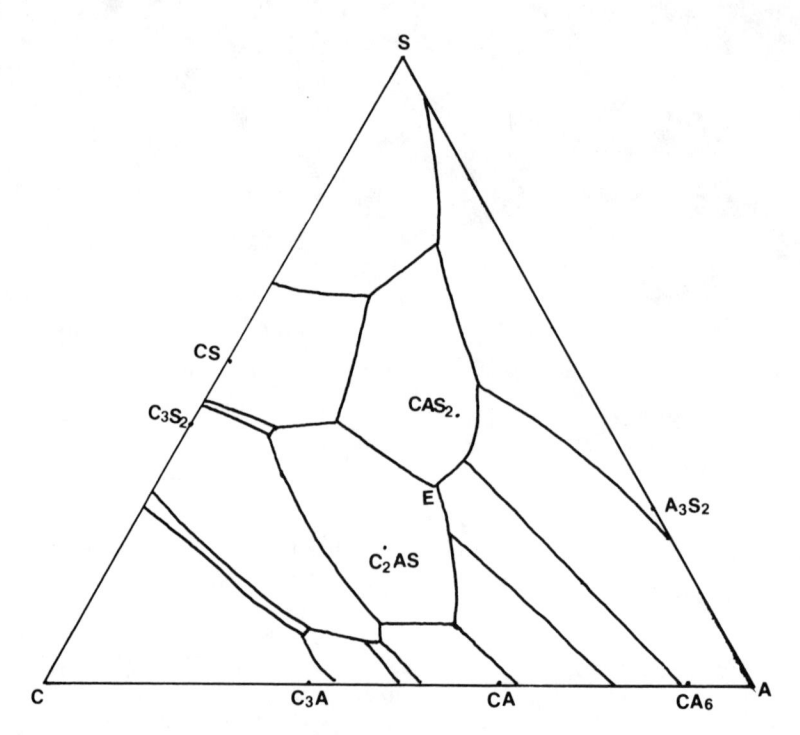

Fig. 7. Al_2O_3-CaO-SiO_2 equilibrium phase diagram. The bulk composition of McDanel 998 is in the Al_2O_3 corner; the measured composition of the glass is indicated. The glass corresponds to an unexpected eutectic composition (E), which may result from either low processing temperature or nonequilibrium cooling.

is also described by Kingery and may result either from nonequilibrium cooling or else from processing below the temperature required for the higher eutectic reaction (1545°C).

The formation of twin boundaries during grain growth is well-known in recrystallized metals but is much less familiar in ceramics. The formal geometry of these twins in alumina is discussed in these proceedings by Morrissey and Carter.[9]

Since $(11\bar{2}0)$ is a mirror plane in alumina, it cannot be a twinning plane. The fact that a large number of our observations lie on $(11\bar{2}0)$ shows that these are not mechanical twins, which are expected to lie on K_1. The $\frac{1}{3}\langle10\bar{1}0\rangle$ Burgers vectors are observed as partials in single crystals[10] and are translation vectors of the DSC lattice for this twin as well;[9] they have been reported by Hockey[11] in deformation-induced basal twins, and the propagation of the coherent twin boundary has been explained by their motion. The presence of dislocations in twin boundaries in liquid-phase-sintered material is important because it indicates that the twin boundaries cannot be wet by glass, since the amorphous film would destroy the DSCL in which the Burgers circuit defining the SGBD's is performed. The basal twin faults only the aluminum sublattice and makes a poor paradigm for twinning in ceramics of other crystal structures; nonetheless, it seems from Table I to be the single microstructural feature most strongly discriminative between specimens undergoing normal and exaggerated grain growth.

The rhombohedral twins are more common in hot-pressed than in sintered materials. Again, they are recognizable by their small interfacial energy. They lie on $(\bar{1}012)$ planes, which are both the most dense planes of the CSL and K_1 as well, so that these may originate from deformation.

It is to be expected that differences in the microstructures of these boundaries will be reflected in differences in their properties, including both interfacial energy and mobility. It has already been shown (Ref. 3 and the present study) that the surface tensions of special boundaries are markedly (roughly a factor of 10, if one assumes force balance at triple junctions) lower than those of neighboring nonspecial boundaries. This reduction is sufficiently large to prevent the capillary rise of the otherwise widely distributed glass up the twins in liquid-phase-sintered materials. A similar anisotropy in surface tension must drive the faceting of alumina/glass boundaries.

It is likely that both these structural anisotropies produce anisotropic boundary mobilities as well. Yan[12] has shown that variations in boundary mobility can lead to exaggerated growth of "grains" in computer simulation studies and has mentioned both glass distribution and formation of special boundaries as potential sources of that variation. Indeed, such variations in structure and mobility are well-known in metals, where glassy boundary phases are known not to form. Rutter and Aust[13] showed that boundaries in lead near special coincidence orientations have higher mobilities than random boundaries and that the mobilities of the special boundaries are less effectively reduced by solid solution than those of the random ones. Seigel et al.[14] likewise showed that coherent annealing twins in gold are largely ineffective as sinks for quenched-in vacancies, while random boundaries were highly effective; thus, random boundaries produced zones denuded of stacking fault tetrahedra on annealing while twin boundaries did not. Since polycrystalline alumina is known for its poor performance in thermal shock, the present irradiation experiment was designed as a test (without quenching) of sink efficiency for twins in alumina.

The metallurgical literature also offers a plausible explanation for the inconclusive result of the irradiation experiment. The chemical driving force produced here by point defect supersaturation should be several orders of magnitude greater (roughly 6 orders; following Ref. 15) than that available from grain boundary curvature to drive exaggerated grain growth. It is possible that in the face of these driving forces the subtleties of crystallographic anisotropy become negligible factors in controlling boundary mobility.

Another possible difficulty in interpreting this result stems from the presence of widely dispersed glass in the sample. It appears that void formation at the random boundary occurred primarily in the boundary glass. In this case, we hypothesize that during our irradiation the random boundary was saturated early by damage in the glass and was subsequently unable to produce the expected denuded zone.

Conclusions

Commerical liquid-phase-sintered aluminas are typically rich in both faceted crystal/glass boundaries and in basal annealing twins. They are also prone to exaggerated grain growth. A laboratory hot-pressed material apparently without glassy phase showed both basal and rhombohedral twins and a propensity to exaggerated grain growth. The propensity is a consequence of variation in mobility between special and random boundaries and should be demonstrable in comparative irradiation experiments, although a first round of these experiments was inconclusive.

Acknowledgments

This work was supported by the Division of Materials Sciences of the U.S. Department of Energy under Contract No. DE-AC02-76ER01198. It utilized facilities from the Center for Microanalysis of Materials of the University of Illinois Materials Research Laboratory and from the Tandem/High Voltage Electron Microscope facility of the Argonne National Laboratory. We thank L. K. V. Lou and Frank Gac and D. E. Day, who have contributed materials to this investigation.

References

[1]T. Vasilos and R. M. Spriggs, "Pressure Sintering: Mechanisms and Microstructures for Alumina and Magnesia," *J. Am. Ceram. Soc.*, **46** [10] 493–6 (1963).
[2]R. L. Coble, "Sintering Crystalline Solids: II, Experimental Test of Diffusion Models in Powder Compacts," *J. Appl. Phys.*, **32** [5] 793–9 (1961).
[3]S. C. Hansen and D. S. Phillips, "Grain Boundary Microstructures in a Liquid-Phase Sintered Alumina," *Philos. Mag.*, **47** [2] 209–34 (1983).
[4]D. G. Howitt and T. E. Mitchell, "Electron Irradiation Damage in Alumina," *Philos. Mag.*, [Part A], **44** [1] 229–38 (1981).
[5]L. K. V. Lou, T. E. Mitchell, and A. H. Heuer, "Impurity Phases in Hot-Pressed Si_3N_4," *J. Am. Ceram. Soc.*, **61** [9–10] 392–6 (1978).
[6]D. R. Clarke, "High Resolution Techniques and Application to Nonoxide Ceramics," *J. Am. Ceram. Soc.*, **62** [5–6] 236–46 (1979).
[7]E. M. Levin, C. R. Robbins, and H. F. McMurdle; Phase Diagrams for Ceramists, No. 631, 1964.
[8]W. D. Kingery, H. K. Bowen, and D. R. Uhlmann; p. 317 in Introduction to Ceramics, 2nd ed. Wiley, New York, 1976.
[9]K. Morrissey and C. B. Carter; these proceedings.
[10]A. H. Heuer and J. Castaing; these proceedings.
[11]B. J. Hockey, "Pyramidal Slip on (11$\bar{2}$3) ($\bar{1}$100) and Basal Twinning in Al_2O_3"; pp. 167–80 in Deformation of Ceramic Materials. Plenum, New York, 1975.
[12]M. F. Yan, "Microstructural Control in Processing Electronic Ceramics," *Mater. Sci. Eng.*, **48** [1] 53–72 (1981).

[13]J. W. Rutter and K. T. Aust, "Migration of ⟨100⟩ Tilt Boundaries in Lead," *Acta Metall.*, **13** [3] 181–6 (1965).
[14]R. W. Seigel, S. M. Chang, and R. W. Balluffi, "Vacancy Loss at Grain Boundaries in Quenched Polycrystalline Gold," *Acta Metall.*, **28** [3] 249 (1980).
[15]R. W. Balluffi, "High Angle Grain Boundaries as a Source or Sinks for Point Defects"; pp. 297–328 of Grain Boundary Structure and Properties. American Society for Metals, 1980.

---

*Materials for this study were obtained from a variety of sources. Our representative liquid-phase-sintered material was commercial McDanel 998 containing 99.8% Al_2O_3 (by weight), 0.06% SiO_2, 0.04% CaO, and numerous minor impurities. Our representative hot-pressed material was prepared by Frank Gac of Los Alamos National Laboratory by holding Baikowski Cr-10 powder for 45 min at 1350°C under 4000 psi (2.75 × 10 Pa). Our magnesia saturated material contained 0.15% MgO and was donated by L. K. V. Lou of General Electric Co. Two other biological implant materials were provided by D. E. Day of the University of Missouri — one material (from Feldmuhle) contained detectable glass while another (from McDanel) did not.

†Phillips Electronic Instruments Co., Mount Vernon, NY.

Solute Segregation at Grain Boundaries in Polycrystalline Al$_2$O$_3$

Chien-Wei Li and W. David Kingery

Massachusetts Institute of Technology
Department of Materials Science and Engineering
Cambridge, MA   02139

Polycrystalline alumina samples containing Ca, Si, Ga, Sc, La, V, Zr, Mn, Y, Ni, Mg, Zn, Ti, and Cr have been examined in a dedicated scanning transmission electron microscope to determine the extent of solute segregation at the grain boundaries. In many samples second phases occurred at triple grain junctions or along grain boundaries. For divalent and trivalent solutes, ion misfit strain energy seems to be the main driving force for segregation, and results are well represented by McLean's equilibrium segregation isotherm. Silica also fits this relationship, but there is evidence of a valence effect for tetravalent titanium.

Solute segregation to grain boundaries is a quite general phenomenon widely observed in both metals[1] and oxides[2] and is particularly important in relationship to diffusional processes such as the sintering of ceramics and high-temperature deformation. In metals, the driving force for boundary segregation is usually attributed to solute strain energy being lower at the grain boundary than in the bulk.[3] For ionic solids, the electrostatic interaction between aliovalent solutes and grain boundaries at a potential different from the bulk must also be considered.[2] While substantial segregation of large ions such as yttrium[4,5] and calcium[5] has been reported, the experimental results reported for divalent ions such as magnesium[6] and nickel[4,5,7] have been contradictory in aluminum oxide.

The objective of this study is to report measurements of grain boundary segregation of a number of different solute materials in aluminum oxide and evaluate the relative influence of size misfit strain energy and valence effects on the segregation.

Experimental Methods

Sample Preparation

Three different sets of samples of aluminum oxide with solute additions have been available for study. One group of samples was doped with 0.2 or 0.5 wt% of Ga$_2$O$_3$, Sc$_2$O$_3$, La$_2$O$_3$, V$_2$O$_5$, ZrO$_2$, MnO$_2$, or TiO$_2$. These samples were prepared by mixing the solutes in solution with a slurry of alumina powder* in glass vessels, calcined at 1000°C, pressed, and fired at 1775°C in oxygen for 2 h. These samples were withdrawn from the furnace and rapidly air-cooled. In addition to the added solutes, some of these samples contain about 0.2 to 0.5 wt% silica. One TiO$_2$-containing sample was fired in air and another in hydrogen, which reduced the titania to Ti$_2$O$_3$.[8]

A second set of samples was prepared by adding 0.2 or 0.5 wt% NiO, Y$_2$O$_3$, MgO, ZnO, or TiO$_2$ to a slurry of alumina* mixed in plastic containers and fired

96 h in oxygen at 1575°C. These samples were withdrawn from the furnace and immediately quenched into liquid nitrogen.

A third set of samples were hot-pressed 15 min in graphite dyes at 1500°C (with 0.7 wt% Cr_2O_3) and 1800°C (with 1.0 wt% Y_2O_3) to form dense samples. These were prepared at the Ceramics Laboratory, École Polytechnique de Lausanne.

STEM Measurements

The polycrystalline samples were mechanically thinned to a thickness of 50–100 μm. Ion milling was then used to thin the samples to a thickness suitable for electron microscope observation, 500–2000 Å (50–200 nm). A carbon coating was applied to each foil to eliminate charging, prior to measurements in the scanning transmission electron microscope.[†]

The grain boundaries studied were parallel to the electron beam and had an apparent boundary width no more than 20 Å (2 nm). The electron beam was manually moved along a direction perpendicular to the boundary and X-ray counts were collected and analyzed. The grain-boundary segregation data are obtained in terms of the intensity of X-ray radiation from the solute to the intensity of the aluminum signal. Typical data showing large (La, Ti) and moderate (Sc, Ni) boundary segregation are illustrated in Fig. 1. In cases where silicon was present, boundary segregation of the silicon was always observed and calcium segregation occurred even when the bulk concentration was below detection limits. Experimental data for all the samples are collected in Table I.

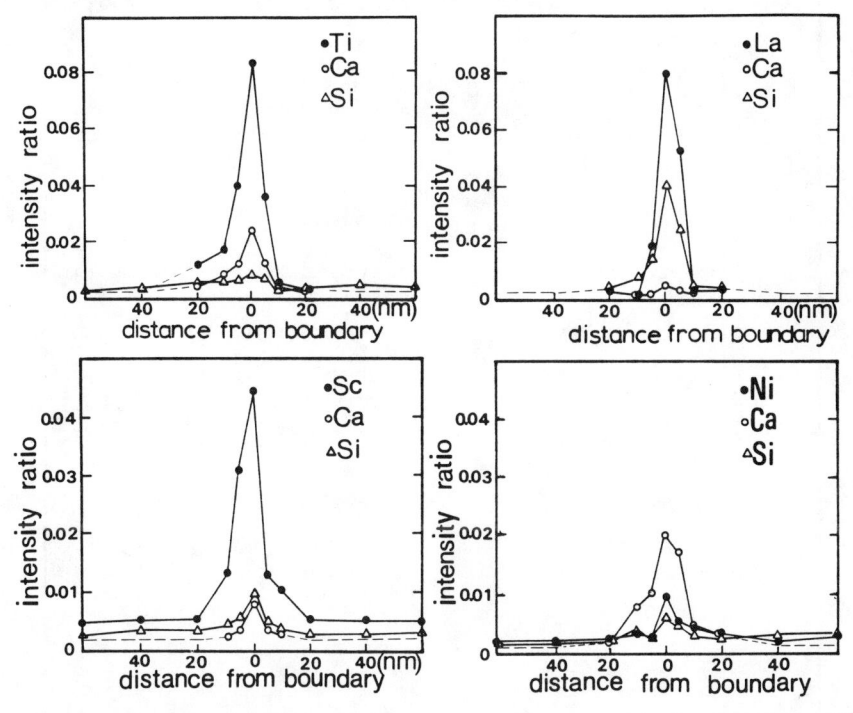

Fig. 1. X-ray count ratio data (I_{solute}/I_{Al}) for samples of Al_2O_3 containing Ti, La, Sc, Ni, Si, and Ca as solutes.

Table I. Experimental Data for the Samples Measured

SiO_2^* (wt%)	Solute (wt%)	Firing condition	$\left(\dfrac{I_{Sol}}{I_{Al}}\right)_{gb}$	$\left(\dfrac{I_{Sol}}{I_{Al}}\right)_{l}$	$\left(\dfrac{I_{Si}}{I_{Al}}\right)_{gb}$	$\left(\dfrac{I_{Si}}{I_{Al}}\right)_{l}$	$\left(\dfrac{I_{Ca}}{I_{Al}}\right)_{gb}$	Second phase at boundary	$\left(\dfrac{I_{Sol}}{I_{Al}}\right)_{gb} \Big/ \left(\dfrac{I_{Sol}}{I_{Al}}\right)_{l}$ [†]
0.46	0.5 La_2O_3	O_2, 1775°C	0.080	not detected	0.040	not detected	0.005	no	~100±
0.3	0.5 ZrO_2	O_2, 1775°C	0.019	not detected	0.003	not detected	0.004	no	
0.2	0.5 MnO_2	O_2, 1775°C	0.020	0.002	0.014	not detected	not detected	yes	10
0.1	0.2 TiO_2 (1) liquid layer indicated	O_2, 1525°C	0.17	not detected	0.06	0.004	0.15	yes	~150±
	(2) "clean" boundary		0.085	not detected	0.009	0.004	0.025	no	~80±
0.1	0.5 TiO_2	air, 1775°C	0.062	not detected	0.025	0.009	0.01	no	~60±
0.1	0.5 TiO_2	H_2, 1775°C	0.0105	0.005	not detected	not detected	0.014	no	2.1
<0.01	0.2 Ga_2O_3	O_2, 1775°C	0.0036	0.0030	0.008	0.003	0.021	no	1.2
<0.001	0.2 Sc_2O_3	O_2, 1775°C	0.045	0.005	0.010	0.003	0.009	no	9
0.02	0.5 V_2O_5	O_2, 1775°C	0.010	0.0025	0.007	0.002	0.003	probably	4
<0.01	0.2 Y_2O_3	O_2, 1575°C	0.024	0.0009	0.009	0.003	0.004	no	27
0.03	0.5 NiO	O_2, 1575°C	0.009	0.0025	0.006	0.002	0.02	no	3.6
<0.01	0.5 ZnO	O_2, 1575°C	0.0023	0.0006	not detected	not detected	0.0032	no	3.8
<0.01	0.5 MgO	O_2, 1575°C	not detected	not detected	not detected	not detected	0.01	no	
Hot-Pressed:									
<0.01	0.7 Cr_2O_3	graphite, 1500°C	0.0071	0.0079	not detected	not detected	not detected	no	1.0
<0.07	1.0 Y_2O_3	graphite, 1800°C	0.014	0.0025	0.004	0.001	0.012	no	5.6

*Atomic absorption spectroscopy.
[†]The actual segregation values are estimated to be about 4 times these experimental apparent segregation values.

Data Analysis

Within the range of studied sample thicknesses and with the mass absorption coefficients for the characteristic X rays, absorption corrections for transforming intensity ratios to concentration ratios are unnecessary, and we have used the simple Cliff–Lorimer relationship, $C_A/C_B = K_{AB}I_A/I_B$,[9] where C_A and C_B are the weight fractions of element A and B and I_A and I_B are the characteristic X-ray intensities of element A and B above background. K_{AB} is the Cliff–Lorimer constant.

While the diameter of the electron beam used is small (15–25 Å), it is larger than the thickness of the segregation at the grain boundary (10–20 Å) and there is electron beam spreading during transmission through the sample. As a result, X rays are generated from a volume larger than that of the grain boundary region and some correction is necessary to convert the measured X-ray intensity ratio into a "boundary concentration." Using a simple electron single-scattering model, Bender et al.[4] developed a relationship for the ratio of the true solute boundary concentration to the apparent value, which depends on the grain boundary enrichment width, the electron beam diameter, and the beam scattering characteristics. For a beam diameter of 20 Å, an enriched segregation width of 10 Å, and an aluminum oxide sample thickness of 1000 Å, the true boundary concentration is about 5 times the apparent value. Depending on the exact beam diameter, boundary enrichment thickness and sample foil thickness, this ratio may vary between 2.5 and 8. More precise Monte Carlo calculations indicate that multiple scattering leads to somewhat less variation than this,[10] as experimentally found by Hall et al.[11] It has been pointed out by various authors[12,13] that the true solute grain boundary concentration can only be deconvoluted by Monte Carlo calculations for each case. However, as an approximation to indicate the order of magnitude of the concentrations associated with grain boundary segregation, we will use the relationship derived from Bender et al.:[4]

$$\beta = \left(\frac{C_{gb}}{C_l}\right)_{true} = A\left(\frac{C_{gb}}{C_l}\right)_{app} + 1 - A \quad (1)$$

with $A = 4$, where C_{gb} and C_l are the solute concentrations in the grain boundary and grain lattice, respectively, and β is the enrichment factor.

Results and Discussion

Second Phases at the Boundary

The width of boundary observed in scanning transmission electron microscopy is about 20 Å. Without further more powerful techniques such as lattice imaging, it is not possible to say with certainty from this observation that no second-phase boundary film occurs. However, when a second phase is present with a finite contact angle (Fig. 2) such as the spinel grains in magnesia-doped alumina, this provides good assurance that there is no continuous second phase along boundaries. Nickel-doped and zinc-doped samples also have spinel precipitates, which also have dihedral angles approaching 90°. In contrast, for the MnO_2/SiO_2-doped sample, faceted grain boundaries contain a continuous manganese and silica-rich second phase, as shown in Fig. 3. In this system, no boundaries were observed that were assuredly free of a second phase.

For some boundaries in which $TiO_2 + SiO_2$ was present in the system, defocusing of the STEM image led to observation of fringes corresponding to a continuous thin wetting second phase between grains[14] (Fig. 4). In other areas of

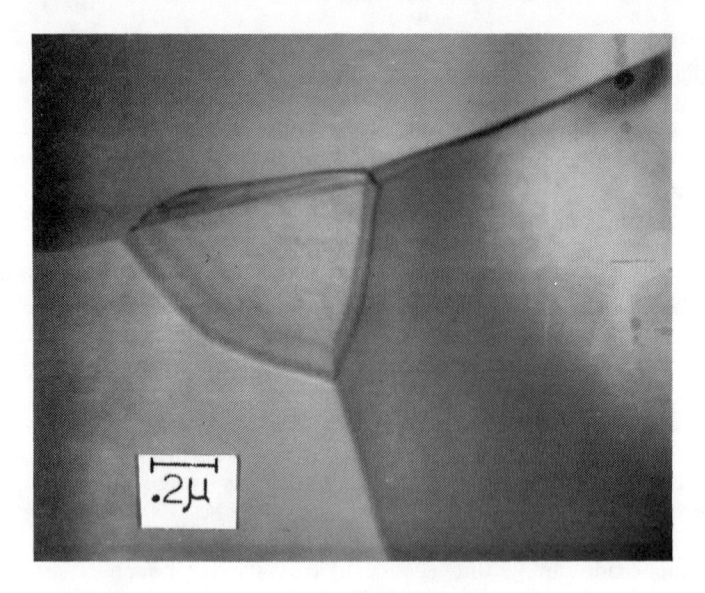

Fig. 2. MgAl$_2$O$_4$ spinel particles at triple grain junctions have a high contact angle.

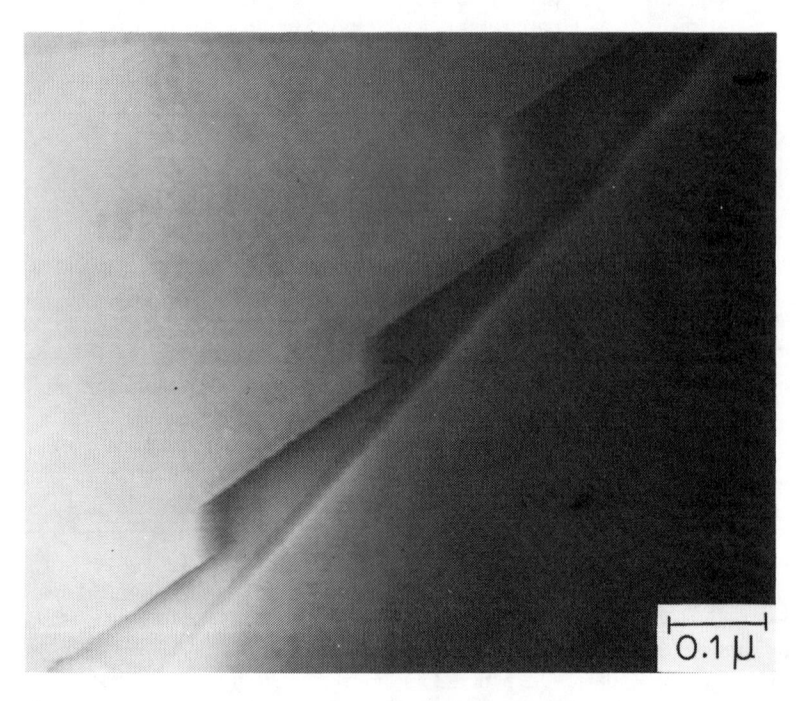

Fig. 3. Grain boundaries in Mn Si-doped Al$_2$O$_3$ consist of a manganese- and silica-rich phase separating faceted crystals.

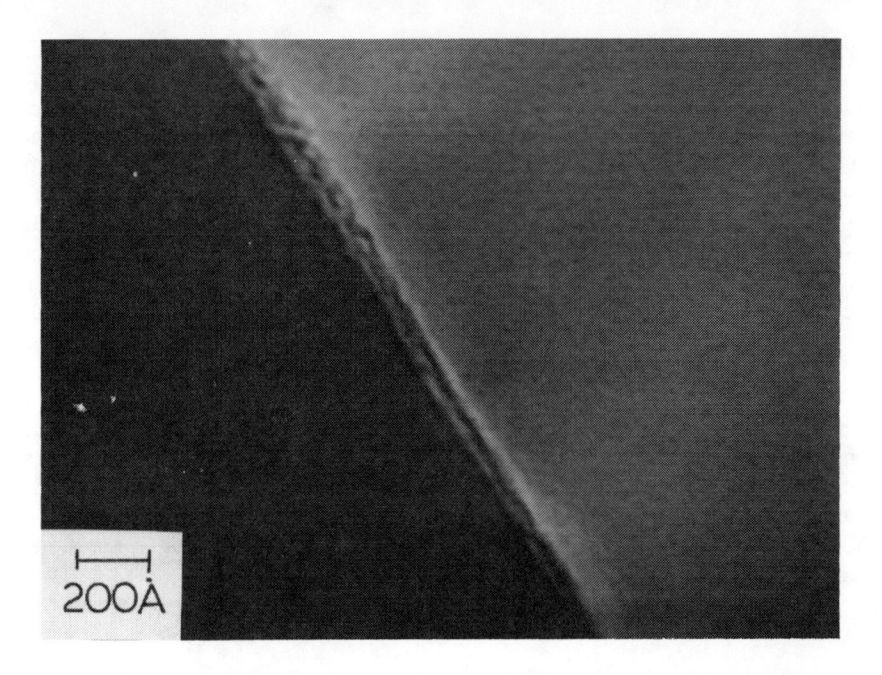

Fig. 4. Defocused STEM image of a grain boundary in Al$_2$O$_3$ doped with TiO$_2$ plus SiO$_2$ and fired in oxygen.

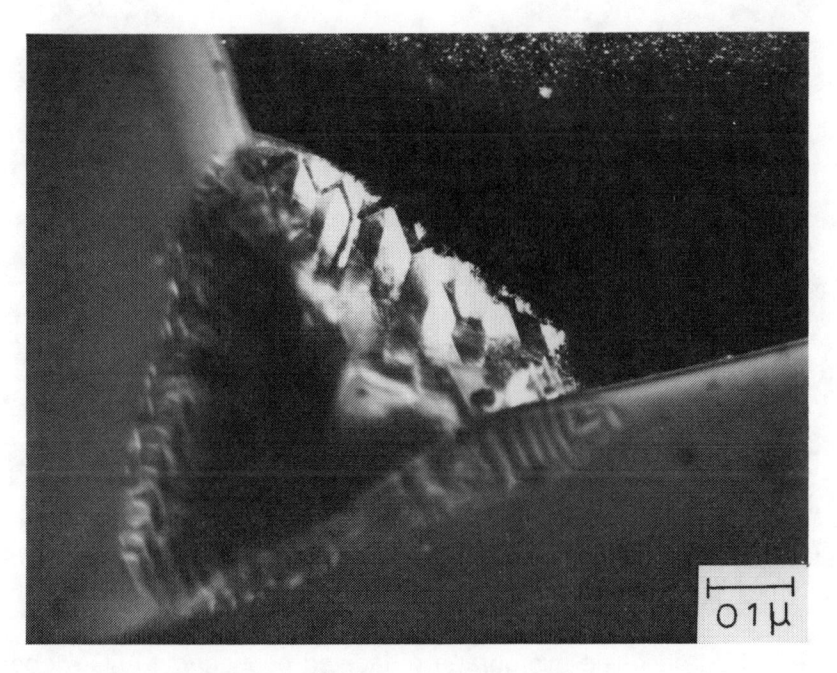

Fig. 5. TEM dark-field micrograph of twinned ZrO$_2$ phase present at triple grain junction in Al$_2$O$_3$ doped with ZrO$_2$ plus SiO$_2$.

the sample, dislocations without precipitates or second phase were observed at the boundary. Our observations indicated that a liquid phase penetrates some, but not all, of the boundaries between grains. This seems to be in accord with the observation of Hanson and Phillips,[15] who found that nearly all boundaries in a commercial 99.8% sintered alumina ceramic were wetted by an amorphous film, with the observation of Carter and Morrissey[16] that many grain boundaries were free of a second phase, and with the lattice imaging observations of Krivanek, Harmer, and Geiss[17] that titania-doped boundaries were free of a second phase. We know that in the system Al_2O_3-TiO_2-SiO_2-CaO the solidus temperature of the Al_2O_3 primary field is less than 1500°C;[18] thus, a liquid phase would have been present during the sintering process.

In the sample containing ZrO_2 and SiO_2, an internally twinned second phase was observed at triple grain junctions, as would be expected for zirconium oxide (Fig. 5). Some other boundaries in this system are faceted and on those boundaries one orientation of the facets shows a silica-rich precipitate, while the other orientation appears to be free of precipitates (Fig. 6). These results are in accord with the phase equilibrium diagram,[19] indicating that ZrO_2, Al_2O_3, and mullite should be the phases present, with the ternary eutectic temperature above the sintering temperature.

In the sample containing vanadium oxide, there was an indication of a silicon-rich second phase forming continuously along some boundaries, as shown in

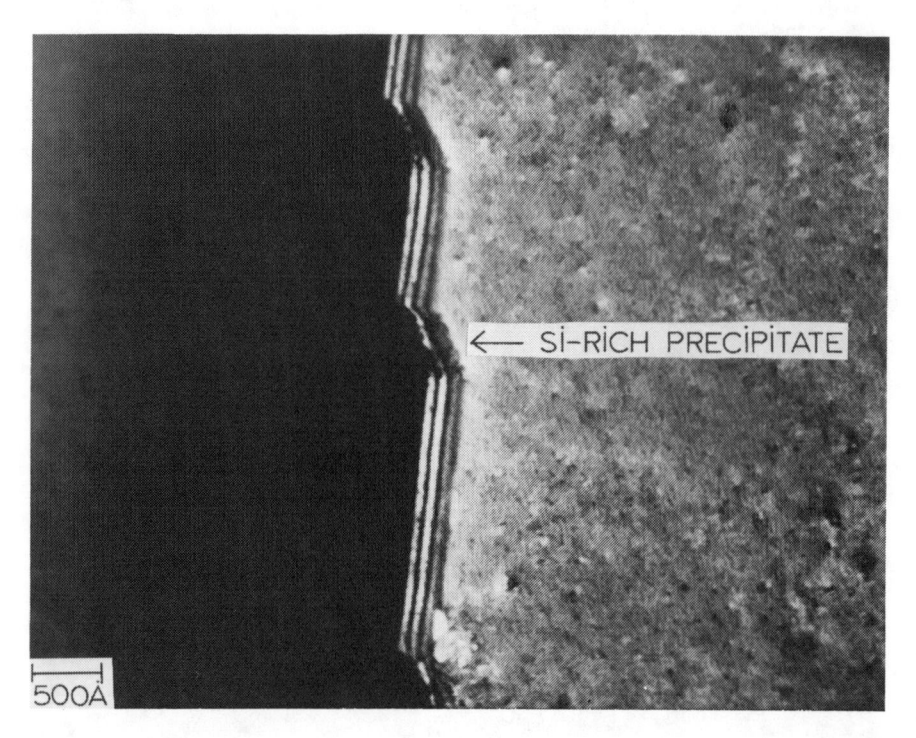

Fig. 6. TEM dark-field micrograph of faceted boundary in Al_2O_3 doped with ZrO_2 and SiO_2. Polycrystals of a silica-rich phase (presumably mullite) occur along one of the facet orientations.

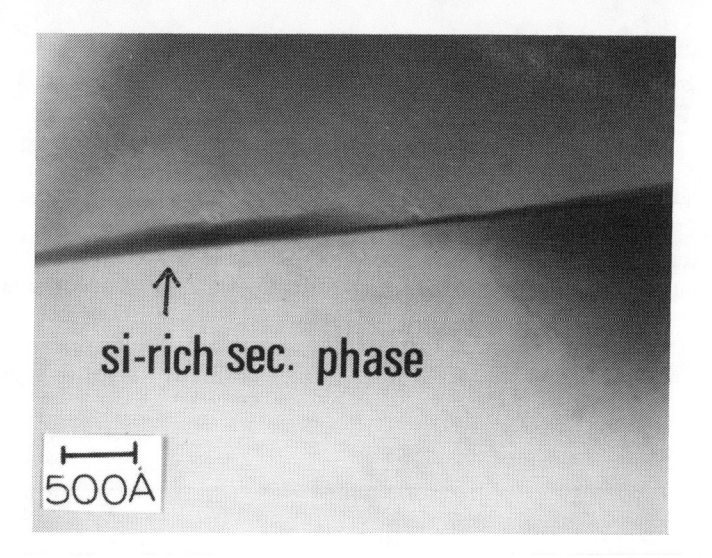

Fig. 7. Electron micrograph of V_2O_5-doped Al_2O_3 containing 0.02% SiO_2 which has area of Si-rich second phase along boundaries.

Fig. 7. We searched for but found no indication of a second phase in the samples containing gadolinium, scandium, lanthanum, yttrium, and chromium.

In summary, then, the dihedral angle in the samples containing Ni, Mg, and Zn additives assure that we are observing grain boundaries free of a second phase. In samples containing vanadium and manganese, we are doubtful of having observed clean boundaries. In samples containing Zr and Ti, precipitates are observed, but areas believed to be clean boundaries are also seen, while in Ga, Sc, La, Y, and Cr we believe, but cannot absolutely prove, that clean boundaries are being observed.

Segregation on "Clean" Grain Boundaries

As long as the concentration of solute at the grain boundary is not too large, the ratio of the grain boundary to bulk concentration is expected to be[3]

$$\beta = \frac{C_{gb}}{C_l} \cong \exp\left(\frac{\Delta G}{kT}\right) \quad (2)$$

where ΔG is free energy change due to segregation; k and T have their usual meanings. One contribution to the decrease in energy with grain boundary segregation is the strain energy resulting from solute misfit in the lattice. McLean has derived a relationship for this energy:[3]

$$W = \frac{24\pi K G r_1 (r_2 - r_1)^2}{3K + 4G\left(\frac{r_1}{r_2}\right)} \quad (3)$$

where K is the bulk modulus of the solute, G is the shear modulus of the solvent, r_1 and r_2 are the ionic radii of the solvent and solute ions. If strain energy reduction is the major contributor to the segregation energy, then by combining Eqs. (2) and

(3) we obtain:

$$\ln \frac{C_{gb}}{C_l} = \frac{\Delta S}{k} + \frac{W}{kT} = \frac{\Delta S}{k} + \frac{24\pi KGr_1^3}{3K + 4G\left(\frac{r_1}{r_2}\right)} \left(\frac{r_2 - r_1}{r_1}\right)^2 \frac{1}{kT} \qquad (4)$$

where ΔS is the entropy change related to solute segregation at the boundary.

For solutes not too far different in size from the solvent ion, the logarithm of the segregation ratio should be nearly proportional to the square of the induced strain, i.e., $\ln C_{gb}/C_l \sim (r_2 - r_1/r_1)^2$. We have plotted our data for divalent and trivalent solutes this way in Fig. 8, using ionic radii data of Belov and Bokii[20] and $T = 1900$ K. The data fall on a straight line, in accord with the expectation that a decrease of misfit strain energy is the principal driving force for segregation.

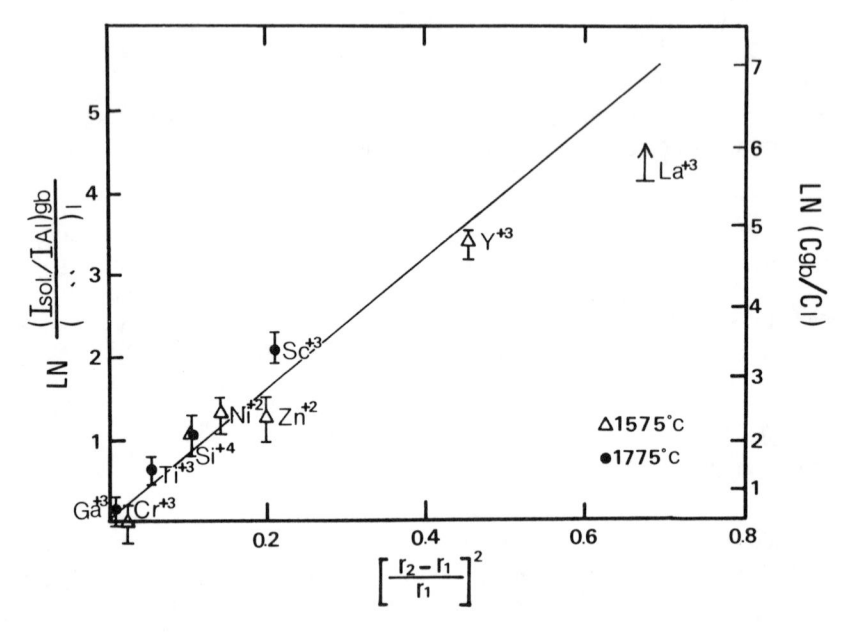

Fig. 8. Plot of logarithm of $(I_{solute}/I_{Al})_{gb}/(I_{solute}/I_{Al})_l$ vs $(r_2 - r_1/r_1)^2$ for divalent and trivalent solutes in Al_2O_3. The scale on the right side of the graph is calculated from Eq. (1), assuming $A = 4$.

At 1600°C, Young's modulus for Al_2O_3 is 40×10^6 psi[21] (28×10^4 MPa) with a Poisson's ratio of 0.25, and we calculate the shear modulus to be 16×10^6 psi (11×10^4 MPa). From the slope of 8.0 from Fig. 8, we calculate from Eq. (4) that the bulk modulus of the solute material should be 15×10^6 psi (10×10^4 MPa), according to the McLean equation. From the limited data available,[22] the bulk modulus of solute oxides at this temperature is in the range $(10-20) \times 10^6$ psi $[(7-14) \times 10^4$ MPa$]$. Considering everything, the data are in quite good accord with the model proposed by McLean, in which strain energy is the principal contributor to boundary segregation.

For the isoelectric solutes, Ga, Cr, Ti, Sc, Y, La, this is surely anticipated. Mackrodt's[23] calculations suggest that magnesium goes into solid solution in alumina in a locally self-compensating mode with both Mg'_{Al} and $Mg_i^{..}$ lattice defects present such that the electrical charge associated with these solutes cancels out. As a result, strain energy would be the major driving force for boundary segregation. Presumably this is also true for Zn and Ni.

The bulk concentration of zirconium in solid solution is too low for calculation of a segregation coefficient. Segregation of vanadium is greater than that expected solely from strain energy considerations, but a liquid phase may be present. The hot-pressed sample with yttria has less segregation than the sintered sample suggesting that equilibrium was not achieved in the 15-min hot-pressing. Silicon has an enrichment factor falling right on the strain energy line as indicated in Fig. 8. We presume that silicon goes into solid solution on a tetrahedral site with strongly associated vacancies (similar to the structure of mullite) to give charge neutrality such that strain energy is the major driving force for silicon segregation.

For the titanium-doped samples, Krivanek et al.[17] found an apparent 10-fold boundary segregation using a 100-Å diameter probe in a sample for which lattice imaging indicated no second phase at the boundary. In the boundaries of our TiO_2 sample in which a liquid phase was present, we found a highly enhanced segregation of titanium, calcium and silicon, in accord with the results of Hansen and Phillips.[15] However, in boundaries with observed dislocation fringes and other boundaries micrographically free of a second phase, we observed segregation of titanium to a boundary estimated to be about 60–80 times the bulk concentration (the X-ray intensity in the bulk is less than the detection limit so that these values are only approximations). These values are in general accord with those of Krivanek et al.[17] when the beam size effect is considered and corresponds to a level of segregation far above that expected from strain energy alone. This result is in accord with a model of substitutional titanium solution with charge compensation by increased concentrations of negatively charged aluminum vacancies or oxygen interstitials giving rise to a negative charge at the boundary and additional titanium segregated in the space charge layer.

Summary

The grain boundary segregation of divalent solutes, trivalent solutes, and silicon in aluminum oxide corresponds to a level resulting from solute misfit strain energy. This result suggests that silicon and divalent solutes go into solution with local self-compensation of electrical charge and lends support to the defect calculations of Mackrodt.

Boundaries of alumina containing TiO_2 as a solute, along with small amounts of silica and CaO, show many boundaries with an amorphous phase present. Boundaries that are free of this amorphous phase show a segregation of titanium greater than can be accounted for by strain energy alone, suggesting that substitution solution of Ti on aluminum sites leads to a negatively charged grain boundary with excess titanium present in the adjacent space charge layer.

Acknowledgment

The authors gratefully acknowledge Masayuki Fujimoto for the TEM micrographs, Alain Mocellin for the hot-pressed samples, and support from the U.S. Department of Energy under Contract No. DE-AC02-76ER02390.

References

[1] E. D. Hondros and M. P. Seah, *Int. Metall. Rev.*, **22**, 262 (1977).
[2] W. D. Kingery, *J. Am. Ceram. Soc.*, **57**, 1 (1974); **57**, 74 (1974).
[3] D. McLean; Grain Boundaries in Metals. Oxford University Press, London, 1957.
[4] B. Bender, D. B. Williams, and M. R. Notis, *J. Am. Ceram. Soc.*, **63**, 542 (1980).
[5] W. C. Johnson, D. F. Stein, and R. W. Rice; p. 261 in Grain Boundaries in Engineering Materials. Edited by J. L. Walter, J. H. Westbrook, and D. A. Woodford. Claitor's, 1975.
[6] W. C. Johnson, *Met. Trans.*, **8A**, 1413 (1977).
[7] D. J. Jorgenson and J. H. Westbrook, *J. Am. Ceram. Soc.*, **47**, 332 (1964).
[8] S. K. Roy and R. L. Coble, *J. Am. Ceram. Soc.*, **51**, 1 (1968).
[9] G. Cliff and G. W. Lorimer, *J. Microsc.*, **103**, 203 (1975).
[10] G. Cliff and G. W. Lorimer; p. 47 in Quantitative Microanalysis with High Spatial Resolution. Edited by G. W. Lorimer, M. H. Jacobs, and P. Doig. The Metal Society, London, 1981.
[11] E. L. Hall, D. Imeson, and J. B. Vander Sande, *Philos. Mag.*, **43**, 1569 (1981).
[12] J. I. Goldstein and D. B. Williams; p. 5 in Ref. 10.
[13] I. P. Jones and A. W. Nicholls, *J. Microsc.*, **130**, 155 (1983).
[14] D. R. Clarke, *J. Am. Ceram. Sci.*, **63**, 104 (1980).
[15] S. C. Hansen and D. S. Phillips, *Phil. Mag., [Part A]*, **A47**, 209 (1983).
[16] C. B. Carter and K. J. Morrissey, "Grain Boundary Structure in Al_2O_3"; these proceedings.
[17] O. L. Krivanek, M. Harmer, and R. Geiss; p. 414 in Ninth International Congress on Electron Microscopy, Toronto. Microscopical Society of Canada, 1978.
[18] J. White, *Glass Ceram. Bull.*, **23**, 43 (1976).
[19] E. M. Levin, C. R. Robbins, and H. F. McMurdle; Phase Diagrams for Ceramists, 1964. Edited by M. K. Reser. The American Ceramic Society, Columbus, OH. Fig. 772.
[20] The Oxide Handbook, 2nd ed. Edited by G. V. Samsonov. IFI/Plenum, New York, 1982.
[21] J. B. Wachtman and D. G. Lam, Jr., *J. Am. Ceram. Soc.*, **42**, 254 (1959).
[22] (a) "Engineering Properties of Selected Ceramic Materials." Compiled by Battelle Memorial Institute. The American Ceramic Society, Columbus, OH, 1966. (b) W. R. Manning and O. Hunter, Jr., *J. Am. Ceram. Soc.*, **52**, 492 (1969). (c) S. L. Dole, O. Hunter, Jr., and F. W. Calderwood, *J. Am. Ceram. Soc.*, **60**, 167 (1977).
[23] W. C. Mackrodt, "Calculated Point Defect Formation, Association, and Migration Energies in MgO and $\alpha\text{-}Al_2O_3$"; these proceedings.

---

*Linde A, Linde Div., Union Carbide Corp., New York, NY.
†Model VG HB5, Vacuum Generator Microscopes Ltd., East Grinstead, U.K. A few samples were examined with a transmission electron microscope, Model 200-CX, JEOL Co., Tokyo, Japan.

Oxygen Diffusion in MgO and Al$_2$O$_3$

Y. OISHI AND KEN ANDO
Kyushu University
Department of Nuclear Engineering
Fukuoka 812, Japan

Mechanisms of oxygen diffusion in α-Al$_2$O$_3$ and MgO are discussed on the basis of self-diffusion coefficients determined for single crystals. Extrinsic oxygen diffusion reported for crushed Al$_2$O$_3$ and MgO samples was proved to be not impurity sensitive but sensitive to structural defects related to dislocations. Oxygen diffusion was insensitive to the Fe concentration in MgO. The results are interpreted on the basis of a vacancy pair mechanism. The effect of sample preparation on the determination of diffusion coefficients by the isotope exchange technique was discussed.

With respect to the oxygen ion sublattice, MgO and α-Al$_2$O$_3$ have similar crystal structures, in which oxygen ions, the larger species, are close packed and expected to diffuse by the vacancy mechanism due to Schottky-type vacancies.

If MgO and Al$_2$O$_3$ are ideal pure single crystals, they should exhibit intrinsic oxygen diffusion where the concentration of oxygen vacancies is temperature dependent and the experimental activation energy for self-diffusion involves the migration and formation energies of the oxygen vacancy. However, the presence of aliovalent cation impurities may cause extrinsic oxygen diffusion in MgO and Al$_2$O$_3$.

In the case of substitutional solid solution of low-valence cation impurities, the oxygen vacancies introduced are composition dependent and temperature independent. Consequently, the experimental activation energy for diffusion is only the migration energy of the oxygen vacancy, and the diffusion coefficient is higher than that for the intrinsic diffusion in the pure oxide.

In the case of high-valence cation impurities, the composition-dependent cation vacancies that are introduced control the Schottky equilibrium, resulting in a decreased concentration of oxygen vacancies. In this case, if the free vacancy mechanism is predominant, the oxygen self-diffusion should exhibit a higher activation energy and a lower diffusion coefficient than in the intrinsic diffusion. However, if the concentration of any defect associations is higher than that of free vacancies, as is the case for the vacancy pair in MgO,[1] a mechanism other than the free vacancy mechanism may become predominant for oxygen diffusion.

If the presence of dislocations enhances oxygen diffusion in a crystal, the oxygen diffusion may become an extrinsic process where the experimental activation energy is the migration energy of the oxygen ion through dislocations.

Self-diffusion coefficients of oxygen ions in single-crystal Al$_2$O$_3$[2] and MgO[3] were first determined by Oishi and Kingery using an isotope exchange technique. In their determinations, Al$_2$O$_3$ exhibited two diffusion mechanisms with differing activation energies, high in the high-temperature regime and low in the low-temperature regime, while MgO exhibited only one diffusion mechanism similar to that of the low-temperature regime of Al_2O_3, which is interpreted to be extrinsic diffusion.

Since crushed particles with irregular shapes were employed as diffusion samples in those determinations, the absolute magnitudes of the self-diffusion coefficients could not be expected to be determined with high accuracy. It was also not clarified whether extrinsic diffusion observed was due to an impurity effect or was structure sensitive. To clarify those questions, we redetermined the oxygen self-diffusion in Al_2O_3[4,5] and MgO[6] by using plate samples with more carefully prepared surface conditions.

Holt and Condit[7] determined an oxygen self-diffusion coefficient in MgO at one temperature by using a proton activation analysis. Proton activation analysis was employed also in Reddy and Cooper's determinations for Al_2O_3[8] and MgO.[9] Reed and Wuensch[10] conducted the determination for Al_2O_3 by using an ion probe technique.

The self-diffusion coefficients of oxygen ions are lower than those of cations in single-crystal Al_2O_3 and MgO, respectively. The oxygen diffusion is enhanced by the presence of grain boundaries in both Al_2O_3[2] and MgO.[11]

Oxygen Diffusion in Al_2O_3

Oishi and Kingery's[2] determination of the oxygen self-diffusion coefficients in single-crystal α-Al_2O_3 was conducted by using an isotope exchange technique in which the increase of the concentration of ^{18}O tracer in the solid sample was determined after diffusion anneal by means of a mass spectrometer. Samples in their work were crushed particles (200–230 mesh) and spheres (0.8-mm diameter) prepared from flame-fused sapphire.* Both samples exhibited different temperature dependences of the oxygen self-diffusion in the high-temperature and low-temperature regimes. The oxygen self-diffusion coefficients determined for crushed particles are shown by squares in Fig. 1. The high-temperature regime was represented by $D = 1.9 \times 10^3 \exp[-636 \text{ (kJ/mol)}/RT]$ (cm^2/s) and was interpreted to be intrinsic diffusion. The low-temperature regime was interpreted as impurity-sensitive or structure-sensitive extrinsic diffusion.

If the extrinsic diffusion observed by Oishi and Kingery is impurity controlled, it should be confirmed by a determination for a sample with different impurity contents; a sample with reduced content of low-valence cation impurities relative to Al should exhibit a lower diffusion coefficient in the extrinsic diffusion regime. Triangles in Fig. 1 show oxygen self-diffusion coefficients determined for crushed particles of a high-purity single-crystal Al_2O_3 prepared by a chemical vapor deposition technique.[†,12] Impurity contents of a similar NBS crystal are listed in Table I.[13] Similar extrinsic diffusion exhibited by two samples of differing impurity contents suggests that the present extrinsic diffusion is not impurity controlled but is due to the enhancement by the secondary defects introduced in the crushed samples.

If the extrinsic diffusion observed with crushed samples is due to secondary defects, a sample with a reduced dislocation density should exhibit a reduced oxygen diffusion coefficient in the extrinsic regime, or extrinsic diffusion may not be observed. Figure 2 shows oxygen self-diffusion coefficients determined for several samples prepared differently from flame-fused single-crystal Al_2O_3. The dotted line represents the high-temperature regime previously determined for crushed particles.[2] The thin solid line designated as "diamond-paste finished" represents Linde single crystal finished with a 1-μm diamond paste.[4] (The impurity contents are listed in Table I.) This diamond-paste finished sample exhibited only

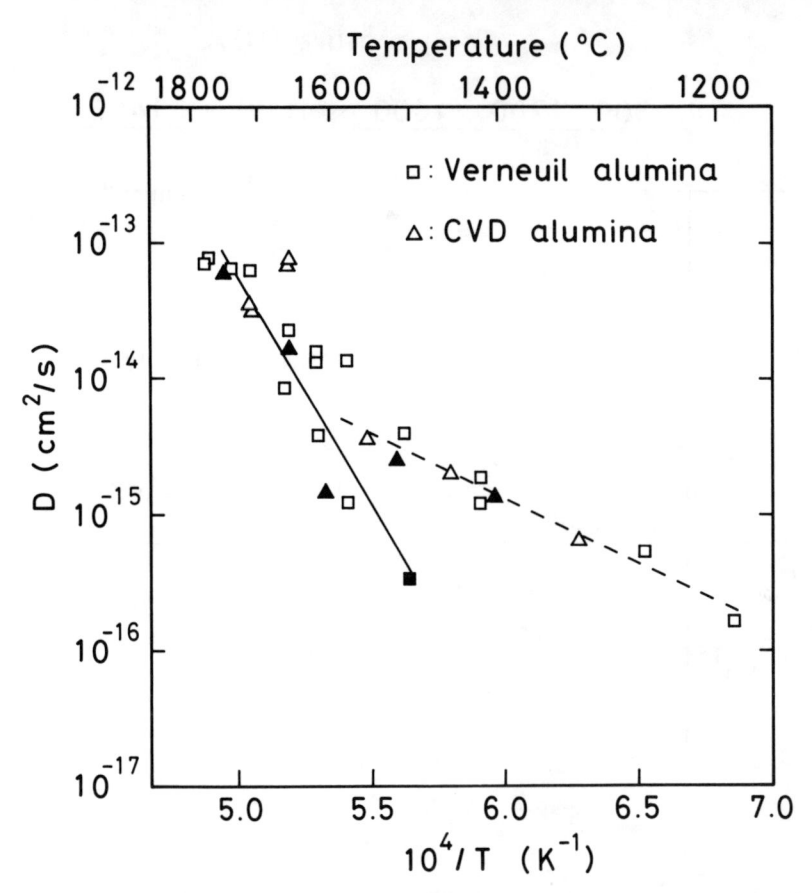

Fig. 1. Comparison of oxygen self-diffusion coefficients in two crushed Al$_2$O$_3$ single crystals of differing purities.

Table I. Impurity Contents in Alumina Samples*

Element	Verneuil alumina	CVD alumina[†]
Fe	90	
Ti	100	
Si	20	
Ca	40	
Mn	≪10	≈0.0005
Mg	≪10	
Ni	≪10	
Na	≪10	
Cu	≪10	0.027–0.038
V	≪10	
Ir		0.0005–0.002
Sc		≈0.015

*Level (ppm wt).
[†]Reference 13.

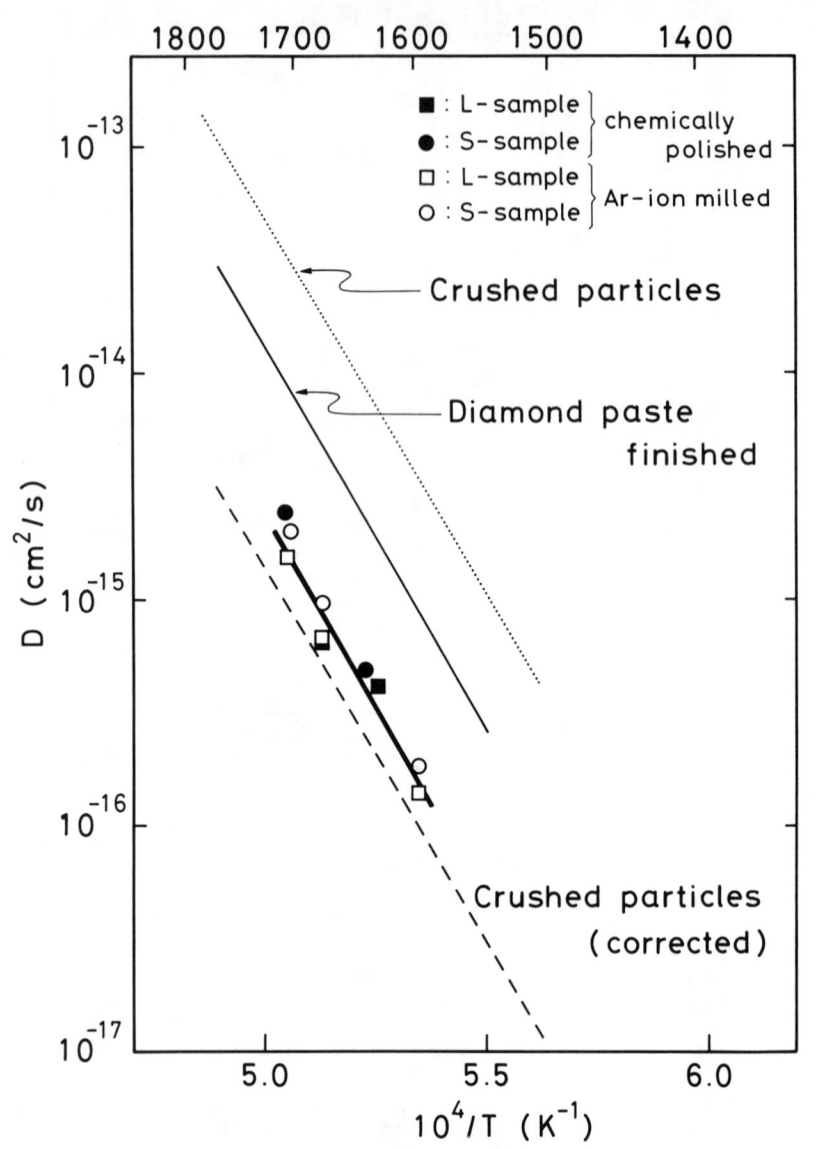

Fig. 2. Oxygen self-diffusion coefficients in crushed, diamond paste finished, chemically polished, and Ar ion milled Al_2O_3 samples.

the high-temperature regime represented by Eq. (1):

$$D = 1.12 \times 10^3 \exp[-649 \text{ (kJ/mol)}/RT] \quad (\text{cm}^2/\text{s}) \quad (1)$$

but did not exhibit the extrinsic diffusion regime. The results confirm that the extrinsic diffusion exhibited by only crushed particles is not impurity sensitive but structure sensitive and related to dislocations introduced in crushed samples.

Figure 2 shows the results also for two single crystals from the Linde Co. (designated as L-sample) and from Shinkosha Co., Japan (designated as S-sample).[5] These crystals, after being polished with a 1-μm diamond paste, were chemically polished with H_3PO_4 or $H_3PO_4:H_2SO_4 = 1:1$ and separately milled with Ar ions at 6 kV to eliminate 10 μm from surface layers, in which dislocations might have been previously introduced by polishing with a 1-μm diamond paste. As shown in Fig. 2, all these samples exhibited similar oxygen self-diffusion coefficients and indicated no significant differences due to differences of samples and polishing techniques. The temperature dependence inclusive of the oxygen self-diffusion coefficient for all the determinations is represented by Eq. (2):

$$D = 562 \exp[-665 \text{ (kJ/mol)}/RT] \quad (\text{cm}^2/\text{s}) \tag{2}$$

As seen in the comparison of Fig. 2, the activation energies of oxygen diffusion are similar for the chemically finished or Ar ion milled sample (665 kJ/mol), the diamond paste finished sample (649 kJ/mol), and crushed samples in the high-temperature regime (636 kJ/mol), though magnitudes of diffusion coefficients are different. The determined activation energies may be essentially intrinsic to oxygen self-diffusion in single-crystal Al_2O_3.

The high magnitudes of the diffusion coefficients exhibited by the diamond-paste finished sample, relative to those for chemically polished and Ar ion milled samples, are presumably due to surface-emergent dislocations and/or microcracks introduced by diamond grains during polishing. Surface-emergent dislocations and microcracks contribute to the effective surface area for diffusion. The diffusion amount is generally described as a function of $S(Dt)^{1/2}$ as in Eq. (3):

$$\frac{M_t}{M_\infty} = f[S(Dt)^{1/2}] \tag{3}$$

where M_t and M_∞ are diffusion amounts at times t and infinity, respectively, and S is the surface area:volume ratio of the sample. The presence of surface-emergent dislocations and microcracks contributes to S. However, since S is constant, increase in S increases only the magnitude of the determined diffusion coefficient but does not change the temperature dependence.

Because of the geometrical irregularity of crushed particles, high accuracy of the absolute magnitude should not be expected in the diffusion coefficients determined with crushed particles. However, the relatively high magnitudes of oxygen self-diffusion coefficients in Oishi and Kingery's determination[2] were found to be due to the surface area:volume ratio being underestimated by the photomicrographic method, which tends to underestimate the surface area and overestimate the volume of irregularly shaped particles.[5] Underestimation of S results in an overestimated D, when calculated by Eq. (3) from a given M_t/M_∞.

When the specific surface area of crushed particles of an Al_2O_3 single crystal was determined by using a BET technique, oxygen self-diffusion coefficients were recalculated for the high-temperature regime in Oishi and Kingery's determination, as given by Eq. (4):

$$D = 56 \exp[-636 \text{ (kJ/mol)}/RT] \quad (\text{cm}^2/\text{s}) \tag{4}$$

The results are close to those determined for the chemically polished and Ar ion milled samples, as shown in Fig. 2. (For the same reason, the low-temperature regime in Fig. 1 is recalculated as $D = 1.8 \times 10^{-11} \exp[-180 \text{ kJ/mol})/RT]$ cm^2/s).

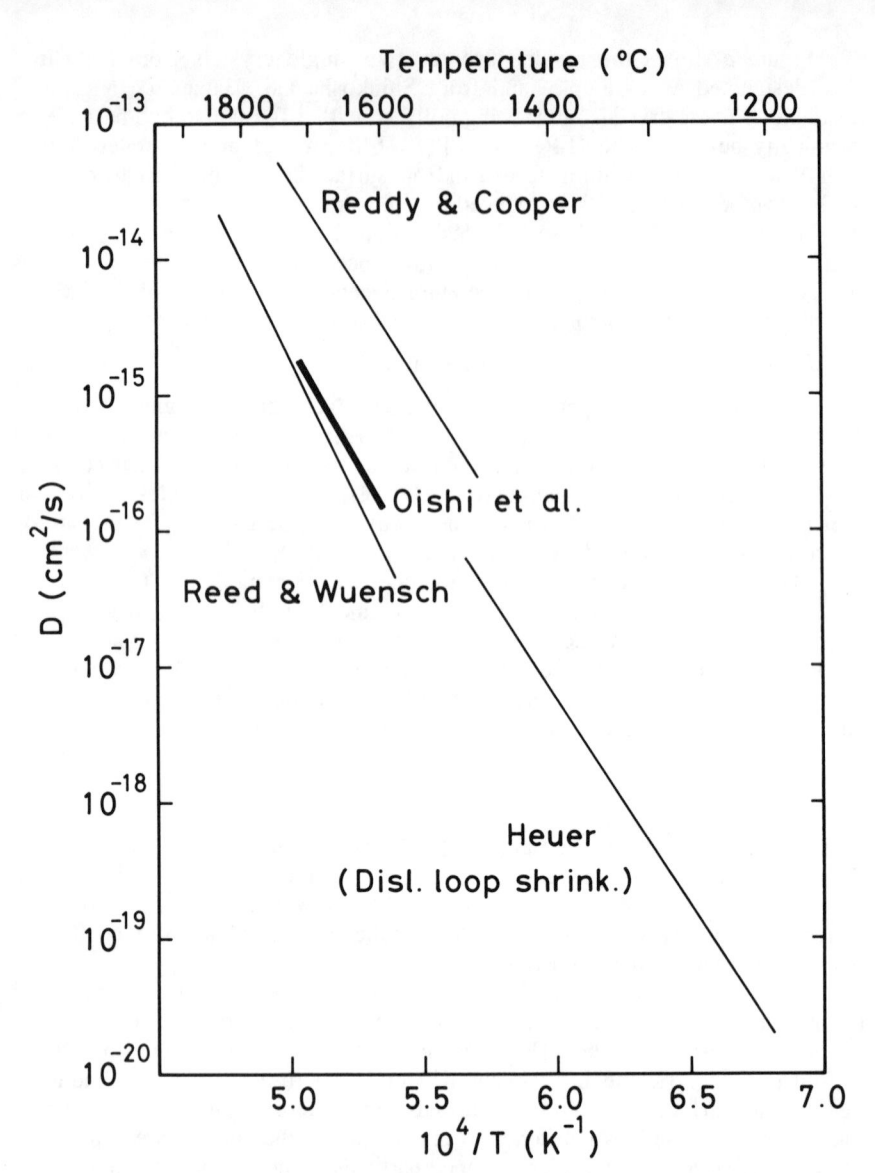

Fig. 3. Comparison of oxygen self-diffusion coefficients and diffusion coefficient calculated from dislocation loop shrinkage in Al_2O_3.

As compared in Fig. 3, Reed and Wuensch's determination for Czochralski alumina by the ion probe technique[10] gave oxygen self-diffusion coefficients (with an activation energy of 787 kJ/mol) that are close in magnitude to the results for chemically polished and Ar ion milled samples. Reddy and Cooper's determination by proton activation analysis[8] gave a similar activation energy (615 kJ/mol) but higher magnitudes than our results.

The diffusion coefficients calculated by Heuer et al.[14] from the shrinkage rates of dislocation loops in Al_2O_3 are close to the oxygen self-diffusion coefficients.

Table II. Impurity Contents in the Norton and ORNL MgO Single Crystals*

	Al	Si	K	Ca	Cr	Fe	Cu	Na	V	Mn	Ni	Pb	Bi	P
Norton	37	<3	<4	210	4	70	<1	<2	<1	7	3	<1	<4	46
ORNL[†]	15	22	<3	50	1	7	<0.1	<1	<2	<1	<3	<2	<0.5	13

*Cationic ppm.
[†]Reference 15.

This agreement implies that the shrinkage rate is controlled by diffusion of the oxygen ion, which is the slower diffusing species in Al_2O_3.

Oxygen Diffusion in MgO

In Oishi and Kingery's work, the oxygen self-diffusion coefficient in MgO was determined by using crushed particles from a single crystal.[3] The results were represented by Eq. (5):

$$D = 2.5 \times 10^{-6} \exp[-261 \text{ (kJ/mol)}/RT] \quad (\text{cm}^2/\text{s}) \quad (5)$$

With comparison of the diffusion parameters previously determined for Al_2O_3, this diffusion was interpreted to be impurity-controlled or structure-sensitive extrinsic diffusion.

In order to elucidate whether the mechanism of the extrinsic diffusion observed with crushed particles is impurity sensitive or structure sensitive, we determined oxygen self-diffusion coefficients for two single crystals with different impurity contents.[‡,6] The latter was purer than the former, as compared in impurity contents listed in Table II, where major impurities other than Ca are high-valence cations relative to Mg. To confirm the results, the determination was conducted also for Fe-doped MgO single crystals.[16] In order to elucidate the effect of dislocation density, oxygen self-diffusion coefficients were compared in three samples prepared by crushing, cleaving, and chemical polishing.[6]

By analogy with oxygen diffusion in Al_2O_3,[4] appearance of the "high-temperature regime" other than the extrinsic regime was expected to carefully prepared samples with low dislocation densities.

Chemically Polished Samples

Diffusion samples were prepared by cleaving Norton crystal and by cutting ORNL crystal with a diamond wire saw. Subsequently, both plate samples were chemically polished with phosphoric acid to remove 50 μm from surface layers, which might have been damaged by cleaving or cutting. As shown in Fig. 4, both chemically polished Norton and ORNL samples exhibited similar oxygen self-diffusion, despite their differing impurity contents;[6] the temperature dependence changed at 1500°C, giving higher activation energies in the high-temperature range, and the temperature dependence of the low-temperature regime was similar to that of the extrinsic diffusion observed by Oishi and Kingery[3] with crushed samples.

The high-temperature regime of Norton sample is represented by Eq. (6):

$$D = 6.76 \exp[-536 \text{ (kJ/mol)}/RT] \quad (\text{cm}^2/\text{s}) \quad (6)$$

and the low-temperature regime by Eq. (7):

$$D = 2.2 \times 10^{-9} \exp[-213 \text{ (kJ/mol)}/RT] \quad (\text{cm}^2/\text{s}) \quad (7)$$

Evidently, the magnitude of the activation energy in the high-temperature regime involves the energy of defect formation.

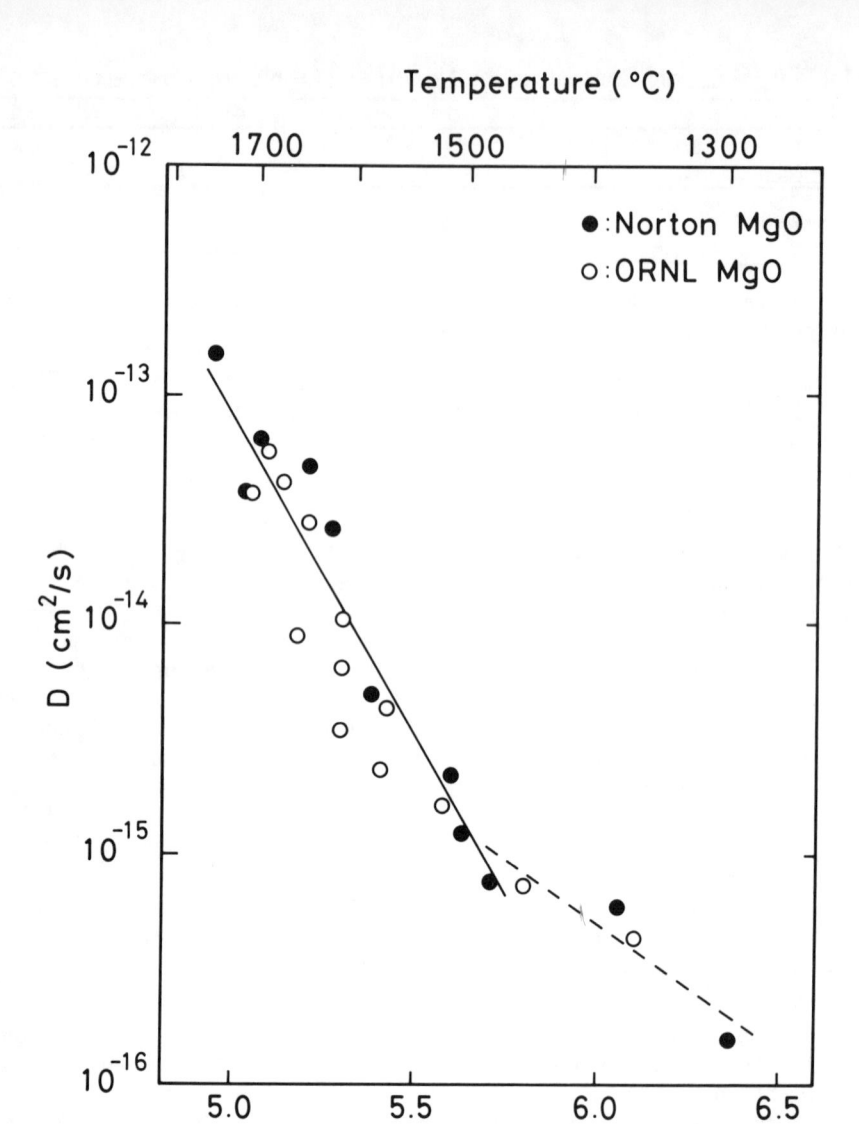

Fig. 4. Oxygen self-diffusion coefficients in two chemically polished MgO single crystals.

As compared in Fig. 5, the oxygen self-diffusion coefficient in MgO determined by Holt and Condit[7] using proton activation analysis is close to the result determined for the chemically polished sample. Reddy and Cooper's determinations[9] are close in magnitude to the present results, but with a relatively small preexponential term (1.9×10^{-4} cm^2/s) and activation energy (370 kJ/mol).

Diffusion coefficients calculated by Narayan and Washburn[17] from shrinkage rates of dislocation loops agree with the oxygen self-diffusion coefficients. This

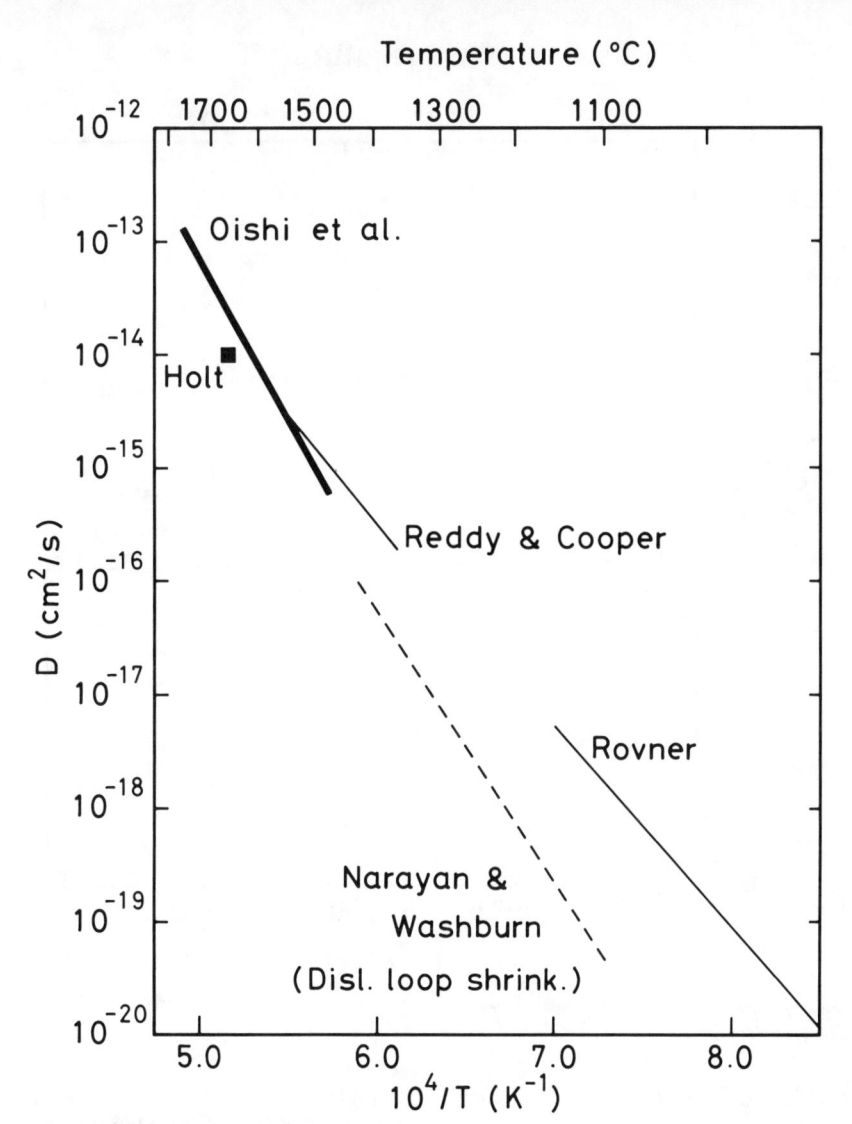

Fig. 5. Comparison of oxygen self-diffusion coefficients and diffusion coefficient calculated from dislocation loop shrinkage in MgO.

agreement confirms that the shrinkage rate is controlled by diffusion of oxygen ions, the slower diffusing species in MgO, as was the case in Al_2O_3.

The oxygen self-diffusion coefficients in single-crystal NiO determined by Dubois et al.[18] are represented by $D = 50 \exp[-539 \text{ (kJ/mol)}/RT]$ (cm^2/s). The magnitude and activation energy are close to those found for single-crystal MgO as given by Eq. (6), suggesting similar mechanisms for oxygen diffusion in MgO and NiO, which have the same crystal structure.

Crushed Particles and As-Cleaved Samples

Figure 6 shows oxygen self-diffusion coefficients determined for crushed

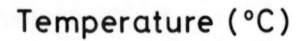

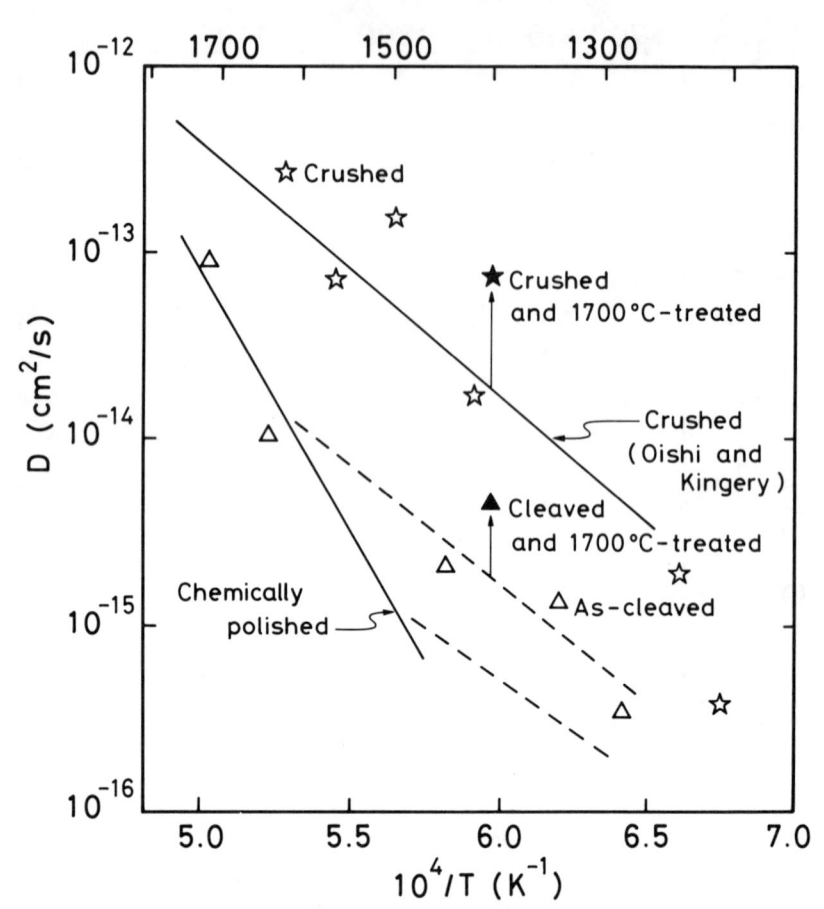

Fig. 6. Oxygen self-diffusion coefficients in crushed, cleaved, and chemically polished single-crystal MgO.

particles (stars) and as-cleaved samples (triangles)[6] prepared from Norton MgO compared with the results for the chemically polished Norton MgO and for Oishi and Kingery's crushed particles.[3] The two crushed samples gave similar oxygen self-diffusion coefficients. The temperature dependence of the as-cleaved sample is similar to those of the crushed particles and of the chemically polished sample in the low-temperature regime but with magnitudes of diffusion coefficients higher than those of the chemically polished sample. (The 1717°C data for the as-cleaved sample are already in the high-temperature regime.)

Figure 6 illustrates that samples prepared in different ways from the same MgO crystal give similar temperature dependences but differing magnitudes of the diffusion coefficient. This indicates that observed extrinsic diffusion is not impurity controlled but structure sensitive, being closely related to dislocation densities and dependent on sample preparation.

The enhancement of oxygen diffusion by the treatment at 1700°C for the crushed sample and the as-cleaved sample is presumably due to the dislocation network or subgrain boundaries resulting from rearrangement of dislocations or polygonization. A similar enhancement of oxygen diffusion was observed also in heat-treated Al_2O_3.[4] These results imply that enhancement by subgrain boundaries formed by rearrangement of dislocations is higher than the enhancement by the original dislocations before rearrangement.

Fe-Doped MgO Single Crystal

The Mg self-diffusion coefficients so far reported by several investigators[19-22] have been interpreted to be due to extrinsic diffusion by free Mg vacancies, which

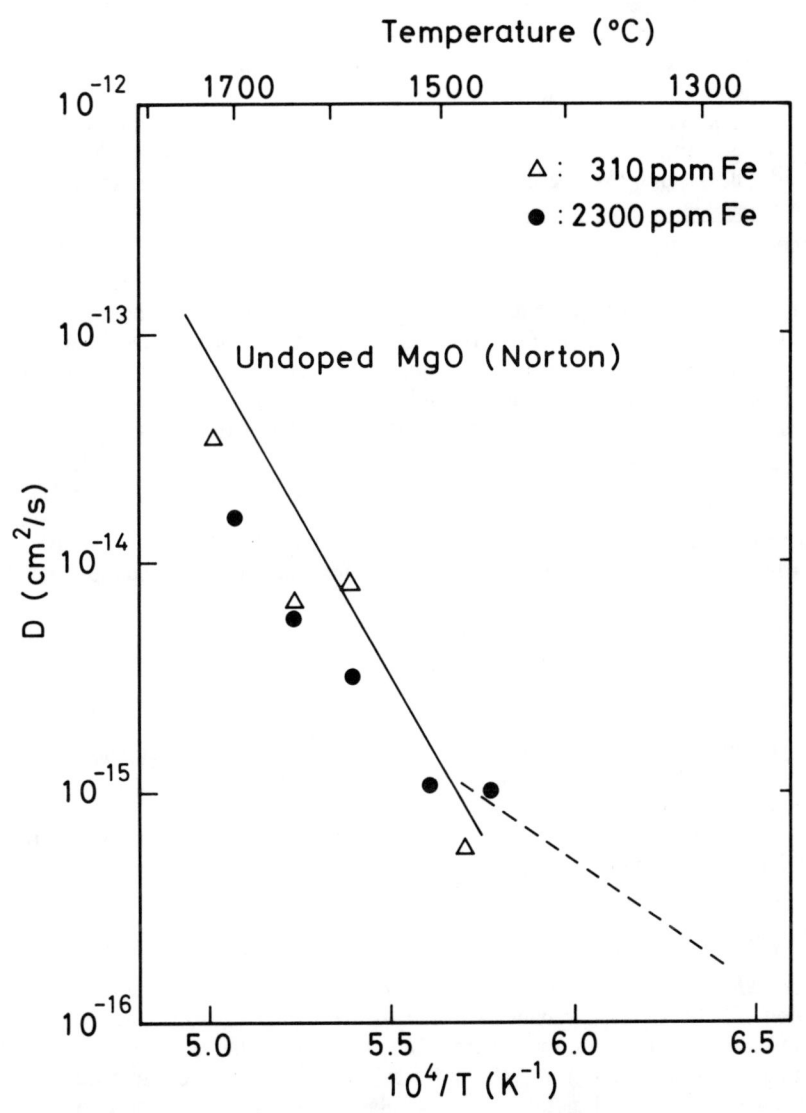

Fig. 7. Oxygen self-diffusion coefficients for Fe-doped MgO single crystals in comparison with those for undoped single crystal.

are caused by the presence of cationic impurities with high valences relative to Mg.[22,23] If the oxygen ion diffuses by the free vacancy mechanism, the oxygen diffusion should be also affected by the presence of those high-valence cations. However, as mentioned earlier, two MgO single crystals with different contents of high-valence cationic impurities did not exhibit significant difference of the oxygen self-diffusion coefficients in the high-temperature regime. Those results suggest that the oxygen diffusion is not due to a simple free vacancy mechanism but due to an impurity-insensitive mechanism. The possibility of diffusion insensitive to the high-valence dopant was further confirmed by using Fe-doped MgO.

Figure 7 shows oxygen self-diffusion coefficients determined for two MgO single crystals doped with 310 and 2300 ppm of Fe, respectively.[8,16] (The basic purity of those crystals is 99.99%, with 70 Si, 30 Ca, and 10 Cr in ppm.) Samples were chemically polished with phosphoric acid, prior to the diffusion experiment, to eliminate subsurface damage.

As compared in Fig. 7, the oxygen self-diffusion coefficients in Fe-doped crystals are similar to or somewhat lower than those in the undoped Norton MgO (solid lines), and two Fe-doped crystals exhibit no significant Fe concentration dependence.

Mechanism of Oxygen Diffusion in MgO

The experimental results illlustrated in Fig. 7, that the oxygen diffusion is somewhat slower in Fe-doped MgO than in undoped MgO, may be interpreted as due to the valence effect of Fe^{3+} ion on the free oxygen vacancy mechanism. This can be checked by theoretical calculation of oxygen self-diffusion coefficients on the basis of assumed diffusion mechanisms.

Free oxygen vacancy mechanism: The concentration of free oxygen vacancies in pure MgO is given by Eq. (8):

$$[V_O^{\cdot\cdot}]_{pure} = [V_{Mg}'']_{pure} = K_s^{1/2} \tag{8}$$

where K_s is the Schottky equilibrium constant. When Fe is substitutionally doped in MgO, formed $Fe_{Mg}^{\cdot}$ determines the concentration of Mg vacancies and consequently controls the Schottky equilibrium where the concentration of oxygen vacancies is given by Eq. (9):

$$[V_O^{\cdot\cdot}]_{doped} = \frac{2}{[Fe_{Mg}^{\cdot}]} K_s \tag{9}$$

If the free oxygen vacancy mechanism is the case, the oxygen self-diffusion coefficient, which is proportional to the oxygen vacancy concentration, should decrease with increasing $[Fe_{Mg}^{\cdot}]$, as given by Eq. (9). The ratio of $[V_O^{\cdot\cdot}]_{doped}$ in Eq. (9) and $[V_O^{\cdot\cdot}]_{pure}$ in Eq. (8) gives a relative oxygen self-diffusion coefficient for doped MgO against that for pure MgO, as given by Eq. (10):

$$\frac{D_{O(doped)}}{D_{O(pure)}} = \frac{2}{[Fe_{Mg}^{\cdot}]} K_s^{1/2} = \frac{2}{[Fe_{Mg}^{\cdot}]} \exp\left(\frac{\Delta S_s}{2R}\right) \exp\left(-\frac{\Delta H_s}{2RT}\right) \tag{10}$$

where ΔH_s and ΔS_s are the enthalpy and entropy, respectively, required for formation of Schottky-type vacancies and D_O is the oxygen self-diffusion coefficient.

With the assumption of $\exp(\Delta S_s/2R) = 1$ to 10, the relative D_O at 1500°C calculated by Eq. (10) for $\Delta H_s = 5.5$ eV and $\Delta H_s = 7.5$ eV, respectively, is shown in Fig. 8 as a function of $[Fe_{Mg}^{\cdot}]$ and is compared with the experimental results.[16] The $[Fe_{Mg}^{\cdot}]$ formed in Fe-doped MgO was taken from the Gourdin et al. experimental results.[24] The discrepancy by factors of 10^2 to 10^7 of

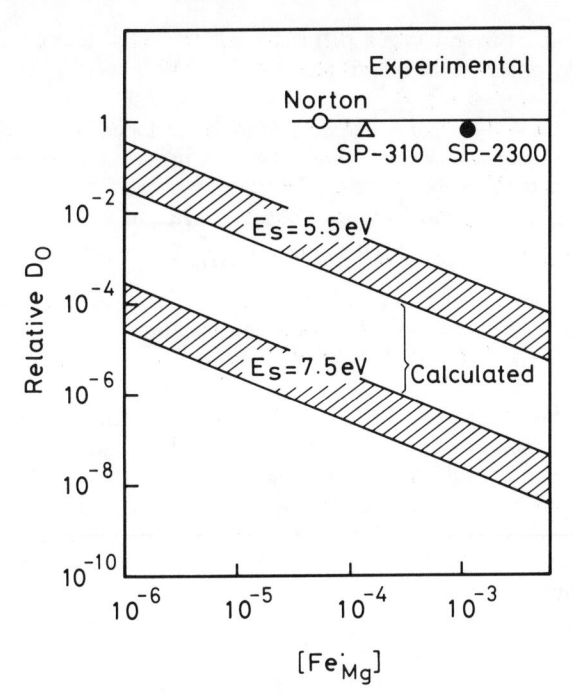

Fig. 8. D_O in Fe-doped MgO relative to D_O in pure MgO at 1500°C calculated for the free oxygen vacancy mechanism.

the experimental results from the calculated relative D_O indicates that the free oxygen vacancy mechanism is not the predominant mechanism in the present oxygen diffusion.

The free defect mechanism due to Frenkel-type oxygen defects is also not the case, since the formation energy of the Frenkel pair in MgO is high (12.1 eV).[25] *Vacancy pair mechanisms:* The small difference of oxygen self-diffusion coefficients in Fe-doped and undoped MgO crystals was not properly explained on the basis of the free oxygen vacancy mechanism. This dopant-insensitive oxygen diffusion can be explained by the vacancy pair mechanism.[16]

Formation of the vacancy pair from Schottky-type free vacancies is described by Eq. (11):

$$V_O^{\cdot\cdot} + V_{Mg}'' = (V_O^{\cdot\cdot} V_{Mg}'') \tag{11}$$

Consequently, the concentration of the vacancy pair is given by Eq. (12):

$$[(V_O^{\cdot\cdot} V_{Mg}'')] = K_s K_b \approx z \exp[-(\Delta H_s + \Delta H_b)/RT] \tag{12}$$

where K_b is the equilibrium constant for Eq. (11), ΔH_b is the corresponding enthalpy change, z is the number of orientations with respect to the vacancy pair, and the entropy term of the order of 1 is assumed.

The diffusion coefficient due to the vacancy pair mechanism should be independent of the dopant concentration, as the vacancy pair concentration is constant, as given by Eq. (12). This explains the experimental results for the Fe-insensitive oxygen diffusion. (Assumption of the 4-vacancy or a higher vacancy associate instead of the vacancy pair leads similarly to dopant-independent oxygen diffusion.) Dopant-insensitive oxygen diffusion was recently determined also for a Sc-doped single crystal in the Henriksen et al. work (2.9×10^{-15} cm^2/s at 1500°C for 1382 ppm Sc-doped MgO).[26]

If the oxygen self-diffusion determined in the undoped Norton MgO is due to the vacancy pair mechanism, the activation energy in the high-temperature regime, 536 kJ/mol = 5.6 eV, is interpreted to be the sum of the formation and migration energies for the vacancy pair, as described by Eq. (13):

$$Q = \text{formation energy } (\Delta H_s + \Delta H_b) + \Delta H_m \tag{13}$$

where Q is the experimental activation energy and ΔH_m the migration energy of the vacancy pair. The value of ΔH_b has been theoretically calculated by Gourdin and Kingery[25] and ΔH_m by Mackrodt.[27] Substitution of $\Delta H_b = -2.67$ eV, $\Delta H_m = 2.47$ eV, and $Q = 5.6$ eV in Eq. (13) gives the formation energy of Schottky-type free vacancies as $\Delta H_s = 5.8$ eV. This value is smaller than the theoretical ΔH_s calculated for pure MgO, $7.4 - 7.9$ eV[28] and 7.7 eV.[25]

Predominance of the vacancy pair mechanism relative to the free vacancy mechanism requires that the concentration of the vacancy pair be higher than that of the free oxygen vacancy, since the migration energy of the vacancy pair is higher than or similar to that of the free oxygen vacancy. As calculated by Eq. (14), which is derived from Eqs. (9) and (12):

$$\frac{[(V''_{Mg}V^{\cdot\cdot}_O)]}{[V^{\cdot\cdot}_O]_{doped}} \approx z \frac{[Fe^{\cdot}_{Mg}]}{2} \exp\left(-\frac{\Delta H_b}{RT}\right) \tag{14}$$

the concentration ratio becomes unity for $[Fe^{\cdot}_{Mg}] \approx 2 \times 10^{-8}$ at 1600°C and 4×10^{-7} at 2000°C, respectively; the presence of Fe^{3+} ions higher than the respective concentrations makes the concentration of the vancancy pair higher than that of the free oxygen vacancy at the respective temperatures.

Acknowledgments

The authors acknowledge W. D. Kingery's discussion and encouragement and the authors' former students, K. Matsuhiro, Y. Kubota, N. Suga, Y. Hiro, H. Kurokawa, and M. Yamakura, who have contributed to this work.

References

[1] W. D. Kingery, H. K. Bowen, and D. R. Uhlmann; p. 151 in Introduction to Ceramics. Wiley, New York, 1976.
[2] Y. Oishi and W. D. Kingery, "Self-Diffusion of Oxygen in Single Crystal and Polycrystalline Aluminum Oxide," J. Chem. Phys., 33 [2] 480–6 (1960).
[3] Y. Oishi and W. D. Kingery, "Oxygen Diffusion in Periclase Crystals," J. Chem. Phys., 33 [3] 905–6 (1960).
[4] Y. Oishi, K. Ando, and Y. Kubota, "Self-Diffusion of Oxygen in Single Crystal Alumina," J. Chem. Phys., 73 [3] 1410–2 (1980).
[5] Y. Oishi, K. Ando, N. Suga, and W. D. Kingery, "Effect of Surface Condition on Oxygen Self-Diffusion Coefficients for Single-Crystal Al$_2$O$_3$," J. Am. Ceram. Soc., 66 [8] C-130–C-131 (1983).
[6] Y. Oishi, K. Ando, H. Kurokawa, and Y. Hiro, "Oxygen Self-Diffusion in MgO Single Crystals," J. Am. Ceram. Soc., 66 [4] C-60–C-62 (1983).
[7] J. B. Holt and R. H. Condit; pp. 13–29 in Materials Science Research, Vol. 3. Edited by W. W. Kriegel and H. Palmour III, Plenum, New York, 1966.
[8] K. P. R. Reddy and A. R. Cooper, "Oxygen Diffusion in Sapphire," J. Am. Ceram. Soc., 65 [12] 634–8 (1982).
[9] K. P. R. Reddy and A. R. Cooper; private communication.
[10] D. J. Reed and B. J. Wuensch, "Ion-Probe Measurement of Oxygen Self-Diffusion in Single-Crystal Al$_2$O$_3$," J. Am. Ceram. Soc., 63 [1–2] 88–92 (1980).

[11] S. Shirasaki and Y. Oishi, "Role of Grain Boundaries in Oxygen Self-Diffusion in Polycrystalline MgO," *Jpn. J. Appl. Phys.,* **10** [8] 1109–10 (1971).

[12] Y. Oishi, K. Ando, and K. Matsuhiro, "Self-Diffusion Coefficient of Oxygen in Vapor-Grown Single Crystal Alumina," *Yogyo-Kyokai-Shi,* **85** [10] 522–4 (1977).

[13] H. S. Parker and C. A. Harding, "Vapor Growth of Al_2O_3 Bicrystals," *J. Am. Ceram. Soc.,* **53** [11] 583–5 (1970).

[14] A. H. Heuer et al.; private communication.

[15] M. M. Abraham, C. T. Butler, and Y. Chen, "Growth of High-Purity and Doped Alkaline Earth Oxides: I," *J. Chem. Phys.,* **55** [8] 3752–6 (1971).

[16] K. Ando, H. Kurokawa, and Y. Oishi, "Oxygen Self-Diffusion in Fe-doped MgO Single Crystals," *J. Chem. Phys.,* **78** [11] 6890–2 (1983).

[17] J. Narayan and J. Washburn, "Self-Diffusion in Magnesium Oxide," *Acta Metall.,* **21** [5] 533–8 (1973).

[18] C. Dubois, C. Monty, and J. Philibert, "Oxygen Self-Diffusion in NiO Single Crystals," *Philos. Mag.,* **46** [3] 419–33 (1982).

[19] R. Lindner and G. D. Parfitt, "Diffusion of Radioactive Magnesium in Magnesium Oxide Crystals," *J. Chem. Phys.,* **26**, 182–5 (1957).

[20] B. C. Harding, D. M. Price, and A. J. Mortlock, "Cation Self-Diffusion in Single Crystal MgO," *Philos. Mag.,* **23**, 399–408 (1971).

[21] B. C. Harding and D. M. Price, "Cation Self-Diffusion in MgO up to 2350°C," *Philos. Mag.,* **26**, 253–60 (1972).

[22] B. J. Wuensch, W. C. Steele, and T. Vasilos, "Cation Self-Diffusion in Single Crystal MgO," *J. Chem. Phys.,* **58**, 5258–66 (1973).

[23] D. R. Sempolinski and W. D. Kingery, "Ionic Conductivity and Magnesium Vacancy Mobility in Magnesium Oxide," *J. Am. Ceram. Soc.,* **63** [11–12] 664–9 (1980).

[24] W. H. Gourdin, W. D. Kingery, and J. Driear, "The Defect Structure of MgO Containing Trivalent Cation Solutes: The Oxidation-Reduction Behavior of Iron," *J. Mater. Sci.,* **14**, 2074–82 (1979).

[25] W. H. Gourdin and W. D. Kingery, "The Defect Structure of MgO Containing Trivalent Cation Solutes: Shell Model Calculations," *J. Mater. Sci.,* **14**, 2053–73 (1979).

[26] A. F. Henriksen, Y. M. Chiang, W. D. Kingery, and W. T. Petuskey, "Enhanced Oxygen Diffusion at 1400°C in Deformed Single-Crystal Magnesium Oxide," *J. Am. Ceram. Soc.,* **66** [8] C-144–C-146 (1983).

[27] W. C. Mackrodt; private communication.

[28] C. R. A. Catlow, I. D. Faux, and M. J. Norgett, "Shell and Breathing Shell Model Calculations for Defect Formation Energies and Volumes in Magnesium Oxide," *J. Phys. C: Solid State Phys.,* **9**, 419–29 (1976).

---

*Linde Div., Union Carbide Co., New York, NY.
†Prepared at the National Bureau of Standards.
‡One from the Norton Co. and the other from the Oak Ridge National Laboratory, Oak Ridge, TN.
§W & C Spicer Ltd., England.

Secondary Ion Mass Spectrometric Analysis of Oxygen Self-Diffusion in Single-Crystal MgO

Han-Ill Yoo and B. J. Wuensch

Massachusetts Institute of Technology
Department of Materials Science and Engineering
Cambridge, MA   02139

W. T. Petuskey*

University of Illinois at Urbana—Champaign
Department of Ceramic Engineering
Urbana, IL   61801

Single-crystal layers of Mg^{18}O were grown epitaxially on substrates of normal MgO through chemical transport with HCl. Exchange between ^{18}O in the crystals and the ^{16}O present in air was produced by annealings in the temperature range of 1000–1650°C, and diffusion coefficients were determined from concentration gradients measured with the aid of secondary ion mass spectrometry. The temperature dependence of the oxygen self-diffusion coefficients so obtained may be represented by an activation energy of 3.24 ± 0.13 eV and D_0 of $(1.8^{+2.9}_{-1.1}) \times 10^{-6}$ cm^2/s. The magnitude of the diffusivities is lower than that first obtained by exchange measurements but agrees well with recent values obtained from gradients established through proton activation analysis. The present activation energy is in the middle of the range reported in previous studies. This value is too large to represent an enthalpy for anion vacancy migration and is far too small to represent the combination of defect migration and formation enthalpies to be expected if, as presently believed, the defect structure of MgO were dominated by cation impurities.

Approximately a dozen studies of oxygen self-diffusion in single-crystal MgO have been reported subsequent to the first exchange measurements[1] performed by Oishi and Kingery in 1960. Although recent advances in theory have provided reliable estimates of the enthalpies for defect formation, migration, and association in MgO,[2–4] the various activation energies for anion self-diffusion provided by experiment are not in especially good agreement with either theory or each other. Moreover, the magnitudes of the diffusion coefficients that were obtained range over more than 2 orders of magnitude although this difference, of itself, would not be unexpected if extrinsic, impurity-controlled transport were involved.

The difference between the activation energies of previous studies may be at least partly due to the small temperature range over which experiments have been conducted. Each data set extends over a temperature range of 450 K at most. This attaches considerable uncertainty to the reported activation energy. The significance of differences between two values cannot be established unless the errors associated with the various physical measurements are assessed and combined to provide a realistic standard deviation. Results may be especially misleading if the measurements extend over a temperature range associated with a change in diffusion mechanism. If insufficient data are available to resolve a change in slope in an Arrhenius plot of the diffusion coefficients, the apparent activation energy will be some value intermediate to those for the two mechanisms.

Oxygen self-diffusion coefficients were measured over a wider temperature range, 1000–1700°C, in the present work. The predominant isotope in air, ^{16}O, was exchanged with ^{18}O in an isotopically enriched layer of $Mg^{18}O$, which had been grown epitaxially on a substrate of normal MgO. Diffusion coefficients were evaluated from concentration gradients that were established with secondary ion mass spectrometry (SIMS). Attention was given to the propagation of error in the measurements to facilitate meaningful comparison of the present results with earlier data.

Experimental Procedures

Preparation of Samples

Diffusion coefficients were determined in an ^{18}O-enriched single-crystal layer of MgO. This material was grown epitaxially on a substrate of commercially available single crystals[†] by a chemical transport method. A powder of $Mg^{18}O$ was prepared by burning magnesium metal ribbon[‡] in an atmosphere that had a nominal isotopic enrichment of 95 to ~99% $^{18}O_2$.[§] The $Mg^{18}O$ powder was cold-pressed into a 6.35-mm diameter × 2 to ~3 mm pellet for use as the source material in a closed-system transport apparatus.

The chemical transport apparatus followed closely a device described by Gruber.[5] Crystal growth was achieved through transport by HCl gas. The substrate crystal was supported over the pellet of source material by a Pt ring, and this assembly was enclosed in a Pt crucible at 1000 to ~1100°C and suspended, in turn, in a fused-silica chamber. A small temperature gradient was established between the source pellet and substrate by heating the lower portion of the Pt crucible by a radio frequency induction field. The rate of epitaxial growth on the substrate depends on temperature, the temperature gradient, the partial pressure of HCl in the growth chamber, and the source-to-substrate separation. With reduction of the latter distance to a few millimeters, growth rates up to 100 μm/h were achieved.

The surface of the substrate crystal had been prepared, prior to deposition, by cleavage on (100). A layer of approximately 100 μm was then removed by etching in hot phosphoric acid to eliminate damage caused by cleaving. The substrate was then annealed in situ in the transport system for 10 h at the growth temperature. The epitaxial isotopic layers were grown to a thickness of ~3 to ~50 μm. The surfaces of these layers were observed to bear an irregular distribution of pits but were used directly in the as-grown state for the exchange experiments in order to avoid mechanical damage or contamination by impurities. Gruber[5] had shown that dislocations remaining in the substrate grow out in the first few micrometers of deposit and also that the purity of the layer is superior or at least equal to that of the source material for most impurities. In particular, incorporation of Cl from the transport agent is less than 10 ppm. The samples were cut by a commercial low-speed saw[¶] along 100 planes normal to the growth surface to provide up to 9 specimens bearing an epitaxial isotope deposit of about 4 × 4 mm area.

Diffusion Annealing

Substrate crystals bearing the as-grown isotope deposit were annealed in an air atmosphere at temperatures in the range 1000–1650°C. Specimens were placed in a Pt envelope with an open end to minimize the possibility of contamination during the annealing. The duration of the annealing was selected to produce a concentration gradient on the order of 1 μm in depth for subsequent in-depth profiling with SIMS.

Unlike the conventional oxygen exchange technique, the minor isotope ^{18}O (0.2% natural abundance) diffused *out* of the sample to be replaced by the predominant ^{16}O isotope contained in air. The gas reservoir from which exchange occurs with the sample is thus essentially infinite, and problems of change in isotopic composition of the atmosphere as exchange progresses are not encountered. As a concentration gradient in the sample is directly determined by SIMS analysis, instead of monitoring the isotopic composition of the atmosphere, the low exchange rate at reduced temperature is a less constraining influence. It is not necessary to contain a small volume of atmosphere at a fixed composition of special isotope. Long diffusion annealing may be employed without difficulty. (With 6 months as an arbitrary upper limit and a diffusion depth of 500 Å (50.0 nm) as desirable to establish a well-defined diffusion gradient, diffusivities as low as 10^{-19} cm^2/s may be easily measured. Values to 4×10^{-19} cm^2/s were obtained in the present work.) A unique aspect of the procedure is that two diffusion coefficients may be determined from each sample: one from the exchange gradient at the surface of the isotopic layer (constant surface concentration boundary conditions) and one from the gradient produced by interdiffusion between the isotopic epitaxial layer and substrate crystal (the times and temperatures are sufficiently low that negligible interdiffusion occurs during deposition and semiinfinite source initial conditions apply). One can thus compare exchange diffusion coefficients and tracer interdiffusion coefficients for the same sample and thereby assess the possible influence of surface reaction, sample vaporization, or surface flaws on diffusion coefficients measured by exchange. Although an interdiffusion gradient is indeed available for measurement in each of the samples that were prepared, analysis of these portions of the specimens has not yet been completed due to the time-consuming necessity of sputtering through the entire thickness of the epitaxial layer. A comparison of results will be reported elsewhere. Preliminary examination of a few samples indicates, however, that the diffusion coefficients provided by the two gradients are in very good agreement.

Measurement of Concentration Gradients

The distribution of isotopes in the annealed specimens was determined by depth-profiling sputtering with an ion microprobe.** The primary ion beam was ^{40}Ar$^+$-accelerated to 9.645 $\pm$ 0.008 keV. The intensity of the positive secondary ions ^{16}O$^+$, ^{18}O$^+$, and ^{24}Mg^{2+} was measured. The primary beam was rastered over a 250 $\times$ 250 μm area during irradiation, but a mechanical aperture was used to restrict the area that contributed to the measured intensity to a central area of approximately 150 μm diameter in order to minimize artifacts arising from edge effects and redeposition of material.[6] The problems caused by charging of the insulator surface under bombardment by the incident ion beam were almost completely eliminated in the present work by depositing 300 Å (30.0 nm) of gold film on the sample surface and by flooding the sputtered area of the surface with an appropriate number of electrons. Determination of the proper electron flux and maintenance of the flux at the correct location were quite laborious. An interpretation of the charge neutralization process has been provided by Hunt et al.[7] and Müller.[8]

An estimate of the sputtering rate was used to establish the time necessary to remove a layer of desired depth. This thickness (10–80 Å) (1–8 nm), in turn, was based on the total anticipated extent of the concentration gradient. The intensity accumulated during this time interval was stored in a multichannel analyzer. After completion of the analysis, the total depth, d, of the sputtered crater was measured with a profilometer.[††] Most of the crater depths could be measured to within a 10% relative error at the 100% confidence level. A few measurements, however, were subject to an error as high as 50%. This was usually due to the inadvertent selection for analysis of a locally rough area of the original specimen surface. As the sputtering rate remained constant within reasonable uncertainty for a prefixed primary accelerating voltage, an attempt was made to average the uncertainty in crater depth over the entire set of analyses. This was done by plotting for each sample the measured crater depth, d, per unit primary ion beam current,[‡‡] I_p, as a function of sputtering time. Figure 1 shows the relation is linear, and the results may be satisfactorily represented by the equation

$$d/I_p = (1.3 \pm 0.2) \times 10^{-10}t + (3.2 \pm 0.7) \times 10^{-7} \quad \text{(cm/nA)} \quad (1)$$

Since the primary beam current, I_p, had been well stabilized and the sputtering time, t, was subject to a negligible uncertainty, Eq. (1) rather than the measured value was used to evaluate a corrected total crater depth, d_0, for a given experi-

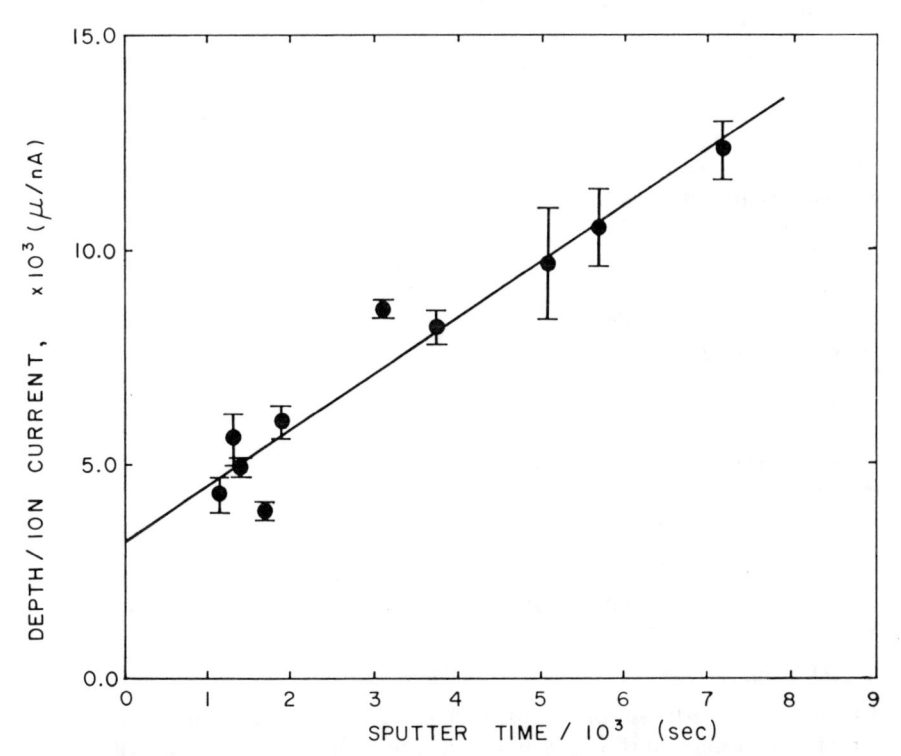

Fig. 1. Crater depth per unit primary ion current as a function of sputtering time for MgO irradiated by a 9.65-kV [40]Ar[+] primary ion beam.

ment. The thickness of each layer sputtered from the sample was thus provided by

$$\Delta X = d_0/N \tag{2}$$

The surface of the ^{18}O-enriched crystal was maintained at constant surface concentration $^{16}C_s$ (atomic fraction Mg^{16}O) during the course of the diffusion annealing. The anticipated distribution of ^{16}O is accordingly

$$(^{16}C - {}^{16}C_0) = (^{16}C_s - {}^{16}C_0)[1 - \text{erf } X(4Dt)^{-1/2}] \tag{3}$$

where penetration into the sample, X, is ΔX times the channel number, according to Eq. (2); $^{16}C_0$ is the uniform atomic fraction of Mg^{16}O initially present in the sample (experimentally found to be 0.5–0.9 in the samples examined), and erf is the Gaussian error function. The isotopic ratio [^{18}O]/[^{16}O] was used to establish the atomic fraction Mg^{18}O, the species with lowest natural abundance. With relations of the form $^{18}C = 1 - {}^{16}C$ (the concentration of a third stable isotope, ^{17}O, is negligible, as its natural abundance is 0.00037), Eq. (3) may be recast in the form

$$(^{18}C - {}^{18}C_s)/(^{18}C_0 - C_s) = \text{erf } X(4Dt)^{-1/2} \tag{4}$$

If the gas exchange reaction at the free surface of the epitaxial layer is sufficiently rapid, $^{18}C_s$ is expected to be close to the natural abundance of ^{18}O (0.00204). If this were not found to be the case experimentally, the gas exchange reaction should be taken into account as, e.g., a first-order quasi-chemical reaction, and Eq. (4) should be replaced by the more appropriate representation[9]

$$\{(^{18}C - {}^{18}C_s)/(^{18}C_0 - {}^{18}C_s) - \exp(k^2t/D - kX/D) \text{ erfc } [X(4Dt)^{-1/2} - k(t/D)^{1/2}]\}$$

$$= \text{erf } X(4Dt)^{-1/2} \tag{5}$$

where k is a reaction constant. As will be seen, use of Eq. (4) with the present data proved highly satisfactory, correlation coefficients between X and erf^{-1} $(^{18}C - {}^{18}C_s)/(^{18}C_0 - {}^{18}C_s)$ being always greater than 0.98. This implies that the second term on the left of Eq. (5) is unimportant within experimental error.

Use of Eq. (4) is subject to another qualification as the concentrations of ^{18}O and ^{16}O are comparable throughout much of the diffusion gradient, and neither is at vanishingly small "tracer" levels. The diffusion coefficient, D, must therefore be of the Nernst–Planck type[10]

$$D = {}^{18}D^{16}D/(^{18}C^{18}D + {}^{16}C^{16}D) \tag{6}$$

where ^{18}D and ^{16}D are the diffusion coefficients of ^{18}O and ^{16}O, respectively. ^{16}D is expected to be the larger, because of the isotope effect,[11] and is given by

$$^{16}D = (1 + \alpha)^{18}D \tag{7}$$

where $\alpha = f((^{18}M/^{16}M)^{1/2} - 1)$. The symbol f is the correlation factor, and ^{16}M and ^{18}M are isotopic masses of ^{16}O and ^{18}O, respectively; α is no greater than 0.05. Substitution of Eq. (7) in Eq. (6) thus provides

$$D = {}^{18}D(1 + \alpha)/(1 + {}^{16}C\alpha) \approx {}^{18}D(1 + {}^{18}C\alpha) \tag{8}$$

The diffusion coefficient of Eq. (4) is thus a function of concentration, and the error function solution is obviously not rigorously applicable. The contribution of the term $^{18}C\alpha$ to D amounts to only a few percent, however, and is negligible in comparison to other experimental uncertainties.

Results

A typical plot of atomic fraction of Mg^{18}O as a function of channel number or penetration X (d_0/N times channel number) is presented in Fig. 2. The surface concentration in all such gradients was sufficiently close to the natural abundance of Mg^{18}O to justify the use of Eq. (3) rather than the surface reaction limited expression, Eq. (6). The former predicts that a plot of erf^{-1} ($^{18}C - {}^{18}C_s$)/($^{18}C_0 - {}^{18}C_s$) as a function of penetration should be linear with a slope A equal to $(4Dt)^{-1/2}$.

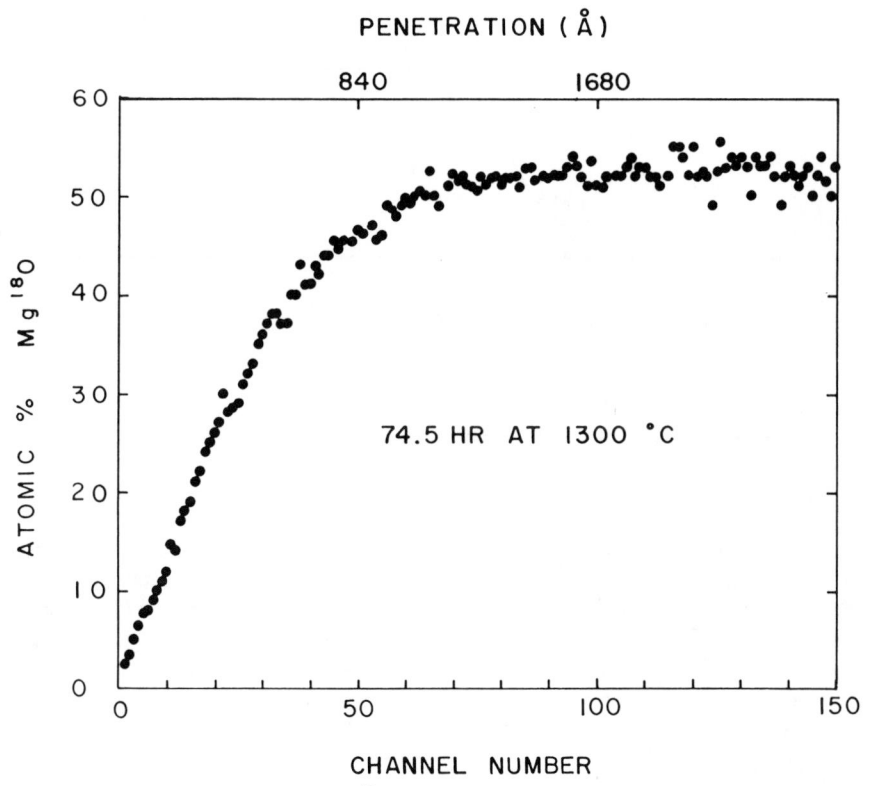

Fig. 2. Atomic fraction of Mg^{18}O as a function of penetration in a layer of isotopically enriched single crystal Mg^{18}O after exchange for 74.5 h at 1300°C with the 0.9976 $^{16}O_2$ content of air.

Figure 3 presents such a plot for the depth profile of Fig. 2. The result is linear and extrapolates to zero at zero penetration. A value for the diffusion coefficient was determined from a least-squares fit to the data up to erf^{-1} <0.6 or 0.7 to minimize the propagation of error contained in the measurement of C and the uncertainty in C_0. The standard deviation in the slope was found to vary between 1% and a maximum of 20% but, as previously noted, with a correlation coefficient always ≥0.98. When multiple analyses were performed on a given sample, the resulting diffusion coefficients were combined into an average, weighted according to the individual standard deviations as described below.

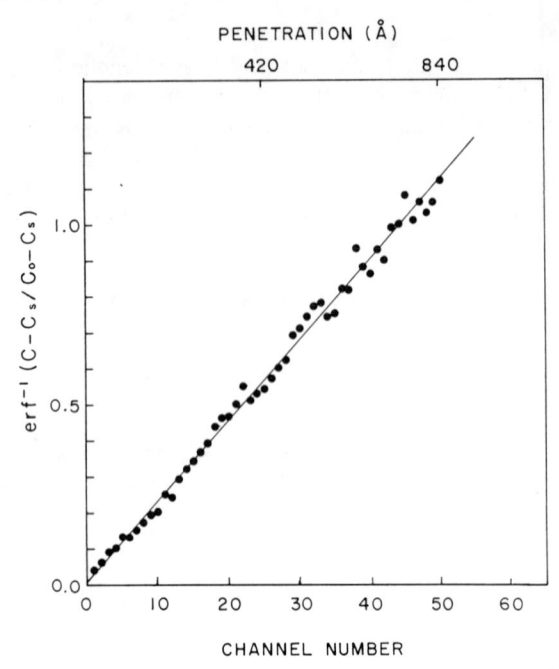

Fig. 3. Plot as a function of penetration of the inverse error function of the ratio of atomic fraction Mg^{18}O less surface concentration to initial atomic fraction less surface concentration.

The average diffusion coefficients determined in the present work are listed in Table I along with the conditions under which these values were obtained. A plot of the logarithm of the diffusion coefficient as a function of reciprocal temperature is presented in Fig. 4; the error bars, however uncosmetic, represent the standard deviations of Table I, based on realistic assessment of error and not merely the standard deviation of the least-squares fit to plots such as that in Fig. 3. As the standard deviations in the values of D are not only comparable ($\sigma_D/D \approx 0.4$) but

Table I. Oxygen Self-Diffusion Coefficients in Single-Crystal MgO

T (°C)	Time (h)	Extent of gradient* ($\times 10^3$ Å)	D (cm^2/s)	σ_D/D
1650	2.0	1.2	$(5.2\pm2.7)\times10^{-15}$	0.52
1600	3.0	1.6	$(5.9\pm2.3)\times10^{-15}$	0.39
1550	4.7	1.0	$(1.4\pm0.8)\times10^{-15}$	0.57
1500	4.0	1.0	$(1.6\pm0.7)\times10^{-15}$	0.44
1400	12.0	0.8	$(3.7\pm1.6)\times10^{-16}$	0.43
1300	74.5	0.8	$(5.6\pm2.4)\times10^{-17}$	0.43
1200	360.0	0.7	$(1.0\pm0.4)\times10^{-17}$	0.40
1100	480.0	0.4	$(1.9\pm0.5)\times10^{-18}$	0.26
1000	960.0	0.3	$(4.5\pm1.0)\times10^{-19}$	0.22

*Approximate penetration to a value of erf^{-1} of ca. 1.0 (1.2×10^3 Å$=0.12\times10^4$ nm).

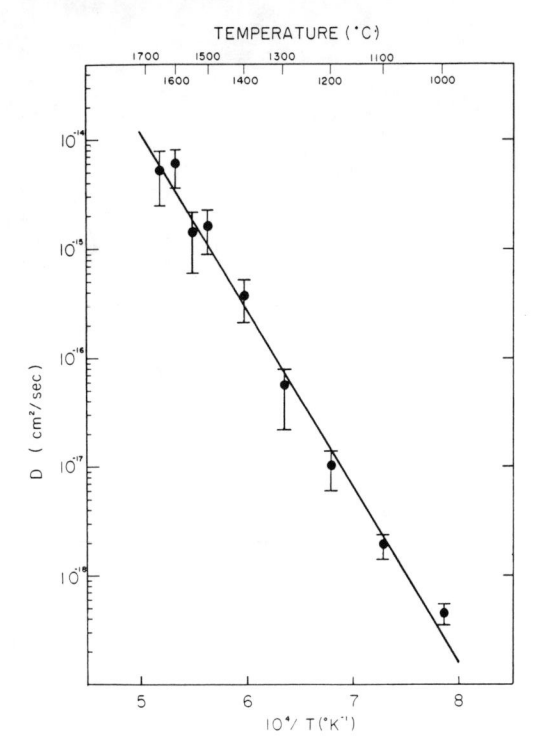

Fig. 4. Plot of the logarithm of the diffusion coefficient for ^{18}O in single-crystal MgO as a function of reciprocal temperature.

large compared to the uncertainty in the measurement of temperature, equal weights were assigned to each measurement in determining a least-squares fit to the data of Fig. 4. The result may be expressed by

$$D = (1.8^{+2.9}_{-1.1}) \times 10^{-6} \exp\left(-\frac{3.24 \pm 0.13 \text{ (eV)}}{kT}\right) \quad (\text{cm}^2/\text{s}) \qquad (9)$$

over a temperature range 1000°C ≤ T < 1700°C. The correlation coefficient of the set $\{D, 1/T\}$ is -0.99.

Discussion

The results of the present analyses are compared in Fig. 5 with the oxygen self-diffusion coefficients reported in previous studies.[1,12-17] The magnitudes of the present diffusion coefficients are 2 orders of magnitude smaller than those obtained by early exchange measurements,[1,12] but Oishi et al.[17] showed that chemical polishing to remove the surface damage introduced in crushed samples considerably lowered the apparent diffusivity. In addition, certain heat treatments could increase the apparent diffusivity by creating etch pits, which increased the area available for exchange.

In contrast, the present results are in good agreement with those of Rovner[15] and agree in magnitude, if not activation energy, with results obtained by Reddy[14] through proton activation measurements of the gradients produced by ^{18}O ex-

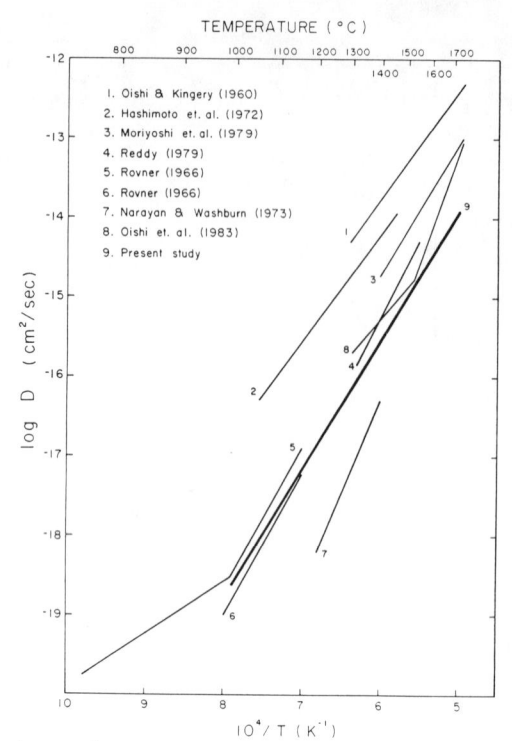

Fig. 5. Comparison of the oxygen self-diffusion coefficients of the present work with previously reported values.

change. The latter agreement is of particular interest as discrepancies that could not be satisfactorily explained were present between the results of SIMS analysis[18] and proton activation analysis[19] of anion self-diffusion in Al_2O_3. Also included in Fig. 5 are results of an indirect determination of diffusion coefficients based upon measurement of dislocation loop shrinkage with the aid of transmission electron microscopy.[16] The diffusivities so obtained are smaller than any directly measured coefficients. The values hinge, however, on the model assumed for loop shrinkage.

The activation energies reported in previous studies vary from 2.6 to 3.8 eV for oxygen diffusion measured with gas exchange techniques. The activation energy obtained in the present work, 3.24 ± 0.13 eV, is squarely in the middle of this range. It is difficult to decide whether individual differences are truly significant in view of the limited temperature range and uncertain error limits of most studies.

An activation energy on the order of 3.2 eV is very difficult to interpret in terms of the available theoretical estimates of energies relevant to the defect structure of MgO and the prevailing interpretation of cation transport mechanisms. By virtue of good agreement between the calculated and experimental energies for cation vacancy migration as well as the magnitudes of ionic electrical conductivity and cation self-diffusion, it seems well established that cationic transport in MgO occurs by a vacancy mechanism and that vacancy concentrations are determined by aliovalent impurities.[20] If such is the case and if Schottky equilibrium is maintained, anion vacancy concentrations and anion self-diffusion coefficients should be depressed—and probably much more so than the ca. 3 orders of magnitude difference presently observed between the measured anion and cation self-diffusion coefficients. Moreover, the activation energy anticipated for anion self-diffusion in a crystal whose defect structure is controlled by cation impurities should be given by $H_m + H_f$, where H_m and H_f are the enthalpies for anion vacancy motion and Schottky vacancy-pair formation, respectively. The sum of theoretical estimates for these enthalpies, namely, 2.38[2] or 2.11[3] eV plus 7.5[2] or 7.72[3] eV, is clearly incompatible with the experimental values. Similarly, the difference between experiment and H_m seems too great for all investigations to date to provide a satisfactory interpretation. Consideration of cation–anion vacancy association leads to the activation energy of $H_a + H'_m + H_f$ for an extrinsic crystal of cationic origin, where H_a and H'_m are the enthalpies for divacancy formation and migration, respectively. H_a has been theoretically estimated to be -2.55 eV,[2] and the H'_m has been shown to be comparable with or to be greater than the activation energy for single vacancy movement.[21] The activation energy is, thus, at least $-2.55 + 2.38 + 7.5$ eV or 7.33 eV, which is far greater than the observed 3.21 eV. The nature of the anion transport mechanism accordingly remains unclear and is not satisfactorily interpretable in terms of any obvious model for defect structure.

Acknowledgments

The authors are grateful to N. Shimizu of the M.I.T. Department of Earth and Planetary Sciences and D. A. Reed and J. E. Baker of the Materials Research Center at the University of Illinois for valuable discussions and assistance with SIMS analysis. This work was supported under Contract No. DE-AC02-76ER02923 with the Office of Basic Energy Sciences of the U.S. Department of Energy. The SIMS analyses were performed at the Center for Microanalysis of Materials in the Materials Research Laboratory at the University of Illinois, a national facility supported by the U.S. Department of Energy under Contract No. DE-AC02-76ER01198.

Appendix: Assessment of Errors

In order to meaningfully compare experimental diffusivities and the diffusion parameters derived therefrom, it is necessary to assign realistic standard deviations to the measured diffusion coefficients. This must include not only the standard deviation in a least-squares fit to the concentration gradient, but also the uncertainty in each of the measured parameters that are necessary to evaluate D.

Equation (1) expressed a relation between sputtering time and depth per unit ion beam current and was used to average measured crater depths over the entire data set. As the primary beam current is well stabilized and sputtering time is subject to negligible uncertainty, the error in corrected crater depth is given by the uncertainty in the parameters that describe the least-squares fit to the sputtering data. Thus

$$\delta d_0 = I_p[(2 \times 10^{-11})t + 7 \times 10^{-8}] \quad \text{(cm)} \tag{A1}$$

Although the typical value of $\delta d/d$ for an individual measurement was ca. 10%, use of Eq. (A1) resulted in an error on the order of 30%. The error propagated to the evaluation of the thickness of each layer removed from the specimen, Eq. (2), is thus

$$\delta(\Delta X) = (1/N)\delta d_0 \tag{A2}$$

Solute penetration X is given by ΔX times channel number, so that a plot of the inverse error function as a function of channel number, following Eq. (4), has slope A given by $d_0(4Dt)^{-1}$. The diffusion coefficient is thus provided by $d_0^2/(4A^2t)$, where t is the duration of the diffusion annealing, and

$$\delta D/D = \left\{4\left(\frac{\delta A}{A}\right)^2 + 4\left(\frac{\delta(\Delta X)}{\Delta X}\right)^2 + \left(\frac{\delta t}{t}\right)^2\right\}^{1/2} \quad (A3)$$

Equation (A3) is, of course, based on the assumption that measurements of the slope, A, channel thickness, ΔX, and annealing time, t, are independent and contain only random errors. In spite of the uncertainties in annealing time associated with heating and cooling of the sample, $\delta t/t$ is small compared with $\delta A/A$ and $\delta(\Delta X)/\Delta X$. Moreover, with very few exceptions, $\delta d_0/d_0 \gg \delta A/A$ for the analyses performed in the present work. The largest contribution to σ_D is uncertainty in the measured final depth of the sputtered crater—created, in turn, by roughness in the as-grown surface of the epitaxial layers which were examined.

Equation (A3) was used to evaluate the standard deviations of the diffusion coefficients reported in Table I. As more than two analyses were performed for each specimen, a weighted average was then taken on the basis of the best estimate[22]

$$\overline{D} = \sum \sigma_{D_i}^{-2} D_i \bigg/ \sum \sigma_{D_i}^{-2} \quad (A4)$$

with

$$\overline{\sigma}_D = \left(\sum \sigma_{D_i}^{-2}\right)^{-1/2} \quad (A5)$$

References

[1]Y. Oishi and W. D. Kingery, "Oxygen Diffusion in Periclase Crystals," *J. Chem. Phys.*, **33** [3] 905–6 (1960).
[2]W. C. Mackrodt and R. F. Stewart, "Defect Properties of Ionic Solids: III. The Calculation of the Point-Defect Structure of the Alkaline-Earth Oxides and CdO," *J. Phys. C.*, **12** [23] 5015–36 (1979).
[3]M. J. L. Sangster and D. K. Rowell, "Calculation of Defect Energies and Volumes in Some Oxides," *Philos. Mag.*, [Part A], **44** [3] 613–24 (1981).
[4]E. A. Colbourn and W. C. Mackrodt, "The Influence of Impurities on the Migration Energy of Cation Vacancies in MgO," *Ceramurgia Int.*, **8** [3] 90–2 (1982).
[5]P. E. Gruber, "Growth of High Purity Magnesium Oxide Single Crystals by Chemical Vapor Transport Techniques," *J. Cryst. Growth*, **18**, 94–8 (1973).
[6]H. W. Werner, "Quantitative Secondary Ion Mass Spectrometry: A Review," *Surf. Interf. Anal.*, **2** [2] 56–74 (1980).
[7]C. P. Hunt, C. T. H. Stoddart, and M. P. Seah, "The Surface Analysis of Insulators by SIMS: Charge Neutralization and Stabilization of the Surface Potential," *Surf. Interf. Anal.*, **3** [4] 157–60 (1981).
[8]G. Müller, "Surface Analysis of Insulating Materials by Secondary Ion Mass Spectrometry (SIMS)," *J. Appl. Phys.*, **10** [4] 317–24 (1976).
[9]J. Crank; The Mathematics of Diffusion, 2nd ed. Oxford University Press, London, 1975.
[10]H. Schmalzried; Solid State Reactions, 2nd ed. Verlag Chemie Gmbh, Weinheim, West Germany, 1981.
[11]J. N. Coles and J. V. P. Long, "An Ion-Microprobe Study of the Self-Diffusion of Li$^+$ of Lithium Fluoride," *Philos. Mag.*, **29** [3] 457–71 (1974).
[12]H. Hashimoto, M. Hama, and S. Shirasaki, "Preferential Diffusion of Oxygen along Grain Boundaries in Polycrystalline MgO," *J. Appl. Phys.*, **43** [11] 4828–9 (1972).
[13]Y. Moriyoshi, T. Ikegami, S. Matsuda, and S. Shirasaki, "The Formation of Subgrain Boundaries in MgO Single Crystal," *Z. Phys. Chem. (Wiesbaden)*, **118**, 187–95 (1979).
[14]K. P. R. Reddy, "Oxygen Diffusion in Close-Packed Oxides"; Ph.D. Thesis, Case Western Reserve University, Cleveland, OH, 1979.
[15]L. H. Rovner, "Diffusion of Oxygen in Magnesium Oxide"; Ph.D. Thesis, Department of Physics, Cornell University, Ithaca, NY, 1966.

[16]J. Narayan and J. Washburn, "Self-Diffusion in Magnesium Oxide," *Acta Metall.*, **21** [5] 533–8 (1973).
[17]Y. Oishi, K. Ando, H. Kurokawa, and Y. Hiro, "Oxygen Self-Diffusion in MgO Single Crystals," *J. Am. Ceram. Soc.*, **66** [4] C60–C62 (1983).
[18]D. J. Reed and B. J. Wuensch, "Ion-Probe Measurement of Oxygen Self-Diffusion in Single-Crystal Al_2O_3," *J. Am. Ceram. Soc.*, **63** [1–2] 88–92 (1980).
[19]K. P. R. Reddy and A. R. Cooper, "Oxygen Diffusion in Sapphire," *J. Am. Ceram. Soc.*, **65** [12] 634–8 (1982).
[20]B. J. Wuensch, "Diffusion in Stoichiometric Close-Packed Oxides"; pp. 353–76 in Mass Transport in Solids. Edited by F. Beniere and C. R. A. Catlow. Plenum, New York, 1983.
[21]K. Tharmalingam and A. B. Lidiard, "Mobility of Vacancy Pairs in Ionic Crystals," *Philos. Mag.*, **6** [69] 1157–62, 1961.
[22]J. R. Taylor; An Introduction to Error Analysis. University Science Books, Mill Valley, CA, 1982.

---

*Present address: Department of Chemistry, Arizona State University, Tempe, AZ.
†Norton Research Corp. (Canada) Ltd., Niagara Falls, Ontario, Canada.
‡Matheson Coleman & Bell, Norwood, OH.
§Mound Laboratory, Monsanto Research Corp., Miamisburg, OH.
˙Isomet, Springfield, VA.
**Cameca ims-3f, Cameca S. A., Paris, France.
††Sloan Dektak Surface Profile Measuring System.
‡‡Normalization by the total current rather than by current density is justified, as the irradiated sample area was kept constant in all analyses.

Grain-Boundary and Lattice Diffusion of ^{51}Cr in Alumina and Spinel

V. S. STUBICAN AND J. W. OSENBACH

The Pennsylvania State University
Department of Materials Science and Engineering
University Park, PA 16802

The grain-boundary diffusion parameter, $\alpha D'\delta$, which is the product of the segregation coefficient, the grain-boundary diffusion coefficient, and the boundary width has been measured in the temperature region 1200–1600°C along low-angle and high-angle grain boundaries of Al_2O_3 and $MgAl_2O_4$. The activation energies for ^{51}Cr grain-boundary diffusion in Al_2O_3 and $MgAl_2O_4$ are 170 ± 10 and 185 ± 10 kJ/mol, respectively, which is ~0.55 times the activation energy for the lattice diffusion. The grain-boundary diffusivity was strongly dependent on the grain-boundary orientation. The ratio of $D'_{\|}$ (parallel) to $D'_{\perp}$ (perpendicular) to the growth direction, $D'_{\|}/D'_{\perp}$, is ~23 for a 6°(01$\bar{1}$2) Al_2O_3 tilt boundary and ~2 for a 13°(01$\bar{1}$2) Al_2O_3 tilt boundary. The same ratio is ~87 for a 4.7°(100) $MgAl_2O_4$ tilt boundary and ~4 for a 23°(100) $MgAl_2O_4$ tilt boundary. These results and previously obtained results with MgO indicate that the mechanism of diffusion in grain boundaries of oxides is a dislocation pipe mechanism. It was found that the temperature dependence of the lattice diffusion coefficients of ^{51}Cr in Al_2O_3 and $MgAl_2O_4$ can be expressed as $D_{Cr} = (2.58 \pm 1.5) \times 10^{-3} \exp-(306 \pm 58 \text{ (kJ/mol)}/RT)$ cm^2/s and $D_{Cr} = (2.4512) \times 10^{-2} \exp-(337 \pm 45 \text{ (kJ/mol)}/RT)$ cm^2/s, respectively.

The kinetics of many chemical transport phenomena in ceramic systems such as solid-state reactions, sintering, grain growth, and creep are controlled by the rate of solid-state diffusion. Mechanisms of lattice diffusion in ionic systems are reasonably well understood theoretically, and the large volume of experimental data is basically consistent with the theory. However, diffusion along grain boundaries in ionic materials is not well understood. There are several uncertainties concerning the mechanism of diffusion along grain boundaries in ionic materials. A space charge is expected to exist at all interfaces in ionic materials.[1-3] The effect that the space charge has on grain-boundary diffusion has been a point discussed in several papers. Kingery[4] reviewed the experimental data and suggested that the space charge in the grain-boundary core may affect grain-boundary diffusion in ionic systems. Tiku and Kröger[5] have developed a theory that takes into account the effect that the grain-boundary space charge has on the electrical conductivity of polycrystalline Al_2O_3. They found that the grain-boundary space charge has a negligible effect on the overall conductivity. Yan et al.[6] have determined that the grain-boundary space charge also has a negligible effect on the grain-boundary diffusivity of ionic materials.

Another point of uncertainty is what effect grain-boundary segregation or precipitation has on the grain-boundary diffusivity in ionic materials. Wuensch and Vasilos[7] have reviewed the early literature on grain-boundary diffusion in oxides and have concluded that enhanced diffusion along grain boundaries is found only if there is evidence of solute segregation or precipitation. Recently, Atkinson and

Taylor[8] measured the diffusion of ^{63}Ni along grain boundaries in polycrystalline NiO. No evidence of solute segregation was found, and a large enhancement of diffusion along the grain boundaries was observed. Chen and Peterson[9] also found a strong enhancement of ^{51}Cr and ^{60}Co diffusion along the grain boundaries of NiO bicrystals of high purity.

In a previous paper, it was reported that the grain-boundary diffusion of ^{51}Cr in MgO and Cr-doped MgO bicrystals was several orders of magnitude faster than the lattice diffusion.[10] A large degree of anisotropy in the grain-boundary diffusion in MgO was found. This indicated that the grain-boundary diffusion in MgO is connected to the orientation of dislocations, and the mechanism is one of dislocation pipe diffusion, which is similar to the grain-boundary diffusion in metallic systems.

The purpose of this investigation was to determine the grain-boundary diffusivity at different temperatures of ^{51}Cr in Al_2O_3 and $MgAl_2O_4$ as a function of grain-boundary orientation. It was hoped that the obtained results would contribute to the general understanding of the grain-boundary structure in oxide systems.

To evaluate the grain-boundary diffusion coefficients, the lattice diffusion coefficients for the diffusion of ^{51}Cr in Al_2O_3 and $MgAl_2O_4$ had to be determined.

Experimental Procedures

Grain-boundary diffusion of ^{51}Cr was studied in Al_2O_3 and stoichiometric $MgAl_2O_4$. Bicrystals (~5 × 3 × 3 mm) were cut out of the group of crystals with a diamond saw. The Al_2O_3 crystals were grown by the Czochralski method, and the $MgAl_2O_4$ crystals were grown by the fusion cast method. The bicrystals were oriented with a Laue back-reflection camera. All of the bicrystals used in this study had tilt boundaries that were approximately normal to the surface. Spectrographic analysis of the Al_2O_3 bicrystals indicated the following impurities: Si, 300–500 ppm; and Ca, 200–300 ppm. The stoichiometric $MgAl_2O_4$ bicrystals contained the following impurities: Si, 150–200 ppm; Ca, 400–500 ppm; Fe, <10 ppm; Cu, <5 ppm by weight.

The bicrystals were polished to a 1-μm finish with SiC paper and diamond paste. They were then annealed at 1450°C for ~7 days to ensure that the vacancy and dislocation substructures were equilibrated. The radioactive isotope ^{51}Cr was applied to the surface of the bicrystal by drying a 50-μL drop of ^{51}Cr$_2$(SO$_4$)$_3$ solution. Two bicrystals were then placed together with their active faces in common and wrapped with Pt wire. The couple was then annealed at the temperature of interest in a verticle Mo-wound resistance furnace controlled to ±1°C. After the diffusion anneal, both bicrystals were edge- and back-sectioned to a depth of >10 Dt to eliminate edge effects. A serial sectioning technique was used to determine the concentration profiles. The 0.32-MeV ^{51}Cr γ ray was counted with a NaI (Tl) crystal and multichannel analyzer. The residual activity in the bicrystal was counted after a section of a known thickness was removed. To optimize counting statistics, only activities >10 times the background were used.

Results and Discussion

Several authors gave solutions that can be used to calculate the grain-boundary diffusion coefficient.[11-14]

In this work Whipple's infinite source solution

$$D'\delta = \left(\frac{\partial \ln \bar{c}}{\partial y^{6/5}}\right)^{-5/3} \left(\frac{4D}{t}\right)^{1/2} (0.78)^{5/3} \tag{1}$$

was used. D' is the grain-boundary diffusion coefficient, δ is the effective grain-boundary thickness, $\bar{c}$ is the average specific activity of tracer in section Δy thick from $-\infty \leq x \leq \infty$ at a distance y from the original surface, D is the lattice (tracer) diffusion coefficient, and t is the time of anneal. Equation (1) is correct if the ratio $D'\delta/2D(Dt)^{1/2}$ is >10. For solute diffusion (impurity diffusion), a solute segregation factor, α, should be considered, so that δ should be replaced by $\alpha\delta$.[15]

In Fig. 1 the difference in recorded tracer activity at y and y' ($\bar{c}$ at y') vs $y^{6/5}$ for a 4.8°(100) MgAl$_2$O$_4$ bicrystal tilt boundary is shown. There are two regions in this curve: a high-activity region, which is mainly due to lattice diffusion, and a low-activity region (the straight line section), which is mainly due to grain-boundary diffusion.

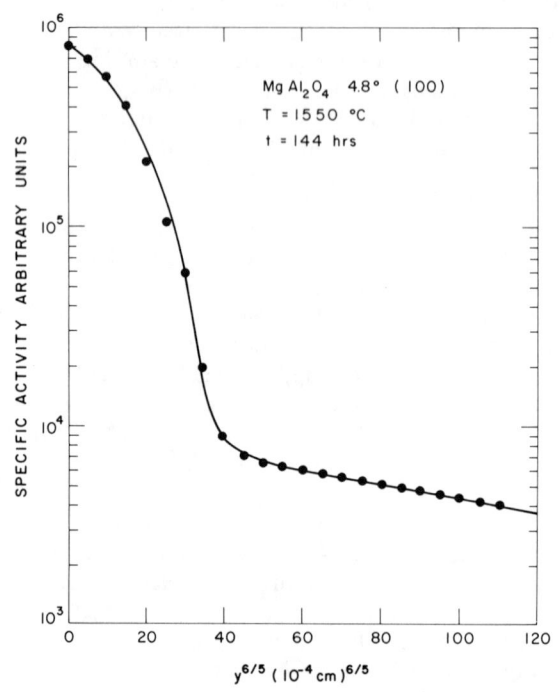

Fig. 1. Plot of specific activity of ^{51}Cr as a function of (penetration depth)$^{6/5}$ for a MgAl$_2$O$_4$ bicrystal annealed for 144 h at 1550°C. Tilt angle $\theta = 4.8°(100)$; diffusion is parallel to the growth direction.

With use of a technique described in a previous paper,[10] the lattice and the grain-boundary diffusion contributions were separated. Figures 2 and 3 show the temperature dependence of the lattice diffusion coefficients of ^{51}Cr in Al$_2$O$_3$ and MgAl$_2$O$_4$, respectively. The error bars represent a 95% confidence interval on the data, and the solid lines represent the best fit regression line on the data; $R^2 > 0.99$ for Al$_2$O$_3$ and for MgAl$_2$O$_4$.* The temperature dependence of the lattice diffusion coefficients of ^{51}Cr in Al$_2$O$_3$ and MgAl$_2$O$_4$ can be expressed as $D_{Cr} = (2.58 \pm 1.5) \times 10^{-3} \exp{-(306 \pm 58 \text{ (kJ/mol)}/RT)}$ cm^2/s and $D_{Cr} = (2.45 \pm 2) \times 10^{-2} \exp{-(337 \pm 45 \text{ (kJ/mol)}/RT)}$ cm^2/s, respectively.

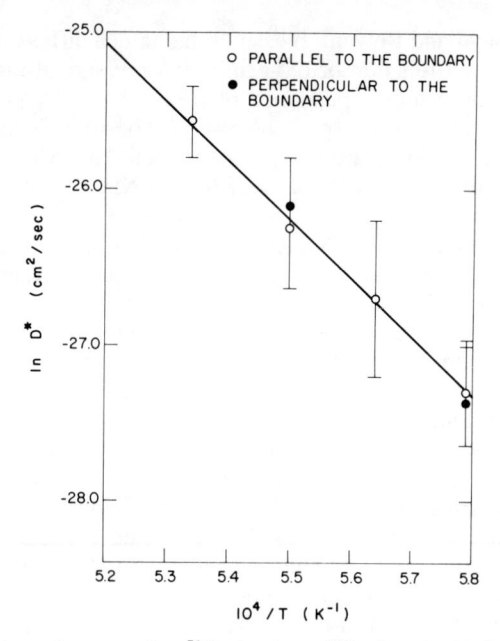

Fig. 2. Arrhenius diagram for ^{51}Cr lattice diffusion in Al$_2$O$_3$. The open circles are the data measured normal to a 6°(01$\bar{1}$2) plane, and the closed circles are the data measured parallel to this plane.

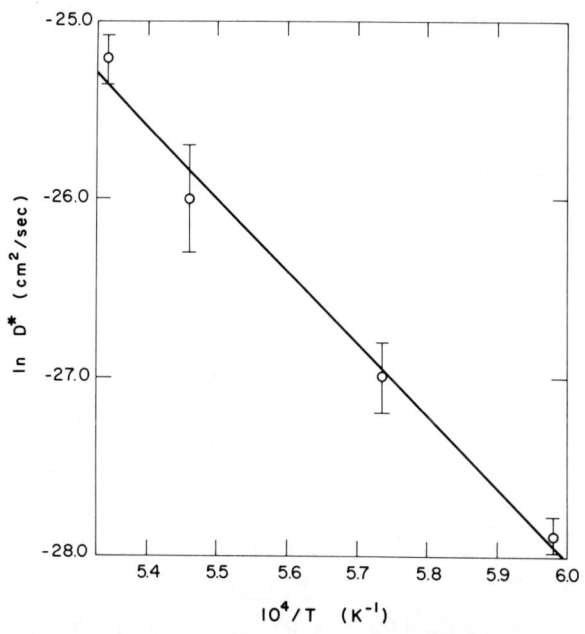

Fig. 3. Arrhenius diagram for ^{51}Cr lattice diffusion in MgAl$_2$O$_4$.

Recently, Lloyd and Bowen[16] measured the lattice diffusion of ^{59}Fe in single crystals of Al_2O_3 as a function of oxygen partial pressure at three different temperatures. The temperature dependence of their data at $P_{O_2} \approx 0.21$ can be expressed by $D_{Fe} = 1.4 \times 10^{-2} \exp{-(318 \pm 67 \text{ (kJ/mol)}/RT)}$ cm^2/s. Our data on ^{51}Cr lattice diffusion in Al_2O_3 are fairly close to their data. Minford and Stubican[17] measured the interdiffusion coefficients of Al^{3+} in $NiAl_2O_4$. An activation energy of 326 ± 41 kJ/mol characterized the temperature dependence of the interdiffusion of Al^{3+} for all of the Ni/Al ratios studied. Our results on ^{51}Cr lattice diffusion in $MgAl_2O_4$ are also close to their interdiffusion results.

In Fig. 2, the solid circles are the lattice diffusion coefficients measured parallel to a 6°($01\bar{1}2$) plane of Al_2O_3, and the open circles are the coefficients measured normal to this plane. As shown within experimental error, no anisotropy in lattice diffusion was detected. Since the oxygen sublattice of Al_2O_3 has a close-packed arrangement, large anisotropy effects should not be found.

The dependence of the grain-boundary diffusion parameter ($\alpha D'\delta$) on grain-boundary orientation was determined in bicrystals of Al_2O_3 and $MgAl_2O_4$. Bicrystals of Al_2O_3 with tilt angles of ~6°, 13°, and 23°($01\bar{1}2$) and $MgAl_2O_3$ with tilt angles of ~5° and 23°(100) were used in this study. The value of $\alpha D'\delta$ was determined both parallel ($\alpha D'\delta_\parallel$) and perpendicular ($\alpha D'\delta_\perp$) to the growth direction of the bicrystals. The temperature dependence of $\alpha D'\delta$ of ^{51}Cr in a 6°($01\bar{1}2$) Al_2O_3 tilt boundary is shown in Fig. 4. The solid lines are the best fit regression

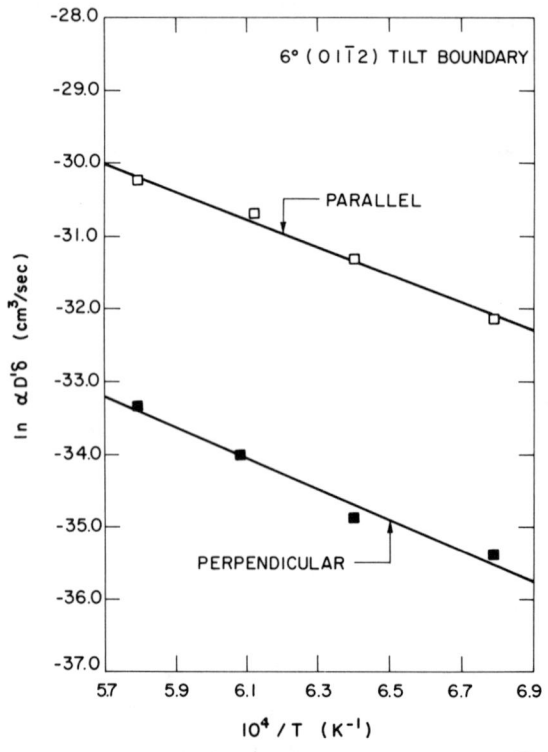

Fig. 4. Arrhenius diagram for ^{51}Cr diffusion along a 6°($01\bar{1}2$) tilt boundary in Al_2O_3.

lines ($R^2 > 0.95$). The two lines are approximately parallel; the activation energy for $\alpha D'\delta_\parallel$ is 164 ± 4 kJ/mol, and for $\alpha D'\delta_\perp$, it is 169 ± 6 kJ/mol. Since both of the activation energies are approximately equal, one expects that the mechanism of diffusion is the same for diffusion in both directions. However, $\alpha D'\delta_\parallel$ is approximately 23 times larger than $\alpha D'\delta_\perp$ over the entire temperature region study. The activation energy did not change significantly when the tilt angle was increased. However, the differences between $\alpha D'\delta_\parallel$ and $\alpha D'\delta_\perp$ decreased. Figure 5 is an Arrhenius plot showing the dependence of $\alpha D'\delta$ on temperature for a 13°(01$\bar{1}$2)

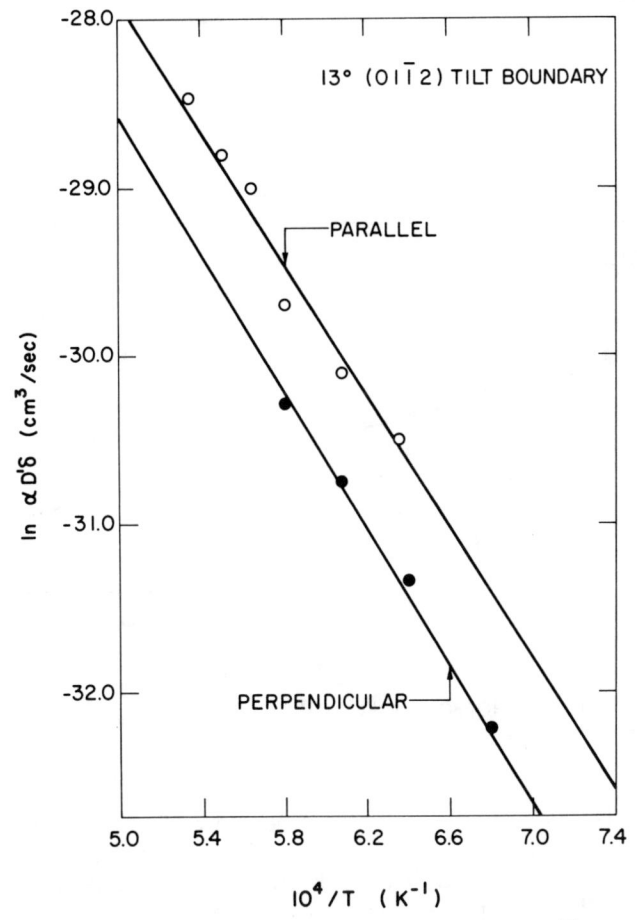

Fig. 5. Arrhenius diagram for ^{51}Cr diffusion along 13°(01$\bar{1}$2) tilt boundary of Al_2O_3.

Al_2O_3 tilt boundary. The solid lines ($R^2 > 0.95$) are approximately parallel, yielding activation energies of 168 ± 5 kJ/mol for $\alpha D'\delta_\parallel$ and 168 ± 7 kJ/mol for $\alpha D'\delta_\perp$. In this boundary, $\alpha D'\delta_\parallel$ is approximately 2 times larger than $\alpha D'\delta_\perp$ over the entire temperature region studied. Results for a 23°(01$\bar{1}$2) Al_2O_3 tilt boundary were essentially identical to those shown in Fig. 5.

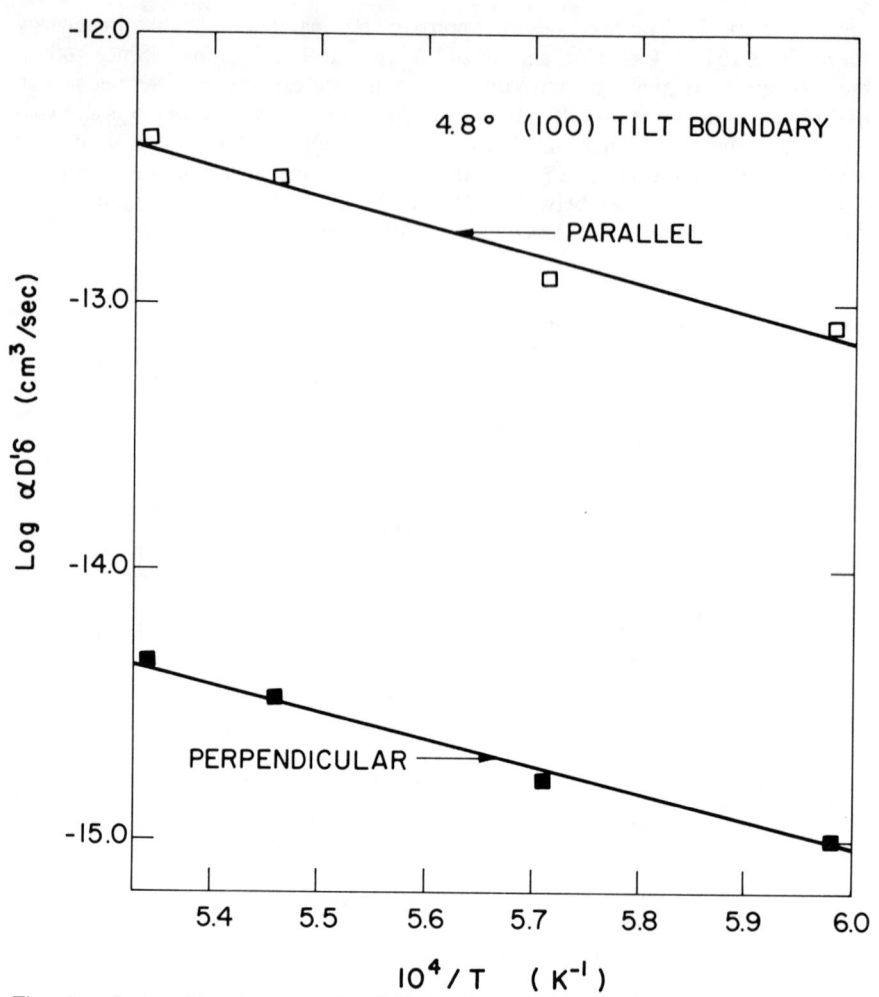

Fig. 6. Arrhenius diagram for ^{51}Cr diffusion along 4.8°(100) tilt boundary in $MgAl_2O_4$.

Similar behavior was found for the dependence of $\alpha D'\delta$ on grain-boundary orientation in the $MgAl_2O_4$ system. Figure 6 is an Arrhenius plot showing the temperature dependence of $\alpha D'\delta$ in a 4.8°(100) $MgAl_2O_4$ tilt boundary. The solid lines ($R^2 > 0.95$) are approximately parallel, yielding activation energies of 184 ± 10 kJ/mol for $\alpha D'\delta_\parallel$ and of 195 ± 9 kJ/mol for $\alpha D'\delta_\perp$. In this boundary, $\alpha D'\delta_\parallel$ is approximately 87 times larger than $\alpha D'\delta_\perp$ over the entire temperature region studied. No significant activation energy change was observed when the tilt angle was increased. However, in a 23°(100) $MgAl_2O_4$ tilt boundary, $\alpha D'\delta_\parallel$ was approximately 4 times larger than $\alpha D'\delta_\perp$ (Fig. 7).

The results can be combined in one plot, Fig. 8, which is a plot of $D'_\parallel/D'_\perp$ vs tilt angle for Al_2O_3, $MgAl_2O_4$, and MgO. The ratio $D'_\parallel/D'_\perp$, which reflects the anisotropy of grain-boundary diffusion, is constant for a tilt angle above ~*10°*. Figure 8 does not reveal, however, which diffusion coefficient is responsible for

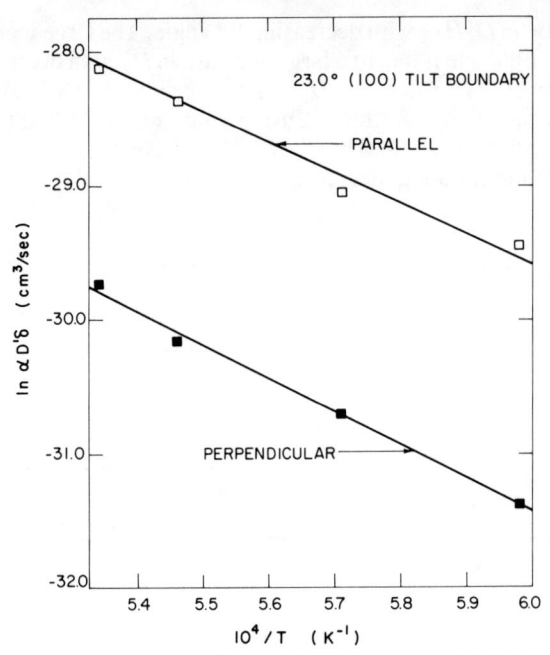

Fig. 7. Arrhenius diagram for ^{51}Cr diffusion along 23°(100) tilt boundary in $MgAl_2O_4$.

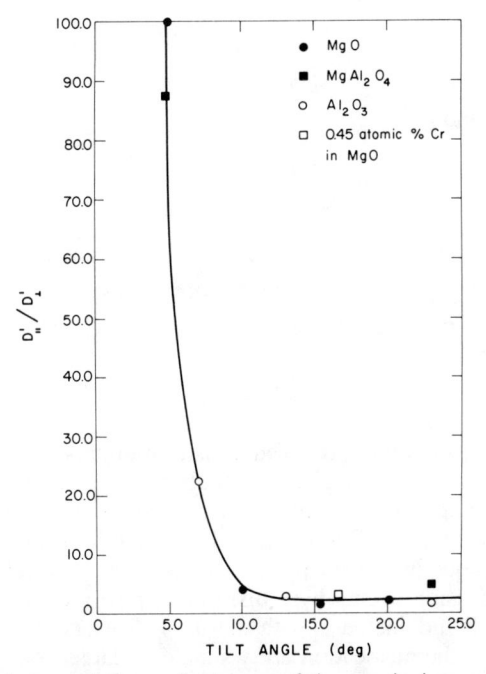

Fig. 8. Dependence of the anisotropy of the grain-boundary diffusion on tilt angle θ for ^{51}Cr diffusion in Al_2O_3, $MgAl_2O_4$, and MgO.

the large increase in $D'_{\parallel}/D'_{\perp}$ with decreasing tilt angle. The large increase in $D'_{\parallel}/D'_{\perp}$ with decreasing tilt angle is due to a large decrease in $D'_{\perp}$ with decreasing tilt angle, as shown in Fig. 9. The value of $\alpha D'\delta_{\perp}$ for diffusion of ^{51}Cr in Al$_2$O$_3$ at 1200°C decreased by a factor of ~27 for diffusion from a 13°(01$\bar{1}$2) tilt boundary to a 6°(01$\bar{1}$2) tilt boundary of Al$_2$O$_3$. However, $\alpha D'\delta_{\parallel}$ decreased only by a factor of ~2 at 1200°C when the tilt angle decreased (Fig. 10).

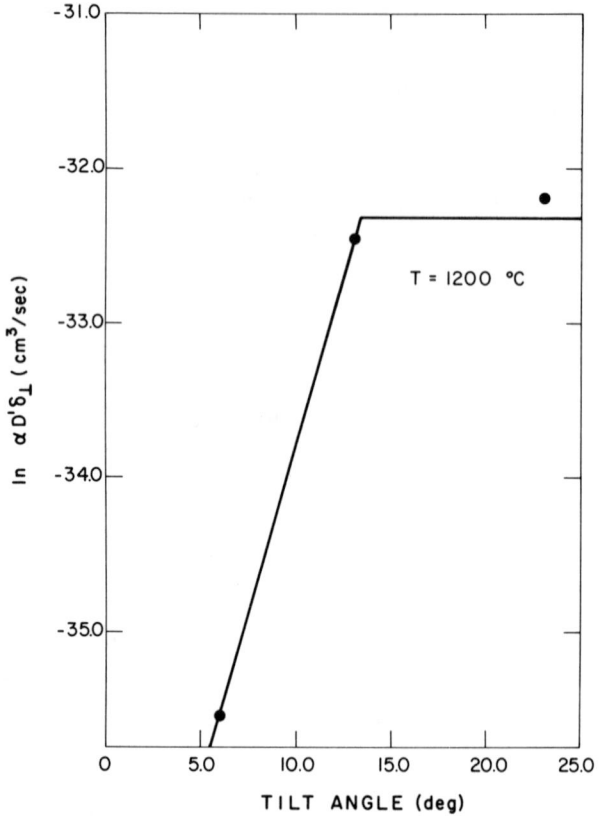

Fig. 9. Dependence of $\alpha D'\delta_{\perp}$ on tilt angle θ for ^{51}Cr diffusion along (01$\bar{1}$2) tilt boundaries of Al$_2$O$_3$.

Turnbull and Hoffmann[18] proposed a dislocation pipe model for diffusion in low-angle grain boundaries. Schober and Balluffi[19] later defined this macroscopic dislocation model with a microscopic atomistic model and extended this model to high-angle grain boundaries. Turnbull and Hoffmann[18] assumed that a tilt boundary is formed by equally spaced parallel dislocations in a relatively unstrained lattice and that diffusion occurs by a dislocation pipe mechanism. According to this model $D'\delta_{\parallel}$ should increase with $\sin(\theta/2)$. This model also predicts that diffusion parallel to the dislocation array should be larger than diffusion perpendicular to the dislocation array because of the difference in dislocation density. Finally, the ratio $D'_{\parallel}/D'_{\perp}$ should go to 1 as the angle increases since the dislocation

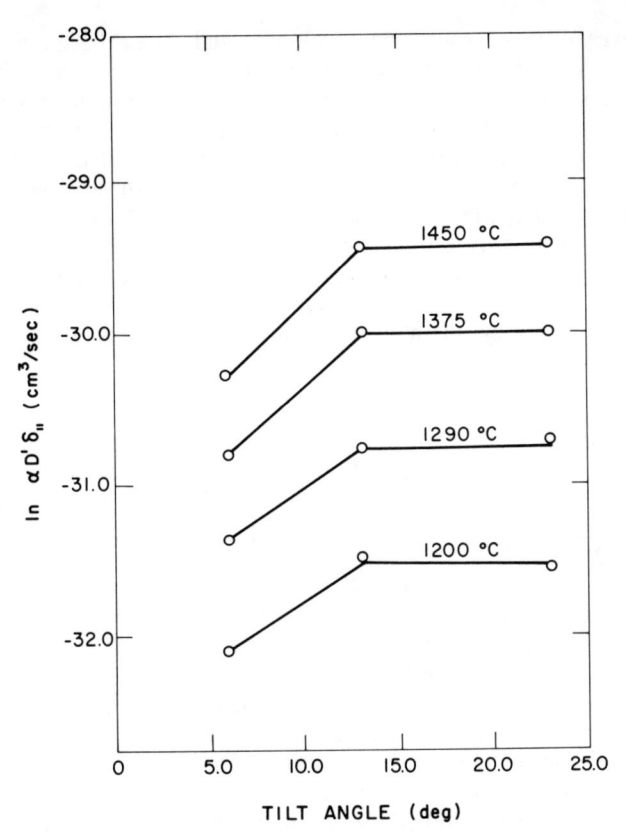

Fig. 10. Dependence of $\alpha D'\delta_\parallel$ on tilt angle θ for ^{51}Cr diffusion along (01$\bar{1}$2) tilt boundaries of Al_2O_3.

array becomes random at high angles. Balluffi's model predicts similar behavior, except the dislocation array never becomes random; therefore, there is always some anisotropy.

The results obtained in this study indicate that the dislocation model is valid for grain-boundary diffusion in Al_2O_3 and $MgAl_2O_4$. Our results indicate that grain boundaries with tilt angles less than ~10° can be considered low-angle grain boundaries. Similar results have recently been obtained for grain-boundary diffusion in (100) MgO tilt boundaries.[10]

Furthermore, Gifkin[20] has shown that an activation energy for grain-boundary diffusion of the order of 0.5–0.8 times the activation energy for lattice diffusion is consistent with the dislocation pipe mechanism. Our results show that the activation energies for ^{51}Cr in Al_2O_3 and $MgAl_2O_4$ are 306 ± 58 and 337 ± 45 kJ/mol, respectively. The activation energies for ^{51}Cr grain-boundary diffusion in Al_2O_3 and $MgAl_2O_4$ are 170 ± 10 and 185 ± 10 kJ/mol, respectively. These results show that the activation energies for the grain-boundary diffusion of ^{51}Cr in Al_2O_3 and $MgAl_2O_4$ are 0.56 and 0.55 times the activation energy for lattice diffusion.

The measured values of $\alpha D'\delta$ for ^{51}Cr diffusion along grain boundaries of Al_2O_3 and $MgAl_2O_4$ may be used to estimate the order of magnitude of the grain-boundary diffusion coefficients. Atkinson and Taylor[8] have estimated, from diffusion of ^{63}Ni along grain boundaries of NiO, δ to be ~0.7 nm. The value of the partition coefficient (segregation coefficient), α, of the solute (tracer) between the boundary and the lattice is equal to 1 for self-diffusion. The results obtained for diffusion of ^{57}Co along grain boundaries of NiO indicated that for this case $\alpha \cong 1$.[21] If we assume that the same is true for ^{51}Cr diffusion along Al_2O_3 and $MgAl_2O_4$ grain boundaries, then the grain-boundary diffusion coefficients are ~10^5–10^6 times the lattice diffusion coefficients.

However, no values for the segregation factor, α, are known, and no theoretical model exists for the possibility that impurities and/or temperature alter the effective grain-boundary width, which makes the quantitative interpretation of results somewhat tentative.

Conclusions

The lattice diffusion coefficient, D, and the grain-boundary diffusion parameter, $\alpha D'\delta$, of ^{51}Cr in Al_2O_3 and $MgAl_2O_4$ were determined. The temperature dependence of the lattice diffusion coefficients of ^{51}Cr in Al_2O_3 and $MgAl_2O_4$ can be expressed as $D_{Cr} = (2.58 \pm 1.5) \times 10^{-3} \exp-(306 \pm 58 \text{ (kJ/mol)}/RT)$ cm^2/s and $D_{Cr} = (2.45 \pm 2) \times 10^{-2} \exp-(337 \pm 45 \text{ (kJ/mol)}/RT)$ cm^2/s, respectively, in the temperature region 1400–1650°C. No anisotropy was found for the lattice diffusion of ^{51}Cr in Al_2O_3, which is probably due to the close packing of the oxygen sublattice. A substantial enhancement in diffusivity of ^{51}Cr along the grain boundaries of Al_2O_3 and $MgAl_2O_4$ was found. The activation energies for grain-boundary diffusion of ^{51}Cr in Al_2O_3 and $MgAl_2O_4$ are 170 ± 10 and 185 ± 10 kJ/mol, respectively. The activation energies for grain-boundary diffusion are 0.56 and 0.55 times the activation energy for lattice diffusion in Al_2O_3 and $MgAl_2O_4$, respectively. A large degree of anisotropy was found in the grain-boundary diffusion along the small angle ($\theta < $ ~10°) tilt boundaries. This phenomenon is consistent with the dislocation pipe grain-boundary diffusion mechanism.

Acknowledgments

This work was supported by the U.S. Department of Energy under Contract No. DE–ACD2–78 ERO 4998.

References

[1] J. Frenkel; p. 36 in Kinetic Theory of Liquids. Oxford University Press, New York, 1946.
[2] K. Lehovec, "Space-Charge Layer and Distribution of Lattice Defects at the Surface of Ionic Crystals," *J. Chem. Phys.*, **21** [7] 1123–8 (1953).
[3] R. B. Poeppel and J. M. Blakely, "Origin of Equilibrium Space-Charge Potentials in Ionic Crystals," *Surf. Sci.*, **15** [3] 507–23 (1969).
[4] W. D. Kingery, "Plausible Concepts Necessary and Sufficient for Interpretation of Ceramic Grain Boundary Phenomena: Part I," *J. Am. Ceram. Soc.*, **57** [1] 1–8 (1974); "Part II," *J. Am. Ceram. Soc.*, **57** [2] 74–83 (1974).
[5] S. K. Tiku and F. A. Kröger, "Effects of Space Charge, Grain-Boundary Segregation, and Mobility Differences Between Grain Boundary and Bulk on the Conductivity of Polycrystalline Al_2O_3," *J. Am. Ceram. Soc.*, **63** [3–4] 183–9 (1980).
[6] M. F. Yan, R. M. Cannon, H. K. Bowen, and R. L. Coble, "Space Charge Contributions to Grain Boundary Diffusion," *J. Am. Ceram. Soc.*, **60** [2–4] 120–7 (1977).
[7] B. J. Wuensch and T. Vasilos, "Origin of Grain Boundary Diffusion in MgO," *J. Am. Ceram. Soc.*, **49** [1] 433–6 (1966).
[8] A. Atkinson and R. I. Taylor, "Diffusion of ^{63}Ni Along Grain Boundaries in Nickel Oxide," *Philos. Mag.*, [Part A], **43** [4] 979–98 (1981).

[9]W. K. Chen and N. L. Peterson, "Grain Boundary Diffusion of ^{60}Co and ^{51}Cr in NiO," *J. Am. Ceram. Soc.*, **60** [9–10] 655–70 (1980).
[10]J. W. Osenbach and V. S. Stubican, "The Grain Boundary Diffusion of ^{51}Cr in MgO and Cr-Doped MgO," *J. Am. Ceram. Soc.*, **63** [3] 191–6 (1983).
[11]J. C. Fisher, "Calculation of Diffusion Penetration Curves for Surface and Grain Boundary Diffusion," *J. Appl. Phys.*, **22** [1] 74–7 (1951).
[12]T. Suzuoka, "Lattice and Grain Boundary Diffusion in Polycrystals," *Trans. Jpn. Inst. Met.*, **2**, 35 (1961).
[13]R. T. Whipple, "Concentration Contours in Grain Boundary Diffusion," *Philos. Mag.*, [Part A], **45** [371] 1220–36 (1954).
[14]A. D. LeClaire, "The Analysis of Grain Boundary Diffusion Measurements," *Br. J. Appl. Phys.*, **14**, 351–6 (1963).
[15]G. B. Gibbs, "Grain Boundary Impurity Diffusion," *Phys. Status Solidi*, **16**, K27–9 (1966).
[16]I. K. Lloyd and H. K. Bowen, "Ion Tracer Diffusion in Aluminum Oxide," *J. Am. Ceram. Soc.*, **64** [2] 744–7 (1981).
[17]W. J. Minford and V. S. Stubican, "Interdiffusion and Association Phenomena in the System NiO-Al$_2$O$_3$," *J. Am. Ceram. Soc.*, **57**, 363–6 (1974).
[18]D. Turnbull and R. Hoffmann, "The Effect of Relative Crystal and Boundary Orientation on Grain Boundary Diffusion Rates," *Acta Metall.*, 419–26 (1954).
[19]T. Schober and R. W. Balluffi, "Quantitative Observation of Misfit Dislocation Arrays in Low and High Angle Twist Grain Boundaries," *Philos. Mag.*, **21** [169] 109–23 (1970); "Dislocation in Symmetric High Angle [001] Tilt Boundaries in Gold," *Phys. Status Solidi B*, **44** [1] 115–26 (1971).
[20]P. G. Gifkin, "Development of Island Model for Grain Boundaries," *Mater. Sci. and Eng.*, **2** [4] 181–92 (1967).
[21]A. Atkinson and R. I. Taylor, "Diffusion of ^{57}Co along Grain Boundaries of NiO," *Philos. Mag.*, **45** [4] 583–92 (1982).

---

*R is the coefficient of determination.

Understanding Defect Structure and Mass Transport in Polycrystalline Al$_2$O$_3$ and MgO via the Study of Diffusional Creep

R. S. GORDON

University of Utah
Department of Materials Science and Engineering
Salt Lake City, UT 84112

Factors that promote diffusional creep deformation in polycrystalline Al$_2$O$_3$ and MgO at elevated temperatures and low stresses will be discussed including the effects of cation dopants in solid solution, oxygen fugacity, and grain size. Well-defined extrinsic defect states can be established in these materials by doping with soluble cations, such as Fe^{2+}, Fe^{3+}, Mn^{2+}, Mn^{3+}, Ti^{3+}, and Ti^{4+}, either separately or in various combinations. All cation dopants, present in the lattice as single impurities (e.g., Fe^{3+} in MgO; Fe^{2+}, Mn^{2+}, or Ti^{4+} in Al$_2$O$_3$) or as ion pairs (e.g., Fe^{2+}–Ti^{4+} or Mn^{2+}–Ti^{4+}), increased diffusional creep rates by enhancing either cation lattice or cation (or anion) grain-boundary diffusion. Examples of Nabarro–Herring creep, which is controlled by cation lattice diffusion, and Coble creep, which is rate limited by either anion or cation grain-boundary diffusion, have been identified in these oxide systems. Diffusion constants extracted from these measurements are the primary sources of transport data in extrinsic Al$_2$O$_3$ and MgO. Most direct measurements of diffusion coefficients are conducted on nominally pure but still probably extrinsic materials with ill-defined defect states. Divalent impurities such as Fe^{2+} and Mn^{2+} considerably enhance cation grain-boundary diffusion in polycrystalline Al$_2$O$_3$. Anion grain-boundary diffusion is very fast in both polycrystalline Al$_2$O$_3$ and MgO and is seldom rate limiting. Under conditions of enhanced cation lattice diffusion and relatively coarse grain sizes, oxygen grain-boundary diffusion has been found to be partially rate limiting in the diffusional creep of iron-doped polycrystalline MgO. Diffusional creep can also be used to determine conditions under which mass transport is rate limited by interfacial defect reactions at grain boundaries. These include appropriate combinations of grain size, enhanced lattice diffusivity, and depressed grain-boundary diffusivity.

For a number of years, we have studied diffusional creep phenomena in polycrystalline MgO[1-3] and Al$_2$O$_3$.[4,5] The principal objective of this research was to establish dominant extrinsic defect states in these materials by doping with soluble aliovalent cation impurities (e.g., Fe, Mn, Ti, Fe–Ti, Mn–Ti). Well-defined examples of diffusional creep have been identified under certain conditions, leading to useful information concerning the effects of dopants (and hence defect concentrations) on both lattice and grain-boundary diffusion processes. In this paper, the effects of dopant type, dopant concentration, oxygen fugacity, and temperature on both lattice and grain-boundary diffusivities will be reviewed. All of the data, except for a few tracer measurements on single crystals, have been interpreted from diffusional creep data obtained in the temperature range between

1150 and 1550°C. It will be demonstrated that the study of diffusional creep on well-characterized, dense polycrystalline material is an effective method for studying lattice and grain-boundary diffusion in compounds such as Al_2O_3 and MgO. Lattice defect concentrations in these materials can be readily controlled by soluble dopants because of the relatively large band gaps and high formation energies for intrinsic lattice defects. This approach to the study of extrinsic diffusion phenomena should be readily extended to other compounds having a strong contribution of ionic bonding (e.g., TiO_2, $CaTiO_3$, ZrO_2, ThO_2, CaF_2, Cr_2O_3, and Y_2O_3).

The objective of this paper is to analyze and compare the effects of variables such as temperature, dopant concentration, and oxygen fugacity (P_{O_2}), which control cation (and in some cases anion) lattice and grain-boundary diffusion processes in polycrystalline Al_2O_3 and MgO. The role of these processes on defect creation and/or annihilation reactions at grain boundaries will also be covered.

The reader is referred elsewhere[1-8] to the massive experimental support for diffusional creep in polycrystalline Al_2O_3 and MgO over certain ranges of stresses, grain sizes, temperatures, and dopant concentrations. All results, which will be reported herein, were derived from steady-state and viscous creep data taken under conditions of dead-load, four-point bending on polycrystalline specimens with densities in excess of 98% of theoretical density.

Diffusional Creep Theory

In general the diffusional creep of a polycrystalline ceramic solid, $A_\alpha B_\beta$, follows a relation of the following form:[9,10]

$$\dot{\varepsilon} = \frac{44\Omega_V \sigma}{kT(GS)^3} D_{complex} \tag{1}$$

in which $\dot{\varepsilon}$ is the steady-state strain rate, σ is the stress, (GS) is the grain size, Ω_V is the molecular volume of $A_\alpha B_\beta$, and $D_{complex}$ is a complex mass transport parameter.

Depending on the mechanism of mass transport, the apparent grain-size exponent [i.e., m in $\dot{\varepsilon}\alpha(GS)^{-m}$] can take on values between 1 and 3. In Table I, a summary is given for the various mechanisms of diffusional creep. In the general situation, three different processes can control diffusional creep in a polycrystalline compound, $A_\alpha B_\beta$, in which the transport of the cation is coupled with that of the anion: (1) lattice diffusion, (2) grain-boundary diffusion, and (3) interfacial defect reactions at grain boundaries. Assuming that the rates of the interfacial defect reactions vary linearly with stress and that they act in series only with lattice

Table I. Grain Size Dependencies in Diffusional Creep

Grain size exponent (m)	Controlling mechanism
1	Defect creation–annihilation at grain boundaries
2	Cation and/or anion lattice diffusion
3	Cation and/or anion grain-boundary diffusion
3	"Pressure-solution" transport in an intergranular liquid, glassy, or molten phase*

*See Ref. 11.

diffusion processes, Ikuma and Gordon[6] derived the following equation for $D_{complex}$:

$$\frac{1}{D_{complex}} = \frac{\alpha\left[\frac{(GS)D_A^l}{\pi} + \frac{(GS)^2 K_A}{44}\right]}{\frac{(GS)^3 D_A^l K_A}{44\pi} + (\delta_A D_A^b)\left[\frac{(GS)D_A^l}{\pi} + \frac{(GS)^2 K_A}{44}\right]} + \frac{\beta\left[\frac{(GS)D_B^l}{\pi} + \frac{(GS)^2 K_B}{44}\right]}{\frac{(GS)^3 D_B^l K_B}{44\pi} + (\delta_B D_B^b)\left[\frac{(GS)D_B^l}{\pi} + \frac{(GS)^2 K_B}{44}\right]} \quad (2)$$

Here, D_A^l and D_A^b are the $A^{\beta+}$ ion lattice and grain-boundary diffusivities, D_B^l and D_B^b are the $B^{\alpha-}$ ion lattice and grain-boundary diffusivities, δ_A and δ_B are the effective widths of the regions of enhanced diffusion near the grain boundaries for $A^{\beta+}$ and $B^{\alpha-}$ ions, and K_A and K_B are the interfacial rate constants for anion and cation vacancy (or interstitial) creation or annihilation, respectively, at the grain boundaries.

In general, it is clear that the diffusion coefficients, which are extracted from diffusional creep data, are complex quantities that involve several basic mass transport parameters. Soluble impurities that are present can significantly alter lattice and grain-boundary diffusivities and hence influence the value of $D_{complex}$. Furthermore, $D_{complex}$ is a strong function of the grain size.

Lattice and/or Grain-Boundary Diffusion-Controlled Kinetics

Frequently,[3,5,7] the apparent grain size exponent lies between values characteristic of Nabarro–Herring[12] and Coble[13] creep. If anion grain-boundary diffusion is rapid (typically observed in polycrystalline MgO and Al_2O_3), then diffusional creep will be rate limited by a combination of cation grain-boundary and lattice diffusion with $2 < m < 3$; i.e.,

$$\dot{\varepsilon} = \frac{44\Omega_V \sigma D_A^l}{\pi k T \alpha (GS)^2} + \frac{44\Omega_V \sigma \delta_A D_A^b}{k T \alpha (GS)^3} \quad (3)$$

Note that at small grain sizes and low temperatures, Eq. (3) predicts that Coble creep ($m = 3$) will be dominant and controlled by cation grain-boundary diffusion. Coble creep has been reported by Terwilliger et al.[1] for the deformation of Fe-doped MgO (0.05–0.27 cation%) for grain sizes between 7 and 23 µm at temperatures ≤1300°C. In this case, diffusional creep rates were controlled by cation (magnesium) grain-boundary diffusion. The only other well-defined case of Coble creep was observed in Fe-doped (2 cation%) Al_2O_3 (17–100 µm) tested under reducing conditions ($P_{O_2} \sim 10^{-1}$ Pa) where a larger concentration of the iron was in the divalent state.[5] Any one of three conditions is important for the enhancement of diffusional contributions due to cation grain-boundary diffusion: (1) small grain size, (2) low temperatures, and (3) specific dopants (e.g., F^- in MgO, Fe^{2+} in Al_2O_3, Mn^{2+} in Al_2O_3).

At larger grain sizes and high temperatures $(GS/\pi)D_A^l \gg \delta_A D_A^b$, Nabarro–Herring creep ($m = 2$), which is rate limited by cation lattice diffusion, should be dominant. Abundant examples of Nabarro–Herring creep controlled by cation lattice diffusion have been observed[2,3,5,8] in polycrystalline MgO and Al_2O_3.

When the cation lattice diffusivity is sufficiently enhanced by appropriate dopants (e.g., Fe^{3+} in MgO), the grain size is reasonably coarse, and the temperature is sufficiently high, diffusional creep kinetics can be controlled in part by anion grain-boundary diffusion.[2] The conditions appropriate for this situation are as follows: $(GS/\pi)D_A^l \gg \delta_A D_A^b$ and $(GS/\pi)D_B^l \ll \delta_B D_B^b$. The general creep equation then becomes

$$\dot{\varepsilon} = \frac{44\Omega_V \sigma}{\alpha \pi k T (GS)^2} \left[\frac{D_A^l}{1 + \frac{\beta(GS)D_A^l}{\alpha \pi \delta_B D_B^b}} \right] \quad (4)$$

Note that in the limit of large grain size and high cation lattice diffusivity, Coble creep ($m = 3$) is dominant and controlled by anion grain-boundary diffusion. Mixed kinetics ($2 < m < 3$) according to Eq. (4) have been observed[2] in Fe-doped MgO (2.65–5.30%). In this system, cation lattice diffusion is significantly enhanced by the presence of trivalent iron in solid solution through the creation of cation lattice vacancies. If cation lattice diffusion becomes too fast, then it no longer is entirely rate limiting and oxygen grain-boundary diffusion becomes rate limiting in part. In the case of Al_2O_3 doped with iron or manganese, while enhancing cation lattice diffusion to some degree, doping significantly enhanced cation grain-boundary diffusion.[7,8] Oxygen grain-boundary diffusion was always too fast to be rate limiting. In this case Eq. (3), not Eq. (4), described the mixed diffusional creep kinetics.

Role of Interfacial Reactions

Let us briefly examine a few situations in which interfacial reactions may be important. First, in the limit of very rapid anion grain-boundary diffusion (good assumption for the model materials analyzed in this paper), the following relation can be written for $D_{complex}$:

$$D_{complex} = \frac{(GS)^3 D_A^l K_A / 44\pi\alpha}{\frac{(GS)D_A^l}{\pi} + \frac{(GS)^2 K_A}{44}} + \delta_A D_A^b / \alpha \quad (5)$$

Equation (5) has several interesting limits. First, in the limit of a rapid interfacial reaction

$$D_{complex} = \frac{1}{\alpha} \left[\frac{(GS)D_A^l}{\pi} + \delta_A D_A^b \right] \quad (6)$$

and diffusional creep is just a competition between cation lattice and cation grain-boundary diffusion. Second, in the limit of slow interfacial kinetics

$$D_{complex} = \frac{1}{\alpha} \left[\frac{(GS)^2 K_A}{44} + \delta_A D_A^b \right] \quad (7)$$

and diffusional creep is controlled by the fastest of either cation grain-boundary diffusion or the cation interfacial reaction.

Finally, in the limit of slow cation grain-boundary diffusion

$$\dot{\varepsilon} = \frac{\Omega_V \sigma}{\pi \alpha k T (GS)} \frac{D_A^l K_A}{\left[\frac{D_A^l}{\pi} + \frac{(GS)K_A}{44} \right]} \quad (8)$$

At small grain sizes and/or high cation lattice diffusivities, creep rates are controlled by the cation interfacial reaction according to

$$\dot{\varepsilon} = \frac{\Omega_V \sigma K_A}{\alpha k T (GS)} \tag{9}$$

The important point to be made here is that interface-controlled creep kinetics are not restricted entirely to the small grain size limit. They can be controlling in intermediate to large grain size materials providing cation *lattice diffusion is very large* and cation *grain-boundary diffusion is slow*. The creep of Ti-doped $Al_2O_3^6$ apparently satisfied these conditions (i.e., intermediate grain size, very high D_{Al}^l, and apparently low $\delta_{Al} D_{Al}^b$). At these compositions, the grain size dependence was much smaller ($m \sim 1$). The titanium dopant was so effective in enhancing aluminum lattice diffusion that the interfacial defect reactions at the grain boundaries became rate limiting. When Fe or Mn was present in combination with Ti, cation grain-boundary diffusion was significantly enhanced and the interfacial reactions were no longer a limiting problem. It is possible that the interfacial rate constant for each diffusing species depends on the magnitude of the corresponding grain-boundary diffusion coefficient.

Mass Transport in Polycrystalline MgO

Lattice Diffusion

In Fig. 1 magnesium ion lattice diffusion coefficients for undoped and iron-doped material* are compared with tracer measurements[14] on nominally pure

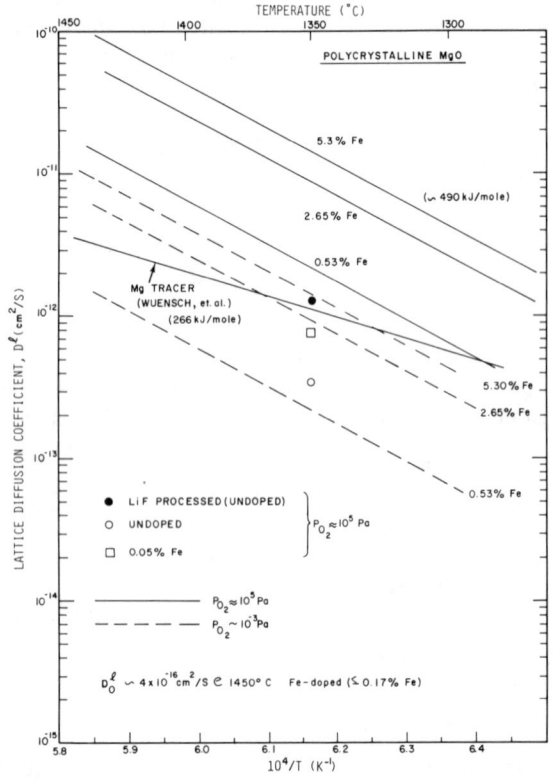

Fig. 1. Cation lattice diffusion in polycrystalline MgO.

(~0.01% cation lattice defects) single crystals. From the effects of iron dopant concentration and oxygen fugacity, it is apparent that trivalent iron enhances magnesium lattice diffusion according to a quasi-chemical defect reaction of the type

$$V''_{Mg} + O_O + 2Fe^{\cdot}_{Mg} = 2Fe^{\times}_{Mg} + 1/2O_2(g) \tag{10}$$

In Eq. (10) V''_{Mg} is a cation vacancy, $Fe^{\cdot}_{Mg}$ is a substitutional trivalent iron, and $Fe^{\times}_{Mg}$ is a substitutional divalent iron. The concentration of trivalent iron establishes the extrinsic vacancy concentration and, hence, the value of the cation lattice diffusivity. As is to be expected, cation lattice diffusion in iron-doped MgO approaches tracer values at low dopant concentrations and/or low oxygen fugacities. Under these conditions, the concentration of divalent iron is maximized. The higher apparent activation energy for cation diffusion in iron-doped material is due to the enthalpy associated with formation of cation vacancies caused by the oxidation of divalent iron in the lattice.[15]

Diffusion coefficients calculated from creep data are consistent with kinetics controlled by cation lattice diffusion. Oxygen lattice diffusion is approximately 4 orders of magnitude slower.[16] This result implies a rapid or short circuit diffusion of oxygen in the grain boundaries for grain sizes up to at least 100 μm (i.e., $\delta_0 D_0^b > 10^{-18}$ cm^3/s at 1450°C[†]).

Finally, it is of interest to note that nominally pure (99.98%) MgO, which was hot-pressed by using an LiF sintering aid, has a significantly higher (~3–4 times) cation lattice diffusivity than material prepared without the hot-pressing additive.[3] Residual fluorine in substitutional solid solution on the anion sublattice (i.e., $F_O^{\cdot}$) could lead to an enhanced cation vacancy concentration and higher cation diffusivities.

Equation (10) predicts that the magnesium diffusion coefficient (D^l_{Mg}) should be proportional to $P_{O_2}^{1/6}$. In Fig. 2, diffusion data at three iron dopant concentrations

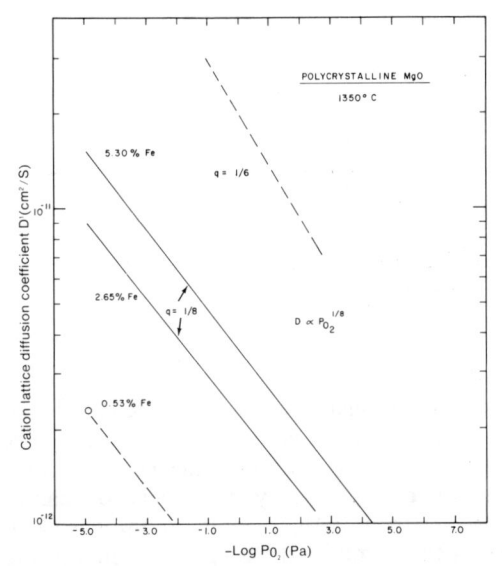

Fig. 2. Effect of oxygen fugacity on cation lattice diffusion in iron-doped polycrystalline MgO.

(0.53, 2.65, and 5.3 cation%) indicate a slightly different dependence (i.e., $D^l_{Mg} \propto P^{1/8}_{O_2}$).[‡] It is probable that the simple defect reaction assuming unassociated lattice vacancies is an oversimplification. At high dopant concentrations, substantial complexing between the cation vacancy and the trivalent iron is known to occur.[17] Thus, not all of the lattice vacancies take part in the diffusion process. This conclusion is substantiated from the data in Fig. 3, which indicate that the magnesium diffusivity (actually $D^l - D^l_0$, where D^l_0 is the diffusivity for the undoped base material) is not proportional ($D^l_{Mg} \propto [V''_{Mg}]^p$ with $p = 1$) to the concentration of lattice vacancies.[§] The effect of vacancy concentration is substantially less ($p = 0.65$) than that predicted from the simple unassociated defect model.

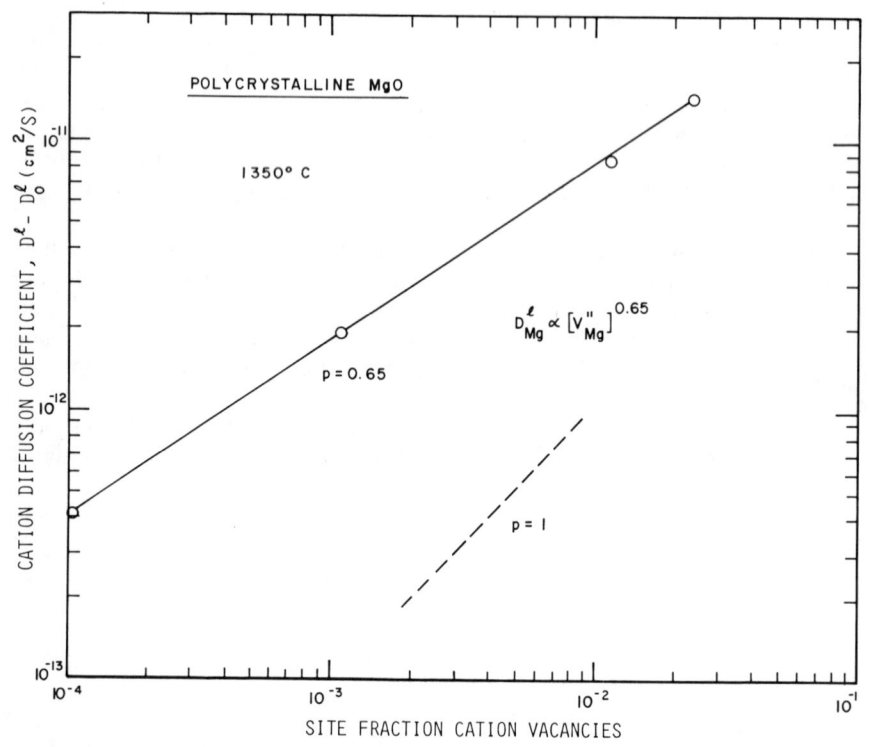

Fig. 3. Correlation between cation lattice diffusivity and cation vacancy concentration in polycrystalilne MgO.

Grain-Boundary Diffusion

Two examples of diffusional creep, which were controlled in part by grain-boundary diffusion, have been observed for iron-doped polycrystalline MgO. First, at small grain sizes (<25 μm) and low temperatures (<1400°C), Coble creep controlled by magnesium grain-boundary diffusion is dominant.[1] In Fig. 4, values of $\delta_{Mg} D^b_{Mg}$ are plotted for the 0.05% Fe composition. It is clear that iron doping substantially enhances cation transport through the grain boundaries. Magnesium grain-boundary diffusion in undoped material is at least 3 orders of magnitude lower. The transition grain size (i.e., $GS \approx \pi \delta_{Mg} D^b_{Mg}/D^l_{Mg}$) between Coble and

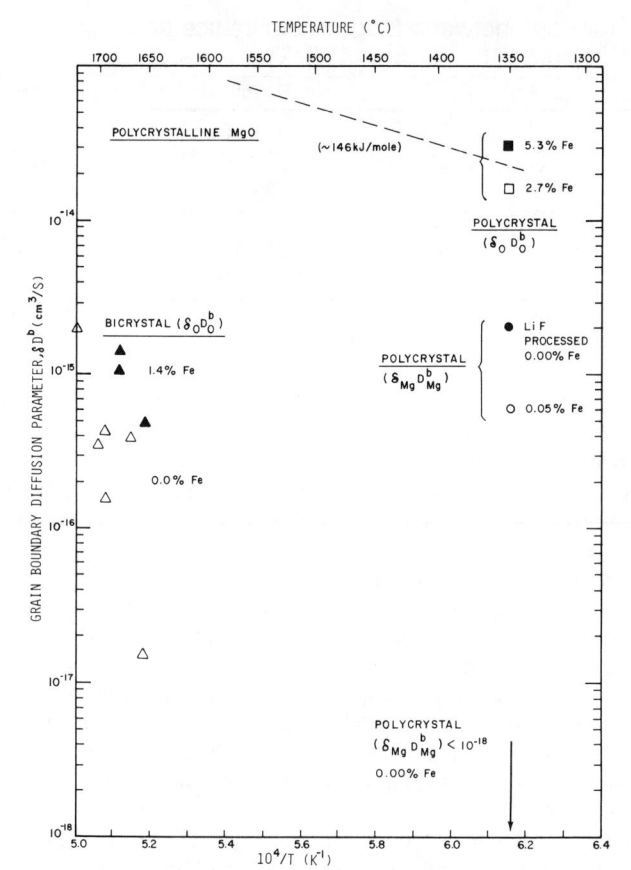

Fig. 4. Cation and anion grain-boundary diffusivities in polycrystalline MgO.

Nabarro–Herring creep at the 0.05% dopant level is about 25 μm at 1350°C. At higher dopant levels where cation lattice diffusion is enhanced even further, Nabarro–Herring creep is expected to be dominant, as has been observed.[2]

Nominally pure polycrystalline MgO, which was processed with the LiF hot-pressed aid, also exhibited enhanced magnesium grain-boundary transport comparable to low levels of iron doping.[3]

As mentioned in the previous section, increased concentrations of trivalent iron (higher dopant concentrations and/or high oxygen fugacities) enhance the magnesium lattice diffusivity to the point that the kinetics become controlled in part by oxygen grain-boundary diffusion. This situation was encountered at the 2.65% and 5.3% Fe compositions tested in oxidizing atmospheres ($m = 2.38$).[2] Values of $\delta_O D_O^b$ extracted from an analysis of the data using Eq. (4) are plotted in Fig. 4 and compared with data taken from proton activation experiments[19] on MgO bicrystals (pure and doped with iron). It is clear that oxygen grain-boundary transport in polycrystalline MgO ($\delta_O D_O^b \sim (2\text{--}3) \times 10^{-14}$ cm^3/s) is substantially enhanced by iron doping. The grain sizes for equal contributions of magnesium lattice and oxygen grain-boundary diffusion were estimated from the relation

Table II. Transition between Magnesium Lattice and Oxygen Grain-Boundary Diffusion (1350°C)

Cation (%Fe)	P_{O_2} (Pa)	D'_{Mg} (cm²/s)	$\delta_O D^b_O$ (cm³/s)	Transition grain size (μm)
0.05	~10^5	7.7×10^{-12}	~1.6×10^{-14}*	~650
0.53	~10^5	2.3×10^{-12}	~1.6×10^{-14}*	~220
2.65	~10^5	9.0×10^{-12}	1.6×10^{-14}	57
2.65	10^{-3}	9.6×10^{-13}	~1.6×10^{-14}*	~520
5.30	~10^5	15×10^{-12}	3.2×10^{-14}	67
5.30	10^{-3}	15×10^{-13}	~3.2×10^{-14}	~670

*Estimated values—upper limit $\delta_O D^b_O$ was assumed to be independent of oxygen fugacity.

$GS = (\delta_O D^b_O)\pi/D'_{Mg}$. These estimates at 1350°C are summarized in Table II. For dopant levels ≤0.53%, Nabarro–Herring creep ($m = 2$) should only be observed in atmospheres with high oxygen fugacities, since the practical upper limit for the grain size is 100 μm. Also for a given dopant concentration and temperature, a decrease in the oxygen fugacity (i.e., decrease in D'_{Mg}) will lead to an increase in the transition grain size.

Oxygen boundary transport in polycrystalline MgO is substantially enhanced by iron doping. Minimum values (~10^{-17}–10^{-18} cm³/s) of $\delta_O D'_O$ in nominally pure MgO were computed from oxygen tracer measurements ($\approx (GS)D'_O/\pi$) on polycrystalline particulates.[20] In these O^{18} exchange experiments, oxygen diffusion along the grain boundaries was so rapid that the grain sizes, instead of the overall particle sizes, had to be used to normalize the data. The normalized diffusion data agreed very well with measurements on single crystals. These minimum values at least put lower limits on oxygen grain-boundary diffusion in nominally pure polycrystalline MgO.

An estimate for the activation energy of oxygen grain-boundary diffusion in polycrystalline, iron-doped MgO was made (~146 kJ/mol) from a careful analysis of the temperature dependence of the apparent creep activation energy, which varied from about 473 kJ/mol at 1350°C to about 293 kJ/mol at 1450–1500°C.[21] Equation (4) and an activation energy of 490 kJ/mol for magnesium lattice diffusion were used to make the estimate.

Finally, the relative importance of oxygen grain-boundary diffusion at large iron dopant concentration, oxygen fugacities, and grain sizes is shown in Fig. 5. In this limit, the diffusional creep rate at high oxygen fugacities is relatively insensitive to changes in the oxygen partial pressure. The transition between Nabarro–Herring creep behavior (i.e., $m = 2$) and that influenced by oxygen grain-boundary diffusion (i.e., $2 < m < 3$) occurs at lower oxygen fugacities as the iron dopant concentration and/or the grain size are increased.

Mass Transport in Polycrystalline Al_2O_3

Lattice Diffusion

In Fig. 6, cation lattice diffusion coefficients ($P_{O_2} \sim 10^5$ Pa) in polycrystalline Al_2O_3 are given for undoped material and Al_2O_3 doped with Mg, Mn, Fe, Ti, Fe–Ti, and Mn–Ti. Aluminum tracer measurements are shown on the figure for comparison.[22] The Cannon composite[23] represents an analysis of several previous studies on the creep of polycrystalline Al_2O_3 saturated with MgO. It is clear that lattice diffusion coefficients inferred from the diffusional creep of poly-

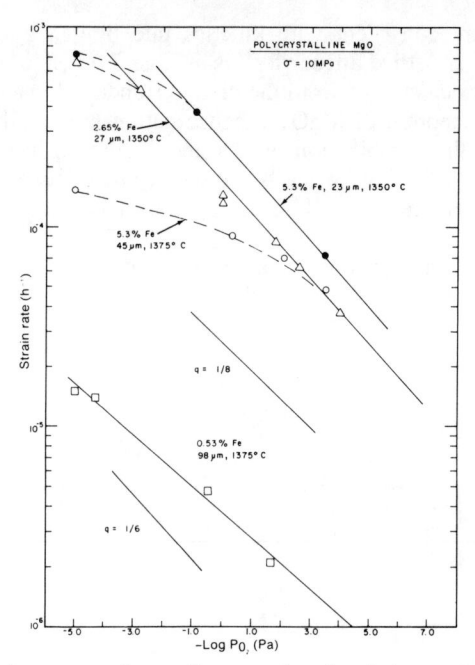

Fig. 5. Effect of oxygen fugacity on steady-state creep of iron-doped polycrystalline MgO.

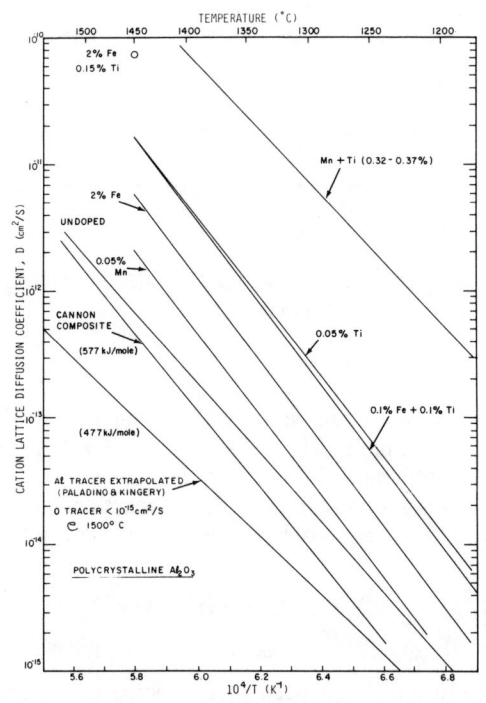

Fig. 6. Cation lattice diffusion in polycrystalline Al_2O_3.

crystalline Al₂O₃ are consistent with kinetics rate limited by aluminum lattice diffusion. The oxygen lattice diffusivity[24] is at least 3 orders of magnitude slower than the diffusivities calculated from the creep of undoped material. All dopants, with perhaps the exception of MgO, enhance cation lattice diffusion. The most effective dopant is the combination of Mn and Ti, which in small amounts enhances cation lattice diffusion by nearly 3 orders of magnitude. Small amounts of Mn or Ti are more effective in enhancing cation lattice transport than is Fe as a single dopant.

In Fig. 7 the effect of oxygen fugacity on the diffusional creep of transition-metal-doped polycrystalline Al₂O₃ is shown at several dopant concentrations.

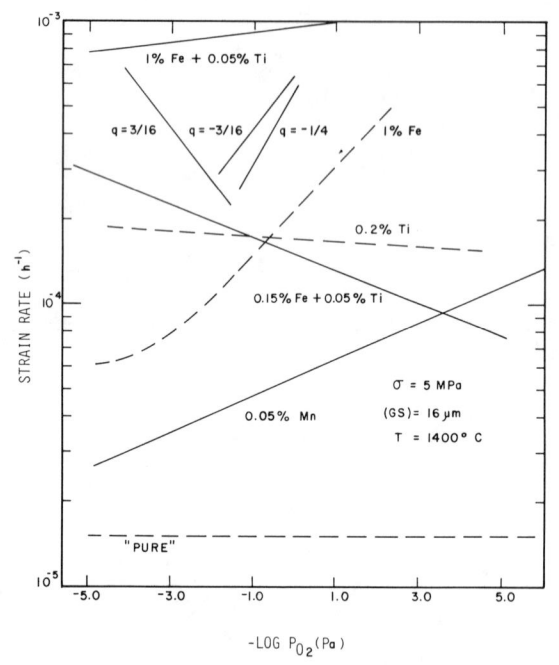

Fig. 7. Effect of oxygen fugacity on steady-state creep of polycrystalline Al₂O₃ doped with transition-metal oxides.

Several important conclusions can be drawn from the data in Fig. 7:

(1) Quadravalent titanium ($Ti^{\cdot}_{Al}$), divalent iron (Fe'_{Al}), and divalent manganese (Mn'_{Al}) in substitutional solid solution enhance aluminum lattice diffusion according to the following quasi-chemical defect reactions: Ti-doping (Eq. (11)); Fe(Mn)-doping (Schottky compensation) (Eq. (12)); Fe(Mn)–Frenkel compensation (Eq. (13)).

$$6Ti^{*}_{Al} + 3/2O_2(g) = 2V'''_{Al} + 3O^{*}_O + 6Ti^{\cdot}_{Al} \quad q = 3/16 \quad (11)$$

$$O^{*}_O + 2Fe^{*}_{Al} = 2Fe'_{Al} + V^{\cdot\cdot}_O + 1/2O_2(g) \quad q = -1/4 \quad (12)$$

Note: Cation diffusion is enhanced by the increased concentration of aluminum interstitials (a minority lattice defect) through the cation Frenkel equilibrium.[4]

$$3O_O^\times + 2Al_{Al}^\times + 6Fe_{Al}^\times + 2V_I = 6Fe'_{Al} + 2Al_I^{\cdot\cdot\cdot} + 3/2 O_2(g) \quad q = -3/16 \tag{13}$$

In Eqs. (11–13), V'''_{Al} is an aluminum ion vacancy, $Ti_{Al}^\times$ and $Fe_{Al}^\times$ are substitutional trivalent titanium and iron, respectively, $V_O^{\cdot\cdot}$ is an oxygen ion vacancy, Al_I''' is an aluminum ion interstitial, and V_I is a vacant interstitial site.

Theoretical dependencies (i.e., $\dot{\varepsilon}\alpha P_{O_2}^q$) representing cation transport controlled by a vacancy mechanism ($q = 3/16$) and an interstitial type mechanism ($q = -3/16$ or $-1/4$) are shown for points of reference. Only the data for Fe-doped material approximate predicted behavior ($q \sim -1/4$).

(2) For doping with titanium only, the creep rate is essentially insensitive to a reduction in oxygen fugacity. This effect is probably related in part to the presence of a second phase and interfacial controlled creep. In the case of titanium dopants, the creep rate varies inversely with the grain size.[6] However, even under these conditions, creep rates are substantially enchanced by the titanium dopant.

(3) Small amounts of titanium (up to about 0.2 cation%), either by themselves or in combination with iron (0.05–1.0%) are significantly more effective in enhancing aluminum lattice diffusion than larger amounts (0.2–1%) of iron present as a single dopant.

(4) The effect of titanium doping in compositions with iron dopant concentrations below about 2 cation% suggests that the Ti^{4+}/Ti^{3+} ion ratio is significantly larger than the Fe^{2+}/Fe^{3+} ion ratio in oxidizing atmospheres.[8]

(5) Small amounts of manganese are more effective than comparable amounts of iron in enhancing aluminum lattice diffusion, suggesting that the Mn^{2+}/Mn^{3+} ion ratio is also larger than the Fe^{2+}/Fe^{3+} ion ratio.[7]

The dominance of Ti^{4+} at 1450°C in mixed Fe–Ti compositions at Fe concentrations below about 2 cation% is readily seen from the data in Fig. 8 in which the

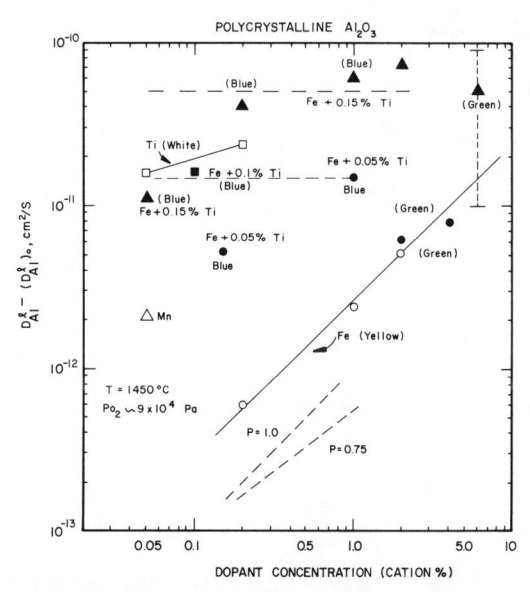

Fig. 8. Effect of transition-metal dopants on cation lattice diffusion in polycrystalline Al_2O_3.

aluminum lattice diffusion coefficient[¶] is plotted vs dopant concentration. In mixed Fe–Ti compositions when Ti^{4+} is dominant, the color of the specimen varies from light to dark blue.[8] For Fe–Ti compositions with iron contents ≥2%, Fe^{2+} is dominant and the specimen color is green.[8] At lower concentrations, the color is yellow. In the case where Fe^{2+} dominates the defect structure, the aluminum ion diffusivity varies with the dopant concentration in a manner predictable** from Eqs. (12) and (13) (i.e., $D^l_{Al} \propto [Fe_T]^p$ with $p = 1.0$ (Schottky) or $p = 0.75$ (Frenkel)). It is clear from the data in Fig. 8 that small amounts of Mn and $Ti^{††}$ present as single dopants also significantly enhance aluminum lattice diffusion. The data (lower bounds) for the single titanium dopants were obtained from creep data on coarse grain size material where interfacial effects are minimized.[6] The cation lattice diffusivity at 1450°C for the mixed Mn–Ti compositions is in excess of 10^{-10} cm^2/s (Fig. 6). The synergistic effect of these two dopants in enhancing cation diffusion to such high levels is not completely understood at this time.

Grain-Boundary Diffusion

In Fig. 9, aluminum grain-boundary diffusion characteristics in pure and transition-metal-doped polycrystalline Al_2O_3 are shown for a range of tempera-

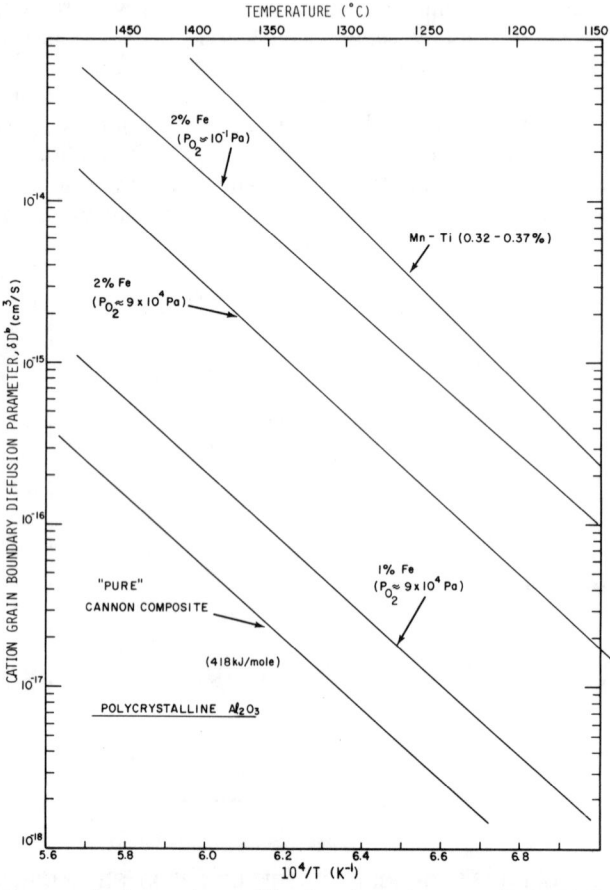

Fig. 9. Cation grain-boundary diffusion in polycrystalline Al_2O_3.

tures. The Cannon composite[23] represents a summary of several studies on polycrystalline Al$_2$O$_3$ saturated with MgO. It is clear that iron (probably divalent iron) and possibly manganese enhance cation grain-boundary transport in polycrystalline alumina. As the concentration of divalent iron is increased, either by increasing the total dopant concentration or by reducing the oxygen fugacity, the contribution to diffusional creep of cation grain-boundary diffusion increases. In fact, at the 2% Fe composition tested in an atmosphere with $P_{O_2} \sim 10^{-1}$ Pa, Coble creep ($m = 3$) controlled entirely by aluminum grain-boundary diffusion was observed.[5]

In all of the creep studies on polycrystalline Al$_2$O$_3$, mass transport has been interpreted in terms of cation lattice and cation grain-boundary diffusion or combinations of both (i.e., Eq. (3)). As a consequence, oxygen grain-boundary diffusion must be very fast, particularly in the doped aluminas. Values of $\delta_O D_O^b$ at 1450°C are probably in excess of 10^{-13} cm^3/s. Enhanced transport of oxygen along the grain boundaries has been postulated by others.[25]

Interfacial Defect Reactions

As discussed earlier, interfacial-controlled kinetics can be promoted by several factors in polycrystalline alumina including (1) a small grain size, (2) a high cation lattice diffusivity, and (3) a slow or depressed cation grain-boundary diffusivity. Interfacial effects may also be promoted by the presence of a second phase at grain boundaries that inhibits the annihilation or creation of defects at the boundary sources.[26] It is believed that most of these conditions are in effect when alumina is singly doped with titanium. Creep rates while enhanced by the titanium dopant reach a limiting value in the fine grain size limit (~ 10 μm).[8] The grain size exponent ($m = 1$) and the weak effect of changes in the oxygen fugacity are both consistent with kinetics controlled by interfacial defect reactions at grain boundaries.

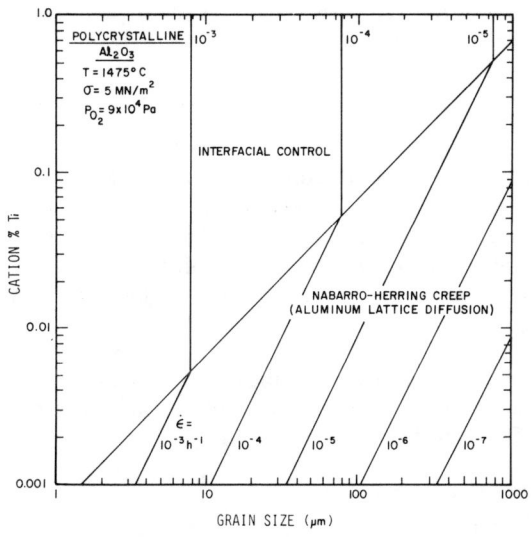

Fig. 10. Relative contributions of cation lattice diffusion and interfacial kinetics at grain boundaries on diffusional creep of titanium-doped polycrystalline Al$_2$O$_3$.

To demonstrate the relative effects of the interfacial reaction and aluminum lattice diffusion on the creep of polycrystalline alumina doped with titanium, a creep deformation map was constructed involving grain size and dopant concentration as the major variables. To construct this map, two simplifying assumptions were made: (1) the interfacial rate constant (K_{Al}) is independent of dopant concentration, and (2) the aluminum ion diffusion coefficient (D'_{Al}) is proportional to the total titanium concentration. Two mechanisms are expected to be important in the diffusional creep of polycrystalline alumina doped with titanium: (1) at small grain sizes and high concentrations of titanium, interfacial kinetics will be dominant, and (2) at larger grain sizes and low concentrations of titanium, creep rates will be controlled by aluminum lattice diffusion. Values of K_{Al} at 1475°C were estimated from the 0.05% (4.20 × 10^{-8} cm/s) and 0.2% (5.64 × 10^{-8} cm/s) compositions, using Eq. (9). The aluminum lattice diffusion coefficient (2.6 × 10^{-11} cm^2/s) at 1475°C for the 0.05% Ti composition was estimated from the Nabarro–Herring equation ($m = 2$) in the limit of coarse grain size (~100 μm). Using this value of D'_{Al} and an average value of K_{Al} (4.97 × 10^{-8} cm/s), we can construct the creep map in Fig. 10. The relatively weak grain size dependence of interfacial reaction controlled creep is indicated by the wide separation between the isostrain rate lines. To avoid any interfacial effects in the diffusional creep of polycrystalline alumina doped with titanium, dopant concentrations less than about 0.01% (~100 ppm) are required.

Since D'_{Al} is about 40 times faster in Ti-doped alumina compared to undoped material, interfacial kinetics would not become important in undoped alumina at 1475°C until the grain size dropped below about 2 μm (assuming K_{Al} to be relatively concentration independent).

When Fe^{2+} or Mn^{2+} are present together with Ti^{4+}, even though cation lattice diffusion is enhanced via the dominance of Ti^{4+} (e.g., Fe–Ti), the divalent dopant enhances $\delta_{Al}D^b_{Al}$ and prevents an interfacial kinetics limitation to occur. Thus, mass transport kinetics are controlled by appropriate combinations of cation lattice and grain-boundary diffusion (i.e., $2 \le m \le 3$), even though cation lattice transport is substantially enhanced by doping.

Discussion

Lattice Diffusion

In Table III, cation lattice diffusion in MgO and Al_2O_3 at 1350°C is compared for different dopants and dopant concentrations. In general, cation diffusivities are 1–2 orders of magnitude faster in MgO than in Al_2O_3 with the exception of polycrystalline Al_2O_3 doped simultaneously with Mn and Ti.

Estimates were made of the defect diffusivities by using a relation of the form

$$D'_A \approx [D]D'_d \tag{14}$$

in which D'_d is the lattice diffusivity of the defect (vacancy or interstitial) and [D] is the cation defect concentration. Cation vacancy and interstitial concentrations were calculated from the following relations:[‡‡]

$$[Fe^{\cdot}_{Mg}] = 2[V''_{Mg}] \tag{15}$$

$$[Fe'_{Al}] = 3[Al^{\cdot\cdot\cdot}_i] \tag{16}$$

$$[Ti^{\cdot}_{Al}] = 3[V'''_{Al}] \tag{17}$$

The concentration of trivalent iron in MgO was measured by chemical analysis.

Table III. Cation Lattice Diffusivity in MgO and Al$_2$O$_3$ (1350°C)

Oxide	Composition (cation%)	Oxygen fugacity (Pa)	Diffusivity (10^{-14} cm^2/s)	Cation defect concentration (%)	Defect diffusivity (10^{-10} cm^2/s)	Ref.
MgO	Tracer (single crystal)	Argon	115	~0.01	~100	Wuensch et al.
MgO	Undoped	~10^4	36	~0.01	~40	This work
MgO	Undoped (LiF process)	~10^4	130			This work
MgO	0.05 Fe	~10^4	77	~0.01	~80	This work
MgO	0.53 Fe	~10^5	230	0.12	20	This work
MgO	0.53 Fe	~10^{-3}	24			This work
MgO	2.65 Fe	~10^5	900	1.13	8	This work
MgO	2.65 Fe	~10^{-3}	96			This work
MgO	5.30 Fe	~10^5	1500	2.38	6	This work
Al$_2$O$_3$	Tracer (single crystal)	~10^4	1.4	<0.01	>1	Paladino and Kingery
Al$_2$O$_3$	Undoped	~10^4	7.0	<0.001	>70	This work
Al$_2$O$_3$	MgO (sat'd)	~10^4	3.8			Cannon and Coble
Al$_2$O$_3$	0.05 Mn	~10^5	14			This work
Al$_2$O$_3$	0.2 Fe	~10^5	13	~0.001	~100	This work
Al$_2$O$_3$	2.0 Fe	~10^5	38	~0.01	~40	This work
Al$_2$O$_3$	0.1 Fe + 0.1 Ti	~10^5	110	~0.027	~40	This work
Al$_2$O$_3$	0.05 Ti	~10^5	120	~0.013	~90	This work
Al$_2$O$_3$	Mn + Ti (~0.35)	~10^5	2400			This work

The concentration of divalent iron in Al$_2$O$_3$ was assumed to be approximately 1.3% of the total iron concentration.[8] A majority (80%) of the titanium was assumed to be quadrivalent.[8] In both MgO and Al$_2$O$_3$, defect diffusivities were in the range of approximately 10^{-9}–10^{-8} cm^2/s at 1350°C. The defect diffusivities in MgO doped with large iron concentrations ($\geq$2.65% Fe) are probably underestimated, since only a fraction of the cation vacancies are mobile.

The activation energies for lattice diffusion were ~490 kJ/mol in Fe-doped MgO and 544–628 kJ/mol in Al$_2$O$_3$, pure and doped with various transition-metal dopants. These values were significantly higher than those obtained from tracer measurements on nominally pure single crystals, i.e., 266 kJ/mol for MgO and 477 kJ/mol for Al$_2$O$_3$.

Grain-Boundary Diffusion

In Table IV, grain-boundary diffusion parameters (δD^b) at 1350°C are compared between polycrystalline MgO and Al$_2$O$_3$. Estimates of grain-boundary diffusivities were made by assuming boundary widths of 10 and 1000 Å (1.0 and 100.0 nm).

Several interesting observations can be made from the data in Table IV. (1) Enhanced oxygen boundary diffusion in MgO is comparable to enhanced aluminum grain-boundary diffusion in Al$_2$O$_3$ doped simultaneously with small amounts of Mn and Ti. (2) The assumption of a grain-boundary thickness of 10 Å (1.0 nm) leads to cation grain-boundary diffusivities in both MgO and Al$_2$O$_3$ that are comparable to defect diffusivities in the lattice. (3) Oxygen grain-boundary diffusion in polycrystalline Al$_2$O$_3$ is very fast at 1350°C and is probably in excess of 10^{-16}–10^{-13} cm^3/s, depending on the dopant and its concentration. (4) Divalent iron and manganese enhance aluminum grain-boundary diffusion in Al$_2$O$_3$, while

Table IV. Grain-Boundary Diffusivity in MgO and Al$_2$O$_3$ (1350°C)

Oxide	Composition (cation%)	Oxygen fugacity (Pa)	Ion	δD^b (cm^3/s)	D^b (cm^2/s) (δ = 10 Å)[†]	D^b (cm^2/s) (δ = 1000 Å)
MgO	Undoped	~10^4	Mg	<10^{-18}	<10^{-11}	<10^{-13}
MgO	0.05 Fe	~10^4	Mg	6×10^{-16}	6×10^{-9}	6×10^{-11}
MgO	Undoped (LiF process)	~10^4	Mg	2×10^{-15}	2×10^{-8}	2×10^{-10}
MgO	Undoped		O	>10^{-17}	>10^{-10}	>10^{-12}
MgO	2.65 Fe	~10^5	O	1.6×10^{-14}	1.6×10^{-7}	1.6×10^{-9}
MgO	5.30 Fe	~10^5	O	3.2×10^{-14}	3.2×10^{-7}	3.2×10^{-9}
Al$_2$O$_3$	MgO (sat'd)*	~10^4	Al	2.5×10^{-17}	2.5×10^{-10}	2.5×10^{-12}
Al$_2$O$_3$	1.0 Fe	~10^5	Al	9.6×10^{-17}	9.6×10^{-10}	9.6×10^{-12}
Al$_2$O$_3$	2.0 Fe	~10^5	Al	1.4×10^{-15}	1.4×10^{-8}	1.4×10^{-10}
Al$_2$O$_3$	2.0 Fe	~10^{-1}	Al	6.6×10^{-15}	6.6×10^{-8}	6.6×10^{-10}
Al$_2$O$_3$	Mn–Ti (~0.35)	~10^5	Al	2.5×10^{-14}	2.5×10^{-7}	2.5×10^{-9}

*Cannon & Coble.
[†]δ = 1.0 nm.

iron (trivalent?) enhances both magnesium and oxygen grain-boundary diffusion in MgO.

Finally, the activation energy for oxygen grain-boundary diffusion in MgO was relatively low (~146 kJ/mol) compared to aluminum grain-boundary diffusion in Al$_2$O$_3$ (420–498 kJ/mol).

Effects of Second Phases

Some comments should be made concerning the solubilities of the various dopants in MgO and Al$_2$O$_3$ at the temperatures of interest. In most cases, the concentrations were within the solubility limits, with the exception of alumina doped with titania. In the latter, depending upon the concentration and the presence of a codopant, some second phase was present at the grain boundaries.

In the case of the iron dopant in MgO, FeO is completely miscible with MgO at temperatures of 1200–1450°C and oxygen fugacities (down to 10^{-5} Pa) of interest. For trivalent iron, the solubility of ferric oxide at temperatures of 1200°C and higher exceeds 5 cation% Fe.[27] Electron microprobe analysis of polycrystalline specimens revealed no segregation of iron at the grain boundaries, except at the highest dopant concentration (5.3 cation%). In this case, small amounts (under 5%) of an iron-rich second phase were observed in some samples. The microprobe examination was conducted on specimens that were slowly cooled to room temperature, such that precipitation might be expected during cooling, particularly at high dopant concentrations. However, on the basis of the reported solubility of ferric oxide in MgO, it is unlikely that any grain-boundary second phase existed at the creep testing temperatures (1275–1450°C).

For the iron dopant in alumina, the solubility[28] of trivalent iron at 1450°C is about 10 cation%, well in excess of the dopant concentrations that have been studied. In oxidizing atmospheres, less than 2% of the iron in solid solution is in the divalent state.[8] For divalent iron, the solubility in air at 1450°C is much lower (i.e., 0.7–1.4 cation%).[29] Hence, at low oxygen fugacities (i.e., less than about 10^{-2} Pa), the solubility of total iron at 1450°C is about 1 cation%, and second phases at grain boundaries may become a problem in highly reducing atmospheres. Thus, for most of the conditions studied, the iron dopant is in substitutional solid solution.

Very little agreement exists in the literature concerning the solubility of quadravalent titanium in oxidizing atmospheres. Estimates between 0.13 and 0.3 cation% at 1300–1400°C have been made by various investigators.[30-33] Little if any second phase has been observed by microprobe analysis in polycrystals doped simultaneously with iron and titanium.[8] It is possible that double doping with iron and titanium enhances the solubility of titanium, as has been reported for the Mg–Ti couple.[31] In reducing atmospheres, as might be expected, the solubility of trivalent titanium is substantially higher.[30,31] An isolated second phase at the grain boundaries was readily detected in specimens (quenched in air to room temperature) doped with 0.2 cation% titanium[6] and doped simultaneously with manganese and titanium.[7] At these compositions, it is likely that the dopants are present in solution at the solubility limit.

The presence of a second phase at grain boundaries could be responsible in part for the interfacial controlled creep that was observed in polycrystalline alumina doped with titanium. Grain-boundary precipitates could alter the kinetics of defect creation or annihilation at the boundaries.[26] The presence of isolated second-phase particles on grain boundaries should have a slight retarding effect on the diffusional creep rate.[34] If precipitate formation was responsible for interfacial controlled deformation, then the creep rate in titanium-doped alumina should be no greater than that observed for undoped material. However, significantly enhanced creep rates were observed, suggesting that some solubility of quadravalent titanium exists at the creep testing temperature, even if some second phase exists.

In summary, even though small amounts of isolated second phase were present in polycrystalline alumina doped either with titanium or simultaneously with titanium and manganese, substantial evidence exists (grain size, atmosphere, dopant, and temperature effects) to support the hypothesis that these dopants in solid solution, albeit at the solubility limit, play a significant role in the high-temperature creep of polycrystalline alumina. Whether or not dopant segregation is responsible for the enhanced grain-boundary transport that was observed in polycrystalline alumina doped with iron at low oxygen fugacities or simultaneously with titanium and manganese remains to be determined in future work.

Summary

Diffusional creep is an excellent tool by which to obtain information on diffusion processes in polycrystalline ceramics, particularly those associated with grain boundaries. Because of their wide band gaps and high intrinsic defect formation energies, extrinsic defect states that are well-defined can be readily produced in compounds like MgO and Al_2O_3. These can be used to study extrinsic lattice and grain-boundary diffusion processes.

Rapid oxygen grain-boundary diffusion is common to both MgO and Al_2O_3. Transition-metal dopants such as Fe and Mn enhance cation grain-boundary diffusion and perhaps oxygen boundary diffusion as well. A theoretical understanding of this enhancement (e.g., dopant segregation) needs to be developed.

Trivalent iron enhances magnesium lattice diffusion in MgO, while divalent iron, divalent manganese, and quadravalent titanium enhance aluminum lattice diffusion in Al_2O_3.

Interfacial effects related to the annihilation or creation of defects at grain-boundary sinks and sources are not necessarily restricted to small grain sizes. They depend significantly on the magnitudes of the lattice and grain-boundary diffusivities for the diffusing species.

Acknowledgments

This work was supported by the U.S. Department of Energy under Contract No. EY-76-S-02-1591. The author acknowledges the contributions of his former students: H. K. Bowen, R. A. Giddings, J. D. Hodge, G. W. Hollenberg, Y. Ikuma, P. A. Lessing, G. R. Terwilliger, and R. T. Tremper.

References

[1] G. R, Terwilliger, H. K. Bowen, and R. S. Gordon, "Creep of Polycrystalline MgO–Fe$_2$O$_3$ Solid Solutions at High Temperatures," *J. Am. Ceram. Soc.*, **53** [5] 241–51 (1970).

[2] R. T. Tremper, R. A. Giddings, J. D. Hodge, and R. S. Gordon, "Creep of Polycrystalline MgO–FeO–Fe$_2$O$_3$ Solid Solutions," *J. Am. Ceram. Soc.*, **57** [10] 421–8 (1974).

[3] J. D. Hodge and R. S. Gordon, "Grain Growth and Creep in Polycrystalline Magnesium Oxide Fabricated with and without a LiF Additive," *Ceramurgia Int.*, **4** [1] 17–20 (1978).

[4] G. W. Hollenberg and R. S. Gordon, "Effect of Oxygen Partial Pressure on the Creep of Polycrystalline Al$_2$O$_3$ Doped with Cr, Fe, or Ti," *J. Am. Ceram. Soc.*, **56** [3] 140–7 (1973).

[5] P. A. Lessing and R. S. Gordon, "Creep of Polycrystalline Alumina, Pure and Doped with Transition Metal Impurities," *J. Mater. Sci.*, **12** [11] 2291–302 (1977).

[6] Y. Ikuma and R. S. Gordon, "Role of Interfacial Defect Creation–Annihilation Processes at Grain Boundaries on the Diffusional Creep of Polycrystalline Alumina"; pp. 283–94 in Surfaces and Interfaces in Ceramic and Ceramic Metal Systems. Edited by J. Pask and A. Evans. Plenum Press, New York, 1981.

[7] Y. Ikuma and R. S. Gordon, "Enhancement of the Diffusional Creep of Polycrystalline Al$_2$O$_3$ by Simultaneous Doping with Manganese and Titanium," *J. Mater. Sci.*, **17**, 2961–7 (1982).

[8] Y. Ikuma and R. S. Gordon, "Effect of Doping Simultaneously with Iron and Titanium on Creep of Polycrystalline Al$_2$O$_3$," *J. Am. Ceram. Soc.*, **66** [2] 139–47 (1983).

[9] R. S. Gordon, "Mass Transport in the Diffusional Creep of Ionic Solids," *J. Am. Ceram. Soc.*, **56** [3] 147–52 (1973).

[10] R. S. Gordon, "Ambipolar Diffusion and Its Application to Diffusion Creep"; pp. 445–64 in Mass Transport in Ceramics. Edited by A. R. Cooper and A. H. Heuer. Plenum Press, New York, 1975.

[11] B. Burton, "Diffusional Creep of Polycrystalline Materials"; pp. 106–9 in Diffusion and Defect Monograph Series, No. 5. Trans Tech Publications, Bay Village, OH, 1977.

[12] (a) F. R. N. Nabarro; 75–90 in the Report of the Conference on Strength of Solids. Physical Society, London, 1948.
(b) C. Herring, "Diffusional Viscosity of a Polycrystalline Solid," *J. Appl. Phys.*, **21** [5] 437–45 (1950).

[13] R. L. Coble, "Model for Boundary Diffusion Controlled Creep in Polycrystalline Materials," *J. Appl. Phys.*, **34** [6] 1679–82 (1963).

[14] B. J. Wuensch, W. C. Steele, and T. Vasilos, "Cation Self Diffusion in Single Crystal MgO," *J. Chem. Phys.*, **58** [12] 5258–66 (1973).

[15] G. W. Hollenberg and Ronald S. Gordon, "Origin of Anomalously High Activation Energies in the Sintering and Creep of Impure Refractory Oxides," *J. Am. Ceram. Soc.*, **56** [3] 109–10 (1973).

[16] (a) Y. Oishi, Ken Ando, H. Kurokawa, and Y. Hiro, "Oxygen Self-Diffusion in MgO Single Crystals"; to be published in *J. Am. Ceram. Soc.* (b) Ken Ando, Y. Kurokawa, and Y. Oishi, "Oxygen Self-Diffusion in Fe-Doped MgO Single Crystals"; to be published in *J. Chem. Phys.*

[17] C. B. Alcock and G. N. K. Iyengar, "A Study of the Oxidation-Reduction Equilibria of Dilute Magnesiowustites," *Trans. Proc. Br. Ceram. Soc.*, **8**, 219–29 (1967).

[18] Jakob Chi-Kang Soong and Ivan B. Cutler, "Diffusion in Polycrystalline Magnesiowustite Determined by Reduction Kinetics," *J. Solid State Chem.*, **53** [7] 339–406 (1970).

[19] D. R. McKenzie, A. W. Searcy, J. B. Holt, and R. H. Condit, "Oxygen Grain Boundary Diffusion in MgO," *J. Am. Ceram. Soc.*, **54** [4] 188–90 (1971).

[20] (a) S. Shirasaki and Y. Oishi, "Role of Grain Boundaries in Oxygen Self-Diffusion in Polycrystalline MgO," *Jpn. J. Appl. Phys.*, **10** [8] 109–10 (1971). (b) H. Hashimoto, M. Hana, and S. Shirasaki, "Preferential Diffusion of Oxygen Along Grain Boundaries in Polycrystalline MgO," *J. Appl. Phys.*, **43** [11] 4828–9 (1972). (c) S. Shirasaki and M. Hana, "Oxygen-Diffusion Characteristics of Loosely Sintered Polycrystalline MgO," *Chem. Phys. Lett.*, **20** [4] 361–6 (1973).

[21] J. D. Hodge, P. A. Lessing, and R. S. Gordon, "Activation Energies in the Diffusional Creep of Polycrystalline Ceramics," *J. Am. Ceram. Soc.*, **60** [7] 318-20 (1977).

[22] A. E. Paladino and W. D. Kingery, "Aluminum Ion Diffusion in Aluminum Oxide," *J. Chem. Phys.*, **37** [5] 957-62 (1962).

[23] R. M. Cannon and R. L. Coble, "Review of Diffusional Creep of Al_2O_3"; pp. 61-100 in Deformation of Ceramic Materials. Edited by R. C. Bradt and R. W. Tressler. Plenum, New York, 1975.

[24] (a) Y. Oishi and W. D. Kingery, "Self-Diffusion of Oxygen in Single Crystal and Polycrystalline Aluminum Oxide," *J. Chem. Phys.*, **33** [2] 480-6 (1960).

(b) K. P. R. Reddy and A. R. Cooper, "Oxygen Diffusion in Sapphire," *J. Am. Ceram. Soc.*, **65** [12] 634-8 (1982).

[25] A. E. Paladino and R. L. Coble, "Effect of Grain Boundaries on Diffusion-Controlled Processes in Aluminum Oxide," *J. Am. Ceram. Soc.*, **46** [3] 133-6 (1963).

[26] B. Burton, "On the Mechanisms of the Inhibition of Diffusional Creep by Second Phase Particles," *Mater. Sci. Eng.*, **11**, 337-43 (1973).

[27] B. Phillips, S. Somiya, and A. Muan, "Melting Relations of Magnesium Oxide-Iron Oxide Mixtures in Air," *J. Am. Ceram. Soc.*, **44** [4] 167-70 (1961).

[28] A. Muan, "On the Stability of the Phase Fe_2O_3-Al_2O_3," *Am. J. Sci.*, **256** [6] 413-22 (1958).

[29] I. A. Novokhatskii, B. F. Velov, A. V. Gorokh, and A. A. Savinskaya, "The Phase Diagram for the System Ferrous-Oxide-Alumina," *Russ. J. Phys. Chem.*, **39** [11] 1498-9 (1965).

[30] W. D. McKee, Jr. and E. Aleshin, "Aluminum Oxide-Titanium Oxide Solid Solution," *J. Am. Ceram. Soc.*, **46** [1] 54-8 (1963).

[31] S. K. Roy and R. C. Coble, "Solubilities of Magnesia, Titania, and Magnesium Titanate in Aluminum Oxide," *J. Am. Ceram. Soc.*, **51** [1] 1-6 (1968).

[32] E. R. Winkler, J. F. Sarver, and I. B. Cutler, "Solid Solution of Titanium Dioxide in Aluminum Oxide," *J. Am. Ceram. Soc.*, **49** [12] 634-7 (1966).

[33] B. J. Pletka, T. E. Mitchell, and A. H. Heuer, "Precipitation in Titanium-Doped Sapphire"; pp. 413-21 in The Society of Materials Science, Vol. IV. Tokyo, Japan, 1972.

[34] R. Raj and M. F. Ashby, "On Grain Boundary Sliding and Diffusional Creep," *Metall. Trans.*, **2** [4] 1113-27 (1971).

---

*Calculated from the Nabarro-Herring creep equation.

†This limit is obtained from the inequality $(GS)D'_O/\pi < \delta_O D_O^b$ with $(GS) = 100$ μm and $D'_O \approx 4 \times 10^{-16}$ cm^2/s.

‡Actually, when the data are normalized by plotting $D'-D'_O$ (where D'_O is the diffusion coefficient for undoped MgO) instead of D', the exponent q in Fig. 2 is approximately 7.

§The concentrations of lattice vacancies (one-half of the concentration of trivalent iron) were determined from quantitative analyses of the divalent and total iron concentrations, using a standard analytical technique (Refs. 17 and 18).

¶Actually, the diffusivities have been normalized by subtracting the diffusivity for the base material $(D_{Al})_0$.

**The data are not sufficiently sensitive to distinguish between Frenkel or Schottky disorder in Al_2O_3 (Ref. 8).

††These samples probably have some titanium-rich second phase present in the microstructure.

‡‡Equation (16) will place an upper bound on the defect concentration since V_O may actually compensate divalent iron in substitutional solid solution.

Lattice-, Grain-Boundary-, Surface-, and Gas-Diffusion Constants in Magnesium Oxide

J. M. VIEIRA* AND R. J. BROOK

Department of Ceramics
University of Leeds
Leeds LS2 9JT, U.K.

Reliable data on the transport parameters and on the elastic constants of a material are necessary for the interpretation of processes such as sintering and mechanical deformation. In this connection, a review of the diffusion coefficients and of the processes of vapor transport in MgO is presented. Tracer diffusion coefficients are compared with mass transport coefficients determined by other techniques: electrical conductivity, creep, sintering, and hot-pressing. Possible approaches to the estimation of these constants when experimental data are lacking are discussed. The scatters in the values of the activation energies for lattice diffusion and for surface diffusion in MgO are found to be consistent with the compensation law. Analogies between the grain-boundary diffusivities in MgO and in isomorphous alkali halides are supported by the available data. The high activation energies of some of the surface diffusion data are consistent with the close packing of the crystal structure type of the oxide.

Several sets of material properties are required for the discussion of sintering, hot-pressing, and creep results or for the construction of the corresponding maps.[1-3] These sets of properties include crystallographic and thermal data, elastic constants, surface energies, diffusion coefficients, and vapor partial pressures together with the values of the constants that go into the model equations for the various mechanisms.[1]

Crystallographic data, elastic constants, and surface energies show only small changes within the pressure and temperature ranges commonly found in experiments or used for maps.[1,4] If the exact values of these properties are not known or if they cannot be reliably measured at high temperatures, the errors that result from using estimated or extrapolated values have little effect on the nature of the conclusions or on the form of the maps. The inaccuracy of the model equations for the contributing mechanisms is likely to be more important than errors in these properties. In contrast to this, diffusion coefficients, being thermally activated, are strongly dependent on temperature, if less so on pressure.[5] Therefore, the errors that can occur in the analysis of experimental data or in the construction of maps can become important if one estimates values of the diffusion coefficients outside the temperature ranges where they have been measured.[6]

Many determinations of the tracer bulk diffusion coefficients of the host ions and of impurity ions in MgO[6-8] have been made since the early measurements of Lindner[9] and Oishi.[10] A comprehensive review of the tracer bulk diffusion coefficients in MgO has been given by Wuensch,[7] and the same data have been recently included in more general reviews of the diffusion coefficients of oxides.[6,8] It is the main purpose of the present study to add to these data by calculating effective diffusion coefficients from electrical conductivity and from sintering, hot-pressing, and creep results; these coefficients are then reviewed and compared with the tracer diffusion coefficients.

The experimental data on grain-boundary and surface diffusion in MgO are sparse compared with those for bulk diffusion. Owing to the lack of good-quality data on grain-boundary diffusion, several authors[1,2,11-13] have used estimated values of the grain-boundary diffusion coefficients in the construction of maps. The arguments used for calculating these diffusion coefficients are discussed.

Vapor transport processes are also important in this context since they can produce particle coarsening and grain growth during sintering and hot-pressing at high temperatures. Four main reactions contribute to the vaporization of MgO,[14] the results then being consistent with early observations of vapor transport in MgO. The same four reactions are used as the basis for the calculations of the gas diffusion coefficients of the corresponding reaction products given in the final section of the paper.

Bulk Diffusivities of Magnesium and Oxygen in MgO

Measured values for the tracer diffusion coefficients of magnesium and oxygen in the MgO lattice are presented in Tables I(a) and II(a), respectively. The effective bulk diffusion coefficients that can be calculated from measurements of the electrical conductivity of MgO at high temperatures are given in Table I(b). The values in Table I(c) are the effective diffusion coefficients calculated from creep, hot-pressing, and sintering results when the corresponding rates are considered to be controlled by the diffusive viscous deformation of the MgO lattice. Effective values of the oxygen bulk diffusion coefficient, D_1^O, obtained from the analysis of dislocation movement, creep, hot-pressing, and sintering are shown in Table II(b).

In the compilation of Tables I(a)–II(b), many data have been obtained directly from tables and graphs of diffusion coefficients available in the literature. When necessary, the Nernst–Einstein equation[15] has been used to calculate the diffusion coefficients from electrical conductivity data. The diffusion coefficients represented in the creep rates given by Vasilos,[16] Passmore,[17] and Tagai[18] have been calculated by using the Nabarro–Herring creep equation.[2,19]

For creep controlled by plastic flow resulting from glide/climb of dislocations, two slightly different formulations of the power law[2,20] have been used for comparison; the corresponding differences occurring in the final results are shown by the error bars of the corresponding points in Fig. 2. In the calculation of the diffusion coefficient from Snowden's creep rates[21] using Weertman's equation,[22] the value 1.0×10^{11} m^{-2} has been taken for the dislocation density, as suggested by the author and in good agreement with the direct measurements of Clauer.[23]

To reduce the errors, graphical data have only been used for the calculation of preexponential factors, the activation energies being those explicitly stated by the authors.

Early measurements of the bulk tracer diffusion coefficients of magnesium[9] and oxygen[10] in MgO have been used many times as bases for comparison with the diffusivities calculated from other processes of material transport: creep, sintering, and electrical properties. A common difficulty has appeared in interpretations[16,24-26] based on such comparison between calculated diffusivities and tracer values, namely, that a wide range of values has been observed for the activation energies

Table I(a). Tracer Diffusion Coefficient of Magnesium in the MgO Lattice

Temp (K)	D_o (m²·s⁻¹)	Q (kJ·mol⁻¹)	Process	Ref	Notes
1429–1616	$0.249^{+0.320}_{-0.140} \times 10^{-4}$	330.5	²⁸Mg	7, 9	Norton, M.;* 10⁻⁵ Ca; 10⁻⁶ Al, Fe, Co
1405–1673	$0.54^{+0.64}_{-0.29} \times 10^{-4}$	308.8 ± 9.6	²⁸Mg	7, 108	Monocrystals and Norton, M. air "Precipitation" region
1673–1742	1.6×10^{-9}	173.6	²⁸Mg	7, 108	Ventron, Co. M. air
1673–2015	1.2×10^{-9}	154.4	²⁸Mg	7, 108	Monocrystals Co. M. air
1813–2173	7.48×10^{-10}	150.6 ± 7.7	²⁸Mg	7, 33	Ventron Co. M. Diluted tracer. 200 ppm Fe³⁺, "extrinsic"
2178–2615	$7.43^{+12.16}_{-4.61} \times 10^{-6}$	339.9 ± 18.4	²⁸Mg	7, 33	idem: "intrinsic" region
1723–2673	$4.19^{+2.45}_{-1.55} \times 10^{-8}$	266.1 ± 7.5	²⁶Mg Vapor exchange	7, 109	Spicer, Norton. M. <700 ppm total impurity, argon

*M. = monocrystal.

Table I(b). Effective Magnesium Diffusion Coefficient in the MgO Lattice from Measurements of the Electrical Conductivity

Temp (K)	D_0 (m²·s⁻¹)	Q (kJ·mol⁻¹)	Material*	Ref	Notes
1473–1873	8.5 × 10⁻⁹	220 ± 20	M., doped	114	Al, Sc, Fe; $|I_{Mg}|$
1270–1695	3.0 × 10⁻⁶	287	Polyc., pure	32	wt%: 0.02 Ca, Zn; 0.01 Na; sintering, 1773 K
1370–1770	1.3 × 10⁻⁸	226	M., pure	32	wt%: 0.05 Fe; 0.03 Na; 0.02 Ca
1325–1725	6.3 × 10⁻¹⁰	184	Doped (Fe)	32	"Polyc." + 0.005 wt% Fe; sintering 1773 K
1379–1725	2.4 × 10⁻⁹	193	Doped	32	idem; 0.54 wt% Fe
1475–1725	5.7 × 10⁻⁷	257	Doped	32	idem; 1.94 wt% Fe
1274–1725	1.18 × 10⁻⁷	220	Doped	32	"Polyc." + 0.05 wt% Al₂O₃
1073–1623	2.42 × 10⁻⁷	243	M., pure	32, 110, 111	Norton Co. "Optical"; 100 ppm Fe; $p_{O_2} = 10^{-2}$ atm
1073–1623	4.70 × 10⁻⁷	261	M., pure	110, 112	idem; $p_{O_2} = 10^{-8}$ atm
1314–1667	1.90 × 10⁻⁵	338	M., very pure	110, 112	General Electric Co; ppm: 7.9 Fe; 5 Sr; 3 S
1344–1642	3.28 × 10⁻⁴	378	Polyc., pure	112	wt% 0.10 SiO₂; 0.03 Fe; 0.013 Li; $p_{O_2} = 10^{-5}$ atm
1257–1383	8.44 × 10⁻⁷	258	Polyc., pure	112	idem: $p_{O_2} = 10^{-2}$ atm
1383–1582	1.33 × 10⁻³	504	Polyc., pure	112	idem
973–2073	2.96 × 10⁻⁷	261	Polyc., pure	112, 113	wt% 0.01 total; vacuum

*M. = monocrystal; Polyc. = polycrystals.

Table I(c). Effective Magnesium Diffusion Coefficient in MgO Lattice from Creep, Hot-Pressing, and Sintering

Temp (K)	D_0 (m²·s⁻¹)	Q (kJ·mol⁻¹)	Process*	Ref	Stress (MPa)	G (μm)	Notes
1380–1800	8.76×10^{-3}	402	c, 4PBend, DM	2, 17	6.9–34.5	2.0	99.8% pure, Fe, C, Si, Al, Ca; $n = 1.5$; m—not measured
1380–1800	1.79×10^{-4}	357	c, 4PBend, DM	2, 17	6.9–34.5	3.6	99.8% pure, Fe, C, Si, Al, Ca; $n = 1.5$; m—not measured
1473–1773	1.63×10^{-2}	435	c, 4PBend, DM	2, 18, 115	1.2–3.5	20.0	99.9% pure, $n = 1$; corrected for g size
1523–1723	$0.46^{+4.7}_{-0.41}$	490 ± 42	c, 4PBend, DM	55	2.5–26.5	10–65	99.9% pure + 0.05% Fe; $p_{O_2} = 0.86$ atm; $n = 1$; $m = 1.94$
1453–1533	7.14×10^{-6}	310	c, 4PBend, FT	2, 16	11–19.3	1–3	99.8% pure; $n = 1$; m—not measured
1423–1673	8.03×10^{-4}	326	HP, FT	24	27.6–70	0.8–4.3	Same purity; corrected for g size
1460–1673	3×10^{-1}	490	HP, DM	116	1.5–22.3	0.7–2	Analar, wt%: 0.3 Na, 0.03 Ca; $n = 1$; corrected for g size
1460–1773	3.1	469	S, Int. stg, FT	24, 25		3–16	99.8% pure ppm: 100 Si, Al; <100 Cd, Mn, Zn, Sn
1723–1873	8.7×10^{-3}	427	S, Int. stg, FT	25			
1553–1703	1.06	496	S, FT	46		0.45	wt%: 0.5 Na; 0.05 Ca, 0.01 Fe; constant g size
1380–1800	4.95×10^{-8}	244	c, 4PBend, DM	2, 17	6.9–34.5	5–20	99.8% pure Fe, C, Si, Al, Ca; $n = 1$; $m = 2.5$ mixed creep
1448–1493	1.65×10^{-7}	293	c, Comp, DM	21	55.1–138	12	"High purity" $n = 1.8$; mixed creep

*c, creep; HP, hot-pressing; S, sintering; Int. stg, intermediate stage; 4PBend, 4 points bending; Comp, compression; DM, Dorn method; FT, fixed temperature; g, grain.

of processes which are ostensibly controlled by the same type of diffusivity—Fig. 1. New determinations of the tracer diffusion coefficients (Tables I(*a*) and II(*b*)) have further increased the difficulties of such identification, since different values of the activation energies have become available for the tracer diffusion of the same ions.

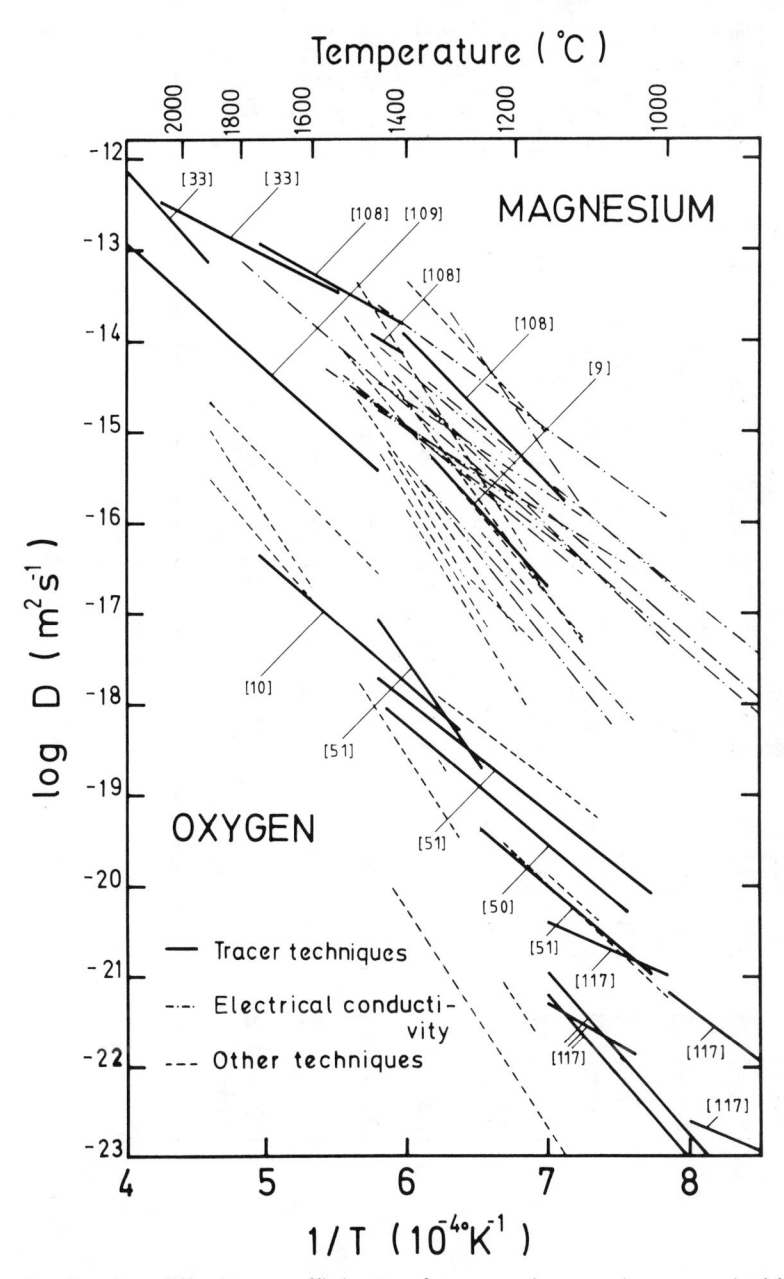

Fig. 1. Lattice diffusion coefficients of magnesium and oxygen in MgO.

Table II(a). Tracer Diffusion Coefficient of Oxygen in the MgO Lattice

Temp (K)	D_0 (m²·s⁻¹)	Q (kJ·mol⁻¹)	Process	Ref	Notes
1573–2023	2.5×10^{-10}	261	¹⁸O, gas exchange	10	200–230 and 60–65 mesh; $p_{O_2} = 150$ torr
1023–1248	4.8×10^{-18}	126 ± 13	¹⁸O, mass spectroscopy	7, 117	Norton Co., $p_{O_2} = 1$ torr (not plotted)
1248–1423	4.3×10^{-9}	344	¹⁸O, mass spectroscopy	7, 117	Norton Co., $p_{O_2} = 1$ torr
1148–1423	2.4×10^{-9}	344 ± 13	¹⁸O, mass spectroscopy	7, 117	Semi elements Co.; $p_{O_2} = 1$ torr
1323–1711	4.5×10^{-11}	252	¹⁸O, gas exchange	50	Sintered polyc.; $p_{O_2} = 40$ torr
1293–1533	1.6×10^{-11}	252 ± 25	¹⁸O, gas exchange	51	Loosely sintered powder; sintering, 1543 K
1533–1723	9.9×10^{-5}	430 ± 42	¹⁸O, gas exchange	51	Idem, "solution-reaction" region
1293–1723	2.4×10^{-11}	234 ± 21	¹⁸O, gas exchange	51	Well-sintered powder; sintering, 1773 K
1107–1273	1.5×10^{-12}	228 ± 13	¹⁸O, gas exchange	7, 117	Doped, 330 ppm Cr; $p_{O_2} = 1$ torr
1273–1423	4.8×10^{-16}	142 ± 16	¹⁸O, gas exchange	7, 117	Idem
1313–1423	1.4×10^{-15}	178 ± 21	¹⁸O, gas exchange	7, 117	Doped, 380 ppm Li; $p_{O_2} = 1$ torr

Table II(b). Effective Oxygen Diffusion Coefficient in MgO Lattice from Dislocation Kinetics (Disl), Creep (C), Hot-Pressing (HP), and Sintering (S)

Temp (K)	D_0 (m²·s⁻¹)	Q (kJ·mol⁻¹)	Process	Ref	Stress (MPa)	G (μm)	Notes
1339–1523	5×10^{-7}	252 ± 15	Disl annealing	61			wt%: 0.06 Al; 0.03 Fe, Ca; "pipe" diffusion
1373–1700	$1.37(\pm 0.26) \times 10^{-6}$	460 ± 18	Disl annealing	118			ppm: 200 Al, Si; 30 Fe
1573–1773	$(0.51–1.57) \times 10^{-4}$	465 ± 50	C compression	2, 119, 20	6.9–44.8	13–68	wt%: 0.04 Si; $n = 3.1 \pm 0.5$
1448–1498	$(3.86–4.93) \times 10^{-12}$	213 ± 21	C compression	2, 20, 56	34.4–137.6	11.8–52	ppm: 75 Li, 40 Ca, 30 Si; $n = 3.3$
1448–1498	1.03×10^{-5}	460	C compression	21, 23	55.1–138	17	Reagent grade "A"-high purity; $n = 7.5$
1596	$(0.88–6.1) \times 10^{-4}$	460 ± 25	C compression	2, 11, 120	50–90	10–14	wt%: 0.1 SiO₂; $n = 3$
1275–1496	1.90×10^{-10}	281	S, initial stage	32		1.25×10^{-2}	wt%: 0.02 Ca, Zn, 0.01 Na; air; ref, "HP700"
1359–1430	1.20×10^{-10}	272	S, initial stage	32		1.6×10^{-2}	idem; ref, "HP100"
1359–1604	1.50×10^{-10}	254	S, initial stage	32		2.25×10^{-2}	wt%: 0.01 Ca, Na; air; ref, "F900"
1728–2166	$(2.5–6.0) \times 10^{-8}$	300 ± 7	HP	4	15.2		Specpure; $n = 3$
1873–2173	$(4.8–10.9) \times 10^{-5}$	451 ± 7	HP	4	13.4		Specpure + 4030 ppm; mol Al₂O₃; $n = 3$
1880–2173	$(0.8–36) \times 10^{-8}$	343 ± 32	HP	4	15.6		Specpure + 284 ppm; mol Al₂O₃

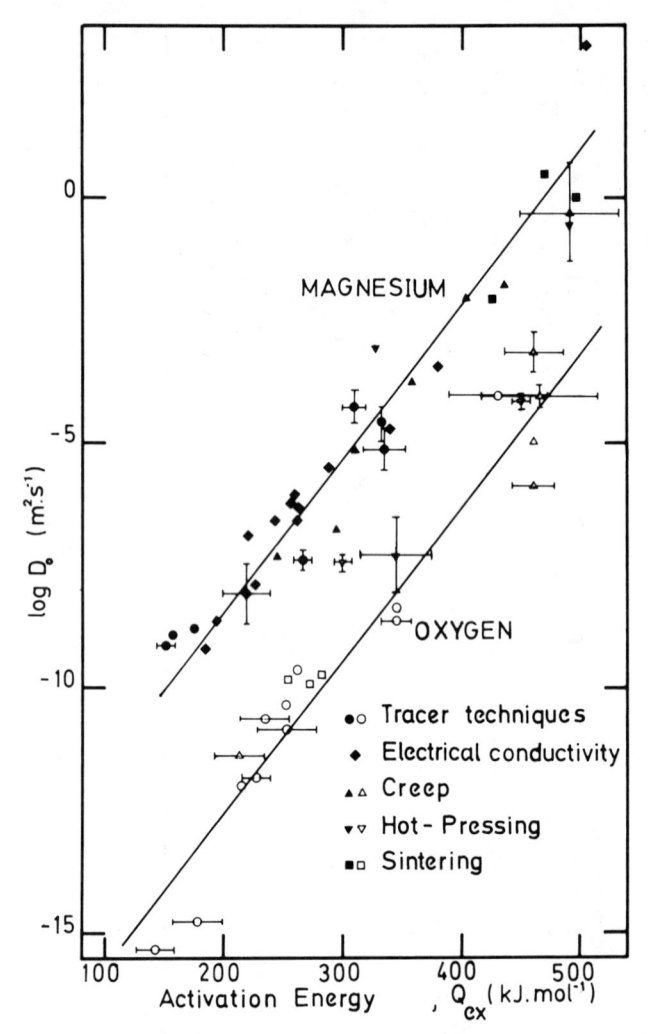

Fig. 2. Compensation law (Eq. (1)) applied to lattice diffusion coefficients of magnesium and oxygen in MgO.

The reasons for the spread of the observed values of the activation energies can be considered. From the review by Wuensch[7] of the available tracer bulk diffusion coefficients in MgO, a few properties stand out as firmly established: (1) The enthalpy of formation of intrinsic defects of the Schottky type, ΔH_S, must be in excess of 4.5 eV (434 kJ·mol^{-1}). The semiempirical correlations of Barr and Dawson[27] and computer calculations based on available models of the interatomic potentials[28–31] fix the value of ΔH_S as lying in between 7 and 8 eV (676–772 kJ·mol^{-1}). Although the existence of Frenkel disorder has on occasion been proposed,[7,32] it emerges as distinctly less likely on the basis of these theoretical calculations. (2) The enthalpy of motion, ΔH_m, for both magnesium and oxygen vacancies in the diffusion process is around 2 eV (193 kJ·mol^{-1}). (3) Contrary to early identification of Lindner's results[9] as corresponding to intrinsic diffusion and to similar claims by Harding,[33] it is improbable that intrinsic diffusion has ever been observed in MgO.

For the observation of intrinsic diffusion below the melting point (T_m = 3125 K),[34] total impurity levels below 10^{-7} are necessary. Such a target is beyond the best current practice in terms of the available crystals and experimental methods. As a consequence, the lattice diffusion coefficients must be considered to be controlled by the impurities in the lattices. Three temperature regions may then be distinguished with increasing temperature, according to the state and quantity of the impurities: (a) For temperatures below 1500°C, the solid solution limits of many impurities in MgO are low.[35–38] The concentrations of the impurities in the MgO lattice are then dependent on solution–precipitation reactions, on grain-boundary segregation, and on association–dissociation of defect complexes in the lattice.[30,39,40] The activation energy for diffusion in this region is then expected to be higher than the enthalpy of motion ΔH_m alone, owing to the simultaneous occurrence of such reactions.[41] (b) Extrinsic diffusion with an activation energy close to ΔH_m becomes more probable for temperatures above 1500–1600°C. (c) This second region is expected to be followed by a wide region[42] of transition ($\Delta T > 400°C$), leading in the limit to a region of intrinsic diffusion. This third region (region c) is characterized by an increase of the activation energy with increasing temperature from ΔH_m (=190 kJ·mol^{-1}) to $\Delta H_m + \Delta H_s/2$ (=560 kJ·mol^{-1}). The limits of these regions in any particular set of experiments will be strongly dependent on the impurities and on previous thermal treatments. The few measurements of diffusion coefficients that have been made above 1500–1600°C are not enough to allow clear separation between these different regions.

The compensation law[42,43] holds in the transition region (region c). It states that the logarithm of the preexponential factor of the diffusion coefficient, D_0, measured over narrow ranges of temperatures, is proportional to the experimentally determined value of the corresponding activation energy, Q_{ex}. The slope of the compensation law for this case[42,43] is $d \ln (D_0)/dQ_{ex} = (1/RT) - (2/\Delta H_s)$. Owing to the high value of ΔH_s in MgO (see above), the second term in the slope is some 2 orders of magnitude smaller than $1/RT$: the compensation law is then properly approximated by

$$\ln (D_0) = \frac{Q_{ex}}{RT} + \text{constant} \tag{1}$$

A second cause for the linearity (Eq.(1)) is purely statistical, resulting from the relation that exists between the estimated values of D_0 and Q_{ex} that arise when an Arrhenius plot is fitted to the measured diffusion coefficients over a narrow range of temperatures.[43] Finally, the compensation law (Eq. (1)), with the term $1/RT$ dominating the slope, is also observed when several thermally activated processes contribute simultaneously to the observed kinetic data.

The value of the compensation law as a tool to investigate the consistency of data for the tracer diffusion coefficients in MgO emanating from different authors has been demonstrated.[43] As shown in Tables I(a)–II(b) and Fig. 2, such consistency is also found in the diffusion data derived from other material transport phenomena in MgO.

What may be the coincidental fitting of some of the data in Tables I(c) and II(b) to the lines in Fig. 2 deserves mention and introduces a cautionary note. The data of Passmore[17] and of Snowden,[21] both with low activation energy, are discussed below as representing grain-boundary diffusion processes; however, they also come into close fit with the magnesium line of Fig. 2 if the results are interpreted as Nabarro–Herring creep[17] or as mixed creep with one of the components being Nabarro–Herring creep.[21] The good fit of Dequenne's data[32] for initial sintering in MgO to the oxygen line in Fig. 2 is also rather unexpected, since it contradicts the observations of enhanced oxygen diffusion through the grain boundaries in this material (discussed below). A similar (indicating lattice diffusion) exponent of the time dependence of shrinkage during sintering has been observed in early work by Quirck,[44] although grain-boundary diffusion has been reported as controlling the initial stage of sintering for similar MgO powders at slightly lower temperatures;[45] sintering has also been recognized as subject to the effect of higher water vapor pressures.[46,47] Ambiguity of this type results from the limited ability of models for initial-stage sintering to discriminate between different sintering mechanisms on the basis of log–log plots of shrinkage vs time.[48,49]

A few conclusions are possible from the data of Tables I(a) and II(b) and from Figs. 1 and 2: (i) The compensation law (Eq. (1)) is a useful tool for investigating the consistency of diffusion data from different sources. (ii) The hot-pressing and creep of MgO with particle sizes in the range of 1 μm are controlled by the diffusion of the cation when viscous deformation governed by lattice diffusion is encountered (anion grain-boundary diffusion is possible for coarser powders or in strongly oxidizing atmospheres), and they are controlled by the bulk diffusion of the anion if plastic deformation by motion of the dislocations is dominant. (iii) The data from creep and sintering with high activation energies do not correspond to "nearly" intrinsic diffusion; rather, the low-temperature ranges at which these data have been obtained suggest that simultaneous reactions might have interfered with the determination of the activation energies.

Grain-Boundary Diffusivities

The evidence for the enhanced grain-boundary diffusion of oxygen in MgO is reviewed and followed by discussion of the conditions under which grain-boundary diffusion of the cation may become rate controlling.

Tracer measurements of the oxygen diffusion coefficient in polycrystalline MgO[50,51] have demonstrated that, for the range of temperatures investigated (1020–1450°C), the grain boundaries act as paths where the diffusivity of oxygen is much higher than that in the lattice. This agrees with the observations of Holt and Condit[52] at higher temperatures, 1600–1670°C, and is analogous to the observation of enhanced grain-boundary diffusion for chlorine in NaCl.[53] The quantitative measurement of the ratio, $B = \delta_{gb} D_{gb}^O/(D_l^O)^{1/2}$, by McKenzie[54] has shown an increase for pure MgO from 1×10^{-14} to 1×10^{-13} m$^2 \cdot$s$^{-1/2}$ in the range of temperatures 1621–1743°C. Such an increase of B with T implies that the activation energies for the grain-boundary diffusion of oxygen, Q_{gb}^O, and for the lattice diffusion of oxygen are in the relation $Q_{gb}^O > 1/2 Q_l^O$. The term B is approximately constant and equal to 1.0×10^{-12} m$^2 \cdot$s$^{-1/2}$ for iron-doped samples.

Additional information on the grain-boundary diffusion of oxygen can be obtained from a few creep studies.[17,21,55,56] Terwilliger's results[56] for the viscous creep of iron-doped MgO (5–15 μm) at 1300°C have shown good correlation between the strain rate and G^{-3}, where G is the average particle size. The values of $\delta_{gb} D_{gb}$ calculated by using the Coble creep equation are in good agreement with the values calculated by using the values of B extrapolated from McKenzie's results and from values of D_l^O taken from Ref. 10, as shown in Fig. 3 and

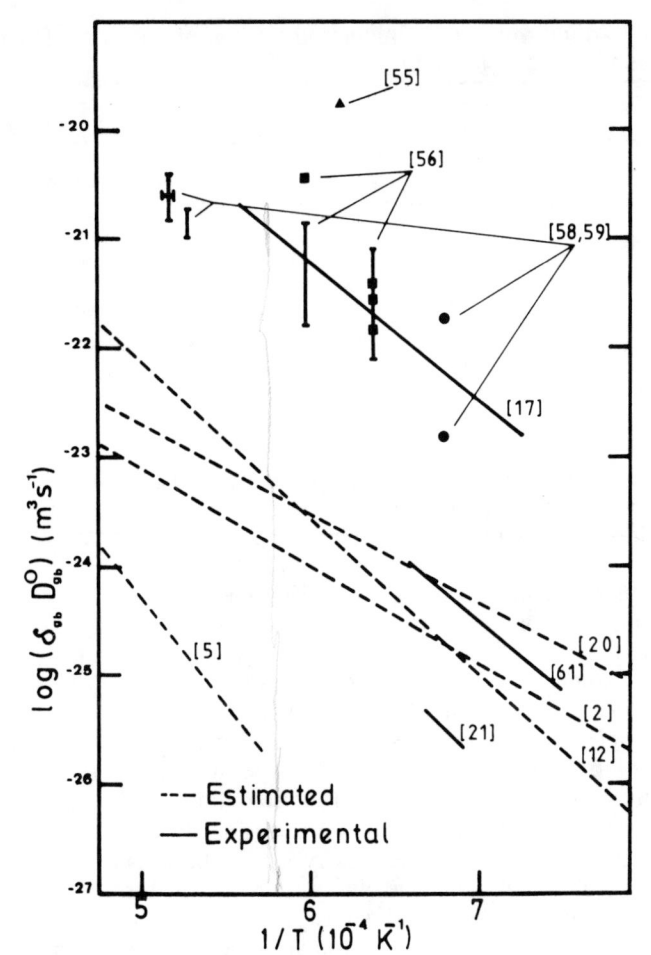

Fig. 3. Grain-boundary diffusion of oxygen in MgO.

Table III. At 1400°C, the creep rates correlate better with G^{-2}, and the values of $\delta_{gb}D_{gb}$ calculated by using the Coble equation deviate from those calculated by using B. Passmore's creep results[17] with 99.8% pure MgO at similar stresses and temperatures (1110–1530°C) showed a decrease in activation energy from 377 kJ·mol^{-1} for samples of 2-μm particle size to 226 kJ·mol^{-1} for samples with a particle size above 5 μm. The creep rates in the region of low activation energies are found to be proportional to $G^{-2.5}$. When this decrease of the activation energies with increasing grain size is interpreted as the result of a transition from Nabarro–Herring creep to Coble creep,[2,20,57] the value of $\delta_{gb}D_{gb}$ calculated at 1700 K and $G = 18$ μm on the basis of Coble's equation is in close agreement with Terwilliger's results.[56] A similar transition from Nabarro–Herring to Coble creep with a corresponding decrease in the activation energies from 497 ± 42 to 339 ± 21 kJ·mol^{-1} has also been observed by Tremper[55] at similar temperatures (1350°C) in Fe-doped samples in oxidizing atmospheres. The value of $\delta_{gb}D_{gb}^{O}$ calculated[55] at 1350°C by using Gordon's equation[58,59] for mixed mechanisms of

Table III. Experimental and Calculated Values of the Grain-Boundary Diffusivities of Oxygen in MgO

Temp (K)	$(D_O)_{gb}$ (m²·s⁻¹)	$\delta_{gb}(D_O)_{gb}$ (m³·s⁻¹)	Q_{gb} (kJ·mol⁻¹)	Ref	Notes
1380–1800		2.62 × 10⁻¹⁴	244	17	Creep, $n = 1$; $m = 2.5$, 6.9–34.5 MPa, 5–20 μm
1448–1493	8.12 × 10⁻⁷	8.12 × 10⁻¹⁶	293	21	Creep, $n = 1.8$, 55.1–138 MPa, 12 μm
1339–1523	5.0 × 10⁻⁷	5.0 × 10⁻¹⁶	252 ± 15	61	Dislocations, "pipe" diffusion
	2.5 × 10⁻¹⁰	2.5 × 10⁻¹⁹	172	2	Calculated D_l^o from Ref. 10
	2.5 × 10⁻¹⁰	2.5 × 10⁻¹⁹	157	20	Idem
	1.37 × 10⁻⁶	1.37 × 10⁻¹⁵	277	12	Idem, D_l^o from Ref. 118
		4.7 × 10⁻¹⁵	382	5	Archetypal Eq.

creep is 1 order of magnitude higher than those of Passmore and Terwilliger discussed above (Fig. 3). Snowden's creep tests[21] at 1200°C and at 55–138 MPa using a high-purity MgO of middle grain size (12 μm) show a stress exponent of $n = 1.8$. The corresponding activation energy, 293 kJ·mol⁻¹, is again lower than the value thought to correspond to volume diffusion of oxygen, 460 kJ·mol⁻¹, and obtained in the creep of preannealed samples, as discussed in the previous section. As shown in Fig. 3, the grain-boundary diffusion coefficient calculated by using Langdon's equation[60] for creep by grain-boundary sliding through the glide/climb of dislocations along the grain boundary is in good agreement with the theoretical calculations of D_{gb}^O that are discussed next; it is also in agreement with the diffusion coefficient of oxygen along the cores of dislocations, D_c, as determined by Narayan[61] from the direct observation of the self-climb of dislocation loops.

Simultaneous enhancement of oxygen diffusion along dislocations and along grain boundaries in MgO has been observed.[52] If the enhanced grain-boundary diffusivity can all be attributed to the grain-boundary "core,"[62] the excess of oxygen ions expected at the grain-boundary core of MgO owing to the negative sign of the grain-boundary electrical potential[63] suggests a dislocation pipe mechanism for grain-boundary diffusion,[39,63] with rapid migration of the excess ion occurring along the grain-boundary dislocation cores. D_{gb}^O and D_c^O must then be essentially the same diffusion coefficient (Fig. 3).

As in the case of the anion in MgO and in the alkali halides, many metals show enhanced grain-boundary diffusion.[2,5,64] The following relations are commonly observed in metals: (i) the preexponential factors of D_{gb} and of D_l are approximately equal to one another; (ii) the activation energies of grain-boundary diffusion, Q_{gb}, have values in the range 0.53–0.68 of the corresponding activation energies for lattice diffusion. These relations, with $Q_{gb} = 0.6Q_l$, have been used to calculate grain-boundary diffusivities for the published deformation maps of MgO.[2,12,20] The corresponding values of $\delta_{gb}D_{gb}^O$ are compared in Fig. 3 with the others discussed above.

In the calculation of $\delta_{gb}D_{gb}^O$ from D_{gb}^O in Fig. 3, $\delta_{gb} = 1$ nm has been used. The separation between the two sets of values in Fig. 3 can be narrowed by using a larger value for δ_{gb},[39] namely, $10 < \delta_{gb} < 10^2$ nm. Such thick grain boundaries may result from the effects of space charges, from solute segregation,[65] from precipitates, or from the insufficient annealing of mechanical deformation.[62,64,66,67] A 2 order of magnitude increase in the grain-boundary diffusion product, $\delta_{gb}D_{gb}$, for Cr^{3+} diffusion along 15° (100) tilt boundaries in Cr-doped MgO owing to increased Cr^{3+} concentration in the samples has been reported.[39] The magnitude of the gap between the lower band of data in Fig. 3 and the upper band, which corresponds to heavily doped (Fe) or to less pure MgO samples, is approximately equal to the increase observed in Cr-doped MgO. The separation thus reflects differences of purity between the samples in the two bands.

No study of the tracer diffusion of Mg in the grain boundaries of MgO is available. From analogous studies of the diffusion of several other cations in bicrystals and polycrystals of MgO[33,68-70] and by analogy with the diffusion of Na in NaCl,[53] no enhanced grain-boundary diffusion of the cation is expected in precipitate-free grain boundaries in MgO. This is corroborated by the negligible effect of grain boundaries in electrical conductivity measurements on bicrystals and polycrystals of MgO.[71]

For the cation, the use of the relations, $Q_{gb} = 0.6Q_1$ and $(D_0)_{gb} = (D_0)_1$ in the calculation of D_{gb}^{Mg} from D_1^{Mg} leads to an enhanced grain-boundary diffusion of the magnesium ion, as it did for the oxygen ion. It results in an overestimate of the extent of the magnesium grain-boundary diffusion-controlled regions in deformation maps.[11,12,20]

The replacement of the relation $(D_0)_{gb} = (D_0)_1$ by an equality between the grain boundary and bulk diffusion coefficients at the melting point, $D_{gb}^{Mg} = D_1^{Mg}$ at T_m, and the retention of the relation $Q_{gb}^{Mg} = 0.6Q_1^{Mg}$ reduces D_{gb}^{Mg} by 2–3 orders of magnitude. This revised procedure gives a value in better agreement with the lack of enhanced grain-boundary diffusion for the cation. It also still accounts for an enhanced cation grain-boundary diffusion at low temperatures,[33,69,70] resulting from short-circuiting paths due to second phase nucleation and precipitation in the grain boundaries. Though apparently arbitrary, the equality between D_{gb} and D_1 at the melting point has also been reported as valid for many metals.[64] The lack of enhanced grain-boundary diffusion of the cation means that deformation controlled by D_{gb}^{Mg} is only possible with fine powders and at low temperatures.[12,33]

Surface Structure and Surface Diffusion in MgO

Surface transport is generally less easy to characterize than bulk behavior, since it is highly dependent on the surface structure and on the cleanliness of the crystal surfaces,[72] the adsorption and segregation of impurities not only changing the concentrations of the surface diffusing defects but also, at impurity levels well below the corresponding bulk solubility limits, inducing surface structure transformations.[72-74]

Surface Structure

MgO has a highly anisotropic surface energy,[28,75] γ_s, with the minimum value corresponding[76,77] to the (100) faces. These are also the planes of cleavage. In the ionic, fcc, NaCl type crystals, the Coulombic interactions prevent the (111) planes, which consist of a single type of ion, from being the planes of lowest surface energy.

The equilibrium shape of a crystal is derivable from a knowledge of the γ plot[76,77] by Wulff's construction. The appearance of a given face depends on the value of the corresponding minimum or "cusp" and on the local curvature of the γ plot, the most commonly observed faces being those corresponding to absolute minima. Theoretical calculations for NaCl predicted a very deep cusp in the γ plot for the (100) faces.[28,75] Also, the calculations in Table IV show that $\gamma_s(100)$ is much lower than $\gamma_s(110)$ or $\gamma_s(111)$ in MgO, a result supported by the fact that

Table IV. Experimental and Calculated Values of Surface Energy of MgO, γ_s, as a Function of the Orientation of the Surfaces

Orientation	Experimental (J·m⁻²)	Calculated (J·m⁻²)
(100)	1.04,*,†,‡ 1.2,*,†,‡ 1.15*,†,‡	1.17,† 1.45,* 1.3*,‡ 1.44,‡ 1.23,‡ 1.36‡
(110)		2.98,† 3.89‡
(111)		2.6 ± 0.3

*Ref. 75.
†Ref. 28.
‡Ref. 121.

the faces of the cube, the (100) planes, are the most commonly observed free surfaces in MgO crystals.[18,25,78–81] The faces of the octahedron, the (111) planes, have also been observed,[81,82] but they disappear with increasing temperature;[82] this suggests that they are associated with a local minimum in the γ plot at the (111) orientations.

The theoretical calculations of Mackrodt[28] using the HADES package and allowing for relaxations of the ions at the surface[28,33] give values of $\gamma_s(100)$ that are particularly close to the experimental ones (Table IV). These numerical methods are not suitable for the calculation of $\gamma_s(111)$ because the (111) surfaces do not constitute neutral planes. In Table IV, we present a semiempirical estimation of $\gamma_s(111)$ calculated from the anisotropy of the wetting angle of liquid Cu, Co, and Fe on the (100), (110), and (111) surfaces of MgO[84] and using the values of the pair $\gamma_s(100)$ and $\gamma_s(110)$ as calculated by Mackrodt.[28] The value of $\gamma_s(111)$ is slightly smaller than the value of $\gamma_s(110)$.

Since it corresponds to a deep cusp in the γ plot, $\gamma_s(100)$ will bring proportionately less weight to the average surface energy, $\bar{\gamma}_s$,[85] of a MgO powder of isotropic spherelike particles, which is closer[86] to the values of $\gamma_s(110)$ and $\gamma_s(111)$ than to the value of $\gamma_s(100)$. A powder compact consisting of cubelike particles has an average surface energy close to $\gamma_s(100)$ and may therefore show reduced sinterability owing to the lower value of $\bar{\gamma}_s$.

As a final point, the terms for thermal expansion and for vibrational energy and entropy become more important in the surface energy with increasing temperature and attenuate the anisotropy of γ_s, which is mostly due to differences of the potential energy of the surfaces at 0 K.[28] Impurities have a complex effect on the surface energy but may also be active in modifying the anisotropy of γ_s.[72,87,88] If selectively adsorbed or segregated at specific surface sites, the impurities can either reduce or increase the extent of anisotropy.[4,72]

Surface Diffusion

The surface diffusion coefficient, D_s, defined as the excess mass transfer diffusivity in a surface layer of thickness δ_s in comparison to the diffusivity in the bulk of the crystal, has been measured in MgO by grain-boundary grooving[74,90–92] and by single-scratch decay.[91] The corresponding values of D_s are collected in Table V. McAllister's results, using a more elaborate experimental procedure at 1500 and 1600°C, have served to confirm the results of preceding authors.[90,91] Below 1500°C, the agreement between the results of the different authors for nondoped MgO is good.[74,93] Cr^{3+} additions below the bulk solubility limit reduce the values of D_s by 1 order of magnitude at intermediate Cr^{3+}

Table V. Surface Diffusion of MgO

Orientation	Temp (K)	D_{os} (m²·s⁻¹)	Q_s (kJ·mol⁻¹)	Method	Ref	Notes
Polycrystals	1473–1773	8.0	370 ± 63	gb grooving	91	<0.5 wt% total cations
(100)	1373–1573	2.1 × 10⁴	445	scratch decay	91	<0.5 wt% total cations
(100)			509	calc (HADES)	28	
≃(110)	1443–1773	2.3 × 10¹	377 ± 42	gb grooving	90	0.04 wt% total cations
(110)	1373–1573	2.3 × 10¹	361	scratch decay	91	<0.5 wt% total cations
(110)			325	calc (HADES)	28	
$\theta < 17°$(100)	1373–1673	3.4 × 10⁻⁵	210 ± 20	gb grooving	74	Norton Co, 0.00 at% Cr Si + Ca + Fe + Al = 700 ppm
$\theta < 17°$(100)	1373–1673	3.7 × 10⁻⁴	239 ± 5	gb grooving	74	PSU, 0.00 at% Cr Si + Ca + Fe + Al = 800 ppm
$\theta < 17°$(100)	1373–1673	5.5 × 10⁻⁴	237 ± 5	gb grooving	74	PSU, 0.038 at% Cr Si + Ca + Fe + Al + Mn = 1900 ppm
$\theta < 17°$(100)	1373–1673	1.9 × 10⁻⁶	178 ± 8	gb grooving	74	PSU, 0.254 at% Cr Si + Ca + Fe + Al = 750 ppm
$\theta < 17°$(100)	1373–1673	4.7 × 10⁻⁶	178 ± 20	gb grooving	74	PSU, 0.376 at% Cr Si + Ca + Fe + Al = 1700 ppm
$\theta < 17°$(100)	1373–1673	1.6 × 10⁻⁶	167 ± 20	gb grooving	74	PSU, 0.442 at% Cr Si + Ca + Fe + Al = 750 ppm

concentrations—0.254 at.% Cr^{3+}. On further increase in Cr^{3+} concentration, the values of D_s are brought back to the level of D_s in nondoped samples. The values of Q_s, the activation energy of D_s, are also changed by doping with Cr^{3+}. In a similar way, small concentrations of Al^{3+} in solid solution (135 ppm) have been found more effective in reducing surface transport during the sintering of high purity MgO at high temperatures[4] than have higher Al^{3+} concentrations (1000 ppm), the changes of surface transport being followed by measurement of the grain growth/densification ratio. Above 1500°C, Robertson has found that the exponent of the time dependence of the groove width changes from ¼, characteristic of grain-boundary grooving by surface diffusion, to ⅓, corresponding to grooving by either bulk diffusion or diffusion through the gas phase. Strong faceting[72] of the grain-boundary grooves can be associated with measurements on (100) surfaces and has hindered the determination of D_s, as observed by Robertson.[90] This has not been reported by any of the other authors.[74,91,92] The only observable form of anisotropy of D_s has been the slight change of value from one type of surface to another.[90,91,93]

The assumption that the experimental results of Table V indeed correspond to surface diffusion in MgO has been questioned.[92,94] Identification of the kinetics of grain-boundary grooving and scratch decay as being controlled by surface diffusion relies on the finding of the value ¼ for the slope of the plot of the logarithm of the groove width against the logarithm of the corrected time,[90,91] t. The time correction is necessary because Mullin's equation is only applicable when the shape of the grooves is close to the steady-state condition.[89,91,93] The curvature in the log–log plot is, however, then dependent on the value chosen for the time correction. It has been claimed[94] that the groove widths as determined by Robertson,[90] if plotted against $t^{1/4}$, extrapolate to a negative unphysical width at zero time, whereas, if plotted against $t^{1/3}$, the widths at zero time become positive and significant. In this case, where mixed kinetics of grain-boundary grooving may well be present, if one were to subtract the bulk diffusion contribution from the measured grain-boundary width, the corrected width could still increase in proportion to $t^{1/4}$. The hypothesis that the slope ⅓ observed above 1500°C[90] is due to grooving brought about by vapor diffusion is unlikely in view of the temperatures, total gas pressures, and type of atmospheres (air) used.

The consistency of the surface diffusion data in Table V can be tested by using the compensation law (Eq.(1)). It can be seen (Fig. 4) that a good fit of the compensation law to the data is obtained ($T^* = 1573$ K is the midpoint temperature[42] for the set of data in Table V). Cr^{3+}-doped samples have lower Q_s and lower $(D_0)_s$ values, but the corresponding points remain close to the compensation law line.

The relationship between D_s and D_1^{Mg} may be seen in Fig. 4: if the preexponential factors of both diffusion coefficients are assumed to be approximately equal, $(D_0)_s = (D_0)_1^{Mg}$, then $Q_s/Q_1^{Mg} = 0.6$ at low values of Q_s ($Q_s = 167$ kJ·mol^{-1}). An important characteristic of the D_s data as given in Table V is the simultaneously high values of Q_s and $(D_0)_s$ of some of the experimental data. By comparison with D_1 in Fig. 4, for the same value of the activation energy, $(D_0)_s$ is approximately 4 orders of magnitude higher than the corresponding preexponential factor of D_1^{Mg} and even higher if the comparison is made with D_1^O. The high values of Q_s are contrary to the conventional picture that diffusion coefficients are generally in the sequence $D_1 < D_{gb} < D_s$ with activation energies in the reverse order $Q_s < Q_{gb} < Q_1$.[8,96] The observation of linearity over the wide range

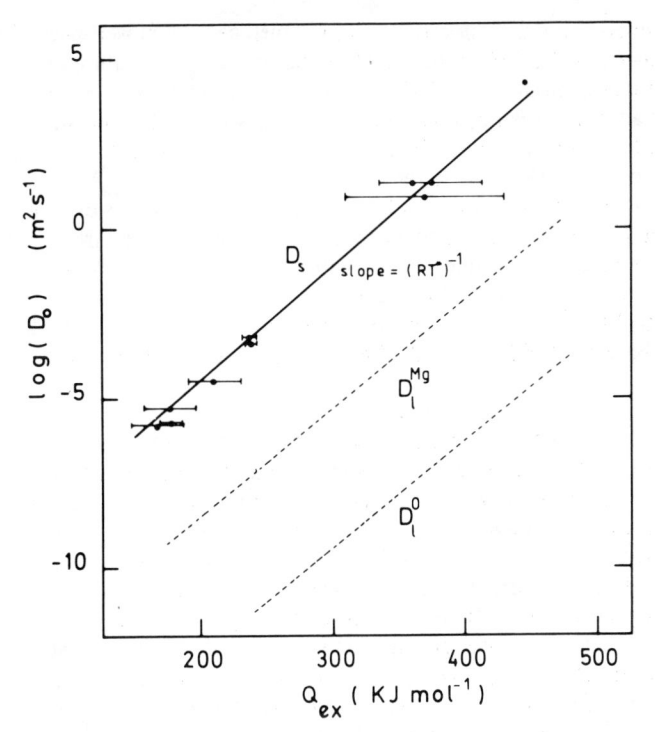

Fig. 4. Consistency of the surface diffusion data of MgO; $T^* = 1573$ K.

of values of Q_s shown in Fig. 4 ($167 < Q_s < 445$ kJ·mol^{-1}) is indicative that the set of D_s data corresponds to mixed behavior,[43] with two or more mechanisms of surface diffusion competing together. Surface diffusion in Al_2O_3 is better documented[43,95,97] than in MgO, and a good fit of the compensation law (Eq.(1)) to the D_s data of Al_2O_3 has also been observed,[43] with Q_s values ranging from 234 to 556 kJ·mol^{-1}. Two different surface diffusion mechanisms have been considered to yield the mixed behavior in Al_2O_3[97] and are consistent with the linear fitting observed: for $T > 0.6T_m$, a high-temperature mechanism with simultaneously high values of Q_s and $(D_0)_s$ holds; below this temperature, a low-temperature mechanism with low values of Q_s and $(D_0)_s$ is dominant. The same two mechanisms of surface diffusion have also been observed in fcc metals,[72,95] with the values of Q_s for the low-temperature mechanism being half those of the other mechanism.

The existing models of surface diffusion have been developed mainly to explain the observations on such elemental solids as the metals.[72,76,77,95] Some care is needed[83,95] in extending these models to ionic solids: (i) there are at least two ions diffusing simultaneously in the case of ionic crystals; (ii) the singular faces are different (see previous section) and so is the nature of nearest neighbors; (iii) the space charge of the surface interacts with the populations of charged defects in the surface region. However, those conclusions that proceed simply from the close packing of the structure may still retain some validity.

From the balance of energy needed to create an adsorbed atom on a singular surface and to move it through the saddle points of the potential of the surface, the activation energy for diffusion on the (100) surface of a closed-packed metal[98] is estimated as $Q_s = 1/2\Delta H_v$, where ΔH_v is the heat of sublimation of the metal. In a more elaborate model[89,95] (TLK), the surfaces near a singular orientation, the vicinal surfaces, are seen as composed of terraces parallel to the singular surface. The energy to create an adatom from kinks in the ledges of the terraces and to move it through the saddle points on the terraces is $Q_s = 2/3\Delta H_v$. As vicinal surfaces with different orientations are built from different combinations of similar terraces, Q_s must then be nearly independent of the orientation of the surfaces. The use of pairwise potentials and of interactions with higher order neighbors further widens the range of the values of Q_s to $0.42\Delta H_v < Q_s < 0.60\Delta H_v$. Since ΔH_v for MgO is around 610 kJ·mol^{-1},[99] the observed high values of $Q_s = 350$–380 kJ·mol^{-1} are in fact between $0.6\Delta H_v$ and $2/3\Delta H_v$, in agreement with this type of model for surface diffusion.

Shell model calculations by the HADES method[28] give 68 and 116 kJ·mol^{-1} as the enthalpies for the migration of vacancies in the (100) and (110) surfaces of MgO, respectively. These values are smaller than that of the enthalpy of vacancy migration in the bulk of MgO, 230 kJ·mol^{-1}, in accordance with the simplified argument presented above. The enthalpy of formation of the vacancies in these surfaces, however, is high and, for the (100) plane, is even higher than in the bulk. The observation of intrinsic vacancy surface diffusion in the (100) surfaces is consequently rather improbable, owing to the high value of Q_s for this surface (Table V); it cannot, however, be ruled out for the (110) surfaces, since the corresponding value of Q_s is comparable to the observed values of Q_1 for extrinsic bulk diffusion. The measured activation energies for surface diffusion in the (100) and (110) surfaces are not equal,[91] Table V being closer to the corresponding theoretical values[28] than they are to each other. Such comparisons are based on too few experimental points to allow firm conclusions.

In summary, either type of model — the TLK model or the vacancy model based on a transition toward intrinsic vacancy diffusion — can explain the high activation energy observed in some of the measurements of D_s in MgO. The intrinsic vacancy model, however, cannot explain the corresponding high values of $(D_0)_s$.

From the theory of diffusion, the preexponential factor of the diffusion coefficient is[95,96]

$$(D_0)_s = \alpha_d \lambda^2 \nu_s \exp[(\Delta S_f + \Delta S_m)/R] \qquad (2)$$

where α_d is a constant approximately equal to unity, λ is the atom jump distance (equal to an atomic spacing for vacancy diffusion), ν_s is the vibrational frequency, ΔS_f and ΔS_m are the entropies of formation and motion of the defect, respectively, and R is the gas constant. For diffusion by a pure vacancy mechanism, ν_s, the vibrational frequency at the surface is expected to be similar to that in the bulk, as are the respective entropies.[93] If an adatom has two vibrational degrees of freedom at the saddle point,[72,95] the vibrations being strongly anharmonic (the Debye–Waller factor for the surface is 2–3 times higher than for the bulk[100]), then there must be a large dispersion of frequencies, and the actual frequency may be 2–3 orders of magnitude higher than that of the purely harmonic vibrations. If, instead, the adatom at the saddle point has two translational degrees of freedom, then[72,95] a reasonable jump distance of four to five interatomic distances is enough to explain the higher values of $(D_0)_s$.

If the population of adatoms has two translational degrees of freedom, it constitutes a two-dimensional gas; and then $\Delta H_m = \Delta H_v/6$ is much higher than for diffusion by way of localized states. Since $\Delta H_v/6 = 100$ kJ·mol^{-1} in MgO, the anticipated value of ΔH_m (70–120 kJ·mol^{-1}) is much higher than $RT = 15$ kJ·mol^{-1} and the adatoms will spend most of the time in vibrational (bound) states; the condition for the continuous existence of a two-dimensional gas is accordingly not fulfilled in MgO. At any event, the motion of an atom on the surface, in contrast to its motion in the bulk, involves little cooperative motion of the neighbor surface atoms.[76] Once an atom acquires sufficient thermal energy to overcome the potential barriers and to escape, it may then travel a significant number of lattice spacings before being adsorbed into another vacant site. When the excited species in a two-dimensional gas are molecules or dimers, the deactivation to the bound state may be made difficult by bond length transitions, resulting in long jump distances in the excited state.[95] In oxides and carbides, the high bond strength makes molecular dissociation into individual ions unlikely on clean surfaces (see also next section). Impurities in the gas phase (H$_2$O, C) can, however, form activated complexes with the surface ions, and they may increase the concentration of adatoms in the excited state, thus promoting surface diffusion by the adatom mechanism.[72,74,95,101] It is not possible to establish a clear separation between the vibrational and the translational models for adatom defects on the surfaces. Both may account for the significant increase of $(D_0)_s$ at the surface as compared with the preexponential factors for bulk diffusion.

In the calculation of D_s from grain-boundary grooving or scratch-decay measurements, δ_s is commonly set equal to one interatomic spacing.[72,74,76,89–92] This value of δ_s is too small to account for the lattice distortion near the surfaces[28,76,83,102] — which in NaCl goes as deep as four atomic layers — and for the distribution of space charge in the surface region of ionic crystals, the width of which may extend several tens of nanometers into the bulk of the crystal.[100] $(D_0)_s$ is overestimated by this convention.

In summary, despite the reservations noted above, the experimental values for the surface diffusion coefficient of MgO reported in the literature are analogous to the general properties of surface diffusion that have been observed in other materials with close-packed structures, namely, the fcc metals and Al$_2$O$_3$. In the absence of better information on surface diffusion in MgO, the values of the coefficients currently available can be interpreted as representing surface diffusion by a number of mechanisms, one being extrinsic vacancy surface diffusion and the other being adatom diffusion.

Vapor Transport in MgO

The effect of vapor transport on sintering and hot-pressing has already been mentioned above. The rate of vapor transport under vacuum is limited by the rate of evaporation–condensation, whereas at higher pressures it is controlled by diffusion through the gas phase.

The easy volatilization of magnesia in contact with carbon above 1600–1700°C, with the formation of white smoke, is a well-known phenomenon.[103] On the other hand, dry hydrogen seems to have no influence on the volatilization of MgO up to 2500°C. From the Ellingham diagrams,[104] carbon acts to reduce MgO only above 2100°C so the easy volatilization at lower temperature may be due to the presence of water vapor in the gaseous products formed by the combustion of carbonaceous fuels. It is known[14] that MgO reacts with water vapor to form volatile hydroxides. Under vacuum, magnesia is more volatile than alumina[105] but has good stability in neutral or oxidizing atmospheres. The four reactions of vaporization

Table VI. Reactions of Vaporization of MgO; ΔG (J·mol^{-1})

Vaporization reaction	$K_r = \exp(-\Delta G/RT)$	Ref
(I) MgO(s) → Mg(g) + ½O$_2$(g)	K_I = 6.429 × 10^{10} exp(−734,473/RT)	14
(II) MgO(s) → MgO(g)	K_{II} = 6.763 × 10^8 exp(−576,275/RT)	99
(III) MgO(s) + ½H$_2$O(g) → Mg(OH)(g) + ¼O$_2$	K_{III} = 8.550 × 10^6 exp(−491,954/RT)	99
(IV) MgO(s) + H$_2$O(g) → Mg(OH)$_2$(g)	K_{IV} = 8.072 × 10^2 exp(−277,884/RT)	14

listed in Table VI can be used to account for the different effects of the type of atmosphere on the volatility of MgO. The corresponding equilibrium constants are represented graphically in Fig. 5.

The values of K_I and K_{IV} are from experimental studies[14] of the free energy changes, ΔG, of the corresponding reactions; K_{II} and K_{III} have been calculated from JANAF tables.[99] The values of the equilibrium constants in Fig. 5 show that the dissociative vaporization in reaction I is only important at very high tem-

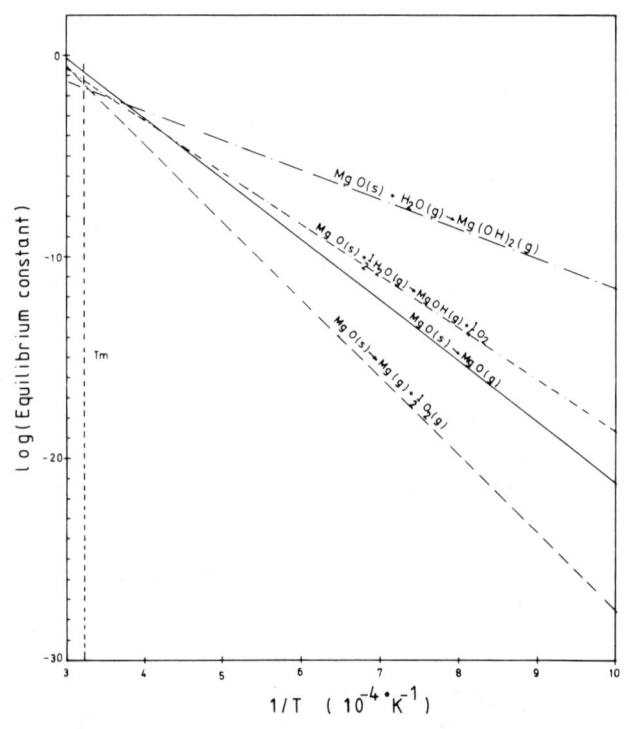

Fig. 5. Equilibrium constants of the main reactions of vaporization of MgO.

Table VII. Proportionality Constants, C_A and C'_A for Gas Diffusion

Gas	ϕ_A (Å)	m_A (kg·mol⁻¹)	$C_A \times 10^4$ (SI units)	$C'_A \times 10^9$ (SI units)
Mg	3.20*	0.02431	2.599	2.565
MgO	3.65*,†	0.04031	1.562	1.542
MgOH	4.05*,†	0.04132	1.379	1.361
Mg(OH)₂	5.50*,†	0.05833	0.7423	0.7326
O₂	3.64‡	0.03200	1.874	1.849
"Air"	3.80‡	0.02882	1.944	1.919

*Ref. 107.
†Ref. 99.
‡Ref. 106.

peratures or in extremely reducing atmospheres, while reaction IV is the dominant one in wet air.[14] Reaction IV may be important even at relatively low temperatures owing to the low value of the corresponding heat of reaction. The congruent vaporization reaction (reaction II) is the more probable process over a broad range of conditions, namely, from moderate vacuum up to atmospheric pressure or above, in neutral or oxidizing atmospheres, and at modest moisture content. Owing to the high value of the corresponding heat of reaction, however, the rates of vaporization by this process are low for the temperatures below 2000°C[103] at which most of the sintering and hot-pressing studies with MgO are done.

As referred to at the outset of this section, diffusion through the gas is the rate-limiting step in vapor transport at high total-gas pressures. The diffusion coefficient of a gas A: $D_{g,A}$ with partial pressure p_A, through another gas is[106]

$$D_{g,A} = \frac{(2R/\pi)^{3/2}(m_A m_B/(m_A + m_B))^{1/2}}{3N_a(1 - p_A/p_t)((\phi_A + \phi_B)/2)^2}\left(\frac{T^{3/2}}{p_t}\right) \quad (3)$$

where p_t is the total gas pressure, N_a is the Avogadro number, m_A and m_B are the molecular weights, and ϕ_A and ϕ_B are the diameters of the molecules of the gases. If $p_t \gg p_A$, $D_{g,A}$ is proportional to $T^{3/2}/p_t$. Values of the constant of proportionality, C_A, in $D_{g,A} = C_A T^{3/2}/p_t$ are given in Table VII for the diffusion in air of the vapors appearing in reactions I–IV. A second constant, $C'_A = C_A/p_t$, for the fixed pressure of $p_t = 1$ atm (1×10^5 Pa) is also given. The values for ϕ_i are taken from measurements of the diffusivities of the gases[106] or are calculated, with tabulated atomic radii,[107] from data on the shape of the molecules and their bond lengths,[99] defining the diameter ϕ_i as the maximum linear dimension of the molecule.

The important feature from Table VII is that all these gases and vapors have approximately the same diffusivity at a given temperature and total pressure, in spite of the differences in their molecular weights and cross sections. Therefore, it is the partial pressure of each one that is the key factor in defining the relative contribution of each vapor species to the overall process of vapor transport.

Acknowledgments

The support of this work by JNIC-INVOTAN (N.A.T.O.) Lisbon under award 1977–80 is gratefully acknowledged.

References

[1] H. J. Frost and M. F. Ashby, Deformation Mechanism Maps, Pergamon, Oxford, 1982.
[2] M. F. Ashby, "A First Report on Deformation-Mechanism Maps," Acta Metall., 20, 887–97 (1972).
[3] M. F. Ashby, "A First Report on Sintering Diagrams," Acta Metall., 22, 275–89 (1974).
[4] J. M. Vieira, "The Hot-Pressing and Fast-Firing of High Purity Magnesium Oxide"; Ph.D. Thesis, Department of Ceramics, The University of Leeds, Leeds, U.K., 1981.
[5] A. M. Brown and M. F. Ashby, "Correlations for Diffusion Constants," Acta Metall., 28, 1085–101 (1980).
[6] R. Freer, "Bibliography Self-Diffusion and Impurity Diffusion in Oxides," J. Mater. Sci., 15, 803–24 (1980).
[7] B. J. Wuensch, "On the Interpretation of Lattice Diffusion in Magnesium Oxide"; pp. 211–31 in Material Science Research, Vol. 9. Plenum Press, New York, 1975.
[8] S. Mrowec, "Defects and Diffusion in Solids an Introduction"; pp. 100–19, 223–9, 275–83, 434–5 in Materials Science Monographs, Vol. 5. Elsevier, Amsterdam, 1980.
[9] R. Lindner and G. D. Parfit, "Diffusion of Radioactive Magnesium in Magnesium Oxide Crystals," J. Chem. Phys., 26, 182–5 (1957).
[10] Y. Oishi and W. Kingery, "Oxygen Diffusion in Periclase Crystals," J. Chem. Phys., 33, 905–6 (1960).
[11] T. G. Langdon, "Deformation Mechanism Maps for Ceramics," J. Mater. Sci., 11, 317–27 (1976).
[12] T. G. Langdon, "Deformation Mechanism Maps for Applications at High Temperatures," Ceramurgia Int., 6 [1] 11–18 (1980).
[13] J. D. Hodge, P. A. Lessing, and R. S. Gordon, "Creep Mapping in a Polycrystalline Ceramic: Application to Magnesium Oxide and Magnesiowustite," J. Mater. Sci., 12, 1598–1604 (1977).
[14] T. Sata, T. Sasamoto, H. L. Lee, and E. Maeda, "Vaporization Processes from Magnesia Materials," Rev. Int. Hautes Temp. Refract., 15, 237–48 (1978).
[15] R. F. Brebrick, "Part 9. Atom Movements Diffusion"; pp. 591–607 in Solid State Chemistry and Physics, Vol. 2. Edited by P. F. Weller. Marcel Dekker, New York, 1974.
[16] T. Vasilos, J. B. Mitchell, and R. M. Spriggs, "Creep of Polycrystalline Magnesia," J. Am. Ceram. Soc., 47 [4] 203–4 (1964).
[17] E. M. Passmore, "Creep of Dense, Polycrystalline Magnesium Oxide," J. Am. Ceram. Soc., 49 [11] 594–600 (1966).
[18] H. Tagai and T. Zisner, "High Temperature Creep of Polycrystalline Magnesia: I, Effect of Simultaneous Grain Growth," J. Am. Ceram. Soc., 51 [6] 303–10 (1968).
[19] F. R. N. Nabarro, "Deformation of Crystals by Motion of Single Ions"; pp. 75–90 in the Report on the Conference of the Strength of Solids. University of Bristol, July 1947, 1948.
C. Herring, "Diffusional Viscosity of Polycrystalline Solid," J. Appl. Phys., 21 [5] 437–45 (1950).
[20] T. G. Langdon, "Grain Boundary Deformation Processes"; pp. 101–26 in Deformation of Ceramic Materials. Edited by R. C. Bradt and R. E. Tressler. Plenum Press, New York, 1975.
[21] W. E. Snowden and J. A. Pask, "High Temperature Deformation of Magnesium Oxide," Philos. Mag., 29, 441–5 (1974).
[22] J. Weertman, "Steady State Creep Through Dislocation Climb," J. Appl. Phys., 28, 362–4 (1957).
[23] A. H. Clauer, "High Temperature Tensile Creep of Magnesium Oxide Single Crystals," J. Am. Ceram. Soc., 59, 89–96 (1976).
[24] T. Vasilos and R. M. Spriggs, "Pressure Sintering: Mechanisms and Microstructures for Alumina and Magnesia," J. Am. Ceram. Soc., 46, 493–6 (1963).
[25] T. K. Gupta, "Sintering of MgO: Densification and Grain Growth," J. Mater. Sci., 6, 25–32 (1971).
[26] E. Yasuda, M. Ootsuda, S. Kimura, and H. Tagai, "Compression Creep of Polycrystalline Magnesia at High Temperature," Bull. Tokyo Inst. Techn., 108, 113–21 (1972).
[27] L. W. Barr and D. K. Dawson, "Correlations Between Melting Temperatures and Formation and Cation Motion Energies of Schottky Defects in Ionic Crystals," Proc. Br. Ceram. Soc., 19, 151–60 (1971).
[28] W. C. Mackrodt and R. F. Stwart, "Defect Properties of Ionic Solids: Point Defects at the Surfaces of Face-Centred Cubic Crystals," J. Phys. C: Sol. State Phys., 10 [9] 1431–46 (1977).
[29] C. R. A. Catlow, I. D. Faux, and M. J. Norgett, "Shell and Breathing Shell Model Calculations for Defect Formation Energies and Volumes in Magnesium Oxide," J. Phys. C: Sol. State Phys., 9, 419–29 (1976).
[30] W. H. Gourdin and W. D. Kingery, "Defect Structure of MgO Containing Trivalent Cation Solutes: Shell Model Calculations," J. Mater. Sci., 14, 2053–73 (1979).
W. H. Gourdin, W. D. Kingery, and J. Driear, "Defect Structure of MgO Containing Trivalent Cation Solutes: The Oxidation Reduction of Iron," J. Mater. Sci., 14, 2074–82 (1979).
[31] E. A. Colburn and W. C. Mackrodt, "The Calculated Defect Structure of Bulk and (001) Surface Cation Dopants in MgO," J. Mater. Sci., 17, 3021–38 (1982).

[32] J. Dequenne, "Condutibilite Electrique et Frittage de l'Oxide de Magnesium," *Rev. Int. Hautes Temp. Refract.*, **10** [3] 141–53 (1973).

[33] B. C. Harding and D. M. Price, "Cation Self-Diffusion in MgO up to 2350°C," *Philos. Mag.*, **26**, 253–60 (1972).

[34] R. C. Weast; pp. B-134 in the CRC Handbook of Chemistry and Physics, 59th edition. CRC Press, Florida, 1978/1979.

[35] A. F. Henriksen and W. D. Kingery, "The Solid Solubility of Sc_2O_3, Al_2O_3, Cr_2O_3, SiO_2, and ZrO_2 in MgO," *Ceramurgia Int.*, **5**, 11–7 (1979).

[36] J. R. H. Black and W. D. Kingery, "Segregation of Aliovalent Solutes to Adjacent Surfaces in MgO," *J. Am. Ceram. Soc.*, **62**, 176–8 (1979).

[37] T. Mitamura, E. L. Hall, W. D. Kingery, and J. B. Vander Sande, "Grain Boundary Segregation of Iron in Polycrystalline Magnesium Oxide Observed by STEM," *Ceramurgia Int.*, **5** [4] 131–6 (1979).

[38] R. C. McCune and P. Wynblatt, "Calcium Segregation to a Magnesium Oxide (100) Surface," *J. Am. Ceram. Soc.*, **66** [2] 111–7 (1983).

[39] J. W. Osenbach and V. S. Stubican, "Grain-Boundary Diffusion of ^{51}Cr in MgO and Cr-Doped MgO," *J. Am. Ceram. Soc.*, **66** [3] 191–5 (1983).

[40] T. A. Yager and W. D. Kingery, "Kinetic Reactions in Quenched MgO:Fe^{3+} Crystals," *J. Am. Ceram. Soc.*, **65** [1] 12–5 (1982).

[41] J. W. Hollenberg, "Origin of Anomalously High Activation Energies in Sintering of Impure Refractory Oxides," *J. Am. Ceram. Soc.*, **56**, 109–10 (1973).

[42] T. Dosdale and R. J. Brook, "Cationic Conduction and Diffusion and the Compensation Law," *J. Mater. Sci.*, **13**, 167–72 (1978).

[43] T. Dosdale and R. J. Brook, "The Comparison of Diffusion Data and of Activation Energies," *J. Am. Ceram. Soc.*, **66** [6] 392–95 (1983).

[44] J. F. Quirck, "Factors Affecting Sinterability of Oxide Powders: BeO and MgO," *J. Am. Ceram. Soc.*, **42** [4] 178–81 (1959).

[45] F. R. Wermuth and W. J. Knapp, "Initial Sintering of MgO and LiF-Doped MgO," *J. Am. Ceram. Soc.*, **56**, 401 (1973).

[46] B. Wong and J. A. Pask, "Models for Kinetics of Solid State Sintering" and "Experimental Analysis of Sintering of MgO Compacts," *J. Am. Ceram. Soc.*, **62**, 138–46 (1979).

[47] P. F. Eastman, "Effect of Water Vapour on Initial Sintering of Magnesia," *J. Am. Ceram. Soc.*, **49** [10] 526–30 (1966).

[48] R. L. Coble and R. M. Cannon, "Current Paradigms in Powder Processing"; pp. 151–70 in Material Science Research, Vol. 11. Plenum Press, New York, 1978.

[49] J. A. Varela and O. J. Whittemore, "Structural Rearrangement During the Sintering of MgO," *J. Am. Ceram. Soc.*, **66** [1] 77–82 (1983).

[50] H. Hashimoto, "Preferential Diffusion of Oxygen Along Grain Boundaries in Polycrystalline MgO," *J. Appl. Phys.*, **43** [11] 4828–9 (1972).

[51] S. Shirasaki and M. Hama, "Oxygen Diffusion Characteristics of Loosely Sintered Polycrystalline MgO," *Chem. Phys. Lett.*, **20** [4] 361–5 (1973).

[52] J. B. Holt and R. H. Condit, "Oxygen-18 Diffusion in Surface Defects on MgO as Revealed by Proton Activation"; pp. 13–29 in Material Science Research, Vol. 3. Plenum Press, New York, 1966.

[53] J. F. Laurent and J. Benard, "Self-Diffusion of Ions in Polycrystalline Alkali Halides," *J. Phys. Chem. Solids*, **7**, 218–27 (1958).

[54] D. R. McKenzie, A. W. Searcy, J. B. Holt, and R. H. Condit, "Oxygen Grain-Boundary Diffusion in MgO," *J. Am. Ceram. Soc.*, **54**, 188–90 (1971).

[55] R. T. Tremper, "Creep of Polycrystalline MgO-FeO-Fe_2O_3 Solid Solution," *J. Am. Ceram. Soc.*, **57** [10] 421–8 (1974).

[56] G. R. Terwilliger, H. K. Bowen, and R. S. Gordon, "Creep of Polycrystalline MgO and MgO-Fe_2O_3 Solid Solutions at High Temperatures," *J. Am. Ceram. Soc.*, **53**, 241–51 (1970).

[57] R. L. Coble, "A Model for Boundary Diffusion Controlled Creep in Polycrystalline Materials," *J. Appl. Phys.*, **34**, 1679–82 (1963).

[58] R. S. Gordon and G. R. Terwilliger, "Transient Creep in Fe-Doped Polycrystalline MgO," *J. Am. Ceram. Soc.*, **55**, 450–5 (1972).

[59] R. S. Gordon, "Mass Transport in the Diffusional Creep of Ionic Solids," *J. Am. Ceram. Soc.*, **56**, 147–52 (1973).

[60] T. G. Langdon, "Grain Boundary Sliding as a Deformation Mechanism During Creep," *Philos. Mag.*, **22**, 689–700 (1970).

[61] J. Narayan and J. Washburn, "Self-Climb of Dislocation Loops in Magnesium Oxide," *Philos. Mag.*, **26**, 1179–90 (1972).

[62] M. F. Yan, R. M. Cannon, H. K. Bowen, and R. L. Coble, "Space Charge Contribution to Grain-Boundary Diffusion," *J. Am. Ceram. Soc.*, **60** [3–4] 120–7 (1977).

[63] W. D. Kingery, "Plausible Concepts Necessary and Sufficient for Interpretation of Grain-Boundary Phenomena: I, Grain-Boundary Characteristics, Structure, and Electrostatic Potential. II, Solute Segregation, Grain-Boundary Diffusion, and General Discussion," *J. Am. Ceram. Soc.*, **57** [1] 1–8; **57**, [2] 74–83 (1974).

[64] D. W. James and G. M. Leak, "Grain Boundary Diffusion of Iron, Cobalt, and Nickel in

Alpha-Iron and of Iron in Gamma-Iron," *Philos. Mag.*, **12**, 491–503 (1965).

[65]Y. M. Chiang, A. F. Henriksen, and W. D. Kingery, "Characterization of Grain Boundary Segregation in MgO," *J. Am. Ceram. Soc.*, **64** [7] 385–9 (1981).

[66]L. W. Barr, I. M. Hoodless, J. A. Morrison, and R. Rudham, "Effects of Gross Imperfections on Chloride Ion Diffusion in Crystals of Sodium Chloride and Potassium Chloride," *Trans. Faraday Soc.*, **56** [449] 697–708 (1960).

[67]A. E. Paladino and R. L. Coble, "Effects of Grain-Boundaries on Diffusion Controlled Processes in Aluminum Oxide," *J. Am. Ceram. Soc.*, **46**, 133–6 (1963).

[68]I. Zaplatynsky, "Grain Boundary Diffusion in Oxides," *J. Appl. Phys.*, **35** [4] 1358 (1964).

[69]B. J. Wuensch and T. Vasilos, "Origin of Grain-Boundary Diffusion in MgO," *J. Am. Ceram. Soc.*, **49**, 433–6 (1966).

[70]J. Minkes, "Diffusion of Ni^{2+} in MgO," *J. Am. Ceram. Soc.*, **54**, 65–6 (1971).

[71]C. M. Osburn and R. W. West, "Electrical Properties of Single Crystals, Bicrystals and Polycrystals of MgO," *J. Am. Ceram. Soc.*, **54**, 428–35 (1971).

[72]N. A. Gjostein and W. L. Winterbottom, "The Structure and Properties of Metal Surfaces"; pp. 42–74 in Fundamentals of Gas-Surface Interactions. Edited by H. Satsburg. Academic Press, New York, 1967.

[73]J. Nowotny, I. Sikora, and J. B. Wagner, Jr., "Segregation and Near-Surface Diffusion for Undoped and Cr-Doped CoO," *J. Am. Ceram. Soc.*, **65** [4] 192–6 (1982).

[74]S. A. Lytle and V. S. Stubican, "Surface Diffusion in MgO and Cr-Doped MgO," *J. Am. Ceram. Soc.*, **65** [4] 210–2 (1982).

[75]M. P. Tosi, "Cohesion of Ionic Solids in the Born Model"; pp. 1–120 in Advanced Research in Applied Solid State Physics, Vol. 16. Edited by F. Seitz. Academic Press, New York, 1964.

[76]J. M. Blakely; Introduction to the Properties of Crystal Surfaces. Pergamon Press, Oxford, 1973.

[77]W. W. Mullins, "Solid Surfaces Morphologies Governed by Capilarity"; pp. 17–66 in Metals Surfaces. Edited by the American Society for Metals, Metals Park, OH, 1963.

[78]R. K. Stringer, C. E. Warble, and L. S. Williams, "Phenomenological Observations During Solid Reactions"; pp. 53–95 in Material Science Research, Vol. 4. Plenum Press, New York, 1969.

[79]H. D. Oel, "Crystal Growth in Ceramic Powders"; pp. 249–72 in Material Science Research, Vol. 4. Plenum Press, New York, 1969.

[80]E. Smethurst and D. W. Budworth, "The Preparation of Transparent Magnesia Bodies I. by Hot-Pressing. II. by Sintering," *Trans. Br. Ceram. Soc.*, **71**, 45–53 (1972).

[81]W. M. Rhodes, "Relation Between Precursor and Microstructure in MgO," *J. Am. Ceram. Soc.*, **56**, 495–6 (1973).

[82]A. F. Moodie, C. E. Warble, and L. S. Williams, "Magnesia Precursor and Sintering Study," *J. Am. Ceram. Soc.*, **49**, 676–7 (1966).

[83]C. G. Kinniburg, "A LEED Study of MgO (100): II. Theory at Normal Incidence," *J. Phys. C: Sol. State Phys.*, **8**, 2382–94 (1975).

[84]W. D. Kingery, H. K. Bowen, and D. R. Uhlman; pp. 210 in the Introduction to Ceramics, 2nd edition. Wiley, New York, 1976.

[85]J. M. Blakely and J. C. Shelton, "Part 4. Equilibrium Adsorption and Segregation"; pp. 189–239 in Surface Physics of Materials, Vol. I. Edited by J. M. Blakely. Material Science Series, Academic Press, New York, 1975.

[86]R. W. Davidge; pp. 75–103 in Mechanical Behavior of Ceramics. Cambridge University Press, Cambridge, 1979.

[87]P. G. Shewmon and W. M. Robertson, "Variation of Surface Tension with Orientation"; pp. 67–98 in Metals Surfaces. Edited by the American Society for Metals, 1963.

[88]B. Mutaftschiev, "Adsorption and Crystal Growth"; pp. 73–86 in Chemistry and Physics of Solid Surfaces, Vol. 1. Edited by R. Vanselow. CRC Press, Inc., Florida, 1977.

[89]N. A. Gjostein, "Surface Self-Diffusion"; pp. 99–154 in Metals Surfaces. Edited by the American Society for Metals, 1963.

[90]W. M. Robertson, "Kinetics of Grain Boundary Grooving in Magnesium Oxide"; pp. 215–32 in Sintering and Related Phenomena. Edited by G. C. Kuczynski. Gordon and Breach, New York, 1967.

[91]J. Henney and J. W. S. Jones, "Surface Diffusion on UO_2 and MgO," *J. Mater. Sci.*, **3**, 158–64 (1968).

[92]P. V. McAllister, "Thermal Grooving of MgO and Al_2O_3," *J. Am. Ceram. Soc.*, **55**, 351–4 (1972).

[93]W. M. Robertson, "Surface Diffusion in Oxides," *J. Nucl. Mater.*, **30**, 36–49 (1969).

[94]P. V. McAllister and I. B. Cutler, "Evaluation of Thermal Grooving Data," *J. Am. Ceram. Soc.*, **52** [6] 348–9 (1969).

[95]H. P. Bonzel, "Part 6. Transport of Matter at Surfaces"; pp. 279–338 in Surface Physics of Materials, Vol. II. Edited by J. M. Blakely. Material Science Series, Academic Press, New York, 1975.

[96]P. Kofstad; pp. 121–7 in Nonstoichiometry, Electrical Conductivity, and Diffusion in Binary Metal Oxides. Wiley, New York, 1972.

[97] J. M. Dynys, R. L. Coble, and W. S. Coblenz, "Mechanisms of Atom Transport During Initial Stage Sintering of Al$_2$O$_3$"; pp. 391–404 in Material Science Research, Vol. 13. Plenum Press, New York, 1980.

[98] J. P. Hirth and G. M. Pound, "Evaporation of Metal Crystals," J. Chem. Phys., **26**, 1216–24 (1957).

[99] D. R. Stull; JANAF Thermochemical Tables, 2nd ed. Dow Chemical Co., No. NSRDS-NBS 37, (1971).

[100] G. A. Somorjai, "Surface Chemistry"; pp. 668–720 in Solid State Chemistry and Physics. Edited by P. F. Weller. Marcel Dekker, New York, 1974.

[101] Tomoyasy Ito, "The Initial Sintering of Alkaline Earth Oxides in Water Vapour and Hydrogen Gas"; Chem. Abstr., **95**, 124 191u (1981).

[102] G. C. Benson and T. A. Claxton, "Application of Shell Model to the Calculation of the Surface Distortion in Alkali Halide Crystals," J. Chem. Phys., **48** [3] 1356–60 (1968).

[103] E. Ryshkewitch; pp. 275–317 in Oxide Ceramics. Academic Press, New York, 1960.

[104] A. W. Searcy; pp. 27, 33–43 in Chemical and Mechanical Behaviour of Inorganic Materials. Wiley, New York, 1970.

[105] M. S. Chandrasekharaiah, "Volatilities of Refractory Inorganic Compounds"; pp. 498–507 in The Characterization of High Temperature Vapors. Edited by J. L. Margrave. Wiley, New York, 1967.

[106] S. Glasstone; pp. 274–83 in the Textbook of Physical Chemistry, 2nd ed. McMillan and Co., London, 1968.

[107] R. M. Tennent; Science Data Book. Oliver and Boyd, Edinburgh, 1976.

[108] B. C. Harding, D. M. Price, and A. J. Mortlock, "Cation Self-Diffusion in Single Crystal MgO," Philos. Mag., **23**, 399–408 (1971).

[109] B. J. Wuensch, "Cation Self-Diffusion in Single Crystal MgO," J. Chem. Phys., **58** [12] 5258–66 (1973).

[110] S. P. Mittof, "Electrical Conductivity of Single Crystals of MgO," J. Chem. Phys., **31** [5] 1261–9 (1959).

[111] S. P. Mittof, "Comments on Doctor Schmalzreid's Letter," J. Chem. Phys., **33**, 941 (1960).

[112] M. O. Davies, "Transport Phenomena in Pure and Doped MgO," J. Chem. Phys., **38** [9] 2047–55 (1963).

[113] W. Weigelt and G. Haase, "Der Elektrische Widerstand von Hochvakuum-gesintertem Magnesium Oxyd," Ber. Dtsch. Keram. Ges., **31**, 45–54 (1954).

[114] D. R. Sempolinski and W. D. Kingery, "Ionic Conductivity and Magnesium Vacancy Mobility in Magnesium Oxide," J. Am. Ceram. Soc., **63** [11–12] 664–9 (1980).

[115] A. G. Evans and T. G. Langdon, "Structural Ceramics," Prog. Mater. Sci., **21** [2] 171–441 (1976).

[116] W. Beere, "Diffusional Flow and Hot-Pressing: a Study on MgO," J. Mater. Sci., **10**, 1434–40 (1975).

[117] L. H. Rovner, "Diffusion of Oxygen in Magnesium Oxide"; Ph.D. Thesis, Department of Physics, Cornell University, 1966.

[118] J. Narayan and J. Washburn, "Self-Diffusion in Magnesium Oxide," Acta Metall., **21**, 533–8 (1973).

[119] J. H. Hensler and G. V. Cullen, "Stress Temperature and Strain Rate in Creep of Magnesium Oxide," J. Am. Ceram. Soc., **51** [10] 557–9 (1968).

[120] P. J. Dixon-Stubbs and B. Wilshire, "Part 5. Factors Affecting the Deformation Behaviour During Creep of Magnesium Refractory Raw Materials," Trans. J. Br. Ceram. Soc., **79** [1] 21–8 (1980).

[121] H. Conrad, "Mechanical Behaviour of Ceramic Materials"; pp. 378 in Chemical and Mechanical Behaviour of Inorganic Materials. Edited by A. W. Searcy. Wiley, New York, 1970.

---

*Presently with the Department of Ceramics and Glass, University of Aveiro, 3800 Aveiro, Portugal.

Oxygen Surface Diffusion and Surface Layer Thickness in MgO

YASURO IKUMA and WAZO KOMATSU

Ikutoku Technical University
Kanagawa 243-02, Japan

An oxygen isotope exchange reaction between $^{18}O_2$ gas and MgO powder was monitored with a microbalance. The weight change of the specimen was plotted as a function of the square root of time, which resulted in a curve having two straight lines with a transition region between them. The first steep line was due to the oxygen exchange reaction within the surface layer, and the less steep line was probably due to the oxygen lattice diffusion. From the first steep lines at various temperatures (600–900°C), the oxygen surface diffusion coefficients were calculated as $D_{s,O} = 9.14 \times 10^{-19} \exp(-107(kJ/mol)/RT)$ m^2/s. The surface diffusion measured by the current method was diffusion through a surface layer rather than diffusion along a surface. From the weight gain at the intersection of the two lines, the surface layer thickness of MgO has been estimated. The surface layer thickness was 0.3–0.5 nm over the temperature range of 600–900°C and had a small positive temperature dependence (8.8 kJ/mol). The method could be used to measure the oxygen surface diffusion coefficient and the surface layer thickness of any other oxides in powder form.

It is well-known that the surface diffusion coefficient and surface layer thickness* are some of the important parameters in understanding the kinetic processes of powder, such as sintering and solid-state reactions. In the past, several methods have been used to determine these parameters related to the surface. The product of the surface diffusion coefficient and surface layer thickness appears in the kinetic equation of combined sintering and in the solution of the differential equation of diffusion involving the surface diffusion. Therefore, the product can be obtained by studying the sintering[1] or by the well-prepared diffusion experiment.[2] The limitation of these methods is that only the product of the quantities can be determined and the estimation of the individual quantity is not possible. Thermal grooving and scratch smoothing methods, which are based on the theory developed by Mullins and others,[3-6] are other ways to determine the prarameters related to the surface. In these methods, only the surface diffusion coefficient of single crystals and sintered bodies can be measured and no surface diffusion coefficient for powder can be determined. The estimation of surface layer thickness is not possible in these methods.

In our laboratory, the surface layer thickness of powder has been measured by making use of the solid-state reactions between two kinds of powder and the reaction between solids and gases (kinetic method).[7-9] The disadvantage of the method is that there are not always any appropriate solid reactions to be utilized for the measurement of surface layer thickness of all the oxides. Another important fact concerning the kinetic method is that the surface layer thickness is measured at a very special moment when the products are just formed. The product at this moment has generally a disturbed structure[10-12] where the diffusion is enhanced. Consequently, the surface layer thickness measured by this method might be overestimated. Recently, the authors[13,14] have developed a new method in which the surface diffusion coefficient and the surface layer thickness can be measured concurrently and independently (isotope exchange method). In this method, we follow the oxygen isotope exchange reaction between $^{18}O_2$ gas and oxide powder by measuring the weight of the specimen. From the measurement of weight gain, both the surface diffusion coefficient and the surface layer thickness can be evaluated independently. Since we can record the diffusion process continuously with a microbalance, we are able to gain important information that is sometimes not detected in the conventional diffusion experiment. The method was successfully applied to NiO and Y_2O_3 powders.[14,15] In the present study, the surface diffusion coefficient and surface layer thickness of MgO powder were obtained by an isotope exchange method.

Experimental Procedures

Nominally pure MgO[†] was annealed in a platinum crucible at the temperature of 1150°C and the oxygen pressure of 6.7×10^3 Pa for 3 h. The surface area (18.35 m^2/g) of the specimen was determined by the BET method. Assuming that the specimen consists of spherical powder, we calculated the average particle size of 46 nm from the BET surface area. The specimen of ~200 mg was measured precisely and placed in a platinum bucket (pure Pt or Pt + 20% Rh). First the specimen was annealed in an $^{16}O_2$ atmosphere (pressure, 6.7×10^3 Pa) over the temperature range of 600–900°C for 5–8 h. In the last 3–6 h of annealing, no weight change of the specimen was observed. The specimen was quenched to room temperature, and the $^{16}O_2$ gas was replaced by the $^{18}O_2$ gas. The specimen was then reheated to the temperature where the previous annealing was performed. An exchange reaction during the diffusion annealing can be expressed by Eq. (1), and $^{18}O_2$ diffuses into the specimen.

$$Mg^{16}O(s) + \tfrac{1}{2}\,^{18}O_2(g) \rightarrow Mg^{18}O(s) + \tfrac{1}{2}\,^{16}O_2(g) \tag{1}$$

In order to follow the diffusion process, a weight increase due to the exchange reaction was monitored continuously with a microbalance.[‡] The specimen was heated with an infrared image furnace,[§] and the heating rate was 50°C/min. The reaction chamber made of quartz glass is shown schematically in Fig. 1. To avoid the exchange reaction between $^{18}O_2$ gas and quartz glass, the glass was cooled with water.

Results and Discussion

Typical results (weight gain, Δw) are plotted in Fig. 2 as a function of $t^{1/2}$. At all the temperatures, the curve resulted in two straight lines with a transition region between them. If we take an imaginary spherical material in which the diffusion coefficient is D_s (surface diffusion coefficient) from $r = 0$ to $r = a$ where a is the radius of the sphere, the Δw–$t^{1/2}$ curve will be line I shown in Fig. 3. However, in the real specimen, the diffusion coefficient is a function of r. To demonstrate the Δw–$t^{1/2}$ curve for the real material, we assume that the diffusion coefficient is D_l (lattice diffusion coefficient) at $0 \leq r \leq a - \delta$ and is D_s at $a - \delta < r \leq a$, where δ is the surface layer thickness ($a \gg \delta$ and $D_s \gg D_l$). Then at the beginning of the diffusion annealing, the diffusion distance is very short and Δw–$t^{1/2}$ curve will follow line I (Fig. 3). As the diffusion distance becomes longer than δ, the Δw–$t^{1/2}$ curve will gradually deviate from line I (line

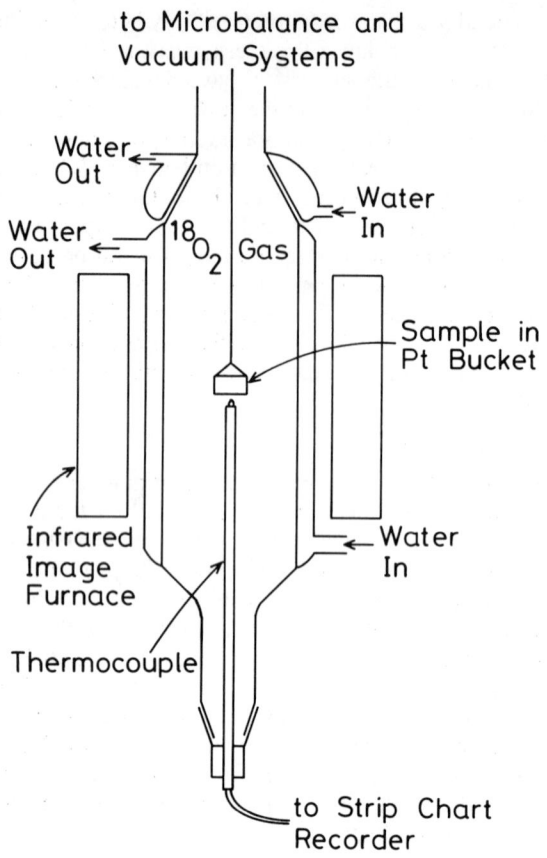

Fig. 1. Schematic of exchange reaction chamber.

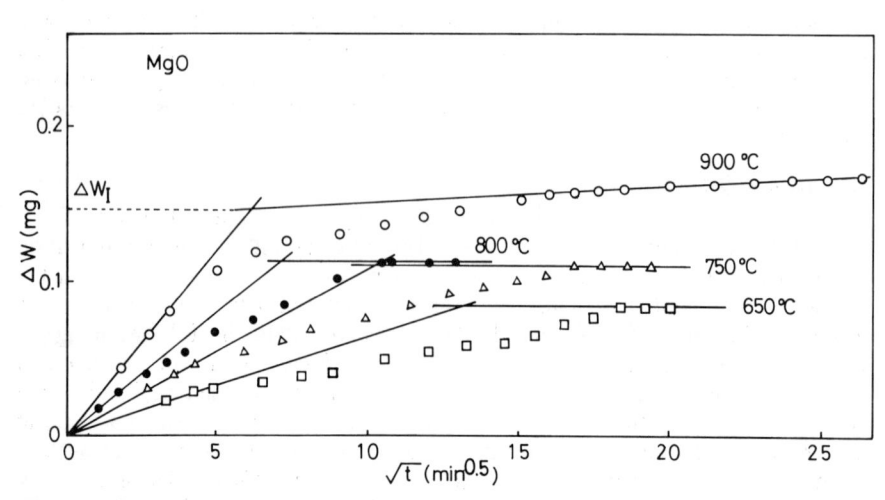

Fig. 2. Some typical weight gains of MgO during the diffusional annealing plotted as a function of $t^{1/2}$.

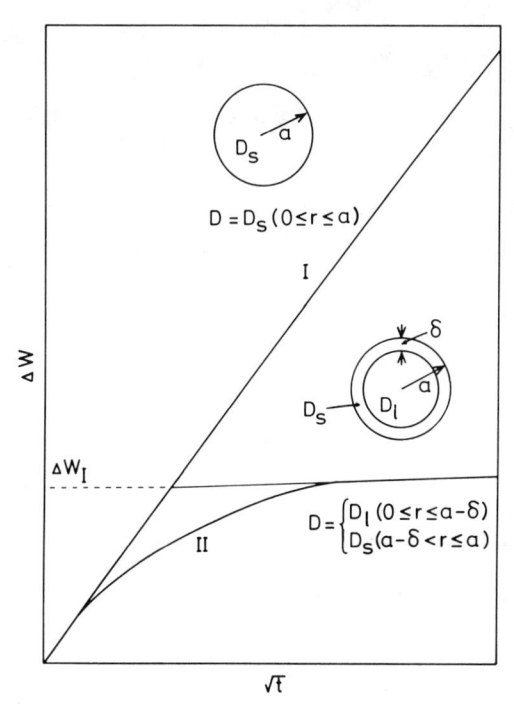

Fig. 3. Model curves of the weight gain (Δw) vs $t^{1/2}$ for a particle with two different diffusion coefficients.

II in Fig. 3). This is the transition region. When the exchange reaction within the surface layer has been completed, the $\Delta w - t^{1/2}$ curve will be approximately linear again (plateau or the less steep line). In this region, the lattice diffusion is the only diffusion that gives rise to the weight change of the specimen. We can calculate the surface diffusion coefficient, therefore, from the first steep line of the experiment by using Eq. (2):[16]

$$\frac{M_t}{M_\infty} = 1 - \sum_{n=1}^{\infty} \left(\frac{6\alpha(\alpha + 1) \exp(-D_s q_n^2 t/a^2)}{9 + 9\alpha + q_n^2 \alpha^2} \right) \quad (2)$$

where M_t is the total amount of $^{18}O_2$ in the specimen at time t, M_∞ is the corresponding quantity after infinite time, the q_n's are the nonzero roots of $\tan q_n = 3q_n/(3 + \alpha q_n^2)$ and α is an experimental constant determined by the initial atomic ratio of ^{18}O in the gas phase to ^{18}O in the solid. In the present experiments α was about 1.2.

The oxygen surface diffusion coefficients calculated in this way were plotted in Fig. 4 along with the data (D_s and D_l) obtained by other workers.[1,18-22] The current result was expressed as

$$D_{s,0} = 9.14 \times 10^{-19} \exp(-107(\text{kJ/mol})/RT) \quad (\text{m}^2/\text{s}) \quad (3)$$

The activation energy for the oxygen surface diffusion is in good agreement with the energy (87 kJ/mol) for anion vacancy migration at the {001} surface calculated by Colbourn and Mackrodt.[17] Robertson[18] has measured the D_s of MgO bicrystals by thermal grooving. Moriyoshi and Komatsu[1] have estimated the surface diffusion

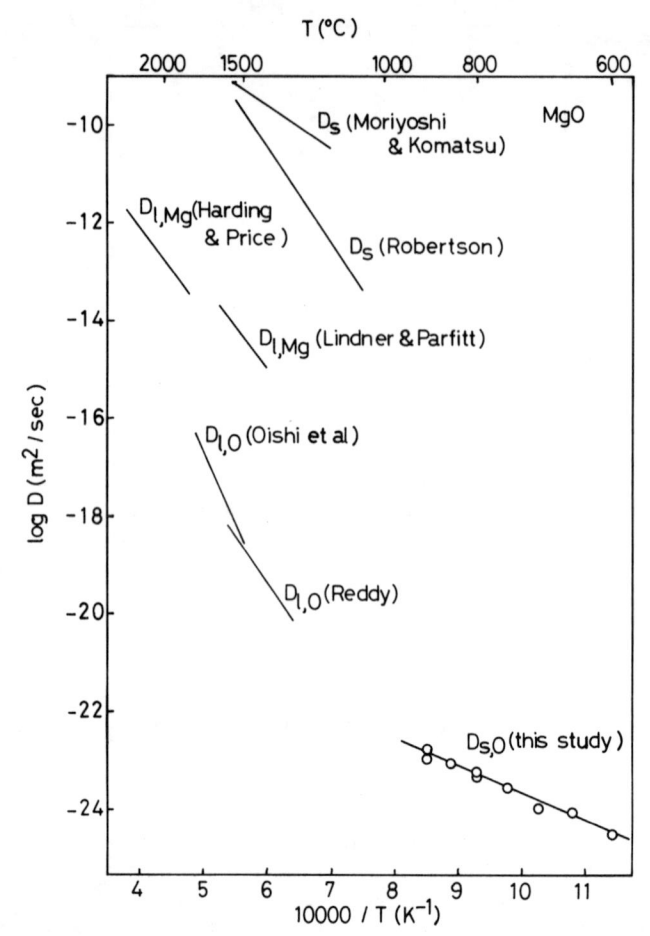

Fig. 4. Surface and lattice diffusion coefficients in MgO.

coefficient of MgO powder from the study of combined sintering. In these results, the diffusing species (cation or anion) that controls the rate processes is not known.

The lattice diffusion coefficient of Mg in MgO ($D_{l,Mg}$) was studied by Harding and Price[19] and Lindner and Parfitt.[20] The lattice diffusion coefficient of oxygen in MgO ($D_{l,O}$) was investigated by Oishi et al.[21] and Reddy.[22] They are all shown in Fig. 4. It can be noticed that the $D_{s,O}$ of this study is higher than oxygen lattice diffusion coefficients of both Oishi et al. and Reddy over the temperature range investigated in this study. However, the oxygen surface diffusion coefficient at 1200°C is almost identical with $D_{l,O}$.

The activation energy of surface diffusion is about ¼ to ⅓ of the activation energy of lattice diffusion (370–536 kJ/mol). A similar trend for the activation energies is also observed in the metallic systems.[23]

The oxygen surface diffusion coefficient measured in this study is smaller than that obtained by sintering[1] and thermal grooving.[18] The explanation to this discrepancy is as follows: in the solid-state reaction between two oxides, for example, the spinel formation reaction, if the product–reactant boundary is in contact with air, the transport of oxygen is through the gas phase. Consequently, the diffusion of cation in the product layer is rate limiting.[24] A similar mechanism is also operable for the transport of oxygen in thermal grooving and sintering controlled by surface diffusion. Since the experiments of both Robertson and Moriyoshi and Komatsu were performed in air, the oxygen could diffuse through the gas phase. Then it is more probable that the cation surface diffusion coefficient has been determined in their studies. The difference between the cation surface diffusion coefficient obtained from these methods and the anion surface diffusion coefficient measured in this study agrees qualitatively with the relative difference between cation lattice and anion lattice diffusion coefficients; i.e., in both cases the diffusion of the cation is faster than the anion. Another important difference in the results of these methods is that the diffusion processes that were observed by Robertson and Moriyoshi and Komatsu are probably the diffusion along the surface, whereas the diffusion in this study is, by the nature of investigation, diffusion perpendicular to the surface. It is generally expected that the diffusion coefficient in the surface layer is a function of radius, r. As shown in Fig. 5, at $r = a$ (surface) the diffusion coefficient is very high. As r decreases, the diffusion

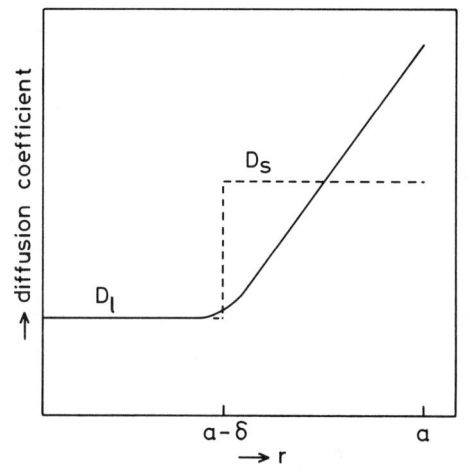

Fig. 5. Profile of the diffusion coefficient in a spherical model material that is close to the real material. In the present study, we assumed that the diffusion coefficient near the surface layer is represented by the dashed line.

coefficient decreases gradually to the value in the bulk, and finally at $r \leq a - \delta$, the diffusion coefficient is equal to the lattice diffusion coefficient. However, for mathematical simplicity, we assume that within the surface layer of the thickness δ, the property of material is homogeneous and that a diffusion coefficient D_s that is actually the average of varying diffusion coefficients is constant within the surface layer. Consequently, it is not surprising to find that the D_s obtained in this study is closer to the lattice diffusion coefficient than the value measured by other methods.

From the less steep line at 900°C in Fig. 2, a diffusion coefficient (presumably the lattice diffusion coefficient) can be calculated. It is 2.2×10^{-26} m^2/s, which is approximately on the line extended from the results of Oishi et al.[21] and Reddy[22] (Fig. 4).

If we assume that the exchange reaction in the surface layer has been completed at the intersection of two lines (Fig. 2), the surface layer thickness can be estimated by substituting the weight change,¶ Δw_I, at the intersection into the following equation:

$$\delta = a\left\{1 - \left(1 - \frac{\Delta w_\delta M_{MgO}}{2wf}\right)^{1/3}\right\} \tag{4}$$

where w is the weight of the specimen, M_{MgO} is the formula weight of MgO, and f is a constant determined by the experimental condition. In the present experiment, the value of f is approximately equal to unity. Equation (4) was derived under the assumption that a spherical specimen with a radius a has the surface layer thickness of δ.

The values of δ obtained from Eq. (4) are shown in Fig. 6. They can be represented by $\delta = 1.14 \exp(-8.8 \text{ (kJ/mol)}/RT)$ nm. The values of δ measured

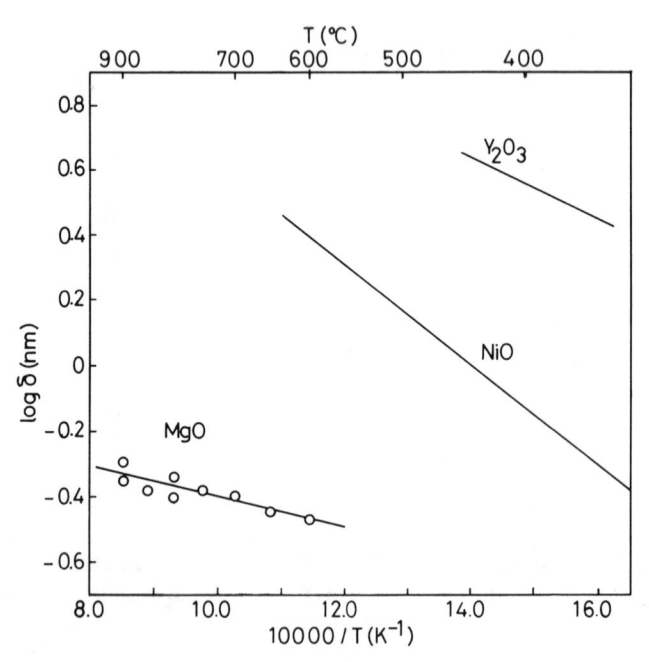

Fig. 6. Surface layer thicknesses of oxides obtained by the isotope exchange reaction.

in other oxide systems are also shown in the figure. It can be noticed that, among the surface layer thicknesses obtained by the isotope exchange reaction, δ of MgO is the smallest. Additional experiments are needed to find the cause of this difference. In Table I are shown the surface layer thicknesses obtained by both the kinetic method and isotope exchange reaction. Admittedly, there is not a sufficient amount of data. Especially the measurements of δ in one oxide using both the isotope exchange reaction and the kinetic method are needed. However, there is

Table I. Comparison of Surface Layer Thicknesses in Oxides

Oxides	Temp (°C)	Surface layer thickness (nm)	Temp dependence (kJ/mol)	Method of measurement
CoO	253–723	4.0–60	54	Kinetic*
Cu_2O	280–380	7.5–92	110	Kinetic[†]
TiO_2 (rutile)	883–965	4.5–20	130	Kinetic[‡]
α-Al_2O_3	900–1000	3.0–13	180	Kinetic[†]
NiO	300–600	0.3–2.7	29	Isotope exchange[§]
Y_2O_3	350–450	2.8–4.6	17	Isotope exchange[¶]
MgO	600–900	0.3–0.5	8.8	Isotope exchange

*Refs. 7 and 9.
[†]Ref. 25.
[‡]Ref. 8.
[§]Ref. 14.
[¶]Ref. 15.

a general trend that the surface layer thicknesses obtained by the kinetic method are much larger than those obtained by the isotope exchange reaction. The difference is partially due to the fact that in the kinetic method the surface layer thickness is measured during the formation of product layer, which has a disturbed structure.[10-12] Consequently, the kinetic method gives a thicker surface layer.

Since the surface layer thickness of MgO measured in this study is 0.3–0.5 nm (lattice parameter of MgO is ~0.4 nm), there might be some concern about the nature of the oxygen ions that have been exchanged with ^{18}O; i.e., most of the weight change could be caused by the oxygen that has been chemically absorbed on to the surface of MgO. To test this possibility, a specimen was first annealed at 650°C in an oxygen pressure of 1.3×10^2 Pa. After reaching an apparent equilibrium, we increased the oxygen pressure to 7.3×10^2 Pa and the weight gain due to the chemical absorption was measured. From the measurement, it was found that weight change of 0.01 mg during the oxygen isotope exchange reaction is caused by oxygen that is chemically absorbed on to the surface of MgO. This experiment suggests that ~4% of oxygen which had been exchanged with ^{18}O was chemisorbed oxygen.

The exchange reactions similar to that shown in Eq. (1) are applicable not only to oxide systems but also to nitride systems. Therefore, the method presented in this study is a general method and can be used to determine the surface layer thickness and the surface diffusion coefficient of almost all the oxides and nitrides, provided that there is no experimental or inherent limitation. One of the limitations is that the weight change other than exchange reaction should not take place during the diffusion annealing.

Conclusions

The oxygen surface diffusion coefficient and the surface layer thickness in MgO have been determined by recording the weight of the specimen during the oxygen isotope (^{18}O) exchange reaction. In the temperature range 600–900°C, the diffusion coefficient can be represented by

$$D_{s,O} = 9.14 \times 10^{-19} \exp(-107 \text{ (kJ/mol)}/RT) \quad (m^2/s)$$

and the surface layer thickness by

$$\delta = 1.14 \exp(-8.8 \text{ (kJ/mol)}/RT) \quad (nm)$$

The oxygen surface diffusion measured by the current method is diffusion through a surface layer rather than diffusion along a surface. This can be one of the reasons why the surface diffusion coefficients obtained by other methods (thermal grooving and sintering) are higher than the current results. Another possible reason was also discussed.

In general the surface layer thicknesses obtained by the oxygen isotope exchange reaction are smaller than those determined by the kinetic method.

The method could be used in principle to determine both the surface diffusion coefficients and the surface layer thicknesses of any oxides and nitrides.

Acknowledgments

The authors thank Y. Hirabayashi for obtaining some of the data. This study was supported in part by the Science Foundation of the Japanese Ministry of Education.

References

[1] Y. Morioshi and W. Komatsu, "Analysis of Initial Combined Sintering," *Yogyo-Kyokai-Shi*, **81** [1] 102–7 (1973).

[2] K. Hirota and W. Komatsu, "Concurrent Measurement of Volume, Grain-Boundary, and Surface Diffusion Coefficients in the System NiO-Al$_2$O$_3$," *J. Am. Ceram. Soc.*, **60** [3–4] 105–7 (1977).

[3] W. W. Mullins, "Theory of Thermal Grooving," *J. Appl. Phys.*, **28** [3] 333–9 (1957).

[4] W. W Mullins, "Grain Boundary Grooving by Volume Diffusion," *Trans. Metal. AIME*, **218**, 354–61 (1960).

[5] R. T. King and W. W. Mullins, "Theory of the Decay of a Surface Scratch to Flatness," *Acta Metall.* **10**, 601–6 (1962).

[6] F. A. Nichols and W. W. Mullins, "Morphological Changes of a Surface of Revolution due to Capillarity-Induced Surface Diffusion, *J. Appl. Phys.*, **36** [6] 1826–35 (1965).

[7] T. Maruyama, K. Takada, and W. Komatsu, "Measurement of Surface Layer Thickness by a Kinetic Method," *Z. Phys. Chem. Neue Folge*, **102**, 221–30 (1976).

[8] M. Yamashita, T. Maruyama, and W. Komatsu, "Surface Layer Thickness of Titanium Dioxide by a Kinetic Method," *Z. Phys. Chem. Neue Folge,* **105**, 187–96 (1977).

[9] W. Komatsu, Y. Chida, and T. Maruyama, "Direct Observations and Effect of Grinding on the Surface Layer Thickness of CoO"; pp. 430–35 in Reactivity of Solids. Edited by K. Dyrek, J. Haber, and J. Nowotny. Elsevier, Amsterdam, 1982.

[10] W. Jander and K. F. Weitendorf, "Processes of Solid State Reaction," *Z. Elektrochem. Angew. Phys. Chem.*, **47** [7] 435–44 (1935).

[11] G. F. Hüttig, "Active State during Chemical Reactions of Two Metallic Oxides," *Z. Elektrochem. Angew. Phys. Chem.*, **41** [7] 527–38 (1935).

[12] (a) C. Kröger, and G. Ziegler, "On the Rate of Glass Forming Reactions, II," *Glastech. Ber.*, **26** [11] 346–53 (1953).
(b) C. Kröger and G. Zeigler, "On the Rate of Glass Forming Reactions, III," *Glastech. Ber.*, **27** [6] 199–212 (1954).

[13] W. Komatsu, Y. Ikuma, K. Uematsu, Y. Kawashima, S. Yamaguchi, and H. Takamura, "Concurrent Measurement of Surface Layer Thickness and Surface Diffusion Coefficient of Oxides by ^{18}O-Exchange Reaction"; pp. 669–704 in Ceramic Powders. Edited by P. Vincenzini. Elsevier, Amsterdam, 1983.

[14] W. Komatsu and Y. Ikuma, "Concurrent Measurement of Surface Layer Thickness and Surface Diffusion Coefficient of Oxides by ^{18}O-Exchange Reaction," *Z. Phys. Chem. Neue Folge,* **131**, 79–88 (1982).

[15] W. Komatsu and Y. Ikuma; unpublished work.

[16] J. Crank, p. 88 in The Mathematics of Diffusion. Oxford University Press, London, 1956.

[17] E. A. Colbourn and W. C. Mackrodt, "Point Defects and Chemisorption at the {001} Surface of MgO"; these proceedings.

[18] W. M. Robertson, "Kinetics of Grain Boundary Grooving on Magnesium Oxide"; pp. 215–32 in Sintering and Related Phenomena. Edited by G. C. Kuczynski, N. A. Hooton, and C. F. Gibbon. Gordon and Breach, New York, 1967.

[19] B. C. Harding and D. M. Price, "Cation Self-Diffusion in MgO up to 2350°C," *Philos. Mag.*, **26**, 253–60 (1972).

[20] R. Lindner and G. D. Parfitt, "Diffusion of Radioactive Magnesium in Magnesium Oxide Crystals," *J. Chem. Phys.*, **26** [1] 182–5 (1957).

[21] Y. Oishi, K. Ando, H. Kurokawa, and Y. Hiro, "Oxygen Self-Diffusion in MgO Single Crystals," *J. Am. Ceram. Soc.*, **66** [4] C60–62 (1983).

[22] K. P. R. Reddy, "Oxygen Diffusion in Close Packed Oxides"; Ph.D. Dissertation, Case Western Reserve University, 1979.
[23] J. H. Brophy, R. M. Rose, and J. Wulff; "Thermodynamics of Structure." Wiley, New York, 1964.
[24] H. Schmalzried; p. 105 in Solid State Reactions. Verlag Chemie, Weinheim, Germany, 1981.
[25] T. Maruyama and W. Komatsu; unpublished work.

---

*We define the thickness of surface layer to be the width of region where the diffusion is enhanced due to a disturbed structure.

†After being annealed at 1150°C for 3 h, the impurity content was measured by emission spectroscopy. The major impurities were Si (~200 ppm), Fe (~60 ppm), Mn (10–60 ppm), and P (10–60 ppm). Rare Metallic Co.,

‡Shimadzu RMB-50V, Shimadzu Seisakusho Ltd., Kyoto, Japan.

§Shinkuriko RHL-E45.

¶Actually, the exchange reaction is taking place also before the specimen was heated to the experimental temperature. The weight gain during this period is added to Δw_i. Therefore, the weight gain shown in Fig. 2 is part of the weight gain (Δw_δ) that used to calculate δ.

Oxygen Diffusion in Undoped and Impurity-Doped Polycrystalline MgO

S. Shirasaki, S. Matsuda, H. Yamamura, and H. Haneda

National Institute for Research in Inorganic Materials
Sakura-Mura, Niihari-Gun, Ibaraki, Japan

Oxygen self-diffusion coefficients in polycrystalline MgO both undoped and doped with small amounts of Si and Ca were measured at an ambient oxygen pressure of about 40 torr (5.33 × 10³ Pa) over the temperature range 1000–1570°C by a solid–gas exchange technique using ^{18}O as a tracer. The volume diffusion level in the extrinsic region of Si-doped materials increased considerably with increasing sintering temperature of the materials. This fact is interpreted in terms of the metastable occurrence of oxygen vacancies formed as a result of thermal decomposition of the oxygen sublattice during sintering. Such a metastability may be caused by trapping electrons around Si''_{Mg} centers, in which difficulty in the elimination through oxidation of the thermally formed oxygen vacancies arises on cooling the Si-doped materials. No sintering temperature dependence of the extrinsic diffusion level was found in Ca-doped materials because of lack of such trapping centers. The "intrinsic" region was always found in impurity-doped and undoped specimens, and both the diffusion level and activation energy were strongly influenced by small amounts of impurities. In this view, even this region can also be regarded as structure sensitive.

Magnesium oxide has generally been regarded as a model material, which could serve as a paradigm for other more complex materials. Our present understanding of the defect structure and diffusion characteristics of MgO and metal oxides in general is not necessarily satisfactory. Diffusion has intrinsic and extrinsic regions in general. It is considered that the extrinsic region is directly controlled by small amounts of impurities whereas the intrinsic region is not. The purpose of the present study is to show the essential character of intrinsic and extrinsic oxygen diffusion in MgO. To achieve this important purpose, a number of oxygen diffusion coefficients of extremely pure, undoped, and Si- and Ca-doped polycrystalline MgO were measured.

Previous data on volume diffusion coefficients in many oxides were restricted down to values greater than 10^{-14} cm²/s.[1] Because of this limitation, one often cannot detect an extrinsic region of behavior that might appear at low temperatures. The solid–gas exchange technique using ^{18}O as a tracer has been used to study the oxygen diffusion in many single and polycrystalline oxides.[2] If the solid–gas exchange experiments are done by using polycrystalline particles where the grain size is small enough and the grain-boundary diffusion is faster than that of the volume diffusion, the exchange amount increases. This permits measurements of a low level of volume diffusivities. In the present study, a new method[3,4] of obtaining volume diffusion coefficients (D_1) from the exchange rate between polycrystalline particles and oxygen gas enriched with ^{18}O was applied, and the reliability is examined and discussed.

Experimental Procedure

Materials

Undoped polycrystalline MgO and impurity-doped materials for the present diffusion study were prepared as follows: Pure magnesium chloride* (total impurities <75 ppm) was used as a starting material. The chloride solution was added to a pure sodium carbonate solution to get a magnesium hydroxide precipitate. The resultant precipitate was repeatedly washed with distilled water. The washed and then dried precipitate was calcined at about 900°C for 1 h to get magnesium oxide powder. Atomic adsorption spectrometric analysis showed that the Na content in the calcined powder was very low (≈ 10 ppm). The resultant powder was pelletized under a pressure of 1 t/cm^2 (9.0718 × 10^2 kg/cm^2) and then sintered at desired temperatures for 4 h in air. The sintered material was crushed into particles, and then the particles were screened by sieves to collect 14–16, 28–48, 48–80, and 200–325 mesh particles, respectively.

To get Si- and Ca-doped polycrystalline MgO, desired amounts of water glass and calcium chloride solutions were added, respectively, to the host magnesium chloride solution. Details of other procedures were essentially the same as those of the undoped case.

To examine the oxygen diffusion characteristics of an extremely pure polycrystalline MgO, MgO powder was purchased commercially.[†] According to the certificate of analysis provided by the manufacturer, impurity levels detected were [Bi] = 1 ppm, [Ca] = 1 ppm, and [Li] = 1 ppm. As-purchased powder was pressed, sintered at 1450°C, and then crushed into particles.

A different type of an extremely pure polycrystalline MgO was also prepared in an effort to eliminate dislocation and microcracks introduced into the particles during crushing of the sintered tablets. As-purchased MgO powder was pressed at 1 t/cm^2 (9.0718 × 10^2 kg/cm^2) and then fired at a very low temperature, e.g., 800°C. Very porous and fragile tablets resulted in this process. The tablets in this state were softly crushed, and small aggregates passing through a 325-mesh sieve (0.0044 cm in diameter) were removed. The resultant aggregates were fired at 1450°C for 4 h and carefully screened by sieves to collect various sizes of particles. During these processes, one might expect an introduction of a lower level of dislocations in the specimen compared with those in the "crushed" specimen.

Measurements of Oxygen Diffusion Coefficients

The self-diffusion coefficients of oxygen were determined by measuring the exchange rate of oxygen between a gas phase and the heated polycrystalline "mesh" particles over the temperature range 1000–1570°C. Oxygen gas enriched with about 20% ^{18}O was used as the tracer. The amount of oxygen exchange was measured by monitoring gas composition with a mass spectrometer. The reaction chamber was made of transparent silica glass whose outer surfaces were cooled by circulating water to minimize the exchange between the constituent oxygen of the chamber and the enriched oxygen gas. Other details of the experimental setup and procedure have been described elsewhere.[2]

The Crank relation[5] was used to calculate diffusion coefficients of oxygen:

$$\frac{M_t}{M_\infty} = 1 - \sum_{n=1}^{\infty} \frac{6\alpha(\alpha + 1)\exp(-Dq_n^2 t/a^2)}{9 + 9\alpha + q_n^2\alpha^2} \tag{1}$$

where the q_n's are the nonzero roots of $\tan q_n = 3q_n/(3 + \alpha q_n^2)$, a is the solid sphere radius, α is the gram atom ratio of oxygen present in the solid sphere particles to that in the gas phase, and M_t/M_∞ is the total amount of solute in the sphere after time t as a function of the corresponding quantity after infinite time.

Results and Discussion

Characterization of Specimens

Table I shows characteristic data of specimens studied. Specimens with relative densities more than 81.0% contained closed pores alone. Analysis by EPMA[‡] showed that in the Si-doped material sintered at 1550°C, precipitation and/or segregation of Si was observed along grain boundaries. The detailed phase of the precipitate was not known. One might expect homogeneous dissolution of silica throughout the bulk because the specimen was prepared through a chemical co-precipitation method.

Table I. Characteristic Data of Studied Specimens*

Specimen	Grain diameter (μm)	Relative density (%)
High purity-1450	0.3	50.7
Undoped-1250	0.12	68.2
Undoped-1400	2.1	73.4
Si-1250	0.26	81.0
Si-1450	6.7	88.0
Si-1550	13.4	95.6
Ca-1250	0.12	66.0
Ca-1550	3.9	88.0

*For example, high purity-1450, undoped-1250, and Si-1450 denote an extremely pure specimen sintered at 1450°C, undoped specimen sintered at 1250°C, and Si-doped specimen sintered at 1450°C, respectively.

Determination of D_1[3,4]

Figure 1 shows typical plots of time t vs the dimensionless quantity $D_t t/a^2$ determined by using Eq. (1). The plots show straight lines passing through the origin. This suggests that the surface exchange reaction is not important. One can calculate apparent diffusion coefficients, D_a, by using either the grain radius, a_g, or the particle radius, a_p, for the value of a in Eq. (1). Figure 2 indicates Arrhenius plots of D_a for undoped polycrystalline particles of different sizes prepared through sintering at 1400°C; D_a was calculated by taking the grain radius (≈ 1.05 μm) for a value. As seen there, the apparent oxygen diffusion coefficients vary with the size of particles used. The magnitude of the coefficient also varies with particle radius as the value of a; this case is not shown in Fig. 2. These facts indicate that the oxygen exchange is influenced by grain-boundary diffusion besides the volume diffusion in this specimen.

Data in Fig. 2 can be used to plot the relation between log (D_a) and log (particle diameter). Figure 3 shows the typical plots for the undoped specimen sintered at 1400°C. The upper plot is for the calculations of D_a as $a = a_p$, and the lower, of D_a as $a = a_g$.

The particle size dependence characteristics of D_a in polycrystalline particles may be divided into three categories: $D_g \gg D_1$, $D_g = D_1$, and $D_g > D_1$. In the case of $D_g \gg D_1$, D_g is so large that the concentration of ^{18}O along grain boundaries is nearly equal to that in a gas phase throughout a diffusion anneal. In this case,

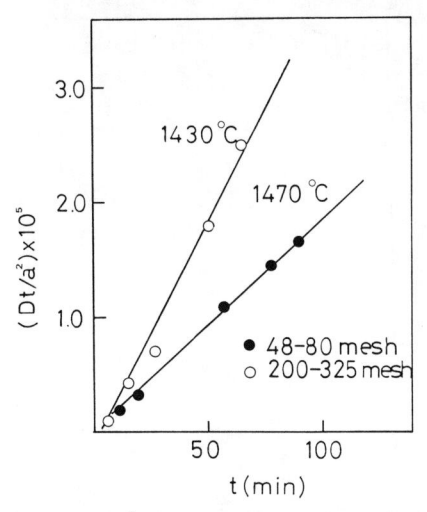

Fig. 1. Typical plots of D_t/a^2 vs time for undoped polycrystalline MgO particles prepared through sintering at 1400°C. The calculations of D were made as a = (grain radius).

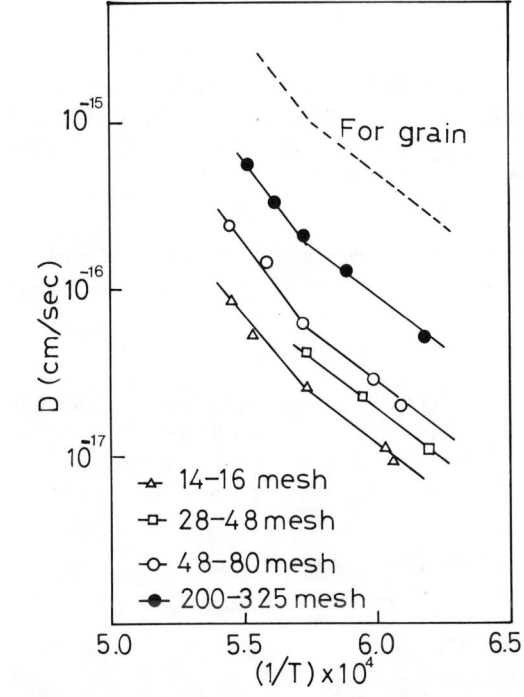

Fig. 2. Temperature dependence of apparent diffusion coefficient, (D_a) of oxygen for undoped polycrystalline MgO particles with different sizes, prepared through sintering at 1400°C. The calculations of D_a were made with a = (grain radius). Calculated volume diffusion coefficients (D_1) are also indicated.

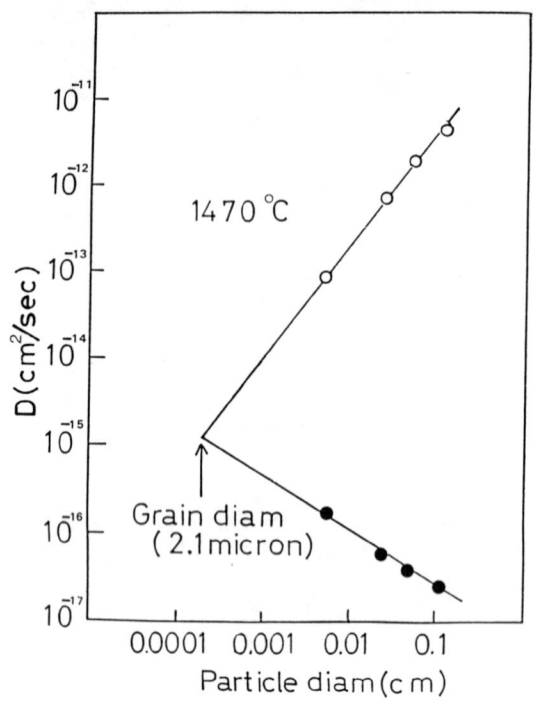

Fig. 3. Plots of log (apparent D) vs log (particle size) for undoped polycrystalline MgO particles prepared through sintering at 1400°C. The calculations of D_a were made with a = (grain radius) (closed circles) and with a = (particle radius) (open circles).

Eq. (2) holds at a constant temperature, irrespective of particles with different sizes, if the capacity of grain boundaries is negligibly small with respect to that of the volume

$$D_l/a_g^2 = C \qquad (2)$$

where C is a constant. This case was found in oxygen diffusion in polycrystalline MgO,[6] Al$_2$O$_3$,[7] and La-doped BaTiO$_3$,[8] and nitrogen diffusion in polycrystalline Si$_3$N$_4$.[9] If the calculations of D_a are done in this case by taking a_p instead of a_g in Eq. (2), one can get the following equation including particle radius (or particle diameter):

$$D_a = a_p^2 C \quad \text{or} \quad \log(D_a) = 2 \log(a_p) + C' \qquad (3)$$

In the case of $D_g = D_l$, on the other hand, Eq. (4) holds for particles with different sizes:

$$D_l/a_p^2 = f(a_p) \qquad (4)$$

This case was encountered in oxygen diffusion in Y$_2$O$_3$-stabilized zirconia.[10] If the calculations of D_a are made by taking a_g instead of a_p here, D_a is expressed by

$$D_a = f(a_p)a_g^2 = D_l a_g^2/a_p^2 = C''/a_p^2 \quad \text{or} \quad \log(D_a) = C''' - 2 \log(a_p) \qquad (5)$$

Emphasis is placed on the fact that, in these two extreme cases, linear relations hold between log (apparent diffusivity) and log (particle radius) (or diameter). This situation is schematically shown in Fig. 4.

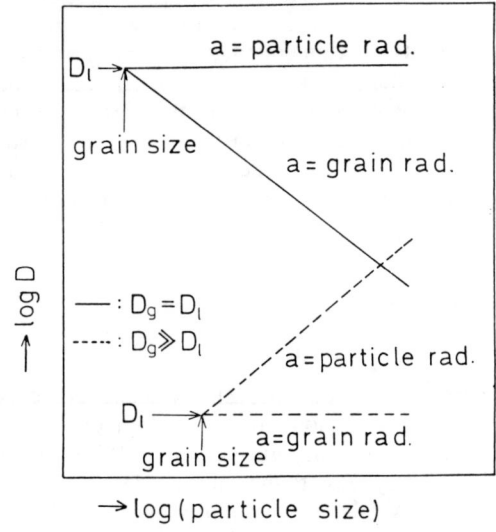

Fig. 4. Schematic illustration of the relations between log D (D_a and D_1) vs log (particle size) for the two extreme cases of $D_g \gg D_1$ and $D_g = D_1$, where D_1 is the volume diffusion coefficient and D_g is the grain-boundary diffusion coefficient.

It is found that the present case (Fig. 3) falls between the two extremes, $D_g \gg D_1$ and $D_g = D_1$, hereafter denoted as a category where $D_g > D_1$. Plots in Fig. 3 consist of two lines, one with negative slope for the calculations of D_a as $a = a_g$ and one with positive slope of D_a as $a = a_p$. If these two straight lines can be extrapolated to the value of the grain diameter (=2.1 μm), D_a at this point should correspond to D_1. It is emphasized that the extrapolated curves using a_g and a_p do intersect at a grain diameter that is identical to that of the experimentally measured grain diameter. Such a situation could also be seen in other specimens in the present study. This is strong support for the model used in this paper.

Remarks may be necessary for the surface geometry of the grains and particles of this specimen and other specimens in the present study. The observation of microstructure of these sintered specimens suggests that there is no problem in the assumption of spheres for the actual grains. However, problems may arise in the assumption of spheres for the particles. The crushed particles have irregular surfaces in general, in which a high estimation of D_1 may result. This is true when single-crystal particles were used for the diffusion measurements. Fortunately, many cases in the present study belong to categories where $D_g \gg D_1$ and $D_g > D_1$. In these cases, the effect of surface geometry of particles on the oxygen exchange may be less important. If the effect in the calculations of D_a is considerable, plots with open circles in Fig. 3 should lead to a high estimate. Data in Fig. 3, however, show the two extrapolated curves do intersect at an experimentally measured grain size, as was described. This strongly suggests that the surface irregularities of particles are not important in the present study.

Another remark may be necessary for the grain-boundary width of the specimen and other specimens in the present study. Throughout the calculations of D_a, the capacity of grain boundaries is assumed to be negligibly small with respect to the volume. Almost all specimens studied belong to categories where $D_g \gg D_1$ and $D_g > D_1$. In these cases, one might observe a fast exchange in an early stage of the exchange, if the capacity of grain boundaries was relatively large. Actually, we observed a slightly fast exchange in several cases where the exchange temperature was very low; however, this was not usually the case. This fact permits us to assume relatively small capacities of grain boundaries in the specimens studied.

The Arrhenius plot of the D_1 thus calculated is also shown in Fig. 2. One can find two diffusion regions, and they can be formulated, respectively, as follows:

$$D_1 = 3.5 \times 10^{-3} \exp\left[-\frac{384.9(\text{kJ/mol})}{RT}\right] \quad 1466°C \leq T < 1570°C \quad (6)$$

$$D_1 = 3.0 \times 10^{-7} \exp\left[-\frac{253.2(\text{kJ/mol})}{RT}\right] \quad 1209°C < T < 1466°C \quad (7)$$

The present method of calculating D_1 by using polycrystalline particles has been applied in oxygen diffusion studies of Mg_2TiO_4,[3] $MgFe_2O_4$,[11] and other oxides.[12] In at least these materials, the values of D_1 obtained by using these polycrystals were always very close to the values of D_1 directly measured by using the corresponding single crystals. The reliability of this method will be discussed later again.

D_1 of Undoped MgO

A different type of undoped polycrystalline MgO was prepared also through sintering, at 1250°C ($a_g = 0.06~\mu$m). The D_1 of this type of particles was calculated in the foregoing way, and the Arrhenius plot is shown in Fig. 5 together with that of the undoped material sintered at 1400°C. They are expressed

$$D_1 = 1.2 \times 10^{-3} \exp\left[-\frac{409.6(\text{kJ/mol})}{RT}\right] \quad 1136°C \leq T < 1330°C \quad (8)$$

$$D_1 = 1.1 \times 10^{-8} \exp\left[-\frac{282.4(\text{kJ/mol})}{RT}\right] \quad 1043°C < T < 1136°C \quad (9)$$

It is necessary to remark that no measurable grain growth occurred during diffusion anneals of this 1250°C-sintered specimen at temperatures up to about 1350°C. Accordingly, the measurements were done at temperatures <1350°C. Such care was similarly taken with other specimens.

As seen in Fig. 5, D_1 is influenced by the sintering temperature of two types of undoped polycrystals. The first important fact is that the sintering temperature dependence is not observed in the "intrinsic" region within experimental error. Such an agreement of "intrinsic" diffusion levels, irrespective of the use of particles with considerably different grain sizes, also indicates this method of obtaining D_1 by using polycrystals is reliable. As will be shown later, such an agreement in "intrinsic" diffusion was confirmed not only in the present undoped case but in Si and Ca-doped specimens as well.

Another important fact can be seen in the region of extrinsic diffusion in the two undoped materials. The two activation energies are close to each other and also agree with that of single-crystal MgO, reported by Oishi and Kingery.[7] However, the discrepancy of extrinsic diffusion levels between the two undoped specimens

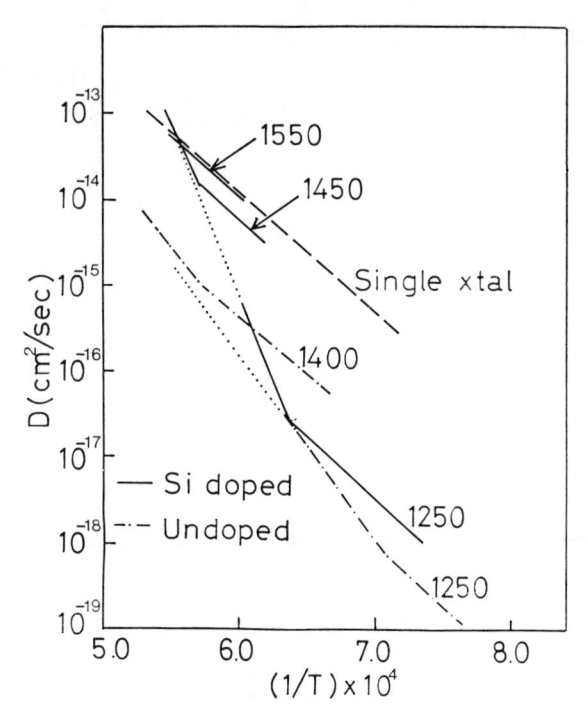

Fig. 5. Arrhenius plots of volume diffusion coefficients of oxygen in undoped and Si-doped polycrystalline MgO particles. Numbers on curves show the sintering temperature of these materials. Data of single-crystal MgO are also shown for comparison (from Oishi and Kingery).

is considerable. This discrepancy may not directly be attributed to a difference in their impurity levels between the two because the origins of the two specimens are the same.

It seems difficult to understand the origins of a series of new facts seen in extrinsic and intrinsic diffusion on the basis of the present data alone. The following study was therefore made in an effort to clarify the origins.

D_1 of an Extremely Pure MgO

It seems important to study oxygen diffusion characteristics of an extremely pure polycrystalline MgO in relation to those of undoped and impurity-doped polycrystalline MgO. Figure 6 shows an Arrhenius plot of D_a of this specimen (total impurity $\cong$ 3 ppm) sintered at 1450°C. No particle size dependence of D_a was observed in this specimen, when the calculations of D_a were done by using $a = a_g$. This result indicates this material to be in the category of $D_g \gg D_1$. Accordingly, the observed values of D_a are directly regarded as D_1. They can be expressed by

$$D_1 = 1.5 \times 10^{-5} \exp\left[-\frac{334.7(\text{kJ/mol})}{RT}\right] \quad 1156°C \leq T < 1513°C \quad (10)$$

$$D_1 = 1.2 \times 10^{-8} \exp\left[-\frac{2565.5(\text{kJ/mol})}{RT}\right] \quad 1043°C < T < 1156°C \quad (11)$$

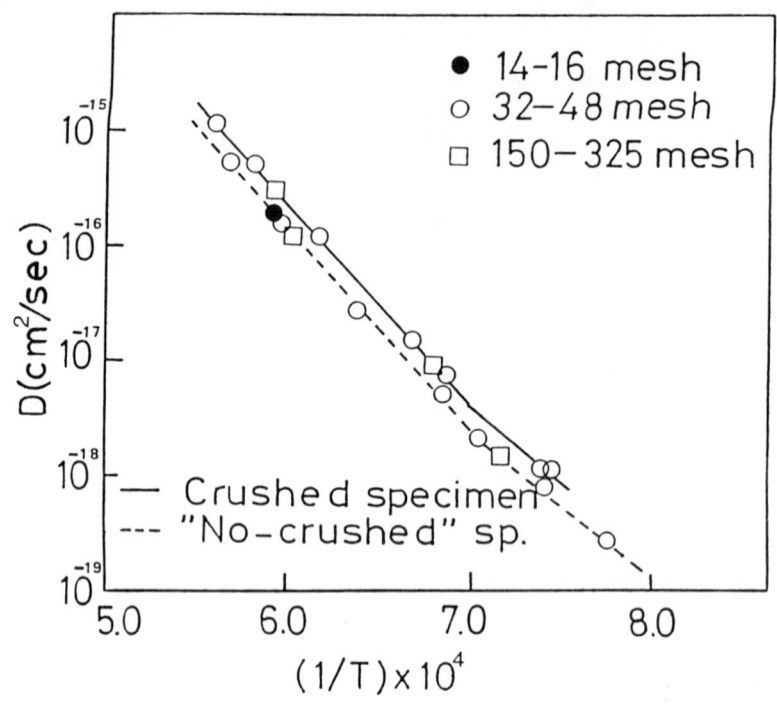

Fig. 6. Arrhenius plots of volume diffusion coefficient of oxygen for two types of extremely pure polycrystalline MgO particles, i.e., "crushed" and "uncrushed" particles.

Equation (10) was an expected one; the Arrhenius plot was completed with the help of the "uncrushed" specimen, as will be described later (see Fig. 6).

It is noted that both the diffusion level and activation energy in the "intrinsic" region are low with respect to those of the undoped materials. This indicates that even "intrinsic" diffusion is also sensitive to impurities, as has been found for extrinsic diffusion in general.

In the extrinsic region of the extremely pure MgO, the diffusion level is very low, but the activation energy is roughly the same as were the undoped data. One might speculate on a variety of extrinsic diffusion levels in terms of microcrack and dislocation densities. To check this, a different type of an extremely pure polycrystalline MgO particle was prepared, in which dislocation and microcracks levels are expected to be lower than those of the foregoing "crushed" specimen. The Arrhenius plot of D_1 of this type of extremely pure specimen is also shown in Fig. 6. No measurable difference is observed between the crushed and uncrushed specimens with extremely high purity. It seems unlikely that a variety of extrinsic diffusion levels in undoped and extremely pure polycrystals can be interpreted in terms of contribution of dislocation and microcracks. In relation to the dislocation effect on diffusion, it seems important to take care in interpreting data from undoped specimens (Fig. 5). If the discrepancy of the extrinsic diffusion levels of the undoped specimens sintered at 1250 and 1400°C is due to a difference in dislocation densities of the two, why are not the "intrinsic" diffusion levels influenced by dislocation? It seems very difficult to answer this.

D_1 of Si- and Ca-Doped MgO

We have at first considered Si as a dopant, which is the principal impurity in MgO. Three types of Si-doped particles were prepared by varying the sintering temperatures, i.e., 1250, 1450, and 1550°C, respectively; other procedures were essentially the same. Spectrophotometric analysis indicated that Si content of the doped powders calcined at 900°C was ≈2000 ppm. The Arrhenius plots of D_1 of these doped materials are also shown in Fig. 5. They are expressed as follows for Si-doped MgO sintered at 1550°C:

$$D_1 = 5.0 \times 10^{-5} \exp\left[-\frac{263.6(\text{kJ/mol})}{RT}\right] \quad 1380°C < T < 1540°C \quad (12)$$

for Si-doped MgO sintered at 1450°C:

$$D_1 = 9.8 \times 10^5 \exp\left[-\frac{669.4(\text{kJ/mol})}{RT}\right] \quad 1466°C \leq T < 1540°C \quad (13)$$

$$D_1 = 4.4 \times 10^{-5} \exp\left[-\frac{279.4(\text{kJ/mol})}{RT}\right] \quad 1340°C < T < 1466°C \quad (14)$$

for Si-doped MgO sintered at 1250°C:

$$D_1 = 2.0 \times 10^8 \exp\left[-\frac{753.1(\text{kJ/mol})}{RT}\right] \quad 1330°C \leq T < 1370°C \quad (15)$$

A considerable variation of D_1 of the Si-doped materials with respect to the undoped ones indicates possible dissolution of Si into the volume of MgO, although segregation and/or precipitation of Si more or less occurred along grain boundaries. On the basis of the present fact of considerably enhanced oxygen volume diffusion in Si-doped materials, it seems that the Si segregated layer along grain boundaries plays an important role in preferential oxygen diffusion along the boundaries. As seen in Fig. 5, the extrinsic diffusion level also increases considerably with increasing sintering temperature without any considerable change in their activation energies, as was seen in the extrinsic diffusion in undoped specimens. These results may not be directly attributed to differences in their impurity levels and dislocation densities containing in the bulk of specimens concerned as was already suggested.

A possible interpretation of the origin of these facts may be given in terms of the metastable occurrence of oxygen vacancies. If oxygen vacancies form as a result of preferential thermal decomposition of the oxygen sublattice at elevated temperatures during sintering, this process can be written as

$$O_O = V_O^x + \tfrac{1}{2}O_2 = V_O^{\cdot} + e + \tfrac{1}{2}O_2 = V_O^{\cdot\cdot} + 2e + \tfrac{1}{2}O_2 \quad (16)$$

When these materials are cooled, the oxygen vacancies that form at elevated temperatures should generally be reduced in concentration if the vacancy concentration of Eq. (16) represents equilibrium. Difficulty in the reduction of oxygen vacancy concentrations may arise on cooling after sintering the materials, if electrons necessary for disappearance of the vacancies by oxidation cannot be supplied. One may consider $Si_{Mg}^{\cdot\cdot}$ centers in Si-doped material that can trap electrons. Thus, oxygen vacancies formed at elevated temperatures can hold metastably on cooling the Si-doped material. The higher the sintering temperature, the higher the metastable oxygen vacancy level. This idea makes it possible to understand clearly various diffusion levels in the extrinsic region, dependent on the sintering temperature. It is emphasized that such a sintering temperature effect seems difficult to understand on the basis of the occurrence of simple Schottky defects. Of course, the formation of Mg vacancies by thermal dissociation of the Mg sublattice at still higher temperatures should also be considered. However, as far as the present temperature range studied is concerned, a preferential formation of oxygen vacancies seems important, partly because MgO has essentially an *n*-type character of oxygen deficiency. Emphasis is placed on the fact that high extrinsic diffusion levels were similarly found in Al-doped polycrystalline MgO sintered at a high temperature in which Al'_{Mg} centers can also form for electron trapping.[12]

To further check the reliability of this model, measurements of D_1 of the Ca-doped specimen will be useful. One might not expect the metastable existence of oxygen vacancies in this specimen, because Ca^x_{Mg} cannot trap any electrons. Atomic absorption spectrometric analysis showed that the Ca content in the powdered materials after calcining is ≈ 100 ppm. Figure 7 shows Arrhenius plots of D_1 of two types of Ca-doped MgO sintered at 1250 and 1550°C, respectively. The diffusion data are expressed as follows for Ca-doped MgO sintered at 1550°C:

$$D_1 = 6.9 \times 10^9 \exp\left[-\frac{769.9(kJ/mol)}{RT}\right] \quad 1250°C < T < 1451°C \quad (17)$$

for Ca-doped MgO sintered at 1250°C:

$$D_1 = 2.1 \times 10^9 \exp\left[-\frac{779.0(kJ/mol)}{RT}\right] \quad 1242°C \leq T < 1340°C \quad (18)$$

$$D_1 = 2.0 \times 10^{-7} \exp\left[-\frac{283.3(kJ/mol)}{RT}\right] \quad 1116°C < T < 1242°C \quad (19)$$

As seen in Fig. 7, a high extrinsic diffusion level is not found even for the Ca-doped specimen sintered at 1550°C, implying that a high level of metastable oxygen vacancies does not occur.

One might correlate such a high extrinsic diffusion level with an increase of solubility of Si into the bulk at elevated temperatures. One may consider two types of defect equilibria for the dissolution of Si into the host lattice:

$$SiO_2 \rightleftharpoons Si^{...}_{Mg} + V''_{Mg} + 2O_O \quad (20)$$
$$SiO_2 + V^{..}_O \rightleftharpoons Si^{...}_{Mg} + 2O_O \quad (21)$$

In both cases, an increased solubility of Si into the host with a temperature elevation does not cause the formation of extrinsic oxygen vacancies.

We now have a number of "intrinsic" diffusion data in MgO, i.e., extremely pure, undoped, and Si- and Ca-doped specimens. These data all imply that not only the "intrinsic" diffusion level but also the activation energy are sensitive to impurities. "Intrinsic" diffusion has been considered to be insensitive to small amounts of impurities. A typical example was reported by Mapother et al. in their study of cation diffusion in the Cd-doped NaCl system.[13] This is not the case in the present study. Therefore, designation of "intrinsic" diffusion is not appropriate in the present study.

It is surprising that even the doping, with Ca with the same valency as the host cation, results in a fast "intrinsic" diffusion with a high activation energy with respect to those of undoped MgO, as were also found in Si-doped materials. It seems difficult to explain unexpectedly high activation energies in the "intrinsic" region of Si- and Ca-doped materials in terms of simple Schottky defects. However

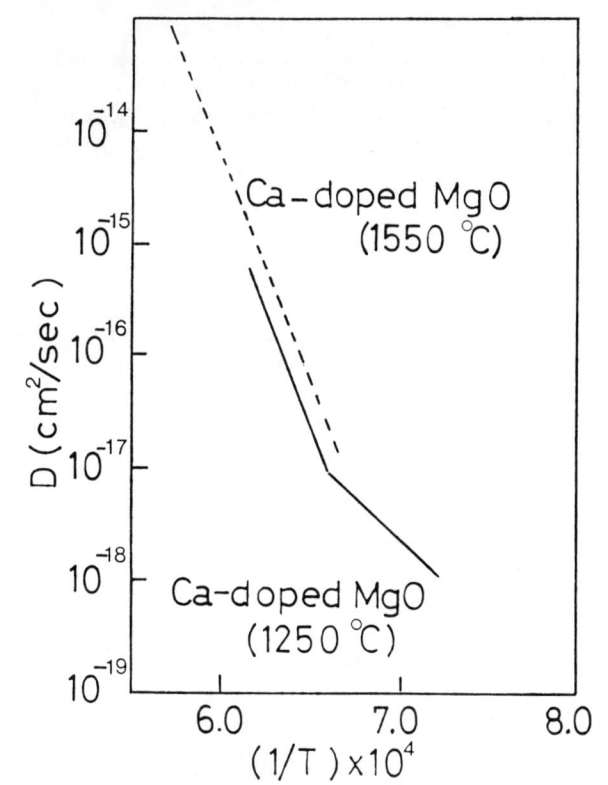

Fig. 7. Arrhenius plots of the volume diffusion coefficient of oxygen for two types of Ca-doped polycrystalline MgO particles prepared through sintering at 1250 and 1550°C.

if Eq. (16) based on reduction/oxidation is considered, a wide variation of the activation energies dependent on impurities may be understood. It has actually been known that oxidation/reduction kinetics of oxides in general is considerably influenced by impurities. In our studies[9,12] of semiconducting La-doped $BaTiO_3$, the activation energy in the intrinsic oxygen diffusion fell between 242 and 418 kJ/mol, dependent on La concentration. These energies no doubt are associated with the formation energy of oxygen vacancies by thermal decomposition of the constituent oxygen sublattice, by which semiconducting behavior resulted.

It is noted that all the dopants studied until now, i.e., Si, Ca, Al,[12] and Fe,[12] the latter two being outside the present study, resulted in high "intrinsic" diffusion levels. On the basis of these results, it is speculated that an introduction of small amounts of impurities into the host and/or accompanied formation of defects (for example, cation vacancies expected in the Si doping) may cause lattice distortion in which thermal (vibrational) entropy and/or configurational entropy increase. An increase of entropies may require a high activation energy of oxygen vacancy formation and/or a high intrinsic diffusion level. Such a compensation effect between activation energy and a frequency factor including entropy terms has often been observed in many chemical reaction processes. It seems to us that diffusion kinetics should also be studied from that point of view in the future.

Summary

It seems that extrinsic oxygen vacancies in undoped and doped materials all occur as a result of the metastable occurrence of oxygen vacancies formed at elevated temperatures. Both the diffusion level and activation energy in the "intrinsic" diffusion (diffusion by thermally formed oxygen vacancies) is appreciably influenced by small amounts of impurities.

Appendix

A series of original Arrhenius plots of D_a for various particles with different mesh sizes of specimens studied are summarized here. All the calculations were made with $a = a_g$. The resultant oxygen volume diffusion coefficients calculated by an extrapolation method are also shown (Figs. 8–13).

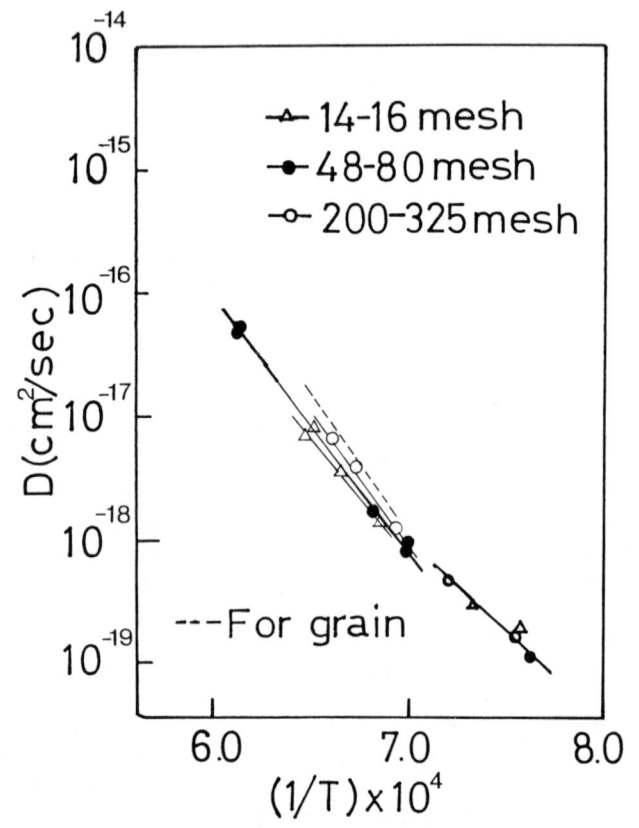

Fig. 8. Temperature dependence of apparent diffusion coefficient (D_a) of oxygen for undoped polycrystalline MgO particles with different sizes, prepared through sintering at 1250°C. The calculations of D_a were made with $a = a_g$. Calculated volume diffusion coefficients (D_1) are also indicated.

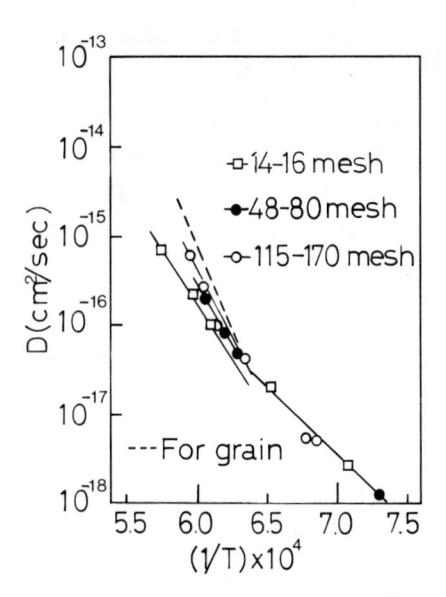

Fig. 9. Temperature dependence of apparent diffusion coefficient (D_a) of oxygen for Si-doped polycrystalline MgO particles with different sizes, prepared through sintering at 1250°C. The calculations of D_a were made with $a = a_g$. Calculated volume diffusion coefficients (D_1) are also indicated.

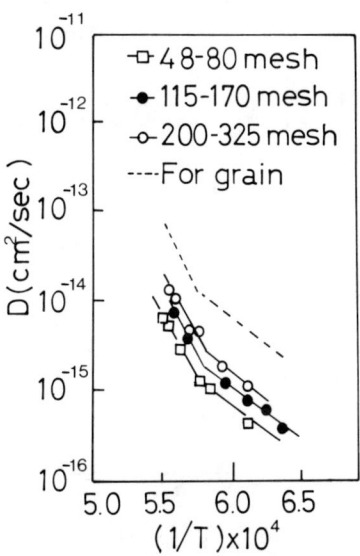

Fig. 10. Temperature dependence of apparent diffusion coefficient (D_a) of oxygen for Si-doped polycrystalline MgO particles with different sizes, prepared through sintering at 1450°C. The calculations of D_a were made with $a = a_g$. Calculated volume diffusion coefficients (D_1) are also indicated.

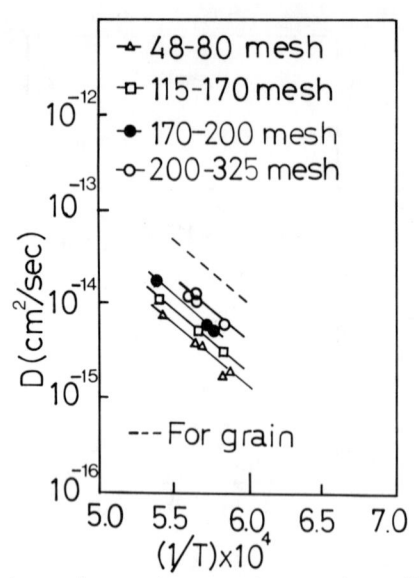

Fig. 11. Temperature dependence of apparent diffusion coefficient (D_a) of oxygen for Si-doped polycrystalline MgO particles with different sizes, prepared through sintering at 1550°C. The calculations of D_a were made with $a = a_g$. Calculated volume diffusion coefficients (D_1) are also indicated.

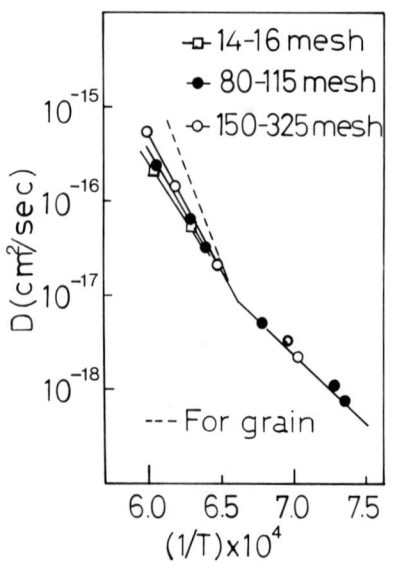

Fig. 12. Temperature dependence of apparent diffusion coefficient (D_a) of oxygen for Ca-doped polycrystalline MgO particles with different sizes, prepared through sintering at 1250°C. The calculations of D_a were made with $a = a_g$. Calculated volume diffusion coefficients (D_1) are also indicated.

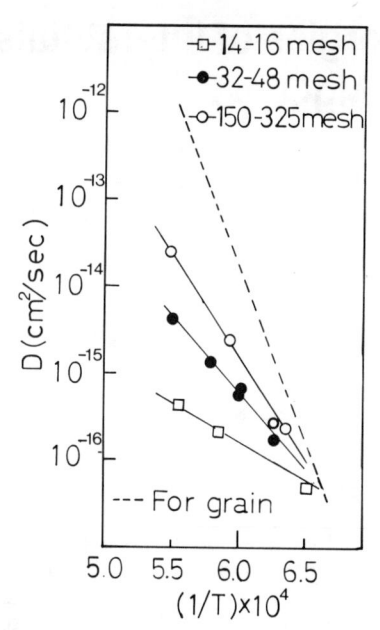

Fig. 13. Temperature dependence of apparent diffusion coefficient (D_a) of oxygen for Ca-doped polycrystalline MgO particles with different sizes, prepared through sintering at 1550°C. The calculations of D_a were made with $a = a_g$. Calculated volume diffusion coefficients (D_1) are also indicated.

References

[1] W. D. Kingery, H. K. Bowen, and D. R. Uhlman; p. 223 in Introduction to Ceramics, 2nd ed. Wiley, New York, 1976.
[2] R. Haul and D. Just, "Disorder and Oxygen Transport in Cadmium Oxide," *J. Appl. Phys.*, **33** [1] 489–93 (1962).
[3] S. Shirasaki, I. Shindo, and H. Haneda, "Relationships Between Oxygen Diffusion Characteristics of Polycrystalline and Single Crystal 2MgO·TiO$_2$," *Chem. Phys. Lett.*, **50** [3] 459–62 (1977).
[4] S. Shirasaki, Y. Moriyoshi, H. Yamamura, H. Haneda, and K. Kakegawa, "Oxygen Diffusion and Defect Structure of ZnO Particles," *Zairyō*, **31** [348] 850–4 (1982).
[5] J. Crank; p. 88 in Mathematics of Diffusion, 1st ed. Clarendon Press, London, 1956.
[6] H. Hashimoto, M. Hama, and S. Shirasaki, "Oxygen Diffusion in Polycrystalline MgO," *J. Appl. Phys.*, **43** [11] 4828–9 (1971).
[7] W. D. Kingery, "Plausible Concepts Necessary and Sufficient for Interpretation of Ceramic Grain-Boundary Phenomena: II," *J. Am. Ceram. Soc.*, **57** [2] 74–83 (1974).
[8] K. Kijima and S. Shirasaki, "Nitrogen Diffusion in Polycrystalline Si$_3$N$_4$," *J. Chem. Phys.*, **65** [7] 2668–73 (1976).
[9] S. Shirasaki, H. Yamamura, H. Haneda, K. Kakegawa, and J. Moori, "Oxygen Diffusion and Defect Structures of Undoped and La-Doped Polycrystalline Barium Titanates," *J. Chem. Phys.*, **73** [9] 4640–5 (1980).
[10] M. Akiyama, T. Ando, and Y. Oshi, "Oxygen Diffusion in Y$_2$O$_3$ Stabilized Zirconia"; Report on Design and Preparation of Functional Ceramics, Vol. 8. Ministry of Education, 1983.
[11] H. Haneda et al.; unpublished work.
[12] S. Shirasaki et al.; unpublished work.
[13] D. E. Mapother, H. N. Crooks, and R. J. Maurer, "Cation Diffusion in Cd-Doped Single Crystal NaCl," *J. Chem. Phys.*, **18**, 1231 (1950).

*Wako Pure Chemical Industry, Ltd., Osaka, Japan.
†Johnson Matthey Co., New York, NY.
‡Model JXA-5A, Nikon Denki Co., Tokyo, Japan.

Migration Energies of Interstitials and Vacancies in MgO

CHIKEN KINOSHITA AND KAZUNORI HAYASHI

Kyushu University
Department of Nuclear Engineering
Fukuoka 812, Japan

T. E. MITCHELL

Case Western Reserve University
Department of Metallurgy and Materials Science
Cleveland, OH 44106

The migration energies of interstitials and vacancies in MgO have been determined through a study of the nucleation and growth of dislocation loops under electron irradiation in a high-voltage electron microscope. Interstitial type dislocation loops nucleate at the beginning of irradiation, and they grow in proportion to the irradiation time to a power smaller than unity at lower temperatures where only interstitials are mobile. The dependence of the volume density and size of loops on irradiation time and electron flux is analyzed by using a theory based on a nucleation mechanism controlled by interstitial motion, in which one or two pairs of Mg and O interstitials correspond to the stable nuclei of loops. The irradiation temperature dependence of the volume density gives the migration energy of interstitials. Dislocation loops grow linearly with irradiation time at higher temperatures where vacancies as well as interstitials are mobile. The difference between the products of the mobility and concentration for the interstitials and vacancies explains the linear growth and the observed square root dependence of the growth rate on electron flux. The dependence of growth rate on irradiation temperature gives the migration energy of vacancies as ~1.9 eV. Considering the fact that the interstitial loops are stoichiometric, these values are interpreted as the activation energies for migration of the slower interstitial and vacancy species in MgO.

There has been increased interest in recent years in studying radiation effects in ceramics, because of their potential use for fusion reactors and nuclear waste applications. However, little definitive experimental work has been done on the migration energies of interstitials and vacancies, which are the most essential parameters to understand radiation effects in solids. Ceramics most frequently are ionic and have their own particular characteristics, such as polyatomicity, the necessity to preserve electrical neutrality, the reciprocity between defect structure and stoichiometry, the existence of a charge state on point defects, and the radiation-induced bias. These characteristics together with their high impurity contents make radiation phenomena in ceramics more complicated than in metals.[1]

High-voltage electron microscopy (HVEM) makes it possible to observe directly and continuously the change in structure of defect aggregates under electron irradiation, and it has been recognized to be a powerful technique to solve problems not only on radiation damage itself but also on the kinetics of point defect processes in metals and other materials.[2-16] The migration characteristics of interstitials and vacancies in metals have been successfully determined by Kiritani and his coworkers from systematic experiments and their analysis.[6,7]

It is now hoped that these experimental and analytical procedures can be applied to ceramics and other practical materials.

Magnesium oxide has generally been regarded as a model material that serves as a paradigm for other more interesting ceramic materials, because of its simple crystal structure and better characterization. Because of this, the general features of nucleation and growth processes in MgO under electron irradiation have been studied and described in terms of the migration energies of interstitials and vacancies on the basis of various mechanisms. In the present paper, an attempt is made to determine the migration energies of interstitials and vacancies in MgO through observations of defect aggregation during irradiation in an HVEM.

Experimental Procedures

Single crystals of MgO were obtained from Norton Co., Tateho Co., and Oak Ridge National Laboratory (ORNL). Their impurity contents are listed in Table I.[17]

Table I. Impurity Contents of MgO Single Crystals*

Crystal	Al	Si	K	Ca	Cr	Fe	Cu
Norton MgO	280	100		140		130	
ORNL MgO (at. ppm)	15	22	<3	50	1	7	<0.1
Tateho MgO	<22	<14	<6	<40	<8	<36	

Crystal	Na	V	Mn	Ni	Pb	Bi	P	Ti
Norton MgO								80
ORNL MgO (at. ppm)	<2	<2	<1	<3	<2	<0.5	13	
Tateho MgO	<6	<2	<3	<2			<3	<2

*Obtained commercially from Norton and Tateho and from ORNL.

They were cleaved along {100} planes and cut to get thin sections 3 mm in diameter and 0.1 mm in thickness. Some thin sections of the Norton MgO were annealed at ~1770 K for 3–5 h in purified oxygen to reduce their carbon contents and are denoted as purified MgO or P-Norton MgO. The thin sections were chemically polished to electron transparency by a jet polishing technique in hot orthophosphoric acid at 400–~410 K and subsequently rinsed in distilled water at 300 K for a few seconds.

Electron irradiation and microscopy were simultaneously performed in a JEM-1000 HVEM operated at 1000 kV. A single tilt, side entry holder was used up to its maximum temperature of ~1270 K. The irradiation was always made along a direction a few degrees away from $\langle 001 \rangle$ to minimize electron diffraction channeling. Bright-field and weak-beam dark-field micrographs were taken during irradiation at the 800 Bragg position in the 200 systematic condition without strong extra reflections. The electron flux was measured with a Faraday cage located between the projection lens and the fluorescent screen. The fluctuation of the electron flux during irradiation was within ±3%. The specimen temperature was measured with a fine thermocouple attached to the specimen holder. Electron beam heating of the irradiated area was confirmed to be within 20 K by the critical voltage method. The radiated heat from the objective aperture caused a temperature fluctuation of a few degrees. Thickness of specimens was estimated from thickness fringes of dynamical and/or weak-beam dark-field micrographs. The position of defect aggregates was determined as a function of distance from the specimen surface by stereomicroscopy.

Nucleation and Growth Kinetics of Dislocation Loops

Electron irradiation produced a high density of defect aggregates at temperatures lower than ~900 K. As the irradiation temperature was increased, the density of these defect aggregates decreased and they grew more rapidly into well-defined interstitial-type dislocation loops with ½⟨101⟩ Burgers vectors lying on {101} planes.[11,18] Irradiation at higher temperatures also caused voids outside the irradiated area.[18] The nucleation and growth processes of dislocation loops were extensively observed and analyzed on the basis of pseudochemical rate theory. Some of the results have been published.[18] Their outline pertinent to determining the migration energies of interstitials and vacancies will be described in this section.

Figures 1 and 2 are typical sequences for the ORNL MgO irradiated at temperatures below and above ~900 K, respectively. Note that the loops in Figs. 1 and 2 are elongated along ⟨010⟩ as described previously;[11,18] this was also observed in the Tateho MgO but not the Norton MgO, where the loops were circular, as discussed later. The nucleation of dislocation loops is complete at the beginning of irradiation. The growth rate of loops, on the other hand, gradually slows down for the lower temperature case, while it stays almost constant for the higher temperature case.

The volume density and size of dislocation loops were examined at temperatures from 300 to 1250 K. The volume density is plotted as a function of irradiation time in Fig. 3, which shows examples of the constant density with time for irradiation of the ORNL MgO at an electron flux of 2×10^{23} e/m^2·s. The saturated volume density is plotted logarithmically in Fig. 4 as a function of electron flux for the ORNL MgO irradiated at 655 K. The slope of the line is ~½.

The size of dislocation loops was also followed as a function of time under irradiation at various temperatures, and examples are shown in Fig. 5 for individual loops in the ORNL MgO together with the as-received Norton MgO. On the assumption that loops grow with a relationship proportional to some power of the irradiation time, the power n was determined and is shown in Fig. 6 for all types of MgO including the as-received Norton MgO and the Tateho MgO. The value of n for the as-received Norton MgO is ~⅔ at ~800 K, in contrast to ~⅓ for the Tateho MgO.[18] The value increases and approaches 1 at ~900 K and again decreases with increasing irradiation temperature. In the cases of the ORNL and purified Norton MgO irradiated at temperatures higher than ~900 K, the initial parabolic growth is followed by linear growth. Therefore, these are assigned a value of $n = 1$ above ~900 K in Fig. 6. The dependence of the growth rate of loops on electron flux was traced by changing electron flux in a stepwise fashion at temperatures above ~900 K. The growth rate is shown as a function of electron flux in Fig. 7 for the ORNL MgO, and it is nearly proportional to the square root of electron flux.

If one summarizes the experimental results at the irradiation temperature T on the volume density C_L and the diameter d of loops as a function of the electron flux

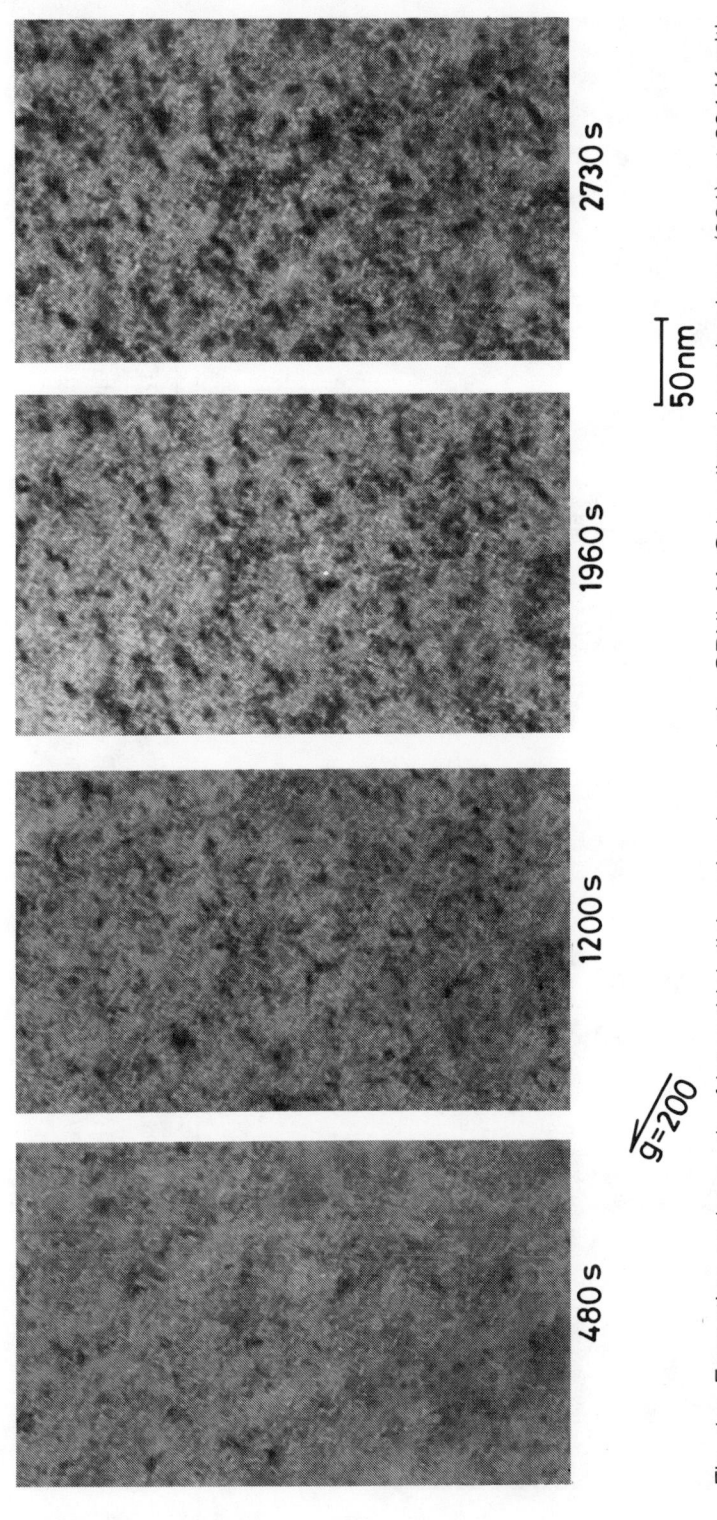

Fig. 1. Formation and growth of interstitial dislocation loops in the ORNL MgO irradiated nearly along ⟨001⟩ at 604 K with 1-MeV electrons of 2×10^{23} e/m$^2\cdot$s.

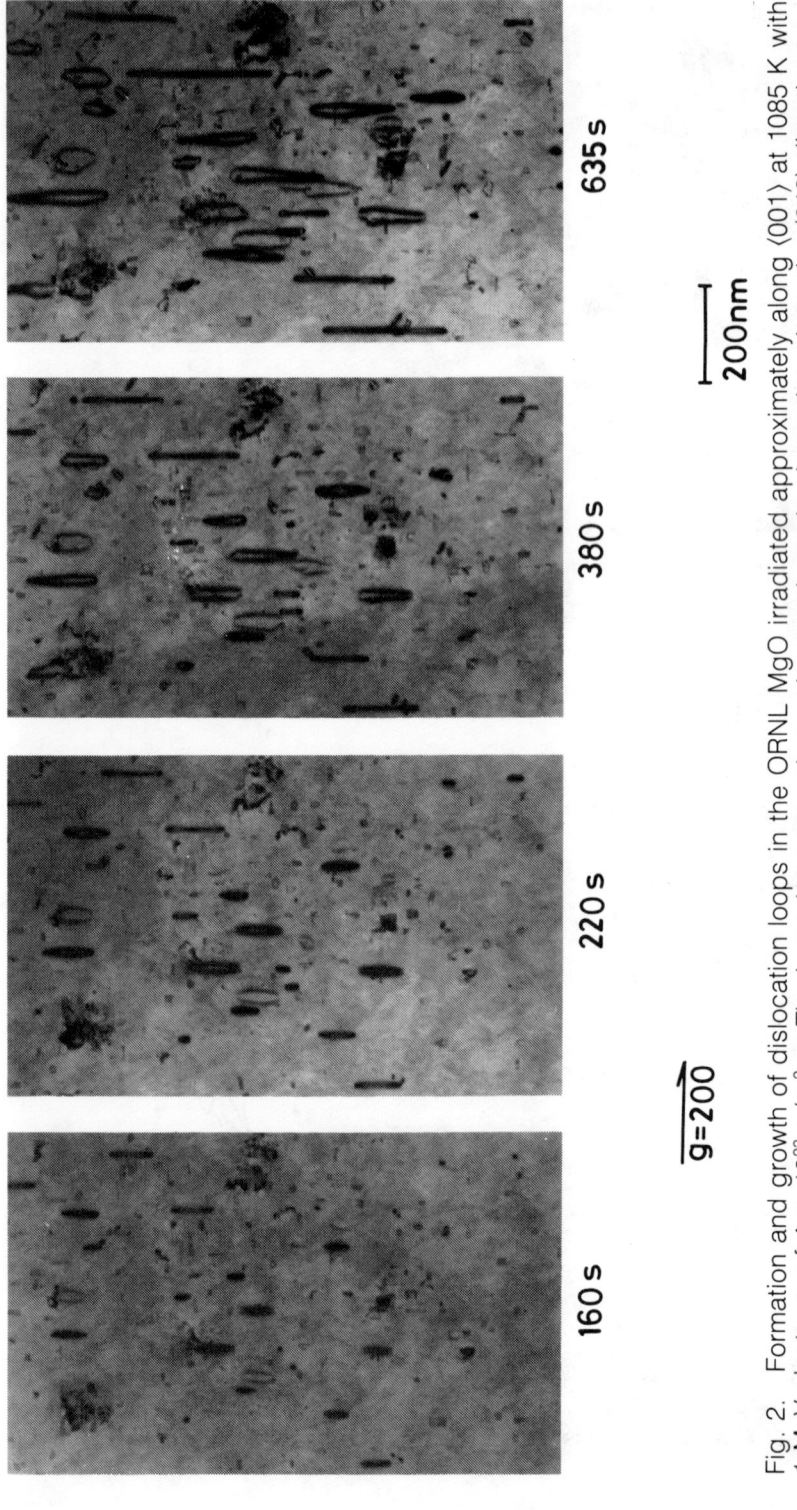

Fig. 2. Formation and growth of dislocation loops in the ORNL MgO irradiated approximately along ⟨001⟩ at 1085 K with 1-MeV electrons of 4 × 10²³ e/m²·s. The loops have an anisotropic growth mode elongating along the ⟨010⟩ directions.

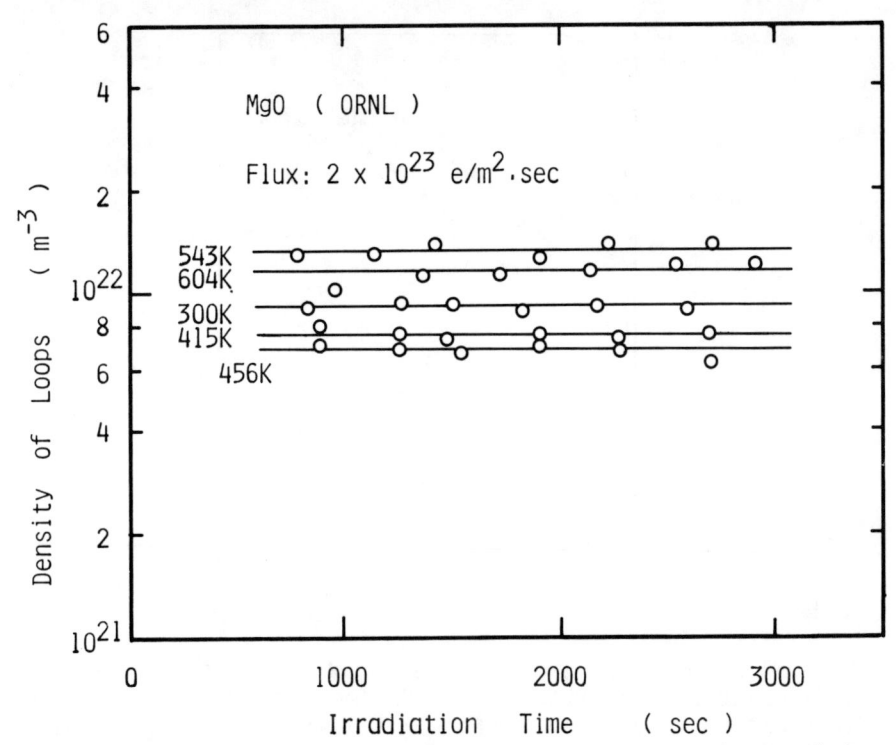

Fig. 3. Time dependence of the volume density of loops during irradiation of the ORNL MgO with 1-MeV electrons of 2×10^{23} e/m²·s. Irradiation temperatures are shown in the figure.

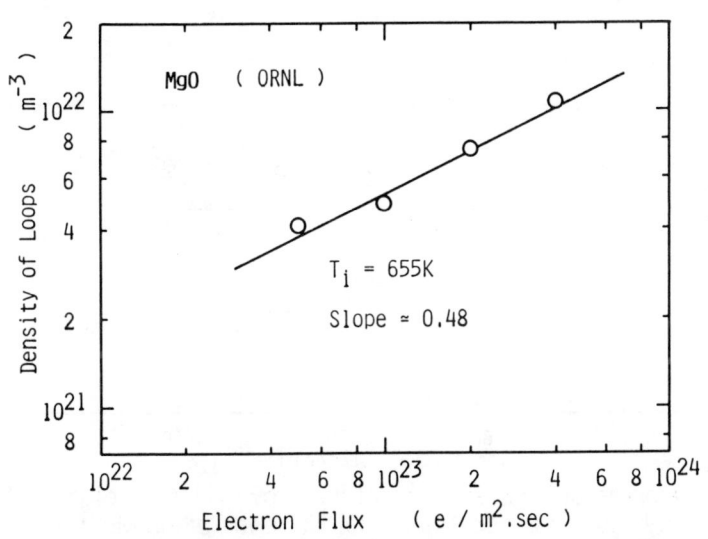

Fig. 4. Electron flux dependence of the saturated volume density of loops in ORNL MgO irradiated at 655 K.

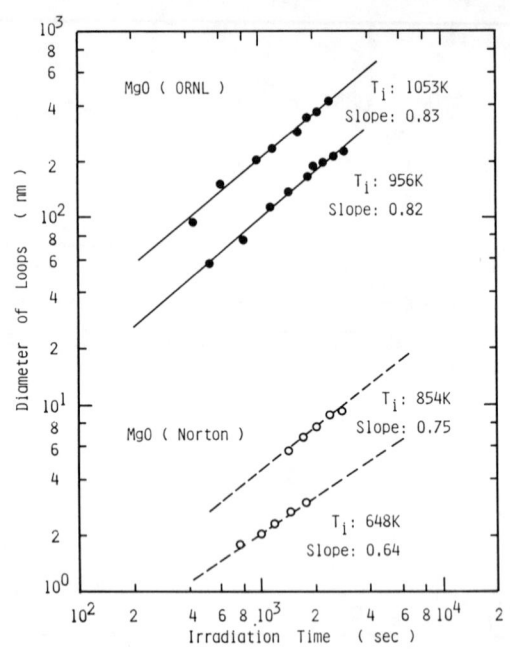

Fig. 5. Typical plots of loop diameter vs irradiation time for the ORNL MgO and the as-received Norton MgO. Irradiation temperatures are shown in the figure. The loops grow with the relation $d \propto t^n$, where d is the loop diameter, t the irradiation time, and n a constant independent of t.

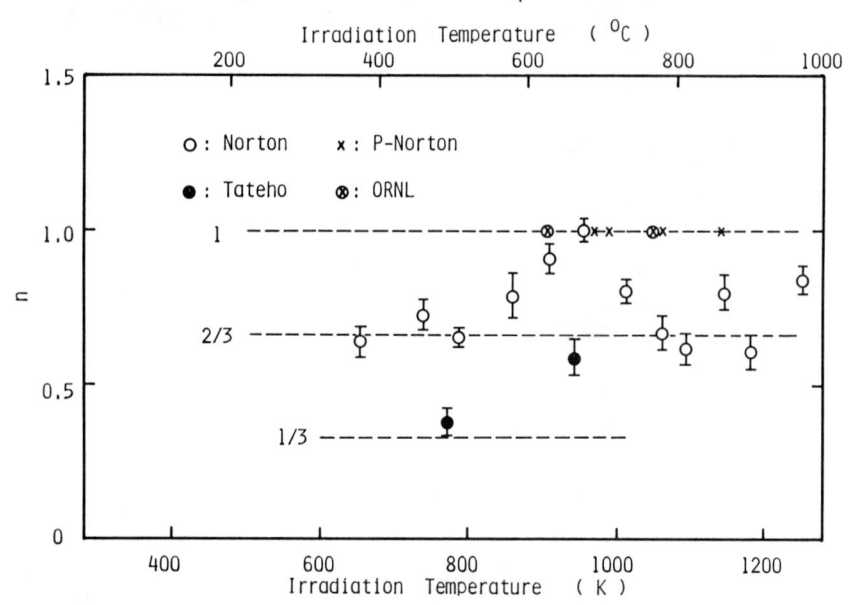

Fig. 6. Temperature dependence of the growth kinetics of loops for the as-received Norton, purified Norton, ORNL, and Tateho MgO. The values of n were determined on the assumption that the loops grow with the relation, $d \propto t^n$.

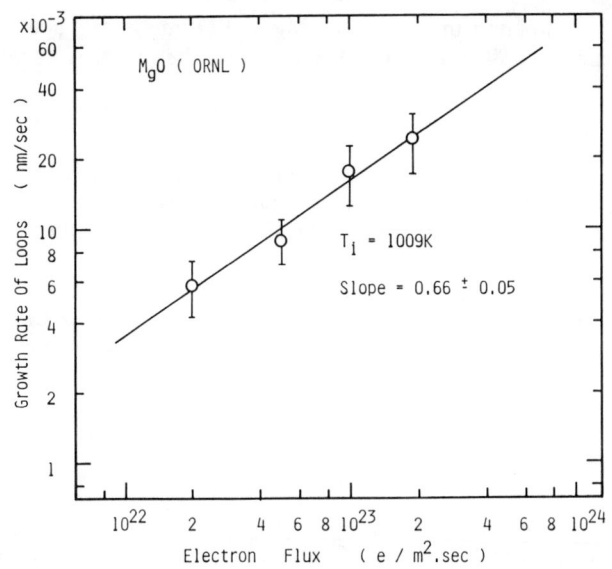

Fig. 7. Dependence of the growth rate of loops on electron flux for the ORNL MgO irradiated at 1009 K.

ϕ and irradiation time t, the empirical relations are

$$C_L \propto \phi^{-1/2}t^0 \quad \text{and} \quad d \propto t^{-1/3-2/3} \quad \text{at} \quad T < \sim 900 \text{ K} \tag{1}$$

$$C_L \propto t^0 \quad \text{and} \quad d \propto \phi^{-1/2}t^{-1} \quad \text{at} \quad T > \sim 900 \text{ K} \tag{2}$$

The kinetics and properties of loops in the ORNL and the purified Norton MgO are not so much different from each other except that loops in the purified Norton MgO are circular in shape, in marked contrast to the loops elongated along $\langle 010 \rangle$ in the ORNL MgO.

The kinetic behavior of interstitial loops should be directly related to the behavior of interstitials and vacancies. Characteristics of the kinetics of dislocation loops are classified in terms of the mobility of interstitials and vacancies, foil thickness, and nucleation sites of loops.[2-7]

In pure metals diinterstitials are the stable nuclei of interstitial loops. At low temperatures where only interstitials are mobile with mobility M_I, the density of loops saturates at the beginning of irradiation, and C_L and d follow

$$C_L \propto \phi^{1/2}t^0 M_I^{-1/2} \quad d \propto t^{1/3} \tag{3}$$

At high temperatures where vacancies are mobile, interstitials and vacancies establish their mutually balanced steady state just after irradiation, and d is expressed by ϕ, t, and the vacancy mobility M_V as

$$d \propto \phi^{1/2}t M_V^{1/2} \tag{4}$$

For diatomic oxides such as MgO, it is unlikely that displacement rates are identical for both anions and cations, nor in general will defect mobilities for each ion be the same. Consequently, these circumstances have to be taken into account for understanding the kinetics of secondary defects in MgO. Considering the fact that the defect aggregates maintain their stoichiometry, we have proposed a model based on the assumption that a pair of Mg and O interstitials acts as the stable nucleus of each loop. The loop kinetics based on the model is expressed by the following equations:

$$C_L \propto \phi^{1/2} t^0/(M_J^I)^{-1/2} \quad \text{and} \quad d \propto t^{1/3} \tag{5}$$

where M_J^I (J stands for Mg or O) is the mobility of J interstitials, corresponding to the slower species. If each dislocation loop nucleates with two pairs of Mg and O interstitials, the power dependence of ϕ on C_L and that of t on d in Eq. (5) become ~½–1 and ~⅔, respectively.[18,19]

At high temperatures where vacancies are also mobile, interstitials and vacancies establish their mutually balanced steady state just after irradiation. The growth rate of interstitial loops is given by the difference between the absorption rates of interstitials and vacancies. Considering the defect aggregates to be stoichiometric, we have

$$\frac{dd}{dt} = a(Z_J^{IL} - Z_J^{VL})\left(\frac{\sigma_J C_J \phi M_J^V}{Z_J^{IV}}\right)^{1/2} \quad (J = \text{Mg or O}) \tag{6}$$

where a is the change in the diameter of dislocation loops by adding interstitials to all available sites on the loops, Z_J^{IL} and Z_J^{VL} are the numbers of capture sites around one atomic site on the dislocation for J interstitials and J vacancies, respectively, Z_J^{IV} is the number of sites around a defect where mutual recombination will spontaneously occur, σ_J is the displacement cross section of J ions, C_J is the concentration of J ions, and M_J^V is the mobility of J vacancies.[18,19] The growth rate should be controlled by J vacancies with the lower value of Eq. (6).

Migration Energies of Interstitials and Vacancies

The simplified model given in the preceding section is based on one or two pairs of Mg and O interstitials as the stable nuclei of loops and on the free migration of interstitials; it reproduces the experimental results on C_L and d at temperatures below ~900 K in a satisfactory manner. If one expresses M_J^I in Eq. (5) in terms of the migration energy of the slower interstitials E_M^I as

$$M_J^I = \nu \exp(-E_M^I/kT) \tag{7}$$

Eqs. (5) and (7) allow us to determine the value of E_M^I from the Arrhenius plot of $\log C_L$ vs $1/T$. This is represented in Fig. 8 for the various MgO specimens. The volume density of loops decreases with increasing irradiation temperature at a rate that is faster above 900 K, where interstitials and vacancies are mobile, than below 900 K, where only interstitials are mobile. There is an anomalous increase around 600 K. Therefore, the correct value of E_M^I is uncertain, but the slope below ~500 corresponds to a value of ~0.05 eV for the ORNL MgO.

At high temperatures, on the other hand, where vacancies are also mobile, the simplified model gives Eq. (6), which satisfactorily reproduces the observed flux dependence of the growth rate of loops in both purified Norton and ORNL MgO under irradiation at temperatures above ~900 K. The explicit form of M_J^V in Eq. (6) is

$$M_J^V = \nu \exp(-E_M^V/kT) \tag{8}$$

where E_M^V is the migration energy of J vacancies. From Eqs. (6) and (8), the Arrhenius plot of $\log (dd/dt)$ vs $1/T$ gives the value of E_M^V.

Figure 9 is an example of sequences showing the growth of individual loops in thick specimens (~1 μm) of the purified Norton MgO for stepwise changes of

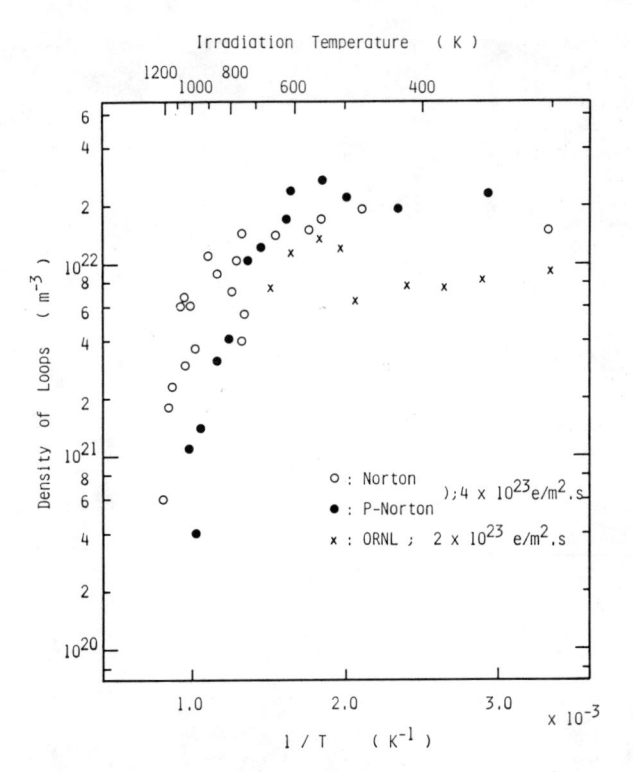

Fig. 8. Temperature dependence of the saturated volume density of loops in the as-received and purified Norton MgO and the ORNL MgO under irradiation with 1-MeV electrons with 4×10^{23} and 2×10^{23} e/m$^2 \cdot$s, respectively.

specimen temperature during irradiation for an electron flux of 1×10^{23} e/m$^2 \cdot$s. The diameter of individual loops was traced and is plotted as a function of irradiation time in Fig. 10. The temperature dependence of the growth rate of loops gives rise to the Arrhenius plot of Fig. 11. The slope of the lines for individual loops should be directly proportional to the migration energy of vacancies. The values of E_M^V for 40 loops were plotted as a function of the distance from the nearest-neighbor loop, but the data were extremely scattered. The basic equation Eq. (6) is applicable only for loops that are isolated from the other loops and are not near the specimen surfaces. However, there was no correlation between E_M^V and the distance from the nearest-neighbor loop. The statistics give 1.5 eV for 40 loops.

Interstitial loops in MgO lie mainly on {101} planes and change their habit planes slightly during irradiation. Inclined loops, therefore, may introduce a larger error to the diameter measurements due to the contrast effects. Figure 12 shows the statistical errors of E_M^V for edge-on loops and inclined loops. As expected, inclined loops show rather large statistical errors. The experimental values of E_M^V for edge-on loops and inclined loops are shown in Fig. 13. The values for inclined loops are extremely small. The reason is not clear at present, but it is believed that the reason is not always due to the contrast effect but also due to some effects of the displacement process and/or kinetic process.

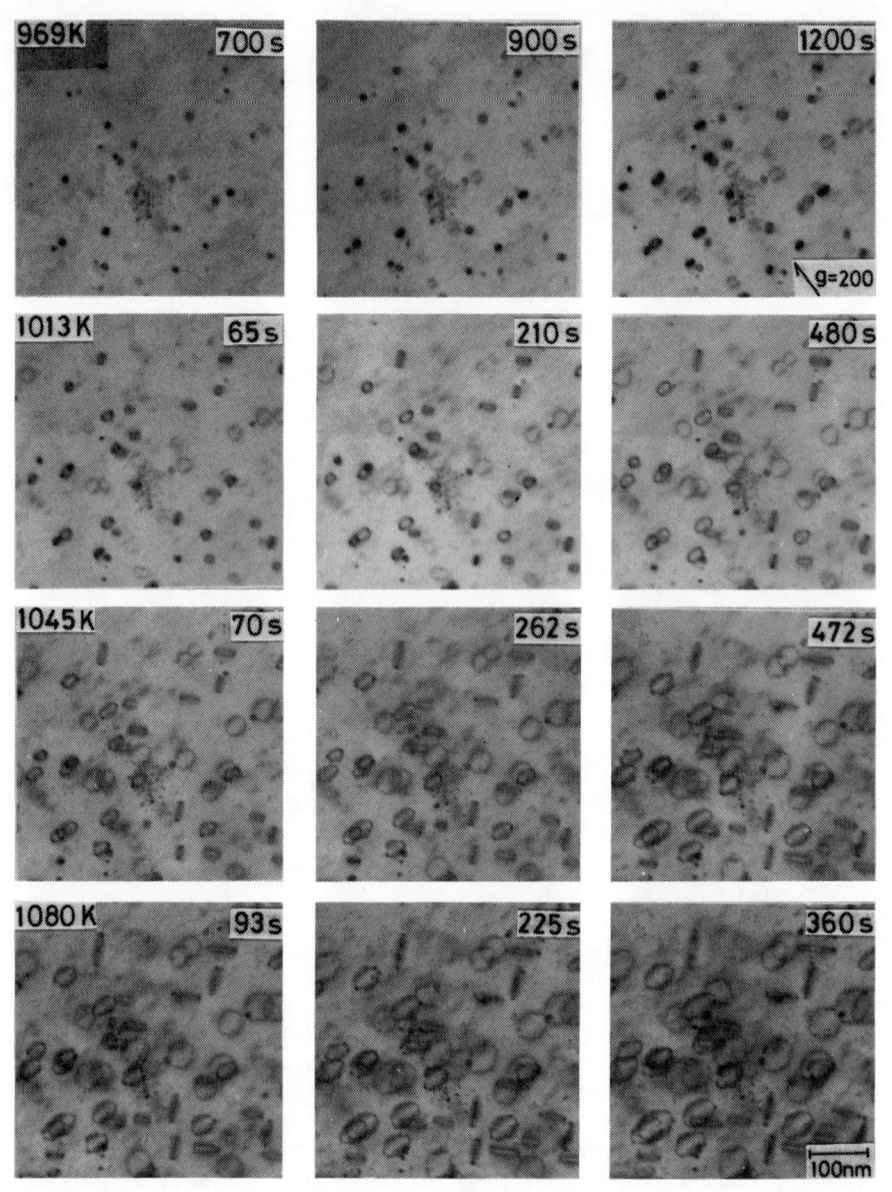

Fig. 9. Growth of the dislocation loops in the purified Norton MgO for stepwise changes of specimen temperature during irradiation for an electron flux of 1×10^{23} e/m$^2 \cdot$s.

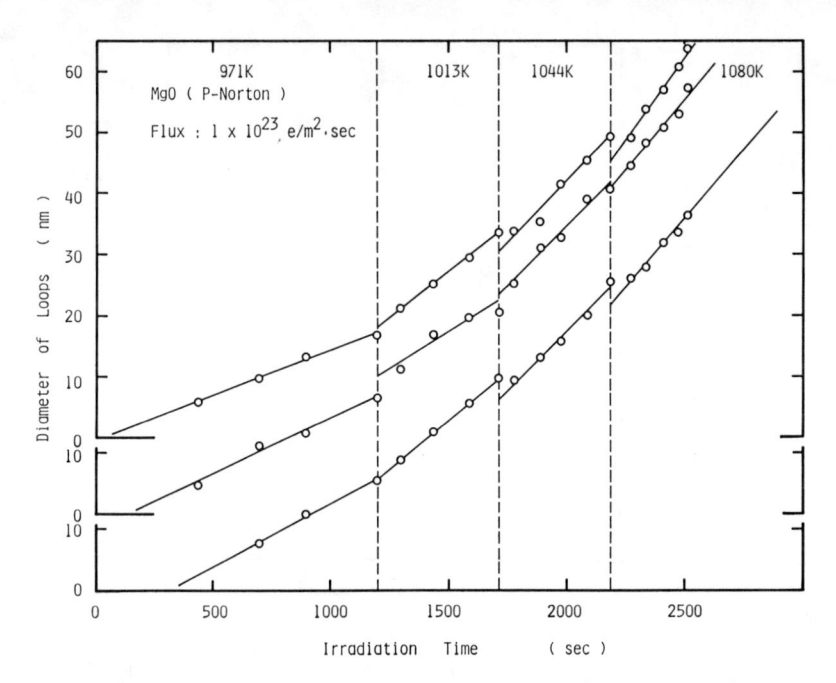

Fig. 10. Time dependence of the diameter of individual loops in the purified Norton MgO for stepwise changes of specimen temperature during irradiation for an electron flux of 1×10^{23} e/m$^2 \cdot$s.

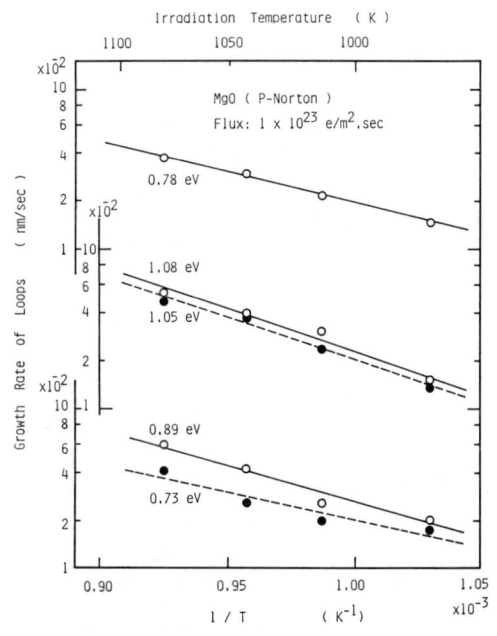

Fig. 11. Arrhenius plot of log growth rate vs reciprocal temperature for the purified Norton MgO.

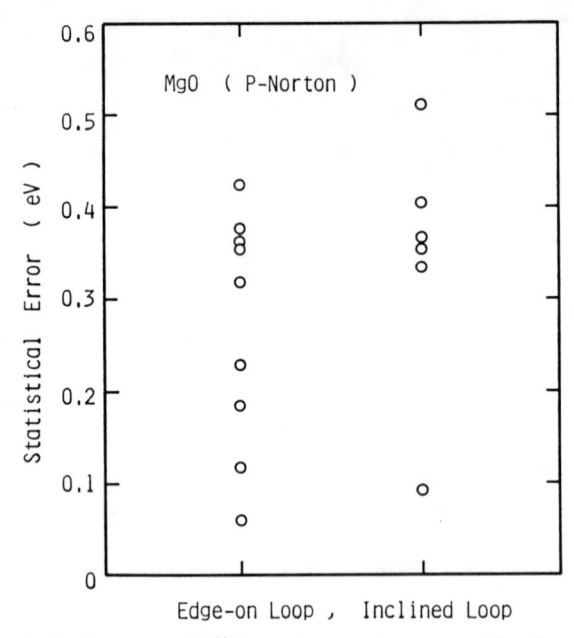

Fig. 12. Statistical errors of E_M^V for edge-on loops and inclined loops in the purified Norton MgO.

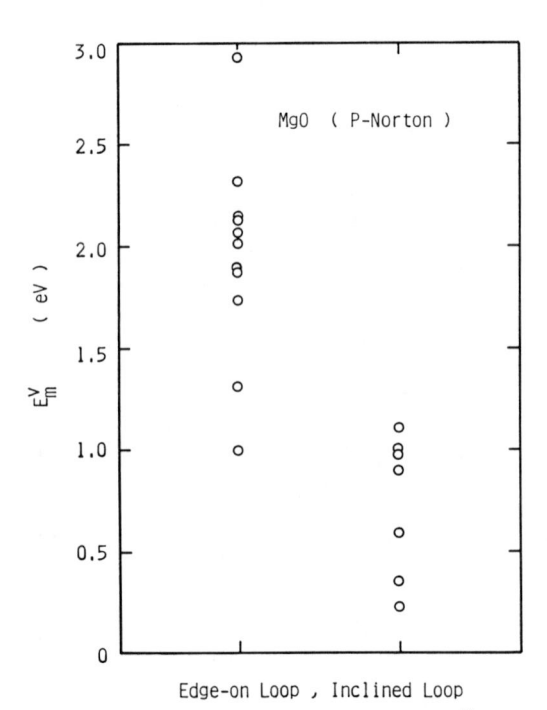

Fig. 13. Apparent migration energies of vacancies E_M^V of individual edge-on loops and inclined loops in the purified Norton MgO.

The values of E_M^V for edge-on loops are plotted in Fig. 14 as a function of distance from the specimen surface. There is no definite relationship between E_M^V and the distance from the specimen surface, and one gets 1.95 ± 0.2 eV on average as the migration energy of vacancies for the purified Norton MgO. Following the same procedure with the long axis of the elongated loops, one gets Fig. 15 for the ORNL MgO. The average value is 1.9 ± 0.2 eV.

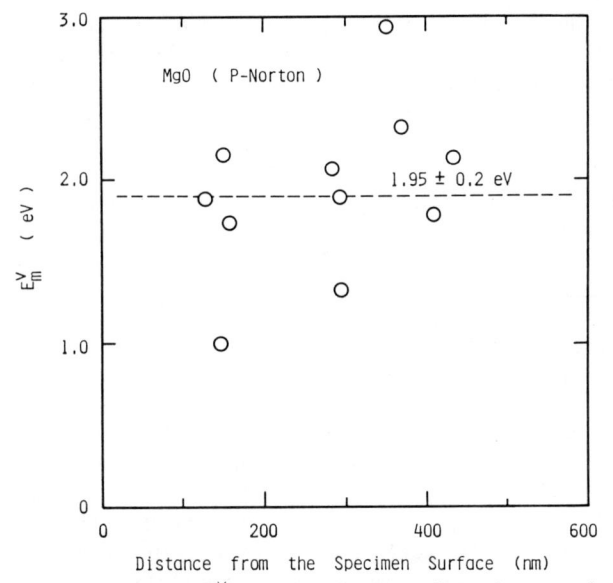

Fig. 14. Relation between E_M^V and the distance from the specimen surface for the purified Norton MgO. One gets 1.95 ± 0.2 eV on average as the migration energy of vacancies.

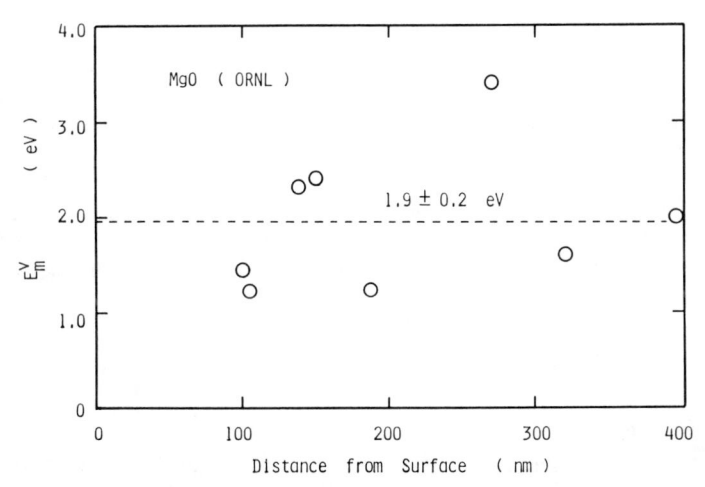

Fig. 15. Same as that in Fig. 14 but for the ORNL MgO. The averaged value of the migration energy of vacancies is 1.9 ± 0.2 eV.

Discussion

Equation (6), which is based on the free migration of Mg and O defects, allows one to determine the migration energy of the species that controls the loop growth rate. The displacement cross sections of Mg and O ions are almost the same as for MgO irradiated with 1000-keV electrons;[10] that is, all the parameters for Mg and O species are the same except the mobility of the vacancy species. Therefore, the migration energy of 1.9 eV corresponds to the slower vacancy species, which is probably oxygen.

In the model described before, neither correlation between Mg defects and O defects nor electrical neutrality in the matrix are considered. Actually, however, the same magnitude of electrostatic attraction exists between Mg and O interstitials and between Mg and O vacancies. The loop kinetics may be controlled by the diffusion via neutral species as in pure metals; in this case the diffusion relates to movement of an associated MgO interstitial or MgO vacancy molecule. In the event that the di-interstitial molecules are the stable nuclei of interstitial loops, the loop kinetics is analogous to that in pure metals, as in Eqs. (3) and (4), but M_I and M_V should be read respectively as the mobilities of MgO interstitial and MgO vacancy molecules.

As noted earlier, the $½\langle 101\rangle\{101\}$ dislocation loops grow in an anisotropic, elongated mode along $\langle 010\rangle$ directions in the ORNL MgO, in contrast to the circular loops in the purified Norton MgO. Furthermore, the growth rate of loops in the ORNL MgO is 1 order higher than in the purified Norton MgO. The difference in growth rate between both types of MgO is due to the bias factor in Eq. (6), because E_M^V is almost the same for each. The elongated growth mode may also be explained by a difference in the point defect–dislocation interactions (i.e., the bias factor in Eq. (6)) between jogs formed on the $\langle 10\bar{1}\rangle$ and $\langle 010\rangle$ sides of the edge dislocations. The details of the atomic configuration that lead to this difference are not understood. In the case of the less pure "purified" Norton MgO, it is suggested that the large amounts of cation impurities make all the sites on the perimeter of the loop equivalent, and therefore no preferential growth occurs in the $\langle 010\rangle$ directions. It is planned to investigate the precise mechanism by using well-characterized, doped MgO.

Conclusion

Electron irradiation causes the rapid nucleation and growth of interstitial dislocation loops in MgO. The kinetic behavior of loops below ~900 K, where only Mg and O interstitials are mobile, implies that the loops nucleate with one or two pairs of Mg and O interstitials or two pairs of MgO interstitial molecules. The growth kinetics of loops above ~900 K, where both interstitials and vacancies are mobile, is analyzed with a model based on a steady-state concentration of interstitials and vacancies with high mobilities, yielding a migration energy for the slower vacancy species as 1.9 ± 0.2 eV.

Acknowledgments

The authors thank R. A. Youngman for his interest and discussion and Y. Oishi for the ORNL MgO. This work was performed by using a high-voltage electron microscope JEM-1000 of the HVEM laboratory, Kyushu University.

References

[1] L. W. Hobbs, *J. Phys.*, **37**, C7-3 (1976).
[2] M. J. Makin, *Philos. Mag.*, **20**, 1133 (1969).
[3] L. M. Brown, A. Kelly, and R. M. Mayer, *Philos. Mag.*, **19**, 721 (1969).
[4] D. I. R. Norris, *Philos. Mag.*, **22**, 1273 (1970).
[5] M. J. Goringe, *Rad. Eff.*, **10**, 169 (1971).
[6] N. Yoshida and M. Kiritani, *J. Phys. Soc. Jpn.*, **35**, 1418 (1973).
[7] M. Kiritani, N. Yoshida, H. Takata, and Y. Maehara, *J. Phys. Soc. Jpn.*, **38**, 1677 (1975).
[8] R. Drosd, T. Kosel, and J. Washburn, *J. Nucl. Mater.*, **69, 70**, 801 (1978).
[9] K. Urban and N. Yoshida, *Philos. Mag.*, [Part A], **A44**, 1193 (1981).
[10] J. W. Sharp and D. Rumsby, *Rad. Eff.*, **17**, 65 (1973).
[11] R. W. Youngman, L. W. Hobbs, and T. E. Mitchell, *J. Phys.*, **41**, C6-227 (1980).
[12] C. Kinoshita, T. E. Mitchell, K. Nakai, and S. Kitajima; p. 465 in the Proceedings of the Tenth International Conference on Electron Microscopy. Hamburg, 1982.
[13] G. P. Pells, *Rad. Eff.*, **64**, 71 (1982).
[14] G. P. Pells and D. C. Phillips, *J. Nucl. Mater.*, **80**, 207 (1979).
[15] D. G. Howitt and T. E. Mitchell, *Philos. Mag.*, [Part A], **A44**, 299 (1981).
[16] T. Yoshiie, H. Iwanaga, H. Shibata, M. Ichihara, and S. Takeuchi, *Philos. Mag.*, [Part A], **A40**, 297 (1979).
[17] M. M. Abraham, C. T. Butler, and Y. Chen, *J. Chem. Phys.*, **55**, 3752 (1971).
[18] C. Kinoshita, K. Hayashi, and S. Kitajima; Proceedings of the Second International Conference on Radiation Effects in Insulators; unpublished work.
[19] C. Kinoshita and T. Yamamoto; unpublished work.

Sintering of Shock-Conditioned Materials

H. Palmour III, T. M. Hare, A. D. Batchelor, K. Y. Kim, K. L. More, and T. T. Fang

North Carolina State University
Department of Materials Engineering
Raleigh, NC 27695

Enhanced sinterability of shock-conditioned ceramic particulates is generally associated with "activation" of the powder attributable to shock-induced defects, treated in detail in a companion paper. Recent research based on precision dilatometric sintering studies of various shocked aluminas, as compared to unshocked analogues, has shown the relationship to be complex and often confounded by shock-induced morphological alteration and/or aggregation of the powders. In this paper, the experimental findings are reviewed for one alumina material, the adverse effects of excessive shock compaction on the densification kinetics and microstructural development are considered, competing thermally activated, rate-sensitive processes (recrystallization, grain growth) are identified, and some examples of enhanced sinterability due to shock conditioning per se (reasonably free of confounding morphological effects) are presented. Factors affecting the optimization of (1) shock conditioning and (2) intermediate (postshock) processing of the powders as well as (3) subsequent densification during sintering are also discussed.

A companion paper[1] has reviewed the shock wave alteration of ceramic particulates, with emphasis on the generation of point and line defects. In 1966, the pioneering study of shock-conditioned Al_2O_3 and other ceramic powders by Bergmann and Barrington[2] demonstrated the enhanced sinterability of such powders. Shock conditioning has also been extensively investigated overseas, e.g., by Dremin and coworkers in Russia[3] and by Sawaoka and coworkers in Japan.[4]

Shock wave processing of ceramic powders, whether carried out directly (by dynamic compaction to final density) or indirectly (by shock activation, followed by some more conventional densification step) is an inherently complex and expensive undertaking.[5] It is most likely to be justifiable — in technological and economic senses — for those situations where (1) the material in question is inadequately responsive to conventional densification processes, (2) special needs exist for unusual sizes or shapes, or retention of metastable phases etc., not readily attainable by conventional processing methods, or (3) specific needs exist for unique combinations of density, microstructure, and resultant properties made possible by shock processing.

Clearly, alumina is already known to sinter well, and thus under these criteria, it does not qualify as a prime candidate per se for shock conditioning. Rather, alumina is being investigated as a convenient, familiar, and available model material to aid in identifying the main directions and interrelationships involved, prior to undertaking shock conditioning and sintering studies with other less tractable materials (e.g., TiC, TiB_2, AlN, or SiC).

Background

The direct dynamic compaction and the shock conditioning of metal and ceramic powders have been extensively reviewed.[5] The historical background and the principal scientific and technical issues involved in shock-activated sintering of ceramic particulates have also been reviewed.[6] Research on the sintering behavior of shock-conditioned, fine, pure alumina powders has been underway at North Carolina State University since early 1982. Earlier findings from that study have been given elsewhere.[7-12]

Materials and Experimental Procedures

The principal experiments were carried out with a 99.98% pure alumina* with an as-received specific surface of 11.3 m^2/g and a typical particle size of ~0.13 μm. With procedures documented in detail elsewhere,[7-11] the powder was characterized, dispersed by dry ball milling, and remilled dry to incorporate fugitive binder and lubricant additives. The prepared powder was pressed uniaxially in a cylindrical die, yielding 7.62-cm diameter by 1.01-cm thick compacts. Forming pressures of 6.82, 68.2, and 206.8 MPa were employed to yield compacts having constant dimensions but differing in the resultant binder-free fractional green densities: 0.571, 0.605, and 0.62, respectively. After binder removal, the compacts, each capable of yielding >100 g of powder, were forwarded to Battelle Columbus Laboratories for encapsulation (in individual welded steel recovery packages), thermal outgassing to 723 K in vacuo, and explosive shocking.[7,13]

Figure 1 shows the "mousetrap" shocking fixture, in which an explosively driven steel flyer plate is dynamically bent and accelerated downward to cause

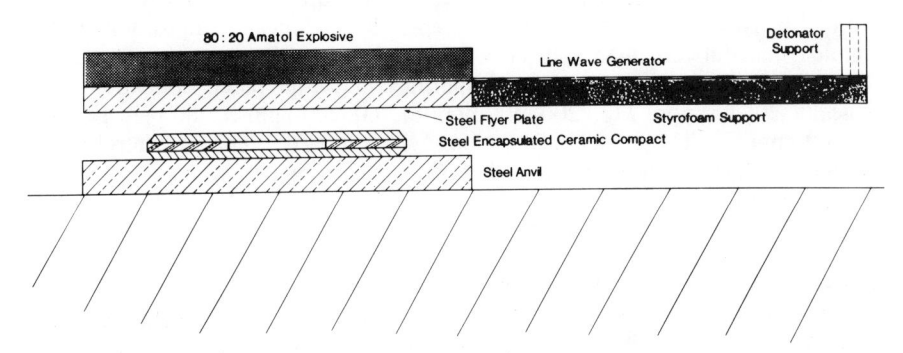

Fig. 1. Arrangement for shock conditioning of powder compact with an explosively driven flyer plate (Refs. 11 and 13).

planar, high-velocity impact impinging on the upper surface of the recovery package containing the specimen. Figure 2 diagrams the results of a preplanned sequence of five such shots. On its vertical axis, the fractional green densities were maintained at ~0.605, and the nominal shock pressure† was varied over the range 51–107 kbar (5.1–10.7 GPa; ~7.5 × 10^5–1.57 × 10^6 psi). On its horizontal axis, the shock pressure was maintained at 74 kbar (7.4 GPa; 1.09 × 10^6 psi) while the fractional density prior to shocking was varied over the range 0.57–0.62.

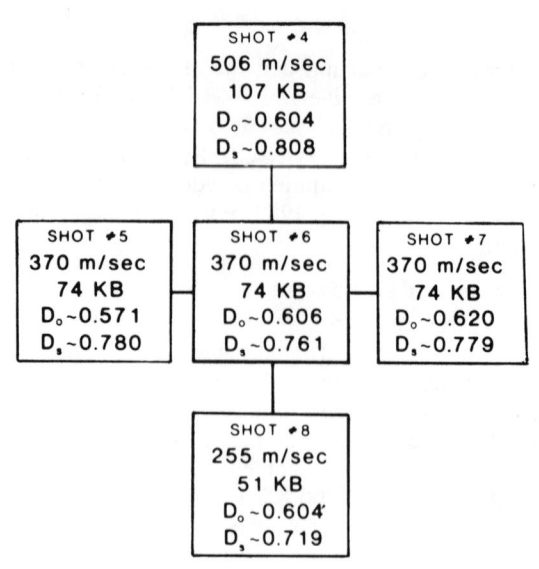

```
        ┌─────────────────┐
        │    SHOT #4      │
        │   506 m/sec     │
        │    107 KB       │
        │  D₀~0.604       │
        │  Dₛ~0.808       │
        └─────────────────┘
                │
┌──────────┬────┴────┬──────────┐
│ SHOT #5  │ SHOT #6 │ SHOT #7  │
│370 m/sec │370 m/sec│370 m/sec │
│  74 KB   │  74 KB  │  74 KB   │
│D₀~0.571  │D₀~0.606 │D₀~0.620  │
│Dₛ~0.780  │Dₛ~0.761 │Dₛ~0.779  │
└──────────┴────┬────┴──────────┘
                │
        ┌─────────────────┐
        │    SHOT #8      │
        │   255 m/sec     │
        │     51 KB       │
        │  D₀~0.604'      │
        │  Dₛ~0.719       │
        └─────────────────┘
```

Fig. 2. Main sequence shock conditioning experiments (Refs. 7, 8, 11).

After recovery, the shock-conditioned powder from each shot was returned to NCSU where it was magnetically deironed, remilled with binder/lubricant additives, and reconstituted by dry-pressing at 137.9 MPa (20 kpsi) to yield small 1.27-cm diameter cylindrical compacts having binder-free green densities ≥0.62.[8,11] Through all these process steps, the chemical composition of the alumina material remained substantially unchanged. ICP analyses (courtesy of R. Meisenheimer, LLNL) showed impurities in the prepared compacts to be typically 20–100 ppm Mg, 200–300 ppm Si, 60–100 ppm K, 60 ppm Fe, and 35–100 ppm Ca. These specimens, along with similarly prepared control specimens of unshocked alumina, were then sintered in flowing, filtered air under a variety of firing conditions, making use of a precision digital dilatometer.[16]

Results

Shock-Induced Alterations of the Powders

A variety of techniques have been employed in characterizing shocked powders after remilling, including residual strain measurements by X-ray line broadening (Fig. 3) and particle size distributions by sedimentation technique (Fig. 4), and in reconstituted compacts, pore size distributions by Hg intrusion porosimetry (Fig. 5). Other methods employed have included BET,[7,8,11] DTA,[8,11] optical microscopy,[12] SEM,[7–12] and TEM.[7,8,11]

The results obtained as a function of shock pressure (vertical axis, Fig. 2) have been reported in part elsewhere.[8–12] As expected, the shocked powders showed increasing residual strain with increasing shock pressure, as a consequence of localized plastic flow (creation and movement of dislocations) occurring under shock conditions. However, they also displayed a concomitant (and unexpected) coarsening of the particles through some shock-induced strong aggregation process, together with reductions in the mean sizes of pores, and a broadening of the pore size distributions. The data presented here (Figs. 3–5) show generally similar effects, but in this instance, the observed differences are attributable only to

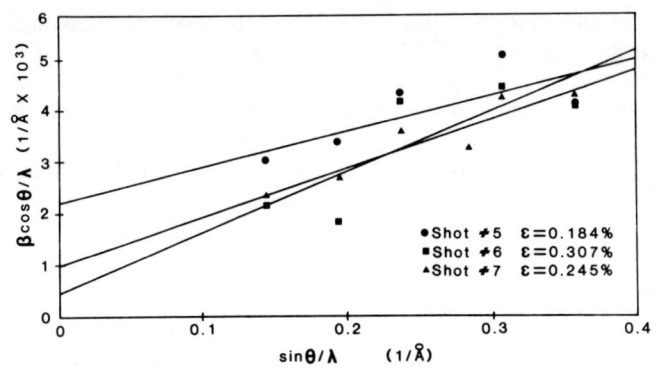

Fig. 3. Hall–Williamson plots of X-ray line broadening data for alumina powders shocked at 7.4 GPa.

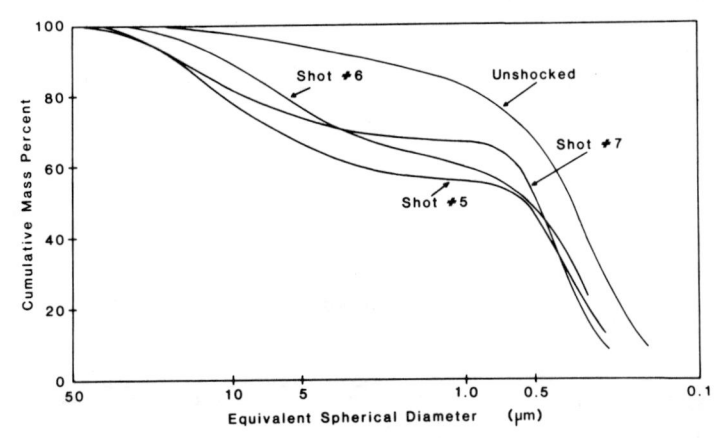

Fig. 4. Particle size distributions comparing milled alumina powders shock conditioned at 7.4 GPa to unshocked, milled master batch. Sedigraph data from Micromeritics Instrument Corp., Norcross, Ga.

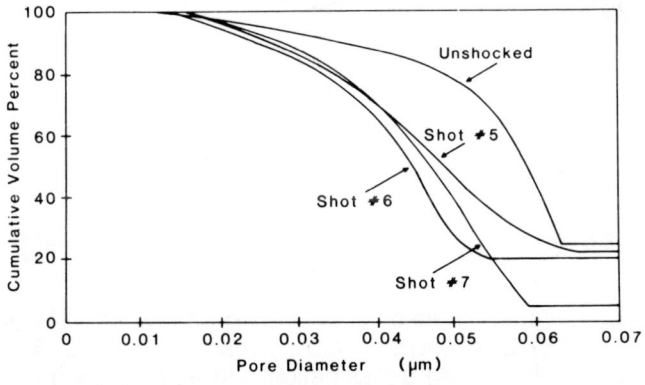

Fig. 5. Pore size distributions comparing reconstituted alumina compacts shock conditioned at 7.4 GPa to compacts of unshocked master batch.

deliberate variations in the compact density prior to shock conditioning at a constant nominal shock pressure of 74 kbar (horizontal axis, Fig. 2). Apart from some shock-induced comminution (primarily affecting the low end of the size distribution), the available microscopic evidence indicates that, though the shapes and contact (neck) areas were altered, the sizes of the individual, fine particles of alumina remained substantially unaffected. Rather, it is primarily the strong reaggregation of those particles by the shock waves that has caused the size distributions and surface areas to have been extensively modified by the various shock treatments.

Sintering Behavior

As discussed elsewhere,[8-12] all of the materials that had been shock conditioned in the main sequence experiments shown in Fig. 2 were found to be *overcompacted* ("overshocked," in the sense that excessive shock-induced thermodynamic heating and resultant strong particle–particle bonding at contact points had occurred). In such overcompacted powders, anomalous sintering behavior was consistently displayed. In Fig. 6, those findings are summarized for one of the

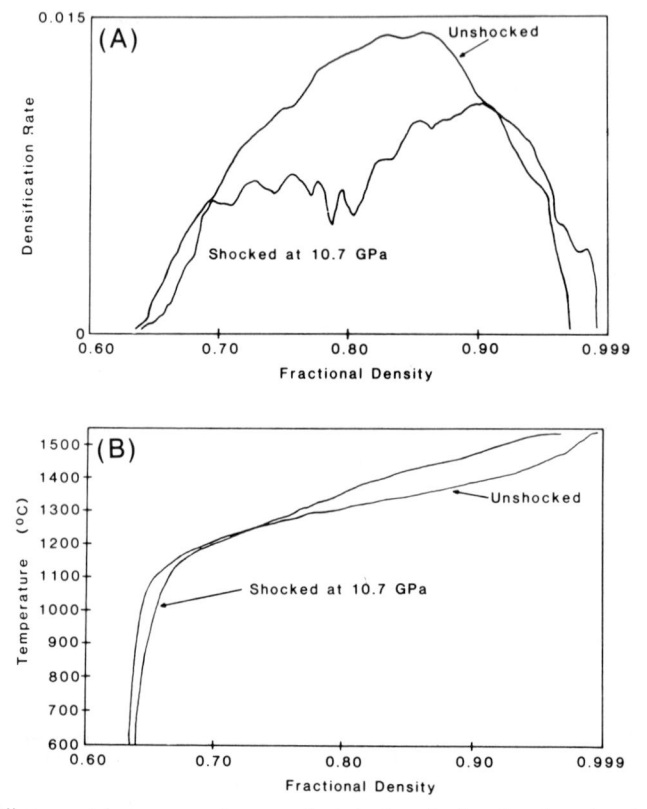

Fig. 6. Dilatometric comparisons of sintering behavior for alumina shock conditioned at 10.7 GPa and unshocked master batch (CTS: 15 K/min to 1803 K, 30-min hold): (*A*) densification rate as a function of fractional density; (*B*) temperatures required to attain given fractional densities (Refs. 8 and 11).

materials precompacted to $D = 0.605$, shock conditioned at the highest pressure, 10.7 GPa (see vertical axis, Fig. 2).[8,11] On this axis, highly bimodal grain size distributions were observed in the final microstructure. Both strain-related recrystallization effects (in the intermediate stage of sintering) and morphologically dictated grain growth effects (at the transition to final-stage sintering) appeared to be involved.[8,12]

The kinds of kinetic effects observed in these unshocked and shocked (overcompacted) alumina powders are compared in Fig. 7. The unshocked alumina

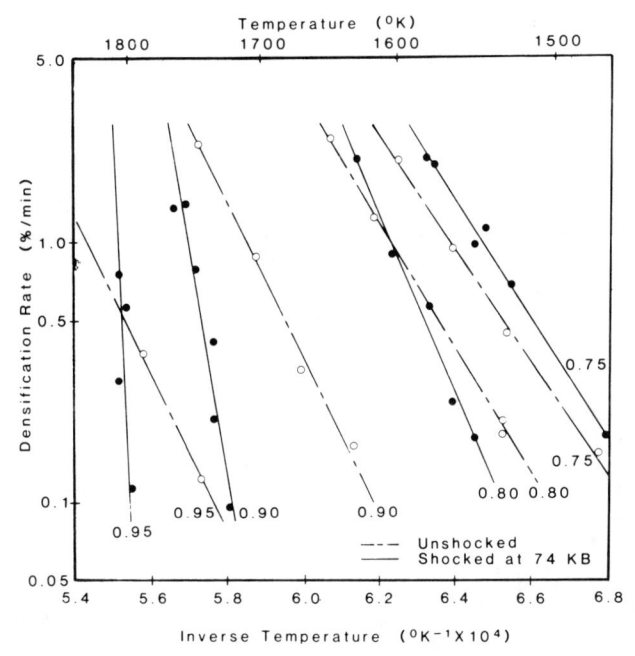

Fig. 7. Apparent densification kinetics for alumina shocked at 7.4 GPa (shot No. 6) and unshocked master batch (Ref. 12).

(dashed lines) yields (1) essentially constant slopes (proportional to the apparent activation energy) and (2) rather regular spacings of the successive isodensity lines over the whole range of densities investigated, in very good agreement with earlier findings[17] for a similar fine, high-purity alumina. Initally (e.g., at $D = 0.75$), the shocked material displays enhanced sinterability, as evidenced by (1) lower temperatures and (2) a slightly steeper slope. As densification proceeds, the slopes of the isodensity lines progressively rotate to steeper angles for the shocked material, and their spacings, particularly at low densification rates, become more widely separated. The adverse consequences of overcompaction are most evident in comparisons of the shocked and unshocked data at $D = 0.90$ and 0.95, especially when densified at slow rates. The expectation from Fig. 7 that sintering of such overcompacted powders at high densification rates would be advantageous has been in part confirmed by earlier experiments in the rate-controlled (RCS) and CTS sintering modes.[8,11]

Figure 8 shows similar dilatometric plots for specimens taken from shock-conditioned powders along the horizontal, variable density axis of Fig. 2. An unusual double peak has been found to be characteristic of overcompacted, shock-conditioned alumina: it occurs at the transition from intermediate to final stage densification. The second peak, resulting from relative reacceleration of the rate

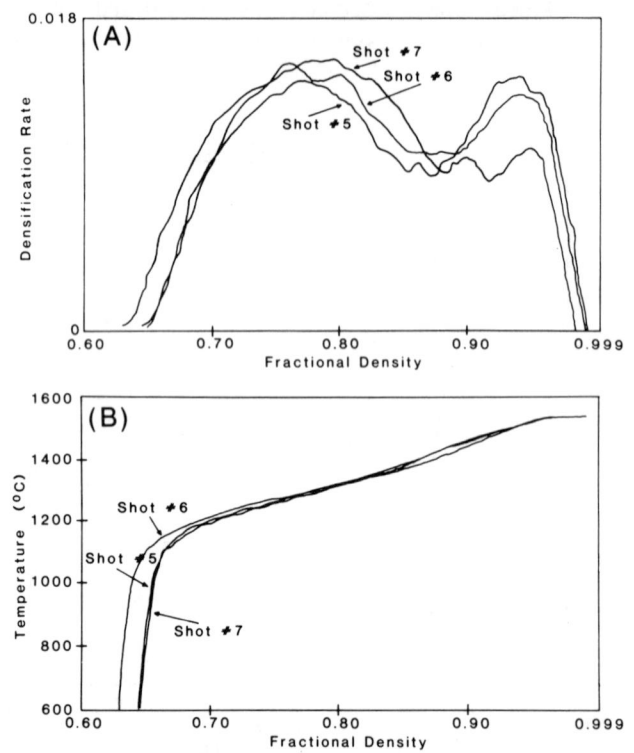

Fig. 8. Dilatometric comparisons of sintering behavior for alumina shock conditioned at 7.4 GPa: (A) densification rate as a function of fractional density; (B) temperatures required to attain given fractional densities.

at that stage, remains essentially invariant with sintered fractional density ($D = 0.93$–0.95). However, its magnitude is inversely dependent on the fractional green density of the initial powder compact prior to shock conditioning. The higher the initial compact density, the more the late-stage peak is suppressed during subsequent sintering. Figure 9 illustrates resultant final microstructures along this density axis with the grain growth being coarsest (and most uniform) for the lowest initial density (shot No. 5). Various degrees of uncontrolled secondary grain growth effects are shown, generally comparable to those previously observed along the vertical shock pressure axis.[8-12]

The amount of waste energy (thermodynamic heating) created during a shock wave event in a powder compact at any given shock pressure is known to increase as the initial compact density decreases (e.g., see Fig. 4 of Ref. 1). Both the magnitude of the late-stage rate peak and the coarseness of the final microstructure of these overcompacted, sintered powders are considered to have resulted from

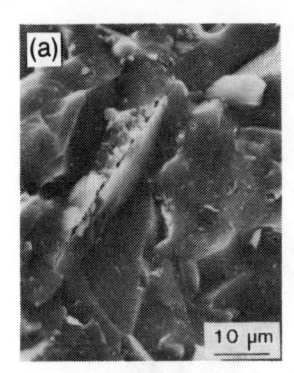

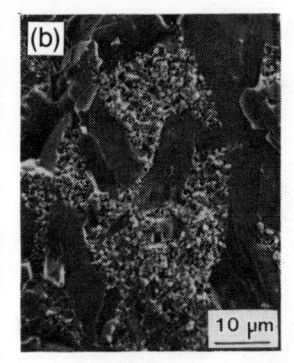

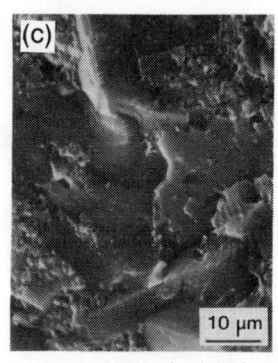

Fig. 9. Sintered microstructures for reconstituted compacts of alumina shock conditioned at 7.4 GPa (CTS: 15 K/min to 1803 K, 30-min hold): (a) 0.571 (shot No. 5); (b) 0.605 (shot No. 6); (c) 0.62 (shot No. 7).

morphological changes attributable to thermally activated strong bonding at alumina grain–grain contact points during shocking at pressures ≥51 kbar.

It recently has been shown that these shock-related sintering anomalies, especially the observed late-stage reacceleration of densification rate (i.e., in the second peak) coupled with discontinuous growth of very large grains in the final microstructure, are entirely morphological–topological in nature.[‡12] The second peak effect observed in alumina apparently is characteristic of inhomogeneously distributed agglomerate structures, regardless of their origins. It can be closely reproduced in preshocked but fully annealed material, as well as in unshocked but presintered, crushed, remilled, and recompacted material.[12]

In earlier experiments, somewhat different compact densities, thicknesses, encapsulation procedures, etc. had been employed to produce alumina powder shock conditioned at 56 kbar (nominal),[7,13] which sintered very well along a single peaked dD/dt vs D curve and displayed both a high final density and a fine microstructure, superior to that of unshocked controls.[7] That preliminary experiment (shot No. 2) clearly served as an example of beneficial *shock conditioning*[2] (i.e., case 2: yielding friable, shocked particulates[1,9]).

On the other hand, later experiments shown in Fig. 2 were found to be *overcompacted* (i.e., to have entered the *dynamic compaction regime*[5] where thermodynamic heating effects have become pronounced). Especially when accentuated by release-wave cracking of the shocked powder compact (see Fig. 5 of Ref. 1), the strong bonding processes that accompany overcompaction must inevitably lead to highly bimodal size distributions in the aggregated, only partially comminuted powders.[9] In these experiments, the overcompacted aggregates were found to be quite resistant to subsequent comminution (Fig. 10).[8,9] These shock-induced strong aggregation effects have been shown to yield an undesirable, overcompacted powder (i.e., case 3: locally strongly bonded, but still partly friable, not readily recomminutable, and hence irrevocably bimodally distributed[1,9]) that thereafter displays consistently poor sintering characteristics and excessive grain growth.

To demonstrate this point, another compact having the same nominal dimensions, fractional density, etc. was shocked in the same type of recovery fixture but at a lower shock pressure, 42 kbar (4.2 GPa, 6.17 × 10^5 psi). The results from this

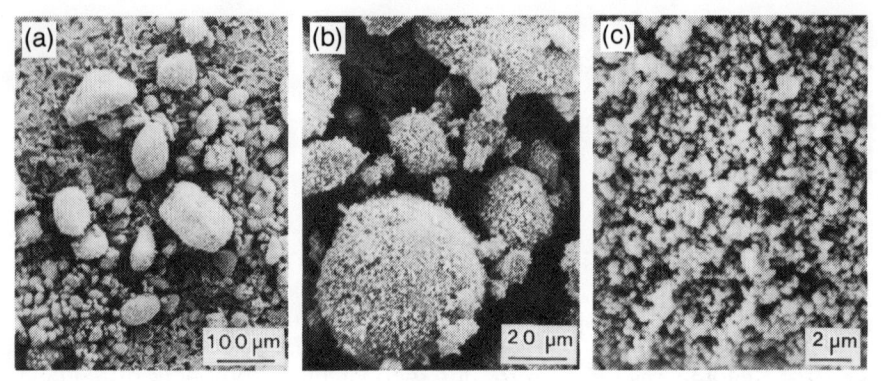

Fig. 10. Strongly aggregated, overcompacted particles from shot No. 6 (7.4 GPa) at increasing magnifications (a), (b), and (c). Note that the strongly bonded aggregates persist in all size ranges, even after intensive dry ball milling (18 h).

experiment (identified as shot No. 8), compared with unshocked control specimens and with overcompacted material from shot No. 6, are presented in Figs. 11–14. Although the nominal shock pressure for shot No. 8 had been ~43% less than that of shot No. 6 (74 kbar) and only about 18% less than that of shot No. 8A (51 kbar, Fig. 2), the results suggest that significant differences had occurred between these samples in the way shock wave effects (including reflections and interactions resulting in step-wise "ringing-up" of the pressure) had been accommodated within the powder. Presumably, the level of localized thermodynamic heating had been substantially reduced at the 42 kbar level: fragmentation and/or deagglomeration, rather than extensive strong aggregation, appears to have occurred, as evidenced by the quite narrow particle size and pore size distributions shown for shot No. 8 in Figs. 11 and 12, respectively. This interpretation is also supported by microscopic observations of the powders and by equivalent spherical

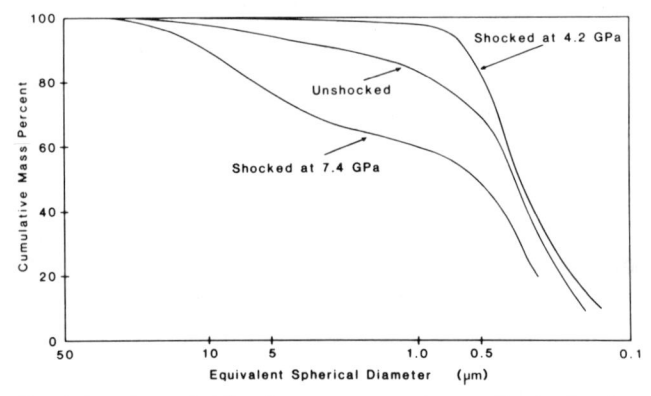

Fig. 11. Particle size distributions comparing milled alumina powders shock conditioned at 4.2 GPa (shot No. 8) and 7.4 GPa (shot No. 6) to unshocked master batch.

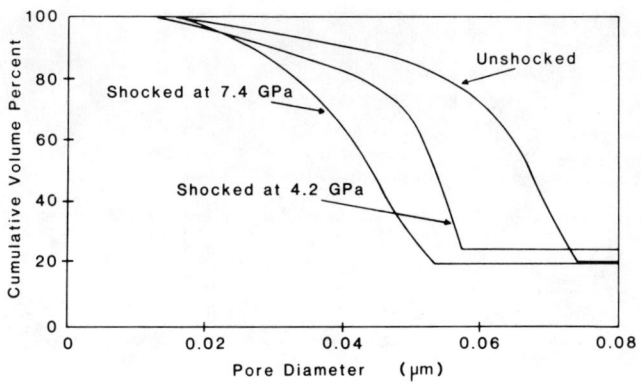

Fig. 12. Pore size distributions comparing reconstituted alumina compacts shock conditioned at 4.2 and 7.4 GPa to unshocked master batch.

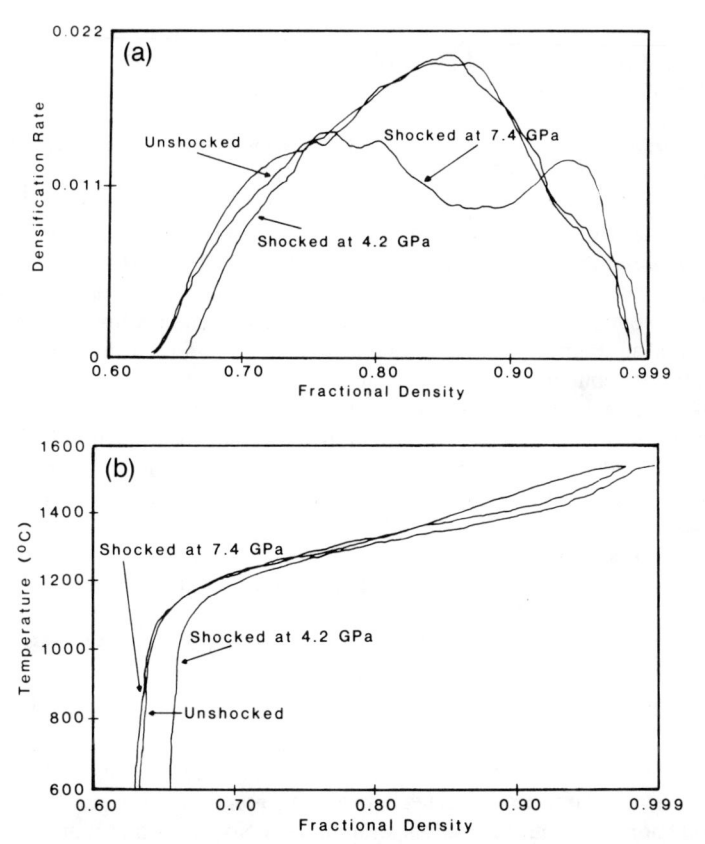

Fig. 13. Dilatometric comparisons of sintering behavior for alumina shock conditioned at 4.2 and 7.4 GPa, and unshocked master batch (CTS: 15 K/min to 1803 K, 30-min hold): (a) densification rate as a function of fractional density; (b) temperatures required to attain given fractional densities.

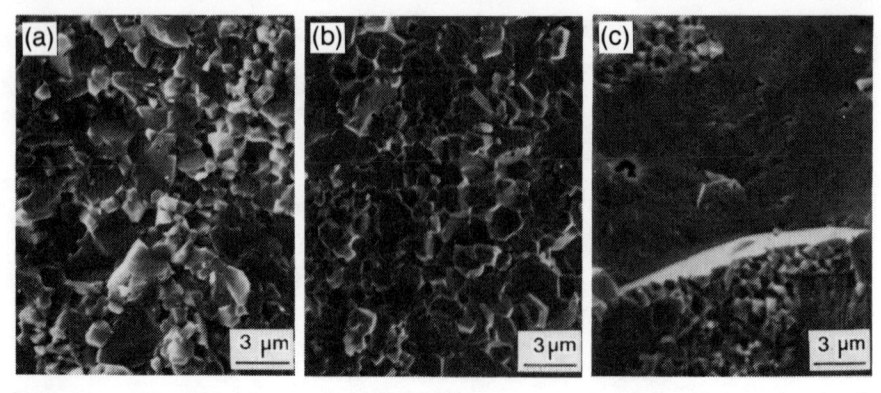

Fig. 14. Sintered microstructures for compacts of alumina (a) unshocked, (b) shock conditioned at 4.2 GPa, and (c) shock conditioned at 7.4 GPa (CTS: 15 K/min to 1803 K, 30-min hold).

diameters calculated from BET specific surface data (unshocked, 12.5 m^2/g, ~0.120 μm; 42 kbar, 11.45 m^2/g, ~0.132 μm; 74 kbar, 10.6 m^2/g, ~0.142 μm).

The sintering curves in Fig. 13 make it evident that the shot No. 8 42-kbar material sinters in the constant rate of heating (CRH) mode along a *single-peaked* rate profile, which is generally similar to that of the unshocked control. It does *not* display the *double-peaked* response (involving both the retardation of rate at $D > $ ~0.75 and the reacceleration of rate at $D \sim 0.90$) characteristically shown by shot No. 6 and by all other overcompacted aluminas examined to date.[8-10] In Fig. 13(a), note that up to $D \sim 0.75$, both the shocked specimens display more acceleration of sintering rates than the unshocked control (in this plot, shown as $[d(dD/dt)]/dD$, but in dD/dt vs t space, it is equivalent to dD^2/d^2t). The shot No. 8 42-kbar material retains its maximum densification rate, ~2%/min, to a higher fractional density, $D \sim 0.88$, than does the unshocked control, $D \sim 0.85$. In Fig. 13(b), note that, to reach an equivalent density, the shot No. 8 42-kbar material consistently requires a lower temperature than does the unshocked control, whereas the overshocked shot No. 6 74-kbar material requires higher temperature, particularly above $D \sim 0.80$. Figure 14 illustrates the resultant final microstructures. It shows that the properly shock-conditioned powder from shot No. 8 yields a finer and more uniform microstructure than does the unshocked control. The overshocked material from shot No. 6 yields a less dense compact having highly bimodal grain size distributions.

Annealing Studies

To assist in distinguishing the possible effects of shock-induced lattice strain-producing defects from other shock-induced morphological–topological changes in the powder assemblage that materially affect sinterability, a separate study of the isothermal annealing out of residual strain in shot No. 6 74-kbar material has been completed.[18] To avoid as much as possible the topological aspects of the sintering of particulate masses (as well as isothermal lag times), the shocked powders were loosely and thinly spread over a shallow Pt foil dish in an alumina refractory boat. The results, summarized in Fig. 15 and Table I, indicate that both the annealing effects and sintering kinetics are strongly, but differently, rate dependent; the

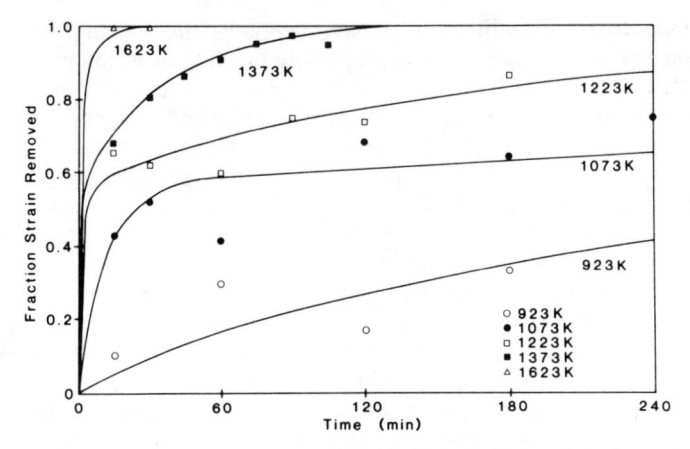

Fig. 15. Temperature–time-dependent release of shock-induced residual strain in alumina shock conditioned at 7.4 GPa (Refs. 12 and 18).

higher the heating rate, the greater the residual strain remaining at a given fractional density during sintering.[12,18] For example, these calculations predict that at 20 K/min, ~28% of the original strain would still be present at $D = 0.70$, but at 4 K/min, only 4% would remain at that density level.[12]

Table I. Calculated Fractional Residual Strain in Shot No. 6 74-kbar Alumina Powder at Increasing Levels of Fractional Density for Heating Rates of 5 K/min and 20 K/min, Respectively*

	5 K/min		20 K/min	
Density	Temp (K)	Remaining strain	Temp (K)	Remaining strain
0.65	1273	0.29	1277	0.42
0.70	1426	0.04	1501	0.28
0.75	1490	0.01	1572	0.19
0.80	1553	<0.01	1621	0.12
0.85	1622	0.00	1673	0.06

*References 12 and 18.

Shock Conditioning

From the results obtained with this particular combination of material, preprocessing, recovery package design, and planar shock configuration, it is evident that proper shock conditioning (yielding case 2 powders[1,9]) will be enhanced by ensuring the attainment of high fractional green density in the precompacted powder (e.g., $D_0 \geq 0.65$). This would yield a high mean coordination number, n_c, which would also be relatively narrowly distributed.[7,19] Highly coordinated particles would thus be better able to resist extensive rearrangement by shearing displacements, even under the influence of highly dynamic shock pressures. From overall waste energy considerations, as well as locally intense frictional heating effects at sliding contact points,[1] one of the principal benefits of increasing the initial compact density will be to reduce thermodynamic heating and the concommitant risk of undesirable particle–particle bonding or welding.

As demonstrated by Figs. 4, 5, and 8, even modest increases in D_0, and thus in n_c, cause a substantial reduction in the late-stage second peak of the rate profile. This phenomenon, first observed in overcompacted shocked powders,[8] has been unambiguously correlated with the presence of coarse, strong aggregates, distributed bimodally among a matrix of finer, dispersed (but also aggregated) particles.[12]

The adverse effects of overcompaction (yielding case 3 powders[1]) have been consistently observed in these experiments with CR-10 alumina at nominal shock pressures $\geq$50 kbar. Note, for example, the marked differences between *shock conditioning* at 42 kbar (shot No. 8, yielding case 2 powders[1]) and *overcompaction* at 51 kbar (shot No. 8A, resulting in case 3 powders[1]). Shock propagation is wavelike in nature, including complex reflections, interactions, etc. Hence, shock pressure buildup is discrete and stepwise, rather than being smoothly continuous, so it is not particularly surprising that shock pressure dependent responses during subsequent sintering should also be selectively discrete. Clearly, small increments in the nominal shock pressure might result in large differences in the character of shocked powders.

There is a real need for knowledge about actual pressure levels attained within powders during shocking. It must come from computer simulations as well as direct experimentation and should be carried out in appropriate shock conditioning modes rather than in the geometries and loadings normally used for direct dynamic compaction (in the latter case, high shear displacements and associated thermodynamic heating effects are usually required to ensure proper bonding of particles[5]).

From the sinterer's viewpoint, then, the message to the powder shocker has become increasingly clear: *Strain the grains, but don't aggregate (or overheat) the particles!*[9]

Benefits of Shock Activation

The benefits during subsequent densification attributable to shock activation of a variety of ceramic powders have been discussed elsewhere.[2-11,14] Results obtained in this study of the sintering of undoped CR-10 alumina powder are summarized in Fig. 16, which compares the final densities obtained in a series of CTS firings (15°C/min to 1400, 1450, 1500, or 1530°C, respectively, with a 30-min hold at temperature) for three different initial powder conditions: unshocked (master batch); shock conditioned (4.2 GPa, shot No. 8); overcompacted (7.4 GPa, shot No. 6).[11] All compacts had been carefully prepared under similar conditions; however, they did display some shock history dependent variations in fractional green density.

As one would expect from Fig. 7, the unshocked control material shows a smooth, almost linear progression of density with increasing temperature of sintering. The properly shock-conditioned material (4.2 GPa) consistently sinters better, attaining given levels of fractional density at temperatures ~65–75°C lower than did the unshocked control material. These findings clearly support Bergmann and Barrington's original observations of beneficial shock conditioning in alumina.[2] By contrast, during sintering, the overcompacted material (7.4 GPa) lags far behind the unshocked control at temperatures below ~1500°C and reaches only very slightly higher densities at 1500 and 1530°C. The resultant microstructures (shown for the 1530°C firing in Fig. 14) show the properly shock-conditioned

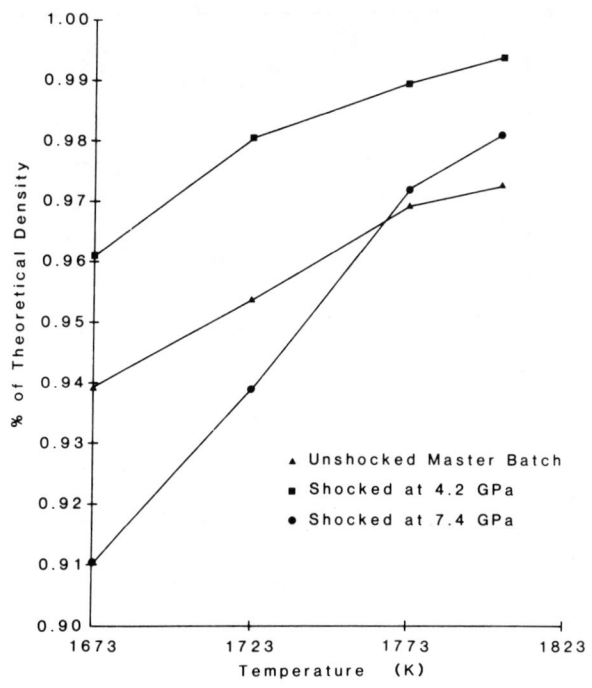

Fig. 16. Temperature dependence of fractional density for sintered CR-10 alumina compacts, shocked and unshocked. Sintered to a given temperature at 15°C/min, with a 30-min hold (Ref. 11).

material (4.2 GPa) to be finer and more uniform than the unshocked control. Obviously, both are far superior to the badly bimodal, coarse-grained textures that resulted from overcompaction at 7.4 GPa.

Particle Packing and Sintering

Sintering and microstructure development are known to be very dependent on specific morphological–topological conditions.[20,21] These issues are, for example, implicit in the well-known MIT paradigm for successful sintering (pure, fine, uniform, well packed, etc.[22]). They are treated in a more fundamental way in Kuczynski's statistical theory of sintering.[23] In that theory, the principal morphological–topological parameter is expressed in terms of the width of the pore size distribution: the narrower the initial pore size distribution, the better the sintering, and the finer and more uniform the final grain size. Though no attempt has been made to "fit" the data from these experiments to the Kuczynski model per se, pore size distribution has indeed been shown to be an excellent predictor for sinterability (e.g., compare Figs. 5 and 8 and 12 and 13). It has also shown that purely morphological–topological effects can profoundly alter the apparent densification kinetics (see Figs. 6, 7, 8, and 13).[8–12]

From the sinterer's viewpoint, then, the message to those who would study kinetics has also become clear: *Avoid overinterpreting the apparent densification kinetics in terms of mechanisms unless (or until) the morphology is closely controlled and well understood!*

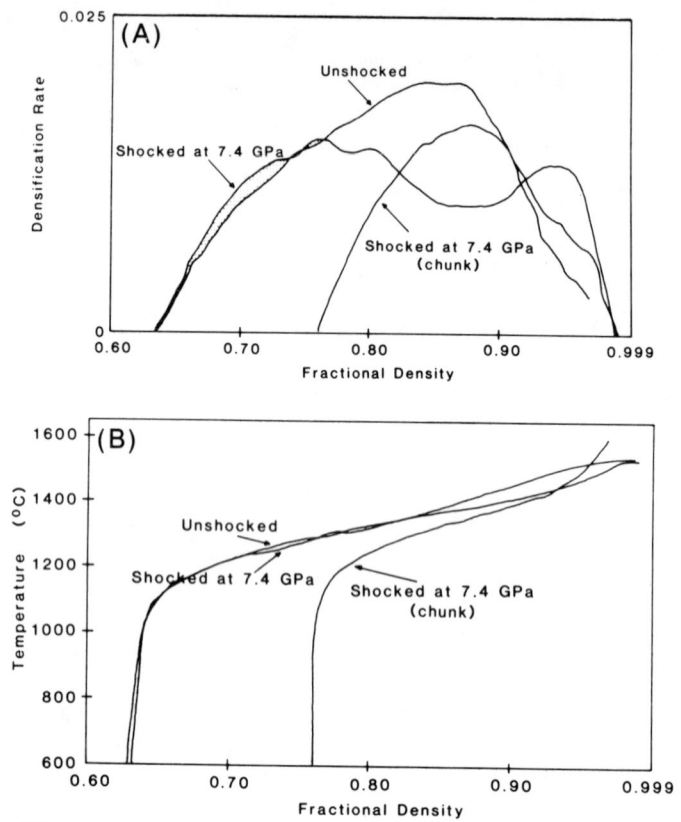

Fig. 17. Dilatometric comparisons of sintering behavior for remilled alumina shock conditioned at 7.4 GPa, a recovered chunk shock conditioned at 7.4 GPa, and unshocked master batch (CTS: 15 K/min to 1803 K, 30 min hold): (A) densification rate as a function of fractional density; (B) temperatures required to reach given densities.

Energetic vs Morphological Effects

One of the most interesting questions, of course, is the degree to which shock conditioning of a ceramic powder can produce any strain- and/or defect-related effects on subsequent sinterability that are separable and distinct from other known effects attributable to concomitant shock-induced morphological changes in the particles and/or particle assemblages. Figure 17 compares the sinterability of CR-10 alumina in shocked and unshocked states under closely controlled and highly comparable conditions. It shows dD/dt vs D and T vs D plots, respectively, for two dilatometric firings carried out for unshocked and shocked (shot No. 6, 74 kbar) materials at 15 K/min to 1803 K (1530°C) with a 30-min hold. Though the two paths are prior history dependent and significantly different from one another, the beginning and end points, D_0 and D_f, are almost identical. For illustrative purposes, data from a 20 K/min firing of a "chunk" specimen (i.e., case 1, recovered intact and substantially free of gross flaws[1]) from shot No. 6 has also been superposed.

There are several important points to be made in connection with the data given in Fig. 17. First, it is important to recognize that even from high shock pressures, undisturbed case 1 materials (as represented by this "chunk," $D_0 \sim 0.76$) sinter easily; they follow a *single-peaked* rate profile and yield fine textures via *normal* grain growth patterns. It is only when the powder is imperfectly dispersed and reconstituted in compacts having unavoidably bimodal size distributions that the *double-peaked* profile so characteristically displayed in this study by overcompacted materials is observed. For such case 3 powders, the final broad size distributions are thought to result not only from the initial strong bonding stage that accompanies the initial dynamic compression but also from a later fracturing stage that must accompany the necessary intense tensile pressure release wave. Individually, recovered "chunks" of overcompacted material are found to be rather uniformly compacted to a high intermediate level of fractional density (see Figs. 2 and 17). However, because in bulk powders the recovered fragments are individually already quite strongly bonded, very randomly distributed, and quite comminution resistant (see Fig. 10), no practical way has been discovered to reconstitute them successfully and uniformly into an easily sintered compact.

Second, the densification rates attained at given levels of fractional density (to $D \leq 0.90$) by the chunk material are appreciably lower than those attained in unshocked material. This is in keeping with the greater resistance to movement resulting from the very high initial density (~ 0.76) and coordination number (>10) of the shock-compacted chunk material.[7]

Third, there appears to be a causal relationship between the onset of sintering of embedded strong aggregates (i.e., small chunks) and the unusual deceleration of densification in the reconstituted material. Presumably, as the fine matrix begins to sinter, the coarse aggregates initially act only as inert inclusions (roughly equivalent to gravel in a concrete mix). At higher temperatures, when the denser coarse aggregates begin to sinter (but at relatively slower rates), the observed overall densification rates, as well as the internal balance of interparticle forces, become significantly altered.

These phenomena are considered in principle to be related to the differential sintering recently discussed by F. F. Lange et al.[24–26] Those authors define the differential strain developed after some sintering between agglomerates (subscript *a*) and the matrix (subscript *m*) as

$$\Delta \varepsilon = \varepsilon_m - \varepsilon_a = (\rho_{oa}/\rho_a)^{1/3} - (\rho_{om}/\rho_m)^{1/3} \tag{1}$$

where ρ_{oa} and ρ_{om} are green densities of the agglomerates and matrix, respectively. For this overcompacted material, where the observed sizes of the fine (ultimate) particles (but not necessarily the neck sizes at their contact points) remain almost invariant through prior shock treatment, the magnitudes of the stresses developed by the differential strain are thought to depend on the stress-strain and strain-rate sensitivity, which in turn depends on the green density and densification rate(s) of the coarser agglomerates (strong aggregates) and the finer (but still somewhat aggregated) matrix. In the case of overshocked material, from the microstructures of Fig. 10 and from Fig. 17, it is clear that the green density of the matrix must be less than that of the hard aggregates.[7,9] The matrix material initially densifies more rapidly, so it will exert a compression strain on the agglomerates, as mentioned in Ref. 25. This case is considered to be different from the usual one[27,28] in which the agglomerates densify more quickly; thus, reduction of densification rate in the initial stage of sintering is not observed. However, it is believed that the anomalous sintering kinetics observed over the range ~0.75–0.90 should be attributed in part to the effect of stress field variation.[24-26]

Finally, attention is directed to the appreciable acceleration in densification rate displayed in Fig. 17 by the shocked material over the density range ~0.66–0.74, corresponding to $T \sim$ 1403–1563 K. The shaded region is considered to be proportional to energetic shock enhancement effects on the densification kinetics, in generally good agreement with discernible energy releases over the same range of temperatures shown by DTA curves for first anneals of similarly shocked aluminas.[8]

It is in this part of the densification regime, i.e., in the initial and/or early intermediate stages, where it was long ago predicted that the sintering consequences of annealable excess energy (the "little q" effect) would be most likely to be observed.[29] It is also consistent with recent determinations of the kinetics of the annealing out of residual strain in this same shocked material,[18] which showed that, at these heating rates, a significant amount of residual strain remained at 1273 K for times up to 60 min or more. At 1623 K or above, all X-ray evidences of residual strain had been annealed out in times on the order of 5 min (see Table I and Fig. 15).

From other evidences presented in this conference,[30,31] it is clear that the presence of work-induced dislocations in oxide ceramics like Al_2O_3 can result in order-of-magnitude enhancements of apparent diffusivities when compared to the intrinsic, essentially dislocation-free state. Here it has been shown (Table I and Fig. 15) that at the higher heating rates, shock-induced residual strains (i.e., resulting from dislocations) do persist to approximately 1623 K. Thus, they remain until the compact is well into the intermediate stage of sintering.[18] It has also been shown that over this same region, and in comparison to unshocked material, the shock-conditioned alumina demonstrates accelerated densification rates (Fig. 17).

Though this particular shocked powder has been morphologically confounded, the effect of the coarsened, thermally bonded aggregates on the early stages of sintering is minimal: for reasons already discussed, they must be considered as quite inert relative to the finer, discrete particles present. Indeed, in the early part of sintering, these coarser, denser aggregates must be acting as drag components relative to the finer, less dense matrix material, such that they would tend to offset, rather than to aid, any energetic acceleration effects. Even so, shock enhancement effects were observed in the early stages of sintering (shaded regions). This attribution of that observed acceleration to shock-induced energetic (little q) effects[29] seems well founded.

In fine alumina powders, the energetic effect is subtle, requiring careful and closely controlled experimentation to separate it from morphological effects. Obviously, it should become much easier to demonstrate and utilize productively when heavily strained, but not overcompacted, ceramic powders become available.

Further study of sintering behavior and elucidation of microscopic details to determine the mechanisms by which this shock-induced enhancement of sintering proceeds now seems well warranted. Ultimately, these continuing studies should aid in the development of better models for understanding and predicting these kinds of defect-dependent sintering phenomena.

The shock enhancement effect is considered real, being both well documented and strategically located in the right part of the overall densification regime. If properly understood and wisely managed, it can profoundly influence the early development of initial and intermediate stage grain, neck, and pore morphologies, which thereafter will become dominant influences in the development of final stage microstructures.

Summary and Conclusions

Recovered shock-compacted powders from the principal shock experiments (51–107 kbar) were generally friable but also contained strong aggregates (chunks); shocked density, D_s, ranged from ~0.68 to 0.808, typically increasing with nominal shock pressure but decreasing slightly with increasing initial fractional density, D_0.

Residual strain (from X-ray line broadening) was found to increase with increasing shock pressure and, in general, to decrease with increasing fractional density of the initial compact; however, at low compact density, the residual strain was also reduced, presumably by annealing effects resulting from the increased level of shock-induced heating.

In these experiments, all alumina materials shock conditioned at nominal shock pressures ⩾51 kbar were found to be *overcompacted*, as evidenced by bimodal distributions of particle sizes in the milled powders and of pore sizes in the reconstituted compacts.

After being redispersed by ball milling and reconstituted in the form of small cylindrical compacts ($D_0 \sim 0.63$), the shock-conditioned powders, together with unshocked but similarly milled and processed control specimens, were characterized and sintered by precision digital dilatometric methods.

In comparison to unshocked controls, which displayed single-peaked rate vs density profiles, *overcompacted* alumina powders were found to display characteristic densification rate anomalies during sintering under constant-rate-of-heating conditions, consisting of the following: (a) at $D > 0.75$, an abnormally low densification rate; (b) at $D \sim 0.90$–0.95, a short-lived reacceleration of the densification rate, thus creating a second peak in the rate vs density profile.

During sintering, the late-stage second peak characteristics of overcompacted aluminas tended to increase in intensity with increasing nominal shock pressure and to decrease with increasing fractional green density of the initial powder compact.

The observed late-stage second peak of the rate vs density profile is considered to be entirely morphological–topological in origin, stemming from the distinctively bimodal character of the pore size distribution and local density variation induced by the strongly bonded, heavily aggregated (in essence, nondispersable) nature of the overcompacted fine particles.

This type of morphology-dependent, late-stage behavior has also been observed in other overcompacted but fully annealed, shocked aluminas, as well as unshocked specimens that were presintered, recomminuted, reconstituted, and resintered.

The late-stage second peak of the rate vs density profile in overcompacted alumina was found to correspond to the rapid uncontrolled growth of very large grains within a much finer grained matrix; in the final microstructure, the ratio of coarse to fine grains decreased with increases in the initial green density of the compact (i.e., to decrease with *decreases* in shock-induced thermodynamic heating).

Rapid densification paths tended to suppress the adverse effects of shock overcompaction on the final sintered microstructures.

The rate of heating affected the annealing out of residual strains; at high rates, significant levels of residual strain persisted to $D > 0.70$, i.e., well into the active sintering range.

Within the defect-annealing range ($\sim 0.63 < D < 0.75$), compacts reconstituted from shocked powders densified more easily than did unshocked controls.

In an iteration experiment (with $D_0 = 0.618$), the nominal shock pressure was reduced to just 42 kbar (4.2 GPa), yielding $D_s = 0.682$; the observed residual strain was lowered somewhat ($\sim 58\%$ of that previously found at 51 kbar), and the sintering behavior was found to be normal (single peaked) to $D \sim 0.99$, yielding a fine, uniform sintered microstructure.

The slightly lower nominal shock pressure chosen (42 kbar) yields a properly shock-conditioned powder that was apparently free of any adverse morphological conditions; over the range 1673–1803 K (30-min soak), reconstituted compacts of that powder sintered to given densities at temperatures 65–75 K lower than did unshocked controls.

Acknowledgments

This work was supported by the U.S. Army under Contract No. DAAK11-83-K-0002. G. L. Moss (BRL/APG) served as Technical Monitor. Other activities undertaken in a related DARPA-sponsored program under LLNL Subcontract No. 6853501 have contributed substantially to the overall level of experimental and interpretive effort under way within our laboratory. The contributions of that program to the analyses and interpretations reported here are also gratefully acknowledged. We also acknowledge the many contributions of research colleagues at NCSU and elsewhere, and laboratory assistance by M. Bridges, L. Freed, T. G. Goudey, E. M. Gregory, G. S. McGaughey, M. J. Paisley, H. H. Stadelmaier, R. Russell, and M. Ganoza.

References

[1] R. F. Davis, Y. Horie, R. O. Scattergood, and H. Palmour III, "Defects Produced by Shock Conditioning: An Overview"; this volume.

[2] O. R. Bergmann and J. Barrington, "Effect of Explosive Shock Waves on Ceramic Powders," J. Am. Ceram. Soc., 49 [9] 502 (1966).

[3] A. V. Anan'in, O. N. Breusov, A. N. Dremin, V. B. Ivanova, S. V. Pershin, V. F. Tatsii, and F. A. Fekhretdinov; pp. 28–34 in Action of Shock Waves on Refractory Compounds. Translation in Sandia National Laboratories, Rept. SAND80-6009, April 1980.

[4] (a) A. Sawaoka, T. Soma, and S. Saito, "Structure Determination of Boron Nitride Transformed by Shock Compression," Jpn. J. Appl. Phys., 13, 891–2 (1974). (b) A. Sawaoka, S. Saito, and M. Araki, New Type of Boron Nitride Sintered Under Very High Pressure"; Proceedings of the International Conference on High Pressure (AIRAPT VI), 1977. (c) A. Sawaoka, K. Kondo, N. Hashimoto, and S. Saito, "Very High Pressure Sintering of Covalent Compounds"; Symposium of Oxide and Nonoxide Ceramics. Edited by S. Somiya. Tokyo Institute of Technology, Tokyo, Japan, 1978.

[5] Dynamic Compaction of Metal and Ceramic Powders. National Materials Advisory Committee Rept. NMAB-394, National Research Council, March 1983.

[6] E. K. Beauchamp and R. A. Graham (Sandia National Laboratory), "Shock Activated Sintering"; unpublished work.

[7] H. Palmour III, K. Y. Kim, K. L. More, and R. C. Motley, "Sintering and Characterization Studies of Shock-Conditioned and Dynamically Compacted Powders"; pp. 331–72 in First Quarterly Report, DARPA Dynamic Synthesis and Consolidation Program. Edited by C. F. Cline. Lawrence Livermore National Laboratory Rept. UCID-19663, Aug. 1982.

[8] K. Y. Kim, A. D. Batchelor, K. L. More, and H. Palmour III, "Rate Controlled Sintering of Explosively Shocked Alumina Powders," pp. 749–64 in Emergent Process Methods for High Technology Ceramics. Plenum, New York, 1984. Edited by R. F. Davis, H. Palmour III, and R. L. Porter.

[9] Y. Horie, H. Palmour III, and J. K. Whitfield, "Shock Compaction and Sintering Behavior of Selected Ceramic Powders"; pp. 130–216 in Second Quarterly Report, DARPA Dynamic

Synthesis and Consolidation Program. Edited by C. F. Cline. Lawrence Livermore National Laboratory Rept. UCID-19663-83-1, March 1983.

[10]K. L. More, "Morphological Effects on the Sinterability of Shock-Conditioned Alumina"; unpublished work.

[11]K. L. Kim, "Densification of Shock-Conditioned Ceramic Powder by Rate Controlled Sintering"; Ph.D. Dissertation, Department of Materials Engineering, North Carolina State University, Raleigh, NC, 1983 (under the direction of H. Palmour III).

[12]T. M. Hare, K. L. More, A. D. Batchelor, and H. Palmour III, "Sintering Behavior of Overcompacted Shock-Conditioned Alumina Powder"; pp. 265–80 in Sintering and Heterogeneous Catalysis. Edited by G. C. Kuczynski, A. E. Miller, and G. A. Sargent. Plenum, New York, 1984.

[13]R. R. Wills, J. H. Adair, and V. D. Linse (Battelle Columbus Laboratories), First Progress Report on Work Conducted in the DARPA Dynamic Synthesis and Consolidation Technology Program; pp. 1–79 in Ref. 7.

[14]R. A. Graham, E. K. Beauchamp, B. Morosin, E. Venturini, D. M. Webb, and L. W. Davison (Sandia National Laboratories), "Shock Activated Sintering"; pp. 258–97 in Ref. 7.

[15]M. L. Wilkins and C. F. Cline (Lawrence Livermore National Laboratory), "Computer Simulation of Dynamic Compaction"; pp. 163–89 in Ref. 7.

[16]A. D. Batchelor, M. J. Paisley, T. M. Hare, and H. Palmour III, "Precision Digital Dilatometry: A Microcomputer-Based Approach to Sintering Studies"; pp. 233–51 in Ref. 8.

[17]H. Palmour III, M. L. Huckabee, and T. M. Hare, Microstructural Development During Optimized Rate Controlled Sintering; pp. 308–19 in Ceramic Microstructures '76. Edited by R. M. Fulrath and J. A. Pask. Westview Press, Boulder, CO, 1977.

[18]K. L. More, "The Release of Shock-Induced Strain Energy and Its Relationship to the Sintering Behavior of Overcompacted Alumina"; unpublished work.

[19]T. M. Hare, "Statistics of Early Rearrangement by Computer Simulation"; pp. 77–93 in Material Science Research, Vol. 13. Edited by G. C. Kuczynski. Plenum Press, New York, 1980.

[20]R. T. DeHoff, "A Cell Model for Microstructural Evolution in Sintering"; in Ref. 12.

[21]F. N. Rhines and R. T. DeHoff, "Channel Network Decay in Sintering"; in Ref. 12.

[22]K. Uematsu, R. M. Cannon, R. D. Bagley, M. F. Yan, U. Chowdry, and H. K. Bowen, "Microstructure Evolution Controlled by Dopants and Pores at Grain Boundaries"; pp. 190–205 in Ref. 4(c).

[23]G. C. Kuczynski, "Statistical Theory of Sintering and Microstructure Evolution"; pp. 37–44 in Sintering-Theory and Practice, Materials Science Monographs, Vol. 14. Edited by D. Kolar, S. Pejovnik, and M. M. Ristic. Elsevier, Amsterdam, 1982.

[24]F. F. Lange, "Processing-Related Fracture Origins: I, Observations in Sintered and Isostatically Hot-Pressed Al_2O_3/ZrO_2 Composites," *J. Am. Ceram. Soc.*, **66** [6] 396–8 (1983).

[25]F. F. Lange and M. Metcalf, "Processing-Related Fracture Origins: II, Agglomerate Motion and Cracklike Internal Surfaces Caused by Differential Sintering"; pp. 398–406 in Ref. 24.

[26]F. F. Lange, B. I. Davis, and I. A. Aksay, "Processing-Related Fracture Origins: III, Differential Sintering of ZrO_2 Agglomerates in Al_2O_3/ZrO_2 Composites"; pp. 406–8 in Ref. 24.

[27]M. D. Sacks and J. S. Pask, "Sintering of Mullite-Containing Materials, II, Effect of Agglomeration," *J. Am. Ceram. Soc.*, **65** [2] 70–7 (1982).

[28]W. H. Rhodes, "Agglomerates and Particle Size Effects on Sintering Yttria-Stabilized Zirconia," *J. Am. Ceram. Soc.*, **64** [1] 20–2 (1981).

[29]H. Palmour III, R. A. Bradley, and D. Ray Johnson, "A Reconsideration of Stress and Other Factors in the Kinetics of Densification"; pp. 392–407 in Materials Science Research, Vol. 4. Edited by T. J. Gray and V. D. Frechette. Plenum Press, New York, 1969.

[30]Y. Oishi and K. Ando, "Oxygen Diffusion in MgO and Al_2O_3"; this volume.

[31]A. H. Heuer and J. Castaing, "Dislocations in α-Al_2O_3"; this volume.

---

*Grade CR-10 Al_2O_3, a product of Baikowski International, Charlotte, NC.

†The nominal shock pressure is that calculated at the steel-to-steel flyer plate/recovery package interface. Very complex reflections, attenuations, reinforcements, etc. that occur as the shock wave traverses the steel container and the encapsulated powder compact cause the pressure to increase in stepwise jumps ("ringing-up") and make direct estimating of actual pressures in the powder quite difficult. To establish the details of pressure distributions, shear displacements, adiabatic heating effects, etc. computer code simulations must be employed (Refs. 8, 14, 15).

‡These experiments were carried out with a high-purity, undoped alumina, i.e., in a material where late-stage grain growth effects are certainly not unexpected. However, similar experiments carried out with shocked (overcompacted) MgO-doped Bayer-process alumina produced essentially identical results, including double-peaked sintering rate curves, very coarse final microstructures, et al. (Ref. 9). It is evident that the morphological–topological observations resulting from overcompaction during shocking dominate the subsequent sintering behavior, such that the usual grain growth inhibition effects attributed to MgO doping of alumina are in essence negated.

Role of Al₂O₃ in Sintering of Submicrometer Yttria-Stabilized ZrO₂ Powders

R. C. Buchanan and D. M. Wilson

University of Illinois
Department of Ceramic Engineering
Urbana—Champaign, IL 61801

The use of Al₂O₃ (up to ~3.5 wt%) as a sintering aid to promote rapid densification of precipitated yttria-stabilized (8 wt%) zirconia (YSZ) powders in the range 1100–1350°C was investigated. The Al₂O₃ was added as hydrated Al(OH)₃ and dispersed by milling in a 60:40 alcohol:water solution, followed by pressing at 205 MPa. Significantly increased densification was obtained with Al₂O₃, even below 1200°C, and optimum densification (>99.0% of theoretical density) occurred at 1350°C/h with 0.325 wt% Al₂O₃. These samples exhibited enhanced electrical conductivity and larger grain size (0.3–0.5 μm). TEM microstructural observations and densification kinetic data indicated a liquid-phase-assisted sintering mechanism. Solid-state doping of the ZrO₂ by Al was inferred from the electrical conductivity data.

The beneficial effect of Al₂O₃ on the sintering of stabilized zirconia has been noted by several investigators.[1,2] However, the mechanism for the observed densification increases has yet to be adequately explained. An additive such as Al₂O₃ can be accommodated in a host material in one of three distinct ways: as a solid-solution dopant, as a grain-boundary segregant, or as a discrete second phase. Combinations of these mechanisms, as determined by the thermodynamics and kinetics of the system, are also possible.

With many additive systems, densification is affected by the formation of an intergranular liquid phase. This contributes to particle rearrangement through grain-boundary sliding, assists in the dissolution of particle–particle contacts and in some cases provides a pathway for rapid mass transport during sintering. Significant enhancement in the densification behavior of ceramic systems has been observed with liquid contents ≤1.0 vol%[3,4] and capillary forces are largest for small liquid contents.[5] The effectiveness of the intergranular phase is strongly dependent on its composition, since the liquid-phase kinetics is determined by the solution of the solid particles in the melt phase. The presence of an intergranular phase in the fired body can, however, be detrimental to such properties as high-temperature strength[6] and electrical conductivity,[7,8] both of critical importance for electrolyte applications.

Other additives may enhance sintering without the formation of an intergranular liquid phase. Dopants soluble in the host lattice can enhance densification by increasing the defect concentration of the diffusing species. Thus, Harmer et al.[9] attributed increased sintering of high-purity Al₂O₃ (Al^{3+} lattice diffusion controlled) doped with Ti^{4+} to an increase in the aluminum vacancy concentration. Conversely, Mg^{2+} additions promoted sintering by the formation of Al^{3+} interstitials.

Densification enhancement for segregated dopants can be attributed to such effects as decreased grain boundary to surface energy ratios[10] or to reduced grain-boundary mobility due to the presence of discrete solid second phases, pores, or segregated solutes. In those systems with a preference for grain-boundary diffusion, sintering may also be affected by changes in the amount and nature of the boundary impurities. Segregated impurities have been shown to significantly reduce grain-boundary electrical conductivity in YSZ,[8,11,12] possibly due to trapping[7] or to occupation of interstitial sites.[50] Parallel effects may also exist for cationic diffusion.

In stabilized zirconias, the stabilizing oxide may be enriched on the grain boundary due to its affinity for a liquid phase[14] or for a segregated additive. This is particularly likely in calcia-stabilized zirconia (CSZ) due to the high affinity of Ca for most grain-boundary impurities. Thus, the distribution of an additive may be partially determined by existing impurities and solutes as well as by sintering temperature.

Alumina, although only 0.1 mol% soluble in YSZ at 1300°C,[13] can be dissolved up to 1–2 mol% at 1700°C.[12] Bernard[13] reported low grain-boundary conductivities in samples cooled slowly to room temperature but conductivities equivalent to bulk values after rapid quenching.

Radford et al.[2] (YSZ and CSZ), Mallinckrodt[3] (CSZ), and Takagi[16] (CSZ) all have attributed enhanced sintering with Al_2O_3 to a liquid phase formed by the dopant, the stabilizing oxide and existing impurities such as MgO, SiO_2, and CaO. Assuming a sufficient impurity level, this is a plausible interpretation, since numerous eutectics could be formed with the above-mentioned oxides below 1500°C, and alumina has been found to be concentrated on the grain boundaries of CSZ along with associated Ca, Mg, and Si impurities.[8,16] Sintering temperatures investigated by the three investigators were in the range 1480–1800°C.

Radford noted a densification enhancement in CSZ + Al_2O_3 both for nuclear grade (99.7% pure) and lower purity (~97%) technical grade samples. Mallinckrodt noted a density decrease in Al_2O_3-doped samples for the highest firing temperatures (1800°C). Although all zirconia sintering additives such as Al_2O_3, Fe_2O_3, TiO_2, and SiO_2 decreased conductivity,[17,2] Radford reported that the decrease with Al_2O_3 additions was relatively small, especially for the lower purity samples. Takagi noted substantial grain growth with Al_2O_3 additions and Radford noted a decrease in grain size. This difference suggests that the effect of the Al_2O_3 on densification is dependent on sample purity and preparation as well as on firing conditions.

Liquid-phase sintering was first refuted as a possible sintering mechanism by Bernard.[13] The beneficial effect of Al_2O_3 additions was found to be strong as low as 1100°C, well below temperatures where the liquid phase would normally be expected. Microstructural examination indicated that Al_2O_3 was present mainly as second-phase inclusions and the grain boundaries were free of liquid. AC impedance spectroscopy indicated a diminution of intergranular resistance with Al_2O_3 additions, and a net increase in conductivity was reported. This was attributed to increased grain size, which reduced the high-resistivity grain-boundary area. The Al_2O_3 additive level could be varied between 0.44 and 1.70 mol% without affecting electrical or densification behavior. Butler et al.[18] supported these conclusions and proposed that the Al_2O_3 particles acted as scavengers for intergranular SiO_2. As the grain boundaries moved past the (assumed stationary) Al_2O_3 inclusions, intergranular SiO_2 diffused rapidly to the Al_2O_3 particles due to the greater thermodynamic stability of mullite (3Al$_2$O$_3$·2SiO$_2$) compared to zircon (ZrO$_2$·SiO$_2$), forming silica-rich cusps on the Al$_2$O$_3$ inclusions. Densification was attributed to grain-boundary pinning by the Al$_2$O$_3$ inclusions. An increase in conductivity could be assumed due to the removal of amorphous second phases from the grain boundary. The presence of Al$_2$O$_3$ in YSZ as discrete inclusions was also reported by Rao et al.[19] Silica was present at triple points and in <20 nm thick films on grain boundaries, which were depleted in yttrium.

The presence of a continuous, segregated grain-boundary phase was also reported by Verkerk et al. in a study of the electrical behavior of YSZ. Impurities such as Ca, Ti, and sintering aids such as Fe$_2$O$_3$ and Al$_2$O$_3$, which were considered to be enriched on the grain boundaries, reduced boundary conductivity significantly.[20]

Early studies on the kinetics of sintering of >1 μm zirconia showed shrinkage time exponents to be near 0.50, which would indicate bulk diffusion control.[21,22] Creep data has supported this conclusion.[23] Nevertheless, Rhodes and Carter, while observing bulk diffusion control during sintering, found boundary diffusivities to be up to 10^5 times as high as bulk values.[22] Shrinkage exponents near 0.3 in a low-temperature study using Cr$_2$O$_3$-stabilized powders indicated the dominance of grain-boundary diffusion.[24] Young et al. also report grain-boundary diffusion control in zyttrite (~100 Å particle size YSZ),[25] while Wirth noted grain-boundary control in submicrometer CSZ.[26] Changes in sintering pathway from lattice to grain-boundary control as grain size decreases has been observed in both MgO and Al$_2$O$_3$.[27] A grain-boundary sintering mechanism in submicrometer YSZ would suggest considerable sensitivity to segregated impurity ions and to amorphous or second phases.

In stabilized zirconias with large oxygen vacancy concentration, the cation diffusion would be rate controlling. Usually, cation vacancies are assumed, although zirconium interstitials have been identified, though only under high-temperature (1800°C) conditions.[28] As pointed out by Brook,[29] compressive creep data[30] indicated a significant maximum creep rate in CSZ at 15 mol% CaO. This is very close to the point at which full stabilization is achieved, conductivity is at a maximum, and the free oxygen vacancy concentration is greatest. A diffusion rate maximum for zirconium under these conditions would be more consistent with zirconium interstitial control than with vacancy control.

On the basis of the above, the object of this investigation was to examine, mechanistically, the role of Al$_2$O$_3$ as a sintering aid for submicrometer stabilized ZrO$_2$ (YSZ) powders with a view to achieving lower densification temperatures and times, as well as improved optical and mechanical properties.

Experimental Procedures

The powder used in this study was commercially available submicrometer yttria-stabilized (8 wt%; ~9.1 mol% YO$_{1.5}$) zirconia (YSZ). Typical lot analysis and physical properties for the powders used are given in Table I.

Residual chlorine, shown by Scott and Reed[31] to inhibit densification, was removed by washing in distilled water. Dilute suspensions (1.0 vol%) were subjected to ultrasonic vibrations for 15 min, followed by centrifuging and decanting of the liquid. Chemical analysis, carried out by atomic absorption and by Hg titration for Cl, indicated a reduction in Cl content from 0.80 wt% to ≤0.04 wt%. Alumina, obtained from fine-grained aluminum hydroxide (Al(OH)$_3$·3H$_2$O), 99.9% pure, was added as a sintering aid. The additive level of Al$_2$O$_3$ was varied from 0–3.24 wt% (3.92 mol%).

Table I. Typical Lot Analysis for Yttria-Stabilized Zirconia (YSZ) Powders*

Composition (wt%)			
Constituent	wt%	Constituent	wt%
ZrO_2	~90	NiO	0.03
Y_2O_3	8.0	Fe_2O_3	0.01
HfO_2	1.6	SiO_2	0.10
Al_2O_3	0.04	TiO_2	0.06
CaO	0.30	Na_2O	0.20
BaO	0.03	K_2O	0.02
MgO	0.01	Cl	~1.0

Physical properties	
Crystalline phase	cubic
Crystalline size	0.02–0.03 μm
Surface area (BET)	50 m^2/g

*Zircar Corp., Florida, NY.

The as-received powders were ball milled for 12 h with ZrO_2 balls in polyethylene jars to reduce agglomeration. Figure 1 shows the considerable reduction in agglomerate size distribution from ~10–15 μm for the as-received to ~0.3–0.5 μm for the milled powder. Figure 1(B) also shows, from the enhanced fine structure, crystallite sizes in the range 0.02–0.03 μm. The milled suspensions, with 1.0 wt% carbowax 4000 and 1.0 wt% PVA added as binders, were spray dried* and pellets of 1.6 cm diameter and ~0.15 cm thickness were pressed uniaxially at 220 MPa. Firing was carried out on Pt foil on ZrO_2 setters in a $MoSi_2$ furnace in the range 1100–1350°C from 1 min to 24 h.

Sintered densities were determined by the Archimedes technique. The theoretical density of YSZ (8.0 wt%) was calculated to be 6.022 g/cm^3, using the lattice parameter data of Tuohig.[32] Densities calculated by using a series mixing formula decreased progressively with added Al_2O_3, the value for 0.65 wt% alumina being 6.00 g/cm^3.

Fig. 1. SEM photomicrographs of YSZ powders: (A) as-received and (B) after milling for 12 h in 2:1 2-propanol:H_2O solution.

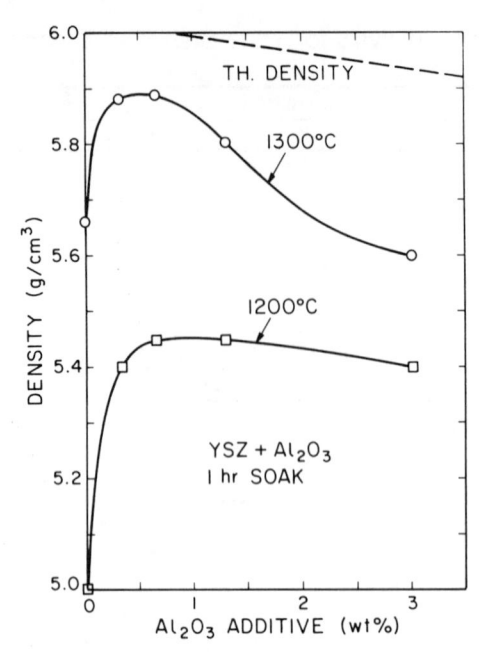

Fig. 2. Plot of fired density for YSZ powder at 1200 and 1300°C for 1 h showing sintering enhancement with Al$_2$O$_3$ additions (wt%).

Microstructures were analyzed by SEM, TEM, and EDS microanalysis techniques. Grain sizes were determined from SEM photomicrographs of polished and thermally etched sections, using the line intersection technique of Mendelsohn.[33] TEM samples were prepared with a ball-cratering device[†] followed by <10 h ion milling, thereby assuring a minimal of milling artifacts. DC electrical resistivity was measured by using a universal bridge.[‡] Specimens were polished plane parallel and provided with Pt paste electrodes, which were fired at 800°C in air. Measurements were made, in air ambient, up to 900°C.

Results and Discussion

Figure 2 shows the effect of Al$_2$O$_3$ additions (0–3.25 wt%) on the fired densities of precipitated YSZ powders at 1200 and 1300°C for 1 h. Densification at 1200°C was significantly enhanced by Al$_2$O$_3$ additions ≤0.65 wt%, with a similar effect noted at 1300°C, where higher overall densities were achieved. Above 0.65 wt% Al$_2$O$_3$ additive content, a relative decrease in densification, more pronounced at 1300°C, was observed. This suggested the optimum Al$_2$O$_3$ additive content to YSZ to be in the range 0.3–0.65 wt%.

Figure 3 compares, for even shorter soak times (0.5 h), the relative densities achieved for YSZ and YSZ + 0.325 wt% Al$_2$O$_3$ samples at sintering temperatures between 1100 and 1350°C. The difference in density between the two samples was seen to increase as the sintering temperature was increased. However, both samples achieved >90% of theoretical density at 1350°C/0.5 h. Figure 4 compares the shrinkage behavior (log $\Delta L/L_0$) as a function of time for the two samples in Fig. 3. The shrinkage data shown corresponded to relative density values in the range 65–92%. Shrinkage for the Al$_2$O$_3$-doped sample was higher than that for the YSZ

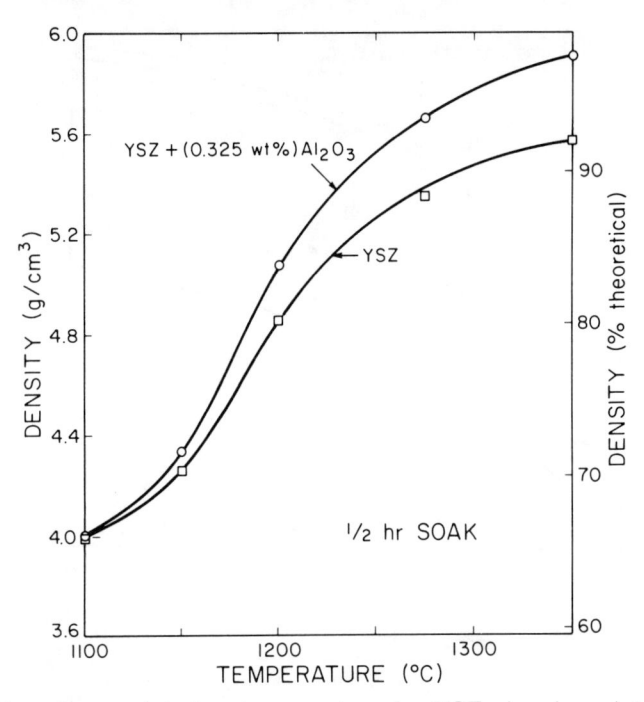

Fig. 3. Density vs sintering temperature for YSZ showing sintering enhancement with temperature for Al_2O_3 additions.

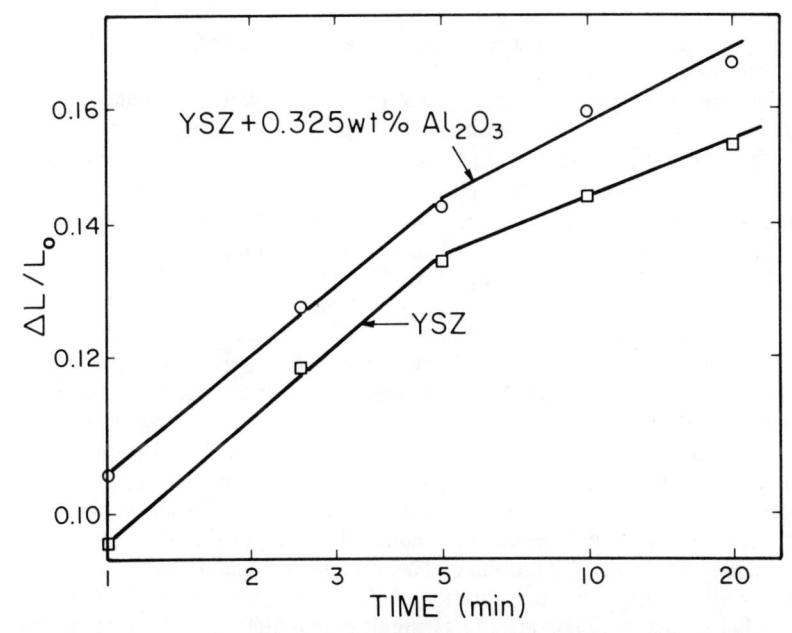

Fig. 4. Plot of shrinkage ($\Delta L/L_0$) at 1275°C as a function of soak time for YSZ and YSZ + 0.325 wt% Al_2O_3 samples.

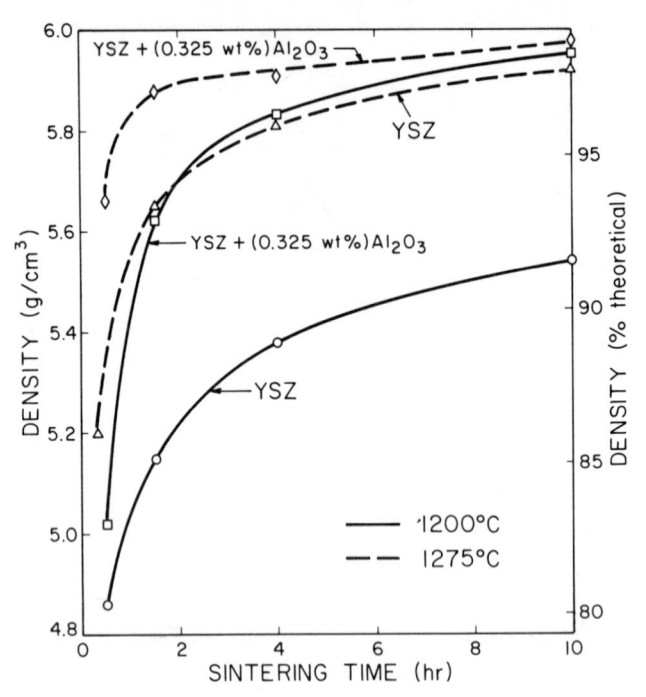

Fig. 5. Plot of density vs sintering time for YSZ at 1200 and 1275°C showing increased densification with Al_2O_3 additions.

sample, in line with the higher densification rate, but the parallel shrinkage curves indicated a similar densification mechanism.

The nonlinearity of the sintering curves in Fig. 3 would indicate liquid-phase densification with different amounts of liquid present. Likewise, the two slopes for the shrinkage curves coupled with the rapid densification rate would classically be interpreted as evidence for liquid-phase sintering. Particle rearrangement would be predominant in the initial stages followed by a solution precipitation mechanism at longer sintering times. This behavior is evident from Fig. 5, which shows the changes in density with sintering time for the YSZ and Al_2O_3-doped samples sintered at 1200 and 1275°C. For short soak times (~0.5 h), these density differences were much greater at 1275°C than at 1200°C, in line with the data presented in Fig. 3. In contrast, after a 10-h soak, the YSZ sample had achieved nearly equivalent density to the Al_2O_3-doped sample at 1275°C, but substantially lower density at 1200°C.

These data suggest that at 1200°C, insufficient liquid was present in either sample to cause significant rearrangement, at least in the YSZ sample, and that subsequent densification could primarily be attributed to solution precipitation and grain-boundary sliding. The higher densification rate for the Al_2O_3-doped sample must, therefore, reflect the presence of a more reactive and perhaps lower viscosity intergranular phase with incorporation of Al_2O_3. Conversely, with the higher expected liquid-phase content at 1275°C, significant initial densification occurred and subsequent densification mechanisms became relatively less important, at least for the Al_2O_3-doped samples.

Table II. Sintered Densities of YSZ and YSZ + Al$_2$O$_3$ Samples for Different Soak Temperatures and Times

Samples	Soak time (h)	Sintered densities (% theoretical)* 1200°C	1275°C	1350°C
YSZ	0.5	80.5	88.0	93.1
(6.02 g/cm^3)†	4.0	89.2	96.3	99.0
	24.0	94.0	99.2	99.7
YSZ + 0.325 wt%	0.5	81.1	94.1	98.3
Al$_2$O$_3$	4.0	97.0	98.5	99.3
(6.01 g/cm^3)†	24.0	99.5	99.3	99.3
YSZ + 0.65 wt%	0.5	83.8	96.5	98.2
Al$_2$O$_3$	4.0	96.9	98.8	99.0
(5.99 g/cm^3)†	24.0	99.0	99.0	99.1
YSZ + 1.30 wt%	0.5	80.7	96.0	97.3
Al$_2$O$_3$	4.0	95.8	97.8	97.6
(5.96 g/cm^3)†	24.0	98.2	98.0	98.3

*Accuracy: ±0.1%.
†Calculated theoretical densities.

Fired density data are given in Table II for YSZ with different concentrations of Al$_2$O$_3$ additive at sintering temperatures of 1200, 1275, and 1350°C and for soak times of 0.5, 4.0, and 24.0 h. The trends in the data are as illustrated in Figs. 2 and 5; that is, densities generally increased with sintering temperature and soak time for all samples. The slightly lower ultimate density achieved by the samples containing 0.325 wt% Al$_2$O$_3$ (99.3% relative density compared to 99.7% for the undoped YSZ) at 1350°C was attributed to Al$_2$O$_3$ inclusions present in the sample. However, equivalent densities could generally be achieved at lower temperatures and for shorter soak times with Al$_2$O$_3$ doping.

Figure 6 shows SEM photomicrographs of polished and thermally etched sections for the samples in Table II that were fired at 1275°C/4 h. No second phases were evident from the photomicrographs presented except for few intergranular pores present in the YSZ sample. The YSZ sample also showed evidence of stacking faults in the grains, indicative of lattice strain in the sample. Stacking faults were not very evident in the YSZ samples which were doped with Al$_2$O$_3$. This would be consistent with the existence of a liquid phase and the dissolution of Al in the ZrO$_2$ lattice. Indications of exaggerated grain growth were also present for the 1.3 wt% sample especially at higher sintering temperatures.

Figure 6 shows an increase in the fired grain sizes as Al$_2$O$_3$ was added. Average grain sizes determined were approximately 0.36, 0.40, 0.38, and 0.37 μm for samples A, B, C, and D. These differences may be only marginally significant, but measurements were made on several samples. In any event, this change roughly parallels the sintering behavior (Fig. 2) and measured densities shown in Table II, where the optimum effect on densification occurred at an Al$_2$O$_3$ additive content of 0.325 wt%. This concentration (0.325 wt%, 0.392 mol%) should nominally represent the solubility for Al$_2$O$_3$ in the YSZ structure. However, some Al$_2$O$_3$ was present as discrete particles and also dissolved in the intergranular phase. In addition, the presence of some Al$_2$O$_3$ interstitially in the YSZ structure

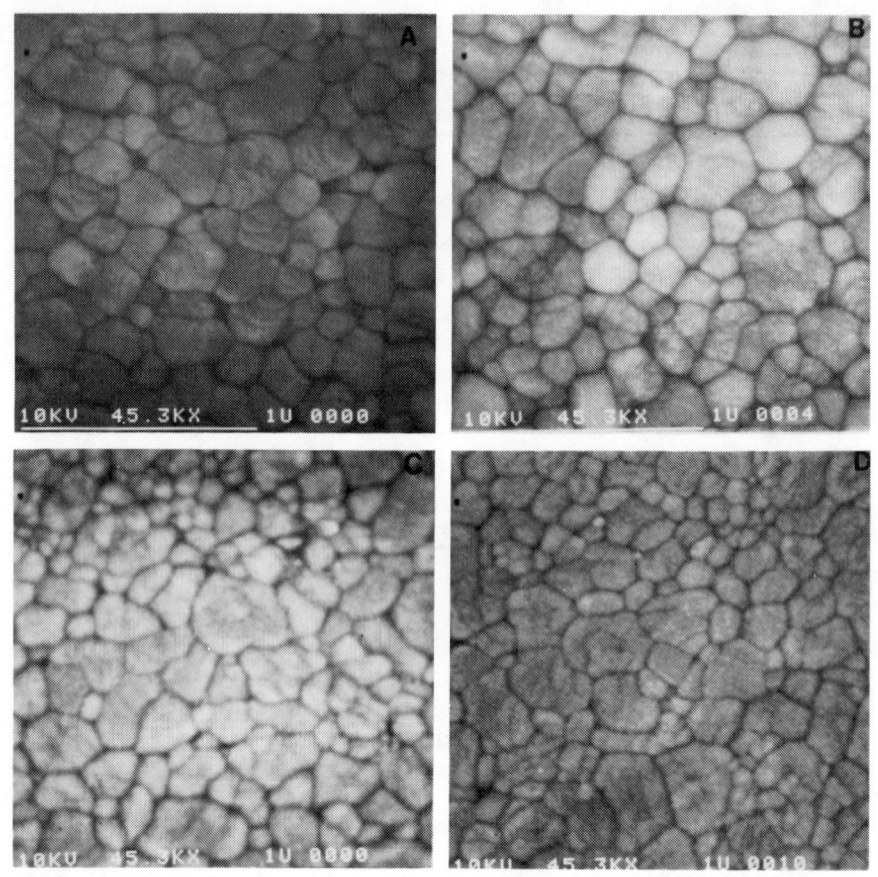

Fig. 6. SEM photomicrographs of polished and thermally etched sections of YSZ and YSZ + Al$_2$O$_3$ samples sintered at 1275°C for 4 h. (A) YSZ, (B) YSZ + 0.325 wt% Al$_2$O$_3$, (C) YSZ + 0.65 wt% Al$_2$O$_3$, and (D) YSZ + 1.3 wt% Al$_2$O$_3$.

might also be expected. The true Al$_2$O$_3$ solubility, therefore, might well be closer to that reported by Bernard (0.1 mol%).[13]

As indicated, Al$_2$O$_3$ inclusions could be found in the Al$_2$O$_3$-doped samples. These were manifested as darker areas, which EDS analysis showed to be rich in Al. Closer examination revealed these to be apparently undissolved Al$_2$O$_3$ grains or inclusions which were considerably larger than the matrix YSZ grains. These inclusions were occasionally associated with porosity, and a perturbation of the microstructure surrounding the inclusion was also observed. The presence of Si could not definitely be identified in the perturbed region from EDS analysis, although its presence would be expected from the work of Butler and Drennan.[18] The frequency of these inclusions increased with added Al$_2$O$_3$ content above 0.325 wt%, but a few observations were made even in the undoped YSZ sample, which would indicate the presence of existing Al$_2$O$_3$ impurities.

Figure 7 shows a plot of grain size vs soak time for the YSZ and YSZ + 0.325 wt% Al$_2$O$_3$ samples. The grain size results were obtained from samples sintered at 1275°C for soak times up to 24 h. Density data for these

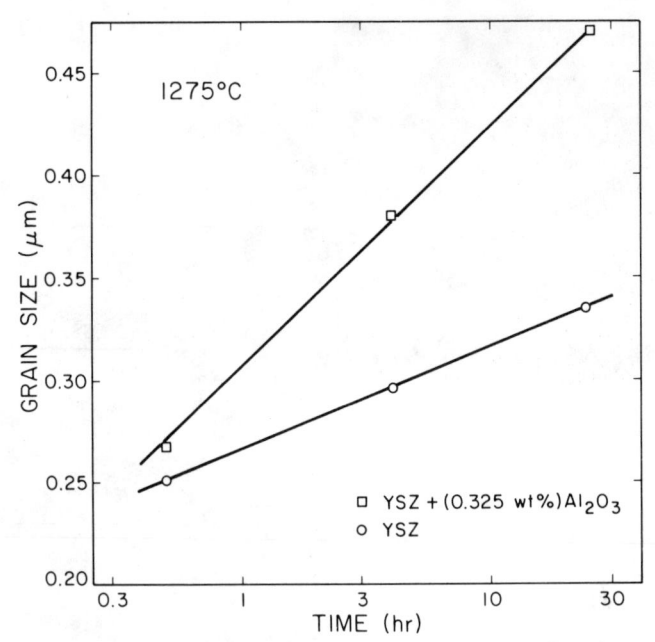

Fig. 7. Grain size vs. sintering time for YSZ samples showing enhanced grain growth with Al$_2$O$_3$ additions.

samples are included in Table II. As indicated previously, the Al$_2$O$_3$ additive samples showed larger grain sizes under all conditions of equivalent densities. However, only moderate grain growth was observed between 93 and 99% relative density. Significant increases occurred only as near complete densification was achieved, and this was accompanied by exaggerated grain growth for the higher Al$_2$O$_3$-doped samples. This circumstance would locate the residual porosity mainly on the grain boundaries and at grain intersections, where they would be eliminated during the final grain-coarsening phase. For samples showing exaggerated grain growth, lower final densities were also achieved, but this would be associated with trapped intergranular porosity, since no pore phase was detected within the grains of the sintered samples.

Figure 8 shows TEM photomicrographs of grain intersections for the YSZ and YSZ + 3.25 wt% Al$_2$O$_3$ samples. Figures 8(A) and 8(B) represent bright-field images of the respective samples. Figure 8(A) (YSZ) shows clearly the existence of a liquid (X-ray amorphous) phase at the triple points and along the grain boundaries. The thin, relatively flat grain boundaries suggested a low concentration of a wetting liquid at the sintering temperature (1350°C). In contrast, grain boundaries for the Al$_2$O$_3$-additive samples were more rounded (Fig. 8(B)) and were also wider, as shown in Fig. 8(C) (dark-field image), indicative of a higher liquid-phase content. Figure 8(B) also showed the existence of an inclusion adjacent to the grain boundary, but the boundaries otherwise appeared free of discrete second phases or inclusions.

Within the grains of the sintered samples, second phases were also not observed, although some tetragonal inclusions might have been expected, considering the low yttria content of the YSZ powder (4.5 mol%). X-ray diffraction studies on the powder and fired samples likewise did not indicate the presence of

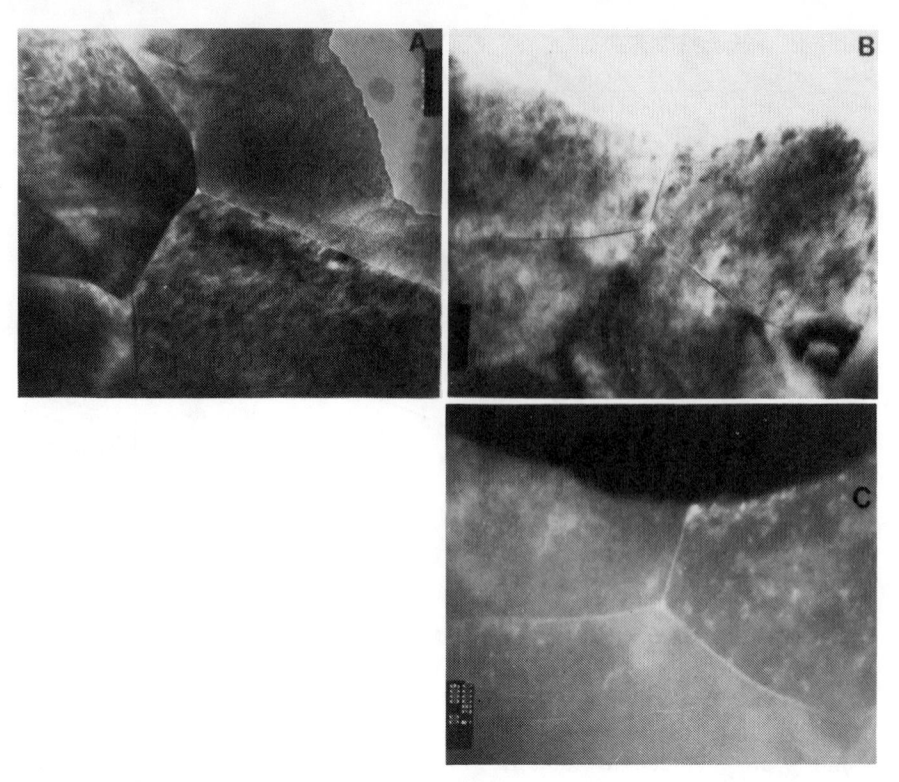

Fig. 8. TEM photomicrographs showing bright-field images of grain boundaries and triple points for (A) YSZ and (B) YSZ + 0.325 wt% Al_2O_3 samples fired at 1350°C for 1 h and 4 h, respectively. Dark-field image of (B) is shown in (C).

a tetragonal phase, but this is normally difficult to distinguish from the cubic phase.[34] Only the cubic YSZ phase was identified in the samples studied, and no crystalline intergranular phases were found.

EDS spectra were obtained from the TEM samples. These were taken in the grain centers and at the triple points for the YSZ sample. These data are presented in Table III and show only Al and Si as significant impurities. The Si and Al average concentration in the YSZ grains were higher than the chemical analysis in Table I would indicate. This may reflect possible (Si) contamination during TEM sample preparation and also likely errors in the EDS analysis. Noteworthy points from the data in Table III are (1) the Al and Si enrichment of the triple-point regions, (2) the higher overall Al concentration within grains and in intergranular regions for the Al_2O_3-doped samples, and (c) the significant increase in Y concentration in the triple-point regions. The concentration of Al and Si at the triple points is consistent with the formation of a liquid boundary phase which aids in sintering. Moderate alumina enrichment of this phase would likely cause increased fluidity and enhanced sintering. Higher concentrations of yttria at the boundary phase might be expected to destabilize the YSZ structure, but as indicated, this was not observed. The existence of a Y-, Si-, Al-, and Ca-rich boundary phase was noted also by Moghadam et al. for a similar YSZ powder.[34]

Table III. Elemental Analysis of YSZ and YSZ + 0.325 wt% Al$_2$O$_3$ Samples by EDS Technique

Sample*	Average concentration (wt%)				Location
	Al	Si	Y	Zr	
YSZ	0.28	0.99	7.42	91.09	Gr center
YSZ + Al$_2$O$_3$	0.56	0.97	7.69	89.51	Gr center
YSZ	0.36	4.56	9.77	84.86	Triple pt
YSZ + Al$_2$O$_3$	0.65	4.43	10.30	86.22	Triple pt

*Sintering temp: YSZ, 1350°C/4 h; YSZ + Al$_2$O$_3$, 1350°C/1 h.

Figure 9 shows a plot of electrical conductivity vs reciprocal absolute temperature for the YSZ and Al$_2$O$_3$-doped samples. The conductivity shows a significant increase with Al$_2$O$_3$ additions, though the activation energy (0.97 eV) remained unchanged. A conductivity maximum at 0.325 wt% was observed, a trend similar to that noted previously for the grain size and densification behavior.

From the TEM and sintering kinetic data presented, the existence of an intergranular phase in the samples studied would seem to be well established. Densification in the submicrometer YSZ powders, with or without Al$_2$O$_3$ doping, can, therefore, be attributed primarily to liquid-phase-assisted sintering, as discussed. The liquid would be formed from impurities present in the YSZ powders,

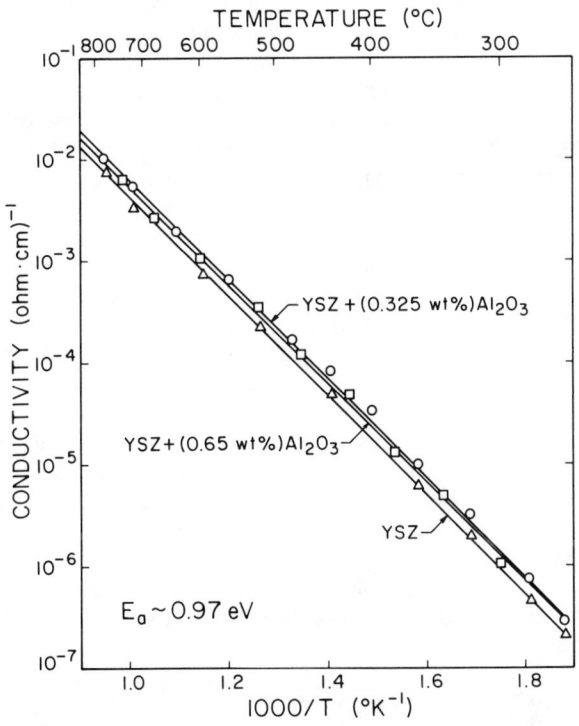

Fig. 9. Plot of DC conductivity vs temperature for YSZ and Al$_2$O$_3$-doped samples at optimum densities.

which have been shown to be concentrated in the intergranular regions. It should be noted, moreover, that the impurities constituted ~0.8 wt% of the YSZ powder and were comprised primarily of such glass forming oxides as SiO_2, Na_2O, K_2O, MgO, CaO, BaO, and Al_2O_3. If converted into a glassy phase, this would constitute ~1.5% of the sample volume, an amount of liquid sufficient to show significant effects of liquid-phase sintering.

The primary role of the added Al_2O_3 as a densification aid appears to be an enhancement in the amount and reactivity of the liquid phase at equivalent temperatures, which causes an increase in the densification rate. Within the glassy phase that might be formed from the impurity oxides present, perhaps 10–20 wt% Al_2O_3 could be dissolved, with beneficial effects on the fluidity and reactivity of the melt phase.[36] On this basis, less than 0.3 wt% of the Al_2O_3 would be present in the intergranular or boundary phase. Higher Al_2O_3 contents would lead to a more viscous and, therefore, a less reactive melt phase. A maximum in the reactivity of the intergranular liquid as its Al_2O_3 content was varied would also explain the observed maximum in densification kinetics at the optimal (0.325 wt%) Al_2O_3 additive level. As fired grain sizes are known to be enhanced by the presence of a reactive liquid phase,[35] the observed similarities between grain growth behavior and densification kinetics with varying Al_2O_3 additions become evident.

The role attributed to Al_2O_3 in the above discussion as a densification aid for submicrometer YSZ powders, is in general agreement with work reported by Radford and Bratton,[2] Mallinckrodt et al.,[3] and Takagi.[16] The authors attribute increased densification with Al_2O_3 additions to YSZ and CSZ powders to formation of an intergranular liquid phase with existing impurities, particularly SiO_2. The Al_2O_3 was reported to be present mostly in the grain-boundary phase. Radford and Bratton noted an apparent grain growth inhibition with added Al_2O_3 at 1480°C, but Takagi reported a substantial increase in grain size at higher temperatures.

In contrast to the above, Bernard,[13] Butler and Drennan,[18] and Rao and Schreiber[19] all reported the presence of Al_2O_3 as discrete inclusions in zirconia. Enhanced densification was observed with the added Al_2O_3, but the effect of the Al_2O_3 was considered by Butler and Drennan to be the scavenging of intergranular SiO_2 by Al_2O_3 inclusions with subsequent grain-boundary pinning. The present study has identified Al_2O_3 inclusions in the YSZ samples, but their effect on densification, at least at low concentrations, was minor.

Increased electrical conductivity was noted for YSZ samples with a low concentration (<0.65 wt%) of added Al_2O_3. Bernard and Verkerk et al.[7] noted this increase and attributed it to increased grain size and consequent reduction in the more resistive grain-boundary area. As grain size and conductivity behavior both reached a maximum at the 0.325 wt% additive level, this mechanism undoubtedly accounts for part of the observed increase. However, some contribution to the conduction process from the defect substitution of Al^{3+} into the YSZ lattice might also be expected.

As pointed out by Wilhelm and Howarth[37] in connection with the incorporation of Fe_2O_3 into YSZ, the trivalent cation can be accommodated into the lattice both interstitially and by direct substitution for Zr^{4+} as follows:

$$Al_2O_3 + 3V_O^{\cdot\cdot} \xrightarrow{ZrO_2} 2Al_i^{\cdot\cdot\cdot} + 3O_O \tag{1}$$

$$2Al_i^{\cdot\cdot\cdot} \rightarrow 2Al'_{Zr} + V_O^{\cdot\cdot} \tag{2}$$

Reaction (1) represents the incorporation of alumina whereby the Al^{3+} ions would be accommodated interstitially with the suppression of existing oxygen vacancies. The cation defects would not contribute in any significant way to the conduction process, but cation mobilites would be affected. This mechanism may explain the ready dissolution of Al_2O_3 into the glassy phase. The decrease in the oxygen vancancy concentration would lower the conductivity, however, depending on the magnitude of the effect.

Reaction (2) indicates substitution of Al^{3+} ions on Zr^{4+} sites with the expected formation of oxygen vacancies. These vacancies would contribute to the conductivity, although with the initial substitution of Y^{3+} into the ZrO_2 lattice, significant oxygen vacancies would already exist.

The magnitude of these substitution effects with Al^{3+} is not known, but with consideration of the size disparity between the Al^{3+} ($r = 0.53$ Å) and Zr^{4+} ($r = 0.84$ Å) ions and the fact that the conductivity does increase, lattice substitution of the Al would seem to be the dominant effect.

Conclusions

Sintering studies carried out on submicrometer YSZ powders with Al_2O_3 additives showed a significant enhancement in densification rate above 1150°C. Grain sizes, which were slightly increased by the Al_2O_3 additions, were in the range 0.3–0.5 μm. Densification and grain growth decreased relatively at Al_2O_3 additive levels >0.6 wt%. Nearly complete densification (>99% relative density) was achieved at 1350°C in 1 and 4 h soak time for the 0.325 wt% Al_2O_3 additive and undoped YSZ, respectively. Microstructural observations and time–temperature sintering kinetics indicated that densification occurred by a liquid-phase mechanism, with enhanced densification in Al_2O_3-additive samples resulting from an increased Al melt content. Conductivity was increased 1.5 times by 0.325 wt% Al_2O_3 additions due partly to increased grain growth. Relative decreases in both grain growth and conducitvity occurred at higher Al_2O_3 additive levels due to a decrease in boundary diffusion kinetics.

Acknowledgments

The assistance of D. S. Phillips of the Materials Research Laboratory and Department of Metallurgy in carrying out the TEM studies is gratefully acknowledged. This work was supported by the Office of Naval Research under Contract No. N-000014-80-K-0969 and in part by the National Science Foundation under MRL Grant DMR-80-20250.

References

[1] H. Yanagida, M. Takata, and M. Nagai, "Fabrication of Transluscent ZrO_2 Film by a Modified Doctor Blade Method," *J. Am. Ceram. Soc.*, **65** [2] C-34–C-35 (1981).
[2] K. C. Radford and R. J. Bratton, "Zirconia Electrolyte Cells," *J. Mater. Sci.*, **14** [1] 59–65 (1979).
[3] D. V. Mallinckrodt, P. Reynan, and C. Zografou, "The Effect of Impurities on Sintering and Stabilization of ZrO_2 (CaO)," *InterCeram*, **31** [2] 126–29 (1982).
[4] G. K. Layden and M. C. McQuarrie, "Effect of Minor Additions on Sintering of MgO," *J. Am. Ceram. Soc.*, **42** [2] 89–92 (1959).
[5] W. J. Huppman, "Sintering in the Presence of Liquid Phase"; pp. 359–78 in Materials Science Research, Vol. 10. Edited by G. C. Kuczynski. Plenum, New York and London, 1975.
[6] R. C. Garvie; pp. 117–66 in High Temperature Oxides, Part II. Edited by A. M. Alper. Academic Press, New York and London, 1970.
[7] M. J. Verkerk, B. J. Middelhuis, and A. J. Burggraaf, "Effect of Grain Boundaries on the Conductivity of High-Purity ZrO_2-Y_2O_3 Ceramics," *Sol. State Ionics*, **6** [2] 159–70 (1982).
[8] N. M. Beekmans and L. Heyne, "Correlation Between Impedance Microstructure and Composition of Calcia-Stabilized Zirconia," *Electrochim. Acta*, **21** [4] 303–10 (1976).
[9] M. Harmer, E. W. Roberts, and R. J. Brook, "Rapid Sintering of Pure and Doped α-Al_2O_3," *Trans. Br. Ceram. Soc.*, **78** [1] 22–5 (1979).

[10] S. Prochazka and R. M. Scanlan, "Effect of Boron and Carbon on Sintering of SiC," *J. Am. Ceram. Soc.*, **58** [1–2] 72 (1975).

[11] S. H. Chu and M. A. Seitz, "The Electrical Behavior of Polycrystalline ZrO_2-CaO," *J. Solid-State Chem.*, **23** [3–4] 297–314 (1978).

[12] J. E. Bauerle, "Study of Solid Electrolyte Polarization by a Complex Admittance Method," *J. Phys. Chem. Solids*, **30** [12] 2657–70 (1969).

[13] H. Bernard, "Microstructure et Conductivite de l'Zirone Stabilisee Frittee"; Ph.D. Thesis, Institut National Polytechnique de Grenoble, 1980.

[14] R. C. Buchanan and A. Sircar, "Densification of Calcia-Stabilized Zironia with Borates," *J. Am. Ceram. Soc.*, **66** [2] C20–C21 (1983).

[15] M. J. Bannister, "Development of The $SIRO_2$ Oxygen Sensor: Sub-Solidus Phase Equilibria in the System ZrO_2-Al_2O_3-Y_2O_3," *J. Aust. Ceram. Soc.*, **18** [1] 6–9 (1982).

[16] H. Takagi, S. Kuwabara, H. Matsumoto, "Effects of Alumina on Sintering of Zirconia Stabilized with Calcia," *Sprechsaal*, **107** [13] 584–88 (1974).

[17] K. C. Radford and R. J. Bratton, "Zirconia Electrolyte Cells; 2, Electrical Properties," *J. Mater. Sci.*, **14** [1] 66–69 (1979).

[18] E. P. Butler and J. Drennan, "Microstructural Analysis of Sintered High-Conductivity Zirconia with Al_2O_3 Additions," *J. Am. Ceram. Soc.*, **65** [10] 474–78 (1982).

[19] B. V. Narasimha Rao and T. P. Schreiber, "Scanning Transmission Electron Microscope Analysis of Solute Partitioning in a Partially Stabilized Zirconia," *J. Am. Ceram. Soc.*, **65** [3] C44–C45 (1982).

[20] M. J. Verkerk, A. J. A. Winnubst, and A. J. Burggraaf, "Effect of Impurities on Sintering and Conductivity of Yttria-Stabilized Zirconia," *J. Mater. Sci.*, **17** [12] 3113–22 (1982).

[21] P. J. Jorgenson, "Diffusion Controlled Sintering in Oxides"; pp. 401–22 in Sintering and Related Phenomena. Edited by G. C. Kuczynski, N. A. Hooton, and C. F. Gibbon. Gordon and Breech, NY, 1967.

[22] W. H. Rhodes and R. E. Carter, "Cationic Self Diffusion in Calcia Stabilized Zirconia," *J. Am. Ceram. Soc.*, **49** [5] 244–49 (1966).

[23] M. S. Selzer and P. K. Talty, "High-Temperature Creep of Y_2O_3-Stabilized ZrO_2," *J. Am. Ceram. Soc.*, **58** [3–4] 124–30 (1975).

[24] M. Heughebaert-Therasse, "Contribution a L'etude de L'evolution et du Frittage de la Zircone Stabilisee par Differents Oxydes, a des Tempertures Inferieures a 1300°C," *Ann. Chim.*, **2** [4] 229–43 (1977).

[25] W. S. Young and I. B. Cutler, "Initial Sintering with Constant Rates of Heating," *J. Am. Ceram. Soc.*, **53** [12] 659–63 (1970).

[26] D. Wirth, Jr., "Sintering Kinetics of Ultrafine Calcia Stabilized Zirconia"; Ph.D. Thesis, University of Illinois, Champaign—Urbana, 1967.

[27] R. L. Coble and R. M. Cannon, "Current Paradigms in Powder Processing"; pp. 151–70 in Materials Science Research, Vol. II. Edited by H. Palmour III, R. F. Davis, and T. M. Hare. Plenum, New York, 1977.

[28] A. M. Diness and R. Roy, "Experimental Confirmation of Major Change of Defect Type with Temperature and Composition in Ionic Solids," *Solid-State Comm.*, **3** [6] 123–25 (1965).

[29] R. J. Brook, "Preparation and Electrical Behavior of Zirconia Ceramics"; pp. 272–85 in Advances in Ceramics, Vol. 3. Edited by A. H. Heuer and L. W. Hobbs. The American Ceramic Society, Columbus, OH, 1981.

[30] R. G. St. Jacques and R. Angers, "The Effect of CaO-Concentration on the Creep of CaO-Stabilized ZrO_2," *Trans. Br. Ceram. Soc.*, **72** [6] 285–89 (1973).

[31] C. E. Scott and J. S. Reed, "Effect of Laundering and Milling on the Sintering Behavior of Stabilized ZrO_2 Powders," *J. Am. Ceram. Soc.*, **58** [6] 587–90 (1979).

[32] W. D. Tuohig and T. Y. Tien, "Subsolidus Phase Equilibria in the System ZrO_2-Y_2O_3-Al_2O_3," *J. Am. Ceram. Soc.*, **63** [9–10] 595–96 (1980).

[33] M. I. Mendelsohn, "Average Grain Size in Polycrystalline Ceramics," *J. Am. Ceram. Soc.*, **52** [8] 443–46 (1969).

[34] F. K. Moghadam, T. Yamashita, and D. A. Stevenson, "Characterization of Yttria-Stabilized Zirconia Oxygen Solid Electrolytes"; pp. 364–79 in Advances in Ceramics, Vol. 3. Edited by A. H. Heuer and L. W. Hobbs. The American Ceramic Society, Columbus, OH, 1981.

[35] W. D. Kingery, "Plausible Concepts Necessary and Sufficient for Interpretation of Ceramic Grain-Boundary Phenomena: II, Solute Segregation, Grain Boundary Diffusion, and General Discussion," *J. Am. Ceram. Soc.*, **57** [2] 74–83 (1973).

[36] B. Locsei; Molten Silicates and Their Properties, Chapters 3–6. Chemical Publishing, New York, 1970.

[37] R. V. Wilhelm, Jr. and D. S. Howarth, "Iron Oxide-Doped Yttria-Stabilized Zirconia Ceramic: Iron Solubility and Electrical Conductivity," *Am. Ceram. Soc. Bull.*, **58** [2] 228–32 (1979).

---

*Buchi Laboratory Spray Dryer, Brinkman Instruments, New Jersey.
†VSZ Ball Cratering Instrument, The Technology Shop, Inc., Sudbury, Mass.
‡Model 4260A, Hewlett-Packard.

Vaporization from Magnesia and Alumina Materials

Toshiyuki Sata and Tadashi Sasamoto*
Kumamoto Institute of Technology
Kumamoto 860, Japan

Langmuir vaporizations from MgO and Al_2O_3 were conducted under vacuum, and their vaporization coefficients were obtained with different temperatures. Effect of porosity brought increases of both vaporization rate and vaporization coefficient. Vacuum vaporizations of impurities in magnesia and alumina were approximately proportional to their vapor pressures but showed certain characteristic difficulties for some impurities by interaction with the matrix. The vapor pressures of MgO(g) were measured in oxygen atmospheres; the vapor pressures over MgO and Al_2O_3 were calculated in atmospheres with various P_{O_2}. Vaporization of MgO was accelerated in humid atmosphere by formations of hydroxide vapor species, and the vapor pressure of $Mg(OH)_2(g)$ was obtained by the transpiration method. Thus, vapor pressures over MgO in atmospheres with both P_{O_2} and P_{H_2O} were calculated for four species of Mg(g), MgO(g), MgOH(g), and $Mg(OH)_2(g)$.

Magnesia and alumina are the most popular materials among oxide ceramics and have high melting temperatures over 2000°C. They are used in general as high-performance refractories (for instance, in the steel industry) and as high-temperature electrical insulators (in MHD channels or for substrates for IC) in various atmospheres. MgO is basic and Al_2O_3 is neutral. These properties are applied against corrosion atmospheres in contact with other kinds of materials in various furnaces.

In recent years, these materials have been sintered or used in various atmospheres from reducing to oxidizing with various oxygen partial pressures. In particular, it is well-known that vaporization loss of the MgO component is significant in the reducing atmosphere at high temperature. In this report, we treated the vaporizations of MgO and Al_2O_3 materials containing impurities in various atmospheres including not only oxygen atmospheres but humid atmospheres.

Langmuir Vaporization Rates

Single crystals of MgO and Al_2O_3 were heated under vacuum (10^{-2}–10^{-3} Pa) at 1400–1950°C, and their weight changes were obtained over all the vaporization rates, which were constant during the vaporization time used.[1,2] Results are shown in Table I. Differences between the vaporization rates of MgO and Al_2O_3 under vacuum are approximately 3–4 orders of magnitude. SEM observations of the MgO single-crystal surface (100) during the vaporization showed that the polished flat surface became rough first at 1200°C due to inhomogeneous vaporization along flaws of polishing, followed by an increase again in flatness at 1400°C due to the larger vaporization rate. This flatness became rough again at 1500°C, and then it decreased at 1600°C.

Table I. Langmuir's Vaporization Rates of MgO and Al_2O_3 Single Crystals

	Oxide temperature (°C)	Vaporization rate ($\times 10^8$ g·cm^{-2}·s^{-1})
MgO	1400	0.53
	1500	4.66
	1550	14.9
	1600	38.9
	1640	79.3
	1700	242
	1800	1260
	1900	1390
Al_2O_3	1850	0.773
	1900	2.02
	1955	7.10

Vaporization Coefficient

The vaporization coefficient, α, is defined as the ratio of the Langmuir vaporization rate to the Knudsen effusion rate (=equilibrium vaporization rate). The Knudsen rate of MgO was calculated by using the data cited from thermochemical data.[3] The Knudsen rate of Al_2O_3 was obtained from the target collection of effused mass from an orifice of the Knudsen cell of W[4] as follows:

Temperature (°C)	Effusion rate (g·cm^{-2}·s^{-1})
1812	1.3×10^{-8}
1865	3.2×10^{-8}
1910	8.7×10^{-8}
1942	1.6×10^{-7}
1967	2.3×10^{-7}

The vaporization coefficients calculated at various temperature levels are shown in Fig. 1 using a nondimensional abscissa of T_m/T.[1,5] T_m is the melting point (K) of MgO or Al_2O_3. It was found that the vaporization coefficients increased with increasing temperatures and tended to unity at their melting temperatures.[5,6] This behavior was also confirmed for some oxides — Al_2O_3, ThO_2, Y_2O_3, Yb_2O_3, Cr_2O_3, and $MgCr_2O_4$ — by recalculation of published Langmuir data.[7–12]

Effect of Porosity on the Langmuir Vaporization Rate

Samples with porosities up to 30% were fabricated by sintering green bodies that were obtained by pressing 53–73 μm of powder of fused magnesia at several kinds of molding pressures (0.10–3.65 tons/cm^2). Sinterings were performed at 1650°C for 1 h.

Overall vaporization rates were measured by weight loss under vacuum from 1400 to 1640°C. Results are plotted against $1/T$ in Fig. 2 and against porosity in Fig. 3. Zero porosity corresponds to the value for the single crystal described before. The effect of porosity on the vaporization rate was fairly small, about double for 30% porosity vs 0%, that of the single crystal. This behavior may be due to that pore with a cell orifice on the surface, acting as a Knudsen effusion cell rather than an increase in vaporization area and in kink or ledge points in the pores. Vaporization coefficients α became higher with porosities and showed the same trend toward unity at their melting points.

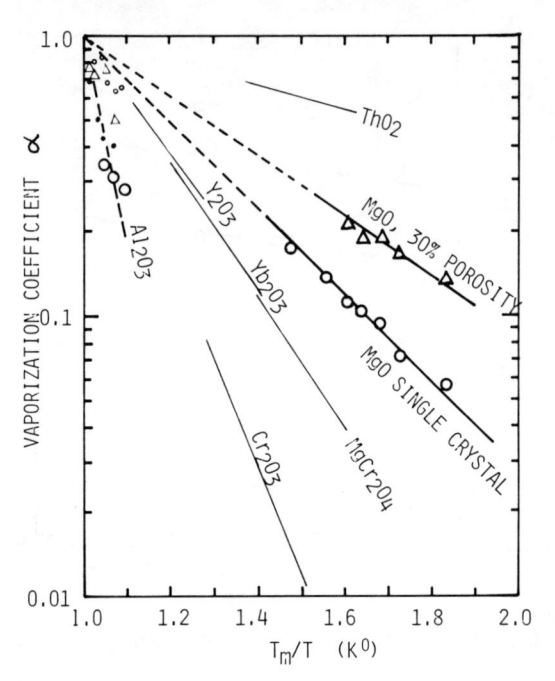

Fig. 1. Vaporization coefficients for MgO and Al$_2$O$_3$; T_m is the melting temperature.

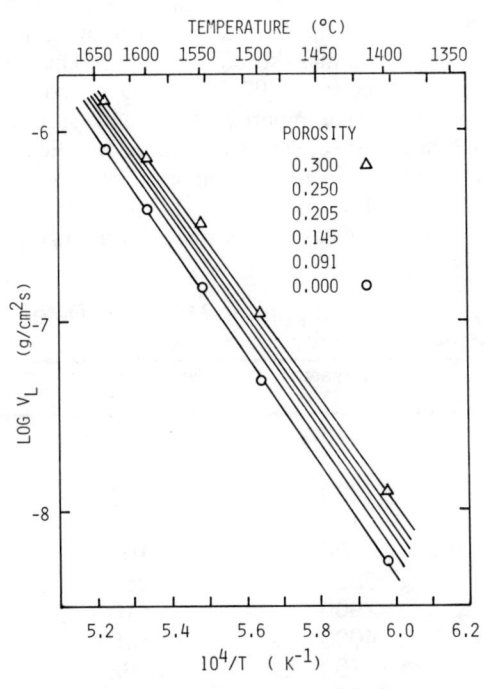

Fig. 2. Langmuir's vaporization rate V_L of MgO with various porosities.

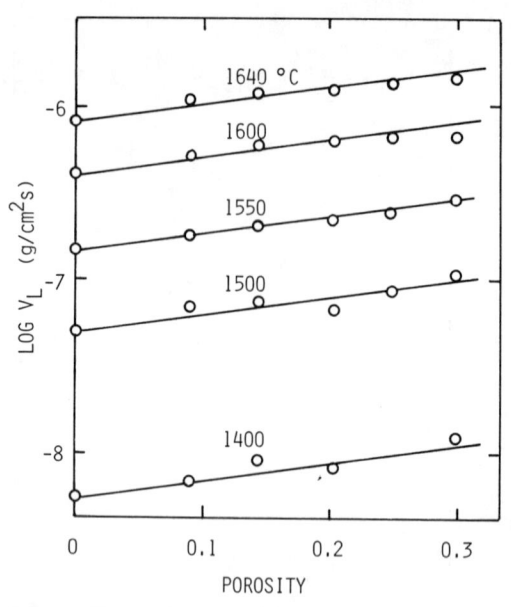

Fig. 3. Effect of porosity on Langmuir's vaporization of MgO.

Vaporization of Impurities

Commercial magnesia clinkers containing impurities shown in Table II were heated at 1800°C and then 2000°C for 2 h each under vacuum in a tungsten resistance furnace. The original and heated specimens were analyzed for 10 impurity elements by spark-source mass spectrometry. The change of impurity contents in sample No. 1 with purity of 97.21% MgO is shown in Fig. 4, as an example, taking vapor pressures of impurity oxides in the abscissa.[13] The analyzed values for original and heated specimens are included in a field whose values and width in content decreased with increasing vapor pressure of the impurity oxides, as indicated by solid and dashed lines in Fig. 4. Thus, the impurity oxides with higher vapor pressures are more easily vaporized during high-temperature (about

Table II. Impurity Contents (wt ppm) in Magnesia Determined by Spark Source Mass Spectrometry

Magnesia sample content (%)	No. 1 sintered (97.21)	No. 2 sintered (99.68)	No. 3 fused (98.44)
K_2O	5.2	5.0	4.1
Na_2O	80	16	20
B_2O_3	1200	84	170
MnO	130	37	570
Fe_2O_3	2100	1100	1750
Cr_2O_3	6300	260	430
SiO_2	2900	410	2600
CaO	14000	1100	1900
TiO_2	250	62	10
Al_2O_3	940	120	190

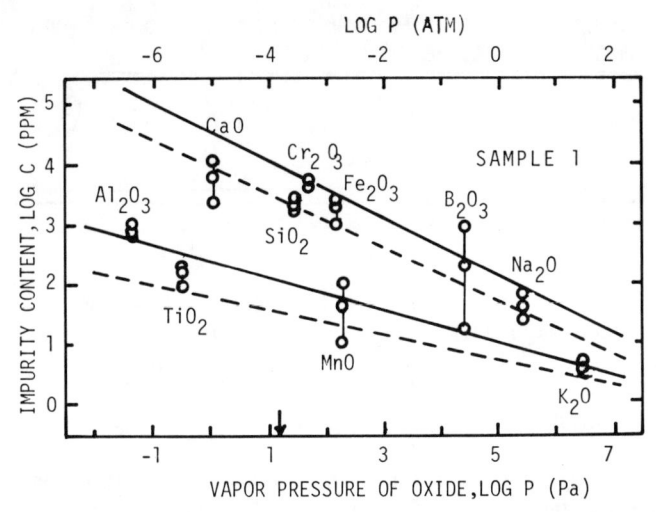

Fig. 4. Changes in impurity contents in sintered magnesia (sample No. 1) at 1800 and then at 2000°C for 2 h each, plotted against vapor pressures of impurities.

1800°C) firing in a rotary kiln. Similar relations were obtained for sample Nos. 2 and 3.

If the remaining ratio C_2/C_1 of content is taken against C_1, where C_1 and C_2 are the contents of an impurity oxide before and after heat treatment, respectively, the vaporizing process for each impurity is represented as shown in Fig. 5. Slopes

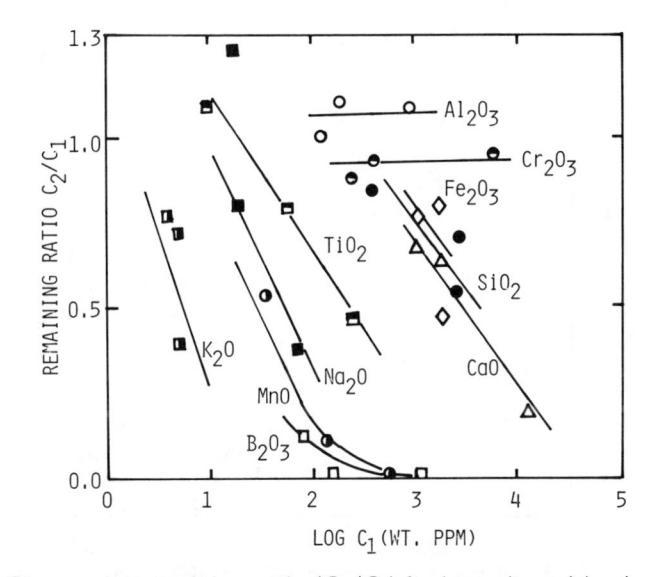

Fig. 5. Change in remaining ratio (C_2/C_1) for impurity oxides in magnesia sample Nos. 1–3 after heat treatments at 1800 and 2000°C for 2 h each. C_1 and C_2 are contents before and after a heat treatment, respectively.

Table III. Minimum Limits of Remaining Impurity Content (wt ppm) in Magnesia after 2000°C Heat Treatments for 2 h

K_2O	1	TiO_2	20	Al_2O_3	NP*
B_2O_3	3	CaO	150	Cr_2O_3	NP*
MnO	4	Fe_2O_3	250		
Na_2O	8	SiO_2	250		

*The minimum limits of Al_2O_3 and Cr_2O_3 are not predicted (NP).

in the linear portion of the remaining ratio against log C_1 decrease in the order of K_2O, B_2O_3, MnO, Na_2O, TiO_2, CaO, SiO_2, and Fe_2O_3. This shows that an impurity tends to remain easily with a decrease in its content. But this behavior is very characteristic for Al_2O_3 and Cr_2O_3 as shown in Fig. 5, indicating very difficult vaporization. This may be because these oxides dissolve in MgO to form refractory Mg spinels.

The remaining ratios have trends taking values over unity. This is due to the fairly large vaporization of the MgO matrix itself at these temperatures. Since the content of Ca is comparable to or larger than that of other impurity oxides, these impurities may combine with the more basic CaO. From these considerations may be predicted the minimum impurity contents in magnesia after heat treatments, as shown in Table III.

In the same experiment for commercial alumina, similar behavior was also observed in spite of less accuracy. In this case, contents of Cr_2O_3 and Fe_2O_3, which dissolved in Al_2O_3, showed different behaviors, as shown in Fig. 6.

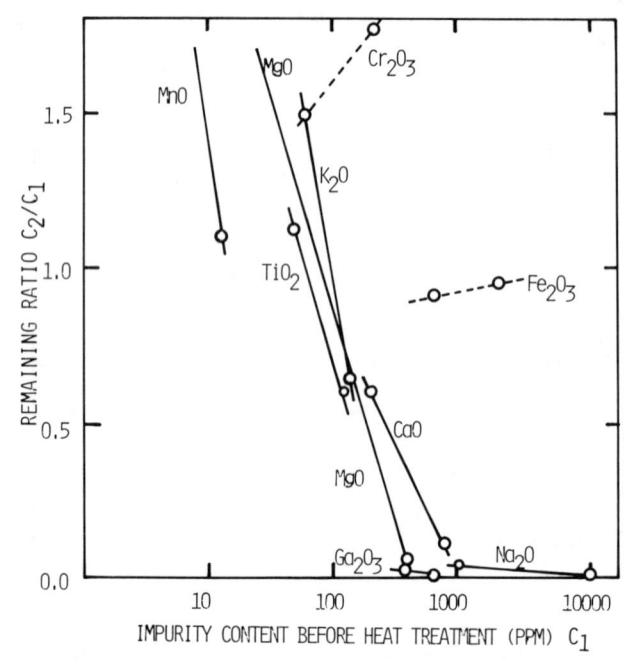

Fig. 6. Change in remaining ratio (C_2/C_1) for impurity oxides in alumina. C_1 and C_2 are contents before and after a heat treatment, respectively.

Vaporization in Atmospheres with Various Oxygen Pressures

It is known that vaporization of MgO takes place by the following reactions in a wide range of atmospheres.

$$MgO(s) = Mg(g) + \tfrac{1}{2}O_2 \qquad (1)$$

$$MgO(s) = MgO(g) \qquad (2)$$

Reaction 1 is the decomposition to the element gaseous species, and it receives the effect of P_{O_2}, while reaction 2 is the simple vaporization without the effect of P_{O_2}. Results of experiments for reaction 1 have been mentioned in the second paragraph of this paper as Langmuir vaporizations under vacuum.

Here reaction 2 was confirmed in an atmosphere of pure oxygen, which can suppress reaction 1 by using the transpiration method as described in the following paragraph. $\Delta H^{\circ}_{298} = 600.8 \pm 10.0$ kJ·mol^{-1} for reaction 2 was obtained from 10 experiments conducted from 1825 to 1977 K for 3–51 h and agreed well with JANAF data.[14] Pressure of MgO(g) was given as follows:

$$\log(P_{MgO}(\text{Pa})) = -(30\,100 \pm 2\,300)/T + (13.82 \pm 2.56) \qquad (3)$$

With the JANAF table,[14] influences of P_{O_2} on the vapor pressures of Mg(g) and MgO(g) were calculated at 1600 and 1900 K as shown in Fig. 7. At lower pressures of oxygen, the vapor pressure of Mg(g) is considerably high, as experienced in reduced atmospheres. At higher pressures of oxygen, the main species is the MgO(g) molecule where the pressure becomes higher than that of Mg(g).

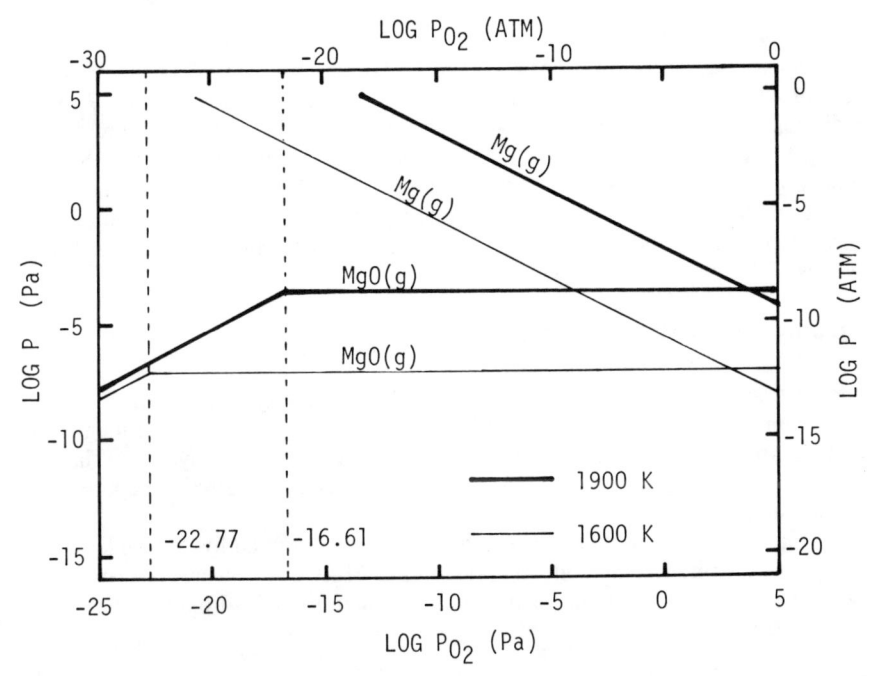

Fig. 7. Vapor pressures of MgO in atmospheres with various P_{O_2} at 1600 and 1900 K.

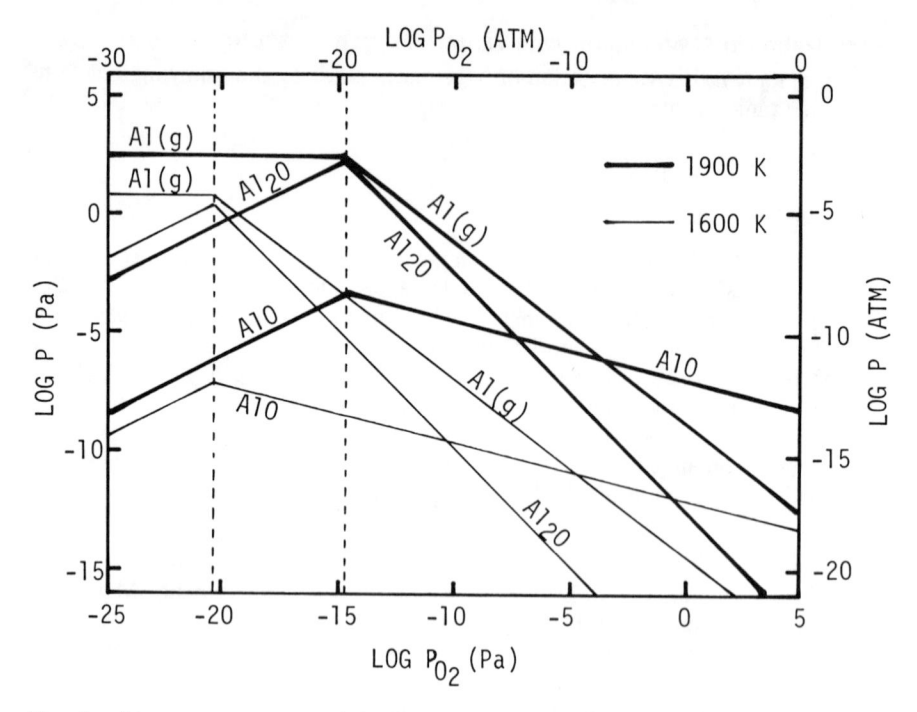

Fig. 8. Vapor pressures of Al$_2$O$_3$ in atmospheres with various P_{O_2} at 1600 and 1900 K.

For the vaporization of Al$_2$O$_3$, a similar diagram is shown in Fig. 8. In the case of Al$_2$O$_3$, vapor species of Al$_2$O$_3$(g) is neglected. As shown in Fig. 9, which indicates highest pressures of species from MgO and Al$_2$O$_3$, the pressure of Al(g) + AlO(g) at 10^{-3} Pa of P_{O_2} is 7 orders of magnitude less at 1600 K and 6 orders of magnitude less at 1900 K than those of MgO. So with Langmuir vaporization of Mg–Al spinel, Mg(g) vaporized preferentially and a corundum layer formed on the surface. This layer lowers the vaporization rate by a diffusion process through the dense layer.

Vaporization in a Humid Atmosphere

In high-temperature industrial processes, a humid atmosphere is usual in combustion of carbonaceous fuel by normal air and oxygen. Alkali or alkaline earth elements usually make hydroxide vapor species. The presence of magnesium monohydroxide vapor species MgOH(g) has been confirmed in hydrogen flame by optical absorption spectroscopy. Al$_2$O$_3$ does not make a hydroxide vapor species in the humid atmosphere.

Vaporization of MgO in humid atmosphere was studied by Alexander et al.[15] using the transpiration method, and pressures of Mg(OH)$_2$(g) were confirmed. This was reexamined by a similar technique.[16] Formations of hydroxide vapor follow these two reactions:

$$MgO(s) + \tfrac{1}{2}H_2O(g) = MgOH(g) + \tfrac{1}{4}O_2(g) \tag{4}$$

$$MgO(s) + H_2O(g) = Mg(OH)_2(g) \tag{5}$$

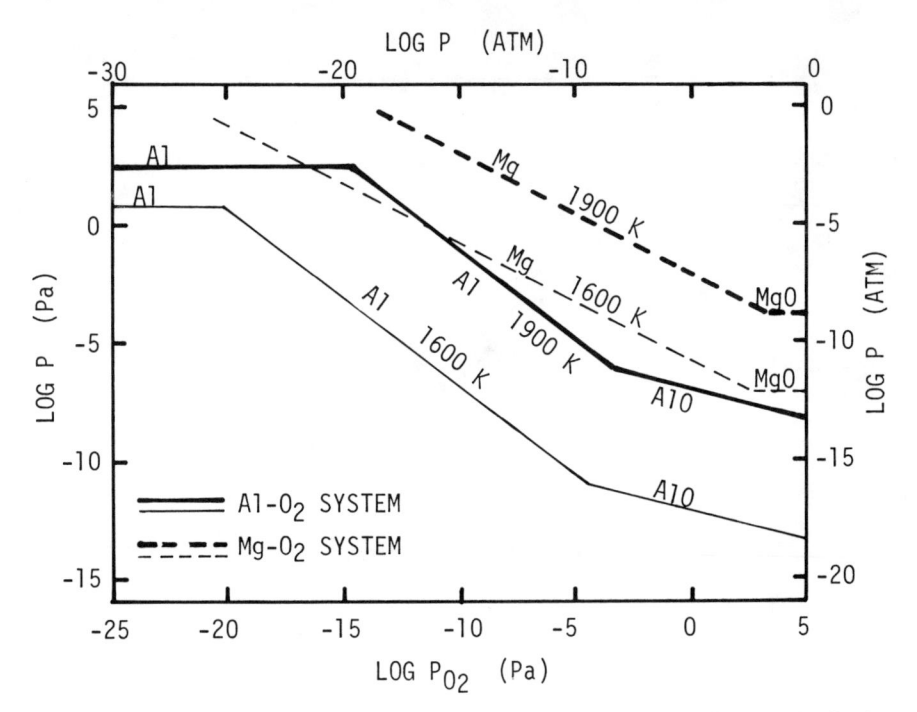

Fig. 9. Maximum vapor pressures for species from MgO and Al$_2$O$_3$ in atmospheres with various P_{O_2}.

These two reactions are difficult to measure at the same time by the transpiration method. Reaction 5 only was examined by the transpiration experiment using H$_2$O-containing oxygen, which suppressed reaction 4. The cell (12 mm in diameter), made of Pt–20%Rh alloy, was set in an alumina tube. Water vapor was introduced into the cell through a Pt–Rh tube as saturated steam produced over water kept at constant temperatures. Outlet gas saturated by Mg-containing molecules goes to a Pt condenser tube through a fine (1-mm ID) pipe of the cell. The condensate was removed into nitric acid solution, and Mg was analyzed by atomic absorption spectrometry.

Experiments were performed in conditions of 1670–2000 K and 2.7×10^3–1.9×10^4 Pa of P_{H_2O}. At first, it was confirmed from linearity with a slope of about unity in the log P_{H_2O} vs log (pressure of Mg(OH)$_2$ vapor) relation that reaction 5 mainly occurred in the oxygen atmosphere in this condition. Vaporization was continued up to about 20 h, and equilibrium constants of reaction 5 were calculated as shown in Fig. 10 by using the following equation.

$$K_P = -(14\,500 \pm 600)/T + (2.91 \pm 0.35) \tag{6}$$

With the data of Δc_p cited from the JANAF tables, the second law values for Mg(OH)$_2$(g) were

$$\Delta H^\circ_{298} = 283.3 \pm 11.3 \text{ kJ} \cdot \text{mol}^{-1}, \quad \Delta S^\circ_{298} = 48.1 \pm 8.4 \text{ J} \cdot \text{mol}^{-1} \cdot \text{K}^{-1}$$

and the third law value was

$$\Delta H^\circ_{298} = 273.6 \pm 2.5 \text{ kJ} \cdot \text{mol}^{-1}$$

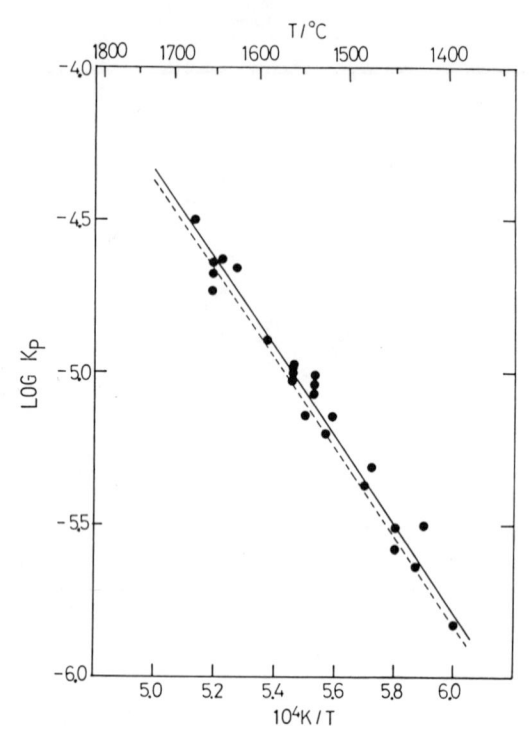

Fig. 10. Equilibrium constant for the reaction: MgO(s) + H$_2$O(g) = Mg(OH)$_2$(g).

with fef from the JANAF table.[13] Alexander's K_p is 10% lower than the present value, the H_r° agrees well with this.

Figure 11 shows calculated pressures of Mg(g), MgO(g), MgOH(g), and Mg(OH)$_2$(g) over MgO(s) under four typical conditions of P_{O_2} and P_{H_2O} in the temperature range from 1400 to 1700°C. In the respective conditions, the highest pressure appeared from the respective vapor species. In usual combustion gases, Mg(g) and MgOH(g) are main species in a reducing atmosphere and Mg(OH)$_2$(g) in an oxidizing atmosphere. In dry oxygen, MgO(g) is the main species.

Summary

This paper treats the vaporization from MgO(s) and Al$_2$O$_3$(s), which are very popular as industrial refractory materials, basic and neutral, respectively. The vaporization from both oxides is fairly large in reducing atmospheres. Basic oxides as MgO make stable hydroxide vapors that are very important in industrial processes using combustion of fuel. MgOH(g) exists in a reducing atmosphere and Mg(OH)$_2$(g) in an oxidizing atmosphere.

Since impurities in MgO and Al$_2$O$_3$ materials are vaporized in vacuum, they may contaminate other neighbor materials or may be purified themselves by high-temperature treatments. Vaporization loss decreases with decreasing contents of impurity during the vaporization process. These losses are proportional to their vapor pressures, but an interaction with the matrix suppresses the impurity vaporization.

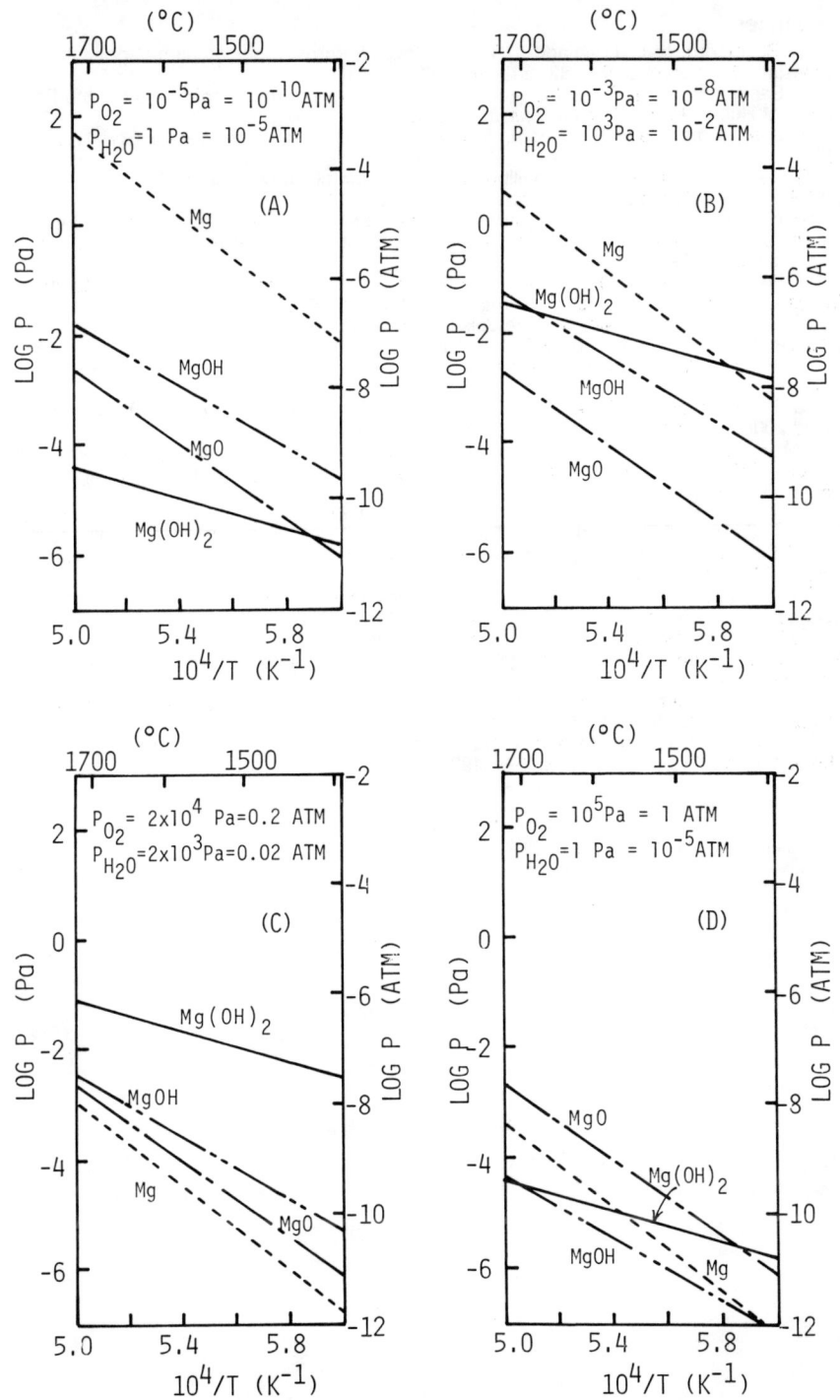

Fig. 11. Vapor pressures of magnesium-bearing species over MgO under four atmosphere conditions.

References

[1] T. Sasamoto, H. L. Lee, and T. Sata, "Effect of Porosity on Vacuum-Vaporization of Magnesia," *Yogyo Kyokai Shi*, **82**, 603–10 (1974).

[2] T. Sasamoto and T. Sata, "Langmuir's Free Vaporization from Single Crystals and Polycrystals of Ruby," *Kogyo Kagaku Zasshi*, **74**, 832–9 (1971).

[3] R. J. Ackermann and R. J. Thorn, "Vaporization of Oxides," *Prog. Ceram. Sci.*, **1**, 39–88 (1961).

[4] T. Sasamoto and T. Sata, "Equilibrium Vaporization of Chromic Oxide from Al_2O_3-Cr_2O_3 Solid Solutions," *Kogyo Kagaku Zasshi*, **74**, 578–86 (1971).

[5] H. L. Lee and T. Sata, "Temperature Dependence of Vaporization Coefficients for Refractory Oxides"; pp. 103–19 in Report of Research Laboratory of Engineering Materials, Vol. 2. Tokyo Institute of Technology, 1977.

[6] T. Sata, T. Sasamoto, H. L. Lee, and E. Maeda, "Vaporization Processes from Magnesia Materials," *Rev. Int. Hautes Temp. Refract.*, **15**, 237–48 (1978).

[7] E. G. Wolff and C. B. Alcock, "The Volatilization of High Temperature Materials in Vacuo," *Trans. Br. Ceram. Soc.*, **61** [10] 667–87 (1962).

[8] M. Pollock, "Vaporization Rate from an Alumina Single Crystal," *Trans. Br. Ceram. Soc.*, **61**, 684–6 (1962).

[9] C. B. Alcock and M. Peleg, "Vaporization Kinetics of Ceramic Oxides at Temperatures around 2000°C," *Trans. Br. Ceram. Soc.*, **66** [5] 217 (1967).

[10] M. Pelleg and C. B. Alcock, "The Mechanism of Vaporization and the Morphological Changes of Single Crystals of Alumina and Magnesia at High Temperature," *High Temp. Sci.*, **6**, 52–63 (1974).

[11] H. L. Lee, T. Sasamoto, and T. Sata, "Effect of Porosity on Vacuum-Vaporization of Chromium Oxide and Thermodynamic Treatments," *Yogyo Kyokai Shi*, **84** [11] 578–83 (1976).

[12] T. Sata and H. L. Lee, "Vacuum Vaporization in the System MgO-Cr_2O_3," *J. Am. Ceram. Soc.*, **61** [7–8] 326–9 (1978).

[13] T. Sasamoto, H. Hara, and T. Sata, "Change in Impurity Content in Magnesia after Vaporization in Vacuum," *Yogyo Kyokai Shi*, **84**, 444–7 (1976).

[14] JANAF Thermochemical Tables, 2nd ed. Edited by D. R. Stull and H. Prophet. U.S. Government Printing Office, Washington, D.C., 1971 (supplements 1974, 1975, and 1978; edited by Chase et al.).

[15] C. A. Alexander, J. S. Ogden, and A. Levy, "Transpiration Study of Magnesium Oxide," *J. Chem. Phys.*, **39**, 3057–60 (1963).

[16] E. Maeda, T. Sasamoto, and T. Sata, "Vaporization from Magnesia in O_2-H_2O Atmospheres," *Yogyo Kyokai Shi*, **86**, 491–9 (1978).

---

*Tokyo National Technical College, Hachioji 192, Japan.

Vaporization of Magnesium Oxide in Hydrogen

V. Bheemineni and D. W. Readey

The Ohio State University
Columbus, OH 43210

In general, gas–solid reaction kinetics can be controlled by the reaction at the interface or by gaseous diffusion of the reactant or product gases through a gaseous boundary layer. In the case of surface reaction control, there are many possible rate-determining steps, which include adsorption of the reactants, surface diffusion of the reactants or products, reactant decomposition at specific surface sites such as kinks, reaction of the adsorbed reactant with the surface, removal of the product species from the reactive site, and desorption. Surface reaction kinetics can indeed be complex even in the simple case of evaporation.[1] As a result, there have been very few studies carried out on nonmetallic solids. On the other hand, gas–solid reaction kinetics are often controlled by gas-phase diffusion. The gas can be usually considered ideal so its thermodynamic and transport properties can be calculated from the kinetic theory of gases and the fluid dynamics of the system modeled with a fair degree of confidence. If the thermodynamics of the gas–solid reaction are known, it should be possible to accurately model the reaction kinetics. Therefore, the purpose of this study was to examine the hydrogen reduction kinetics of MgO single crystals to determine under what conditions gaseous diffusion kinetics were rate controlling and to see how accurately the kinetics could be modeled.

Gaseous Diffusion Kinetics

Thermodynamics

For the reduction of MgO in hydrogen, there are three most likely reactions:[2]

$$MgO(s) + H_2(g) = Mg(g) + H_2O(g)$$

with $\Delta G° = 54.8$ kcal/mol, $K_e = 1.0 \times 10^{-7}$, and $\Delta H° = 115.0$ kcal/mol;

$$2MgO(s) + H_2(g) = Mg(OH)_2(g) + Mg(g)$$

with $\Delta G° = 98.1$ kcal/mol, $K_e = 2.9 \times 10^{-13}$, and $\Delta H° = 178.9$ kcal/mol;

$$2MgO(s) + H_2(g) = 2MgOH(g)$$

with $\Delta G° = 90.4$ kcal/mol, $K_e = 2.8 \times 10^{-12}$, and $\Delta H° = 173.3$ kcal/mol at 1700 K where K_e is the equilibrium constant for the reaction at that temperature. From these data the first reaction is the most likely one, and the equilibrium partial pressures of the product gases will be on the order of 10^{-3}–10^{-4} atm if pure hydrogen is used as the reactant gas.

Boundary Layers

A small plate-shaped crystal undergoing reaction in either a stagnant or flowing gas stream of large extent compared to the size of the crystal will

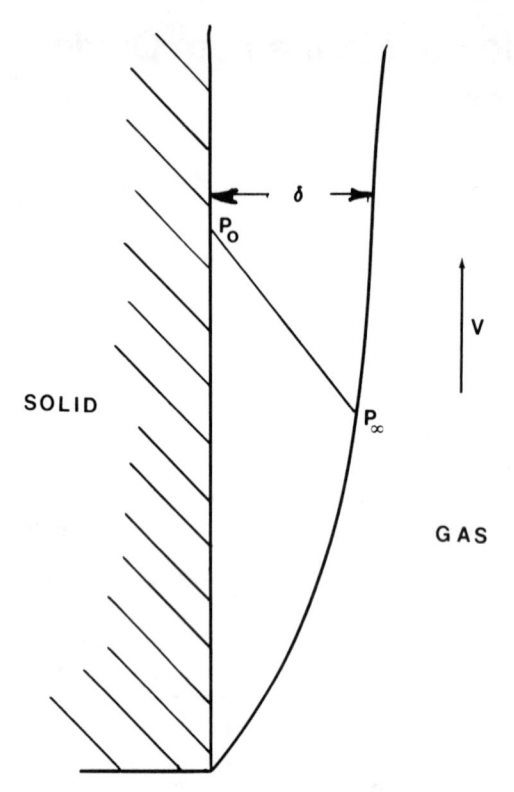

Fig. 1. Diffusion through a gaseous boundary layer of thickness δ. The equilibrium partial pressure at the gas–solid interface is p_0 and p_∞ is the pressure in the flowing gas of velocity v.

be surrounded by a diffusion boundary layer of thickness δ shown in Fig. 1 and given by[3]

$$\delta = \delta_0 + 3.09 Re^{-1/2} Sc^{-1/3} \qquad (1)$$

where Re = Reynold's number = Lv_∞/μ, Sc = Schmidt coefficient = μ/D, L = the distance along the plate, μ = the kinematic gas viscosity, v_∞ = the stream velocity parallel to the plate far from the surface, δ_0 = a term independent of stream velocity, which takes into account that diffusion is taking place into an infinite medium, and D = an appropriate gaseous diffusion coefficient. The diffusion flux density is given by

$$J = -D\frac{\partial C}{\partial x} = -\frac{D}{\delta RT}(p_\infty - p_0) \qquad (2)$$

where C = molar concentration, p_∞ = partial pressure of the diffusing species far from the surface, and p_0 = equilibrium partial pressure of the diffusing species at the interface. Therefore, as the gas velocity is increased, the boundary layer thickness decreases and the rate of reaction increases with the square root of the gas velocity. A commonly used test to determine whether surface reaction or diffusion kinetics are rate controlling is to determine the rate as a function of the fluid velocity. If the rate increases as the fluid velocity increases, the reaction is assumed to be controlled by diffusion kinetics.

Multicomponent Gaseous Diffusion

In general, the total molar flux density of a species A relative to a fixed coordinate system, N_A, is given by[4]

$$N_A = J_A + C_A v \tag{3}$$

where J_A is the diffusion flux density relative to the moving stream, C_A is the molar concentration of A, and v is the molar average velocity of the bulk stream now parallel to the diffusion flux. The stream velocity depends on the fluxes of the diffusing gas species through

$$v = (1/C)(N_A + N_B + \ldots + N_i) \tag{4}$$

This gives for the flux density

$$N_A = -CD_{AM}\frac{dX_A}{dx} + X_A(N_A + N_B + \ldots + N_i) \tag{5}$$

where X_A is the mole fraction of A, C is the total molar concentration, and D_{AM} is a multicomponent diffusion coefficient determined from the Stefan–Maxwell equation[5] and is in terms of the individual gas binary diffusion coefficients of the gases in the system. In this particular case, they are $D_{H_2-H_2O}$, D_{Mg-H_2}, and D_{Mg-H_2O}. These binary diffusion coefficients are relatively insensitive to composition and can be obtained from tabulated data, by comparison to tabulated data for similar gaseous molecules, or can be calculated from the Chapman–Enskog equation easily to within 10%.[5] As was pointed out above, the equilibrium constants for all the reactions of interest in the MgO–H$_2$ system are quite small, resulting in low product gas pressures. In this case, the multicomponent diffusion coefficient for magnesium is $D_{Mg} \cong D_{Mg-H_2}$, and that for water vapor is $D_{H_2O} \cong D_{H_2O-H_2}$.

Kinetics

For the most likely reaction

$$MgO(s) + H_2(g) = Mg(g) + H_2O(g)$$

the equilibrium constant is given by

$$K_e = \frac{p(Mg)p(H_2O)}{p(H_2)} \tag{6}$$

If the reactant hydrogen gas is very dry so that $p_\infty(H_2O) \cong 0$, then, for the product gases, $p_0(H_2O) \cong p_0(Mg)$. Likewise, if the initial hydrogen gas pressure is significantly higher than the product gas pressures, $p_0(H_2) \cong p_\infty(H_2) = p(H_2)$, then

$$p_0(Mg) = p_0(H_2O) = p(H_2)^{1/2}K_e^{1/2} \tag{7}$$

In this case the rate of the reaction is given by

$$N_{Mg} = N_{H_2O} = \frac{D_{Mg}}{\delta RT}K_e^{1/2}p(H_2)^{1/2} \tag{8}$$

The only exponentially temperature-dependent term in the above is K_e, so the apparent activation energy for the reaction is $\Delta H°/2$. On the other hand, if $p_\infty(H_2O) \gg p_0(Mg)$ or the hydrogen is wet to begin with, then

$$p_0(\text{Mg}) = K_e\left(\frac{p(\text{H}_2)}{p_\infty(\text{H}_2\text{O})}\right) \tag{9}$$

with the rate of reaction now given by

$$N_{\text{Mg}} = \frac{D_{\text{Mg}} K_e p(\text{H}_2)}{\delta RT p_\infty(\text{H}_2\text{O})} \tag{10}$$

In this case, the rate of reaction is inversely proportional to the water vapor pressure in the system, and the apparent activation energy is simply $\Delta H°$ or the standard enthalpy for the reaction.

Thus, there are many experimental checks that can be made to determine whether or not gaseous diffusion is rate controlling in this system. (1) The rate of reaction should increase as the square root of the gas stream velocity. (2) In very dry hydrogen, the apparent activation energy is one-half of the enthalpy for the reaction. (3) In wet hydrogen, the apparent activation energy should equal the enthalpy for the reaction, and the rate should be inversely proportional to the water vapor pressure. (4) The rate of reaction should vary linearly or should vary as the square root of the hydrogen pressure, depending on whether the gas is wet or dry, respectively. (5) With a change of the carrier gas, the gaseous diffusion coefficient can be varied. (6) Finally, the rates should be calculable from thermodynamic and gas property data.

Experimental Procedure

Small plate-like high-purity single crystals of MgO (obtained from Y. Chen at the Oak Ridge National Laboratory) weighing initially about 100 mg were suspended in pure flowing hydrogen gas at 1 atm (9.8×10^4 Pa) with a molybdenum wire. The sample weight loss was monitored as a function of time with a microbalance having microgram sensitivity. For a given set of experimental conditions, the rate of weight loss was essentially constant during the loss of several milligrams. Therefore, a single sample could be used over a wide range of experimental conditions. The effects of temperature, gas flow rate, and water content of the hydrogen gas were investigated. The H_2O content was varied by passing the gas over a temperature-regulated water bath, and the water vapor pressure was monitored at the exit end of the furnace with an electronic hygrometer.

Results and Discussion

Figure 2 shows the effect of gas velocity on the rate of the reaction calculated from the initial sample size and rate of weight loss at a constant temperature and water vapor content. The data fit reasonably well a straight line when plotted vs the square root of the velocity. As discussed above, since the reaction rate is quite dependent of the stream velocity and varies as the square root of the velocity, this is strong evidence that the rate of reaction is controlled by gaseous diffusion.

Figure 3 gives typical results for the effect of water vapor pressure in the hydrogen at a constant gas velocity and temperature. The rate of reaction shows the expected decrease but not as rapid as the inverse water vapor pressure dependence predicted above. For all the samples run, the average slope of log–log plots of reaction rate vs H_2O partial pressure was -0.84 ± 0.16 rather than the predicted value of -1. In part, this discrepancy can be rationalized on the basis that, at the lower water vapor pressures, the equilibrium water vapor pressure (p_0) is only 2–3 times the water vapor pressure in the gas stream, and the latter really cannot be neglected as has been done in Fig. 3 and similar plots. This has the effect

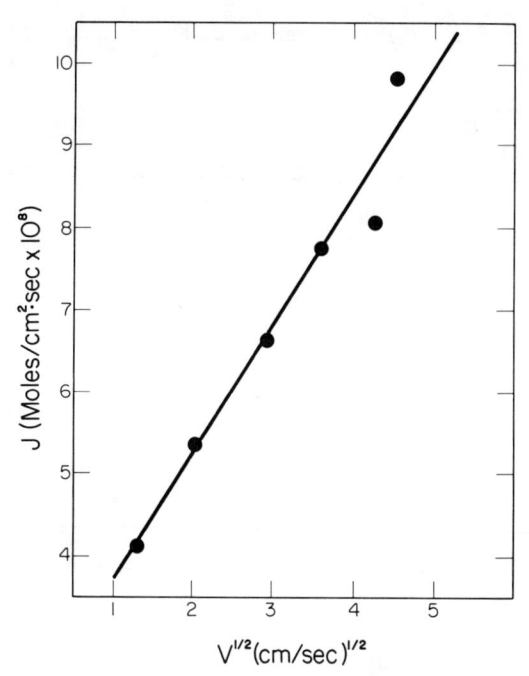

Fig. 2. Effect of the gas velocity on the reaction rate. Temperature and water vapor pressure are constant.

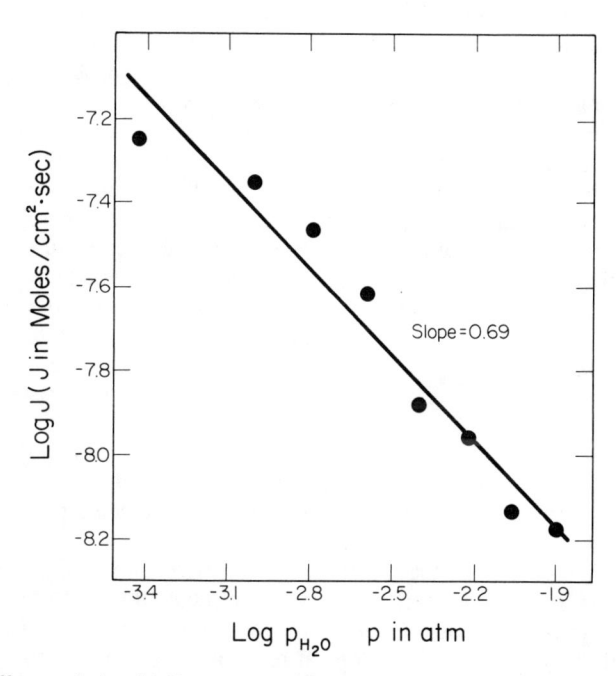

Fig. 3. Effect of the H₂O content of the hydrogen on the reaction rate.

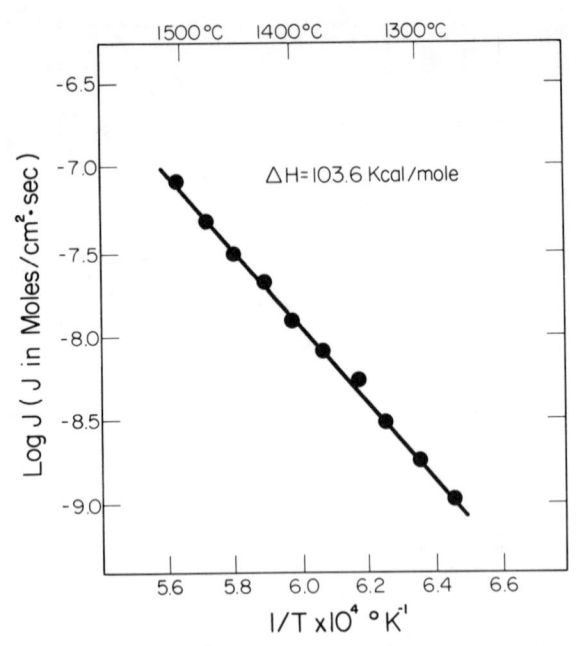

Fig. 4. Temperature dependence of the reaction rate. Water vapor pressure and gas velocity are constant.

of lowering the points at the lower pressures in the log–log plots with a large effect on the apparent slope. Nevertheless, the data do come close to the inverse $p(H_2O)$ dependence predicted by the model of diffusion of the product gases.

In Fig. 4 are plotted typical data showing the effect of temperature on the reaction rate at a constant gas velocity and water vapor pressure. For this particular sample, the apparent activation energy is 103.6 kcal/mol, and for all the samples measured, the average activation energy is 96.31 ± 11.6 kcal/mol. This is to be compared with the activation energy of $\Delta H° = 115$ kcal/mol predicted by the model when the ambient water vapor content of the gas is comparable in magnitude to that produced in the reaction. Again, at the higher temperatures, the $p(H_2O)$ produced in the reaction really cannot be neglected, as has been done in this figure. This has the effect of making the log J vs $1/T$ plot have a smaller slope than it would if the water vapor produced in the reaction could be taken into account. This can partly explain the difference between the expected and measured values of the apparent activation energies. In any event, they still are reasonably close, and the thermodynamic data from the literature may be in error by 10 kcal/mol or more. With the insertion of reasonable approximate values in the rate equations given earlier, the calculated rates are roughly within a factor of 2 of the observed rates, giving further credence to the mechanism being gaseous diffusion of the products.

Up to this point, all of the weight loss data, if not providing exact quantitative corroboration of the model, certainly are in qualitative and semiquantitative agreement. If gaseous diffusion were rate controlling, then uniform attack of the surface would be expected since no local reaction sites should exist. During the early stages of the reaction, after only a small amount of the surface has been

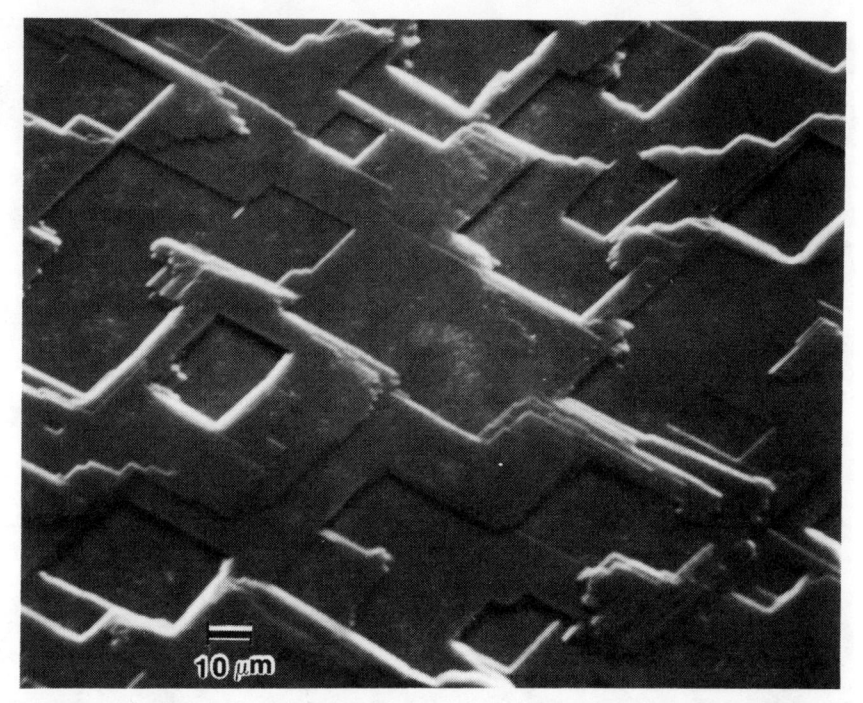

Fig. 5. Surface of a (100) cleavage face in the early stages of reaction showing shallow surface features.

reacted, that is indeed what is found, as shown in Fig. 5. The surface shows more or less uniform attack with slightly raised regions with crystallographic step features. However, after the reaction has proceeded longer, uniform attack virtually ceases at some point, and all further reaction is localized at square pits in the surface, as shown in Fig. 6. Little further reaction occurs at the surface, but weight loss apparently continues at the pit sites. As a result, the reaction occurs below the surface and produces a porous crystal. Figure 7 is a micrograph of a sample that has been cleaved after reaction, which shows the surface and the pits at the bottom and how the attack has proceeded below the surface on the cleaved face at the top of the micrograph.

A great deal of effort has been spent to determine if there is any correlation between the pits or sites for local attack and dislocations, either grown in or produced by scratches or other means of deformation. None could be found. In general, when the surface morphology is followed by examining crystals after a certain degree of reaction and reacting again, etc., the only conclusion that could be reached was that pits would form anywhere on the crystal surface and seemed to be stable when they reached a certain size, which depended on the extent of the reaction. This behavior strongly suggests the possibility that the change in surface morphology may be due to some kinetic instability inherent in the process rather than related to artifacts or impurities. These experimental observations are being made more quantitative, and surface analysis for impurity segregation during reaction is underway to determine more precisely the cause of the apparent local reaction sites.

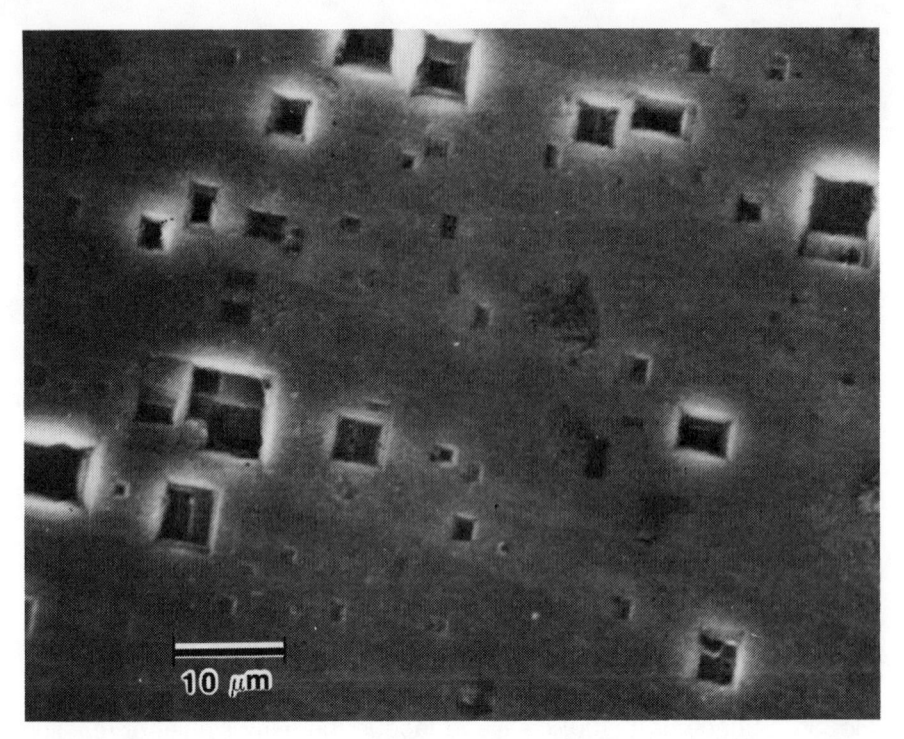

Fig. 6. Pits formed on the surface during the intermediate stages of the reaction.

It might be pointed out, that, although the reaction becomes localized at pit sites on the surface, the overall reaction kinetics still fit the gaseous diffusion model. This is not unreasonable if it is assumed that the gas emerging from the pits is very near to thermodynamic equilibrium, as it must be since surface reaction virtually stops. The pit separation is only on the order of tens of micrometers or less while the boundary layer thickness for gas velocities used in these experiments is on the order of hundreds or thousands of micrometers. As a result, the surface gas concentration is essentially uniform compared to the boundary layer thickness, and gaseous diffusion control would be expected.

Conclusions

The kinetics of reduction of MgO single crystals in pure hydrogen have been studied by monitoring weight loss as a function of time. The rate of reaction varies approximately as the square root of the gas velocity, it is roughly inversely proportional to the water vapor content of the gas, the apparent activation energy is close to the standard enthalpy for the reaction, and the calculated rates are within a factor of 2 of the experimental rates if a gaseous diffusion model is assumed. However, the crystal surfaces show strongly localized attack that cannot be correlated to dislocations and is thought to represent the onset of some fundamental surface instability.

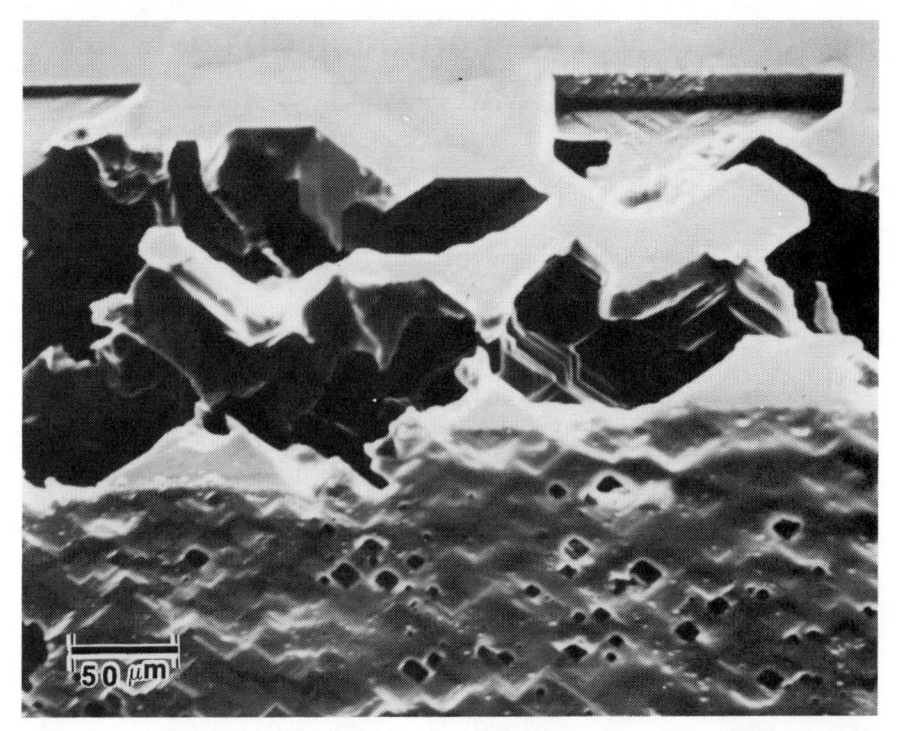

Fig. 7. Reacted surface in the later stages showing pits in the lower half of the micrograph. The upper half is a surface cleaved perpendicular to the original surface after the reaction, which graphically shows the extent of subsurface reaction.

Acknowledgments

The authors thank Y. Chen of the Oak Ridge National Laboratory for supplying the high-purity MgO crystals used in this research. The work was supported by the National Science Foundation.

References

[1] J. P. Hirth, "Evaporation and Sublimation Mechanisms"; p. 453 in The Characterization of High Temperature Vapors. Edited by J. L. Margrave. Wiley, New York, 1967.

[2] D. R. Stull and H. Prophet, JANAF Thermochemical Tables, 2nd ed. National Bureau of Standards, Rept. No. NSRDS–NB537, June 1971.

[3] A. R. Cooper, Jr., and W. D. Kingery, "Corrosion of Refractories by Liquid Slags and Glasses"; p. 85 in Kinetics of High Temperature Processes. Edited by W. D. Kingery. M. I. T., Cambridge, MA, 1959.

[4] C. J. Geankoplis, Mass Transport Phenomena. Columbus, OH, 1977.

[5] R. B. Bird, W. E. Stewart, E. N. Lightfoot; Transport Phenomena. Wiley, New York, 1960.

Influence of MgO Vaporization on the Final-Stage Sintering of MgO–Al$_2$O$_3$ Spinel

Minoru Matsui, Tomonori Takahashi, and Isao Oda

NGK Insulators, Ltd.
Research and Development Laboratory
Mizuho, Nagoya, Japan

The influence of MgO vaporization on the final stage sintering of MgO–Al$_2$O$_3$ spinel was studied. Spinel bodies with 1% porosity were fired at 1800°C in a MgO vaporizing atmosphere and an atmosphere with no MgO vaporization. In the MgO vaporizing atmosphere, MgO vaporization at the surface resulted in an Al$_2$O$_3$-rich spinel solid solution and a gradient of Al$_2$O$_3$/MgO concentration in the specimen. This [Al$_2$O$_3$]/[MgO] gradient enhanced mass transport and was effective in pore removal. The mechanism of the final-stage sintering is discussed in terms of the [Al$_2$O$_3$]/[MgO] gradient effect on the flux of vacancies diffusing from pores to the surface. The flux for the specimen with [Al$_2$O$_3$]/[MgO] gradient is estimated to be about 5000 times as large as the flux in the specimen without MgO vaporization.

Transparent MgO–Al$_2$O$_3$ spinel sintered bodies, as shown in Fig. 1, were obtained when they were finally sintered at 1800°C in hydrogen gas. It was necessary for pore removal that MgO was vaporizing from the spinel bodies during the sintering.

Spinel bodies of 1% porosity were prepared by sintering at 1600°C and subsequently fired at 1800°C under two conditions, one with MgO vaporization and the other with little MgO vaporization. On the basis of the microstructure and the Al$_2$O$_3$ and MgO composition distributions across the specimens, the mechanism of MgO vaporization effects on the pore removal was studied.

Experimental Procedures

Sample Preparation

MgO–Al$_2$O$_3$ spinel powder prepared by hydrolysis and calcination of magnesium and aluminum isopropoxides was formed into the pellet of 11 mm diameter × 1.2 mm thick and pressed isostatically at 260 MPa. The properties and the microstructure of the powder are shown in Table I and Fig. 2, respectively. The formed bodies were sintered at 1600°C in hydrogen gas. The properties of the spinel bodies are given in Table II, and their microstructure is shown in Fig. 3. During sintering at 1600°C, the spinel bodies had no weight losses and exhibited the homogeneous distributions of Al$_2$O$_3$ and MgO compositions, as determined by an electron-probe microanalyzer (EPMA). As shown in Fig. 3, all pores were closed and located on the grain boundaries in the spinel bodies. The porosity analyzed from the optical micrographs was about 1%.

Firing

The prepared spinel bodies were fired at 1800°C for 12 and 24 h in hydrogen atmospheres under two conditions, one with MgO vaporization in a Mo crucible and one with little vaporization in a spinel powder bed. The [Al$_2$O$_3$]/[MgO] ratio

Fig. 1. Transparent spinel body sintered at 1800°C for 24 h.

of the specimens were calculated from chemical analyses and weight losses during firing. The crystal phases were determined by powder X-ray diffraction. Their microstructures were observed by optical microscopy. Al_2O_3 and MgO composition profiles from the surface to the center of the specimens were analyzed by EPMA, and $[Al_2O_3]/[MgO]$ molar ratio profiles were calculated.

Results

Al_2O_3 and MgO compositions and weight losses during firing of the specimens at 1800°C are given in Table III. The $[Al_2O_3]/[MgO]$ ratio changes of the specimens with the firing time calculated from the results of the chemical analysis agreed with those calculated from weight losses with the assumption of MgO vaporization. MgO vaporized from the specimens in Mo crucibles while vaporization from those in spinel powder beds was not noticeable. The crystal phase of the specimens fired at 1800°C was a single phase of spinel with an Al_2O_3-rich

Table I. Properties of Spinel Powder*

Chemical composition	
Al_2O_3	72.32 wt%
MgO	26.75 wt%
$[Al_2O_3]/[MgO]$	1.07 mol/mol
SiO_2	30 ppm
TiO_2	10 ppm
Fe_2O_3	70 ppm
CaO	<10 ppm
K_2O	10 ppm
Na_2O	20 ppm
ignition loss	1.05 wt%
total	100.13 wt%
Powder X-ray diffraction study	
Crystal phase	Spinel
Lattice constant	8.080 Å[†]
Crystallite size	290 Å

*Specific surface area: 50 m^2/g.
[†]8.080 Å is 0.8080 nm.

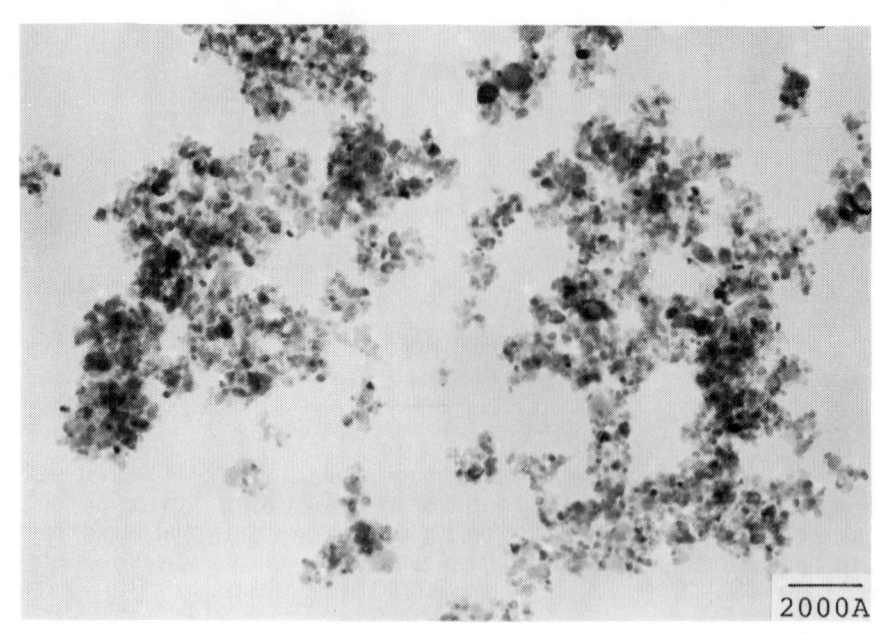

Fig. 2. Electron micrograph of the spinel powder.

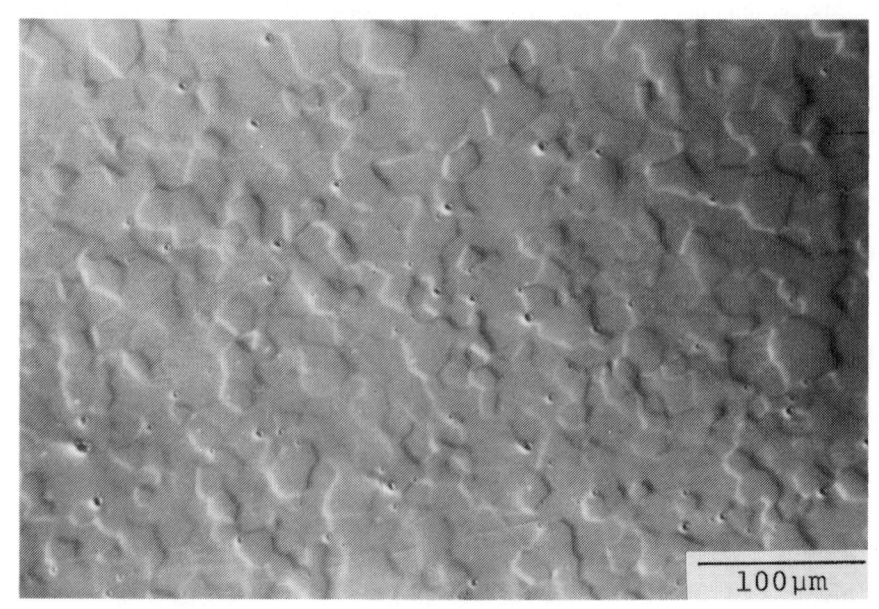

Fig. 3. Microstructure of the spinel body sintered at 1600°C for 5 h.

composition. Since the compositions given in Table III are in the composition range of Al_2O_3-spinel solid solution as shown by the MgO-Al_2O_3 phase diagram in Fig. 4, Al_2O_3-rich solid solution should be produced with MgO vaporization.

Table II. Properties of the Specimen Sintered at 1600°C for 5 h

Dimensions	11 mm (diam) × 1.2 mm (thick)
Crystal phase	Spinel
Chemical composition	
Al_2O_3	73.02 wt%
MgO	26.93 wt%
$[Al_2O_3]/[MgO]$	1.07 mol/mol
Impurities	<200 ppm
Density	3.58 g/cm³
Porosity	~1%

Table III. Chemical Compositions and Weight Losses during Firing of the Specimens

	Sintered at 1600°C for 5 h	Fired at 1800°C in Mo crucibles for 12 h	Fired at 1800°C in Mo crucibles for 24 h	Fired at 1800°C in spinel powder beds for 12 h	Fired at 1800°C in spinel powder beds for 24 h
Chemical composition					
Al_2O_3 (wt%)	73.02	74.40	77.11	73.30	73.49
MgO (wt%)	26.98	25.89	23.17	26.93	26.33
$[Al_2O_3]/[MgO]$	1.07	1.14	1.32	1.08	1.11
Weight loss (%)	0	1.8	5.2	0.4	1.0

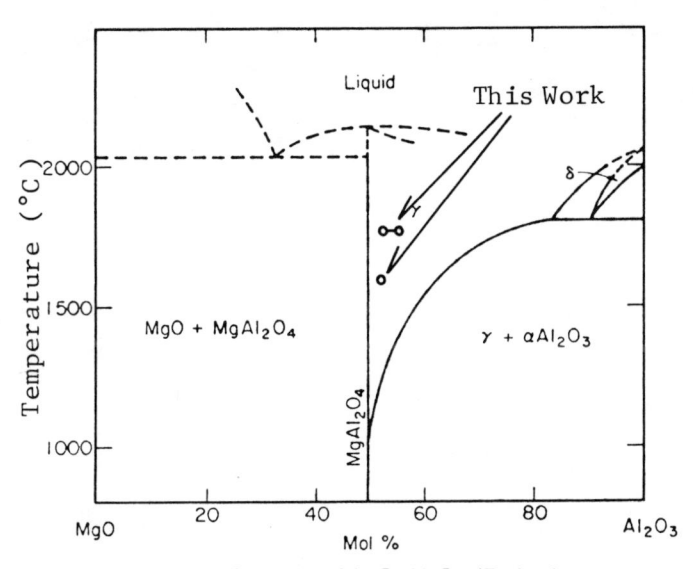

Fig. 4. Phase diagram of system $MgO-Al_2O_3$ (Ref. 1).

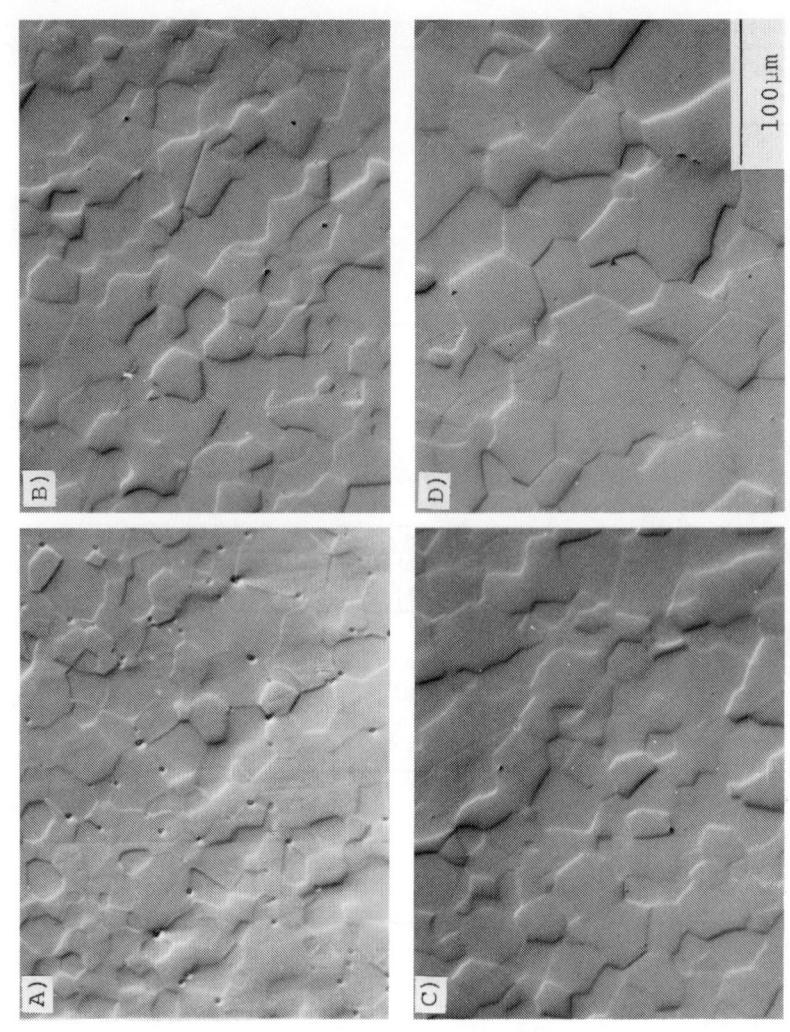

Fig. 5. The microstructures of the specimens fired at 1800°C in Mo crucibles: (A) for 12 h, the center of the specimen, (B) for 12 h, near the surface, (C) for 24 h, the center of the specimen, and (D) for 24 h, near the surface.

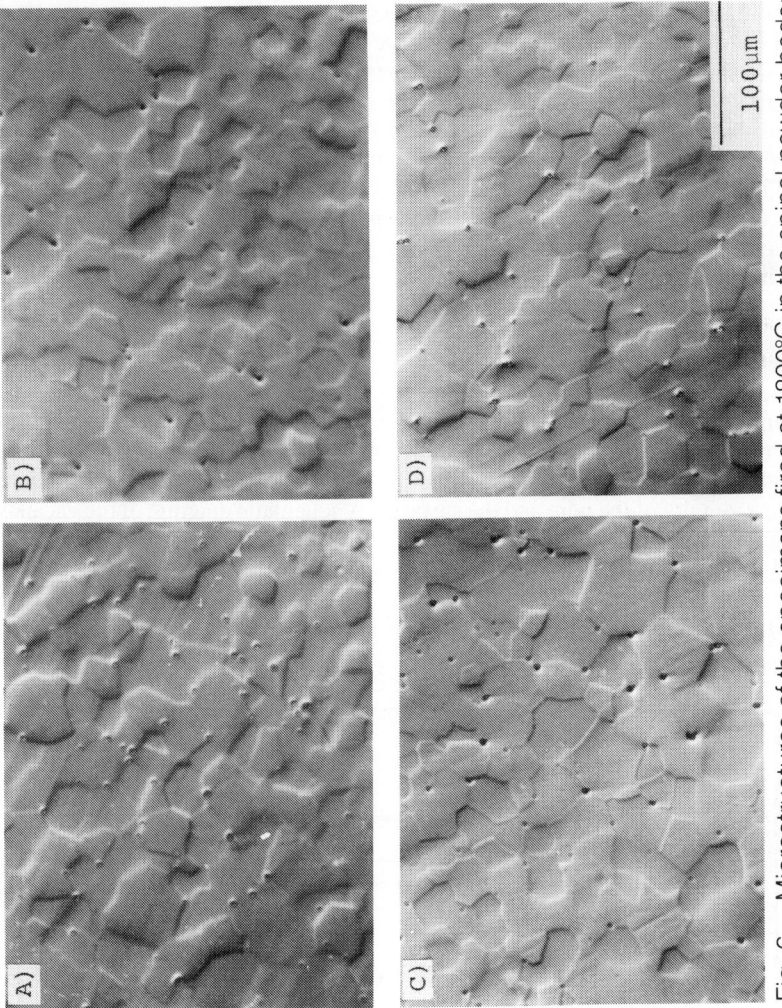

Fig. 6. Microstructures of the specimens fired at 1800°C in the spinel powder beds: (A) for 12 h, the center of the specimen, (B) for 12 h, near the surface, (C) for 24 h, the center of the specimen, and (D) for 24 h, near the surface.

The optical micrographs of the specimens are shown in Figs. 5 and 6, and [Al$_2$O$_3$]/[MgO] ratio profiles from the surface to the center calculated from EPMA results are shown in Fig. 7. In a domain near the surface of the specimen fired for 12 h in the Mo crucible, pores were removed and an [Al$_2$O$_3$]/[MgO] ratio gradient occurred, but in the inner domain pores remained and the gradient did not occur, as shown in Figs. 5 and 7. In the specimen fired for 24 h in the Mo crucible, pores were almost removed and the gradient occurred both in the inner part and near the surface, as shown in Figs. 5 and 7. In both specimens fired in spinel powder beds for 12 and 24 h, pores remained and [Al$_2$O$_3$]/[MgO] gradients did not occur, as shown in Figs. 6 and 7.

Discussion

Mechanism of the Final-Stage Sintering

The mechanism of pore removal during firing at 1800°C will be of the final stage sintering process, because the specimens before firing were dense and pores were closed as shown in Fig. 3. In the final-stage sintering porcess, pores are removed by the vacancy diffusion caused by the vacancy concentration difference between a vicinity of the pore and the surface.[2] A schematic diagram of the vacancy diffusion mechanism in the final-stage sintering process is shown in Fig. 8. The vacancy concentrations near the pore, C_p, and at the surface, C_s, are given by

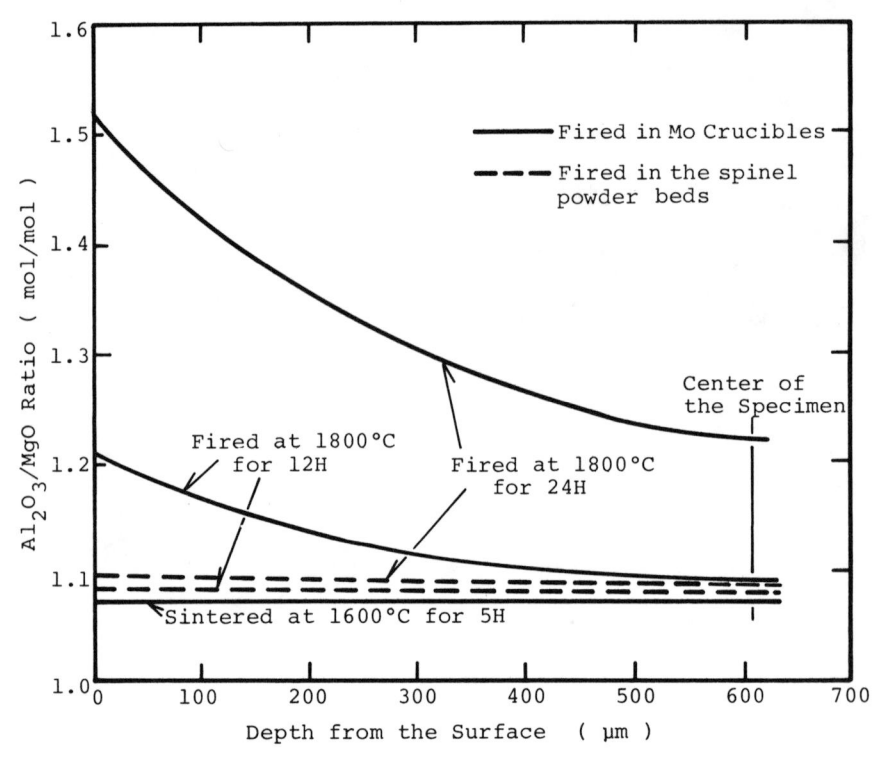

Fig. 7. [Al$_2$O$_3$]/[MgO] ratio profiles across the specimens.

Eqs. (1) and (2), respectively:

$$C_p = C_0 + \frac{\gamma a^3 C_0}{kTr} \tag{1}$$

$$C_s = C_0 + \frac{\gamma a^3 C_0}{kTR} \tag{2}$$

where C_0 is vacancy concentration of a plane surface, γ is surface energy, r is pore radius, R is the effective radius of the surface, a is ionic radius, k is Boltzmann constant, and T is absolute temperature. The flux of vacancies, diffusing from the pore to the surface, is given by Eq. (3):

$$J = -D\frac{dC}{dX} = -D\frac{C_s - C_p}{X} = -\frac{D\gamma a^3 C_0}{XkT}\left(\frac{1}{R} - \frac{1}{r}\right) \tag{3}$$

where X is the distance from the surface to the pore and D is the diffusion coefficient of the vacancy.

Increase of the Flux of the Vacancies by [Al$_2$O$_3$]/[MgO] Gradient

The acceleration of pore removal observed in the spinel bodies fired at 1800°C in Mo crucibles means an increase of the flux of the vacancies diffusing from the pore to the surface given by Eq. (3). MgO vaporization is a phenomenon occurring only outside of the specimens. The enhancement of the vacancy diffusion in the specimens may be related to the [Al$_2$O$_3$]/[MgO] ratio gradient. In each parameter of Eq. (3), the vacancy concentration of a plane surface, C_0, and surface energy, γ, can be influenced by the [Al$_2$O$_3$]/[MgO] ratio gradient. When C_0 and γ are C_{0p} and γ_p in the vicinity of the pore and C_{0s} and γ_s at the surface, the vacancy

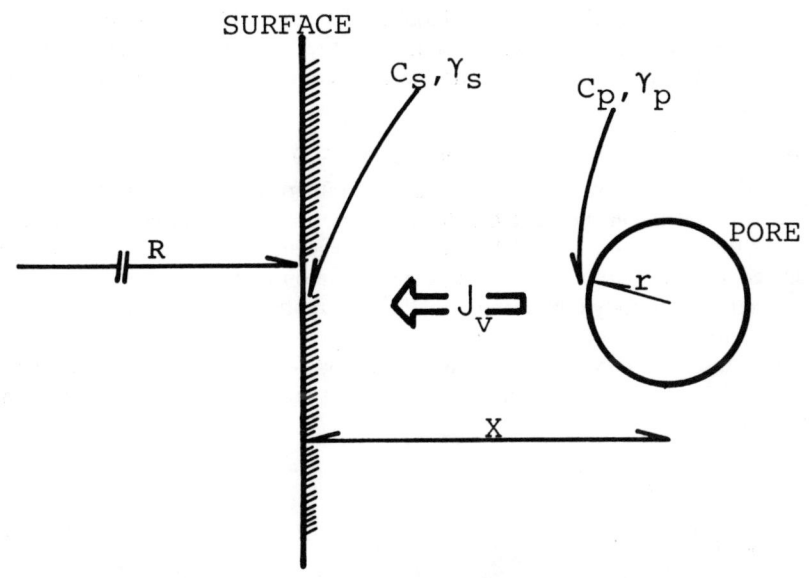

Fig. 8. Schematic diagram of the vacancy diffusion mechanism of the final-stage sintering.

concentration in the vicinity of the pore, C_p, and that at the surface, C_s, are given by Eqs. (4) and (5), respectively:

$$C_p = C_{0p} + \frac{\gamma_p a^3 C_{0p}}{kTr} \qquad (4)$$

$$C_s = C_{0s} + \frac{\gamma_s a^3 C_{0s}}{kTR} \qquad (5)$$

The flux of vacancies diffusing from the pore to the surface is given by Eq. (6):

$$J = -\frac{D}{X}\left(C_{0s} - C_{0p} + \frac{a^3}{kT}\left(\frac{\gamma_s C_{0s}}{R} - \frac{\gamma_p C_{0p}}{r}\right)\right) \qquad (6)$$

The flux, J, was calculated for two conditions: (i) vacancy concentration of a plane surface, C_0, is different between the vicinity of the pore and the surface ($C_{0p} \ne C_{0s}$) and (ii) surface energy, γ, is different between them ($\gamma_p \ne \gamma_s$). The calculated results are given in Table IV. In condition (i), the flux J_1 was calculated by substituting C for the difference between C_{0p} and C_{0s}, C_0 for C_{0p} and γ for γ_p and γ_s ($C_{0p} = C_0$, $C_{0s} = C_0 + \Delta C$, $\gamma_p = \gamma_s = \gamma$).

In condition (ii), the flux J_2 was calculated by substituting $\Delta\gamma$ for the difference between γ_p and γ_s, γ for γ_p, and C_0 for C_{0p} and C_{0s} ($C_{0p} = C_{0s} = C_0$, $\gamma_p = \gamma$, $\gamma_s = \gamma + \Delta\gamma$). In order to compare J_1 and J_2 with the flux J_0 of the homogeneous specimen ($C_{0p} = C_{0s} = C_0$, $\gamma_p = \gamma_s = \gamma$), $(J_1 - J_0)/J_0$ and $(J_2 - J_0)/J_0$ were calculated, and a numerical estimation was done by substituting the value given for each parameter in Table IV. In condition (i) with different C_0, the increase of the flux by the difference in C_0 is $-(\Delta C/C_0) \times 10^4$, and in condition (ii) with different γ, the increase of the flux by the difference in γ is $-(\Delta\gamma/\gamma) \times 10^{-3}$.

Since $\Delta\gamma/\gamma$ should not be as large as 10^3, the effect of the difference in surface energy should be very small. In the next section, the effect of [Al$_2$O$_3$]/[MgO] ratio gradient on the vacancy concentration difference, $\Delta C/C_0$, is discussed.

Oxygen Ion Vacancy Concentration Change along [Al₂O₃]/[MgO] Ratio

In MgO–Al$_2$O$_3$ spinel, the mass transport rate must be determined by oxygen ion vacancy diffusion, because the diffusion coefficient of oxygen ion vacancies is lower than those of Al and Mg ions. The vacancies, the flux of which was discussed in the foregoing section, are the oxygen ion vacancies, and consequently the oxygen ion vacancy concentration, $[V_O^{\cdot\cdot}]$, should be substituted for the vacancy concentration of a plane surface, C_0.

The intrinsic oxygen ion vacancy in MgO–Al$_2$O$_3$ spinel is presumably Schottky defects. Its formation and equilibrium are given by Eqs. (7) and (8), respectively:

$$\text{null} = nV_{\text{metal}}^{m(\prime)} + V_O^{\cdot\cdot} \qquad (7)$$

$$[V_{\text{metal}}^{m(\prime)}]^n [V_O^{\cdot\cdot}] = K_s \qquad (8)$$

where n is between ⅔ and 1, $n = 1$ when vacancies are produced at Mg sites, and $n = ⅔$ when vacancies are produced at Al sites.

Table IV. Changes of the Vacancy Flux*

	No difference	Different C_0	Different γ
$C_{0,p}$	C_0	C_0	C_0
γ_p	γ	γ	γ
$C_{0,s}$	C_0	$C_0 + \Delta C$	C_0
γ_s	γ	γ	$\gamma + \Delta\gamma$

Flux

$$J_0 = -\frac{Da^3\gamma C_0}{XkT}\left(\frac{1}{R} - \frac{1}{r}\right) \qquad J_1 = -\frac{D}{X}\left[\Delta C + \frac{a^3\gamma}{kT}\left(\frac{C_0 + \Delta C}{R} - \frac{C_0}{r}\right)\right] \qquad J_2 = -\frac{Da^3 C_0}{XkT}\left(\frac{\gamma + \Delta\gamma}{R} - \frac{\gamma}{r}\right)$$

$$(J_1 - J_0)/J_0 \sim -\left(\frac{\Delta C}{C_0}\right)\left(\frac{kTr}{\gamma a^3}\right) \qquad (J_2 - J_0)/J_0 = \left(\frac{\Delta\gamma}{\gamma}\right)\left(\frac{r}{r-R}\right)$$

$$(J_1 - J_0)/J_0 \sim -\left(\frac{\Delta C}{C_0}\right) \times 10^4 \qquad (J_2 - J_0)/J_0 \sim -\left(\frac{\Delta\gamma}{\gamma}\right) \times 10^{-3}$$

*$k = 1.38 \times 10^{-16}$ erg/K; $\gamma = 1000$ erg/cm^2; $a = 1.4 \times 10^{-8}$ cm; $r = 1 \times 10^{-4}$ cm; $R = 1 \times 10^{-1}$ cm; $T = 2073$ K.

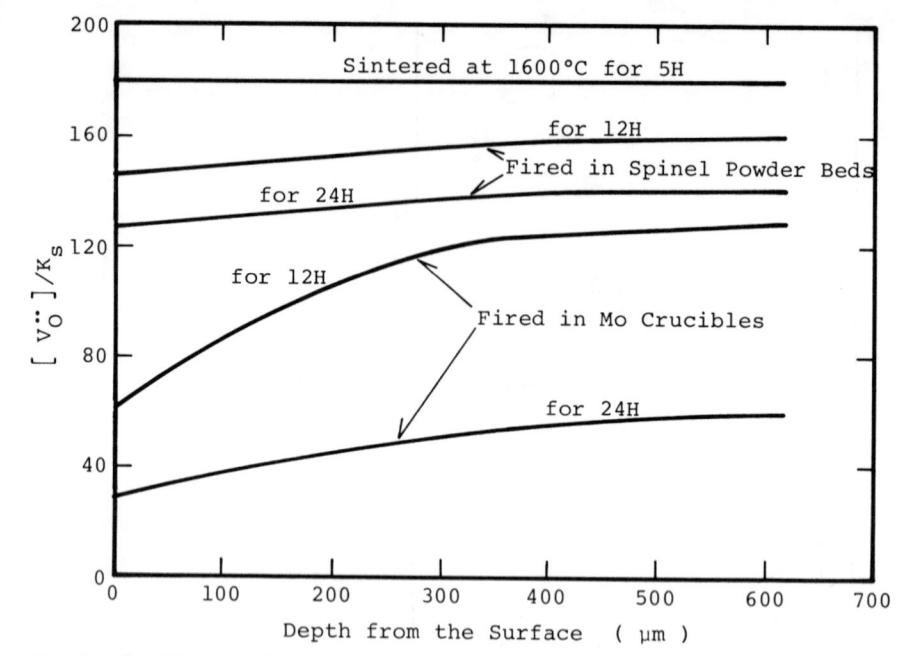

Fig. 9. $[V_{\ddot{O}}]/K_s$ profiles calculated from $[Al_2O_3]/[MgO]$ ratio profiles.

MgO vaporization results in the formation of Al_2O_3-rich spinel solid solution by the reaction given by Eq. (9):

$$p(Al_2O_3) + MgAl_2O_4 = (Mg_{Mg}Al'_{Mgq}V''_{Mg(3p/4)-q}) \times$$
$$(Al_{Alp-(q/2)+1}V'''_{Al(q/2)-(p/4)})_2O_{3p+4} \quad (9)$$

Mg-site and/or Al-site vacancies are produced in the reaction.[3]
When the $[Al_2O_3]/[MgO]$ molar ratio is y and the extrinsic cation vacancies are predominant, the cation vacancy concentration is given by $[V'''^{(\prime)}_{metal}] = (y - 1)/(9y + 3)$. As y is larger than 1.07, $[V'''^{(\prime)}_{metal}]$ is larger than 0.0055, and the extrinsic cation vacancies are predominant. On the assumption that the cation vacancies are positioned at Mg sites, n is 1, and from Eq. (8), the oxygen ion vacancy concentration is given by $[V_{\ddot{O}}] = K_s(9y + 3)/(y - 1)$. $[V_{\ddot{O}}]/K_s$ profiles calculated from $[Al_2O_3]/[MgO]$ ratio profiles in Fig. 7 are shown in Fig. 9.

$\Delta C/C_0$ in the specimen fired at 1800°C for 12 h in the Mo crucible is the largest, as shown in Fig. 9, and $[V_{\ddot{O}}]/K_s$ is 129 at the center and 63 at the surface. Then for $\Delta C/C_0 \simeq -50\%$, the flux J_1 is estimated to be 5000 times as large as the flux J_0 for the homogenous specimen. When the cation vacancies are positioned at Al sites, n is 2/3, and $[V_{\ddot{O}}]/K_s$ is 26 at the center and 15 at the surface. For $\Delta C/C_0 \simeq -40\%$, the flux J_1 is estimated to be 4000 times as large as the flux J_0.

As two mechanisms were discussed above, the change in the plane surface vacancy concentration is effective, while the surface energy change effect is very small.

Summary

Dense transparent $MgO-Al_2O_3$ spinel bodies were obtained only when MgO was vaporized from the specimens during the final sintering. The $[Al_2O_3]/[MgO]$

ratio gradient resulting from MgO vaporization was effective in the pore removal in the final densification.

Two mechanisms of the effects of the $[Al_2O_3]/[MgO]$ ratio gradient on the pore removal, the vacancy concentration change of a plane surface and the surface energy change, were discussed. Due to the effect of the vacancy concentration change of a plane surface, the flux of the vacancies diffusing from the pore to the surface is estimated to be 5000 times as large as the flux for the specimen without the $[Al_2O_3]/[MgO]$ gradient, while the surface energy change effect is very small.

The mechanism of the effect is discussed as follows: (1) An oxygen ion vacancy concentration gradient occurs along the $[Al_2O_3]/[MgO]$ ratio gradient, because Al_2O_3-rich spinel has a high concentration of extrinsic cation vacancies and a low concentration of oxygen ion vacancies for the Schottky equilibrium. (2) Diffusion of oxygen ion vacancies from the pore to the surface is enhanced by the oxygen ion vacancy concentration gradient, and the pore shrinkage rate is accelerated.

Acknowledgments

The authors acknowledge Y. Oishi's helpful advice.

References

[1]F. Colin, "Contribution to the Study of Phases Obtained during the Reduction of Some n-Al_2O_3·MO Oxides," *Rev. Int. Hautes Temp. Refract.*, **5** [4] 267–83 (1968).

[2]R. L. Coble, "Sintering Crystalline Solids. I. Intermediate and Final State Diffusion Models," *J. Appl. Phys.*, **32** [5] 787–92 (1961).

[3]H. Jagodzinski and H. Saalfeld, "Unmixing in Al_2O_3 Supersaturated Mg–Al Spinels," *Z. Kristallogr.*, **109** [2] 87–109 (1957).

Solid-State Sintering: The Attainment of High Density

Suxing Wu, E. Gilbart, and R. J. Brook

University of Leeds
Department of Ceramics
Leeds LS2 9JT, U.K.

Alumina and, to a lesser extent, magnesia have acted as the principal models for work on the densification of ceramic powders. Major difficulties in using this work for the prediction of appropriate processing conditions for other systems have resulted from the complexity and number of the mechanisms that are available for microstructural change during sintering. Nonetheless, it is argued that substantial benefit can stem from regarding processing as a rivalry between densification and coarsening and from adjusting the processing conditions to enhance the former. The effects of temperature, pressure, particle size, and additive are surveyed from this point of view, and the resulting conclusions are applied to zirconium oxide.

Since 1949 and the work of Kuczynski,[1] the study of the densification of powder compacts by diffusion mechanisms has been one of the central concerns in ceramics processing; alumina and to a lesser extent magnesia have served as model materials on which to test the many theoretical contributions to the understanding of these processes, and the progress has been considerable. However, much of the progress has taken the form of an increased recognition of the complexity of microstructural change in powder compacts, and the ability to use the alumina and magnesia experience as a predictive guide for the processing of other systems has been confined to relatively general criteria. In the following, these criteria are reviewed with emphasis on the alumina experience; the results of applying them to a different system, namely zirconia, are then discussed.

Densification and Coarsening

When predictive information from the understanding of densification in alumina and magnesia is sought, the objective is to reach a position where an informed and successful choice can be made of the processing conditions for a given system; in broad terms these conditions are five, namely, the processing temperature, the processing duration, the particle size, the applied pressure (in the event that hot-pressing is used), and the composition, including the use of controlled atmospheres for adjustment of the degree of nonstoichiometry and of additives for such benefits as they may bring. There are many other variables that affect the result of a sintering program, but on the basis that the starting point is a homogeneous structure formed from the random close packing of uniform-sized spherical particles, these five variables remain to be chosen. (It should be noted that one persuasive view[2] in ceramics processing is that the real difficulty is to reach this ideal starting point and that, provided this point can be reached, the selection of conditions for subsequent processing becomes less critical. It seems likely that attention to both aspects will emerge as the most rewarding approach.)

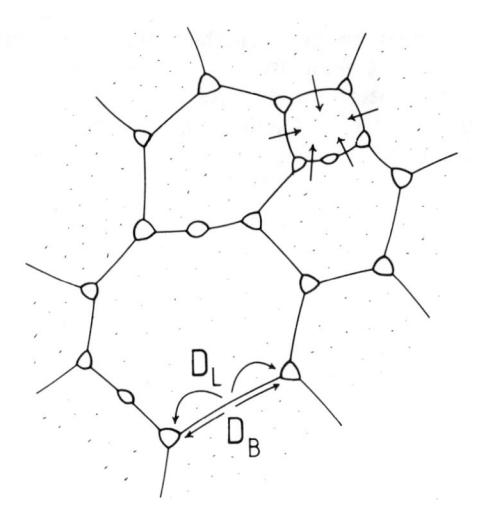

Fig. 1. Late-stage sintering of a powder compact (Ref. 3). The two contributions to surface energy reduction are densification (below) where the arrows show the direction of atom flow by lattice, D_L, or boundary, D_B, paths and coarsening or grain growth (above) where the arrows show the direction of boundary movement.

The difficulty in selecting these conditions lies in the recognition of the complexity of the sintering process that has emerged: the first complication lies in the fact that the powder can reduce its excess surface energy either by densification, i.e., replacement of solid/gas interface by solid/solid interface, or by coarsening (or grain growth), i.e., reduction in the extent of solid/gas interface or solid/solid interface by change in the scale of the microstructure. These processes not only occur simultaneously but actively interact in the sense that the rate of the one depends upon the extent of the other. It is one of the main targets in studying sintering to separate these two processes so that their dependence on the processing variables can be unambiguously studied.

The second complication lies in that each of the two primary processes can take place by many different mechanisms, e.g., coarsening by surface diffusion or vapor-phase transport or densification by lattice or grain boundary diffusion (Fig. 1) and that the relative importance of these mechanisms itself depends on both the processing variables and the instantaneous condition of the microstructure, e.g., grain size. The range of possible mechanisms that may be active, particularly where a many component system is concerned such as $BaTiO_3$, where mass transport mechanisms for each of the components must operate, is then so wide that the identification of the controlling mechanism is both very difficult and also specific to the particular conditions employed. The validity of a general claim for a controlling mechanism in solid-state sintering is, in some contrast to liquid-phase sintering or viscous flow sintering, very much open to question.

Approaches to Attainment of High Density

In the light of this position that has been recognized very much as a consequence of the work on model systems of which alumina has been the main one, the best approach for new systems is perhaps that of recognizing the general features that lie behind the complexity and attempting to use these. In this connection, two targets have been seen as linked to the attainment of high density, namely, first, the enhancement of the densification alternative relative to that of coarsening, and second, the avoidance in the later stages of the process of any separation of porosity from the grain boundary atom sources, i.e., avoidance of abnormal grain growth.

Densification/Coarsening Ratio

The first of these targets, namely, the enhancement of the densification/coarsening ratio, is perhaps so logical a feature of a search for high density that it can be criticized as obvious; however, it has won increasing recognition as the difficulties of applying more quantitative modeling have been seen; it also has the merit of indicating suitable choice for the processing variables.

If the objective is to enhance densification relative to coarsening, the first step is to identify the mechanisms responsible for the two simultaneous processes. Despite the many studies that have been made of sintering and grain growth and despite the often sophisticated modeling that has been applied to these processes, this remains a difficult task. The direct use of analytical models, e.g., the use of the time exponent of the shrinkage rate as a discriminating indicator, has been to some extent discounted owing to the ambiguities of any experiment where both processes are active. Consequently, efforts have been devoted to studies of densification in the absence of coarsening (this is to some extent achievable by hot-pressing[4]), to studies of grain growth in the absence of densification (this is difficult to achieve except at zero porosity[5]), or to studies of model systems intended to display one mechanism, e.g., tracer diffusion and boundary grooving. For each of these cases, the difficulty arises of then applying the information from the isolated conditions to the eventual mix of processes representative of the microstructural changes in practical sintering; in the latter, for example, the occurrence of grain growth during processing may well take the grain size beyond the range for which the model experiments have been completed.

The present status of this approach for alumina is that the mechanisms of densification are believed to be lattice diffusion of the cation or at larger grain sizes boundary diffusion of the anion.[6] This result stems from a wide range of sintering, creep, and hot-pressing results. The mechanisms of coarsening are not known with confidence, though surface diffusion is commonly assumed to be active.[7] (Since both surface diffusion and vapor-phase transport are likely to be heavily influenced by the exact chemistry of the system, this is an area where general conclusions are much open to question.) For magnesia, this picture is less clear, and for other systems such as $BaTiO_3$, even lattice diffusion data are (as for the titanium ion) unknown. In summary, therefore, the study of the single mechanisms remains, with the exception of alumina and such isolated examples as UO_2, in its infancy. The conclusions that may be drawn at this stage are therefore distinctly qualitative, and the nature of predictive work is influenced by this.

Temperature and the Ratio

An example of this approach is the influence of temperature. In the attempt to raise the densification/coarsening ratio, temperatures can be chosen on the basis of the relative activation energies for the two mechanisms involved. For lattice diffusion of the cation (densification), this is from direct diffusion and from creep data in the region of 575 kJ·mol^{-1}.[8] For surface diffusion (coarsening), activation

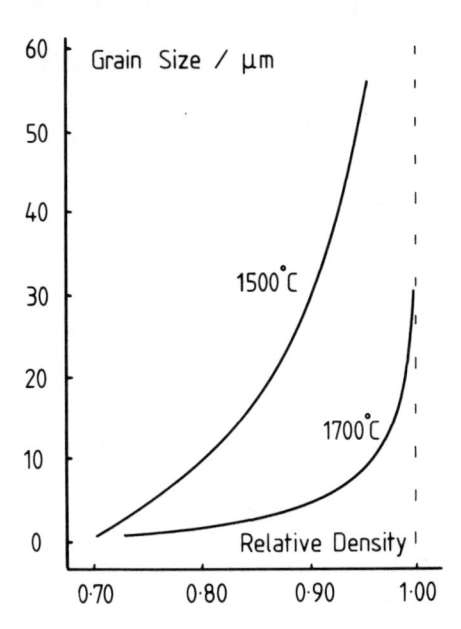

Fig. 2. Prediction of the influence of firing temperature on the microstructure development of pure Al_2O_3 samples. The initial grain size is taken as 0.3 μm and the initial relative density as 0.7 (Figure courtesy of S. J. Jamil, Ref. 3).

energy values range from 263 to 548 kJ·mol⁻¹ [9] (this range of values is not uncommon and is one of the principal problems facing the making of predictions relating to temperature change). However, the highest known surface diffusion value is less than the lattice diffusion value, so the relative rates of densification and coarsening are moved in a favorable direction by firing at high temperature. This is shown on the basis of simple kinetic models in Fig. 2, where it can be seen[3] that high firing temperature allows the attainment of a given density with less grain growth. Experimental data (Fig. 3) are consistent[10] with the expectation.

The picture for MgO in respect to temperature selection is ambiguous, since the range of surface diffusion energies (360–445 kJ·mol⁻¹) is overlapped[11] by the cation lattice diffusion energies (150–504 kJ·mol⁻¹). Under these conditions, the choice of a favorable temperature range will depend on whether the powder being considered corresponds in terms of its chemistry and character to one end or the other of the two spans. In work with $MgCl_2$-derived MgO powder of high purity, Vieira[11] has shown that high-temperature firing is not an advantage for this material.

When this concept is extended to zirconia, the densification energy (lattice diffusion of the cation) is both by tracer[12] and by sintering studies[13] in the region of 440 kJ·mol⁻¹. On the basis of grain growth data,[14] an activation energy for coarsening is 330 kJ·mol⁻¹; preliminary surface diffusion data from boundary grooving[15] suggest 340 kJ·mol⁻¹. As a consequence, the use of a high-temperature firing cycle again appears to be advantageous, and data are again consistent with the expectation (Fig. 4).

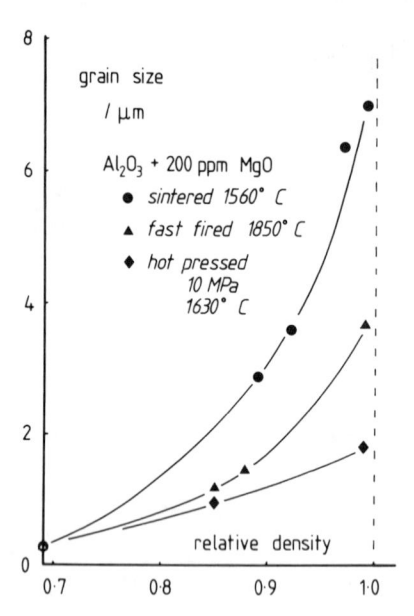

Fig. 3. Experimental data showing the ability of high-temperature firing and hot-pressing to decrease the degree of grain growth for a given degree of coarsening in alumina.

In summary, the influence of temperature as represented in results can be interpreted on the basis of the densification/coarsening ratio. The predictive use of this influence for the processing of other systems is limited first by the lack of data for the separate mechanisms and second by the uncertainty concerning the actual controlling mechanisms themselves. Since the experiment itself is readily performed (comparison of structure following different temperature treatments), this is generally a more effective procedure than attempting prediction.

Particle Size and the Ratio

The influence of particle size has been modeled in terms of scaling laws,[16] which indicate, for each densification and coarsening process, the dependence of the rate on size. Determination of the size dependence under conditions where a single process is active has been performed for alumina,[4] and the validity of such scaling laws for single processes has been indicated. This understanding can then be valuable in selecting the optimum size range for the attainment of densification; if, for example, densification occurs by lattice diffusion (rate $\propto G^{-3}$) and coarsening by vapor transport (evaporation/condensation rate $\propto G^{-2}$), small particle sizes are advantageous.[17,18] The application of such concepts to the densification of other systems has proved helpful.[19]

Pressure and the Ratio

The influence of pressure, P_a, may be understood from its direct action[20] on the driving force for densification $[(a\gamma)/r + P_a/\rho]$ and its lack of consequence in respect to the coarsening rate. Hot-pressing will therefore always act favorably in enhancing fired densities, a result that is well-known for alumina (e.g., Fig. 3) and

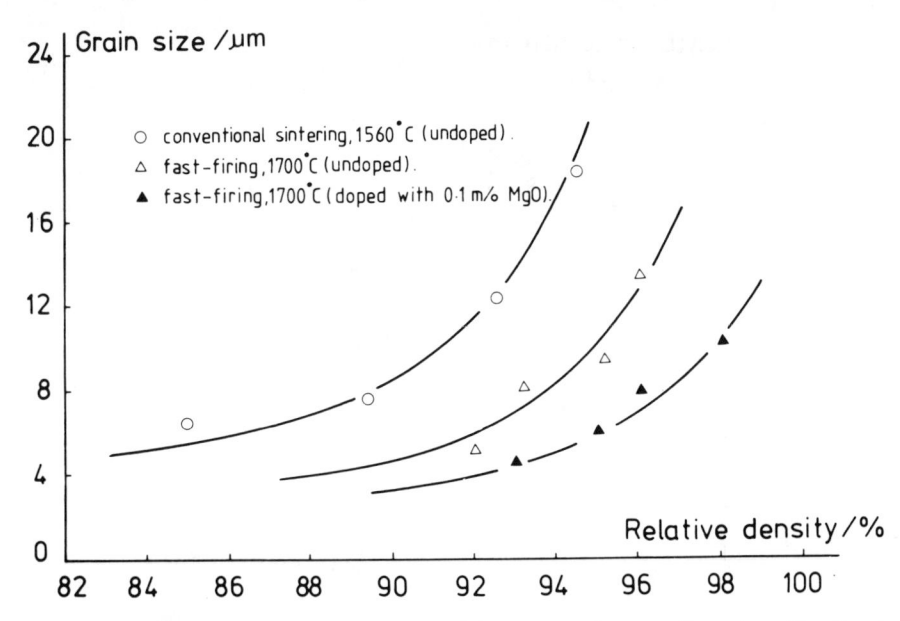

Fig. 4. Ability of high-temperature firing to enhance the densification/coarsening ratio in calcia-stabilized zirconia.

magnesia and for many other systems. This is one of the most confidently held of the predictive conclusions.

Additives and the Ratio: Sintering of Zirconia

The remaining issue is that of the composition and notably the additive. This has been one of the main influences of work on alumina since the benefits of magnesia additions[21] have been so widely recognized. The many possible roles of the additive are discussed elsewhere in the book, but a brief reference to the application of the ideas suggested by the alumina system to the sintering of the zirconia system may be helpful in indicating the use of the densification/coarsening ratio argument.

It is believed[4] that MgO accelerates the densification process in alumina by modifying the lattice diffusion coefficient; there is also evidence[22] to suggest that the rate of surface diffusion is reduced by this additive. There are other consequences of magnesia addition, but these two are already a substantial influence in respect to the value of the densification/coarsening ratio. On this basis, work on zirconia has been directed to the determination of an additive capable of maximizing the densification rate and of suppressing the coarsening rate. Work on the influence of CaO additions has shown[23] the value of 12 mol% CaO concentration in raising the densification rate (Fig. 5), and work on the boundary grooving[15] of stabilized zirconia bicrystals has suggested (Fig. 6) that MgO is effective in lowering the surface diffusion rate. The use of combined additives is accordingly suggested, and Fig. 7 shows the result of MgO additions to 12 mol% CaO stabilized zirconia.[24] The benefits are less dramatic than those with MgO in Al_2O_3, but they are nonetheless real and significant in raising the fired density well into the range of impermeability.

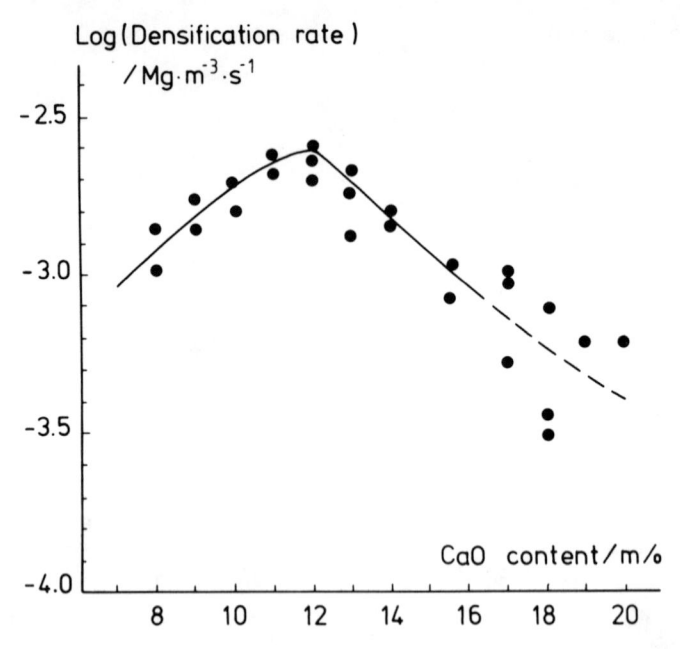

Fig. 5. The influence of CaO content on the kinetics of hot-pressing (1200°C; 20 MPa) of zirconia.

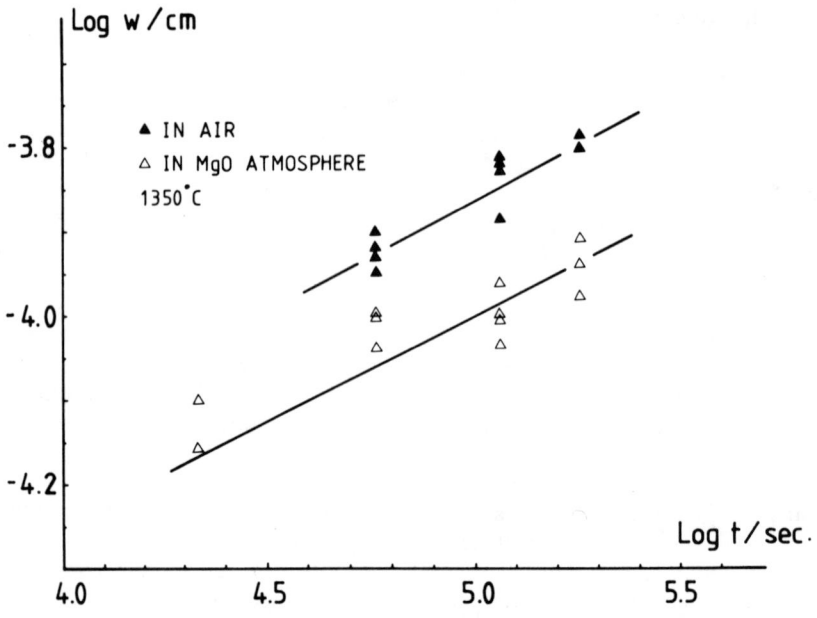

Fig. 6. Growth of grain-boundary grooves on yttria-stabilized zirconia bicrystals. The ability of magnesia to suppress the rate of groove formation is indicated. The lines show the theoretical slope for surface-diffusion controlled growth.

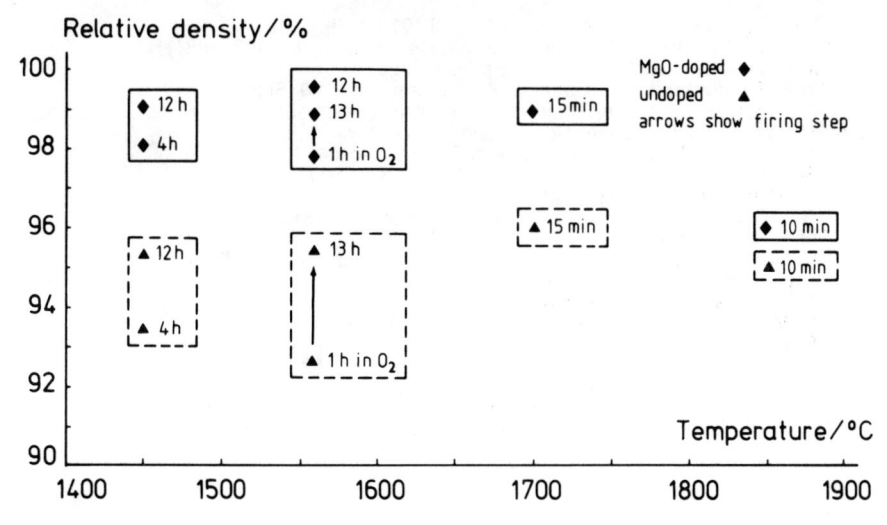

Fig. 7. Final densities achieved by firing calcia-stabilized zirconia under different time/temperature conditions. The arrows indicate firing steps made in sequence.

The additive question is peculiarly complex and really comprises the remaining general issue in the sintering of ceramics where a broad measure of uncertainty exists. The above results do, however, suggest that, as in the case of temperature, the basic consideration of the densification/coarsening ratio can be helpful.

Abnormal Grain Growth

The final issue concerns the importance of the abnormal grain growth question. It can be argued[3] that, with sufficient attention to the enhancement of the rate ratio, the problem of abnormal growth is curtailed and the data on temperature and pressure effects are broadly supportive of this view. However, the growing awareness of boundary-additive interactions,[25] the dramatic role of MgO additions in limiting abnormal growth in alumina,[21] the finding that MgO suppresses the rate of boundary migration[5] in fully dense Al_2O_3, and the difficulty in detecting a direct effect of MgO on the ratio itself (measurements of surface area during densification[26] — the early results have been confirmed for compositions below the solid solution limit[27]) all suggest a degree of caution in relying uniquely on the ratio in the search for understanding of additive function. This will remain an area of active research, and as before, alumina will have to act as one of the surest guides to progress.

References

[1] G. C. Kuczynski, Trans. AIME, **185**, 169 (1949).
[2] E. R. Barringer and H. K. Bowen, J. Am. Ceram. Soc., **65** [12] C-199 (1982).
[3] R. J. Brook, Proc. Br. Ceram. Soc., **32**, 7 (1982).
[4] M. P. Harmer and R. J. Brook, J. Mater. Sci., **15**, 3017 (1980).
[5] S. J. Bennison and M. P. Harmer, J. Am. Ceram. Soc., **66** [5] C-90 (1983).
[6] A. E. Paladino and R. L. Coble, J. Am. Ceram. Soc., **46** [3] 133 (1963).
[7] A. H. Heuer, J. Am. Ceram. Soc., **62** [5–6] 317 (1979).
[8] R. M. Cannon and R. L. Coble; p. 61 in Deformation of Ceramic Materials. Edited by R. C. Bradt and R. E. Tressler. Plenum, New York, 1975.

[9] T. K. Gupta, *J. Am. Ceram. Soc.*, **61** [5–6] 191 (1978).
[10] M. P. Harmer and R. J. Brook, *Trans. Br. Ceram. Soc.*, **80**, 147 (1981).
[11] J. M. Vieira and R. J. Brook; these proceedings.
[12] W. H. Rhodes and R. E. Carter, *J. Am. Ceram. Soc.*, **49** [5] 244 (1966).
[13] J. Jorgensen; p. 401 in Sintering and Related Phenomena. Edited by G. C. Kuczynski, N. A. Hooton, and C. F. Gibbon. Gordon and Breach, New York, 1967.
[14] T. Y. Tien and E. C. Subbarao, *J. Am. Ceram. Soc.*, **46** [10] 489 (1963).
[15] Suxing Wu and R. J. Brook; *Sci. Ceram.*, **12**, 371–80 (1983).
[16] C. Herring, *J. Appl. Phys.*, **21**, 301 (1950).
[17] R. L. Coble and R. M. Cannon, *Mater. Sci. Res.*, **11**, 151 (1978).
[18] M. F. Yan, *Mater. Sci. Eng.*, **48**, 53 (1981).
[19] C. Greskovich and J. H. Rossolowski, *J. Am. Ceram. Soc.*, **59** [7–8] 336 (1976).
[20] R. L. Coble, *J. Appl. Phys.*, **41**, 4798 (1970).
[21] R. L. Coble, *J. Appl. Phys.*, **32**, 793 (1961).
[22] C. Monty and J. le Duigon, *High T-High P,* **14**, 709–16 (1982).
[23] Suxing Wu and R. J. Brook, *Solid State Ionics,* **14** (1984).
[24] Suxing Wu and R. J. Brook, *Trans. Br. Ceram. Soc.*, **82** [6] 200–205 (1983).
[25] W. D. Kingery, *J. Am. Ceram. Soc.*, **57** [1] 1 (1974).
[26] J. E. Burke, K. W. Lay, and S. Prochazka, *Mater. Sci. Res.*, **13**, 417 (1980).
[27] Nancy J. Shaw; British Ceramic Society—Seminar, Swansea, U.K., 1983.

Initial Sintering of MgO in Several Water Vapor Pressures

O. J. Whittemore

University of Washington
Materials Science and Engineering Department
Seattle, WA 98195

J. A. Varela

Instituto de Quimica, UNESP
14800 Araraquara
SP, Brazil

The progress of structural development during the initial stages of sintering MgO indicates several processes occurring simultaneously, viz., rearrangement, grain growth, surface diffusion, and coalescence. Evolution during sintering in varying water vapor pressures was measured by mercury porosimetry, surface area measurements were measured by nitrogen adsorption, and crystal growth was measured by X-ray line broadening. Large changes in the first minute indicate rearrangement as the first controlling process so that rezeroing of shrinkage data cannot be justified. Crystal growth at 1100°C showed a rate constant proportional to water vapor pressure to the 1.5 power, and growth cubed was linearly related to time at from 2.3 to 101 kPa. Very little crystal growth and only pore shrinkage occurred in dry argon up to 10 h at 1300°C. Above 2.3 kPa, pores always grew with time and temperature.

Many studies have been conducted on the effect of water vapor on the sintering of MgO,[1-10] most showing increased sintering rates as judged by the parameters measured. Nearly all studies measured only one property. Both loose powder and compacts have been studied. Bulk density and linear shrinkage were measured on compacts,[1-3,7] surface areas by gas adsorption were measured on powders,[4,8] and crystal size was measured on powders.[5,6] From analysis of the data, sintering mechanisms have been proposed, activation energies calculated, and mechanisms proposed for adsorption of water on the surface of MgO.

Lack of agreement has been found among many of these studies. Although atmosphere, temperature, and time are the most important variables, sintering can also be affected by initial compact density, microstructure, particle morphology, extent of aggregation, impurities, etc.

In this work, magnesia compacts were sintered isothermally in several water vapor pressures and at several temperatures and times. The parameters measured were pore size distribution, crystal size, surface area, shrinkage, and density. Some of the work has been previously reported.[9-11]

Experimental Procedures

High-purity $Mg(OH)_2^+$ was ground in a mortar and dry ball milled for 2 h with alumina balls. No evidence of Al over a detectability of 200 ppm was noted by energy dispersive analysis of X-rays during SEM. The powder was calcined in air at 900°C for 4 h. Immediately after being calcined, the MgO powder was stored in a dessicator, as were all compacts subsequently prepared. MgO compacts were pressed in steel dies at 15 MPa and repressed isostatically in rubber sacks at 210 MPa. The compact density was 1.81 Mg/m^3, or 51% of theoretical density.

Compacts were sintered in an alumina tube with one closed end that was heated in an electric furnace. Isothermal studies were conducted in argon flowing at a rate of 1 L/min that had been bubbled through a gas washer to add a specific water vapor content. Water vapor pressures were confirmed by measuring weight gain on zeolite molecular sieves from a measured gas volume.

The compacts were placed on a MgO powder bed in an alumina boat. The boat was pushed into a zone in the alumina tube where the temperature was about 650°C. After 30 min of equilibration at 650°C in the desired atmosphere, the boat was pushed into the hot zone. The sample reached the desired temperature in 3–5 min, and zero time was taken at that point. After being sintered for the desired time, the sample was withdrawn from the furnace.

Bulk densities were measured by mercury displacement and by measuring the total pore volume by mercury porosimetry. Pore size distribution (PSD) curves were determined by mercury porosimetry. The midpore diameter, defined as the pore size after half of the pore volume has been intruded by the mercury,[12] was determined from the PSD. As the midpore diameter is used later to relate sintering, it should be noted that the pore volume of the initial compacts consisted of 80% of pores between 40 and 50 nm and < 4% of pores smaller than 30 nm. This narrow PSD allows use of the midpore diameter in calculations where a wide PSD would not be applicable.

Surface areas of the sintered compacts were determined from PSD curves, using the relationship[13]

$$S = \int_0^{V_P} Pdv/\gamma_m \cos\theta \tag{1}$$

where S is the total surface area, V the cumulative volume of mercury intruded at pressure P, γ_m the surface tension of Hg, and θ the effective contact angle between Hg and sample. Surface areas were also determined by nitrogen adsorption, and good agreement was observed.

The crystal size of the sintered compacts was measured from X-ray line broadening by using the integral-breadth method.[14]

The characteristics of compacts before sintering as measured by the above methods were the following: crystal size, 43.5 nm; surface area by nitrogen adsorption, 26.0 m^2/g; surface area by mercury intrusion, 25.2 m^2/g; midpore diameter by mercury porosimetry, 45.3 nm.

Results and Discussion

Pore Size Distribution vs Volumetric Shrinkage

During the initial stages of sintering, considerable pore growth of MgO had been measured after sintering in atmospheres containing water vapor as low as a normal room atmosphere of 2.3 kPa water vapor pressure.[10] However, in dry argon, only pore shrinkage was observed. These pore changes have more meaning if related to shrinkage and to the expected uniform shrinkage from a model. To relate the variation of pore sizes, the relative midpore diameter cubed is plotted in Fig. 1 vs the relative volumetric shrinkage. For comparison, the straight line on the

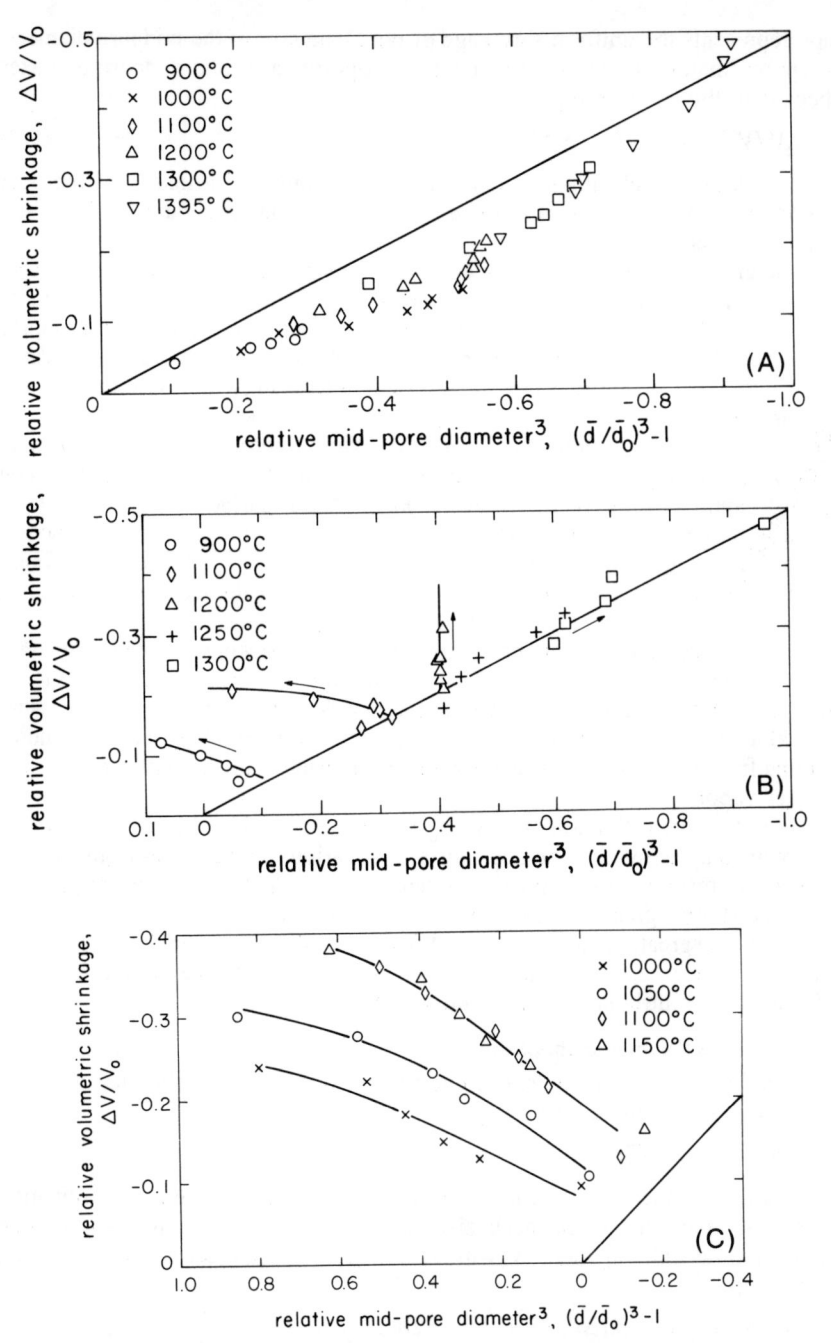

Fig. 1. Relative volumetric shrinkage vs cube of relative midpore diameter for MgO compacts sintered in (A) dry argon, (B) argon with 2.3 kPa of water vapor, and (C) argon with 22.7 kPa of water vapor for 1–600 min at temperatures indicated; solid line represents homogeneous pore shrinkage.

graph represents the uniform shrinkage of pores the size of the midpore diameter and approximately the midpore volume as proportional to the midpore diameter cubed, with this relationship:

$$\Delta V/V_0 = \lambda[(\bar{d}/\bar{d}_0)^3 - 1] \qquad (2)$$

where V_0 is the initial sample volume, ΔV the volumetric shrinkage, λ the pore fraction (0.49 for these compacts), $\bar{d}$ the midpore diameter, and $\bar{d}_0$ the initial midpore diameter.

The great effect of water vapor on pore kinetics is shown in the changes from Fig. 1(A) through (B) to (C). In dry argon (Fig. 1(A)), pores and the sample volume both shrink. The pore shrinkage rate is greater than the volume shrinkage rate up to 15% volume shrinkage. A faster decrease in large pores is indicated at first, contrary to the expected faster decrease of small pores, inferring that rearrangement is the controlling process in the initial stage.

However, when sintered in 2.3 kPa of water vapor in argon as shown in Fig. 1(B), pores grew with time at 900 and 1100°C. Comparison between Fig. 1(A) and (B) suggests that water vapor increases surface diffusion. At temperatures above 1200°C, pores shrank nearly at the rate predicted by the model, probably due to the small effect of surface diffusion. Note that the lowest point at 1250°C was after 1 min of sintering, yet 20% of volumetric shrinkage had occurred.

In 22.7 kPa of water vapor in argon, only pore growth was noted from 1000 to 1150°C and is plotted in Fig. 1(C). Time of sintering proceeds from 1 min on the right to 4 h at the left. Of practical interest is that this water vapor pressure (22.7 kPa) is very close to the water vapor pressure in combustion atmospheres resulting from the stoichiometrical burning of methane with air containing 2.3 kPa of water vapor.

The effective activation energy for densification in 22.7 kPa of water vapor was obtained by estimating the rate constants at various temperatures from the time necessary to reach the same amount of densification. Arrhenius plots are given in Fig. 2 and show higher activation energies for low values of relative linear shrinkage, and all are greater than that obtained from crystal growth (238 kJ/mol). These higher values may also indicate rearrangement, which is temperature dependent, and should have little effect on crystal growth.

Surface Area vs Volumetric Shrinkage

If reduction of surface area is due to neck growth, the following equation relates neck size with surface area:

$$\Delta S/S_0 = N_c(x/2r)^2 \qquad (3)$$

where $\Delta S/S_0$ is the relative change in surface area, N_c the mean coordination number per particle, x the neck size, and r the particle radius. Combining with equations relating density with shrinkage, we can derive the following relationship:[10]

$$\Delta S/S_0 = [16\rho_0/\rho_t(\Delta L/L_0 + 1)^3 - 2](\Delta L/L_0) \qquad (4)$$

where ρ_0 is the initial density, ρ_t is the theoretical density, and $\Delta L/L_0$ is the relative linear shrinkage.

This relationship is plotted as a solid line in Fig. 3(A), and a dashed line is shown for the mass being redistributed uniformly on the surface of the spheres if interpenetration of two spheres occurs during sintering. In sintering real powder

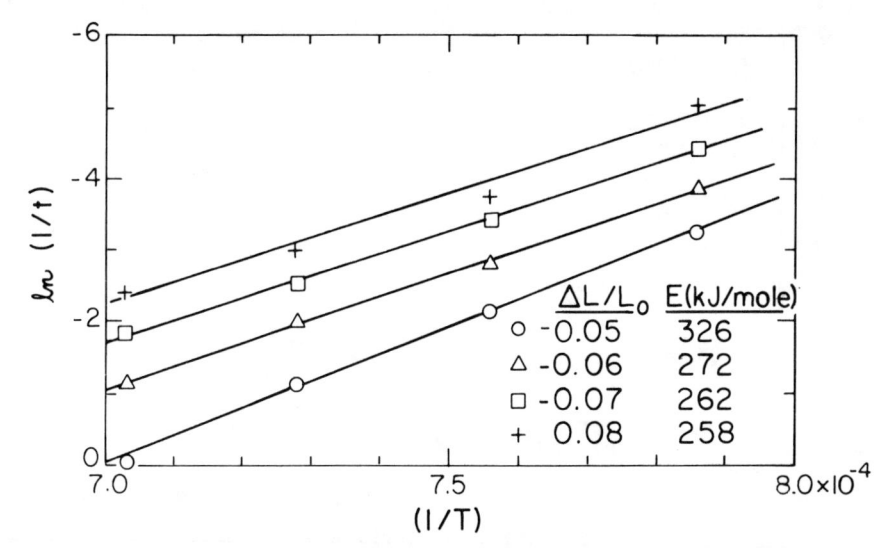

Fig. 2. Arrhenius plots of linear shrinkage of MgO compacts in 22.7 kPa of water vapor and at different stages of shrinkage; $\bar{t}$ is the time necessary to reach the same degree of sintering.

compacts, an intermediate curve between the two lines should be expected.

Shrinkage can be seen to progress at a more rapid rate than loss of surface when sintering is conducted in dry argon (Fig. 3(A)). Rearrangement should have little effect on surface area yet could cause densification and thus is indicated as a main factor in the early and initial stages of sintering these samples.

When the atmosphere contains water vapor, the relation between linear shrinkage and loss of surface is changed greatly (see Fig. 3(B) and (C)). At 2.3 kPa of water (which is the water vapor content of a normal room atmosphere and thus what would be obtained in an electric furnace open to the air), the data fortuitously follow the relationship of Eq. (4). However, in an atmosphere of 22.7 kPA of water vapor (see Fig. 3(C)), the loss of surface proceeds much more rapidly than does linear shrinkage.

Crystal Growth

MgO compacts sintered in dry argon showed very slight crystal growth at temperatures up to 1200°C. From an original 43.5 nm, crystals grew only to 47 nm after 10 h at 1200°C. Even after 10 h at 1395°C, crystals grew only to 79.6 nm, yet the bulk density reached 3.51 Mg/m³ or 98% of theoretical density.

In atmospheres containing water vapor, crystal growth varied with water vapor pressure and with time by

$$G^3 - G_0^3 = kt \qquad (5)$$

where G_0 is the initial crystal size, G is the crystal size at time t, and k is the rate constant for crystal growth. The relationship of Eq. (5) is plotted in Fig. 4 for compacts sintered in argon with 22.7 kPa of water vapor at several temperatures and also in Fig. 5 for compacts sintered at 1100°C in several water vapor pressures in argon. Rate constants (k) were obtained from Fig. 5 and, when related to water

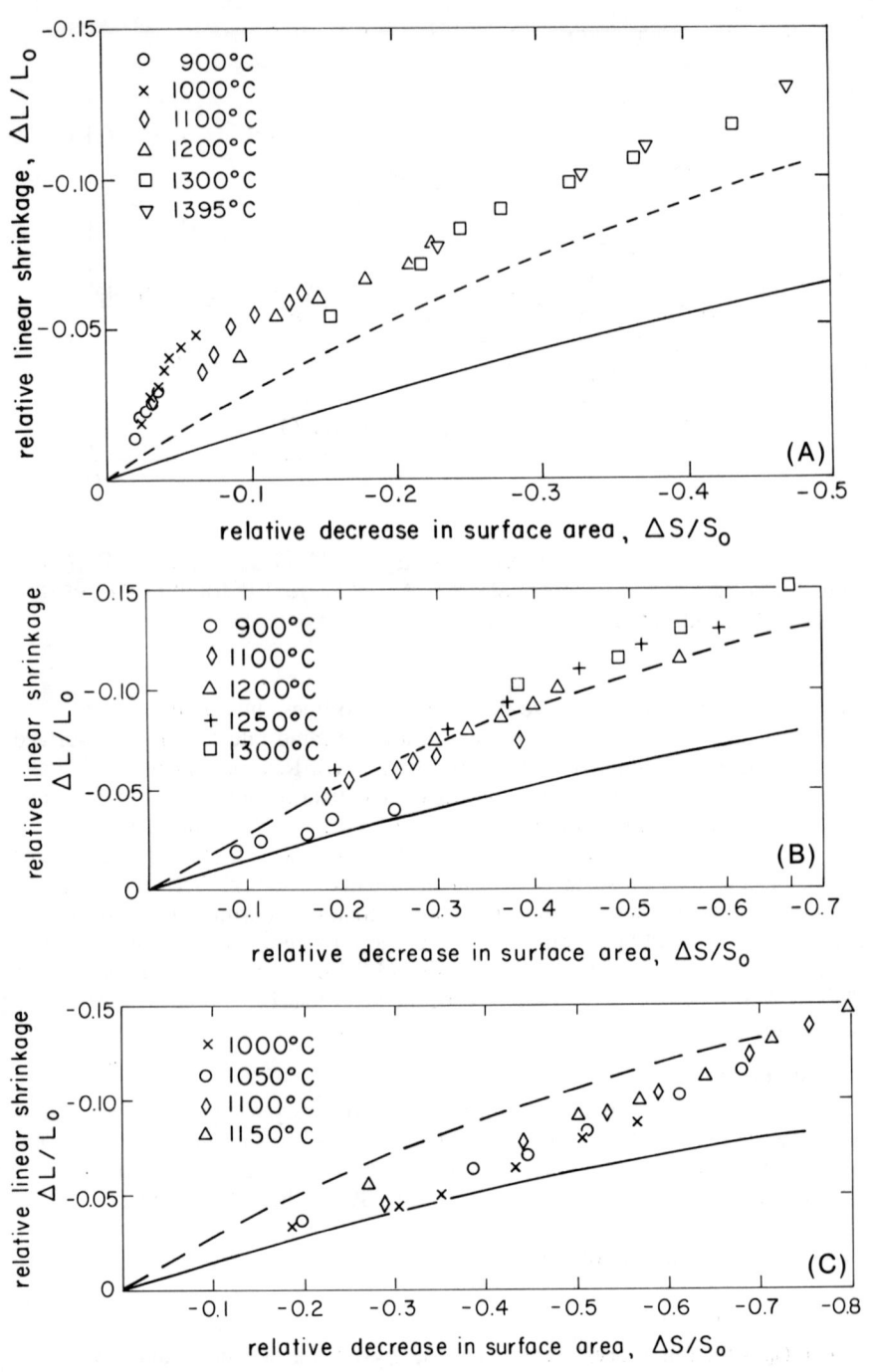

Fig. 3. Relative linear shrinkage vs relative decrease in surface area of MgO compacts sintered in (A) dry argon, (B) argon with 2.3 kPa of water vapor, and (C) argon with 22.7 kPa of water vapor for 1–600 min at temperatures indicated; solid line represents surface loss due to neck growth.

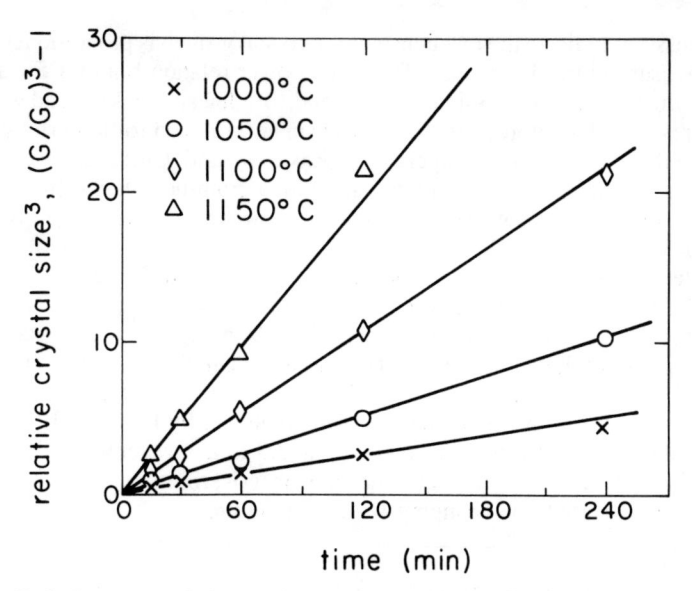

Fig. 4. Relative crystal size cubed vs time for MgO compacts sintered at various temperatures in 22.7 kPa of water vapor in argon.

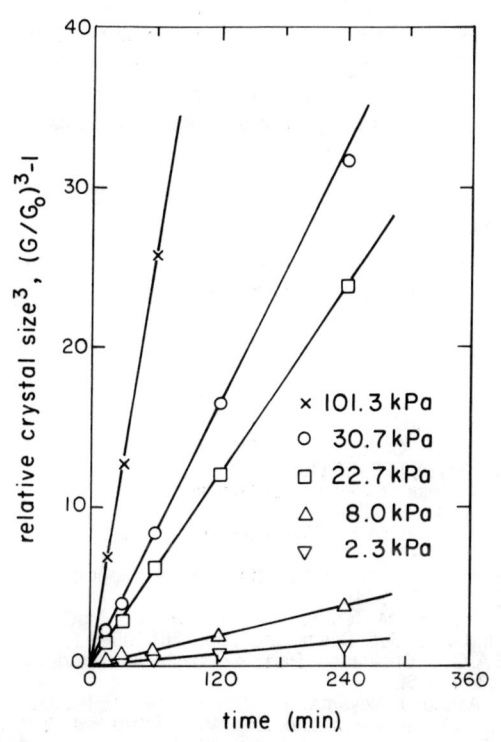

Fig. 5. Relative crystal size cubed vs time for MgO compacts sintered at 1100°C in various water vapor partial pressures.

vapor pressure, indicate the rate constant for crystal growth is proportional to water vapor pressure to the 1.5 power. The same power relationship of 1.5 was found between the time required to sinter to 8, 9, and 10% linear shrinkage and with water vapor pressure. Therefore, the same mechanism(s) is controlling crystal growth and densification at this temperature. Eastman and Cutler also obtained a 1.5 power of water vapor pressure to vary with a grain-boundary diffusion coefficient determined from isothermal shrinkage.[1] However, Hamano et al.[7] obtained a power of 0.5.

Rate constants for crystal growth from 1000 to 1150°C in 22.7 kPa of water vapor in argon were estimated from the time necessary to reach $\Delta G/G_0$ values of 0.4, 0.5, and 0.8 at various temperatures, ΔG being the net crystal growth. Arrhenius plots of these rate constants permitted calculation of an activation energy of 230 kJ/mol for each $\Delta G/G_0$ value. This value agrees with the activation energy for surface diffusion in MgO single crystals obtained by Lytle et al.[15]

Similar calculations for crystal growth at 2.3 kPa of water vapor did not result in a single activation energy. The mechanism governing crystal growth in this atmosphere appears to be changing with temperature.

Conclusions

Without water vapor present, magnesia sinters slowly, but both the mass and the pores shrink with little grain growth occurring. Water vapor increases surface diffusion, and in high water vapor pressures, pores always grow with time and temperature together with densification. At 1100°C the shrinkage rate of MgO compacts is proportional to water vapor pressure to the 1.5 power. The crystal growth rate is also proportional to the same power of water vapor pressure, indicating the same controlling mechanism. The activation energy calculated for crystal growth at high water vapor pressures is 238 kJ/mol, which is in agreement with the activation energy for oxygen self-diffusion in the extrinsic region and also for surface diffusion in a MgO single crystal.

Acknowledgments

The authors gratefully acknowledge the support of the National Science Foundation, U.S.A., grant No. DMR 8,111,111 and of CNPq of Brazil.

References

[1] P. F. Eastman and I. B. Cutler, "Effect of Water Vapor on Initial Sintering of Magnesia," J. Am. Ceram. Soc., **49** [10] 526–30 (1966).

[2] B. Wong and J. A. Pask, "Experimental Analysis of Sintering of MgO Compacts," J. Am. Ceram. Soc., **62** [3–4] 141–6 (1979).

[3] F. R. Wermuth and W. J. Knapp, "Initial Sintering of MgO and LiF-Doped MgO," J. Am. Ceram. Soc., **56** [7] 401 (1973).

[4] P. J. Anderson and P. L. Morgan, "Effects of Water Vapor on Sintering of MgO," Trans. Faraday Soc., **60** [5] 930–7 (1964).

[5] Y. Kotera, T. Saito, and M. Terada, "Crystal Growth of MgO Prepared by the Thermal Decomposition of Mg(OH)$_2$," Bull. Chem. Soc. Jpn., **36**, 195–9 (1963).

[6] K. Aihara and A. C. D. Chaklader, "Particle Growth Characteristics of Active MgO," Acta Metall., **23** [7] 855–64 (1975).

[7] K. Hamano, K. Asano, J. Akiyama, and Z. Nakagawa, "Effects of Water Vapor Pressure on Sintering of Magnesia," Rep. Res. Lab. Eng. Mat., Tokyo Inst. Tech., **4**, 59–68 (1979).

[8] R. I. Resouk, R. S. Mokhail, and J. Regai, "Surface Properties of Calcined Magnesia: Effects of Presence of Water Vapor on Surface Areas and Pore Sizes," J. Appl. Chem. Biotechnol., **23**, 51–61 (1973).

[9] J. A. Varela and O. J. Whittemore, "Grain and Pore Growth During the Sintering of MgO at Different Water Vapor Pressures"; pp. 439–45 in Sintering—Theory and Practice. Material Science, No. 14, 1982.

[10] J. A. Varela and O. J. Whittemore, "Structural Rearrangement During the Sintering of MgO," *J. Am. Ceram. Soc.*, **66** [1] 77–82 (1983).

[11] J. A. Varela, "The Initial Stage of Sintering MgO"; Ph.D. Dissertation, University of Washington, Seattle, 1981.

[12] O. J. Whittemore and J. A. Varela, "Pore Distribution and Pore Growth During the Initial Stages of Sintering"; pp. 51–60 in Sintering Processes, Material Science Research, Vol. 13. Edited by G. C. Kuczynski. Plenum, New York, 1980.

[13] H. M. Rootare and C. F. Prenzlow, "Surface Areas from Mercury Porosimetry Measurements," *J. Phys. Chem.*, **71** [8] 2734 (1967).

[14] C. N. J. Wagner, "Analysis of the Broadening and Changes in Position of Peaks in an X-Ray Powder Pattern"; pp. 219–69 in Local Atomic Arrangements by X-Ray Diffraction. Edited by J. B. Cohen and J. E. Hilliard. Gordon and Breach, New York, 1966.

[15] S. A. Lytle and V. S. Stubican, "Surface Diffusion in MgO and Cr-Doped MgO," *J. Am. Ceram. Soc.*, **65** [4] 210–12 (1982).

---

*Kanto Chemical Co., Tokyo, Japan.

Model of Interactions between Magnesia and Water

E. Longo
Universidade Federal de São Carlos
Department of Chemistry
São Carlos, SP, Brazil

J. A. Varela
Instituto de Quimica de Araraquara
Department of Physical Chemistry
Araraquara, SP, Brazil

C. V. Santilli
Universidade Federal de São Carlos
Department of Materials Engineering
São Carlos, SP, Brazil

O. J. Whittemore
University of Washington
Materials Science and Engineering Department
Seattle, WA 98195

Water vapor in the sintering atmosphere directly affects the kinetics of the process, and the interaction mechanism of the water with the MgO surface is not well-known. A quantum mechanical study was conducted by using the CNDO/2 method to verify the interaction of water with Mg. Two distinct models were studied: (1) the interaction study of Mg with water and (2) the possibility of magnesium vacancy formation. Theoretical calculations indicated that water molecules interact with Mg. Mg vacancy formation is favored by this water interaction which dissociates, creating protonated vacancies and Mg(OH)$_2$.

The sintering mechanisms of magnesium oxide in a controlled atmosphere of water vapor have not been clearly resolved. This is due to the complexity of the interaction between MgO and the water molecule. However, this effect has been studied by several authors.[1-13] Of these works, the results of Eastman and Cutler,[11] Hamano et al.,[10] and Varela and Whittemore[13] are most notable.

According to Eastman and Cutler,[11] water vapor is chemically adsorbed on the surface of MgO, resulting in a magnesium vacancy. In this process a water molecule interacts with a MgO molecule, resulting in magnesium hydroxide. This can be represented by the following quasi-chemical reaction:

$$Mg^{2+} + O^{2-} + H_2O \rightleftarrows Mg^{2+} + 2OH^- + V_{Mg} \qquad (1)$$

From this the water–magnesium vacancy is directly proportional to the water vapor pressure to the ⅓ power:

$$[V_{Mg}] \propto P_{H_2O}^{1/3} \qquad (2)$$

On the other hand, Hamano et al.[10] proposed another mechanism where the water molecule collides with the MgO surface, forming a hydroxide group outside the lattice together with a Mg vacancy on the surface. This vacancy can be filled by a Mg ion from the lattice. After a short time of retention, the water molecules leave the surface, leaving oxygen vacancies in the lattice. Schematically, the following equation represents the phenomena:

$$Mg^{2+} + O^{2-} + H_2O\downarrow \rightarrow Mg^{2+} + 2(OH^-) + V_{Mg} \rightarrow Mg^{2+} + O^{2-} + H_2O\uparrow + V_{Mg} + V_O \qquad (3)$$

The experimental results show, for low water vapor pressure, that the rate of sintering is proportional to water vapor pressure to a power that ranges from 0.3 to 0.5. The mechanisms proposed by Eqs. (1 and 3) reasonably agree with these results. However, these mechanisms cannot explain the experimental results for high water vapor pressures where the rate of sintering is proportional to water vapor pressures to the ½ power.[11,13]

The objective of this work is to use quantum mechanics to study the interaction of water with MgO at the molecular level and to analyze the possibility of vacancy formation on the MgO surface.

Model and Method

Two distinct models were considered in this work: one in which the interaction of magnesium with the water was studied and another in which the possibility of magnesium vacancy formation on the MgO surface was analyzed.

The hexahydrated form was considered for the interaction of water with Mg^{2+} and Mg^0. This form consists of a regular octahedron with six water molecules arranged on the corners and the magnesium (ion or neutral) in the center (Fig. 1).

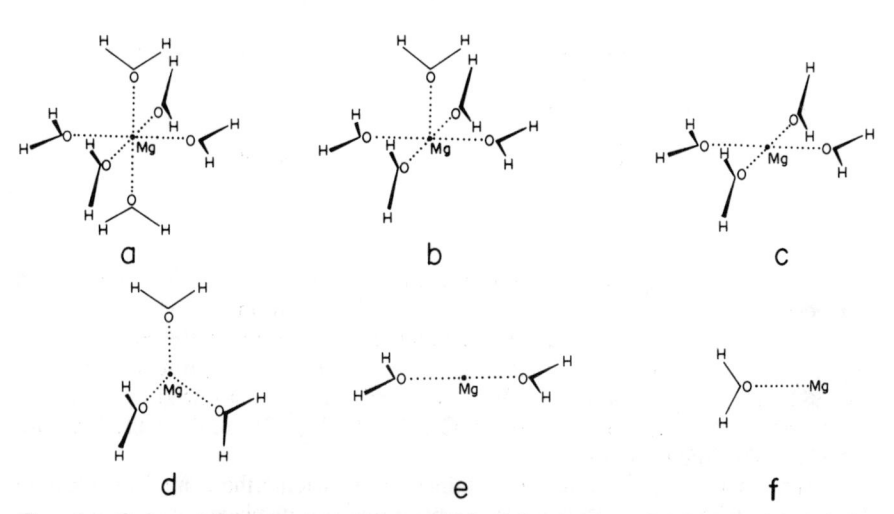

Fig. 1. Schematic representation of MgO hydration: (a) $6H_2O$; (b) $5H_2O$; (c) $4H_2O$; (d) $3H_2O$; (e) $2H_2O$; (f) $1H_2O$.

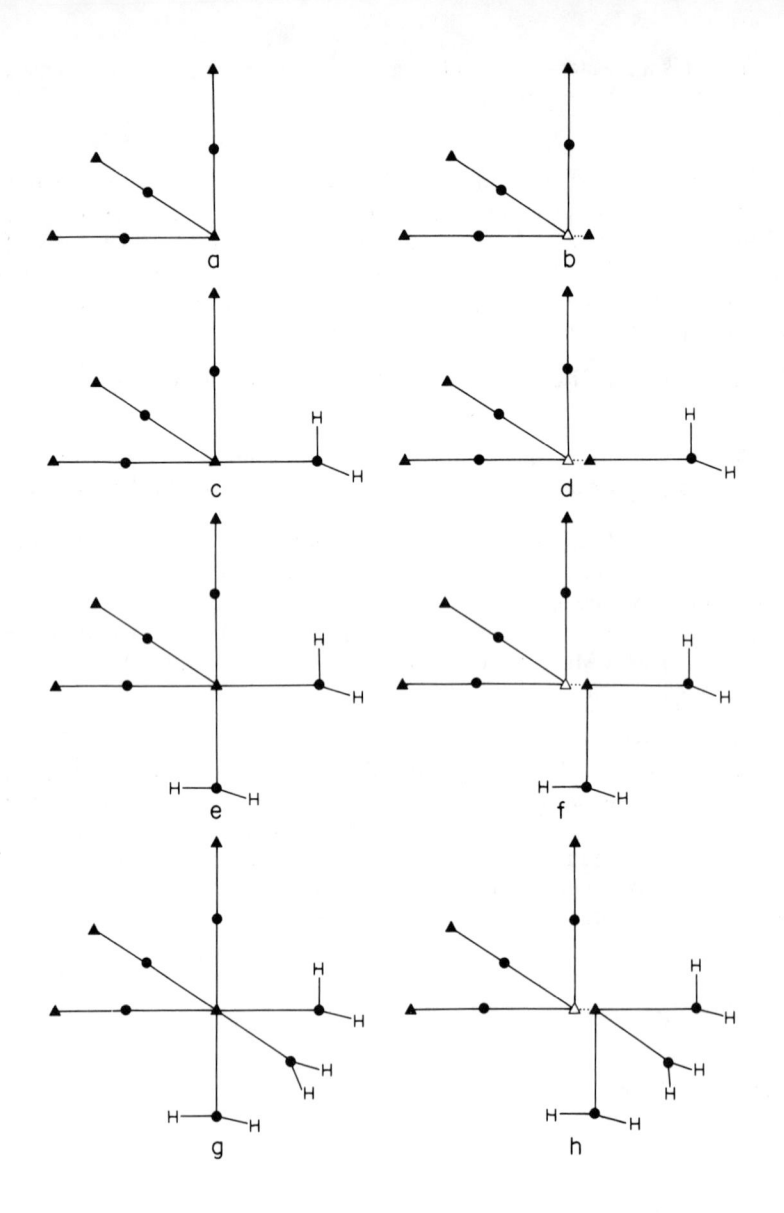

In order to study the magnesium vacancy formation, a simple model was considered instead of the unit cell of MgO. This three-dimensional model for a corner consists of a magnesium atom bonded by three mutually perpendicular MgO molecules, (MgO)$_3$Mg. The edge and the face were represented, respectively, by (MgO)$_4$Mg and (MgO)$_5$Mg. The structure was optimized by obtaining the value of 1.99 Å (0.199 nm) for MgO and 2.17 Å (0.217 nm) for the Mg with respect to the MgO groups.

The magnesium vacancy was obtained by displacing the central magnesium by 0.5 Å (0.05 nm) from its equilibrium position. This displacement was analyzed in several ways, considering whether or not the decomposition of the water during its interaction with the supermolecule (Fig. 2).

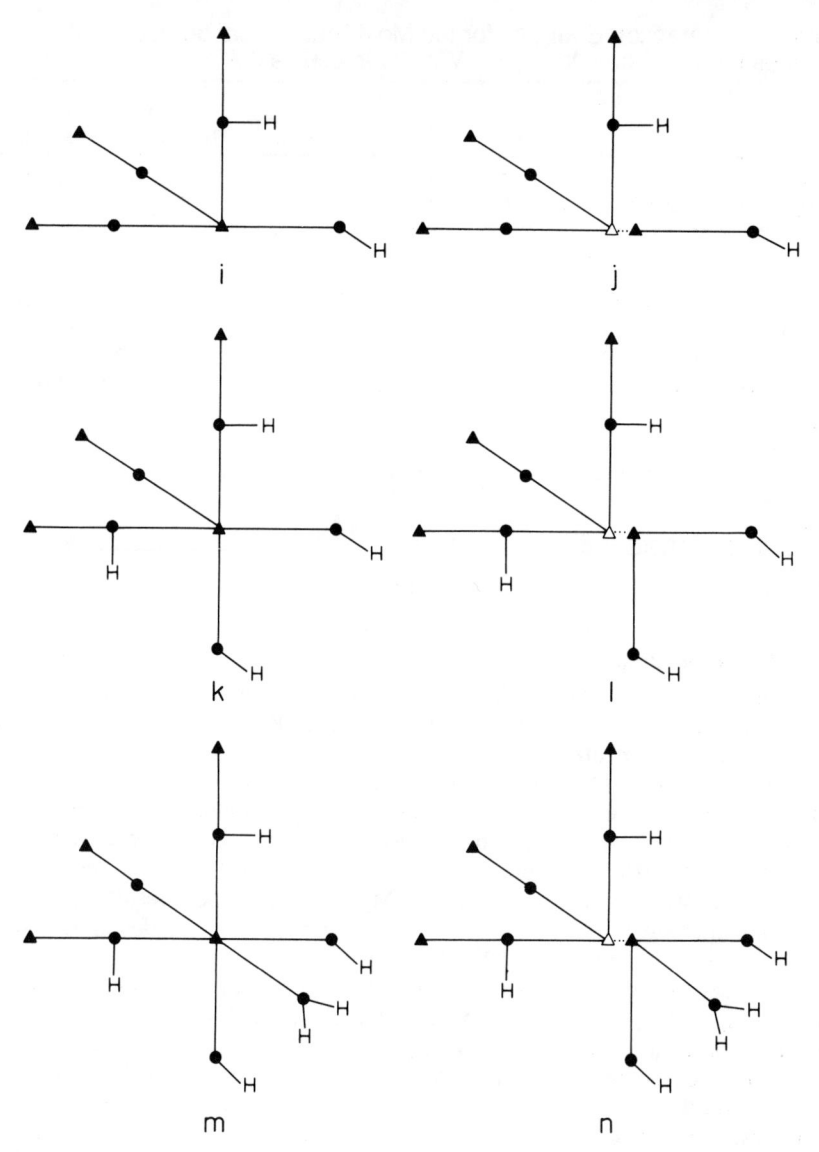

Fig. 2. Crystal model for the corner: (a) magnesium in the normal position; (b) magnesium vacancy; (c) crystal with a water molecule; (d) magnesium vacancy with the help of a water molecule; (e) crystal with two water molecules; (f) magnesium vacancy with the help of two water molecules; (g) crystal with three water molecules; (h) magnesium vacancy with the help of three water molecules; (i) crystal with a dissociated water molecule; (j) magnesium vacancy with the help of a dissociated water molecule; (k) crystal with two dissociated water molecules; (l) magnesium vacancy with the help of two dissociated water molecules; (m) crystal with two dissociated and one undissociated water molecules; (n) magnesium vacancy with the help of two dissociated and one undissociated water molecules.

Table I. Interaction Energies for the Most Stable Configuration of the Magnesium with from One to Six Water Molecules (kJ/mol)

	1H$_2$O	2H$_2$O Linear	3H$_2$O Trigonal plane	4H$_2$O Square plane	5H$_2$O Tetragonal	6H$_2$O Octahedral
Mg^0	−258.6	−244.8	−240.6	−222.2	−213.4	−196.2
Mg^{2+}	−470.7	−466.4	−445.1	−440.7	−428.4	−418.8

The quantum mechanics "complete neglect of differential overlap" (CNDO/2) method, developed by Pople et al.,[14] was used to make the theoretical calculations. These were conducted by considering the models presented above as supermolecules. This approach for interaction studies has been intensively used in quantum mechanics.[15]

The interaction energies were computed by considering the following equation:

$$\Delta E(\text{interaction}) = E^+(\text{supermolecule}) - \sum_i E_i^+(\text{monomers}) \qquad (4)$$

where E^+ and E_i^+ are the optimized energies of the supermolecule and of monomers i.

Results and Discussion

Table I shows the calculated values for the interaction energies between the magnesium neutral or magnesium ion with several water molecules by using Eq. (4). The experimental value for the hydration enthalpy is 320 kJ/mol[15] for the Mg^{2+} hexahydrated. This value is lower than that theoretically calculated. This difference can be due to the fact that the CNDO/2 method underestimates the interaction energies when the ion is charged. However, it was observed that a reasonable interaction of water with Mg^{2+} and Mg^0 exists that would result in the adsorption of water on the surface of the MgO crystals. These results agree well with the experimental data of infrared spectroscopy, where the deuterated water (D$_2$O) was adsorbed on the MgO crystal surfaces.[3]

In order to confirm the above data, the interaction of water with the crystal model ((MgO)$_3$Mg), edge ((MgO)$_4$Mg), and face ((MgO)$_5$Mg) was calculated. The results show the adsorption of water by the MgO crystal (Table II). The comparison of the models for corner, edge, and face indicates that water interacts more strongly with the corner. It is also observed that, as the number of the molecules increases, the interaction energy between the water and the supermolecule decreases. These results are compatible with the interaction of water with Mg^0 and Mg^{2+} (Table I). However, the comparison of Tables I and II shows that the effect of the three molecules of MgO is to stabilize the interaction of water with the magnesium.

Table III shows the calculation of the total energy and the energy necessary for formation of magnesium vacancies ($\Delta E(V_{Mg})$). This energy was defined as the energy required to displace the magnesium atom a distance of 0.5 Å (0.05 nm). With consideration of the magnesium atom as the origin of the coordinate system, the calculation indicates that formation of V_{Mg} occurs preferentially by the displacement of the Mg along one of the system axes. When the magnesium of the system (MgO)$_3$Mg is displaced by 0.5 Å (0.05 nm) in the x axis, there is an increase of 261 kJ/mol, whereas with the same displacement in the diagonal of the plane xy,

Table II. Total Energy, E_T, and Interaction Energy, ΔE_{Int}, for the Model $(MgO)_3Mg$ and Water

Supermolecule	$-E_T$ (eV)	$-\Delta E_{Int}$ (kJ/mol)
$(MgO)_3MgH_2O$	2146.93	251
$(MgO)_3Mg2H_2O$	2690.12	247
$(MgO)_3Mg3H_2O$	3233.22	243
$(MgO)_4MgH_2O$	2671.73	220
$(MgO)_4Mg2H_2O$	3214.63	218
$(MgO)_5MgH_2O$	3195.84	196

Table III. Total Energy, E_T, and Energy Necessary for the Formation of a Magnesium Vacancy, $\Delta E(V_{Mg})$

Supermolecule	$-E_T$ (eV)	$\Delta E(V_{Mg})$ (kJ/mol)
$(MgO)_3Mg$	1603.66	
$(MgO)_3\cdots Mg^*$	1600.95	261
$(MgO)_3\cdots Mg^\dagger$	1599.87	365
$(MgO)_3\cdots Mg^\ddagger$	1599.14	435
$(MgO)_3Mg\cdot H_2O$	2146.93	
$(MgO)_3\cdots Mg\cdot H_2O$	2144.45	239
$(MgO)_3Mg\cdot 2H_2O$	2690.12	
$(MgO)_3\cdots Mg\cdot 2H_2O^*$	2688.24	181
$(MgO)_3\cdots Mg\cdot 2H_2O^\dagger$	2686.88	312
$(MgO)_3\cdots Mg\cdot 2H_2O^\ddagger$	2686.24	374
$(MgO)_3Mg\cdot 3H_2O$	3233.22	
$(MgO)_3\cdots Mg\cdot 3H_2O$	3231.18	196

*Direction of x axis.
†Direction of the diagonal of plane xy.
‡Direction normal to the plane xyz.

the increase is 365 kJ/mol, and in the normal to the plane xyz, the increase is 435 kJ/mol. The same effect was observed in the displacement of a magnesium bonded to two waters.

Water helps the formation of V_{Mg}, as observed in Table III. The "activation" energy barrier decreases by 22, 80, and 65 kJ/mol for one, two, and three molecules of water, respectively. As a first approximation, magnesium would be displaced more easily with the interaction of two water molecules.

Anderson et al.[3] proposed a simple model for water chemisorption on the MgO crystal on the basis of infrared spectroscopy. They observed two distinct surface OH vibrations and assigned the sharp bond at 3752 cm^{-1} to the upper layer of "free" OH groups adsorbed on Mg^{2+} ions and the broader band at 3610 cm^{-1} to the other type of OH groups. This fact suggests the decomposition of the water molecule before the formation of V_{Mg}. On the other hand, studies of hydrogen adsorption on MgO[16,17] show the existence of V_{Mg} and of hydrogen adsorbed on the nearest oxygen. These active sites that hold protons are deactivated under vacuum at high temperatures. However, the same sites are reactivated after heating in the presence of water vapor. This would indicate the decomposition of water on the MgO surface and the location of the proton on the oxygen near the vacancy.

Table IV. Total Energy, E_T, and Energy Necessary for the Dissociation of the Water Molecule in the Region of Active Center, ΔE_{Dec}

Supermolecule	$-E_T$ (eV)	ΔE_{Dec} (kJ/mol)
(MgO)$_3$Mg·H$_2$O	2146.93	
(MgO)$_3$HMgOH	2147.63	−70.3
(MgO)$_3$Mg·2H$_2$O	2690.12	
(MgO)$_3$2HMg(OH)$_2$	2689.80	+30.8
(MgO)$_3$Mg·3H$_2$O	3233.22	
(MgO)$_3$2HMg(OH)$_2$·H$_2$O	3232.34	+84.8

Table V. Total Energy, E_T, and Energy Necessary for the Formation of Magnesium Vacancy, $\Delta E(V_{Mg})$

Supermolecule	$-E_T$ (eV)	$\Delta E(V_{Mg})$ (kJ/mol)
(MgO)$_3$Mg·H$_2$O	2146.93	
(MgO)$_3$H···MgOH	2146.01	88.6*
(MgO)$_3$Mg·2H$_2$O	2690.12	
(MgO)$_3$2H···Mg(OH)$_2$	2689.03	47.3*
(MgO)$_3$Mg·3H$_2$O	3233.22	
(MgO)$_3$2H···Mg(OH)$_2$·H$_2$O	3232.43	76.2*
(MgO)$_3$HMgOH	2147.66	
(MgO)$_3$H···MgOH	2146.01	159†
(MgO)$_3$2HMg(OH)$_2$	2689.80	
(MgO)$_3$2H···Mg(OH)$_2$	2689.63	32.8†
(MgO)$_3$2HMg(OH)$_2$·H$_2$O	3232.34	
(MgO)$_3$2H···Mg(OH)$_2$·H$_2$O	3232.43	−8.7†

*Decomposition of the water and displacement of magnesium.
†Displacement of magnesium.

The experimental results described above suggested a study of the water decomposition on the crystal model (MgO)$_3$Mg, and the calculated results are listed in Table IV. According to this table, the first water molecule dissociates spontaneously, releasing 70.3 kJ/mol of energy. The second water molecule needs 30.8 kJ/mol of energy for its dissociation. This is due to the repulsive electrostatic effect originated by the hydroxide groups bonded to the magnesium. The proton located on the oxygen near the magnesium also contributes to the repulsive field but in a lower scale. As expected, a third water molecule increases the "activation" energy to 84.8 kJ/mol for a decomposition of two water molecules.

The above results suggest that magnesium vacancy formation should be accompanied by water molecule dissociation on the surface. Table V shows the calculation of the total energy (E_T) and the energy necessary for magnesium vacancy formation ($\Delta E(V_{Mg})$). This table shows that, for formation of a magnesium vacancy with the help of one water molecule that dissociates spontaneously, there is an energy barrier of 159 kJ/mol for the displacement of magnesium after the dissociation. However, when the magnesium is associated with two water molecules, this barrier decreases to 32.8 kJ/mol. The association of the magnesium with two dissociated water molecules and a third undissociated water molecule leads to a spontaneous displacement of magnesium (−8.7 kJ/mol).

In a general analysis, the results suggest that the water molecule dissociates at the active site of the MgO crystal surface. The energy released by the first water molecule can help to dissociate a second water molecule. The phenomena under this prism results in a change in reaction kinetics, resulting in the formation of the vacancy by a low-energy path. In this way a third undecomposed water molecule can also help the displacement of the magnesium. As a consequence, three equations can be proposed for the kinetics of vacancy formation with a decomposition of one or two molecules:

$$Mg(MgO)_3 + H_2O \rightleftarrows (MgO)_3H + MgOH + V_{Mg} \tag{5}$$

$$Mg(MgO)_3 + 2H_2O \rightleftarrows (MgO)_3 \cdot 2H + Mg(OH)_2 + V_{Mg} \tag{6}$$

$$Mg(MgO)_3 + 3H_2O \rightleftarrows (MgO)_3 \cdot 2H + Mg(OH)_2 \cdot H_2O + V_{Mg} \tag{7}$$

where the equilibrium constants have the following forms:

$$K_1 = [Mg(OH)][V_{Mg}]/P_{H_2O} \tag{8}$$

or

$$[V_{Mg}] \propto P_{H_2O}^{1/2} \tag{9}$$

$$K_2 = [Mg(OH)_2][V_{Mg}]/P_{H_2O}^2 \tag{10}$$

or

$$[V_{Mg}] \propto P_{H_2O} \tag{11}$$

$$K_3 = [Mg(OH)_2 \cdot H_2O][V_{Mg}]/P_{H_2O}^3 \tag{12}$$

or

$$[V_{Mg}] \propto P_{H_2O}^{3/2} \tag{13}$$

The mechanism of vacancy creation proposed by the Eq. (5) leads to a relation between vacancy concentration and water vapor partial pressure given by Eq. (9). If one assumes that the rate of sintering is proportional to the magnesium vacancy concentration, then the rate of sintering is proportional to $P_{H_2O}^{1/2}$. The experimental results of Eastman and Cutler[11] and of Hamano et al.[10] obtained for water vapor pressures lower than 5 mm Hg (6.45 × 10^2 Pa) agree reasonably well with this mechanism. On the other hand, the mechanism proposed by Eq. (7) leading to a dependence of the rate of sintering proportional to $P_{H_2O}^{3/2}$ agrees well with experimental results obtained by Eastman and Cutler[11] and by Varela and Whittemore,[13] for high water vapor partial pressures.

The mechanism proposed by Eq. (5) is possible for low water vapor pressure because of the low concentration of water molecules and of the small retention of the water on the MgO surface. When the concentration of water molecules increases, the number of collisions with the MgO surface also increases and the probability of the interaction of two or three water molecules with the active site is higher. In this way the mechanisms proposed by Eqs. (6 and 7) are favored as the temperature and the water vapor pressure increase.

Conclusions

The theoretical calculations show that the formation of a magnesium vacancy is favored by the presence of water that can be dissociated, originating in a vacancy with a proton and formation of magnesium hydroxide. The proposed mechanism agrees well with the experimental data for low and high water vapor pressures.

Acknowledgments

The authors acknowledge the Universidade Federal de São Carlos for computer facilities, the Conselho Nacional de Desenvolvimento Científico e Tecnologico (CNPq) and the Fundacão de Amparo à Pesquisa do Estado de São Paulo (FAPESP) for financial support (Grant No. 82/1491-8), and the National Science Foundation (Grant No. DMR 8, 111, 111).

References

[1] W. R. Eubank, "Calcination Studies of Magnesium Oxide," *J. Am. Ceram. Soc.*, **34** [8] 225 (1951).
[2] P. J. Anderson and P. L. Morgan, "Effects of Water Vapour on Sintering of MgO," *Trans. Faraday Soc.*, **55** [12] 2203 (1959).
[3] P. J. Anderson, R. E. Harlock, and J. F. Oliver, "Interaction of Water with the Magnesium Oxide Surface," *Trans. Faraday Soc.*, **61** [12] 2754 (1965).
[4] R. I. Razouk and R. Sh. Mickhail, "Sorption of Water Vapor on Magnesium Oxide," *J. Phys. Chem.*, **59** [7] 636 (1955).
[5] J. Chown and R. F. Dracon, "Hydration of Magnesia by Water Vapor," *Trans. Br. Ceram. Soc.*, **63** [2] 91 (1964).
[6] H. B. Johnson, O. W. Johnson, and I. B. Cutler, "Effects of Water Vapor on Dielectric Loss in MgO," *J. Am. Ceram. Soc.*, **49** [7] 390 (1966).
[7] P. J. Anderson and P. L. Morgan, "Effects of Water Vapour on Sintering of MgO," *Trans. Faraday Soc.*, **60** [5] 930 (1964).
[8] Y. Kotera, T. Saito, and M. Terada, "Crystal Growth of MgO Prepared by the Thermal Decomposition of Mg(OH)$_2$," *Bull. Chem. Soc. Jpn.*, **36**, 195 (1963).
[9] K. Aihara and A. C. D. Chaklader, "Particle Growth Characteristics of Active Magnesium Oxide," *Acta Metal.*, **23** [7] 855 (1975).
[10] K. Hamano, K. Asano, J. Akiyama, and Z. Nakagawa, "Effects of Water Vapor on Sintering of Magnesia," *Rep. Res. Lab. Eng. Materials, Tokyo, Inst. Tech.*, **4**, 59 (1979).
[11] P. F. Eastman and I. B. Cutler, "Effects of Water Vapor on Initial Sintering of Magnesia," *J. Am. Ceram. Soc.*, **49** [10] 526 (1966).
[12] B. Wong and J. A. Pask, "Experimental Analysis of Sintering of MgO Compacts," *J. Am. Ceram. Soc.*, **62** [3–4] 141 (1979).
[13] J. A. Varela and O. J. Whittemore, "Grain and Pore Growth During the Sintering of MgO at Different Water Vapor Partial Pressures"; p. 439 in Sintering-Theory and Practice, Material Science Monographs, Vol. 14. Edited by D. Kolar, S. Pejovnik, and M. M. Ristic. Elsevier, Amsterdam, 1982.
[14] J. A. Pople, D. P. Santry, and G. A. Segal, "Approximate Self-Consistent Molecular Orbital Theory II — Calculations with Complete Neglect of Differential Overlap," *J. Chem. Phys.*, **43**, 5136 (1965).
[15] B. Pullman, A. Pullman, and H. Berthod, "SCF ab-Initio Study of the Through Water versus Direct Binding of the Na$^+$ and Mg^{2+} Cations to the Phosphate Anion," *Int. J. Quant. Chem. Quant. Biol. Sym.*, **5**, 79 (1978).
[16] J. M. Andre, E. G. Derouane, J. G. Fripiat, and D. P. Vercauteren, "Quantum Mechanical Approach to the Chemisorption of Molecular Hydrogen on Defect Magnesium Oxide Surfaces," *Theor. Chim. Acta (Biol.)*, **43**, 239 (1977).
[17] M. Boudart, A. Delbouille, E. G. Derouane, V. Indocina, and A. B. Walters, "Activation of Hydrogen at 78 K on Paramagnetic Centers of Magnesium Oxide," *J. Am. Chem. Soc.*, **94** [19] 6622 (1972).
[18] E. Longo, J. A. Varela, L. R. Assis, and C. V. Santilli, "Interacão da Água com o Óxido de Magnésio"; unpublished work.

Effects of Crystallinity in Initial-Stage Sintering

J. G. Dash

University of Washington
Department of Physics
Seattle, WA 98195

Richard Stone

University of Washington
Department of Materials Science and Engineering
Seattle, WA 98195

Sintering of amorphous and highly disordered materials is attributed to neck growth motivated by surface tension. In crystalline materials, neck growth is inhibited by crystal morphology and by grain boundary energy; in the absence of compensating mechanisms, sintering involves grain growth without neck growth. A possible important compensation is due to compression, which produces highly disordered intergranular contact regions. When heated to moderate temperatures, these regions form crystallization nuclei for subsequent neck growth at normal sintering temperatures. An experimental program, which distinguishes between connectivity and grain growth in MgO and other crystalline powders, is described in brief.

It is generally understood that the principal motivation for sintering is surface energy. Variations in curvature of the surfaces of the grains in a compacted powder produce variations in surface energy and chemical potential. When the compact is heated, diffusion is toward the regions of lower chemical potential, with a resulting growth of necks between grains in contact. This general picture of the mechanism for neck growth has formed the basis for detailed mathematical descriptions of the growth process.[1] The analysis of simple geometrical shapes has permitted quantitative prediction of growth rate as functions of particle size and neck area. The theory is fairly successful in describing the sintering behavior of macroscopic glass spheres and other simple rounded shapes, but it is quite unsuccessful for typical ceramic powders. Several authors have noted that the failure may be due to the neglect of crystallinity in the actual powders, and some have proposed more realistic models.[2] This paper is offered in the same spirit but differs considerably from previous work in our attention at the outset to the fundamental properties of crystallinity. Here we present a qualitative view of our ideas, with comments on experimental implications. A brief description is given of an experimental study that has been designed to test the theory.

Theory

Grain Growth and Necking in Two Ideal Models

It is instructive to examine the necking process as it might occur in two contrasting ideal models. The first is the familiar model of two contacting spheres

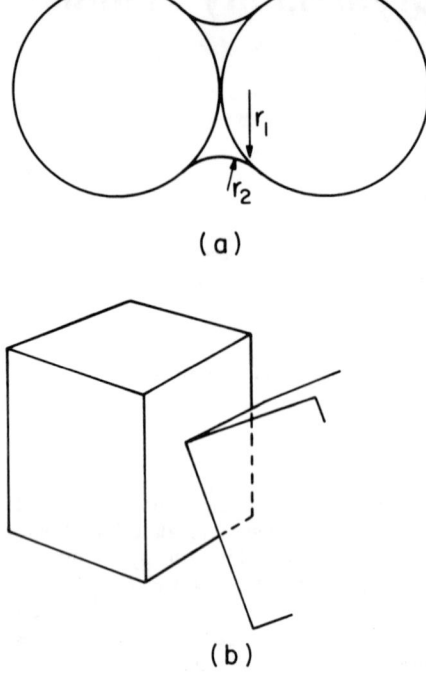

Fig. 1. Two ideal models of powder particles: (a) uniform and isotropic spheres; (b) cubic crystals with plane crystalline faces.

of isotropic uniform media. The second is cubic crystals in point-face contact. The two models are illustrated in Fig. 1.

We analyze both models by first assuming that the principal motivation for sintering is a reduction in the surface energy and that the variation in chemical potential due to shape is given by the familiar equation[3,4]

$$\Delta\mu_\gamma = \gamma v\left(\frac{1}{r_1} + \frac{1}{r_2}\right) \tag{1}$$

where γ is the surface tension, v is the molecular volume, r_1 and r_2 are the principal radii of curvature, and μ is the chemical potential.

In the two-sphere model $\Delta\mu_\gamma < 0$ in the contact region, since r_2, which is negative, is of smaller magnitude than r_1. Therefore, mass transport is toward the contact region, and a neck grows until $r_2 = r_1$.

In the two-cube model, arbitrary contact angles between the crystals yield no large distinction between the principal radii: in general, they will be comparable in magnitude and opposite in sign. However, even in regions of narrow contact angle, where surface tension forces tend to drive mass transport toward the contact region, a second factor intervenes to prevent neck formation. This is the grain boundary energy, which for simple tilt boundaries inclined at angle θ, causes a chemical potential shift[5]

$$\Delta\mu_{gb} \simeq \frac{c_{44} v \theta}{4\pi(1-\nu)}[A - \ln\theta] \tag{2}$$

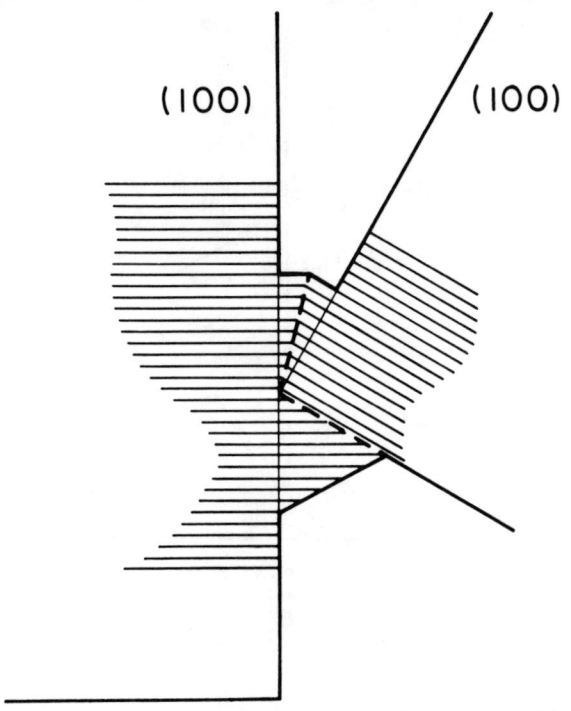

Fig. 2. Illustration of a simple crystalline neck between cubic crystals with (100) faces, showing the necessity for grain boundaries, high index faces, and/or positive curvature surfaces.

where c_{44} is the shear modulus, v is Poisson's ratio, v is molecular volume, and $0.15 \leq A \leq 0.55$ for angles in the range $15° \leq \theta \leq 45°$. The grain-boundary energy contribution is in general unavoidable, since at least one grain boundary must appear in a neck between crystals at arbitrary orientation. For typical materials of interest, the grain-boundary energy per unit area is comparable to the surface tension. Therefore, for contact angles greater than a small value, $\Delta\mu_{gb}$ due to grain-boundary energy will be much larger than that due to surface tension, the total $\Delta\mu > 0$ and consequently a neck will not form. But diffusion will tend to cause grain growth, leading to a net transfer of material from small to large crystals. Thus, in such an ideal powder, there will be surface area reduction without neck growth. Figure 2 illustrates a crystalline neck as described here.

Sintering in Disordered Crystalline Compacts

More realistic models of actual compacts must include crystalline disorder. By this term we mean to include all forms of chemical and crystalline imperfection: dislocations, strains, grain boundaries, point defects, surface imperfections, chemical impurities, and deviations from stoichiometry. A certain amount of disorder is present in the parent material, and an additional amount is usually introduced in the preparation of the uncompressed powder, through milling or other process. Further disorder is introduced in the region of contact when the powder is compressed. These and the following features are illustrated schematically in Fig. 3.

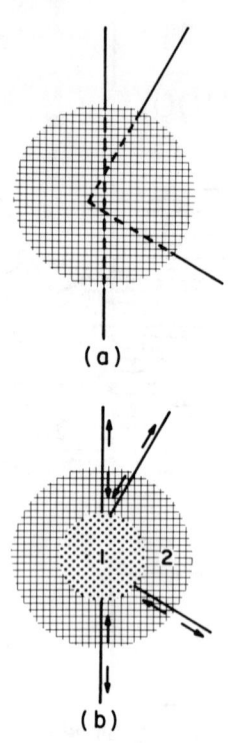

Fig. 3. Schematic of compressed cubic crystals: (a) as pressed, with a deformed contact region composed of fractured, plastically strained, and highly disordered material; (b) upon heating, the most disordered region 1 forms "new crystal" nuclei by nearly diffusionless transformation. Material from the surrounding, less disordered region 2 diffuses both toward the nuclei and toward less damaged crystal.

It is well-known that the magnitude of the work of compression can be quite large. An estimate of the order of magnitude of the local increases indicates that the changes in chemical potential may be comparable to or larger than those due to surface tension on surfaces with curvatures as small as 10 atomic radii. More importantly, the compression energy is concentrated in the regions of contact between the particles. The direction of the chemical potential gradient is therefore such as to drive mobile atoms away from rather than toward the regions of contact. However, these local excesses can, by virtue of their large magnitudes, open an alternate channel for neck formation and growth. This channel is the nearly diffusionless restructuring of the most disordered regions into small "new crystal" nuclei. The motivation for the restructuring approaches, in the limit of atomically disordered contacts, the latent heat of fusion of the material. To be sure, the chemical potential of the nuclei is higher than that of a perfect large crystal because of size effect, but since the interfacial tension between crystal and amorphous solid is relatively small, the increase is moderate. Because the formation of new crystal requires atomic rearrangements on a scale of only a few lattice parameters, the new crystal can form at ambient temperature or after only mild heating. It is this process that we believe accounts for the frequent observation of "presintering."

We now consider the various processes occurring during the sintering of our powder model.

As the temperature is raised, the earliest processes to occur will be those activated by the largest local gradients and excesses of chemical potential. These include the restructuring of the contact regions as described in the previous paragraph.

As the temperature is further increased, diffusion moves material from closely neighboring disordered regions to the neck nuclei. This "fine-grained sintering," which begins as soon as the neck nuclei are formed, is driven by the total gradient of chemical potential, which is composed of all three contributions — those due to variations of local radius, grain-boundary energy, and residual structural disorder.

Competing processes tend to slow the continued growth of the necking regions. Annealing and growth on the exposed surfaces and in the interiors of the powder particles improve their crystalline perfection, thus reducing the chemical potential gradients that could drive material toward the necks. The rate is also slowed by the greater distances for diffusion between contact areas and the partially annealed neighboring regions. Since diffusion and restructuring mechanisms are fundamentally different processes, they proceed at different rates. Therefore, it is possible that at some stage the chemical potential in the noncontact regions may become lower than that in the necking regions. If this occurs, material will then begin to flow away from the bridges: neck growth is halted and then reversed. With still higher temperature and longer time, "sintering" would entail grain growth but little or no neck growth.

Our qualitative model involves three general observations: (1) Compression plays an essential role in the production of local disorder that is required for the nucleation and growth of necks. (2) The initial disorder of the surfaces and interiors of the powder particles, which results from the milling or other preparation methods, is an important factor tending to facilitate neck growth. (3) Grain growth and neck growth are distinct processes, both occurring and competing during sintering.

It is important to note here that we are neglecting any consideration of the effects of active chemical ingredients of the powder or of the sintering atmosphere, which might provide additional channels for reducing chemical potential. Since impurities and additives are known to be extremely effective in certain cases, as e.g., water vapor for MgO,[6] it is a significant omission. It may be possible to include such effects in some future elaboration of the theory, but at the present stage we confine our attention to pure materials.

These observations are the motivation for an experimental study, described in the next section. Since the experiment is now in an early stage, we will describe the techniques to be used but will not report results.

Experimental Procedures

An experimental study has been undertaken to explore grain and neck growth, using techniques that can distinguish between reduction of surface area and connectivity. For surface area measurement, we used physical adsorption. This method has been commonly used to determine surface areas of powders and sintered compacts. In the present study, we will also employ it as a gage of surface quality, using the results of modern studies of physical adsorption on highly uniform surfaces.[7]

For sintering material, we have chosen MgO. The choice was made for two reasons: it is possible to prepare highly uniform fine powders of cubic particles, thus presenting a physical realization of the simple ideal shape depicted in Fig. 1, and because sintered MgO is an important industrial product.

Surface Characterization

Physical adsorption can reveal considerably detailed information about the nature of monolayer films and the solid surfaces on which they are adsorbed. It is now possible to determine, in addition to surface area and molecule–substrate binding energy, the nature of film phases and their transitions, the atomic corrugations of substrate potential, and the positions of the atoms with respect to the substrate lattice. Of greater relevance here, modern studies can yield quantitative gages of the sizes of uniform domains on the exposed crystal facets and the variations of binding energy due to subsurface defects. Film properties are sensitive to such heterogeneity, particularly in the neighborhood of phase transitions. We illustrate this sensitivity in Fig. 4, which shows a series of adsorption isotherms of Kr on MgO smoke particles. In an earlier study[8] of this system, it had been expected that the smoke, which is known to form cube particles with (100) exposed facets, would provide an exceptionally uniform adsorbate, on which the first-order phase transitions of an adsorbed monolayer would be signaled by coverage-independent vapor pressure. The particles did indeed have a high proportion of cube particles according to examination by scanning electron microscopy. However, the isotherms had the smooth sigmoid "BET" shape of typically strongly heterogeneous adsorption, and therefore, the adsorption isotherms were sensitive to forms of heterogeneity beyond the resolution of SEM. Although it is not certain what these are, their thermodynamic significance and their important effect on the chemical potential of the substrate are evident. That such defects can be reduced by certain thermal and processing techniques was shown by a series of isotherms on powders subjected to annealing under vacuum. We found that successively higher temperature anneals produced increasingly uniform surfaces, indicated by the progressively longer and more vertical isotherm steps. A more recent study[9] has succeeded in producing isotherms still closer to ideal, by particular attention to atmosphere control in burning and the avoidance of agglomeration. Examples of isotherms taken from the recent study are shown in Fig. 4.

Quantitative measures of heterogeneity can be obtained from the analysis of such isotherms,[10] and they will be applied in the present study of sintering but will not be described here.

Direct Measurement of Neck Growth

The distinction between surface area reduction and the development of interparticle necks requires a technique that is particularly sensitive to the overall connectivity of the powder. Aside from mechanical measurements such as modulus of fracture, none of the common probes of sintering can satisfy this need. However, a suitable technique does exist: measurement of thermal conductivity or diffusivity.

The thermal conductivity of a loose powder under vacuum is extremely low at moderate and low temperatures, being due to radiation transport only. If the powder is compressed, interparticle connections are produced, which, as we have noted above, are highly disordered and hence have poor thermal conductivities. If the compression is insufficient to produce a number of continuous connected paths through the entire sample, the conductivity remains essentially at its low original value, but if the "percolation transition" is exceeded, the conductivity is increased. This threshold value establishes the lower limit of compressions for subsequent examinations of sintering behavior. Sample conductivity can then be studied as a function of compression, time, and temperature, as a gage of crystallization and

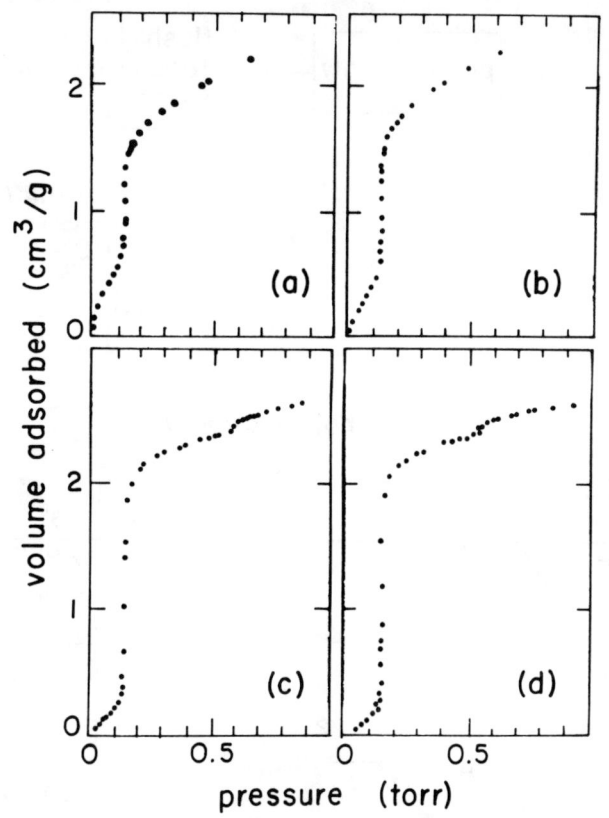

Fig. 4. Vapor pressure isotherms of Kr on uncompressed MgO smoke powder, illustrating the improvement in substrate uniformity, as indicated by increasing height and verticality of the isotherms, caused by modifications in the preparation of the smoke. Preparation parameters that were varied included oxygen concentration in the ignition chamber, subsequent vacuum anneal temperature, and protection from water vapor.

growth of the neck regions. Since the conductivity is a complicated function of the geometry, particle size, and neck crystallinity, it will be necessary to supplement the conductivity measurements with other gages of sintering, such as fracture strength, as well as adsorption isotherms.

Our thermal measurements are carried out by the light flash method. This technique, which has been used by several groups to study solid minerals and ceramics, is particularly suitable for thin disks and slabs, i.e., shapes that are convenient for sintering studies of powders. The principles of the method are illustrated in Fig. 5. The thermal diffusivity α, which is the ratio of conductivity to heat capacity, is obtained directly from the time $t_{1/2}$, the delay between the heating pulse and the time for the temperature on the back face to rise to one-half of its final value. The relation between the experimental parameters is[11]

$$\alpha = \frac{\kappa}{\rho c} = \frac{1.38 L^2}{\pi^2 t_{1/2}} \qquad (3)$$

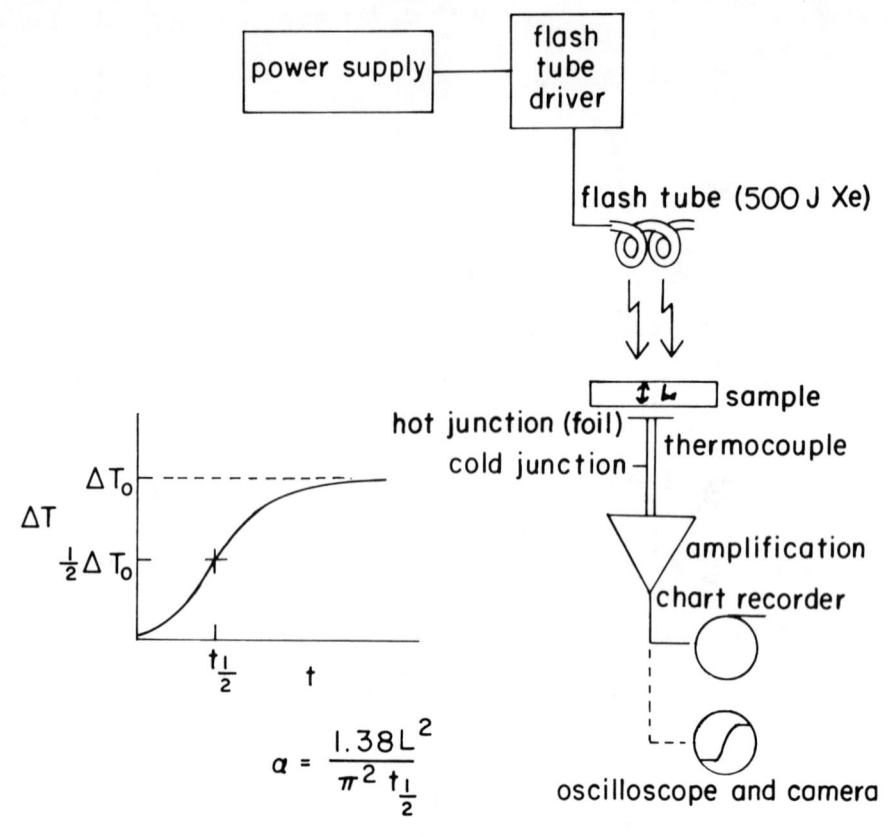

Fig. 5. Schematic arrangement for thermal diffusivity measurements by the flash technique.

$$\alpha = \frac{1.38 L^2}{\pi^2 t_{\frac{1}{2}}}$$

The method and the necessary apparatus are relatively uncomplicated. Our apparatus uses a 500-J xenon flash tube as source and a thermocouple as detector. The thermocouple signal is amplified and then directly displayed on a chart recorder. Initial tests with quartz and olivine samples have yielded thermal conductivities in good agreement with published results.

Acknowledgments

We are grateful to several people for suggestions and criticisms during the development of this study. Our interest in sintering questions was stimulated by the participation (of J.G.D.) in a continuing program of research in ceramic materials coordinated by J.I. Mueller of the Department of Ceramic Engineering at The University of Washington. O.J. Whittemore provides continuing information and advice on the technical aspects of ceramic processing and sintering. We have benefited greatly from the advice and encouragement of O.E. Vilches, whose continuing studies of adsorption and the preparation of uniform MgO have been essential in the design of this project. In the matter of surface tension forces and grain growth, we are grateful to M. Bienfait and B. Mutaftschiev, of Université d'Aix-Marseille. Their critical questions were important in the development of our understanding of the competition between these two components of the chemical potential. Support for this project is obtained from the Department of Physics of The University of Washington and from the National Aeronautics and Space Administration, under Grant NAGW 199.

References
[1] G. Kuczinski, *Sci. Sintering,* **9**, 243 (1977).
[2] R. L. Porter, "Sintering Processes," *Mater. Sci. Res.,* **13**, 129 (1979).
[3] W. D. Kingery, H. K. Bowen, and D. R. Uhlmann; Introduction to Ceramics, Wiley, New York.
[4] L. D. Landau and E. M. Lifshitz; Statistical Physics, Pergamon, London.
[5] J. Friedel; Theory of Dislocations, Pergamon, London.
[6] J. A. Varela and O. J. Whittemore; Proc. 5th Int. Conf. on Sintering, Potoroz, Yugoslavia, 1981.
[7] J. G. Dash; Films on Solid Surfaces, Academic Press, New York.
[8] J. G. Dash, R. E. Ecke, O. E. Vilches, and O. J. Whittemore, *J. Chem. Phys.,* **82**, 1450 (1978).
[9] J.-P. Coulomb, T. Sullivan, and O. E. Vilches; unpublished work.
[10] J. G. Dash and R. D. Puff, *Phys. Rev. B: Condens. Matter,* **B24**, 295 (1981).
[11] W. J. Parker, R. J. Jenkins, C. P. Butler, and G. L. Abbot, *J. Appl. Phys.,* **32**, 1679 (1961).

Effect of Magnesium Chloride on Sintering of Magnesia

Kenya Hamano, Zenbe-e Nakagawa, and Hideo Watanabe

Tokyo Institute of Technology
Research Laboratory of Engineering Materials
Yokohama 227, Japan

Increasing amounts of magnesium chloride added to magnesium hydroxide decreased the lattice parameter of the periclase prepared by calcination of the mixtures, increased its apparent crystallite size, and decreased its specific surface area. Bulk densities of their green compacts and specimens fired at 1500°C for 4 h increased with increase of the chloride addition and reached a maximum value for powder calcined at 700°C. The addition of magnesium chloride eliminated relics of brucite, increased crystallinity of the periclase, and led to a slight amount of grain growth during calcination. The slight growth and uniform size of periclase grains markedly improved compactness of the calcined powders. The high density of green compacts and the slow and successive progress of densification of compacts owing to lower activity and uniform grain size of periclase made the fired compacts very high density. In this study, cation-impurity-free magnesia ceramics of about 99% of theoretical density could be obtained, for firing temperatures as low as 1500°C.

Grain size, shape, crystallinity, and other powder characteristics are the most important factors to obtain dense fine ceramics. These characteristics might be controlled during the calcination of powder precursors. Magnesia is an interesting material for the study of these problems.

Brucite is the most commonly used raw material for preparation of magnesia. To prepare magnesia from brucite, there is an optimum temperature for its calcination, i.e., usually 900°C. If it is calcined at higher temperature, the crystallinity of the formed magnesia becomes too high to retain its activity and aggregates often form. If the temperature is lower, relics of the brucite remain in the calcined powder that decrease the compactness of green specimens and sometimes prevent direct contact of each grain. So in both cases, the magnesia cannot be well densified.[1-3]

If small amounts of certain magnesium compounds are added to the brucite, the properties of the calcined mixtures are different according to the kind of the added compounds, because they behave in various ways during their decomposition and, thus, cause different effects of the calcining powders. Further, the addition of any magnesium compound that finally converts to magnesia does not introduce any cationic impurities into the calcined magnesia.[4,5] From the previous study on the effect of such magnesium compounds on sintering of magnesia, it could be concluded that magnesium chloride is the most promising additive for this purpose.[4] In this paper, effects of the chloride addition were examined in more detail, i.e., effects of amounts of chloride addition, calcining temperature, and chlorine atmosphere.

Table I. Properties of Calcined Powders*

Amount of chloride addition (%)[†]	Phases	Periclase Lattice parameter (Å)[‡]	Crystallite size (Å)	Specific surface area (m^2/g)
0	Periclase	4.213$_1$	540	31
4	Periclase	4.211$_6$	1780	8

*900°C, 1 h.
[†]Calculated as MgO.
[‡]4.2131 Å is 0.42131 nm.

Previous Studies on the Effect of Magnesium Chloride on Sintering of Magnesia Prepared from Magnesium Hydroxide

For the starting material of the host oxide, magnesia, the purest grade of magnesium hydroxide* (MgO > 99.9%) was used. Magnesium chloride was also the extra pure grade chemical.[†] The magnesium hydroxide and chloride were mixed in an agate mortar with ethanol. The amount of chloride addition was 0 or 4 wt%, calculated as magnesia. The mixtures were calcined at 900°C for 1 h in an electric furnace. The calcined powders were examined by X-ray powder diffraction. The lattice parameters and apparent crystallite size of periclase were calculated from its diffraction peak of (420) compared with that of (321) of quartz, used as an internal standard. The specific surface area of the calcined powders was also measured by means of BET. The results obtained are shown in Table I. Addition of magnesium chloride decreased the lattice parameter and increased the apparent crystallite size markedly, that is, increased its crystallinity. Addition of the chloride also decreased its specific surface area.

From the calcined powder, small tablets of 10 mm in diameter were formed by an isostatic compression with pressure of 1000 kg/cm^2 and fired at 1400°C for 4 h in an electric furnace. Bulk density of the specimens was measured by the mercury displacement technique. Table II shows green and fired bulk density of these specimens. Bulk densities of the green compact and also the fired specimen prepared from a mixture containing the chloride are higher than those of ones prepared from the magnesium hydroxide alone.

Figure 1 shows transmission electron micrographs of these two calcined magnesia specimens and the magnesium hydroxide powder. Magnesia prepared from brucite alone shows large hexagonal relics of the brucite, in which very fine (about 0.05 μm) periclase grains are observed. On the other hand, the magnesia prepared from a mixture of brucite and magnesium chloride is uniform, with rather large, clear grains and no relics. These might be reasons for its high green and fired

Table II. Bulk Density of Green and Fired Specimens

Magnesia specimens— amount of chloride addition (%)	Bulk density of Green compact* (g/cm^3)	Fired specimen[†] (g/cm^3)
0	1.52 ±0.03	3.43
4	1.95±0.02	3.52

*1000 kg/cm^2.
[†]1400°C, 4 h.

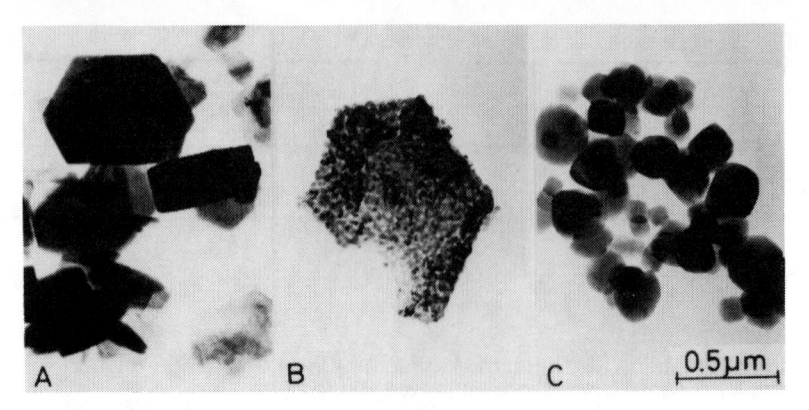

Fig. 1. Transmission electron micrographs of brucite and two calcined magnesia specimens: (A) raw material, brucite; (B) magnesia prepared from brucite calcined at 900°C; (C) magnesia prepared from brucite and 4 wt% MgCl$_2$ calcined at 900°C.

densities. From these results, it became clear that addition of the magnesium chloride made relics of the brucite disappear, improved the crystallinity of periclase, lowered activity, and increased compactness of the calcined powder. As a result of these effects, sintering of the calcined powder proceeded at a moderate rate to high density, and the periclase grains grew slowly without much pore entrapment.

Effect of Calcining Temperature and Amount of Addition of Magnesium Chloride on Sintering of Magnesia

As mentioned above, the magnesium chloride was so effective in promoting densification of magnesia that the effect of the chloride was examined in more detail. In the previous studies, all experiments were carried out with specimens mixed with a chloride concentration (calculated as magnesia) of "4 wt%." The amounts for this study were 0, 1, 2, 4, 8, and 16 wt%, also calculated as magnesia. The magnesium hydroxide and chloride were mixed in an agate mortar with ethanol and calcined at 600, 900, and 1100°C for 1 h. From these calcined powders, small tablets of 10-mm diameter and 3-mm thick or rectangular specimens of 4 × 4 × 40 mm were formed by isostatic pressing, and their properties were measured by the same procedure described in the previous chapter.

Figure 2 shows the relations between the amount of chloride and lattice parameter, apparent crystallite size of periclase, and specific surface area of the calcined powders. With an increase in the amount of chloride, the lattice parameter of periclase decreased, its crystallite size increased, and the specific surface area decreased. These tendencies were more extensive in the specimens calcined at a lower temperature and particularly notable in samples with 1 wt% chloride.

Nearly the same tendency could be observed in the case of the green density of the compact; i.e., with an increase in the amount of chloride, the density increased, as shown in Fig. 3.

Then, these specimens were fired at 1500°C for 4 h, and bulk density of the fired specimens was measured by the mercury displacement method. The results obtained are plotted against the amount of chloride as shown in Fig. 4. Addition of the 1 wt% magnesium chloride markedly increased the density, but its addition

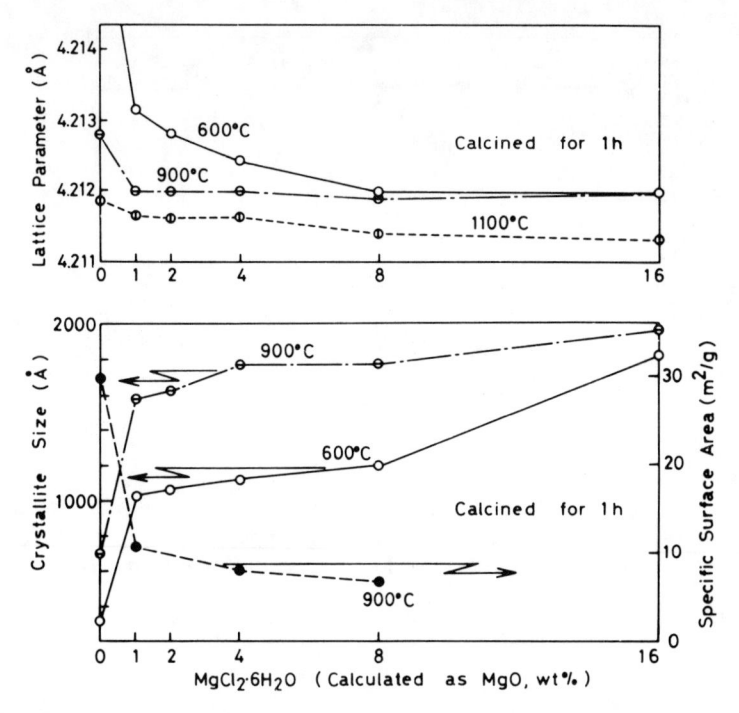

Fig. 2. Relations between the amount of chloride and lattice parameter, apparent crystallite size of periclase, and specific surface area of the calcined matters.

beyond that showed only little or no increase of the density. From the results, it is concluded that addition of the 1–8 wt% magnesium chloride effectively increases density of the fired magnesia specimens.

In the next study, the effect of the calcination temperature was examined with the samples of "no additive" and 4 wt% chloride mixture. Figure 5 shows relations

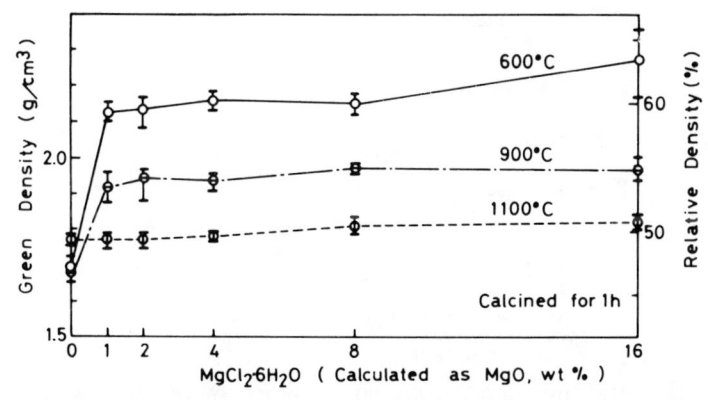

Fig. 3. Relation between green density of the compact and the amount of chloride.

613

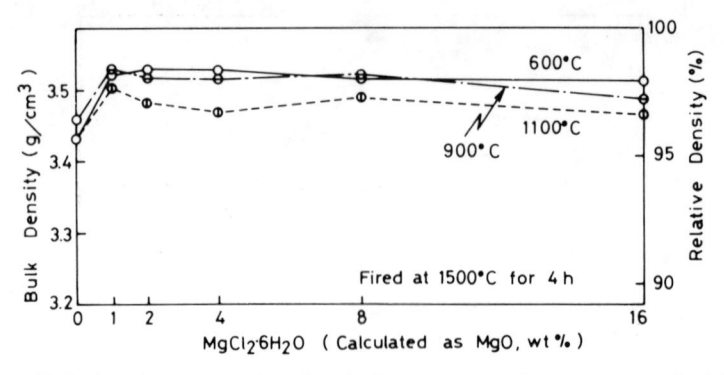

Fig. 4. Relation between the fired density and the amount of chloride.

between the calcining temperature and the lattice parameter and apparent crystallite size of periclase. Results of the measurement of their specific surface areas are also in Fig. 5. With increasing calcination temperature, the lattice parameter and the specific surface area decreased and crystallite size increased. These ten-

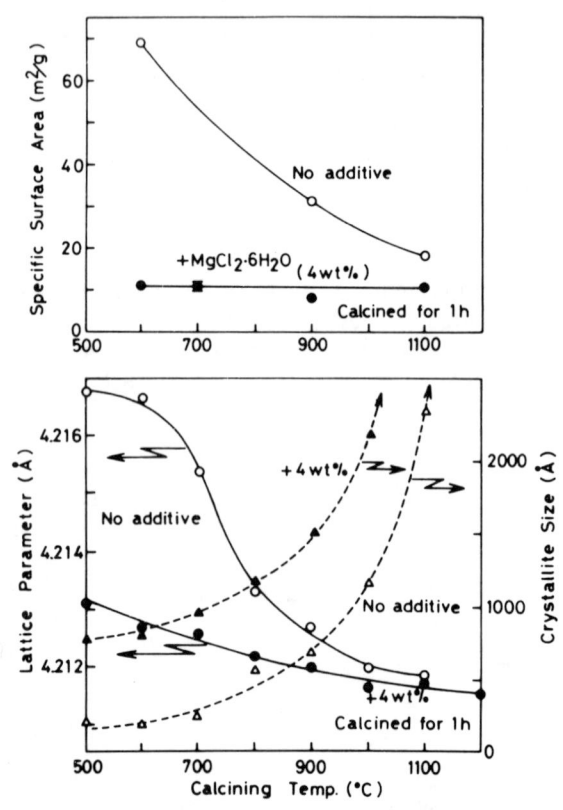

Fig. 5. Relations between the calcining temperature and the lattice parameter, the apparent crystallite size of periclase, and the specific surface area of calcined matters.

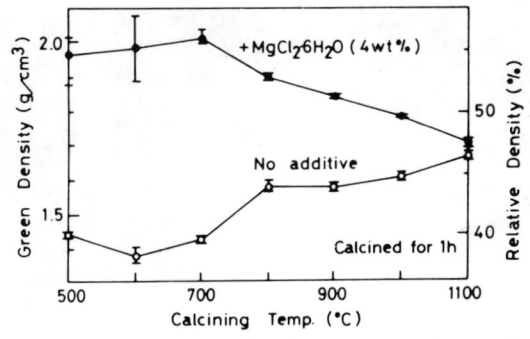

Fig. 6. Relation between the calcining temperature and green density of compacts.

dencies were extensive for the specimens mixed with 4 wt% chloride, and the difference between those values for samples of "no additive" and 4 wt% chloride mixture was more pronounced for those calcined at lower temperature.

Figure 6 shows the relation between the calcination temperature and the green bulk density of compacts formed from the calcined powders. Addition of the chloride increased the density markedly and showed a maximum with the powders calcined at the temperature from 600 to 700°C. For a calcination temperature greater than 700°C, the density decreased. From these results, it was concluded that the effect of chloride addition was remarkable for the calcined powders at lower temperature, i.e., 600–700°C.

These compacts were then fired at 1500°C for 4 h, and their bulk densities were measured. These data were plotted against their calcining temperatures on Fig. 7. The specimens mixed with chloride have higher fired densities than those of "no additive" for all the calcining temperatures, and those of both fired specimens showed their maxima for powders calcined at 700°C. The maximum value of about 99% of theoretical density was obtained for the specimen mixed with 4 wt% chloride and calcined at 700°C, even though its firing temperature was as low as 1500°C. Figure 8 shows transmission electron micrographs of magnesia specimens calcined at 700 and 1100°C for 1 h. Compared with samples calcined at 1100 and 900°C, shown in Fig. 1(C), magnesia calcined at 700°C shows more

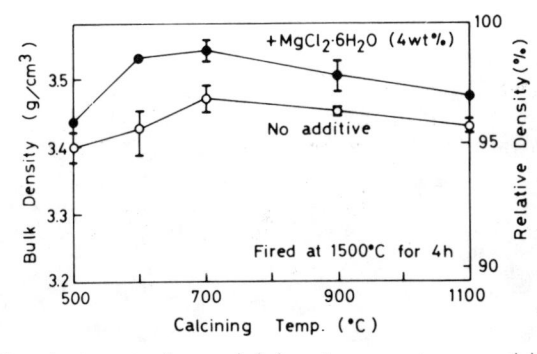

Fig. 7. Relation between the calcining temperature and bulk density of fired specimens.

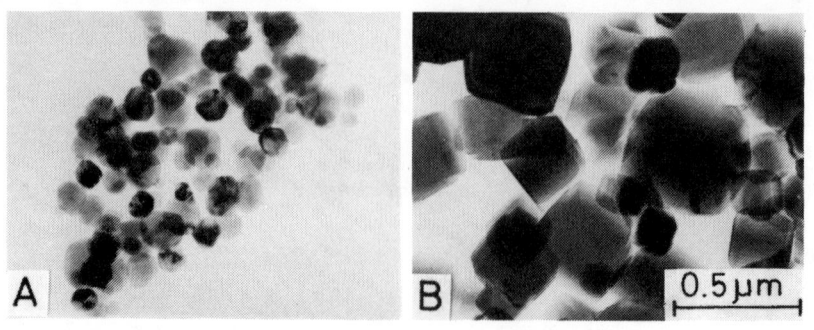

Fig. 8. Transmission electron micrographs of magnesia specimens calcined at 700 and 1100°C for 1 h: (A) 700°C; (B) 1100°C.

uniform, rounder and finer grains. This character of magnesia grains calcined at 700°C is also one reason for the promoting effect of chloride addition on sintering of magnesia. Figure 9 shows reflection microphotographs of a polished and etched surface of specimens fired at 1500°C for 4 h. The specimen mixed with chloride shows smaller and fewer pores.

Effect of Residual Chloride of Calcined Powder and Chlorine Atmosphere

One of the reasons for these remarkable promoting effects of the magnesium chloride might be due to the presence of residual chlorine in the calcined powders, which was determined as 0.08 wt% for the powder calcined at 900°C. The chlorine content of the other calcined powders was also determined by a gravimetric analysis using a silver nitrate solution, and the results are summarized in Fig. 10. The amounts of residual chlorine of the specimens calcined at 900°C are about 0.09 wt%, which is independent of the mixed amount of chloride, and the amount decreased linearly with increasing calcination temperature. But these tendencies did not coincide with the relations between the bulk density of fired specimens and the amounts of mixed chloride. Further to examine the effect of chlorine on sintering of magnesia, compacts of the powder of "no additive" calcined at 900°C were fired at 1400°C in various atmospheres of chlorine and with several firing schedules. Part of the results is shown in Table III. The specimen fired in air densified to 60% relative density, though most of the periclase grains

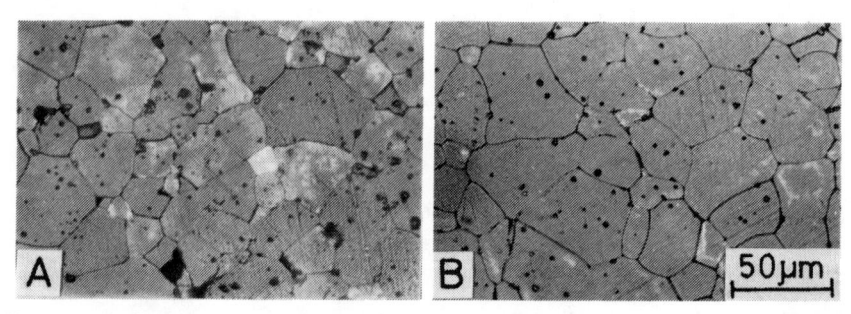

Fig. 9. Reflection microphotographs of polished and etched surface of specimens fired at 1500°C for 4 h: (A) no additive; (B) +4 wt% chloride.

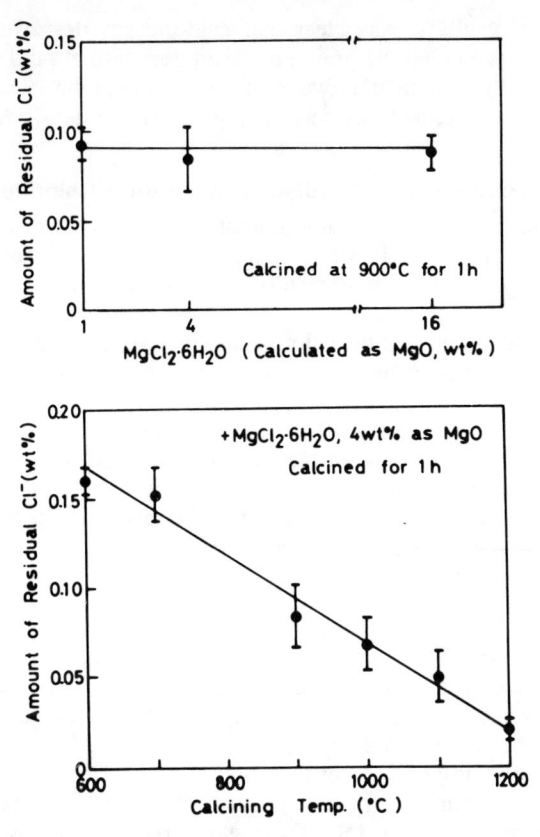

Fig. 10. Residual chlorine of calcined matters.

in it preserved brucite relics of about 0.5 μm. The specimen exposed to chlorine gas atmosphere at the temperature range from 900 to 1100°C did not densify and the one in chlorine gas atmosphere from 1200° to 1400°C also did not densify as much as the specimen fired in air. Chlorine gas destroyed brucite relics during firing and promoted the grain growth of periclase. Size of the periclase grains increased with an increase of temperature at which the specimen was exposed to chlorine gas.

Table III. Relative Density and Particle Size of Specimens

Firing atmosphere*	Relative density (%)	Particle size (μm)
In air	59.8	Brucite relics about 0.5 μm
In chlorine gas[†] from 900 to 1100°C	40.5	Relic free 0.5 to ~1.0 μm
In chlorine gas[†] from 1200 to 1400°C	55.1	2 to ~3

*Fired at 1400°C for 0 h in air and chlorine gas atmosphere; heating rate, 6°C/min; green compact, 40.0 to ~40.3% relative density, calcined powder at 900°C without MgCl$_2$ additive.
[†]Mixing ratio of air and chlorine gas–5:2.

From these results it was clear that chlorine gas destroyed brucite relics during firing and promoted the grain growth of periclase but did not promote the densification of the calcined powder. Thus, it might be concluded that the residual chlorine in calcined powders does not promote densification of the calcined powders.

Results of the Experiment on the Effect of Magnesium Chloride Addition

The chlorine atmosphere did not promote densification of the magnesia compacts. Increases in the bulk density of the green compacts led to a higher density in the fired compacts. From these results, reasons for the promoting effect of the addition of magnesium chloride on sintering of magnesia were inferred as follows. Addition of a small amount of the chloride into magnesium hydroxide makes relics of the hydroxide disappear, increases crystallinity, decreases activity of periclase, and slightly increases growth of the grain of periclase in the calcination stage of the mixture. The disappearance of relics and slight growth of periclase grains markedly improve compactness of the calcined powders. The medium grain size and, especially, lower activity of the periclase cause slow but successive progress of densification. The moderate densification rate and high density of the green compacts lead to high-density fired compacts.

Summary

Addition of a small amount of the magnesium chloride to magnesium hydroxide makes relics of the hydroxide disappear, increases crystallinity, decreases activity, slightly increases growth of the grains of periclase during their calcination, and further improves compactness of the calcined matters. These cause slow but successive progress of densification of dense green compacts, and magnesia ceramics of about 99% of theoretical density could be obtained, even for firing temperatures as low as 1500°C. In general, as in the case of magnesium chloride for magnesia, if a certain suitable compound, having the same cation as that of the host oxide, is selected as an additive agent, at least, dense ceramic bodies free of cationic impurities can be obtained by firing at relatively low temperatures.

Acknowledgments

This work was supported financially in part by the Fund of Science Research given by the Japanese Education Ministry in 1979 (project No. 34705).

References

[1] D. T. Livey, B. M. Wanklyn, H. Hewitt, and P. Murray, *Trans. Br. Ceram. Soc.*, **56**, 217–36 (1957).
[2] Kenya Hamano and Shinichiro Katafuchi, *Taikabutsu*, **32** [5] 243–52 (1980).
[3] Hiroshi Kamizono and Kenya Hamano; 1977 Annual Meeting of the Ceramic Society of Japan, May 11, 1977.
[4] Kenya Hamano, Zenbe-e Nakagawa, and Hideo Watanabe; pp. 159–64 in Sintering—Theory and Practice, Vol. 14. Elsevier, Amsterdam, 1982.
[5] Hideo Watanabe, Zenbe-e Nakagawa, Minoru Hasegawa, and Kenya Hamano; pp. 61–9 in the Rept. RLEMTIT, No. 7, 1981.

---

*MH30, Iwatani Chemical Corp., Osaka, Japan.
†Kanto Chemical Corp., Tokyo, Japan.

Final-Stage Sintering of MgO

C. A. Handwerker, R. M. Cannon,* and R. L. Coble
Massachusetts Institute of Technology
Department of Materials Science and Engineering
Cambridge, MA 02139

The assumptions used for models of simultaneous sintering and grain growth and of pore–boundary interaction during final-stage sintering were tested by using MgO samples of controlled microstructure and purity. The conditions examined were pore locations, sample densities, low background impurity levels, high levels of dopants, and the distributions of dihedral angles among samples. The microstructure evolution was described in terms of the average pore size and grain size with pores attached to boundaries and the transition pore and grain sizes for pore–boundary separation. Pore–boundary separation maps and grain size–pore size trajectories were calculated from literature data and were measured for various low-purity undoped MgO and Al-doped MgO samples. The pore sizes and grain sizes for the low-purity undoped MgO were in good agreement with the calculated transition values for pore–boundary separation. With Al additions, the pore sizes of separation increased relative to the conditions for the undoped MgO. The calculated trajectories for low-purity undoped MgO were in poor agreement with the observed trends; increasing the ratio (assumed) of the diffusivities controlling sintering and grain growth by a factor of 10 brought the calculated and observed values into agreement. The observed surface–boundary dihedral angles of pore–boundary intersections from polished surfaces of hot-pressed and annealed samples were shifted to lower values relative to the angles observed on thermally etched free surfaces. The effects of a distribution of dihedral angles on sintering and pore–boundary separation are discussed.

The microstructural evolution of a material undergoing sintering and grain growth is determined by the diffusivities (surface, lattice, and boundary), the rates of evaporation and condensation, the grain-boundary mobilities, the mobilities of closed pores, the pore–boundary interactions, the surface energy and grain-boundary energy and their variation with orientation, and the initial sintering geometry. The effect of each of these parameters is understood quantitatively when considered separately. The effect of the total set on the microstructural evolution is understood only qualitatively.

The first generation of modeling has described sintering and grain growth separately and as a function of a single variable or a limited set of these variables. The relative contributions of the surface, lattice, and boundary diffusivities and evaporation–condensation can be evaluated for a single sintering geometry to determine whether a material densifies or coarsens during the initial stages of neck growth. For a quantitative analysis of the sintering and grain growth in a powder compact, a mathematical formalism including all parameters is necessary. In this study, models based on larger subsets of the sintering variables are briefly reviewed. These include models of simultaneous sintering and grain growth and models of the transitions from grain growth with pores attached to grain boundaries to grain growth with pores separated from grain boundaries. Particular emphasis has been placed on the effects of dihedral angle distributions on kinetics within a single material.

The microstructural evolution calculated from these models, as defined by the average pore size and grain size and the condition of pore–boundary attachment or separation, is compared with the microstructural evolution observed in MgO compacts of "typical" purity examined in studies by Handwerker,[1] Gupta,[2] and Kamizono and Hamano.[3]

The variation in the observed dihedral angle distribution in MgO compacts undergoing sintering or in initially dense MgO compacts undergoing bloating has been assessed.[1] These distributions of observed dihedral angles in compacts are compared with the distribution of dihedral angles as measured from grooved surfaces. The effect of the width of the dihedral angle distribution on sintering and grain growth is briefly discussed.

Microstructural Evolution

The parameters that determine the sinterability of a ceramic powder compact are the diffusivities D_s, D_l, and D_{gb} of all species, the rates of evaporation and condensation (E/C), the surface and grain-boundary energies γ_s and γ_{gb}, the average particle size and particle size distribution, the particle morphology, and the green density. The values of the diffusivities, the rates of E/C, γ_s, and γ_{gb} are functions of orientation as well as temperature and background impurity concentration. Due to the number of possible combinations of these parameters, significant simplifications are required to model the sintering process. In some cases, either sintering or coarsening is treated. In all cases, anisotropy is not treated. A single value of the dihedral angle, ψ, as defined by $\gamma_{gb}/\gamma_s = 2\cos(\psi/2)$, is sometimes incorporated into the models, but it is generally assumed to be 180°.

The effect of decreasing the dihedral angle is an increase in the radius of curvature and, hence, a decrease in the driving force for sintering. It has been estimated that, with a dihedral angle of ~115°, the sintering time during the initial stage would increase by a factor of 2–50 relative to the sintering time for $\psi = 180°$.[5] The time increase depends on the sintering mechanism and the neck size. However, a dihedral angle $\psi < 180°$ is rarely incorporated into data analyses, since the sintering geometries and the rate equations are approximations and the magnitudes of the diffusivities are uncertain.

Two different but complementary descriptions of the simultaneous densification and grain growth processes and the transition to discontinuous grain growth have been developed in terms of the geometry of the microstructure and material parameters, such as diffusivities. The first treatment was developed by Brook[6] and extended by Yan et al.[7] and Hsueh et al.[8] It involves calculation of the transitions from pore drag controlled boundary migration to solute drag or intrinsic boundary migration controlled growth with the pores either attached to or separated from the boundary as functions of grain size and pore size. In the modeling by Hsueh et al.,[8] the effect of dihedral angle is considered. Pore reattachment after pore–boundary separation has been modeled by Carpay.[9] The second approach involves the calculation of the pore size–grain evolution (trajectories) during sintering, which depends on the relative rates of grain growth, pore coalescence, and densification.[10]

Pore–Boundary Separation

The separation of a boundary from a pore occurs when the boundary velocity, V_b, exceeds the pore velocity, V_p. The velocity, V_i, can be expressed as the product of a mobility M_i and a driving force, F_i. The separation condition, as stated by Brook,[6] is that $V_p < V_b$ for N pores on a unit area of a boundary, which leads to

$$V_p = F_p M_p < V_b = (F_b - NF_p)M_b \tag{1}$$

When $V_p = V_b$, the pores migrate with the boundary and

$$V_b = F_b \frac{(M_p M_b)}{NM_b + M_p} \tag{2}$$

The limiting cases of Eq. (2) are when $NM_b \gg M_p$ and $NM_b \ll M_p$. In the former, $V_b = F_b M_p / N$ and pore mobility determines the boundary velocity. In the latter case, $V_b = F_b M_b$ and the boundary mobility determines the velocity. A "semiquantitative" map of conditions for pore separation was calculated by Brook, who assumed that a single spherical pore is attached to a boundary of length G, the pore moves by surface diffusion, the maximum force on the pore exerted by the boundary is $\pi r \gamma_b$, and the force on the pore-free boundary is due to its curvature and is $2\gamma/G$.

There are three basic types of pore–boundary interactions: pore-drag-controlled migration, boundary-controlled migration with pores attached, and boundary-controlled migration with pores separated. Brook also considered the effect of a segregated impurity on the grain-boundary mobility. The solute drag limited mobility is lower than the intrinsic mobility, allowing larger pores to migrate with the boundary before separation. The "nose" of the separation curve is, thus, moved to larger grain sizes.

The effects of a grain size distribution and solute drag were incorporated into this basic model by Yan et al.[7] The pore–grain geometry assumed in this model (and all models) determines the conditions for pore–boundary separation. The geometry assumed in the model is the tetrakaidecahedron with spherical pores on all corners, and the pore radius, r, is related to the volume fraction pores, v_f, and average grain size, $\bar{G}$.

The transition from pore drag control to pore separation is calculated from Eq. (2) when $NM_b \gg M_p$. The Hillert equation for the force on a boundary due to curvature G in a matrix with a narrow grain size distribution is

$$F_b = (5/2)\gamma_b \left(\frac{1}{G_c} - \frac{1}{G} \right) \tag{3}$$

where G_c is the critical grain size for growth.[11] Equating the maximum restraining force due to N pores on a boundary ($\pi r \gamma_b N$) to the force on the boundary (F_b) gives the condition for separation for $G_c = \bar{G}$:

$$\left| 1 - \frac{G}{\bar{G}} \right| > 2.64 v_f^{1/3} \tag{4}$$

The condition of $G_c = \bar{G}$ indicates that all grains of size G greater than $\bar{G}$ will grow and those of a size less than $\bar{G}$ will shrink. For normal growth (Hillert criterion), $G_{max} = 2\bar{G}$, the critical volume fraction before breakaway, v_f^* is equal to 0.007. For $G_{max} \gg \bar{G}$, $v_f^* = 0.094$. The boundary to which this condition

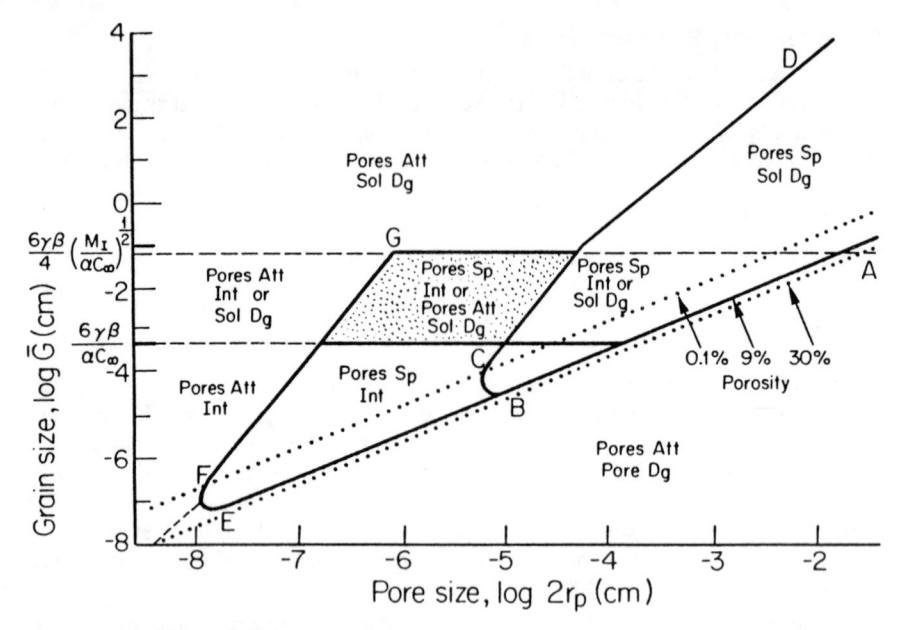

Fig. 1. Schematic of the range of pore–grain boundary interactions during final-stage sintering.

corresponds is shown in Fig. 1 as line $\overline{AE}$. The region to the lower right of the dotted line 30% corresponds to a density of less than 65% below which the pore diameter is greater than the grain facet length.

The critical force for solute drag limited migration with pores attached as calculated from Eq. (2) is

$$F_c = \pi r \gamma_b N + \pi r \gamma_b \frac{A_n}{r^n} \alpha C_\infty = NF_p + F_p \frac{M_p}{M_b} \qquad (5)$$

where $M_b = (\alpha C_\infty)^{-1}$, the solute drag mobility, α is the impurity drag per unit velocity and per unit dopant concentration, C_∞ is the bulk concentration of solute, and the pore mobility M_p is expressed as A_n/r^n. The values of A_n/r^n are dependent on the mechanism of pore migration; these are listed in Table I. The condition for pore and solute separation using Eq. (5) and the geometric relationship among r, v_f, and $\overline{G}$ is

$$5/2 \left| 1 - \frac{\overline{G}}{G} \right| > 6.6 v_f^{1/3} + \frac{\pi 48^{(n-1)/3} A_n \alpha C_\infty}{\overline{G}^{n-2} v_f^{(n-1)/3}} \qquad (6)$$

This determines the "nose" of the curve, line $\overline{BC}$, in Fig. 1.

The pores can remain attached to the boundary after the transition to the intrinsic mobility if the pore mobility is sufficiently high. To determine the range of pore sizes and grain sizes for which the pores remain attached, $1/\alpha C_\infty$ is replaced in Eq. (6) by M_I, the intrinsic mobility. The "nose" of the curve, line $\overline{EF}$, in Fig. 1 is the solution to the pore separation condition. The lines $\overline{CD}$ and $\overline{FG}$ describe pore–boundary separation when the pore velocity is high, i.e., when the second term in Eq. (5) is large. The line $\overline{CD}$ is determined by the solute drag limited boundary velocity, and $\overline{FG}$, by the intrinsic mobility limited velocity.

Table I. Pore Mobilities ($M_p = A_n/r^n$)*

Controlling mechanism	n	A_n
Surface diffusion	4	$\dfrac{3wD_s\Omega}{2\pi kT}$
Lattice diffusion	3	$\dfrac{3D_l\Omega}{2\pi kT}$
Vapor diffusion	3	$\dfrac{3D_g\Omega^2 P_v}{4\pi(kT)^2}$
Evaporation–condensation	2	$\dfrac{3\alpha'\Omega^2 P_v}{4(\pi kT)^{3/2}(2m)^{1/2}}$

*Symbols: w, surface thickness; D_s, surface diffusivity; Ω, molar volume; k, Boltzmann's constant; T, absolute temperature; D_l, lattice diffusivity; D_g, vapor diffusivity; P_v, vapor pressure; α', sticking coefficient; m, molecular weight (from Nichols (1969) (Ref. 4)).

It has been predicted from solute drag theory that an unstable region in the velocity–driving force relationship exists.[12] In this range of driving forces, transitions occur from the solute drag limited velocity to the intrinsic velocity over a range of driving forces. These driving forces can be equated to the curvature-induced driving forces for grain growth (in the absence of other driving forces). The grain size limits that correspond to the driving forces, in the unstable regime, correspond to $(6\gamma\beta)/\alpha C_\infty$ and $[(6\gamma\beta)/4](M_l/\alpha C_\infty)^{1/2}$ in Fig. 1, where β is the maximum boundary velocity for which the solute stays attached (see Ref. 12). In the dotted strippled region in Fig. 1 the pores will stay attached only if the solute stays attached.

An alternate approach has been developed by Hsueh, Evans, and Coble[8] for the motion of pores attached to two-grain interfaces and for the motion and coalescence of pores attached to the faces of a disappearing three-sided grain. A maximum steady-state pore velocity was found that is a function of the dihedral angle. The transition from the steady-state velocity of the pore to a nonsteady state led to grain boundary–pore separation. With comparison of the steady-state pore velocity to the grain-boundary velocity, critical conditions for pore separation were developed for different pore–grain configurations.

A pore size for breakaway was calculated for a pore on one face of a three-sided grain by equating the velocity of the boundary to the maximum steady-state velocity. For a five-sided grain, the condition for pore separation is more analogous to that calculated by Yan et al.[7] and is similar to line $\overline{CD}$ in Fig. 1.

The analysis of Hsueh et al.[8] for stable pore–boundary interactions predicts a limited region of pore size–grain size where pores remain attached. When $V_b > V_p$, a steady-state pore profile was not found for pores on two-grain faces. With decreasing dihedral angle, the maximum pore velocity increased and the critical pore radius that could remain attached to a shrinking three-sided grain increased. The region that is defined by this critical pore radius is similar to that determined by Yan et al. for the transition to boundary mobility controlled boundary migration with pores attached, line $\overline{FG}$ or $\overline{CD}$ in Fig. 1.

The geometrical models assumed for pore–boundary interaction by Hsueh and Yan have limited applicability to actual final-stage microstructural evolution. In the grain growth process, grains below a certain size and a certain number of sides shrink; the others grow. The treatment of Yan et al. assumes that pore separation occurs at an isolated, large grain surrounded by the tetrakaidecahedral "equilibrium" structure, rather than during the disappearance of a grain with fewer sides than that of a tetrakaidecahedron or during a grain switching event.

In contrast, the model by Hsueh describes an isolated pore–boundary separation event involving pores on grain faces of a grain with three sides (in two dimensions). This approach does not describe the stability of a microstructure with pores on four- or three-grain junctions prior to pore–boundary separation.

Therefore, it is more appropriate to use the modeling of Yan et al. to describe a microstructure with a narrow grain size and with a predominance of three- and four-grain junctions and to use the dihedral angle effects on pore mobility determined by Hsueh et al. to modify the pore–boundary separation condition for small pore sizes.

An additional condition for pore–boundary attachment exists in final stage microstructures where pore–boundary separation occurs at one grain size and reattachment occurs at a larger grain size after some amount of grain growth with pores separated.[9] Under these conditions, the pore size and grain size are not related to the volume fraction porosity by a simple expression. Hence, the drag force exerted by a pore is not simply related to the pore size, grain size, and volume fraction porosity, as in Eq. (6). Carpay has calculated the conditions for pore reattachment when the interpore spacing, f, is a constant, and $f < \overline{G}$. As the interpore spacing increases for a given pore size, the grain size for reattachment increases, as shown schematically in Fig. 2.

Simultaneous Coarsening and Densification

Another approach to microstructural evolution during intermediate–final-stage sintering is a dynamic one in which simultaneous sintering and coarsening are considered (Chowdhry, Cannon, and Yan; personal communication). In this treatment, the ratio of the coarsening rate to the densification rate is used to indicate whether a material will densify or coarsen during heat treatment. By substituting geometrical relationships for grain size, pore size, and density into the rates of coarsening and densification, we can express the ratio of these two rates in terms of average microstructural parameters (r_p, G, v_f), and can thus describe the grain size–pore size evolution (trajectory) for a compact of initial (r_{p_0}, G_0, v_{f_0}).

The change in the pore size with time is expressable as the sum of the change in pore size at constant volume fraction porosity and the change in pore size at constant grain size. The grain growth rate is calculated on the basis of control by either boundary mobility or pore drag controlled mobility. The equations for sintering and coarsening are equated through the time derivative in the rate equations, yielding relationships for dG/dr and $d \ln G/(d \ln r)$. Various equations for grain growth and sintering limited by different mechanisms are presented in the Appendix. These equations are based on the tetrakaidecahedral geometry, the final-stage sintering models of Coble, and the pore mobility relations of Nichols, listed in Table I.

The equations shown in the Appendix represent trajectories in pore size–grain size maps. A further modification of this treatment yielded calculations of times to reach a given volume fraction porosity as a function of initial grain size, G_0, initial pore size, r_0, initial green density, v_{f_0}, and the coarsening parameter Γ, which is $3wD_s\gamma_{gb}/(167\delta D_b\gamma_s)$ for the case of grain growth limited by pore mobility determined by surface diffusion control and sintering by grain-boundary diffusion.

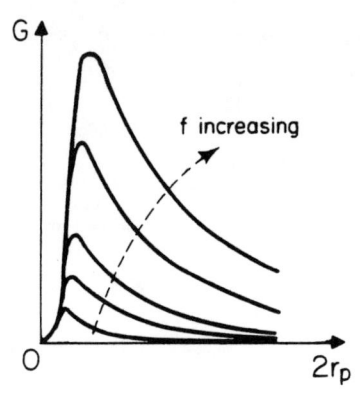

Fig. 2. Pore–boundary separation conditions for constant interpore spacing, f, as derived by Carpay (Ref. 9).

This case is shown schematically in Fig. A1 for the limiting volume fraction porosity for various values of Γ and v_{f_0}. In Fig. A2 is shown the dependence of the time to sinter to a given final density as a function of Γ for a given v_{f_0}.

As with the other models presented here, the restrictive geometry of this model, as well as the assumed forms of the expressions for the sintering and grain

Table II. Data for Pore–Boundary Separation and Sintering-Grain Growth Calculations*

	Diffusivity data	
wD_s (grain boundary grooving in air)	$6.9 \times 10^{-3} \exp(-90 \text{ (kcal)}/RT)$ (cm^3/s)	(Ref. 15)
D_l^{Mg} (tracer in Ar)	$4.19 \times 10^{-4} \exp(-63.5 \text{ (kcal)}/RT)$ (cm^2/s)	(Ref. 16)
+0.53% Fe (creep)	$1.04 \times 10^4 \exp(-117 \text{ (kcal)}/RT)$ (cm^2/s)	(Ref. 17)
+2.65% Fe (creep)	$5.06 \times 10^4 \exp(-117 \text{ (kcal)}/RT)$ (cm^2/s)	(Ref. 18)
D^{Mg} (creep of undoped)	$3.6 \times 10^{-13} \text{ (cm}^2/\text{s)}$ at 1620 K	(Ref. 19)
δD_b^{Mg} (creep of undoped)	$<10^{-18} \text{ (cm}^3/\text{s)}$ at 1620 K	(Ref. 19)
+0.53% Fe (creep)	$2.13 \times 10^{-8} \exp(-55.8 \text{ (kcal)}/RT)$ (cm^2/s)	(Ref. 18)
δD_b^0 + 0.53% Fe (creep)	$1.06 \times 10^{-6} \exp(-55.8 \text{ (kcal)}/RT)$ (cm^2/s)	(Ref. 18)
	Vaporization data	
$MgO(s) \rightarrow Mg(g)$ $+ \frac{1}{2}O_2(g)$	$p_{Mg} = 3.15 \times 10^{-9}$ (atm)* at 1620 K, 0.2 (atm) O_2	(Ref. 20)
	Boundary mobility data	
$M_{intrinsic}$	1.21×10^{-11} (cm·s/dyne·cm^2)	(Ref. 13)
+0.05 cation% Fe	4.64×10^{-14}	(Ref. 21)
+0.25 cation% Fe	6.75×10^{-15}	(Ref. 21)

*3.08×10^4 Pa.

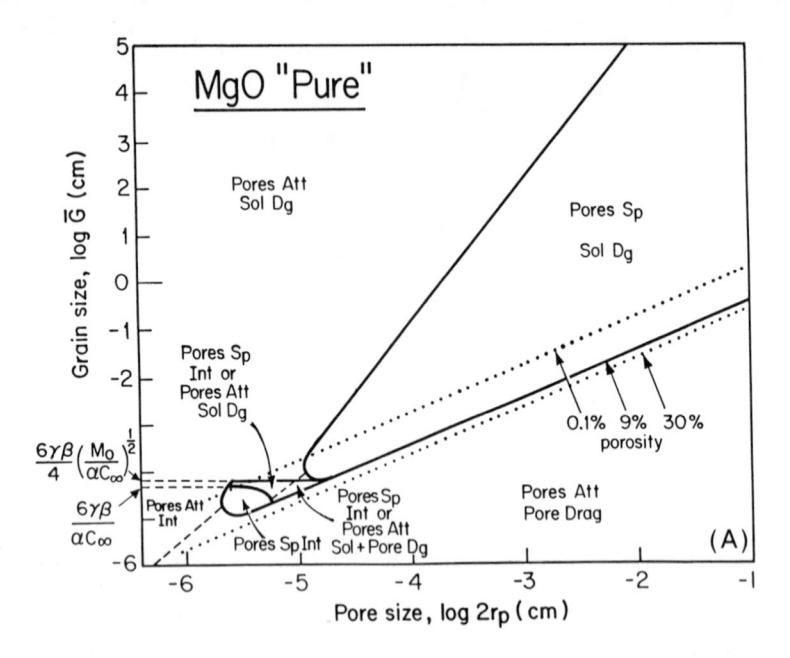

growth rates, makes these relationships useful only as order-of-magnitude indicators of sintering–grain growth trajectories. For the case presented above, the sensitivity of $d \ln G/d \ln r$ to small changes in the diffusivities makes this analysis difficult to apply. Only by comparison with experimental data can the value of Γ be determined for $\Gamma \sim 1$.

Calculation of Microstructure Evolution for MgO

The data base on MgO is the most extensive for ceramics, and the best diffusivity data for various purity levels as a function of temperature are from creep experiments (boundary and lattice diffusivities), tracer diffusion experiments (lattice diffusivities), and grain-boundary grooving experiments (surface diffusivities). Grain-boundary mobility data were derived from the results of grain growth experiments for undoped ("pure") and Fe-doped MgO by Yan et al.[13] The intrinsic boundary mobility was also estimated by Yan et al.,[13] using the simplified Turnbull relation[14] and the boundary diffusivities deduced from creep experiments on Fe-doped MgO. The most extensive data for MgO grain growth and creep are available only near 1623 K. Vapor pressure data for MgO are available over a wide temperature range. Because of inaccuracies in the calculated activation energies for these processes, the calculated values of the coarsening parameter, Γ, and the pore–boundary separation condition may be in error at temperatures where the diffusivity data do not overlap.

Pore size–grain size maps for pure and Fe-doped MgO at 1620 K were calculated by Chowdhry, Cannon, and Yan (personal communication) from the data for grain growth, diffusion, and vapor transport listed in Table II. The data for pure MgO describe material with at least 100 ppm cation impurities. However, the chemical compositions of the MgO samples described in the literature are not generally reported. The calculated maps are shown in Fig. 3(A, B) for $F_b = 3\gamma_b/\overline{G}$ (see Eq. (3)) and in Fig. 3(C) for $F_b = (5/4)\gamma_b 1/\overline{G}$. In the pure MgO map in Fig. 3(A) are seen all the main features described above.

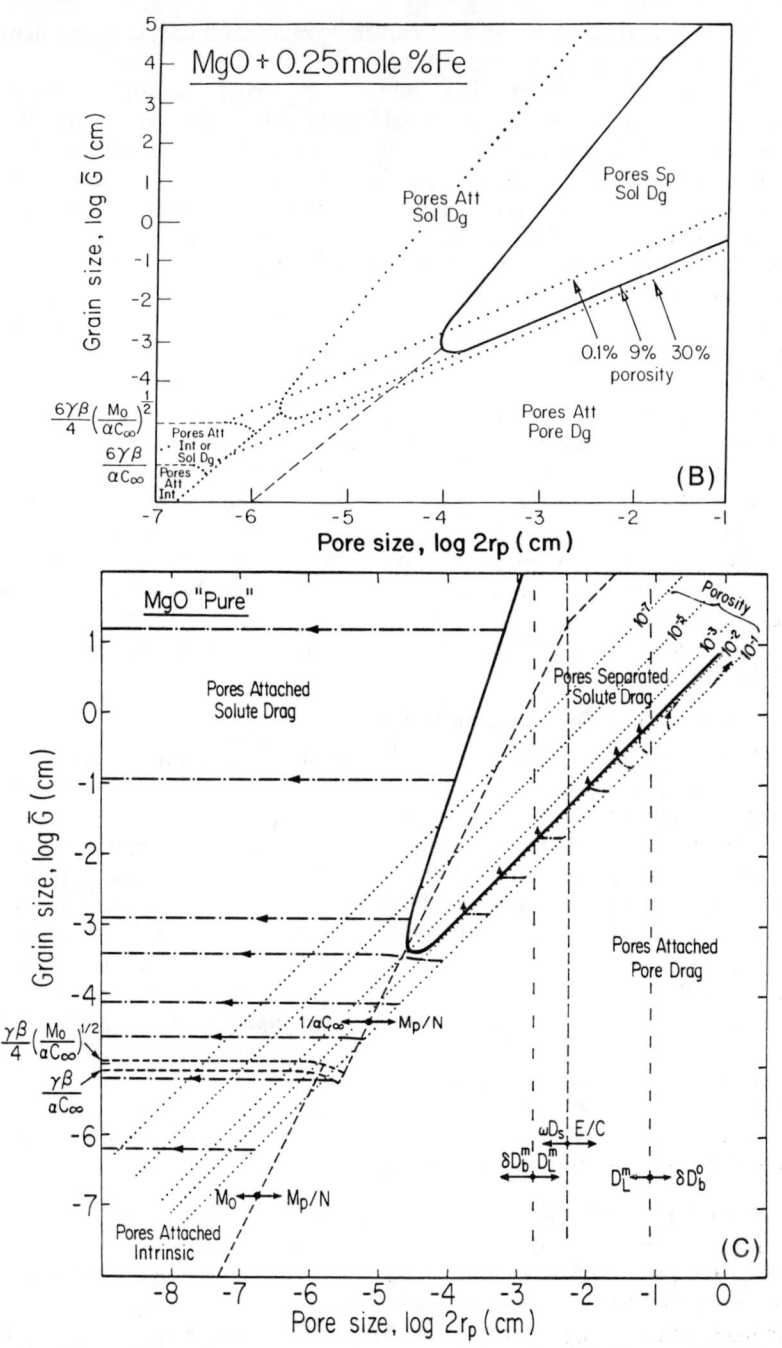

Fig. 3. Pore–boundary separation condition for MgO at 1350°C for different grain size distributions (courtesy R. M. Cannon, M. Yan, and U. Chowdhry): (A) pure MgO with a wide grain size distribution ($G_{max} \gg \overline{G}$); (B) Fe-doped MgO with a wide grain size distribution ($G_{max} \gg \overline{G}$); (C) pure MgO with a narrow grain size distribution ($G_{max} = 2\overline{G}$) in vacus.

The simultaneous sintering and coarsening relations for MgO were calculated by using equations as listed in the Appendix.

With the magnesium boundary and lattice diffusivities at 1623 K, Mg grain-boundary diffusion is predicted to control shrinkage for pore radii less than 10 μm. Surface diffusion dominates the coarsening process at this temperature. At 1623 K, oxygen grain-boundary diffusion is about a factor of 50 greater than magnesium grain-boundary diffusion. Details of the calculations are found elsewhere.[1] At 1623 and 1673 K, the values for the coarsening parameter, Γ, for normal-purity MgO are 0.15 and 0.20, respectively. This implies that the material will sinter without significant coarsening. The times to close 1 μm pores on grain boundaries are 1.3×10^5 s (35 h) and 8.5×10^4 s (28 h) for 1623 and 1673 K, respectively. These values of Γ predict near horizontal trajectories on the $\overline{G}-2r_p$ maps. Thus, for MgO powder of "normal" purity, compacted to high green density with nearly ideal packing, the sintering–coarsening analysis predicts sintering to high density ($\Gamma < 1$). The calculated values of Γ for normal-purity MgO are not much less than 1. Changes in the sintering geometry or in the diffusivities (by a factor of 10) can lead to $\Gamma > 1$, and coarsening will be predicted to dominate microstructure evolution.

The preceding calculations were based on pore-mobility-controlled coarsening. If grain growth–coarsening is controlled instead by the grain-boundary mobility, the microstructure evolution ($\overline{G}$, $2r_p$ trajectory) will be different, and the time to attain a specific density from an initial microstructure (V_0, $\overline{G}_0$, $2r_{p,0}$) will decrease.

Sintering and Grain Growth Experiments

Two experiments to determine densification–coarsening trajectories and pore–boundary separation were performed by using two processing techniques on two powder precursors.

The experiment to determine pore–boundary separation conditions was performed by using hot-pressed and annealed Mg(OH)$_2$-derived undoped and Al-doped MgO samples. A constant valence ion, aluminum, was added to MgO to change the grain-boundary mobility by solute drag. Kingery et al.[23] determined that Al segregates to boundaries in MgO. Chiang[24] determined that Sc^{3+} (an ion similar in charge and solubility to Al^{3+}), Ca^{2+}, and Si^{4+} segregate to MgO grain boundaries up to 3 nm away from the boundaries. (Samples without Ca and Si were not available.

The set of experiments for pore–boundary separation and trajectory determinations used sintered, MgCO$_3$-hydrate-derived MgO samples prepared by the settling of sized powders. Sintering was used to avoid the bloating problems associated with annealing hot-pressed samples.

Experimental Procedures

Magnesium hydroxide was prepared by the hydrothermal conversion of 99.99% Mg metal turnings[†] in saturated steam at a temperature of 620 K and a pressure of 16.3 MPa. This produces hexagonal platelets (<1 μm wide) with a wide distribution of agglomerate sizes. The specific surface area of the Mg(OH)$_2$ was ~11 m^2/g. The hydroxide powder was suspended in methanol and de-agglomerated by placing the hydroxide–methanol mixture in an untrasonic bath for 1 h.

The Al concentrations chosen for this study were ~500 and 1000 ppm, cation basis, which are ⅙ and ⅓ the solubility limit at 1670 K.[25] The Al dopant,

Al(OC$_3$H$_7$)$_3$,[‡] was dissolved in trichloroethylene and added dropwise to the stirred Mg(OH)$_2$–CH$_3$OH suspension. The pure and Al-doped powders were air-dried at 320 K for 24 h and transformed to the oxide at 1270 K in air for 2 h. Phase analyses by X-ray diffraction of the hydroxide and the pure and doped, calcined powders indicated that all were single phase.

The specific surface areas of the pure and Al-doped oxide powders are ~40 and 10 m^2/g, respectively. Chemical composition, both with regard to background and Al dopant levels, varies from sample to sample.

The hydroxide-derived powders were ground by using a mortar and pestle and vacuum hot-pressed at 1673 ± 10 K into disks of ≥90% ρ_{th} and approximately 0.5 cm high in a 5-cm diameter graphite die, which was lined with graphite foil, preannealed under vacuum at 1973 K.

The surfaces of the hot-pressed disks were ground on coarse SiC paper to remove adhering graphite foil and were annealed at 1300 K in air for 24 h to oxidize the surface carbon. The hot-pressed disks were cut into ca. 5-mm cubes. The samples from the center of the disks were annealed in an alumina tube resistance furnace; temperature was controlled to ±10 K. The heat treatments were performed at 1673 K in air or O$_2$ for various times up to 184 h.

The average pore sizes were determined from photographs of the polished but unetched samples with automatic quantitative metallography.[§] For grain size determinations, the polished sections were etched in a solution of HCl–H$_2$O(10:1). The average grain size, $\overline{G}$, was determined from the average line intercept, $\overline{l}$, where $\overline{G} = 1.5\overline{l}$.

The polished surfaces were examined for evidence of pores entrapped within grains. In areas of uniform density, microstructures were classified as either having or not having pores entrapped. Only in uniform regions were average pore sizes and grain sizes categorized according to a "yes–no" classification of entrapment. These data were used to construct breakaway maps.

The second set of experiments on low-purity MgO was performed by using reagent grade, hydrated magnesium carbonate (hydromagnesite). The physical and chemical properties of this powder in the carbonate and oxide forms are described in greater detail in Handwerker et al.[26] and O'Connor.[27] The carbonate was transformed to the oxide by heating in air to 1273 K. The agglomerated oxide powder was suspended in methanol and ultrasonically deagglomerated at 100 W (20 kHz) for 4 min. The powder suspension was allowed to settle for 25 h, and the remaining suspension was decanted. The particle size distribution of the decanted suspension ranged from 0.13 to 0.4 μm. The suspension was allowed to settle in a glass container, and the alcohol was evaporated at 300 K to form a compact of approximately 60% ρ_{th}. The term "top surface" refers to the side of the compact containing the last particles to settle by gravity and the term "bottom surface" to the first part of the compact to form, therefore, containing the largest particle sizes. These designations correspond to particle sizes of 0.13 and 0.4 μm, respectively. The samples were heated at approximately 10 deg/min to 1623, 1943, and 1873 K and annealed for times up to 170 h in O$_2$. The grain sizes and pore sizes on the top and bottom as-fired surfaces of the samples were measured. Fracture surfaces were examined to determine the amount of porosity through the sample thickness. Yes–no decisions of boundary breakaway were also made for these samples on the basis of the presence of entrapped pores on free surfaces and on fracture surfaces.

Results

The as-hot-pressed disks were ≥90% of theoretical density, but all contained

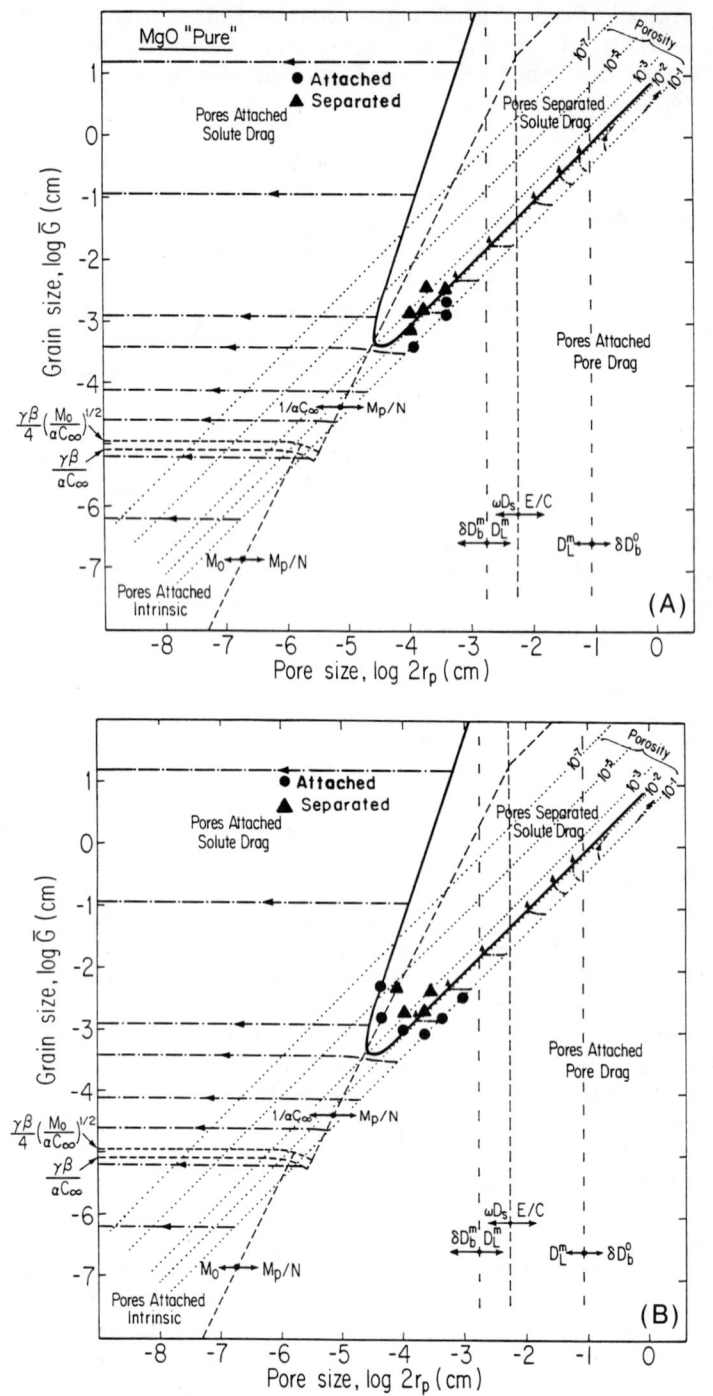

Fig. 4. Grain size–pore size data for boundary–pore attachment and separation plotted on an interaction map for narrow grain size distribution for (A) pure MgO and (B) MgO with 500 ppm Al.

large, well-defined bands of different densities. The lower density bands characterized by open porosity were gray after hot-pressing and remained gray during subsequent annealing at 1673 K. Superimposed on this large, but local density variation was a variation in grain size, pore size, and density due to the action of the die during hot-pressing. After additional heat treatment, these density gradients persisted; they were evident in the microstructures of multiple samples annealed at the same temperature and times. No systematic changes in the mean grain size, density, and number of pores per area for various samples nor in grain size distributions were noted with increasing time at 1673 K. Although the gross density variations precluded any kinetic analysis, local area analysis of pore–boundary breakaway was possible for these pure and Al-doped samples. These boundary breakaway data are presented in Fig. 4(A, B).

For the sintering experiments, the initial particle sizes of the top (0.13 μm) and the bottom (0.4 μm) of the compact as calculated from Stokes equation and the grain and pore sizes after annealing as measured by SEM are listed in Table III. The sample thickness after sintering was approximately 200 μm. Two regimes in pore and grain sizes were observed: small grain and pore sizes with no observable pore separation and large grain sizes and intermediate pore sizes with pores attached predominantly to two-grain intersections. At 1623 and 1743 K for times up to 4 h, no change in microstructure was observable by SEM. At 1623 K for 8 and 24 h, the samples had sintered to the closed pore stage, and the grain sizes were similar: 0.7 μm (top surface) and 0.5 μm (bottom surface). The pore sizes were <0.2 μm for both surfaces. No pore separation from boundaries was observed. At 1743 K for 71 h, the average grain sizes of the compact were 28 (top) and 15 μm (bottom). Entrapped pores and pores at grain intersections were observed on the as-fired bottom surface but not on the as-fired top surface. Examination of the fracture surface revealed entrapped pores through the thickness up to approximately one grain diameter from the top surface. At 1743 K for 170 h, the average

Table III. Grain Size*–Pore Size Relationships in Settled and Sintered MgCO$_3$ Derived MgO

1620 K	t = 0	t = 4 h	t = 8 h	t = 24 h	pores
top	0.13[†]	N.C.[‡]	0.7	N.C.[‡]	r_p < 0.3 μm
bottom	0.4[†]	N.C.[‡]	0.4	N.C.[‡]	

1740 K	t = 0	t = 71 h	t = 170 h	pores
top	0.13	28	35	0.4–3-μm pores attached along grain faces; 0.4–0.8-μm pores within grains
bottom	0.4	15	30	

1870 K	t = 0	t = 6 h	pores
top	0.13	15	0.1–0.3-μm pores attached along grain faces; entrapped pores on top and bottom surfaces
bottom	0.4	15	

*Grain size (μm).
[†]From Sedigraph; in reasonable agreement with SEM.
[‡]N.C. = no observed change from shorter annealing time.

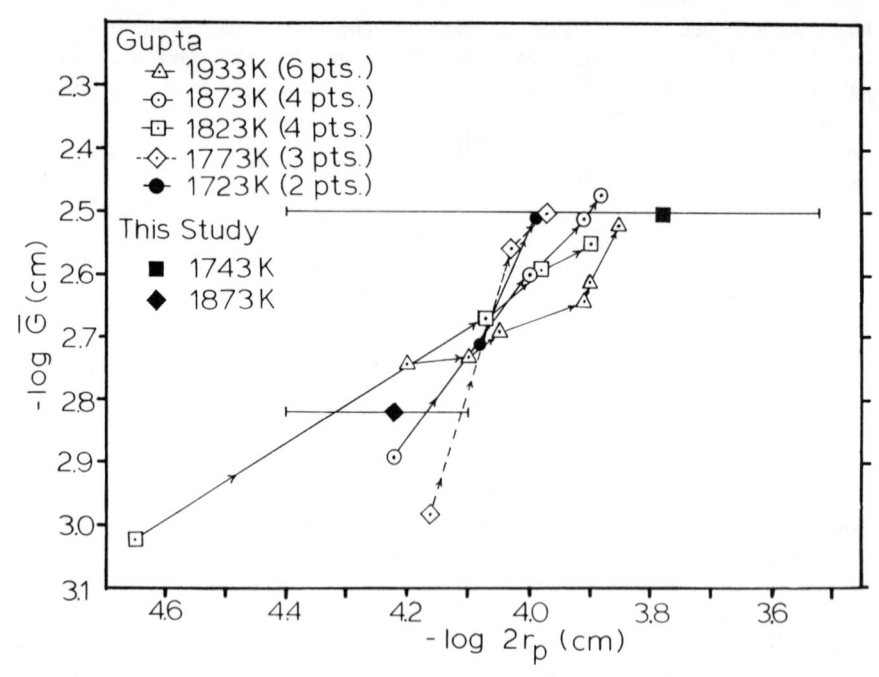

Fig. 5. Comparison of pore reattachment data from this study with data from Gupta (Ref. 2).

grain sizes increased to 35 μm (top surface) and 30 μm (bottom surface). From 71 to 170 h at 1743 K, G_{max}/G at the top surface changed from 2.5 to 4. The bottom surface showed entrapped pores and pores attached to two- and three-grain intersections. Along fracture surfaces, pores ranging from 3 to 0.4 μm were attached to all boundaries (7–25 pores per grain face); pores entrapped inside grains were smaller, 0.8–0.4 μm. At 1873 K for 6 h, the compacts sintered to the closed pore stage. The grain size was uniform through the sample, and entrapped pores were observed in both top and bottom surfaces. Pores of $2r_p \sim 0.3$ μm were also attached to boundaries predominantly along two-grain intersections with approximately 10 pores (>0.1 μm) per grain face.

The microstructural evolution in these samples as described by pore size–grain size trajectories are compared in the following section with the calculated trajectories.

The averages and the ranges of pore sizes and the average grain sizes after grain–pore reattachment are plotted in Fig. 5 with data from the literature.[2]

Discussion

The tetrakaidecahedral model for grain–pore attachment and the results of experiments on the undoped and Al-doped MgO are in good agreement. This may seem surprising in light of the restrictive geometry, the type of data used to construct the maps, and the results by Chiang on the amount of solute segregation and second phase present in these samples. The criteria used to include pore sizes and grain sizes in the attachment–separation plot were chosen to match the geometrical model assumptions and experimental geometries as well as possible. By comparison of the samples used in this study with the samples from which data on wD_s, D_g, and M_b were extracted, the samples are shown to be similar in composition.

The choices of uniform regions for grain size–pore size measurement were made on the basis of the tetrakaidecahedral model with pores at each four-grain intersection (corners). The densities and the pore sizes and grain sizes of the regions examined were required to be uniform for inclusion in the experiment. Regions deviating from the model geometry were due, in part, to the large local variations in density, in pore pinning (variable number and sizes of pores), and grain size. These regions were excluded from the measurements.

The data from which the plots in Fig. 3 were calculated were gathered on samples of high background impurity levels, perhaps similar to the samples examined here. As described in a previous section, the data needed for these analyses are diffusivities (D_g, δD_s, D_l) and grain-boundary mobilities for undoped and doped MgO. In this analysis, it was assumed that oxygen boundary diffusion or oxygen diffusion through a vapor phase is sufficiently fast to make magnesium transport rate limiting. The Mg lattice diffusivities from ^{26}Mg tracer measurements on a single crystal containing at least 300 ppm impurities (possibly up to 2000 ppm) (Wuensch et al.[16]) and from creep of LiF-doped MgO (total impurities not measured) (Hodge and Gordon[19]) were 1.2×10^{-12} and 1.3×10^{-12} cm^2/s, respectively, at 1623 K. The two sets of wD_s data from thermal grooving in air of exposed MgO surfaces of unspecified composition are also similar (Robertson[15]; Henney and Jones[28]). The grain-boundary mobilities for undoped and Fe-doped MgO were calculated from grain growth studies of hot-pressed MgO and magnesiowustite (0.10 and 0.48 at.% Fe$_2$O$_3$) (Gordon et al.[21]). In this study, chemical compositions of the undoped samples prepared from nominally 99.55+% MgO were found to vary from sample to sample but, generally, contained 2500 ppm (by weight) total cation impurity, predominantly Si, Ca, Fe, and Ti.

In a STEM study by Chiang (personal communication) of grain-boundary segregation in the Al-doped MgO samples, significant segregation and high background impurity levels were observed. Typical TEM microstructures of the Al-doped MgO samples are shown in Fig. 6(B,D). Large microstructural and compositional variations were observed that were not seen by optical microscopy of SEM. Some triple points were precipitate and pore free; others contained a glassy phase, a crystalline second phase, or pores. In addition, solute segregation of Al, Si, Ca, and Ti to MgO boundaries was observed at different ratios and different total levels from boundary to boundary, as illustrated in Fig. 6(A, C). These high impurity levels are typical of standard laboratory specimens of MgO.

There exists a window of measurable pore sizes and grain sizes on the attachment–separation map due to the limits of resolution of SEM and optical microscopy and to the size of the uniform density regions. The resolution limit is ~0.5 μm with optical microscopy. However, the practical limit varies from microstructure to microstructure. For example, pores of 0.5 μm adjacent to 1-μm grains are difficult to resolve; well-dispersed 0.5-μm pores adjacent to 10-μm grains are easier to identify and measure. The resolution limit of SEM is much lower (~0.01 μm); however, the good depth of field in SEM leads to a practical limit of ~0.1 μm for pores. At pore sizes less than 0.1 μm and grain sizes less than 0.5 μm, little information on the microstructural evolution can be obtained. At small initial pore sizes and grain sizes, several cycles of normal growth and pore attachment followed by pore separation could occur without detection by optical

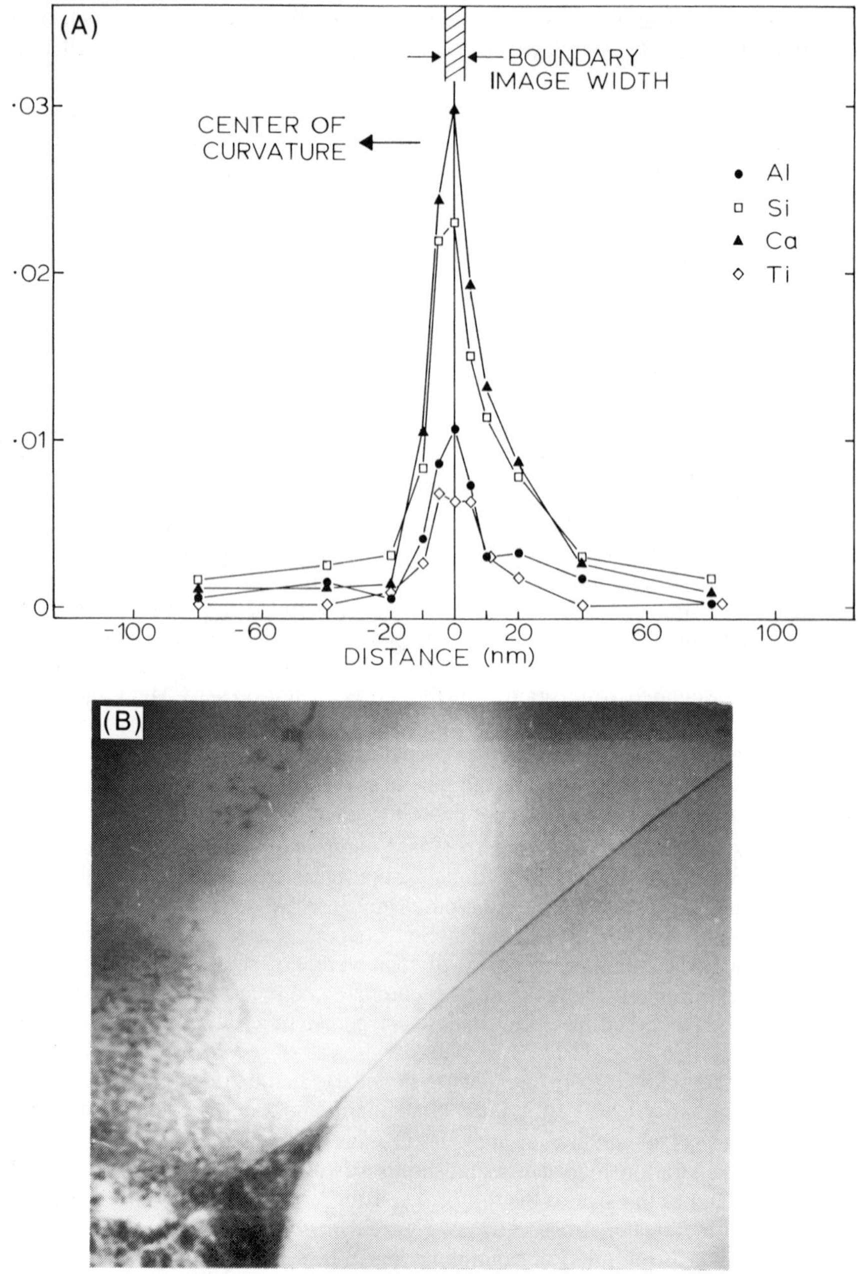
(A)
BOUNDARY IMAGE WIDTH
CENTER OF CURVATURE
Al
Si
Ca
Ti
·03
·02
·01
0
-100
-60
-20
0
20
60
100
DISTANCE (nm)
(B)

microscopy and SEM. Therefore, the data presented here describe only a small part of the G–$2r_p$ space through which microstructures evolve. The upper limits of grain size and pore size in this study were determined by the gross density variation in the samples, caused by the powder type and the fabrication technique. In the undoped samples, the thickness of the high-density lenticular regions vary from 20

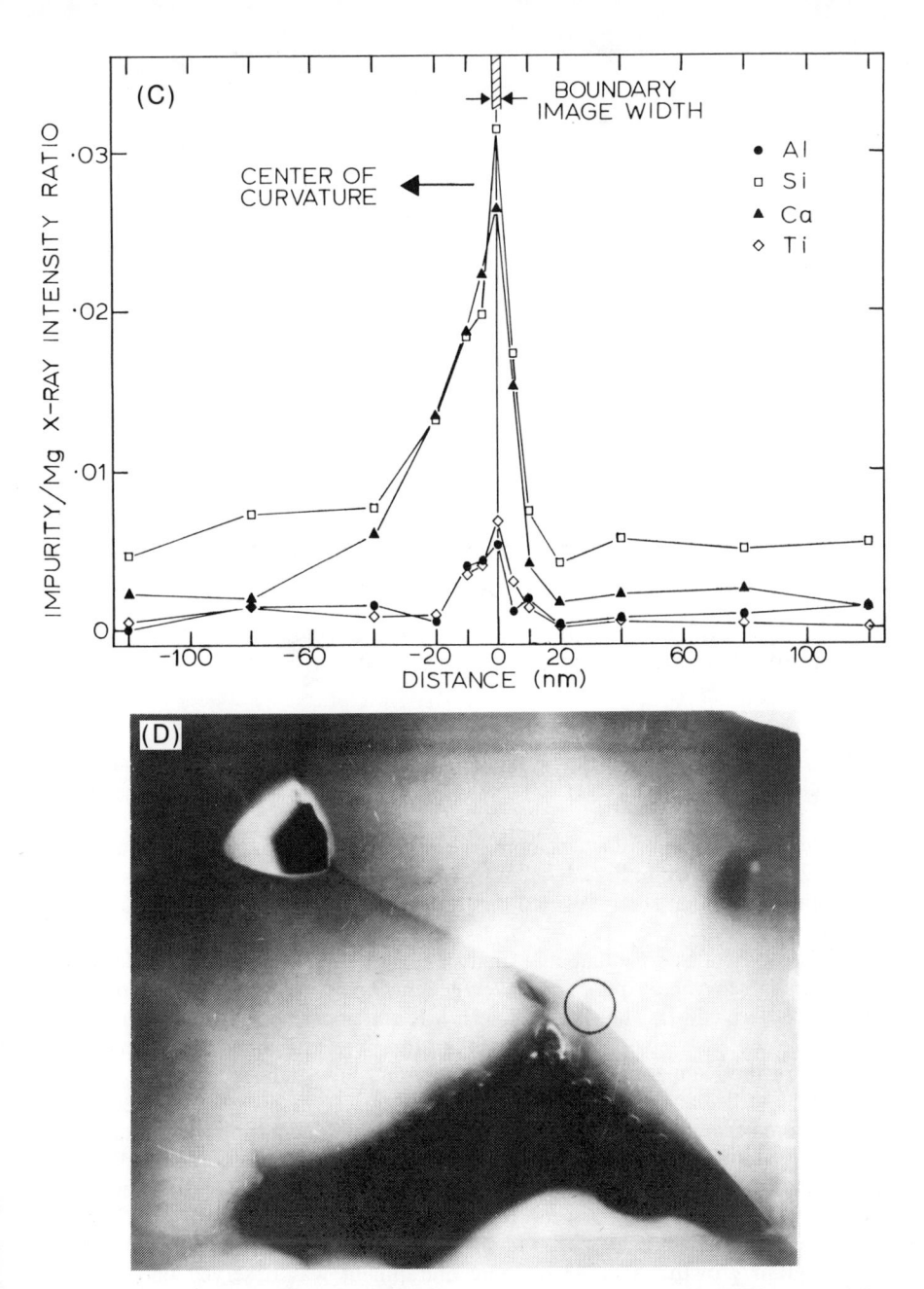

Fig. 6. STEM segregation profiles and bright field micrographs of areas from which profiles were measured. Samples are hot-pressed Al-doped (1000 ppm) MgO appealed at 1673 K for 72 h (courtesy of Y. M. Chiang): (A) STEM profile; (B) boundary area analyzed in (A); (C) STEM profile; (D) boundary area analyzed in (C).

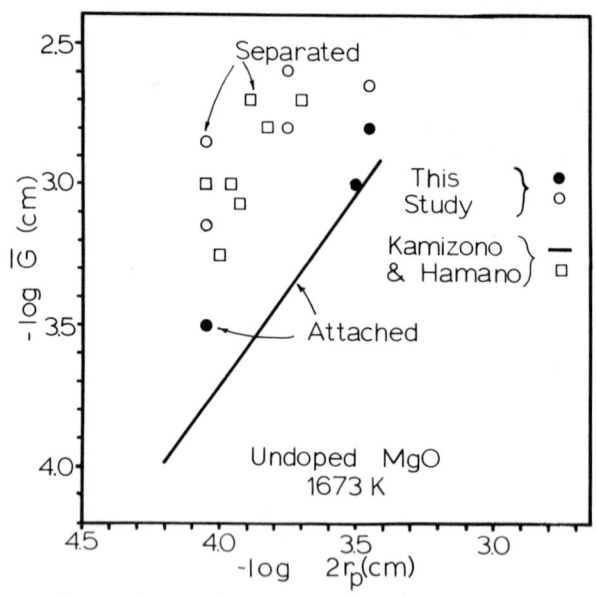

Fig. 7. Comparison of pore–boundary separation and attachment conditions from this study with data from Kamizono and Hamano (Ref. 3).

to 200 μm perpendicular to the pressing direction. In the Al-doped MgO samples, the lenticular regions are larger (40–500 μm) but also occur throughout the entire sample thickness.

The measurement limits for pore size ($2r_p$) and grain size ($\overline{G}$) are drawn on the $\overline{G}-2r_p$ plot shown in Fig. 4(A). When the $\overline{G}-2r_p$ values in a compact sintered to intermediate–final stage fall into this range, the microstructural evolution can be determined.

In a study of intermediate and final stage sintering, Kamizono and Hamano[3] determined the average pore size–average grain size and the pore location for hydroxide-derived MgO hot-pressed at 1173 K and annealed at 1673 K. As shown in Fig. 7, the pore size–grain size ranges for pore attachment and separation overlap for the Kamizono and Hamano study and this study.

In the settled and sintered MgO samples examined in this study, the tendency toward pore separation from boundaries increased with increasing temperature and increasing initial particle size, in agreement with the calculated pore separation maps and pore size–grain size trajectories. At 1873 K, pore entrapment was evident both on as-fired surfaces ($G_0 = 0.13$ μm and $G_0 = 0.40$ μm) and in the bulk. Pore attachment was evident on grain faces, indicating reattachment during grain growth. At 1743 K, pore entrapment was observed only on the ($G_0 = 0.40$ μm) surface. Pore reattachment to grain faces was observed through the sample thickness up to ~10 μm from the top surface ($G_0 = 0.13$ μm). At 1623 K, no pore separation was observed.

As calculated from the simultaneous sintering–coarsening expressions (Appendix), the times at these temperatures to sinter pores associated with 0.13- and 0.40-μm grains are small at 1673 K, $t \sim 1$ s. This is obviously not the situation found for MgO settled compacts. Coarsening is much greater than predicted from

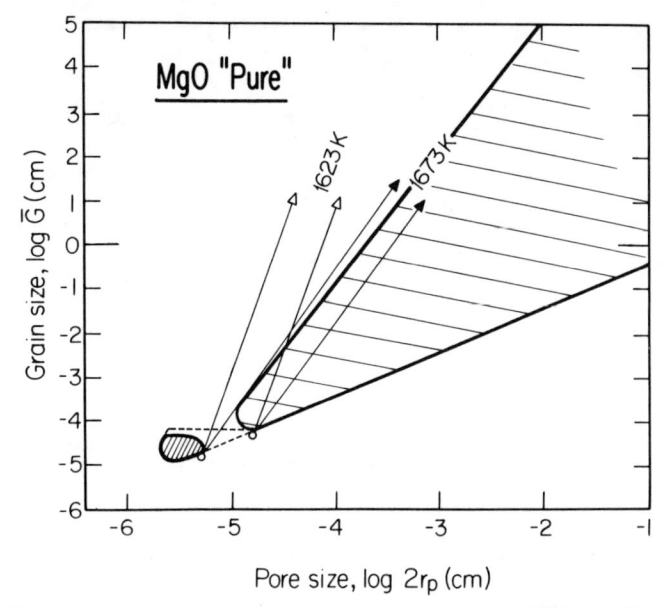

Fig. 8. Trajectories of grain size–pore size during sintering of MgO calculated by using a factor of 10 increase in the coarsening parameter, Γ.

these models and the available diffusivity data. Coarsening can be enhanced by poor particle packing in the green state, by decreasing the dihedral angle, or by increasing wD_s. It is not possible to determine the dominant effect in this set of experiments.

An increase in Γ by a factor of 10, due either to an increase in $wD_s/\delta D_b$ or to an increase in the pore mobility due to a decrease in the dihedral angle, yields $\Gamma > 1$ for 1623, 1743, and 1873 K. The calculated grain size–pore size trajectories are shown in Fig. 8 for the top surface ($G_0 = 0.13$ μm, $r_0 = 0.028$ μm, $v_f = 0.40$) and for the bottom surface ($G_0 = 0.40$ μm, $r_0 = 0.08$ μm, $v_f = 0.40$). The pore separation map is for 1623 K only; the trajectories for 1743 and 1870 K can be qualitatively used to estimate the breakaway conditions. At 1623 K, neither trajectory intersects the primary pore separation region. At 1673 K, the trajectory for the bottom surface intersects the pore separation region, whereas the trajectory for the top surface misses the nose of the curve and does not intersect the separation region. At higher temperatures (not shown), the trajectories for the top surface ($G_0 = 0.13$ μm) intersect the pore separation region at larger grain sizes than the bottom surface ($G_0 = 0.40$ μm) trajectories. At these temperatures, the difference in grain size at which pore separation occurs for $G_0 = 0.13$ μm and $G_0 = 0.40$ μm is small, leading to little difference in grain size through the compact. Pore reattachment at larger grain sizes occurred. This situation is not easily depicted in a breakaway diagram because of the more complex microstructural evolution. These data are plotted separately in Fig. 5 along with pore–grain size data from a study on pore coarsening by Gupta.[2] Gupta determined the average pore size, grain size, and number of attached pores for isostatically pressed and sintered MgO undergoing coarsening-induced bloating. As in samples examined in this study, the microstructures examined by Gupta had most pores on

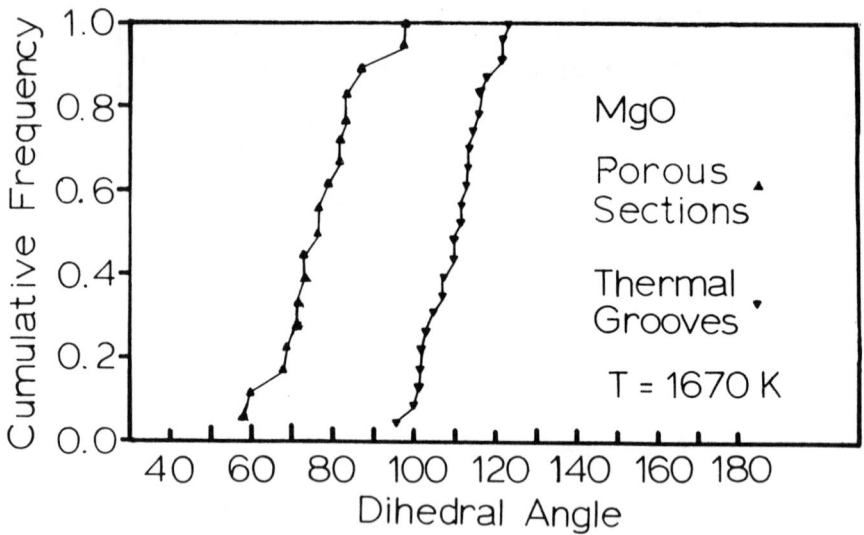

Fig. 9. Comparison of the distribution of dihedral angles from free surface–boundary intersections (thermal grooves) with those from pore–boundary intersection observed on planes of polish.

two grain intersections, and the pore radii varied by a factor of ~10 along a single fracture surface. The smallest average grain size examined by Gupta was 9 μm (1823 K), indicating breakaway at smaller grain sizes. These data are in good agreement with the pore sizes and grain sizes determined in this study for reattachment.

Surface–Boundary Dihedral Angles in MgO

Relative pore stability is determined in part by the grain boundary to surface energy ratio and, hence, the dihedral angle. For four-sided pores in three dimensions, pores with dihedral angles equal to or less than 70.5° will be stable relative to pores with $\psi = 70.5°$.

In porous compacts with wide dihedral angle distributions, stable or growing pores will coexist with shrinking pores.[29] As sintering proceeds, the apparent dihedral angle distribution shifts to smaller dihedral angles due to the stability differences among pores. Even for the case where all four-sided pores have $\psi \geq 70.5°$, there will be stability differences between pores with $\psi = 100°$ and with $\psi = 140°$, leading to differences in sintering rates. For the case of some four-sided pores with $\psi \leq 70.5°$, the pores must become three- or two-sided for sintering to continue; i.e., the pores must separate from one or two boundaries without separation from all boundaries. In poorly packed green compacts (nonideal, typical), n-sided pores, with $n > 4$, frequently occur. As n increases, the critical dihedral angle defining pore stability increases. For these cases, n must decrease for sintering to proceed.

If the pores become sufficiently large to pin the boundaries, no separation and no further densification will occur. Large local variations in porosity and boundary pinning will exist. Some regions will sinter to closed pore stage and undergo grain growth while other regions contain open pores. If the dihedral angle distribution is broad and the fraction of small angles is large, the probability of having a high density, large grain region adjacent to a smaller grain size, porous region increases. The larger grain size region can then grow into the immobile porous region. This is the classic description of discontinuous grain growth.

The results of a previous study of dihedral angle distribution from grain-boundary grooves on free surfaces and in porous compacts[1] are presented in Fig. 9. The dihedral angles from free surfaces, as measured by the metal reference line technique, range from 97° to 123° over the median angle, $\psi_n = 112°$. The apparent dihedral angle distribution from planes of polish of hot-pressed and annealed Al-doped MgO were observed to shift to significantly lower values (56–90°) with $\psi_n = 70°$.

The distribution of apparent dihedral angles for the porous MgO samples supports the predicted microstructural evolution based on dihedral angle variations. At high densities, the dihedral angle distribution is expected to shift to smaller dihedral angles than that of the total possible population due to the lower driving force for densification and higher pore velocity for pores with smaller dihedral angles. As the density increases, the porosity is determined by the pores with low dihedral angles and by entrapped pores. In higher density samples undergoing grain growth, pores that effectively pin the boundaries or that migrate with the boundaries will dominate the distribution. Two-sided pores with the smallest dihedral angles were calculated by Hsueh et al.[8] to more efficiently pin and to migrate faster than pores of equal in-boundary radii but larger dihedral angles.

Conclusions

In hot-pressed and annealed samples of $Mg(OH)_2$ derived MgO, the pore sizes and grain sizes at which boundary separation from pores occurred are in good agreement with the transition values calculated from literature data by using the model of Cannon, Yan, and Chowdhry. With Al additions, the pore sizes at separation increased relative to the separation conditions for undoped MgO, as expected from lower boundary mobilities due to the solute drag of Al.

For normal-purity MgO, sintering to theoretical density with negligible coarsening is predicted from the data at the experimental temperatures. Experimentally, this is not the case. An increase in the surface diffusivity by 10 times or an increase in pore mobility by 10 times would be in better agreement with the observed behavior.

A distribution of dihedral angles in a sintering body will lead to local variations in the driving forces for sintering and in pore mobilities, which determine the pore–boundary separation condition. These two effects of dihedral angles on microstructure evolution are expected to lead to a narrowing distribution during the sintering process with a lower median value than that of the initial distribution.

The observed surface–boundary dihedral angles from polished surfaces of hot-pressed and annealed MgO samples were shifted to lower values relative to the angles observed on thermally etched free surfaces.

Acknowledgments

This research was supported by the U.S. Department of Energy (Contract No. DE-AC02-76ER02390) and the National Science Foundation, Division of Materials Research (Contract No. 77-11585-DMR).

Appendix

Equations for simultaneous densification and coarsening during final-stage sintering have been derived by Yan, Cannon, and Chowdhry (personal communication). The Coble equations for final-stage sintering were used to describe the change in pore size with time at constant grain size.[22] The change in grain size with time was assumed to be controlled by grain growth limited by either pore mobility or grain-boundary mobility. An additional term describing the change in pore size with time is due to grain growth and pore coalescence. This correponds to a change in pore size with time at constant volume fraction porosity. Thus,

$$\frac{d\overline{G}}{dt} = M\frac{\gamma_b}{G} \tag{A-1}$$

where the mobility, M, is either that for grain-boundary mobility or pore mobility, listed in Table I, and

$$\frac{dr}{dt} = \left(\frac{dr}{dt}\right)_G + \left(\frac{dr}{dt}\right)_{v_f} = \left(\frac{dr}{dt}\right)_G + \frac{r}{G}\frac{dG}{dt} \tag{A-2}$$

The term $(dr/dt)_G$ is determined by either lattice or boundary diffusion. The coarsening term $(dr/dt)_{v_f}$ is controlled by the same mechanisms as grain growth. The relationship $d \ln G/(d \ln r)$ is calculated from Eqs. (A–1) and (A–2) through the time derivative. Grain size–pore size trajectories for various combination of mechanisms are presented in Table A1. Examples of the microstructural relationships that can be derived from these equations are shown in Figs. A1 and A2.

Table A1. Densification–Coarsening Equations for Various Rate Controlling Mechanisms*

Densification (boundary diffusion) — coarsening (surface diffusion)

$$\frac{d \ln G}{d \ln r} = \frac{1}{1 - \frac{176}{3}\left(\frac{D_b}{wD_s}\right)\frac{\gamma_s}{\gamma_b}g(v_f)}$$

Densification (lattice diffusion) — coarsening (surface diffusion)

$$\frac{d \ln G}{d \ln r} = \frac{1}{1 - 32\frac{rD_l}{wD_s}\left(\frac{\gamma_s}{\gamma_b}\right)g(v_f)}$$

Densification (lattice diffusion) — coarsening (evaporation–condensation)

$$\frac{d \ln G}{d \ln r} = \frac{1}{1 - \frac{1}{r}\left(\frac{64(2\pi mkT)^{1/2}D_l\gamma_s}{\alpha'p_r\Omega\gamma_b}\right)g(v_f)}$$

Densification (boundary diffusion) — coarsening (boundary mobility controlled)

$$\frac{d \ln G}{d \ln r} = \frac{1}{1 - \frac{11}{3}\left(\frac{\gamma D_b\Omega\gamma_s G^2}{M_b\Omega kT\gamma_b r^4}\right)g(v_f)}$$

*An alternate relation is $d \ln G/(d \ln r) = \Gamma/(\Gamma - 1)$.

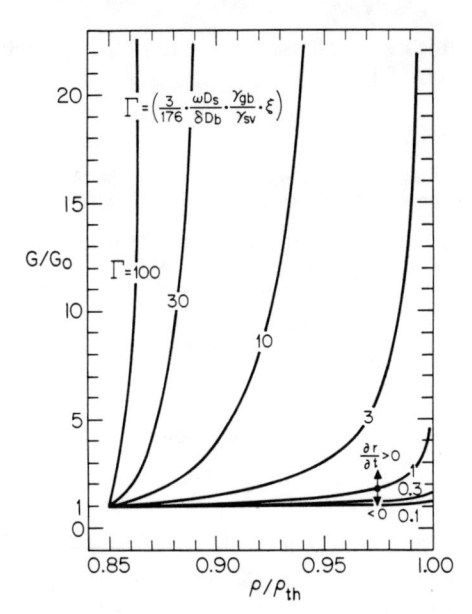

Fig. A1. Grain size–porosity evolution during final-stage sintering as a function of Γ and the starting grain size and porosity.

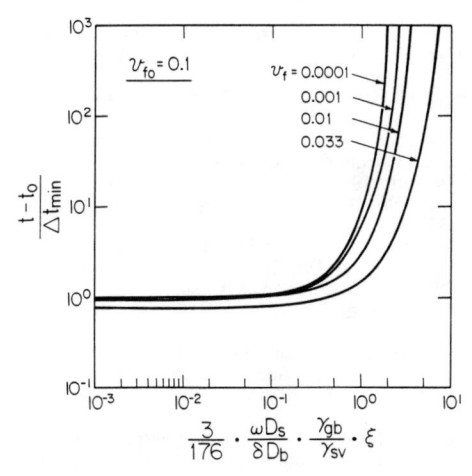

Fig. A2. Time to sinter to various porosities as a function of Γ.

The relationship $d \ln G/(d \ln r)$ can be expressed in general terms as

$$\frac{d \ln G}{d \ln r} = \frac{\Gamma}{\Gamma - 1} \tag{A-3}$$

as seen in Table A1. For the combination of grain boundary diffusion controlled densification and surface diffusion controlled pore mobility, the variation of relative grain size with relative density has been calculated as a function of Γ, as shown in Fig. A2. For the same mechansims, the time to sinter to a given density has been calculated as a function of Γ.

References

[1] C. A. Handwerker, "Sintering and Grain Growth of MgO"; Sc.D. Thesis, M.I.T., 1983.
[2] T. K. Gupta, "Kinetics and Mechanisms of Pore Growth in MgO," *J. Mater. Sci.*, **6**, 989 (1971).
[3] H. Kamizono and K. Hamano, "Relation between Pore Distribution and Grain Size of Magnesia in the Intermediate Stage," *Yogyo-Kyokai-Shi*, **88** [10] 48 (1978).
[4] F. A. Nichols, "Kinetics of Diffusional Motion of Pores in Solids," *J. Nucl. Mater.*, **30**, 148 (1969).
[5] J. M. Dynys, R. L. Coble, W. S. Coblenz, and R. M. Cannon, "Mechanism of Atom Transport during Initial Stage Sintering"; in Sintering Processes. Edited by G. C. Kuczynski. Plenum Press, New York, 1980.
[6] R. J. Brook, "Pore-Grain Boundary Interactions and Grain Growth," *J. Am. Ceram. Soc.*, **52** [1] 56 (1969).
[7] M. Yan, R. M. Cannon, H. K. Bowen, and U. Chowdhry, "Effect of Grain Size Distribution on Sintered Density"; to be published in *Mater. Sci. Eng.*
[8] C. H. Hsueh, A. G. Evans, and R. L. Coble, "Microstructure Development During Final/Intermediate Stage Sintering: I, Pore/Grain Boundary Separation," *Acta Metall.*, **30**, 1269 (1982).
[9] F. M. A. Carpay, "Discontinuous Grain Growth and Pore Drag," *J. Am. Ceram. Soc.*, **60**, 82–3 (1977).
[10] K. Uematsu, R. M. Cannon, R. D. Bagley, M. F. Yan, U. Chowdhry, and H. K. Bowen, "Microstructure Controlled by Dopants and Pores at Grain Boundaries"; in Factors in Densification and Sintering of Oxide and Non-Oxide Ceramics. Edited by S. Somiya and S. Saito. Gakujutsu Bunken Fuku-kai, Tokyo, Japan, 1978.
[11] M. Hillert, "Theory of Normal and Abnormal Grain Growth," *Acta Metall.*, **13** [3] 227 (1965).
[12] J. W. Cahn, "Impurity Drag Effect in Grain Boundary Motion," *Acta Metall.*, **10** [9] 789 (1960).
[13] M. Yan, H. K. Bowen, and R. M. Cannon, "Grain Boundary Migration in Ceramics"; in Ceramic Microstructure '76. Edited by R. M. Fulrath and J. A. Pask. Westview Press, Boulder, CO, 1976.
[14] D. Turnbull, "Theory of Grain Boundary Migration Rates," *Trans. A.I.M.E.*, **191**, 661 (1951).
[15] W. M. Robertson, "Kinetics of Grain Boundary Grooving on Magnesium Oxide"; in Sintering and Related Phenomena. Edited by G. C. Kuczynski, N. A. Hooton, and C. F. Gibbon. Gordon and Breach, New York, 1967.
[16] B. F. Wuensch, W. C. Steele, and T. Vasilos, "Cation Self-Diffusion in Single Crystal MgO," *J. Chem. Phys.*, **55**, 6258 (1973).
[17] R. T. Tremper, R. A. Giddings, J. D. Hodge, and R. S. Gordon, "Creep of Polycrystalline MgO-FeO-Fe$_2$O$_3$ Solid Solutions," *J. Am. Ceram. Soc.*, **55** [8] 421 (1972).
[18] R. S. Gordon, "Ambipolar Diffusion and its Application to Diffusion Creep"; p. 445 in Mass Transport Phenomena in Ceramics. Edited by A. R. Cooper and A. H. Heuer. Plenum Press, New York, 1975.
[19] J. D. Hodge and R. S. Gordon, "Grain Growth and Creep in Polycrystalline Magnesium Oxide Fabricated with and without a LiF Additive," *Ceram. Int.*, **4**, 17 (1978).
[20] R. Altman, "Vaporization of MgO and Its Reaction with Alumina," *J. Phys. Chem.*, **67** [2] 366 (1961).
[21] R. S. Gordon, D. D. Marchant, and G. W. Hollenberg, "Effects of Small Amounts of Porosity on Grain Growth in Hot-Pressed Magnesium Oxide and Magnesiowustite," *J. Am. Ceram. Soc.*, **53** [7] 399 (1970).
[22] R. L. Coble, "Sintering Crystalline Solids: I, Intermediate and Final Stage Diffusion Models," *J. Appl. Phys.*, **22** [5] 787 (1961).
[23] W. D. Kingery, W. L. Robbins, A. F. Henriksen, and C. E. Johnson, "Surface Segregation of Aluminum (Spinel Precipitation) in MgO Crystals," *J. Am. Ceram. Soc.*, **59** [5–6] 239 (1976).
[24] Y. M. Chiang, A. F. Henriksen, W. D. Kingery, and D. Finello, "Characterization of Grain Boundary Segregation in MgO," *J. Am. Ceram. Soc.*, **64** [7] 388 (1981).
[25] A. F. Henriksen, "Precipitation in MgO"; Ph.D. Thesis, M.I.T., 1978.
[26] C. A. Handwerker, M. M. O'Connor, R. M. Cannon, and R. L. Coble, "Preparation and Characterization of High Purity MgO Powder Compacts"; in Processing of Metal and Ceramic Powders. Edited by R. M. German and K. W. Lay. TMS-AIME Special Publication, 1980.
[27] M. M. O'Connor, "Effect of Powder Precursor and Alcohol Dispersant on the Surface and Compact Characteristics of MgO Powder"; S. M. Thesis, M.I.T., 1982.
[28] J. Henney and J. W. S. Jones, "Surface Diffusion Studies on UO$_2$ and MgO, "*J. Mater. Sci.*, **3**, 158 (1968).

[29]W. D. Kingery and B. Francois, "The Sintering of Crystalline Oxides: I, Interaction Between Grain Boundaries and Pores"; p. 471 in Sintering and Related Phenomena. Edited by G. C. Kuczynski, N. Hooton, and C. Gibbon. Gordon and Breach, New York, 1967.

*Present address: University of California, Berkeley, CA.
†Atomergic Chemetals Corp., Plainview, NY.
‡Alfa-Ventron Corp., Danvers, MA, chemical assay 99.9%, metallic assay 6N.
§Magiscan, Joyce-Loebl Ltd., Gateshead, U.K.
Note added in press: The numerical coefficient in Eq. (3) becomes 3 at large $G_{max}/\overline{G}$.

Processing of Narrow Size Distribution Alumina

Todd R. Gattuso

Cabot Corp.
Billerica, MA   01821

H. Kent Bowen

Massachusetts Institute of Technology
Cambridge, MA   02139

XA-139 alumina was dispersed in an organic solvent and classified by settling with the aid of a centrifuge. The tape casting and sintering behavior of a narrow size cut centered at $d = 0.2$ μm were compared with those of unclassified XA-139 alumina. It was found that tapes cast from the classified material were more uniform from top to bottom than tapes cast from unclassified material. Sintering experiments performed in a dilatometer showed that the classified material sintered to higher final densities in shorter times and at lower temperatures than unclassified material.

It is well-known that mixtures of coarse and fine particles can lead to high green densities in ceramic bodies.[1,2] However, such structures do not necessarily produce the best fired ware. Davis et al.[1] studied the effects of particle size distribution on compaction and sintering of alumina. They obtained the highest fired densities in samples that had the lowest green densities. Coble[3] recognized that wide particle size distributions can affect the initial stages of sintering. Large particles in a compact inhibit shrinkage among the small particles. The resulting stresses can lead to tearing of contacts rather than coherent shrinkage. In the later stages of sintering, it is important that pores remain on grain boundaries; abnormal grain growth in particular must be avoided.[4] The criterion commonly used for normal grain growth is the Hillert criterion,[5] which states that the largest grain should be smaller than twice the mean size. In this respect, wide initial particle size distributions have been recognized as predisposing toward abnormal grain growth.[6]

Recent work has demonstrated that monosized, spheroidal particles of SiO_2 and TiO_2 can be consolidated into essentially close-packed arrays with uniform green densities.[7,8] These bodies sinter to high densities (>99% of theoretical density) at shorter times and lower temperatures than those commonly used. There are obvious economic advantages to reduced firing schedules, but other benefits also accrue. Lower firing temperatures allow materials substitutions in cofired systems. Normal grain growth is reduced by reducing firing schedules, leading to bodies with higher strength.

While perfectly uniform particles generally are not available, the advantages with respect to processing of particles with uniform size and shape can be demonstrated with commercial powders by sorting them into narrow size distribution cuts. This report describes the classification, casting, and sintering of a commercial aluminum oxide.

Fig. 1. Unclassified XA-139 alumina.

Dispersion

XA-139* SG alumina was chosen for these experiments because of its fine particle size and high purity. The as-received material was dispersed by vibratory milling in isopropyl alcohol with small additions of p-hydroxybenzoic acid (p-HBA), typically 0.7 wt% solids. Ethanol was used as the solvent in a few experiments. This dispersant system was selected to keep formulations simple and nonaqueous. The common tape casting dispersant system of menhaden fish oil in toluene and trichloroethylene[9] was rejected because of the uncertainty of the structure and composition of fish oil and its tendency to age. Also, the required solvents present toxicity problems. Milling was done in polypropylene jars with high-purity alumina media. Milling times ranged from 4 to 6 h.

Classification

Classification was accomplished by centrifugation. Times and rates of rotation were adjusted to yield particles with diameters in the range 0.2–0.3 μm. Calculated settling rates were based on Stokes' law and spherical particles. Settling rates for alumina particles of this size are too low to do classification by gravitational sedimentation. The as-received XA-139* powder was dispersed as described above in a slurry of about 50 vol% solids. The dispersion was diluted to a 15 vol% slurry and then centrifuged until all particles with a diameter greater than 0.3 μm fell out. The supernatant was siphoned into fresh containers and centrifuged until all particles with a diameter greater than 0.2 μm fell out. The supernatant was siphoned off and discarded, and the remaining cake was dried and removed from the centrifuge bottles.

Figure 1 shows unclassified XA-139* alumina, which is a mixture of large platey particles and smaller more nearly equiaxed particles. The fine cut from a

Fig. 2. Fine cut from a centrifuge run.

centrifuge run is shown in Fig. 2, where the narrow size distribution of the particles can be seen. Particles in this cut also are more nearly uniform in shape than the unclassified particles, although they are not spheroidal.

Fine-cut yields were low, only about 10%. A disadvantage of the centrifugal classification method is that fine particles from near the bottom of the dispersion tend to settle out along with coarse particles from near the top of the dispersion. This behavior partially accounts for the low yield of 0.2–0.3 μm particles in the final size cut. The centrifugal classification scheme is summarized in Table I.

Table I. Summary of Centrifugal Classification Scheme

Classification
XA-139*
0.67 wt% solids *p*-HBA
Isopropyl alcohol (50 wt% solids)
Vibramill[†] 4–6 h
Dilute to 15 vol% solids
Centrifuge (650 *g*'s at r_{max})
Decant—save supernatant
Centrifuge (650 *g*'s at r_{max})
Decant—save cake

*Alcoa, Pittsburgh, PA.
[†]Vibramill, Sweco, Inc., Los Angeles, CA.

Tape Casting

The tape casting process is an important part of the semiconductor packaging industry. The production of thin, very smooth insulating layers is vital to producing higher device density. Narrow size distribution powders should permit reaching this goal more readily than material of broad particle size range. Slips were prepared and cast to compare classified and unclassified material.

The as-received XA-139* alumina was dispersed as described above to produce slips with about 40 vol% solids. Because only a small amount of classified material was available, the centrifuged powder was redispersed ultrasonically. Experiments showed that a binder was necessary to allow handling of thin, cast tapes. Poly(vinylbutyral)[†] was used because of its compatibility with isopropyl alcohol and ethanol and because of its moderate plasticity. Concentrations of about 1.3 wt% solids were used with the binder added at the final stage of milling.

Tapes were cast with a laboratory-scale caster[‡] onto either acetate or parafilm sheets. The casting rate was 60 cm/min, and the dry tape thickness typically was 40 μm. Tapes cast onto acetate were peeled off, and bisque was fired at 900°C in air to remove organics and establish some strength. Tapes cast onto parafilm were similarly fired but without being peeled off the substrate. The wax was burned off with the other organics. The tape casting procedure is summarized in Table II.

Figure 3(A) shows the top surface of a tape cast from unclassified XA-139* alumina, and Fig. 3(B) shows the bottom surface of the same tape. The segregation of particles with respect to size from top to bottom is striking. This structure would not densify without curling. A much smoother bottom surface compared to the top surface would also be expected after firing.

In contrast, Fig. 4(A, B) shows the top and bottom surfaces, respectively, of a tape cast with classified material. Because the size distribution in the starting material is narrow, size segregation from top to bottom in the tape cast is substantially reduced compared to the unclassified tape. Packing is quite good in both top and bottom surfaces.

Sintering

Sintering experiments were performed in air in a vertical alumina muffle furnace with platinum windings and a programmable temperature controller. Real time shrinkage was monitored by means of an LVDT attached to a sapphire rod in contact with the specimen. Thermal expansion of the furnace assembly was auto-

Table II. Summary of Tape Casting Procedure

Tape casting
XA-139*
0.67 wt% solids *p*-HBA
Isopropyl alcohol/Ethanol (40 vol% solids)
Vibramill[‡] 4 h
1.3 wt% solids Butvar 76[†]
Vibramill 30 min
Cast onto parafilm (t = 0.002 in.)
Bisque fire

*Alcoa, Pittsburgh, PA.
[†]Butvar, Monsanto Co., St. Louis, MO.
[‡]Vibramill, Sweco, Inc., Los Angeles, CA.

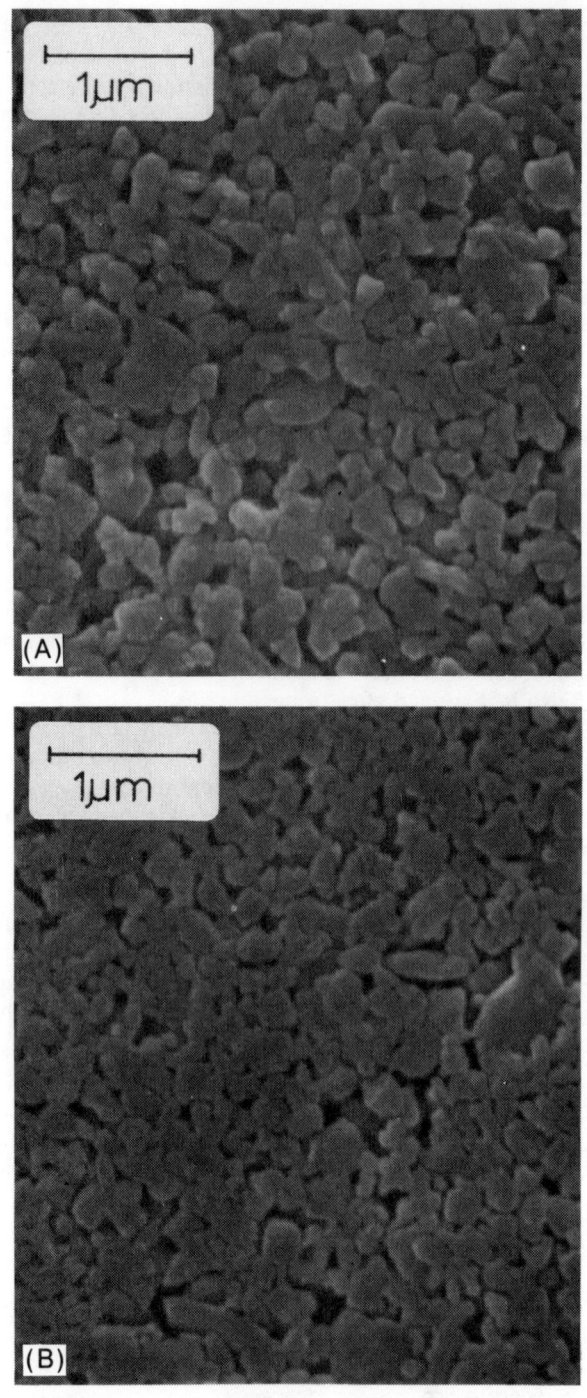

Fig. 3. Top surface of a tape cast from unclassified XA-139 alumina, (A). Bottom surface of the tape shown in Fig. 3(A), (B).

Fig. 4. Top surface of a tape cast from classified XA-139 alumina, (A). Bottom surface of the tape shown in Fig. 4(A), (B).

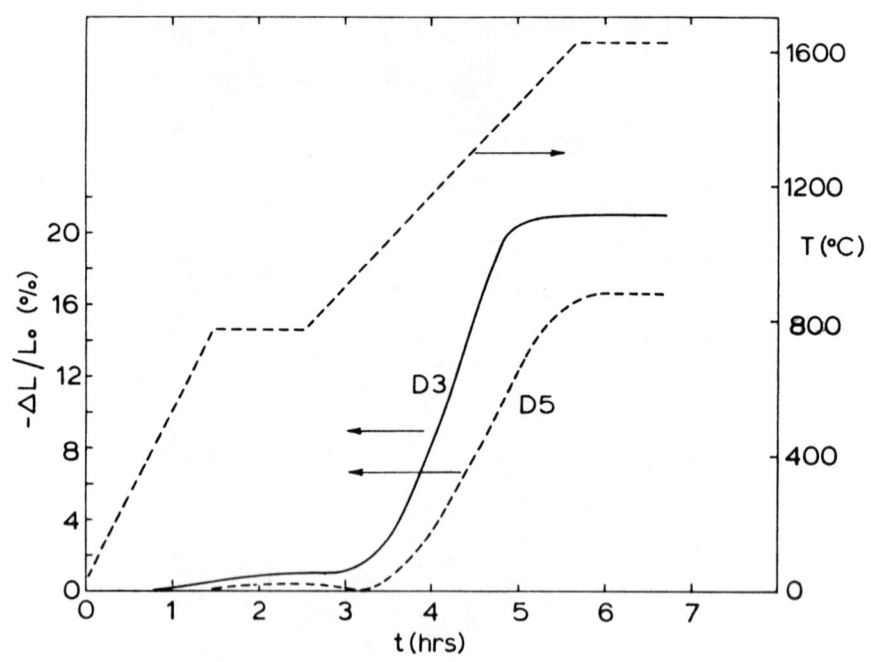

Fig. 5. Temperature and shrinkage vs time for classified (D3) and unclassified (D5) alumina.

matically compensated for with a second LVDT attached to a sapphire rod in contact with the specimen stage. Samples of classified XA-139* alumina were broken out of centrifuge cakes, and two opposite faces were made flat and parallel by abrasion against a high-purity alumina slab. Firing schedules typically included a 1 h soak at 800°C to allow burnout of organics.

Figure 5 shows temperature and shrinkage vs time for centrifugally classified and cast material (sample D3) and unclassified material (sample D5). The samples were heated at 500°C/h to 800°C, held there for 1 h, and then heated at 265°C/h to 1630°C, where they were held for 1 h. The green structure of the classified sample is shown in Fig. 2. Sample D3 was centrifugally cast at a radial acceleration of 650 g's and had a green density of 0.49 theoretical density. The green structure of the unclassified sample is shown in Fig. 1. Sample D5 was prepared by dispersing the as-received XA-139* in ethanol and p-HBA at high volume fraction solids and then allowing the solvent to evaporate. The green density of D5 was 0.58 theoretical density.

The shrinkage of D3 at $T < 900°C$ is typical of the centrifugally classified and cast samples. This shrinkage could be due to the homogeneous approach of particle centers as surface layers are desorbed. Low-temperature shrinkage occurred to a much smaller extent in the unclassified samples. The lower shrinkage in this case is probably a consequence of the complex green structure where particle rearrangement and homogeneous approach are unlikely. The dip in the shrinkage curve of D5 at 1000°C is not real and probably reflects an error in the thermal expansion correction of the dilatometer.

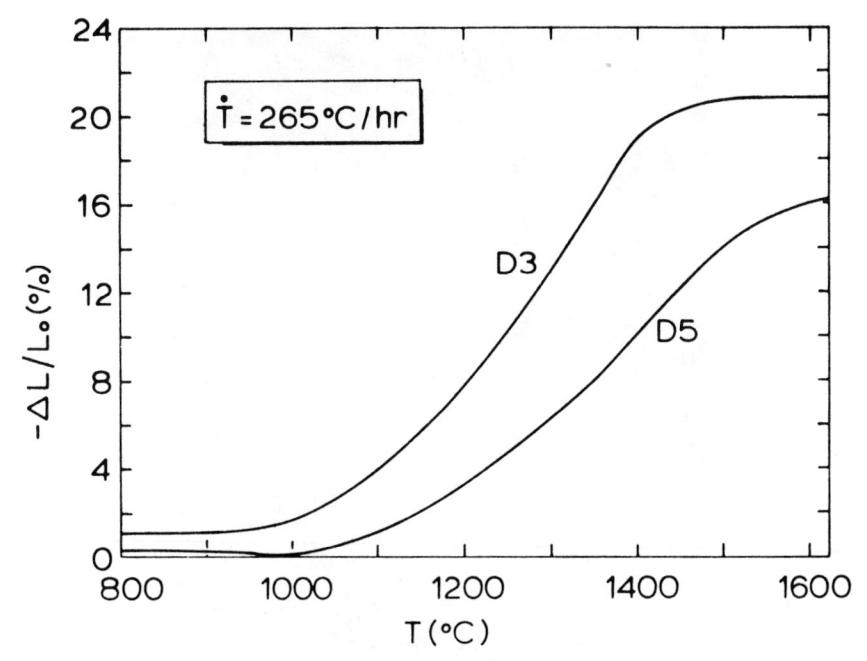

Fig. 6. Shrinkage vs temperature during constant heating rate portion of the firing schedule for samples D3 (classified) and D5 (unclassified).

The narrow size distribution sample D3 reached a final density greater than 0.99 theoretical density. The broader size distribution sample D5 reached a final density of only 0.98 theoretical density. Both samples had stopped shrinking by the end of the firing cycle. Sample D3 had effectively stopped shrinking before the 1630°C soak temperature was reached. This is shown more clearly in Fig. 6, which shows shrinkage vs temperature during heating at 265°C/h between 800 and 1630°C. The difference in the rates of approach to final density between the two samples is also apparent. The uniform particle size of D3 results in uniform size and distribution of pores, all of which are eliminated at approximately equal rates. Thus, shrinkage stops more or less abruptly. The broad particle size distribution of D5 results in a wider range of pore sizes. These are eliminated at a wider range of rates; consequently, the approach to final density is more gradual.

Microstructure

Figures 7 and 8 show fracture surfaces of samples D3 and D5, respectively. It should have been expected from the shrinkage data of Fig. 5 that D3 was overfired, and Fig. 7 shows that extensive grain growth occurred. In pure materials, pores pin grain boundaries, and as high densities are reached, boundary migration becomes uninhibited. The pores trapped within grains in Fig. 7 are of 0.1 μm in size; the few larger pores present seem still to be located on boundaries. Figure 8 shows the onset of abnormal grain growth in sample D5, which had a broad initial particle size distribution. The intragranular pores in D5 are larger than those in D3.

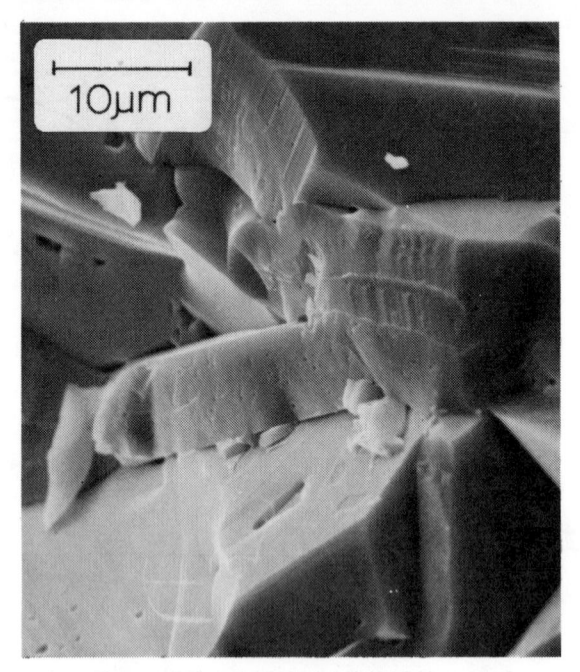

Fig. 7. Fracture surface of fired sample D3.

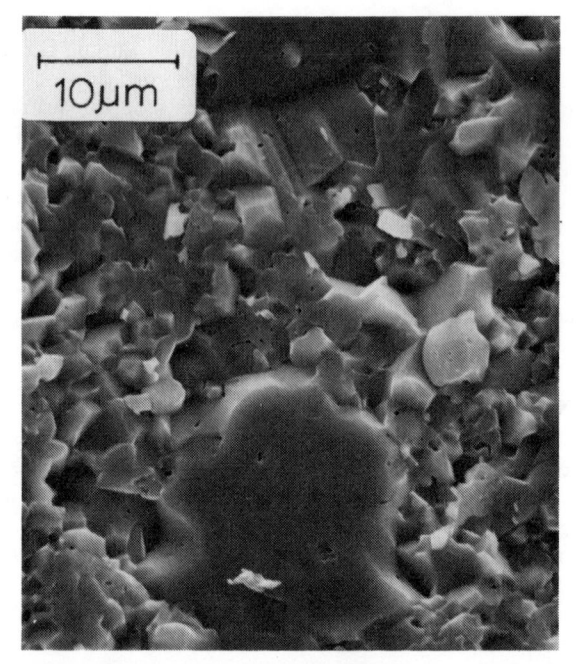

Fig. 8. Fracture surface of fired sample D5.

Fig. 9. Centrifugally classified and cast alumina fired to 1530°C with no soak.

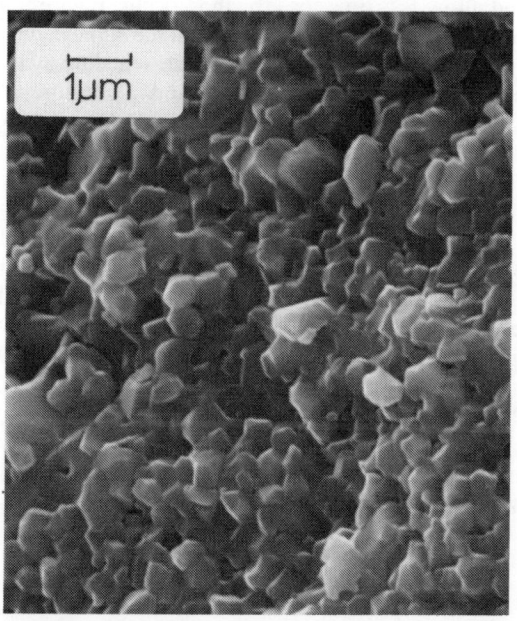

Fig. 10. Centrifugally classified and cast alumina doped with 600 ppm MgO and fired at 1400°C for 1 h.

Fig. 11. Same as that in Fig. 10 except fired for 2 h at 1400°C.

Grain growth can be controlled in at least two ways. The first is to reduce the firing schedule and allow less time for growth to occur. Figure 9 shows a fracture surface of sample D7. Sample D7 was taken from the same centrifuge cake as D3 and fired in the dilatometer according to the same schedule as that of D3 except the temperature was taken to only 1530°C and then down, with no high-temperature soak. Sample D7 shrank 20.2% linearly, slightly less than D3 (20.8%). While some grain growth occurred in sample D7, the distribution remained normal and the microstructure is a tremendous improvement over that of sample D3.

The second method of controlling grain growth is through addition of dopants to act as inhibitors. Centrifugally classified and cast XA-139* alumina was doped with MgO by dissolving magnesium acetate in ethanol and applying drops of the solution to pieces from centrifuge cakes (0.50 theoretical density). These pieces were then fired in air at 950°C to convert the magnesium to MgO. Dopant levels were ~600 ppm MgO.

Figure 10 shows a fracture surface of a sample doped as described above and then sintered by rapidly heating to 1400°C, holding for 1 h, and then rapidly cooling to room temperature. This sample shrank about 18% linearly and the average grain size is less than 1 μm. Figure 11 shows a sample similar to that in Fig. 10, except the soak time at 1400°C was 2 h and the linear shrinkage was nearly 20%.

Summary

XA-139* SG alumina was dispersed in isopropyl alcohol and p-hydroxybenzoic acid and classified by centrifugation. Tapes cast from classified material showed better uniformity from top to bottom than tapes cast from unclassified material. Centrifugally classified and cast material could be sintered to higher densities in shorter times and lower temperatures than unclassified material, despite higher green densities in the unclassified samples. There also was a reduced tendency for abnormal grain growth in the classified material. Normal grain growth in the narrow size distribution material could be controlled by reducing the firing schedule and by MgO additions.

References

[1] J. E. Davis, V. G. Carithers, and D. R. Watson, *Am. Ceram. Soc. Bull.*, **50**, 906–12 (1971).
[2] A. N. Patankas and G. Mendal, *Trans. J. Br. Ceram. Soc.*, **79**, 59–66 (1980).
[3] R. L. Coble, *J. Am. Ceram. Soc.*, **56**, 461–6 (1973).
[4] R. J. Brook; in Ceramic Fabrication Processes, Treatise on Science and Technology, Vol. 9. Academic Press, New York, 1976.
[5] Hillert, *Acta Metall.*, **13**, 227 (1965).
[6] G. Chol, *J. Am. Ceram. Soc.*, **54**, 34–9 (1971).
[7] E. Barringer, N. Jubb, B. Fegley, R. L. Pober, and H. K. Bowen, "Processing Monosized Powders"; Ch. 26 in Ultrastructure Processing of Ceramic Glasses and Composites. Edited by L. L. Hench and D. R. Uhlmann. Wiley & Sons, New York, 1984.
[8] E. A. Barringer and H. K. Bowen, "Formation, Packing, and Sintering of Monodisperse TiO_2 Powders," *J. Am. Ceram. Soc.*, **65** [12] C-199–C-201 (1982).

Research sponsored by MIT's Ceramics Processing Consortium.
*Alcoa, Aluminum Company of America, Pittsburgh, PA.
†Butvar 76, Monsanto Co., St. Louis, MO.
‡Cladan, Inc., San Diego, CA.

Sintering of α-Al$_2$O$_3$ in Gas Plasmas

D. Lynn Johnson, Wayne B. Sanderson, Jennifer M. Knowlton, and Eric L. Kemer

Northwestern University
Department of Materials Science and Engineering
Evanston, IL 60201

Aluminum oxide powders were sintered in gas plasmas generated by RF, microwave, and DC excitation. Rods were passed through the plasmas at a variety of rates up to 6 cm/min, resulting in high densification rates, reasonably high densities, and fine grain sizes. MgO doping increased the densification rate and the final density and decreased the final grain size.

The initial plasma sintering work by Bennett et al.[1,2] indicated that Al$_2$O$_3$ and some other ceramic materials can be sintered successfully and rapidly in plasmas generated by microwave excitation. They observed significantly finer grain sizes and greater densification rates under isothermal conditions than in conventionally sintered materials at the same temperature. Among other things, they hypothesized that the plasma might be enhancing sintering by cleansing the particle surfaces.

The next set of plasma sintering experiments was carried out in a plasma generated in a cylindrical hollow cathode device.[3-5] Little quantitative sintering data were given, except the mention that alumina sintered more rapidly in the plasma than in a conventional furnace at the same temperature. RF induction-coupled plasma (ICP) and rapid pass-through techniques were utilized by Johnson and Rizzo[6] to sinter β''-alumina to high density and fine grain size. In particular, they noted that the conversion of the precursors to the β'' form occurred rapidly and completely during the sintering operation, which took place at a translation rate of up to 2 cm/min. Conventionally sintered tubes using similar starting materials required an extended annealing treatment after sintering to complete the conversion to the β''-alumina form.[7]

Johnson and Kim[8,9] used the same ICP to sinter α-Al$_2$O$_3$. They passed thin-wall tubes and small-diameter rods of both pure and 0.25% MgO-doped powder through an argon plasma at rates up to 6 cm/min for the tubes and 4 cm/min for the rods. The highest density they reported was 99.5% of the X-ray density for MgO-doped tubes translated at 6 cm/min.

Both densification and grain growth rates were extremely high, (1%/s and 0.1 μm/s, respectively), although the final grain size was small (about 5 μm for doped material) because of the short time. Maximum temperatures approached the melting point, although temperatures could not be measured precisely because of the luminosity of the plasma.

Anomalous heating effects were observed. The specimen temperature increased as the rate of translation through the plasma increased. At a given argon pressure and power level, surface melting occurred at sufficiently high translation rates. MgO-doped alumina required higher translation rates to cause melting than did undoped material. If the translation of a specimen was stopped, the temperature dropped approximately 800°C. Finally, a sintered specimen could not be reheated to high temperature upon reinsertion into the plasma. Kramb[10] confirmed these findings and observed further anomalous heating effects. Specimens translated at low rates, i.e., 1 cm/min, and at low argon pressures demonstrated cyclic heating and cooling. A section of tube would achieve high temperature, densify rapidly, and then cool down several hundred degrees. As the specimen moved through the induction coil, the plasma would travel with the sintered portion of the tube. As the plasma reached the bottom of the coil it would suddenly jump upward to the unsintered region, whereupon this region would quickly heat to high temperature, undergo densification, and cool down, and the cycle would be repeated. The resulting tube showed bands of high density separated by bands of low-density material.

To explore the causes of these heating effects, Kramb prepared two types of macroporous specimens by adding organic pore forming materials. In both cases, interconnected porosity existed in a sintered matrix after plasma sintering. The pore sizes were about 100 and <5 μm. In both cases the specimens did not cool down if held stationary in the plasma, and both could be reheated at will upon reinsertion into the plasma. It was therefore demonstrated that the anomalous heating effects were related to the porosity.

Further work has been undertaken in the ICP and in plasmas generated by microwave and DC excitation.

Experimental Methods

Specimens of alum-derived alumina* with specific surface areas of 30 m^2/g (CR30) and 6 m^2/g (CR6) were sintered by translating through these plasma furnaces. The powders were milled 65 h in polypropylene jars in isopropyl alcohol with a single 12-mm diameter high-purity alumina grinding cylinder.[11] The milled powder was dried, mixed with 3% poly(vinylbutyral)† in acetone, dried, and screened through 60 mesh, and 5-mm rods were isostatically pressed at 275 MPa. Powders doped with 0.25% MgO were prepared by addition of magnesium nitrate solutions to the milled powder prior to the addition of binder. The rods were presintered at 600°C to burn out the binder and develop sufficient strength for handling. The magnesium nitrate in the doped rods would decompose to MgO at this time.

A spherical hollow cathode DC plasma furnace was designed and constructed. The inside of the stainless-steel cathode constituted a truncated sphere of 89-mm radius. The anode was spaced 3.2 mm from the cathode on the exterior of the cathode. DC potentials on the order of 600–1500 V, with plasma power of 350–800 W, were applied at gas pressures of 47 Pa. Sintering was attempted in Ar, O$_2$, and CO$_2$.

A microwave plasma sintering furnace has been designed and constructed. A 12-mm or 25-mm air-cooled quartz tube was inserted in a reduced height slotted waveguide that was powered by up to 450 W of microwave energy at 2450 MHz.

Results

It was impossible to heat the alumina rods to high temperatures in Ar in the DC plasma, but the specimens could be heated to the melting point both in O$_2$ and in CO$_2$. Figures 1 and 2 show the relative density of rods translated at 3 cm/min through CO$_2$ plasmas at a variety of power levels. The effect of MgO doping is shown in Fig. 1. The higher density curve for the undoped powder is due primarily to the fact that the temperature of the undoped specimens was greater

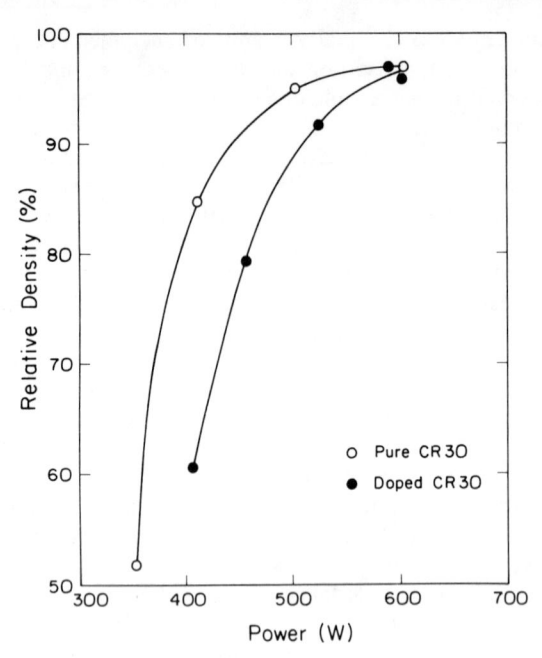

Fig. 1. Relative density of pure and 0.25% MgO-doped alumina translated through the DC CO_2 plasma at 3 cm/min at various power levels.

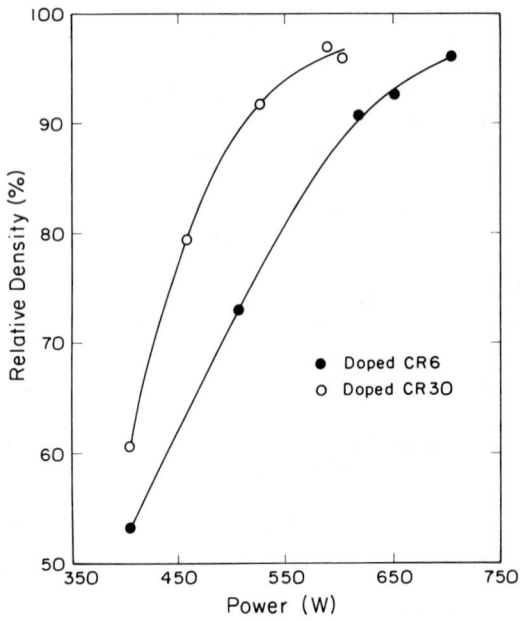

Fig. 2. Relative density as a function of power for rods of differing particle size translated through the DC CO_2 plasma at 3 cm/min.

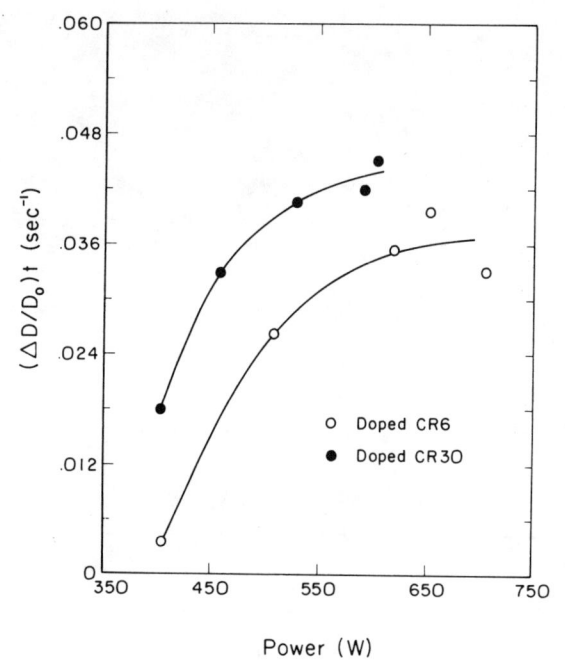

Fig. 3. Average densification rates for rods translated through the DC CO_2 plasma at 3 cm/min at various power levels.

than that of the doped ones. Higher power levels were required to cause surface melting of the MgO-doped specimens compared with the undoped specimens. This is consistent with the ICP result of Kim and Johnson.[9] The reason for this has not been determined.

Average densification rates are shown in Fig. 3. Here the average rate of change of the rod diameter is plotted vs the applied power. The CR30 sintered more rapidly than did the CR6, as expected. Very high linear shrinkage rates are readily apparent.

The anomalous heating effects observed in the ICP were not found in the DC plasma. In fact, solid rods of sintered alumina could be heated readily to the melting point in O_2 and CO_2 plasmas in this apparatus.

Figures 4 and 5 show microstructures of doped and undoped CR30 sintered in oxygen at a pressure of 48 Pa. The power applied was 585 W for the undoped powder and 700 W for the doped material. In both cases the power level was just below that at which surface melting would be observed.

Optimum heating at 450-W input power in the microwave plasma occurred in the 12-mm quartz tube at an Ar pressure of 5300 Pa. Specimens of CR30 translated at 1, 2, or 3 cm/min exhibited approximately the same surface temperature, as measured by optical pyrometry, but the temperature was reduced at 4 and 5 cm/s. Figure 6 shows the fired density as a function of translation rate, and Fig. 7 is a fractograph of the most dense specimen. Spontaneous cooling upon cessation of translation was observed in the argon plasma, but the drop in temperature was only 150°C.

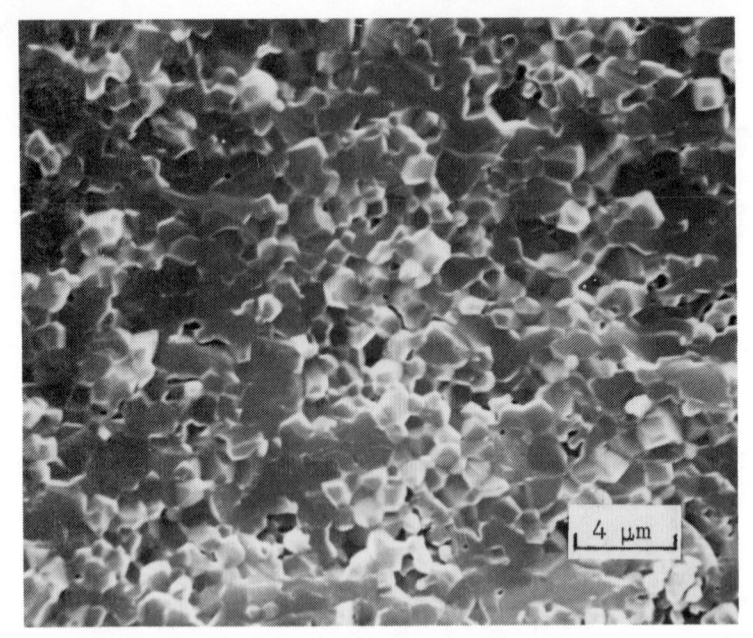

Fig. 4. Fracture surface of MgO-doped CR30 sintered in the DC O_2 plasma at a translation rate of 3 cm/min at 585 W.

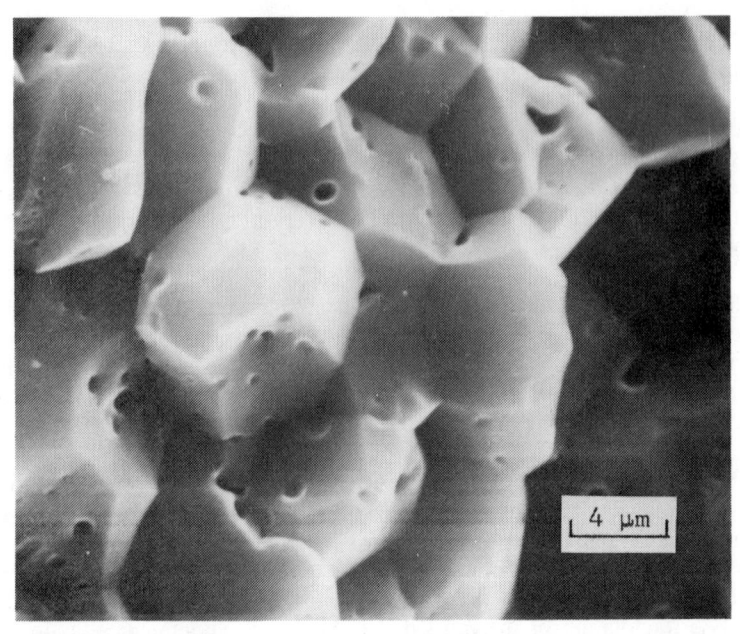

Fig. 5. Fracture surface of undoped CR30 translated through the DC O_2 plasma at 3 cm/min at an applied power of 700 W.

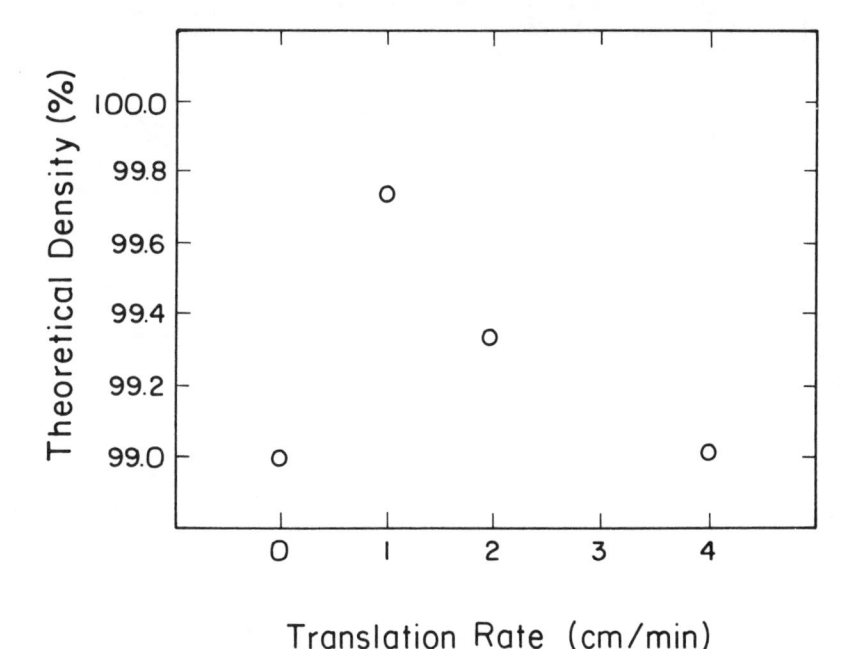

Fig. 6. Relative density of MgO-doped CR30 sintered in the microwave Ar plasma at 450 W at various translation rates.

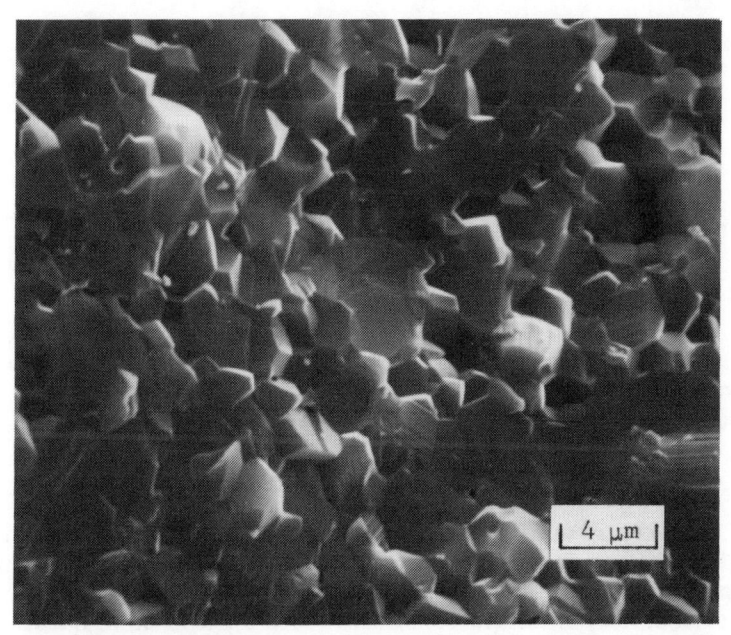

Fig. 7. Fracture surface of the specimen in Fig. 6 that was translated at 1 cm/min.

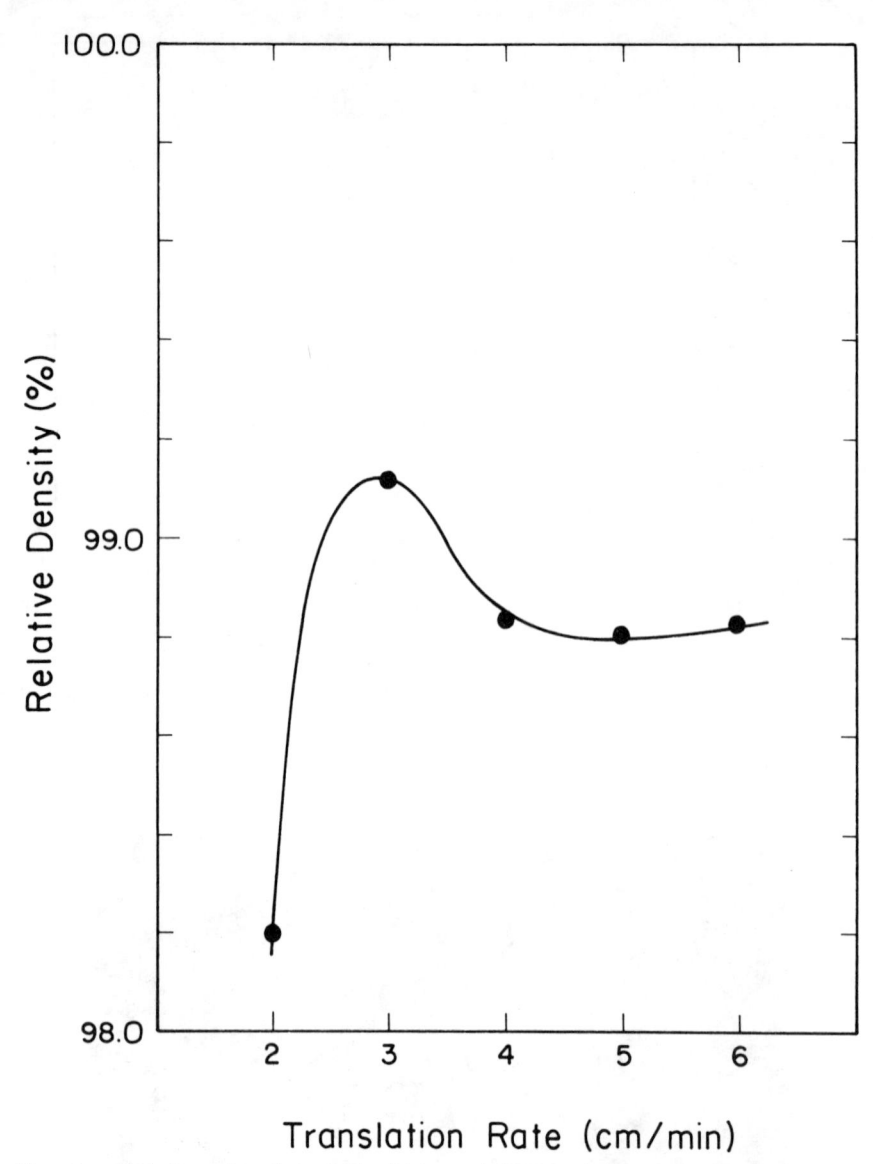

Fig. 8. Relative density of MgO-doped CR30 sintered in the Ar ICP at 75 torr and 5.7 kW and various translation rates.

Rod specimens of CR30 and CR6 were translated through the ICP in argon. The CR6 rods broke up at all combinations of power, pressure, and translation rates attempted, including very slow translation rates. Rods of CR30 translated through the plasma at rates of from 1–6 cm/min showed a maximum in the density at 3 cm/min, with a slightly lower plateau at 4–6 cm/min, as shown in Fig. 8. Figures 9 and 10 are fractographs of 3 and 6 cm/min specimens.

Fig. 9. Fracture surface of the specimen in Fig. 8 that was translated at 3 cm/min.

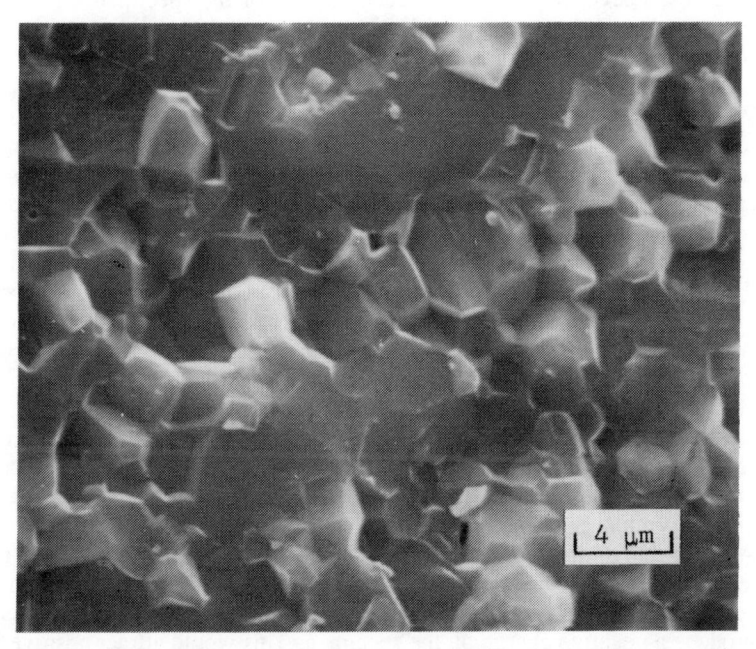

Fig. 10. Fracture surface of the specimen in Fig. 8 that was translated at 6 cm/min.

Discussion

Some similarities and some significant differences exist among the results of sintering in the three plasma furnaces. Alumina rods translated through each of them could be sintered at high rates to >99% of theoretical density with magnesium oxide doping. They each produced specimens with fine grain size. A number of differences were observed, however. For one thing, the pressure range for optimum heating was least for the DC and greatest for the IC plasma. The DC plasma operated best at pressures in the range of 50 Pa. The optimum heating in the microwave plasma was at 5300 Pa, while the ICP could be operated under stable conditions up to atmospheric pressure.

The anomalous heating effects were most pronounced in the argon ICP, where a temperature drop of approximately 800°C was observed when sample translation was stopped. The drop in the case of the microwave argon plasma was only 150°C, while no visually discernable temperature drop was observed in O_2 or CO_2 in the DC plasma.

The shrinkage rate in the DC CO_2 plasma was the highest at a given translation rate, followed by the ICP and, finally, the microwave plasma. This is a manifestation of the steepness of the temperature gradient in each of these plasmas, as governed by the degree of confinement of the plasma. Therefore, the translation rate to produce a given linear shrinkage rate would be greatest for the microwave and lowest for the DC plasma.

The densification rate of MgO-doped alumina in the ICP was observed by Kim and Johnson[9] to be higher than that for undoped material while the opposite was observed in the DC plasma. The temperatures of the magnesia-doped specimens in the DC plasma were clearly significantly lower than those of the undoped powder at the same power level, as manifested both by visual appearance as well as by the power level required to produce melting of the specimen surface. The difference in the grain size in the MgO-doped and undoped powders shown in Figs. 4 and 5 is a manifestation of temperature differences as well as any effect of MgO on grain growth. The cause of this behavior is not known.

While the cause of the anomalous heating effects cannot be clearly defined at the present time, there are some possibilities that might be mentioned. Kramb's[10] results clearly show that the inability to reheat a sintered specimen to the sintering temperature in the ICP is related to the presence of porosity. This effect could arise from at least two causes. If the plasma can be excited within the pores of an unsintered specimen, then the energy transfer from the plasma to the specimen would be significantly enhanced over simple surface contact. This would result in a higher temperature in porous specimens. However, this would necessitate ionization in extremely small volumes, i.e., within the pores of green specimens, which does not seem to be likely. The second cause could be that the heating is related to surface phenomena such as recombination of ions and electrons or surface deexcitation of excited gas atoms or molecules. The amount of heat transferred to the specimen would then be related to the surface area of the specimen, and porous green or partially sintered materials, with their higher surface area, would receive greater energy input and therefore higher temperature.

A factor that may affect the heating behavior is the relative trajectories of the charged particles in the different plasmas. Electrons in the DC plasma, which is generated within a truncated spherical cavity, are focused toward the center and may produce a negative charge on the specimen. This would attract positive ions, which would recombine with the electrons and give at least part of their energy to the specimen. In the other cases, the charged particles move tangent to the surface, either circumferentially in the ICP or longitudinally in the microwave plasma. Heating could be due more to surface roughness. The fact that solid rods can be melted in the DC plasma but not in the others may be an indication that the focused field directs the excited and ionized species more directly to the surface. Further work will be required to make more definitive statements on the heating mechanisms and heating effects.

Conclusions

Sinterable alumina can be sintered readily in gas plasmas generated by DC, radio frequency, and microwave excitation. Linear shrinkage rates in excess of 4%/s can be achieved with final densities above 98% and grain sizes in the range of 2 μm for MgO-doped alumina. Densities in excess of 99% can be achieved at somewhat lower shrinkage rates. Undoped aluminas show larger grain growth, although secondary recrystallization is usually not observed, primarily because of the rapidity with which the temperature excursion takes place. Temperature measurements are difficult, but in both the DC plasma and the ICP, temperatures up to the melting point were achieved.

Acknowledgments

This material is based upon work supported by the National Science Foundation under Grant No. DMR-7918403 and DMR-8216710. This work was conducted in the Ceramics, SEM, and Metallography facilities of Northwestern University's Materials Research Center, supported in part under the NSF-MRL program (Grant No. DMR 79-23573).

References

[1] C. E. G. Bennett, N. A. McKinnon, and L. S. Williams, "Sintering in Gas Discharges," *Nature (London)*, **217**, 1287–8 (1968).
[2] C. E. G. Bennett and N. A. McKinnon, "Glow Discharge Sintering of Alumina"; pp. 408–12 in Kinetics of Reactions in Ionic Systems. Edited by T. J. Gray and V. D. Frechette. Plenum, New York, 1969.
[3] L. G. Cordone and W. E. Martinsen, "Glow-Discharge Apparatus for Rapid Sintering of Al_2O_3," *J. Am. Ceram. Soc.*, **55** [7] 380 (1972).
[4] G. Thomas, J. Freim, and W. E. Martinsen, "Rapid Sintering of UO_2 in a Glow Discharge," *Trans. Am. Nucl. Soc.*, **17**, 177 (1973).
[5] G. Thomas and J. Freim, "Parametric Investigation of the Glow Discharge Technique for Sintering UO_2," *Trans. Am. Nucl. Soc.*, **21**, 182–3 (1975).
[6] D. L. Johnson and R. R. Rizzo, "Plasma Sintering of α Alumina," *Am. Ceram. Soc. Bull.*, **59** [4] 467–72 (1980).
[7] G. E. Youngblood, A. V. Virkar, W. R. Cannon, and R. S. Gordon, "Sintering Processes and Heat Treatment Schedules for Conductive, Lithia-Stabilized β''-Al_2O_3," *Am. Ceram. Soc. Bull.*, **56** [2] 206–10 (1977).
[8] D. L. Johnson and J. S. Kim, "Ultra-Rapid Sintering of Ceramics"; pp. 573–8 in Sintering—Theory and Practice, Material Science Monographs, Vol. 14. Edited by D. Kolar, S. Pejovnik, and M. M. Ristic. Elsevier, Amsterdam, 1982.
[9] Joung Soo Kim and D. Lynn Johnson, "Plasma Sintering of Alumina," *Am. Ceram. Soc. Bull.*, **62** [5] 620–2 (1983).
[10] V. A. Kramb, "Rapid Sintering of α-Alumina in a Radio Frequency Plasma"; M.S. Thesis, Northwestern University, 1983.
[11] R. D. Bagley; private communication, 1983.

---

*Baikowsksi International Corp., Charlotte, NC.
†Butvar-76, Monsanto, St. Louis, MO.

Effect of Magnesia on Grain Growth in Alumina

R. D. Bagley

Corning Glass Works
Research and Development Div.
Corning, NY 14831

D. Lynn Johnson

Northwestern University
Department of Materials Science
Evanston, IL 60201

Small amounts of magnesia added to alumina control grain growth so that alumina can be sintered to a high density and good translucency. The purpose of this research was to examine the complex interrelationship between grain growth and magnesia volatilization during sintering. Magnesia added to alumina is distributed in the alumina grains, as a solid solution at the grain boundaries or as a spinel second phase, typically at trigrain boundaries. During sintering, magnesia volatilizes from the sample surface, progressively depleting the sample of the spinel second phase from the surface inwardly following parabolic kinetics. As the second phase is depleted from the surface of the sample, regular grain growth in the depleted zone increases markedly so that the surface grain size is significantly larger than the grain size in the sample interior where spinel second phase still remains. When all of the second phase volatilizes, spectacular exaggerated grain growth occurs.

The purpose of this study was to examine the interrelationship between grain growth in magnesia-doped alumina and magnesia loss from the alumina by volatilization during sintering.

When amounts of magnesia, as low as 0.025 wt%, are added to alumina, a spinel second phase forms at grain boundaries during sintering. At sintering temperatures, magnesia volatilizes from the sample surface, leaving a zone at the surface that is depleted of spinel second phase. The width of the depleted zone is a function of the amount of magnesia added and the firing time and temperature. The change in depletion zone width with time is essentially parabolic if the grain size is constant.

Many investigators have contributed to the understanding of the role of magnesia as a grain growth inhibitor and as an exaggerated grain growth suppressant in alumina. Several articles and reviews have examined the sintering of alumina or the broader subject of sintering with additives.[1-19]

Experimental Procedure

Sample Preparation

Alumina of a similar type from two different suppliers was used. Linde 0.3 A alumina* was used for most of the magnesia volatility studies, and Baikowski

CR15 AS2 alumina[†] was used for the grain growth experiments discussed in this paper. Both aluminas can be sintered to translucency when doped with magnesia.

Magnesia from 0.025 to 0.8 wt% was added as magnesium nitrate hexahydrate to the alumina powders to provide the desired magnesia dopant levels; 1% sucrose was added as a binder. Pill-shaped samples were isostatically pressed at 138 MPa for magnesia volatility studies and to 69 MPa in a plunger type die with plastic-coated faces for the grain growth studies.

The samples for magnesia volatility and grain growth studies were prefired in air for 12 h at 1050°C and 4 h at 1200°C, respectively, to burn out the binder and to decompose the magnesium nitrate. Samples for the magnesia volatility studies were then fired either to 1600°C for 2 h or to 1850°C for 19 h under vacuum. These samples were then sectioned to expose a surface below the spinel depletion zone. These surfaces were used as references for measuring magnesia volatilization when the samples were used so that the 1600°C prefired samples would undergo significant additional grain growth and the 1850°C prefired samples would undergo little additional grain growth during subsequent firings. The 1600°C, 2-h firing was chosen to produce a small grain sized, nonporous sample that could be cut and ground without absorption of water. Final grinding of all samples was done with distilled water. Figure 1 shows a block diagram of the process used for preparing samples for the magnesia volatility study.

Firings for determining magnesia volatilization losses were mostly made in a tungsten mesh heated vacuum furnace, although some firings were made in hydrogen in a molybdenum wound alumina tube furnace. Magnesia loss results were similar for both hydrogen and vacuum firings.

Firings for grain growth studies described in this paper were done at a later time period than the magnesia loss studies, but the same furnaces were used. Again, most of the firings were done under vacuum. For both magnesia loss and grain growth studies, most samples were suspended from fine molybdenum wires so that all surfaces were exposed.

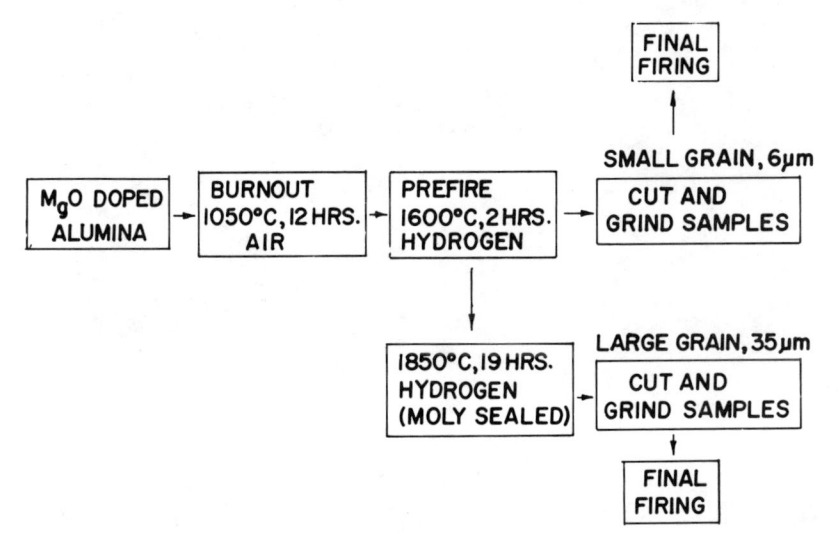

Fig. 1. Flow diagram for preparing samples for magnesia volatility studies.

Magnesia Volatility

Figure 2 shows a typical reflected-light micrograph of polished and thermally etched samples of Linde 0.3 A alumina* containing no magnesia and 0.05 wt% magnesia. No second phase is visually observable, although the second phase can be readily observed by using transmitted light. Figure 3 shows micrographs of samples containing a larger amount of magnesia, 0.4 wt%. The second phase is observable in the reflected-light micrograph. However, since we wanted to examine magnesia dopant levels as low as 0.025% MgO, transmitted light was used. Figure 3 also shows micrographs made with regular transmitted light or phase contrast transmitted light as observed through polished thin section wedges in a matching index oil. Using transmitted-light techniques, we were able to distinguish even small amounts of spinel second phase, corresponding to as little as 0.025 wt% MgO, the smallest amount of MgO added in this study. The phase contrast technique was used for most of the magnesia loss measurements because it provided good visual resolution of the spinel second phase in alumina.

Grain Size Experiments

Micrographs of as-fired surfaces were made to determine surface grain size. Hydrogen-fired samples have clearly visible surface grain boundaries; however, surface grain boundaries for vacuum-fired samples are not as well defined, particularly for short firing times or lower temperatures. Grain sizes were measured by using the line intercept technique and multiplying the result by a factor of 1.5.

To measure the grain size as a function of depth, the surface was progressively ground off, polished, and etched, and grain sizes were measured until the core of the sample containing spinel second phase was reached. Typical depths from the

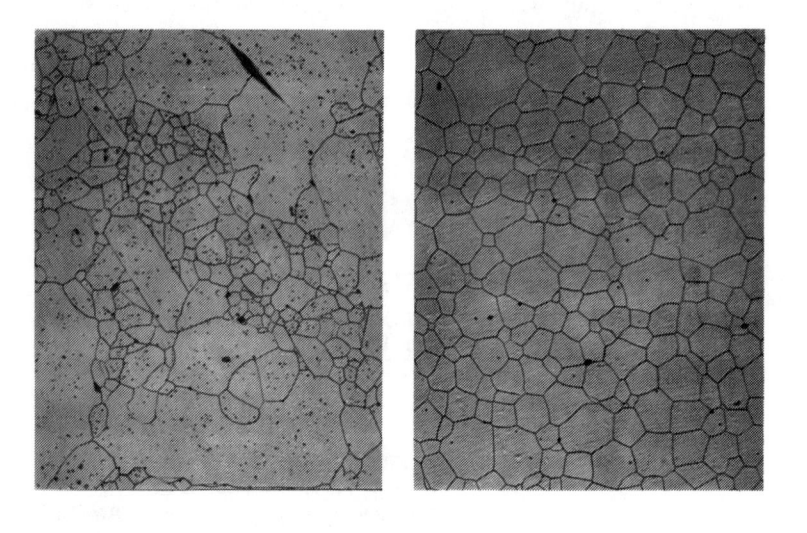

ALUMINA, NO M$_g$O   ALUMINA + 0.05 WT% M$_g$O

100 μm

Fig. 2. Micrographs of alumina with and without magnesia added (1850°C, 20 h under vacuum).

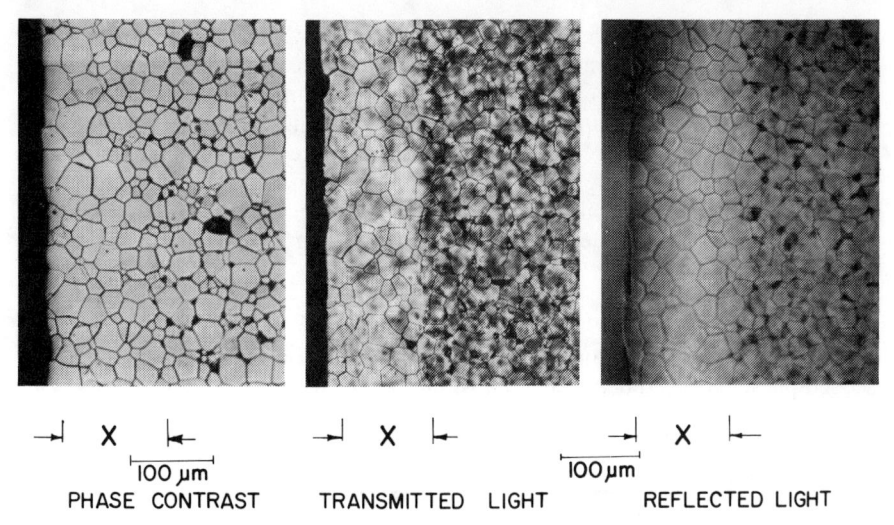

Fig. 3. Micrographs showing spinel second phase in alumina, made with reflected and transmitted light. X is width of the spinel depletion zone in alumina (alumina plus 0.4 wt% MgO; 1850°C; 27 h under vacuum).

surface at which grain sizes were measured were 12.7, 51, 127, 203, 305, 432, and 584 μm. The firing conditions were chosen to produce a spinel depletion zone width wide enough so that several grain size data points could be made before encountering the second-phase zone. Samples of alumina containing 0.05 wt% magnesia were used, since the spinel second phase at this MgO concentration is readily visible in dense samples and the depletion zone could be formed in reasonable times at temperatures of 1700°C and above.

An experiment was made where a porous, as-pressed alumina sample was fired in an enclosed system containing magnesia vapor (see Fig. 4). A molyb-

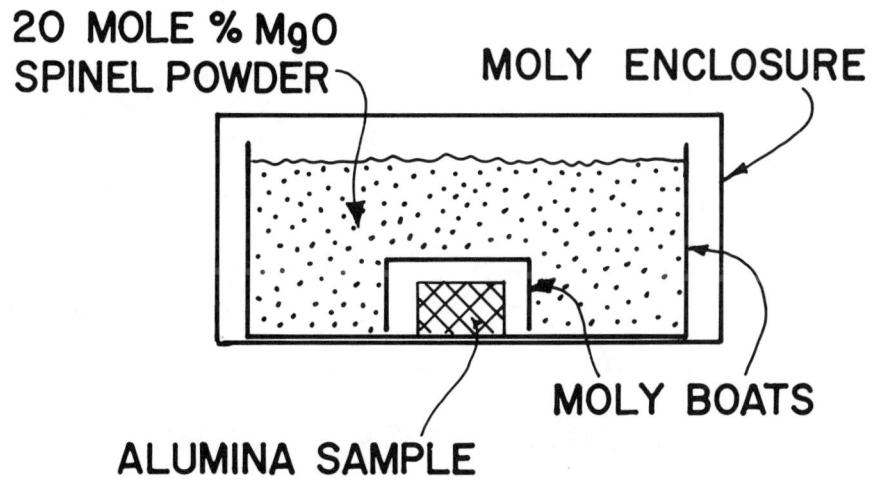

Fig. 4. Setup for saturating alumina with magnesia.

denum cap was placed over the top of the alumina sample to prevent physical contact between the spinel and the alumina. The composition of the spinel powder was chosen so that no spinel second phase would be formed on the outside of the alumina. The spinel encapsulation powder containing 80 mol% alumina and 20 mol% magnesia was prepared by reacting a mixture of ammonium/aluminum alum and magnesium nitrate at 1000°C for 68 h. Johnson and Coble[1] demonstrated, using a similar experimental technique, that a spinel second phase was not necessary to suppress exaggerated grain growth and obtain samples of high density.

Results and Discussion

Magnesia Loss Studies

Figure 3 shows the zone next to the surface of the sample from which the second phase has been depleted. The width "X" of the depleted zone varies with

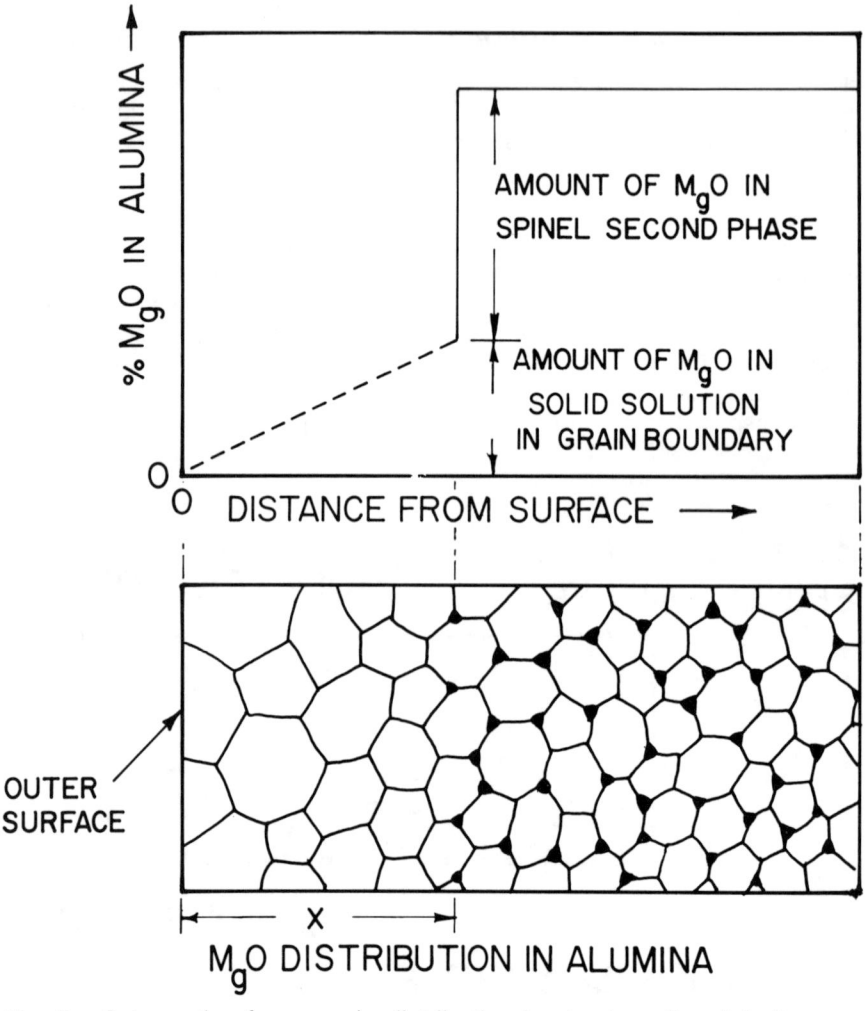

Fig. 5. Schematic of magnesia distribution in alumina after sintering.

firing time and temperature and with the amount of magnesia added to the alumina. Figure 5 shows in schematic detail how the magnesia is distributed in the polycrystalline alumina body after firing at high temperatures where magnesia has volatilized from the surface. The small concentration of magnesia in the grain interior is ignored in this figure.

Figure 6 shows the variation of the depletion zone width with time and with the amount of magnesia added to alumina for an 1800°C firing where the prefire temperature was 1600°C for 2 h. Note that the slope of the line for each different percent magnesia is about 0.32. Significant grain growth occurred simultaneously with magnesia volatilization for samples prefired to 1600°C.

Figure 7 shows the variation in the depletion zone width with time and with the amount of magnesia dopant added to the alumina for samples prefired to 1850°C for 19 h so that, during the subsequent heat treatment at 1800°C, only slight grain growth occurred. The slope of the lines on the plot is about 0.42, which is close to a 0.5 slope that would be expected for diffusion-controlled loss of magnesia.

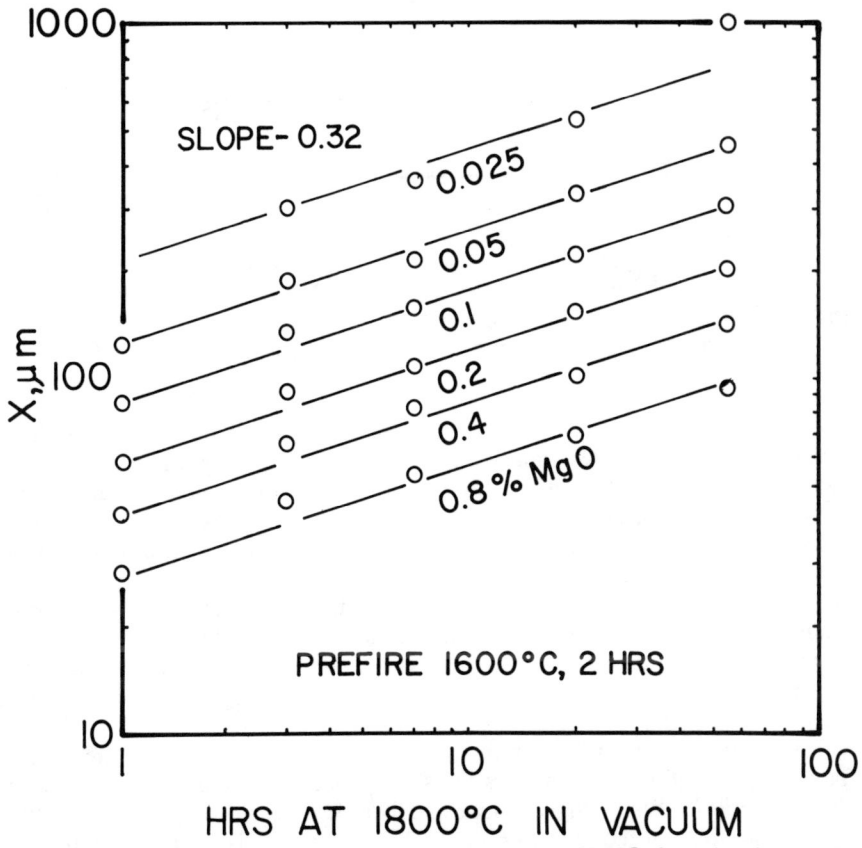

Fig. 6. Spinel depletion distance vs time at 1800°C for aluminas with various magnesia contents and a low-temperature prefire.

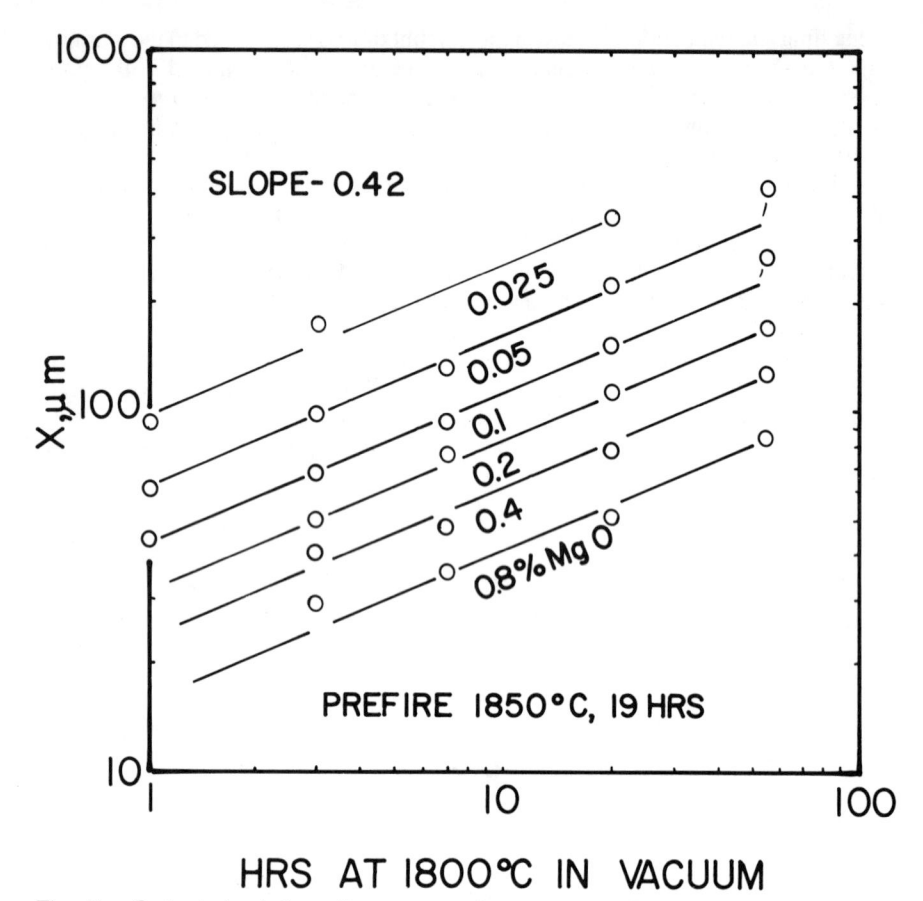

Fig. 7. Spinel depletion distance vs time at 1800°C for alumina with various magnesia contents and a high-temperature prefire.

When Figs. 6 and 7 are compared, it is apparent that the 1600°C prefired samples, which had a finer grain size before the 1800°C magnesia volatilization heat treatment, lost significantly more magnesia for comparable times at 1800°C than the large grain sized samples, which were prefired to 1850°C for 19 h before the 1800°C heat treatment. This indicates that magnesia is lost by diffusion through the alumina grain boundaries. To further test this hypothesis, a thin single-crystal layer was laminated onto the surface of polycrystalline alumina and then tapered to a thin wedge, and high-temperature magnesia volatility experiments were made. No measurable magnesia loss was observed when the polycrystalline alumina surface was covered with a single-crystal sapphire layer as thin as a few micrometers for heat treatments of 25 h at 1850°C under vacuum.

Some alumina volatilizes from the surface during high-temperature firing. The alumina volatilization loss plus the slight amount of grain growth occurring when the 1850°C, 19-h firing was reheated to 1800°C would both tend to slightly decrease the slope of the lines in Fig. 7. The slope would be greater than 0.42 if surface volatilization and the effect of grain growth are considered, which indicates that magnesia loss would be close to parabolic. Since the volatilization rates, as

SURFACE REPLICA   INTERIOR

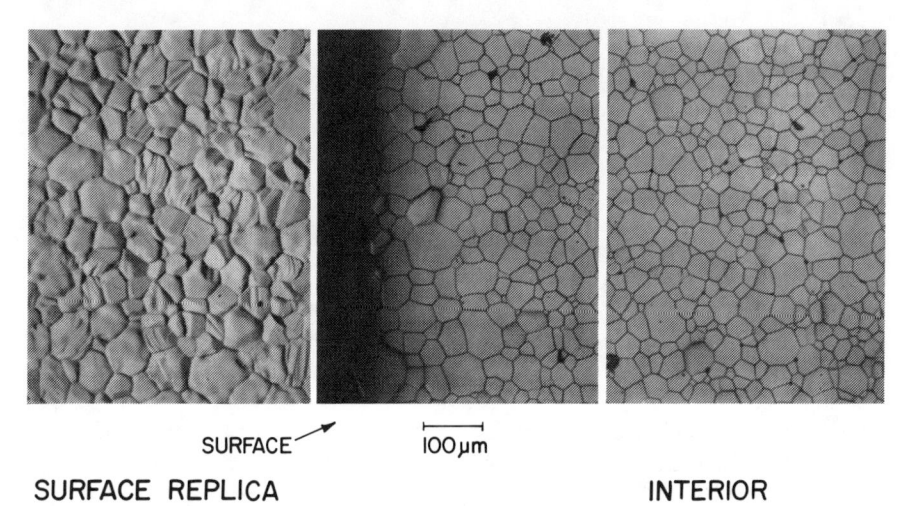

Fig. 8. Micrographs of surface and interior grains (alumina plus 0.05% MgO; 1830°C; 64 h; H$_2$).

indicated by a change in the depletion zone width X, are the same for vacuum and hydrogen and are a function of the grain size, diffusion from the spinel second phase to the surface is probably rate limiting for suspended samples where the MgO leaving the surface can be swept away during firing as under vacuum or in flowing hydrogen. Reduced diffusion rates when the grain size is large and the inhibition of magnesia loss when a thin layer of sapphire single crystal is bonded to the surface of the polycrystalline alumina confirm that magnesia is predominantly lost through the grain boundaries.

Grain Growth

Grain sizes at the surface of magnesia-doped alumina samples where magnesia can readily evaporate were significantly larger than in the interior or core of the sample where second phase spinel was present. Figure 8 shows micrographs of the surface and interior grains and a cross-sectional view from the surface into the sample for a sample containing 0.05% MgO fired for 64 h at 1830°C in hydrogen. The variation in grain size between the surface and interior can be observed visually.

Figure 9 is a plot of 12.7-μm depth (surface) and core grain sizes of alumina samples containing MgO from 0.025 to 0.8 wt% fired for 9 h at 1800°C under vacuum. There was a greater change in grain size with an increase in magnesia content for the surface grains than for the interior grains, where spinel second phase was present. Note that the interior grain sizes did not change greatly even with the magnesia content varying from 0.025 to 0.8 wt%, a factor of 32 increase in MgO. It is also important to note that spinel second phase was not necessary to prevent exaggerated grain growth, even at the low MgO concentration near the surface of the sample.

Figures 10 and 11 show how the grain size changed from the surface to the interior for samples fired at 1800°C in flowing hydrogen or under vacuum. The grain size was essentially constant in the portion of the sample containing spinel second phase. The concentration of magnesia at the grain boundaries was probably

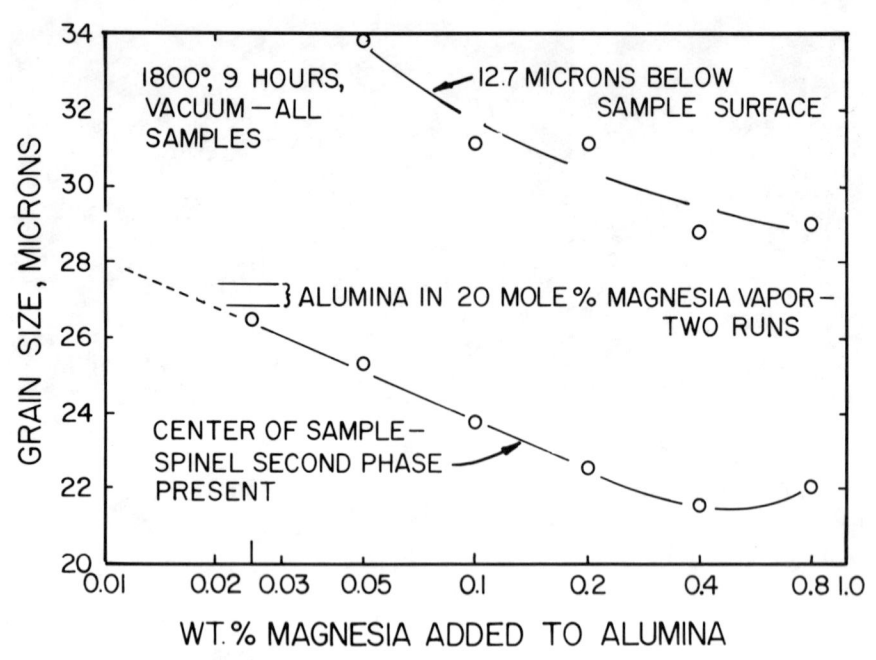

Fig. 9. Effect of magnesia content on grain size of sintered alumina.

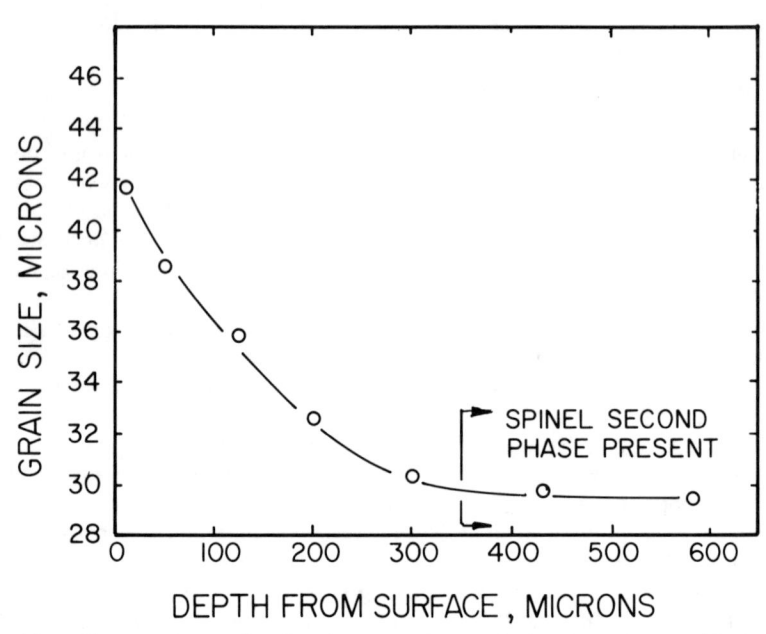

Fig. 10. Change in grain size from surface of alumina sintered in hydrogen (1800°C; H_2; 20 h; alumina plus 0.05 wt% MgO).

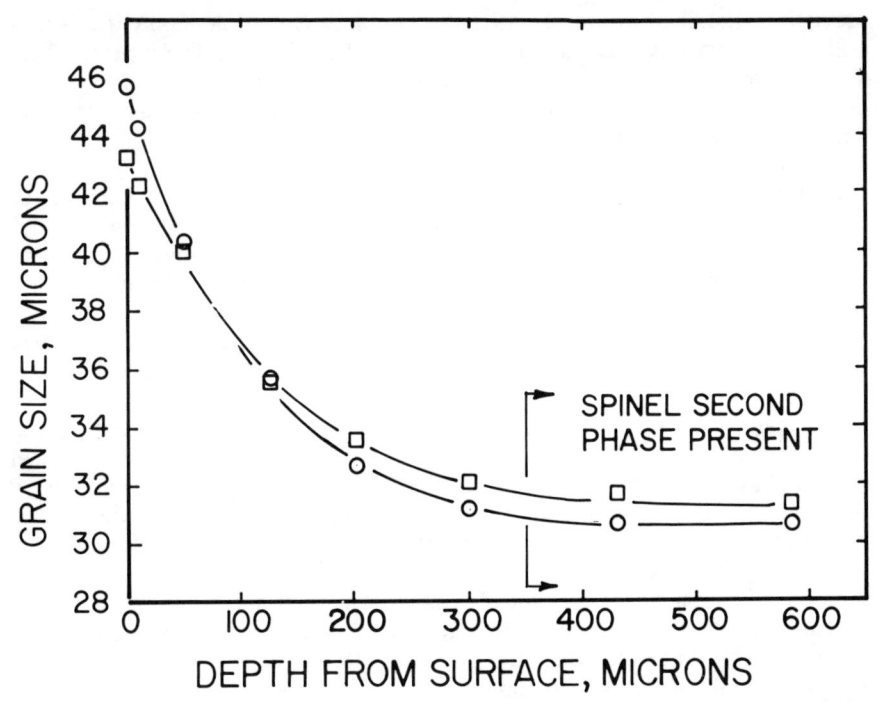

Fig. 11. Change in grain size from surface of alumina sintered under vacuum (1800°C; under vacuum; 20 h; two runs; alumina plus 0.05 wt% MgO).

near zero at the exterior surface in samples fired at high temperatures in flowing hydrogen or under vacuum. This would produce a concentration gradient in magnesia extending from the portion of the sample containing spinel second phase to the surface, as shown in Fig. 5. Thus, grain growth for alumina is controlled by at least two factors where no spinel second phase is present.

First, the rate of change in width of the zone, which is depleted of visibly observable second phase, begins rapidly at first and then becomes progressively slower as the magnesia diffusion distances from the second phase to the surface become greater. Since the grain growth rate is essentially constant in the area containing spinel second phase, grains begin to grow more rapidly only after the spinel second phase disappears. Second, there is a concentration gradient of magnesia in solid solution at the grain boundaries between the area containing spinel second phase and the sample surface, as shown schematically in Fig. 5. The solid solution at the grain boundaries could also control grain growth as a function of the magnesia concentration in solid solution at the grain boundary, with higher concentrations of magnesia at the grain boundaries retarding grain growth to a greater extent. The maximum magnesia solubility at the boundary would occur in the region where there is spinel second phase, and then the amount of magnesia at the grain boundaries would decrease toward the surface.

To further explore the effect of MgO solid solution at the grain boundaries with no spinel second phase present, porous, as-pressed alumina samples with no magnesia dopant or binder were enclosed in a magnesia-containing atmosphere, as previously described in Fig. 4. After the sample was fired at 1800°C for 9 h, a polished section was made through the center of the sample. The interior of the sample contained large exaggerated grains, but the outer rim of the sample was dense and translucent and contained grains with a normal growth habit. Johnson and Coble observed similar results.[1]

In Fig. 9, the normal grain sizes of two different samples fired in 20 mol% magnesia spinel vapor are compared with the surface and interior grain sizes of samples containing from 0.025 to 0.8 wt% magnesia fired for the same time and temperature. Note that the grain size of the samples fired in the spinel atmosphere are slightly larger than that of the interior grains of the samples doped with 0.025 wt% MgO, the smallest amount of MgO used in this study.

The sample fired in the enclosed magnesia atmosphere probably absorbed magnesia preferentially at the grain boundaries until they were saturated. This would be the same condition that exists at most of the grain boundaries in alumina that contain spinel second phase, since the second phase is usually segregated at trigrain boundaries as discrete spinel grains. Grain boundaries next to the spinel second-phase particulates would be saturated with magnesia in solid solution.

The increase in the amount of spinel second phase reduced both the surface and interior grain sizes as shown in Fig. 9. This was in addition to the grain size reduction by magnesia in solid solution at the grain boundaries (Figs. 10 and 11). An approximation of the amount of magnesia in solid solution when the grain boundaries are saturated with magnesia in a spinel atmosphere can be obtained from Fig. 9 by observing where the grain sizes for the samples fired in the spinel atmosphere fall on the extrapolation of the plot of the grain sizes where spinel is present. A solubility of about 0.015–0.02 wt% magnesia was indicated.[19] This would represent magnesia both in solid solution at the grain boundaries and in the bulk. The amount of magnesia absorbed by the alumina sample in the spinel atmosphere would depend on the grain size and temperature if samples were held at temperatures long enough to achieve equilibrium.

Although the spinel second phase decreased the grain size, the magnesia in solid solution had a dominant effect in controlling both normal and exaggerated grain growth. This would also explain why increasing the magnesia content from 0.025 to 0.8 wt% did not decrease the grain size markedly. The spinel second phase does have some effect on reducing the grain size, but it is not necessary for the spinel second phase to be present to prevent exaggerated grain growth. The spinel second phase acts primarily as a reservoir to provide magnesia to the grain boundaries. To further demonstrate this, a sample was prepared with a thickness so that, when it was fired for a certain time and temperature, all of the spinel second phase would be depleted by evaporation. The time required for this to happen can be predicted from a plot such as those in Figs. 6 or 7. The thickness of the sample would need to be less than 2 times the depleted zone width X shown on the plot for a particular time and temperature. When the last of the spinel second phase was depleted and magnesia in solid solution at the grain boundaries evaporated, large grains suddenly grew throughout the sample (see Fig. 12). The size of these grains was unpredictable and nonuniform, but typically was up to several hundred micrometers in diameter. These "postmagnesia" exaggerated grains were typically much larger and contained less porosity than exaggerated grains that grew in sintered alumina which did not contain magnesia. The reason for the lower porosity in the postmagnesia exaggerated grains was due to the period of time the grains grew under normal growth conditions where porosity could be removed at grain boundaries.

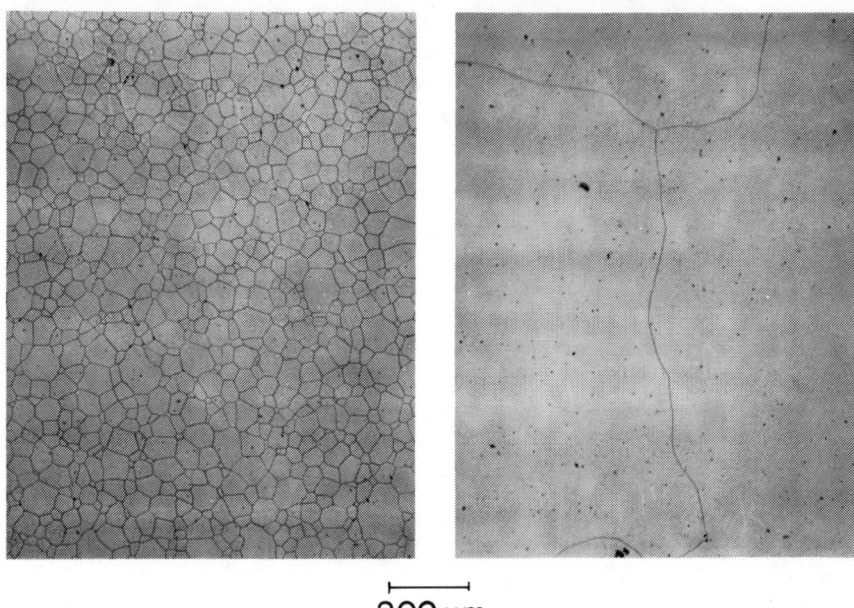

THICKER THAN 2X   THINNER THAN 2X

Fig. 12. Micrographs of alumina samples showing the effect of complete spinel second-phase loss (alumina plus 0.05 wt% MgO; 25 h; 1850°C; under vacuum).

Grains as large as 1 cm in length have been observed in postmagnesia exaggerated grain growth. The reason for grains larger in this than in undoped alumina may be due to the nature of the magnesia depletion front as the last of the spinel second phase in the center of the sample is absorbed by the adjacent grain boundaries and diffuses to the surface and evaporates. A minimum amount of magnesia in solid solution at the boundary is probably necessary for suppressing exaggerated grain growth. When the magnesia content at the grain boundary falls below this level, exaggerated grains can grow. If a single large grain begins to grow in an exaggerated mode, it will follow the minimum magnesia front through the sample, sweeping through the normal grains. Typically, several large grains are nucleated and eventually impinge on each other or in some cases trap islands of normal growth within the sample.

Conclusions

MgO preferentially diffuses at grain boundaries in alumina. Magnesia in solid solution at grain boundaries in alumina can control both normal and exaggerated grain growth without the presence of a spinel second phase. Increased amounts of spinel second phase decrease the grain size moderately, in addition to the magnesia solid solution effect. The spinel second phase acts as a magnesia-containing reservoir to supply magnesia to grain boundaries as magnesia is lost from the sample surface by evaporation. Surface grain growth is more rapid than grain growth where spinel second phase is present. Grain sizes are similar for samples containing low concentrations of magnesia, about 0.025 wt%, and those fired in a magnesia-containing atmosphere. Less than 0.025 wt% (250 ppm) magnesia is required to saturate alumina grain boundaries for coarse-grained samples. When the spinel second phase is totally depleted from the sample, spectacular exaggerated grain growth occurs.

References

[1] W. C. Johnson and R. L. Coble, "A Test of the Second-Phase and Impurity-Segregation Models for MgO-Enhanced Densification of Sintered Alumina," *J. Am. Ceram. Soc.*, **61** [3–4] 110–4 (1978).

[2] H. P. Cahoon and C. J. Christensen, "Sintering and Grain Growth of Alpha-Alumina," *J. Am. Ceram. Soc.*, **39** [10] 337–44 (1956).

[3] R. L. Coble and J. E. Burke, "Sintering in Ceramics"; pp. 197–251 in Progress in Ceramic Science, Vol. 3. Edited by J. E. Burke. Pergamon, New York, 1964.

[4] P. J. Jorgensen and J. H. Westbrook, "Role of Solute Segregation at Grain Boundaries during Final-Stage Sintering of Alumina," *J. Am. Ceram. Soc.*, **47** [7] 332–8 (1964).

[5] P. J. Jorgensen, "Modification of Sintering Kinetics by Solute Segregation in Al_2O_3," *J. Am. Ceram. Soc.*, **48** [4] 207–10 (1965).

[6] S. K. Roy and R. L. Coble, "Solubilities of Magnesia, Titania, and Magnesium Titanate in Aluminum Oxide," *J. Am. Ceram. Soc.*, **51** [1] 1–6 (1968).

[7] H. L. Marcus and M. E. Fine, "Grain Boundary Segregation in MgO-Doped Al_2O_3," *J. Am. Ceram. Soc.*, **55** [11] 568–70 (1972).

[8] G. Rossi and J. E. Burke, "Influence of Additives on the Microstructure of Sintered Al_2O_3," *J. Am. Ceram. Soc.*, **56** [2] 654–9 (1973).

[9] A. Mocellin and W. D. Kingery, "Microstructural Changes During Heat Treatment of Sintered Al_2O_3," *J. Am. Ceram. Soc.*, **56** [6] 309–14 (1973).

[10] R. I. Taylor, J. P. Coad, and R. J. Brook, "Grain Boundary Segregation in Al_2O_3," *J. Am. Ceram. Soc.*, **57** [12] 539–40 (1974).

[11] W. C. Johnson and D. F. Stein, "Additive and Impurity Distributions at Grain Boundaries in Sintered Alumina," *J. Am. Ceram. Soc.*, **58** [11–12] 485–8 (1975).

[12] J. G. J. Peelen, "Influence of MgO on the Evolution of Microstructure of Alumina," *Mater. Sci. Res.*, **10**, 443–53 (1975).

[13] P. Nanni, C. T. H. Stoddart, and E. D. Hondros, "Grain Boundary Segregation and Sintering in Alumina," *Mater. Chem.*, **1**, 297–320 (1976).

[14] R. I. Taylor, J. P. Coad, and A. E. Hughes, "Grain Boundary Segregation in MgO-Doped Al_2O_3," *J. Am. Ceram. Soc.*, **59** [7–8] 374–5 (1976).

[15] W. C. Johnson, "Mg Distributions at Grain Boundaries in Sintered Alumina Containing $MgAl_2O_4$ Precipitates," *J. Am. Ceram. Soc.*, **61** [5–6] 234–7 (1978).

[16] A. H. Heuer, "The Role of MgO in the Sintering of Alumina," *J. Am. Ceram. Soc.*, **62** [5–6] 317–9 (1979).

[17] F. Yan, "Microstructural Control in the Processing of Electronic Ceramics," *Mater. Sci. Eng.*, **48**, 63–72 (1981).

[18] R. J. Brook, "Fabrication Principles for the Production of Ceramics with Superior Mechanical Properties," *Proc. Br. Ceram. Soc.*, **32**, 7–24 (1982).

[19] J. E. Burke, K. W. Lay, and S. Prochazka, "The Effect of MgO on the Mobility of Grain Boundaries and Pores in Aluminum Oxides," *Mater. Sci. Res.*, **13**, 417–25 (1980).

---

*Linde Div., Union Carbide Co., New York, NY.
†Baikowski International Corp., Charlotte, NC.

Use of Solid-Solution Additives in Ceramic Processing

M. P. Harmer

Lehigh University
Materials Research Center
Bethlehem, PA 18015

The use of solid-solution additives continues to show promise as a practical means for controlling microstructure development in ceramic systems. In this paper the solid-solution additive approach to sintering is reviewed with the aid of a new form of final-stage sintering diagram. The system Al_2O_3-MgO is used as a case study for testing the principles involved and for highlighting useful areas for future work.

There has developed an increasing tendency to approach sintering from two very different viewpoints. The first approach is classical in nature and may best be described as "medicinal." With this approach it is recognized that most conventionally processed* powder compacts will, if untreated, develop "ailments" during the firing stage (pore breakaway from moving grain boundaries and abnormal grain growth being the two most commonly occurring "diseases" of sintering (Fig. 1)). A cure is then sought in the form of a sintering additive, which gives the system a high degree of immunity from such effects. In the case of Al_2O_3 sintering, MgO doping has proved to be a most effective additive cure in this sense (Fig. 2).

The additive approach is medicinal in every sense; some systems are more prone to "disease" than others, different systems can respond quite differently to the same treatment, and more encouragingly, a sound scientific basis for finding effective additive cures for new systems is beginning to emerge.[1-3] Finally, there exists among the supporters of this approach, the firm belief that an effective additive cure can be found for the majority of host systems.

The second approach has arisen out of despair for the limited number of systems for which suitable host–additive combinations have been found to work (the system Al_2O_3-MgO has proved to be a rather exceptional case). This second approach may best be described as "infallible." Here great care is taken to form a green compact so perfect[4] (monosized powders, close packed, agglomerate free, etc.) that a good result after firing is practically ensured. Hot-pressing also falls into this category where, in this case the application of high external pressure (brute force!) is used to guarantee a good result. The problem with the infallible approaches is that they tend to be very expensive, less convenient, and more restrictive in terms of the range of sample geometries that can be fabricated.

In this paper the additive approach to sintering is reviewed with emphasis on the use of solid-solution dopants. A new form of final-stage sintering diagram is used to demonstrate the principles of additive effects on microstructure development. The analysis is helpful in establishing general guidelines for the prediction of dopant effects and in lifting some of the mystique associated with the use of

679

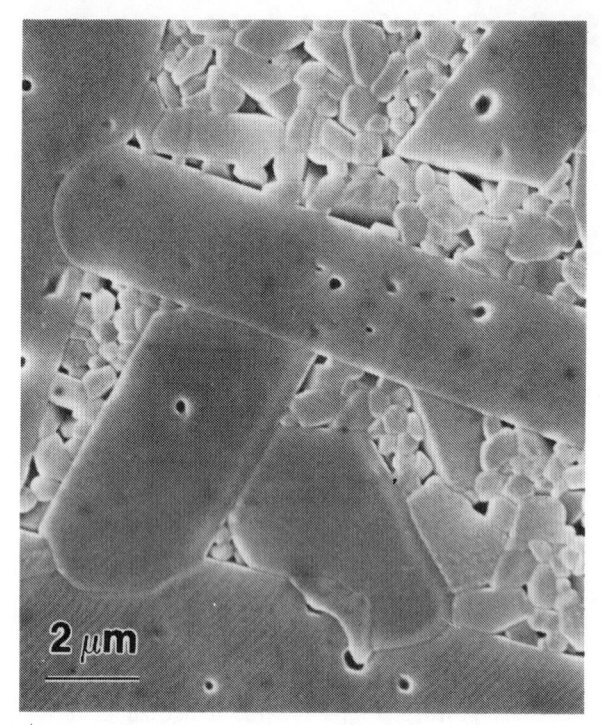

Fig. 1. Abnormal grain growth and pore entrapment in an undoped alumina ceramic.

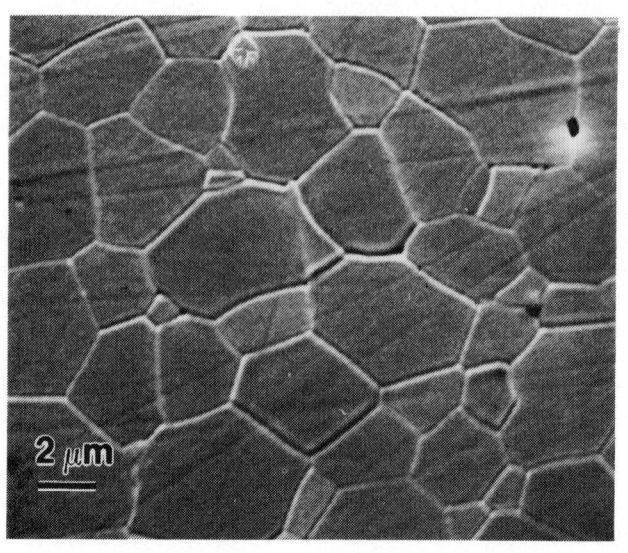

Fig. 2. Uniform fine grain microstructure of a sintered MgO-doped (200 ppm) alumina ceramic.

solid-solution additives. The system Al_2O_3-MgO is used as a case study to demonstrate these principles and to highlight important areas for future work.

Principles

The effect of an additive on microstructure development is felt most strongly through its influence on the various diffusivity and mobility terms: $D_L, D_b'', D_s,$ and M_b, where D_L corresponds to lattice diffusion, D_b'' to diffusion parallel to the grain boundary (diffusion width δ_b''), D_s to surface diffusion (diffusion width δ_s), and M_b to grain-boundary mobility. Various approaches have been used to predict the effect of process variables such as these on microstructure development, the most helpful of which are the mapping approaches, some of which are reviewed briefly below.

Temperature–Density Maps

In this approach developed by Ashby,[5] diagrams are constructed that identify, at a given temperature, grain size, and neck size, the dominant mechanism and show the rate of sintering that all the mechanisms, acting together, produce. The results are plotted in the form of a map of temperature vs density (the ratio of the neck to particle radius), defining areas in which a single mechanism predominates. Since this type of map is restricted to conditions of negligible grain growth, however, it is difficult to use it to follow the development of a microstructure with time.

Surface Area–Density Plots

Plots of surface area vs density have been proposed by Burke et al.[6] as a helpful way to look at the effect of process variables on microstructure development. The approach has been developed only qualitatively thus far, although it does lend itself readily to quantification in a manner similar to that to be described in Grain Size–Density Plots. In this approach sintering is represented by curved trajectories on a plot of surface area vs density. The more convex the trajectory (with respect to the density axes), the greater the contribution from coarsening (see Fig. 3). Since surface area is not easily measured during the final closed pore stage of sintering, this approach is restricted to studying the initial and intermediate stages of sintering.

Grain Size–Pore Size Maps

For the case of idealized microstructures and regularly spaced spherical pores, Brook[7] developed a diagram showing the types of pore-boundary interaction that can occur under different conditions of pore size and grain size (see Fig. 4). The important feature of the diagram is the prediction of a range of pore size and grain size values within which pores will separate from grain boundaries to become isolated within the grains, thus preventing the attainment of full density. Additives that reduce the area of the separation region minimize the risk of developing pore breakaway and abnormal grain growth. If the pores move by surface diffusion, for example, a favorable additive will be one that acts to raise D_s or lower M_b. Modified versions of the Brook diagram have been developed to consider the effects of grain size distribution[8] and a constant interpore spacing.[9] Evans et al.[10–12] have recently conducted a more rigorous analysis of the breakaway problem by considering the actual pore and grain-boundary shape changes that occur during pore-boundary motion. Evans found that the separation conditions for five-sided grain configurations differed from the predictions of the earlier "phenomenological descriptions" by less than a factor of 2. His analysis for the three-sided grain configuration predicts a different form of pore-breakaway map.

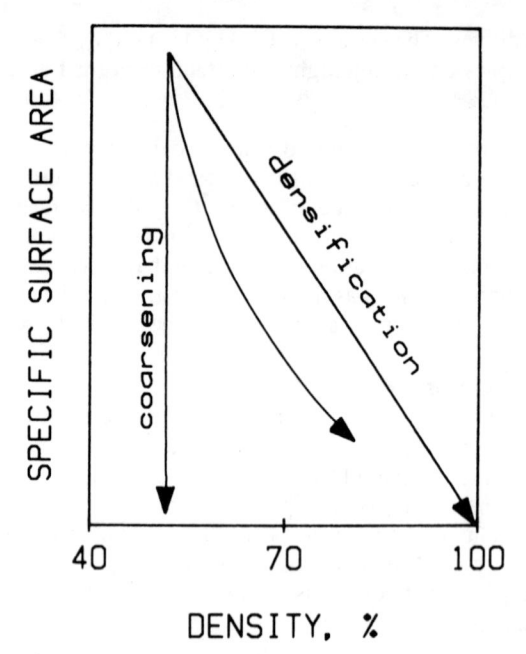

Fig. 3. Schematic representation of microstructure development as seen here in terms of a plot of surface area vs density after Burke et al. (Ref. 6).

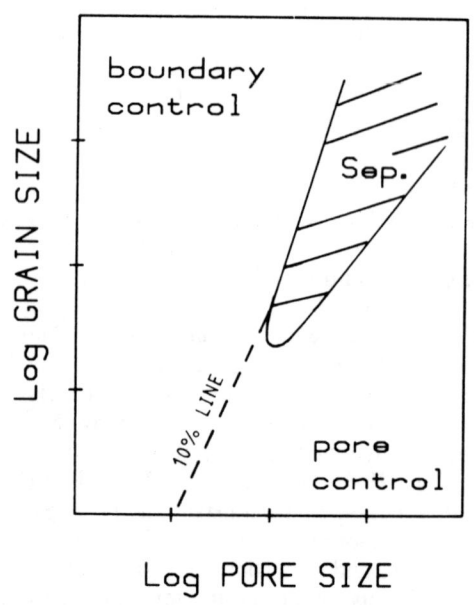

Fig. 4. Schematic representation of the dependence of the type of pore-boundary interaction on pore size and grain size for the case where pores move by surface diffusion, after Brook (Ref. 7).

Grain Size–Density Plots

In this approach, proposed by Yan et al.,[2] a model of simultaneous densification and normal grain growth is used to construct density–grain size trajectories during sintering. The procedure is to obtain expressions for the relative coarsening rate, $d \ln G/dt$, and the relative densification rate, $d \ln \rho/dt$, and to construct a plot of grain size (G) vs density (ρ), based on the ratio of these two terms, i.e., $d \ln G/d \ln \rho$ for a given set of starting conditions (initial grain size G_0, density ρ_0).

For pure densification $d \ln G$ is zero, corresponding to a completely flat trajectory on the G–ρ plot. For pure coarsening $d \ln \rho$ is zero, corresponding to an infinitely steep trajectory on the G–ρ plot. Under normal sintering conditions ($0 < dG/d\rho < \infty$), the actual trajectory will lie somewhere between these two extreme cases (see Fig. 5). The effect of dopants is seen through their influence on the various diffusivity terms included in the relative densification rate–coarsening rate ratio. To take an example, for densification controlled by lattice diffusion in the final stages of sintering[13]

$$\frac{d \ln \rho}{dt} = \text{constant}\left[\left(\frac{1}{\rho}\right)\frac{D_L \gamma_{s/v} \Omega}{G^3 RT}\right] \tag{1}$$

(here $\gamma_{s/v}$ is the solid–vapor interfacial energy and Ω is the molar volume) and for normal grain growth controlled by pore drag with the pores moving by surface diffusion[14]

$$\frac{d \ln G}{dt} = \text{constant}'\left[\left(\frac{D_s \delta_s}{RTG^4}\right)\frac{\gamma_{gb}\Omega}{(1-\rho)^{4/3}}\right] \tag{2}$$

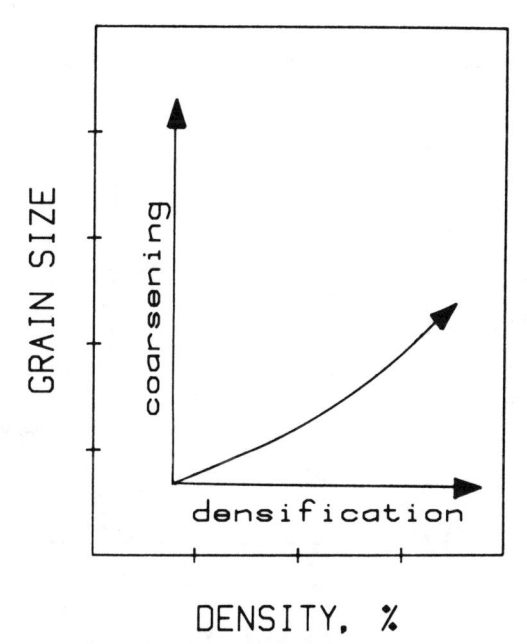

Fig. 5. Schematic representation of microstructure development as seen here in terms of a plot of grain size vs density, after Yan et al. (Ref. 2).

where γ_{gb} is the grain-boundary energy. The ratio term is given as

$$\frac{d \ln G}{d \ln \rho} = \text{constant}'' \left[\frac{D_s \delta_s \gamma_{gb}}{D_L \gamma_{s/v}} \left(\frac{1}{G} \right) \frac{\rho}{(1-\rho)^{4/3}} \right] \tag{3}$$

A favorable additive would be, for these circumstances therefore, one that raises D_L and/or lowers D_s.

Of the four approaches just described, the latter two (grain size–pore size maps and grain size–density plots) are the most useful for considering the effects of additives on microstructure development during late-stage sintering. Recognizing, however, that what may be good for flattening the trajectory on a $G-\rho$ plot (lowering D_s for example) may not be good for averting breakaway on a $G-r$ plot, we developed a new diagram combining both of these approaches.

Microstructure Development Maps

A microstructure development map consists of a plot of grain size vs density with the pore-breakaway area and sintering trajectories superimposed. A hypothetical diagram is shown in Fig. 6. The map is divided into regions of different simultaneous densification, coarsening mechanism pairs with the respective mechanisms indicated by a diffusion coefficient pair, $D_{dens}:D_{coars}$. To take some examples, $D_L:D_s$ corresponds to lattice-diffusion-controlled densification and surface-diffusion-controlled coarsening (pore drag), $D_b'':D_L$ corresponds to boundary-diffusion-controlled densification and lattice-diffusion-controlled coarsening (pore drag), $D_L:D_b^*$ corresponds to lattice-diffusion-controlled densification and boundary-diffusion-controlled coarsening (solute drag), etc. The separation region appears as a shaded area on the far right-hand side of the diagram.

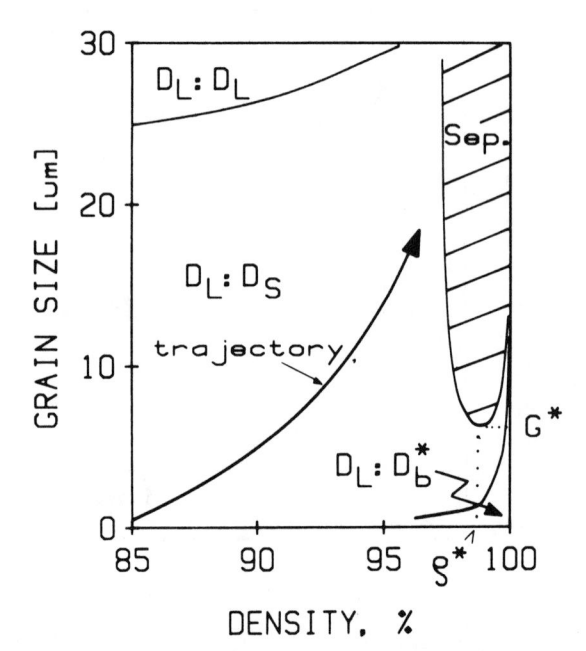

Fig. 6. Schematic illustration of a microstructure development map.

Table I. Data Used in the Construction of Microstructure Development Maps for Undoped Al_2O_3

$D_s = 10^{-10}$ m²·s⁻¹	$M_b = 2 \times 10^{-14}$ m³·N⁻¹·s⁻¹
$\delta_s = 3 \times 10^{-10}$ m	$k = 1.38 \times 10^{-23}$ J·K⁻¹
$\gamma_{s/v} = 0.9$ J·m⁻²	$T = 1873$ K
$\gamma_b = 0.45$ J·m⁻²	$\overline{G}/G_{max} = 0.2$
$\Omega = 2.11 \times 10^{-29}$ m³	$\eta = 5/4$

Pore breakaway during sintering will be avoided if the grain size–density trajectory passes below some critical grain size value, G^*, at some critical density, ρ^*.

It is helpful to use this type of map to illustrate how an additive might affect the course of sintering through its influence on diffusion parameters. To demonstrate the effects, we have constructed maps of semiquantitative validity for undoped Al_2O_3 using values listed in Table I. The modified Brook analysis of Uematsu et al.[8] was used to determine the conditions of density and grain size defining the separation region and the 10% pore control line. The condition for pore separation is given for the case when pores move by surface diffusion as

$$\left(\frac{D_s \delta_s \Omega}{M_b k T r^3}\right) \overline{G}^2 - 2\eta(1 - \overline{G}/G_{max})\overline{G} + 24r = 0 \tag{4}$$

where $\overline{G}$ is the average grain size, G_{max} is the size of the largest grain, and η is a geometrical factor (5/4 for narrow grain size distributions and 9/4 for very wide grain size distributions). The critical grain size, G^*, and density, ρ^*, are given by the expressions

$$G^* = \left(\frac{8192 D_s \delta_s \Omega}{M_b k T \eta^4 (1 - \overline{G}/G_{max})^4}\right)^{1/2} \tag{5}$$

$$\rho^* = 1 - 0.0117 \eta^3 (1 - \overline{G}/G_{max})^3 \tag{6}$$

Grain size–density trajectories were computed by iteration of the following equations for densification and grain growth expressed in integral form as given by Brook:[3]

$$\rho - \rho_0 = \frac{287 D_L \gamma_{s/v} \Omega t}{G^3 k T} \tag{7}$$

For grain growth controlled by surface-diffusion-controlled pore drag:[14]

$$G^4 - G_0^4 = \frac{440 D_s \delta_s \gamma_{gb} \Omega t}{k T (1 - \rho)^{4/3}} \tag{8}$$

The effects of additives on microstructure development are considered with the aid of these diagrams in the following section.

Effect of Additives on D_L and D_b'': The effect of raising D_L by a factor of 10 is, as can be seen from Fig. 7, to cause a significant flattening out of the G–ρ trajectory while leaving the separation region unaffected. The trajectory is altered sufficiently, in the present case, to undercut the separation region, thereby enabling full density to be attained. A similar response would be produced by raising D_b'' if it were controlling densification rather than D_L. The recommendation is, therefore, to raise D_L and/or D_b''.

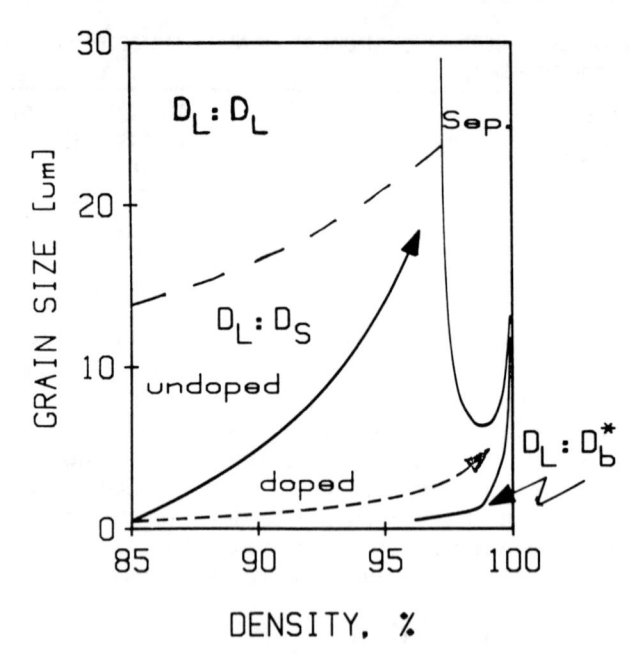

Fig. 7. Microstructure development map for Al_2O_3 showing the effect of raising the lattice diffusion coefficient by a factor of 10.

The effect of an additive on D_L can be predicted by using the principles of defect chemistry.[15] The same principles can be used to predict an additive influence on D_b'''[16] in the near boundary region provided that the solute distribution in the boundary region is known. No adequate theory is available for predicting the effect of an additive on diffusion along the grain-boundary core.

Effect of Additives on M_b: The effect of lowering M_b by a factor of 10 is, as can be seen from Fig. 8, to raise G^* by a factor of $10^{1/2}$ and to extend the region of solute drag control to a wider range of G and ρ values. The trajectory is affected (flattened in this case) only when it enters the solute drag control regime ($D_L:D_b^*$). Lowering of M_b is, therefore, to be recommended.

The theories available[17] for predicting the effects of additives on M_b remain to be convincingly demonstrated for oxide systems.

Effect of Additives on D_s: Lowering D_s by a factor of 10 can be seen, from Fig. 9, to have the favorable effect of flattening the trajectory and the undesirable effect of lowering G^*. Due to these competing effects, lowering D_s does not, in this case, make it possible to reach full density. Raising D_s increases G^*, but the $G-\rho$ trajectory is simultaneously steepened to an intolerably high level (see Fig. 10). The safest recommendations for altering D_s are, therefore, to (i) lower D_s and compensate for the decrease in G^* by lowering M_b simultaneously or (ii) raise D_s (and therefore G^*) and compensate for the steepened trajectory by simultaneously raising D_L. The choice depends on the particular set of circumstances involved (the

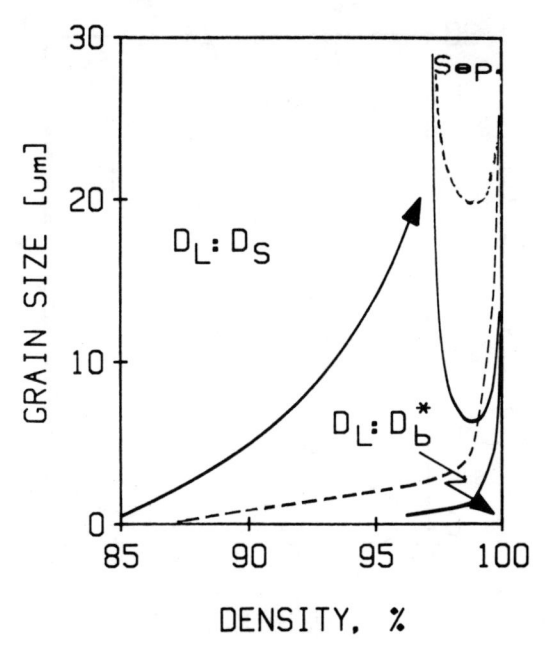

Fig. 8. Microstructure development map for Al_2O_3 showing the effect of lowering the boundary mobility by a factor of 10.

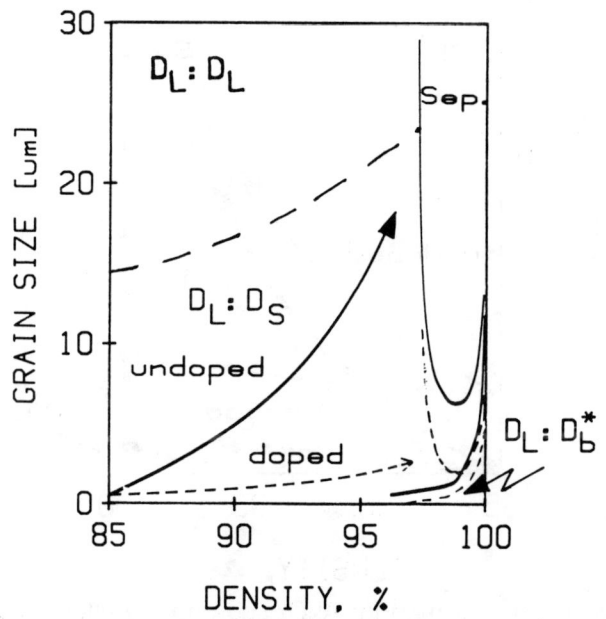

Fig. 9. Microstructure development map for Al_2O_3 showing the effect of lowering the surface diffusion coefficient by a factor of 10.

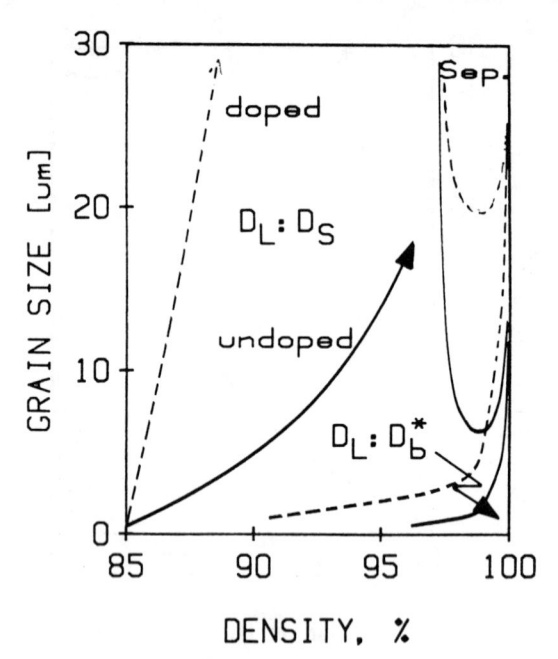

Fig. 10. Microstructure development map for Al_2O_3 showing the effect of raising the surface diffusion coefficient by a factor of 10.

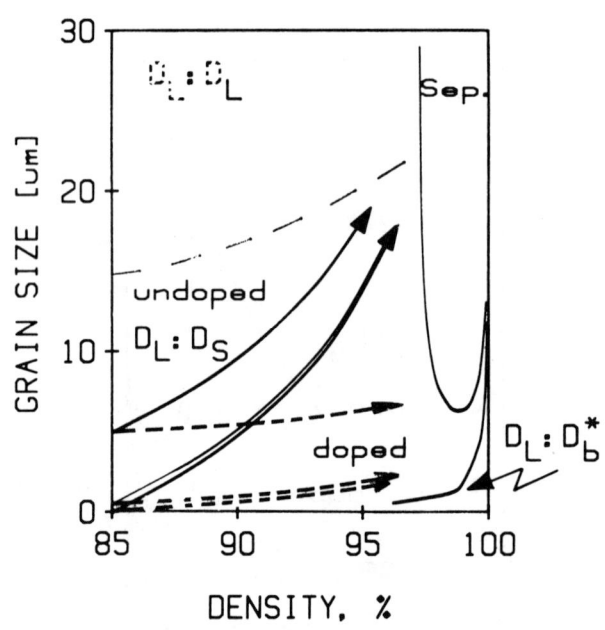

Fig. 11. Microstructure map for Al_2O_3 showing the effect of varying the initial particle size (G_0 values of 5.0, 0.5, and 0.05 μm). The effect of raising D_L by a factor of 10 is also shown.

respective values of G_0, G^*, D_s, D_L, M_b, etc.). Under most sintering circumstances, however, analysis indicates that it is generally more advisable to go with lowering D_s and M_b simultaneously, as it would be in the present case of Al_2O_3.

There is presently no adequate theoretical framework for predicting the effects of additives on surface diffusion in oxide systems.

Finally, for completeness, the effects of the initial particle size and the particle size distribution are considered:

Effect of the Initial Particle Size: Microstructure development maps for powders with three different initial particle sizes (5, 0.5, and 0.05 μm) are shown in Fig. 11. Trajectories have been drawn for both the undoped and doped ($D_L \times 10$ increase) case. The tendency is to observe much less relative grain growth with coarser starting powders, consistent with the predictions of Herrings scaling laws[18] (see companion paper by Brook in this volume). For the present case of lattice-diffusion-controlled densification and surface-diffusion-controlled pore drag, it can be seen that there is little to benefit from the use of ultrafine (<0.5 μm) starting powders other than faster microstructural change.

Effect of Particle Size Distribution: The present simplistic analysis considers the effect of the particle size distribution on the location of the separation region. Microstructure maps for wide and narrow size distributions are shown in Fig. 12. The critical grain size and density are maximized for very narrow size distributions. The critical density can be as high as 99.3% for the narrowest distribution $\overline{G}/G_{max} = 0.5$) and as low as 91% for the widest distribution ($\overline{G}/G_{max} = 0$).[8]

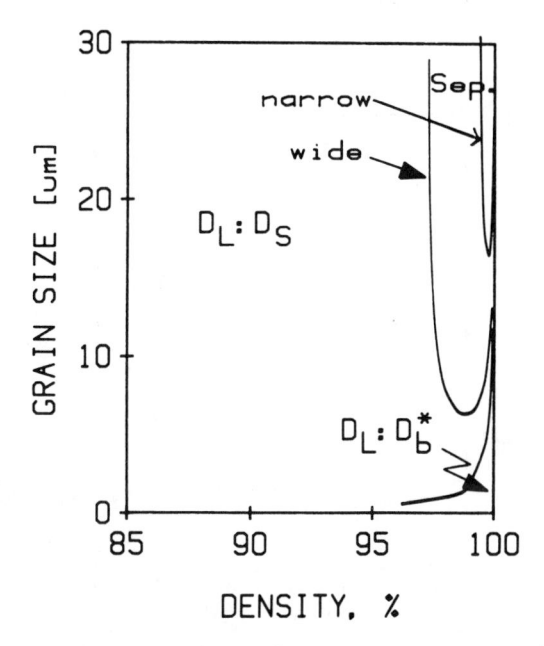

Fig. 12. Microstructure map for Al_2O_3 demonstrating the effect of changing the particle size distribution. One case corresponds to a wide ($G_{max} = 5\overline{G}$) distribution and the other to a very narrow ($G_{max} = 2\overline{G}$) size distribution.

System Al$_2$O$_3$-MgO — A Case Study

In the light of the preceding arguments, it is useful to, first, consider the role of MgO in the sintering of Al$_2$O$_3$ solely from the viewpoint of the known effects of MgO doping on the various diffusivity terms and, second, consider the mechanistic details involved.

Macroscopic View of the Role of MgO

Recent available data on the effect of MgO solute on the various diffusion coefficient terms indicate that (a) MgO raises D_L by a factor of 3,[19] (b) MgO lowers M_b by a factor of 5,[20] and (c) MgO either leaves D_s unchanged[6,21] or lowers it by a factor of 10.[22] It is immediately apparent from these data that MgO affects all of the diffusivity and mobility terms in the most favorable way, as predicted by the previous analysis. Microstructure development maps showing the combined effects of MgO doping for the two possible cases of (i) D_s unchanged and (ii) D_s lowered (times 10) are shown in Figs. 13 and 14, respectively. Both combinations are shown to be effective, on the basis of these diagrams, in enabling Al$_2$O$_3$ to sinter to full density. If the effects of MgO are considered separately, i.e., if MgO only raises D_L, only lowers M_b, or only lowers D_s, the diagrams show that no one single effect is totally effective in enabling full density to be reached. The implication from this is that MgO can only work through the reinforcing action of several contributory mechanisms.

To allow a comparison between theory and experiment to be made, experimentally measured grain size–density plots for MgO-doped (200 ppm) Al$_2$O$_3$ are compared against the computed profiles in Fig. 15. From this it can be seen that a reasonably good correlation exists between the experimental data and the com-

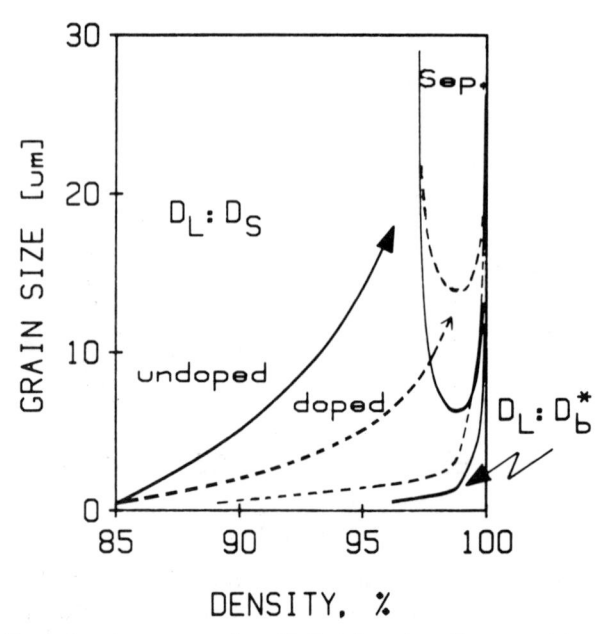

Fig. 13. Microstructure map for Al$_2$O$_3$ showing the combined effects of raising D_L (times 3) and lowering M_b (times 5).

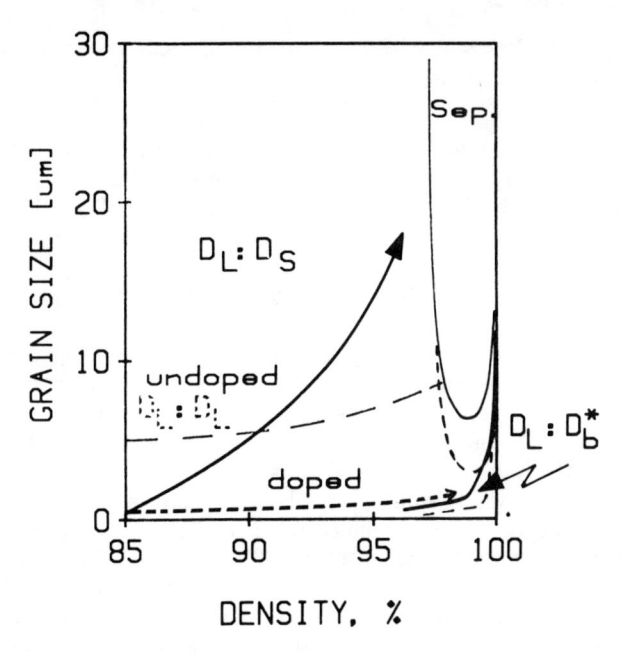

Fig. 14. Microstructure map for Al_2O_3 showing the combined effects of raising D_L (times 3), lowering M_b (times 5), and lowering D_s (times 10).

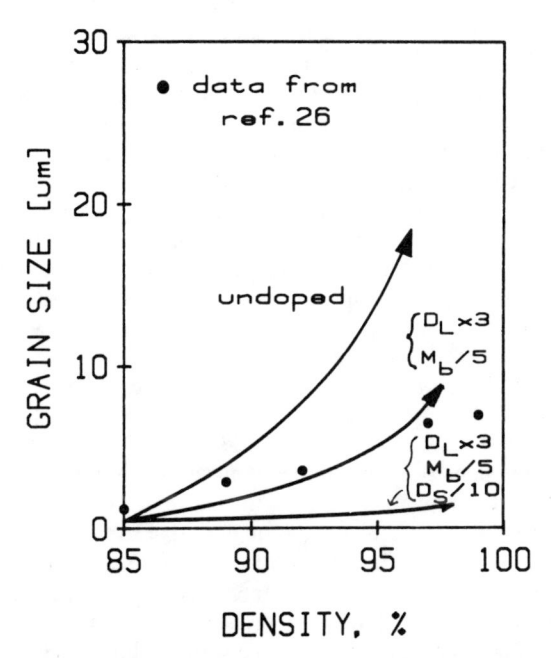

Fig. 15. Grain size–density plots for undoped and doped Al_2O_3 comparing computed profiles with experimental results for a MgO-doped (200 ppm) Al_2O_3 fired at 1560°C (Ref. 26).

puted profiles for the case where MgO acts to simultaneously raise D_L (times 3) and lower M_b (times 5).

Mechanistic Aspects

While the effects of MgO doping are quite well understood from a macroscopic viewpoint, many of the detailed mechanisms still have to be worked out. In particular, the mechanisms governing the action of MgO on M_b and D_s are far from clear. (The effect of MgO on D_L is, by comparison, more clearly understood.[19])

Models have been proposed to explain the effect of MgO solute on M_b on the basis of solute segregation to grain boundaries, the most recent of which involves the preferential segregation of magnesium and calcium ions to specific boundary types.[20] Sufficient experimental data are not yet available, however, to confirm these theories. Another viewpoint supports the notion that grain-boundary structure is controlling M_b.[23] The question of whether the boundary structure determines the solute content, which in turn controls M_b, or whether the solute content and type control the structure, which in turn controls M_b, is indeed difficult to answer.

Another factor to consider is the possible presence of liquid phases in these systems. We have observed large anisotropic flat-sided grains in a recent study of grain growth using high-purity (99.98%) Al_2O_3 starting powders.[20] We have since confirmed, however, that some of these large anisotropic grains are surrounded by a thin intergranular phase (~5 nm thick). The intergranular phase is amorphous in nature, as confirmed by diffuse dark-field imaging and defocus imaging[24] (see Figs. 16 and 17). X-ray microanalysis shows that the grain-boundary phase is rich in Al_2O_3, Na_2O, SiO_2, and CaO. The remarkable feature of MgO doping is the ability of MgO to transform these large-grain anisotropic structures into uniform fine-grain microstructures. These observations raise some new and intriguing possibilities for the role of MgO. Perhaps it simply alters the solid–liquid interfacial energy to promote nonwetting or, alternately, it may act as a scavenger by drawing counterimpurity species such as Si^{4+} into solid solution from the impurity phase. Further studies on the possible role of liquid phases are recommended.

The issue concerning the effect of MgO on D_s needs to be settled. Implications from sintering studies[21,6] are that MgO has no effect on D_s, whereas recent studies using the technique of groove annealing on single-crystal surfaces of special orientation have shown that MgO lowers D_s by a factor of 10 at 1650°C.[22] Recognizing that the surface state during sintering studies is likely to be very different from that during single-crystal studies, great care must be taken when arguing from one to the other. A promising new technique for studying surface diffusion is being developed in our laboratory. The procedure involves (i) vacuum hot-pressing Al_2O_3 powders to full density, (ii) desintering the fully dense disks by air annealing to form uniform grain-boundary pore networks,[25] (iii) reannealing under reducing conditions while the density is maintained constant by applying external pressure, and (iv) measuring grain growth kinetics in porous samples at constant density. This method is potentially useful for the quantitative evaluation of the effects of solid-solution additives on surface-diffusion-controlled pore mobility.

Much of the previous discussion has focused on the conditions necessary for achieving high density by avoiding pore breakaway, and less attention has been given to the related problem of abnormal grain growth. It is important to recognize, however, that, when abnormal grain growth is dealt with, pore breakaway represents only one of many possible causes.[27] Great care should, therefore, be taken when the cause of abnormal grain growth is diagnosed, especially when the

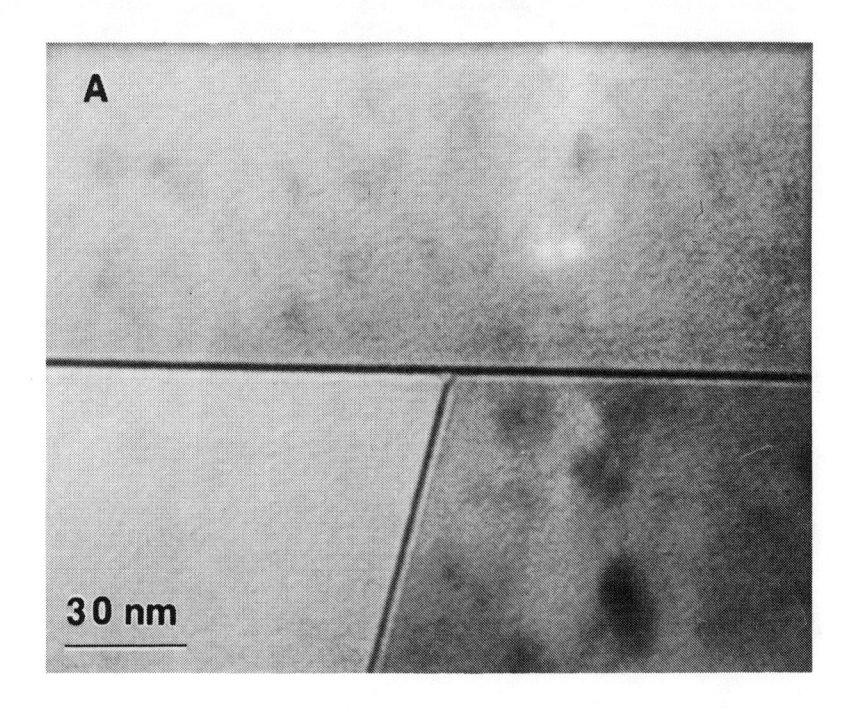

Fig. 16. Bright-field image (A) and diffuse dark-field image (B) of a grain-boundary junction in a pure (99.98%) undoped Al$_2$O$_3$ showing the presence of a thin intergranular amorphous phase.

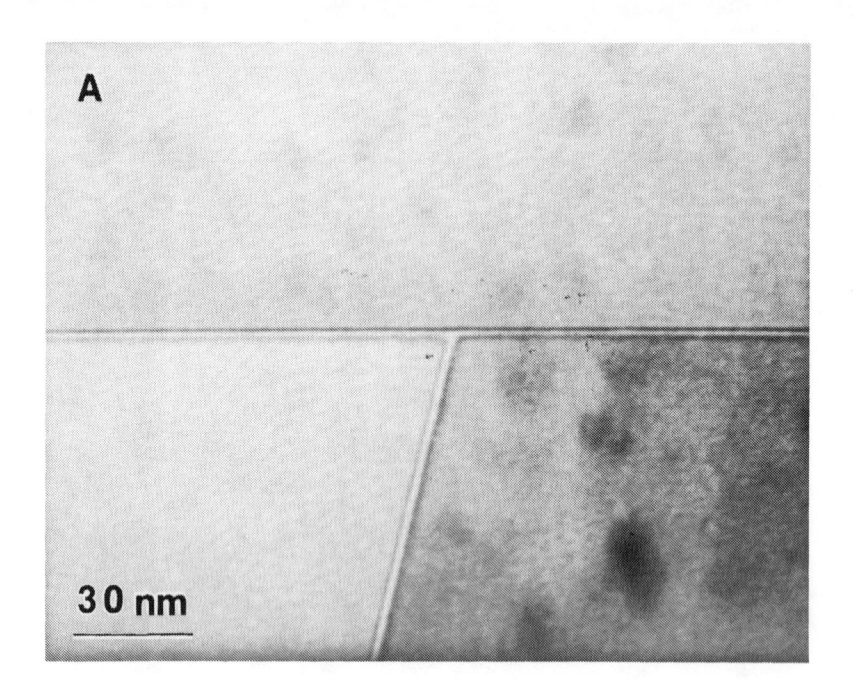

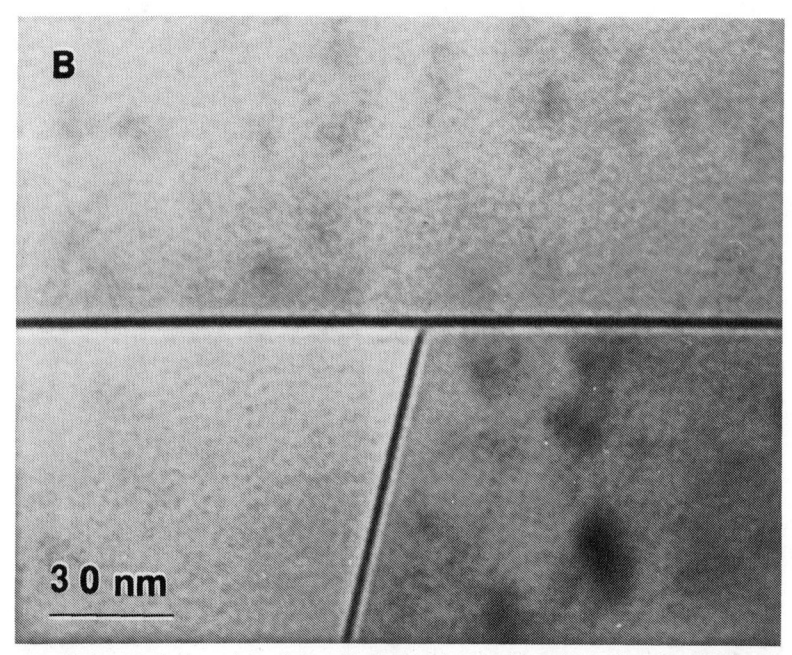

Fig. 17. Through-focal series of a grain-boundary junction in Al$_2$O$_3$ showing contrast reversal in going from the under-focus condition (A) to the over-focus condition (B). The contrast reversal indicates that the grain boundary contains a thin intergranular phase (Ref. 24).

analysis is based on the appearance of the final microstructure, which, for many of the different possible causes, can look the same.

Conclusions and Implications

The use of solid-solution additives continues to show promise as a practical means for controlling microstructure development in ceramic systems. A new form of final-stage sintering diagram has been developed to demonstrate the principles of additive effects on microstructure development during sintering. The role of MgO in the sintering of Al_2O_3 can be adequately understood, solely on the basis of the known effects of MgO solute on diffusion parameters in Al_2O_3; many of the details still need to be worked out. Liquid phases in high-purity ceramics may play an important role in sintering and grain growth. The effects of liquid phases on pore-boundary interactions need to be modeled. Improved methods for studying and predicting the effects of additives on surface diffusion in ceramic systems need to be developed. The idea[3] of using additive combinations for reinforcement purposes needs to be pursued further.

Acknowledgments

The author is grateful to the National Science Foundation for financial support of this work under Grant No. DMR-8116865. Discussions with R. J. Brook and R. M. Cannon and contributions from research students at Lehigh University (S. J. Bennison, M. Lynch, and K. Rumsey) are also gratefully acknowledged.

References

[1] D. W. Budworth, *Mineral Mag.*, **37**, 833 (1970).
[2] (a) M. F. Yan, *Mater. Sci. Eng.*, **48**, 53 (1981); (b) M. F. Yan, R. M. Cannon, U. Chowdhry, and H. K. Bowen, *Am. Ceram. Soc. Bull.*, **56** [3] 291 (1977).
[3] R. J. Brook, *Proc. Br. Ceram. Soc.*, **32**, 7 (1982).
[4] E. A. Barringer and H. K. Bowen, *J. Am. Ceram. Soc.*, **65**, C199 (1982).
[5] M. F. Ashby, *Acta Metall.*, **20**, 887 (1972).
[6] J. E. Burke, K. W. Lay, and S. Prochazka, *Mater. Sci. Res.*, **13**, 417 (1980).
[7] R. J. Brook, *J. Am. Ceram. Soc.*, **52**, 56 (1969).
[8] K. Uematsu, R. M. Cannon, R. D. Bagley, M. F. Yan, U. Chowdhry, and H. K. Bowen; pp. 109–205 in Proceedings of the International Symposium of Factors in Densification and Sintering of Oxide and Non-Oxide Ceramics. Edited by Shigeyuki Sōmiya and Shinroku Saito. Gakujutsu Bunken Fukyu-Kai, Tokyo, Japan, 1979.
[9] F. M. A. Carpay; pp. 261–75 in Ceramic Microstructures 1976. Edited by R. M. Fulrath and J. A. Pask. Westview Press, Boulder, CO, 1977.
[10] C. M. Hsueh, A. G. Evans, and R. L. Coble, *Acta Metall.*, **30**, 1269 (1982).
[11] M. A. Spears and A. G. Evans, *Acta Metall.*, **30**, 1281 (1982).
[12] C. M. Hsueh and A. G. Evans, *Acta Metall.*, **31**, 189 (1983).
[13] R. L. Coble, *J. Appl. Phys.*, **32**, 787 (1961).
[14] S. P. Howlett, "Grain Growth and Grain Boundary Segregation in Doped Cobalt Monoxide"; Ph.D. Thesis, University of Leeds, 1981.
[15] R. J. Brook; pp. 179–267 in Electrical Conductivity in Ceramics and Glass, Part A. Edited by N. M. Tallan. Marcel Dekker, New York, 1974.
[16] M. F. Yan, R. M. Cannon, H. K. Bowen, and R. L. Coble, *J. Am. Ceram. Soc.*, **60**, 120 (1977).
[17] M. F. Yan, R. M. Cannon, and M. K. Bowen; pp. 276–307 in Ceramic Microstructures 1976. Edited by R. M. Fulrath and J. A. Pask. Westview Press, Boulder, CO, 1977.
[18] C. Herring, *J. Appl. Phys.*, **21**, 301 (1950).
[19] M. P. Harmer and R. J. Brook, *J. Mater. Sci.*, **15**, 3017 (1980).
[20] S. J. Bennison and M. P. Harmer, *J. Am. Ceram. Soc.*, **66**, C90 (1983).
[21] J. M. Dynys, R. L. Coble, W. S. Loblenz, and R. M. Cannon, *Mater. Sci. Res.*, **13**, 391 (1980).
[22] C. Monty and J. LeDuigou, "Symposium on Chemical Matter at High Temperature"; to be published in *High Temp.-High Press.*
[23] C. B. Carter; private communication.
[24] D. R. Clarke, *Ultramicroscopy*, **4**, 33 (1979).

[25] S. J. Bennison and M. P. Harmer; to be published in *Advances in Ceramics*.
[26] M. P. Harmer and R. J. Brook, *Trans. J. Br. Ceram. Soc.*, **80**, 147 (1981).
[27] S. J. Bennison and M. P. Harmer; pp. 928–9 in Ceramic Powders. Edited by P. Vincenzini. Elsevier, Amsterdam, The Netherlands, 1983.

---

*Fine (0.1–1 μm) nonspherical particles with a narrow size distribution (monomodal but not monosized) free of hard agglomerates, uniformly compacted.

Mechanical Behavior of Alumina: A Model Anisotropic Brittle Solid

A. G. Evans and Y. Fu

University of California
Materials and Molecular Research Division
Lawrence Berkeley Laboratory and
Department of Materials Science and Engineering
Berkeley, CA 94720

This article describes the mechanical behavior of alumina at both low and high temperatures. The brittle behavior at low temperatures is classified into two regions: one due to microcrack coalescence at large grain sizes and the other due to the growth of preexistent flaws at smaller grain size. The microcracking is predicated on the existence of residual stress due to thermal expansion anisotropy. Microcracks result in R-curve behavior and a very complex dependence of strength on microstructure. Failure at small grain sizes is more straightforward and relatively well understood. At high temperatures, rupture is time dependent and failure involves the nucleation and growth of cracks accompanied by creep cavitation. Eventual failure is determined by the coalescence of blunted creep cracks. Each of these failure processes is discussed and related to microstructure.

The mechanical properties of Al_2O_3 have been extensively investigated during the last decade.[1-4] The behavioral trends thus elucidated provide appreciable insight into the mechanical characteristics of other noncubic brittle solids. The research endeavors on this material thus provide a generic basis for assessing the mechanical behavior of a broad range of technologically important ceramic polycrystals.

The general trends in mechanical behavior can be conveniently set within the framework summarized in Fig. 1. Specifically, at temperatures $\leq T_m/2$ (where T_m is the melting temperature) the material is brittle, as characterized by an essentially linear load/deflection curve (Fig. 1(A)). In this range, the strength is relatively temperature and time insensitive. At higher temperatures, appreciable permanent deformation can precede failure and the strength is strongly temperature and rate sensitive. Failure in this temperature range occurs by a creep rupture process similar to that encountered in metals. In the brittle range, two dominant behaviors occur, depending on the grain size. In fine-grained materials, the strength is relatively grain size insensitive (Fig. 1(B)) and fracture occurs from defects with dimensions that appreciably exceed the grain size. These defects include fabrication flaws such as inclusions, voids, and shrinkage flaws, as well as surface cracks introduced by machining. In large-grained materials, microcrack formation precedes final failure and the failure process embraces a microcrack coalescence phenomenon (Fig. 1(B)). The failure stress in this grain size range tends to scale inversely with the square root of the grain size.

The intent of the present article is to examine the processes that induce fracture in these various failure regimes and, equally as important, to examine the microstructural conditions that lead to transitions between regions. Thereafter, the

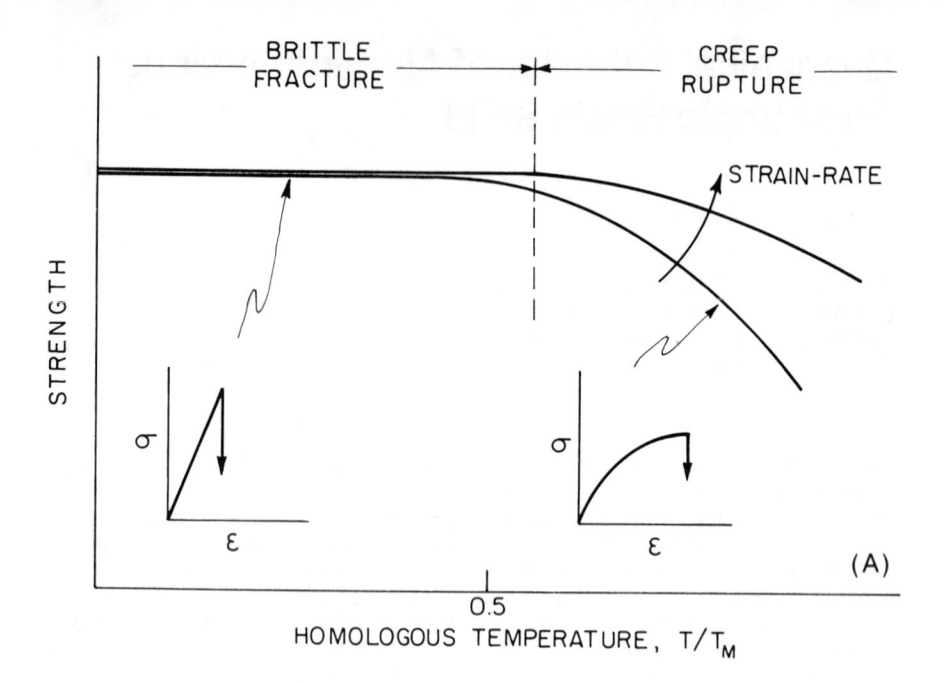

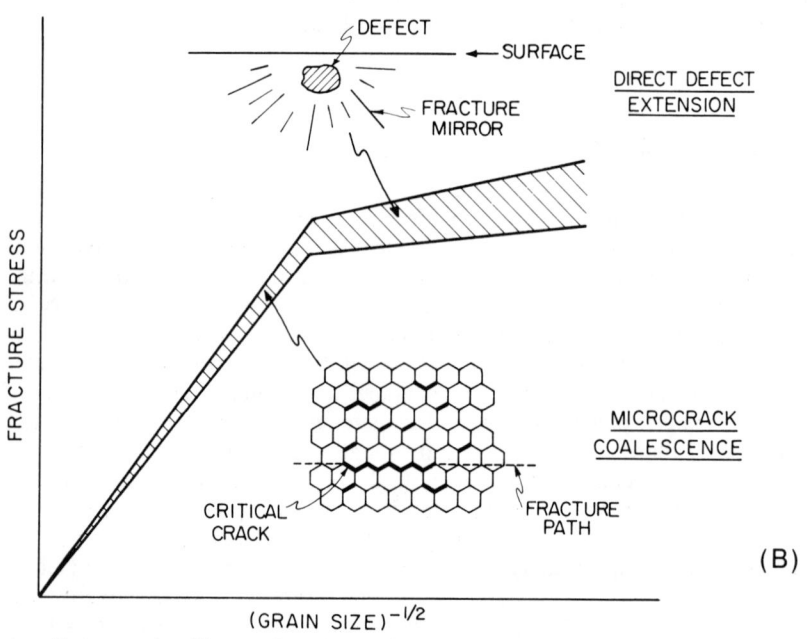

Fig. 1. Schematics illustrating the effects of (A) temperature and (B) grain size on fracture.

implications of the present comprehension of failure in alumina to the interpretation and prediction of failure in other noncubic ceramic polycrystals can be discussed.

The article is arranged such that the microstructural conditions pertinent to microcracking are first examined. With this insight, the brittle fracture process and its grain size dependence can be comprehensively addressed. Thereafter, the mechanisms of failure that obtain at elevated temperatures are presented, and transitions between brittle fracture and creep rupture are considered.

Brittle Fracture: Stress-Induced Microcracking

Initial Microcracking

Microcracks in single-phase brittle polycrystals are a direct consequence of localized residual stresses.[5-7] In materials resistant to dislocation glide and mechanical twinning, these stresses derive from anisotropy in the thermal expansion properties of the material. Such anisotropy is typical of noncubic materials. Thermal expansion anisotropy results in residual stresses with a wavelength of the order of the grain facet length, which oscillate between tension and compression along adjoining facets (Fig. 2(a)). The stress amplitudes can be computed by using Eshelby procedures[7] (Fig. 2(b)) coupled with a grain orientation assumption, e.g., predicated on randomness. The maximum stresses at grain facet centers occur for orthogonally oriented grains[7] (A, B, C, and D in Fig. 2(a)) and vary along individual facets such that a logarithmic singularity exists at the grain corners

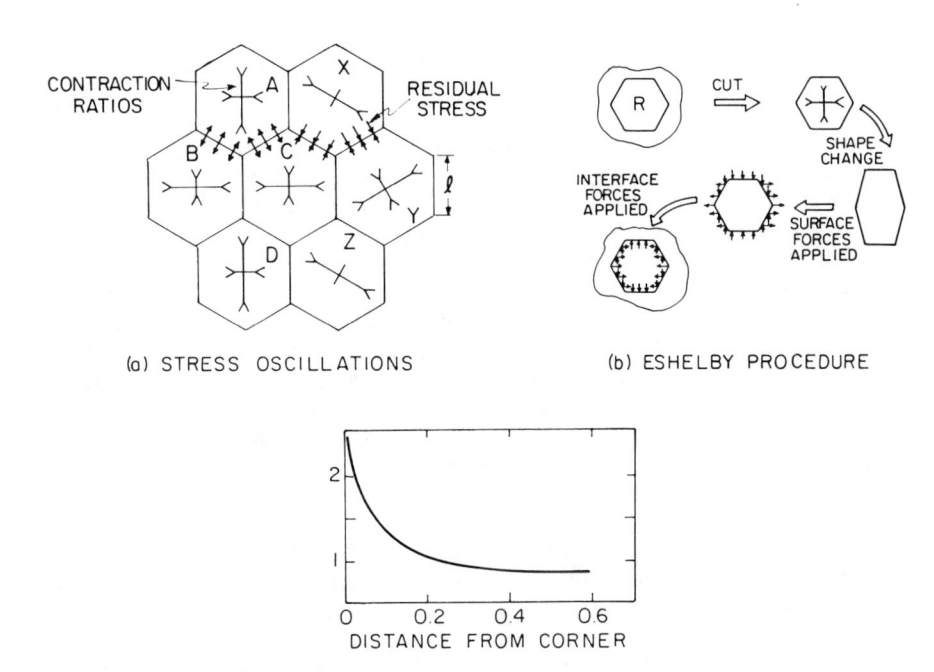

Fig. 2. Thermal-expansion-induced residual stresses (Ref. 7): (a) oscillating residual stress on neighboring facets caused by thermal contraction anisotropy; (b) schematic of the Eshelby method for calculating the residual stresses on grain boundaries; (c) stress variation from the grain corner showing the logarithmic singularity.

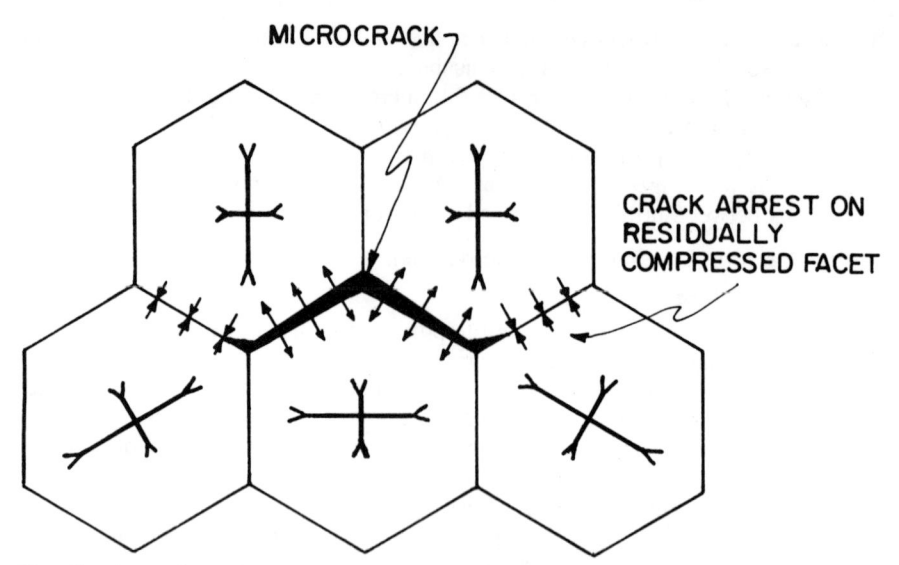

Fig. 3. Location of microcracks with respect to the residual stress on the grain facets.

(Fig. 2(c)). In general, therefore, the stress has the following approximate form:
$$\sigma(x) \approx E\Delta\alpha\Delta T[1 + a \ln (l/x)]F(\theta,\phi) \quad (1)$$
where E is Young's modulus of the polycrystal, $\Delta\alpha$ is the anisotropy in thermal expansion, ΔT is the cooling range (which increases as the grain size increases), l is the grain facet length, x is the distance from the grain corner, F is a function that depends on the orientations (θ,ϕ) of the adjoining grains, and a is a constant.

An essential feature of this stress field with regard to the incidence of microcracking is the scale effect associated with the singular term. This scaling, as $\ln (l/x)$, allows the stress to be sustained over a larger area of grain facet as the grain size increases. Consequently, microcracking becomes more probable as the grain size enlarges. A formal analysis of microcracking within the singular field, in terms of the stress intensity at corner-located heterogeneities,[7,8] reveals the existence of a "critical" grain size,[5] l_c, above which spontaneous grain facet fracture is likely to occur upon cooling, given by*
$$l_c = 3.1(K_b/E\Delta\alpha\Delta T)^2 \quad (2)$$
where K_b is the fracture resistance of the grain boundary. The microfracture propensity increases rapidly as the grain facet size increases above l_c.[6,8] However, the probabilistic aspects of this process have not been appreciably explored.

Microcracking in the presence of applied loads has been examined by a direct extension of the preceding concept.[8] A uniform applied stress can simply be superposed on the residual stress (Eq. (1)) and the stress intensity at corner heterogeneities reevaluated. Such analyses allow the density of microcracked facets to be computed as a function of the applied load. For this purpose, it is generally assumed that the microcracks formed at prior loads do not appreciably perturb the local stress field. The basis for this premise is the appreciation that the microcracks arrest in regions of residual compression and, hence, that the influence

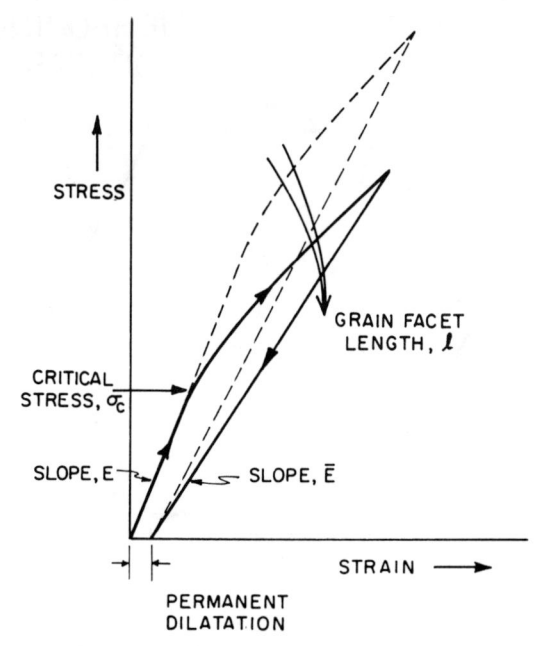

Fig. 4. Stress–strain curve that occurs in a material subject to microcracking. Note the hysteresis due to the modulus change and the dilatation accompanying microcracking.

zone of the microcrack does not extend into the neighboring regions of residual tension (Fig. 3). With a noninteraction analysis of this type the fraction of facets microcracked, f, is found to vary with uniaxial applied stress,[†] σ_∞, as ($f \ll 1$),

$$f \approx m(\sigma_\infty - \sigma_c)/E\Delta\alpha\Delta T \tag{3}$$

where σ_c is the critical stress needed to initiate the first microcrack and m is a constant $\approx \frac{1}{4}$. The critical stress varies with grain facet size l as

$$\sigma_c \approx E\Delta\alpha\Delta T[(l_c/l)^{1/2} - 1] \tag{4}$$

The presence of the microcrack induces a nonlinear material response, by virtue of the effect of the microcracks on the elastic modulus[9]

$$\bar{E}/E \approx 1 - 16(1 - \bar{\nu}^2)(10 - 3\nu)f[45(2 - \bar{\nu}^2)]^{-1} \tag{5}$$

where $\bar{E}$ and $\bar{\nu}$ are the Young's modulus and Poisson's ratio of the microcracked material, respectively. The stress–strain curve on loading is thus predicted to have the form[8] depicted in Fig. 4. Unloading, which is determined by the modulus of the microcracked material $\bar{E}$, occurs elastically. Hysteresis thus results (Fig. 4). Furthermore, a permanent dilatation also develops (Fig. 4) because the microcracks relieve the residual stress on those facets subject to residual tension (Fig. 3).

Process Zone Initiation: Microcrack Transition

An important condition with regard to the onset of a microcrack influence on brittle fracture relates to the ability of discrete microcracks to initiate at the tip of

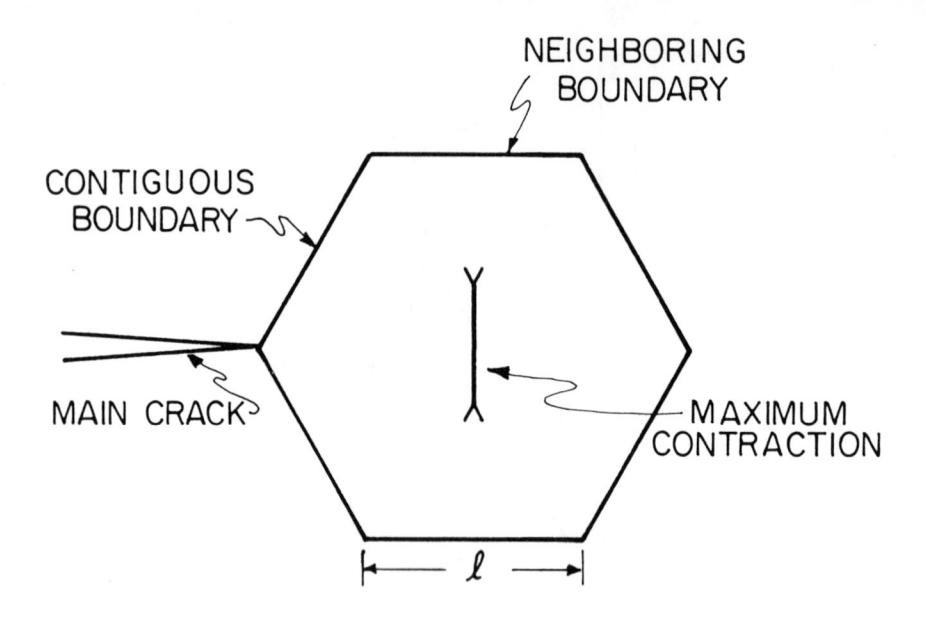

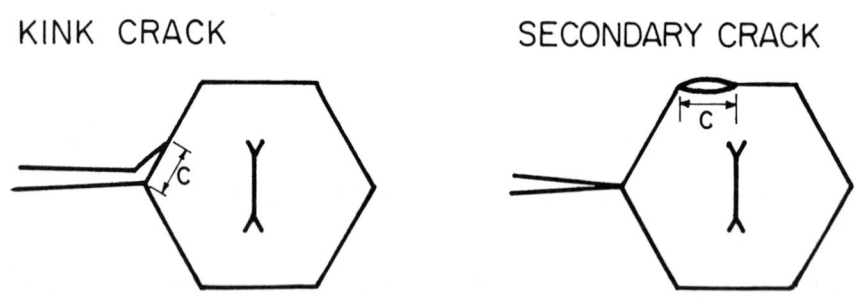

Fig. 5. Schematic basis for evaluating the conditions that permit the nucleation of discrete microcracks ahead of the primary crack.

a major crack (Fig. 5).[10] Some appreciation of this condition has been achieved by calculating the grain size at which a discrete, detached microcrack can form in preference to a microcrack contiguous with the primary crack (Fig. 5). The calculation is again cognizant of the residual stress in the material. Specifically, it is recognized that discrete microcrack formation can only occur when the grain facet contiguous with the primary crack is subject to residual compression, while the facet subject to microcracking is under residual tension (Fig. 5). Analysis of the condition for microcrack initiation on the adjoining tensile facet compared with the condition for kink growth along the contiguous facet (Fig. 5) reveals the relatively simple result[10] that discrete microcracks form when the facet length $l \gtrsim 2l_c/5$.

A probable consequence of discrete microcracking is the onset of microcrack toughening and failure by microcrack coalescence. Conversely, when $l \lesssim 2l_c/5$, failure presumably involves the progressive growth of a dominant macrocrack,

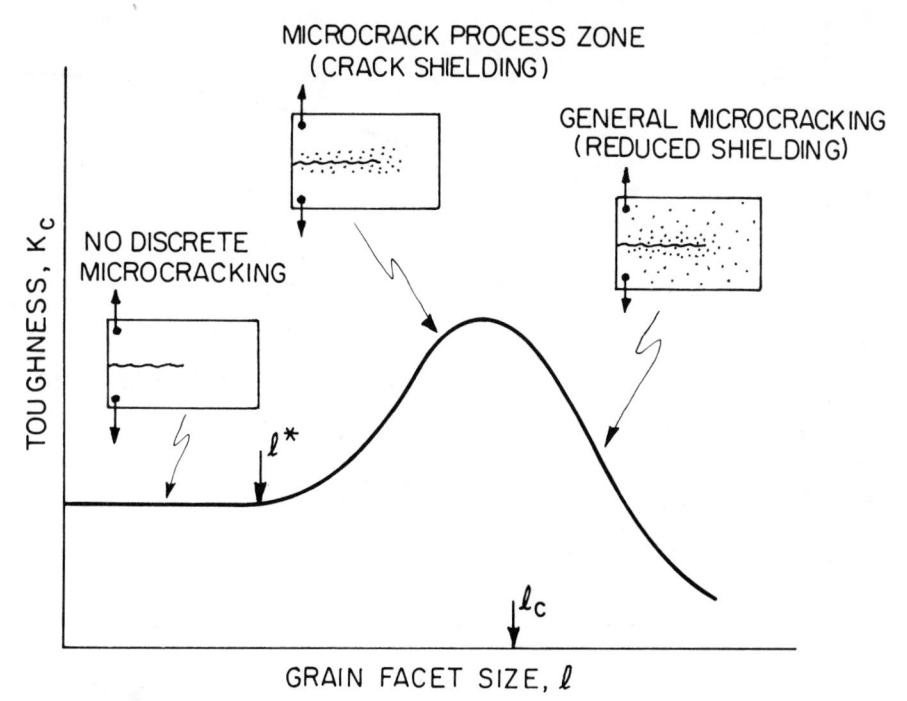

Fig. 6. Effects of grain size on the toughness showing the transition grain size, l^*, between microcrack toughening and deflection toughening and the size, l_c, for thermal microcracking.

e.g., as influenced by a crack deflection mechanism. It is presumed, therefore, that the condition $l = 2l_c/5$ coincides with the transition in Fig. 1(B) between failure from preexistent flaws to a mode of failure involving microcrack coalescence. Hence, for present purposes we define the *transition* size, l^*, specified in Fig. 6, as

$$l^* \approx 1.2(K_b/E\Delta\alpha\Delta T) \qquad (6)$$

Some justification for this choice will be presented in the following section.

Further discussion of brittle fracture first considers the behavior at $l > l^*$, wherein microcrack toughening and microcrack coalescence occurs. Then, in the subsequent section, fracture at $l < l^*$ is addressed, as dominated by preexisting flaws and by crack deflection toughening.

Microcrack Toughening

A discrete microcrack process zone can result in crack shielding and, consequently, a material with an enhanced fracture toughness.[11] Microcrack toughening is invariably characterized by an R curve[11-14] (Fig. 7(a)). Notably, a frontal process zone provides a substantially smaller effect on toughness than a fully extended zone over the crack surface (Fig. 7(a)). The origin of the R curve can be appreciated first by recognizing that the material in the frontal zone is not subject to unloading prior to crack advance. Hence, the path-independent J integral dictates the crack tip field,[11,15] in accord with the relation

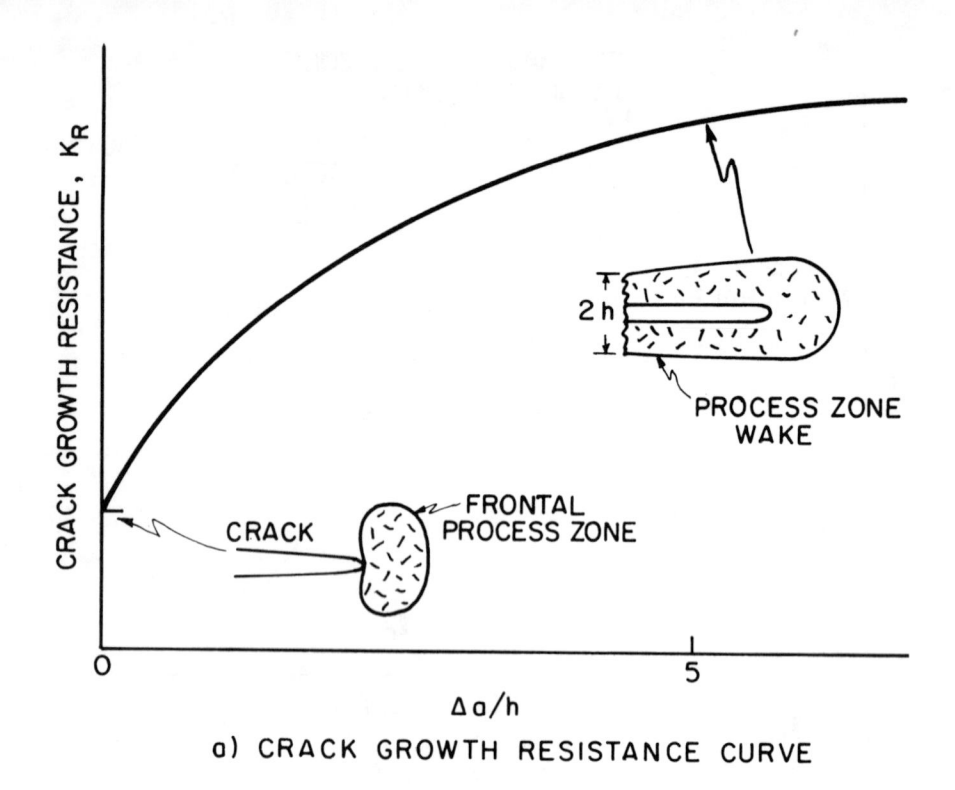

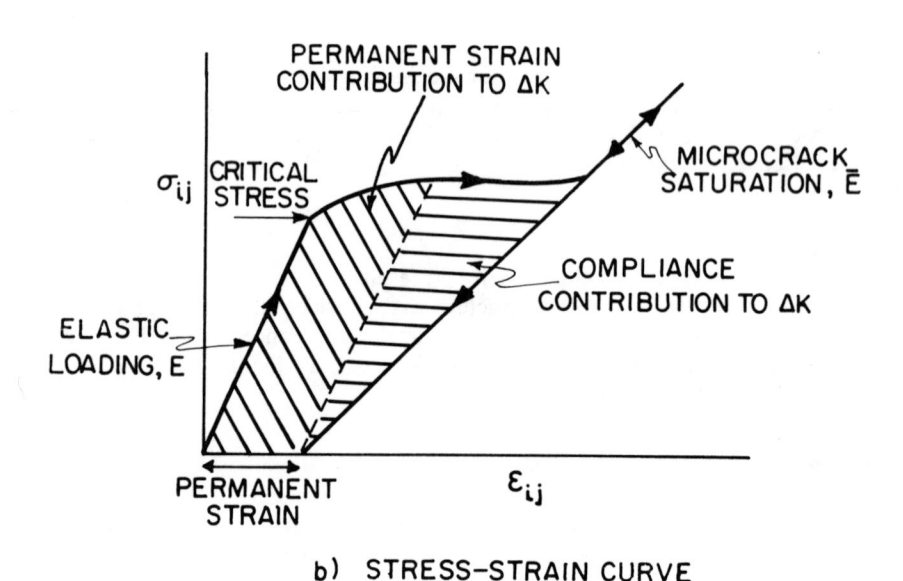

Fig. 7. Schematic R curve typical of polycrystalline Al$_2$O$_3$ (a) and the corresponding stress–strain curve, indicating saturation (b).

$$J \equiv K_\infty^2(1 - \nu^2)/E = K_l^2(1 - \overline{\nu}^2)/\overline{E} \tag{7}$$

where K_∞ is the stress intensity associated with the applied loads and K_l is the local stress intensity at the crack tip.[‡] The magnitude of $\overline{E}/E$ (Eq. (4)) is such that crack shielding, $K_l < K_\infty$, is induced by the microcrack zone. However, there is a concommitant degradation of the material ahead of the crack tip given approximately by[11,16]

$$K_m \approx K_0(1 - f_s) \tag{8}$$

where K_m is the degraded crack growth resistance, K_0 is the intrinsic crack growth resistance of the alumina, and f_s is the saturation density of microcracks. Consequently, since crack growth occurs when $K_l = K_m$, the net effect of the frontal microcrack zone on toughness is negligible, because the shielding (Eq. (7)) is completely counteracted by the degradation[6,11] (Eq. (8)).

More substantial shielding obtains when the process zone encompasses the crack surfaces. For a steady-state zone of uniform width, recognition that the material in the remote wake experiences complete unloading prohibits the use of J but allows the pertinent energy balance integral to be derived as[15]

$$K_\infty^2(1 - \nu^2)/E = K_l^2(1 - \overline{\nu}^2)/\overline{E} + 2\int_0^h U(y)\,dy \tag{9}$$

where h is the microcrack zone width and $U(y)$ is the residual energy density in the wake. The quantity $U(y)$ is the area under the stress–strain curve depicted in Fig. 7(b). Additional crack shielding derives from the wake contribution to the energy balance, and an enhanced toughness invariably ensues. Specifically, with the stress–strain curve evaluated in Process Zone Initiation: Microcrack Transition, the steady-state toughness can be evaluated as[8,11]

$$K_c \approx F[E\theta f_s h^{1/2}/K_0 f_s] \tag{10}$$

where θ is the microcrack-induced dilatational strain and F is the function plotted in Fig. 8. Typically, K_c/K_0 ranges between 1.5 and 2.5 for pertinent choices of the material variables.[8]

More generally, h increases as the crack advances, and the crack growth resistance, K_R, continues to rise with crack advance, Δa[14] (Fig. 7(a)). The measured value of the toughness then depends on the growth requirement, $K = K_R$, and the instability condition, $dK/d\Delta a > dK_R/d\Delta a$, pertinent to the test geometry.[11] In particular, it is important to appreciate that notched-beam tests, with no initial wake, yield instability for small Δa and, hence, give a measured toughness similar to the intrinsic toughness, K_0. However, cantilever tests (such as the DCB test), in which an initial wake has been created during precracking, yield values of toughness comparable to the steady-state value, K_c (Eq. (10)). Specimen geometry effects of this type are invariably encountered in materials that exhibit R-curve behavior.[13]

The higher toughnesses that obtain with a fully developed wake do not normally coincide with a larger breaking strength,[17] at least in single-phase alumina, because of stress-induced flaws generated by microcrack coalescence. Strength and toughness increases can exist, however, in certain two-phase alumina-based systems.[18]

The incidence of microcracking depends on grain size and temperature, as reflected in the magnitude of the zone width, h, and the critical facet size, l_c. The specific grain size dependence predicted from Eq. (10) has the form depicted in Fig. 6. Evidently, toughening can only initiate when $l > l^*$. Additionally, when

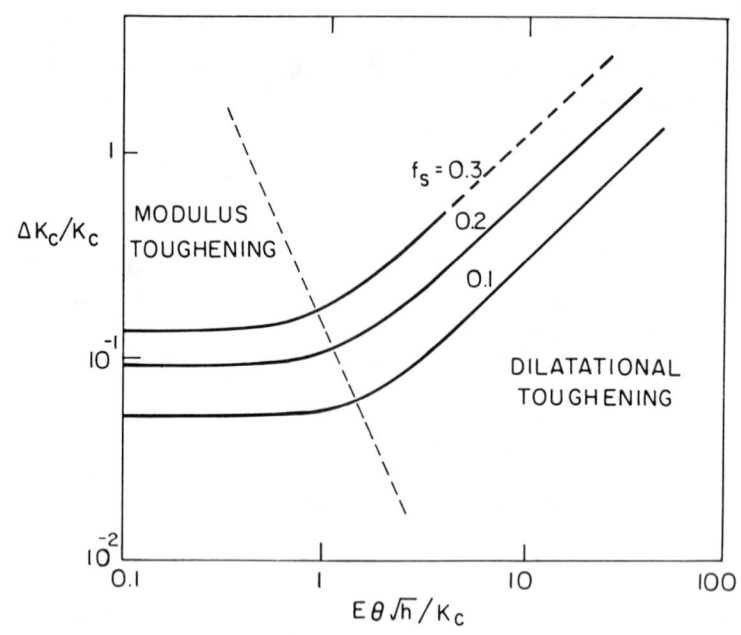

Fig. 8. Trends in toughness with the permanent strain θ and the saturation microcrack density, f_s (Ref. 11).

$l \to l_c$, the stress–strain hysteresis approaches zero, because the *initial* modulus approaches the saturation value $\bar{E}$, due to thermal microcracking, and significant toughening is excluded. The maximum toughness thus obtains at a grain size between l^* and l_c, reaching values about twice that for unmicrocracked material under optimum conditions. Behavior of this type has been substantiated by toughness data obtained by using DCB tests[13,17] (Fig. 9).

The toughness of the material with the optimum grain size decreases with an increase in temperature due to the increase in l_c and, hence, the decrease in h, as the temperature is raised. This behavior is similar to the temperature-dependent toughness observed in transformation toughening. It is also noted that the grain size trends noted above are analogous to particle size effects in transformation toughening.[19]

Microcrack Coalescence

The process of microcrack coalescence is not well understood. However, it is appreciated that coalescence is encouraged by the generation of coplanar microcrack arrays, by concentrating the applied stress onto the residually compressed, uncracked ligaments. A thorough analysis of this phenomenon has not yet been developed. The best available analysis[20] considers that fracture proceeds when sufficient microcracks occur on contiguous grain boundaries that a microcrack of critical size ensues (Fig. 1(B)). This analysis neglects the details of coalescence across residually compressed ligaments and does not consider the R-curve behavior

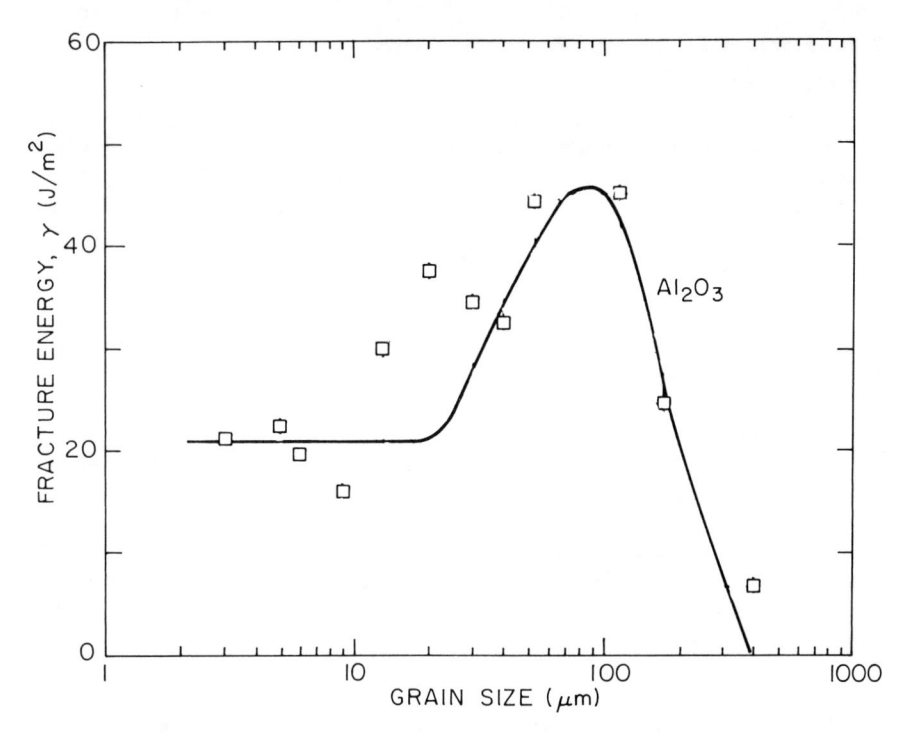

Fig. 9. Effects of grain size on the fracture toughness of alumina (Ref. 17).

that characterizes macrocrack growth. Nevertheless, the analysis provides some useful physical insights. The basis for the analysis is the assumption that the microcracks act independently[§] and have a cumulative failure probability, p. Then, the probability of failure, Φ, as determined by the probability of forming c/l contiguous cracks (where c is the critical crack size for unstable fracture) becomes[8]

$$\Phi \approx (2V/l^3)p^{(c/l+1/2)} \tag{11}$$

where V is the volume of material subject to stress. Noting that p is simply the fraction of microcracked facets, we can combine Eqs. (3) and (11) to yield the failure stress, S, at the median probability level [$\Phi = 1/2$]

$$S = \frac{\pi K_c}{2l^{1/2}} \left\{ \frac{\ln\left[4(S - \sigma_c)/E\Delta\alpha\Delta T\right]}{\ln\left[l^3/4V\right]} \right\}^{1/2} \tag{12}$$

For $S \gg \sigma_c$, the predominant grain size dependence emerges as $S \sim K_c l^{-1/2}$. This reciprocal square root dependence is typical of strength data in the large-grained region (Fig. 1(B)). This dependence derives simply from the condition that fewer contiguous microcracked facets are needed to comprise a critical crack as the grain size increases.

Brittle Fracture: Failure from Preexistent Flaws
Dominant Flaws

In materials with a value of grain facet length less than the transition value, l^*, discrete microcracks cannot be supported, and failure occurs by the growth of a predominant flaw, significantly larger than the grain size.[21] The flaws derive either from processing or from machining. The machining flaws can be particularly deleterious. Such flaws have a size, c, dictated by the load P acting on the diamonds in the grinding wheel, such that[22]

$$c^{3/2} = \lambda(P/K_c)(E/H)^{1/2} \tag{13}$$

where H is the hardness and λ is a material-independent constant. Furthermore, the cracks are subject to residual stress due to the plastic surface layer. Consequently, the breaking process involves the stable extension of cracks, prior to eventual failure, e.g., as monitored by acoustic waves[22] (Fig. 10). Analysis of this stable growth process reveals that the fracture instability occurs when the dominant flaw has the dimensions[22]

$$(a_1 a_2)^{1/2} \approx 5c \tag{14}$$

where a_1 and a_2 are the flaw dimensions at fracture and c is the depth of the machining flaws (Fig. 10(b)). On the basis of this relation, the breaking strength, S, has been deduced to vary as

$$S \approx (3\pi K_c/4)(a_1 a_2)^{-1/4} \tag{15}$$

Inclusions with a smaller thermal expansion than that of the alumina matrix also become deleterious flaws, because of residual tangential tension in the matrix and the attendant crack formation and growth. The cracks are again subject to residual stress, and the cracks grow stably under increasing load, in much the same manner as that observed for machining flaws. Analysis of the stable crack growth process reveals that the breaking strength is given by

$$S = \left(\frac{27\pi}{4}\right)^{1/3} \frac{K_c^{4/3}}{8} \left[\frac{1-2\nu}{Eb^2(\alpha_m - \alpha_i)\Delta T}\right]^{1/3} \tag{16}$$

where α_i and α_m are the thermal expansion coefficients of the inclusion and matrix, respectively, and b is the inclusion radius. Note that, because of the residual stress, the strength varies more strongly with inclusion size, $S \sim b^{-2/3}$, than with the usual dependence for a cracklike flaw, $S \sim a^{-1/2}$. Furthermore, the strength is substantially less than that expected from a Griffith crack of equivalent size.

Most other flaws are less deleterious, for the equivalent size, than either machining flaws or low expansion inclusions. Typically, flaws such as high expansion inclusions and voids exhibit a strength[21]

$$S \gtrsim (\pi/2)K_c b^{-1/2} \tag{17}$$

The extent to which the strength exceeds the equality in Eq. (17) is dictated by the presence of cracklike flaws in the immediate vicinity of the dominant flaws. Such flaws are not well understood. However, it is appreciated that small machining flaws, for example, interact with larger subsurface voids, leading to breaking strengths comparable to that calculated in Eq. (17).

Fracture Toughness

The fracture toughness of single-phase alumina in the absence of a microcrack process zone, $l < l^*$, is dictated by crack deflecting mechanisms.[23] Such mecha-

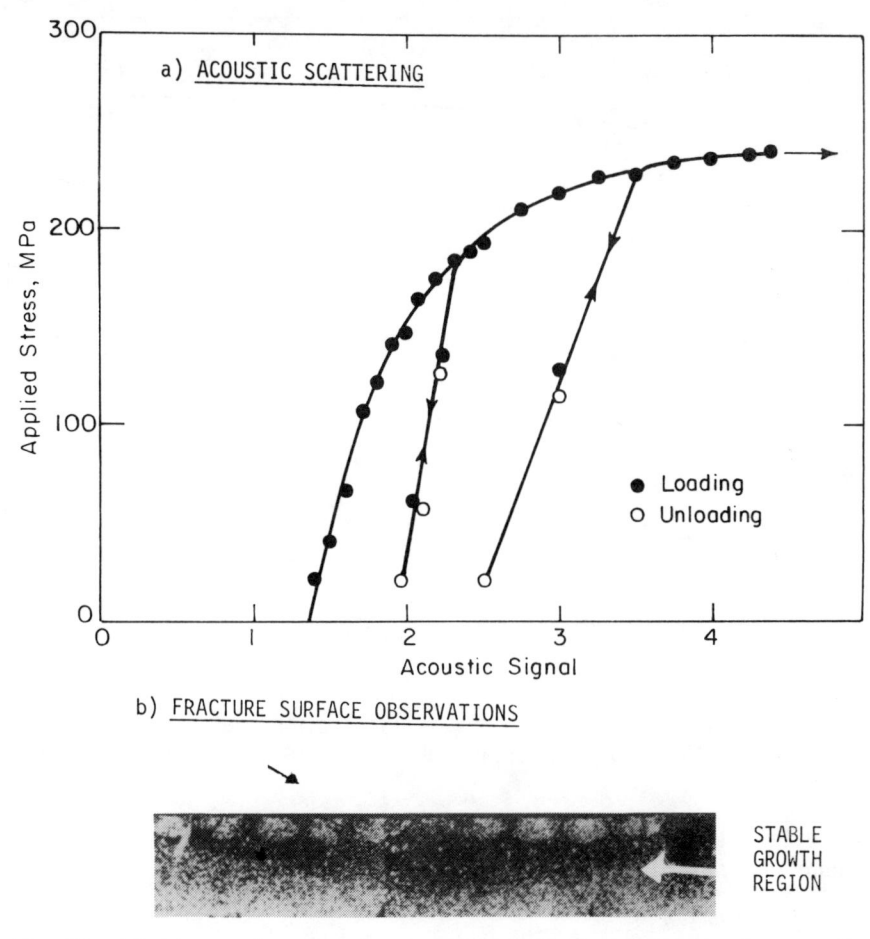

Fig. 10. Stable crack growth in a machined Al_2O_3: (a) change in acoustic scattering; (b) fracture surface illustrating the initial cracks and the stable growth region.

nisms induce a toughness that depends exclusively on the grain shape. Hence, the toughness is relatively insensitive to the grain size (Fig. 6) and temperature, in marked contrast to the strong effects of size and temperature on microcrack toughening.

Equiaxed grains yield a toughness of ~3 MPa·m$^{1/2}$.[17] However, there is some evidence that rod-shaped grains result in a toughness of ~5–6 MPa·m$^{1/2}$.[¶,24] Toughness increases of this magnitude are consistent with crack deflection, by twisting around the rod-shaped grains (Fig. 11).[23] The incidence of deflection is predicated on the development of a cleavage-resistant plane,[4] [0001], normal to the axis of the rod-shaped grains, that resists coplanar crack extension through the grains.

High-Temperature Failure

The process of creep rupture in alumina at elevated temperatures involves the nucleation, growth, and coalescence of cracks, as shown by the sequence depicted

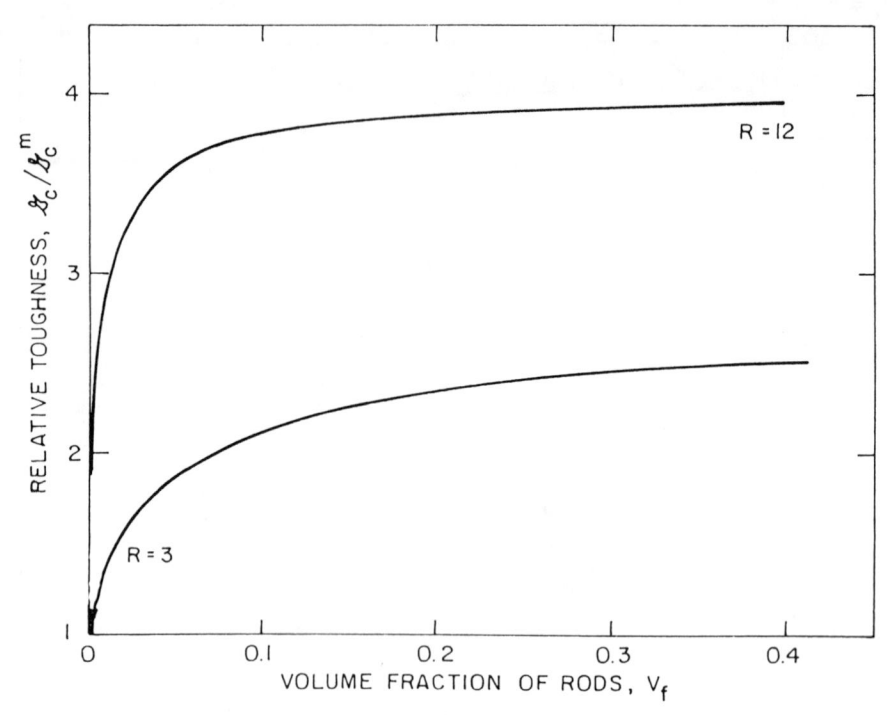

Fig. 11. Crack-twist-induced toughening magnitudes associated with rod-shaped deflecting particles; R is the aspect ratio of the rods (Ref. 23).

in Fig. 12.[3] Inhomogeneities are of particular importance in the failure process. The failure time, t_f, is found to vary as

$$t_f \dot{\varepsilon}_\infty = (\sigma_0/\sigma_\infty)^{-6} \qquad (18)$$

where $\dot{\varepsilon}_\infty$ is the applied strain rate and σ_0 is a constant. A detailed comprehension of this stress dependence had not yet been obtained. However, a certain understanding of the constituent phenomena has been achieved, as discussed in the following sections.

Crack Nucleation

Cracks are observed to nucleate at microstructural and chemical heterogeneities (Fig. 13). An important microstructural inhomogeneity is the large-grained region shown in Fig. 13. These regions are subject to an enhanced stress, about twice the applied stress,[25] due to the higher creep viscosity of the large grains deforming by diffusive creep. The larger stress permits relatively rapid cavity coalescence and crack nucleation. The cracks extend rapidly across the region but then extend more slowly into the fine-grained matrix.

Chemical inhomogeneities are more prevalent.[3] These defects are manifest on the surface as regions that thermally etch during the creep test (Fig. 13). Cracks nucleate within these regions. There is some evidence that regions of this type contain SiO_2. The presence of SiO_2 accelerates crack growth by allowing viscous flow and solution–reprecipitation in the resultant amorphous phase. Furthermore, propagation of the crack away from the inhomogeneity depletes the crack tip of

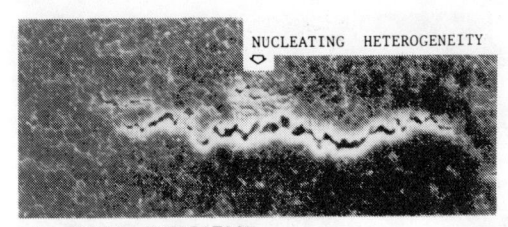

A) CRACK NUCLEATION

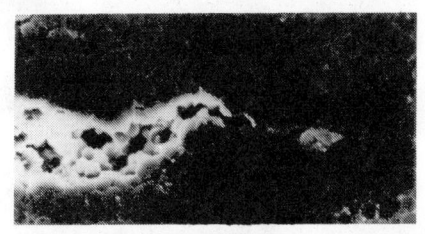

B) CRACK PROPAGATION

C) CRACK BLUNTING

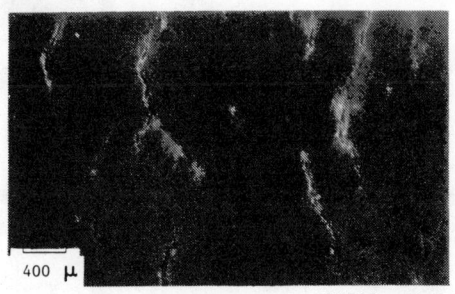

D) CRACK COALESCENCE AT SHEAR BANDS

Fig. 12. Failure sequence in alumina at elevated temperatures showing the nucleation, growth, blunting, and coalescence phases.

amorphous material, resulting in crack arrest and blunting. Behavior of this type is consistent with the observations discussed in the following section.

Models of crack nucleation generally predict a Monkman–Grant relation for the nucleation times, $t_n \dot{\varepsilon}_\infty \approx$ constant. Such behavior is not normally consistent with measured failure times (Eq. (18)), which exhibit highly nonlinear behavior. Consequently, the nucleation time is not regarded as the major time-consuming stage in the failure process. This conclusion is consistent with the observations that cracks nucleate well before eventual failure.

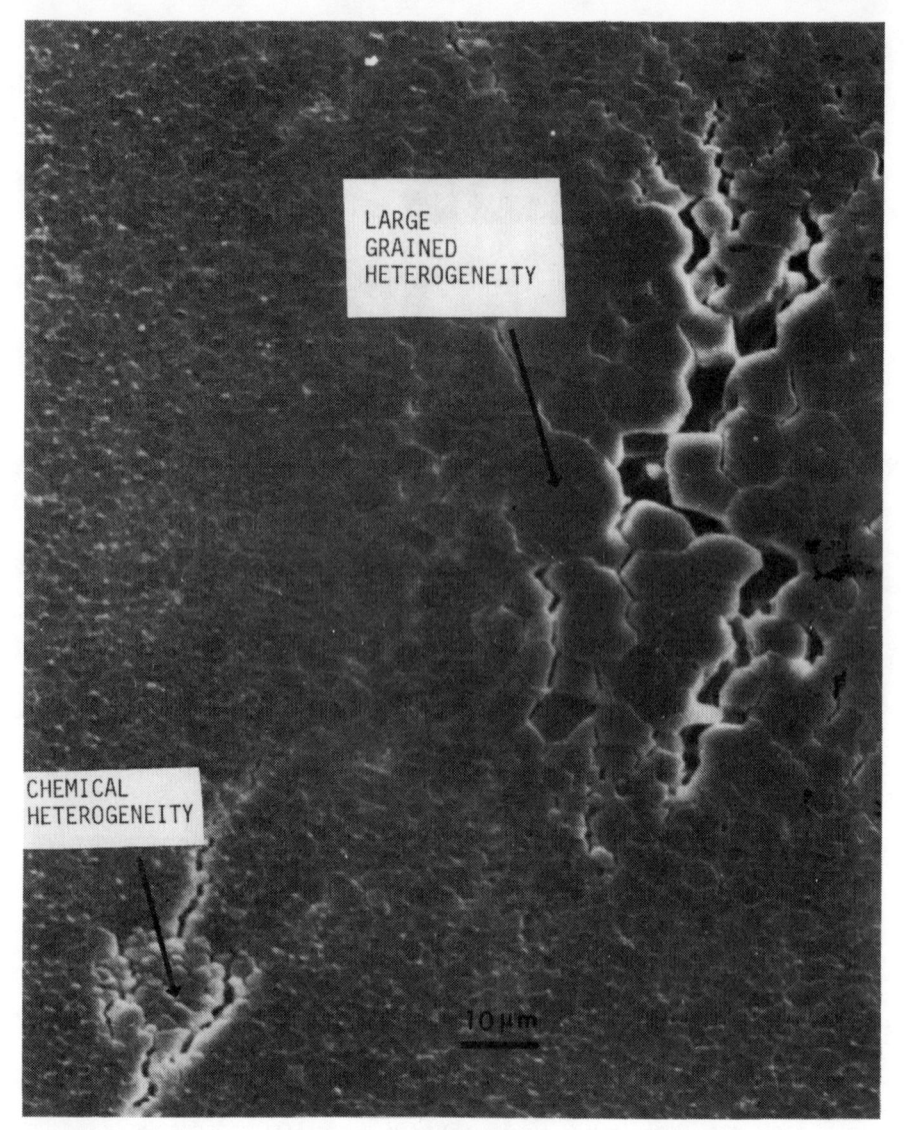

Fig. 13. Scanning electron micrographs of crack nucleation under creep conditions showing a large-grained region and a chemical heterogeneity.

Crack Propagation and Blunting

The general creep crack propagation characteristics observed for alumina include regions of slow crack growth, blunting, and crack healing,[3,26] as depicted in Fig. 14. Slow crack growth occurs with relatively small crack opening displacements (Fig. 12(B)) and can be characterized by a relation between the crack growth rate, $\dot{a}$, and stress intensity factor, K, of the form[27]

$$a \approx \dot{a}_0 (K/K_c)^n \exp(-Q/RT) \tag{19}$$

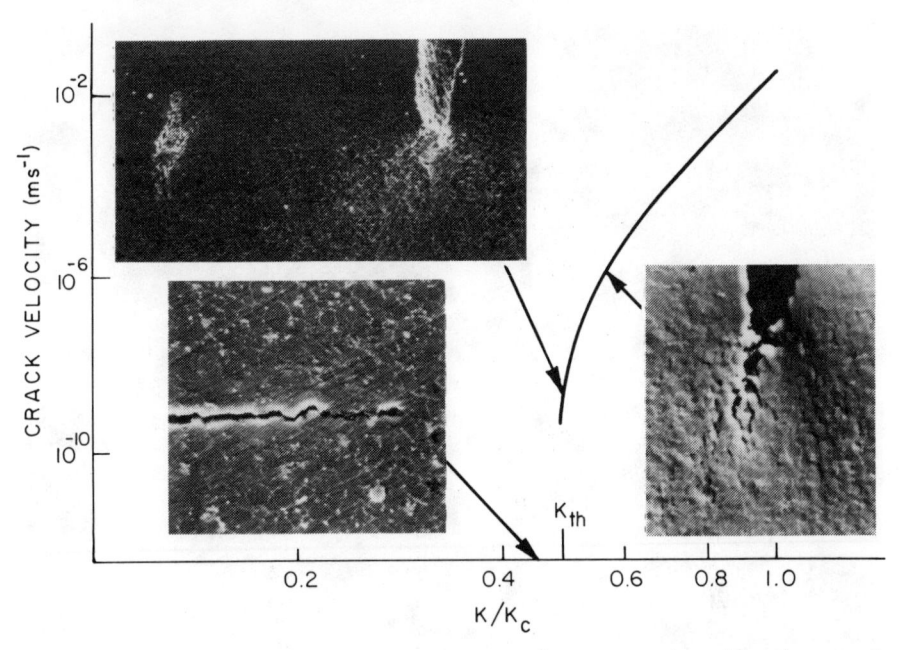

Fig. 14. Schematic showing the slow crack propagation, blunting, and healing regions.

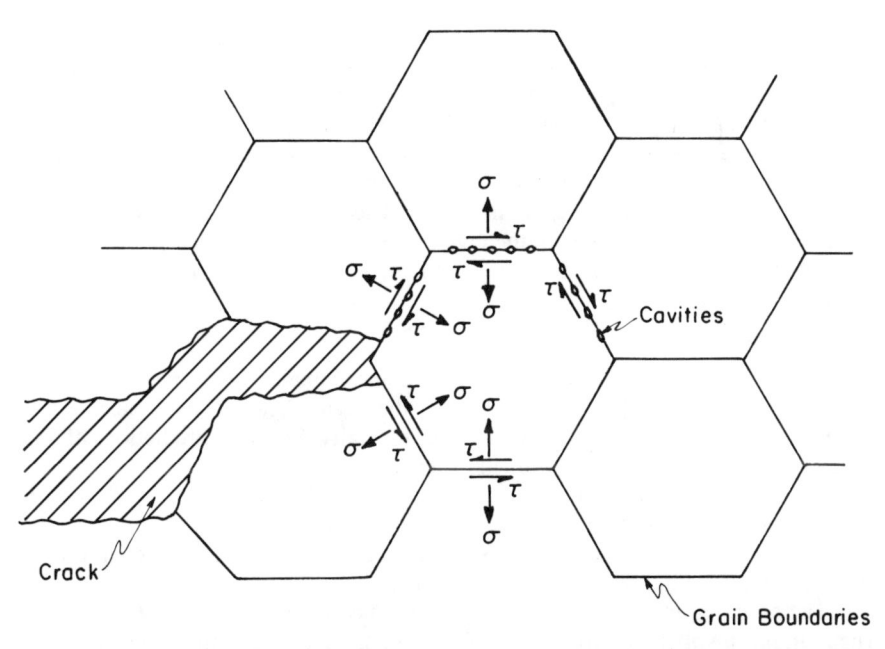

Fig. 15. Characteristics of creep crack growth by a damage zone process showing the damage zone and the state of stress.

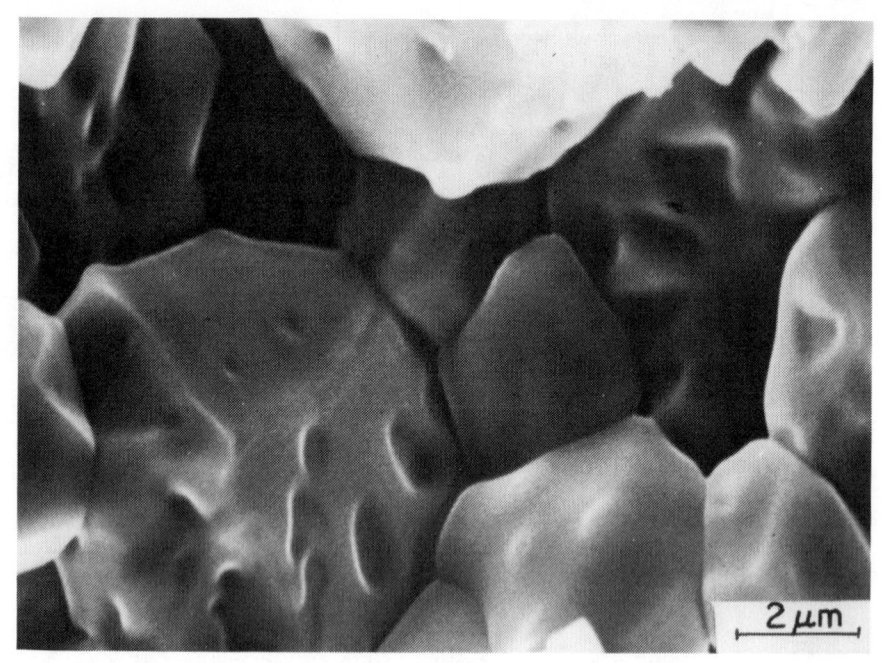

Fig. 16. Cavities on two grain interfaces observed within the damage zone.

where $\dot{a}_0$ is a constant, Q is the activation energy for creep, and n is the creep exponent (~1.8 for fine-grained alumina). The mechanism of growth applicable to extrinsic cracks involves the nucleation, growth, and coalescence of cavities, by diffusion, in a damage zone ahead of the crack, as schematically illustrated in Fig. 15. Observations of the fracture surface reveal the presence of cavities on two grain interfaces (Fig. 16), such that the density of grain facets containing cavities diminishes as the driving force, K, increases. The process has been modeled[27] by embedding a cavitating region in the crack tip stress field and allowing the cavity volume change rate to relax the stress in the damage zone. The model predicts a crack growth rate

$$\dot{a}\eta = (Kl^{1/2})F(\lambda/l, z, A_f) \quad (20)$$

where λ is the spacing between cavities in the damage zone, η is the creep viscosity, z is the damage zone length, A_f is the cavity area at coalescence, and F is a function. For the spacing observed experimentally (Fig. 16), predictions of this model coincide quite closely with the measured crack growth rates. However, superposed on this crack growth process are an acceleration, as $K \to K_c$, and a deceleration as the blunting threshold is approached. The acceleration is accompanied by an increase in the density of noncavitating facets and reflects a mixed brittle fracture–cavitation mode of crack advance.

The blunting of extrinsic cracks occurs in fine grained alumina at $K \approx 0.4K_c$. The blunting involves the formation of damage in side lobes emanating from the crack tip[3] (Fig. 17). The damage consists of many individual and coalesced cavities. The damage develops in a regime subject to both shear and normal stresses,

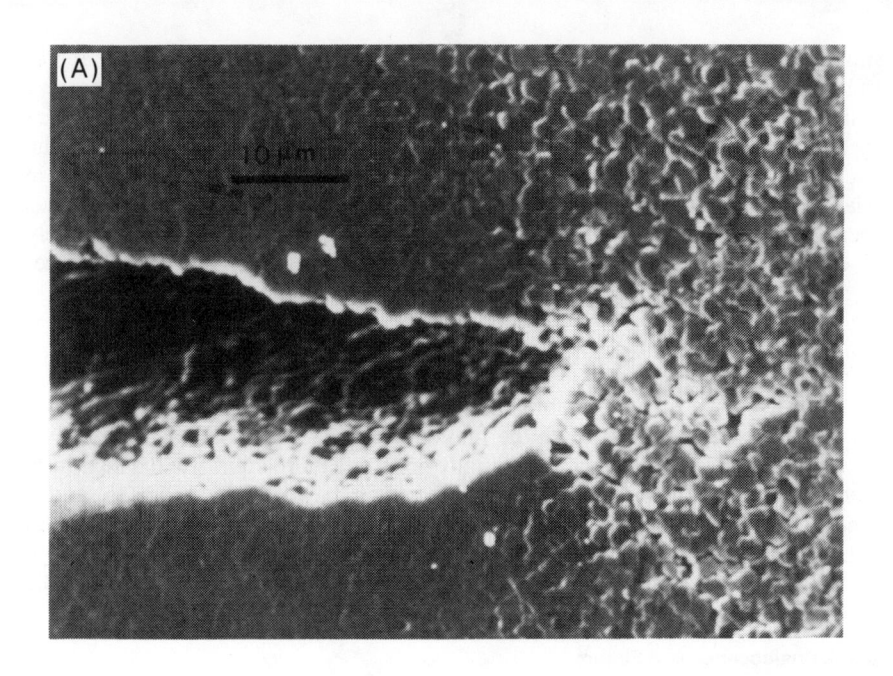

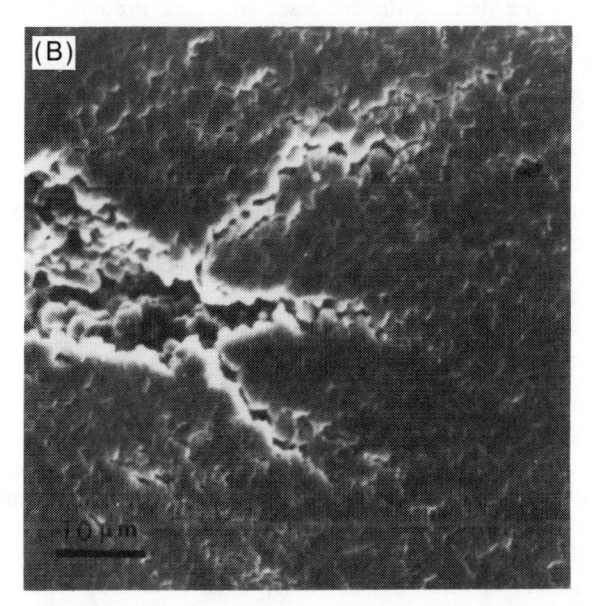

Fig. 17. Morphologies of the crack tip in the blunting range: (A) tip blunting; (B) damage coalescence behind the crack tip.

similar in shape to the plane strain plastic zone in metals. The presence of appreciable shear thus emerges as a necessary participant in damage zone development in the blunting range. Furthermore, since unconstrained cavity growth is governed by the normal stress, it is presumed that the shear stress is involved either in the nucleation of the cavities (e.g., due to grain-boundary sliding transients)[28] or in relaxing constraint on the damage zone.[29] In fact, consideration of the continuous shear displacements around the crack allows the blunting threshold to be tentatively modeled, by assuming that the side lobe damage initiates when the shear displacement beyond the first grain attains the critical value needed to nucleate cavities. On the basis of this premise, the threshold is anticipated[3] to be temperature independent and to vary with grain facet size as

$$K_{th}/K_c \approx (K_c l^{3/2})^{1/(n-1)} \qquad (21)$$

These trends are consistent with the limited data presently available.

Intrinsically nucleated cracks initially grow at a more rapid rate than extrinsic cracks. However, the cracks arrest and blunt when they extend in radius to several times the radius of the nucleating heterogeneity (Fig. 12(C)). This behavior, as noted above, is thought to be due to the presence of amorphous material (SiO_2) at the nucleation site. The blunting of these cracks is accompanied by the same side lobe damage morphology observed for extrinsic cracks.

Crack Coalescence and Failure

Final failure appears to involve coalescence along shear bands between arrested intrinsic cracks (Fig. 12). The macrocrack thus formed grows by the coplanar damage zone mechanism discussed in Crack Propagation and Blunting, whenever the crack attains a sufficient size to allow the stress intensity K to exceed the threshold value K_{th}. The growth time after the threshold condition has been exceeded[3] varies as, $t_p \sim \sigma_\infty^{-2}$. Hence, the final growth of the macrocrack to the critical size is probably not a controlling phase.

The coalescence stage of failure thus appears to emerge as the dominant time-consuming process during the high-temperature failure of alumina. Substantial nonlinearity, of a probabilistic nature, characterizes macrocrack formation by the coalescence of blunt intrinsic cracks[3] (at least when the intrinsic cracks continue to nucleate with time and exhibit a formation probability that increases with the applied stress). However, continuous crack nucleation and the associated stress dependence have not been experimentally investigated, and hence, the preceding emphasis on coalescence should be regarded as tentative.

Creep Rupture Transition

The transition between brittle fracture and creep rupture should coincide with the condition wherein the steady-state creep stress, at constant applied strain rate, becomes less than the brittle fracture stress, S. Hence, the transition can be emplaced, for example, on an isothermal creep map, with grain size and stress as coordinates (Fig. 18). In essence, the creep map is truncated at the brittle fracture stress, such that a creep rupture region emerges at lower strain rates in fine-grained Al_2O_3, while brittle fracture prevails at higher strain rates or in larger grained material. The brittle failure level, S, has been placed on the map, based on data for a single-phase Al_2O_3, with the microcrack transition (to the behavior $S \sim l^{-1/2}$) predicated on the value of l^* determined in Process Zone Initiation: Microcrack Transition. Within the creep rupture region, the dominant creep regimes can be identified and the associated strain rate specified, as indicated on Fig. 18.

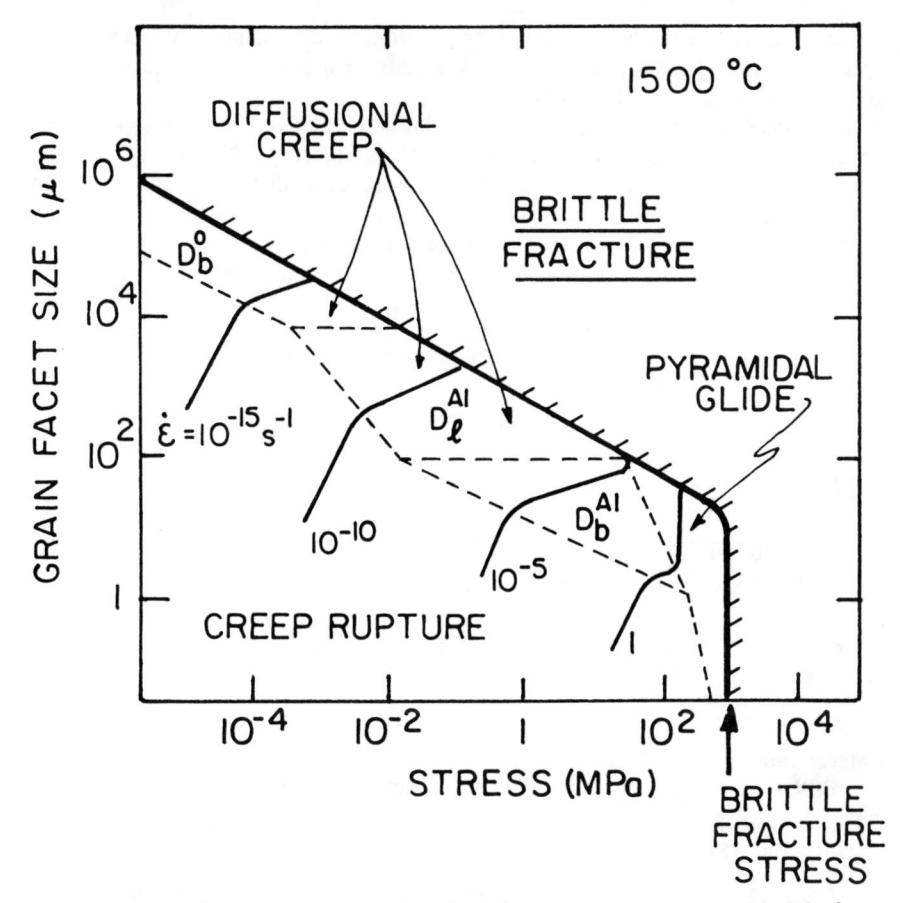

Fig. 18. Modified creep map showing the creep rupture and brittle fracture regimes.

Concluding Remarks

The present article reveals that the current understanding of the brittle failure of alumina at temperatures less than about half the melting temperature is well advanced. The understanding is predicated on knowledge of flaw mechanics, of effects of grain size and shape on the fracture toughness, and of the incidence of microcracking. Residual stresses, both around flaws and within grains, emerge as a dominant influence on the brittle failure process. Furthermore, the residual stresses within grains have been shown to be the source of a transition in behavior from a strongly grain size dependent strength at large grain size to a grain size insensitive strength at small grain size. This transition is related to the incidence of profuse microcracking.

The processes that dictate failure at elevated temperatures are at a more elementary stage of development. The contributing stages have been identified. These include the nucleation, growth, and blunting of intrinsic cracks (formed from heterogeneities), followed by crack coalescence across shear bands and eventual failure by growth of the coalesced intrinsic cracks, via a damage zone mechanism.

However, a general mode of failure, incorporating the dominant stages, has not yet been developed. Furthermore, appreciable additional research will be needed to address this relatively complex problem.

The studies of mechanical failure in alumina undoubtedly have implications for failure in other noncubic polycrystals, such as Si_3N_4, α-SiC, and $BaTiO_3$. The mechanisms of brittle fracture should be the same, with the specific trends in strength and toughness between materials being contingent upon $\Delta\alpha\Delta T$, the cleavage resistance and the dominant processing flaws. In fact, present evidence already indicates that the same strength and toughness trends pertain in Si_3N_4,[1-3] including the effects on toughness of grain shape and of microcracking.

The high-temperature failure of other polycrystals is also known to involve crack nucleation at heterogeneities and creep crack growth, qualitatively similar to that encountered with Al_2O_3. However, additional complexity is likely in the nonoxides, due to effects of oxidation on both crack nucleation and crack growth.[30] Insufficient information is available on this topic to make further comment at this juncture.

Acknowledgment

This work was supported by the Director, Office of Energy Research, Office of Basic Energy Sciences, Materials Sciences Division of the U.S. Department of Energy, under Contract No. DE-AC03-76SF00098, and by the Office of Naval Research under Contract No. N00014-81-K-0362.

References

[1] R. W. Rice, "Microstructural Dependence of Mechanical Behavior of Ceramics," *Treatise Mater. Sci. Technol.*, **11**, 199 (1977).
[2] A. H. Heuer, N. J. Tighe, and R. M. Cannon, "Plastic Deformation of Fine-Grained Alumina (Al_2O_3): II. Basal Slip and Nonaccommodated Grain-Boundary Sliding," *J. Am. Ceram. Soc.*, **63**, 53 (1980).
[3] A. G. Evans, W. Blumenthal, and M. Thouless; Deformation of Ceramics; unpublished work.
[4] S. M. Wiederhorn, "Fracture of Sapphire," *J. Am. Ceram. Soc.*, **52**, 485 (1969).
[5] R. W. Davidge and T. J. Green, "The Strength of Two-Phase Ceramic/Glass Materials," *J. Mater. Sci.*, **3**, 629 (1968).
[6] R. W. Davidge, "Cracking at Grain Boundaries in Polycrystalline Brittle Materials," *Acta Metall.*, **29**, 1695 (1981).
[7] A. G. Evans, "Microfracture from Thermal Expansion Anisotropy—I. Single Phase-Systems," *Acta Metall.*, **26**, 1845 (1978).
[8] Y. Fu and A. G. Evans; unpublished work.
[9] B. Budiansky and J. O'Connell, "Elastic Moduli of a Cracked Solid," *Int. J. Solids Struct.*, **12**, 81 (1976).
[10] Y. Fu and A. G. Evans, "Microcrack Zone Formation in Single Phase Polycrystals," *Acta Metall.*, **30**, 1619 (1982).
[11] A. G. Evans and K. T. Faber, *J. Am. Ceram. Soc.*, **67**, 255 (1984).
[12] H. Hubner and J. Jillek, "Sub-critical Crack Extension and Crack Resistance in Polycrystalline Alumina," *J. Mater. Sci.*, **12**, 117 (1977).
[13] N. Claussen, B. Mussler, and M. V. Swain, "Discussion of Grain Size Dependence of Fracture Energy in Ceramics," *J. Am. Ceram. Soc.*, **65**, C-14 (1982).
[14] R. Krehans and R. Steinbrech, "Memory Effect of Crack Resistance during Slow Crack Growth in Notched Al_2O_3 Bend Specimens," *J. Mater. Sci. Lett.*, **1**, 327 (1982).
[15] B. Budiansky, J. Hutchinson, and J. Lampropolous, "Continuum Theory of Dilatant Transformation Toughening in Ceramics," *Int. J. Solids Struct.*, **19**, 337 (1983).
[16] J. Zwissler and M. Adams; Fracture Mechanics of Ceramics, Vol. 6, 1983; p. 211.
[17] R. W. Rice, S. W. Freiman, and P. F. Becher, "Grain-Size Dependence of Fracture Energy in Ceramics: I, Experiment," *J. Am. Ceram. Soc.*, **64**, 345 (1981).
[18] F. F. Lange, "Transformation Toughening," *J. Mater. Sci.*, **17**, 225 (1982).
[19] D. L. Porter and A. H. Heuer, "Mechanisms of Toughening Partially Stabilized Zirconia (PSZ)," *J. Am. Ceram. Soc.*, **60**, 183 (1977).

[20] F. A. McClintock; p. 93 in Fracture Mechanics of Ceramics, Vol. 1. Edited by R. C. Bradt, D. P. H. Hasselman, and F. F. Lange. Plenum, New York, 1974.
[21] A. G. Evans, "Structural Reliability: A Processing-Dependent Phenomenon," *J. Am. Ceram. Soc.*, **65**, 127 (1982).
[22] D. B. Marshall, A. G. Evans, B. T. Khuri-Yakub, and G. S. Kino, "The Nature of Machining Damage in Brittle Materials," *Proc. R. Soc. (London), [Ser. A]*, **A385**, 461 (1983).
[23] K. T. Faber and A. G. Evans, "Crack Deflection Processes, II. Experiment," *Acta Metall.*, **31**, 577 (1983).
[24] R. W. Davidge and G. Tappin, "The Effect of Temperature and Environment on the Strength of Two Polycrystalline Aluminas," *Proc. Br. Ceram. Soc.*, **15**, (1970).
[25] S. M. Johnson; Ph.D. Thesis, University of California, Berkeley, CA, 1983.
[26] W. Blumenthal and A. G. Evans, Fracture Mechanisms of Ceramics; unpublished work.
[27] M. D. Thouless, C. H. Hsueh, and A. G. Evans, "A Damage Model of Creep Crack Growth in Polycrystals"; to be published in *Acta Metall.*
[28] A. S. Argon, "Intergranular Cavitation in Creeping Alloys," *Scr. Metall.*, **17**, 5 (1983).
[29] J. W. Hutchinson; to be published in *Acta Metall.*
[30] S. M. Wiederhorn and N. J. Tighe, "Proof-Testing of Hot-Pressed Silicon Nitride," *J. Mater. Sci.*, **13**, 1781 (1978).

---

*It is cautioned that this result is predicated upon the invariant existence of a critical sized heterogeneity at each facet corner. The nature of these heterogeneities and their spacial/severity characteristics are not understood. Deviations from Eq. (2) could thus emerge in certain cases that would not be explicable in terms of current comprehension.

†Larger microcrack densities develop for equibiaxial stress, because of a more effective sampling of the facets subject to residual tension (Ref. 8), such that $m \approx \frac{1}{2}$.

‡The crack tip field is characterized by K_I whenever the microcracking saturates, as shown in Fig. 7(b), resulting in a linear stress–strain curve at very large strains. The validity of the linearity premise in single-phase materials is not well understood. However, if it is assumed that the residually *compressed* facets are not susceptible to discrete microcracking (Process Zone Initiation: Microcrack Transition), saturation occurs when 50% of the crack tip facets are microcracked.

§The independence assumption is probably more reasonable than might first be appreciated, because the residually compressed ligaments do not transmit significant stresses to the neighboring facets subject to residual tension.

¶Elongated grains are often formed during liquid-phase sintering, and a residual silicate remains at the grain boundaries. It is presumed, without justification, that the glass phase has no direct influence on the toughness.

Note added in press: The ordinate, "Grain Facet Size," in Fig. 18 is incorrect; the values should be reduced by a factor of 10.

Effects of Temperature and Stress on Grain-Boundary Behavior in Fine-Grained Alumina

J. D. Fridez, C. Carry, and A. Mocellin

Ecole Polytechnique Féderale
Laboratoire de Céramique
CH-1007 Lausanne

Normal grain growth and compressive creep deformation with concurrent grain size change have been studied jointly on identical fine-grained Mg-doped aluminas at 1450°C. The time dependence of the grain growth kinetics is not affected by the application of stress, but the rate constants are increased by a factor of up to ~4 at stresses not exceeding 30 MPa. Intergranular alkali containing platelike precipitates are thought to have limited grain growth but do not seem to have influenced strain rates up to total strains of ~ −0.2 when cavitation sets in. Deformation is mostly governed by grain-boundary diffusion and follows a course similar to that observed in previous studies. Large strains up to −0.50 have been reached by superplastic-type grain rearrangements and are quantitatively analyzed by assuming that quasi-steady state had prevailed during creep. It is argued that grain growth and boundary diffusion accommodating grain-boundary sliding are strongly coupled processes. When their respective rates are not governed by the same mechanism at the atomic level, incompatibilities will build up at the microscopic level, thus impairing the microstructural quality (e.g., by cavitation). Conversely, when both controlling mechanisms are identical, as can be the case in sufficiently pure materials, substantial superplastic strains are accessible with concurrent grain growth, as previously demonstrated with Al_2O_3. The argument is thought to apply to other fine-grained ceramics.

Under appropriate temperature and/or stress conditions, grain boundaries of polycrystalline materials are set to motion, thus bringing changes in microstructures and properties. An element of grain-boundary surface that does not meet singularities (e.g., grain edges or corners) then has a well-defined velocity in general consisting of both normal and tangential components, respectively. The former drives the element of grain-boundary surface toward its center of curvature and for the polycrystal as a whole is usually considered to be in relation with grain growth. The latter component in turn may be considered as describing the relative displacement along the boundary of either crystals adjacent to it, which is usually referred to as grain-boundary sliding. When viewed at the microscopic level then, grain-boundary sliding and grain-boundary migration leading to grain growth appear to be intimately related or coupled as was already pointed out some years ago.[1]

The literature provides a substantial body of information on average grain-boundary mobilities derived from grain growth experiments, particularly in Al_2O_3 and MgO,[2] and so does it on diffusional creep,[3,4] which is often considered in association with grain-boundary sliding.[5] For obvious simplicity reasons, in many studies some sort of steady state (e.g., time-independent distribution of grain sizes relative to the average during grain growth or constant grain size during deformation) was either assumed or aimed at through "microstructure stabilizing pretreatments." To the authors' knowledge, however, transient regimes in normal grain growth have rarely been experimentally investigated, and relatively little information is available on deformation with concurrent grain size changes.

Concerning the latter, either they have at times been claimed inconsequential[6] or the resulting transients were interpreted in terms of steady-state rate equations, a posteriori incorporating the time dependences of grain sizes[7] whereas said equations had been derived under a constant grain size assumption.

Such procedures certainly may be acceptable in some circumstances. But as the above introductory remarks on grain-boundary behavior suggest, it may happen for the grain growth kinetics to be stress (or strain) dependent, particularly so when concurrent with deformation via grain-boundary sliding. This effect, if it occurs,[8–10] should also be considered in theoretical attempts to analyze the deformation of polycrystals. It was the purpose of the present work to obtain information on the average behaviors of grain boundaries in fine-grained alumina as influenced by temperature and stress. Experimental and testing conditions were a priori selected in order that grain-boundary sliding could be a dominant deformation mode and that structural observations be purposely made during transients. It was thought that the effects of coupled mechanisms would thus be more readily brought to evidence and their magnitudes evaluated.

Experimental Material and Procedures

A commercially available alum-derived Al_2O_3 powder* was magnesia doped to a nominal concentration of 500 ppm MgO by weight. According to a typical analysis from the supplier, the residual impurity levels (in ppm by weight) are the following: Na = 50, K = 120, Ca ≤ 5, Fe = 15, Ga = 3–30, Pb = 2, Si = 30, S < 100, and others (Cu, Mn, Mo, Ni, V, B, Ti) <5 ppm each. As the alkali appear in concentrations that might be in excess of their solubility limits,[11] they were independently analyzed by radioactivation techniques (Table I). Sulfur also was independently analyzed by coulometry and found present in the starting powder at the 15 ± 1 ppm level.

Disks with a 5-cm diameter by ~1.2–1.5-cm thickness were vacuum hot-pressed in a tungsten resistance furnace for 15 min at 1450°C, using graphite molds and pistons. The nominal hot-pressing load was 45 MPa. Figure 1 schematically shows the positions and orientations of samples that were obtained from the hot-pressed disks by diamond cutting and grinding. Two sets such as those sketched on Fig. 1 could be prepared from every disk, each consisting of the as-pressed piece ("I") used to characterize the initial microstructure, the 33 × 11 × 11 mm³ bar ("D") to be hot deformed, and the reference ("T") piece

Table I. Impurity Levels in Various Al_2O_3 Samples (ppm wt)

Matrix	Na	K
As-received powder	100 ± 20	150 ± 20
As hot-pressed (1450°C; 15 min)	150 ± 20	290 ± 40
Deformed (30 MPa; −0.5)	60 ± 20	150 ± 40

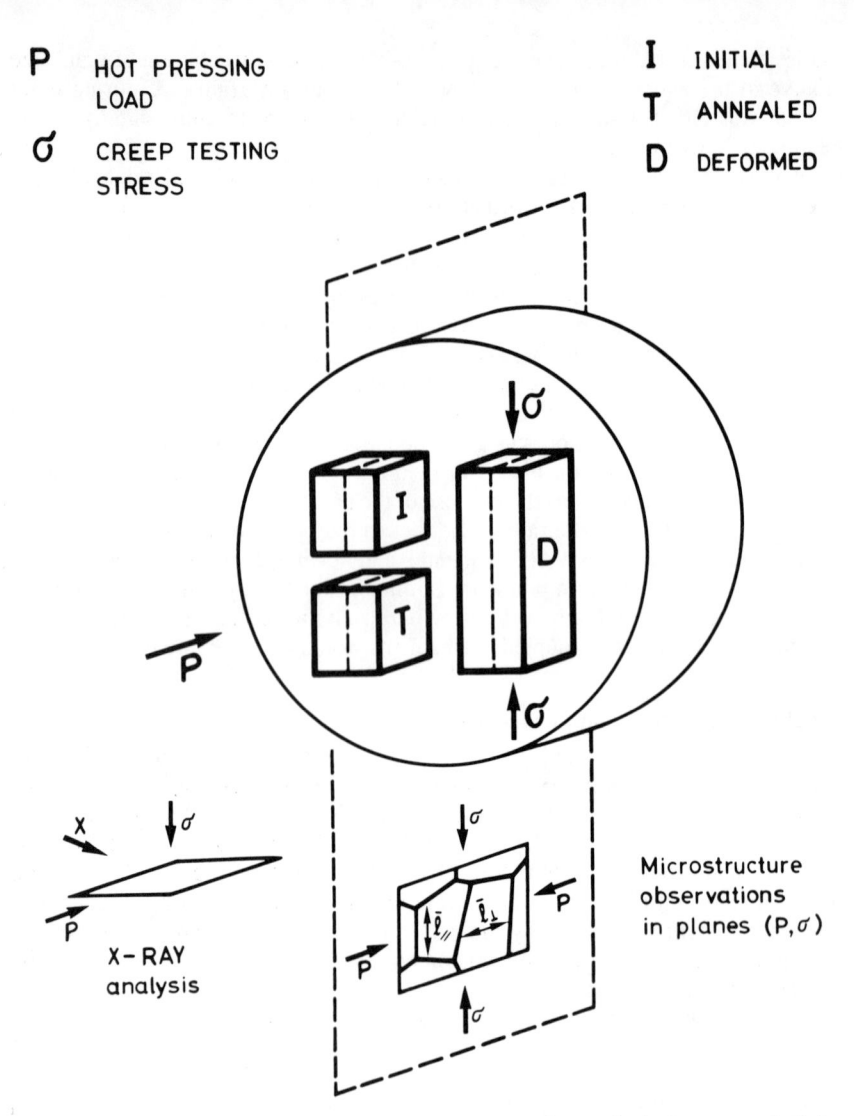

Fig. 1. Positions of samples and observation planes in hot-pressed disks.

to be placed in the high-temperature furnace next to the sample being deformed, so as to receive the same heat treatment as its deformed counterpart but in a stress-free condition.

Also shown on Fig. 1 are the relative orientations of median planes, diamond polished and thermally etched, from which structural information was obtained (~5 min at 1250–1300°C in flowing H_2). For evaluation of average grain section areas and intercept lengths (parallel and perpendicular to the hot-pressing and deformation stress directions, respectively), 700–800 grains were randomly sampled in every case. Bulk densities were measured on all hot-pressed, annealed, and deformed samples with the standard Archimedes method of immersion in water. Results are accurate within ±0.005 g/cm³ absolute specific gravity. In

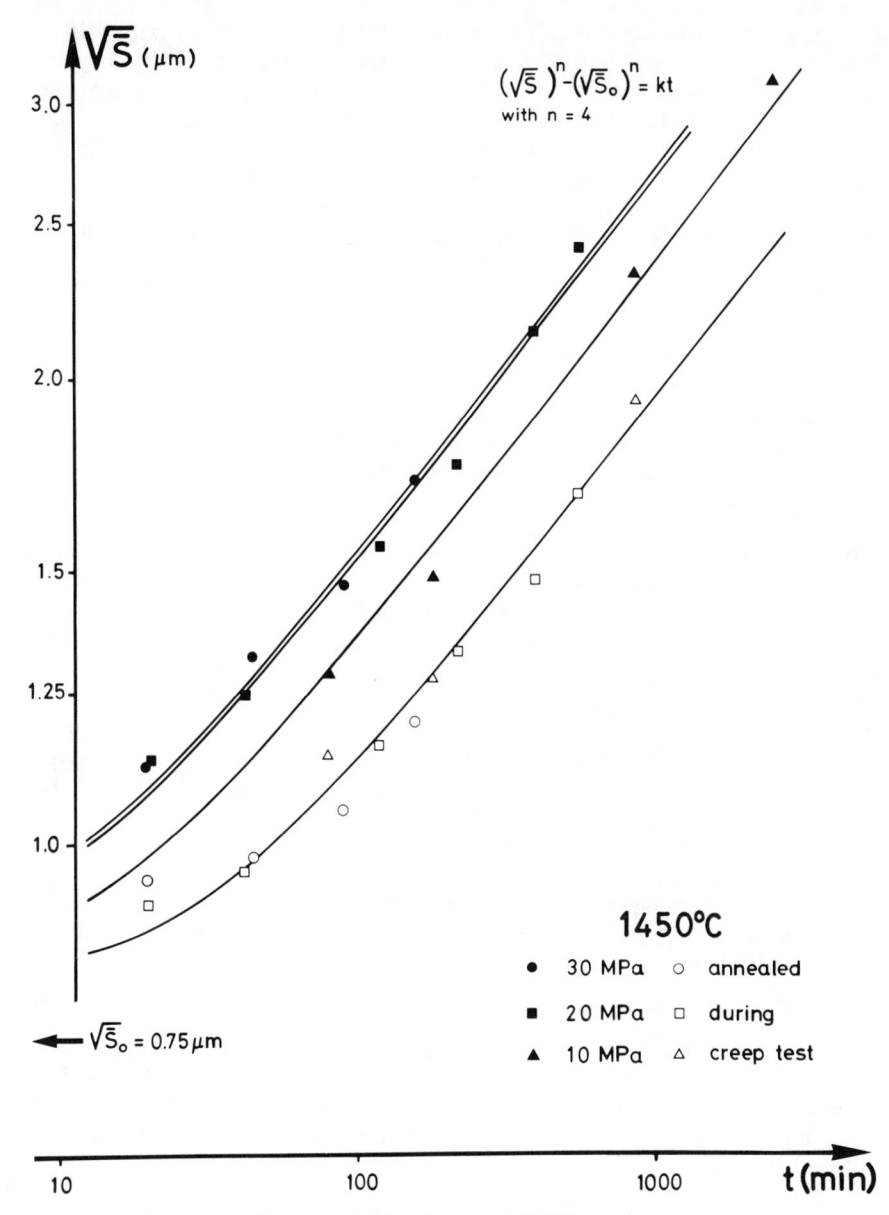

Fig. 2. Normal grain growth kinetics at 1450°C.

$(\sqrt{\bar{S}})^n - (\sqrt{\bar{S}_o})^n = kt$ with n = 4

$\sqrt{\bar{S}_o} = 0.75\,\mu m$

1450°C
- ● 30 MPa ○ annealed
- ■ 20 MPa □ during
- ▲ 10 MPa △ creep test

some instances also TEM observations were made at 100 kV on diamond-ground and ion-thinned slices with the applied stress direction well identified.

Compressive creep tests under constant stresses, mostly 10, 20, and 30 MPa, with some runs at 40 or 50 MPa, were made under flowing purified argon (room-temperature P_{O_2} at the gas inlet was $\sim 10^{-14}$ atm (9.80×10^{-10} Pa)). The total load applied to the sample during deformation was manually adjusted from time to time and permanently controlled within ±20 kg. Test durations ranged from 20 to ~2500 min. The same furnace as for hot-pressing was used for the joint grain growth–creep experiments. High-quality silicon carbide slugs and thin graphite sheets were used to transfer loads from the graphite pistons to the sample. This setup together with the piece to be stress-free annealed was positioned within a graphite sleeve in the furnace. Most runs were made at 1450°C with sheathed 5–26% W–Re thermocouples being used for temperature measurements. True strains $\varepsilon = \ln h/h_0$ (where h_0 and h refer to initial and present sample heights, respectively) and strain rates $\dot{\varepsilon} = d\varepsilon/dt$ were evaluated from continuous recordings of relative piston displacements, as delivered by an external LVDT. Calibration was achieved by a posteriori measuring the overall sample height reduction with a caliper. No barreling was noted even after 50% strain, but in some cases the samples were found slightly twisted after deformation. This effect will be neglected in the following discussion. Finally, stress sensitivity exponents and activation energies of strain rates were evaluated according to the standard method of instantaneous changes in stress (±5 MPa) and temperature (±50°C), respectively.

Results and Discussion

Structural Observations

The normal grain growth kinetics at 1450°C are shown on Fig. 2 as $\log(\bar{s}^{1/2})$ vs log time for both annealed and deformed sample types. Actual times are given in minutes, and linear grain dimensions are taken as the square root of the average grain intercept area. The initial grain size, averaged over several as-pressed microstructures, is also indicated as $\bar{s}_0^{1/2}$. It clearly appears that at all times and stresses the average grain sizes of deformed samples have grown bigger than in their stress-free annealed counterparts. Attempts at fitting standard equations of the type $(\bar{s}^{1/2})^n - (\bar{s}_0^{1/2})^n = K_n t$ were made with a priori assumed n values of 2–4. The best, although not perfect fit[†] was obtained for $n = 4$ at all stress levels. The solid lines on Fig. 2 were drawn for $n = 4$, and numerical values for the rate constants $K_4(\sigma)$ are given in Table II.

The analysis suggests that the effect of stress (or strain) is primarily felt on the rate constant. At this stage then, the observed overall grain growth kinetics can be summarized as

$$(\bar{s}^{1/2})^4 - (\bar{s}_0^{1/2})^4 = K_4(\sigma)t \tag{1}$$

where $K_4(\sigma)$ is some function of the applied stress. It is also of interest to note that the relative influence of stress on grain growth appears more important at low stresses (0–20 MPa) than at higher levels (20–30 MPa). This trend is confirmed by results obtained with 40- and 50-MPa applied stresses, not to be further discussed here.

Figure 3 shows a similar log–log plot of average grain intercept lengths in directions normal and parallel to the applied stress or hot-pressing directions,

Table II. Grain Growth Rate Constants at 1450°C

Applied stress (MPa)	0	10	20	30
Rate constant K_4 (10^{-28} m^4/s)	2.4	5.4	9.0	9.5

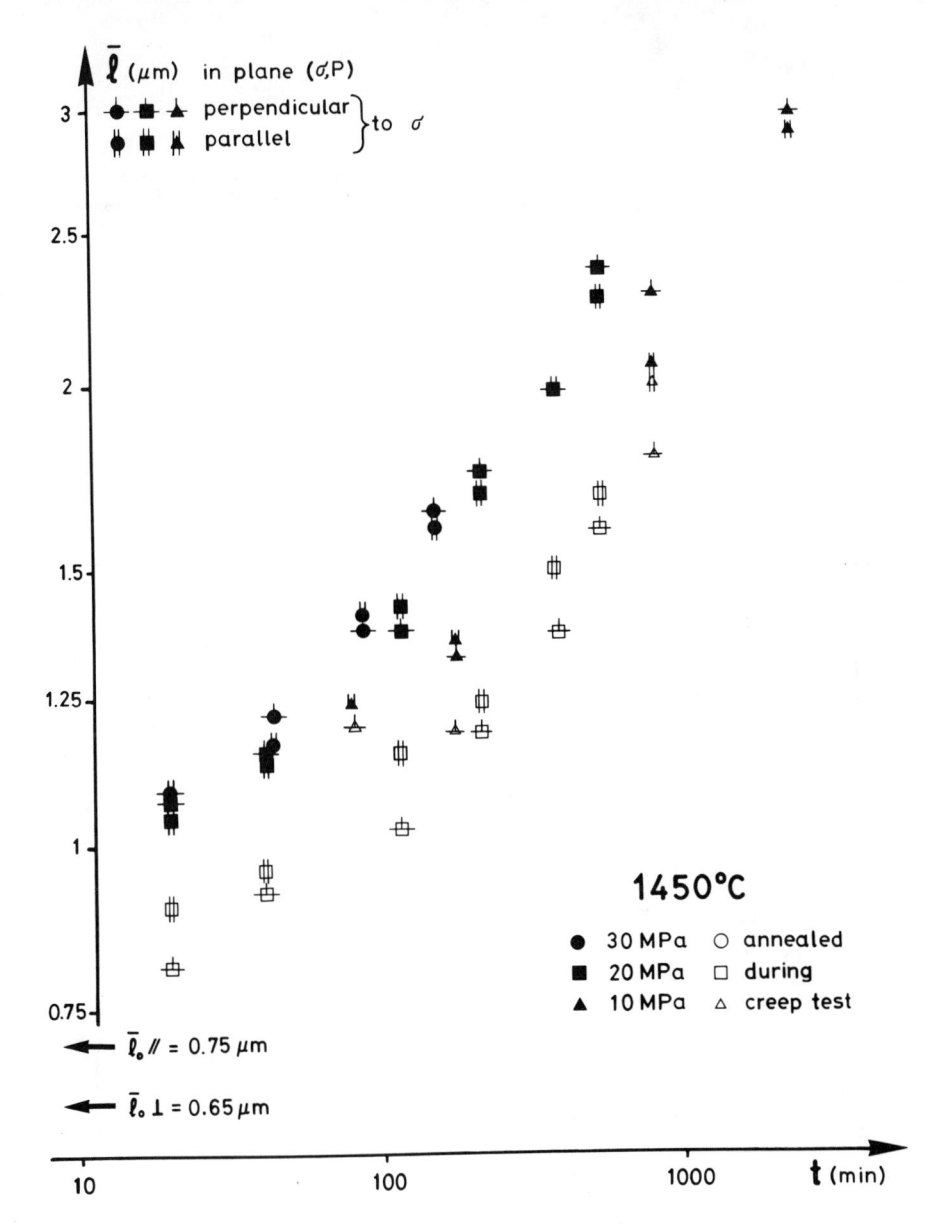

Fig. 3. Grain anisometry changes during growth.

respectively (cf. Fig. 1). It brings confirmation of the stress-enhanced grain growth. It also shows a slight average grain anisometry in the as-pressed condition, which is retained during stress-free annealing, but is rapidly reduced in the stressed samples that maintain an equiaxed structure throughout most of their deformation.

The results of bulk density measurements are shown on Fig. 4 vs final strain. Also shown are corresponding reference values for stress-free annealed samples. The latter are seen to have lost a typical 0.2% relative density, presumably due to

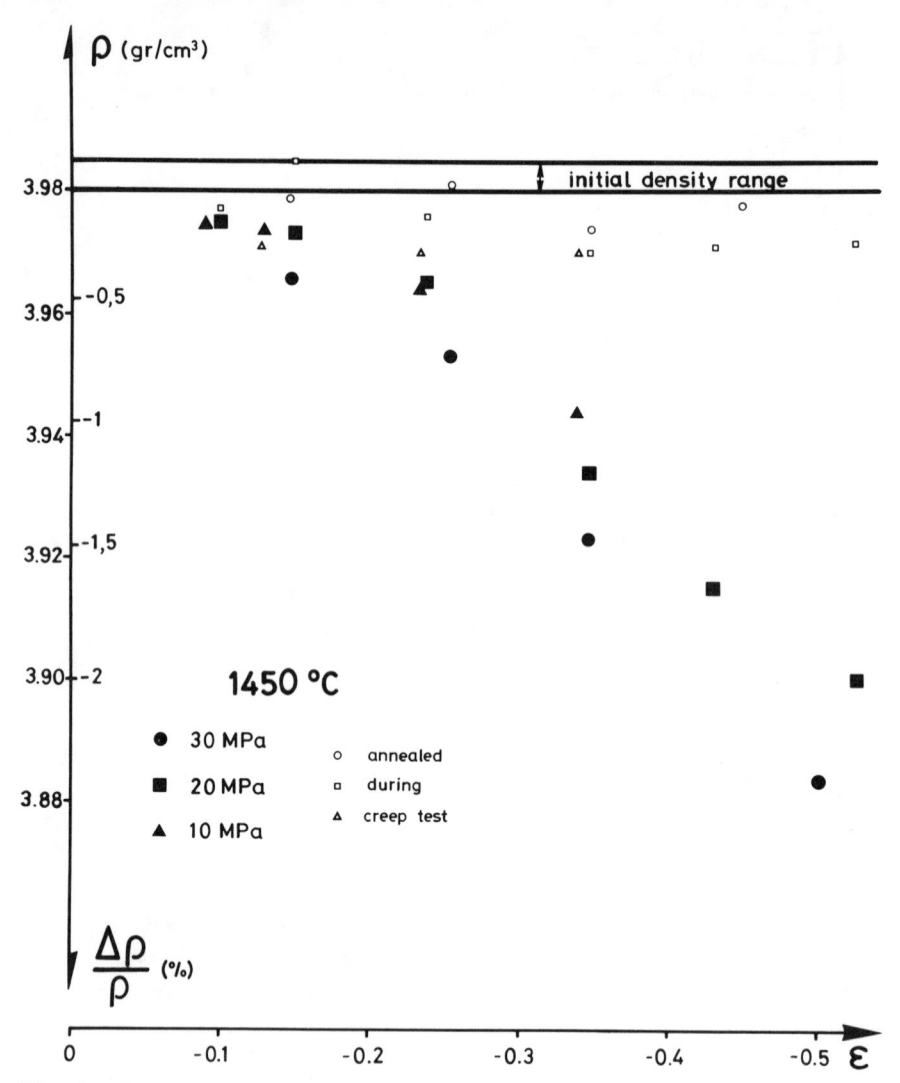

Fig. 4. Density variations during annealing and creep.

internal impurity effects or to exchanges with the furnace environment in zones near the external sample surface. The former lose ~0.5% relative density up to −0.25 strain under 10 or 20 MPa, which may be considered rather small. Density decreases however are more pronounced under a 30-MPa stress.

This difference between samples deformed under 30 MPa and the other sets may also be noted from the results of X-ray analyses. Figure 5 shows the (116) pole diagrams for the −0.50 strained sample and the other two related ones from the same hot-pressed disk — the initial state and the stress-free annealed state. The latter two appear rather similar to each other but somewhat different from the former in which an incipient basal texture can be noted. This trend may already be seen after −0.25 strain on powder patterns analyzed as previously described.[3c]

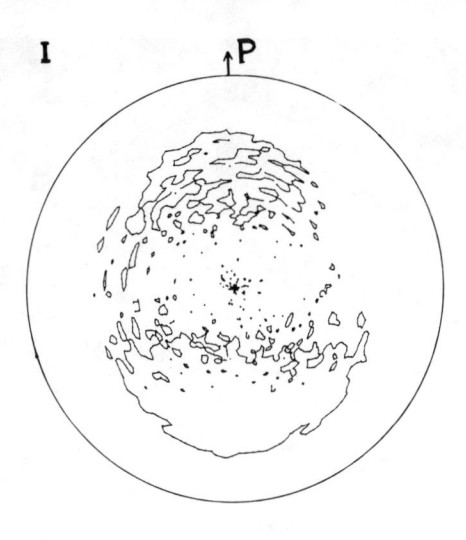

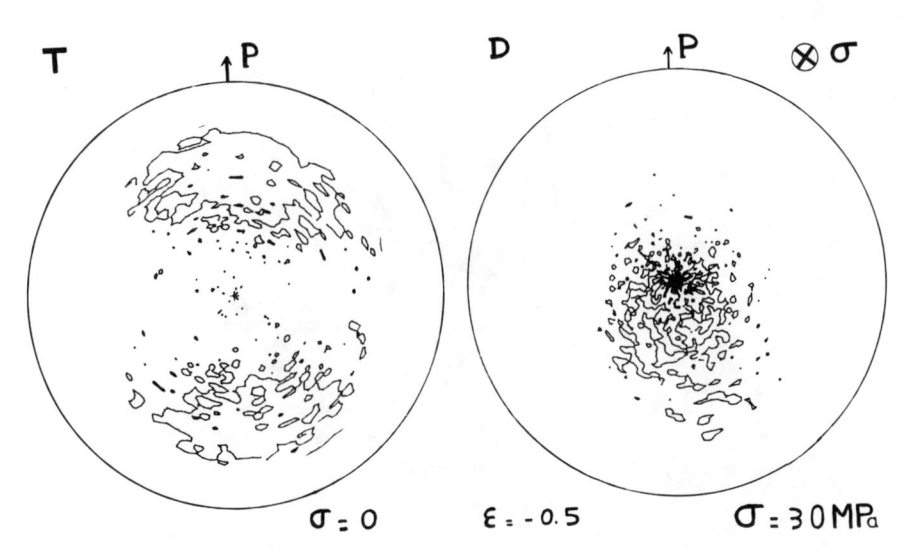

Fig. 5. (116) pole figures: *I*, as-pressed material, level 500; *T*, annealed ~3 h at 1450°C, level 500; *D*, deformed (−0.5; 30 MPa) during same time, level 600.

Direct microstructural observations (Figs. 6–8) are of particular interest in several respects. When they are made on as-pressed, stress-free annealed and deformed samples, respectively, they first confirm the overall homogeneity of microstructures: no evidence for abnormal grain growth was detected throughout this study (Figs. 6–8). They also show well-distributed intergranular platelike

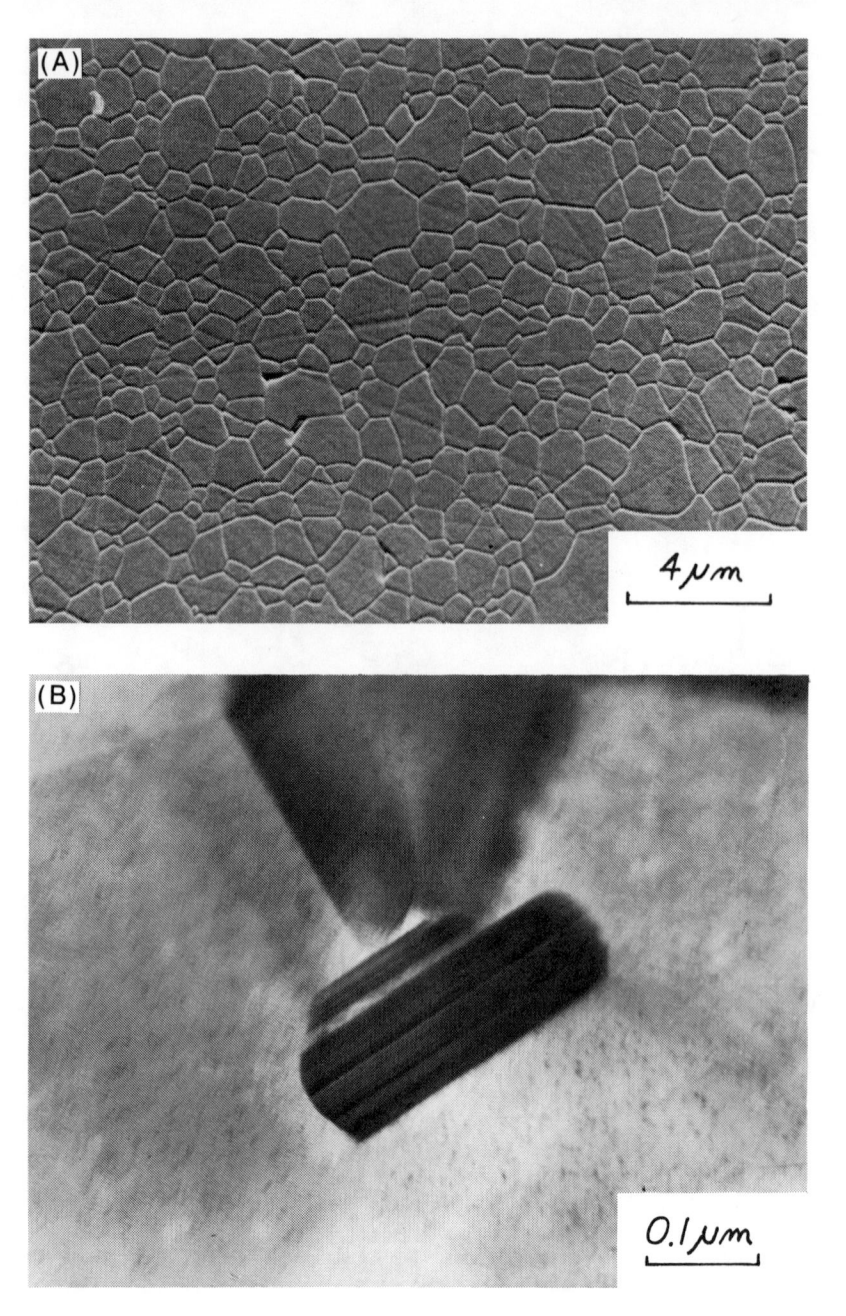

Fig. 6. Microstructures of as-pressed material: (A) SEM of initial microstructure; (B) TEM showing example of intergranular plate.

Fig. 7. Microstructure of stress-free annealed sample with precipitate relicts.

precipitates, presumably of a β-Al$_2$O$_3$ type (Table I and Ref. 11). These are present already in the as-pressed material (Fig. 6(B)), and their average sizes have a tendency to increase upon subsequent heat treatment either with or without stress (Fig. 8). But no systematic quantification was made here of precipitate volume fraction or growth kinetics. Similarly, the absence of intergranular cavities or microcracks was noted in the as-pressed and annealed materials (Figs. 6 and 7) whereas they are clearly visible in varying concentrations, shapes, and sizes after ~15–20% strain (Fig. 8). Often they appear associated with platelike precipitates (Fig. 8). Finally, no dislocations could be seen within the alumina grains except in the most extensively deformed samples under a 30-MPa stress. There small amounts of dislocations, the detailed characteristics of which have not yet been analyzed, were to be found in a few individual crystals (Fig. 8(D)).

Mechanical Behavior

The experimental information pertaining to the deformed samples may be summarized as follows. Figure 9 shows the evolution of recorded strain rates vs total strains at 1450°C for the different stress levels. It should be pointed out that all data points are arithmetic averages of up to six different deformation runs. The run multiplicities are indicated on the figure. It is clear from Fig. 9 that steady state has barely if at all been reached after a respectable deformation to ~50% strain. This of course was to be expected in view of the previously described structural evolution, but the usual information on grain size and stress sensitivity of the strain rate is made more difficult to generate and perhaps more open to question. It seems however possible to approach the influence of grain size through a plot of the recorded terminal strain rates, i.e., those recorded just before the release of stress and furnace shutdown. Figure 10 shows such a plot of log $\dot{\varepsilon}_{\text{terminal}}$ vs log $\bar{s}^{1/2}$, where

(A)

10 μm

(B)

1 μm

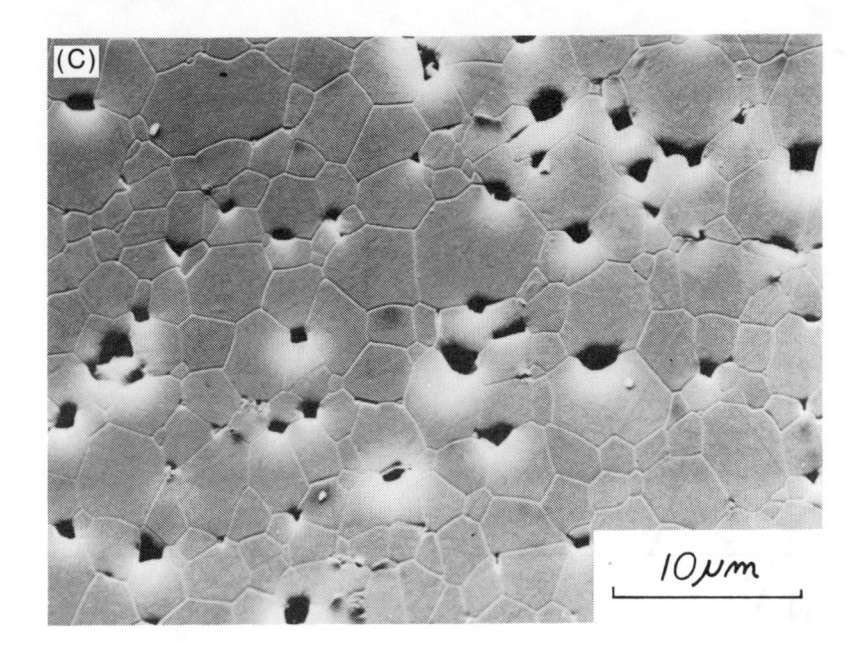

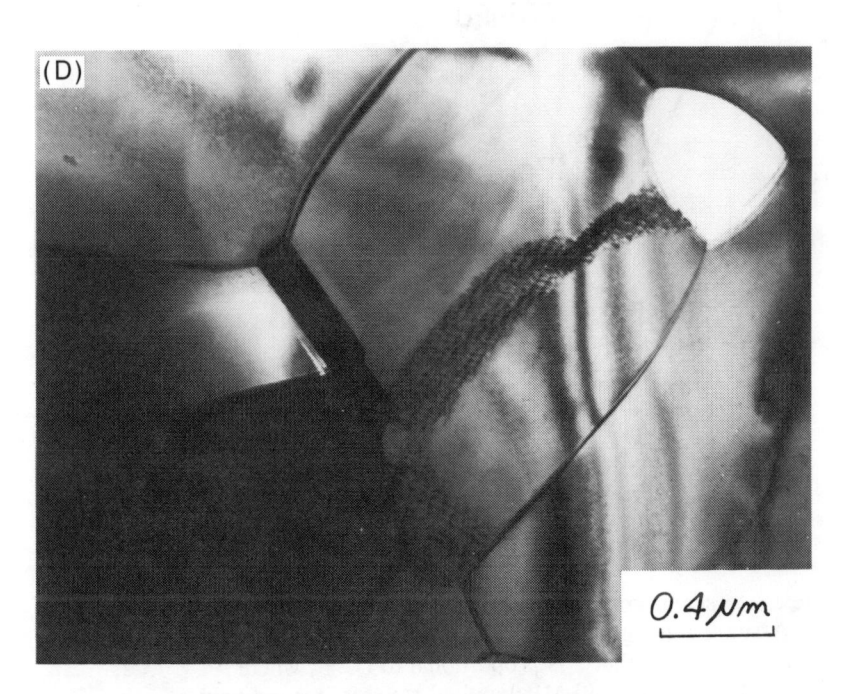

Fig. 8. Microstructures of deformed samples: (A) onset of cavitation ($\varepsilon = -0.24$; $\sigma = 10$ MPa); (B) cavities and occasional dislocations ($\varepsilon = -0.24$; $\sigma = 30$ MPa); (C) cavities distributed at large strains ($\varepsilon = -0.53$; $\sigma = 20$ MPa); (D) dislocations, cavity, and intergranular precipitate ($\varepsilon = -0.50$; $\sigma = 30$ MPa).

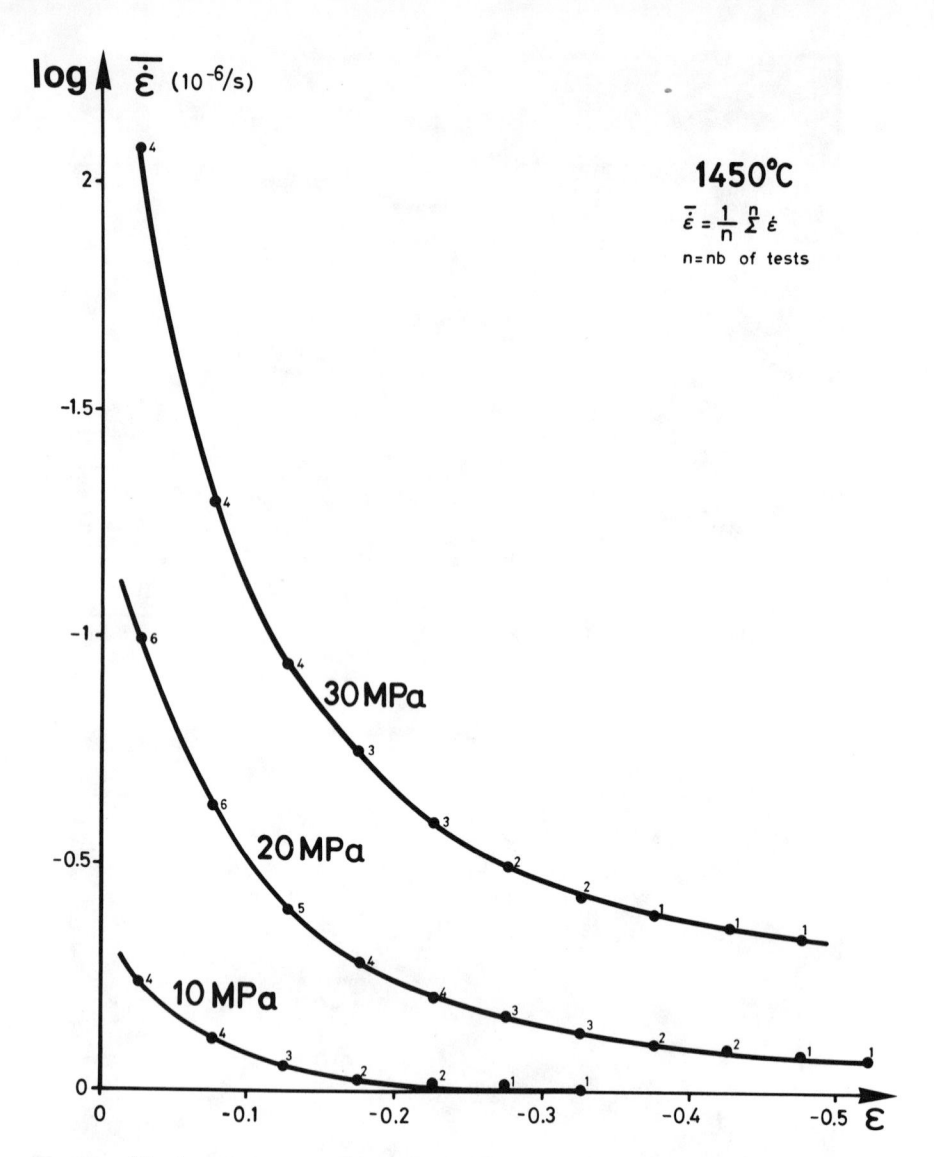

Fig. 9. Strain rates vs strain at 1450°C.

$\bar{s}^{1/2}$ refers to actual grain size measurements obtained on the corresponding samples. Assuming no grain size change during cooling and within experimental accuracy, it appears that $\dot{\varepsilon}$ is proportional to $(\bar{s}^{1/2})^p$, where $p = -3.2, -2.8$, and -2.5 for 10-, 20-, and 30-MPa applied stresses, respectively.

Another perhaps less satisfactory way to evaluate the grain size exponent consists in combining information from Fig. 9 with kinetic Eq. (1), phenomenologically describing the experimental grain growth results. Values computed in this fashion are $p = -2.6, -2.8$, and -3.1 for $\sigma = 10, 20$, and 30 MPa, respectively. It so appears that a common $p = -3$ value does not provide

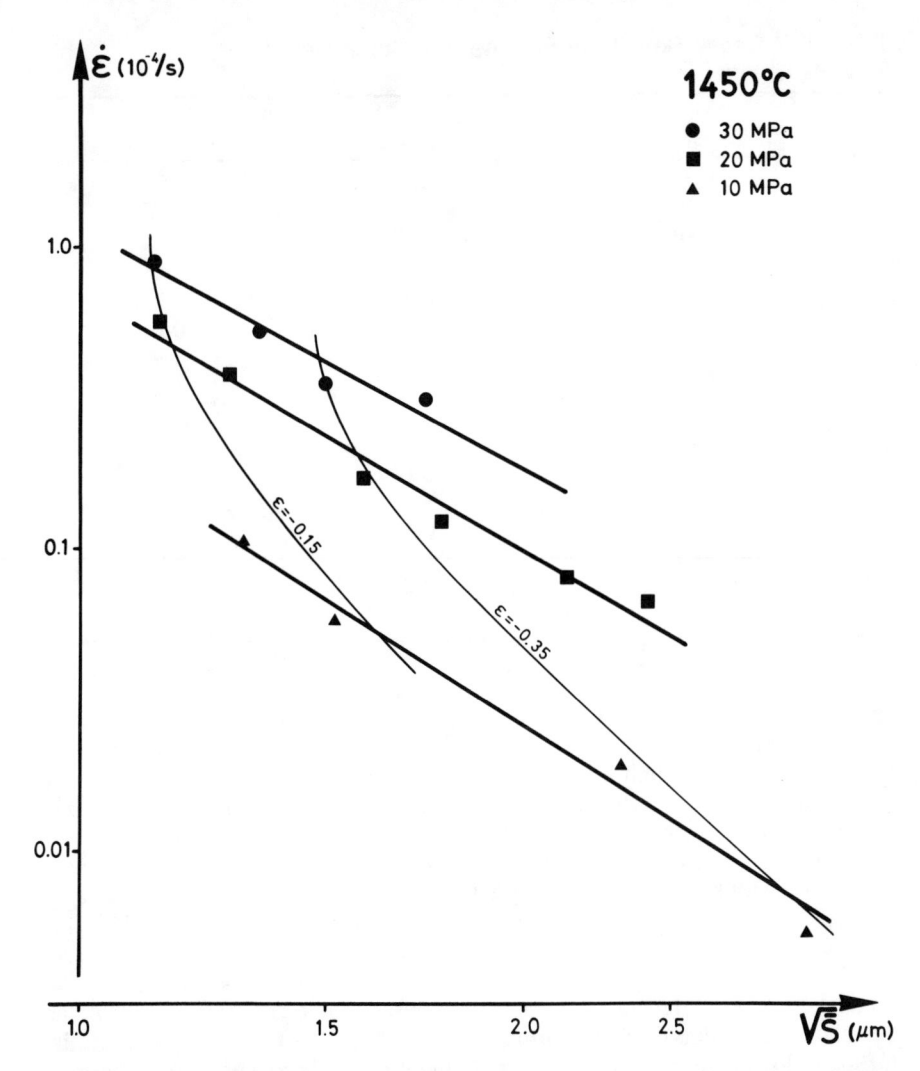

Fig. 10. Measured grain size dependence of strain rate under non-steady conditions.

too unrealistic a picture for the effect of (varying) grain size on the observed strain rates.

The stress sensitivity "m" exponent in the phenomenological $\dot{\varepsilon} \propto \sigma^{1/m}$ relationship was measured at 1450°C. Instantaneous stress changes from 27.5 to 32.5 MPa and back yielded $m \simeq 0.50$ strain. Similarly, when the stress jumps were set between 15 and 20 MPa, a series of different m values were obtained and are given in Table III with the corresponding state of strain reached in the sample.[‡]

Interpretation and Discussion

The observed kinetics and the presence of intergranular precipitates in the as-pressed material and throughout the thermal treatments suggest that grain

Table III. Stress Sensitivity Exponent of Strain Rate under a 20-MPa Stress

Total strain	Stress change (MPa) 20 → 15	Stress change (MPa) 15 → 20
−0.091	0.40	
−0.109		0.67
−0.138	0.74	
−0.156		1.06
−0.175	0.53	
−0.189		0.86
−0.205	0.63	
−0.217		1.20
−0.233	0.72	
−0.241		0.98
−0.258	0.66	
−0.269		0.82
−0.293	0.56	
−0.301		0.63
−0.326	0.49	
−0.333		0.61

growth was to a large extent controlled by this dispersed phase. Although it was not measured directly, the precipitate volume fraction is likely to have been of the order of ~1%, which is sufficient to effectively slow down grain-boundary migration.[11] Voids or cavities, which usually were smaller in amounts and sizes and appeared at a later stage only in the stressed samples, are not believed to have had much influence on grain growth. Such a $t^{1/4}$ kinetic law may result[12] from the viscous drag of intergranular precipitates with material transport governed by interfacial diffusion[13] and/or from precipitate coarsening via grain-boundary diffusion.[14] The gradual dissolution of grain-boundary anchoring precipitates also could yield this time dependence.[12] This latter possibility should be taken into account in view of results of chemical analyses (Table I), suggesting that the overall alkali contents decrease during prolonged heat treatment.[8] It is not essential for the following discussion, however, to more precisely identify the exact grain growth rate-controlling mechanism. But it should be realized that the chemical compositions at or near the grain boundaries may well have been modified during deformation and/or grain growth.

It is possible to further analyze the overall rate constant in Eq. (1). If we assume it may be written as

$$K_4 = K_0' \exp\left[-\frac{Q_G - \sigma V^*}{RT}\right]$$

where Q_G is the grain growth activation energy without stress, σ is the nominally applied stress (constant for each sample set), and V^* is an ad hoc activation volume, then from Table II, $V^* = 1.93$ and 1.57×10^{-27} m^3/"molecule" for $\sigma = 10$ and 20 MPa, respectively. Taking these values as identical within experimental uncertainties finally yields $V^* \approx 40\,\Omega$, where Ω is the Al$_2$O$_3$ molecular volume. It may also be commented that although small (~4–5%) in comparison with typical activation energies in alumina, the mechanical energy term σV^* obviously does play a definite role. And it is of interest to note that the $\exp[\sigma V^*/RT]$ factor may be rewritten as $\exp[\sigma/\sigma^*]$, where σ^* is merely defined as $\sigma^* \equiv RT/V^*$. From Table II, $\sigma^* = 12$ and 15 MPa for $\sigma = 10$ and 20 MPa, respectively. It may be argued finally that the absence of further enhancement of grain growth when the applied stress is changed from 20 to 30 MPa (or beyond) is due to other mechanisms beginning to operate.

With respect to the deformation behavior, both the present and previously[10] published results clearly show that diffusional creep dominates, presumably as a means to accommodate grain rearrangements in structural superplasticity.[15] It is however to be suspected that at least one other mechanism, involving dislocations, may also contribute to the overall deformation for the 30-MPa stress level. Also, considering our working grain sizes and stresses, the present results are qualitatively in agreement with predictions of a recently published[3c] deformation map for polycrystalline MgO-doped Al_2O_3, although our tests were conducted to substantially higher homogeneous strains and under a priori nonsteady conditions, with concurrent grain growth. According to this map, which was redrawn to outline our conditions (Fig. 11),¶ provided that a quasi-steady state prevailed throughout our experiments (an assumption which from now on we shall make), then the following results are observed:

(i) Under a 10-MPa Stress: Deformation appears to take place via grain-boundary diffusion with interface-reaction control in the early stages of deformation, gradually changing to control by matter transport along the boundaries. Since the transition between these two rate-limiting steps in series spreads over most of the grain size range covered here, then our measured strain rate should in fact be written as $1/\dot{\varepsilon} = 1/\dot{\varepsilon}_S + 1/\dot{\varepsilon}_D$, where $\dot{\varepsilon}_s$ refers to the rate when interface reaction is controlling, whereas $\dot{\varepsilon}_D$ corresponds to diffusion control.[16] Thus, even with simple constitutive laws for $\dot{\varepsilon}_S$ and $\dot{\varepsilon}_D$, the apparent stress and grain size sensitivity exponents should deviate from standard values and probably be time dependent. The experimental data now available are felt insufficient to separate each contribution in a precise unambiguous way. In particular, the existence of a threshold stress in the interface control regime is still a matter of controversy with estimates ranging from below 1–2 MPa[17] to ~13.8 MPa[3b] for grain boundaries presumably free from all secondary phases and chemically "stabilized".** The present evidence for an enhancement of grain-boundary migration might also be considered as an alternative to experimental threshold estimates.

(ii) Under a 30-MPa Stress: Deformation besides diffusional creep appears to also involve some dislocation activity, presumably basal glide. This is confirmed here by TEM observations and the development of a basal texture observed in the more highly strained samples. Under such conditions then the overall strain rate should again be a combination of both these mechanisms acting in parallel: $\dot{\varepsilon} = \dot{\varepsilon}_D + \dot{\varepsilon}_G$, where $\dot{\varepsilon}_G$ now refers to the basal glide contribution. Here also, the mixed deformation mode complicates the issue, not to mention the added difficulty arising from the development of cavities that may well offset the further stress enhancement of grain-boundary migration beyond 20 MPa.

(iii) Under a 20-MPa Stress: Finally, things perhaps should be more simple, since most of the investigated grain size range belongs to a single deformation mechanism field: diffusional creep controlled by transport along the grain boundaries. Stress sensitivity values (Table III) bring support to this conclusion except at small overall strains when interface reaction might still be limiting and at large strains

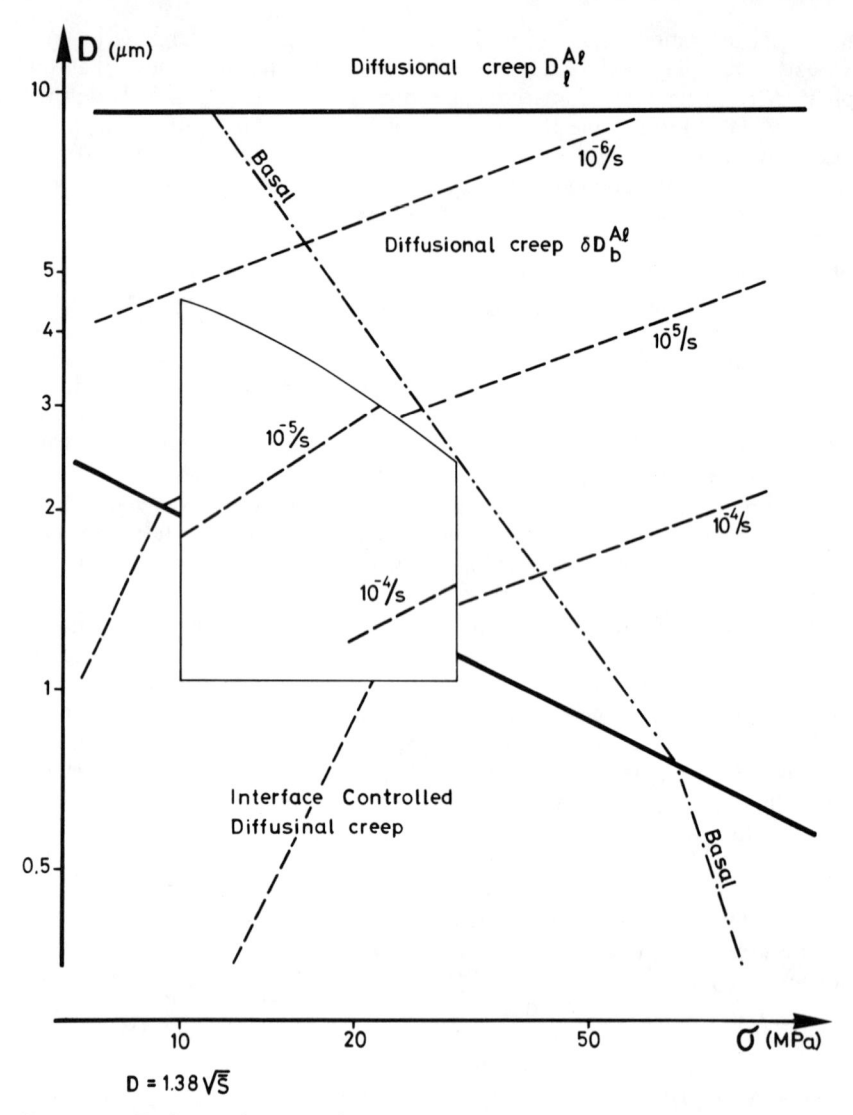

Fig. 11. Deformation mechanism map (Ref. 3c) showing range of present data.

when another mechanism, e.g., cavitation, may intervene. With the present set of data also, it is possible to test the validity of our previous assumption that deformation took place in quasi-steady-state conditions, which has also been made by other authors.[7] According to this assumption the general instantaneous strain rate equation

$$\dot{\varepsilon} = A\frac{\sigma^{1/m}}{(\overline{s}^{1/2})^p} \qquad (2)$$

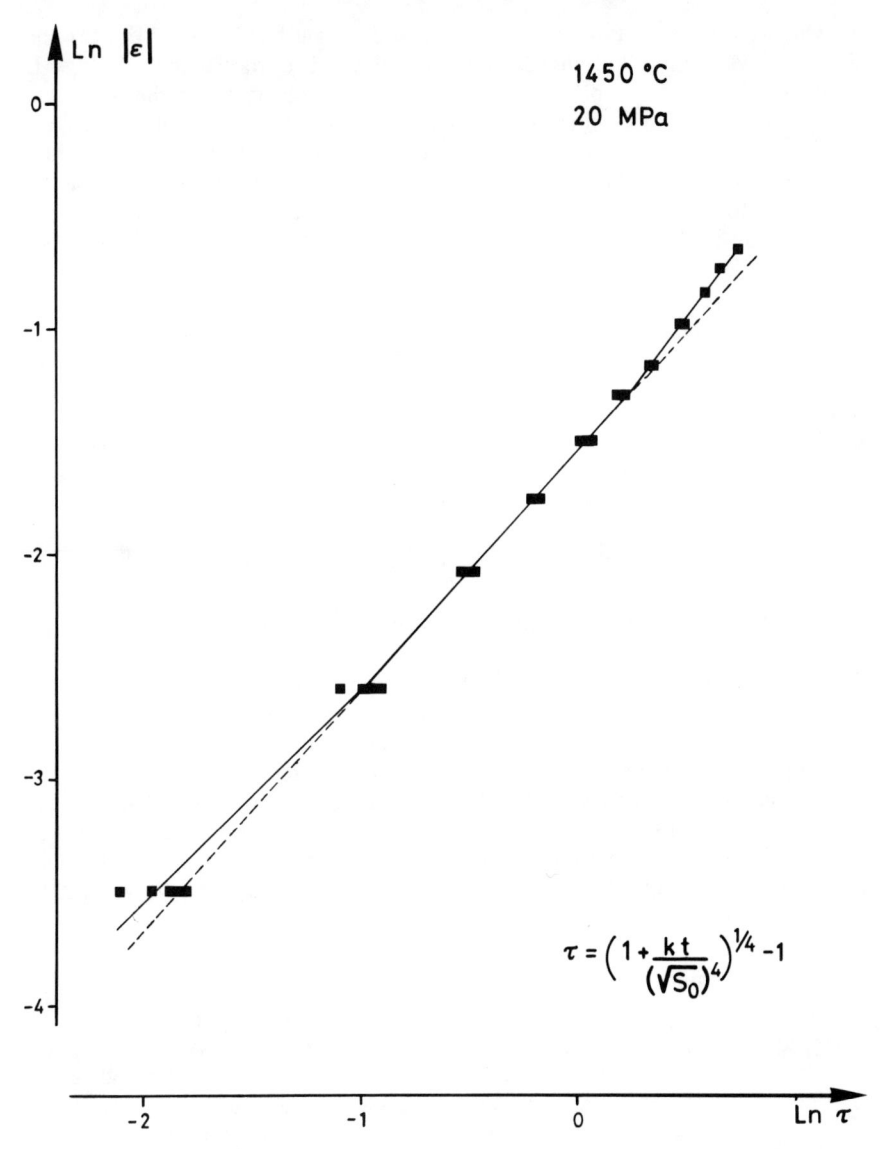

Fig. 12. Strain vs reduced time plot to test quasi-steady-state assumption.

$$\tau = \left(1 + \frac{kt}{(\sqrt{s_0})^4}\right)^{1/4} - 1$$

where A is a temperature-dependent factor, σ is the applied stress, and $\bar{s}^{1/2}$ is the average grain size, should be followed at all times, even when some parameters (e.g., grain size) are time dependent. In the framework of the assumption, Eq. (1) above may be substituted in Eq. (2) and integration with respect to time yields for $m = 1$ and $p = 3$, respectively

$$\frac{\varepsilon K_4}{4A\sigma \bar{s}_0^{1/2}} = \left(1 + \frac{K_4}{(\bar{s}_0^{1/2})^4} t\right)^{1/4} - 1 \tag{3}$$

Introducing a reduced time τ as identical to the right-hand side of Eq. (3) and plotting ε vs τ on a log–log diagram should yield a straight line of slope 1. Figure 12 shows such a plot obtained from the independent experimental measurements of strain and grain size increases with time. The agreement with Eq. (3) is rather satisfactory and, as was to be expected, better in the range where m values are closest to 1. It is also possible with Fig. 12 to numerically evaluate A appearing in Eq. (2), from the condition $\tau = 1$. Setting this condition and replacing σ, K_4, and $\bar{s}_0^{1/2}$ with the corresponding numbers yields $A = 8.36 \times 10^{-30}$ (MKS units). Assuming now that[3b,c]

$$A = \frac{7\pi\Omega}{kT}(\delta D_b) \tag{4}$$

where Ω is the molecular volume of Al_2O_3 (4.276×10^{-29} m^3) and δD_b is the appropriate grain-boundary diffusivity term a numerical value, $\delta D_b \approx 2 \times 10^{-22}$ m^3/s, is obtained from the present experiments at 1450°C, which is in remarkable agreement with that resulting from a previously published expression pertaining to aluminum ion diffusivity (Ref. 3a):

$$(\delta D_b^{Al}) = 8.6 \times 10^{-4} \exp-\left[\frac{100(\text{kcal/mol})}{RT}\right] \quad (\text{cm}^3/\text{s}) \tag{5}$$

It is worth pointing out that the stress-enhanced grain growth rate constant was used in the calculations, namely, $K_4 = 9 \times 10^{-28}$ m^4/s. It could also be argued that Eq. (4) should be valid only for small overall strains and that, for large superplastic strains such as those observed in this study, the 7π factor should be replaced by $7\pi \times \alpha$, where $1.7 \leq \alpha \leq 5.4$ when the process is controlled by grain-boundary diffusion.[15] If this latter course is followed, then the stress-free grain growth rate constant should be introduced in the above calculations of the effective diffusivity, since the evaluations of bounds for α have explicitly postulated the existence of an internal stress field superimposed to the externally applied stress.[15] Under such conditions and with the above calculated δD_b, we would find $\alpha = 3.7$, well within the stated bounds.

Concluding Remarks

The present results are believed to show that normal grain growth can be enhanced by stress and/or deformation when the latter primarily involves grain rearrangements of the type that has been invoked to account for structural superplasticity in single-phase materials. When stress accommodation at grain junctions is effected through matter transport along the grain boundaries, as in this study, it is not clear whether it should be considered that concurrent grain growth favors grain rearrangements hence deformation or whether deformation when large enough should of necessity be accompanied by grain growth if the integrity of the material is to be retained. In fact, both points of view could be claimed equivalent on the basis of considerations concerning the overall topological stability of the polycrystalline structure.[19] Indeed, when such a stability is to be maintained, a definite ratio should exist on the average between the specific amounts of both the elementary topological transformations: cell neighbor interchanges and the balance of triangular cell appearances and disappearances in a plane section.

The basic identification and evaluation of grain-boundary-dominated phenomena at the atomic level in ceramics perhaps can benefit from the experimental approach followed here whenever it is experimentally feasible, since information can be obtained not only on such quantities as activation energies but also on activation volumes. In the present set of fine-grained aluminas containing a priori substantial amounts of residual alkaline impurities, it appears that the atomistic rate-controlling mechanisms are not the same for grain growth as for creep deformation. In particular, the presence of intergranular platelike precipitates did not seem to markedly affect the creep rate up to ~20% strain in samples tested under a 20-MPa stress. The exact reason for it remains unclear. However, it is reasonable to suspect that during testing each of the deformation and grain growth processes tended to set its own pace, which from an overall kinetic point of view at least are not compatible in the long run. The onset of cavitation in the more strained samples may also be viewed as a direct consequence of this incompatibility buildup. In better purified materials, devoid of intergranular precipitates, large superplastic strains should be accessible as theoretically predicted[20] and experimentally observed.[10] A better knowledge and control of the chemistry at and near grain boundaries once again then appears to be of major importance not only for alumina, but presumably for many other ceramics as well.

It may be added finally that the approach outlined here is somewhat at variance with the frequently held view that superplastic deformation necessitates an ultrafine grain size that should be kept so by means of dispersed particles. Certainly a fine grain size is a prerequisite for grain-boundary sliding to dominate deformation. This has been the case for a number of other ceramic systems where substantial strains could be reached, including MgO,[4] ZrO_2,[21] UO_2,[22] magnetic oxides,[23] and also SiC.[24] When intragranular deformation modes do not operate, grain growth then is probably necessary for maintaining microstructural homogeneity.

Acknowledgments

Thanks are due to Dr. Friedli (EPFL) for help with the activation analyses, Prof. J. J. Heizmann (Université de Metz, France) for construction of the pole figures, and the Swiss National Science Foundation for financial support (Contract No. 2.050-0.81).

References

[1] M. F. Ashby, "Boundary Defects and Atomistic Aspects of Boundary Sliding and Diffusional Creep," *Surf. Sci.*, **31**, 498–542 (1972).

[2] M. F. Yan, R. M. Cannon, and H. K. Bowen, "Grain Boundary Migration in Ceramics"; pp. 276–307 in Ceramic Microstructures 76. Edited by R. M. Fulrath and J. A. Pask. Westview, Boulder, CO, 1977.

[3] (a) R. M. Cannon and R. L. Coble, "Review of Diffusional Creep of Al_2O_3"; pp. 61–100 in Deformation of Ceramic Materials. Edited by R. C. Bradt and R. E. Tressler. Plenum, New York, 1975. (b) R. M. Cannon, W. H. Rhodes, and A. H. Heuer, "Plastic Deformation of Fine-Grained Alumina (Al_2O_3) I. Interface Controlled Diffusional Creep," *J. Am. Ceram. Soc.*, **63** [1–2] 46–52 (1980). (c) A. H. Heuer, N. J. Tighe, and R. M. Cannon; pp. 53–8 in Ref. 3b. (d) W. R. Cannon and O. D. Sherby, "Creep Behavior and Grain Boundary Sliding in Polycrystalline Al_2O_3," *J. Am. Ceram. Soc.*, **60** [1–2] 44–7 (1977). (e) P. A. Lessing and R. S. Gordon, "Creep of Polycrystalline Alumina Pure and Doped with Transition Metals Impurities," *J. Mater. Sci.*, **12**, 2291–2302 (1977).

[4] (a) Tuvia Zisner and Hideo Tagai, "High Temperature Creep of Polycrystalline Magnesia II. Effects of Additives," *J. Am. Ceram. Soc.*, **51** [6] 310–4 (1968). (b) R. S. Gordon, "Mass Transport in the Diffusional Creep of Ionic Solids," *J. Am. Ceram. Soc.*, **56** [3] 147–152 (1973). (c) J. Crampon, "The Creep Microstructure of Ultrafine-Grained MgO Polycrystals," *Acta Metall.*, **28**, 123–8 (1980). (d) J. Crampon and B. Escaig, "High Temperature Creep of Ultrafine-Grained Fe Doped MgO Polycrystals," *J. Mater. Sci.*, **13**, 2619–26 (1978).

[5](a) R. Cannon, "The Contribution of Grain Boundary Sliding to Axial Strain during Diffusion Creep," *Philos. Mag.*, **25**, 1489–97 (1972). (b) R. N. Stevens, "Grain Boundary Sliding and Diffusion Creep," *Surf. Sci.*, **31**, 543–65 (1972). (c) R. C. Gifkins, T. Langdon, and D. McLean, "Grain Boundary Sliding and Axial Strain during Diffusional Creep," *Met. Sci.*, **9**, 141–4 (1975). (d) O. A. Ruano and O. D. Sherby, "Low Stress Creep of Fine-Grained Materials at Intermediate Temperatures: Diffusional Creep or Grain Boundary Sliding," *Mater. Sci. Eng.*, **56**, 167–75 (1982).

[6]Tadaaki Sugita and J. A. Pask, "Creep of Doped Polycrystalline Al_2O_3," *J. Am. Ceram. Soc.*, **53**, [11] 609 (1970).

[7]G. R. Terwilliger, H. K. Bowen, and R. S. Gordon, "Creep of Polycrystalline MgO and $MgO-Fe_2O_3$ Solid Solutions at High Temperature," *J. Am. Ceram. Soc.*, **53** [11] 241–51 (1970).

[8]M. A. Clark and T. H. Alden, "Deformation Enhanced Grain Growth in a Superplastic Sn—1% Bi Alloy," *Acta Metall.*, **21**, 1195–1206 (1973).

[9]R. M. Cannon and W. H. Rhodes, "Deformation Processes in Forging Ceramics," Summary Report, Contract NASW 1914, 15, 1970.

[10]C. Carry and A. Mocellin, "Superplastic Forming of Alumina—Proceeding of 3rd Symposium on Fabrication Science," *Proc. Br. Ceram. Soc.*, **33**, 101–15 (1983).

[11](a) M. Blanc, A. Mocellin, and J. L. Strudel, "Observation of Potassium β'''-Al_2O_3 in Sintered Alumina," *J. Am. Ceram. Soc.*, **60**, 403–9 (1977). (b) M. Blanc and A. Mocellin, "Observation of a Complex Grain Growth Mechanism in a Sintered Al_2O_3"; pp. 437–48 in Sintering Processes. Edited by G. C. Kuczynski. Plenum, New York, 1980.

[12]A. Mocellin and W. D. Kingery, "Microstructural Changes during the Heat-Treatment of Sintered Alumina," *J. Am. Ceram. Soc.*, **56**, 309–14 (1973).

[13]F. A. Nichols, "On the Diffusional Mobilities of Particles, Pores and Loops," *Acta Metall.*, **20**, 207–14 (1972).

[14]N. V. Speight, "Growth Kinetics of Grain-boundary Precipitates," *Acta Metall.*, **16**, 133–5 (1968).

[15]M. F. Ashby, G. H. Edward, J. Davenport, and R. A. Verrall, "Application of Bound Theorems for Creeping Solids and their Application to Large Strain Diffusional Flow," *Acta Metall.*, **26**, 1379–88 (1978).

[16]J. H. Schneibel and P. M. Hazzledine, "The Role of Coble Creep and Interface Control in Superplastic Sn-Pb Alloys," *J. Mater. Sci.*, **18**, 562–70 (1983).

[17]Y. Ikuma and R. S. Gordon; pp. 283–93 in Surfaces and Interfaces in Ceramic and Ceramic-Metal Systems. Edited by J. A. Pask and A. G. Evans. Plenum, New York, 1981.

[18]B. Burton and W. B. Beere, "Influence of Second Phase Particles on Diffusional Creep," *Met. Sci.*, 71–6 (Feb. 1978).

[19]E. Carnal and A. Mocellin, "A Topological Model for Plane Sections of Polycrystals," *Acta Metall.*, **29**, 135–43 (1981).

[20]A. G. Evans, J. R. Rice, and J. P. Hirth, "Suppression of Cavity Formation in Ceramics: Prospects for Superplasticity," *J. Am. Ceram. Soc.*, **63**, 368–75 (1980).

[21]P. E. Evans, "Creep in Y_2O_3 and Sc_2O_3-Stabilized Zirconia," *J. Am. Ceram. Soc.*, **53**, 365–9 (1970).

[22]T. E. Chung and J. T. Davies, "The Superplastic Creep of Uranium Dioxide," *J. Nucl. Mater.*, **79**, 143–53 (1979).

[23]M. H. Hodge, W. R. Bitler, and R. C. Bradt, "Deformation Texture and Magnetic Properties of the Magnetoplumbite Ferites"; pp. 483–96 in Ref. 3a.

[24]C. Carry and A. Mocellin, "High Temperature Creep of Dense Fine Grained Silicon Carbides"; pp. 391–403 in Deformation of Ceramics, II. Edited by R. E. Tressler and R. C. Bradt. Plenum, New York, 1984.

---

*Grade A15Z, Rubis synthétique des Alpes, Jarrie/Grenoble (France).

†A more rigorous regression analysis would yield n values between 3.5 and 4.

‡Apparent activation energies for the strain rates also were measured by imposing temperature changes between 1400 and 1450°C, yielding $Q \approx 430$ kJ/mol for a 20-MPa stress. The atomistic significance of such results will not be discussed here.

§The decrease is presumably as a result of fast grain-boundary diffusion and elimination at the outer surface of the samples where color changes can be noted.

¶Isostrain rate lines were plotted at 1450°C on the basis of experimental evidence, and our $\bar{s}^{1/2}$ grain sizes were corrected (times 1.38) for consistency.

**Taking $\sigma_0 = G_b V_v / r$ (Ref. 18) with $V_v = 10^{-2}$, $r = 0.3$ μm, $b = 4.85$ Å (0.485 nm), and $G = 140 \times 10^9$ Pa yields $\sigma_0 \approx 2$ MPa.

High Creep Ductility in Alumina Containing Compensating Additives

W. Roger Cannon

Rutgers University
Department of Ceramics
Piscataway, NJ 08854

Aluminum oxide containing the compensating aliovalent additives, 2 mol% TiO_2 and 2 mol% CuO, exhibits a number of interesting properties. When a reactive alumina* is used, the material sinters to theoretical density at 1200°C. The creep rate is greatly enhanced, 4 orders of magnitude, over that of "pure" alumina, and the creep ductility becomes very large. A number of samples were tested from 1000°C to 1200°C in both bending and compression, and none failed to the limit of our testing ability, which was 33% bending strain. It is proposed that superplastic deformation occurs by an easily deformable mantle allowing the grains to slide past one another. It is also proposed that samples are being tested below the critical void nucleation stress for cavity and crack growth.

In spite of the optimism of the 1960s that structural polycrystalline ceramics might be ductile at high temperatures, very few results of high ductility in polycrystalline ceramics have been reported in the literature. Table I contains a list of all known studies reporting greater than 20% deformation for polycrystalline ceramics without glassy phases. These are divided into three categories: dislocation, diffusional, and transformation superplasticity. Those listed under dislocation creep are cubic materials tested at high temperatures where five independent slip systems are believed to be operating. High ductilities have also been predicted[10] for Al_2O_3 by dislocation deformation but very near its melting temperature where it is difficult to test. Under diffusional creep a number of recent examples have been cited. These are all fine-grained polycrystalline materials whose grain size is 3 μm or less. In each case the authors have proposed that these materials fall in the category of structural superplasticity. The final group transformational superplasticity came from a single research group who have studied Bi_2O_3 compounds strained through an allotropic phase transformation to obtain high deformations.

In a recent paper Evans et al.[11] discussed the circumstances for achieving superplasticity in the diffusional creep regime by avoiding cavity nucleation and growth. On the basis of nucleation and growth equations, they suggested a number of possible ways of suppressing cavitation, among them "additives which encourage grain-boundary diffusion but do not form amorphous phases or grain-boundary precipitates (i.e., solid solution alloys)." In this paper ductilities of alumina with compensating additives is described since these are shown to greatly enhance the diffusion rate.[12]

Compensating Additives

The mechanisms of diffusion enhancement by compensating additives, in Al_2O_3 ($FeO-TiO_2$ and $MnO-TiO_2$), have been discussed in an earlier paper by

Table I. Summary of High-Creep Ductilities Found in the Literature

Author (year)	Material (grain size (μm))	Temperature (°C)	Failure strain (%)
	Dislocation		
Day and Stokes* (1966)	MgO	>1800	40 (T)¶¶
Gentilman et al.[†] (1977)	MgO·2Al$_2$O$_3$	>1750	>20 (B)
Various authors	Alkali halides	~T_m	
	Diffusional		
Crampon and Escaig[‡] (1980)	MgO (0.1)	>1050	⩾20 (C)
Chung and Davies[§] (1979)	UO$_2$ (2)	1400	69 (C)
R. M. Cannon[¶]	Al$_2$O$_3$ (1–3)	1400–1500	35 (C)
			20 (biaxial B)
Fridez et al.**	Al$_2$O$_3$ (1–3)	1450	50 (C)
	Transformational superplasticity		
Johnson et al.[††] (1975)	Bi$_2$O$_3$, Bi$_2$SmO$_3$	730	>100 (C)
Smyth et al.[‡‡] (1975, 1977)	Bi$_2$O$_3$-Sm$_2$O$_3$		

*Reference 1.
[†]Reference 2.
[‡]Reference 3.
[§]Reference 4.
[¶]Reference 5.
**Reference 6.
[††]Reference 7.
[‡‡]References 8 and 9.
¶¶T represents tension, C represents compression, and B represents bending.

Gordon.[12] In the Al$_2$O$_3$-FeO-TiO$_2$ system, for instance, aluminum ion diffusion is enhanced either by V_{Al}''' due to Ti$_{Al}^{\cdot}$ defects or by Al$_i^{\cdot\cdot\cdot}$ due to Fe$_{Al}'$ defects, depending on which of these two species dominates. Ikuma and Gordon[13] observed the largest enhancement in creep rate at equimolar Fe-Ti. This is an unexpected result since the Al$_i^{\cdot\cdot\cdot}$ and V_{Al}''' should annihilate one another, but as they explained, the Ti$_{Al}^{\cdot}$ dominates at almost all concentrations and $\dot{\varepsilon}$ increases with concentration; at very high concentration, interface reactions control and $\dot{\varepsilon}$ slows down. Thus, the equimolar maximum is considered merely coincidental. Similar results, however, were noted by Cutler et al.[14] in their 1957 sintering paper, using two other compensating additives. The highest densities were achieved in each case by equimolar additions. In addition to the earlier suggestion, it may be that a higher degree of solid solution is present at the equimolar composition.[15]

Procedure

One of the compositions studied by Cutler et al.,[14] Al$_2$O$_3$-2 mol% CuO-2 mol% TiO$_2$, was studied here. For this composition an oxidizing atmosphere encourages both the Cu into the 2+ valence and Ti into the 4+ valence. Thus, an oxygen atmosphere was used in most cases.

Powder for sintering was prepared by coprecipitating the Cu$_2$O and TiO$_2$ additives onto a slurry of alumina powder.* Cu$_2$O was added as Cu(NO$_3$)$_2$·3H$_2$O and precipitated with a basis solution. The slurry was then dried and calcined before mixing it into a measured solution of tetraisopropyl orthotitanate in ethanol. The latter solution was then decomposed by the addition of water while stirring.

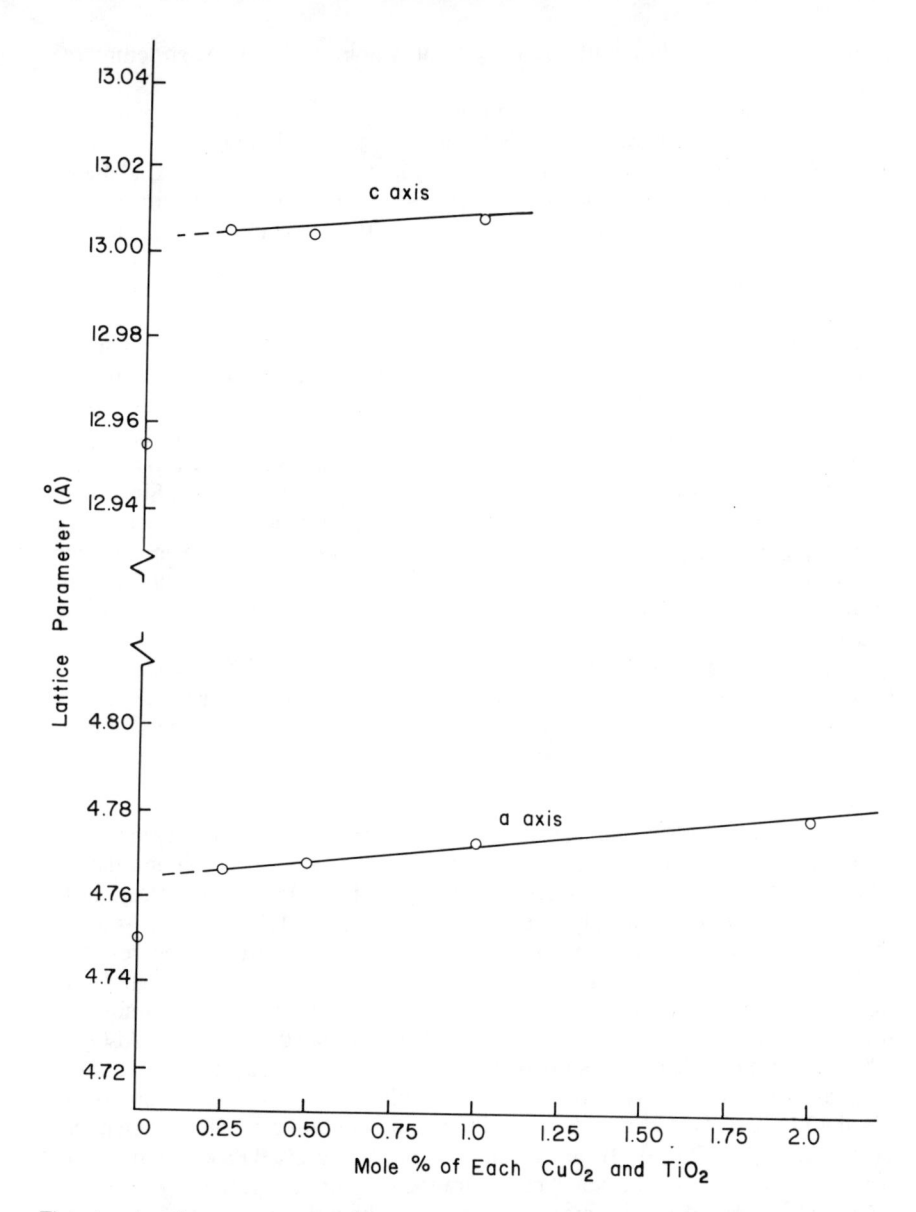

Fig. 1. Lattice parameters for equimolar additions Cu$_2$O and TiO$_2$ to aluminum oxide. The lines indicate an apparent two phase region and are sloped because they are not along a tie line.

The slurry was again dried and calcined and the resulting powder milled overnight. Powder was dry-pressed at 103 MPa (15 000 psi) and isostatically pressed at 206 MPa (30 000 psi) in preparation for sintering. Small pellets were sintered at 1200 and 1300°C for 1 h in flowing O$_2$. Final densities were 3.95 and 3.97 g/cm^3, respectively. Material sintered at 1200°C was used in subsequent studies. Average grain size was 1.5 μm with only a slight amount of duplex structure (see Fig. 3(A, B)).

X-ray analysis revealed no second-phase peaks but did show an approximate 2% shift in the lattice parameters, indicating some solid solution. Roy and Coble[15] were unable to detect lattice parameters changes in $Al_2O_3 + TiO_2$, and Gadalla and White[16] indicated no solid solution between Al_2O_3 and Cu_2O. The shift in lattice parameters noted here indicates increased solubility with compensating additives. Figure 1 shows the relationship between lattice parameters and composition along the equimolar composition line. Results indicate that there is less than 0.25% solid solution.

It has not been determined whether the second phase is at the grain boundary of the alumina grains or not, since no transmission microscopy was performed. Optical micrographs of the Al_2O_3-2% Cu_2O-2% TiO_2 compound, however, show a brown and white speckled appearance in which the fine grains are brown and larger grains white. The interpretation is that grains are surrounded by a brown grain-boundary phase that is more clearly seen among the fine grains. Since X-ray analysis cannot detect the second phase, it is unlikely that the larger number of brown grains correspond to the second phase. Specimens etched in hot phosphoric acid etched very nonuniformly and only along some grain boundaries, indicating perhaps that the second phase was not distributed uniformly along grain boundaries.

Phase diagrams are not available for the Al_2O_3-CuO-TiO_2 system, but in the binary[16] Al_2O_3-Cu_2O, a liquid phase appears at about 1250°C, depending on the P_{O_2}. In the Al_2O_3-TiO_2 binary, liquid first appears at 1800°C.[17] Thus, there is insufficient evidence to decide whether the second phase is liquid or not.

Creep Tests

A limited number of creep tests were performed at several temperatures. Those at 1000°C were performed in air in compression (4–7 MPa) and those at 1200°C in 10^5 Pa (1 atm) in oxygen in bending. The stress exponent, n, in the equation $\dot{\varepsilon} = A\sigma^n d^p$ for steady-state creep at 1000°C was 1.6 ($m = 1/n = 0.62$). Steady state creep results are compared in Fig. 2 with extrapolated results of R. M. Cannon et al.[18] measured in the same grain size range but at a higher stress range and with some results of Ikuma and Gordon[13] for double-doped alumina. The present results are about 4 orders of magnitude faster than the ¼% MgO-Al_2O_3 and 100 times greater than the double-doped Al_2O_3. In the double-doped alumina, bulk diffusion is believed to be rate controlling. The particular composition 1% Fe-0.05% Ti was chosen because Ikuma and Gordon estimated a defect concentration, $V'''_{Al} \cong 0.02\%$. If the dopant level of 2% Cu-2% Ti used in the present paper merely increased the defect concentration proportionally, the expected defect concentration based on the creep results would have been $V'''_{Al} \cong 2\%$, which is undoubtedly too high. It is more likely that the high creep rate is related to rapid diffusion through the boundary containing the second phase. The results of R. M. Cannon et al. are believed to be boundary diffusion controlled. Our stress exponent and activation energy were similar to theirs. Thus, it is suggested that the enhanced boundary diffusion is 4 orders of magnitude faster. The rate of the interface reaction must correspondingly also increase.

Creep Ductility

Seven samples were tested in bending between 1150 and 1200°C to strains ranging from 11 to 33%. Only one sample failed, which appeared to be as a result

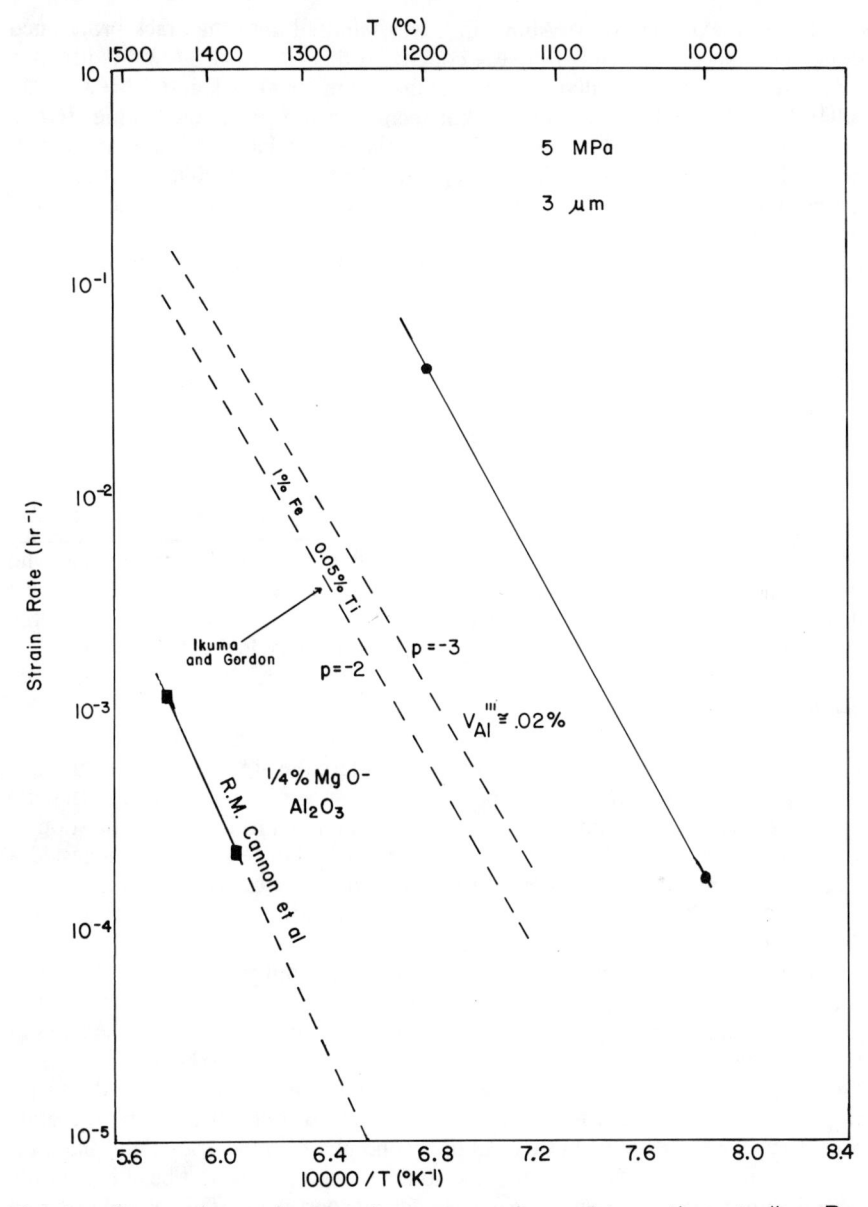

Fig. 2. Comparison of present creep results and two other studies. Results of R. M. Cannon et al. were extrapolated from 20 to 5 MPa and from 2.7 to 3 μm grain size by using $m = 0.65$ and $p = -2.5$. Results of Ikuma and Gordon were extrapolated from a grain size of 13–3 μm by using $p = -2$ and $p = -3$ and from 1450°C by using an activation energy of 406 kJ/mol. The current results were extrapolated at 1200°C from 0.78 to 5 MPa, and the strain rate was calculated from the (total strain/time) × 2.5 to account for primary creep since the creep test was performed without continually monitoring strain.

of a large preexisting crack. About 15 h was required until the crack propagated through the specimen, and so it was concluded that the crack grew by diffusive crack growth. (The estimated K_I value for the 1-mm long crack at 0.8 MPa is only 0.045 N/m$^{3/2}$.) High strain rates did not seem to cause premature failure. In one experiment, a stress on the order of 7 MPa was used at 1100°C. The sample crept in bending to a strain limited by the apparatus (~5%) in less than 1 min but still did not fail. It is possible that at higher strain rates, crack tip stresses are relieved by diffusion.

Strains were measured in two ways. In the first, the total strain was given by

$$\varepsilon = \frac{r_1 - r_2}{r_2} \qquad (1)$$

where r_1 is the outer radius of curvature and r_2 was the neutral axis radius of curvature. In the second, the cross-section dimensions of the deformed specimen were taken under a microscope and the strain was calculated from

$$\varepsilon = \frac{1}{\nu}\left(\frac{d_1 - d_2}{d_2}\right) \qquad (2)$$

where ν is Poisson's ratio (0.25), d_1 is the specimen width at the tensile edge, and d_2 is the specimen width at the neutral axis. This latter method ensured that the deformation was tensile rather than shear strain. Results on the most highly strained specimen are 33 and 32%, by the two methods, respectively.

Superplastic Deformation

A detailed model of the mechanism of deformation is not presented here, but some observations based on the superplasticity literature are made. Although there is not general agreement on the mechanism of superplasticity, there is general agreement that grain-boundary sliding plays an important roll. The key question is: "How does accommodation at grain triple points occur?" Gifkins[20] suggested a macroscopic model of grain-boundary sliding that may be useful in interpreting the present results. He proposed that a grain contain a central core that does not deform and a mantle around the outside of the grain that deforms easily as grains roll over each other. A simple calculation of the minimum possible width of mantle is based on a uniform set of hexagonal grain which may move past each other when the mantles are defined as the areas swept out between the apices and the sides of the hexagons as they rotate. In this case the mantle is required to be only 0.07 times the grain diameter.[21] For the alumina alloy studied here, this mantle could, at least in part, be supplemented by an easily deformable second phase along the grain boundary. For the expected low level of second phase, some grain deformation is also expected as part of the mantle, and so the second phase must be able to easily accept aluminum and oxygen vacancies and transport them. This model does not suggest a liquid phase at the grain boundary, however, since the liquid phases generally contribute to cavitation and early failure.

Cavitation

Figure 3(A, B) compares the microstructures at the tensile edge of the specimen before and after 33% creep strain in bending. The degree of cavitation varied from area to area of the microstructure, but in areas shown in Fig. 3(A, B), which were dense areas, little cavitation is seen. Figure 3(B) shows that nucleation of cavities at triple points was not excessive during creep if it occurred at all.

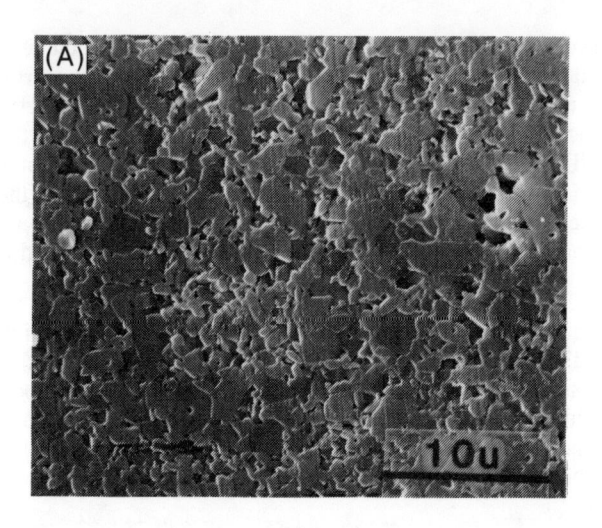

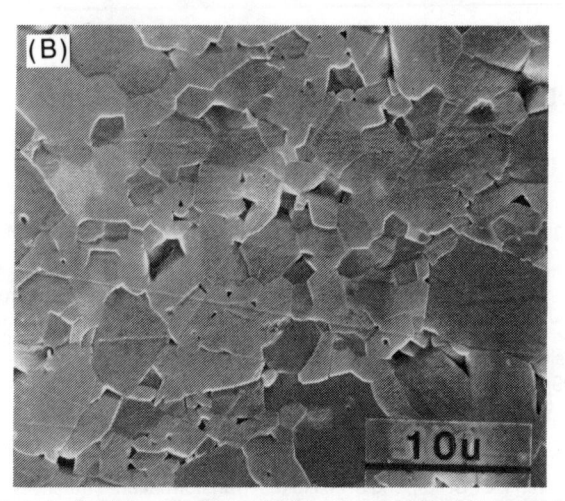

Fig. 3. As-sintered microstructure (etched in boiling phosphoric acid) (A). The overetched appearance is believed to be due to preferential etching of the grain-boundary phase. Microstructure after 33% strain (tensile edge of the bend specimen) (B).

The analysis of cavitation described by Evans, Rice, and Hirth[11] was useful in suggesting the effect of greatly enhanced grain-boundary diffusivities. Their results indicate an exponential relationship between pore nucleation rate and stress, and so there is an effective threshold stress below which nucleation does not occur. The diffusion coefficient lies outside the exponential term and so does not greatly affect the threshold stress. Nevertheless, additives that greatly enhance the creep rate while maintaining a fine grain size allow a large amount of deformation in a reasonable time at a low stress below which cavity nucleation occurs.

Since residual cavities are usually present from sintering, if not from stress enhanced nucleation, it may be contended that cavity growth and diffusive crack growth are more important than nucleation. Evans et al.[11] conclude that the cavity growth rate depends on the ratio of boundary diffusion to surface diffusion, $(D_b \delta_b/D_s \delta_s)$, and it is not easy to predict in our case what this effect is. It can, however, be noted that cavity growth rate is lower at lower stresses that accompany the use of additives, and so in this way additives have an additional effect.

A final effect of additives is the effect on crack growth since they allow tests to be performed at very low stresses where the Griffith crack length is not easily reached, so that ductility is likely to be greater than that in nondoped material.

Additional Comments

The alumina alloy described in this paper has a higher creep rate than any of the alumina alloys discussed by Gordon.[12] At the same time it exhibits excellent ductility at reasonably low temperatures and so may have some potential for hot forging. In addition, the fracture strength of three samples, which were tested in three point bending, averaged 380 MPa (50 000 psi). Apparently, the additives did not degrade the fracture properties.

Acknowledgments

The author thanks Sheldon Wiederhorn at the National Bureau of Standards for performing the 1000°C creep tests. The author also thanks J. B. Wachtman and M. G. McLaren for their help on the project and Susan Forte for the X-ray measurements.

References

[1] R. B. Day and R. J. Stokes, "Mechanical Behavior of Polycrystalline MgO at High Temperatures," *J. Am. Ceram. Soc.*, **49** [7] 345–54 (1966).
[2] R. Gentileman, E. Maguire, and J. Pappis, "High Durability Missile Domes"; Interim Technical Rept., Contract No. N00014-76-C-0635. Office of Naval Research, April 1, 1976 through Sept., 30, 1977.
[3] J. Crampon and B. Escaig, "Mechanical Properties of Fine-Grained Magnesium Oxide at Large Compressive Strains," *J. Am. Ceram. Soc.*, **63** [11] 680–6.
[4] T. E. Chang and T. J. Davies, "The Low-Stress Creep of Fine-Grain Uranium Dioxide," *Acta Metall.*, **27** [4] 627–35 (1979).
[5] R. M. Cannon; private communications.
[6] J. D. Fridez, C. Carry, and A. Mocellin, "Effects of Temperature and Stress on Grain-Boundary Behavior in Fine-Grained Alumina"; these proceedings.
[7] C. A. Johnson, R. C. Bradt, and J. H. Hoke, "Transformational Plasticity in Bi_2O_3," *J. Am. Ceram. Soc.*, **58** [1] 37–40 (1975). C. A. Johnson, J. R. Smyth, R. C. Bradt, and J. H. Hoke, "Transformational Superplasticity in Pure Bi_2O_3 and Bi_2O_3-Sm_2O_3 Eutectoid System"; pp. 443–53 in Deformation of Ceramic Materials. Edited by R. C. Bradt and R. E. Tressler. Plenum Press, New York, 1975.
[8] J. R. Smyth, R. C. Bradt, and J. H. Hoke, "Transformational Superplasticity in the Bi_2O_3-Sm_2O_3 Eutectoid System," *J. Am. Ceram. Soc.*, **58** [9–10] 381–4 (1975).
[9] J. R. Smyth, R. C. Bradt, and J. H. Hoke, "Isothermal Deformation in B_2O_3-Sm_2O_3 Eutectoid System," *J. Mater. Sci.*, **12** [7] 1495 (1977).
[10] C. Gandhi and M. F. Ashby, "Fracture-Mechanism Maps for Materials which Cleave: FCC, BCC, and HCP Metals and Ceramics," *Acta Metall.*, **27** [10] 1565–1602 (1979).
[11] A. G. Evans, J. R. Rice, and J. P. Hirth, "Suppression of Cavity Formation in Ceramics: Prospects for Superplasticity," *J. Am. Ceram. Soc.*, **63** [7–8] 368–75 (1980).
[12] R. S. Gordon, "An Understanding of Defect Structure and Mass Transport in Polycrystalline Al_2O_3 and MgO via the Study of Diffusional Creep"; these proceedings.
[13] Y. Ikuma and R. S. Gordon, "Effect of Doping Simultaneously with Iron and Titanium on the Diffusional Creep of Polycrystalline Al_2O_2," *J. Am. Ceram. Soc.*, **66** [2] 139–47 (1983).
[14] I. B. Cutler, C. Bradshaw, C. J. Christensen, and E. P. Hyatt, "Sintering Alumina at Temperatures of 1400°C and Below," *J. Am. Ceram. Soc.*, **40** [4] 134–9 (1957).
[15] S. K. Roy and R. L. Coble, "Solubilities of Magnesia, Titania, and Magnesium Titanate in Aluminum Oxide," *J. Am. Ceram. Soc.*, **51** [1] 1–6 (1968).
[16] A. M. M. Gadalla and J. White, "Equilibrium Relationships in the System CuO-Cu_2O-Al_2O_3," *Trans. Br. Ceram. Soc.*, **63** [1] 39–62 (1964).

[17] K. T. Jacob and C. B. Alcock, "Thermodynamics of $CuAl_2O$ and $CuAl_2O_4$ and Phase Equilibria in the System $Cu_2O\text{-}CuO\text{-}Al_2O_3$," *J. Am. Ceram. Soc.*, **58** [5–6] 192–5 (1975).

[18] A. M. Lejus, D. Goldberg, and A. Revcolevschi, *C. R. Hebd. Seances Acad. Sci., Ser. C*, **263** [201] 1223 (1966).

[19] R. M. Cannon, W. H. Rhodes, and A. H. Heuer, "Plastic Deformation of Fine-Grained Alumina: I, Interface-Controlled Diffusional Creep," *J. Am. Ceram. Soc.*, **63** [1–2] 46–53 (1980).

[20] R. C. Gifkins, "Grain-Boundary Sliding and its Accommodation During Creep and Superplasticity," *Metall. Trans., A*, **7A** [8] 1225–32 (1976).

[21] R. C. Gifkins and T. G. Langdon, "Comments on Theories of Structural Superplasticity," *Mater. Sci. and Eng.*, **36**, 27–33 (1978).

---

*Alcoa A16SG, Aluminum Co. of America, Pittsburgh, PA.

Crack Healing in Al$_2$O$_3$, MgO, and Related Materials

T. K. Gupta

Westinghouse Research & Development Center
Pittsburgh, PA 15235

The crack healing in ceramics occurs in several stages by distinct morphological changes in the crack and appears to be surface-diffusion controlled. Similar morphological changes also occur with a liquid or a solid as an inclusion in the material. The significance of such changes occurring to the inclusions, in a number of technologically important materials including Al$_2$O$_3$ and MgO, is reviewed in this paper.

Crack healing in ceramics[1-5] is a phenomenon wherein the deleterious effects of surface microcracks on strength are progressively reduced by heat treatment. Associated with this, there is a significant morphological change[6,7] of the crack during crack healing. The phenomenon has been demonstrated in a number of materials, e.g., Al$_2$O$_3$,[1,2,5,8] MgO,[3] ZnO,[4] UO$_2$,[9-11] and in several carbides[12,13] and halides.[14,15] Other techniques are also available for the removal of surface microcracks, e.g., annealing,[16] flame polishing,[17-19] chemical polishing,[20] and gaseous etching,[21,22] but the study of the crack healing by heat treatment is of special interest, because it allows a systematic investigation of the morphological change that takes place in a crack as the healing proceeds to completion. The significance of the above observations is that such morphological change is also known to occur with a liquid or a solid as an inclusion in the materials. The understanding of this morphological change and its effect on material properties for several technologically important areas of interest are the subject of this review.

Crack Morphology during Crack Healing

The accumulated evidence from detailed microstructural studies on Al$_2$O$_3$[1,6-8] and MgO[2,3] indicates that the crack healing is accompanied by a significant morphological change in the crack during heat treatment. When these studies are combined with studies on other ceramics,[12-15] a model for crack healing emerges. Figure 1 illustrates this model, which depicts the sequential change that occurs to a crack as it goes through various stages of crack healing. The reduction of surface energy is the driving force for this change. In the initial stage of crack healing, there is a continuous regression of crack[23] at the crack tip or a discontinuous pinching[1,24] of crack along the length of the crack. This is followed by the cylinderization[1,6-8] of crack in the second stage of crack healing. The third and the most important stage of crack healing is the breakup of the cylinder by ovulation or spheroidization.[6-8] In the final stage of crack healing,[1,3,5] the spherical voids either disappear or grow. This last stage of crack healing is exactly analogous to pore shrinkage and pore growth during sintering[25] and will depend on the materials and experimental conditions. Since the sintering phenomena has been studied extensively in the literature,[26-28] we will limit our discussion only to the first

Crack Morphology

```
                    Surface Crack
            ┌───────────┴───────────┐
      Continuous                Discontinuous
    Crack Regression            Crack Pinching
            └───────────┬───────────┘
                   Cylinderization
            ┌───────────┴───────────┐
        Ovulation               Spheroidization
            └───────────┬───────────┘
                   Void Elimination
                       And/Or
                    Void Growth
```

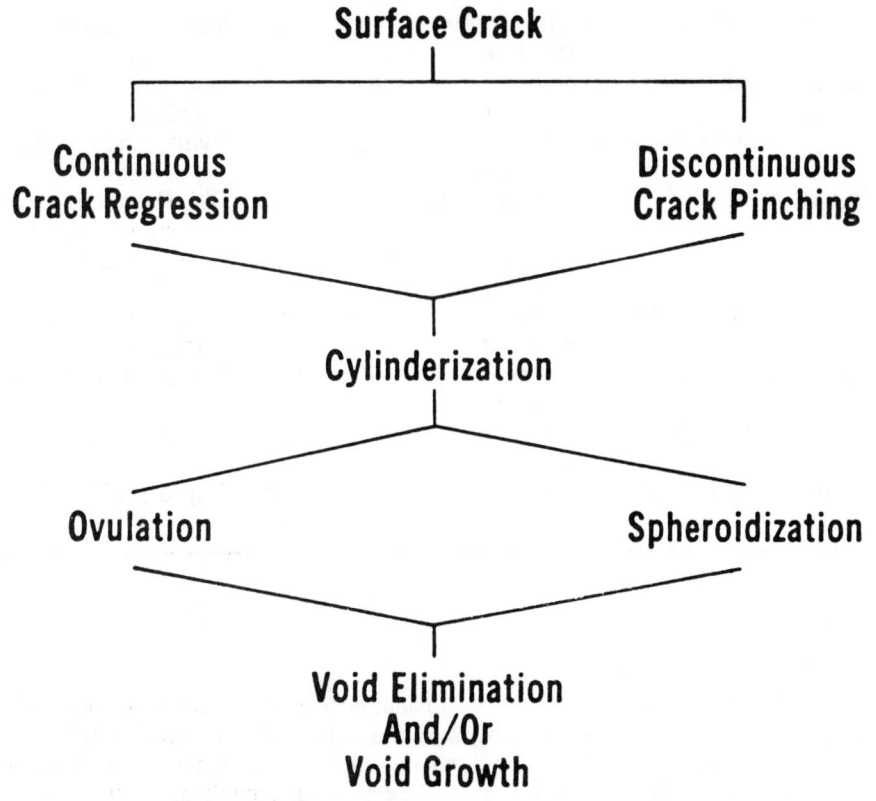

Fig. 1. Morphological change of a crack during crack healing.

three stages of crack healing. The discussion of sintering is beyond the scope of this paper.

Initial Stage of Crack Healing

Most crack healing experiments were conducted by deliberately introducing surface microcracks in ceramics; techniques used were stressing of precracks,[23] indenting,[29] scratching,[8] impacting,[15] and thermal shocking;[1-5] crack morphology was studied by examining the microstructures upon heating the cracked ceramics. Depending upon the method of introduction of the crack, the crack healing started either by the continuous regression of the crack at the crack tip or by the discontinuous pinching of the crack, combined with the reduction of crack width. Table I summarizes the mode of initial crack healing in oxides and halides as described in the literature. In general, crack regression occurred when the crack was introduced by scratching, indenting, or stressing. Impacting and thermal shocking produced cracks that started to heal by crack pinching. This distinction is clearly

Table I. Initial Stage of Crack Healing of Various Oxides and Halides

Material	Technique of crack introduction	Initiation of crack healing by C.R.*	Initiation of crack healing by C.P.†	Reference
Single crystal Al_2O_3	Inscription by WC tool	Yes		Yen and Coble (1972)
Single crystal Al_2O_3	Diamond indentor	Yes		Hockey and Lawn (1975)
Polycrystalline Al_2O_3	Stressing of precracks	Yes		Evans and Charles (1977)
Single crystal LiF	Cleavage crack	Yes		Raj, Pavinich, and Ahlquist (1975)
Single crystal NaCl	Impacting by sharp needle	Yes	Yes	Park and O'Boyle (1977)
Polycrystalline ZnO	Thermal shocking		Yes	Lange and Gupta (1970)
Polycrystalline UO_2	Thermal shocking		Yes	Roberts and Wrona (1973)
Polycrystalline MgO	Thermal shocking		Yes	Gupta (1975)
Polycrystalline Al_2O_3	Thermal shocking		Yes	Gupta (1977)

*C.R. = crack regression.
†C.P. = crack pinching.

evident in Al_2O_3. The NaCl, on the other hand, displayed both types of crack healing when the crack was introduced by impacting. These observations can be rationalized by stating that, if the material remains stress free after the introduction of the crack, crack regression will be predominant; on the other hand, if the material retains stored energy or residual stress after the introduction of the crack, crack pinching will be pronounced.

Cylinderization

The second stage of crack healing is the cylinderization of the flattened inclusion. The cylinderization is most rapid when the inclusion is a void, e.g., a crack, and is progressively less rapid when the inclusion is a liquid and a solid, respectively. Extensive microstructural studies during crack healing of Al_2O_3[6-8] and MgO[3] by the present author revealed that the process of "cylinderization" of a "flat" crack is extremely rapid and cannot be documented for analysis. However, when the inclusion is a liquid or a solid, microstructural documentation was possible. Lange and Clarke[30] have recently observed that, when $MgAl_2O_4$ (spinel) containing LiF was heated at 1500°C, an intergranular second phase, present as a liquid film in a large grain of $MgAl_2O_4$, progressively reduced its area by first forming cylinders, which then broke into liquid-filled spheres. The process of cylinderization was still very rapid for quantitative analysis. However, the cylinderization of a solid inclusion was relatively less rapid and permitted Murty, Morral, Kattamis, and Mehrabian[31] to analyze the data and determine the transport mechanism. The solid inclusion consisted of the sulfides (Fe–Mn–S) in AISI 4340 low-alloy steel, which underwent morphological change during homogenization of the hot-rolled alloy at 1310°C for various lengths of time. During the early stage of homogenization, the flattened and elongated sulfide plates in the as-rolled alloy coarsened and became cylindrical. It can be concluded from these studies that cylinderization of flattened inclusions is the most readily attainable morphology during thermal anneal, as the surface area is reduced by such a change.

Instability of Cylinder

The third stage of crack healing is the classical breakup of the cylinders. This phenomenon is known to occur in a variety of materials, irrespective of whether the inclusion is comprised of a void, a liquid, or a solid.

It is well-known that a cylindrical inclusion is stable when the length L is equal to its circumference; i.e., $L = 2\pi R_0$, where R_0 is the radius of the cylinder. Any longitudinal or circumferential perturbation can make the cylinder unstable;[32] it narrows at one point and bulges at another, resulting in an unduloid that soon breaks into two or more spheres.[6,7]

The breakup of a cylinder has been theoretically understood by assuming that any disturbance or perturbation to a cylinder takes the form of a wave. For a longitudinal perturbation, Nichols and Mullins,[32] from a theoretical study of the instability of a semiinfinite cylinder, defined a characteristic wavelength of perturbation λ_0, where $\lambda_0 = 2\pi R_0$, and related that wavelength to the stability of the cylinder. When the wavelength of perturbation is $<\lambda_0$, the cylinder is stable. This means that a small disturbance (short wave) readjusts itself with time. However, when the wavelength of perturbation is $>\lambda_0$, the cylinder becomes unstable. The breakup of an unstable infinite cylinder is schematically illustrated in Fig. 2.[7] For a finite cylinder, Nichols[33] showed that there are two modes of transformation: ovulation and spheroidization. Ovulation is the breakup of a cylinder into a string of spheres, whereas spheroidization is the direct conversion of a cylinder into a sphere. The mode of transformation depends on the aspect ratio (length to diameter, L/D) of the initial cylinder. Nichols[33] theoretically showed that the cylinders will ovulate with $L/D > 7.2$ and spheroidize with $L/D < 7.2$, although, in practice, the deviation from the above critical aspect ratio has been observed. These modes of transformation are shown schematically in Fig. 3.[7]

As seen, both processes originate with cylindrical configurations A and then undergo three configurational changes. Of these, configurations B (peanut- or dumbbell-shaped structures) and D (sphere) are similar in the ovulation and spheroidization processes. The only configuration that allows a distinction between the two modes is C. For ovulation, C is like two touching tops, and for spheroidization, it is egg shaped. The reason for this difference is that as the undulation proceeds, the two bulging ends recede toward the center of the cylinder, causing λ to decrease.[32] In the case of ovulation, λ is still $>\lambda_0$ at C, so that breakup into voids can occur; in the case of spheroidization, λ is suddenly $<\lambda_0$ during undulation, so the curvature in the middle suddenly changes to comply with the new boundary conditions.[33] In either case, the end results are voids that eventually approach a spherical shape. The only difference in the final state is that there are two voids per ovulation and one void per spheroidization.

For anisotropic materials, a further change in morphology may occur due to the anisotropic nature of surface energy.[34] The spheroids are known to convert to faceted morphology as a result of surface energy anisotropy.

Fig. 2. Schematic of an infinite cylinder under perturbation and final breakup.

$\lambda > \lambda_0$

$\lambda_0 = 2\pi R_0$

$\lambda < \lambda_0$

λ = Wave length of perturbation

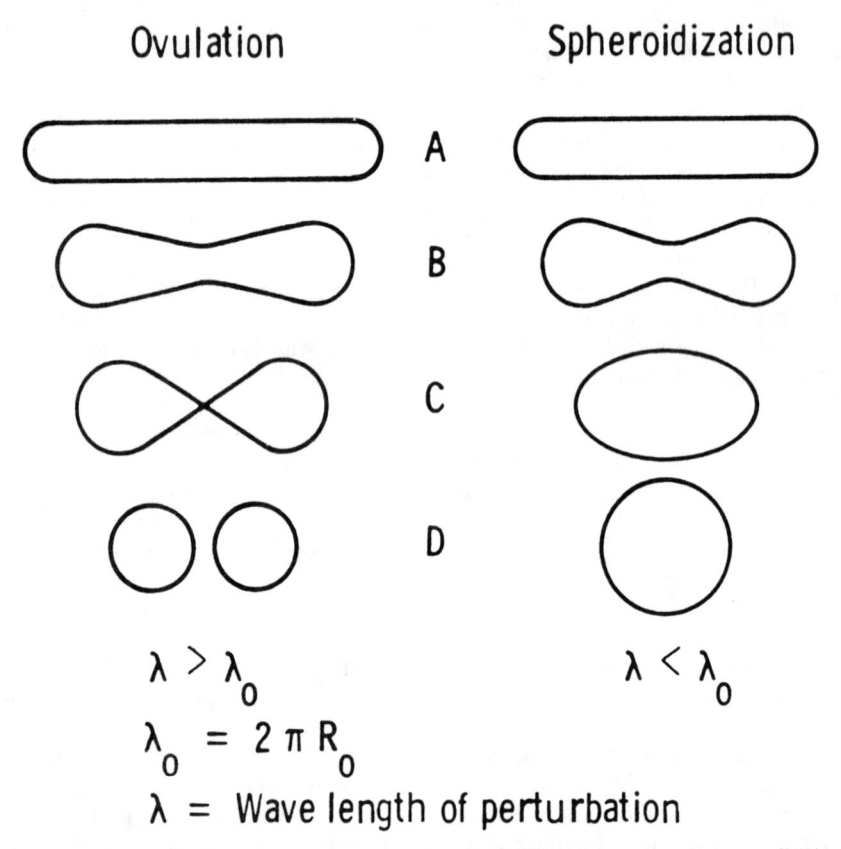

Fig. 3. Schematic of ovulation and spheroidization of a finite cylinder: λ = wavelength of perturbation and $\lambda_0 = 2\pi R_0$ (see text).

Microstructural Observation of Crack Morphology

Various oxides and nonoxides were studied to examine the crack morphology during crack healing. We present here the examples of the various stages of crack morphology from our previous investigations on Al_2O_3[1,2] and MgO.[3] In both ceramics, deep surface cracks were first introduced by thermally shocking the dense polycrystalline ceramics by quenching in water from high temperature. The cracked ceramics were then thermally annealed at 1500–1700°C for Al_2O_3 and 1400–1650°C for MgO. The voids were observed by examining the fractured surface of the ceramics with a scanning electron microscope. All the fracture surfaces examined were intergranular in nature. Figures 4 and 5 show photomicrographs of Al_2O_3 and MgO, respectively, describing the various stages of morphological changes that are discussed previously.

In Fig. 4(A), typical pinching of thermally shocked cracks in Al_2O_3 is illustrated. Pinching occurs at several points along the length of the crack. The crack width also decreases simultaneously. As stated before, crack regression has not been observed in thermally shocked Al_2O_3; however, Evans and Charles[23] reported crack regression in Al_2O_3 where the crack was introduced by stressing of a precrack.

Figure 4(B) illustrates the configurational changes that are schematically described in Figs. 2 and 3. It is seen that many cylinders are in a state of perturbation. Of special interest is the long cylinder (marked by an arrow) which is clearly in a state of strong perturbation, with the neck region indicating a condition prior to a breakup. The peanut-shaped, egg-shaped, and spherical voids are also evident. Examples of cylindrical and spherical voids are illustrated in Figs. 4(C, D). In Fig. 4(C), some of the cylinders have ovulated and/or spheroidized, while others are still present as cylinders. This difference arises because thinner cylinders ovulate more rapidly than the thicker ones.[7] Additionally, for a given cylindrical length, the L/D is favorable to the critical aspect ratio of ~7.2 when the cylinder radius is small. In Fig. 4(D), all the voids are of spherical or near-spherical shape, suggesting complete breakup of the cylinders.

Finally, because of its anisotropic crystalline structure, Al_2O_3 sometimes exhibited faceted morphology instead of spheroids. This is illustrated in Fig. 4(E).

Figure 5 shows the examples of stages of crack evolution obtained from different areas of the fractured surface of MgO specimens, which were annealed at 1500°C for 7 h. Figure 5(A) shows the typical pinching of an initial grain-boundary crack; Fig. 5(B) illustrates a mixture of cylindrical and spherical voids whereas Fig. 5(C) shows extensive spheroidization of voids.

In the final stage of crack healing, the voids are believed to eliminate from the structure, causing densification and strengthening of the cracked ceramics. As in sintering, the grain boundaries seem to play an important role in eliminating the voids. During the study of the crack healing of Al_2O_3, it was repeatedly found that the spherical voids tended to terminate in adjacent grain boundaries that acted as sinks for voids. This phenomenon is illustrated by the scanning electron micrograph of Fig. 6, which depicts the spherical void configuration on an intergranular surface of a nearly healed specimen. It is seen that the regions adjacent to grain boundaries are depleted of voids, suggesting migration and disappearance of voids in the neighboring grain boundaries, while some residual voids are still present in the middle of the grain. Moreover, a trail of voids (marked by an arrow) is also seen approaching the lower grain boundary. Indications are thus strong that there is a diffusional migration of voids toward grain boundaries that acted as sinks for voids. This conclusion is consistent with the sintering theory.

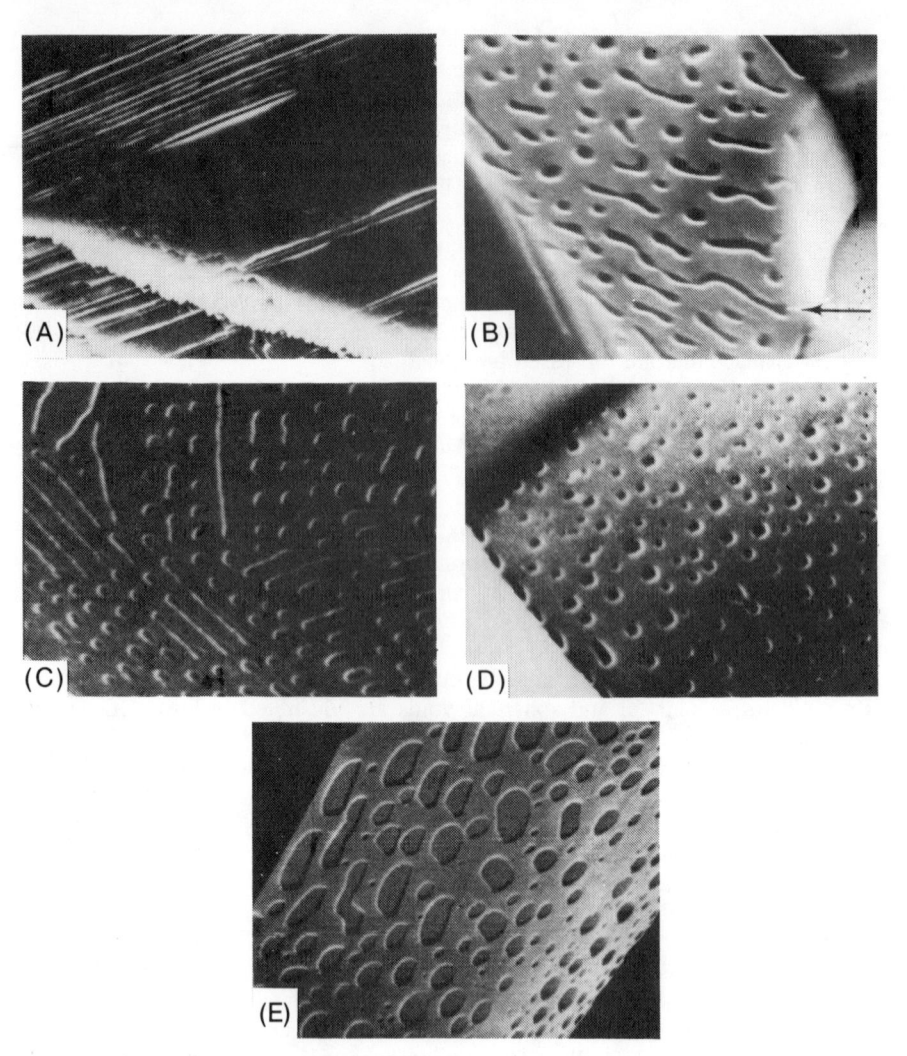

Fig. 4. Sequence of crack morphology during crack healing in Al_2O_3: (A) typical pinching of thermally shocked cracks; (B) cylinders under perturbation; (C) and (D) partial and complete breakup of cylindrical voids; (E) faceted void inclusions.

Kinetics of Morphological Change

Although the kinetics of crack "pinching" has not been investigated because of the rapidity with which it occurs, Evans and Charles[23] studied the kinetics of primary crack regression in Al_2O_3. From the activation energy of the crack regression data, they concluded that the crack regression was controlled by the surface-diffusion mechanism.

The kinetics of cylinderization was studied[31] during homogenization of hot-rolled AISI 4340 low-alloy steel. Longitudinal and transverse sections of the manganese sulfide inclusions prior to and during isothermal treatment at 1310°C

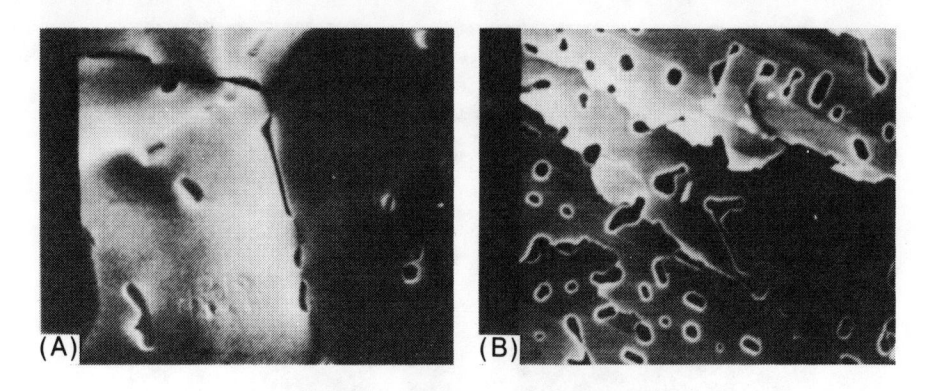

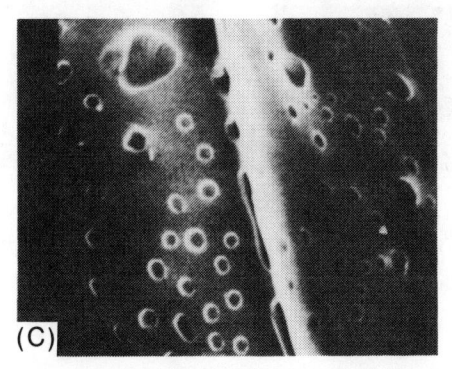

Fig. 5. Crack morphology during annealing of thermally shocked MgO showing (A) crack pinching, (B) cylindrical voids at breakup, and (C) spherical voids.

were measured. It was found that the flattened inclusions developed circular cross sections within the homogenization time of 10 h, while their lengths remained approximately unchanged. After that time, the ovulation became perceptible. The kinetics of the cylinderization process was described by the theory of Nichols and Mullins[32] for the decay of circumferential perturbation on an infinite rod, and the rate of cylinderization was controlled by surface diffusion.

The kinetics of the breakup of cylindrical voids in Al_2O_3 by either ovulation or spheroidization was studied in detail by Yen and Coble[8] and Gupta.[7] In both studies, the controlling transport mechanisms were judged by observing the spacings between two voids (λ_m) and the final spherical void diameter (d) on breakup. The data on Al_2O_3 confirmed the relation

$$\lambda_m = 2.43d \tag{1}$$

which is valid for a surface-diffusion-controlled breakup mechanism.[8,7,46] Following the analysis of Nichols and Mullins,[32] an estimate of the surface-diffusion coefficient (D_s) was made for Al_2O_3 by using the expression[7]

$$D_s = \frac{0.584 r^4 kT}{\tau \gamma \Omega^2 \nu} \tag{2}$$

where τ is the time for the first ovulation, γ the surface energy, ν the number of

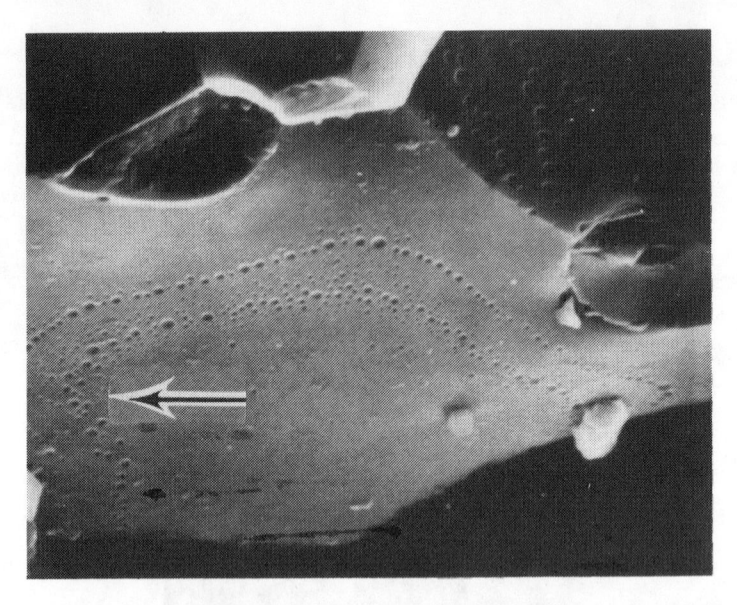

Fig. 6. Void elimination in grain boundaries in Al$_2$O$_3$. Note that the area adjacent to grain boundaries is depleted of voids (annealing conditions: 1700°C, 110 min).

diffusing·atoms/unit surface area, Ω the atomic volume, r the radius of the sphere, k the Boltzmann constant, and T the temperature. Figure 7 compares these diffusion coefficients with other literature values[35-41] for surface-diffusion coefficients in Al$_2$O$_3$ whereas the preexponential factor D_0 and the activation energy Q_s in the diffusion equation $D_s = D_0 \exp(-Q_s/RT)$ are compared in Table II. The data indicate that, depending on the experimental technique and the material, a range of values is obtained for the surface-diffusion coefficient in Al$_2$O$_3$, but the values are within 1–3 orders of magnitude of each other at the temperatures studied. The activation energy is also within the range previously reported in the literature. Considering the uncertainties involved in various experiments, this spread in diffusion coefficient and activation energy was considered well within experimental error. Surface diffusion was concluded to be the predominant mechanism of void brakup in alumina on thermal anneal, but the actual diffusing species was not identified.

The kinetics of the various stages of crack healing is summarized in Table III for Al$_2$O$_3$, MgO, and other materials reported in the literature. It is seen that the surface-diffusion coefficient is the predominant mechanism for the transport of material during the first three stages of crack healing. Transport by volume diffusion becomes predominant only in the last stage of crack healing, which, as stated previously, is analogous to sintering of powders. Coble[42] determined volume diffusion as the rate-controlling mechanism for the densification (void elimination) of Al$_2$O$_3$, whereas Gupta found volume diffusion as the rate-controlling mechanism for the densification[43] of MgO and both surface and volume diffusion for de-densification[44] (void growth) in MgO. An analogous phenomena for void growth

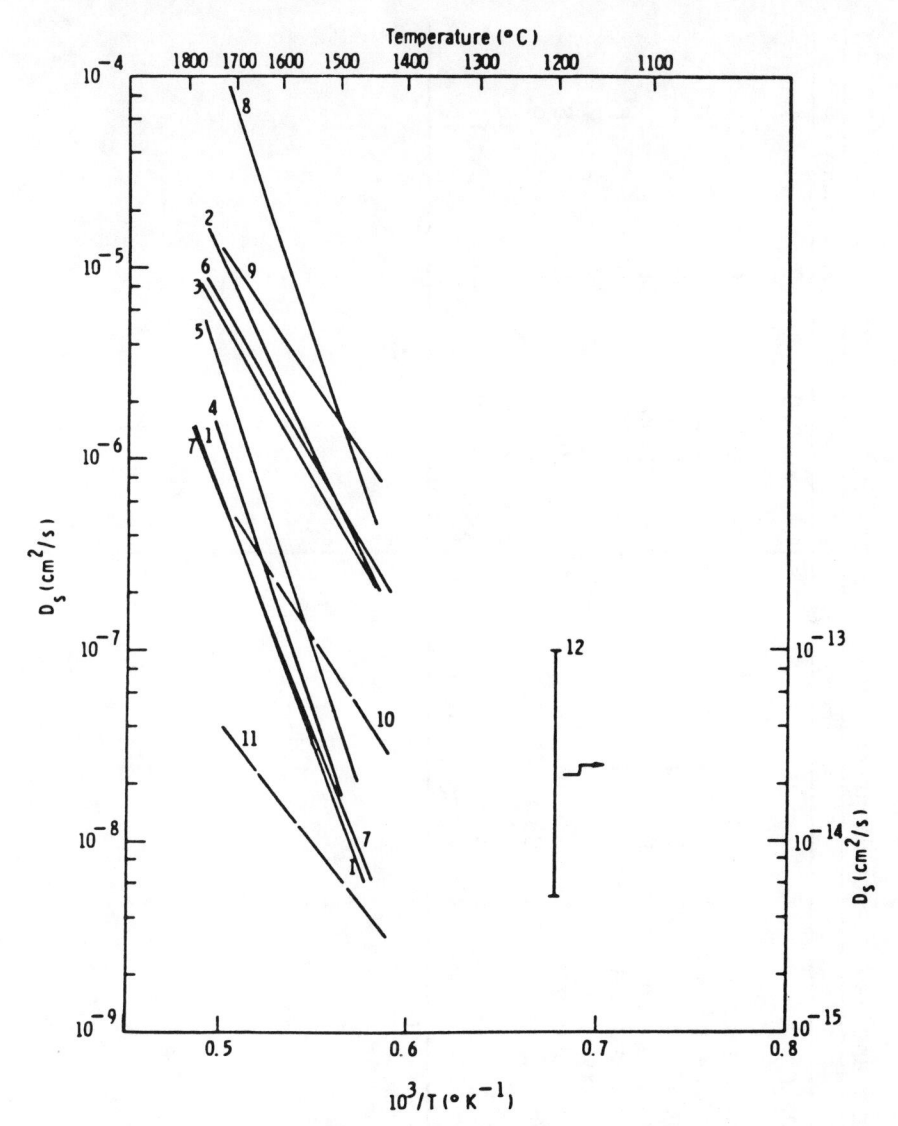

Fig. 7. Comparison of surface diffusion coefficients obtained for Al_2O_3 from crack healing and other techniques. See Table II for identification of curves.

is the coarsening of solid inclusion, and this has been observed with sulfide inclusions in the low-alloy steel, when the latter was homogenized for a prolonged period of time. Murty, Kattamis, Mehrabian, and Flemings[45] have established that the mass transport during coarsening of sulfides at 1310°C was by surface diffusion of sulfur along the sulfide/matrix interface. Further homogenization caused faceting of the inclusion.

Table II. Preexponential Factor and the Activation Energy for Surface Diffusion in Alumina ($D_s = D_0 \exp(-Q_s/RT)$)

Experiment	D_0 (cm²/s)	Q_s (cal/mol)	Ref	Curve no. in Fig. 7
Void breakup by crack healing in polycrystalline alumina[†]	4.2×10^6	118 000	7	1
Void breakup by crack healing in sapphire	2.38×10^3	95 000	8	2
Thermal grooving in polycrystalline alumina	7.0×10^2	75 000	35	3
Thermal grooving in commercial* alumina	1.0×10^8	128 000	36	4
Thermal grooving in commercial[†] alumina	8.0×10^8	133 000	36	5
Thermal grooving in commercial[‡] alumina	9.0×10^2	75 000	36	6
Thermal grooving in alumina bicrystal	5.0×10^5	110 000	37	7
Scratch smoothing in single crystal alumina	1.5×10^{10}	130 000	38	8
Initial sintering in Al_2O_3	1.06×10^2	63 800	39	9
Initial sintering in Al_2O_3	4.8×10^{-2}	56 000	40	10
Initial sintering in Al_2O_3	11.5	67 000	40	11
Initial sintering in Al_2O_3	$D_s (1200°C) \approx 10^{-15} - 10^{-13}$		41	12

*Linde, Union Carbide Corp., New York, NY.
[†]Lucalox, General Electric Co., Schenectady, NY.
[‡]Morganite, Morganite Inc., Dunn, NC.

Table III. Kinetics of the Change of Crack Morphology

Morphology	Material	Kinetics
Crack regression		
Evans and Charles (Ref. 23)	Al$_2$O$_3$	Surface diffusion
Crack pinching	Al$_2$O$_3$	Surface diffusion (anticipated)
Cylinderizing		
Murty et al. (Ref. 31)	Steel (Fe–Mn–S)	Surface diffusion
Ovulation and spheroidization		
Yen and Coble (Ref. 8)	Al$_2$O$_3$	Surface diffusion
Gupta (Ref. 7)	Al$_2$O$_3$	Surface diffusion
Lange and Clarke (Ref. 30)	Spinel–LiF	Surface diffusion
Moon and Koo (Ref. 46)	W–gas	Surface diffusion
Voss, Butler, and Mitchell (Ref. 47)	α-Fe$_2$O$_3$	Surface diffusion
Void elimination and/or void growth		
Coble (Ref. 42)	Al$_2$O$_3$	Volume diffusion
Gupta (Ref. 43, 44)	MgO	Surface and volume diffusion
Coarsening (solid analog of void growth)		
Murty et al. (Ref. 45)	Steel–MnS	Surface diffusion

Application of Crack Healing Phenomena in Several Technological Areas of Interest

The morphological change of the inclusion that occurs during crack healing appears to be widespread in different areas of ceramics and metals technology. The purpose of this section is to briefly review these areas and to point out the effect of these morphological changes on the properties of the materials.

Sintering

All stages of crack healing may be associated with the entire process of sintering wherein the pore shape changes continuously.[28] The initial stage of sintering can be associated with the cylinderization of voids at the grain-boundary interface. The intermediate stage of sintering, where there is a significant transition from a continuous cylindrical pore phase to isolated porosity, can be associated with the breakup of cylindrical voids. The final stage of sintering is analogous to the final stage of crack healing. The voids either disappear or grow, depending on whether the voids are on the grain boundary or are trapped within the grains.

For liquid-phase sintering, the morphological changes of an intergranular thin film in a polycrystalline spinel have been demonstrated by Lange and Clarke.[30] During heat treatment at 1500°C, they observed cylinderization, ovulation, spheroidization, and also an area reduction of the liquid phase (phase elimination) during the final stage.

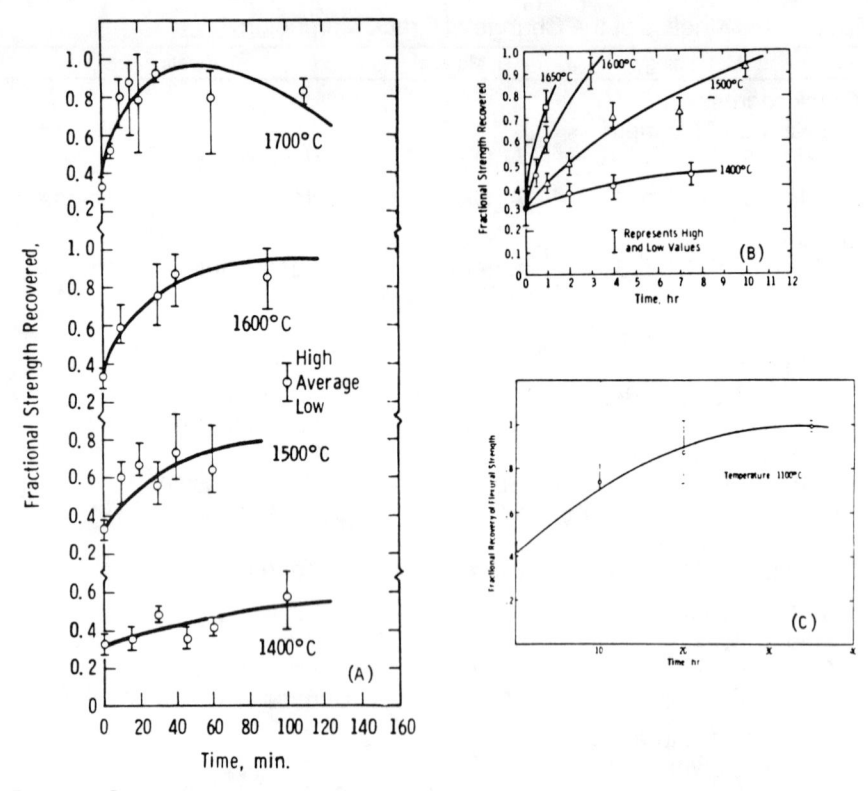

Fig. 8. Strength recovery data as a function of crack healing temperature and time for (A) Al₂O₃, (B) MgO, and (C) ZnO.

Strengthening of Ceramics

The most dramatic effect of crack healing was observed in the improvement of strength of the cracked ceramics, e.g., Al$_2$O$_3$,[1] MgO,[3] and ZnO.[4] The normalized strength recovery data of thermally shocked, reheated oxides are illustrated in Fig. 8. The recovery is strong at the early stage of strengthening; the rate, however, decreases with time as the strengthening continues to completion. In the case of alumina, the stage of complete recovery is followed by a stage of "relapse" where the strength decreases with time. The strengthening mechanism can be visualized in terms of crack healing as follows: cylindrical voids that are initially formed during crack healing break up into a multitude of spherical voids; strengthening of the polycrystalline body occurs as the spherical voids continue to shrink and disappear at the grain boundary (see Fig. 6). When the voids cannot disappear at the grain boundary, they may coalesce and grow instead and may cause the strength to decrease as these large pores may now act as critical flaws in the material.

Strengthening of Metals

The sulfide inclusions in steel have been known to greatly influence the properties of the finished products, and this had led to numerous investigations[31,45] on the effect of sulfide formation on strength. There is now adequate experimental evidence to suggest that the morphology of sulfide inclusions is greatly altered by heat treatment and, as a result, the mechanical properties are significantly improved. This improvement in mechanical properties has been attributed[45] to the cylinderization and spheroidization of sulfide inclusions that were observed during homogenization of flattened and elongated sulfide plates.

Sag-Resistant Tungsten Filament

To promote the life of the incandescent lamp, the tungsten filament is required to have high sag resistance at the operating temperature of 2000–3000°C. For nearly half a century, the high sag resistance has been imparted to the tungsten wire by doping with small amounts of K, Al, and Si. It was shown by Moon and Koo[46] that the sag resistance is developed in the tungsten wire due to the presence of a string of dopant-filled voids of several hundred angstroms in diameter, which are formed during annealing of the wire by the crack healing mechanism.

Growth of Single Crystal

The crack healing mechanism has been recently observed to control the growth of single-crystal hematite (α-Fe_2O_3) in an unusual way. Micrometer-sized blades of α-Fe_2O_3 were observed to grow out of the oxidized surface of iron at high temperatures. Microscopic observations by Voss, Butler, and Mitchell[47] showed a hollow cylinder (tunnel) up the axis of each blade. As the growth proceeded, the cylinder became unstable and broke up into a string of voids, which stopped further growth of the blade. The mechanism of blade growth has also been determined to be surface diffusion controlled.

Composites

Finally, the crack healing mechanism seems to play a vital role in determining the stability of fiber-reinforced composites. Studies were carried out on several different composite systems, and in each system there were signs of cylinderization, ovulation, and spheroidization of the inclusions that greatly affected the properties of the composites.

McLean[48] and coworkers[49] studied the behavior of liquid lead inclusion in aluminum at 450–620°C and observed cylinderization and spheroidization of the inclusion. They further noted that ovulation occurred with $L/D > 8$, whereas the shorter cylinders directly turned into spheres. However, they concluded that the spheroidization of the ~ 10-μm diameter cylinders was controlled by volume diffusion.

Stapley and Beevers[50] studied the stability of sapphire whiskers in nickel in the temperature range of 1100–1400°C. They observed both ovulation and spheroidization as the principal physical processes that led to the degeneration of sapphire whiskers. They gave a very rough estimate of the critical aspect ratio as $L/D \sim 6$. In addition to ovulation and spheroidization, these authors also noted "wasting" of the whisker (configuration B of Fig. 3) from surface undulation and Ostwald ripening, which contributed significantly to spheroidization.

Finally, an interesting observation was recently made with W fiber/Cu composite. It was found that the mechanical strength of the above composite degraded on thermal cycling. Careful studies of W fiber/Cu composite by Yoda, Takahashi, Wakashima, and Umekawa[51] revealed that repetitive thermal cycling caused preferential crack of the matrix along the reinforcing fibers, the amount of which increased with the number of thermal cycles. They further noted that as the cycling proceeded, round and isolated pores were formed that soon increased in numbers and coalesced to form cavities with the resultant decrease in mechanical strength. These authors did not comment on the mechanism of void formation during thermal cycling, but in view of our knowledge regarding the morphological changes during crack healing, it can be stated with a high degree of certainty that the void formation during prolonged thermal cycling was caused by all the morphological changes that were observed during crack healing. Similar degradation phenomenon may also explain the observed thermally induced degradation in other composites.[52]

Summary

In this review of crack healing of Al_2O_3 and MgO, the morphological change of a crack upon heat treatment has been emphasized. From an extensive review of the literature, four stages of morphological change have been identified. The crack healing commences with a continuous regression of the crack at the crack tip or a discontinuous pinching of the crack, combined with the reduction of crack width. This is quickly followed by the cylinderization of the crack. In the third stage of crack healing, the cylinders ovulate or spheroidize, depending on the aspect ratio of the cylinder. In the final stage, the spherical voids may either disappear or grow. Since a crack can be considered as an inclusion in the solid, the above morphological changes are also known to occur when the inclusion is a liquid or a solid. The surface diffusion has been determined to be the dominant transport mechanism for all stages of crack healing except for the final stage, where volume diffusion is also known to occur significantly. The review of literature indicates that morphological change of the inclusion also occurs in several technologically important areas.

The most obvious analogy is between the phenomenon of crack healing and that of sintering. In the latter, the pore shape changes continuously in a fashion similar to that of crack morphology during crack healing. The strengthening of cracked ceramics is a direct consequence of crack healing. Similarly, the alloy steel is known to become mechanically improved due to the presence of sulfide inclusion, which undergoes shape change during homogenization similar to that in crack healing. The sag-resistant tungsten filament contains gas-filled spheres that are the direct result of ovulation and spheroidization upon annealing of the elongated cylinder formed during working of the ingot to wire. The growth of the single crystal of α-Fe_2O_3 has been shown to be controlled by the internal cylindrical void and its subsequent breakup. Finally, degradation of the fibers by ovulation and spheroidization has been reported in fiber-reinforced composites. Additionally, spherical voids, resulting from the cracks at the fiber/matrix interface, have also been identified in several technologically important composites, which may cause a further degradation of mechanical properties. These spherical voids are believed to have been formed by the crack healing mechanism as described in this review.

References

[1]T. K. Gupta, "Crack Healing and Strengthening of Thermally Shocked Alumina," *J. Am. Ceram. Soc.*, **58** [5–6] (1976).
[2]T. K. Gupta, "Kinetics of Strengthening of Thermally Shocked MgO and Al_2O_3," *J. Am. Ceram. Soc.*, **59** [9–10] 448–9 (1976).
[3]T. K. Gupta, "Crack Healing in Thermally Shocked MgO," *J. Am. Ceram. Soc.*, **58** [3–4] (1975).
[4]F. F. Lange and T. K. Gupta, "Crack Healing by Heat Treatment," *J. Am. Ceram. Soc.*, **53** [1] 54–5 (1970).
[5]F. F. Lange and K. C. Radford, "Healing of Surface Cracks in Polycrystalline Al_2O_3," *J. Am. Ceram. Soc.*, **53** [7] 420–1 (1970).

[6]T. K. Gupta, "Alteration of Cylindrical Voids During Crack Healing in Alumina"; pp. 354–65 in 6th International Material Symposium on Ceramic Microstructure. University of California, Berkeley, CA, 1967.
[7]T. K. Gupta, "Instability of Cylindrical Voids in Alumina," J. Am. Ceram. Soc., 61 [5–6] 191–5 (1978).
[8]C. F. Yen and R. L. Coble, "Spheroidization of Tubular Voids in Al_2O_3 Crystals in High Temperatures," J. Am. Ceram. Soc., 55 [10] 507–9 (1972).
[9]J. T. A. Roberts and B. J. Wrona, "Crack Healing in UO_2," J. Am. Ceram. Soc., 56 [6] 297–9 (1973).
[10]G. Bandyopadhyay and J. T. A. Roberts, "Crack Healing and Strength Recovery in UO_2," J. Am. Ceram. Soc., 59 [9–10] 415–9 (1976).
[11]G. Bandyopadhyay and C. R. Kennedy, "Isothermal Crack Healing and Strength Recovery in UO_2 Subjected to Varying Degrees of Thermal Shock," J. Am. Ceram. Soc., 60 [1–2] 48–50 (1977).
[12]F. F. Lange, "Healing of Surface Cracks in SiC by Oxidation," J. Am. Ceram. Soc., 53 [5] (1970).
[13]R. N. Singh and J. L. Routbort, "Fracture and Crack Healing in (U,Pu)C," J. Am. Ceram. Soc., 62 [3–4] 128–33 (1979).
[14]R. Raj, W. Pavinich, and C. N. Ahlquist, "On the Sintering Rate of Cleavage Cracks," Acta Metall., 23 [3] 399–403 (1975).
[15]S. M. Park and D. R. O'Boyle, "Observation of Crack Healing in Sodium Chloride Single Crystals at Low Temperature," J. Mater. Sci., 12, 840–1 (1977).
[16]A. H. Heuer and A. P. Roberts, "Influence of Annealing on the Strength of Corundum Crystals," Proc. Br. Ceram. Soc., 6, 17–27 (1966). M. J. Noone and A. H. Heuer, "Improvements in the Surface Finish of Ceramics by Flame Polishing and Annealing Technique"; pp. 213–32 in Science of Ceramic Machining and Surface Finishing. NBS Publ. No. 348, 1972.
[17]F. P. Mallinder and B. A. Proctor, "Preparation of High Strength Sapphire Crystals," Proc. Br. Ceram. Soc., 6, 9–16 (1966).
[18]P. F. Becher and R. W. Rice, "Flame Polishing of Flat Oxide Bars"; pp. 237–45 in Science of Ceramic Machining and Surface Finishing. NBS Publ. No. 348, 1972.
[19]J. T. A. Pollock, "Continuous Flame Polishing of Sapphire Filament"; pp. 247–57 in Science of Ceramic Machining and Surface Finishing. NBS Publ. No. 348, 1972.
[20]H. P. Kirchner, R. M. Gruver, D. R. Platts, P. A. Rishel, and R. E. Walker, "Chemical Strengthening of Ceramic Materials"; Summary Rep., Contract N00019-68-C-0142, Naval Air Systems, 1969.
[21]W. A. Schmidt and J. E. Davey, "Preparation of Smooth, Crystalline, Damage-Free Sapphire Surfaces by Gaseous Etching"; pp. 259–65 in Science of Ceramic Machining and Surface Finishing. NBS Publ. No. 348, 1972.
[22]R. W. Rice, P. F. Becher, and W. A. Schmidt, "Strength of Gas Polished Sapphire and Rutile"; pp. 267–9 in Ref. 21.
[23]A. G. Evans and E. A. Charles, "Strength Recovery by Diffusive Crack Healing," Acta Metall., 25 [8] 919–29 (1977).
[24]T. K. Gupta, "Effect of Crack Healing on Thermal Stress Fracture"; pp. 365–80 in Thermal Stresses in Severe Environments. Edited by D. P. H. Hasselman and R. A. Heller. Plenum, New York, 1980.
[25]W. D. Kingery, H. K. Bowen, and D. R. Uhlman; Chapter 10 in Introduction to Ceramics. Wiley, New York, 1976.
[26]G. C. Kuczynski, N. A. Hooton, and C. F. Gibson (Editors); Sintering and Related Phenomena. Gordon and Breach, New York, 1967.
[27]G. C. Kuczynski (Editor), "Sintering and Related Phenomena"; Materials Science Research, Vol. 6. Plenum, New York, 1973.
[28]R. L. Coble and J. E. Burke; pp. 197–249 in Progress in Ceramic Science, Vol. III. Edited by J. E. Burke. Pergamon, New York, 1963.
[29]B. J. Hockey and B. R. Lawn, "Electron Microscopy of Microcracking about Indentations in Aluminum Oxide and Silicon Carbide," J. Mater. Sci., 10, 1275–84 (1975).
[30]F. F. Lange and D. R. Clarke, "Morphological Changes of an Intergranular Thin Film in a Polycrystalline Spinel," J. Am. Ceram. Soc., 65 [10] 502 (1982).
[31]Y. V. Murty, J. E. Morral, T. Z. Kattamis, and R. Mehrabian, "Initial Coarsening of Manganese Sulphide Inclusions in Rolled Steel During Homogenization," Metall. Trans., A, 6A, 2031–5 (1975).
[32]F. A. Nichols and W. M. Mullins, "Morphological Changes of a Surface of Revolution Due to Capillarity-Induced Surface Diffusion," J. Appl. Phys., 36 [6] 1826–35 (1965).
[33]F. A. Nichols, "Spheroidization of Rod-Shaped Particles of Finite Length," J. Mater. Sci., 11 [6] 1077–82 (1976).
[34]A. D. Brailsford and N. A. Gjostein, "Influence of Surface Energy Anisotropy on Morphological Changes Occurring by Surface Diffusion," J. Appl. Phys., 46 [6] 2390 (1975).
[35]W. M. Robertson and R. Chang, "Role of Grain Boundaries and Surfaces in Ceramics"; pp. 49–60 in Materials Science Research, Vol. 3. Edited by W. W. Kriegel and Hayne Palmour, III. Plenum, New York, 1966.

[36]W. M. Robertson and F. E. Ekstrom, "Kinetics and Reactions in Ionic Systems"; pp. 271–81 in Materials Science Research, Vol. 4. Edited by T. J. Gray and V. D. Frechette. Plenum, New York, 1969.

[37]J. F. Shackelford and W. D. Scott, "Relative Energies of ($\bar{1}$100) Tilt Boundaries in Aluminum Oxide," J. Am. Ceram. Soc., **51** [12] 688–92 (1968).

[38]T. Maruyama and W. Komatsu, "Surface Diffusion of Single-Crystal Al_2O_3 by Scratch Smoothing Method," J. Am. Ceram. Soc., **58** [7–8] 338–9 (1975).

[39]Y. Moriyoshi and W. Komatsu, "Kinetics of Initial Combined Sintering," Yogyo Kyokai Shi, **81** [3] 102–7 (1973).

[40]S. Prochazka and R. L. Coble, "Surface Diffusion in the Initial Sintering of Alumina: I," Phys. Sint., **2** [1] 1–18 (1970); "II," Phys. Sint., [2] 1–14; "III," Phys. Sint., [3] 15–34.

[41]W. R. Rao and I. B. Cutler, "Initial Sintering and Surface Diffusion in Al_2O_3," J. Am. Ceram. Soc., **55** [3] 170–1 (1972).

[42]R. L. Coble, "Sintering Crystalline Solids," J. Appl. Phys., **32** [5] 787–92 (1961).

[43]T. K. Gupta, "Sintering of MgO: Densification and Grain Growth," J. Mater. Sci., **6**, 25–32 (1971).

[44]T. K. Gupta, "Kinetics and Mechanisms of Pore Growth in MgO," J. Mater. Sci., **6**, 989–97 (1971).

[45]Y. V. Murty, T. Z. Kattamis, R. Mehrabian, and M. C. Flemings, "Behavior of Sulfide Inclusion During Thermomechanical Processing of AISI 4340 Steel," Metall. Trans., A, **8A**, 1275 (1977).

[46]D. M. Moon and R. C. Koo, "Mechanism and Kinetics of Bubble Formation in Doped Tungsten," Metall. Trans., **2** [8] 2115–22 (1971).

[47]D. A. Voss, E. P. Butler, and T. E. Mitchell, "The Growth of Hematite Blades During the High Temperature Oxidation of Iron," Metall. Trans., A, **13A**, 929 (1982).

[48]M. McLean, "Kinetics of Spheroidization of Lead Inclusions in Aluminum," Philos. Mag., **27** [6] 1253–66 (1973).

[49]M. McLean and M. S. Loveday, "In Situ Observations of the Annealing of Liquid Lead Inclusions Entrained in an Aluminum Matrix," Philos. Mag., **9** [7] 1104–14 (1974).

[50]A. J. Stapley and C. J. Beevers, "Stability of Sapphire Whiskers in Nickel at Elevated Temperatures: II," J. Mater. Sci., **8** [9] 1296–306 (1973).

[51]S. Yoda, R. Takahashi, K. Wakashima, and S. Umekawa, "Fiber/Matrix Interface Porosity in Tungsten Fiber/Copper Composites on Thermal Cycling," Metall. Trans., A, **10A**, 1796 (1979).

[52]K. K. Chawla, "Thermal Fatigue Damage in Borsic-Al (6061) Composites," J. Mater. Sci., **11**, 1567 (1976).

Fracture Toughness of Single-Crystal Alumina

M. Iwasa* AND R. C. Bradt
University of Washington
Department of Materials Science and Engineering
Seattle, WA 98195

Fracture toughnesses of the (0001), (10$\bar{1}$2), (10$\bar{1}$0), and (11$\bar{2}$0) planes of single-crystal alumina were measured from room temperature through 1400°C in air. All of the planes exhibited low-temperature regions of toughness decreases with increasing temperature, while the (0001) and (10$\bar{1}$2) also exhibited increasing toughness at elevated temperatures, indicative of crack tip plastic flow. The (10$\bar{1}$2) cleavage plane is preferred at room temperature; however, alumina exhibits a transition to a (0001) cleavage plane above 700°C. On the basis of the fracture toughness trends with increasing temperature, an explanation is proposed for the high-temperature strength minimum of alumina single crystals.

For nearly half a century, the fracture of single crystals of alumina as corundum or in the sapphire or ruby forms has been of continuing interest. It has been subjected to intense study by numerous mineralogists, ceramists, and other scientists. In addition to the purely academic aspect of the fracture process, it has direct commercial implications in the breakage of single-crystal grains of alumina abrasives, the wear and polishing of crystals, the performance of jewel bearings, the failure of vapor arc lamp envelopes, the cracking of microcircuit substrates, and the design and failure of metal–matrix composites using single-crystal alumina whiskers or fibers for reinforcement. Of course, the fine milling of alumina powders for subsequent consolidation into polycrystalline ceramic bodies also depends on the fracture or cleavage of individual crystallites once the powders are sufficiently reduced in size. It is obvious that fracture is a primary concern whenever alumina is utilized.

Studies of the fracture of single crystals of alumina can be conveniently grouped into four categories. These include the following: (i) mineralogical investigations such as those of Shappel,[1] Winchell,[2] and Yamaguchi, Kubo, and Ogawa;[3] (ii) conventional strength studies such as those of Wachtman and Maxwell,[4] Brenner,[5] and more recently Noda;[6] (iii) the fracture mechanics studies of Wiederhorn,[7] Wieland,[8] and Smith and Pletka;[9] (iv) a number of different types of general observational efforts including those of Palmour et al.,[10] Mallinder and Proctor,[11] and Anderson.[12] While all of this research must achieve a high degree of consistency within the overall results, the mineralogical and the fracture mechanics investigations are probably the most appropriate ones to initially review to provide the background for this study.

Although numerous statements exist in the general literature concerning the conchoidal fracture of alumina, similar to that often accepted to occur for crystalline quartz, whenever fracture traces of alumina single crystals have been closely examined, such as by Yamaguchi et al.,[3] it has become readily apparent that a number of distinct crystallographic planes do cleave, as the bonds across some planes are broken more readily than others. However, at room temperature, no plane cleaves more easily than the rhombohedral R plane, the ($10\bar{1}2$). Partings, which are planar-like fractures that are approximately parallel to individual planes but are not usually accepted as true cleavage, have been reported to occur on the (0001), ($10\bar{1}0$), ($11\bar{2}0$), ($10\bar{1}1$), ($11\bar{2}3$), and the ($11\bar{2}6$) planes. One must conclude that crack propagation in single-crystal alumina can and does readily occur on specific crystallographic planes.

The three studies that have concluded that the rhombohedral (1012) plane is the preferred room-temperature cleavage plane, those of Shappel,[1] Wiederhorn,[7] and Iwasa et al.,[13] also agree that the basal or C plane, the (0001), is very difficult to cleave at room temperature. Shappel[1] assigned an ease of cleavability index of 12.4 to the ($10\bar{1}2$) and only 9.5 to the (0001), while Wiederhorn[7] measured a surface energy of 6 J/m^2 for the ($10\bar{1}2$) and estimated that of the (0001) to be >40 J/m^2. Iwasa et al.[13] reported K_{Ic} values of 2.38 ± 0.14 and 4.54 ± 0.32 MPa·m$^{1/2}$ for the ($10\bar{1}2$) and (0001), respectively. Several other features of the difficulty of basal plane cleavage at room temperature are noteworthy. In his double cantilever beam specimens, Wiederhorn[7] was not able to propagate a basal plane crack the full thickness of the specimen web. He attributed this difficulty to a lack of charge neutrality on the basal plane. Noda[6] and Becher[14] also confirm the difficulties with room-temperature basal plane cleavage in their strength studies. Becher's calculations yielded a discontinuity that could not be deciphered for fracture on the (0001). Thus, the reasons for the lack of basal plane cleavage at room temperature remain unclarified. However, the reason for ($10\bar{1}2$) rhombohedral plane cleavage is generally attributed to the extensive array of vacant cation lattice sites parallel to the plane.

In spite of the aforementioned difficulties with room-temperature basal plane cleavage, there are a number of well-documented reports of moderate to extensive basal plane fracture. Becher[14] shows a micrograph of a flat (0001) region on a room-temperature fracture, while Firestone and Heuer[15] show a similar (0001) region in an 1800°C fracture originating during creep of 0° crystals at that temperature. Abdel-Latif et al.[16] describe extensive flat mirror regions of basal plane fractures for c-axis sapphire crystals tested in tension, and Brenner[5] also reports smooth basal plane fractures. However, the most outstanding report of consistent basal plane cleavage is that of Anderson[12] for sapphire vapor arc lamp envelopes. Anderson very frequently observed (0001) basal plane cleavage of the single-crystal envelopes to occur at ≈1100°C. Castaing et al.[17] have observed well-defined basal plane cleavage at high pressures, the equivalent to an elevated temperature observation. Thus, (0001) basal plane cleavage does occur in alumina, and it apparently occurs more easily at elevated temperatures. The statement of Wiederhorn et al.[18] that cleavage "planes preferred at low temperatures might not be preferred at high temperatures" might be applicable to basal plane cleavage. That intriguing possibility was one of the primary reasons for measuring the K_{Ic} for fracture on the (0001), ($10\bar{1}2$), ($10\bar{1}0$), and ($11\bar{2}0$) planes of single-crystal alumina at elevated temperatures in this study.

Experimental Procedures

The single crystals measured in this study were commercial synthetic Czochralski-grown transparent boules approximately 5 cm in diameter. To obtain suitable test specimens perpendicular to the (0001), ($01\bar{1}0$), ($11\bar{2}0$), and ($10\bar{1}2$)

planes, the boules were oriented by the back-reflection Laue technique and then diamond sawed and diamond ground to bar specimens 0.2 × 0.2 × 2.5 cm. The final surface finish was by a 600-grit diamond wheel, after which the specimen edges were hand beveled by using diamond paste to ensure fracture initiation from the Knoop-indenter-induced microflaws.

Fracture toughness, which will be referred to as K_{Ic} in this paper, was measured by using the technique popularized by Petrovic and Mendiratta.[19] After the specimens were ground, microflaws were introduced on the (0001), (10$\bar{1}$0), (11$\bar{2}$0), and (10$\bar{1}$2) planes in the [11$\bar{2}$0], [0001], [0001], and [12$\bar{1}$0] directions, respectively, by using a Knoop diamond indenter with a 3600-g load. Residual stresses were removed by grinding an additional 25 μm from the surface. These precracked specimens were then broken in air in three-point loading over a span of 1.99 cm at a crosshead speed of 0.254 cm/in. The fracture toughness was calculated from the equation

$$K_{Ic} = 2.06\sigma_f\left(\frac{a}{\pi}\right)^{1/2} \qquad (1)$$

where σ_f is the fracture stress in bending and a is the characteristic flaw dimension, which was measured after cleavage with a calibrated microscope.

Elevated temperature fracture toughnesses were measured in air at 200°C intervals through 1400°C. Individual specimens were tested by inserting them into a preheated vertical tube Pt furnace after positioning the sample on the two bottom knife edges. Fractures were affected immediately upon reaching temperature after inserting the specimen and knife edges into the furnace. After fracture, the specimen surfaces were examined by optical and scanning electron microscopy to determine the flaw dimensions and generally observe the fracture surface features. Different numbers of specimens were tested at the different temperatures, sometimes as many as 15; however, the only data reported are those for which the flaw dimensions could be determined with a high degree of confidence. This latter point is important, for often it is not possible to positively identify the critical flaw dimensions.

Results and Discussion

Room-Temperature Cleavage

The room-temperature fracture toughnesses for the four previously mentioned planes are listed along with the Shappel[1] ease of cleavability estimates and the fracture surface energies of Wiederhorn[7] in Table I. It is evident from these that the rhombohedral or R plane, (10$\bar{1}$2), is the preferred cleavage plane at room temperature and that the basal or C plane, (0001), is considerably more difficult to

Table I. Cleavage Planes of Alumina at Room Temperature

Plane	Shappel Cleavability*	γ_f (J/m²)†	K_{Ic} (MPa·m$^{1/2}$)‡
C (0001)	9.5	>40	4.54 ± 0.32
M (10$\bar{1}$0)	10.2	7.3	3.14 ± 0.30
R (10$\bar{1}$2)	12.4	6.0	2.38 ± 0.14
A (11$\bar{2}$0)	9.2		2.43 ± 0.26

*Reference 1. Higher values represent easier cleavage.
†Reference 7.
‡This study.

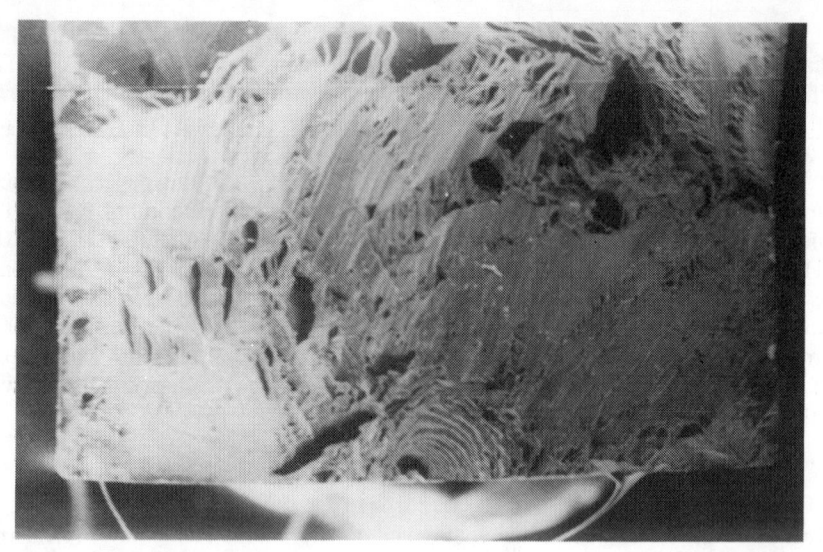

Fig. 1. Surface of a (0001) specimen fractured at room temperature. Note that only near the microflaw at the bottom center of the specimen is there any basal cleavage.

fracture. Although these three sets of results are the most comprehensive, it is of interest to discuss a number of other related fracture observations for they generally confirm these results. Stofel and Conrad[20] observed numerous rhombohedral cleavage fractures in bend tests of 0° sapphire rods, while Wieland[8] observed the R plane to cleave much more easily than the M plane, although expressed on a ratio of loads that are nearly equal to the K_{Ic} ratios of Table I. Smith and Pletka[9] examined the microindentation toughnesses of the $(10\bar{1}0)$ and $(11\bar{2}0)$ planes and observed crack extension on the $(10\bar{1}0)$ to be more difficult than on the $(11\bar{2}0)$ plane, also in general agreement with Table I. Kirchner and Gruver[21] in examining slow crack growth in polycrystalline aluminas have estimated and assigned K_{Ic} values to various crystallographic planes of the corundum structure. Their values agree in principle with the order expressed in Table I. In addition they also estimated a K_{Ic} of 4.3 MPa·m$^{1/2}$ for the $(11\bar{2}6)$ plane, which might be expected to be a rather difficult plane to cleave. Noda[4] made extensive strength measurements on polished single-crystal specimens of various orientations including different angles of tension on all four of the planes listed in Table I and concluded that the R plane, $(10\bar{1}2)$, was indeed the dominant room-temperature cleavage plane for sapphire. For alumina, it must be concluded that the rhombohedral or R plane, $(10\bar{1}2)$, is the preferred cleavage plane at room temperature and that the basal or C plane, (0001), is one of the most difficult planes to cleave at room temperature.

It was also confirmed in this study, by using microflaws, that extensive crack propagation at room temperature on the basal plane is quite difficult. This confirmed the larger crack studies using the double cantilever beam specimen by Wiederhorn[7] regarding extensive crack propagation on the (0001) at room temperature. The specimens oriented and precracked for basal plane cleavage in this study showed only very limited basal plane crack extension at room temperature.

Shortly after the initiation of fracture, the microcracks that were introduced on the basal plane changed their propagation direction and reoriented onto the familiar (10$\bar{1}$2) rhombohedral planes characterized by their easily recognized stairlike checkerboard cleavage fracture patterns. Figure 1 illustrates one of the room-temperature basal (0001) C plane specimens. Palmour et al.[10] have extensively described these fracture surface patterns.

The high fracture toughness of the (0001) C plane and the lower toughnesses of the R, M, and A planes suggest that these differences may be related to the elastic moduli normal to those planes. Becher[14] has previously attempted a similar line of reasoning. Combining the linear elastic fracture mechanics equation for the fracture toughness[22]

$$K_{Ic}^2 = \frac{2E\gamma_f}{(1-\nu^2)} \quad (2)$$

with that for the fracture surface energy[23]

$$\gamma_f = \frac{E}{a_0}\left(\frac{\lambda}{\pi}\right)^2 \quad (3)$$

yields

$$K_{Ic} = \frac{\lambda 2^{1/2} E}{\pi a_0^{1/2}(1-\nu^2)} \quad (4)$$

suggesting that a graph of K_{Ic} vs E, where E is the Young's elastic modulus normal to the plane, should yield a straight line, providing that the interplanar spacing, a_0, the relaxation distance for the interatomic forces, λ, and Poisson's ratio, ν, are similar. Figure 2 illustrates such a plot of fracture toughness vs Young's elastic modulus applying the single-crystal elastic constants after Wachtman et al.[24] to calculate the Young's elastic moduli. With the exception of the M plane, there is reasonably good prediction, although the Young's modulus normal to the rhombohedral cleavage plane exceeds that of both the A and M planes. Since both the M and A planes have the same Young's elastic modulus, $1/s_{11}$ or 425 GPa, cleavage of the M plane, (10$\bar{1}$0), must initiate some additional energy dissipation processes. The most likely of these is twinning, as has been extensively discussed by Stofel and Conrad[20] and more recently by Scott and Orr.[25]

As previously noted, Becher[14] was not overwhelmingly successful in applying a γ_f vs E analysis to sapphire similar to the analysis presented in Fig. 2. These problems raise questions as to the purely elastic aspects of the fracture process. This is especially true since more satisfactory K_{Ic} vs E plots have been observed. A good example is that for $MgAl_2O_4$.[26] The reason for the failure of the crystallographic cleavage of alumina to be dominated by the Young's elastic modulus variation is not completely understood; however, it may be related to the lack of any substantial degree of elastic anisotropy within the alumina structure. Although many crystals exhibit Young's elastic moduli differences of a factor of 2 or more for different crystallographic directions, the difference between the minimum and maximum E in single-crystal alumina is only about 10%. With this relatively small degree of elastic anisotropy, it must be concluded that other structural features probably dominate the ease of cleavage of specific planes in the alumina structure. In the case of (10$\bar{1}$2) rhombohedral cleavage, the vacant cation lattice sites must dominate, whereas the difficult (0001) basal plane cleavage may be attributed to the lack of charge neutrality as suggested by Wiederhorn[7] or perhaps the strong

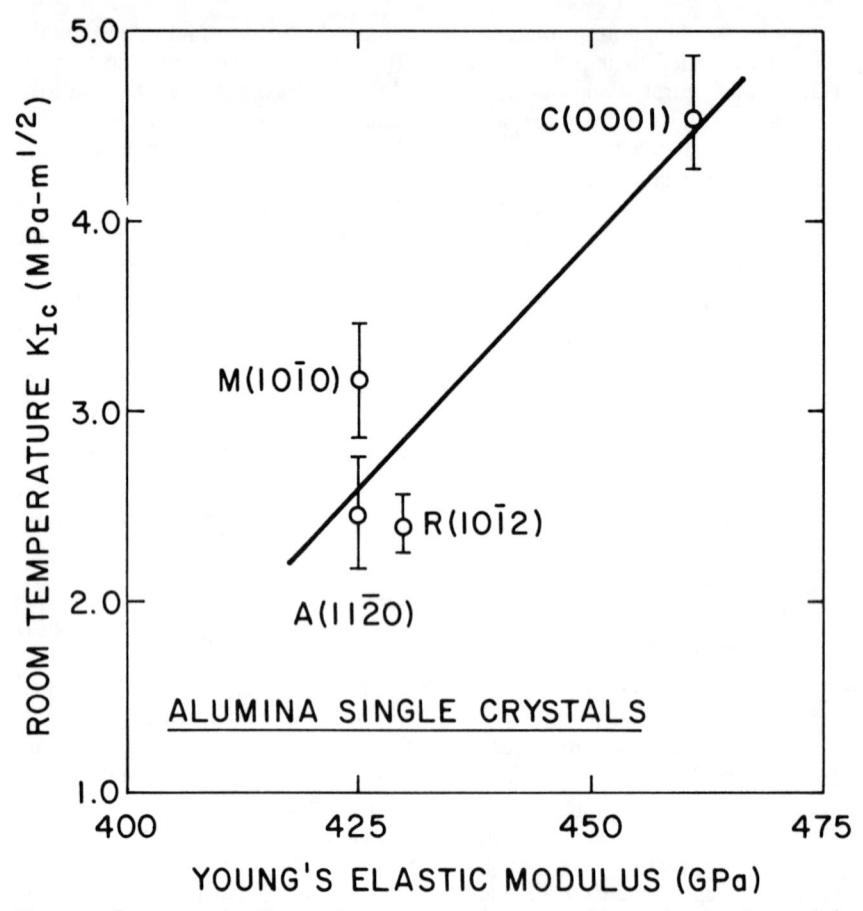

Fig. 2. Room-temperature fracture toughness vs Young's elastic modulus diagram for alumina.

cation–cation interaction in the [0001] as a consequence of the face sharing of octahedra along that direction.

Elevated-Temperature Cleavage

The effects of temperature on the K_{Ic} values of the $(10\bar{1}2)$, (0001), $(10\bar{1}0)$, and $(11\bar{2}0)$ were also studied. Even at 1400°C, the load-displacement curves of the precracked microflaw containing specimens appeared totally linear elastic to fracture. All of the orientations exhibited a distinct decrease of K_{Ic} with increasing temperature at lower temperatures, while the $(10\bar{1}2)$ and (0001) also exhibited increases of K_{Ic} above about 800–1000°C, indicative of the brittle to ductile transition but, as previously noted, without macroscopically noticeable plastic flow. Wiederhorn et al.[18] also studied the effect of temperature on the K_{Ic} of the $(10\bar{1}0)$, M plane, and observed it to decrease from about 2.5–1.8 MPa·m$^{1/2}$ between room temperature and 600°C. Wieland[8] compared the M plane and the R plane but not in direct K_{Ic} form. He reported decreases with increasing temperature for both, with the R plane toughness perhaps decreasing slightly more

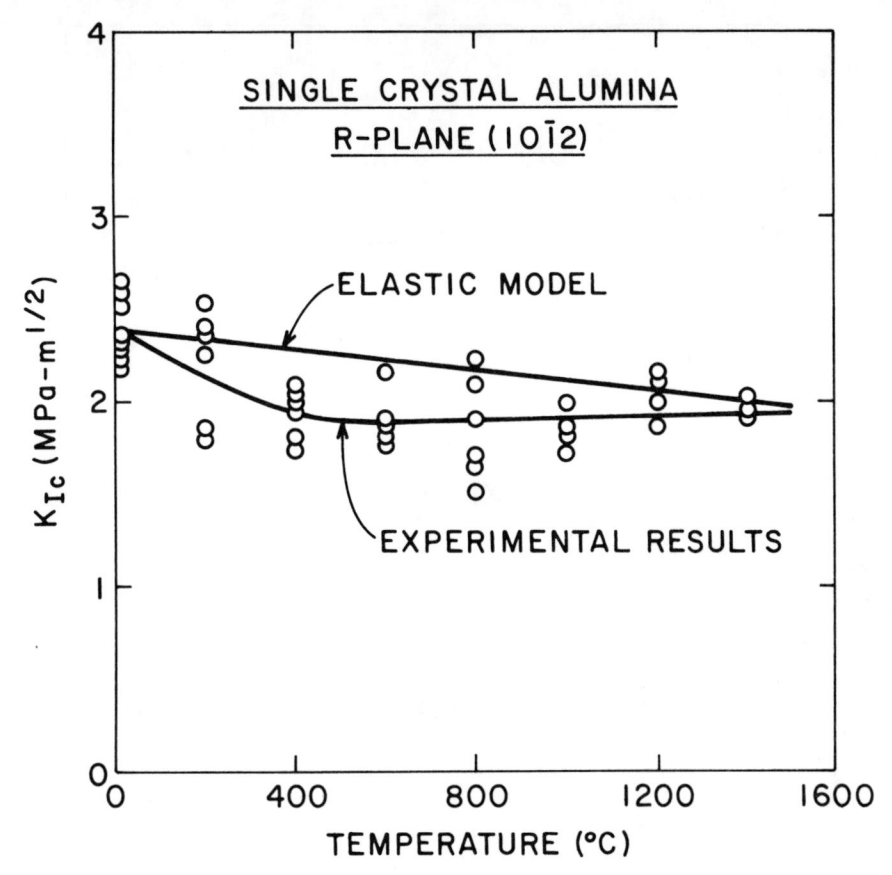

Fig. 3. Experimentally measured fracture toughness variation with temperature on the (10$\bar{1}$2) and the calculated elastic change.

rapidly. Figures 3 and 4 illustrate the K_{Ic} trends with increasing temperature for the (10$\bar{1}$2) and the (0001) planes, while Fig. 5 illustrates the composite temperature trends of all four of the planes that were measured. Above 800°C, the (10$\bar{1}$0) M plane specimens did not fracture from the preinduced microflaws and thus could not be measured. Wiederhorn et al.[18] reported a similar difficulty of high-temperature crack propagation for that plane. Before addressing the implications of Fig. 5, it is appropriate to initially consider the low-temperature fracture toughness decreases.

If the cleavage fracture of single-crystal alumina is considered to be elastic in the low-temperature region, a reasonable assumption since dislocation plastic flow processes should be expected to increase K_{Ic} with increasing temperature, then an estimate of the slope of (dK_{Ic}/dT) can be achieved by differentiating Eq. (4) and rearranging it to yield

$$\left(\frac{dK_{Ic}}{dT}\right) = K_{Ic}\left(\frac{1}{E}\frac{dE}{dT} - 1/2\alpha\right) \qquad (5)$$

by treating $(d\lambda/dT)$ and $(d\nu/dt)$ as negligible, for λ and ν are nearly constant.

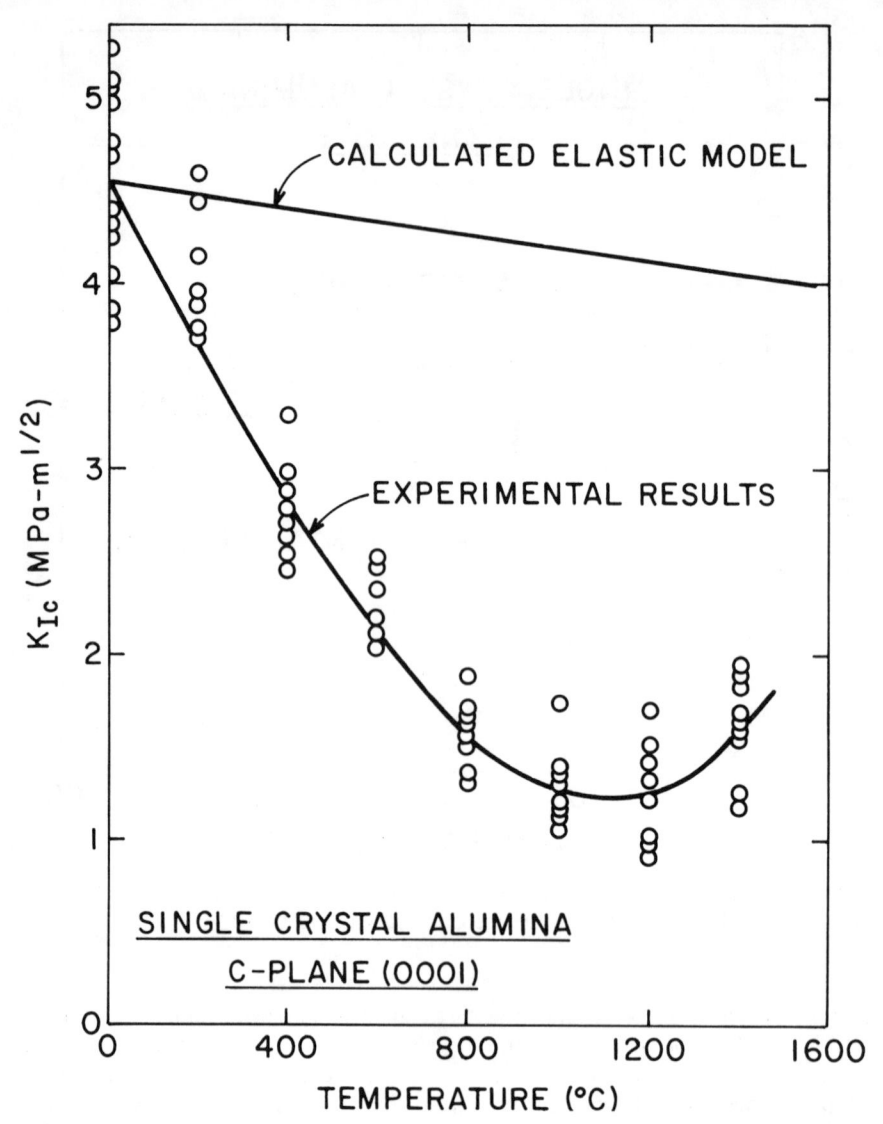

Fig. 4. Fracture toughness on the basal (0001) plane and the calculated elastic change.

Stewart and Bradt[27] and deWith[28] have successfully applied this approach to single-crystal spinel and polycrystalline boron carbide, respectively. Values for (dK_{Ic}/dT) in Eq. (5) can be readily calculated from the temperature dependence of the elastic constants after Tefft[29] and the directional coefficients of thermal expansion after Wachtman et al.[30] The calculated elastic model K_{Ic} vs temperature lines are dashed on Figs. 3 and 4 for the $(10\bar{1}2)$ and (0001) cleavages, respectively. The results of these elastic calculations yield negative (dK_{Ic}/dT) slopes with values of 10^{-3}–10^{-4} (MPa·m$^{1/2}$/°C), in agreement with the measurements of Wiederhorn

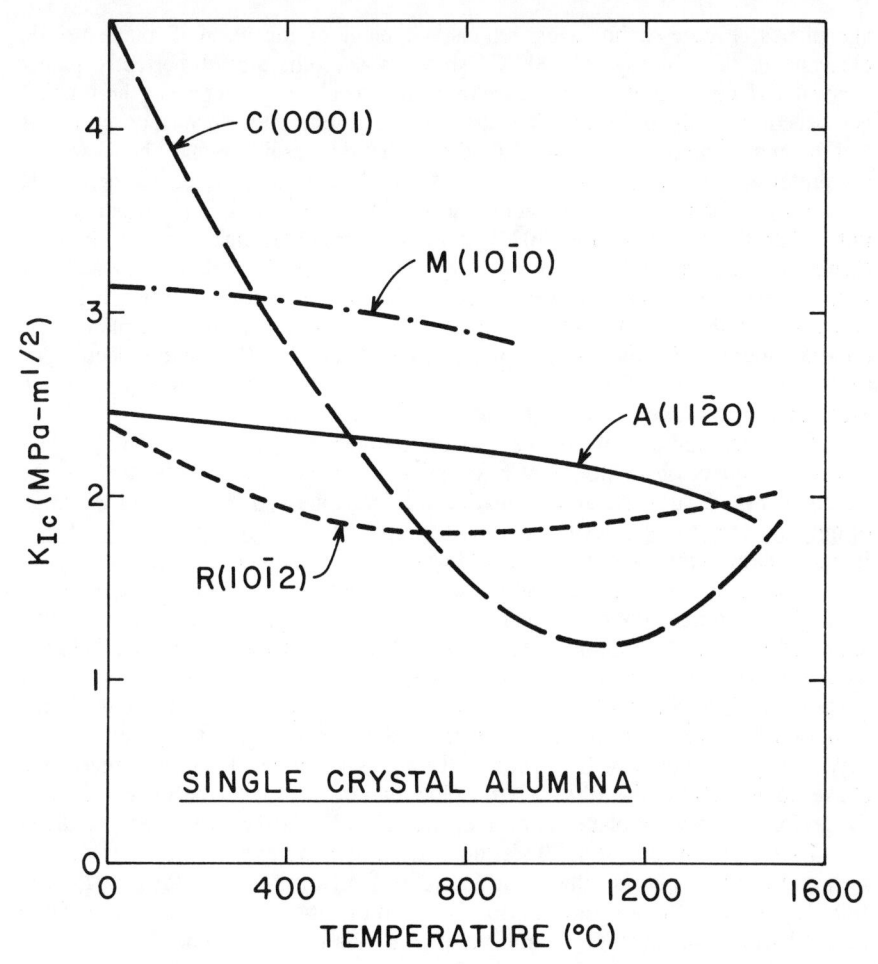

Fig. 5. Composite of the experimental fracture toughnesses of the planes of interest from room temperature through 1400°C.

et al.[7] and Wieland.[8] The R plane measurements in this study are described quite well by the simple elastic model; however, those for the C plane are not. The failure of this simple elastic model for the (0001) basal plane cleavage is surprising, for it has been successfully applied to $MgAl_2O_4$,[27] to B_4C,[28] and also to vitreous fused silica.[31] The reason for the significant difference in this single instance is not understood at this time. However, fractographic evidence and other researchers' results both support the rapid toughness decrease that is observed experimentally for the (0001) as opposed to the more gradually predicted theoretical elastic calculations based on Eq. (5).

Figure 1 illustrated the room-temperature (0001) fracture specimen's surface. After the crack initiated from the original microflaw, it rapidly tended to seek the lower toughness R plane, and the remaining fracture surface areas are just a recurring series of the characteristic stairlike checkerboard pattern of $(10\bar{1}2)$ cleavage planes. Figure 6, however, which is typical of the basal C plane (0001) cleavage in the vicinity of 1000°C, shows a smooth, almost perfectly planar cleaved surface without any evidence of numerous cleavage steps, for the fracture is proceeding on the preferred cleavage plane. These fracture surface observations and the experimental measurements indicate that the basal C plane, the (0001), is the preferred cleavage plane of single-crystal alumina at elevated temperatures. Figure 5 clearly illustrates that above about 700°C the basal C plane, or (0001) rather than the rhombohedral ($10\bar{1}2$), is the preferred cleavage plane of alumina. However, the form of K_{Ic} vs temperature for the (0001) C plane very strongly suggests that, perhaps above about 1700°C, the rhombohedral R plane, ($10\bar{1}2$), or perhaps even the ($11\bar{2}0$) A plane may become the preferred cleavage plane. The rapid K_{Ic} increase for the basal C plane above about 1100°C probably occurs due to dislocation plastic flow processes in the crack tip region, as suggested by Wachtman and Maxwell[4] and by Wiederhorn et al.[18]

This transition to preferred basal plane cleavage explains Anderson's[12] extensive and frequent observations of basal plane cleavage at 1100°C. Extending the results of Fig. 5 suggests an explanation for Firestone and Heuer's[15] observation of only a partial basal cleavage at 1800°C, similar to room-temperature fractures. It is only natural that Anderson[12] should observe very frequent and quite extensive basal cleavage in the vapor arc lamp envelopes fractured at 1100°C, for the (0001) basal plane is indeed the preferred cleavage plane of single-crystal alumina at that temperature. The observation of Firestone and Heuer[15] at 1800°C of a rapid change from a planar basal fracture to cleavage steps of another crystallographic variety is very similar to the fracture surfaces observed by Wachtman and Maxwell[4] and Abdel-Latif et al.[26] for room-temperature fractures. All three of the papers contain very similar micrographs. Extension of the K_{Ic} vs T curves of Fig. 5 suggests that above about 1700°C the (0001) basal plane would no longer be expected to be the preferred cleavage plane, but rather the ($11\bar{2}0$) or ($10\bar{1}2$) would have a lower K_{Ic}. Thus, for fracture at 1800°C, the crack might be expected to reorient and seek one of those planes rather than the (0001), if it initiates on the (0001). The 1800°C fracture pattern should closely resemble that for a room-temperature basal (0001) plane fracture. That type of behavior is exactly what Firestone and Heuer[15] observed.

Temperature Dependence of Strength

Since Jackman and Roberts[32] first reported a minimum for the bend strengths of single-crystal corundum with increasing temperature, there have been numerous studies of the phenomenon including those of Charles and Shaw,[33] Heuer and Roberts,[34] Davies,[35] Bayer and Cooper,[36] and most recently Shahinian.[37] The latter was a tensile strength study that clearly substantiated the strength minimum that had been observed to occur between 400 and 900°C, its exact location depending on the growth process of the crystals, their purity, orientation, surface conditioning, annealing, and the particular test conditions. The cause of this strength minimum with increasing temperature has not been specified with any degree of certainty. It has variously been attributed to different phenomena, including the relief of residual strains developed during crystal growth, alteration of the surface condition or surface structure of the specimens, the role of twinning in crack nucleation, various stress corrosion processes, although Brenner[5] suggests that they are not a factor, and the onset of a limited amount of ductility in the form of dislocation processes in the vicinity of stress concentrations. This latter mechanism has been applied to describe the lower temperature weakening by dislocation

Fig. 6. Fracture surface typical of those observed for basal cleavage in the temperature range of 800–1200°C.

plastic flow nucleation of cracks and also the elevated-temperature strengthening by a crack tip blunting process. Certainly, each of the aforementioned processes can be envisioned as a potential contributor to one aspect or another of the strength minimum. The composite fracture toughness results depicted in Fig. 5 can equally well, perhaps even better, explain the strength minimum with increasing temperature.

The strength of a material is related to the fracture toughness, K_{Ic}, and the flaw size, c, though the Griffith equation:

$$\sigma_f = K_{Ic} Y c^{-1/2} \qquad (6)$$

where σ_f is the strength and Y is a flaw geometry parameter. During stressing, fracture can be expected to occur for the most serious situation, that of the largest flaw and lowest toughness. The combined form of the K_{Ic} values for the $(10\bar{1}2)$ and the (0001) planes in Fig. 5 clearly suggests a strength minimum, the exact temperature of which is determined by the critical flaw size relative to the $(10\bar{1}2)$ or (0001) planes and perhaps the residual stress at the crack tips as well. A minimum can readily occur if the lower temperature strength decrease is dominated by the toughness decrease on the $(10\bar{1}2)$, which eventually transfers control to the (0001) toughness, as it assumes a major role in determining the minimum as well as the subsequent increase at the higher temperatures. Since various flaw configurations relative to the $(10\bar{1}2)$ and (0001) planes are highly probable, as discussed by Becher,[14] the gradual nature of the often observed strength minimum is easily visualized as being controlled by the minimum toughness in the temperature regime of interest.

Summary and Conclusions

Single-crystal alumina specimens were diamond cut from oriented Czochralski boules and precracked by using the Knoop indenter method to yield fractures on the (10$\bar{1}$2), (0001), (10$\bar{1}$0), and (11$\bar{2}$0) planes. The fracture toughnesses were measured in air from room temperature to 1400°C, and the fracture surfaces were examined. At room temperature the cleavage plane is the (10$\bar{1}$2), which has a fracture toughness of only 2.38 ± 0.14 MPa·m$^{1/2}$. The C plane (0001) has a much higher toughness, confirming the strength reports and general observations of other investigators. A room-temperature elastic model to explain the cleavage planes was only moderately successful in ordering their fracture toughnesses.

The toughnesses of all four of the planes decreased with increasing temperature at moderately elevated temperatures, while the (10$\bar{1}$2) and (0001) also increased in toughness at much higher temperatures, indicative of plastic flow processes in the crack tip region. At about 700°C, there exists a cleavage plane transition from the (10$\bar{1}$2) to the (0001). The basal plane remains the preferred cleavage plane through 1400°C; however, the trends of both the (10$\bar{1}$2) and (11$\bar{2}$0) relative to those of the (0001) suggest that above about 1700°C, the basal plane may no longer be the preferred cleavage plane. Other published research is in agreement with the 700°C basal plane cleavage transition and also supports the occurrence of the second cleavage plane transition at higher temperatures.

The combined (10$\bar{1}$2) and (0001) cleavage plane fracture toughness trends also offer an explanation of the elevated-temperature strength minimum that has been reported for alumina single crystals. The overall toughness trend of those two planes clearly predicts a strength minimum without resorting to the somewhat more speculative mechanisms such as those of residual strain relief and stress corrosion processes.

Acknowledgments

The authors are grateful to the Alcoa Foundation for a grant to obtain some of the single crystals measured in this study and to P. Warren and B. J. Sedlak for recommendations on the alumina crystals. The assistance of H. Palmour and S. M. Wiederhorn for supplying reports and discussion relative to unpublished research is also appreciated.

References

[1] M. D. Shappel, "Cleavage of Ionic Minerals," *Am. Mineral.*, **21** [2] 75–102 (1936).
[2] H. Winchell, "Orientation of Synthetic Corundum for Jewel Bearings," *Am. Mineral.*, **29** [11–12] 399–414 (1944).
[3] G. Yamaguchi, Y. Kubo, and H. Ogawa, "Fracture in Thin-Sectioned Corundum Crystals," *Bull. Chem. Soc. Jpn.*, **39** [2] 287–99 (1966).
[4] J. B. Wachtman and L. H. Maxwell, "Plastic Deformation of Ceramic-Oxide Single Crystals," *J. Am. Ceram. Soc.*, **37** [7] 291–5 (1954).
[5] S. S. Brenner, "Mechanical Behavior of Sapphire Whiskers at Elevated Temperatures," *J. Appl. Phys.*, **11** [1] 33–9 (1962).
[6] K. Noda, "Fracture Properties of Sapphire Single Crystals"; M. S. Thesis, Osaka University, Osaka, Japan, 1979.
[7] S. M. Wiederhorn, "Fracture of Sapphire," *J. Am. Ceram. Soc.*, **52** [9] 485–91 (1969).
[8] G. Wieland, "Investigations on the Temperature Dependence of Dislocation Morphology During Fracture of Sapphire"; Diploma Thesis, University of Erlangen-Nurnberg, 1976.
[9] S. S. Smith and B. J. Pletka, "Indentation Fracture of Polycrystalline Alumina"; pp. 189–210 in Fracture Mechanics of Ceramics, Vol. 6. Edited by R. C. Bradt et al. Plenum, New York, 1983.
[10] H. Palmour, III, D. R. Johnson, C. S. Kim, and C. E. Zimmer, "Fractographic and Thermal Analyses of Shocked Alumina"; ONR Tech. Rept. 69-5, Contract No. N00014-68-A-0187, Aug, 1971.
[11] F. P. Mallinder and B. A. Proctor, "Preparation of High-Strength Sapphire Crystals," *Proc. Br. Ceram. Soc.*, **6**, 9–16 (1966).
[12] N. C. Anderson, "Basal Plane Cleavage Cracking of Synthetic Sapphire Arc Lamp Envelopes," *J. Am. Ceram. Soc.*, **62** [1–2] 108–9 (1979).

[13] M. Iwasa, T. Ueno, and R. C. Bradt, "Fracture Toughness of Quartz and Sapphire Crystal at Room Temperature," *J. Jpn. Soc. Mater. Sci.,* **20** [33] 1001–4 (1981).
[14] P. F. Becher, "Fracture Strength Anisotropy of Sapphire," *J. Am. Ceram. Soc.,* **59** [1–2] 59–61 (1976).
[15] R. F. Firestone and A. H. Heuer, "Creep Deformation of 0° Sapphire," *J. Am. Ceram. Soc.,* **59** [1–2] 24–9 (1976).
[16] A. I. A. Abdel-Latif, R. E. Tressler, and R. C. Bradt, "Fracture Mirror Formation in Single Crystal Alumina"; pp. 933–9 in Fracture 1977, Vol. 3. Proceedings of the ICF-4, Waterloo, Canada.
[17] J. Castaing, J. Cadoz, and S. H. Kirby, "Deformation of Al_2O_3 Single Crystals between 25°C and 1800°C, Basal and Prismatic Slip," *J. Phys. (Paris),* **42** [6] C3-43–C3-47 (1981).
[18] S. M. Wiederhorn, B. J. Hockey, and D. E. Roberts, "Effect of Temperature on the Fracture of Sapphire," *Philos. Mag.,* **28**, 783–96 (1973).
[19] J. J. Petrovic and M. G. Mendiratta, "Fracture from Controlled Surface Flaws"; pp. 83–102 in ASTM-STP 678, Fracture Mechanics Applied to Brittle Materials. Edited by S. W. Freiman. ASTM, Philadelphia, PA, 1979.
[20] E. Stofel and H. Conrad, "Fracture and Twinning in Sapphire," *Trans. AIME,* **227** [5] 1053–60 (1963).
[21] H. P. Kirchner and R. M. Gruver, "Fractographic Criteria for Subcritical Crack Growth Boundaries in 95% Al_2O_3," *J. Am. Ceram. Soc.,* **63** [3–4] 169–74 (1980).
[22] R. W. Hertzberg; p. 273 in Deformation and Fracture Mechanics of Engineering Materials. Wiley, New York, 1976.
[23] A. Kelley; p. 3 in Strong Solids. Clarendon Press, London, 1966.
[24] J. B. Wachtman, W. E. Tefft, D. C. Jam, Jr., and R. P. Stenchfield, "Elastic Constants of Synthetic Corundum at Room Temperature," *J. Res. Natl. Bur. Stand.,* **64A** [3] 213–28 (1960).
[25] W. D. Scott and K. K. Orr, "Rhombohedral Twinning in Alumina," *J. Am. Ceram. Soc.,* **66** [1] 27–32 (1983).
[26] R. L. Stewart, M. Iwasa, and R. C. Bradt, "Room Temperature K_{Ic} Values for Single-Crystal and Polycrystalline $MgAl_2O_4$," *J. Am. Ceram. Soc.,* **64** [2] C-22 (1981).
[27] R. L. Stewart and R. C. Bradt, "Fracture of Single Crystal Spinel," *J. Mater. Sci.,* **15** [1] 67–72 (1980).
[28] G. deWith, "High Temperature Fracture of Boron Carbide," unpublished work.
[29] W. E. Tefft, "Elastic Constants of Synthetic Single Crystal Corundum," *J. Res. Natl. Bur. Stand.,* **70A**, 277–80 (1966).
[30] J. B. Wachtman, T. G. Scuderi, and G. W. Cleek, "Linear Thermal Expansion of Aluminum Oxide and Thorium Oxide from 100 to 1100°K," *J. Am. Ceram. Soc.,* **45** [7] 319–23 (1962).
[31] N. Shinkai, R. C. Bradt, and G. E. Rindone, "Fracture Toughness of Fused SiO_2 and Float Glass at Elevated Temperatures," *J. Am. Ceram. Soc.,* **64** [7] 426–30 (1981).
[32] E. A. Jackman and J. P. Roberts, "On the Strength of Polycrystalline and Single Crystal Corundum," *Trans. Br. Ceram. Soc.,* **54**, 389–98 (1955).
[33] R. J. Charles and R. R. Shaw, "Delayed Failure of Polycrystalline and Single Crystal Alumina"; General Electric, Rept. No. 62-RL-308M, 1962.
[34] A. H. Heuer and J. P. Roberts, "The Influence of Annealing on the Strength of Corundum Crystals," *Proc. Br. Ceram. Soc.,* **6**, 17–27 (1966).
[35] L. M. Davies, "The Effect of Heat Treatment on the Tensile Strength of Sapphire," *Proc. Br. Ceram. Soc.,* **6**, 29–35 (1966).
[36] P. B. Bayer and R. E. Cooper, "Tensile Strength of Sapphire Whiskers at Elevated Temperatures," *J. Mater. Sci.,* **4** [1] 15–20 (1969).
[37] P. Shahinian, "High Temperature Strength of Sapphire Filament," *J. Am. Ceram. Soc.,* **54** [1] 67–8 (1971).

---

*Currently with the Government Industrial Research Institute, Osaka, Ikeda-shi, 563, Japan.
†Union Carbide Corp., San Diego, CA.

Effect of MgO Dopant Dispersing Method on Density and Microstructure of Alumina Ceramics

ABRAHAM COHEN, C. P. VAN DER MERWE, AND A. I. KINGON

CSIR
Division of Ceramics, Glass, and Phase Studies
National Institute for Materials Research
Pretoria 0001, South Africa

It is generally accepted that MgO additions to Al_2O_3, below the solid solubility limit at the sintering temperature, promote the densification rate and the rate of normal grain growth. A number of studies, using various techniques for distributing the MgO, reported some differences in the results. Distribution of low concentrations of the MgO second phase is difficult and critical. The effect of MgO dispersing methods on the properties of the Al_2O_3 ceramics is discussed. Eight distributing methods were compared, including mechanical and chemical distribution of various MgO precursors. Commercial alumina, containing insignificant MgO impurities, was used. The dopant level was 0.03 wt% MgO. It has been found that the method used for dispersion has an effect on sintered densities, densification rates, grain sizes, grain growth rates, and porosity values and type. Microstructural differences are also observed clearly.

It has been known for many years that the addition of a small amount ($\leq 0.25\%$) of MgO to alumina acts as a densification aid and allows the preparation of Al_2O_3 ceramic materials with near-theoretical density.[1] The precise mechanisms, whereby the MgO acts,[2-7] remain debatable. However, it is generally accepted that MgO additions to Al_2O_3 below the solid solubility limit at the sintering temperature (e.g., 250 ppm at 1630°C under vacuum[6]) enhance both the densification and the normal grain growth rates.[8] Above the solid solubility limit, when a second phase is present, MgO apparently acts as a grain growth inhibitor and the densification is also influenced negatively.[8] In order to prevent effects resulting from the formation of a second phase, it is essential to limit the amount of MgO additive to its solubility in Al_2O_3 and to disperse it thoroughly.

In various studies on the effect of MgO additions to Al_2O_3, a number of methods have been used for dispersing (distributing) the MgO. Clearly, because of the low additive concentration, the MgO dispersion in alumina is difficult and critical. The different methods used by various researchers for distributing the MgO, therefore, create difficulties when their results are compared.

The present work compares the effects when several common means of doping are used (i.e., for dispersing a small amount, 0.03 wt%, of MgO) on the sintered density and the microstructural evolution of the Al_2O_3 ceramics.

Experimental Procedure

The present work was carried out on commercial alumina.* It is a high-purity material (>99.5% Al_2O_3) with a median particle size of ~3.5 μm. Impurity specifications supplied by the manufacturer include the following (wt%): Na$_2$O, 0.08; SiO$_2$, 0.03; Fe$_2$O$_3$, 0.015; CaO, 0.01; MgO, <0.01. The MgO content was measured by atomic absorption spectrophotometry and found to be 36 ± 14 ppm.

Previous sintering tests showed that small amounts of absorbed moisture could significantly affect the sintering process. The starting alumina powder was therefore dried at 250°C for 16 h and stored under vacuum over P$_2$O$_5$.

All samples were wet ball-milled in polyethylene containers for 54 h, using ZrO$_2$ balls in a mass ratio of 1:5 (powder:balls) and absolute methanol. The median particle size after milling was ~2.2 μm, measured by the sedimentation method. The particle size distribution of this alumina, as-received and after 54 h of wet ball-milling, is shown in Fig. 1.

The alumina samples were doped with 0.03 wt% MgO or one of the following precursors: magnesium stearate, magnesium acetate, or magnesium nitrate. All were analytical grade reagents. Dispersing methods and conditions were the following: (a) MgO dopant was mixed into alumina by wet ball-milling for 15 min, 1.5 h, 9 h, or 54 h, using absolute methanol as the liquid medium. MgO was therefore added either 15 min, 1.5 h, 9 h, or 54 h before the completion of the 54-h milling period. (b) Magnesium stearate (Mg(C$_{18}$H$_{35}$O$_2$)$_2$) was mixed into alumina by wet ball-milling for 9 h, using absolute methanol as the liquid medium. The molecular weights of MgO and magnesium stearate are 40.31 and 591.27 g/mol, respectively; both are insoluble in methanol. (c) A methanolic solution of Mg(NO$_3$)$_2$ was mixed into a methanolic slurry of alumina, using a commercial mixer[†] for 10 min. (d) An aqueous solution of Mg(NO$_3$)$_2$ was mixed into an aqueous slurry of alumina, using a commercial mixer[†] for 10 min. (e) A methanolic solution of magnesium acetate (Mg(CH$_3$CO$_2$)$_2$) was mixed into a methanolic slurry of alumina, using a commercial mixer[†] for 10 min.

When dispersing methods (a) and (b) were applied, the dopant was added and mixed in the course of the ball-milling operation. For dispersion by methods (c), (d), and (e), a commercial mixer[†] was used for blending the aqueous/methanolic suspension of dopant precursor in alumina for 10 min. This mixing method did not cause significant particle size reduction, and it followed 54 h of wet ball-

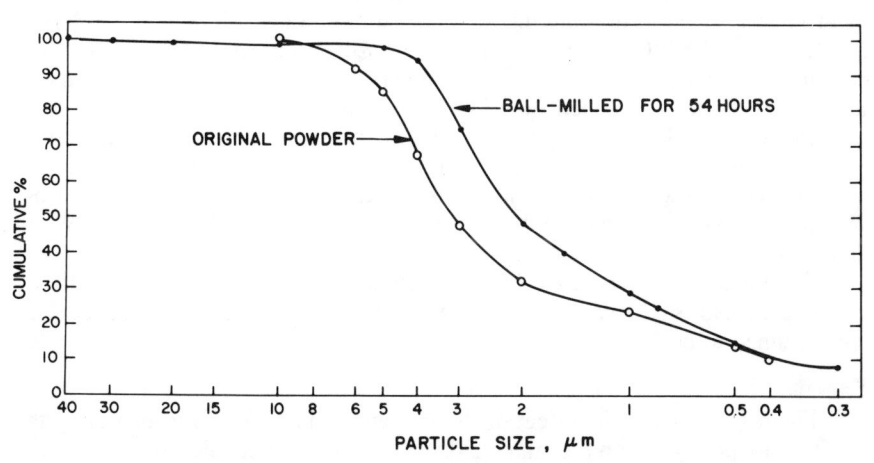

Fig. 1. Particle size distribution for commercial aluminas.*

Table I. Green Density (g/cm³) of Alumina,* Wet Ball-Milled for 54 h and Doped with 0.03 wt% MgO by Using Different Dispersing Methods

Dispersing method	Average of 8 pellets
Wet ball-milling of:	
MgO powder for 15 min	2.546 ± 0.003
MgO powder for 1.5 h	2.540
MgO powder for 9 h	2.550
MgO powder for 54 h	2.583
Magnesium stearate powder for 9 h	2.600
Using a mixer[†] for 10 min to disperse:	
Methanolic soln of magnesium acetate	2.547
Methanolic soln of Mg(NO$_3$)$_2$	2.599
Aqueous soln of Mg(NO$_3$)$_2$	2.575

*Alcoa A-17, Aluminum Co. of America, Pittsburgh, PA.
[†]Silverson.

milling. All samples were dried by evaporation at 80°C and ground by hand in a porcelain mortar.

Samples obtained by using mixing methods (b), (c), (d), and (e) underwent calcination at 900°C for 2 h to decompose the stearate, acetate, and nitrate to oxide. These samples were ground by hand in a porcelain mortar and sieved through a 106-μm sieve to improve the dispersion of MgO.

Polyethylene glycol (4 wt%), diluted in absolute methanol, was added to all samples as a binder and mixed by hand.

The samples, after being dried and ground, were formed in a metal die by using a uniaxial pressure of 92 MPa (13 400 psi). The green pellets had a diameter of 17.0 mm and a thickness of ~5.7 mm. The binder was burnt out at 400°C for 2 h. Green densities (Table I) were in the range 63.9–65.2% of theoretical (assumed to be 3.986 g·cm^{-3} after Peelen[8]) density.

Sintering was carried out in a commercial resistance rapid-heated furnace[‡] at 1500, 1600, and 1650°C for 15, 30, 60, and 120 min. The firings were carried out in an air atmosphere. The sintering program selected was the following: heating at 100°C/min to 1000°C, 50°C/min to 1200°C, 20°C/min to 1400°C, and 10°C/min up to the sintering temperature; cooling at 50°C/min from sintering temperature to 1200°C and then quenching in air to room temperature.

Density was determined to a precision of ±0.15% (Table II) by measuring mass and dimensions. Grain size measurements were made on scanning electron photomicrographs of polished and thermally etched (at 1400°C for 1 h) cross sections of sintered pellets, by the linear-intercept technique described by Wurst and Nelson.[9]

Open porosity was measured with a mercury porosimeter[§] up to a pressure of 60 000 psi (4.13 × 10^8 Pa). The contact angle between mercury and the surface was assumed to be 140°.

Results

Figures 2–4 show the effect of the various methods used for dispersing the MgO, or its precursor, on the sintered density of commercial alumina.* Sintering temperatures were 1500, 1600, and 1650°C, respectively.

Table II. Density (g/cm^3) of Alumina,* Wet Ball-Milled for 54 h and Doped with 0.03 wt% MgO by Using Different Dispersing Methods, Fired at 1600°C for Different Times

Dispersing method	\multicolumn{2}{c}{Firing time (min)}						
	15		30		60		120	
Wet ball-milling of:								
MgO powder for 15 min	3.172	3.176	3.335	3.332	3.433	3.432	3.590	3.591
MgO powder for 1.5 h	3.177	3.181	3.342	3.345	3.452	3.453	3.612	3.616
MgO powder for 9 h	3.195	3.196	3.350	3.353	3.486	3.484	3.627	3.624
MgO powder for 54 h	3.195	3.196	3.351	3.350	3.484	3.487	3.624	3.628
Magnesium stearate powder for 9 h	3.303	3.308	3.473	3.474	3.601	3.602	3.727	3.731
Using a mixer† for 10 min to disperse:								
Methanolic soln of Mg(CH$_3$CO$_2$)$_2$	3.344	3.349	3.494	3.501	3.640	3.641	3.786	3.785
Methanolic soln of Mg(NO$_3$)$_2$	3.359	3.357	3.509	3.505	3.652	3.650	3.782	3.780
Aqueous soln of Mg(NO$_3$)$_2$	3.373	3.371	3.511	3.512	3.653	3.652	3.771	3.773

*Alcoa A-17, Aluminum Co. of America, Pittsburgh, PA.
†Silverson.

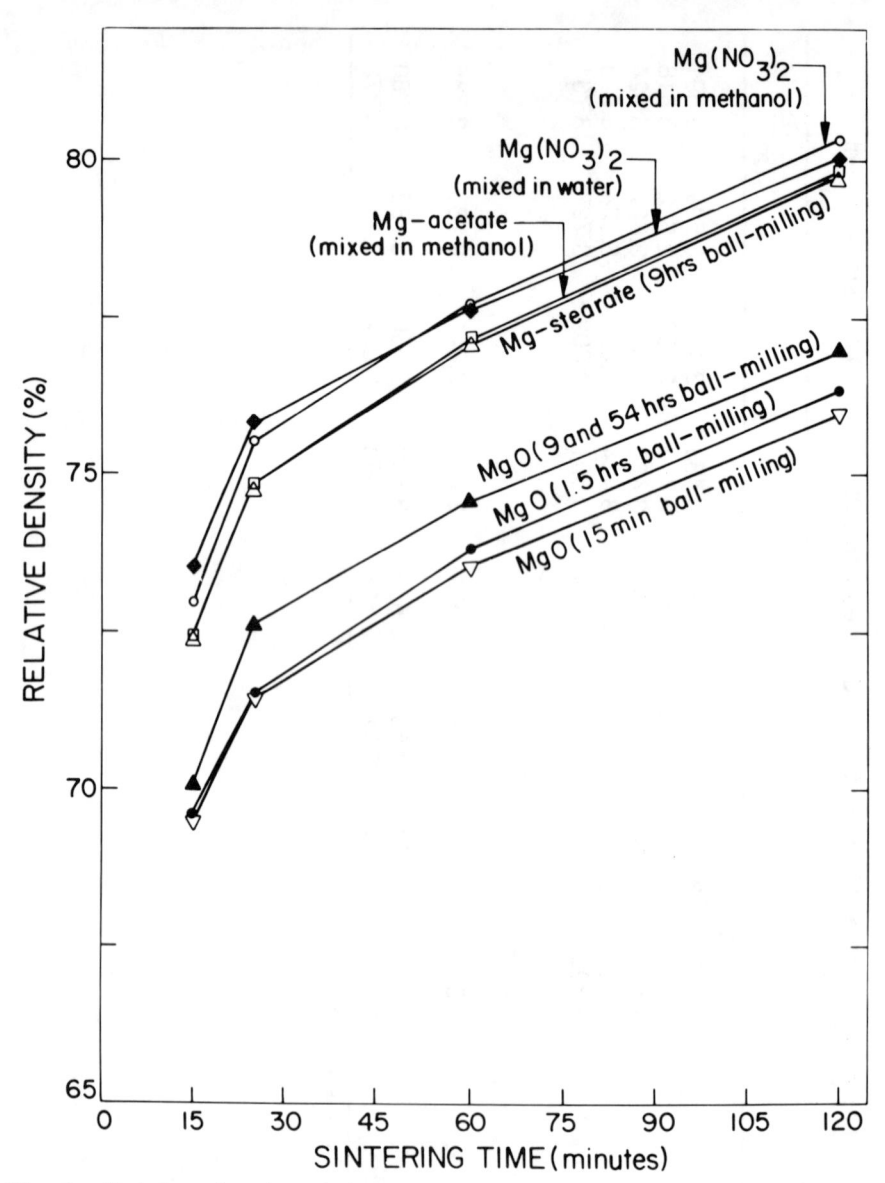

Fig. 2. Relative density against sintering time for alumina* doped with MgO with different dispersing methods, sintered at 1500°C.

It is clear that the dispersing method has a substantial effect on sintered densities. Higher fired densities were measured for aluminas mixed with MgO by wet ball-milling for longer times. The optimal mixing time was 9 h. No significant increase in sintered density was detected after longer times. Mixing by wet ball-milling for 9 h with magnesium stearate as a precursor for MgO showed a significant increase in density above the use of MgO. An additional increase in fired

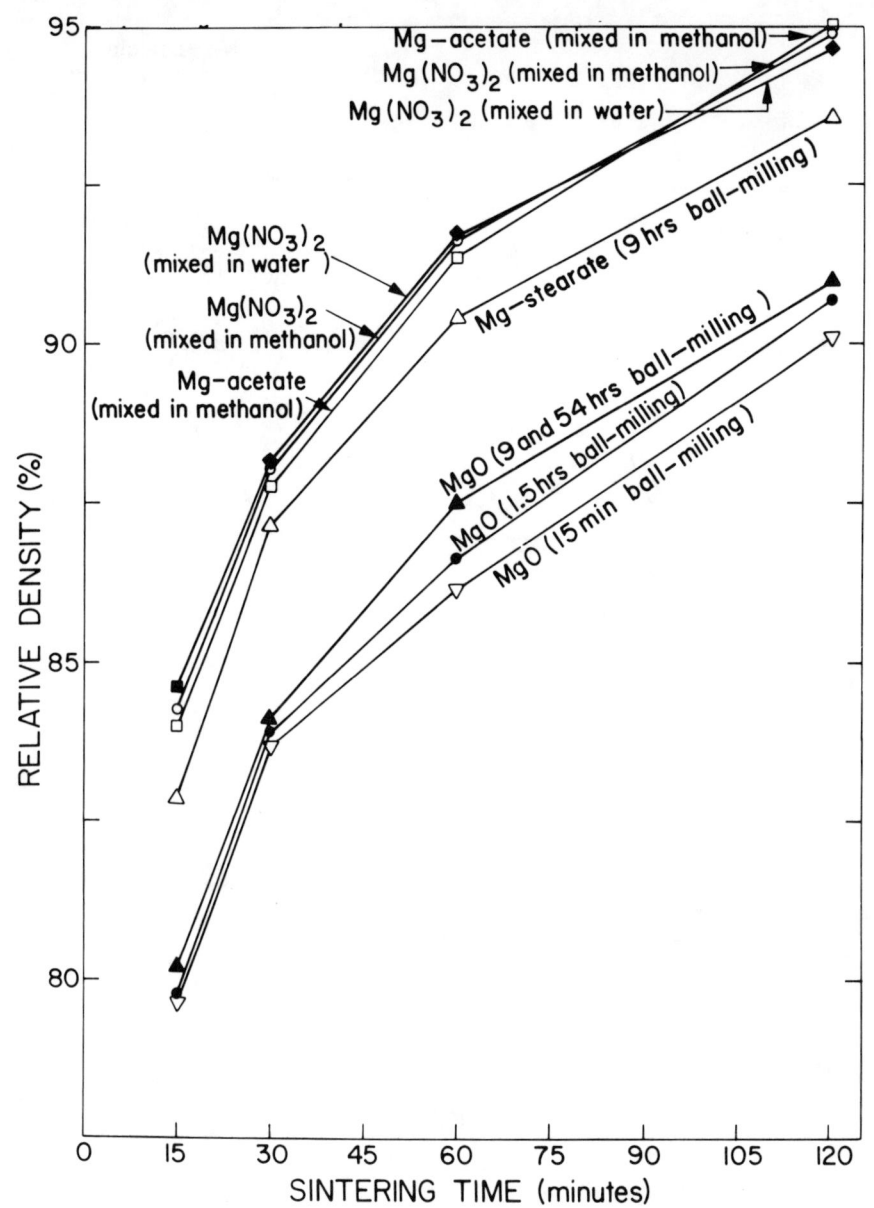

Fig. 3. Relative density against sintering time for alumina* doped with MgO with different dispersing methods, sintered at 1600°C.

density was observed when alumina samples were doped by mixing alcoholic or aqueous solutions of magnesium acetate or Mg(NO$_3$)$_2$ with a commercial mixer[†] for 10 min.

Representative types of sintered alumina samples were chosen for microstructural examination. Figure 5 shows the influence of using different methods for

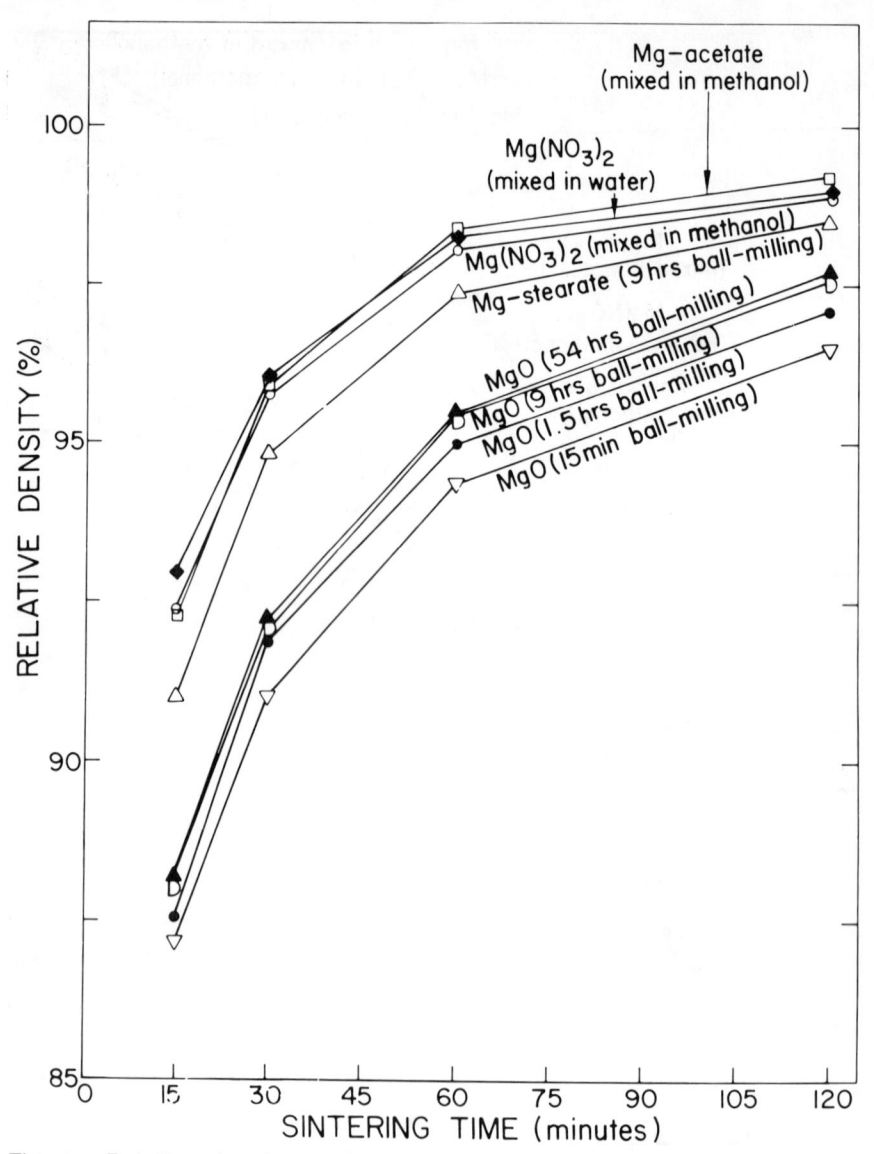

Fig. 4. Relative density against sintering time for alumina* doped with MgO with different dispersing methods, sintered at 1650°C.

distributing MgO dopant (or its precursor) on the microstructural homogeneity (pore number, size, and distribution). It is apparent, when one moves from the least effective dispersing method (a) to the most effective one (d) that there is a decrease in pore number and size together with a significant increase in microstructural homogeneity. Note that samples with poor MgO distribution, especially samples shown in Figs. 5(A, B), contain a large number of pores that are substantially greater than the mean grain size.

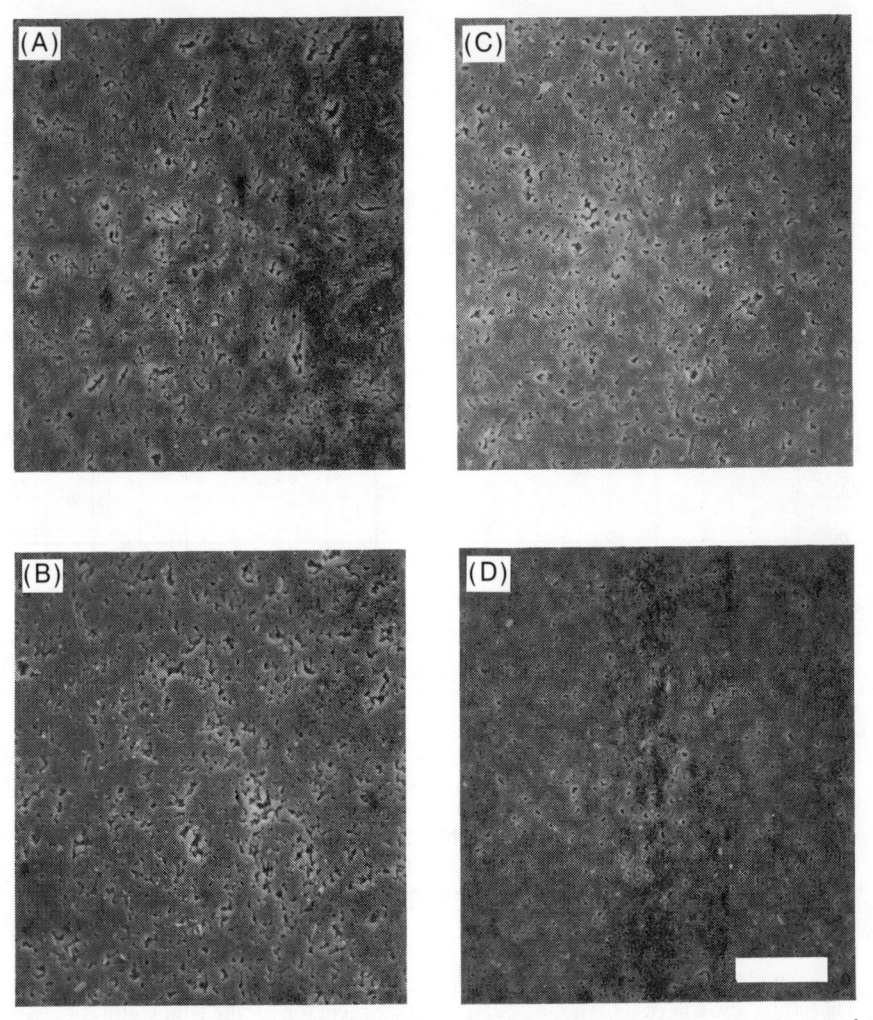

Fig. 5. Scanning electron micrographs of polished cross sections of alumina* doped with 0.03 wt% MgO and sintered at 1650°C for 15 min. The dopant added as (A) MgO by ball-milling for 15 min, (B) MgO by ball-milling for 9 h, (C) magnesium stearate by ball-milling for 9 h, and (D) aqueous solution of Mg(NO$_3$)$_2$ by mixing with a mixer.

Open, closed, and total porosity values are given in Table III for commercial alumina* doped with 0.03 wt% MgO or its precursor, using different dispersing methods, sintered at 1500°C for 1 h. Figure 6 shows the closed porosity as a percentage of the total porosity. As expected, there is a decrease in total porosity values with increasing dispersing method efficiency. In addition, it is evident that there is a decrease in the percentage of the closed porosity (out of the total) with the increase of microstructural homogeneity.

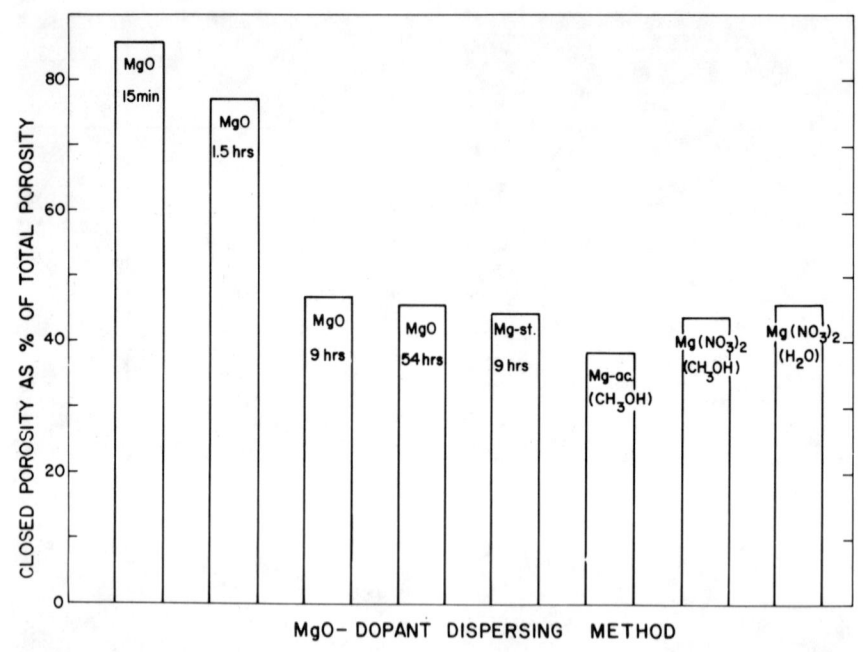

Fig. 6. Closed porosity as a percentage of total porosity, in alumina sintered at 1500°C for 1 h, for different MgO dopant dispersing methods.

Discussion

The present work shows that sintered density varies directly with ball-milling mixing time of MgO dopant (Figs. 2–4). On the other hand, no significant increase in density was measured after 9 h of ball-milling mixing time. Mixing, by using the same method for 9 h, with magnesium stearate as the precursor for MgO, shows a significant increase in density. This result is attributed to the 15 times higher molecular weight of magnesium stearate in comparison with that of MgO. Therefore, the use of a high molecular weight precursor for the MgO allows better dopant distribution, and consequently, higher density is obtained. Note that both MgO and magnesium stearate are insoluble in methanol, the liquid medium used for ball-milling. Therefore, they mix mechanically in the solid state on a particle scale.

An additional increase in sintered density is observed when alumina samples are doped by mixing alcoholic or aqueous solutions of magnesium acetate or $Mg(NO_3)_2$ with alcoholic or aqueous alumina slurries, using a commercial mixer[†] for 10 min. The use of aqueous or alcoholic solutions of the dopant provides intimate mixing of MgO precursor with alumina particles. This mixing technique leads to improved homogeneity and, as a result, to the highest attainable density.

Slight differences in sintered density values are observed between the various solution mixing techniques. It seems that there is a slight preference for using magnesium acetate in methanolic solution to using $Mg(NO_3)_2$, especially for sintering at high temperatures and over longer periods. In addition, no significant difference is detected between the use of aqueous or methanolic solution of $Mg(NO_3)_2$.

Table III. Porosity Values (%) for Alumina,* Wet Ball-Milled for 54 h and Doped with 0.03 wt% MgO by Using Different Dispersing Methods, Sintered at 1500°C for 1 h

Dispersing method	Total porosity[†] (P_t)	Open porosity[‡] (P_o)	Closed porosity[§] (P_c)	100(P_c/P_t)
Wet ball-milling of:				
MgO powder for 15 min	26.5	3.8	22.7	85.7
MgO powder for 1.5 h	26.2	6.0	20.2	77.1
MgO powder for 9 h	25.5	13.5	12.0	47.0
MgO powder for 54 h	25.4	13.8	11.6	45.7
Magnesium stearate powder for 9 h	22.9	12.7	10.2	44.5
Using a mixer[¶] for 10 min to disperse:				
Methanolic soln of magnesium acetate	22.9	14.1	8.8	38.4
Methanolic soln of Mg(NO$_3$)$_2$	22.3	12.5	9.8	44.0
Aqueous soln of Mg(NO$_3$)$_2$	22.4	12.1	10.3	46.0

*Alcoa A-17, Aluminum Co. of America, Pittsburgh, PA.
[†]Calculated from density values.
[‡]Measured by using a Quantachrome mercury porosimeter up to a pressure of 60 000 psi (4.13 × 10^8 Pa).
[§]Calculated by difference.
[¶]Silverson.

The microstructure examination supports the density measurement findings. Figure 5 shows that the more uniform the dispersion of MgO, the more uniform and less porous the microstructure and, therefore, the denser the resulting sintered body. It must be emphasized that the trend in sintered densities is an indication of a more serious microstructural problem that has already been pointed out in the results (with reference to Fig. 5). Poor MgO distribution results in inhomogeneous densification, with a consequent increase in porosity and pore size distribution. The largest pores must be considered as flaws. These are thermodynamically stable and will not be removed, whatever the sintering schedule. Moreover, they act as high-stress regions during brittle fracture and thus affect the strength of the ceramics. A dependence of strength upon the MgO dispersing method is thus to be expected.

Open porosity values, measured for sintered alumina samples doped with MgO or its precursor by using different dispersing methods, form an additional tool for homogeneity evaluation. For ideal alumina compacts, including one with perfect MgO dopant dispersion, the porosity is open until the density reaches ~90% of its theoretical value. In this case, all porosity is expected to move from an open state to a closed state while the density increases from ~93 to ~96% of its theoretical value during sintering. The extent of deviation from this behavior may be taken as a measure of the deviation from the ideal state. In nonideal, inhomogeneous compacts, local centers of high MgO concentration are formed together with others lacking MgO. As a result, the densification kinetics differ significantly in different regions of the compact. Accordingly, some of these regions reach high densities, and their porosity is consequently closed, while the bulk density is still low. Therefore, for sintered alumina samples with densities well below 90% of the theoretical value, a decrease in the percentage of the closed porosity (out of the total) may be interpreted as an increase in the microstructural homogeneity.

The closed porosity as a percentage of the total porosity in alumina samples doped with MgO or its precursor, using different dispersing methods and sintered at 1500°C for 1 h, shows (Fig. 6) a clear trend of decreasing with increasing dispersing method efficiency, and as a result, improving the microstructural homogeneity. Using this measurement as a criterion for homogeneity leads to the conclusion that doping with magnesium acetate in methanolic solution is the best dispersing method used. In addition, the use of $Mg(NO_3)_2$ in methanolic solution is slightly preferable to the use of $Mg(NO_3)_2$ in aqueous solution.

Conclusions

Final density and microstructure are significantly influenced by the MgO dopant dispersing method. The use of a mechanical mixing technique, such as wet ball-milling, to dope alumina powder to a low level (0.03 wt%) of MgO additive leads to inhomogeneities in composition and, as a result, to lower final density and less uniform microstructure. Longer mixing times of MgO dopant in alumina by wet ball-milling improve the homogeneity of the compact up to an optimal time of 9 h. No significant improvement in homogeneity is detected after much longer mixing times. The use of a high molecular weight precursor for the MgO dopant allows better dopant dispersion, when wet ball-milling is used as the dispersing method. In order to achieve improved homogeneity, aqueous or alcoholic solution techniques should be used for dispersing the additive, thus forming a thin layer of MgO precursor on alumina particles. There is a slight preference for using magnesium acetate ($Mg(CH_3CO_2)_2$) in methanolic solution rather than using magnesium nitrate ($Mg(NO_3)_2$) as a precursor for MgO dopant. Closed porosity values, as a percentage of the total porosity, can be used as a criterion for homogeneity evaluation.

References

[1] R. L. Coble, *J. Appl. Phys.*, **32**, 793 (1961).
[2] I. B. Culter, *Am. Ceram. Soc. Bull.*, **57**, 316 (1978).
[3] W. C. Johnson and R. L. Coble, *J. Am. Ceram. Soc.*, **61**, 110 (1978).
[4] W. D. Kingery and B. Francois; pp. 471–98 in Sintering and Related Phenomena. Edited by G. C. Kuczynski, N. A. Hooten, and C. F. Gibbon. Gordon and Breach, New York, 1967.
[5] D. L. Johnson; p. 137 in Materials Science Research, Vol. 11. Edited by H. Palmour III, R. F. Davis, and T. M. Hare. Plenum, New York, 1978.
[6] S. K. Roy and R. L. Coble, *J. Am. Ceram. Soc.*, **51**, 1 (1968).
[7] A. H. Heuer, *J. Am. Ceram. Soc.*, **62**, 317 (1979).
[8] J. G. J. Peelen; p. 443 in Materials Science Research, Vol. 10. Edited by G. C. Kuczynski. Plenum, New York, 1975.
[9] J. C. Wurst and J. A. Nelson, *J. Am. Ceram. Soc.*, **55**, 109 (1972).

---

*Alcoa A-17, Aluminum Company of America, Pittsburgh, PA.
†Silverson L2R.
‡Kanthal Corp., Bethel, CT.
§Quantachrome Corp., Greenvale, NJ.

Overview of Current Understanding of MgO and Al_2O_3

Defects in MgO and Al$_2$O$_3$

J. H. CRAWFORD, JR.

University of North Carolina
Department of Physics and Astronomy
Chapel Hill, NC 27514

This paper focuses on the basic experimental probes available to us for detecting and identifying intrinsic defects in oxides produced thermally or by irradiation. Equally important for understanding the role that these defects play in controlling the electronic and chemical behavior of their host is their charge state and the manner in which this is altered by ambient conditions. Therefore, to be effective, the probes must be able to detect and identify not only the physical structure of the defects but their charge states as well. We who work with insulators such as refractory oxides have a great advantage over those who work with metals because we can use a variety of probes which are denied them by the free electrons; e.g., metals are opaque to photons and magnetic fields. Therefore, the very convenient nondestructive probes of optical absorption (or emission) and electron paramagnetic resonance are excluded. Furthermore, because of the high, temperature-independent, electron concentrations, changes in electronic behavior of metals can be detected only through modification of electronic scattering mechanisms which are relatively insensitive compared to photoconductivity, thermally stimulated conductivity, and polarization processes in insulators. Therefore, one might conclude that defect studies in insulators should present few if any problems to the experimenter. Clearly, this is not the case. Although we may be relatively better off with regard to defect probes than those studying metals, the processes of interest to us involving defects are more complex in binary and ternary oxides and, indeed, this complexity arises directly from the basic character of insulators which also gives us the convenient probes for defects, namely, the fact that electrons are localized to form ions and, where ions are missing, electrons and holes can be trapped. However, we have a more serious problem than complexity: There is a great disparity in the effectiveness of probes for different types of defects. Although we do quite well for lattice vacancies,[1-4] performance for interstitials is comparatively poor. In fact, clearly identified optical absorption or emission bands associated with isolated interstitial cations and anions, even in their simplest complexes, have yet to be confirmed. Evans and Stapelbroek[5] speculated on rather indirect evidence that the absorption band at 4.1 eV with a zero phonon band at 3.957 eV in neutron-bombarded α-Al$_2$O$_3$ is possibly associated with a complex involving an Al interstitial. Electron paramagnetic resonance (EPR) signals associated with interstitials combined with lattice atoms in special environments have been reported: Halliburton and Kappers[6] detected EPR spectra which they attribute to oxygen molecular ions in fast-neutron-damaged MgO, which appear to be caused by oxygen interstitials combined with lattice ions near cation vacancies, and Cox and Herve[7] reported at least two types of Al pairs in fast-neutron-bombarded α-Al$_2$O$_3$ crystals, which originate from Al interstitials.

One possible reason for the paucity of experimental evidence for interstitial cations or anions in irradiated oxide crystals is that as such they may be unstable.

In other words, their migration rates at all irradiation temperatures thus far investigated may be so large that either they are trapped in some stable form or they recombine. Therefore, the difference in vacancy concentrations for the two constituent species, cations and anions, results from the chemical behavior of the respective interstitials, which of course conditions their storage. For example, formation of molecular ions by interacting O^- interstitials (O^{2-} interstitials are expected to be unstable, see below) provides a storage mechanism, allowing a buildup of anion vacancies which manifest themselves as F^+ centers (one trapped electron) and F centers (two trapped electrons) through their optical absorption and emission bands. The final yield of these anion vacancy centers then depends on competing reaction paths for the migrating oxygen interstitials.

Because of intense Coulomb repulsion, interaction between Al^{3+} interstitials is not expected. Hence, any storage of this defect species apparently involves other structural irregularities, which could account for the very small enhancement of aluminum vacancy centers (V-type centers in which one or more holes are trapped by adjacent O^{2-} ions) at low damage levels in fast-neutron or electron-bombarded specimens.[8] In fact, the interstitial Al complexes whose EPR signals Cox observed in fast-neutron-bombarded Al_2O_3 saturate at rather low concentrations, suggesting that impurity ions may be necessary to facilitate the formation of a stable complex involving the interstitial Al^{3+}.

It should be pointed out, however, that, depending on circumstances (temperature, ionization intensity, and radiation dose rate), interstitial Al^{3+} does accumulate in appreciable quantities. Pells and coworkers[9-12] used electrons in a high-voltage electron microscope (HVEM) to damage Al_2O_3. Using the growth of the V-band absorption (holes trapped at cation vacancies), they showed that the displacement energy for cations was much less (17 eV) than for anions (75 eV) where the latter were detected via the F absorption band. On irradiation above 900 K to several dpa (displacements per atom) at 1 MeV incident electron energy, an energy capable of displacing both cations and anions (though the cation displacement rate is ≈ 14 times as great), transmission electron microscopy (TEM) exhibited, in addition to dislocation loops, aluminum metal precipitates as well as irregular voids. Evidently, under these experimental conditions, Al^{3+} interstitials aggregate to form metal precipitates and vacancies of both species combine to form voids as well as the interstitials from both sublattices aggregating to form interstitial dislocation loops. If irradiations are carried out below 800 K, no resolvable features are visible other than a general darkening of the TEM image. If irradiations were carried out with incident electron energies sufficient to displace Al^{3+} ions but not O^{2-}, i.e., 300 keV electrons above 800 K, stoichiometric interstitial dislocation loops are formed. Pells[12] interpreted this surprising result as indicating that interstitial Al^{3+} ions can extract O^{2-} ions from adjacent planes, thus forming loops; in the process O^{2-} vacancies are injected into the crystal. Hence, there is ample evidence for the formation of cation Frenkel defects by knock-on type collisions. However, it appears that high damage rates coupled with intense background ionization at an elevated temperature are required for them to be readily manifested. Prolonged fast-neutron bombardment (doses to several dpa) also indicates that cation interstitials play an important role in the structure of the residual damage through the formation of stoichiometric dislocation loops.[13] Here, however, it is not unreasonable to conclude that the lattice damage in the form of displacement cascades can provide trapping sites for Al^{3+} interstitials, especially at high fluences when these begin to overlap.

To recapitulate, experimental probes available to us for identifying and detecting anion vacancies in their neutral (two trapped electrons, i.e., the F center) and single positive (one trapped electron, the F^+ center) charge states are quite adequate. Both absorption and emission spectra are able to distinguish these centers in Al_2O_3 (Table I) — F^+ luminescence is very sensitive; a few 10^{13} cm^{-3} F^+ centers can be detected in Al_2O_3 via the 3.8 eV photon. The 3.0 eV F-center emission is also quite sensitive in thermochemically colored Al_2O_3 crystals, although the F-center luminescence is quite weak in neutron-bombarded crystals because of concentration quenching in displacement cascades.[14] In MgO (Table I) both F^+ and F centers absorb at nearly the same wavelength, which requires EPR signal amplitude to distinguish the concentrations of the two centers. Emission spectra are definite signatures of these centers also. By contrast there is no optical absorption, optical emission, or EPR spectrum which has been identified as arising from free interstitials of either type in MgO or Al_2O_3, although EPR resonances due to (1) simple complexes of oxygen interstitials have been reported for neutron-bombarded MgO[6] and those due to (2) Al interstitials paired with lattice cations have been observed in neutron-bombarded Al_2O_3.[7]

We now come to our main concern: How can the insights that we have gained about the behavior of defects in magnesia, alumina, and spinel be applied fruitfully to other ionic oxides and related materials? First, it is worth noting that identification of anion vacancy defects (the F^+ and F centers) was made easy by the groundwork laid in color-center research in alkali halides. It is interesting and significant that these basic defect structures span uni–univalent, di–divalent, and tri–divalent to ternary compounds. This certainly gives us hope that a similar

Table I. Anion Vacancy Centers in Oxides (Photon Energies in eV)

Center	Absorption	Emission	EPR
	MgO		
F^+	4.9	3.1	yes
F	5.0	2.3	?
F_2	2.1 (Vibronic)	?	yes
	CaO		
F^+	3.7 ←(3.50 ZPL)*→	3.3	yes
F	3.1	2.1 (2.16 ZPL)	yes (ODMR)
F_2	2.35 ZPL		yes (ODMR)
	α-Al_2O_3		
F^+	6.3 (?)	3.8	yes
	5.4		
	4.8		
	4.1 (?)		
	$MgAl_2O_4$		
F	6.0	3.0	
F_2	3.45 ←(3.36 ZPL)*→	3.30	
F^+	4.8	none	
F	5.3	none	

*Zero phonon line.

projection will hold for other oxides, sulfides, etc., with less well-behaved cations and with ternary and higher order compounds, provided of course that ionic character dominates.

In this regard it should be mentioned that F^+ centers have been detected in neutron-irradiated BeO and electron-bombarded ZnO, both of which crystallize in the hexagonal wurtzite structure. DuVarney and coworkers[15] used both EPR and ENDOR measurements to identify F^+ in BeO. Smith and Vehse[16] identified the EPR spectrum of the F^+ center in ZnO. There was no well-defined optical absorption band associated with the F^+ center in ZnO; only a shoulder on the absorption edge which grows with electron bombardment. Using this shoulder as an index of the introduction of lattice damage, Vehse and coworkers[17] determined the threshold energy for damage to be 0.55 MeV for incident electrons. It is not clear whether F centers are also introduced along with the F^+ or whether the threshold measurements apply to displacement of cations or anions. F^+ and F centers have also been identified in $MgAl_2O_4$, a typical ternary ionic oxide (see Table I).

Little work involving color centers in ZnO_2, CeO_2, and similar oxides has been reported. One problem in these materials is the inavailability of single crystals which are helpful in interpreting any optical or EPR spectra. These dioxide systems, along with certain of the sesquioxides like Y_2O_3, La_2O_3, etc., are of considerable interest because of their refractory character. They should be susceptible to study by means of optical transitions of native point defects. Optical and EPR absorption would also be of interest in sulfides and selenides as well as in ternary oxides of other structures and should parallel experience in MgO and $\alpha\text{-}Al_2O_3$ provided these materials are sufficiently ionic and the cations are of fixed valence.

What can we expect to learn about F-type centers and V-type centers in these more complex oxides, sulfides, etc.? In every case, provided ionic character is the controlling element, we will still be dealing with oxygen: its vacancy and the consequences of removing the oxygen atom or ion on the electronic states associated with the vacancy on the one hand, and the cation vacancy and the accommodation of the excess negative charge associated with it by the surrounding shell of O^{2-} ions on the other. It seems safe to predict that, as has been pointed out,[18,19] since removing the oxygen atom or O^- ion from an oxide ion site leaves the first electron more tightly bound than when the site was occupied, nude or empty anion vacancies will not exist. This situation also applies to sulfides and selenides, provided the latter are sufficiently ionic. In this connection, it should be pointed out that there are only two stable charge states possible for an oxide vacancy: neutral (the F center) and single positive (the F^+ center). The nude oxide vacancy corresponding to a doubly positive charge state would remove an electron from an O^{2-} in a neighboring shell, thus releasing a hole into the lattice. Therefore, one would expect the intrinsic thermal defect production to be strongly affected. In MgO, for example, the Schottky pair would not be V''_{Mg} and $V^{..}_O$, but rather V'_{Mg} and $V^{.}_O$ (the F^+ center). Similarly, the Frenkel pair on the anion sublattice would be $V^{.}_O$ and O'_i. In either case, the energy gained by putting an electron on the $V^{.}_O$ and the resultant hole on the V'''_{Mg} would lower the energy of formation of the Schottky by a substantial amount with a similar consequence for the $V^{.}_O$, O'_i pair. An interesting situation arises in $\alpha\text{-}Al_2O_3$: Normal creation of Schottky defects requires that the defect source (a jog on a dislocation) provide two V'''_{Al} for every three $V^{..}_O$. With the Choi postulate taken into account, charge neutrality would require one V''_{Al} for every two $V^{.}_O$. Hence, there would soon develop a bottleneck at the jog source, inhibiting further Schottky defect formation. This would tend to favor $V_O^{\cdot}$, O_I' pair formation. It is interesting to note that electron microscopy studies of dislocation loop formation and dissolution at high temperatures suggests that thermally created oxygen interstitials are involved.[21] Finally, the lower charge on the intrinsic defects would also result in a decrease in the motion energy of the defect below the presently available calculated values, which may help to explain why the experimental diffusion activation energies are consistently ≈ 2 eV less than the calculated values.[22]

One would expect V-type centers to be somewhat the same in the unexplored systems as in the MgO and Al_2O_3 structures. The reason, of course, is that one is observing the transitions of an O^- ion perturbed by a large electric field gradient due to the adjacent cation vacancy. In MgO, even partial compensation by other positive charges does not shift the transition appreciably. Therefore, V-type centers in all ionic oxides should be pretty much the same with the absorption band very broad because of its strong coupling to the lattice and its position determined by a small dependence upon lattice parameter and charge on the cation. The binding energy of the hole to an otherwise uncompensated cation vacancy, as indicated above, can be substantial. Tench and Duck[23] report a value of 1.6 eV for MgO. Studies using thermoluminescence of V-type centers in irradiated α-Al_2O_3 indicate that the single trapped hole center (V^{2-} or V_{Al}''') is stable to 500 K, which indicates a trap depth of ~ 1.5 eV, whereas the V^- (V_{Al}') center (two trapped holes) is stable to 390 K (≈ 1.2 eV trap depth). The V absorption band maxima are at 2.3 eV photon energy in MgO and 3.0 eV in Al_2O_3.

In summary, our probes for intrinsic defects vary widely in their effectiveness. Optical and, for the most part, paramagnetic resonance probes are quite good for anion vacancy defects in MgO, Al_2O_3, and spinel, and the V-type absorptions are well established in MgO and Al_2O_3. However, interstitials, except when they are bound in some complex, are virtually invisible, and even the bound interstitials are detected only through EPR. One expects similar vacancy probes to be effective in studying more complex ionic oxides, since one is dealing with the levels in an O^{2-} vacancy on the one hand and transitions of an O^- ion adjacent to a cation vacancy on the other. The F^+ center model of Choi raises the interesting question of the identity of Schottky defects in ionic oxides. Since the nude O^{2-} vacancy does not appear to exist and the O^{2-} vacancy containing a single electron (F^+ center) is the most stable species, Schottky defects would seem to be F^+ centers and V^- centers. The consequences of this possibility for calculations of energies of formation and migration should be considerable.

The author wishes to acknowledge support from the U.S. Department of Energy under Contract DE-A505-78ER05866.

References

[1] For a review of anion vacancy centers in MgO, see: B. Henderson, *CRC Crit. Rev. Solid State Mater. Sci.*, **9**, 1 (1980).

[2] For a review of V centers in MgO, see: Y. Chen and M. M. Abraham, *New Phys. (Seoul)*, **15**, 47 (1975).

[3] For a review of radiation defects in Al_2O_3, see: J. H. Crawford, Jr., *Semicond. Insul.*, **5**, 599 (1983).

[4] For a discussion of defects in $MgAl_2O_4$, see: G. P. Summers, G. S. White, K. H. Lee, and J. H. Crawford, Jr., *Phys. Rev. B*, **21**, 2578 (1980).

[5] B. D. Evans and M. Stapelbroek, *Phys. Rev. B*, **18**, 7089 (1978).

[6] L. E. Halliburton and L. A. Kappers, *Solid State Commun.*, **26**, 111 (1978).

[7] R. T. Cox and A. Herve, *C. R. Hebd. Seances Acad. Sci.,* **261**, 5080 (1965); R. T. Cox, *Phys. Lett.,* **21**, 503 (1966).
[8] J. H. Crawford, Jr., *J. Nucl. Mater.,* **108/109**, 644 (1982).
[9] G. P. Pells and D. C. Phillips, *J. Nucl. Mater.,* **80**, 207 (1979).
[10] G. P. Pells and D. C. Phillips, *J. Nucl. Mater.,* **80**, 215 (1979).
[11] T. Shikarna and G. P. Pells, *Philos. Mag.,* **47**, 369 (1983).
[12] A. Y. Stathopoulos and G. P. Pells, *Philos. Mag.,* **47**, 381 (1983).
[13] F. W. Clinard, G. F. Hurley, and L. W. Hobbs, *J. Nucl. Mater.,* **108/109**, 655 (1970).
[14] B. Jeffries, G. P. Summers, and J. H. Crawford, Jr., *J. Appl. Phys.,* **51**, 3984 (1980).
[15] R. C. DuVarney, A. K. Garrison, and R. H. Thorland, *Phys. Rev.,* **188**, 657 (1969).
[16] J. M. Smith and W. E. Vehse, *Phys. Lett.,* **31A**, 147 (1970).
[17] W. E. Vehse, W. A. Sibley, F. J. Keller, and Y. Chen, *Phys. Rev.,* **167**, 828 (1968).
[18] S.-I. Choi and T. Takeuchi, *Phys. Rev. Lett.,* **50**, 1474 (1983).
[19] S.-I. Choi; this conference.
[20] H. A. Wang, C. H. Lee, F. A. Kröger, and R. T. Cox, *Phys. Rev. B,* **27**, 3821 (1983).
[21] M. G. Blanchin, J. Castaing, G. Fontaine, A. H. Heuer, L. W. Hobbs, and T. E. Mitchell, *J. Phys. (Paris),* **42**, C3-L (1981).
[22] For example, see: Y. Oishi and K. Ando; this conference.
[23] A. J. Tench and M. J. Duck, *J. Phys. C,* **6**, 1134 (1973).

Dislocations in Ceramics

A. H. Heuer
Case Western Reserve University
Department of Metallurgy and Materials Science
Cleveland, OH 44106

The state of knowledge of dislocations in Al_2O_3 and MgO is at a relatively advanced level and can serve as a guide for studying dislocation properties and behavior in other ceramics. However, a number of issues still need resolution:

(1) The issue of glide vs climb dissociation has been of concern for some time. What is now clear is that given a choice, i.e., at high temperatures, a dissociated dislocation will adopt a climb configuration when at rest, although such climb-dissociated dislocations are sessile. (The elastic repulsion of two climb-dissociated partial dislocations is smaller than two glide-dissociated partials, and the diffusion required to effect dissociation is very short range.)

The orientation dependence of the stacking fault energy in Al_2O_3 and spinel ($[MgO]:[Al_2O_3] \geq 1$) is modest; is this true for all ceramics because of the high temperature at which climb dissociation must necessarily occur? To date, dissociation has not been identified in other ceramics where dislocation behavior has been studied — UO_2, TiO_2, MgO, etc. Although structural reasons permit rationalization of the lack of dissociation in MgO and other rock salt oxides, it is not known if the lack of observed dissociation in other materials is due to the poor TEM resolution in the studies performed to date or to a sufficiently high stacking fault energy that any dissociation is confined to the core of the dislocation. In this regard, it is worth noting that in both Al_2O_3 and spinel, the observed dissociation involves only the cation sublattice; disruptions of the oxygen sublattice must produce very high energy stacking faults.

(2) The core structures of dislocations in oxides are essentially unexplored. Calculations have been done for MgO. They should be repeated for Al_2O_3 and other oxides and compared with experimental determination of the core structure. The latter is now possible in modern high-resolution electron microscopes by comparing images of end-on dislocations in thin foils with calculated images assuming different core configurations; no such studies have been reported for oxides, although spectacular success has been achieved with Si and Ge.

(3) Twinning is an important low-temperature deformation mode in Al_2O_3 and some other oxides. Whether or not twin growth requires motion of twinning dislocations in Al_2O_3 is not known; likewise, the dislocations that are known to exist within some twin matrix interfaces in Al_2O_3 have not been adequately characterized. The situation with regard to other ceramics is even worse.

(4) The temperature dependence of the critical resolved shear stress in Al_2O_3 is marked, follows a logarithmic law over the temperature range 200–2000 K, and is sufficiently less marked for prism plane than for basal slip, so that the former is the preferred slip system at temperatures less than ~900 K. No theory exists to explain these experimental observations, nor why MgO and other rock salt oxides show such different behavior — they can be deformed in single-crystal form down to helium temperatures.

(5) The issue of slip anisotropy is as troubling in other oxides as it is for Al_2O_3. While the preference for {110} slip over {100} slip in rock salt structure oxides was rationalized on structural grounds, evidence exists in the isostructural alkali halides suggesting that the preference for {110} slip has a chemical rather than a structural origin.

(6) Other examples are known where slip anisotropy is virtually absent (spinel) or else varies systematically with nonstoichiometry, as in UO_{2+x}, where {111} slip is favored for oxygen-excess samples (although the CRSS does not vary with the O:U ratio). A convincing theoretical explanation for such aberrant behavior is lacking, as are explanations for the variations of the plastic deformation of other oxides with changes in stoichiometry, e.g., Al-rich nonstoichiometric Mg–Al spinel.

(7) As already noted, rock salt oxides have a low brittle → ductile transition temperature ($T_{B→D}$), even under ambient confining pressures, while oxides with the corundum structure require elevated temperatures for appreciable plasticity. The situation with regard to other oxide families, e.g., those with the fluorite structure, is murky. For example, $T_{B→D}$ is <500 K for UO_{2+x} but ~1500 K for stabilized ZrO_2. These specific differences, as well as the difference between different oxides, have not been satisfactorily explained. One constant worry is that sufficiently pure crystals have not yet been prepared to permit study of intrinsic dislocation behavior in any system.

(8) It has recently been recognized that the low-temperature fracture behavior of brittle polycrystalline ceramics often depends crucially on residual stresses introduced by elastic/plastic incompatibility due to limited near-surface plastic deformation under contact loading. The near-surface deformed regions in Al_2O_3 have been studied, but more work is needed; virtually nothing has been reported on other oxides.

Structure of MgO and Al$_2$O$_3$ Surfaces

Victor E. Henrich

Yale University
Department of Applied Physics
New Haven, CT   06520

Compared to bulk properties such as diffusion, point defects, or impurity states, our understanding of the surface geometric and electronic structure of MgO and Al$_2$O$_3$ ceramics is meager. This is largely due to the relatively recent development of both the experimental and theoretical tools necessary to address the properties of surfaces. Although rather detailed pictures of metal and semiconductor surfaces have been developed in the last few years, the study of ceramic surfaces is still in its infancy.

Of the two oxides on which this conference has primarily focused, the surface properties of MgO are much better understood. MgO has one extremely stable face, (100), along which it cleaves well. Low-energy-electron diffraction (LEED) experiments and multiple scattering calculations indicate that the atomic geometry of the MgO (100) surface is very nearly a truncation of the bulk crystal structure. However, the electronic bandgap is narrowed by about 2 eV at the surface due to the large electric fields that arise from the abrupt termination of the highly ionic lattice. The perfect MgO (100) surface is quite inert with respect to the chemisorption of most molecular species.

Point defects on the MgO (100) surface have been treated both theoretically and experimentally, and theoretical work has begun on extended defects such as steps, kinks, and corners. Lattice defect calculations have predicted the type and magnitude of the atomic relaxation in the vicinity of various types of defects, and while the perfect (100) surface is only slightly relaxed, relatively large atomic displacements are predicted at steps, corners, etc. Such calculations have also been applied to the chemisorption of CO and H$_2$ at defect sites. Virtually no experimental confirmation of these calculations is yet available. The excited electronic states of some point defects have been observed experimentally and compared with calculations, but the agreement is not wholly satisfactory. Measurements of the segregation of Ca atoms to the MgO (100) surface have been made, and ordered arrangements of Ca atoms have been observed. The observed impurity segregation is in general agreement with theoretical predictions.

The surface properties of Al$_2$O$_3$ on an atomic scale are almost completely unknown. Due to the large unit cell in the corundum lattice, theoretical calculations have not been performed for any Al$_2$O$_3$ surface, of which several are relatively stable. No LEED measurements have been performed on any Al$_2$O$_3$ surface; thus, nothing is known about relaxation, reconstruction, etc. The few surface-sensitive measurements that have been made of the electronic structure of Al$_2$O$_3$ do not show any major deviations from the bulk structure for the filled valence band, but the width of the bandgap between the valence and conduction bands at the surface has not been investigated. Calcium segregation on Al$_2$O$_3$ has been studied by using surface-sensitive techniques, but problems were encountered that limited the accuracy of the results. Nothing at all is known about the properties of extended defects in Al_2O_3.

In short, the surface properties of MgO and Al_2O_3, while extremely important in a wide range of phenomena, constitute one of the least studied areas of ceramics. The theoretical and experimental techniques necessary to address the properties of ceramic surfaces are now quite well established, and all that is lacking is the motivation to apply them. Hopefully, the next 10 years will bring about a veritable explosion in the state of our understanding of ceramic surfaces, an explosion whose effects will be felt in all areas of ceramics.

Grain Boundaries in MgO and Al$_2$O$_3$

W. D. KINGERY

Massachusetts Institute of Technology
Department of Materials Science and Engineering
Cambridge, MA 02139

The progress that has been made in developing viable models for grain-boundary structures and chemistry in the last decade is gratifying indeed. The utility of the generalized coincidence site lattice and Bollmann O-lattice concepts for describing special grain boundaries in both MgO, which has the rock salt structure, and Al$_2$O$_3$ with the corundum structure seems well established. Results for these paradigm materials can be extended with some confidence to other oxide ceramics. In this respect the similarity in results found for NiO and MgO by Duffy and Tasker is particularly welcome.

Results have been reported at this conference on experiments with computer calculations of the structure of both tilt and twist boundaries, transmission electron microscopy, electron diffraction, boundary grooving, and scanning transmission electron microscopy. Except for the boundary grooving experiments, these techniques were unavailable until a decade or so ago and have developed into sophisticated tools only in the last few years. The results from quite different points of view and with very different techniques lead to a generally converging picture.

Both calculations and experiments indicate that coincident boundaries are low in energy, more open in structure than the corresponding boundaries in metals, quite different from one another in structure and openness, have many low-energy sites for solute segregation, and are anisotropic in their structure. The preference for coincidence orientations is indicated by the prevalence of faceting in both rock salt and corundum structure boundaries. In both MgO and Al$_2$O$_3$, the presence of small amounts of eutectic liquids — Ca, Si, and Ti are the most common offenders — can lead to liquid phases which penetrate along boundaries separating the grains in a way important for both processing and properties.

It seems that the foundation has been laid to attack much more poorly understood problems relating to the structure of high-angle boundaries, the influence of nonstoichiometry and segregation on boundary structure, and the influence of boundary structure and composition on the physical and chemical *properties* of grain boundaries, which are necessary to understand and develop ceramics. Phillips' observation of the absence of twin orientation boundaries in MgO-doped Al$_2$O$_3$ is suggestive of the practical results that may be anticipated.

Early determinations of crystal structures were necessary before theories of crystalline solutions and lattice defects could be developed. In the same way, as we develop a better understanding of boundary structure, we build a foundation for beginning to understand the role of solutes in boundaries and "defects" in boundaries.

For studies of grain-boundary structure, chemistry, and properties, MgO and Al$_2$O$_3$ seem to have served us well as paradigm oxides. And on the basis of history, we can confidently predict exciting and technologically important developments as the grain-boundary story unfolds.

Electrical and Optical Properties of MgO and Al₂O₃: An Overview

HARRY L. TULLER

Department of Materials Science and Engineering
Massachusetts Institute of Technology
Cambridge, MA   02139

A major theme of these proceedings has been concerned with the question of how effectively the refractory oxides MgO and Al₂O₃ serve as models for other ceramic systems. Given the subject of this overview, one must be particularly concerned with the utility that electrical and optical measurement techniques provide in revealing the basic properties of these compounds. In the following, a summary is provided which outlines:

(a) The features of these measurements which make them particularly suited for determining thermodynamic and kinetic data.

(b) Recent notable success in obtaining such data for MgO and Al₂O₃ by these techniques.

(c) Limitations to be aware of when applying these techniques.

Electrical Measurements

Charge transport in ionic compounds proceeds via the simultaneous motion of ionic and electronic charge carriers, each with its characteristic charge $Z_j q$. The nature and concentration, n_j, of these charged defects are generally sensitive functions of the crystal structure, temperature, atmosphere, and purity, whereas the carrier mobilities, μ_j, depend primarily on structure (atomic and electronic) and temperature. These factors are combined in the electrical conductivity, given by

$$\sigma = \sum_j n_j Z_j q \mu_j \tag{1}$$

It therefore follows that the measurement and appropriate interpretation of the electrical conductivity can, in principle, provide insight into both the thermodynamic and kinetic factors necessary for the physicochemical understanding of ionic solids. Table I lists a number of important parameters obtainable for ionic solids in this way.

The versatility of a particular measurement technique is nearly as important as the nature of the information that can be derived from it. In this regard, electrical measurements are particularly outstanding. Although other experimental techniques are sometimes more direct in their interpretation (e.g., tracer diffusion, thermogravimetric analysis, electron spin resonance), they are generally difficult to apply over an extensive range of experimental conditions, particularly those most critical for refractory oxides, i.e., elevated temperatures and controlled atmospheres. Electrical measurements, on the other hand, may be applied quickly, routinely, and in situ over a wide range of experimental conditions, as summarized in Table II.

Aside from versatility, there are few techniques which possess both the sensitivity and dynamic range of electrical conductivity measurements. Instrumen-

Table I. Parameters Obtainable for Ionic Solids

Thermodynamic:
- Intrinsic defect disorder, e.g. Frenkel and Schottky products, thermal band gap
- Redox equilibria—nonstoichiometry
- Stability limits
- Order–disorder

Kinetic:
- Ion diffusion, electron mobility
- Redox kinetics

Interfacial:
- Electrode kinetics
- Space charge barrier characteristics

---

tation is available which can measure conductivities over approximately 25 orders of magnitude with the lower limits corresponding to electron densities as low as 1 electron/cm^{-3}! (See Table II.)

What is gained in terms of versatility and sensitivity is, however, often lost in terms of complexity. As is apparent from Eq. (1), the experimentally determined parameter, σ, is a superposition of contributions from a variety of defects, each with its own characteristic mobility. Furthermore, processes occurring at interfaces (electrodes, grain boundaries, precipitates, etc.) may further complicate the electrical response of the solid.

Fortunately, techniques have been developed, and are being refined, which now enable investigators to deconvolute these various factors and thereby individually characterize bulk carrier densities, carrier mobilities, and dispersion phenomena originating at grain boundaries and other interfaces. Table III lists some of the other electrical techniques which are often used to complement the classical DC electrical conductivity measurements. Examples, taken from the recent literature, are used below to illustrate how they may be applied most effectively.

Perhaps as important as the techniques listed in Table III is the application of defect chemical modeling (discussed in detail by Kröger[1] elsewhere in these proceedings) toward interpretation of the electrical conductivity. Here, the unique atmosphere and dopant dependencies predicted for each of the defects are used to identify the defect dominating the electrical conductivity under given experimental

Table II. Hallmarks of Electrical Measurements

Versatility:
    Temperature: $10^{-3} < T < 2 \times 10^3$ K
    Atmosphere: $10^{-30} < P_{O_2} < 10^{+3}$ atm
    Time: $10^{-9} < t < 10^8$ s

Dynamic range
    Resistance: $10^{-6} < R < 10^{20}$ Ω

Sensitivity: 1 electron/cm^3*

*For $\sigma = 10^{-15}$ $(\Omega \cdot \text{cm})^{-1}$ and $\mu = 6 \times 10^3$ cm^2/V·s

Table III. Important Electrical Measurement Techniques

Technique	Utility
Impedance spectroscopy	Separation of bulk from interfacial phenomena
Open circuit-concentration cell measurements	Separation of ionic and electronic contributions
Dilatocoulometry	Separation of cationic and anion contributions
Thermoelectric power and Hall effect	Give sign and concentration of majority carrier (most generally applied to electronic conductors)

conditions. Recently, notable success has been achieved in applying this approach to defect and transport interpretation in MgO. This is summarized in the following.

Defect Interpretation in MgO — A Model System

Magnesia is a highly stable oxide, as reflected by its high melting temperature and thermal band gap. Correspondingly high Schottky and Frenkel energies are likewise expected, as recently confirmed by computer simulation.[2] Consequently, transport behavior can be expected to depend on defects formed in response to impurities and deviations from stoichiometry rather than intrinsic defects. Further, since both the Mg and O ions are stable in their respective valence states, it is safe to assume that impurities will control the defect equilibria, except for the highest-purity specimens.

The above considerations have several important consequences. First, defect concentrations are expected to be low, thereby satisfying the dilute solution approximation. Second, control of the defect equilibria by impurities markedly narrows the likely defect models which must be considered. The significance of these simplifying factors in interpreting defect equilibria in MgO are illustrated in the following.

The three major mechanisms for defect formation in donor-doped MgO, i.e., (a) thermally induced intrinsic, (b) redox-induced, and (c) impurity-induced defects are listed in Table IV, along with the appropriate electroneutrality relation.

Table IV. Defect Reactions: Donor-Doped MgO

Schottky equilibria:
$$\text{Null} \rightarrow V''_{Mg} + V^{\cdot\cdot}_{O}; \quad K_s(T) \tag{2}$$

Electron-hole equilibria:
$$\text{Null} \rightarrow e' + h^{\cdot}; \quad K_e(T) \tag{3}$$

Redox equilibria:
$$\tfrac{1}{2} O_2 \rightarrow V''_{Mg} + 2h^{\cdot} + O_O; \quad K_{ox}(T) \tag{4}$$

Impurity Incorporation:
$$D_2O_3 \xrightarrow{MgO} 2D^{\cdot}_{Mg} + V''_{Mg} + 3O_O \tag{5}$$

Electroneutrality:
$$n + 2[V''_{Mg}] = p + 2[V^{\cdot\cdot}_{O}] + [D^{\cdot}_{Mg}] \tag{6}$$

Application of the standard piecewise solution to the simultaneous equations of Table IV (Brouwer approximation[3]) gives results like those illustrated in Fig. 1. Here, the defect densities, normalized by their respective mobilities (see Eq. (1)), are plotted as a function of oxygen partial pressure for a given temperature T. In each of the four defect regimes, designated by the applicable simplified electroneutrality relation, the defects exhibit a different characteristic P_{O_2} dependence, which may be used to identify the applicable defect regime experimentally.

Because of the large inequality which normally exists between ion and electron mobilities (i.e., $\mu_{ion} \ll \mu_{electron}$), analysis of the total conductivity is markedly simplified. Thus in the three defect regimes in which an ionic defect is balanced by an electronic defect, in the simplified electroneutrality relations (see Fig. 1), only the predominant *electronic* defect will contribute measurably to the conductivity. Under circumstances where the electronic conductivity is fixed by dopants and is therefore P_{O_2}-independent (e.g., $n = [D_{Mg}^{\cdot}]$), one may readily extract the electron mobility, i.e.,

$$\mu_e = \sigma/q[D_{Mg}^{\cdot}] \tag{7}$$

as recently demonstrated in the derivations of μ_e in U-doped CeO_2,[4] and μ_h in

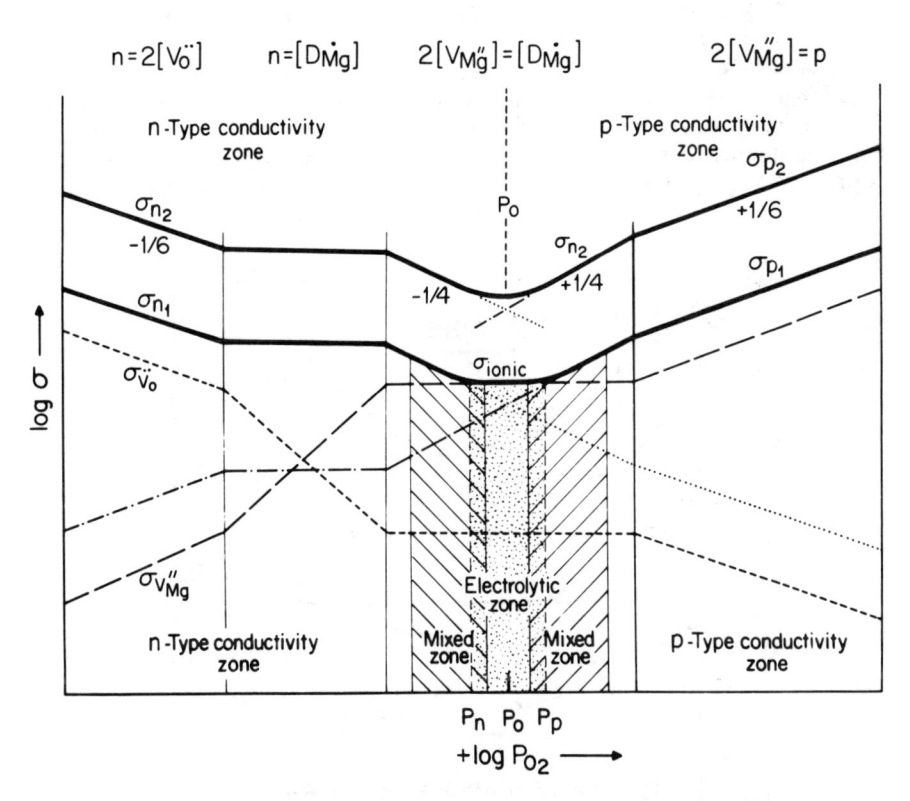

Fig. 1. Plot of the partial ionic and electronic conductivities for donor-doped MgO as a function of P_{O_2}. Solid curves represent the total conductivity with the upper and lower curves satisfying the following conditions, respectively, $\mu_{V_{Mg}} \ll \mu_e, \mu_h$; $\mu_{V_{Mg}} < \mu_e, \mu_h$.

Y-doped UO_2.[5] On the other hand, when the electron conductivity is P_{O_2}-dependent, as in the left and right extremes of Fig. 1, measurements of σ as a function of T can be used to extract values of some of the equilibrium constants listed in Table IV. This approach has been applied successfully in many semiconducting oxide systems, including (for example) CoO,[6] and CeO_2,[7] as well as many others.[8]

As it happens, all of the experimental data obtained to date for MgO appear to be limited to the third defect regime of Fig. 1 characterized by the neutrality relation

$$2[V''_{Mg}] = [D^{\cdot}_{Mg}] \tag{8}$$

Since both majority defects in this region are ionic and notably higher than the electron and hole densities, it becomes possible for σ to become predominantly ionic even with the mobility asymmetry being ionic and electronic carriers. This possibility is reflected in the lower of the two solid curves in Fig. 1. On the other hand, if $[V''_{Mg}] > n, p$ is insufficient to offset μ_e, $\mu_h \gg \mu_{Mg}$, σ will remain electronic at all atmospheres, as reflected in the upper solid curve. Note that, in the former case, a P_{O_2}-independent ionic conductivity plateau is bounded by two P_{O_2}-dependent segments with opposite slopes ($\sigma \propto P_{O_2}^{\pm 1/4}$), whereas in the latter a minimum is obtained where these segments intersect. The intersection point is designated by P_0 and corresponds to the condition at which intrinsic electronic disorder is obtained,* i.e., $n = p$.

Both of the above possibilities have been observed experimentally in donor-doped MgO with the lower curve characteristic of lower T and higher $[D^{\cdot}_{Mg}]$. These features are illustrated in Figs. 2 and 3, in which the electrical conductivity and ionic transference data of Sempolinski and Kingery[9] for high-purity and scandia-doped (1500 ppm) MgO are plotted as a function of P_{O_2} at a series of temperatures.

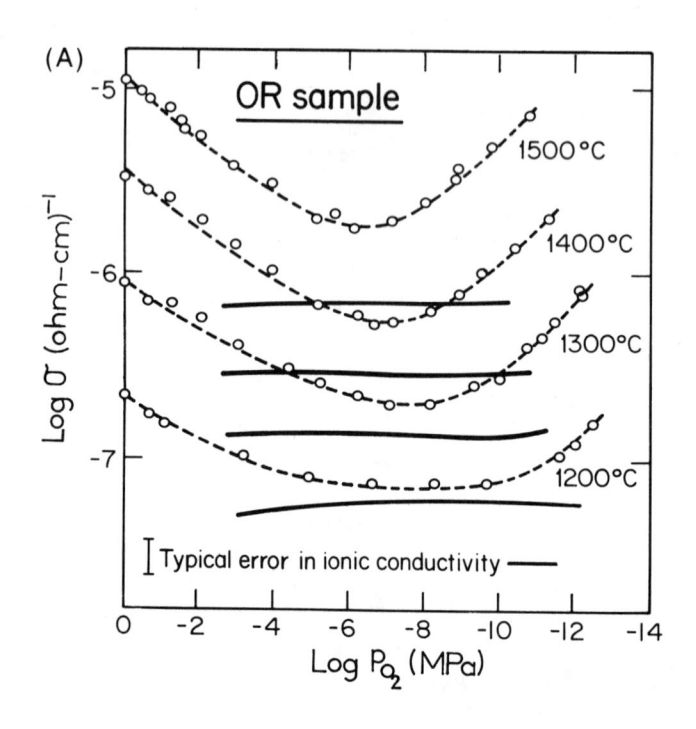

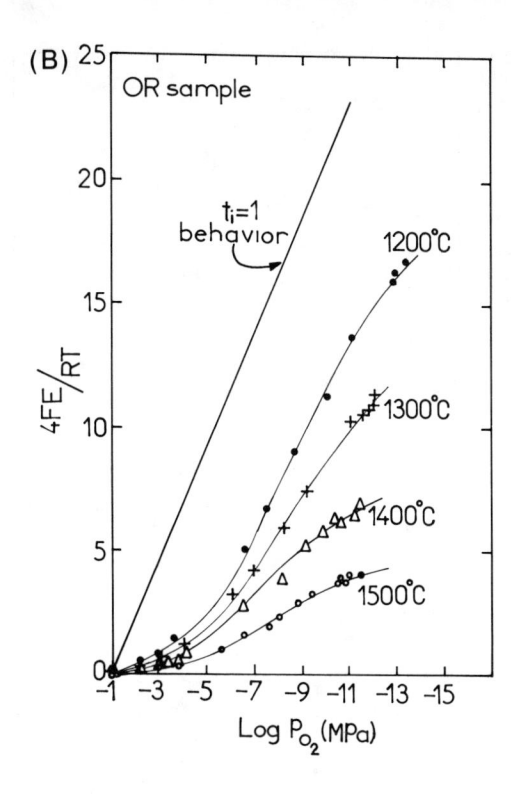

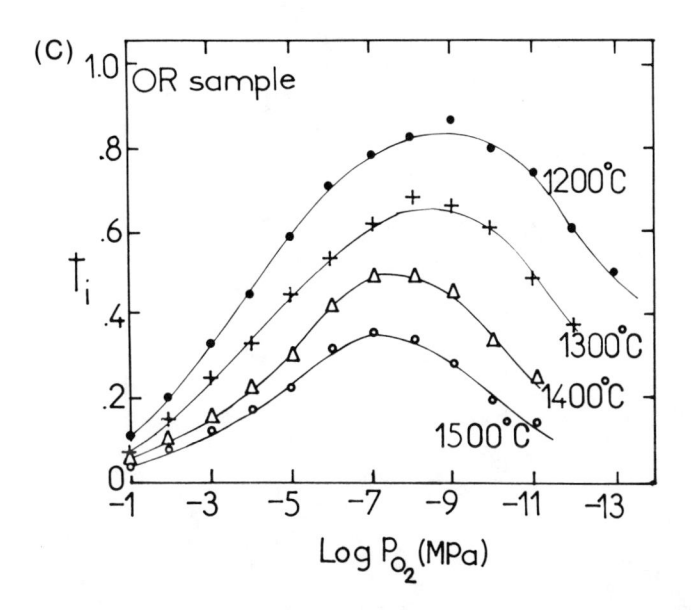

Fig. 2. Electrical data for high-purity MgO. Shown as functions of P_{O_2} are (A) total and ionic (solid curves) conductivity; (B) emf; and (C) ionic transference number (from Ref. 9).

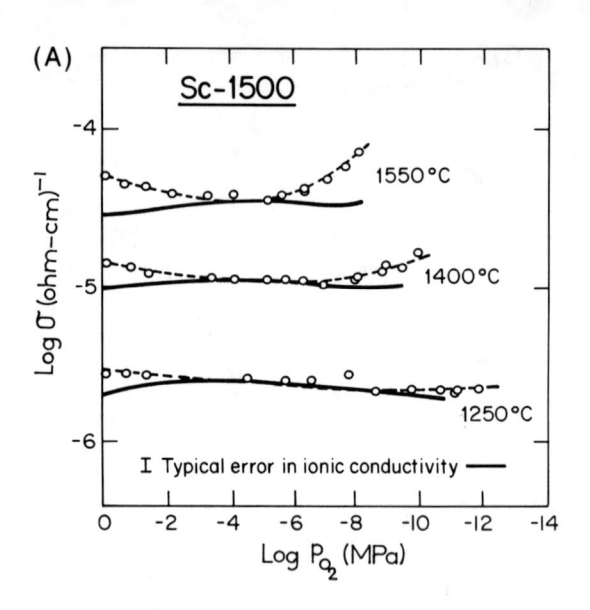

(A)

Sc-1500

1550°C

1400°C

1250°C

I Typical error in ionic conductivity

Log σ (ohm-cm)$^{-1}$

Log P_{O_2} (MPa)

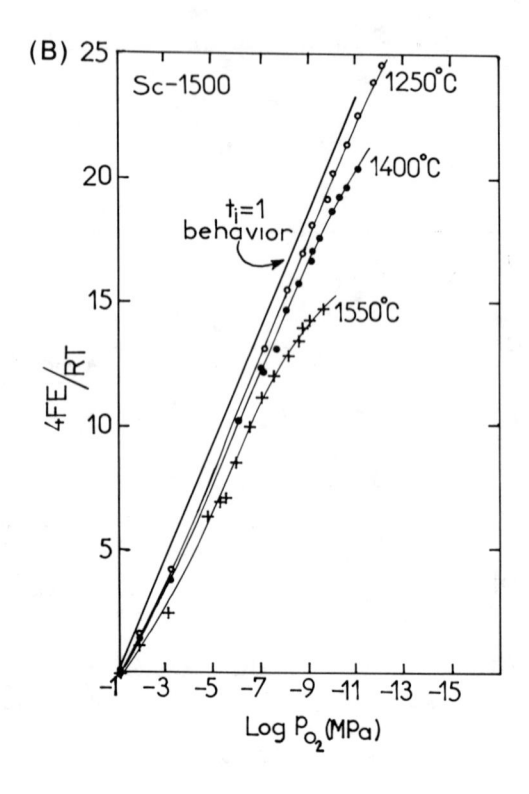

(B)

Sc-1500

1250°C
1400°C
1550°C

$t_i=1$ behavior

4FE/RT

Log P_{O_2} (MPa)

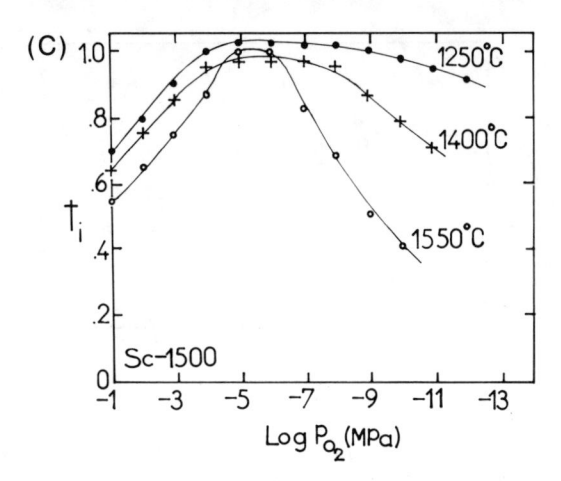

Fig. 3. Electrical data as in Fig. 2, but for doped MgO (from Ref. 9).

The ionic conductivity in the defect regime under discussion is given by

$$\sigma_{ion} = [V''_{Mg}]2_q(\mu_0/T)\exp(-E_m/kT) \qquad (9)$$

where E_m is the migration energy of magnesium vacancies and μ_0 is the ionic mobility preexponential term which includes the jump distance, attempt frequency, and a number of other constants. Since $[V''_{Mg}]$ is fixed by the donor density $[D^{\bullet}_{Mg}]$, one may readily solve for the ion mobility given by

$$\mu_{V_{Mg}} = (\mu_0/T)\exp(-E_m/k_T) = \sigma_{ion}/q[D^{\bullet}_{Mg}] \qquad (10)$$

Further employing the Nernst-Einstein relation

$$D_{V_{Mg}} = \frac{kT}{2q}\mu_{V_{Mg}} = \frac{kT}{2q^2}\frac{\sigma_{ion}}{[D^{\bullet}_{Mg}]} \qquad (11)$$

one also obtains values for the defect diffusivity.

In the study by Sempolinski and Kingery quoted above, the donor density (Sc, Al, or Fe) was controlled over nearly two orders of magnitude. Figure 4 shows that σ_{ion} follows a near-linear dependence on donor density, as expected from Eq. (9). The temperature dependence of the defect mobility and diffusivity, derived from measurement of σ_{ion} vs $1/T$ is shown in Fig. 5, and expressions defining the best fits to these data are given by

$$\mu_{V_{Mg}} = \frac{8800}{T}\exp\left(-\frac{220\text{ kJ/mol}}{RT}\right)\text{ cm}^2/\text{V}\cdot\text{s} \qquad (12)$$

$$D_{V_{Mg}} = 0.38\exp\left(\frac{-220\text{ kJ/mol}}{RT}\right)\text{ cm}^2/\text{s} \qquad (13)$$

One commonly assumes the nature of the ionic defect subsequent to an examination of the crystal structure. Thus a magnesium vacancy was chosen above, rather than an oxygen interstitial, given the close packing of the rock salt structure. Such choices may, however, be confirmed experimentally, as was done recently by Duclot and Deportes,[10] who measured the cationic transport number t_{Mg} using dilatocoulometry. This technique establishes the magnitude of metal electro-

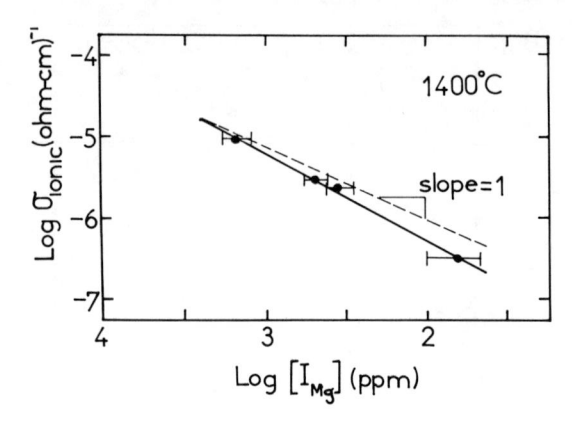

Fig. 4. Plot of log σ_{ion} vs log $[D_{Mg}]$, showing ionic conductivity vs total impurity concentration at 1400°C (from Ref. 9).

migration by monitoring the displaced volume for a given level of charge passed through the specimen during a period of time, t. No displacement is expected for electronic conduction nor for oxygen ion transport since the atmosphere serves as a source and sink for atoms.

The dilatocoulometric measurements showed that t_{Mg} decreases from ≈ 0.6 at 1100° to ≈ 0.2 at 1305°C, clearly supporting a mixed conduction model in which

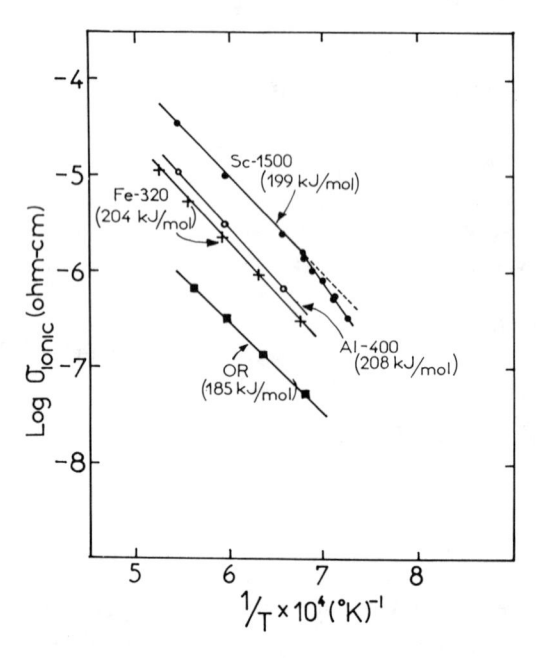

Fig. 5. Ionic conductivity as a function of temperature for MgO doped with a variety of donors. Number adjacent to elements corresponds to dopant levels in terms of parts per million (from Ref. 9).

the mobile ion is Mg. A plot of $\sigma_{Mg} = t_{Mg}\sigma_{total}$ vs $1/T$ gave an activation energy for ionic conduction of 2.2 eV (213 kJ/mol), in excellent agreement with Eq. (12).

At this stage it is worth noting that only after the above ionic conductivity studies were completed was a reliable and consistent picture of magnesium diffusion obtained.[9] This was true in spite of a previous series of Mg tracer studies.

We next shift our attention to the electronic component of conduction. The ionic transference measurements quoted above show that MgO becomes progressively more electronic with increased T and exhibits p-type conduction at high P_{O_2} and n type at low P_{O_2}, as predicted in Fig. 1. More specifically, $\sigma_e = \sigma_{total}(1 - t_{ion})$ is observed to pass through a $p \rightarrow n$ transition at $\approx 10^{-8}$ MPa with P_O shifting weakly to lower P_{O_2} with increasing T (see Fig. 2).

Sempolinski et al.[11] showed that the minimum in electronic conductivity is given by

$$\sigma_{min} = 2q(N_c N_v \mu_e \mu_h)^{1/2} \exp(-E_g/2kT) \tag{14}$$

in which N_c and N_v are, respectively, the conduction and valence band effective densities of states. Analysis of the temperature dependence of σ_{min} in MgO in terms of Eq. (14) allowed for the first realizable experimental determination of MgO's thermal band gap energy of 6.8 ± 0.5 eV (657 ± 48 kJ/mol). Further, assuming a large polaron electronic mobility mechanism, values for the electronic mobilities ($\mu_h \approx 5$ cm^2/V·s, $\mu_e \approx 15$ cm^2/V·s, 1673 K) were estimated from the pre-exponential term of the above equation.

With expressions for all relevant defect mobilities in hand, it is next possible to derive expressions for a number of the equilibrium constants. For example, the hole conductivity at fixed P_{O_2} is given by

$$\sigma_p = \frac{K_{ox}^{1/2} P_{O_2}^{1/4} e \mu_h}{1/2 [D'_{Mg}]} \tag{15}$$

Since the donor density may be derived from the ionic conductivity (see Fig. 4) and μ_h and P_{O_2} are known, one may plot σ_p vs $1/T$ to obtain K_{ox}. Following this approach, Sempolinski et al.[11] derived the following:

$$K_{ox}(T) = 7 \times 10^{63} \exp(-605 \pm 77 \text{ kJ/mol}) \text{cm}^{-9} \text{ atm}^{1/2} \tag{16}$$

Incorporation of multiple-valent dopants such as Fe (2+, 3+) requires some modification of the above approach. An additional defect reaction which describes the Fe ionization reaction

$$Fe_{Mg} \rightarrow Fe^{\cdot}_{Mg} + e' : K_{Fe}(T) \tag{17}$$

and an Fe mass balance equation, i.e.,

$$Fe_{total} = [Fe^{x}_{Mg}] + [Fe^{\cdot}_{Mg}] \tag{18}$$

must also be considered. One finds that for conditions for which $Fe_{total} \simeq [Fe^{x}_{Mg}]$, both ionic and hole conductivities become P_{O_2}-dependent approaching $P_{O_2}^{+1/6}$. Applying these relations, Sempolinski et al.[11] derived the following expression for $K_{Fe}(T)$:

$$K_{Fe}(T) = 1.7 \times 10^{21} \exp(-3.3 \pm 0.5 \text{ eV}/kT) \text{ cm}^{-3} \tag{19}$$

showing that the Fe energy level lies 3.3 eV (319 kJ/mol) below the conduction band edge or nearly mid-gap.

In summary, electrical measurements, including electrical conductivity, dilatocoulometry, and oxygen concentration cell emfs, when combined with a defect chemical approach, have been successfully applied to MgO in deriving both thermodynamic and kinetic data, much of it either unavailable or suspect prior to these studies. Several key features allowed for the unambiguous determination of the ionic defect density $[V''_{Mg}]$ and the ionic mobility in MgO. First, single crystals with relatively low concentrations of background impurities were available, to which relatively high concentrations of donor dopants could be intentionally added. Second, intrinsic defect generation could safely be assumed to be negligible. In this way, the ionic and electronic defects could be systematically controlled by doping. This is less true for Al_2O_3, where the solubility limit of aliovalent dopants tends to be lower.

Defect Interpretation in Al_2O_3 — Some Comments

The defect structure and electrical properties of Al_2O_3 are analyzed in detail by Kröger[1] elsewhere in this volume. In the following we only attempt to compare some of the characteristics of Al_2O_3 in relation to those of MgO and show how they impact the interpretation of electrical measurements.

In general, the system Al_2O_3 is more difficult to interpret. This is a consequence of a number of factors, including very low diffusivities, generally low dopant solubilities, very high resistivities, and very large defect formation energies. The first feature makes it difficult to obtain equilibrium data of the form illustrated in Figs. 2 and 3 for MgO, except for temperatures of 1500°C and above. This implies additional experimental difficulties as well as reliance on nonequilibrium data which are more difficult to interpret.

The second feature, i.e., low dopant solubilities, implies a difficulty in fixing the defect equilibria with any one dopant. It then becomes possible for a superposition of background impurities to compensate for the intentional dopant. When this is true, it becomes difficult or impossible, for example, to derive the ionic diffusivity as done previously for donor-doped MgO.

Kröger and coworkers[12-14] showed, however, that the major features of the defect equilibria of Al_2O_3 can be understood. In general, the electrical properties support a picture in which a transition occurs from electronic to ionic conduction as the P_{O_2} range is traversed isothermally. Experimentally, this is seen as a transition from a $P_{O_2}^{+1/x}$ to a $P_{O_2}^{-1/x}$ dependence in going from high to low P_{O_2}, with a minimum in conductivity observed at the transition. This type of minimum in σ should not be confused with the minimum in electronic conduction given by Eq. (14). Nevertheless, many more assumptions must be made in interpreting the electrical data for Al_2O_3 than for MgO.

To overcome some of the limitations listed above for Al_2O_3, Kröger and others have turned to complementary techniques including optical absorption, electron spin resonance, nonequilibrium electrical measurements, etc. Even so, many fundamental thermodynamic and kinetic parameters still remain unknown or unclear in Al_2O_3, including such fundamental factors as the nature of the predominant ionic defects, a variety of defect formation energies including the thermal bandgap, and most of the defect mobilities (see Ref. 15 for summary of reported data). Further clarification of the defect and electrical properties of Al_2O_3 remains an important challenge to future investigators.

Polarization Studies

AC conductivity (complex impedance) measurements were mentioned previously in regard to electrically deconvoluting interfacial from bulk phenomena. Such studies may however also be applied toward improved characterization of defect complexes in solids, as discussed by Crawford[16] in this volume.

Such studies become particularly important at reduced temperatures where defects of opposite charge tend to associate into defect complexes due to Coulombic attraction. For defect complexes with substantial defect association enthalpies, this implies nearly total association of mobile defects, e.g., V''_{Mg} in MgO up to rather high temperatures.

Because the dielectric loss peak at low temperatures corresponds to dipole reorientation, it can be used to (a) estimate the concentration of such defect pairs and (b) estimate the reorientation or the migration energy. Now if the DC conductivity is measured as a function of T, the sum of the migration and one-half the defect association enthalpy is obtained. Thus the combination of the two measurements allows the two components to be isolated.

Another important feature of polarization studies is their ability to distinguish between different defect complexes, since the frequency of the dielectric loss peak is characteristic of the specific dipole.

Surprisingly few polarization studies have been performed on oxides. Only relatively recently have Nowick and coworkers[17] shown their true versatility toward understanding ionic conduction in the fluorite oxides. Elements of this work and the recent studies on Li-doped MgO are reviewed by Crawford[16] in his chapter. Given such versatility, polarization studies as applied to oxides deserve much greater attention in the future.

Optical Studies — Some Comments

The field of optics is at least as broad as that of electrical phenomena. Since the author is much less expert in this field than the other contributors to this area, i.e., Summers[18] and Burns and Burns,[19] only some comments concerning their utility toward defect interpretation will be attempted.

Perhaps the most important attribute of optical over that of electrical measurements is that defects generally possess unique optical signatures, which may correspond to either unique absorption or emission frequencies. For example, many substitutional transition metal or rare earth dopants in MgO or Al_2O_3 can be identified readily by their absorption or emission signatures. Further, different valence states of the same dopant can also be distinguished by their different absorption spectra. This, for example, can be very valuable in confirming or establishing the nature of a defect model based on variable valent impurities.[12]

Not only do optical studies allow one to identify a defect or impurity, but they also allow one to place them in energy relative to a conduction or valence band edge. Such information can sometimes be used directly in the derivation of some equilibrium constants, e.g., Eq. (19).

Finally, details of the structure of defect complexes may be determined, e.g., the complex symmetry, by utilizing polarized light together with oriented single crystals. Such studies are relatively well developed, for example, in the field of color centers in alkali halides.

As with all techniques, these also suffer some limitations. Perhaps most important is the inability to do many detailed and controlled studies at high temperatures, where defects are in equilibrium with their environment. Instead, the vast majority of optical studies are performed at room temperature or below where one examines only the quenched-in defect structure which is generally considerably more complex than that at elevated T due to defect association and ordering.

Part of the difficulty in performing high-temperature optical studies is instrumental. Here sources of difficulty include competing blackbody radiation from

Table V. Model Systems: Characteristics of MgO and Al_2O_3

+ Availability of single crystals and relatively high-purity powders
+ Large band gaps—ideal optical hosts for transition and rare earth metal impurities
+ Simple structures and well-defined stoichiometries
− Difficult or impossible to observe intrinsic phenomena
− Slow kinetics limit equilibrium measurements to very high temperatures
− Extremely sensitive to thermal history effects
− High resistivities

furnace and sample holder, need for high-temperature windows, and temperature-insensitive spectrometers. Progress along these lines is being made, however, as increasing use is made of high pass filters, chopped light, and laser heaters instead of furnaces.[20]

The difficulty not amenable to direct improvement, however, is the feature of thermal broadening, in which an optical peak either broadens beyond recognition or overlaps with other spectra to form a near continuum. Similar problems arise due to overlapping absorption by free carriers thermalized during heating.

Summary and Recommendations

MgO and Al_2O_3 do serve as model systems in many important ways, as summarized in Table V. The positive features no doubt have contributed, over the last decade, to a markedly improved understanding of MgO. The negative features, on the other hand, often lead to difficulties in obtaining reproducible data for refractory materials; e.g., as in the case of Al_2O_3. Some of these difficulties lead to different degrees of nonequilibria at, for example, interfaces such as grain boundaries and electrodes. This is an area which can now be much better treated by AC electrical measurements rather than the DC measurements conventionally utilized. AC impedance spectroscopy is therefore bound to be utilized more extensively for these and other materials in the future, and can be expected to lead to an improved understanding of interfacial as well as bulk phenomena.

In general, there is a need for additional complementary techniques which allow for the deconvolution of many effects, e.g., ionic vs electronic, n vs p, bulk vs interfacial, etc. Techniques such as those listed in Table III must be used more frequently and in concert to ensure correct interpretation of experimental phenomena.

Modeling of the phenomena in terms of defect chemical models further ensures correct interpretation of data as well as allowing for the derivation of critical thermodynamic and kinetic parameters. The defect chemical approach is proving its versatility in ever greater numbers of oxide systems[21] and is therefore likely to become more actively used in future studies.

Optical spectroscopists have made major strides in characterizing the nonequilibrium defect structures of oxides at reduced temperatures, while ceramists are doing the same for those at elevated temperatures. Unfortunately, the spectroscopists and the ceramists or solid-state chemists use very different defect nomenclatures and apparently rarely read each others' literature. Consequently, proposals by one group are sometimes made which are in clear disagreement with models already clearly proved by the other group. More communication must be encouraged between the various groups of investigators. This meeting was only one small step in the right direction. In future meetings, an urgent matter for discussion must be the development of a common defect notation.

In closing, it may be noted that as we progress in any field we come closer to the truth but are likely never to attain the full truth. The same may be said about our model systems — significant progress in understanding of these systems has been made in the last several decades. In that sense they have served well as model systems, *but* we still have much to learn.

Acknowledgment

The continuing support by the National Science Foundation under Contract No. 82-03697 DMR for our research is gratefully acknowledged.

References

[1] F. A. Kröger, "Experimental and Calculated Values of Defect Parameters and Defect Structure of α-Al$_2$O$_3$: Electrical Properties of α-Al$_2$O$_3$"; this proceedings.
[2] W. C. Mackrodt, "Calculated Point Defect Formation, Association and Migration Energies in MgO and α-Al$_2$O$_3$"; this proceedings.
[3] G. Brouwer, Philips Res. Rept. **9**, 366 (1954).
[4] H. L. Tuller and T. G. Stratton, "Defect Structure and Transport in Oxygen Excess Cerium Oxide-Uranium Oxide Solid Solutions," *in* Proc. 3rd Int. Conf. on Transport in Nonstoichiometric Compds., Penn. State Univ., June 10-16, 1984; to be published.
[5] N. J. Dudney, R. L. Coble, and H. L. Tuller, "Electrical Conductivity of Pure and Yttria-Doped Uranium Oxide," *J. Am. Ceram. Soc.*, **64**, 627-32 (1981).
[6] B. Fisher and D. S. Tannhauser, "Electrical Properties of Cobalt Monoxide," *J. Chem. Phys.*, **44**, 1663 (1966).
[7] H. L. Tuller and A. S. Nowick, "Defect Structure and Electrical Properties of Nonstoichiometric CeO$_2$ Single Crystals," *J. Electrochem. Soc.*, **126**, 209-17 (1979).
[8] P. Kofstad, Nonstoichiometry, Diffusion and Electrical Conductivity in Binary Metal Oxides. Wiley-Interscience, New York, 1972.
[9] D. R. Sempolinski and W. D. Kingery, "Ionic Conductivity and Magnesium Vacancy Mobility in Magnesium Oxide," *J. Am. Ceram. Soc.*, **63**, 664 (1980).
[10] M. Duclot and C. Deportes, *J. Solid State Chem.*, **31**, 377 (1980).
[11] D. R. Sempolinski, W. D. Kingery, and H. L. Tuller, "Electronic Conductivity of Single Crystalline Magnesium Oxide," *J. Am. Ceram. Soc.*, **63**, 669 (1980).
[12] B. V. Dutt, J. P. Hurrell, and F. A. Kröger, "High-Temperature Defect Structure of Cobalt-Doped α-Alumina," *J. Am. Ceram. Soc.*, **58**, 420 (1975).
[13] S. K. Mohapatra and F. A. Kröger, "Defect Structure of α-Al$_2$O$_3$ Doped with Magnesium," *J. Am. Ceram. Soc.*, **60**, 141 (1977).
[14] S. K. Mohapatra and F. A. Kröger, "Defect Structure of α-Al$_2$O$_3$ Doped with Titanium," *J. Am. Ceram. Soc.*, **60**, 381 (1977).
[15] S. K. Tiku and F. A. Kröger, "Levels of Donor and Acceptor Dopants on Electron and Hole Mobilities in α-Al$_2$O$_3$," *J. Am. Ceram. Soc.*, **63**, 31 (1980).
[16] J. H. Crawford, "Polarization-Depolarization Studies of Defects in Oxides"; this proceedings.
[17] See, for example, R. Gerhart-Anderson, F. Zamani-Noor, and A. S. Nowick, "Study of Sc$_2$O$_3$-Doped Ceria by Anelastic Relaxation," *Solid State Ionics*, **9–10**, 931 (1983).
[18] G. P. Summers, "Luminescence and Photoconductivity in MgO and α-Al$_2$O$_3$ Crystals; this proceedings.
[19] R. G. Burns and V. M. Burns, "Optical and Mossbauer Spectra of Transition Metal-Doped Corundum and Periclase"; this volume.
[20] R. H. French, H. P. Jenssen and R. L. Coble, "High Temperature VUV Spectrometer"; in Proceedings of an International Conference on Vacuum Ultraviolet Spectroscopy, Jerusalem, 1983.
[21] H. L. Tuller; p. 271 in Nonstoichiometric Oxides. Edited by O. T. Sorensen. Academic Press, New York, 1981.

---

*Assuming $\mu_e = \mu_h$.

Mechanical Properties of MgO and Al$_2$O$_3$

R. M. Cannon

University of California
Berkeley, CA 94720

The demonstration of room-temperature ductility in MgO crystals by Gorum et al.[1] and the pioneering work of Kronberg[2] on Al$_2$O$_3$ stimulated extensive work on MgO and more recently on Al$_2$O$_3$. Due in part to studies of these materials, the understanding of and the ability to control the mechanical properties of ceramics have developed dramatically in the ensuing 25 years. They provide useful models for two classes of materials: (1) Al$_2$O$_3$, as a hard, strong, anisotropic ceramic, and (2) MgO, as a somewhat softer, cubic material.

This paper assesses the state of knowledge regarding the mechanical behavior of these oxides. It outlines the basic behavior, and then focuses on the underlying factors that control properties and areas where current understanding is poor. It should provide insights regarding the behavior of other ceramics.

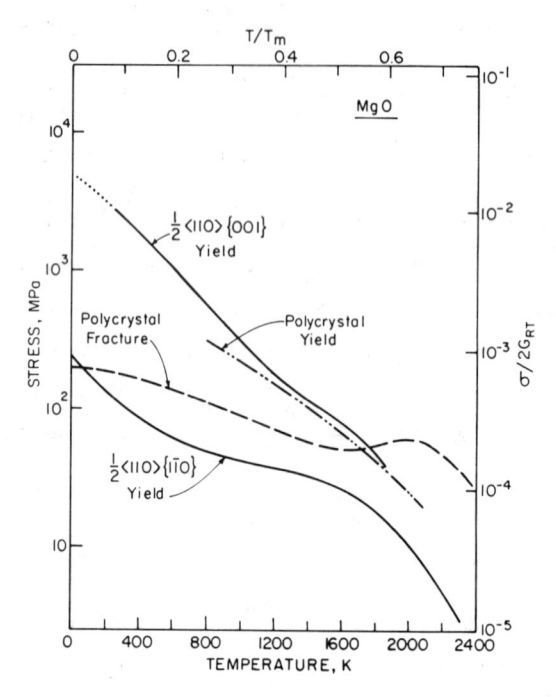

Fig. 1. Schematic of temperature dependence of the tensile yield stresses for primary and secondary slip for periclase and the fracture stress typical of ≈100 μm grain-sized, dense MgO polycrystal.

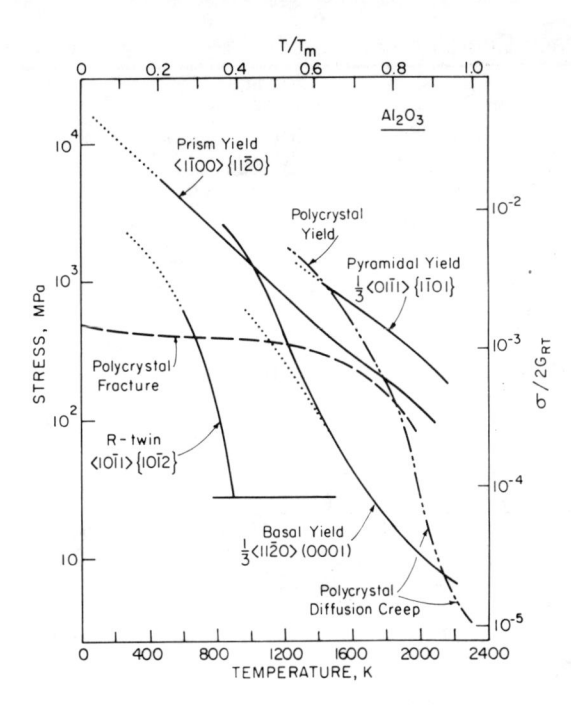

Fig. 2. Schematic of temperature dependence of the tensile yield stresses for primary, secondary, and tertiary slip for sapphire and the fracture stress typical of a 10–20 μm grain-sized, dense Al_2O_3 polycrystal.

Overview

Characteristic behavior of MgO and Al_2O_3 is shown schematically in Figs. 1 and 2. The dominant slip systems have been largely identified for both Al_2O_3 and MgO. For each, slip is highly anisotropic. Particularly for Al_2O_3, high stresses are required to induce general yielding at any temperature. Curves of the yield stresses for the primary and secondary slip systems can be constructed for a significant range of temperatures from various measurements on single crystals, e.g., from the summaries of Heuer et al.,[3] Castaing et al.,[4] Hulse,[5] Copley and Pask,[6] Day and Stokes,[7] and Reppich.[8]

Both rhombohedral, $\langle 10\bar{1}1\rangle \{10\bar{1}2\}$, and basal, $\langle 1\bar{1}00\rangle (0001)$, twinning occur in Al_2O_3, even at low temperatures.[9] Twinning is found more frequently in compression. At intermediate temperatures, <1300°C, it is easier than slip.[10,11]

Cleavage of both oxides is also highly anisotropic. Cleavage energies have been measured for several orientations in Al_2O_3 and for the {001} cleavage plane for MgO (Table I). Cleavage on {001} in MgO is much easier than for Al_2O_3. Tasker and colleagues have done atomistic calculations of surface energies and configurations for several MgO and Al_2O_3 surfaces; some energies are listed in Table I.

Without application of hydrostatic pressure, polycrystals of both materials are brittle at low temperatures. Fracture often results from extrinsic flaw extension, especially for Al_2O_3. Thus, low-temperature strengths usually depend on processing- or machining-related flaws, rather than the critical stresses for yield

Table I. Surface Energies

Al_2O_3	(θ, ϕ)	Wiederhorn*[12,13]	Cleavage (J/m^2) Becher [14]	Iwasa and Bradt [15]	Theory Tasker et al.[§][20]
$\{10\bar{1}0\}$	90, 30	7.3 4 700°C	8.8	12 11 900°C	
$\{1\bar{2}10\}$	90, 0		6.7	7.3 5 1450°C	
$\{10\bar{1}2\}$	57, 30	6.0		7.0 4 700°C	
$\{1\bar{2}13\}$	61, 0		6.8		
$\approx\{1\bar{2}19\}$	~30, 0		20.0		
(0001)	0,	>40	>30	26 2 1100°C	3.0 relaxed 6.5 unrelaxed

MgO	Temp.	Gilman [16]	Cleavage Gutshall and Gross [17]	Westwood and Goldheim [18]	Shockey and Groves [19]	Theory Tasker et al. [21,22]
{001}	<100 K	1.5[†]	1.6[‡]			1.2
	room temperature	≥2.0 R		≥1.6 R	≥1.7	
{031}						1.7
{021}						2.0
{011}						2.9

Measurements at room temperature unless noted otherwise.
*Measurements in dry N$_2$ at room temperature or in vacuo; others from K_c measured in air.
$(\theta, \phi) - \theta$, angle of normal to [0001]; ϕ, angle of basal projection of normal to $\langle1\bar{2}10\rangle$.
[†]Measured at 77 K in liquid N$_2$ and recalculated by Wiederhorn (Ref. 23).
[‡]Measured at 98 K in vacuo and recalculated by Wiederhorn (Ref. 23).
R Measured in ambient air and recalculated as per Wiederhorn (23).
[§]These calculations were for basal planes with half the cations partitioned between the two half-crystals; "unrelaxed" is for the atoms positioned as for perfect crystal; "relaxed" is after polarization and relaxation displacements, but without reconstruction.

or twinning. For MgO with polished surfaces and small internal flaws, inhomogeneous slip can nucleate fracture at relatively low temperatures. For Al$_2$O$_3$, twin-induced microcracking may control the higher strengths, especially in compression.

At higher temperatures, extensive deformation can occur, particularly in favorably oriented single crystals or very-fine-grained polycrystals. Ductility obtains at lower temperatures for single crystals oriented for single or duplex slip. For MgO crystals oriented for general slip or for some clean polycrystals, a gradual transition occurs over a wide temperature range from slip-induced fracture to fully ductile behavior above ≈0.7 T_m. For sapphire oriented for basal slip the transition to ductility is rather abrupt; otherwise ductility is limited. For coarse-grained polycrystals, ductility is usually greater and evident at lower temperatures for MgO. At higher temperatures and stresses, power law or recovery creep is important.

At intermediate to low stresses and higher temperatures, diffusion can con-

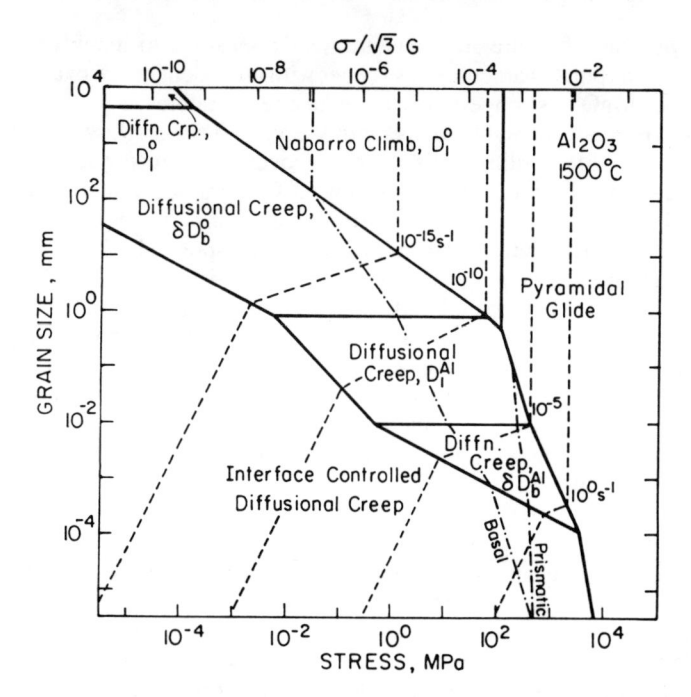

Fig. 3. Deformation map for MgO-doped Al$_2$O$_3$ at 1500°C (Ref. 3). The pyramidal glide field involves general slip on five independent systems. The lines marked "basal" and "prismatic" denote the conditions for which basal or prismatic glide gives strain rates equal to those occurring in the polycrystalline sample where deformation is controlled by diffusional creep.

tribute significant strain. Especially for polycrystals, it can produce appreciable deformation under conditions where general slip is too difficult, even at stresses above those for primary yield. For Al$_2$O$_3$, diffusional creep dominates deformation over a wide range of conditions (Fig. 3).

At high temperatures and lower stresses, diffusional cavitation and crack growth control fracture, although final failure may involve cleavage. Mounting evidence indicates that a different set of heterogeneities can act as flaws at high temperature than at low temperature, and that impurities can significantly lower strength and enhance crack-growth rates at high temperatures.

The ductility depends very much on loading conditions. For high rate, tensile loading, large deformations obtain only at very high temperatures for pure MgO or for sapphire oriented for basal slip. In contrast, appreciable strain can occur at much lower temperatures in compression, especially with an additional hydrostatic stress component. With Al$_2$O$_3$, hydrostatic constraint is necessary for yielding below 0.5 T_m.* Even in multiaxial tension, extensive high-temperature deformation can result for polycrystals of either, if strain rates and maximum stresses are carefully controlled. However, internal cavitation frequently occurs. Thus, at high temperatures, these materials, particularly Al$_2$O$_3$, can present the apparent anomaly of being extensively deformable and having poor creep resistance but exhibiting no useful impact resistance or ductility at high strain rates.

The yield and fracture strengths are highly sensitive to impurities for most temperatures. Impurity hardening is stronger with aliovalent than isovalent cations. Hardening of MgO is strongest at intermediate temperatures, e.g., 0.1 to 0.5 T_m; for Al_2O_3, it is greater at higher temperatures, e.g., $>0.5\ T_m$. Aliovalent impurities can also change diffusivities and so affect recovery and creep rates.

Fabrication advances providing dense, fine-grained material and elimination of flaws have led to room-temperature strengths, σ_f, up to 600–700 MPa ($\sigma_f/E = 0.0016$) for sintered substrate-grade or hot-pressed Al_2O_3. The highest strengths measured in sapphire whiskers or flame-polished rod, 11–15 GPa, give $\sigma_f/E \approx 0.027$–0.038 compared to theoretical[24] values of $\sigma_f/E \approx 0.1$. The softer MgO polycrystals are usually weaker, although strengths of 550 MPa for fine-grained material have comparable σ_f/E, ≈ 0.0019.

Recently, major improvements have been made in the strength and toughness of two-phase Al_2O_3-based alloys, notably Al_2O_3-ZrO_2 materials which are nearly twice as strong as polycrystalline Al_2O_3 and even relatively tougher, $K_c \approx 8$–10 MPa·m$^{1/2}$.

Assessment

Much of the mechanical behavior has been experimentally characterized for some range of conditions. Potentially important mechanisms have been identified and many properties have been modeled, although often only at order-of-magnitude accuracy. Numerous discrepancies exist between experiment and theory or emerge from comparing all existent data, and controlling mechanisms are often disputed. In the following, areas are emphasized where a sense of clarity has been attained or where further work seems particularly desirable.

Deformation

For MgO, slip on $\{1\bar{1}0\}$ is apparently always easier than on $\{001\}$ although the difference becomes small above $\approx 1600°C$, $0.6\ T_m$; both systems are needed to provide five independent slip systems. Dislocation loops and debris causing hardening and creep substructures, often involving subgrains, have been characterized. Hardening and recovery creep are not understood in detail, but are similar to the behavior for alkali halides. Trivalent cation impurities cause potent hardening either as solutes or fine precipitates. The most extensive quantitative analyses are those of Reppich and coworkers[8,25] for solute and precipitate hardening in Fe-doped MgO. The coherent spinel causes slightly more hardening at peak aging (≈ 50 Å) than when in solution, but is nearly stress-free and soft enough to be cut.[25] Other, unidentified impurities can give greater precipitate hardening.[26] Above $\approx 1200°C$, $0.45\ T_m$, hardening from solutes and precipitates diminishes, at least for low impurity contents, and is poorly characterized. Magnesia has not been studied significantly above $\approx 0.7\ T_m$, 1700°C.

For primary slip, theoretical analyses by Reppich[8] and Mitchell and Heuer[27] have shown that the tetragonally distorted elastic field from complexes, such as the trimer of a cation vacancy and two Fe^{3+} ions, causes a strong solute-dislocation interaction, i.e., Fleischer hardening. This reportedly exceeds electrostatic interactions from shearing complexes or between dipoles and charged jogs. Two somewhat different analyses have given plausible agreement between theory and experiment for hardening by the Fleischer interaction.[8,27] For Fe-doped MgO, Reppich[8] found that thermally activated dislocation passage of stationary complexes controlled the yield stress, τ_y, for primary slip below $\approx 450°C$, $0.25\ T_m$. Athermal hardening from an induced Snoek effect involving complex reorientation was considered dominant above this temperature. The temperature-independent plateau in τ_y is not evident in all studies. Within this temperature range, where clustering and precipitation occur, dislocation locking and dynamic aging can also increase τ_y. These results are consistent with recent analyses of yielding in alkali halides below 0.6 T_m by Skrotzki and Haasen[28]; based on alkali halide behavior, primary slip may be Peierls-stress-controlled below $\approx 0.05\ T_m$, depending on the impurity content. However, this interpretation fails to explain the mobility differences between edge and screw dislocations, charged dislocations, or Rebinder effects.

Solute effects on secondary slip have not been determined. Comparison with alkali halide behavior[28] suggests that the Peierls stress controls slip below $\approx 0.25\ T_m$ and that solute hardening dominates at higher temperatures. As the Fleischer interaction is theoretically weaker for secondary than primary slip, electrostatic interactions that depend on the core structure and charge were postulated to be important. The order-of-magnitude higher Peierls stress indicated for secondary slip presumably owes to like-charge repulsion at the half-slipped position.

For Al_2O_3, the primary, secondary, and tertiary slip systems operating at high temperature have been ranked in terms of resistance to slip, Fig. 2. The pyramidal slip plane is somewhat uncertain. This system is required, as the other two provide only two independent slip systems each.[29] At room temperature another secondary system has been found to operate around indentations.[30] Compressive τ_y data[4,31] suggest that basal slip may be more difficult than prism slip at temperatures below 700°C; in contrast, TEM observations[32] after room-temperature indentations indicate that basal slip may occur more frequently. Twinning may have caused excessive basal hardening during low-temperature compression. Above 1300°C, the basal τ_y are confirmed by tensile measurements which avoid twinning.[11] Solute hardening of basal slip is very strong between 1100° and 1600°C, 0.6 and 0.8 T_m, particularly for aliovalent solutes.[33] The room-temperature hardness can be increased mildly by alloying. Pyramidal flow stresses are mildly influenced by some solutes at 0.7–0.9 T_m.[34] Nabarro climb can apparently control deformation of c-axis sapphire at lower stresses in this temperature range.[35]

Impurity effects are complex for basal slip at the high temperatures of interest. Divalent solutes cause potent hardening, as well as strain aging,[33] but also reduce work-hardening rates,[36] presumably because of the enhanced lattice oxygen diffusion. Tetravalent Ti, which reduces the oxygen diffusivity, increases both τ_y and work-hardening rates.[36,37] Theoretical analysis, Pletka et al.,[37] which neglects thermal-activation and solute-mobility effects, purportedly shows that size misfit causes the major interaction for isovalent solutes and that for tetravalent solutes, for which the hardening per ion ($d\tau_y/dC \approx G$) is 20 times greater, the tetragonal distortion from defect complexes causes stronger interaction than does shearing complexes.

Comparison between MgO and Al_2O_3 shows several differences. The τ_y are much lower for MgO below 1600°C or at all comparable T/T_m. The anisotropy in τ_y is much less for MgO at higher temperatures but may be greater at low temperature. Slip involves one Burgers vector, **b**, and cross slip can occur between the primary and secondary systems in MgO. In Al_2O_3, three Burgers vectors are involved; because of the structural cation vacancies, they are all longer. Further, the flow stresses do not correlate with the lengths of **b** for Al_2O_3.[36] Cross slip for the basal (primary) system involves a different plane and is harder than secondary slip; thus, multiplication and slip band spreading may be difficult and even involve climb at higher temperatures.

Heuer, Castaing and coworkers have found that dislocations in sapphire often dissociate by climb, whereas appreciable glide dissociation is less frequently observed.[36] At high temperature, both basal and prism dislocations climb-dissociate into ⅓⟨10$\bar{1}$0⟩ partials (the oxygen repeat vector). As they are immobile without a trailing (or migrating) stacking fault, remobilization must be difficult once dislocations stop and climb dissociation begins. In addition, during prism or pyramidal slip, networks, which have a high fraction of basal dislocations, form rapidly. For prism slip, this can result from decomposition into (⅓)⟨1$\bar{2}$10⟩ dislocations and subsequent reaction.[36] The decomposition products are perfect, but decomposition is out of the glide plane, which inhibits both edge and screw dislocation motion. There is limited evidence of glide dissociation at low temperature for both ⅓⟨1$\bar{2}$10⟩ and ⟨10$\bar{1}$0⟩ dislocations. Lack of indicated slip by ⅓⟨1$\bar{2}$10⟩ dislocations on {1$\bar{1}$00} suggests that glide dissociation is sufficient to prevent this cross slip; if so, all primary and secondary glide may involve ⅓⟨10$\bar{1}$0⟩ partials having small separation. All the frequently observed dissociations involve cation faults, leaving the oxygen lattice perfect.[36] Appreciable dissociation has not been reported for MgO; as the cation sublattice is full, cation faults do not exist, and the ½⟨110⟩ Burgers vector is the same as the ⅓⟨10$\bar{1}$0⟩ partial.

It has been suggested that the Peierls stresses can be important for basal slip below 0.75 T_m and for pyramidal slip up to 0.9 T_m or higher, although a comprehensive, quantitative evaluation of the various glide mechanisms is lacking. For prism slip the dynamics are inconsistent with a simple Peierls mechanism.[31,38] More likely, the high slip resistance also depends on core dissociations and decompositions.[31] Difficult cross slip and climb dissociation may be factors causing high basal yield stresses. For rhombohedral dislocations, it is uncertain whether glide resistance or climb dissociation causes the unexpected result that climb can apparently be easier than glide.

The significance of the twinning stresses for Al_2O_3 are unknown. The lower-temperature stresses would not be inconsistent with Peierls stress resistance to the motion of partial dislocations.

Folweiler's[39] work on Al_2O_3 was among the first affirmations of diffusional creep for any material. Since then, several experiments with Al_2O_3 or MgO have verified the expected stress and grain-size dependences and have given reasonable creep-rate magnitudes, although the grain-size dependences and rates often indicate that a combination of lattice and boundary diffusion is controlling.[40–42] When analyzed self-consistently, the data for Al_2O_3 from many investigators agree well.[43]

Frequent observations of creep with stress exponents of 1–2 have caused confusion. These have three identifiable causes: (1) interface-controlled diffusional creep for relatively fine-grained material; (2) concurrent cavitation and cracking, manifest by strain softening and most apparent with suspected or proven impurity phases at grain boundaries; and (3) simultaneous dislocation slip or climb for coarser grain sizes.

For both oxides, the cation lattice diffusivity far exceeds that of oxygen. Limited tracer data and current interpretation of the creep behavior indicate that boundary diffusion is usually much slower for cations than anions. Typical values of wD_b/bD_1 are $\approx 10^{-4}$–10^{-5} for Al and $\geq 10^{-9}$–10^{-10} for O ions in Al_2O_3. The controlling diffusivity depends on grain size; the sequence illustrated in Fig. 3 for Al_2O_3 is similar for MgO. Oxygen lattice diffusion is unimportant for both oxides (except for climb).[43,44]

Aliovalent impurities significantly affect creep rates, a fact that has been used to study impurity effects on transport. Where aliovalent impurities are considered to be affecting the controlling cation lattice diffusivity, increases of one to two orders of magnitude have been observed. More dramatic increases occur in Al_2O_3 co-doped with divalent and tetravalent impurities, which affect both lattice diffusion and boundary processes.[44,45]

When cation lattice diffusion is shown to be controlling, the impurity effects are better explained. For MgO, the increased rates with Fe^{3+} additions are consistent with Schottky defects being predominant, although detailed explanation depends on partial ionization and defect association.[41,44] For Al_2O_3, extensive data,[43] particularly those of Gordon and coworkers,[44] show that either divalent or tetravalent cations accelerate creep when cation lattice diffusion is apparently controlling. The accepted interpretation, reviewed by Ikuma and Gordon,[44] assumes that either Al vacancies or Al interstitials cause diffusion at appropriate stoichiometries. Although the presumption that cation Frenkel defects are important, even if minority defects may seem to be at variance with the higher cation interstitial energies predicted theoretically, e.g., Mackrodt,[46] inconsistency has not been verified using a suitable defect model that includes association and plausible defect entropies.

For boundary-diffusion control, significant sensitivity to some but not all impurities is indicated, but not understood. For example, Fe^{2+} apparently enhances Al boundary diffusion in Al_2O_3,[45] but Mg does not[40]; Fe^{3+} and F apparently enhance Mg diffusion in MgO.[47] Reinterpretation of older data and recent data for co-doped material by Gordon et al.[44,45] suggests that O boundary diffusivities for Al_2O_3 exceed previous estimates, and that no certain measure of them exists.

Although the creep-rate magnitudes and dependences on dissolved impurities are reasonable, detailed evaluation of theoretical models is precluded by incomplete tracer-diffusion data, unaccounted impurity effects, and inexact understanding of lattice defects. Creep and tracer measurements have not been made on similar materials, which would allow assessment of the factor of 3 uncertainty in the numerical coefficient for the models; this arises from uncertainty about the amount of concurrent grain rearrangement and boundary sliding and the effect of grain-size distribution. Creep rates for undoped MgO are less than predicted using the Nabarro-Herring relation and tracer data, contrary to observations in metals. Creep rates and their activation energies for undoped Al_2O_3 exceed those predicted using tracer data, but that is expected if divalent cations are the dominant dissolved impurities. Appreciable transient strains have been observed; they may result from stress or impurity redistribution but have not been adequately explained.

Diffusional creep in Al_2O_3 can be interface-controlled, but the controlling mechanisms are not identified. With low impurity levels or MgO doping, interface control is manifest at very fine grain sizes by the nonlinear stress dependence at low stresses; in this regime the ion transport is almost entirely by boundary diffusion.[40] With heavy TiO_2 doping to raise the Al lattice diffusivity, the weak grain size and P_{O_2} dependences suggest that creep is interface-controlled.[45] In the latter case, Fe doping seems to accelerate the limiting boundary processes. Where creep is interface-controlled with transport by boundary diffusion, grain-boundary sliding rather than point-defect creation is plausibly rate-controlling. The nonideal behavior has strong implications on crack growth, stress concentrations, and sintering, which differ greatly for the two extremes of difficult grain-boundary sliding or point-defect creation.

Grain-boundary sliding occurs above $\approx 1000°C$; however, little is known about the mechanisms. Bicrystal experiments have shown that sliding is irregular, nonlinear in stress, strongly dependent on the misorientation angle, and often accompanied by fracture.[48] Sliding is greatly facilitated by impurities. Interpretation is uncertain because of the difficulty in distinguishing behavior controlled by sliding, per se, and by the accommodating lattice deformation. Internal friction and bicrystal experiments suggest that sliding in glass-free Al_2O_3 may be much more difficult than in metals.

The data for dense polycrystals, taken in compression or with hydrostatic constraint, indicate that the yield stresses are similar to the secondary stresses for MgO and may be at or above the prismatic or pyramidal yield stresses for Al_2O_3. Impurity and grain-boundary hardening and cracking presumably cause unexplained variabilities. Power-law creep has been observed in each, but the contributions of glide, climb, and diffusion are often unresolved. At lower strain rates, finer-grained materials are often softer owing to diffusion. There are few definitive data on impurity hardening or softening, except numerous ones illustrating that stimulating grain-boundary sliding and cavitation lower the apparent yield stresses.

Fracture

Several factors indicate that cleavage is a complex process involving irreversible losses and perhaps lattice trapping. Among these are a trend in the ratios of the measured cleavage energies, γ_c, to the calculated surface energies, γ_t, which seem to increase on going from close-packed, neutral planes, to nonclose-packed, neutral planes, to planes which are nonneutral without partition of the cations across the fracture plane. For {001} MgO, $\gamma_c/\gamma_t \approx 1.3$ at 100 K or less. For Al_2O_3, γ_c are 6–12 J/m² for the prism and pyramidal cleavage planes, which all seem too high for equilibrium surface energies; if the calculated energies for relaxed, neutral (0001) or for {011} of MgO are a guide, then $\gamma_c/\gamma_t \approx 2$–4 for these nonclose-packed planes. For the Al_2O_3 basal plane at room temperature, cleavage rarely, if ever, occurs, but the apparent γ_c of 26–40 J/m² represent $\gamma_c/\gamma_t \approx 8$–13, comparing with the relaxed neutral surface. These comparisons presume that the calculated surface energies are closer to the equilibrium surface energies, γ_s, than are the cleavage energies.

These results suggest that the excess (dissipated) energy is smallest for close-packed, neutral surfaces, is greater for nonclose-packed surfaces, and is very high for surfaces which would be nonneutral on ideally planar cleavage. Recent calculations suggest that energies for unrelaxed surfaces can be twice those for relaxed ones, for nonclose-packed oxide surfaces. Perhaps the excess cleavage energy results because polarization and relaxation occur several planes deep into the crystal. Thus, as bonds break, the atoms initially unload into an intermediate position because of the constraint of nearby uncracked lattice; as the crack moves away, the energy released by relaxation and converted to phonon energy would be unrecoverable. Where the cations must partition between the two surfaces to produce neutral surfaces, the partition may require time and impede rapid propagation mechanisms. If local, nonneutral patches develop, cleavage energies would be very high.

Thermally activated bond breaking should alleviate some sources of irreversibility. Thermally activated, subcritical crack growth was found along {10$\bar{1}$0} in sapphire in vacuo at 200°–600°C, by Wiederhorn et al.[13] For some measurements, the energies for rapid cleavage in sapphire decrease proportionally faster with heating than does the modulus, particularly the basal cleavage energy measured by Iwasa and Bradt[15]; this also suggests such effects.

There are no boundary cleavage energy data. Duffy and Tasker[22] have calculated γ_b and γ_e for several symmetrical grain boundaries for MgO or NiO, where

$2\gamma_e = 2\gamma_s - \gamma_b$ is the binding energy. The boundary, γ_b, and surface energies are comparable but vary with orientation; thus, the γ_e vary significantly. For [001] tilt or twist boundaries, which naturally give neutral cleaved surfaces, the calculated γ_e range from 0.5 to 1.4 J/m² and are roughly one-half of the equilibrium energies of the surfaces formed. Handwerker et al.[49] found that the dihedral angles formed during annealing vary widely and have median values of 110° to 120° for both MgO and Al_2O_3. Although this information is indirect, it infers $\gamma_s \approx \gamma_b$ and a wide distribution of γ_e, as do the calculations.

Grain-boundary cleavage does not usually produce neutral, close-packed planes; thus, we expect irreversible effects. Further, the grain-boundary calculations illustrate that nonequilibrium surfaces can be produced. When two equilibrium surfaces are bonded to form a boundary, diffusional reconstruction may be required to equilibrate the boundary. Rapid cleavage of such a boundary produces nonequilibrium surfaces. Only for slow-growing cracks at high temperature can equilibrium surfaces be produced directly and with minimal energy dissipation when diffusional reconstruction is required. The calculations of Tasker and Duffy[50] indicate that reconstruction, involving removal of appreciable cations and anions, is required to stabilize (001) twist boundaries in MgO. For the stable $\Sigma = 5$ boundary, the net binding energy would be reduced from γ_e of 0.77 to 0.26 J/m² if equilibrium surfaces could be produced directly. Corrosion, segregation, or entropically induced disorder could ameliorate the need for reconstruction and reduce γ_e. These factors may contribute to the increased intergranular cleavage and strength reduction at high temperature.

The best evidence, i.e., TEM of crack tips, suggests that cracks in Al_2O_3 propagate with sharp tips below 400°C, and that even up to 800°C emitted or nearby dislocations are relatively infrequent; they are usually confined to arrested cracks or cracks being repropagated, Wiederhorn et al.,[13] Lawn et al.[51] TEM evidence indicates that behavior is similar in MgO although the details are not elaborated, Hockey.[52] We suggest that once crack-tip emission becomes possible for moving cracks, a sharp increase in toughness with further heating should ensue. No such increase has been seen for MgO polycrystals[53] below 1600°C or Al_2O_3 polycrystals[54,55] below 1400°C. Only mild increases are indicated for prism or pyramidal planes in sapphire above 700°C. Thus, many, if not all, rapidly moving cracks may be sharp at least to 1400°–1600°C for both Al_2O_3 and MgO.

Toughening by crack-tip shielding can result from motion of dislocations near a crack that reduces the stresses at the tip. For MgO, the {001} cleavage energies at room temperature can be two to three times the surface energy as a result of plastic deformation. Presumably the toughening results from shielding rather than blunting. There is also evidence that polycrystal fracture energies are influenced by plasticity; for example, the room-temperature MgO toughness can be significantly (50%) reduced by irradiation hardening.[56] However, for MgO, between room temperature and 1600°C, the primary τ_y often drops by a factor of four or more, and the secondary τ_y drops more. The basal and prism τ_y for Al_2O_3 decrease by one to two orders of magnitude between 400° and 1400°C. Yet the polycrystalline toughnesses for both materials and most sapphire cleavage energies tend to decrease in this interval, which is counter to expectation for an important shielding contribution.

With shielding, the cleavage energy can be approximated as:

$$\gamma_c = g(\phi)\gamma^0(1 + B) \qquad (1)$$

where γ^0 is the cleavage energy without shielding. The factor $g(\phi)$ is ≥ 1 and depends on crack deflection or twist. If bond breaking involves irreversible processes, γ^0 should decrease with heating or reduction of velocity. The shielding term, B, should be inversely proportional to τ_y because the plastic-zone size increases as the yield stress decreases; thus, B should increase with heating. The shielding is proportional to the crack-tip stresses which depend on γ^0. Thus, the temperature dependence of γ_c represents a balance between the reduction in requisite crack-tip stresses and the effect of shielding, which increases with reductions in τ_y on heating, but decreases as the critical crack-tip stresses decrease. When γ^0 is near the equilibrium γ_s, as for {001} cleavage of MgO, plasticity effects should become apparent on further heating, as is seen on going from 100 K to room temperature for {001} cleavage. Etch-pit evidence revealing that dislocation activity is greater near arrested than moving crack tips for MgO indicates that this competition depends on loading rate.

These results indicate that cleavage energies most closely approach thermodynamic surface energies for cracks in inert environments at low velocities and at temperatures where a minimum energy obtains.[†] At higher temperatures or lower velocities, plasticity may increase the fracture energies. The minimum γ_c at any temperature for rapid fracture of sapphire occur at 600°–1000°C or higher, e.g., Iwasa and Bradt,[15] Fig. 5, and range from 2 to 5 J/m² for different planes, Table I; they would sometimes be lower were low velocity data used. These approach expected equilibrium energies.

Basal cleavage in Al_2O_3 is more complex, e.g., Iwasa and Bradt.[15] It is nearly impossible at room temperature, and does not occur at high speed at 1500°–1800°C. The cleavage energy drops rapidly with temperature, goes through a minimum, and then increases. At intermediate temperatures, it is the lowest cleavage energy for sapphire. It is unknown whether the rapid increase in γ_c, relative to that for other planes, results simply from plasticity. Perhaps adsorption or adsorption-assisted reconstruction facilitates cleavage at intermediate temperatures. Similar behavior is expected for {111} cleavage in MgO.

Unaccounted variations have been observed in the cleavage energies and their apparent temperature dependences for sapphire. These may represent effects of propagation direction. This could affect the ease of bond breaking, basal (or prism) slip, or deflection to an easier cleavage plane, and deflection induced by asymmetric slip and shielding.

A major development has been the realization that polycrystalline fracture energies can depend on crack size and *history* as well as grain size.[57-59] For MgO, the primary effect appears to be that, at small ratios of crack size to grain size, C/G, the fracture energy is that for the appropriate plane or grain boundary, whereas, for cracks with large C/G the energy is higher owing to tortuosity and deflection, e.g., Rice et al.[60] For Al_2O_3, two additional effects arise from the tendency for microcracking caused by the thermal expansion anisotropy stresses. The macrocrack toughness can increase with crack propagation owing to formation of a microcrack zone around the crack[61]; this apparently also stimulates stable branching.[62] Thus, for Al_2O_3, the macrotoughness may increase without limit during extension, whereas for MgO the macrocrack is presumed to have an asymptotic fracture energy at large C/G. The microcracking causes the toughness to depend on grain size. It is very low at large grain sizes (e.g., >100 μm) where spontaneous microcracking occurs, and it seems to reach a maximum value at intermediate grain sizes where stress-induced microcrack shielding and branching are significant.[61,62] The R-curve behavior, i.e., the dependence of γ on prior extension, can cause extensive stable crack growth in Al_2O_3 and accounts for some of the considerable error in toughness measurements which has plagued the field.

Analytic treatment of the toughness of polycrystals is incomplete. Data do not exist for γ_c for all orientations or boundaries. The polycrystalline values[62] for MgO, ≈ 11 J/m^2, and very fine-grained Al$_2$O$_3$, ≈ 18 J/m^2, may be 3–10 times the γ_c for single-crystal cleavage planes. In the absence of microcracking, these high ratios have been attributed to crack deflection and plasticity. These macro-toughnesses tend to be higher than are accountable simply by crack deflection or twist (e.g., a factor of 1.3 for equiaxed polycrystals). One cause may be the higher incidence of cleavage steps than occur with the best single-crystal measurements. For MgO, deflection may induce more dislocation shielding owing to the mode II and III loading. For Al$_2$O$_3$, the cause is less clear, but similar discrepancies exist for the toughness of polycrystals and for the nominal basal fracture which is actually microfacets of rhombohedral cleavage and chonchoidal fracture. The effect of microcrack shielding, but not branching, has been modeled by Evans and Fu.[61] Neither sufficient theory nor data exist to describe the behavior at intermediate C/G confidently.

For fine to medium grain-sized Al$_2$O$_3$, the critical toughness, K_c, usually decreases with heating up to 1400°C, reflecting thermally activated bond breaking and reduced thermal expansion anisotropy-induced microcracking. Subcritical growth is evident at all temperatures in air,[54] but is least near 600°C, below which corrosion presumably dominates the available data. Very rapid subcritical growth at intermediate temperatures, e.g., at 1000°–1300°C, is usually too fast for diffusion to occur concurrently or cause the growth (e.g., $V > D/a$ for surface or boundary diffusion where a is the ion spacing).

At 1400°C, 30 μm Lucalox Al$_2$O$_3$ undergoes little bulk deformation or tip blunting.[55] Taking the average between K for detectable crack growth and detectable healing[54] as representing the thermodynamic balance gives $\gamma \approx 0.9$ J/m^2. In contrast, rapid fracture[54,55] yields $\gamma_c \approx 7$ J/m^2. Even reducing these by 0.7 for deflection effects, the latter is too high for $\gamma_\sigma - \gamma_b/2$, whereas the former may be reasonable at this temperature. This lower energy would give $\gamma_s \approx 2\gamma_e \approx 1.3$ J/m^2, which is between the lowest γ_c measured for sapphire and $\gamma_s = 0.9$ and 0.7 J/m^2 for solid and liquid energies deduced from sessile drop and contact angle experiments[63] on impure materials at higher temperatures.

At higher temperatures, crack growth becomes diffusion-controlled. Although this is only now being studied extensively, several characteristics are evident from work on very-fine-grained Al$_2$O$_3$ in a regime with diffusional crack-tip blunting and bulk diffusional creep, e.g., Evans et al.[61,64] Subcritical crack growth involves cavitation damage ahead of cracks. At intermediate K, cracks are arrested and blunted; however, cavitation damage can occur in shear bands ahead of the crack, and cracks may continue to open. At lower K, cracks appear to heal unevenly. The blunting and damage zones mean that crack shapes, toughness, and subcritical velocities are history-dependent. Second, from comparison of results of several studies we infer that the maximum K occurs at low velocity for diffusively blunted cracks, and that this "K_c" may be nearly twice that expected at higher velocity.

Moisture-assisted slow crack growth occurs in Al$_2$O$_3$. Limited data indicate it occurs, but less aggressively, in MgO. If this depends on reactivity, then corrosive crack growth is inferred to be less severe in the more-reactive MgO. Alternatively, corrosive subcritical growth may be less on neutral, close-packed planes, as may be suggested by the toughening[19] and lack of subcritical crack growth at low K along {001} in MgO in contrast with the occurrence of static fatigue failure in polycrystals.

Strength and Ductility

Several distinct mechanisms for initiation of brittle fracture have been identified: (1) extension of preexistent flaws, (2) crack formation at stress concentrations from heterogeneous strain (e.g., dislocation pile-ups at kink walls or where slip bands interpenetrate), (3) crack formation from dislocation pile-ups at grain boundaries, (4) grain-boundary cracking from boundary sliding-induced stress concentrations at boundary jogs or triple junctions, (5) cracks induced by thermal expansion anisotropy or mismatch around inclusions, (6) cracks induced at twins. All but the first may involve crack nucleation in good material, as well as propagation.

Preexistent flaw extension can be described by the Griffith relation:

$$\sigma_f = yK_c/C^{0.5} = y(2\gamma_c E/C)^{0.5} \tag{2}$$

The geometrical constant, y, is of order unity. When dislocation coalescence is controlling, shorter slip bands give lower stress concentrations and higher strengths. Often the strengths obey the Petch relation:

$$\sigma_f = 2\tau_0 + k/d^{0.5} \tag{3}$$

where τ_0 is the lattice friction stress for slip and d is the grain or cell size controlling the slip band length. For samples with ground surfaces or large internal flaws, extension of preexistent flaws can be dominant to quite high temperatures.

For MgO, the first four mechanisms have been shown to be important,[48,53,56,65,66] although the fourth has not been well characterized or quantified. For Al_2O_3, the first and fifth are considered to be important at low temperatures, although twinning may sometimes be involved.

Many of the important characteristics of the brittle-to-ductile transition for MgO have been elaborated by Stokes and coworkers[7,65,67] and others. For single crystals, the transition occurs in stages with heating; each gives more strain-to-failure. Strength and strain-to-failure depend on sample orientation and shape, which influence the uniformity and multiplicity of slip. For crystals undergoing multiple slip, extensive tensile ductility occurs above $\approx 1700°C$, $\approx 0.65\ T_m$.

The crucial features involve: (1) The primary yield stress drops below that for preexistent flaw extension. (2) Sufficient cross slip, climb, or recovery occurs that slip bands broaden and local stress concentrations cannot nucleate extendable cracks. (3) Slip bands interpenetrate, giving more homogeneous slip with smaller cell sizes. (4) The glide stresses on primary, secondary, and even tertiary planes, e.g., {111}, become low enough that work-hardening can provide resistance to plastic instability, such as necking, without inducing cleavage. Thus, ductility is apparent at lower temperature with single or duplex slip, but uniform elongation is greater with multiple slip (e.g., 100–150%).[7] For general ductility, interpenetration of oblique slip bands is a key factor even in compression; it is difficult because of favorable reactions producing sessile dislocations.

In tension, these conditions obtain only at high enough temperatures that climb, polygonization, and dynamic recrystallization can occur as well as secondary slip.[7] With heating through the transition, the failure mechanism goes from cleavage, cleavage in necks, to ductile rupture, giving large reductions in area and involving intergranular failure at high strains and temperatures. Under some conditions, void formation is a precursor to failure by cleavage or intergranular rupture; presumably, the voids result from dislocation coalescence or boundary sliding stress concentrations.

For MgO polycrystals, the brittle-ductile transition stages are often reflected in the fracture stress-temperature curve. After the initial transition to primary slip-induced failure, σ_f decreases with the primary τ_y, Eq. (3), and is higher for finer grain sizes.[53] At temperatures where cross slip or recovery prevent primary slip-induced failure, the onset of secondary slip can apparently cause fracture. This sometimes causes an inflection in the strength-temperature curve, shown in Fig. 1, where σ_f rises to nearer the secondary slip yield stress than the primary one. At temperatures above that for the strength maximum, failure strains may become high, whereas, at and below it, total strain is very small. Failure often involves intergranular cleavage below the temperature for this maximum and more ductile intergranular separation above it.[67] Grain-boundary sliding and low boundary strengths may also contribute to the strength decreases above $\approx 1000°C$. For polycrystals made from powder, the observed transition temperatures have been much higher, $\approx 2200°C$, $0.8\ T_m$, or above, depending on purity and porosity, and achievable strains are usually small.[67]

A key factor is inducing secondary slip in adjacent grains to relieve stress concentrations from boundary sliding and blocked slip bands without causing fracture. The final transition temperatures are the same for single crystals and for polycrystals made by recrystallizing single crystals. This similarity and the void formation observed in both types of samples[67] suggest that crack-tip blunting from thermally activated dislocation emission perhaps aided by diffusion may be helping to inhibit cleavage failure. The considerable decrease in the transition temperature with reduced strain rate indicates that easier plasticity offsets easier crack nucleation and cleavage.

The conditions controlling strength and ductility are less well characterized for Al_2O_3. For sapphire with only basal slip occurring, a sharp transition to extensive tensile ductility occurs, apparently above the temperature where the strain rate-dependent yield stress drops below that for preexistent flaw extension.[68] This is typically near $1300°C$, $0.7\ T_m$. For sapphire oriented to preclude basal slip, tensile deformation obtains only above $\approx 1700°C$, $0.85\ T_m$, and extension is limited.[34,38] For c-axis sapphire, rapid failure occurs by cleavage[35] even at $1800°C$. The high glide stresses and high-temperature recovery causing stress saturation make Al_2O_3 more sensitive to preexistent flaws and reduce the amount of work-hardening available to inhibit necking. In compression, ductility is apparent at lower temperature, but microcracking, possibly resulting from twin-twin or twin-slip band interactions, is often evident. Extensive tensile ductility by general slip has not been demonstrated for polycrystals; in compression the transition has been observed infrequently. Limited microscopic evidence suggests that, at $\approx 1400°C$, crack or pore nucleation more often results from boundary sliding, although occasionally twins and perhaps slip bands cause microcracking.[3,69] Often polycrystalline ductility depends on diffusion, which can provide extensive, although strain-rate-sensitive deformation; under these conditions high stresses cause cleavage, usually on boundaries, and high local stresses are relieved by cracking, not slip.

The extent to which grain-boundary sliding, slip, twinning, or easier cleavage contribute to the strength decrease at intermediate temperatures (above $\approx 1000°C$) for polycrystals, or the strength minima (at $\approx 600°C$) in sapphire is unresolved. However, none of these factors should be discounted.

Impurities can significantly affect the low-temperature strength of MgO. Cation impurities can raise τ_y and so raise the fracture strength where initial yield controls the strength. In contrast, anions or entrapped gases apparently significantly weaken grain boundaries; they increase the degree of intergranular fracture and can decrease the low-temperature strength by a factor of 3 or more.[70,71]

At high temperature, extensive data for both materials show that impure materials are usually weaker and more susceptible to intergranular failure at high temperature.[6,67,71] Presumably this reflects reduced boundary strengths, easy crack nucleation, and the onset of boundary sliding at lower temperatures where accommodational lattice deformation is more difficult. Local silicate impurities significantly enhance crack-growth rates in Al_2O_3.[64] For MgO polycrystals made from powder, anion impurities may contribute to the higher brittle-ductile transition temperatures and low failure strains.[70] An exception occurs for Al_2O_3 with appreciable glassy phase; a narrow temperature range exists, near the glass-softening temperature, within which moderate strengthening obtains.[72] At higher temperatures, the creep and creep rupture resistance of such materials rapidly diminish.

Although various mechanisms contributing to the strength-temperature-grain size relations have been qualitatively demonstrated, few, if any, cases exist where all the features can be quantitatively explained. Below the temperatures for grain-boundary sliding, Eq. (3) seems to describe MgO samples with well-prepared surfaces and no large pores.[53,56,70] Occasionally, machined samples seem to satisfy Eq. (2) with $C \approx G$. More often, ground samples are better described phenomenologically by Eq. (3), with the constant term inversely dependent on grit size. Presumably, residual stresses from machining and thermal expansion mismatch, which drive small cracks, and the increase in toughness with increasing C/G cause the inverse grain size dependence; however, the effects of both factors need to be better quantified. Internal processing flaws could also give an inverse grain-size dependence for strength; in these cases the stress concentration factors are often uncertain. With most data too many variables, such as grain size, porosity, surface damage, and chemistry, have been changed simultaneously to allow analysis. In addition, characterization of corrosion-induced crack growth and the temperature dependence thereof is inadequate. Generally the behavior is better understood at low than at high temperature. The yield and fracture strengths of single crystals often depend sensitively on surface damage, which provides dislocation sources as well as extendable cracks.

It is often uncertain at what crack size the critical load occurs. Where it occurs after appreciable extension or during microcrack coalescence, interpretations of fractographs or estimates of controlling fracture energies are ambiguous. Many such ambiguities exist, particularly for Al_2O_3. For high-strength materials, criticality may occur early during propagation, with the lower toughnesses controlling. Higher polycrystalline values should control crack arrest and the strength after damage, e.g., after contact damage or thermal shock. Analytical, statistical, and experimental techniques are needed for situations where criticality develops while several cracks are interacting, but have not coalesced. These issues and the effect of R-curve behavior on lifetime predictions and multiaxial stress effects[73] are among the outstanding problems for design and property optimization.

In compression, strengths are much higher, and failure may involve microcrack coalescence. Sometimes extensive damage occurs in shear bands. Strength-temperature curves can have a local minimum and maximum, e.g., Lankford's[74] data for Al_2O_3, suggesting that the onsets of primary and of general slip cause effects similar to those in tension. These inflections occur at lower temperatures, however. For Al_2O_3, twinning causes microcracking at low temperatures, which may be a factor controlling compressive strengths.[74]

Microcrack and pore nucleation are often neglected theoretically but can control mechanical behavior, especially of dense materials. Nucleation-limited situations include dense Al_2O_3-ZrO_2 alloys in which extremely high local stresses, $\sigma/E \approx 0.05$, can exist around fine particles without causing microcracks,[75] microcrack formation in dense, anisotropic polycrystals, and pore nucleation during creep and creep crack growth. If nucleation is thermally activated, then near-equilibrium surface energies may be controlling, which could be much lower than those for subsequent crack extension. For both oxides, the dihedral angles found in pores tend to be much smaller than those on polished and annealed surfaces.[49] This implies that pore or crack nucleation is much easier at particular boundaries or junctions (with low γ_s and high γ_b) than would be anticipated based on average surface and boundary energies. Preexistent or diffusionally delivered segregant can further facilitate nucleation.

Porosity generally reduces the modulus, toughness, strength, and creep resistance because of the reduced cross-sectional area and internal stress concentrations; correlations have been reviewed by Rice.[71] However, several observations indicate that small amounts of intragranular porosity in MgO can raise the fracture toughness, lower the temperature of observable yield, especially in compression, and raise the low-temperature fracture strength, presumably by acting as dislocation sources. Porous materials can creep very rapidly, in part owing to easier grain-boundary sliding. Sometimes creep appears to be controlled by primary slip, which is presumably accompanied by pore or crack growth. Ultimately, the best tensile ductility obtains with pore-free materials.

Three notable results with Al_2O_3-ZrO_2 or Al_2O_3-HfO_2 alloys indicate that significant property improvements can accrue with multiphase materials. The low-temperature strengths and toughness achieved with the materials fabricated from powder are far superior to those achieved with any bulk single-phase ceramics.[75] These improvements can accrue from transformation and/or microcrack-toughening. The high-temperature strengths achieved with the eutectically grown samples,[76] ≈ 600 MPa at 1580°C, are among the highest recorded for any material above 1500°C. Also, these eutectic materials sometimes have extraordinary low-temperature toughnesses,[76] $\gamma \approx 1000$ J/m^2, owing to severe crack deflection.

Concluding Remarks

Major developments in brittle fracture have occurred in several areas. (1) Fracture mechanics has been utilized for materials improvement: Numerous sources of preexistent flaws have been identified; tougher materials have been developed; and the effects of internal stresses on crack extension have been appreciated and are being quantified. (2) The role of dislocations has been clarified. Slip can independently affect initiation and propagation. In addition, toughening by shielding from slip near a sharp crack and from crack-tip blunting have been distinguished; at low temperatures, cracks are considered to be atomically sharp with any toughening resulting from shielding. (3) The realization that the macroscopic toughness for polycrystals can depend on crack size and extension history, as well as grain size, has opened the way to resolving outstanding ambiguities regarding the flaw sizes and fracture energies controlling criticality and microstructure effects on subcritical crack growth. (4) Finally, recent data and theoretical calculations confirm that cleavage is a complex process, except possibly for close-packed, naturally neutral planes; cleavage energies usually exceed surface energies, and thermally activated crack growth can occur.

The necessary slip systems have been identified and flow stresses ranked for MgO and at high temperature for Al_2O_3. Several factors that depend on the structural cation vacancies and trigonal structure may provide a basis for understanding the high yield stresses for Al_2O_3. Several large Burgers vectors are necessary. Thus, Peierls stresses may be high, and noncolinear dissociation, decomposition out of the glide plane, and climb dissociation occur, which may make glide and cross slip difficult even at high temperature. Solute hardening for primary slip in MgO is partially understood, but the exceptional high-temperature hardening for Al_2O_3 is not well explained and may depend on core structures. The discrepancies about the low-temperature slip systems must be resolved and twinning better understood before fracture of Al_2O_3 at high stresses and intermediate temperatures can be understood.

Several important ingredients in the brittle-ductile transitions have been established. Both primary and secondary slip and slip band interpenetration must occur without inducing fracture from stress concentrations at blocked slip bands, and often must relieve stress concentrations from grain-boundary sliding. An effect of crack blunting is suggested but has not been established. The low cleavage energy for {001} in MgO and certain boundaries with particularly low crack nucleation and cleavage energies contribute to the inherent tendency toward brittleness for these oxides. Impurities, including anions, are a major cause of low strength at all temperatures, and poor ductility.

Diffusional creep has been established as important and may be the best-understood deformation mechanism, although theoretical uncertainties remain. Impurities have substantial effects on the creep rate, which are partially understood when lattice diffusion is controlling. The difficulties of interpreting diffusional-creep data when more than one mechanism may be controlling are formidable, especially when impurity-dependent diffusivities and interface-controlled creep are possible. Nevertheless, the boundary diffusivities deduced from such studies are more plausible than existent tracer data for these materials, and the lattice diffusivities have given important insights.

As most important mechanical behavior has been phenomenologically characterized or modeled, a basis exists for more-refined investigation of specific mechanisms and engineering properties, and for more-intense investigation of multiphase materials. Work on defect structures, impurity solubilities, interface structures, and diffusivities in bulk and along interfaces and dislocations is a necessary complement to mechanical-property studies. Modeling of glide and creep improves as more is known about solubilities and defect structures, particularly for complexes. Grain-boundary processes, including point-defect creation, sliding, fracture, and diffusion, and the impurity effects on them are not understood. Investigation is needed for boundaries and surfaces of all orientations, not just low-energy, equilibrium ones. Anion diffusivities and anion impurity-induced defects are poorly understood; however, anions can significantly weaken boundaries and affect controlling diffusivities.

Magnesia and/or Al_2O_3 provide models for different classes of materials. Diffusivities for them are approximately similar or higher for MgO at any given temperature, despite the higher melting point of MgO. Magnesia is appreciably softer and dislocation motion can affect mechanical properties at virtually all temperatures. MgO may be a good model for other softer, cubic materials such as halides and higher molecular weight compound semiconductors. It is uncertain whether the particularly low cleavage energies for rock salt materials make MgO more brittle or whether the greater dislocation glide resistance of covalent materials make them more brittle. Slip is so difficult for Al_2O_3 that many properties of very-fine-grained polycrystals may be unaffected by dislocation motion, except for compression and indentation damage. It may provide a useful model for noncubic ceramics, including nonoxides such as SiC or Si_3N_4. The thermal expansion and elastic anisotropy for Al_2O_3 are small in comparison with many noncubic oxides. However, the microcracking effects from thermal expansion anisotropy seem to be similar, but the "critical" grain sizes for peak toughness or spontaneous cracking shift with expansion anisotropy. For harder cubic ceramics such as ZrO_2, SiC, and mixed oxides, such as spinel, where slip can be extremely difficult, behavior has similarities to both MgO and Al_2O_3. Comparison of H/E may indicate which of Al_2O_3 ($H/E = 0.06$) or MgO ($H/E = 0.02$) would be the better model for behavior that depends on the ease of slip. Finally, neither of these two illustrates the interesting effects of intrinsic nonstoichiometry.

Acknowledgments

Gratitude is expressed to W. D. Kingery for encouragement, to the Office of Naval Research, under contract N00014–81–K–0362, for financial support, to A. H. Heuer and A. G. Evans for critical comments, and to A. M. Glaeser for review of the manuscript. S. M. Wiederhorn, B. J. Dalgleish, B. J. Hockey, P. F. Becher, and P. W. Tasker are gratefully acknowledged for helpful discussions and for early release of unpublished results or for consultation of old records to confirm data.

References

[1] A. E. Gorum, E. R. Parker, and J. A. Pask, "Effect of Surface Conditions on Room-Temperature Ductility of Ionic Crystals," *J. Am. Ceram. Soc.*, **41** [5] 161–64 (1958).

[2] M. L. Kronberg, "Plastic Deformation of Single Crystals of Sapphire: Basal Slip and Twinning," *Acta. Metall.*, **5** [9] 507–24 (1957).

[3] A. H. Heuer, N. J. Tighe, and R. M. Cannon, "Plastic Deformation of Fine-Grained Alumina (α-Al_2O_3): II, Basal Slip and Nonaccommodated Grain-Boundary Sliding," *J. Am. Ceram. Soc.*, **63** [1–2] 53–58 (1980).

[4] J. Castaing, J. Cadoz, and S. H. Kirby, "Deformation of Al_2O_3 Single Crystals Between 25°C and 1800°C: Basal and Prismatic Slip," *J. de Phys. Coll.*, *C-3*, **42** [6] 43–47 (1981).

[5] C. O. Hulse, "Plastic Anisotropy of MgO and Other NaCl-Type Crystals"; pp. 307–19 in Anisotropy in Single-Crystal Refractory Compounds, Vol. 2. Edited by F. W. Vahldiek and S. A. Mersol. Plenum, New York, 1968.

[6] S. M. Copley and J. A. Pask, "Deformation of Polycrystalline MgO at Elevated Temperatures," *J. Am. Ceram. Soc.*, **48** [12] 636–42 (1965).

[7] a. R. B. Day and R. J. Stokes, "Mechanical Behavior of Magnesium Oxide at High Temperatures," *J. Am. Ceram. Soc.*, **47** [10] 493–503 (1964).
b. R. B. Day and R. J. Stokes, "Effect of Crystal Orientation on the Mechanical Properties of Magnesium-Oxide at High Temperatures," *J. Am. Ceram. Soc.*, **49** [2] 72–80 (1966).

[8] B. Reppich, "Solid Solution Hardening in Iron Doped MgO," *Mater. Sci. Eng.*, **22** [1] 71–84 (1976).

[9] A. H. Heuer, "Deformation Twinning in Corundum," *Philos. Mag.*, **13** [122] 379–93 (1966).

[10] W. D. Scott and K. K. Orr, "Rhombohedral Twinning in Alumina," *J. Am. Ceram. Soc.*, **66** [1] 27–32 (1983).

[11] H. Conrad, "Mechanical Behavior of Sapphire," *J. Am. Ceram. Soc.*, **48** [4] 195–201 (1965).

[12] S. M. Wiederhorn, "Fracture of Sapphire," *J. Am. Ceram. Soc.*, **52** [9] 485–91 (1969).
[13] a. S. M. Wiederhorn, B. J. Hockey, and D. E. Roberts, "Effect of Temperature on the Fracture of Sapphire," *Philos. Mag.*, **28** [4] 783–96 (1973).
b. S. M. Wiederhorn, private communication (1984); confirmed that the K_c-T data in Ref. 13(a) were for planes near {10$\bar{1}$0}, National Bureau of Standards.
[14] P. F. Becher, "Fracture-Strength Anisotropy of Sapphire," *J. Am. Ceram. Soc.*, **59** [1–2] 59–61 (1976).
[15] M. Iwasa and R. C. Bradt, "The Fracture Toughness of Single Crystal Alumina"; this proceedings, pp. 767–79.
[16] J. J. Gilman, "Direct Measurements of the Surface Energies of Crystals," *J. Appl. Phys.*, **31** [12] 2208–18 (1960).
[17] P. L. Gutshall and G. E. Gross, "Cleavage Energies of NaCl and MgO in Vacuum," *J. Appl. Phys.*, **36** [8] 2459–60 (1965).
[18] A. R. C. Westwood and D. L. Goldheim, "Cleavage Surface Energy of {100} Magnesium Oxide," *J. Appl. Phys.*, **34** [11] 3335–39 (1963).
[19] D. A. Shockey and G. W. Groves, "Origin of Water-Induced Toughening in MgO Crystals," *J. Am. Ceram. Soc.*, **52** [2] 82–85 (1969).
[20] P. W. Tasker, "Basal Plane Surface Energies of Al_2O_3" private communication (1982); to be published.
[21] E. A. Colbourn, W. C. Mackrodt, and P. W. Tasker, "The Segregation of Calcium Ions At the Surface of Magnesium Oxide: Theory and Calculation," unpublished work.
[22] D. M. Duffy and P. W. Tasker, "The Properties of Grain Boundaries in Rocksalt Structured Oxides"; this proceedings, pp. 275–89.
[23] S. M. Wiederhorn, "Crack Propagation in Polycrystalline Ceramics"; pp. 317–38 in Ultrafine-Grain Ceramics. Edited by J. J Burke, N. L. Reed, and V. Weiss. Syracuse University Press, 1970.
[24] A. Kelly, Strong Solids, Clarendon Press, Oxford, 1966.
[25] a. B. Reppich, "Strengthening Mechanisms in MgO Containing Coherent Stress-Free Precipitation Particles—I. Theory," *Acta Metall.*, **23** [9] 1055–60 (1975).
b. H. Knoch and B. Reppich, "Strengthening Mechanisms in MgO Containing Coherent Stress-Free Precipitation Particles—II. Experiments and Comparison with Theory," *Acta Metall.*, **23** [9] 1061–68 (1975).
[26] R. J. Stokes, "Thermal-Mechanical History and the Strength of Magnesium Oxide Single Crystals: I, Mechanical Tests," *J. Am. Ceram. Soc.*, **48** [2] 60–67 (1965).
[27] T. E. Mitchell and A. H. Heuer, "Solution Hardening by Aliovalent Cations in Ionic Crystals," *Mater. Sci. Eng.*, **28** [1] 81–97 (1977).
[28] W. Skrotzki and P. Haasen, "Hardening Mechanisms of Ionic Crystals on {110} and {100} Slip Planes," *J. de Phys. Coll.*, C-3, **42** [6] 119–48 (1981).
[29] J. D. Snow and A. H. Heuer, "Slip Systems in Al_2O_3," *J. Am. Ceram. Soc.*, **56** [3] 153–57 (1973).
[30] B. J. Hockey, "Pyramidal Slip on {11$\bar{2}$3}⟨$\bar{1}$100⟩ and Basal Twinning in Al_2O_3"; pp. 167–79 in Deformation of Ceramic Materials. Edited by R. C. Bradt and R. E. Tressler. Plenum, New York, 1975.
[31] J. Castaing, J. Cadoz, and S. H. Kirby, "Prismatic Slip of Al_2O_3 Single Crystals Below 1000°C in Compression Under Hydrostatic Pressure," *J. Am. Ceram. Soc.*, **64** [9] 504–11 (1981).
[32] B. J. Hockey, "Plastic Deformation of Aluminum Oxide by Indentation and Abrasion," *J. Am. Ceram. Soc.*, **54** [5] 223–31 (1971).
[33] K. C. Radford and P. L. Pratt, "The Mechanical Properties of Impurity-Doped Alumina Single Crystals," *Proc. Br. Ceram. Soc.*, **15**, 185–202 (1970).
[34] R. E. Tressler and D. J. Michael, "Dynamics of Flow of C-Axis Sapphire"; pp. 195–215 in Deformation of Ceramic Materials. Edited by R. C. Bradt and R. E. Tressler. Plenum, New York, 1975.
[35] R. Firestone and A. H. Heuer, "Creep Deformation of 0° Sapphire," *J. Am. Ceram. Soc.*, **59** [1–2] 24–29; [3–4] 184 (1976).
[36] A. H. Heuer and J. Castaing, "Dislocations in Al_2O_3"; this proceedings, pp. 238–57.
[37] B. J. Pletka, T. E. Mitchell, and A. H. Heuer, "Solid Solution Hardening of Sapphire (α-Al_2O_3)," *Phys. Status Solidi, Sect. A*, **39** [1] 301–11 (1977).
[38] D. M. Kotchick and R. E. Tressler, "Deformation Behavior of Sapphire via the Prismatic Slip System," *J. Am. Ceram. Soc.*, **63** [7–8] 429–34 (1980).
[39] R. C. Folweiler, "Creep Behavior of Pore-Free Polycrystalline Aluminum Oxide," *J. Appl. Phys.*, **32** [5] 773–78 (1961).
[40] R. M. Cannon, W. H. Rhodes, and A. H. Heuer, "Plastic Deformation of Fine-Grained Alumina (Al_2O_3): I, Interface-Controlled Diffusional Creep," *J. Am. Ceram. Soc.*, **63** [1–2] 46–53 (1980).
[41] R. T. Tremper, R. A. Giddings, J. D. Hodge, and R. S. Gordon, "Creep of Polycrystalline MgO-FeO-Fe_2O_3 Solid Solutions," *J. Am. Ceram. Soc.*, **57** [10] 421–28 (1974).
[42] P. A. Lessing and R. S. Gordon, "Creep of Polycrystalline Alumina, Pure and Doped with Transition Metal Impurities," *J. Mater. Sci.*, **12** [11] 2291–2302 (1977).

[43] R. M. Cannon and R. L. Coble, "Review of Diffusional Creep of Al_2O_3"; pp. 61–100 in Deformation of Ceramic Materials. Edited by R. C. Bradt and R. E. Tressler. Plenum, New York, 1975.

[44] Y. Ikuma and R. S. Gordon, "Effect of Doping Simultaneously with Iron and Titanium on the Diffusional Creep of Polycrystalline Al_2O_3," *J. Am. Ceram. Soc.*, **66** [2] 139–47 (1983).

[45] R. S. Gordon, "An Understanding of Defect Structure and Mass Transport in Polycrystalline Al_2O_3 and MgO via the Study of Diffusional Creep"; this proceedings, pp. 418–37.

[46] W. C. Mackrodt, "Calculated Point Defect Formation, Association, and Migration Energies in MgO and α-Al_2O_3"; this proceedings, pp. 62–78.

[47] J. D. Hodge and R. S. Gordon, "Grain Growth and Creep in Polycrystalline Magnesium Oxide Fabricated with and without a LiF Additive," *Ceramurgia Int.*, **4** [1] 17–20 (1978).

[48] A. J. Mountvala and G. T. Murray, "The Role of the Grain Boundary in the Elevated Temperature Fracture Behavior of Magnesia," *Philos. Mag.*, **13** [123] 441–53 (1966).

[49] C. A. Handwerker, R. M. Cannon, and R. L. Coble, "Final-Stage Sintering of MgO"; this proceedings, pp. 619–43.

[50] P. W. Tasker and D. M. Duffy, "On the Structure of Twist Grain Boundaries in Ionic Oxides," *Philos. Mag. A*, **47** [6] L45–L48 (1983).

[51] B. R. Lawn, B. J. Hockey, and S. M. Wiederhorn, "Atomically Sharp Cracks in Brittle Solids: An Electron Microscopy Study," *J. Mater. Sci.*, **15** [5] 1207–23 (1980).

[52] B. J. Hockey, "Observations on Crack Tips in MgO" private communication (1984); to be published.

[53] A. G. Evans, D. Gilling, and R. W. Davidge, "The Temperature-Dependence of the Strength of Polycrystalline MgO," *J. Mater. Sci.*, **5** [3] 187–97 (1970).

[54] A. G. Evans, "High-Temperature Slow Crack Growth in Ceramic Materials"; pp. 373–96 in Ceramics for High-Performance Applications. Edited by J. J. Burke, A. E. Gorum, and R. N. Katz. Brook Hill, Chestnut Hill, Mass., 1974.

[55] P. L. Gutshall and G. E. Gross, "Observations and Mechanisms of Fracture in Polycrystalline Alumina," *Eng. Fract. Mech.*, **1** [3] 463–71 (1969).

[56] A. G. Evans and R. W. Davidge, "The Strength and Fracture of Fully Dense Polycrystalline Magnesium Oxide," *Philos. Mag.*, **20** [164] 373–88 (1969).

[57] B. J. Dalgleish, P. L. Pratt, R. D. Rawlings, and A. Fakhr, "The Fracture Toughness Testing of Ceramics and Acoustic Emission," *Mater. Sci. Eng.*, **45** [1] 9–20 (1980).

[58] H. Hübner and R. W. Jillek, "Sub-Critical Crack Extension and Crack Resistance in Polycrystalline Alumina," *J. Mater. Sci.*, **12** [1] 117–25 (1977).

[59] R. Steinbrech, R. Khehans, and W. Schaarwächter, "Increase of Crack Resistance during Slow Crack Growth in Al_2O_3 Bend Specimens," *J. Mater. Sci.*, **18** [1] 265–70 (1983).

[60] R. W. Rice, S. W. Freiman, and J. J. Mecholsky, Jr., "The Dependence of Strength-Controlling Fracture Energy on the Flaw-Size to Grain-Size Ratio," *J. Am. Ceram. Soc.*, **63** [3–4] 129–36 (1980).

[61] A. G. Evans and Y. Fu, "The Mechanical Behavior of Alumina: A Model System"; this proceedings, pp. 697–719.

[62] R. W. Rice, S. W. Freiman, and P. F. Becher, "Grain-Size Dependence of Fracture Energy in Ceramics: I, Experiment," *J. Am. Ceram. Soc.*, **64** [6] 345–50 (1981).

[63] W. D. Kingery, "Role of Surface Energies and Wetting in Metal-Ceramic Sealing," *Am. Ceram. Soc. Bull.*, **35** [3] 108–12 (1956).

[64] A. G. Evans and W. R. Blumenthal, "High Temperature Failure Mechanisms in Ceramic Polycrystals"; Deformation of Ceramic Materials, Vol. 2. Edited by R. C. Bradt and R. E. Tressler. Plenum, New York; in press.

[65] R. J. Stokes and C. H. Li, "Dislocations and the Tensile Strength of Magnesium Oxide," *J. Am. Ceram. Soc.*, **46** [9] 423–34 (1963).

[66] R. C. Ku and T. L. Johnston, "Fracture Strength of MgO Bicrystals," *Philos. Mag.*, **9** [98] 231–47 (1964).

[67] R. B. Day and R. J. Stokes, "Mechanical Behavior of Polycrystalline Magnesium Oxide at High Temperatures," *J. Am. Ceram. Soc.*, **49** [7] 345–59 (1966).

[68] M. L. Kronberg, "Dynamical Flow Properties of Single Crystals of Sapphire: I," *J. Am. Ceram. Soc.*, **45** [6] 274–79 (1962).

[69] P. F. Becher, "Deformation Substructure in Polycrystalline Alumina," *J. Mater. Sci.*, **6** [4] 275–80 (1971).

[70] R. W. Rice, "Strength and Fracture of Hot-Pressed MgO," *Proc. Br. Ceram. Soc.*, **20**, 329–63 (1972).

[71] R. W. Rice, "Microstructure Dependence of Mechanical Behavior of Ceramics"; pp. 199–381 in Treatise on Materials Science and Technology, Vol. II, Properties and Microstructure. Academic Press, New York, 1977.

[72] R. W. Davidge and G. Tappin, "The Effects of Temperature and Environment on the Strength of Two Polycrystalline Aluminas," *Proc. Br. Ceram. Soc.*, **15**, 47–60 (1970).

[73] B. J. Pletka and S. M. Wiederhorn, "A Comparison of Failure Predictions by Strength and Fracture Mechanics Techniques," *J. Mater. Sci.*, **17** [5] 1247–68 (1982).

[74] J. Lankford, "Temperature-Strain Rate Dependance of Compressive Strength and Damage Mechanisms in Aluminum Oxide," *J. Mater. Sci.*, **16** [6] 1567–78 (1981).

[75]N. Claussen and M. Ruhle, "Design of Transformation-Toughened Ceramics"; pp. 137–63 in Advances In Ceramics, Vol. 3. Edited by A. H. Heuer and L. W. Hobbs. The American Ceramic Society, Columbus, Ohio, 1981.
[76]C. O. Hulse, "Mechanical Properties of Al_2O_3-HfO_2 Eutectic Microstructures"; pp. 903–12 in Fracture Mechanics of Ceramics, Vol. 4. Edited by R. C. Bradt, D. P. H. Hasselman, and F. F. Lange. Plenum, New York, 1978.

---

*The melting points, T_m, are 3125 K for MgO and 2313 K for Al_2O_3.
†If bulk specimen deformation occurs, artificially low γ_c, even below γ_s, may result unless a plastic analysis, such as for J integrals, is used.

Sintering and Grain Growth in Alumina and Magnesia

R. L. Coble and H. Song

Massachusetts Institute of Technology
Cambridge, MA   02139

R. J. Brook

University of Leeds
Department of Ceramics
Leeds, U.K.

C. A. Handwerker

National Bureau of Standards
Gaithersburg, MD   20878

J. M. Dynys

IBM Corporation
Poughkeepsie, NY   12603

The purpose of this conference was to review the sintering behaviors of alumina and magnesia, the "model" ceramics research materials, through studies of which our qualitative and semiquantitative understanding of the sintering behavior of nonmetals has been best developed. Research on these materials was intensified as a result of the finding of magnesia as an additive in alumina to achieve theoretical density by sintering. Subsequent studies have been conducted to identify why magnesia works as a sintering aid in alumina and to determine how to extend those results to other systems.[1]

Dense alumina and magnesia products are important to many modern technologies. Because of the multiple applications of alumina for high-temperature insulation and the availability of high-purity alumina ceramics and sapphires for research, there have been extensive studies of the basic properties, such as defect structure, electrical properties, tracer diffusion coefficients, and creep and fracture behaviors, and of the processing to support the requirements of developing technologies. Magnesia has been a model material in the laboratory due to its simple crystallographic structure and the availability of high-purity single crystals. In general, the sintering studies of magnesia are not applicable to the fabrication of commercial magnesia products, such as magnesia refractories which contain substantial contents of impurities and, frequently, liquid phases.

The paradigms for powder processing and sintering of ceramics have evolved from the empirical sintering studies in alumina, magnesia, and other materials and their interpretation in terms of the first-order modeling of sintering and grain growth. The paradigms, as reviewed in 1978,[2] are presented in Table I. Other important reviews by Brook[3] on the effects of the relative rates of densification and coarsening and of pore mobility and boundary mobility, and by Yan et al.[4] on grain

Table I.   Paradigms for Powder Processing of Compounds (Ref. 2)

	Theory	Practice
Thermodynamic stability	X	Cr_2O_3, ZnO, NaCl
Small particle size	X	General
Narrow particle-size distribution	X	Al_2O_3, ferrites
Uniform high-density compacts	X	General
Narrow pore-size distribution	X	$ZrO_2:Y_2O_3$
Grain-growth-controlling additive	X	$Al_2O_3:MgO$ + many others
Soluble, diffusible atmospheres	X	Al_2O_3 and others
New results, not emphasized or represented there		
Relative rates (Ref. 3)		
Densification:coarsening	X	
Pore mobility:boundary mobility	X	
Possible change to interface control (sol-gel)	X	Use of smaller particle sizes: sol-gel
Pore growth to pore size maximum (limits pore mobility)	X	
Dihedral angle effects and faceting	X	

growth in ceramics provide important modeling and sources to earlier literature. Our purpose here is to bring the status of understanding to date for Al_2O_3 and MgO and to briefly review the results since the paradigms review in 1978 to focus attention on research needed in the near future.

A question posed for the conferees was whether continued efforts were justified on alumina and magnesia as paradigms. To understand sintering and grain growth we need: (1) thorough *characterization* of the powders, compacts, and microstructure evolution; (2) *models* that are suitable for those changes; and (3) *data* for all interfacial energies and transport coefficients for a material under study/development to test the applicability of the models. The status is: (1) Characterization may be adequate for powders but is incomplete for compacts or the microstructure evolution, (2) the models are *idealized*, and (3) the data base is *inadequate*.

The theories for solid-state sintering, hot-pressing, reactive sintering, liquid-phase sintering, and associated grain growth, breakaway grain growth, and the isolation of pores from grain boundaries, are far from complete. We cannot predict the time dependence of densification for a compact regardless of how well the powders and powder compacts are characterized. The particle-size distributions and variable shapes for real powders deviate from the models derived for single-size isotropic spheres. To determine how to improve the models, a material is needed in which all of the interfacial energies and transport coefficients are known. Thus, it is the data base on alumina and magnesia that, in part, warrants continuation of their study. However, the sintering behavior of the covalently bonded materials is sufficiently different that work on them is required as well. Silicon could serve as a "model" covalent material with silicon carbide as the analog compound.

If we were to select a material solely on the requirement for a data base, the best overall choice would probably be gold, and the choice among the ceramics is probably uranium dioxide. However, the sinterability of magnesia to high density

Table II. Materials Emphasized in Sintering and Grain Growth Studies

Realm	Model	Metallic	Ionic	Covalent
Sintering glasses	Many	?	?	SiO_2
Single-phase solid	Au	Fe/Cu/Ni	Al_2O_3	Si/SiC
	Alkali halide		MgO	
			UO_2	
Liquid-phase sintering	W:Ni/Mo:Ni	WC:Co	$MgO:SiO_2$	Si_3N_4:X
		Fe:Cu	Refractories	
			Al_2O_3	
			Insulators	
Reactive sintering	Mo:Ni	Fe:Ni	Porcelains	Sialon
		Fe:Cu	PZT	
Hot-pressing	?	Ni-alloys	Al_2O_3	Si_3N_4

and of alumina to theoretical density allows study of these two materials at all stages of the sintering process. Thus, we conclude that further work is warranted on alumina and magnesia but, as noted in this conference, the data base could be vastly improved.

Liquid-phase sintering and associated (breakaway) grain-growth studies have been focused on the heavy metal alloys because of their applicability and value. Alumina with talc and clay as additives, magnesia with silica as an impurity, and silicon nitride with the various dopants that form silicate liquids also warrant further research. There are also numerous instances in which small amounts of liquid are present in nominally single-phase systems. The need, then, to understand the influence of the composition and contents of liquids due to dopants or uncontrolled impurities on the sintering and grain-growth behavior is obvious. Several model systems and systems which have been applied to practice are summarized in Table II.

Characteristics of Powder Compacts and the Phenomena of Sintering

Powder compacts consist of assemblages of powder particles that are generally irregularly shaped and consist of a distribution of particle (crystal) sizes. In some compacts, tens to thousands of individual crystals are grouped together into agglomerates in which the packed density is higher than the average density for the compacts (50–60%). The pore spaces formed among the imperfectly fitting particles and groups of particles typically have a bimodal or multimodal distribution of sizes. The "structure" of this pair of interpenetrating continuous networks could probably be "defined" on the basis of either of them, but because of the complex geometry, our measurements and topological descriptions are incomplete; partial characterization includes features of both based on conveniently measurable parameters. Complete characterization is intended to be implied by the descriptions in Table III, but the topology is generally ignored. It is recognized that from a known size distribution of particles there are limits to achievable coordination numbers and average densities, but there are simply too many possible local combinations and permutations to consider (and the results depend on fabrication procedures). Hence, only the average values of *some* of these characteristics have been commonly measured; i.e., those that are dependent on the accessible pa-

Table III. Characteristics of Powder Compacts

Load-Bearing Network	
Crystal density (theoretical)	ρ_t
Crystal shapes	"spheres": radius a
Crystal particle-size distribution	$g(PS)$ and $\overline{PS}$
Surface area	$\underline{g}(6/(\rho_t \cdot PS))$
Agglomerate size	d
Agglomerate density	ρ_a
Coordination number of particles	$g(CN)$ and $\overline{CN}$
Neck size between particles	$g(x)$ and $\bar{x}$
Relative density of compact	
Initial value	ρ_0
Final value	ρ_f
Grains	
Size distribution	$g(G)$ and $\overline{G}$
Shapes	equiaxed/tabular
Porosity	
	$1 - \rho$
Size distribution	$g(r)$ and $\bar{r}$
Connectivity/shape	
Open: branches and nodes	Connected "cylinders"
Grain CN/node	
Closed	Isolated "spheres"
Dihedral angles	$g(\psi)$ and $\bar{\psi}$
Pore locations	
On/intersected by GB's	
Number of surrounding grains	
Isolated from GB's	
Composition	
Segregation at grain boundaries, surfaces	
Second solid-phase particles	
Wetting angles	

rameters from quantitative metallography. Until recently, the particle size, compact density (ρ_0), shrinkage, final density (ρ_f), grain size ($\overline{G}$), and product (im)permeability have received the most attention. In addition, attention has now been focused on the pore size and pore-size distribution, and on the dihedral angles of pore:boundary junctions. The fact that characterization has been incomplete originates with the small sizes of important particles and pores. Quantitative metallography of the pores and their shapes (dihedral angles) is quite difficult. The growth of pores was deduced by mercury porosimetry[5] rather than by metallography. Surface area reduction (by BET) is another more easily conducted measurement from which pore growth might be inferred. The phenomena which generally occur during sintering are summarized in Table IV; and Table V lists the atom transport mechanisms.

Models

Until about a decade ago, much of the "model" work on sintering involved measurements of the neck growth between spheres and/or the shrinkage during the

Table IV. The Phenomena of Sintering in Single, Solid-Phase Systems

Neck growth/surface area reduction	
Shrinkage/densification	
$L_0 \to L \to L_f \qquad \rho_0 \to \rho \to \rho_f$	
Isotropic shrinkage	
$\rho = \dfrac{m_1}{L_3} \qquad \rho L^3 = \rho_0 L_0^3 = \rho_f L_f^3$	
Pore growth	$g(r,t)$
Pore continuity change:	Rayleigh instability
Grain growth:	
Normal	$\overline{G_m} = g(t)$
Discontinuous	$G_s = g\left(\dfrac{1}{G_m}t\right)$
Breakaway from pores	$M_b > M_p$
Breakaway and reattachment	

Table V. Atom Transport Mechanisms of Sintering

Neck growth	$D_s \; D_b \; D_l \; P^0$
Shrinkage	$D_b \; D_l$
Pore growth	
Ostwald ripening	$P^0 \; D_s \; D_l \; D_b$
Drag and coalescence	
Pore motion	$D_s \; D_l \; P^0$
Mobility M_p	$g(D_s/r^4), \; (D_l/r^3), \; P^0/r^2$
Grain-boundary motion	$D^*, \; D_b$
Mobility M_b	

initial stage of sintering as a function of the initial particle size, compact density, temperature, and time. The time dependence was then frequently interpreted in terms of mechanisms of atom transport based on the initial-stage sintering models that had been developed starting with Kuczynski[6] but have not ended yet. A critical review of the individual models for the initial stage was reported by Coblenz et al.[7] A significant paper which increased our understanding of the sintering process was by Johnson,[8] who synthesized neck growth and shrinkage curves using the independent models for independent atomic transport paths (mechanisms) in an additive way. The notable result shown was that approximately any time dependence for shrinkage could be generated from the combinations of different mechanisms. The time dependence of shrinkage would be different than that which would be predicted by a shrinkage model alone and the activation energy for shrinkage would also be different with *mixed mechanisms* of sintering than that assumed for the atom transport mechanism responsible for shrinkage. In spite of that finding, papers are still generated with attempts to interpret the time dependence of shrinkage, for example, in terms of a controlling mechanism, or atom transport path.

    There are many of us who believe that, for the technological materials being processed or studied, there are few instances in which a single mechanism of atom

Table VI. Initial-Stage Sintering Models from Critical Review (Ref. 7)

Mechanism (Source → Sink)	Geometry Assumption (Neck radius)	Neck Growth Rate
Evaporation:condensation (total sphere surface → neck)	$\rho \approx \dfrac{x^2}{2a}$	$\left(\dfrac{\dot{x}}{a}\right)_{\varepsilon-c} = \dfrac{\alpha P^\circ \gamma \Omega^2}{\sqrt{2\pi m}\,(RT)^{3/2}} \left(\dfrac{1}{a\rho}\right)$
Surface diffusion (near neck sphere surface → neck)	$\rho \approx 0.58 \left(\dfrac{x}{a}\right)^{2/3} x$	$\left(\dfrac{\dot{x}}{a}\right)_s = \dfrac{0.811 \delta D_s \gamma \Omega}{RT} \left(\dfrac{1}{a\rho^3}\right)$
Grain-boundary diffusion (grain boundary → neck)	$\rho \approx \dfrac{x^2}{4a}$	$\left(\dfrac{\dot{x}}{a}\right)_{G.B.} = \dfrac{8 W D_b \gamma \Omega}{RT} \left(\dfrac{1}{\rho x^3}\right)$
Lattice diffusion (grain boundary → neck)	$\rho \approx \dfrac{x^2}{4a}$	$\left(\dfrac{\dot{x}}{a}\right)_{l-s} = \dfrac{6 D_l \gamma \Omega}{RT} \left(\dfrac{1}{\rho x^2}\right)$
Lattice diffusion (near neck sphere surface → neck)	$\rho \approx 0.58 \left(\dfrac{x}{a}\right)^{2/3} x$	$\left(\dfrac{\dot{x}}{a}\right)_{l-c} = \dfrac{2 D_l \gamma \Omega}{RT} \left(\dfrac{1}{a\rho^2}\right)$

transport is overwhelming. Most materials will have mixtures of mechanisms such that the kinetics will be quite complicated and, at this time, *uninterpretable*. Although there have been no adequate tests of the sintering models, we have taken the view that the set reviewed by Coblenz et al.[7] are appropriate to use in predicting qualitative microstructural changes for the initial-stage sintering; they are reproduced in Table VI.

A need for model improvement arises with mixed mechanism considerations and the desire to predict the size:shrinkage:time:temperature conditions at which control shifts from one mechanism to another. A pair of order-of-magnitude models considered together could give a choice over multiple orders-of-magnitude intervals (in time, for example) in which the transition would occur. For the initial stage, least confidence exists for the lattice diffusion model; an analytical solution or computer simulation is needed.

The best proof of the modeling and of mixed mechanism sintering comes from the coarsening mechanism, evaporation:condensation (EC) that can be *quantitatively* assessed. For oxide ceramics the available thermochemical data are very good. The statement in the paradigms that the material is thermodynamically stable means that the vapor pressure must be low in order to avoid coarsening by EC or, worse, to have the samples evaporate. Sample weight loss is an indication that EC is significant as a competing (coarsening) mechanism to densification. The effects of the oxidation/reduction potential and, frequently, the influences of water vapor and halogen gas pressures can also be quantitatively controlled to give the "effective vapor pressures" of many oxides. Readey (et al.) demonstrated the inhibition of densification by controlled increases in the "effective vapor pressure" of several oxides (Al_2O_3, ZnO, ferrite, MgO) by changes in the oxygen pressure in the atmosphere. That work is taken as quantitative support for the EC model;

for example, see Bheemineni and Readey's paper[9] on MgO:H$_2$ in this volume and those by Whittemore and coworkers.[10,11]

For the several stages of sintering, the processes of densification, grain growth, and pore growth have been modeled independently, and several approaches have been taken to consider them jointly. In the first stage, neck/pore growth and densification are the only important ones to consider because grain growth is assumed to be pinned by the porosity present, i.e., at the interparticle necks. Some pores with large numbers of surrounding grains (packed agglomerates) may exhibit growth only until after grain growth occurs to decrease the numbers of surrounding grains, which gives reversal of the surface curvature and then allows shrinkage. This will occur after the agglomerates have become dense.

Since Burke[12] applied Zener's model for pore:boundary interactions to define the critical condition for breakaway of boundaries in discontinuous grain growth in the final stage, we have assumed qualitatively that porosity controls the grain size during the intermediate stage. Quantitative support for this came from Yan et al.'s[4] exhaustive review of grain growth in ceramics, which showed that the "boundary mobilities" were much lower for porous than for dense ceramics. Assuming that growth simply occurs to the Zener limit during the intermediate stage (makes boundary migration a fast process, by inference) means that the simpler analysis of pore shrinkage and/or coarsening should be sufficient to define the important parameters to maximize densification:coarsening. This calls for maximizing the ratio of D_b/D_s or D_l/D_s and γ_{sv}/γ_{gb}, the same as for the initial stage of sintering.

Kuczynski[13] modeled the second and third stages on a phenomenological, statistical basis and found that the *breadth* of the pore-size distribution was a determining factor for the alternatives of pore growth/shrinkage. Further statistical modeling is needed for the first and second stages of sintering, that would include stress effects in the load-bearing network and the possibilities of rearrangement. It *may* be required that various kinds of "cells" be analyzed before all relationships between the important features can be identified. In the meantime, more thorough characterization of the microstructures and their evolution would help to guide the modeling.

The transition of the continuous pore network to isolated pores on grain boundaries at the final stage has been treated in a semiquantitative manner only. While the breakup of the complex pore network has not been analyzed, the evolution of a cylindrical void into spherical pores has been for surface diffusion and lattice diffusion. The breakup of cylindrical voids in Al$_2$O$_3$ (sapphire) and in LiF-doped polycrystalline MgO follow the model for surface-diffusion-controlled mass transport. In powder compacts, the transition to closed pores begins at approximately 15% pores and is complete at approximately 5%. This density range over which the microstructural transformation takes place is determined by the same factors which affect the earlier sintering stages (pore size, grain size and grain-size distribution, density distribution, anisotropy in γ_s, γ_{gb}, D's, etc.). The continuity of pores is relatively "stabilized" by facets or by dihedral angles at pore intersections with grain boundaries less than 180°.

For the final stage of sintering the critical event has been identified as the breakaway of boundaries from pores, leaving pores trapped within grains during discontinuous grain growth. The important models to apply then are: (1) the forces between pores and boundaries, (2) the pore mobilities by surface and lattice diffusion, (3) pore coarsening (Ostwald ripening by D_b, D_l), (4) pore coarsening by drag and coalescence, and (5) the boundary mobilities (intrinsic or impurity drag). Principal emphasis has been given to the pore:grain boundary mobilities and to the pore-size:grain-size trajectories.

Modeling of the final stages of sintering and, in particular, the conditions for pore-boundary separation are in better agreement with experiment than for any other modeling and experiments described here. Two descriptions of pore-boundary separation based on a single pore in a migrating boundary[14] and on an average 3-D grain-pore assemblage[15] are discussed in a companion paper[16] and will not be repeated here.

Because of the complexity of the process, Brook refocused attention on the information that can be extracted from experimental measurements of the relative rates of densification to coarsening and of the boundary mobility to the pore mobility. Thus, for the influences of initial particle size, compacting pressure, temperature, heating rate, and dopants over which we have control, the experimental measurements can be directed to the relative behaviors of: (1) densification:grain growth, (2) densification:pore growth (or surface area decrease), and (3) grain-size:pore size trajectories, which represent summations from all mechanisms that are operating. For the data, plots of density:grain size, density:pore size, and grain size:pore size at various temperatures and times can be compared for changes in the fabrication variables, heating rate, atmospheres, and dopants to determine how they affect the overall densification:coarsening mechanisms. In the absence of understanding the kinetics, this is the best choice to decide about "improvements" in processing.

We wish to maximize the densification by control of the dopants and conditions, and would like to generate (in fabrication) a high ratio of grain:pore sizes to avert breakaway grain growth. Yan et al.[4] also emphasized the importance of control of the initial grain/particle-size distribution. The pore-size distribution should *also* be narrow to control pore growth.

From the individual models, and by consideration of them jointly, the properties we would like to influence and the relative directions of change of interest for the first, second, and final stage of sintering are:

(1) To maximize densification, maximize the surface energy, lattice, and boundary diffusivities, and minimize surface diffusion.

(2) To minimize grain growth, minimize the boundary energy and boundary mobilities.

(3) To maximize the pore:boundary drag force, the ratio of boundary to surface energy should be large.

(4) To maximize pore mobility in the final stage calls for a maximum in surface diffusivity and/or lattice diffusivity.

(5) To avoid pore detachment, the pore mobility:boundary mobility should be maximized.

We end up with conflicting requirements for the surface diffusivity and for the surface:grain boundary energy ratio as we pass from the earlier to the final stage of the process. The interesting conjecture on the most desirable influence of a dopant (MgO in Al_2O_3) has originated from this difference. A view was that the surface diffusivity might be enhanced by magnesia doping, thereby increasing the pore mobility to reduce the probability of pore detachment. A second view was that pore growth earlier might be decreased by a reduction in D_s. This would change the pore size:grain size trajectory and yield smaller, more mobile pores ($M_p \propto D_s/r_p^4$) and perhaps miss the critical condition for breakaway.

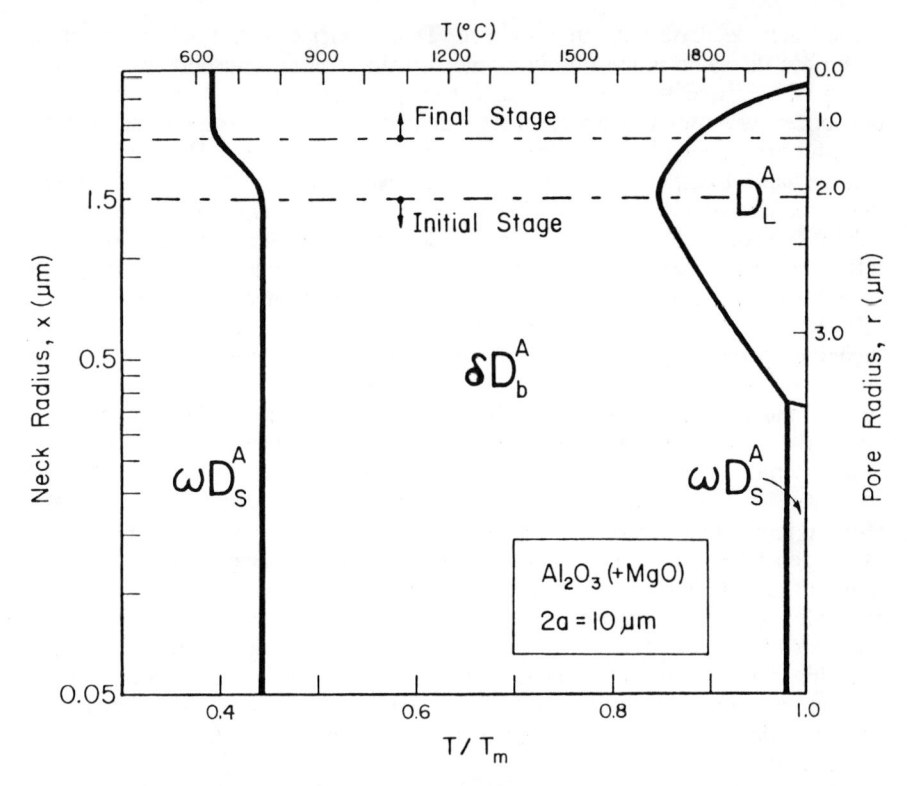

Fig. 1. Best-choice sintering map for Al_2O_3.

Application of Models to the Sintering of Alumina and Magnesia

Ashby introduced the use of sintering maps to illustrate, for example, the behavior of any given material where the dominant mechanisms of atom transport are displayed as fields on a two-dimensional plot of shrinkage vs temperature at constant grain size. These maps have the advantage of condensing large amounts of information that involve the models and the data base for any given material. The disadvantage is that their accuracy depends on both. A map for Al_2O_3 is shown in Fig. 1. The boundaries between fields represent the loci of points at which equal contributions to shrinkage are calculated from the models based on selected data. The initial stage models, reviewed by Coblenz et al.[7] are used here. Isotropic surface energies and diffusivities are assumed, although there is evidence to the contrary.

The data base used for these calculations has also been reviewed by Dynys et al.[17] Because of the multiple sources of samples and the different measurements from tracer diffusion, mass transport by creep, sintering, and surface diffusion or grain-boundary diffusion with different possible effects of impurities, it is obvious that the data base does not represent "an alumina." In this Proceedings, Wuensch covers the diffusion data and questions their accuracy and whether any of them represents intrinsic behavior.

The data selection was constrained by the following observations or assumptions. At low temperatures, the aluminum oxide samples display neck growth or surface area decrease without shrinkage. Thus, either evaporation: condensation or surface diffusion is the dominant mechanism of atom transport. Readey and Kuczynski showed some years ago that evaporation: condensation is important only under very low oxygen pressures for aluminum oxide. The representation here is for a broad range of oxygen pressures not strongly reducing. Thus, we assume that the vapor transport can be neglected and *surface diffusion is therefore dominant at low temperatures*. At higher temperatures, alumina samples consistently exhibit shrinkage, thus requiring a shift to dominance by either boundary or lattice diffusion. The increased sinterability rates with increased heating rates are taken as additional evidence for the transition in the dominant mechanism from surface diffusion to lattice or grain-boundary diffusion at higher temperatures. In the higher temperature regime where densification occurs, the influences of divalent cation doping on creep rates and on the sintering rates and of titanium doping (under oxidizing conditions where the titanium is presumed to be quadravalent) also have been assumed to be evidence for lattice diffusion as the dominant transport mechanism with Frenkel disorder on the aluminum cation sublattice.

From the data base, we have assumed that aluminum ion diffusion in the lattice is more rapid than oxygen and that significantly enhanced diffusive transport of oxygen takes place at grain boundaries. We assume that parallel paths for transport of aluminum through the lattice and oxygen at grain boundaries is the probable mechanism for sintering at larger grain sizes. The fact that breakaway grain growth leaves pores inside grains which are then quite stable is consistent with this view, i.e., when the pores are intersected by grain boundaries the boundaries provide a rapid transport path for oxygen, but when pores are isolated from grain boundaries the slower transport path becomes, of necessity, oxygen diffusion in the lattice.

Finally, the sensitivity of the sintering behavior of alumina to particle size is taken as evidence for the applicability of the diffusion models with a rejection, therefore, of plastic flow as the mechanism of atom transport. This is also supported by the grain size and stress dependences of deformation rate in creep and in hot-pressing.

For this discussion of the atomistic processes, we have rejected the notion that oxygen interstitials are important and have not included possible significance of the electronic disorder (the prospect that electron or hole transport may become controlling for some of these mass transport events). The overall interpretation of the sintering and hot-pressing processes should be regarded as phenomenological. Our greatest confidence can be placed only in comparison with creep rates, grain-boundary grooving rates, and similar sources of mass transport data. There is no adequate basis to force an atomistic interpretation of mass transport control on either of those simpler phenomena or on the sintering process at this time.

There are two prominent facts concerning the sinterability of aluminum oxide. (1) Essentially all of the alumina raw materials with $\rho_0 \approx 50\%$ are sinterable, whether derived from Bayer-process aluminas, chloride-source precipitated hydroxides or nitrate-source aluminum salts, alkoxides, or various others. It seems not to matter what the different species and amounts of impurities are to "sinterability." (2) Magnesium oxide works as an additive controlling breakaway of grain boundaries from pores in the final stage of sintering in all these different raw materials, suggesting that different distributions of impurities do not matter!

MgO

The construction of MgO sintering maps is based on the same initial-stage

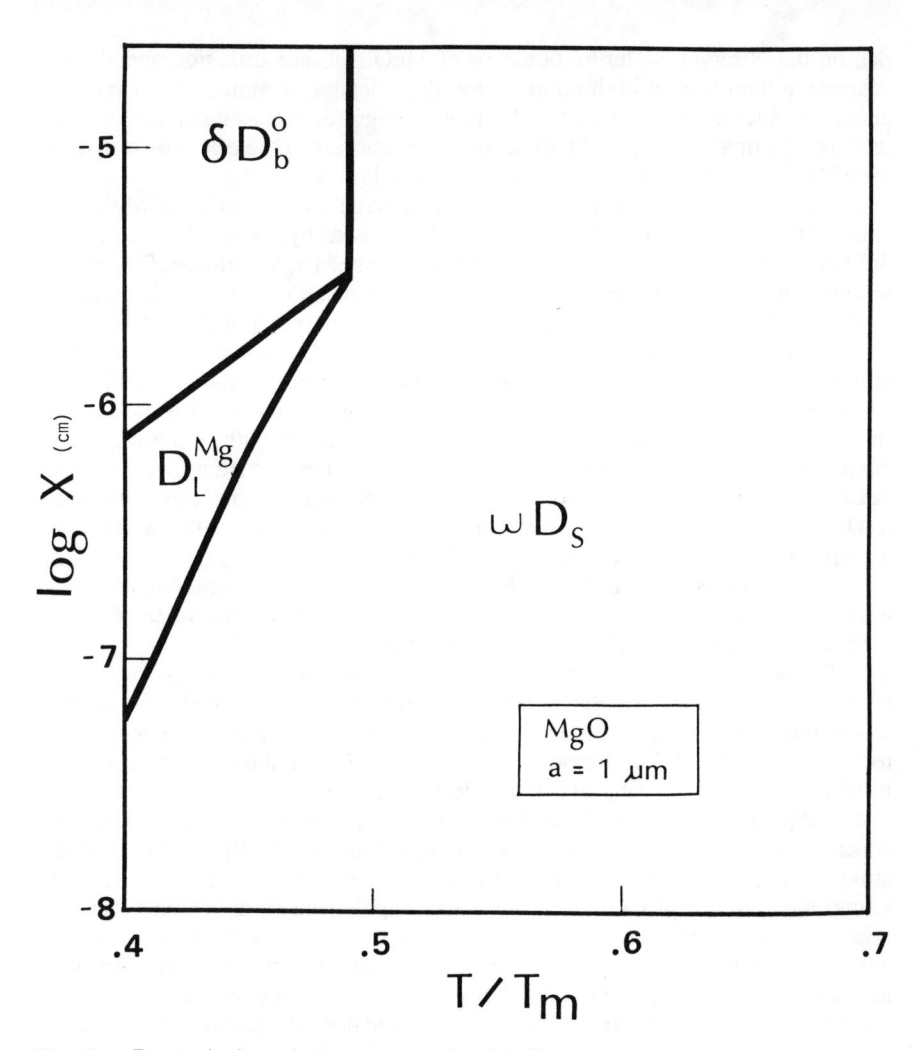

Fig. 2. Best-choice sintering map for MgO.

models as the Al$_2$O$_3$ maps and depends on the particular selections among the data for tracer diffusion, surface diffusion from grain-boundary grooving/surface smoothing, and other effective diffusivities from creep and sintering studies. Maps were constructed using all possible permutations of the mass transport data for MgO. Comparison of the predicted microstructure evolution (shrinkage and neck formation) using the various sets of diffusivities led to a rejection of most maps. Most maps contained no fields in which densification contributed significantly to the sintering process; surface diffusion was predicted to dominate at all temperatures, particle sizes, and neck sizes of interest. Since MgO of typical purity is observed to densify, this indicates that either the measured diffusivities do not apply to the MgO powders used in sintering (or that the initial-stage models do not describe sintering of MgO). It is not now possible to determine the cause of the discrepancy. The sintering maps do agree qualitatively with the effect of heating rate on the observed sintering behavior of MgO. Surface diffusion and, hence, coarsening dominate at high temperature; densification dominates at lower temperatures. Vieira[18] found that fast heating of MgO to high temperature did not increase the final density of MgO samples, in contrast to the effect of fast firing on Al_2O_3, where surface diffusion dominates at low temperatures.

Impurity effects are important in the microstructural evolution of MgO. The sinterability of high-purity MgO or MgO doped with hydroxyl or halide ions is different from the behavior of MgO of "typical purity." High-purity MgO produced as cubic smoke particles from burned magnesium metal does not undergo densification during heat treatment. In the absence of water vapor most boundaries formed between smoke particles are $\Sigma 1$, that is, two particles form a single crystal.[19] In the presence of water vapor, other (100) twist boundaries will form. This result indicates that (100) twist boundaries are unstable with respect to the two (100) surfaces except in the presence of an impurity. Other, more general, boundaries in MgO are expected to be similar. Similarly, magnesium chloride-hexahydrate, decomposed to form cubic oxide particles, compacted, and heat-treated, does not densify.[18] High-purity MgO does not exhibit significant densification at temperatures up to 1600°C.[20,21]

During final-stage sintering of MgO, the calculated microstructure evolution is in good agreement with the measured grain-size–pore-size trajectories[21] and also the grain-size:pore-size ratio for pore-boundary separation.[16,21]

At low temperatures, in calcination, magnesium oxide grain coarsening has been shown to depend on the water vapor pressure. Also, magnesium oxide is much more susceptible to enhanced volatilization when the oxygen pressure is reduced than is Al_2O_3. Furthermore, the uniformity of the grain structure developed in hot-pressing vs sintering and the sintering behaviors of carbonate-derived vs hydroxide-derived powders show that carbon is an important impurity. Thus, we conclude that magnesium oxide is much more susceptible to variations in the atmosphere than is alumina, and that under normal ambient conditions (air sintering) the presence of CO_2 and water vapor supplement complications that may arise from specific impurities in the raw materials. We close with the assertion that more care must be exercised in control of the atmosphere when dealing with magnesium oxide before reproducible results and a self-consistent set of data with respect to independently measured diffusivities and dopant effects can be expected.

MgO Effect in Al_2O_3

A succession of papers have appeared on the mechanism by which magnesium oxide inhibits the breakaway of grain boundaries from pores. A recent set regarded the possible effect of magnesium oxide on the *surface diffusion coefficient*. Several have argued that the surface diffusion coefficient might be increased, thereby increasing the mobility of the pores in the final stage. Alternatively, Brook[3] argued that the surface diffusion coefficient may be reduced by MgO, thereby decreasing the pore-growth rate, improving the prospect of maintaining a higher mobility of the pores, and avoiding breakaway at the critical condition. At this stage it seems best to be conservative and simply review the phenomenological results — Bennison and Harmer[1] showed that MgO reduced the rate and anisotropy of grain growth in Al_2O_3. In this Proceedings, Harmer[22] reports the presence of liquid in the grain boundaries in Al_2O_3. Although we believe in the impurity drag modeling, the existence of liquid at boundaries and an inability to find MgO segregated at boundaries require reconsideration of the whole subject.

Conclusions

For Al$_2$O$_3$ sintering with submicrometer powder under normal heating rates, the initial stage is dominated by coarsening and, from the evidence of pore growth, the intermediate stage is as well. This correlates with the D_s data from boundary grooving. Under high heating rates to high temperature, densification can be completed in minutes, with small final grain sizes. This behavior can be semi-quantitatively justified as being due to grain-boundary diffusion, based on the models and the boundary diffusivities derived from the creep data. The stress and grain-size effects in creep suggest mixed mechanism (diffusion) control by lattice and boundary diffusion. It is inferred that cation Frenkel defects exist from dopant-enhanced creep by M^{2+} and M^{4+}. That this could arise from boundary-diffusion enhancement is a conjecture not now subject to independent testing. The influence of MgO doping is thought to be due mainly to reduced grain-growth rate with a secondary effect of increasing the lattice diffusivity. No effect by MgO on D_s is indicated in boundary grooving or ρ:surface area behaviors. However, the presence of a liquid phase in a high-purity source material has been shown, and a need to document liquid-phase effects is compelling.

For sintering magnesia, boundary diffusion is dominant at low temperatures, yielding to surface diffusion at high temperatures. There are much more important effects due to atmospheric variables — water, carbon dioxide, etc. — than is the case for alumina. It is also more dependent on impurities; pure MgO does not densify.

References

[1] S. J. Bennison and M. P. Harmer, "Effect of MgO Solute on the Kinetics of Grain Growth in Al$_2$O$_3$," *J. Am. Ceram. Soc.*, **66**, C–90 (1983).

[2] R. L. Coble and R. M. Cannon, "Current Paradigms in Powder Processing," *Mater. Sci. Res.*, **11**, 151 (1978).

[3] R. J. Brook, "Controlled Grain Growth"; pp. 331–64 in Treatise on Materials Science and Technology, Vol. 9. Edited by F. F. Y. Wang. Academic Press, New York, 1976.

[4] M. F. Yan, R. M. Cannon, and H. K. Bowen, "Grain Boundary Migration in Ceramics"; pp. 276–307 in Ceramic Microstructures. Edited by R. M. Fulrath and J. A. Pask. Westview Press, Boulder, CO, 1977.

[5] O. J. Whittemore and J. J. Sipe, "Pore Growth During the Initial Stages of Sintering Ceramics," *Powder Technol.*, **9**, 159 (1974).

[6] G. C. Kuczynski, "Self-Diffusion in Sintering of Metallic Particles," *Trans. Am. Inst. Mining Met. Engrs.*, **185**, 169 (1949).

[7] W. S. Coblenz, J. M. Dynys, R. M. Cannon, and R. L. Coble, "Initial Stage Sintering Models, A Critical Analysis and Assessment"; pp. 141–57 in Sintering Processes. Edited by G. C. Kuczynski. Plenum, New York, 1980.

[8] D. L. Johnson, "Solid-State Sintering"; pp. 173–84 in Ultrafine-Grain Ceramics. Edited by J. J. Burke, N. L. Reed, and V. Weiss. Syracuse University Press, 1970.

[9] V. R. Bheemineni and D. W. Readey, "Vaporization of MgO in Hydrogen"; this conference.

[10] O. J. Whittemore and J. A. Varela, "Initial Sintering of MgO in Several Water Vapor Pressures"; this conference.

[11] E. Longo, J. A. Varela, C. A. Santili, and O. J. Whittemore, "Model of Interactions Between MgO and Water"; this conference.

[12] J. E. Burke, "Role of Grain Boundaries in Sintering," *J. Am. Ceram. Soc.*, **40**, 80 (1957).

[13] G. C. Kuczynski, "Pore Shrinkage and Ostwald Ripening"; pp. 217–24 in Sintering and Related Phenomena. Edited by G. C. Kuczynski. Plenum, New York, 1973.

[14] C. H. Hsueh, A. G. Evans, and R. L. Coble, "Microstructure Development During Final/Intermediate Stage Sintering-I. Pore/Grain Boundary Separation," *Acta Metall.*, **30**, 1269 (1982).

[15] M. F. Yan, R. M. Cannon, and U. Chowdhry, "Theory of Grain Growth During Densification"; unpublished work.

[16] C. A. Handwerker, R. L. Coble, R. J. Brook, and R. M. Cannon, "Final Stage Sintering of MgO"; this conference.

[17] J. M. Dynys, R. L. Coble, W. S. Coblenz, and R. M. Cannon, "Mechanism of Atom Transport During Initial Stage Sintering"; pp. 391–404 in Sintering Processes. Edited by G. C. Kuczynski. Plenum, New York, 1980.

[18] R. Vieira, "Fast-Firing and Hot Pressing of MgO"; Ph. D. Thesis, University of Leeds, U. K., 1980.

[19] P. Chaudhari and J. W. Matthews, "Coincidence Twist Boundaries Between Crystalline Smoke Particles," *J. Appl. Phys.*, **42**, 3063 (1971).

[20] R. A. Brown, "Sintering of Very Pure Magnesium Oxide and Magnesium Oxide Containing Vanadium," *J. Am. Ceram. Soc.*, **44**, 483 (1965).

[21] C. A. Handwerker, "Sintering and Grain Growth of MgO"; Sc. D. Thesis, Massachusetts Institute of Technology, 1983.

[22] M. P. Harmer, "The Use of Solid Solution Additives in Ceramic Processing"; this conference.

Author Index

Ando, K. See Oishi, Y.
Bagley, R. D. and Johnson, D. L. Effect of Magnesia on Grain Growth in Alumina, 666
Batchelor, A. D. See Palmour III, H.
Bheemineni, V. and Ready, D. W. Vaporization of Magnesium Oxide in Hydrogen, 553
Bowen, H. K. See Gattuso, T. R.
Bradt, R. C. See Iwasa, M.
Brook, R. J. See Vieira, J. M.
Brook, R. J. See Wu, S.
Buchanan, R. C. and Wilson, D. M. Role of Al_2O_3 in Sintering of Submicrometer Yttria-Stabilized ZrO_2 Powders, 526
Burns, R. C. and Burns, V. M. Optical and Mossbauer Spectra of Transition-Metal-Doped Corundum and Periclase, 46
Burns, V. M. See Burns, R. C.
Cannon, R. M. Mechanical Properties, of MgO and Al_2O_3 818
Cannon, R. M. See Handwerker, C. A.
Cannon, W. R. High Creep Ductility in Alumina Containing Compensating Additives, 741
Carry, C. See Fridez, J. D.
Carter, C. B. and Morrissey, K. J. Grain-Boundary Structure in Al_2O_3, 303
Castaing, J. See Heuer, A. H.
Choi, S. and Takeuchi, T. Semiempirical Analysis of the Electronic Processes of F-Type Centers in α-Al_2O_3, 152
Coble, R. L. Sintering, and Grain Growth in Alumina and Magnesia, 839
Coble, R. L. See Handwerker, C. A.
Cohen, A.; Van der Merwe, C. P.; and Kingon, A. I. Effect of MgO Dopant Dispersing Method on Density and Microstructure of Alumina Ceramics, 780
Colbourn, E. A. and Mackrodt, W. C. Point Defects and Chemisorption at the {001} Surface of MgO, 190
Crawford, J. H., Jr. Polarization-Depolarization Studies of Defects in Oxides, 16
Crawford, J. Defects in MgO and Al_2O_3, 793
Dash, J. G. and Stone, R. Effects of Crystallinity in Initial Stage Sintering, 601
Davis, R. F.; Horie, Y.; Scattergood, R. O.; and Palmour III, H. Defects Produced by Shock Conditioning: An Overview, 157
Duffy, D. M. and Tasker, P. W. Properties of Grain Boundaries in Rock-Salt Structured Oxides, 275
Eastman, J.; Schmuckle, F.; Vaudin, M. D.; and Sass, S. L. Electron Diffraction and Microscopy Studies of the Structure of Grain Boundaries in NiO, 324
Evans, A. G. and Fu, Y. Mechanical Behavior of Alumina: A Model Anisotropic Brittle Solid, 697
Fang, T. T. See Palmour III, H.
Freund, F.; King, B.; Knobel, R.; and Kathrein, H. Low Atomic Number Impurity Atoms in Magnesium Oxide: Hydrogen and Carbon, 119
Fridez, J. D.; Carry, C.; and Mocellin, A. Effects of Temperature and Stress on Grain-Boundary Behavior in Fine-Grained Alumina, 720
Fu, Y. See Evans, A. G.
Gattuso, T. R. and Bowen, H. K. Processing of Narrow Size Distribution Alumina, 644
Gilbart, E. See Wu, S.
Gordon, R. S. Understanding of Defect Structure and Mass Transport in Polycrystalline Al_2O_3 and MgO via the Study of Diffusional Creep, 418
Gupta, T. K. Crack Healing in Al_2O_3, MgO, and Related Materials, 750
Hamano, K.; Nakagawa, Z.; and Watanabe, H. Effect of Magnesium Chloride on Sintering of Magnesia, 610
Handwerker, C. A.; Cannon, R. M.; and Coble, R. L. Final-Stage Sintering of MgO, 619

Haneda, H. See Shirasaki, S.
Hare, T. M. See Palmour III, H.
Harmer, M. P. Use of Solid-Solution Additives in Ceramic Processing, 679
Hayashi, K. See Kinoshita, C.
Henrich, V. E. Surface Electronic Structure of MgO and Al_2O_3 Ceramics, 205
Henrich, V. Structure of MgO and Al_2O_3 Surfaces, 801
Heuer, A. H. Dislocations in Ceramics, 799
Heuer, A. H. and Castaing, J. Dislocations in α-Al_2O_3, 238
Horiai, N. See Kimura, S.
Horie, Y. See Davis, R. F.
Iwasa, M. and Bradt, R. C. Fracture Toughness of Single-Crystal Alumina, 767
Ikuma Y. and Komatsu, W. Oxygen Surface Diffusion and Surface Layer Thickness in MgO, 464
Johnson, D. L.; Sanderson, W. B.; Knowlton, J. M.; and Kemer, E. L. Sintering of α-Al_2O_3 in Gas Plasmas, 656
Johnson, D. L. See Bagley, R. D.
Kathrein, H. See Freund, F.
Kemer, E. L. See Johnson, D. L.
Kim, K. Y. See Palmour III, H.
Kimura, S.; Yasuda, E.; Horiai, N.; and Moriyoshi, Y. Boundary Structure Observed in MgO Bicrystals, 347
King, B. See Freund, F.
Kingery, W. D. Grain Boundaries in MgO and Al_2O_3, 803
Kingery, W. D. See Li, C. W.
Kingery, W. D. See Yager, T. A.
Kingon, A. I. See Cohen, A.
Kinoshita, C.; Hayashi, K.; and Mitchell, T. E. Migration Energies of Interstitials and Vacancies in MgO, 490
Knobel, R. See Freund, F.
Knowlton, J. M. See Johnson, D. L.
Komatsu, W. See Ikuma, Y.
Kröger, F. A. Experimental and Calculated Values of Defect Parameters and the Defect Structure of α-Al_2O_3, 100
Kröger, F. A. Electrical Properties of α-Al_2O_3, 1
Ku, R. C. See McCune, R. C.
Li, C. W. and Kingery, W. D. Solute Segregation at Grain Boundaries in Polycrystalline Al_2O_3, 368
Longo, R.; Varela, A.; Santilli, C. V.; and Whittemore, O. J. Model of Interactions between Magnesia and Water, 592
McCune, R. C. and Ku, R. C. Calcium Segregation to MgO and α-Al_2O_3 Surfaces, 217
Mackrodt, W. C. Calculated Point Defect Formation, Association, and Migration Energies in MgO and α-Al_2O_3, 62
Mackrodt, W. C. See Colbourn, E. A.
Matsuda, S. See Shirasaki, S.
Matsui, M.; Takahashi, T.; and Oda, I. Influence of MgO Vaporization on the Final Stage Sintering of MgO-Al_2O_3 Spinel, 562
Mitchell, T. E. See Kinoshita, C.
Mocellin, A. See Fridez, J. D.
More, K. L. See Palmour III, H.
Moriyoshi, Y. Dislocations in MgO, 258
Moriyoshi, Y. See Kimura, S.
Morrissey, K. J. See Carter, C. B.
Nakagawa, Z. See Hamano, K.
Oda, I. See Matsui, M.
Oishi, Y. and Ando, K. Oxygen Diffusion in MgO and Al_2O_3, 379
Osenbach, J. W. See Stubican, V. S.
Palmour III, H.; Hare, T. M.; Batchelor, A. D.; Kim, K. Y.; More, K. L.; and Fang, T. T. Sintering of Shock-Conditioned Materials, 506

Palmour III, H. See Davis, R. F.
Petuskey, W. T. See Yoo, H. I.
Phillips, D. S. and Shiue, Y. R. Grain-Boundary Microstructures in Alumina Ceramics, 357
Readey, D. W. See Bheemineni, V.
Sanderson, W. B. See Johnson, D. L.
Sangster, M. J. L. See Stoneham, A. M.
Santilli, C. V. See Longo, E.
Sasamoto, T. See Sata, T.
Sass, S. L. See Eastman, J.
Sata, T. and Sasamoto, T. Vaporizations from Magnesia and Alumina Materials, 541
Scattergood, R. O. See Davis, R. F.
Schmuckle, F. See Eastman, J.
Shirasaki, S.; Matsuda, S.; Yamamura, H.; and Haneda, H. Oxygen Diffusion in Undoped and Impurity-Doped Polycrystalline MgO, 474
Shiue, Y. R. See Phillips, D. S.
Stone, R. See Dash, J. G.
Stoneham, A. M.; Sangster, M. J. L.; and Rowell, D. K. Calculations of Entropies and of Absolute Diffusion Rates in Oxides, 79
Stubican, V. S. and Osenbach, J. W. Grain-Boundary and Lattice Diffusion of ^{51}Cr in Alumina and Spinel, 406
Summers, G. P. Luminescence and Photoconductivity in MgO and α-Al_2O_3 Crystals, 25
Takahashi, T. See Matsui, M.
Takeuchi, T. See Choi, S.

Tasker, P. W. Surfaces of Magnesia and Alumina, 176
Tasker, P. W. See Duffy, D. M.
Van der Merwe, C. P. See Cohen, A.
Varela, J. A. See Longo, E.
Varela, J. A. See Whittemore, O. J.
Vaudin, M. D. See Eastman, J.
Vieira, J. M. and Brook, R. J. Lattice, Grain Boundary, Surface, and Gas Diffusion Constants in Magnesium Oxide, 438
Watanabe, H. See Hamano, K.
Whittemore, O. J. and Varela, J. A. Initial Sintering of MgO in Several Water Vapor Pressures, 583
Whittemore, O. J. See Longo, E.
Wilson, D. M. See Buchanan, R. C.
Wolf, D. Comparison of the Calculated Properties of High-Angle (001) Twist Boundaries in MgO, FeO, CoO, and NiO with MgO, 290
Wu, S.; Gilbart, E.; and Brook, R. J. Solid-State Sintering: The Attainment of High Density, 574
Wuensch, B. J. See Yoo, H. I.
Yager, T. A. and Kingery, W. D. Defect Association in MgO, 139
Yamamura, H. See Shirasaki, S.
Yasuda, E. See Kimura, S.
Yoo, H. I.; Wuensch, B. J.; and Petuskey, W. T. Secondary Ion Mass Spectrometric Analysis of Oxygen Self-Diffusion in Single-Crystal MgO, 394

Subject Index

Additives, compensating, high creep ductility in alumina containing, 741
 solid-solution, use of in ceramic processing, 679
Alumina, α-, and MgO, calculated point defect formation, association, and migration energies in, 62
 α-, and MgO crystals, luminescence and photoconductivity in, 25
 α-, electrical properties of, 1
 α-, experimental and calculated values of defect parameters and the defect structure of, 100
 α-Al$_2$O$_3$, dislocations in, 238
 α-, semiempirical analysis of the electronic processes of F-type centers in, 153
 α-, and MgO, calcium segregation to surfaces of, 217
 α-, sintering of in gas plasmas, 656
 and MgO, oxygen diffusion in, 379
 and magnesia materials, vaporizations from, 541
 and magnesia, surfaces of, 176
 and MgO ceramics, surface electronic structure of, 205
 defects in, 793
 and spinel, grain-boundary and lattice diffusion of ^{51}Cr in, 406
 ceramics, effect of MgO dopant dispersing method on density and microstructure of, 780
 ceramics, grain-boundary microstructures in, 357
 containing compensating additives, high creep ductility in, 741
 dislocations in, 799
 effect of magnesia on grain growth in, 666
 electrical and optical properties of, 804
 fine-grained, effects of temperature and stress on grain-boundary behavior in, 720
 grain boundaries in, 803
 lattice defects in, 791
 MgO, and related materials, crack healing in, 750
 MgO-spinel, influence of MgO vaporization on the final stage sintering of, 562
 mechanical behavior of: a model anisotropic brittle solid, 697
 narrow size distribution, processing of, 644
 polycrystalline, and MgO, understanding of defect structure and mass transport in via the study of diffusional creep, 418
 solute segregation at grain boundaries in, 368
 role of in sintering of submicrometer yttria-stabilized ZrO$_2$ powders, 526
 single-crystal, fracture toughness of, 767
 sintering of, 839
 surfaces in, 801
Analysis, secondary ion mass spectrometric, of oxygen self-diffusion in single-crystal MgO, 394
 semiempirical, of the electronic processes of F-type centers in α-Al$_2$O$_3$, 153
Association, defect, in MgO, 139
 formation, and migration energies, calculated point defect, in MgO and α-Al$_2$O$_3$, 62
Atoms, impurity, low atomic number, in magnesium oxide: hydrogen and carbon, 119
Behavior, mechanical, of alumina: a model anisotropic brittle solid, 697
Bicrystals, MgO, boundary structure observed in, 347
Boundaries, grain, in MgO and Al$_2$O$_3$, 803
 in NiO, electron diffraction and microscopy studies of, 324
 in polycrystalline Al$_2$O$_3$, solute segregation at, 368
 in rock-salt structured oxides, properties of, 275
 microstructures, in alumina ceramics, 357
 structure in Al$_2$O$_3$, 303
 high-angle (001) twist, in MgO, FeO, CoO, and NiO, comparison of the calculated properties of with MgO, 290
 structure, observed in MgO bicrystals, 347
Calcium, segregation to MgO and α-Al$_2$O$_3$ surfaces, 217
Carbon and hydrogen, low atomic number impurity atoms in magnesium oxide, 119
Centers, F-type, in α-Al$_2$O$_3$, semiempirical analysis of the electronic processes of, 153
Ceramics
 alumina, effect of MgO dopant dispersing method on method on density and microstructure of, 780
 grain-boundary microstructures in, 357
 MgO and Al$_2$O$_3$, surface electronic structure of, 205
 processing, use of solid-solution additives in, 679
Chemisorption and point defects at the {001} surface of MgO, 190
Chromium -51, grain-boundary and lattice diffusion of in alumina and spinel, 406
Cobalt oxide, MgO, FeO, and NiO, high-angle (001) twist boundaries in, comparison of the calculated properties of with MgO, 290
Conditioning, shock, defects produced by: an overview, 157
Constants, lattice-, grain-boundary, surface-, and gas-diffusion, in magnesium oxide, 438
Corundum and periclase, transition-metal-doped, optical and Mössbauer spectra of, 46
Cracks, healing in Al$_2$O$_3$, MgO, and related materials, 750
Creep, diffusional, understanding of defect structure and mass transport in polycrystalline Al$_2$O$_3$ and MgO via the study of, 418
 high, ductility in alumina containing compensating additives, 741
Crystallinity, effects of in initial stage sintering, 601
Crystals, MgO and α-Al$_2$O$_3$, luminescence and photoconductivity in, 25
Defects, association in MgO, 139
 formation, calculated, point, association, and migration energies in MgO and α-Al$_2$O$_3$, 62
 in MgO and Al$_2$O$_3$, 793
 in oxides, polarization–depolarization studies of, 16
 lattice, in MgO and Al$_2$O$_3$, 791
 parameters, and the defect structure of α-Al$_2$O$_3$, experimental and calculated values of, 100
 point, and chemisorption at the {001} surface of MgO, 190
 produced by shock conditioning: an overview, 157
 structure, and defect parameters, of α-Al$_2$O$_3$, experimental and calculated values of, 100
 and mass transport in polycrystalline Al$_2$O$_3$ and MgO via the study of diffusional creep, 418
Density and microstructure, of alumina ceramics, effect of MgO dopant dispersing method on, 780
 high, the attainment of: solid-state sintering, 574
Depolarization–polarization studies of defects in oxides, 16
Diffraction, electron, and microscopy studies of the structure of grain boundaries in NiO, 324
Diffusion, grain-boundary and lattice, of ^{51}Cr in alumina and spinel, 406
 lattice-, grain-boundary, surface-, and gas, constants in magnesium oxide, 438
 oxygen, in MgO and Al$_2$O$_3$, 379
 in undoped and impurity-doped polycrystalline MgO, 474
 surface, and surface layer thickness in MgO, 464
 rates, absolute, and calculations of entropies, in oxides, 79
 self, secondary ion mass spectrometric analysis of in single-crystal MgO, 394
Dislocations in α-Al$_2$O$_3$, 238
 in MgO, 258

and Al$_2$O$_3$, 799
Dopants, MgO, dispersing method, effect of on density and microstructure of alumina ceramics, 780
Ductility, creep, high, in alumina containing compensating additives, 741
Electron diffraction and microscopy studies of the structure of grain boundaries in NiO, 324
Electronic processes, semiempirical analysis of the F-type centers in α-Al$_2$O$_3$, 153
structure, surface, of MgO and Al$_2$O$_3$ ceramics, 205
Energy migration, association, and calculated point defect formation, in MgO and α-Al$_2$O$_3$, 62
of interstitials and vacancies in MgO, 490
Entropies, calculations of, and of absolute diffusion rates in oxides, 79
F-type centers, semiempirical analysis of the electronic processes of in α-Al$_2$O$_3$, 153
Formation association, and migration energies, calculated point defect, in MgO and α-Al$_2$O$_3$, 62
Fracture toughness of single-crystal alumina, 767
Gas, plasmas, sintering of α-Al$_2$O$_3$ in, 656
Grain boundaries, in MgO and Al$_2$O$_3$, 803
Growth, grain, in alumina, effect of magnesia on, 666
Healing, crack, in Al$_2$O$_3$, MgO, and related materials, 750
Hydrogen and carbon, low atomic number impurity atoms in magnesium oxide, 119
vaporization of magnesium oxide in, 553
Impurity atoms, low atomic number, in magnesium oxide: hydrogen and carbon, 119
Interactions between magnesia and water, model of, 592
Interstitials and vacancies, migration energies of in MgO, 490
Iron oxide, MgO, CoO, and NiO, high-angle (001) twist boundaries in, comparison of the calculated properties of with MgO, 290
Luminescence and photoconductivity in MgO and α-Al$_2$O$_3$ crystals, 25
Magnesia and α-Al$_2$O$_3$, calculated point defect formation, association, and migration energies in, 62
and Al$_2$O$_3$ ceramics, surface electronic structure of, 205
and α-Al$_2$O$_3$ crystals, luminescence and photoconductivity in, 25
and Al$_2$O$_3$, defects in, 793
oxygen diffusion in, 379
and α-Al$_2$O$_3$ surfaces, calcium segregation to, 217
Al$_2$O$_3$, and related materials, crack healing in, 750
-Al$_2$O$_3$ spinel, influence of MgO vaporization on the final stage sintering of, 562
and alumina materials, vaporizations from, 541
surfaces of, 176
and polycrystalline Al$_2$O$_3$, understanding of defect structure and mass transport in via the study of diffusional creep, 418
and water, model of interactions between, 592
bicrystals, boundary structure observed in, 347
comparison of the calculated properties of high-angle (001) twist boundaries in MgO, FeO, CoO, and NiO with, 290
defect association in, 139
dislocations in, 258
dopant, dispersing method, effect of on density and microstructure of alumina ceramics, 780
effect of on grain growth in alumina, 666
electrical and optical properties of, 804
FeO, CoO, and NiO, comparison of the calculated high-angle (001) twist boundaries with MgO, 290
final-stage sintering of, 619
grain boundaries in, 803
initial sintering of in several water vapor pressures, 583
lattice defects in, 791
grain-boundary, surface-, and gas-diffusion constants in, 438
low atomic number impurity atoms in: hydrogen and carbon, 119

mechanical properties of, 818
migration energies of interstitials and vacancies in, 490
oxygen surface diffusion and surface layer thickness in, 464
point defects and chemisorption at the {001} surface of, 190
polycrystalline, undoped and impurity-doped, oxygen diffusion in, 474
single-crystal, secondary ion mass spectrometric analysis of oxygen self-diffusion in, 394
sintering of, 839
effect of magnesium chloride on, 610
surfaces in, 801
vaporization, influence of on the final stage sintering of MgO-Al$_2$O$_3$ spinel, 562
in hydrogen, 553
Magnesium chloride, effect of on sintering of magnesia, 610
Materials, magnesia and alumina, vaporizations from, 541
shock-conditioned, sintering of, 506
Microscopy studies, and electron diffraction, of the structure of grain boundaries in NiO, 324
Microstructure and density, of alumina ceramics, effect of MgO dopant dispersing method on, 780
grain-boundary, in alumina ceramics, 357
Migration energies, calculated point defect formation, and association, in MgO and α-Al$_2$O$_3$, 62
of interstitials and vacancies in MgO, 490
Mössbauer and optical spectra of transition-metal-doped corundum and periclase, 46
Nickel oxide, electron diffraction and microscopy studies of the structure of grain boundaries in, 324
MgO, FeO, and CoO, high-angle (001) twist boundaries in, comparison of the calculated properties of with MgO, 290
Optical and Mössbauer spectra of transition-metal-doped corundum and periclase, 46
Oxides, calculations of entropies and of absolute diffusion rates in, 79
polarization–depolarization studies of defects in, 16
rock-salt structured, properties of grain boundaries in, 275
Oxygen diffusion in MgO and Al$_2$O$_3$, 379
in undoped and impurity-doped polycrystalline MgO, 474
self-diffusion, secondary ion mass spectrometric analysis of in single-crystal MgO, 394
surface diffusion and surface layer thickness in MgO, 464
Parameters, defect, and the defect structure of α-Al$_2$O$_3$, experimental and calculated values of, 100
Periclase and corundum, transition-metal-doped, optical and Mössbauer spectra of, 46
Photoconductivity and luminescence in MgO and α-Al$_2$O$_3$ crystals, 25
Plasmas, gas, sintering of α-Al$_2$O$_3$ in, 656
Polarization–depolarization studies of defects in oxides, 16
Powders, submicrometer yttria-stablized ZrO$_2$, role of Al$_2$O$_3$ in sintering of, 526
Pressures, water vapor, initial sintering of MgO in, 583
Processes, electronic, of F-type centers in α-Al$_2$O$_3$, semiempirical analysis of, 153
Processing, ceramic, use of solid-solution additives in, 679
of narrow size distribution alumina, 644
Properties, calculated, of high-angle (001) twist boundaries in MgO, FeO, CoO, and NiO, comparison of with MgO, 290
electrical, of α-Al$_2$O$_3$, 1
and optical, in MgO and Al$_2$O$_3$, 804
mechanical, in MgO and Al$_2$O$_3$, 818
of grain boundaries in rock-salt structured oxides, 275
Rock salt structured oxides, properties of grain

boundaries in, 275
Segregation, calcium, to MgO and α-Al$_2$O$_3$ surfaces, 217
  solute, at grain boundaries in polycrystalline Al$_2$O$_3$, 368
Shock -conditioned materials, sintering of, 506
  conditioning, defects produced by: an overview, 157
Sintering, final-stage, of MgO, 619
  of MgO-Al$_2$O$_3$ spinel, influence of MgO vaporization on, 562
  initial, of MgO in several water vapor pressures, 583
  effects of crystallinity in, 601
  of α-Al$_2$O$_3$ in gas plasmas, 656
  of magnesia, effect of magnesium chloride on, 610 and Al$_2$O$_3$, 839
  of shock-conditioned materials, 506
  of submicrometer yttria-stabilized ZrO$_2$ powders, role of Al$_2$O$_3$ in, 526
  solid-state: the attainment of high density, 574
Solids, anisotropic brittle: a model for mechanical behavior of alumina, 697
Solute segregation at grain boundaries in polycrystalline Al$_2$O$_3$, 368
Spectra, optical and Mössbauer, of transition-metal-doped corundum and periclase, 46
Spinel, and alumina, grain-boundary and lattice diffusion of ^{51}Cr in, 406
  MgO-Al$_2$O$_3$, influence of MgO vaporization on the final stage sintering of, 562
Stress and temperature, effects of on grain-boundary behavior in fine-grained alumina, 720
Structure, boundary, observed in MgO bicrystals, 347
  defect, and defect parameters, of α-Al$_2$O$_3$, experimental and calculated values of, 100
  and mass transport, understanding of in polycrystalline Al$_2$O$_3$ and MgO via the study of diffusional creep, 418

grain-boundary, in Al$_2$O$_3$, 303
  of grain boundaries in NiO, electron diffraction and microscopy studies of, 324
  surface electronic, of MgO and Al$_2$O$_3$ ceramics, 205
Surface layer thickness and oxygen surface diffusion in MgO, 464
Surfaces, {001}, of MgO, point defects and chemisorption at, 190
  electronic structure, of MgO and Al$_2$O$_3$ ceramics, 205
  in MgO and Al$_2$O$_3$, 801
  MgO and α-Al$_2$O$_3$, calcium segregation to, 217
  of magnesia and alumina, 176
  oxygen, diffusion, and surface layer thickness in MgO, 464
Temperature and stress, effects of on grain boundary behavior in fine-grained alumina, 720
Toughness, fracture, of single-crystal alumina, 767
Transport, mass, and defect structure, understanding of in polycrystalline Al$_2$O$_3$ and MgO via the study of diffusional creep, 418
Vacancies and interstitials, migration energies of in MgO, 490
Vapor, water, pressures, initial sintering of MgO in, 583
Vaporization from magnesia and alumina materials, 541
  MgO, influence of on the final stage sintering of MgO-Al$_2$O$_3$ spinel, 562
  of magnesium oxide in hydrogen, 553
Water and magnesia, model of interactions between, 592
  vapor pressures, initial sintering of MgO in, 583
Yttria—stabilized ZrO$_2$ powders, role of Al$_2$O$_3$ in sintering of, 526
Zirconia powders, submicrometer yttria-stabilized, role of Al$_2$O$_3$ in sintering of, 526